W9-CHK-390

Genetics:
Human Aspects

GENETICS: HUMAN ASPECTS

SECOND EDITION

Arthur P. Mange
UNIVERSITY OF MASSACHUSETTS

Elaine Johansen Mange
AMHERST, MASSACHUSETTS

SINAUER ASSOCIATES, INC.
Publishers
Sunderland, Massachusetts

The Cover:
At Play, original batik 26″ × 44″,
by Ugandan artist Paul Nzalamba.

Courtesy of the artist:
P.O. Box 65036,
Los Angeles, California 90065.

Genetics: Human Aspects, Second Edition

Library of Congress Cataloging-in-Publication Data
Mange, Arthur P.
Genetics : human aspects / Arthur P. Mange, Elaine Johansen Mange.
—2nd ed.
p. cm.
Bibliography: p.
Includes index.
ISBN 0-87893-501-0
1. Human genetics. I. Mange, Elaine Johansen. II. Title.
QH431.M279 1989 89-19662
573.2′1—dc20 CIP

Printed in U.S.A.

5 4 3 2 1

To our nonagenarian parents

Janet Purvin Mange
Clarence E. Mange
Anna Nicolaysen Johansen
Ragnar A. Johansen

Contents

Preface xiii

PART ONE **BASIC CONCEPTS**

1. ***Perspectives*** 3

Science and Scientists 4
A Brief History 5

Mendel 6; Further History of Transmission Genetics and Cytogenetics 7; Molecular and Biochemical Genetics 7; Population Genetics 9; The Nature–Nurture Concept: Quantitative Genetics 10

Human Genetic Literature 10

2. ***Rules of Inheritance for Single Genes*** 12

Autosomal Genes: Mendel's First Law 12

Mendel's Experiments 13; Mendel's First Law 14; Modern Nomenclature 15; A Checkerboard Method for Predicting Offspring 15; Progeny Testing and Matings 16; An Example from Humans 17

X-Linked Genes 17

Sex Determination 18; Genes on Sex Chromosomes 18; Color Blindness 18; Hemophilia 20; X-Linked Lethal Genes and Shifts in Sex Ratio 21

Y-Linked Genes 22
Mitochondrial Genes 22

3. ***Pedigrees*** 26

Data Gathering 27
Pedigree Construction 28

Pedigree and Biochemical Analyses 29; Some Caveats 30

Autosomal Dominant Inheritance: Brachydactyly 30

Autosomal Recessive Inheritance: Cystic Fibrosis 32

Clinical Features and Treatment 33; Pedigrees 33; Biochemistry and Molecular Genetics 35; Detection of Carriers and Homozygotes 35

X-Linked Recessive Inheritance: Duchenne and Becker Muscular Dystrophies 36

Clinical Features 36; Pedigrees 37; Biochemistry and Molecular Genetics 38; Dystrophin 39; Detection of Carriers and Hemizygotes 39

4. ***Chromosomes*** 43

Chromosomes, Genes, and DNA 43

Gross Structure of Chromosomes 44; The Components of Genes and Chromosomes 45; Fine Structure of Chromosomes 46

Human Chromosomes 49

Methods of Chromosome Preparation and Analysis 50; The Normal Human Karyotype 54

5. ***Gametes and Cell Division*** 60

Gametes and Fertilization 60

The Mature Egg and Its Cell Structures 60; The Mature Sperm and Fertilization 63

The Cell Cycle and Mitosis 66

Interphase 66; Mitosis 68

Meiosis and Gametogenesis 71

Meiosis I 71; Meiosis II 74

Cell Cycle Mutants 76

6. ***Genetics of Development*** 81

Embryonic Development 81
Genetics of Embryonic Development 83

Maternal Effects 84; Segmentation and Pattern Formation 85; The Need for Both Maternal and Paternal Genomes in Mammals 88; Future Prospects 90

Usual Sexual Development 90

Prenatal Development: Genetic and Hormonal Factors 91; Postnatal Development: Environmental Factors 93; Postnatal Development: Physiological Factors 93

Some Errors in Sexual Development 94

True Hermaphrodites 94; Pseudohermaphrodites 96; Gender Identity 99

PART TWO **CYTOGENETICS**

7. ***Nondisjunction*** 107

Mistakes in Cell Division 108

Nondisjunction and the Chromosome Theory 108; Meiotic Nondisjunction 109; Mitotic Nondisjunction 110

Nondisjunction of Human Autosomes: Sporadic Down Syndrome 112

General Features 112; Possible Relation to Alzheimer Disease 113; Frequency of Down Syndrome 114; Chromosomal-Genetic Basis for Down Syndrome 115; Search for Factors Altering Nondijunction Rates 116; Down Syndrome Mosaics 118

Other Autosomal Aneuploids 118

Trisomy 18 and Trisomy 13 118; Very Rare Autosomal Aneuploids 118

8. ***Sex Chromosomes and Their Abnormalities*** 123

Y Chromosome Present 124

Klinefelter Syndrome: 47,XXY Males 124; The 47,XYY Karyotype 126

Y Chromosome Absent 128

Turner Syndrome: 45,X Females 128; 47,XXX and 48,XXXX Females 129; Other Poly-X Karyotypes 130

Sex Chromatin and the Lyon Hypothesis 130

Detection of Sex Chromatin in Humans 131; The Inactive-X Hypothesis 131; X Chromosome Inactivation and Reactivation 132

Structure of the Mammalian X and Y Chromosomes 135

The Pseudoautosomal Region of the X and Y Chromosomes 135; XX Males, XY Females, and the Testis Determining Factor 135

The Fragile X Syndrome 136

Fragile X Syndrome in Males and Females 137; Genetics of the Fragile X Syndrome 138

9. ***Other Chromosomal Abnormalities*** 143

Deletions and Duplications 144

Origins and Examples of Deletions 144; Origins and Examples of Duplications 147

Translocations 147

Origins of Translocations 147; Translocations in Meiosis 148; Translocation Down Syndrome 150; Other Translocations 151

Inversions 151

Behavior of Inversions during Meiosis 152 Inversions in Humans; 153

Other Abnormalities 153

Ring Chromosomes 153; Polyploidy 154

Frequencies of Chromosomal Abnormalities 155

Gametes 157; Prenatal Tests 158; Abortions, Stillbirths and Neonatal Deaths 158; Unselected Newborns and Some Adult Groups 159; Conclusions on Fetal Wastage 160

Chromosomal Abnormalities and Gene Mapping 160

Variant Chromosomes and Inversions 160; Deletions, Duplications, and Translocations 161

PART THREE GENE TRANSMISSION

10. ***Allelic Segregation and Probability*** 169

Multiple Alleles 169

Origin of Multiple Allelic Series 169; Combining Alleles into Genotypes 170

The Phenotype of a Heterozygote 170

Levels of Observation 171

The Rules of Probability 172

Definition 173; Five Rules for Combining Probabilities 173

Problem Solving 179

The Binomial 179; Conditional Probability 180

11. ***Recombination of Nonallelic Genes*** 183

The Concept of Recombination 183

Genes on Different Chromosomes: Mendel's Second Law 184

The Proportions of the Gametes 185; Offspring Types for Two Genes 185; Gene × Gene Method 186; Meiosis and the Second Law 187

Genes on the Same Chromosome: Crossing Over 187

The Proportions of the Gametes 189; The Physical Basis of Recombination 189; The Limits of Recombination 192; The Phase Problem 193

Mapping the Human X Chromosome by Family Linkage Studies 194

Green-Shift and Red-Shift 194; Two Clusters of X-Linked Genes 194; Further Considerations 195

Significance of Gene Mapping 197

Building and Using Gene Maps 197; Gene Mapping and Evolution 199

12. ***Complicating Factors*** 204

Variation in Gene Expression 204

Sex Differences 204; Variable Expressivity 206; Incomplete Penetrance 209; One Gene with Several Effects 210; One Phenotype from Different Causes 210

Ascertainment Bias 212

The Always-Missing Families 212; The Sometimes-Missing Families 215

Testing Hypotheses 215

The Chi-Square Method 216; Interpretation 216; Comments 217

Is the Trait Genetic? 218

Kuru 218

PART FOUR GENES, METABOLISM, AND DISEASE

13. ***Gene Structure and Function*** 225

DNA Structure 225

Base Complementarity 227

Replication: DNA → DNA 227

Bubbles and Forks 228

Transcription: DNA → RNA 230

Types of RNA 232

Protein Synthesis: RNA → Protein 233

Amino Acids 234; Translation 235

Gene Organization in Chromosomes 237

Interrupted Genes and mRNA Processing 237; Gene Clusters and Pseudogenes 239; Repetitive DNA 241; Transposable Elements 242

Mitochondrial Genes 243

14. ***Genetic Information and Misinformation*** 247

The Triplet Code 247

Deciphering Specific Codons 249; Properties of the Code 249

Mutations 251

Types of Base Substitutions 251; Spontaneous and Induced Mutations 252

Hemoglobin Mutations 253

Sickle-Cell Anemia 254; Other Amino Acid Substitutions 255; Thalassemias 255

Chemical Mutagenesis 258

Screening Systems 258; The Ames Test 259; Sister Chromatid Exchange 259

Radiation Mutagenesis 261

Human Radiation Exposures 262; Evaluation of Radiation Effects 264

15. ***New Genetic Technologies*** 269

Recombinant DNA 270

Making Recombinant DNA 270; Manufacturing Proteins 272; Public Concerns 275

DNA Manipulations 276

Electrophoresis and Restriction Maps 276; Sequencing 277; Making a Gene Library 281; Screening a Gene Library 283; Southern Blotting and the Molecular Biology of Color Vision 284; Restriction Fragment Length Polymorphism (RFLP) 286; Huntington Disease and Sickle-Cell Anemia 287; DNA Fingerprinting and the Polymerase Chain Reaction 289

Somatic Cell Genetics 292

Selecting and Isolating Hybrid Cells 292; Chromosome Assignment by a Gene's Protein Product 294; Methods for Regional Assignments 295

A Further Note on Mapping 298

16. ***Inborn Errors of Metabolism*** 303

Amino Acid Metabolism 305

Phenylketonuria and Other Hyperphenylalaninemias 306; Albinism 311

Lipid and Lipoprotein Metabolism 314

Tay-Sachs Disease 314; Familial Hypercholesterolemia 317

Purine Metabolism 321

Lesch-Nyhan Syndrome 322

Medical Considerations 325

Detection of Metabolic Disease in Newborns 325; Pharmacogenetics 325

17. ***Genetics of Blood Groups*** 332

Blood Grouping 332

Red Blood Cells 333; Polymorphisms 334; Techniques 335; How Some Systems Were Discovered 337

The ABO Blood Group System 340

Transfusions 341; Biosynthesis of the Antigens 342; The Curious Case of the Bombay Bloods 344

Rh and the Prevention of Hemolytic Disease 345

Rh Genes, Antigens, and Antibodies 345; Incompatibility 346; Protection against Rh-Caused Hemolytic Disease 347

18. ***Genetics of Immunity*** 352

The Immune Response 353

Main Components of the System 353; The B Cell Response 354; The T Cell Response 357

Antibody Structure and Genetics 358

Antibody Structures 358; Antibody Diversity 359; Monoclonal Antibodies 363

Transplantation Antigens 365

Tissue Typing: The HLA System 365; Antigens and Alleles of the *BCA* Loci 367; Transplantation 368

The Immune System and Disease 371

Associations with HLA 371; Allergies 373; Autoimmune Disorders 373; Immune Deficiency Disorders 374

19. ***Genetics of Cancer*** 378

Cancerous Cells 379

Oncogenes and Proto-Oncogenes 380

Oncogenes Carried by Retroviruses 381; Finding Human Oncogenes by Gene Transfer 382; The Proteins of Oncogenes 384; Trangenic Mouse Models 386

Cancer and Cytogenetics 386

Proto-Oncogenes at Translocation Breakpoints 387; Other Specific Chromosomal Abnormalities 390

Cancer and Mendelian Inheritance 392

Dominants: Retinoblastoma and Wilms Tumor 392; Recessives: Chromosome Breakage Syndromes 394

Genetic and Environmental Factors 398

PART FIVE **POPULATION AND QUANTITATIVE GENETICS**

20. ***Population Concepts*** 405

Genotype, Phenotype, and Allele Frequencies 405

Relation between Allele and Genotype Frequencies 406

Mating Frequencies 406

Random Mating 407; Offspring from Random Mating 407

The Hardy-Weinberg Law 408

A Proof 408; The Meaning of the Hardy-Weinberg Law 409; The Array of Genotype Frequencies 410; Another Proof of the Hardy-Weinberg Law 410

Some Applications 412

The Sickle-Cell Allele among American Blacks 412; Alkaptonuria and the Problem of Dominance 412; ABO: A Three-Allele System 413; X-Linked Genes 413

21. ***Inbreeding and Isolates*** 417

Relatives and Their Offspring 417

Cousins 417; Inbreeding 419

Measuring Inbreeding from Pedigrees 421

Alleles Identical by Descent 421; Calculating the Coefficient of Inbreeding 422; More Complex Pedigrees 423

The Effects of Inbreeding 424

Increase in Specific Recessive Traits 425; Health and Mortality Data 426; The Prognosis for a First Cousin Marriage 427

Isolates 428

Recessive Disease in the Old Order Amish 428; Inbreeding in the Hutterites 429

22. ***Processes of Evolution*** 435

Mutation 436

Measuring Spontaneous Mutation Rates 437; The Average Mutation Rate for Human Genes 439

Selection 441

Selection Against a Recessive Lethal Phenotype 442; Selection Against Both Homozygotes 444

Migration and Drift 446

Migration and Gene Flow 447; Random Genetic Drift 448

Differing Views on the Mechanisms of Evolution 449

The Role of Drift 449; The Nature of Variability 450

A Genetic Concept of Race 451

The Reality but Arbitrariness of Races 451; Human Variability 452
Individuality; 454

23. ***Quantitative and Behavioral Traits*** 458

Genetic and Environmental Variation 459

Heritability 460

The Genetic Component 461

An Additive Model 462; Skin Color 464

Twins: Their Usefulness and Limitations 465

The Biology of Twinning 465; Twins in Genetic Research 467

Behavioral Traits 468

A Genetic Component to Human Behavior 469; Intelligence 471; Alcoholism 474

PART SIX **APPLICATIONS OF GENETIC TECHNIQUES**

24. ***Beginnings of Life*** 481

Negative Eugenics 482

Involuntary Sterilization 483; Voluntary Sterilization 483

Positive Eugenics 484

Germinal Choice 484; Artificial Insemination 484

Other Reproductive Technologies 486

In Vitro Fertilization and Embryo Transfer 486; Variations on a Theme 487; Ethical and Legal Issues 488; Reproductive Technologies in Farm Animals; 490

Sex Selection 491

Available Techniques (Postconception) 493; Contemplated Techniques (Preconception) 493

25. ***Genetic Practices and Prospects*** 498

Genetic Counseling 499

The Procedure 499; Ethical Concerns 501

Prenatal Diagnosis 502

Procedures for Obtaining Fetal Cells 502; Indications and Results 503; Sex Prediction 507

Genetic Screening 508

Screening For Early Detection of Genetic Disease 509; Screening For Heterozygotes 511

Treatments for Genetic Disease 512

Current Practices 513; Gene Therapy 516

Some Thoughts on Science and Society 518

Public Influence on Science Policy 518; The Lysenko Experience 519; Conclusion 520

APPENDICES

1. ***Some Common Units of Measurement*** 525
2. ***Some Basic Mathematics*** 526
3. ***Some Basic Chemistry*** 530
4. ***The Human Gene Map*** 536

Bibliography 542
Name Index 568
Subject Index 574

Preface

This book is designed for a one-semester course in genetics—especially human genetics—for students with diverse interests. Readers with a good high school background in biology, chemistry, and algebra should have little difficulty with the material, but an up-to-date college course in biology would provide better preparation. (For readers who may need review, we give some chemical and mathematical background in Appendices 2 and 3.) We present the general principles of genetics, show how they operate in humans, and discuss their implications for individuals and for society. We strive for interest, rigor, and clarity in both text and illustrations. The main text is divided into six parts:

	Part	Number of chapters
One.	Basic Concepts	6
Two.	Cytogenetics	3
Three.	Gene Transmission	3
Four.	Genes, Metabolism, and Disease	7
Five.	Population and Quantitative Genetics	4
Six.	Applications of Genetic Techniques	2

Part One introduces some fundamental aspects of genetics; it is a first circuit of genetic knowledge, touching on readers' prior experience and preparing them for more advanced material in later chapters. Since the book is fairly comprehensive, instructors may wish to leave out some material. For example, Chapters 6 (Genetics of Development), 21 (Inbreeding and Isolates), and 24 (Beginnings of Life) could be omitted without interrupting the flow of the text. Chapters that are rich with examples, such as 16 (Inborn Errors of Metabolism), could be presented in shortened form.

The first edition of this book appeared in 1980. Since then, the molecular biology revolution has changed the way that much genetic research is conducted, and vastly increased our knowledge of how genes work and where they are located on the chromosomes. Indeed, new findings now accumulate so rapidly that it is difficult to stay abreast from week to week. The second edition of this book reflects the new approach throughout. In particular, we include two new Chapters, 15 (New Genetic Technologies) and 19 (Genetics of Cancer), and greatly expand the molecular approach in 16 (Inborn Errors of Metabolism), 18 (Genetics of Immunity), and 25 (Genetic Practices and Prospects).

But science is more than just technologies, facts, and concepts. Every science has an interesting history, and progresses toward an unpredictable future through the collective efforts of many creative people. For these reasons, we provide historical material where appropriate, and plentiful references to past and current scientific literature. The citations in the text—giving the author's name and year of publication—refer the reader to a complete listing of bibliographic data at the back of the book. Each chapter also has a brief guide to reference material, so that readers can easily pursue any topic that they choose.

To aid understanding, we include chapter summaries and many questions—nearly all with answers—and detailed indices. More questions and answers and an additional bibliography are included in a teacher's guide that we have prepared.

Many of our students, colleagues, and teachers have contributed to the development of this book. We particularly thank the following persons who took the time to review chapters of our manuscript—thus making the book a far better account of human genetics. We are indebted to them for their expertise and diligence.

Adelaide T. C. Carpenter, University of California, San Diego
Patricia A. DeLeon, University of Delaware
Robin E. Denell, Kansas State University
Carter Denniston, University of Wisconsin
Richard W. Erbe, Harvard University Medical School
Thomas A. Ferguson, Washington University School of Medicine
Molly Fitzgerald-Hayes, University of Massachusetts
Stanley M. Gartler, University of Washington
Douglas Green, University of Alberta Medical School
Emanuel Hackel, Michigan State University
Carl A. Huether, University of Cincinnati
Richard E. LaFond, University of Massachusetts
Claude J. Migeon, Johns Hopkins Children's Center
T. K. Mohandas, University of California, Los Angeles
Hans Ris, University of Wisconsin
Margretta R. Seashore, Yale University School of Medicine

Earl Seidman, University of Massachusetts
Howard M. Temin, University of Wisconsin Medical School
Irene A. Uchida, McMaster University Medical Centre
J. Bruce Walsh, University of Arizona
Steven A. Williams, Smith College
Doris T. Zallen, Virginia Polytechnic Institute

We also express our deep appreciation to the staff of the University of Massachusetts Libraries—to many anonymous souls who over the years accumulated a superb collection of genetic and medical literature, as well as to all individuals who assisted us expertly and cheerfully on innumerable occasions. Among many in the latter group we thank particularly Alena F. Chadwick, James L. Craig, Virginia W. Craig, and Laurence M. Feldman.

Our friend and colleague Katherine Doktor-Sargent did all of the artwork for the first edition (most of which is retained) and much of the artwork for the second edition. We continue to marvel at her talents. Finally we acknowledge the help and support of the personnel of Sinauer Associates, especially Andrew D. Sinauer, Carol J. Wigg, Joseph Vesely, Jodi L. Simpson, and Dean Scudder. Their professionalism, patience, and attention to fine detail have contributed greatly to the finished product.

Arthur P. Mange
Elaine Johansen Mange

PART ONE BASIC CONCEPTS

1. *Perspectives*
2. *Rules of Inheritance for Single Genes*
3. *Pedigrees*
4. *Chromosomes*
5. *Gametes and Cell Division*
6. *Genetics of Development*

Chapter 1 Perspectives

Science and Scientists

A Brief History
Mendel
Further History of Transmission Genetics and Cytogenetics
Molecular and Biochemical Genetics
Population Genetics
The Nature–Nurture Concept: Quantitative Genetics

Human Genetics Literature

> *The elements of beautiful science are familiar: first the confrontation of the human mind with a natural phenomenon, then its investigation through observations and experiments, the continuing proposal of theories, the testing of predictions, and finally, in best case, the convincing demonstration of the validity of one of the theories through confirmation of its specific predictions. The process can take only a few years and involve only a few scientists or it can span centuries and involve many. The practical consequence may be revolutionary... or it may have little or no use. In either case, a full scientific story, especially one that has been unfolding over historic times, can be a lovely thing, like a classical symphony or a gothic cathedral.*
>
> *David Botstein (1986)*

All people are recognizably human, but no one is exactly like anyone else, not even an identical twin. The similarity and diversity that coexist in all species have puzzled and intrigued people for thousands of years. Only during the past century, as the science of genetics developed new ways to approach such problems, have explanations begun to emerge. Experiments with a wide variety of plants, animals, and microorganisms have yielded detailed knowledge of (1) how traits are passed on from one generation to another, (2) how genetic information within individuals is chemically "decoded" and expressed during their development, and (3) how the genetic variability that arises among the members of populations can account for the gradual evolution of populations. These three major areas of inquiry are known, respectively, as *transmission genetics*, *molecular* (and *biochemical*) *genetics*, and *population genetics*.

Only during the past several decades has the genetics of humans become a flourishing discipline with increasingly sophisticated techniques for the diagnosis and treatment of many inherited conditions. Along with the new possibilities for alleviating human suffering, however, have arisen social, political, and legal problems that will not be easily resolved. For example:

- Should parents have the right to choose the sex of their children? If so, might this have undesirable effects on the population?
- Should certain populations be screened to detect individuals who, although normal themselves, are "carriers" of detrimental genes, such as the sickle-cell gene in Blacks or the Tay-Sachs gene in Jews?
- Should families be informed when a child is born with a genetic abnormality whose effects are not well understood? Will the benefits of studying and perhaps ultimately helping such children be offset by the possibility of stigmatizing them?
- Should the costs of expensive treatments for genetic disease be borne by society? If so, should society have a say in whether or not such affected individuals ought to be born, if this can be prevented by prenatal diagnosis and selective abortion?
- Should some kinds of research, such as that which involves releasing into the environment potentially beneficial but "engineered" organisms, be strictly regulated?
- If it becomes possible to alter our genetic destiny by using techniques such as gene replacement and cloning, how shall these methods be controlled?

Answers to such questions do *not* appear at the back of this book. Indeed, tidy answers do not exist anywhere, however much we all may yearn for them. This is because the necessary "facts" and their interpretations often turn out to be rather slippery, changing as scientists ask new questions and develop new techniques for trying to answer them. Usually, we must make do with the information at hand, imperfect or incomplete as it may be.

For these and other reasons, the formulation of public policy on complex social issues should involve the thoughtful participation of all citizens. As Daniel Callahan (1972) has written:

> The promised benefits are mainly mathematical, helpful to that statistically unknown portion of the population which either needs, or thinks it needs, or would like, the benefits. One can ordinarily feel only slight personal involvement in these problems... But the problems themselves are not ordinary. Even if we leave aside the respect we should (ordinarily) pay the desires and work of scientists, and leave aside the benefits their work may bring for some statistical portion of the population... we cannot leave aside quite so readily the long-term implications for man of the work being done on "New Beginnings in Life." Because we are part of the human community and bear some responsibility for the future of man, what is being done by our generation (even if not by us personally) becomes part of our responsibility. Thus even if it is true that most of us are laymen in these matters and that most of us will personally be neither helped nor harmed appreciably in our lifetime by what develops, we are not thereby excused from deciding or acting on the issues.

SCIENCE AND SCIENTISTS

If lay people are to have a say in developing guidelines for the conduct of certain scientific problems, they should first have a clear sense of what science can be expected to accomplish and what it cannot do. This in turn depends on some understanding of how research is carried on. Unfortunately, a scientific mystique has evolved over the years and has been reinforced by truly spectacular achievements in many fields, leading the public to expect more of science than it can possibly deliver: Too little food? Not enough fuel? Too many people? Not to worry! Science will take care of everything! Then, when science fails to solve all our old problems, and sometimes even creates some new ones, disillusionment and hostility set in. But withdrawing large-scale support for science can be as harmful to public welfare as ignoring the ethical problems that arise when science begins to impinge on certain human rights and values (Handler 1979).

It would help if scientists made a greater effort to explain their everyday work to the public. If gee-whiz-type media announcements of new discoveries also stressed how many areas of ignorance still remain, a more balanced picture would emerge. Instead, science is usually viewed with awe, and perhaps with a little fear or suspicion, as a field that ordinary mortals cannot fathom, let alone aspire to enter. These extravagant stereotypes, hopes, or fears can be dangerous if they affect public policies, including the use and abuse of science (Goodfield 1981). For these reasons, and because this topic is rarely treated in textbooks, we wish to elaborate a bit on the popular images of scientists.

The people who "do" science are, like any other group, quite varied. Yet, like any other group, they are often misunderstood. In cartoons and comic strips, "mad scientist" types abound (Figure 1). In movies and books, scientists are often portrayed as brilliant but out of touch with reality (*The Man in the White Coat*), or as totally objective and unswervingly dedicated to their work (*Microbe Hunters*), or as coldly sinister (*Dr. Strangelove*). Usually they are male. Seldom are they treated as just plain folks capable of pettiness or insecurity. James D. Watson's *The Double Helix* (1968), however, provides a rare glimpse into the less praiseworthy but more human side of scientists, and Anne Sayre's *Rosalind Franklin and DNA* (1975) shows how personal interactions can affect the conduct of scientific research.

Textbooks (including this one) tend to be glib in their accounts of scientific achievements, presenting a smooth continuum of success after success; rarely do readers learn about the intervening false starts, dead ends, misinterpretations, or outright failures. And, as Stephen Jay Gould (1986) points out in a review of Nobel laureate Peter Medawar's engaging autobiography (1986), the same is true for articles in scientific journals, whose standard writing style

> ...misconstrues, even falsifies, the actual doing of science... The epitomized logic of inductivist accounts—from introduction, to materials and methods, results and conclusions—omits the basic human dimensions of hypothesis, confusion, error and collegiality. The false starts are in the wastebasket, not the *Science Citation Index*. By focusing his text on the frustrations, the errors and the bullheaded approaches, until kicked in the pants by data or good advice from colleagues with other perspectives, Medawar has illustrated his favourite themes by honestly discussing his own work.

Truth be told, scientific research quite often yields confusing or negative or equivocal results, even at the hands of the world's greatest scientists (Crick 1988). Furthermore, even successful research involves a lot of tedium —and, except for the hope of finding something exciting now and then, some scientists might at times wish they had chosen another line of work.

Of all the world's scientists who ever existed, most

"Crack out the liquid nitrogen, dumplings . . . we're on our way."

Figure 1 This conception of the potential outcome of recombinant DNA research, although whimsical and amusing, has no basis in reality. Nevertheless, many people may believe that such results are possible. For information on this type of research, see Chapters 13 and 15. (Reprinted courtesy of *The Boston Globe.*)

are alive today. A tour of their laboratories would reveal a wide range of intellects and personalities—from geniuses to plodders, from Renaissance men and women to those with limited interests, from the flamboyant to the very shy, from intensely competitive workers to those with modest aspirations, from generous and caring individuals to selfish boors.

Even among the most intelligent and ambitious, very few will ever be mentioned in a textbook. Indeed, some of their published papers will never be cited anywhere, because scientific reports are now being generated at such a rate that nobody can hope to keep track of more than a tiny fraction of them. Browsing through scientific journals, one finds an astonishing and bewildering array of articles whose titles are so specific that they may sound utterly ridiculous to lay people—especially to those who feel that science should be completely goal-oriented and not concerned with apparently trivial pursuits. But the gradual accumulation of unexciting bits of knowledge and experimental techniques by many workers often sets the stage for the great discoveries or insights by a few, usually in ways that could never have been predicted in advance.

Those who studied human genetics, especially in the earlier decades, encountered some unique difficulties along with the special fascinations. Our species was long considered a poor subject for genetic research because of long generation time, small families, and the occurrence of so many genetically uninformative matings. Yet our biochemistry is better understood than that of any other complex species, and research on humans *has* led to new and important generalizations. For example, insight into the relationship between human genes and enzymes (by Archibald Garrod in the early 1900s) preceded such analyses in other species, and discoveries about the precise effect of mutations on protein structure in humans (by Vernon Ingram in the 1950s) preceded those in bacteria and viruses. Obviously, the use of experimental organisms has provided extensive knowledge of fundamental genetic phenomena and permitted large-scale testing and refinement of concepts. But because we are so curious about ourselves, investigators willingly coped with the special problems of human genetics, or found ways around them. What is difficult today may be less difficult tomorrow with new ideas and new technology.

A BRIEF HISTORY

The following sketch illustrates several features of scientific research. One theme is the *codiscovery* of a principle by several investigators working independently, for example, the rediscovery of Mendelian principles in 1900. In some cases codiscoveries are not surprising, because the exchange of ideas in scientific journals and at meetings may draw many people to work on an interesting problem—especially in "hot" new areas of inquiry. In other cases, and particularly in recent times, the problems have become so complex that solutions require teamwork by several laboratories, each specialized in a different type of research.

A second theme in the history of science is the so-called *premature discovery*, which is ultimately confirmed and accepted by others, but only after a long period of neglect. Gregor Mendel's explanation in 1865 of how hereditary factors operate, Archibald Garrod's suggestion in 1909 that genes make enzymes, Peyton Rous's discovery in 1910 that a virus can induce cancer in chickens, and Oswald Avery's evidence in 1944 that DNA is the genetic material—all are examples of ideas that were too advanced for their time.

A third theme is the critical role that *new technology* plays in the development and testing of scientific ideas. The process is never-ending: better microscopes, novel ways of staining chromosomes, improvements in methods for growing and handling cells, remarkable new techniques for isolating and analyzing the genetic material, increasingly powerful (and accessible) computers, plus many other technological developments, were necessary for the advancement of our knowledge to its current level. These research tools do not, of course, generate, prove, or disprove hypotheses. People do that, often alone but rarely in complete isolation.

This historical sketch begins with seminal discoveries reported around the 1860s in transmission genetics, molecular genetics, and population genetics. Rather than providing detailed explanations of genetic concepts, it is meant as a preview—defining the scope of human genetics while stressing the important experimental role of other species in advancing knowledge of our own. We try to show how answers to scientific problems rest on strata of knowledge accumulated by earlier investigators. These layers do not form at constant rates; indeed, for certain periods in some disciplines, there may be no progress at all. Each age sees itself as the modern age. But in science, as elsewhere, final answers are rarely final.

Figure 2 Gregor Mendel (1822–1884). This Czechoslovakian stamp, issued in 1965, commemorates the centennial of Mendel's report on his experiments with garden peas.

Mendel

The Austrian cleric Gregor Mendel experimented for nine years to confirm theories of inheritance that—by some remarkable insight—he might already have formulated. Far from being a shy recluse who labored in secrecy and obscurity, Mendel was a broadly trained, energetic, and highly respected scientist and teacher (Figure 2). He was also active in agricultural societies organized by local sheep, apple, and grape growers to further their interests in animal and plant breeding and to keep Brunn a thriving commercial center. Mendel tackled the problem of heredity at the suggestion of his superior, Abbot Napp—an enlightened administrator who fostered scientific research at the monastery in Brunn, and who even had a greenhouse built for some of Mendel's peas (Orel 1984). Despite their confidence in Mendel and their interest in the hybridization problem, however, his colleagues failed to grasp the significance and generality of his findings.

Distinctly different from any prior work were the two basic **laws of transmission** that he proposed (Chapters 2 and 11). The First Law states that a simple genetic trait is determined by a pair of separable factors (now called **alleles of a gene**), one inherited from each parent. Mendel's Second Law deals with the inheritance of two or more traits and states that each pair of alleles behaves independently of the other pairs. Thus it is possible to predict, in simple mathematical terms, the various outcomes of matings. In opposition to the popular notions of blending inheritance, Mendel proposed discrete units of heredity—the genes. This idea, increasingly refined by later investigators, has permeated all areas of biology.

Mendel's publication was read before a local scientific group in 1865, and new evidence suggests that it was enthusiastically received. "But the fact remains that no participant saw the far-reaching implications of his theory, and no one tried to repeat the experiments or to carry out similar experiments with other plant species" (Orel 1984). Copies of his paper were sent to 120 libraries in Europe and America as well as to the leading botanists of the time, with whom he corresponded. (Unfortunately, he did not write to Charles Darwin, despite having followed with great interest the controversy surrounding the 1859 publication of *The Origin of Species by Means of Natural Selection.*) Although Mendel's work was cited in numerous reviews of hybridization (including one in the 1881 *Encyclopaedia Britannica*), its significance went unrecognized during his lifetime.

Further History of Transmission Genetics and Cytogenetics

In 1900 three European botanists independently worked out the principles of inheritance; their searches of previous literature then turned up Mendel's 1865 publication. During the intervening 35 years, many details of cell structure and cell division (Chapters 4 and 5) had been worked out. These included the discovery of **chromosomes**, cellular elements contributed by parents to offspring through fertilization of an individual egg by a single sperm. During cell divisions, the chromosomes exhibit very specific patterns of behavior, leading to their orderly distribution among daughter cells.

Soon after 1900 it was noted that both visible chromosomes and abstract genes obeyed the same rules of transmission. This observation provided strong circumstantial evidence that genes were parts of chromosomes. Conclusive proof came later, with the analysis of abnormal patterns of inheritance: mistakes in the distribution of chromosomes to eggs and sperm were precisely matched by corresponding errors in the transmission of genes. One type of mistake, called nondisjunction, is now known to be a major cause of genetic ills (Chapter 7).

The inheritance of human ABO blood groups, discovered in 1900 (Chapter 17), was found to obey Mendel's laws, as did that of many diseases, including hemophilia (Chapter 2). The same rules applied to all plant and animal species that were studied. By the 1920s, Thomas Hunt Morgan and his students at Columbia University had collected many gene variations in fruit flies and analyzed their modes of inheritance. They described how sex was determined by X and Y chromosomes and worked out the pattern of inheritance of genes on the X (Chapter 2). They were also the first to **map genes**, that is, find the relative locations of genes on a given chromosome (Chapter 11), and to describe many errors in chromosomal transmission. By the 1930s, other geneticists had done comparable studies on a wide variety of organisms.

Not so fortunate were those who first tried to investigate the genetics of humans: it took until 1956 just to determine that we have 46 chromosomes (Chapter 4). But by 1960 a number of conditions (including Down syndrome) were found to be associated with abnormalities in chromosome number or structure (Chapters 7–9, 19). The mapping of human genes likewise proved very difficult. Only after 1968, when scientists figured out how to "cross" mammalian cells in laboratory cultures, did it become possible to assign many human genes to specific chromosomes—and in the 1980s this method was augmented by newer techniques for cloning and chemically analyzing human genes (Chapter 15).

Molecular and Biochemical Genetics

In 1871 Friedrich Miescher, a Swiss biochemist and contemporary of Mendel (Figure 3), isolated a phosphorus-rich acidic substance from cell nuclei. He believed that this "nuclein" was essential to cell multiplication, and predicted that "A knowledge of the relationships between nuclear materials, proteins, and their immediate products of metabolism will gradually help to raise the curtain which at present so completely veils the inner process of cellular growth." Although staining reactions later identified this nucleic acid as a major component of chromosomes, its significance was ignored for decades, because the more complex chromosomal proteins seemed a better candidate for the hereditary material. Interest in **deoxyribonucleic acid (DNA)** developed in 1944, when Oswald Avery and colleagues at the Rockefeller Institute reported that this substance, but *not* protein, could permanently transform the heredity state of some bacteria from one type to another. Further experiments with different organisms convinced many skeptics, and in 1953 the American geneticist James Watson and the English physical chemist Francis Crick proposed the now famous double helix structure for DNA (Chapter 13).

What a gene actually does was considered much earlier by Archibald Garrod, an English physician. In 1909 he published studies of four inherited diseases, accurately described as "inborn errors of metabolism" (Chapter 16), in which a specific **enzyme** was either inactive or missing. (An enzyme is a protein molecule that catalyzes a specific biochemical reaction in cells. All proteins are composed of one or more **polypeptide chains**, which are in turn made up of units called **amino acids**.) The idea that genes make enzymes was not seriously reconsidered until the 1940s, when research began on the synthesis of eye pigments in fruit flies and on the nutritional requirements of a bread mold. In both systems, each mutant gene blocked a specific biochemical reaction normally mediated by a specific enzyme. Additional research (Chapters 13–14) has refined this concept, which in its simplified form states that **one gene makes one polypeptide**.

Human hemoglobin, a large protein found in red blood cells, was the touchstone for understanding gene action (Chapter 14). In 1949 Linus Pauling, a chemist at the California Institute of Technology, suggested that the inherited disease sickle-cell anemia was caused by an abnormal hemoglobin. In the 1950s Vernon Ingram, a biochemist at the Massachusetts Institute of Technology, determined the amino acid sequences of the polypeptide chains that make up both normal and sickle-cell hemoglobins. Amazingly, the two hemoglobins differed by only *one* amino acid out of hundreds in each molecule. This observation proved that each different form (**allele**) of a gene makes a slightly different polypeptide. Clearly, the sequence of basic units (**nucleotides**) in the DNA chain encodes a special sequence of amino acids in the polypeptide chain, and one mistake in the DNA leads to one mistake in its polypeptide product. This exact correspondence between the positions of changes in a gene and the amino acid substitutions in the gene product has also been elegantly demonstrated in microorganisms.

Cracking the code, that is, determining exactly which sequence of nucleotides specifies which amino acid, was the most exciting genetic undertaking of the early 1960s. This code is identical in the nuclear DNA of all organisms, and almost identical in the DNA of certain cellular inclusions (Chapters 13–14).

In recent decades, knowledge of how genes work has come from studies of viruses, bacteria, and other microorganisms—all of which yield millions of progeny in a very short time (Lederberg 1987). Especially instructive are viruses that can coexist peacefully with the cells they infect; they accomplish this coexistence by inserting their DNA into their host's DNA. Molecular biologists have learned how to copy this trick by splicing into bacterial DNA the genes from other organisms (including humans); they can then use such cloned genes for research or commerce. (Cuts within specific, short sequences of DNA can be made by enzymes called **restriction nucleases**, whose existence in bacteria was first discovered in 1962. By 1975 scientists had purified these enzymes and devised novel ways of using them as tools

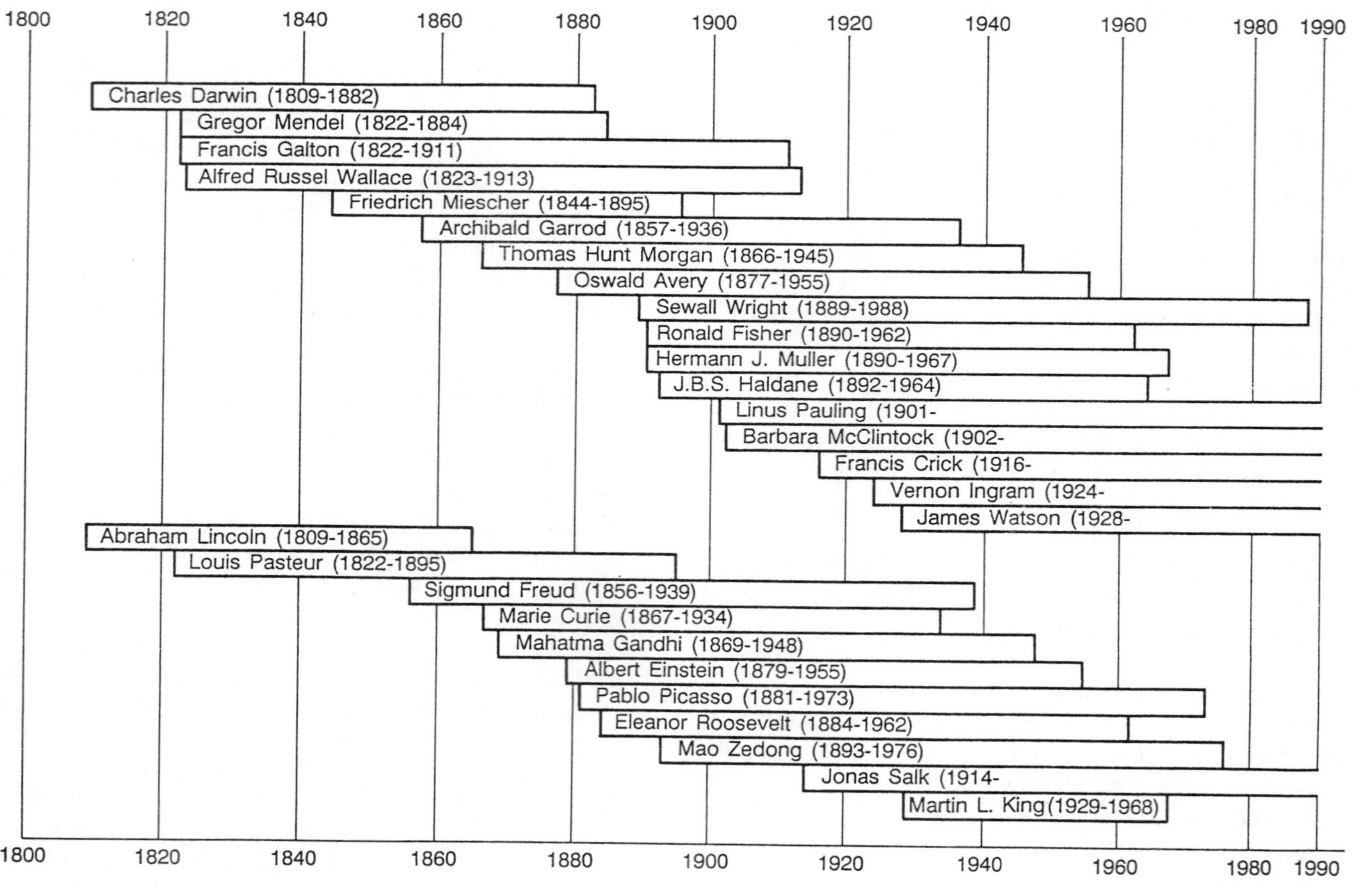

Figure 3 The life spans of most of the scientists mentioned in this chapter (above) and some of their contemporaries (below).

to analyze and restructure DNA molecules.) These sophisticated and powerful **molecular technologies** have revolutionized virtually all branches of genetics (Chapter 15); for example, they allow researchers to isolate and study genes even before their protein products are known (Rowland 1988). Now it is also possible (although still a gargantuan task) to determine the entire nucleotide sequence of human DNA! Even among scientists, this project has aroused considerable debate as to its usefulness and its future impact on human genetics and medicine (Lewin 1986).

Work with bacteria also led to the concept of **regulator genes** that switch other genes on and off. Their presence in higher organisms would help explain how cells become differentiated from each other—how, for example, a nerve cell and a liver cell with identical genes can look and function so differently—one of the major unsolved problems in biology. A breakthrough came in 1984, however, with the discovery that fruit flies, mice, and humans all contain the same gene segment (called a **homeobox**), which during embryonic development regulates the expression of other groups of genes (Chapter 6). Barbara McClintock's work with corn, begun in the 1940s, has provided another possible model for development; she found special classes of mutator genes (called **transposable elements**) that move around within the chromosomes, turning genes on and off. Recent discoveries of similar systems in a wide variety of organisms have aroused a great deal of interest in this phenomenon. Certain other genes, which regulate normal cell growth, have been found to induce cancer when changed into **oncogenes** (Chapter 19).

Population Genetics

Unlike Mendel's and Miescher's works, the entire first printing of Charles Darwin's *The Origin of Species* (1859) sold out in one day. The year before, Darwin and Alfred Wallace had published short essays on evolution, a concept that had been discussed for decades—ever since studies of fossils had pointed to slow but continuous changes in life forms. What Darwin and Wallace proposed was a *mechanism* for such changes: evolution by means of natural selection.

Evidence collected independently, including data from extended world voyages, led both men to suggest that gradual changes in a species occurred over successive generations because better adapted members left more progeny than their less adapted cohorts. The slight variations that made some individuals more successful —at gathering food, or finding shelter, or avoiding predators—were assumed to be inherited. Because these individuals left more progeny, the average characteristics of a population would shift over generations. Ultimately, a species with geographically isolated subgroups might evolve into two or more new species.

The theory was controversial, challenging the belief that all species were created at the same time and never changed thereafter. It also implied that humans, as mere products of evolution, may not differ from other animals with respect to Divine connections.

Unsolved by Darwin were the problems of how the variations were inherited and how they arose; indeed, his ideas on these topics were completely wrong. Mendel, of course, had solved the first problem, but never wrote Darwin about it. An approach to the second problem was suggested in 1900: sudden, small, permanent gene changes (called **mutations**) could provide the variations necessary for evolution.

In the early 1900s mathematical formulae were developed to describe how genes are distributed within large, randomly mating populations (Chapter 20). By the 1930s the Englishmen J. B. S. Haldane and Ronald Fisher and the American Sewall Wright had constructed a sophisticated mathematical framework for population genetics (Chapter 22). Their formulae evaluated the relative importance of selection, mutation, migration, and population size in producing evolutionary changes—and how quickly, or slowly, these changes might come about. Later studies of sickle-cell anemia in African populations illustrated how genetic change can be related to environmental conditions—in this case, why sickle-cell anemia was more frequent in areas where malaria was endemic and less frequent in areas where malaria was absent. More recently, mathematical geneticists also have been studying evolution at the level of molecules (including the theory that some mutations are selectively neutral) and evolution that seems to occur in fits and starts (with relatively rapid changes in certain groups of organisms).

The raw material for evolution is the continuing production of mutations. These rare but permanent changes in genes lead to alterations in the structure or rate of production of their corresponding polypeptides. Science fiction tales notwithstanding, the vast majority of mutations preserved by natural selection lead only to minor changes in body structure or function. Mutations are unpredictable; usually they have no immediate value to the organism, or are downright detrimental. Rarely, however, their presence enables bearers to better adapt to a prevailing or future environment. In the 1920s the American geneticist Hermann Muller developed ingenious ways to measure mutation rates in fruit flies. He showed that, although X rays could dramatically increase this rate, the mutations so produced were no different from those that occurred spontaneously in nature. Later, mutation rates were measured for many species, and other types of radiation as well as many chemicals were found to be mutagenic. Evaluating the genetic effects of both new and old environmental agents continues to be a complex and difficult task.

The Nature–Nurture Concept: Quantitative Genetics

Not all variability within a species is due to gene differences. Environmental differences play a large role, too. Before 1900 the English scientist Francis Galton tried to analyze the relative influence of hereditary and environmental differences on **quantitative traits** like height, mental abilities, and behavioral characteristics. The expression of such traits in a population varies continuously from one extreme to another rather than separating into a few distinct, nonoverlapping forms. Galton laid the foundation for **biometry**, the science of statistics applied to biological observations. He and others at first favored a *blending* type of inheritance for quantitative traits, however, rather than accepting Mendel's *particulate* model. Around 1910 it was shown that the inheritance of quantitative traits such as human skin color could be explained as the joint action of a number of Mendelian genes. Although biometric techniques have led to better understanding and improvement of agricultural practices, neither these nor any other approaches have yet untangled the genetic contributions (nature) from the environmental contributions (nurture) to most human quantitative variability. Controversies about race and IQ center precisely on this problem (Chapter 23). Although studies of human twins have provided some clues, these data are imprecise and limited in usefulness.

Galton coined the term **eugenics** to describe the hoped-for improvement of our own genetic endowment. It reflects a naive optimism about applying methods of artificial selection, so successful in agriculture, to humans: just encourage "superior" individuals to be fruitful and discourage or prevent "inferior" people from reproducing at all. Unfortunately, the eugenics movement attracted demagogues along with idealists, and distortion of the meaning of Darwinian selection contributed to the Holocaust in Nazi Germany. With its emphasis on eliminating the "bad" genes of paupers, criminals, feebleminded individuals, and immigrants, this brand of eugenics fell into disrepute (Kevles 1985).

We now realize that even the most honorable attempts at improving the human gene pool are fraught with difficulty. First of all, the simplest equations of population genetics show that preventing people with rare recessive disorders from reproducing will *not* significantly reduce the frequency of these detrimental genes. Moreover, the criteria used to determine whether reproduction should be encouraged or discouraged are fuzzy at best and highly susceptible to social and political mischief.

But human genetics can be employed humanely at the personal level—as increasingly sophisticated prenatal testing and genetic counseling are able to avert family tragedy or reassure those individuals who turn out not to be at risk for certain hereditary disorders (Chapter 25). Great caution, care, and sensitivity are required, however, when considering or carrying out the large-scale testing of populations that are known to be at increased risk for certain conditions. Even for those who prefer no testing at all, the problem of just maintaining our current genetic endowment remains—when, for example, medical advances preserve detrimental genes that would otherwise be eliminated by the death of these individuals. The possible long-term genetic and social consequences of medical practices will be debated for years to come.

HUMAN GENETICS LITERATURE

Before proceeding further, we should note how readers will be introduced to the literature of human genetics. Pertinent articles and books are mentioned throughout the text, and at the end of each chapter we describe further reading chosen for general interest and clarity of writing. All citations are by author and date of publication. A single alphabetical listing of these works, with bibliographic information, appears at the end of the book. In addition to these specific references, we suggest that the reader browse through library holdings in the following Library of Congress classifications:

HQ 753, Eugenics
QH 431, General and Human Genetics
RB 155, Medical Genetics
RG 626, Birth Defects

Books on the genetics of many species and on the role of genetics in society, at all levels of sophistication, can be found here. There are also numerous reference books and research journals, some of which are devoted primarily to human genetics.

Readings appropriate to Chapter 1 include Nossal's (1975) perceptive description of medical scientists and the world in which they live. Allen (1985a,b), Sapp (1987), Goodfield (1981), Merton (1973), and Zuckerman (1977) also offer interesting approaches to the sociology and history of science. Among the best histories of genetics are those by Dunn (1965), Sturtevant (1965), Judson (1979), and Portugal and Cohen (1977). For biographies of Mendel, see Orel (1984) and Iltis (1924); Sandler and Sandler (1984) have also written on his life and work. Stern's (1965) comments on Mendel's contribution to human genetics are included in an issue of the *Proceedings of the American Philosophical Society* devoted to the significance of Mendel's work. One translation of Mendel's paper is in Stern and Sherwood (1966), which also reprints an interesting statistical analysis of his experimental results. Rosenberg (1984) and Bajema (1976) collect many fascinating articles on eugenics; Boyer (1963) and Schull and Chakraborty (1979) reprint research papers important to the early development of human genetics; Corwin and Jenkins (1976), Srb et al. (1970), Peters (1959), and L. Levine (1971) do the same for general genetics. Articles on the history and development of human genetics include Dunn (1962), McKusick (1975), Motulsky (1978), Jacobs (1982), Caskey (1986), and Bodmer (1986a). Essays on the definition or concept of the gene include those by Stadler (1954), Muller (1947, 1965), and Falk (1984).

Chapter 2 Rules of Inheritance for Single Genes

Autosomal Genes: Mendel's First Law
Mendel's Experiments
Mendel's First Law
Modern Nomenclature
A Checkerboard Method for Predicting Offspring
Progeny Testing and Matings
An Example from Humans

X-Linked Genes
Sex Determination
Genes on Sex Chromosomes
Color Blindness
Hemophilia
X-Linked Lethal Genes and Shifts in Sex Ratio

Y-Linked Genes

Mitochondrial Genes

> *So the crucial questions were, what* is *inherited? and how is it inherited? According to Napp [Abbot of the Augustinian monastery at Brunn], the answers lay in physiological research. He later remarked . . . that a further unknown was the role of chance.*
>
> *Vitezslav Orel (1984)*
>
> *It requires a good deal of courage indeed to undertake such a far-reaching task; however, this seems to be the one correct way of finally reaching the solution to a question whose significance for the evolutionary history of organic forms must not be underestimated.*
>
> *Gregor Mendel (1865)*

How simple Mendel's solution to the problem of heredity seems today, and how surprising that nobody understood his carefully reasoned analysis in 1865. His work, which marked the beginning of quantitative biology, still stands as a model of brilliant experimental design. We cover the first of Mendel's laws in this chapter, and the second in Chapter 11.

AUTOSOMAL GENES: MENDEL'S FIRST LAW

Drawing upon his extensive training in mathematics (including probability), physics, and chemistry, Mendel was among the first to apply statistical principles to biological phenomena. He noted the different classes of offspring that arose from specific matings and how many individuals occurred in each class. He observed ratios among these classes and showed that they fit the terms of simple algebraic expansions. He recognized the value of large-scale experimentation: "The greater the numbers, the more are merely chance effects eliminated" (Mendel 1865).

Mendel designed experiments in such a way that they could test a specific hypothesis about inheritance. Rather than worrying about how *all* traits are inherited in *all* organisms, he concentrated on the inheritance of easily distinguishable alternative forms of a *few* traits in a *single* organism, the garden pea. This plant was chosen for three reasons: (1) It exhibited a number of clearly defined variations such as tall versus short plants, round seeds versus wrinkled seeds, inflated pods versus constricted pods, and white seed coats versus gray seed coats. (2) Its flower structure, tightly enclosing both male and female parts, normally guarantees self-fertilization. But cross-fertilization can be arranged by removing the male (staminate) parts from each flower before the pollen matures and later dusting the female (pistillate) parts with pollen from a different line. (3) The hybrids between two lines are as fertile as the parental lines, so large numbers of progeny can be studied in every generation.

Mendel started with **pure-breeding lines**, wherein

all the progeny of self-fertilization resemble the parents, generation after generation. A pure-breeding line has no genetic variation within it. When making crosses between two different lines, he always set up **reciprocal matings**: females of line A × males of line B, as well as males of line A × females of line B. In this way he could determine whether the inheritance of a trait was influenced by the sex of the parents. Mendel was careful to keep the plant generations separated from each other and to study the descendants of each hybrid line through at least four generations of self-fertilization. He apparently realized the importance of keeping environmental conditions as constant as possible. The necessity for every one of these precautions is obvious today, but Mendel was the first to recognize them.

Mendel's Experiments

With each of seven pairs of traits, he performed the same series of experiments and got the same kinds of results. For the sake of brevity, we present data for only one pair: *tall* plant (about 200 cm, or 6.5 ft) versus *short* plant (about 30 cm, or 1 ft). (We now know that this difference in height is due to the presence or absence of a growth hormone whose synthesis is under genetic control.) All the crosses are diagramed in Figure 1. Each generation represents 1 year's work. First, Mendel made a large number of reciprocal matings between a pure-breeding tall line and a pure-breeding short line. These constitute the **parental generation (P)**. Since the parents differed from each other in just one characteristic, the mating is called a **monohybrid cross**. The progeny of these matings, called **F_1 hybrids**, were all tall. Thus, he reasoned, the tall plant character must be **dominant** in expression to the short plant character, which was designated **recessive**. For the other six pairs of traits, likewise, only one of the alternative forms was expressed in the F_1 hybrid.

Next, Mendel allowed the F_1 hybrid plants to self-fertilize, collected the resulting seeds, and sowed them the following year. Among this **F_2** generation, 787 plants were tall and 277 were short, a ratio of about 3 to 1. Comparable F_2 ratios were obtained for the other six pairs of traits. "No transitional forms were observed in any experiments," Mendel noted; that is, no plants of medium height, or with any characteristic intermediate between the two parental lines, ever appeared.

The third step was to again allow self-fertilization, sow the seeds from individual F_2 plants and look at the F_3 generation. The plants with the recessive trait were always found to be pure-breeding: short plants yielded

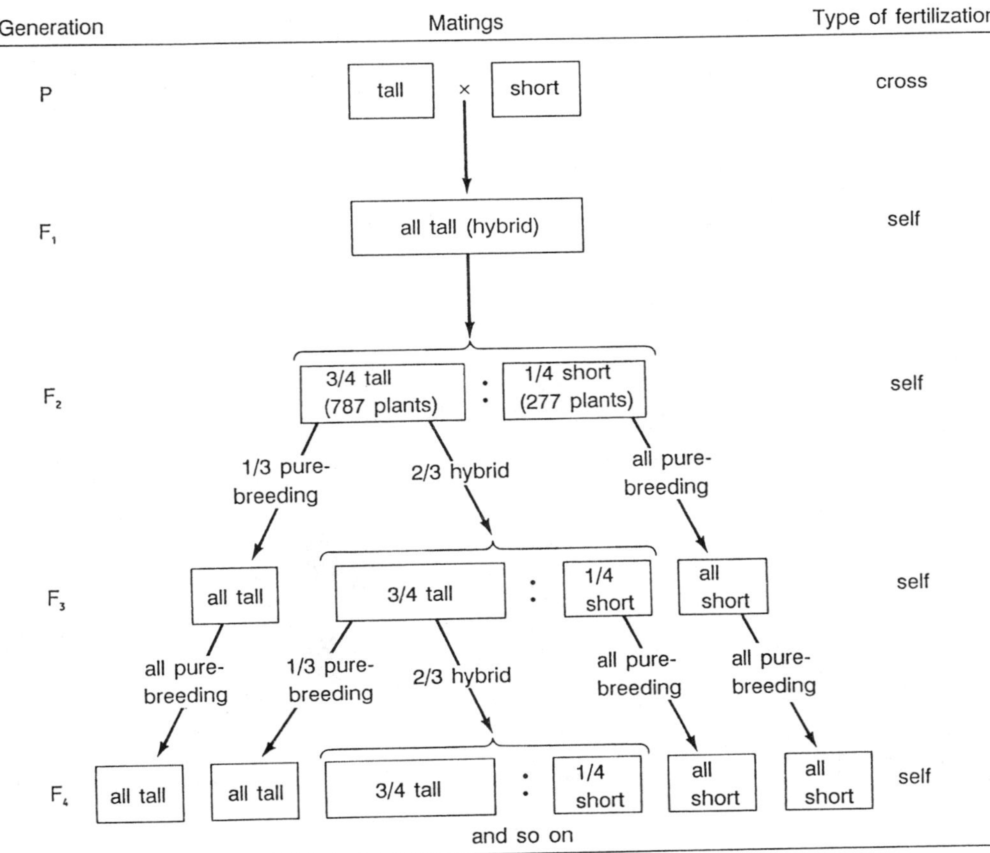

Figure 1 The system of matings used by Mendel for one of his monohybrid crosses. (The use of the term *hybrid* does not imply a cross between two different species, as it often does. Here it means a cross between two previously isolated and pure-breeding lines of the *same* species.)

only short progeny. But the progeny of the F_2 plants with the dominant trait were of two types. About 1/3 of the F_2 tall plants were pure-breeding, giving rise to only tall plants after self-fertilization. But the remaining 2/3 of the tall F_2 plants, after self-fertilization, yielded both tall and short progeny in a ratio of 3 to 1. Thus, Mendel concluded, they must be hybrids like the F_1. As shown in Figure 1, he continued this procedure for several generations and found that this pattern repeated in all cases.

These experimental results did not fit prevailing ideas about heredity, which held that the progeny of two different lines would all be intermediate in appearance as a result of a permanent blending of their characteristics. According to this hypothesis, the F_1 plants should have been all of medium height, rather than all tall. It also followed that the original parental traits (in this case, tall and short) should never be recovered in future generations, having contaminated each other in the hybrids. This outcome was not observed either. Mendel found that each F_1 hybrid line gave rise to F_2 progeny resembling one or the other of the two original parents (i.e., tall or short) but never intermediate in appearance. Thus, no blending or contamination of the two alternative traits occurred while they were combined in the hybrids.

Mendel's First Law

Mendel's interpretation of these data is called the **law of segregation**. From the ratios described above, he reasoned that each observable trait, e.g., plant height, is determined by a pair of cellular factors. During the formation of **gametes** (reproductive cells), *the two members of each pair of factors are separated (segregated),* so an egg cell or a pollen grain contains one or the other, but not both, of the original factors.*

Because reciprocal crosses gave the same results, Mendel assumed that the male and female parents contribute equally to the formation of offspring. Thus, as shown in Figure 2, each F_1 hybrid inherits one "tall" factor (A) from its tall parent plus one "short" factor (a) from its short parent, and its factor composition can be designated as A/a. Although the recessive a factor is not expressed in the A/a hybrid, it remains unchanged and may be expressed in future generations.

When the hybrid A/a plants form gametes, half of them will contain the A factor and the other half will carry the a factor. "It remains, therefore, purely a matter of chance

*Although chromosomes had not yet been discovered, the cell theory propounded by Schleiden and Schwann in 1838–39 was fully accepted in Mendel's time, as were reports that eggs and sperm were single cells. Cytological proof that an egg is fertilized by just *one* sperm did not come until 1879, however.

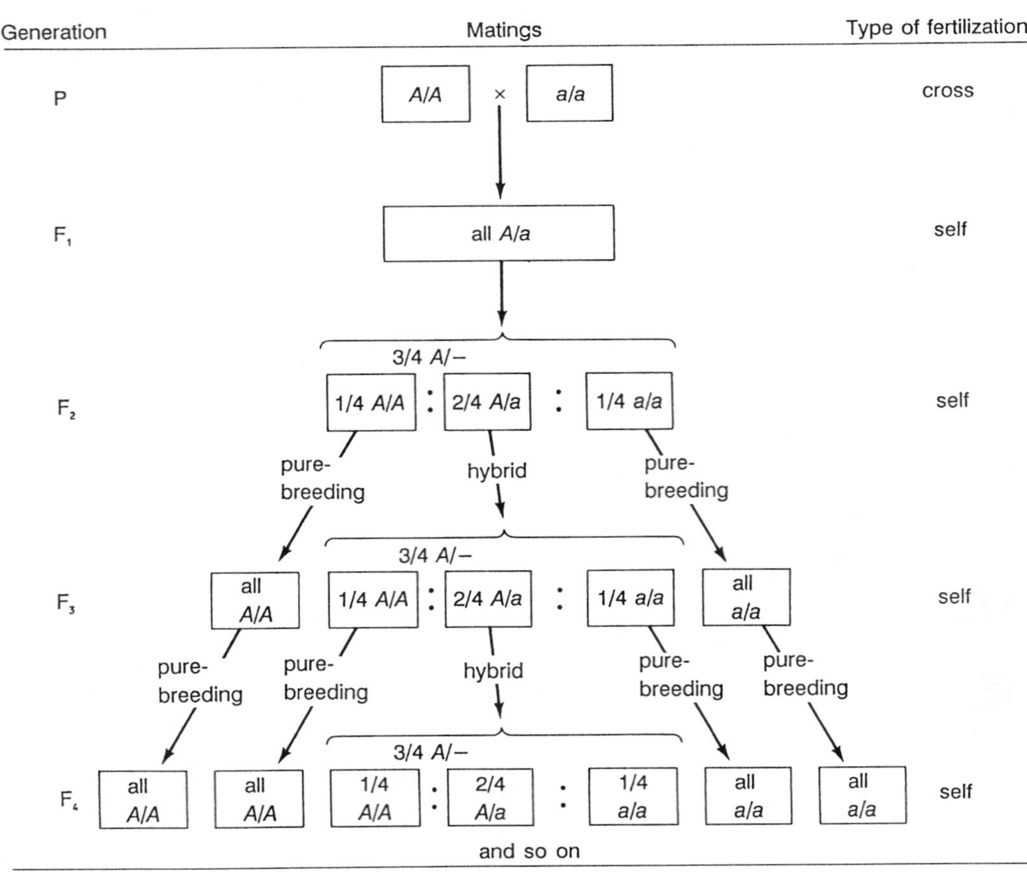

Figure 2 Genetic interpretation of Mendel's monohybrid crosses. Compare with Figure 1.

which of the two sorts of pollen will become united with each separate egg cell. According, however, to the law of probability, it will always happen, on the average of many cases, that each pollen form, *A* and *a*, will unite equally often with each egg cell form, *A* and *a*" (Mendel 1865).

The four equally likely combinations are as follows:

	Type of egg cell		Type of pollen grain	Type of seed and plant		
1/4	*A*	×	*A*	→ *A/A*		3/4 tall
1/4	*A*	×	*a*	→ *A/a*	1/2 *A/a*	
1/4	*a*	×	*A*	→ *a/A*		
1/4	*a*	×	*a*	→ *a/a*		1/4 short

Hence the recovery, in the F_2 generation, of 1/4 *A/A* : 1/2 *A/a* : 1/4 *a/a* plants. Since the tall trait is dominant, however, it is impossible to distinguish the *A/A* plants from the *A/a* plants just by looking at them. What one sees is 3/4 tall plants and 1/4 short plants.

Modern Nomenclature

Several terms were coined after Mendel's time to describe these situations. The factors *A* and *a* are **alleles** of each other; they are alternative forms of a **gene** that influences plant height. We can thus refer to *A* as the dominant allele of the gene determining plant height, and to *a* as its recessive allele. The terms *gene* and *allele* are sometimes used interchangeably, but in most contexts only one term is correct. It is proper to say that we are discussing the gene (not the allele) for plant height that exists chemically in different allelic forms, one form determining taller growth than the other. (Most genes, by the way, exist in more than two allelic forms, but a given individual carries only two alleles.) In later chapters we discuss the nature of the chemical differences between alleles of a particular gene and how they cause the differences that are observed.

Genotype refers to the precise allelic composition of a cell, such as *A/A* or *A/a* or *a/a*. **Phenotype** refers to what is actually observed in an organism, that is, tall plants or short plants. The phenotype *short* is due to the genotype *a/a*. The phenotype *tall* is due to either the genotype *A/A* or the genotype *A/a*. A **homozygous** genotype is one in which the two alleles are the same (*A/A* or *a/a*); in a **heterozygous** genotype, the two alleles are different (*A/a*). When we are not sure whether a phenotypically tall plant is genotypically homozygous (*A/A*) or heterozygous (*A/a*), we write the genotype *A/*−, the dash representing an allele that could be either *A* or *a*. The designation *A/*− can also be used when we are not interested in distinguishing between the dominant homozygote and the heterozygote.

Genes are present on **chromosomes** inside the nucleus of a cell in higher organisms (Chapter 4). Mendel was unaware of the existence of chromosomes. But just as Mendelian factors (alleles) are paired, so too are chromosomes. The two members of a pair of corresponding chromosomes are known as **homologues**. In fact, the two alleles of the same gene occupy the same relative position on homologous chromosomes. The particular site or location where a gene is found along the length of a chromosome is the **locus** of that gene. For example, in a heterozygous tall plant, *A/a*, the allele *A* occupies a particular locus on a particular chromosome. At a corresponding position on the homologous chromosome is found its allele, *a*. In humans, and in many other organisms with separate sexes (unlike peas), all homologous pairs of chromosomes are called **autosomes**, except for one pair called **sex chromosomes.**

A Checkerboard Method for Predicting Offspring

This procedure, although somewhat cumbersome, helps students to visualize simple crosses involving just one or two gene loci. (Better techniques of problem solving, using the rules of probability, will be presented in Chapter 10.) The steps for this method (see Figure 3) are as follows:

1. Label the rows of a checkerboard with the various kinds of eggs that the female parent can make. There should be one row for each *different* kind of egg. The labeling of rows should include, in addition, the fraction represented by each egg type—what we call the *gamete fraction*—and these egg gamete fractions should together add up to 1. If the total is more than 1, it is likely that an egg type has been written down more than once.

2. Label the columns of the checkerboard in a similar fashion with the different kinds of male gametes. Include the gamete fractions of each different kind. If the number of kinds of pollen or sperm is different from the number of kinds of eggs, you will end up with a rectangle rather than a square.

3. Fill in the body of the checkerboard with the *offspring genotypes* by combining the headings of the corresponding row and column. The fraction of each offspring genotype is the *product* of the fractions heading the corresponding row and column (see also Chapters 10 and 11). In Figure 3, for example, 1/2 *A* × 1/2 *A* gives 1/4 *A/A* for the upper left-hand space.

Summing the two identical genotypes on the ascending diagonal squares (lower left + upper right) gives a total of 2/4 *A/a*. Phenotypically, *A/A* and *A/a* are tall (3/4), and *a/a* is short (1/4). In this checkerboard the egg

	pollen grains: 1/2 (A)	pollen grains: 1/2 (a)	sum of column fractions = 1
eggs: 1/2 (A)	1/4 A/A	1/4 A/a	
eggs: 1/2 (a)	1/4 A/a	1/4 a/a	
sum of row fractions = 1			

Figure 3 Checkerboard depicting Mendel's F_1 cross A/a ♀ × A/a ♂. In the squares of the body of the checkerboard (within the bold lines) appear the genotypes of the progeny generation, i.e., the F_2, and the proportion of each genotype (products of the gamete fractions).

and pollen types are the same, because the cross was by self-fertilization, but with other types of crosses this will probably not be true. Also, in this case the fractions heading the rows (or columns) are both 1/2, reflecting Mendel's law of segregation; the alleles separate randomly into the gametes. Below and in Chapter 11 we describe situations in which the different kinds of gametes are not formed in equal proportions.

Progeny Testing and Matings

Mendel was able to determine the genotypes of the tall F_2 plants by **progeny testing**. He took 10 seeds from each self-fertilized F_2 plant, grew them, and observed the F_3 phenotypes (Figure 2). Homozygous tall F_2 plants (A/A) produce only A gametes; thus, when self-fertilized, they yield only tall plants. But heterozygous tall F_2 plants (A/a) produce equal numbers of both A and a gametes (1/2 A and 1/2 a). Like their hybrid F_1 parents, they yield both tall and short progeny after self-fertilization. What Mendel found among the tall F_2 plants, by these breeding tests, was a 1:2 ratio of homozygotes to heterozygotes. This he expressed as 1/3 pure-breeding to 2/3 hybrid. (Recall that a ratio is converted into fractions by dividing each member of the ratio by the sum of the members. This operation is reviewed in Appendix 1.)

Actually, allowing a tall plant to self-fertilize is not the most reliable way of determining whether it is homozygous or heterozygous. To maximize the probability of detecting the recessive allele in a heterozygote, one should cross each tall plant with a short plant ($A/- \times a/a$). As shown in Figure 4A, heterozygous tall plants, when crossed to short plants, will yield a ratio of 1 tall to 1 short progeny, rather than the ratio of 3 tall to 1 short plant recovered after self-fertilization (Figure 3). Note in Figure 4B and C that homozygous tall plants produce all tall progeny, whether mated to short plants or self-fertilized. Note too (in all three checkerboards here) that *when all the gametes from a parent are the same, they are represented by the fraction 1, and by just one row or one column in the checkerboard.*

Consider that Mendel planted only 10 seeds from each self-fertilized F_2 plant. From heterozygous tall plants $A/a \times A/a$), he could expect to recover 2 or 3 short plants among the progeny ($1/4 \times 10 = 2.5$). Of course, as a result of random variation, he observed sometimes more and sometimes fewer than this number of short plants. The occurrence of even 1 short plant among the 10 progeny would be sufficient to identify its parent as a heterozygote. But what if, by chance, $A/a \times A/a$ yielded 10 tall and 0 short progeny, instead of the expected 7–8 tall and 2–3 short plants? In such a case the parent plant would *mistakenly* be called a homozygote rather than a heterozygote. One can even calculate how often this might happen (Chapter 10), and it is indeed likely that Mendel missed a few heterozygotes this way.

Now, if 10 seeds from the cross $A/a \times a/a$ are planted, we would expect to find 5 short plants among the progeny ($10 \times 1/2 = 5$). Here too, one would observe sometimes more and sometimes fewer than 5 short plants because of random fluctuation. But common sense dictates, and simple calculations prove, that the occurrence

(A) *Heterozygous* tall × short: $A/a \times a/a$

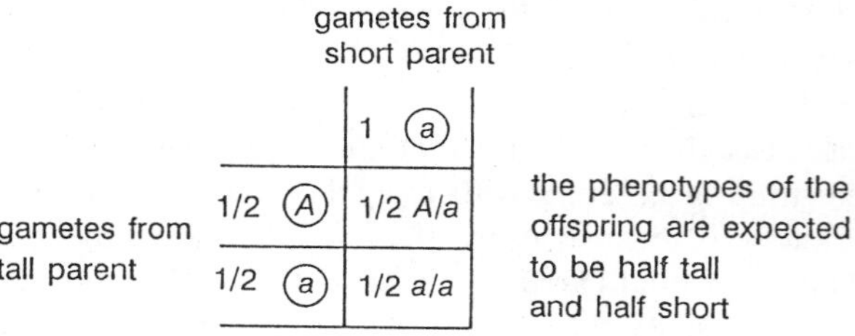

(B) *Homozygous* tall × short: $A/A \times a/a$

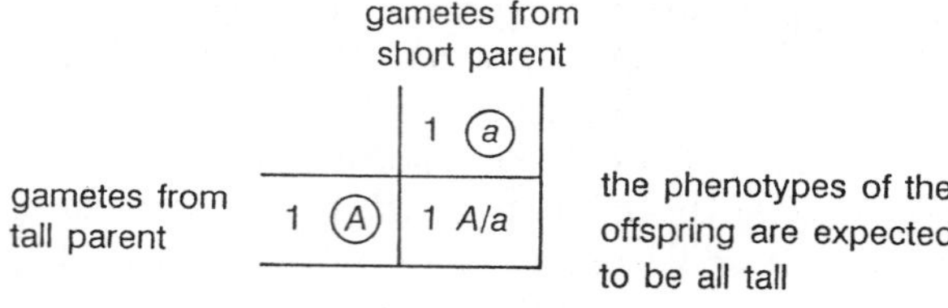

(C) *Homozygous* tall self-fertilized: $A/A \times A/A$

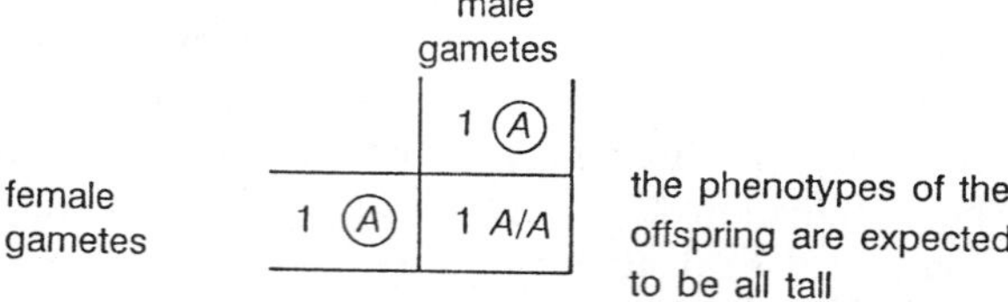

Figure 4 Checkerboards for one gene, two alleles, with A dominant to a. The phenotype of $A/-$ is tall and of a/a, short. (The genotypes of progeny arising from the self-fertilization of a heterozygous tall plant are shown in Figure 3.)

Table 1 Matings for one pair of alleles, *A* and *a*

*Parental genotypes**	*Genotypic ratios among progeny*	*Phenotypic ratios†*
1. *A/A* × *A/A*	all *A/A*	
2. *A/a* × *A/a*	1/4 *A/A*:1/2 *A/a*:1/4 *a/a*	3/4 *A/*–:1/4 *a/a*
3. *a/a* × *a/a*	all *a/a*	
4. *A/A* × *A/a*	1/2 *A/A*:1/2 *A/a*	all *A/*–
5. *a/a* × *A/a*	1/2 *a/a*:1/2 *A/a*	
6. *A/A* × *a/a*	all *A/a*	

*In the first three, the genotypes of the two parents are the same. In the last three, the genotypes of the two parents are different, but here we do not care to specify which sex has which genotype. If reciprocal matings were considered separately, then there would be a total of nine different matings, rather than the six listed here.

†Where different from genotypic ratios and assuming *A* is dominant to *a*.

of 10 tall and 0 short plants from this cross is much less likely than in the case of self-fertilization.

For this reason geneticists will, whenever possible, cross a dominant phenotype to a recessive phenotype in order to determine whether the former is homozygous or heterozygous. Such a mating is called a **testcross**.

A testcross is one of the 6 matings possible for a single pair of alleles, *A* and *a*. That is, there are 6 different ways that the 3 genotypes (*A/A*, *A/a* and *a/a*) can be combined in pairs, as shown in Table 1. In the course of describing Mendel's experiments, all but one of these has been considered. We strongly recommend that you verify for yourself the outcome of each of these matings, using checkerboards if necessary, and then simply commit them to memory as a future time-saving device. They are absolutely basic.

Mendel did many additional experiments to see what happens when more than one pair of characters is considered simultaneously, for example the cross *A/a B/b* × *A/a B/b*. A discussion of these studies, and the resulting law of independent assortment that Mendel proposed, will be deferred until Chapter 11.

An Example from Humans

Mendel's law of segregation is fully applicable to human autosomal genes. **Alkaptonuria**, for example, is one of the "charter group" of heritable metabolic diseases first proposed as Mendelian recessives in the early 1900s (Garrod 1902, 1909). It was correctly suggested that the basic defect in this relatively mild condition is the absence of a specific enzyme that normally converts a substance called *homogentisic acid* to something else (Chapter 16). Homogentisic acid, having nowhere to go biochemically, then accumulates and is excreted in the urine of affected individuals. Because it turns the urine black after exposure to air, alkaptonuria is easily diagnosed. In addition to this striking symptom, observed from infancy on, dark pigmentation of certain connective tissues (the outer layer of the eye, the cartilage of the external ear) appears in young adults. Older patients suffer from arthritis and perhaps exhibit a higher than usual incidence of heart disease (La Du 1978).

The two alleles of this gene can be symbolized *K* and *k*, and the genotypes and phenotypes are as follows:

Genotype	**Phenotype**
K/K	Normal
K/k	Normal
k/k	Affected with alkaptonuria

The disease is rare. Affected children are almost always born to normal parents who are heterozygous (*K/k* × *K/k*) but do not usually know they carry the unusual gene until they have produced a child with alkaptonuria. By Mendelian rules, we can predict that 1/4 of any future children will be expected to be affected. While restoration of the missing enzyme is not yet possible, it has been reported that large doses of ascorbic acid (vitamin C) may inhibit the darkening of connective tissues.

X-LINKED GENES

Mendel found no differences among the progeny from reciprocal matings. The same phenotypic classes occurred, in the same proportions, whether a given trait was inherited through the male or through the female parent. But Mendel's recognition of the *possibility* of such differences is further tribute to his insight. Pea plants, after all, are not distinguishable on the basis of sex. Both male (staminate) and female (pistillate) structures appear in the same organism, a situation that characterizes some animals as well.

Sex Determination

In other plant and animal species, however, the sexes are separate, and the mode of inheritance of some traits does differ with reciprocal matings. For such organisms, the biological basis of sex determination was a mystery until the turn of the century, when research on insects and other animals showed that sex is, with few exceptions, an inherited trait. Specifically implicated was a single **sex chromosome**, called **X**, or two sex chromosomes, called **X** and **Y**. All the remaining chromosomes were designated as autosomes. Humans, for example, have an X–Y pair of sex chromosomes and 22 pairs of autosomes.

How sex is determined varies from one phylogenetic group to another, but a common pattern is the one found in humans. Individuals with two X chromosomes are female, and those with one X chromosome and one Y chromosome are male. As is the case with many organisms, human X and Y chromosomes are structurally and functionally quite different. The medium-sized X is about 2.5 times as long as the tiny Y chromosome; the two also look quite different when stained and viewed under a microscope, because of differences in chromosomal organization that will be discussed more fully in Chapter 4.

Nevertheless, these two chromosomes do possess a short **homologous region** (Goodfellow et al. 1985); and during the cell division that leads to the production of gametes, they form a homologous pair that obeys Mendel's first law. Among the sperm produced by a male, half carry an X chromosome and the other half carry a Y chromosome. All the eggs produced by a female, on the other hand, contain an X chromosome. So it is the father, rather than the mother, whose gamete determines the sex of their offspring. If an X-bearing sperm fertilizes the egg, a female is produced; the union of a Y-bearing sperm with an egg gives rise to a male. This system, diagramed in Figure 5, perpetuates a presumed 1:1 sex ratio at the time of fertilization, generation after generation. Actually, the sex ratio at birth differs slightly from 1:1, perhaps as a result of a differential mortality of the sexes during intrauterine life.

We should emphasize, however, that sexual development is an immensely complex process controlled by many genes and susceptible to alteration by environmental conditions at any stage from before birth through adulthood. In many species (including humans) some sexual characteristics can be modified, or perhaps even reversed, by hormonal treatments, injury, surgery, or many other factors. So chromosomal sex and phenotypic sex are usually, but not always, the same (Chapter 6). Anyway, what concerns us here is the behavior of the sex chromosomes and the genes they carry rather than the development of sex per se.

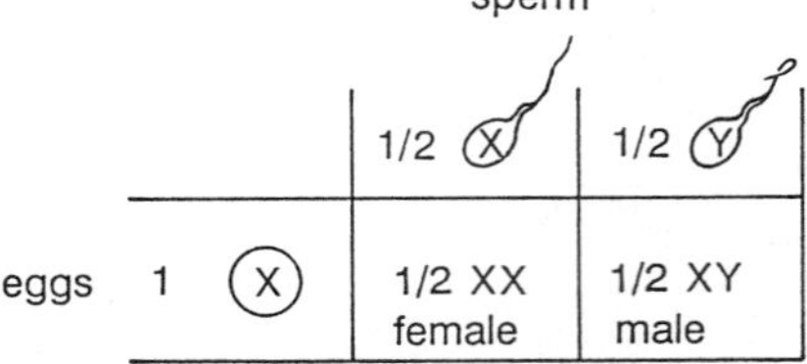

Figure 5 Checkerboard for the chromosomal sex-determining system that operates in humans and many other organisms. All the eggs from an XX female contain an X chromosome; the sperm from an XY male are 1/2 X and 1/2 Y. Thus, a 1:1 sex ratio at fertilization is maintained generation after generation. [All eggs carry an X chromosome, so the corresponding fraction is 1. Since we are not concerned here with genes on the X chromosome (for which the female may be heterozygous), only one row is needed to depict eggs.]

Genes on Sex Chromosomes

The X chromosomes of most species bear roughly as many genes as do autosomes of similar size. These genes are said to be **X-linked** or *sex-linked*, because their inheritance is coupled with that of the X chromosome. Keep in mind, however, that X-linked genes are no different in kind from those found on autosomes. Well over 100 genes have been assigned with certainty to the human X chromosome. They affect characteristics as diverse as blood groups, vision, hearing, the nervous system, the muscular system, teeth, skin, and glucose metabolism. Most X-linked genes have nothing to do with sexual differentiation. Conversely, there are a number of *autosomal* genes known to play a role in sexual development. Thus, *X-linkage is a matter of gene geography, not gene function*.

Color Blindness

The term *color blindness* is a misnomer, because most people with this condition see a range of colors — especially among the blues and yellows. What we really mean is *partial* color blindness or color *defectiveness*, which is mainly a problem of discriminating between reds and greens. Over two centuries ago this condition was known to run in families, including those of some well-known scientists (Dalton 1798). Shortly after the rediscovery of Mendel's paper, and just one year after the assignment of a white-eye gene to the X chromosome of fruit flies, E. B. Wilson (1911) suggested that the gene for color blindness in humans also resides on the X chromosome.*

Later it was found that *two* genes, called *protan* and *deutan*, are involved in red–green color blindness; they

*There are a few truly color blind people, who see only in black and white (and very poorly during the day) because of the absence or nonfunctioning of all color-sensitive cells in the retina. These conditions are extremely rare, however, and are inherited as autosomal recessives.

are located near one end of the X chromosome (Keats 1983; Kalmus 1965). Then in the mid-1980s, investigators at Stanford University (Nathans et al. 1986a,b)—acting on a hunch and using the molecular techniques described in Chapter 15—performed an amazing feat: isolating these two genes and determining their entire DNA sequences!* It turns out that they are 98% identical and lie very close to each other on the chromosome. This similarity, plus the finding that some normal males have more than one copy of the deutan gene, confirmed old theories of color vision and provided new evidence for the widely held idea that many genes evolve through duplication and subsequent divergence (Botstein 1986; Mollon 1986).

Each of the two X-linked color vision loci controls the production of a light-absorbing protein (**opsin**): one making up part of the **red-sensitive** color vision pigment, and the other comprising part of the **green-sensitive** pigment. These two pigment types occur separately in color-sensitive cells (cones) within the retina of the eye, as does a third, blue-sensitive, pigment (not under consideration here). Each of the myriad hues that we see is generated or matched by the intermixture of just three components: the excitation of red-sensitive, green-sensitive, and/or blue-sensitive pigments in the retina. For example, the normal perception of yellow by our eyes results from the joint stimulation of red and green cones. (*Note:* Although their peaks differ, the ranges of wavelengths absorbed by red- and green-sensitive pigments overlap considerably near the middle of the spectrum. For more information, see Rushton [1975] or Hurvich [1981].)

TYPES OF DEFECTS. Four types of X-linked color vision defects are known: at each of the two loci, one mutant allele is associated with an anomalous pigment and the other mutant allele with a missing pigment. (Multiple alleles of a gene are discussed further in Chapter 10.) For both loci, the order of dominance is normal pigment over either mutant and anomalous pigment over absence of pigment.

Altogether, about 8% of males of western European extraction are color blind. The most common defect, **deuteranomaly** (which we call **green-shift**), occurs in about 5% of White males. (For reasons that will become clear later, all these conditions are rare among females.) Affected individuals possess the three pigment types; but their green-sensitive pigment is anomalous, with an absorption spectrum that is shifted toward the longer wavelengths characteristic of the normal red-sensitive pigment. Thus, in certain ranges they see fewer hues than normal; reds are perceived as reddish browns, brighter greens as tans, and olive greens as indistinguishable from browns. Asked to match a yellow light by adjusting the relative proportions of red and green in a diagnostic apparatus called an **anomaloscope**, affected individuals will choose a mixture that looks greenish to normal eyes.

People with **protanomaly** (which we call **red-shift**) also possess three color vision pigments—but here the red one is defective, with an absorption spectrum shifted toward the somewhat shorter wavelengths characteristic of the normal green-sensitive pigment. These individuals, like those with deuteranomaly, can see reds, greens, yellows, blues, blacks, and whites; but they likewise perceive fewer colors and are confused by browns and olive greens. Asked to match a yellow light in the anomaloscope, protanomalous individuals will select a red-green mix that appears reddish to normal eyes. The frequency of red-shifted males among West Europeans is about 1%.

The other two X-linked color defects (each present in about 1% of Caucasian males) are more serious, involving the total absence of a pigment: individuals with **deuteranopia** lack the green-sensitive pigment, whereas those with **protanopia** lack the red-sensitive one. With two rather than three color vision pigments, neither type can tell reds from greens at all, nor can they distinguish either of these from yellow. Basically, what they see are yellows, blues, blacks, whites, and grays. When matching a yellow light in an anomaloscope, both types will indiscriminately select a red or a green light; because protanopes see red so dimly, however, they will choose a red that is about 10 times brighter than that picked by deuteranopes. A simpler but much less exact diagnostic method uses books of colored plates. Those designed by Ishihara (1968) consist of multicolored dots forming a gray background, within which are "embedded" figures that may or may not be recognized by various phenotypes.

GENETICS. Recall that a female inherits two sets of X-linked genes, one from her mother and one from her father. Since the green-shifted allele (call it *g*) is recessive to the normal *G* allele, a female must be homozygous *g*/*g* to express the trait. Heterozygous *G*/*g* females, with one normal allele and one green-shifted allele, are called **carriers**, and their own color vision is usually normal or near normal. Homozygous *G*/*G* females have normal color vision.

A male's single X chromosome can only be inherited from his mother. Whatever genes are carried on his X will be expressed, whether dominant or recessive, because his other sex chromosome is a gene-poor Y received from the father. Thus we cannot use the terms homozygous or heterozygous with respect to X-linked traits in a male. Rather, he is said to be **hemizygous** (Greek *hemi*, "half") for *G* (i.e., phenotypically normal) or hemizygous for *g* (i.e., phenotypically green-shifted). The corresponding genotypic description is *G*/(Y) or *g*/(Y)

*Nathans and his colleagues also isolated two autosomally inherited pigment genes (blue-sensitive opsin from chromosome 7 and rhodopsin from chromosome 3), but these do not concern us here.

Compare the results of reciprocal crosses between normal and green-shifted individuals. First consider the most common such case, a homozygous normal mother and a green-shifted father (Figure 6A). All the mother's eggs (fraction = 1) have a normal *G* gene, so all her sons are normal. The father produces two kinds of sperm; one kind carries the Y chromosome, which is passed on to sons, and the other kind carries the X chromosome with the *g* gene. Thus all daughters will be carriers.

What can be expected from matings between heterozygous normal mothers and green-shifted fathers? As shown in Figure 6B, her eggs are of two types, *G* and *g*. His sperm are of two types, *g* and Y. This leads to four genotypic combinations that are also phenotypically distinct: 1/4 carrier females, 1/4 green-shifted females, 1/4 normal males, and 1/4 green-shifted males. Among the offspring of either sex, half will be normal and half will be green-shifted.

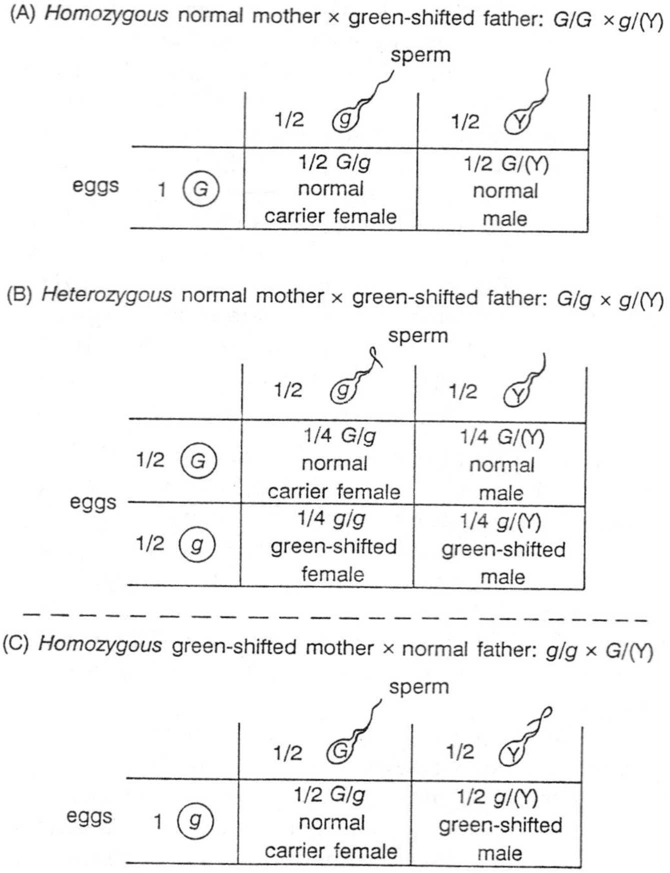

(A) *Homozygous* normal mother × green-shifted father: *G/G* × *g/*(Y)

		sperm 1/2 (g)	sperm 1/2 (Y)
eggs	1 (G)	1/2 *G/g* normal carrier female	1/2 *G/*(Y) normal male

(B) *Heterozygous* normal mother × green-shifted father: *G/g* × *g/*(Y)

		sperm 1/2 (g)	sperm 1/2 (Y)
eggs	1/2 (G)	1/4 *G/g* normal carrier female	1/4 *G/*(Y) normal male
	1/2 (g)	1/4 *g/g* green-shifted female	1/4 *g/*(Y) green-shifted male

(C) *Homozygous* green-shifted mother × normal father: *g/g* × *G/*(Y)

		sperm 1/2 (G)	sperm 1/2 (Y)
eggs	1 (g)	1/2 *G/g* normal carrier female	1/2 *g/*(Y) green-shifted male

Figure 6 Reciprocal matings between normal and green-shifted individuals. Two types of matings, (A) and (B), are possible for a normal mother and a green-shifted father. Only one type of mating, (C), is possible for a green-shifted mother and a normal father.

Finally, consider the reverse of the preceding, in which the father is normal but the mother is green-shifted (Figure 6C). The father must be *G/*(Y) and the mother must be *g/g*. The mother's eggs all carry *g*; of the father's sperm, half carry *G* and half carry the Y chromosome. Thus all sons will be green-shifted and all daughters will be carriers. Note that, in this case, the pattern of color vision in the male and female parents is "criss-crossed" in their offspring.

Now you can see why deuteranomaly occurs much more frequently among males than among females. A male needs only one mutant gene, inherited from either a carrier or a green-shifted mother, to express the trait; the genotype of the father, who contributes only a Y chromosome, is irrelevant. A green-shifted female, however, must have not only a carrier or a green-shifted mother, but a green-shifted father as well. The low probability of such parentage leads to about a 20:1 ratio of green-shifted males to affected females in the population as a whole.

Because one mutant gene guarantees the expression of X-linked recessive traits in males, whereas two are needed for the expression of autosomal *recessive* traits in either sex, it is much easier to detect the former than the latter in pedigree studies. This is not true for *dominant* mutants, which are expressed in single dose, whether X-linked or autosomal. Since most mutants are recessive, however, there is a striking difference between the number of genes assigned to the X chromosome and the number known for a specific autosome. It is not likely that more genes exist on the X chromosome than on autosomes of comparable size. Rather, X-linked mutants are just easier to detect and verify. But with the increased use of molecular genetic techniques, which do not favor X-linked over autosomal DNA, this distinction will disappear.

Hemophilia

Normal blood clotting is an immensely complex process requiring the highly ordered interaction of about 20 substances, most of which are plasma proteins that function as enzymes. Hereditary deficiencies have been reported for nearly all of these blood coagulation factors (McKee 1983; Biggs 1983; Graham et al. 1983), but the best-known are the two X-linked recessive disorders: **factor VIII deficiency** (**hemophilia A**, the classical type), and **factor IX deficiency** (**hemophilia B**, or Christmas disease). Together, their frequency is about 1 per 10,000 in males, with hemophilia A accounting for 80% of all cases. Very few affected females have ever been observed.*

*A third type of hemophilia, called *von Willebrand disease*, is caused by an *autosomal* dominant gene. It occurs equally in males and females, with a total frequency of about 1 in 20,000. Affected individuals lack a large component of factor VIII called VIIIR:WF.

Although hemophilia is rare, its usual mother-to-son pattern of inheritance was recognized over 2,500 years ago. The Talmud states that if a woman produces two "bleeders," any additional sons shall be excused from circumcision. If as many as three sisters bear sons who are bleeders, the sons of any other sisters are not to be circumcised; but no such prohibition applies to the sons of their brothers (Rosner 1977).

Hemophilia A appeared in several interrelated royal families of Europe, apparently arising from a mutation in one of Queen Victoria's parents. Victoria had four sons, one of whom was hemophilic. He survived long enough to produce a carrier daughter who, in turn, had one normal and one hemophilic son. Among Victoria's five daughters, two were carriers who produced a total of three hemophilic sons and three carrier daughters. The latter bore, collectively, five hemophilic sons (great-grandsons of Victoria) and perhaps six carrier daughters. Among all these, the most famous case was that of (Victoria's great-grandson) Tsarevitch Alexis, heir to the throne of Russia—whose illness may have hastened the onset of the Russian Revolution.*

Hemophilias A and B are very serious conditions; and until blood plasma became generally available in the 1950s, affected individuals rarely survived to adolescence. Even with treatments available today, internal bleeding in the joints remains a painful, potentially crippling, and very expensive problem. Added to this burden is the possibility of contracting hepatitis or AIDS (acquired immune deficiency syndrome) from viral contaminants in the plasmas, which are collected from thousands of donors and then highly concentrated.

Biotechnology has come to the rescue, however, by allowing scientists to isolate and "clone" the genes for both hemophilias—and thereby raising hopes for the commercial production of pure, uncontaminated factors VIII and IX (Lawn and Vehar 1986; Lawn et al. 1986; Gitschier et al. 1984; Choo et al. 1982). Equally important, these techniques can reveal a lot about the structure, function, and evolution of the genes and their products and may provide more insight into the clotting process as well (Chapter 15).

Even before the two genes were dissected from their chromosomal surroundings, it was known from pedigree studies that they were situated far apart on the X chromosome. The exact location of the hemophilia B gene is unknown. The gene for hemophilia A (more technically, for antihemophilic globulin factor VIII:C) lies near one tip of the X, close to the two genes for color blindness and a gene called *G6PD* (Chapter 11). There may actually be several alleles at each locus, because both types show considerable variation in the observed amount of antihemophilic factor (0–30%) among the affected population.

*During one of Alexis's many serious bleeding episodes, a mystic (and charlatan) named Rasputin was called in to try to heal the 3-year-old. When their child "recovered," the grateful parents (Czar Nicholas and especially Czarina Alexandra, Victoria's granddaughter) put their faith—and considerable political powers—into the hands of the evil Rasputin. Young Alexis was to suffer and survive many more bleeding episodes before he was finally killed at the age of 14 (along with his family) by Bolshevik revolutionaries—who might have been spurred on partly by Rasputin's outrageous political machinations. For poignant accounts of this bizarre story, see Hartl (1983) and Massie (1985).

Normal carriers, too, show a range in clotting time and expressed amount of the factor—but this could be entirely due to a phenomenon in females called X-inactivation (Chapter 8). Nevertheless, over 70% of carriers can be detected by their less-than-normal levels of clotting factors. Until the genes were isolated, it was not possible to detect hemophilia in a developing fetus without extracting and testing some of its blood—a highly specialized procedure (fetoscopy) that is done in only a few medical centers. Now that geneticists have isolated the two genes, however, hemophilia should become detectable by analyzing the DNA in fetal cells collected by the more standard procedures (amniocentesis or chorionic villus biopsy; see Chapters 15 and 25). And by making use of nearby biochemical marker sites (so-called RFLPs, or restriction fragment length polymorphisms) in maternal DNA as well, scientists can identify carrier females with greater accuracy than by means of clotting times alone (Antonarakis et al. 1985).

X-Linked Lethal Genes and Shifts in Sex Ratio

We have so far discussed genes that produce *normal* phenotypes and genes that cause conditions *detrimental* to the organism. There is a third class of genes, called **lethal mutations**, that disrupt processes absolutely essential to life. Like other genes, lethals can be dominant or recessive and can exist on any chromosome. They can also take effect at various stages of the life cycle, from the earliest embryonic stage through adulthood. Some spontaneous abortions, for example, are due to lethal genes acting during fetal development.

Suppose that a recessive lethal mutation, causing fetal death, occurs on the X chromosome. Any males that inherit it would die before birth. Females who inherit one lethal X chromosome are usually normal, but as shown in Figure 7, half of their sons are not expected to survive. Thus the sex ratio among their live-born offspring will be *2 females to 1 male*, rather than the usual 1 female to 1 male. Of course, in order to distinguish an abnormal sex ratio from random fluctuations about a normal sex ratio, large numbers of progeny from extended human families must be observed. X-linked lethals are rare in human populations, however, and so do not appreciably affect the overall sex ratio.

X-linked recessive lethals have the same effect in fruit

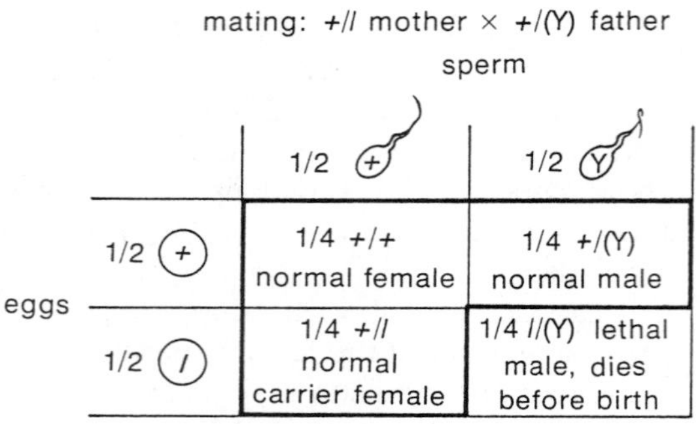

mating: +/*l* mother × +/(Y) father

		sperm: 1/2 (+)	sperm: 1/2 (Y)
eggs	1/2 (+)	1/4 +/+ normal female	1/4 +/(Y) normal male
eggs	1/2 (*l*)	1/4 +/*l* normal carrier female	1/4 *l*/(Y) lethal male, dies before birth

Figure 7 Progeny expected from mating between a female heterozygous for an X-linked recessive lethal and a normal male. When the lethal gene is expressed during fetal development, only half the sons conceived will survive to birth. Thus, *among the live-born progeny* there will be twice as many females as males: 2/3 females and 1/3 males.

Note: It is customary to designate the normal gene as + and its recessive lethal as *l*.

flies, where females are also XX and males XY. In the 1920s H. J. Muller utilized this shift in sex ratio to estimate the frequency of X-linked recessive lethal mutations in natural populations of fruit flies. By comparing this "spontaneous" mutation rate with what happened after X-ray treatments, he was the first to show that X radiation increases the number of lethal mutations. Other types of mutation are also observed more frequently after X irradiation. Evaluating the effect of irradiation in humans by screening large populations for slight changes in sex ratio is incredibly difficult and expensive. One such study attempted to determine whether or not the atomic bombs dropped on Japan in 1945 led to an increase in the mutation rate there (Chapter 14).

Y-LINKED GENES

Very few genes have been detected on the Y chromosome of most species. In humans, the Y is slightly longer than the smallest autosome (which contains 11 confirmed gene loci and 5 provisional loci) and less than half the length of the X chromosome (which contains 139 confirmed and 171 provisional loci). But only three or four loci have been definitively assigned to the Y chromosome, and a certain degree of controversy has surrounded even these (Goodfellow et al. 1985). Besides the few simply inherited traits, to be discussed in Chapter 8, some factors for sperm production, skeletal height, and tooth size have also been provisionally assigned to the Y chromosome (Buhler 1980). The Y is critical to the development of maleness and the prevention of femaleness, since certain rare individuals possessing only one X chromosome, but no Y, are females (Chapter 8). More precisely, a particular region of the Y is needed for male development, and femaleness is what develops in its absence.

One tip of the Y chromosome is actually homologous to a tiny region on one tip of the X chromosome, and this region contains at least two of the simply inherited traits mentioned above. But what concerns us here is the rest of the Y chromosome, the nonhomologous part. Any gene located on this latter part of the Y chromosome has a very striking mode of inheritance: it would be received, and usually expressed, by all the male descendants of an affected male, generation after generation. (And in the other direction it should extend from any male through his paternal grandfather, his grandpaternal great-grandfather, and so on, all the way back to Adam!) Female descendants, lacking the nonhomologous Y region, could neither receive, express, nor transmit such a trait. In the past half-century, claims of Y-linkage have been advanced for over a dozen traits. Upon closer examination of the family histories, however, exceptions to the above requirements were always discovered. The most bizarre such case is described in Chapter 3.

MITOCHONDRIAL GENES

In the 1960s scientists discovered that aerobic (oxygen-using) cells of animals contain two genetic systems: the main one in the nucleus and a very tiny one in the mitochondria, the "powerhouses" of the cell (Chapter 5).* The latter are actually descended from some free-living aerobic bacteria that eons ago invaded and took up permanent residence in primitive anaerobic cells. During this evolutionary process, the mitochondria retained some of their own DNA, which resembles bacterial DNA but now includes just a few genes involved in energy production (Grivell 1983; Chapter 15).

All the mitochondria present in our cells are derived exclusively from our mothers, and none from our fathers. This is because egg cells contain many mitochondria, but sperm cells contain very few mitochondria—and none of the latter survive inside the fertilized egg. Thus, *any trait associated with a mitochondrial gene must be transmitted by the mother to all of her children, both male and female.* In the other direction this lineage should extend from any individual through the maternal grandmother, the grandmaternal great-grandmother, and so on, all the way back to Eve! Indeed, because mutations in mitochondrial genes accumulate at a much higher rate than do those of nuclear genes, geneticists use the former as "tags" to help trace human evolutionary history. One

*Plant cells contain, in addition to these two, a third genetic system in their *chloroplasts*.

study of mitochondrial DNAs taken from 147 people representing five population groups (African, Asian, Australian, Caucasian, and New Guinean) revealed that they all "stem from one woman who is postulated to have lived about 200,000 years ago, probably in Africa. All the populations examined except the African population have multiple origins, implying that each area was colonised repeatedly" (Cann et al. 1987; Wainscoat 1987).

Any mitochondrial mutations that cause diseases might be detected by their maternal pattern of inheritance. They must also be associated with defects in mitochondrial oxidative phosphorylation and sequentially affect those tissues—brain and spinal cord, skeletal muscle, heart, kidney and liver—that have high energy requirements. In humans, good evidence exists for at least two such types of conditions, both quite variable. One group of disorders, called *mitochondrial myopathy* (Greek *mys*, "muscle"; English *pathy*, "disease"), involves muscular weakness or wasting or both (Land et al. 1981; Wallace et al. 1988a). Another, affecting vision, is *Leber hereditary optic neuropathy* (Novotny et al. 1986; Wallace 1986). The latter defect has been traced to a specific enzyme of the mitochondrial membrane involved in the production of the high-energy compound ATP; it is caused by a so-called missense mutation (Chapter 14) of a single nucleotide base (Merz 1988; Wallace et al. 1988b).

SUMMARY

1. Mendel's experiments were carefully planned and executed. He approached the topic of heredity in novel ways designed to obtain clear-cut answers.
2. In modern terminology, Mendel's law of segregation states that the two alleles of a gene separate during the formation of gametes, so each gamete receives at random one of the two alleles.
3. The law of segregation and the randomness of fertilization allow predictions of the genotypic and phenotypic ratios among the offspring of any mating. A convenient way to set up these predictions is to construct a checkerboard whose rows and columns are labeled with the various kinds of eggs and sperm, and with gamete fractions representing their relative abundance.
4. The testcross is the most informative mating for determining whether an organism with the dominant phenotype is homozygous or heterozygous.
5. In humans, the development of the male sexual phenotype is triggered by the presence of a Y chromosome. Normally, XX = female and XY = male.
6. A father transmits his X chromosome to daughters and his Y chromosome to sons; a mother transmits an X chromosome to both sexes. A son receives an X chromosome from his mother and a Y chromosome from his father; a daughter receives an X chromosome from both parents.
7. For X-linked genes, as for autosomal genes, females can be homozygous or heterozygous. Because X-linked genes have no corresponding locus on their Y chromosome, males have but a single allele and are said to be hemizygous.
8. The X chromosome carries many genes, most of which have nothing to do with sexual development. Color blindness and hemophilia, for example, are determined by genes on the X chromosome.
9. X-linked recessive lethal genes that act during fetal development, thereby leading to spontaneous abortions, can alter the sex ratio to favor females.
10. Few genes have been assigned to the Y chromosome, and few human disorders are traceable to the mitochondrial genome.

FURTHER READING

Mendel's paper (1865) is in a collection edited by Peters (1959). Also in this book is Morgan's (1910) paper on X-linked inheritance in fruit flies. For more information on color blindness, see Rushton (1975), Hurvich (1981), G. Jacobs (1981), and Botstein (1986). Good articles on hemophilia include McKusick (1965), Kasper (1982), Brownlee and Rizza (1984), Maddox (1984), and Lawn and Vehar (1986).

QUESTIONS

1. The first edition (1966) of McKusick's catalog, *Mendelian Inheritance in Man*, identifies 506 autosomal loci and 68 X-linked loci, whereas the eighth edition (1988) identifies 2069 autosomal loci and 139 X-linked loci. Thus, although the measured physical length of the X-chromosome comprises 5.1 percent of the total haploid genome, X-linked loci made up 68/574 or 11.8% percent of all confirmed loci in 1966, but 139/2208 or 6.3% percent of all confirmed loci in 1988.

Why are both percentages greater than the actual proportional length of the X chromosome, and what can explain the significant "loss" of X-linked loci between 1966 and 1988?

Classical albinism is caused by the lack of an enzyme necessary for the synthesis of melanin pigments. Enzyme production requires the dominant allele, *C*, so $C/- =$ normal pigmentation and $c/c =$ albino.

2. What progeny are expected, and in what proportions, from a normally pigmented woman who has an albino husband and an albino father?

3. A normally pigmented woman and an albino man have 9 normally pigmented children and 1 albino child. What is the best guess for her genotype? Is any other genotype possible?

4. A normally pigmented woman and an albino man have 10 normally pigmented children and no albinos. What is the best guess for her genotype? Is any other genotype possible?

---

5. Consider a gene with two alleles, *B* and *b*. (a) List all the matings (i.e., the parental genotypes) that could produce a heterozygous child. (b) Which mating in your list gives the greatest proportion of heterozygous offspring?

6. (a) Again considering alleles *B* and *b*, list all the matings whose offspring can be of only one genotype. (b) List all the matings whose offspring can be of two, and only two, genotypes. (c) What one mating is not on either of these lists?

7. Assume that Rh blood types are determined by a gene called *R*, so that $R/- =$ Rh positive and $r/r =$ Rh negative. (The actual genetic system is more complex than this.) Rh positive people have a particular chemical, an antigen, on the surface of their red blood cells; Rh negative people do not have it. List the four possible matings between two persons who are both Rh positive (considering reciprocal matings as separate). What progeny are expected from each mating?

8. For some rare autosomal dominant diseases it is possible that the dominant homozygote has never been seen. One reason is that this genotype may die early in development and be spontaneously aborted. Consider a situation in which

Q/Q dies as an early fetus
Q/q lives to reproduce but is affected
q/q lives and is normal

What genotypes are possible among the live-born children of the mating $Q/q \times Q/q$? Among these, what is the proportion of each genotype?

With regard to the X-linked gene for green-shift, recall that females can have three different genotypes, *G/G, G/g,* or *g/g,* but males can have only two, *G*/(Y) and *g*/(Y). The normal allele *G* is dominant to its green-shifted allele *g*.

9. Systematically list the six possible matings for this X-linked gene, identifying the sex of each parent. For each mating give the expected offspring. List daughters and sons separately.

10. Which of these matings can produce a carrier daughter? In which of these are *all* daughters carriers?

11. Which of these matings is expected to produce some sons with green-shift and some sons with normal color vision? What is the genotype of the mothers in these matings?

12. What offspring phenotypes are possible from a green-shifted mother and a normal father? Identify offspring by sex. (Note that in this particular family there is a crisscross type of inheritance.)

13. A couple has a green-shifted daughter and a son with normal color vision. What are the genotypes of the parents?

---

14. If a male desires to maximize his contribution of genes to future generations, is he better off having daughters or sons? Why?

15. You could not have gotten a sex chromosome from one of your four grandparents. Which grandparent could not have transmitted, via your parents, a sex chromosome to you? [*Note*: The correct answer depends on your sex.]

ANSWERS

[*Note*: ♀ = female, ♂ = male]

1. Humans are not experimental organisms like peas or fruit flies, that can be mated at will and in ways that best reveal their genetic makeup, so geneticists can only identify hereditary traits as they show up in affected individuals. Thus the most easily detectable traits (dominant traits and X-linked traits) are overrepresented, and the least commonly manifested traits (autosomal recessive traits) are underrepresented. But as scientists develop more sophisticated chemical and mathematical techniques for studying human genetics, more autosomal traits are discovered, thereby reducing the proportion of known X-linked traits.

2. *C/c* ♀ × *c/c* ♂ produces 1/2 normal (*C/c*) and 1/2 albino (*c/c*).

3. Presence of the albino child shows that she is heterozygous (*C/c*). No other genotype is possible, barring mutation of *C* to *c* in a *C/C* woman.

4. Most likely she is homozygous (*C/C*), in which case only normally pigmented children are expected. She could, however, be heterozygous (*C/c*), in which case half of her children are expected to be albino. In Chapter 10 we explain how to calculate the probability (which is low, but definitely not zero) that a heterozygote produces no albinos among several children. For 10 children, the probability is $(3/4)^{10} = 0.056$. [*Note:* This situation exactly parallels Mendel's method for progeny testing his tall F_2 plants (see page 16), running about a 5% chance of mislabeling a heterozygote as a homozygote.]

5.

B/b × *B/B* *B/b* × *B/b* *B/b* × *b/b*	50% of progeny expected to be *B/B*
B/B × *b/b*	100% of progeny expected to be *B/b*

6. (a) *B/B* × *B/B*
B/B × *b/b*
b/b × *b/b*

(b) *B/b* × *B/B*
B/b × *b/b*

(c) *B/b* × *B/b*

7. *R/R* ♀ × *R/R* ♂ → all *R/R* (Rh pos)
R/R ♀ × *R/r* ♂ } → 1/2 *R/R*, 1/2 *R/r* (all Rh pos)
R/r ♀ × *R/R* ♂ }
R/r ♀ × *R/r* ♂ → 1/4 *R/R*, 1/2 *R/r*, 1/4 *r/r* (3/4 Rh pos : 1/4 Rh neg)

8. 2/3 *Q/q*, 1/3 *q/q*

9.

	Among daughters	Among sons
G/G ♀ × *G*/(Y) ♂ →	all *G/G*	all *G*/(Y)
G/G ♀ × *g*/(Y) ♂ →	all *G/g*	all *G*/(Y)
G/g ♀ × *G*/(Y) ♂ →	1/2 *G/G*, 1/2 *G/g*	1/2 *G*/(Y), 1/2 *g*/(Y)
G/g ♀ × *g*/(Y) ♂ →	1/2 *G/g*, 1/2 *g/g*	1/2 *G*/(Y), 1/2 *g*/(Y)
g/g ♀ × *G*/(Y) ♂ →	all *G/g*	all *g*/(Y)
g/g ♀ × *g*/(Y) ♂ →	all *g/g*	all *g*/(Y)

10. All but the first and last matings listed in problem 9. All the daughters will be carriers when the mother is homozygous and the father is hemizygous for the *other* allele (second and fifth mating).

11. Those in which the mother is heterozygous. For sons, the genotype of the father is irrelevant with regard to X-linked genes.

12. *g/g* ♀ × *G*/(Y) ♂ → all daughters normal, all sons green-shifted.

13. *G/g* ♀ × *g*/(Y) ♂

14. Daughters. He transmits a gene-rich X chromosome to daughters but a gene-poor Y chromosome to sons.

15. A female receives no sex chromosome from her father's father, whereas a male receives no sex chromosome from his father's mother.

Chapter 3 Pedigrees

Data Gathering

Pedigree Construction
Pedigree and Biochemical Analyses
Some Caveats

Autosomal Dominant Inheritance: Brachydactyly

Autosomal Recessive Inheritance: Cystic Fibrosis
Clinical Features and Treatment
Pedigrees
Biochemistry and Molecular Genetics
Detection of Carriers and Homozygotes

X-Linked Recessive Inheritance: Duchenne and Becker Muscular Dystrophies
Clinical Features
Pedigrees
Biochemistry and Molecular Genetics
Dystrophin
Detection of Carriers and Hemizygotes

A country Labourer, living not far from Euston-Hall in Suffolk, shewed a Boy (his Son) about fourteen Years of Age, having a cuticular Distemper . . . The Father knew of no Accident to account for this Distempered Habit . . . [The] Mother had received no Fright, to his Knowledge, whilst she was with Child; and hath born him many Children, none of which have ever had this, or any other unusual Distemper or Deformity.

John Machin (1732), quoted in Penrose and Stern (1958)

This quote from the minutes of the Royal Society of London is the earliest known description of what came to be called *porcupine men*. Some years later it was reported that Edward Lambert, the affected male, had six children, "all with the same rugged covering as himself: the first appearance whereof in them, as in him, came on in about nine weeks after the birth. Only one of them is now living." In 1756 Lambert and his surviving son were similarly described by a French physician who mentioned, in addition, that the five deceased offspring (all affected) had included both girls and boys. These early accounts were verified two centuries later by the genealogical data that L. S. Penrose and Curt Stern (1958) dug out of old parish registers from the Suffolk area.

But intervening reports on the same Lamberts and their descendants became more and more extravagant. Decades later it was even claimed that whole tribes of porcupine men roamed North America (Figure 1), and in 1833 the following erroneous family history was published by *Lancet*, a distinguished medical journal:

> The great-grandfather of this man was found by a whaler in Davis Strait, in a wild state, similarly covered. He was taken on board, and christened in the captain's name, Lambert. He was ultimately brought to England, where he begat a son, coated in the same manner, who again begat his like, and so the generation went on to the fourth—numerous births of scaly male children occurring, the females being born with proper human skins (Anonymous 1833).

Because of such reports, the Lambert-type condition was for some time thought to be Y-linked. Now called *ichthyosis hystrix gravior*, this trait is actually inherited as an autosomal dominant. Several other types of ichthyosis exist, however, among which one is known to be inherited as an autosomal recessive and another inherited as an X-linked recessive. For more information, see Schnyder (1970).

Although extreme, the case above illustrates an important point: getting complete and accurate family information is a major challenge. But it is essential, because a hypothesis is only as good as the raw data on which it is based. No amount of fancy statistical manipulations can offset uncertainties caused by the sloppy collection of information. Computer workers put it quite simply: "Garbage in, garbage out."

Rather than relying on hearsay of a few family members, careful investigators will personally interview and examine as many living relatives as possible. They may also have to persuade some subjects to part with samples of blood, urine, skin, and so on, for laboratory analyses. This whole process often requires a great deal of travel time and money. Then, because few families keep careful written records, there remains the problem of getting reliable information about deceased relatives. Recollections may be hazy or exaggerated, especially about medical histories; and illegitimate births may go unmentioned.*

Another problem is the reluctance to discuss family members who are in some way abnormal. Indeed, close relatives may not even know about such individuals if they died young or were institutionalized. Stillbirths and miscarriages are even less likely to be remembered. Sometimes concealment may serve to avoid stigmatization, as some people might suffer discrimination if their family background were known—for example, Blacks who are carriers of the sickle-cell trait, or offspring of a person with Huntington disease, or people at risk of developing a serious illness.

One striking case of family secrecy involved a debilitating, autosomal dominant condition among the descendants of a Portugese seaman who settled in California in 1845. He died of a rare neurological disease involving the brain region that controls body movement; symptoms include uncontrollable vomiting, a stumbling gait, slurred speech, and involuntary twitching of the limbs. In every generation, several individuals were afflicted in their 20's and usually died by the age of 40. This condition was variously misdiagnosed by physicians—but family members, who suffered guilt and anxiety about their "stigma," were very reluctant to discuss it among themselves or to seek help from specialists. But in 1975 a family member contacted the National Genetics Foundation, which arranged a reunion for the 120 living members of the Antone Joseph clan. Among those who were examined, 3 new cases were diagnosed and 40 others were found to be at risk of developing what is now called *Azorean neurological disease* (Rosenberg and Fowler 1981). Although no treatment or good diagnostic tests are yet available for this "private" disorder, further research may help.

Aside from lack of knowledge or unwillingness to divulge it because of embarrassment or shame, people may withhold facts just to see whether scientists can figure things out on their own. For all these reasons, investigators must gain the confidence of family members by showing sensitivity, patience, and discretion (Krush and Evans 1984).

All data must be recorded clearly and completely.

Figure 1 Handbill published in London around 1820, advertising a porcupine man. (Courtesy of the Royal College of Surgeons of England.)

*Indeed, based on several family studies of blood groups and other genetic markers, Ashton (1980) estimated that 2.3% of tested children "were probably the result of infidelity, concealed adoption, or another event." Estimates from smaller studies have ranged much higher.

Particularly important are *exact descriptions* of the various phenotypes, so that information from many different studies can be pooled if desired at a later time. Even the *order* in which the data are collected can be important for statistical analyses (Chapter 12). Human geneticists prefer to study large families, in which the capricious effects of small sample size are lessened. Information on at least three generations is desirable—and family records that include all birth and death dates, interrelationships, and complete medical histories are a real find. For consistency of environment and ease of location it helps if the family is not too dispersed. Also, geneticists are interested in the effects of marriage between relatives, which is more likely to occur in families that occupy a restricted area.

Although they rarely encounter the ideal research situation, human geneticists make the most of what they find. The few populations (such as the Amish and the Hutterites) that produce large families, live in well-defined areas, marry almost exclusively among themselves, maintain a uniform life style for generations, and also keep careful records, have provided unique opportunities for genetic study (Chapter 21).

PEDIGREE CONSTRUCTION

One might think that the recent explosion of molecular techniques would make old-fashioned pedigree analyses unnecessary. Not so. Even high-tech geneticists often need to know the exact interrelationships among the sources of DNA molecules they dissect and analyze, while clinicians and genetic counselors must utilize all possible genetic tools, both new and old. Thus they still find it convenient and instructive to summarize, in diagrammatic form, information from family studies. These **pedigrees** are drawn up in accordance with loosely standardized rules, whose symbols are usually described in accompanying legends, so that they can be interpreted by others with a minimum of confusion. Figure 2, for example, represents a hypothetical pedigree in which affected persons possess a dominant autosomal gene.

Separate generations occupy separate horizontal lines, with the most ancestral at the top, and are numbered from top to bottom by Roman numerals. The individuals comprising a given generation are numbered from left to right by Arabic numerals. Note, however, that consecutive individuals may be genetically unrelated, as are II-2 and II-3 in Figure 2. Within each **sibship** (group of brothers and sisters), the birth order, from oldest to youngest, is arranged from left to right.

Males are designated by squares, females by circles. If sex is unknown, a diamond is used; this symbol, with a number enclosed, may also designate unaffected individuals whose sex is irrelevant. Stillbirths, miscarriages, and abortions are indicated by tiny symbols, like that for II-7; deceased individuals are indicated by a symbol with a diagonal slash. An arrow points to the **propositus/ proposita** (or **proband**), the individual through whom the pedigree was discovered (IV-15) and who must be accounted for in statistical analyses (Chapter 12).

Parents who are unaffected and unrelated, such as the mate of III-5, may be omitted from the diagram. (But readers should be informed whether the phenotype of an omitted spouse is normal or unknown.) Likewise, unaffected sibs may be lumped into a single diamond (IV-11 through 13), with a number inside to indicate how many individuals are included. Extensive pedigrees are sometimes drawn in circular or spiral rather than rectilinear form, to make them more compact.

Two parents are joined by a horizontal line, from which drops an inverted T, to which their offspring are attached by short vertical lines. Individuals III-4 through III-6, for example, are sibs, the children of II-3 and II-4. Note that III-2 had two mates, namely III-1 and III-3; thus IV-1 and IV-2 are half-sibs. A single child (such as III-7) is attached by a long vertical line directly to its parents' horizontal "mating line." Twins, such as II-4 and II-5, attach at the same spot along the inverted T if nonidentical. If identical, like III-10 and III-11, they may branch from a short vertical line or be connected by a line. (If the type of twinning is unknown, a question mark appears between them.)

Individuals affected with the trait under consideration (I-1, II-4, II-8) are represented by solid symbols; normal individuals are represented by open symbols. For dominant disorders, affected individuals may be shown as "half-black," because they are usually heterozygotes. Partial expression of a trait may be indicated by lighter shading. A dot inside the symbol is usually used to identify a heterozygous carrier of an X-linked recessive gene; it may sometimes be used to identify an individual who has a gene but does not express it, such as the carrier of a recessive trait, or even of an unexpressed dominant allele (III-12). Alternatively, carriers of autosomal genes, including those showing nonpenetrance, are represented by half-solid symbols. When two or more traits are being studied simultaneously (Figure 8 in Chapter 11), more complex symbolism must be used.

Consanguineous matings (matings between related individuals) are important to note. Individuals IV-8 and IV-9 are related as second cousins, because they have one set of great-grandparents (I-1 and I-2) in common. The mating line between IV-8 and IV-9 (doubled to indicate consanguinity) closes a genetic loop that also passes through ancestral individuals III-6, II-4, II-8, and III-8. The daughter of IV-8 and IV-9, although inbred, is normal with respect to the trait considered in the figure.

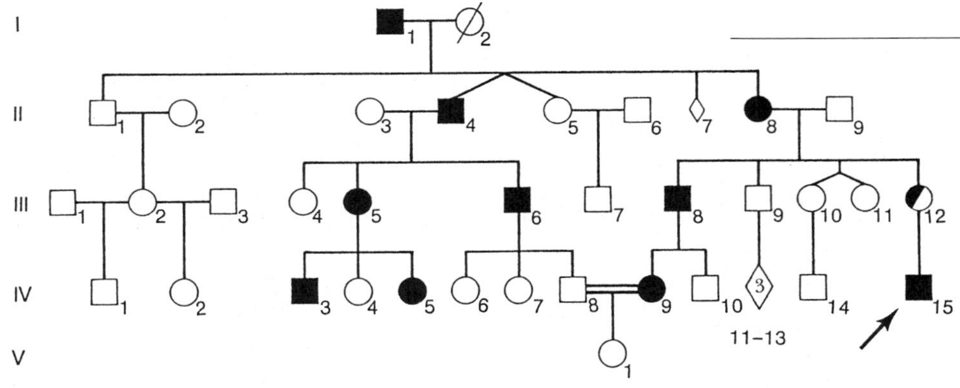

Figure 2 Hypothetical pedigree of a rare autosomal dominant trait, illustrating the symbols most commonly used by American geneticists. The gene is not fully penetrant in individual III-12, who thus does not express the trait.

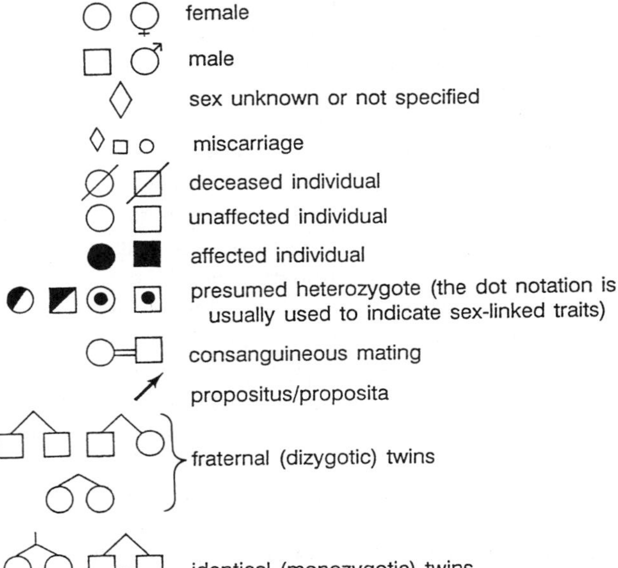

Pedigree and Biochemical Analyses

By studying pedigrees and medical data, geneticists have over a period of decades managed to determine the mode of inheritance for many phenotypic traits. Specifically, the most recent catalogue of Mendelian inheritance in humans (McKusick 1988) includes 1,443 proven autosomal dominant traits, 626 proven autosomal recessive traits, and 139 proven X-linked traits. (In addition, the catalogue lists almost as many phenotypes whose loci are not fully identified or validated.)

Although the gene products for the majority of the disorders listed in this catalogue remain unknown, from various studies taken together an interesting rule of thumb has emerged: *almost all disorders caused by enzyme deficiencies are recessive* (see Chapter 12), whereas those *disorders caused by defects in nonenzymatic, that is, structural, proteins are often (though not always) inherited as dominants.* The reasons given for this distinction are as follows: Although heterozygotes make only half as much of the normal gene product as homozygous normal individuals do, this is usually still enough to get the job done in the case of enzyme reactions, where the catalyst is needed in only small amounts. But in the case of structural proteins, such as collagen, the abnormal gene product may actively interfere with normal cell functions.

To a surprising extent, gene structure and function are *conserved*, not only between closely related groups of organisms, but often between distantly related organisms as well.* Thus researchers often make interspecies and interclass comparisons of phenotypes, genotypes, and genes. "What strikes me more than anything," says cancer researcher Robert Weinberg, "is the complexity of life, the intricacy, the patterns that have been evolving for two billion years, the extraordinary cleverness of biological things. It is still staggering for me to realize how similar we are to flies, earthworms, and the bacteria in our gut" (Langone 1983). For inherited human disorders in particular, animal models are both instructive and practical—allowing many kinds of crosses and experiments that could never be undertaken on people.

And now that traditional pedigree studies can be combined with powerful new techniques for analyzing differences in the structure of DNA itself, a new brand of genetics is emerging. From Mendel's time until the mid-1970s, allelic differences could be detected and studied only if they gave rise to phenotypic variation—such as tall versus short plants, or normal versus hemophilic people—and patterns of inheritance could only be determined by observing what phenotypes, in what proportions, resulted from various types of crosses. This traditional approach, going from the phenotype to the gene product to the gene locus, has been called **forward genetics**. But with the ability to isolate, sequence, and even insert specific mutations into small segments of DNA (Chapter 15) comes the possibility of studying and using gene loci whose allelic variants may show no discernable phenotypic differences whatsoever, whose phe-

*One of the most unusual examples: The gene for *hemoglobin* (the oxygen-carrying molecule of animal blood cells) is found in some *plant* species, prompting the suggestion that one might be able to get blood from a turnip after all (Appleby et al. 1983).

notypes and gene products are not even known. This new approach, starting rather than ending with the gene locus, is called **reverse genetics**.

We begin with the standard approach, however. Summarized below are the key aspects of the three most common types of single-gene inheritance (autosomal dominant, autosomal recessive, and X-linked recessive) that are found by traditional pedigree studies. X-linked dominance does occur, but very few examples are known for certain—partly because it is difficult to distinguish from autosomal dominance. (The kinds of traits determined by the *interaction* of numerous genes defy simple analyses; these will be discussed in later chapters.)

Some Caveats

In trying to determine how a given disorder is inherited, genetic sleuths must keep several things in mind.

1. Although a single pedigree may be used to rule out certain possibilities, it may not suffice as conclusive evidence of a genetic hypothesis. For example, some pedigrees for X-linked recessive inheritance would also fit an autosomal gene acting as a dominant in males and a recessive in females. (The gene for male pattern baldness, discussed in Chapter 12, may act this way.) It may be necessary, therefore, to pool together many studies and analyze the data by more sophisticated mathematical techniques.
2. On the other hand, because mutations at different loci may occasionally give rise to very similar phenotypes, some inherited disorders *do* have multiple modes of transmission. McKusick's catalog of simply inherited human phenotypes (1988) includes about three dozen conditions that are definitely known (and many more that are suspected) to have at least one autosomal and one X-linked variant. In addition, many autosomal disorders are known to have both dominant and recessive forms. In such cases the initial diagnosis of a given disease does not always predict its inheritance, and a very careful clinical and genealogical follow-up is required to determine which form it really is and to provide accurate genetic information to the family.

For these reasons, geneticists who are armchair physicians or physicians who are armchair geneticists should be cautious about dispensing advice to families with suspected genetic conditions. It is better to refer such patients to specialists who are trained in *both* medicine and genetics. These genetic counselors are more likely to be familiar with new developments in diagnosis and treatment and with the various problems— psychosocial as well as medical and genetic—encountered by such families.

AUTOSOMAL DOMINANT INHERITANCE: BRACHYDACTYLY

The pedigree in Figure 3 shows the pattern of inheritance for one form of **brachydactyly** (Type A2) in a Norwegian family descended from individual I-1, who was born in 1764. Affected individuals have shortened index fingers and toes, due to the shortening or near absence of the middle bone in these digits (Greek *brachys*, "short"' *daktylos*, "finger"). Some cases (hatched symbols in the pedigree) are so slightly affected that they might have been classed as normal without careful examination, including X rays. As shown in Figure 4, however, the more severely affected cases (solid symbols in the pedigree) exhibit marked shortening and sometimes crookedness of the index fingers—although never to the extent of causing a real handicap. (In other types of brachydactyly, different digits, or all digits, may be shortened.)

The frequency of brachydactyly in the general population is about one in a million. *For such rare dominant traits, all affected individuals are assumed to be heterozygous, unless their parents were related.* We make this assumption because it is highly unlikely that a second mutant allele would be brought in from outside the family just by chance. In this pedigree, however, individual IV-20 might have been a *homozygote*—because her parents (III-6 and III-7) were first cousins. In their paper the researchers (Mohr and Wriedt 1919) describe her as follows:

> This second daughter, who died one year old, demands special attention. All information is to the effect that she was a cripple, unable to develop; but it has so far been impossible to investigate the exact nature of her malformation . . . through the fact that the family has intentionally not talked about her, regarding the birth of a cripple within the family as a point not to be mentioned. This practice has been followed so carefully that even the daughter of her own sister, when asked, did not know of her existence. The small amount of information resulting from our very elaborate inquiries is limited to the following rather vague statements: Her half-brother . . . states that 'her whole osseous system was in disorder.' Her half-sister knows with certainty that 'her hands and feet, or at any rate her fingers and toes, were entirely absent.'

It is ironic that the family record keeper (II-7) neglected to describe a critical member of this pedigree, his crippled granddaughter (IV-20), whom he outlived by 14 years. Regrettable, but less peculiar, is the lack of infor-

○ normal

● marked shortening of index fingers and toes

⊘ slight shortening of index fingers and toes

⊘ not seen by authors; described as "normal" by some relatives, but probably heterozygous because they produced affected offspring

○? insufficient information to classify phenotype

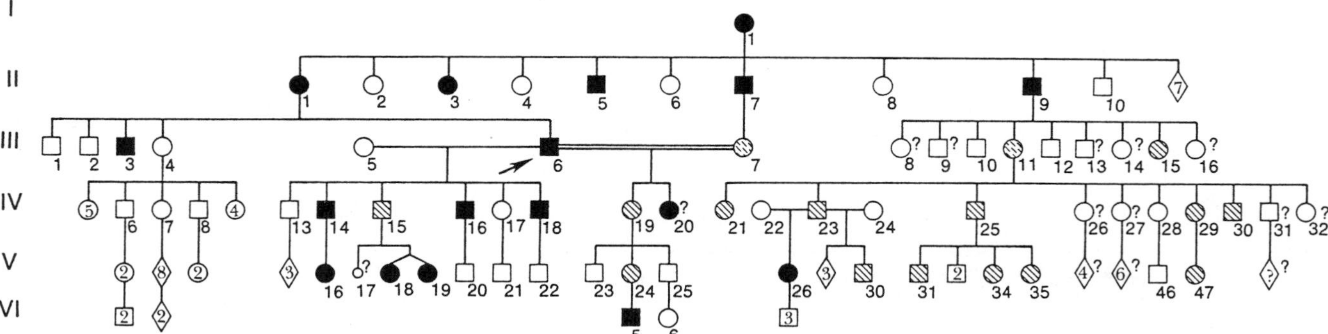

Figure 3 Pedigree of a family that expresses the brachydactylous condition through six generations. Note that individual III-6 married twice, III-7 being his first wife and III-5 the second. Individual IV-20, a cripple who died in infancy, may have been a homozygote. (Redrawn from Mohr and Wriedt 1919.)

mation about his only child (III-7). She was probably the first heterozygote not strikingly affected by the dominant allele.

This pedigree illustrates all the characteristics of autosomal dominant inheritance when a gene is fully, or almost fully, penetrant— that is, expressed phenotypically wherever it occurs (Chapter 12). Note that:

1. *All affected individuals have at least one affected parent.* This is true here—except for two sibships in which one parent (III-7, III-11) was surmised, but not definitely known, to be brachydactylous.

   As a corollary to this rule, *affected individuals are usually found in every generation of a sizable pedigree.* In the six generations here, for example, we find 29 heterozygotes who definitely showed the trait. (The identical twins, having arisen from a single fertilized egg, are counted as one.) In addition, we can count as heterozygotes the aforementioned two individuals whose phenotypes might have overlapped the normal, but who nevertheless had affected children.

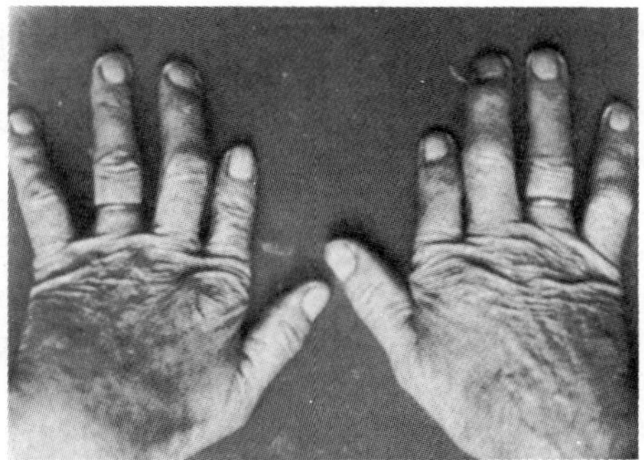

Figure 4 Photograph showing the markedly shortened index fingers of individual III-6 from the pedigree of brachydactyly in Figure 3. (From Mohr and Wriedt 1919.)

2. *Males and females are affected in about equal numbers*, which makes X-linkage less likely and rules out Y-linkage. As shown in Table 1, this pattern is true for both the extremely affected and the slightly affected members of the pedigree.

3. *Both males and females transmit the trait.* Note that female I-1 and male III-6 each produced affected progeny of both sexes. Male-to-male transmission rules out X-linkage, and transmission by a female rules out Y-linkage.

4. *Matings of heterozygous affected × normal produce approximately 1/2 affected and 1/2 normal offspring.* When all the progeny from such matings in this pedigree are added together (Table 2), the totals are 29 known affected: 25 known normal. The phenotypes of 17 additional offspring were unknown.

5. *The trait does not appear in the descendants of two normal parents*, unless the dominant allele is reintroduced by matings with unrelated affected individuals. Note that the allele has disappeared from among the descendants of individual III-4.

6. The trait may be *more severe or extreme in homozygotes* than in heterozygotes, or *it may even be lethal.* Note that individual IV-20, the only candidate for homozygosity in this pedigree, was very severely deformed and lived for only one year.

Table 1 Numbers of heterozygous affected individuals for both the more severe expression and the less severe expression of the brachydactyly gene

	MARKEDLY SHORTENED FINGERS		SLIGHTLY SHORTENED FINGERS		
Generation[a]	*Males*	*Females*	*Males*	*Females*	*Total*
I	0	1	0	0	1
II	3	2	0	0	5
III	2	0	0	3	5
IV	3	0	4	3	10
V	0	3	2	4	9
VI	1	0	0	0	1
Total	9	6	6	10	31

a. Individuals from Figure 3 are summed up by generation. The two probably affected individuals in generation III (7 and 11) are included.

Table 2 Numbers of affected and normal offspring from matings of heterozygote × normal, including sibs whose phenotypes were not determined

Parent[a]	*Affected offspring*	*Normal offspring*	*Offspring of unknown phenotype*	*Total offspring*
I-1	5	5	7	17
II-1	2	3	0	5
II-7[b]	1	0	0	1
II-9[b]	2	2	5	9
III-6b	4	2	0	6
III-11[c]	5	1	4	10
IV-14	1	0	0	1
IV-15	1	0	1	2
IV-16	0	1	0	1
IV-18	0	1	0	1
IV-19	1	2	0	3
IV-23a	1	0	0	1
IV-23b	1	3	0	4
IV-25	3	2	0	5
IV-29	1	0	0	1
V-24	1	0	0	1
V-26	0	3	0	3
Total	29	25	17	71

a. See Figure 3. Omitted from this table are the offspring of III-6 × III-7, because if III-7 was affected, this was a mating between the heterozygotes.

b. Mohr and Wriedt omitted the offspring of II-7 and II-9 from their analysis, thus getting a ratio of 26 affected:23 normal.

c. Phenotype was not definitely known to be affected but was assumed to be heterozygous because affected offspring were produced.

Complicating most pedigree analyses is the fact that phenotypic expression of a trait often varies among individuals (Chapter 12). This can be due to differences in background genotypes (including different modifying genes) as well as to differences in environmental conditions. Here, for example, most of the slightly affected persons would have been missed, had Mohr and Wriedt relied on secondhand information rather than on personal examination of all living family members. When it comes to the "more than questionable memory of individuals now living," looking at *all* the members of a *few* generations is usually better than relying on incomplete or faulty information about generations further back.

AUTOSOMAL RECESSIVE INHERITANCE: CYSTIC FIBROSIS

Written accounts of what we now call **cystic fibrosis (CF)** date back to the eighteenth century, but it was not defined as a specific disease entity until the late 1930s. Now known to be the most common lethal autosomal recessive disorder in Caucasians of European ancestry, CF is a leading cause of childhood deaths. In 1950 few patients survived past infancy; but now the median age of survival is about 24 years, with over 25% of all patients living into their 30s.

Dorothy Anderson (1938) of the Columbia University College of Physicians and Surgeons was among the first to collect and carefully analyze a large number of postmortem cases, among which there were several with affected sibs. Soon it became clear that CF is inherited as an autosomal recessive and that it mainly strikes Caucasians. Frequency estimates vary from about 1 in 500 to 1 in 3,800 among different white populations (M. Thompson 1980); but, overall, about 1 in 2,000 newborns are affected in the United States—and 1 in 20–25 people are thought to be heterozygous for this gene. The incidence of CF in Blacks is about 1 in 17,000 live births, and among Orientals in Hawaii about 1 in 90,000 (Wright and Morton 1968).

One major puzzle is how such a lethal gene can be so common, especially when so few affected individuals reproduce. Perhaps the CF mutations persist because of some small beneficial effect—like a 2% increase in fertility—among normal carriers. Another possibility is a **founder effect**, whereby a few individuals bring an atypical allele into a small, newly formed population, and the atypical allele is maintained among their descendents. (These kinds of phenomena will be discussed in Chapters 20 and 22.) And even if there were two or three different CF loci, as was once suggested but now seems unlikely, this could not possibly account for the tenfold excess of CF among Caucasians (Conneally et al. 1973).

Clinical Features and Treatment

The major triad of symptoms includes serious digestive and respiratory problems, and extremely salty sweat. These and other manifestations are related to decreased chloride and sodium reabsorption in the ducts of all exocrine glands (that is, glands that release their products into externally emptying tubules rather than into blood or lymph vessels). As a result, thick, sticky secretions clog up all the epithelial tubules, causing irreversible damage and malfunctioning of various organ systems (Harris and Super 1987).

Clogged bile ducts may cause cirrhosis of the liver. Pancreatic tubules rupture and spill out the digestive enzymes, repeated incidents that lead to fibrosis (formation of fibrous tissue). Because pancreatic enzymes fail to reach the intestines, the digestion and absorption of proteins and fats is incomplete. In fact, about 10% of CF patients are discovered at birth when a putty-like plug containing undigested protein obstructs the intestine, a condition called *meconium ileus*. CF infants may eat voraciously but still suffer from malnutrition, and their frequent, bulky, foul-smelling stools contain much undigested fat. Intestinal complications are common, as is clubbing of the fingers. Growth is usually (but not always) retarded in children with CF.

The accumulation of thick mucus in the small airways of the lungs results in chronic coughing, impaired breathing, and a greatly increased (and often fatal) susceptibility to bacterial infections. Failure of the ducts of sweat (and salivary) glands to reabsorb sodium and chloride causes the sweat to be extremely salty (about five times normal), and provides the most reliable diagnostic test for CF (di Sant'Agnese et al. 1953).

In adults with CF, respiratory problems become more severe, but digestive symptoms may be controlled by taking replacement pancreatic enzymes by mouth. Adults tend to develop nasal and cardiac problems, massive bleeding from the bronchial tubes, mild diabetes, and emotional difficulties (Davis 1983).* Over 95% of CF males are sterile, because of damaged tubules in the testes; and fertility is greatly reduced in females—probably as a result of abnormal cervical mucus. But many CF adults marry, and most lead active, productive lives until the time of death.

Despite vast improvements in diagnosis and treatment over the last few decades, many CF patients still succumb within a year or two of birth. There is still no cure in sight for this "chronic, relentless disease. For patients and their parents, cystic fibrosis remains a grim sentence, usually prolonged through an unhappy adolescence to a sad, inevitable end" (Littlefield 1981). Digestive problems are treated with a high-protein, low-fat diet plus supplements of digestive enzymes and vitamins, and breathing is somewhat improved by exercise and frequent thumping of the upper body to help drain mucus from the lungs. Psychosocial counseling is also very important.

Pedigrees

The pedigree shown in Figure 5 is taken from a genealogical survey of a small area in northern Brittany where cystic fibrosis is common (1 in 377 live births) and the population is quite isolated and stable (Bois et al. 1978). It exhibits some, but not all, of the features of autosomal recessive inheritance.

1. *Most affected individuals have two normal parents.* This finding indicates that the trait tends to show up erratically and unpredictably. In fact, such a gene is often passed on invisibly for many generations, through normal heterozygotes. Only when two carriers produce an affected child is the condition manifested. In this pedigree, for example, all 14 affected individuals have normal parents, and there are no recorded cases of CF for five to six consecutive generations.

2. *The trait is expressed in both sexes and transmitted by either sex to both male and female offspring* in roughly equal numbers. Here we see 12 affected females and only 2 affected males—clearly an extreme sex ratio, but perhaps due to chance variation, because another study of a larger area in

*Most adults with CF are still treated by pediatricians and hospitalized in pediatric wards. Although 20% of all CF patients in the United States are now over the age of 18, and 20% of these adults were diagnosed after the age of 15, few internists are trained to deal with CF (Davis 1983).

northern Brittany had turned up 16 boys and 18 girls. The trait is obviously carried by heterozygotes of both sexes.

3. *Matings between normal heterozygotes yield both normal and affected offspring* in a ratio of about 3:1. Here we see eight sibships (A–H) produced by matings between known heterozygotes; they contain a total of 32 offspring, of which 14 are affected. At first glance there appear to be far too many CF offspring, that is, 14 rather than 8. But in pedigrees like this the *numbers are always biased in favor of affected offspring*, because we only identify those heterozygous parents who have at least one affected child! Moreover, families having a larger number of affected members are more likely to come to medical attention. As will be explained further in Chapter 12, heterozygous parents who by chance produce *no* affected children will escape detection; such *sibships with heterozygous parents and all normal children are wrongfully excluded* from our CF totals. On the other hand, all CF *homozygotes* are quite readily detectable.

4. *With rare traits, the normal but heterozygous parents of an affected individual are more likely to be related to each other.* When this occurs, the recessive allele that each parent carries has usually originated in an ancestor that they have in common (Chapter 21). In this pedigree, sibship C resulted from a marriage between two distant cousins, each of whom probably inherited a CF allele from the same common ancestor (III-2 or III-3).

5. *Matings of an affected × normal individual will usually yield all normal offspring*; only if the normal parent is heterozygous can affected children be produced. CF pedigrees usually lack such matings, as a result of the sterility of males and the shortened life span of affected individuals, but a few pregnancies of CF females have been reported.

6. *When both parents are affected, all their children are affected.* (This outcome contrasts with that for a dominant autosomal trait, where two affected parents can produce normal children.) For reasons mentioned above, no such matings occur in CF pedigrees.

Figure 5 Autosomal recessive inheritance of cystic fibrosis in eight related families (A–H) from the department of Finistère in northwestern Brittany. By interviewing family members and examining birth, death, and marriage registers, Bois and colleagues (1978) were able to trace these families back six to eight generations, within which there are at least five ancestral couples: I-1 and I-2, II-5 and II-6, III-2 and III-3, III-6 and III-7, and IV-12 and IV-13. Note that there is only one known consanguineous marriage, in generation VII.

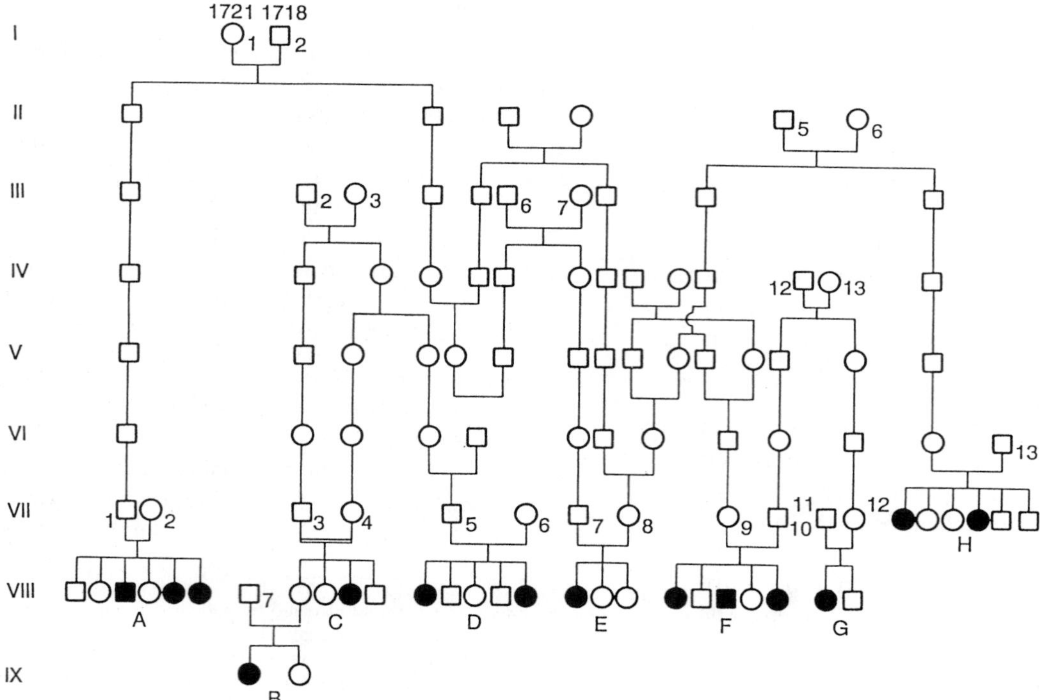

Biochemistry and Molecular Genetics

Scientists still do not know what the cystic fibrosis gene product is or exactly what it does, but they are closing in on answers to these questions. There is some evidence that tissues of fetuses with CF lack a peptidase enzyme that indirectly stimulates exocrine secretion (Gosden and Gosden 1984). Other studies target the basic physiological defect as the *inability of chloride ions to cross the membranes of the specialized epithelial cells* that line all exocrine glands (Quinton and Bijman 1983; Knowles et al. 1983). Other recent work indicates that the chloride channels in cell membranes from CF cells actually function normally—and that the CF defect instead affects the *regulation* or "gating" of these chloride channel proteins, through some pathway that generates a key compound called cyclic AMP (Welsh and Liedtke 1986; Frizzell et al. 1986).

Meanwhile, researchers are also closing in on the highly elusive cystic fibrosis gene *(CF)*. In 1985 three research teams (led by Lap-Chee Tsui in Toronto, Ray White in Salt Lake City, and Robert Williamson in London) reported the discovery of some RFLP* marker sites that were closely linked to the *CF* locus (Knowlton et al. 1985; Roberts 1988a,b) and located on chromosome 7. Also very close to *CF* is the locus for a cancer-promoting gene called *met*. These and other findings have greatly improved the accuracy of prenatal diagnosis and made possible for the first time carrier detection in families with cystic fibrosis. Once the gene product is identified, prospects will also be improved for the development of drugs or antibodies to treat the disease.

Detection of Carriers and Homozygotes

In devising ways of detecting heterozygotes (before or after producing an affected child) and homozygotes (before or after birth) for any disease, researchers must consider several levels of urgency.

1. Top priority is given to high-risk families in which an affected child has already been born. Here, depending on the mode of inheritance, the carrier status of one or both parents has thus been revealed, and the major task is to detect any additional affected offspring before or after they are born.
2. The next priority is given to the close relatives (i.e., sibs, aunts, uncles) of high-risk families, whose carrier status is unknown and who have not yet produced affected offspring. Here the double challenge is to identify carrier parents before they produce affected children and to detect the affected children before birth, if possible.
3. For inherited conditions that are lethal (or very serious) and also relatively common—and especially if an effective treatment is available—the screening of entire populations might be undertaken to detect both carriers and future affected individuals.

Thus far, cystic fibrosis testing has progressed only to the second level, although a few trial screenings have been reported. Several findings have led to the development of reliable tests for detecting CF prenatally in families known to be at risk. David J. H. Brock and coworkers at the University of Edinburgh (1985) reasoned that fetuses with CF and blocked intestines may fail to eliminate some intestinal enzymes which are normally voided (along with fetal meconium) into the amniotic fluid. So they measured the levels of an intestinal enzyme called *alkaline phosphatase (ALP)* in second-trimester amniotic fluid samples from pregnant women known to have a 1-in-4 chance of producing a baby with CF.* Out of 26 cases where a CF pregnancy was later verified, 23 (= 88%) had ALP levels less than half of normal for the amniotic fluid; and in 65 cases where a normal baby was later born, 62 (= 95%) had normal ALP levels. The rate of error is too great and the test is too complicated and expensive to use this ALP assay for mass screening, but it is acceptable for prenatal diagnosis with known high-risk couples (Brock and Van Heyningen 1986; *Prenatal Diagnosis* 1985), especially when molecular techniques for detecting the markers surrounding the *CF* gene are less applicable. Adding tests for two or three other intestinal enzymes reduces the error rate even further (Beaudet and Buffone 1987).

Most genetic centers combine enzyme tests with DNA tests. In families where DNA samples can be obtained from an affected offspring and where both parents are heterozygous for an RFLP marker, molecular techniques for the prenatal detection of affected offspring indirectly through flanking markers are usually at least 98% accurate. Carriers can also be detected in about 80% of families where DNA is available from unaffected sibs as well as from an affected sib (Farrell et al. 1986). After scientists manage to isolate the *CF* gene itself, direct DNA tests should pinpoint its presence or absence with even greater accuracy in both carriers and homozygotes.

One promising approach to *mass screening of newborns* is a test for blood levels of the pancreatic enzyme

*Restriction Fragment Length Polymorphisms. They are biochemically detectable differences in DNA sequence that are not associated with any known phenotypic differences. For details, see Chapter 15.

*A full discussion of the techniques of prenatal diagnosis occurs in Chapter 25. See also *Prenatal Diagnosis* (1985).

trypsin, which tends to be higher in babies with CF—but even this test, the immunoreactive trypsin (IRT) assay, misses a small percentage of the cases (Bowling et al. 1987). Another goal is to find a simple and reliable test for detecting heterozygotes in the general population before they produce affected offspring. Progress has been slow here, too, although some researchers claim that rates of sodium transport are lower in cultured fibroblast cells taken from carriers than in cells from noncarriers (Littlefield 1981).

When the *CF* gene is identified, it may become possible to do more widespread screening for both homozygotes and heterozygotes. But even now, certain DNA markers are known to be associated with a *CF* allele to a greater or lesser extent. Thus, unaffected people who are homozygous for these particular DNA markers can be identified as being at either higher-than-average or lower-than-average risk of carrying the *CF* allele (Estivill et al. 1988).

X-LINKED RECESSIVE INHERITANCE: DUCHENNE AND BECKER MUSCULAR DYSTROPHIES

All muscular dystrophies are characterized by gradual weakening and wasting of muscle tissue (*dys*, "abnormal;" *trophy*, "growth"), and all are inherited—but the clinical manifestations, probable metabolic defects, and modes of inheritance vary considerably.

Clinical Features

The most common and deleterious of these syndromes is **Duchenne muscular dystrophy (DMD)**, which was first described in 1868 by French neurologist Guillaume Benjamin Amand Duchenne. In 1879 W. Gowers reported that affected males inherited this condition through normal mothers, and DMD is now known to be a fully penetrant, X-linked recessive condition occurring in about 1 in 3,600 males at birth (Morton and Chung 1959). Again, geneticists do not understand why DMD, like cystic fibrosis, remains at such a high frequency in the population. About one-third of all cases appear to be new mutations (that is, the mothers are not carriers).

"Usually the onset . . . occurs before age 6 years, and the victim is chairridden by age 12 and dead by age 20" (McKusick 1988). A majority of patients are mildly retarded, although some show normal to superior intelligence. The earliest and often overlooked sign is weakness and gradual wasting of the thigh and pelvic muscles—manifested as unsteadiness, difficulty in walking stairs or rising from chairs, and a cautious, waddling gait. Although weak, the calf muscles are noticeably thickened, mostly as a result of infiltration of fat and connective tissue. As the shoulder, trunk, and back muscles gradually weaken and degenerate, the child develops a "swayback" posture (Figure 6), has trouble maintaining balance, and falls a lot. With gradual contraction of the Achilles tendons (heel cords), he also begins to walk on his toes. Later, tendons connected to the hips, knees, and elbows will also shorten—especially during periods of confinement to bed—and curvature of the spine (scoliosis) will squeeze and interfere with the functioning of internal organs (notably the lungs). Heart problems are the rule, usually including tachycardia (excessively rapid pulse rate) that sometimes leads to sudden death. Fibrosis of the cardiac muscle and enlargement of the heart may occur, too. Patients usually die of respiratory infections or of respiratory failure when diaphragm muscles become affected (Emery 1987).

A similar but rarer X-linked condition is **Becker muscular dystrophy (BMD)**. Although Becker muscular dystrophy appears identical to Duchenne muscular dystrophy with regard to muscle tissue abnormalities, its overall phenotype is more benign and variable—with a tendency to later onset, slower progression, little or no intellectual impairment, and a much longer life span (Emery 1983).

There is no cure for either DMD or BMD, and no highly effective drugs are available. All that can be done is to try to slow the progressive deformity of joints and the loss of muscle strength, thereby prolonging the pa-

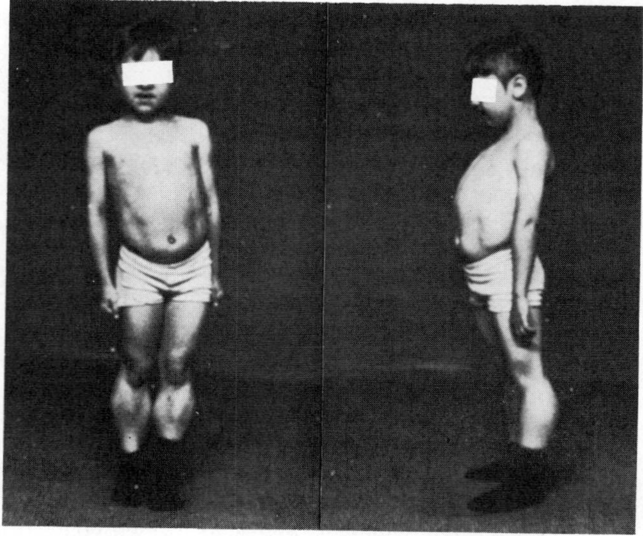

Figure 6 A 7-year-old patient with Duchenne muscular dystrophy. He has thickened calves (left) and a swayback posture (right), protruding shoulder blades, and narrowing of the distance between the lower rib cage and the hips. (From Appel and Roses 1983.)

tient's ability to sit and walk. Treatment includes the use of lightweight splints and braces, physical therapy, pulmonary therapy, and perhaps orthopedic surgery.

Pedigrees

Listed below are the characteristics of X-linked recessive inheritance, many of which are illustrated in Figure 7.

1. *If the gene is not common, many more males than females are affected.* Here, for example, we find five affected males and no affected females. [A few rare cases of females with full-blown DMD are known, but these are associated with abnormalities of the X chromosome (Kolata 1985a) or with the phenomenon of X chromosome inactivation (Chapter 8).]

2. *Father-to-son inheritance is never observed. Instead, affected males always receive their defective gene from their mothers*—who are usually normal but may have affected fathers, brothers, or uncles. Here all five mothers of affected males are normal, and four of them (II-4, II- 8, III-7, and IV-19) have affected brothers and/or uncles. Because males with DMD do not usually reproduce, we cannot say anything about their progeny*— but with other X-linked recessive conditions (red-green color blindness, for example), affected males produce only normal sons and carrier daughters.

3. *Among the sons of carrier mothers, about half are affected and half are normal.* Among live births in this pedigree, known carrier mothers (I-1, II- 4, II-8, III-7, and IV-19) produced a total of six normal and five affected sons. Although these numbers are too small to provide strong evidence, they do fit the expected pattern. [When investigators pool data for genetic analysis, they must take into account the bias introduced by the failure to detect those carrier mothers who produce no affected sons (Chapter 12)].

4. *All the daughters from the mating of a carrier female × normal male are normal, but half will be carriers.* Because carriers can be identified only *after* they have produced an affected son or a carrier daughter, classification of female offspring is iffy, and we cannot even guess about childless females. (This is not true, however, when heterozygotes are identifiable by laboratory tests.) Counting only those females who had sons, we see in Table 3 that among the daughters of known carriers, four were carriers and three were possible noncarriers—about what one might expect, but these numbers are too small to be significant. When one adds the figures from six additional pedigrees collected by Stephens and Tyler, the grand totals are 16 carrier daughters and 19 possible noncarriers.

*Although C. Thompson (1978) reports that one male with DMD fathered two normal children, a son and a daughter.

Table 3 Daughters of mothers carrying Duchenne muscular dystrophy

Carrier mother[a]	*Carrier daughters*	*Possible noncarrier*
I-1	2 (II-4, II-8)	0
II-4	1 (III-7)	3 (III-3, III-5, III-6)
III-7	1 (IV-19)	0
Total:	4	3

a. Carrier mothers are identified by having produced a son with DMD. Among their daughters, only those with at least one son are included in the tabulation; even then, the three noncarrier daughters may be misclassified.

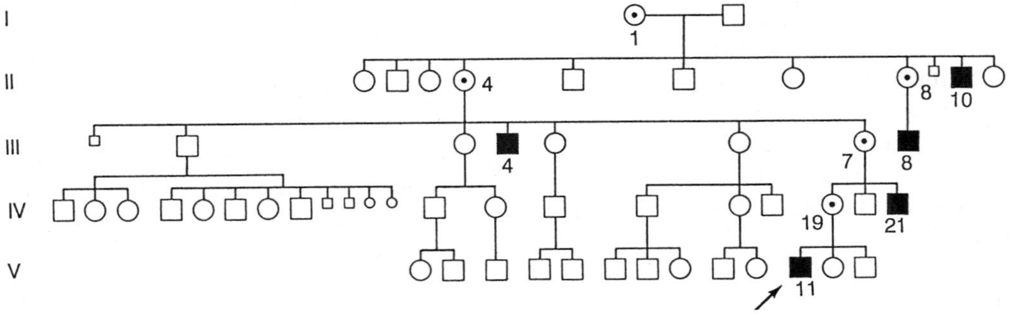

Figure 7 One of 33 Duchenne muscular dystrophy kindreds discovered in Utah by Stephens and Tyler (1951). It illustrates X-linked recessive inheritance, with sterility of affected males. Note, however, that *unaffected* males do reproduce but fail to transmit the disorder to any of their offspring. Instead, carrier females (circles with dots inside) are the sole transmitters. (Redrawn from Stephens and Tyler 1951.)

5. *An affected male × homozygous normal female produces no affected children, but all his daughters will be carriers.* Because males with DMD rarely reproduce, we cannot illustrate this principle here—but with other X-linked conditions, such as hemophilia or red-green color blindness, it has been amply demonstrated.

6. *Occasionally, a carrier female may manifest some symptoms of a recessive X-linked condition.* Indeed, about 20% of carriers show some muscle weakness—due to a phenomenon called **lyonization** (Chapter 8), a normal process whereby one or the other X chromosome in every female cell is rendered inactive. If by chance a female ends up with many cells whose active X chromosome carries the mutant allele (in tissues affected by this disorder), she may be partly or (in very rare cases) fully affected by a recessive condition that usually strikes only males. In fact, *all* females heterozygous for X-linked traits are **mosaics**; their tissues contain various proportions of cells in which one or the other of the two alleles is expressed.

   Investigators can sometimes make use of lyonization to test whether a certain gene coding for an enzyme is X-linked. They do this by taking cell samples (e.g., fibroblasts, or white blood cells) from known heterozygous females and letting the individual cells grow into separate colonies in tissue culture. Then they test each colony for the presence or absence of that enzyme: If some colonies have it while others do not, this gene is definitely X-linked, because one or the other allele was expressed. But if all colonies are phenotypically identical, the gene may or may not be X-linked.

Biochemistry and Molecular Genetics

For decades scientists debated whether the primary *DMD* gene defect was in muscles, nerves, or the blood vessels supplying them. Microscopic examination of Duchenne dystrophic muscles show a significant loss of muscle fibers and variable sizes of those that remain, along with considerable infiltration of fat and connective tissue. Muscle physiology goes awry, and a muscle enzyme called *creatine phosphokinase (CPK)* is found in the serum at levels 20-50 times higher than normal. (Because CPK levels may be normal before birth, however, this diagnostic test is not reliable enough for prenatal use.) CPK is also elevated in about 70% of known female carriers, because of lyonization (Emery 1980). Metabolic disturbances are not limited to muscle tissue, though. Many studies of red blood cells also point to membrane defects that let too much calcium get into the cells—perhaps activating intracellular enzymes called proteases, which digest proteins and cause cell death.

Detection of the *DMD* gene itself became a possibility in the early 1980s (Kedes 1985). Investigators* took three different approaches: studying rare females with DMD, studying affected males who were missing pieces of their X chromosomes, and looking for other genes or markers near the *DMD* locus. In analyzing a few rare cases of *females with full-blown Duchenne muscular dystrophy*, geneticists had found that all of them carried a chromosomal abnormality called a **translocation**: the fusion of parts of two nonhomologous chromosomes (Chapter 9). In all affected females the translocated X chromosome had broken at the same spot, the so-called Xp21 band of the short arm (Chapter 4), but it was joined to a different autosome in each case.† This analysis pinpointed the *DMD* gene to the Xp21 region, which makes up about 20% of the short arm (Worton et al. 1986).

Zeroing in on this new reference point and using human–rodent cell hybrids along with many tricks of molecular biology, scientists set out to find some nearby marker genes. Researchers at Yale University (Lindgren et al. 1984; Francke et al. 1985) determined that the *DMD* gene lies very close to the gene for the enzyme *ornithine transcarbamylase (OTC)*, and also close to a gene for *chronic granulomatous disease (CGD)*. Many *RFLP markers*, some of which closely flanked the *DMD* region, were also being identified by researchers worldwide (Bakker et al. 1985; Monaco et al. 1985).

Although pedigree studies show the *DMD* locus to be transmitted as a simple X-linked recessive, molecular studies show that it is also a huge gene with different defects in different patients (Monaco and Kunkel 1987; Monaco et al. 1987). Rather than being caused by a base-substitution (a point mutation at a single nucleotide) within the DNA sequence that constitutes the gene, in at least 60% of cases a *DMD* mutation results from the **deletion** (loss) of a long stretch of units (bases) within that region of the DNA. (Becker muscular dystrophy patients, too, often show deletions in their DNA.) A particularly lucky break was finding a patient who had Duchenne muscular dystrophy, chronic granulomatous disease, and three other X-linked conditions—as well as a big X chromosome deletion at Xp21 that might span all these loci (Francke et al. 1985). By using a technique called DNA hybridization (Chapter 15) to "line up" his DNA next to normal (i.e., non-deleted) DNA, researchers could identify which region of his X was missing. Then,

*Among the most active teams were those headed by Louis Kunkel at Harvard University, Ronald Worton at the Hospital for Sick Children in Toronto, and Kay Davies at Oxford University.

†Although each of these females also had an unbroken X chromosome, its normal *DMD* allele is not expressed—because whenever an X:autosome translocation occurs in female somatic cells, the normal X chromosome is always found to be inactivated.

by hybridizing the "leftover" tiny piece of normal DNA to DNA from males with deleted X's who had *only* DMD, the part that spanned just the *DMD* deletions was localized.

Cooperation among competing workers contributed to the relative speed with which the *DMD* locus was characterized. By sharing their DNA "probes" as well as information on the DNA of their patients, an international cooperative of 25 research teams then amassed and analyzed data on 1,346 males with Duchenne or Becker muscular dystrophy—among which 88 patients had recognizable deletions. When DNA samples from a subgroup of 57 deletion patients were compared, the deletions turned out to vary considerably in size and location within the Xp21 target area—a few of them showing no overlap whatsoever. This study, published in 1986 by Louis Kunkel of Harvard University and 76 coauthors, showed for the first time that the *DMD* locus is gigantic—about 2 million nucleotide base pairs, over 10 times the size of any other known human gene! They also noted that the size of a *DMD* deletion is not correlated in any obvious way with the severity of disease:

> The profile of deletion breaks indicates that there must be a large segment of DNA which, when disrupted, can yield the phenotype of DMD and/or BMD. The fact that large deletions yield a DMD phenotype similar to that of other DMD boys who have not been demonstrated to bear deletions, indicates that the product of the locus can be either completely absent or aberrant and still yield a similar clinical picture.

Dystrophin

In what has been called a triumph of reverse genetics, Kunkel's team then searched for and found the gene product (Hoffman et al. 1988). To do so, they took advantage of the fact that the X-linked genes of all mammals are very highly conserved and virtually identical (Ohno 1973). After using various snippets of the newly isolated human *DMD/BMD* DNA to "fish out" (by hybridization) comparable sequences of mouse X chromosome DNA, they were able to predict the amino acid sequence and structure of several parts of its unknown protein product. Then they constructed some short protein fragments, got test animals to make antibodies against them, and used the antibodies to isolate the normal DMD/BMD protein from extracts of muscle cells from normal humans and normal mice.

This previously unknown protein, dubbed **dystrophin**, exists at very low levels in the muscle cell membranes of normal individuals. It is thought to act as a strengthener, helping to protect the fibers (especially of skeletal muscle cells) from damage due to repeated contraction (Zubrzycka-Gaarn 1988; Webster et al. 1988). It is present at even lower levels in a dozen other normal tissues (including brain, lung, and kidney), where its function is unknown (Chelly et al. 1988). Dystrophin is missing from the muscles of nearly all DMD patients, however, and present in altered (abnormal) form or amounts in BMD patients (Beam 1988).

A mutant strain of mice carries a mutation called *mdx*. These mice exhibit traits comparable to those of DMD: they have deletions in the same X-linked gene; they lack dystrophin in their muscle cells; and they show the same kind of tissue damage seen in human dystrophic muscle cells. Furthermore, there is great similarity between the DNA sequence of the normal alleles of *DMD* and *mdx* genes—confirming this mutation's status as a genetic counterpart to *DMD* (Hoffman et al. 1987). Yet, for unknown reasons, these mice do not express a DMD-like phenotype; indeed, they are fairly normal and seem to manage quite well without dystrophin (Witkowski 1988). Obviously it would be desirable to find some animals that do develop full-blown DMD, and a few rare cases have been found in dogs (Cooper and Valentine 1988).

> All of this is wonderful and an unbelievable change from the state of ignorance of only a few years ago. But the discovery of the gene and gene product for Duchenne's dystrophy is not the end of the story, because we now face a new set of problems that cannot be solved only by seemingly magical tricks with DNA. Other methods in cell biology, biochemistry, physiology, and developmental biology will be needed to ascertain the function of dystrophin, how it is synthesized and placed in the membrane, how lack of dystrophin causes Becker's or Duchenne's muscular dystrophy, what role dystropin may have in other diseases, how Becker's and Duchenne's dystrophies differ, why the *mdx* mouse escapes disability, and whether this information can be used to define effective therapy (Rowland 1988a).

Detection of Carriers and Hemizygotes

Until 1982, the only way of identifying DMD carriers was by measuring the levels of the enzyme *creatine phosphokinase* (CPK) present in their blood: "about two thirds of definite carriers have levels which exceed the normal 95th percentile" (Emery 1980). In addition, up to 80% of definite carriers show thickening of the calf muscles, about 8% exhibit some muscle weakness, and some show other serum and muscle abnormalities.

Unfortunately, elevated CPK levels do not show up reliably in affected males before birth. Thus, in families known to be at risk for DMD, about the only option before 1982 was to test for sex, and then abort *all* male fetuses—even though 50% would be normal!

By employing batteries of RFLP markers known to surround the *DMD* locus, it is now possible to detect carrier mothers and affected offspring with 95–100% accuracy in most *known* DMD families. But it will never be possible to detect *all* cases of DMD in advance through prenatal testing of carriers' offspring, because about *one-third of all cases represent new mutations*: that is, the mothers of these individuals do not carry the *DMD* gene. Why such a high mutation rate? Nobody knows for sure, but one factor might be the colossal size of the *DMD* gene, which presents a big target for mutation and in addition allows for a great deal of intragene recombination (exchange of homologous parts) during the formation of gametes in carrier mothers. There is also some evidence that at least one small area within the *DMD* locus is a "hot spot," especially prone to breakage and deletions (Van Ommen et al. 1986; Forrest et al. 1987).

It is also possible that some of the "new mutations" are not classical DMD at all—but represent an autosomal recessive mutant whose phenotype mimics X-linked DMD! Emery (1980) estimates that about 5% of all DMD cases are not X-linked. Indeed, this might explain the initial misdiagnosis (by Darras et al. 1987) of two male fetuses in a few families known to be at risk for DMD. These workers conclude:

> In pedigrees without clear evidence of X-linked recessive inheritance of muscular dystrophy... the results of DNA-marker studies will help to establish the true frequency of X-linked versus autosomal recessive forms of DMD... In spite of the discovery of several DNA polymorphisms [markers] within the DMD gene or closely linked to it, prenatal diagnosis and carrier identification are still prone to error... Additional intragenic and flanking DNA polymorphisms that span and bracket the entire region of DNA in which DMD-yielding mutations can occur are needed.

Especially because of its high mutation rate, the testing for DMD in *all* newborn infants (rather than just those in known high-risk families) might be worthwhile if reliable and cost-effective tests were available. It is possible, for example, to detect CPK in the tiny amounts of dried blood (on filter paper) that are routinely obtained from all newborns to test for certain other inherited conditions (Chapter 25). In a screening program undertaken in 1986 in Manitoba, Canada, newborn males were tested for DMD by CPK analysis of these dried blood spots (Greenberg et al. 1988). Among the 18,000 males screened within two years, they found five with DMD, of whom three showed DNA deletions. When the five mothers were tested for elevated serum CPK and for DNA markers or rearrangements, three were considered to be probable carriers. Although no treatment yet exists for the affected babies found through such programs, their families are offered genetic counseling, giving them the information and the choice of what to do about future pregnancies.

SUMMARY

1. Gathering reliable family data is not easy. Interpreting genetic data may be difficult, too, especially when dealing with rare traits.

2. Some inherited conditions show unique patterns of transmission that are recognizable by means of standard pedigrees. In humans, the most common types of simple inheritance are autosomal dominant, autosomal recessive, and X-linked recessive.

3. Autosomal dominant traits that are fully penetrant do not skip generations. They are expressed equally in both sexes and are transmitted by both sexes (including father-to-son transmission).

4. Autosomal recessive traits often do not appear in every generation, and affected individuals usually have normal parents. In cases of very rare traits, the normal parents of an affected child are usually related. The trait is expressed equally in both sexes and transmitted by both sexes (including father-to-son inheritance).

5. X-linked recessive traits skip generations, are more frequent in males than females, and are transmitted differently by the two sexes. Father-to-son inheritance is never observed; instead, males inherit the trait from carrier mothers, who may have affected male relatives and carrier female relatives. Because of the lyonization effect, cultured cells taken from normal carriers may consist of two phenotypic populations, and an occasional carrier female may be partly or fully affected.

6. Some genetic disorders have multiple modes of inheritance or exhibit differences in the molecular basis of their mutations. For this and other reasons, genetic counseling should be done only by those trained in both clinical medicine and genetics.

7. Brachydactyly is a relatively benign autosomal dominant disorder; neither the gene location nor the gene product are known.

8. Cystic fibrosis is a relatively common lethal condition caused by an autosomal recessive allele on the long arm of chromosome 7. Neither the gene nor its product have yet been isolated, but it appears that the defect involves regulation of chloride ion transport through the ducts of exocrine glands.

9. Duchenne muscular dystrophy is caused by a recessive lethal gene located on the short arm of the

X chromosome. About 60% of patients with Duchenne muscular dystrophy have deletions in this gene, which after isolation was found to be huge. The normal gene product, called dystrophin, is absent from the membranes of dystrophic muscle cells. About 5% of all apparent DMD cases may be associated with an autosomal locus rather than the X-linked locus.

FURTHER READING

For simply inherited human traits, the "bible" is a massive catalog compiled and regularly updated by McKusick (8th Ed., 1988); his foreword and appendices are packed with information on human genetics, and the concise descriptions of the 3,300+ phenotypes are fascinating. McKusick (1978) is a collection of studies on the Amish.

Krush and Evans (1984) have prepared a nuts-and-bolts guide on how to design and conduct family studies; although meant for professionals, it is so clear and interesting (especially when discussing personal reactions and interactions) that others would enjoy it, too. Human genetics books written for neophytes include Hartl (1985), Nyhan and Edelson (1976), and Greenblatt (1974); although not up-to-date on recent research, the latter two contain interesting case descriptions and historical sidelights.

Mohr and Wriedt's 1919 monograph on brachydactyly provides an early example of careful genetic sleuthing and reporting. To learn more about cystic fibrosis, consult books written/edited by Harris and Super (1987), Riordan and Buchwald (1988), Hodson et al. (1983), Lloyd-Still (1983), and Taussig (1984), plus articles by Talamo et al. (1983) and McCrae (1983). For information on Duchenne muscular dystrophy, see the book by Emery (1987) as well as articles by Rowland (1988b), Monaco and Kunkel (1987, 1988), Moser (1984), Eméry (1983), Appel and Roses (1983), and Hayden and Nichols (1984).

QUESTIONS

1. (a) Describe the pattern of inheritance of a rare X-linked dominant trait with complete penetrance, and how it can be distinguished from an autosomal dominant trait. (b) Do the same for a rare, fully penetrant X-linked dominant trait that is lethal (before birth) in hemizygous males.

2. How would a Y-linked trait be inherited?

3. (a) What mode(s) of inheritance could explain the pedigree shown here, assuming the trait is *rare*? (b) What mode(s) of inheritance could explain the same pedigree if the trait were *common*?

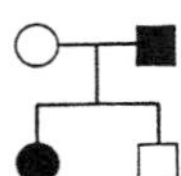

4. Around 1930, several published reports described males who lacked teeth and sweat glands, and were partly bald. This condition had been noted much earlier by Charles Darwin (1875):

> I may give an analogous case . . . of a Hindoo family in Scinde, in which ten men, in the course of four generations, were furnished . . . with only four small and weak incisor teeth and with eight posterior molars. The men thus affected have very little hair on the body, and become bald early in life. They also suffer much during hot weather from excessive dryness of skin. It is remarkable that no instance has occurred of a daughter being affected . . . [T]hough the daughters in the above family are never affected, they transmit the tendency to their sons; and no case has occurred of a son transmitting it to his sons. The affectation thus appears only in alternate generations, or after long intervals.

How is this trait (now called ectodermal dysplasia) inherited?

5. What mode(s) of inheritance could explain the pedigree shown here?

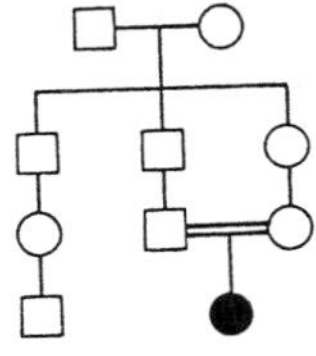

6. Although cystic fibrosis and sickle-cell anemia may be more frequent in certain populations, the most common Mendelian disorder worldwide is familial hypercholesterolemia. It is characterized by elevated levels of serum cholesterol bound to low-density lipoproteins (LDL) and is due to a defect or deficiency of cell membrane receptors for LDL (Goldstein and Brown 1979). Because much LDL and cholesterol are unable to enter cells and be properly metabolized, they accumulate in the plasma and get deposited in tendons and arteries—the latter leading to premature heart disease. From the following pedigrees (Mabuchi et al. 1978), what is the most likely mode of inheritance?

[*Note*: Half-solid symbols = affected; solid symbols = severely affected.]

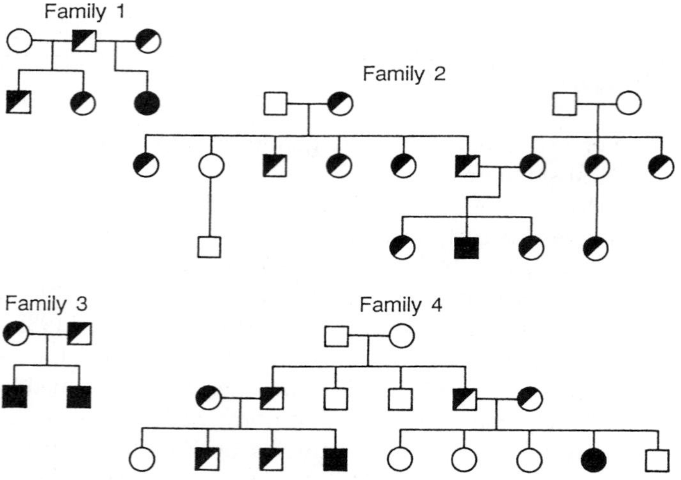

7. (a) Construct, as completely as possible, a pedigree of your own family. (b) Ask as many relatives as possible to eat some asparagus and report to you whether or not they smell a distinctive odor in their urine afterward. If feasible, suggest that they also try to detect this odor in the urine of other family members. Then see if you can determine how this simple and harmless trait is inherited.

ANSWERS

1. (a) Does not skip generations; more females than males are affected, but (heterozygous) females are less severely affected than (hemizygous) males; *affected males × normal females have all affected daughters but no affected sons*, whereas the reciprocal mating yields both affected daughters and affected sons; affected males always have an affected mother. Given the limitations of small family size, this mode of inheritance is very hard to distinguish from autosomal dominant inheritance—but look at the affected males' progeny, among which all daughters should be affected and all sons should be normal. (b) Does not skip generations; expressed only in females and transmitted only through females. Affected females produce only half as many sons as daughters, but all sons are normal. This mode of inheritance is also rare, but one example is oral-facial-digital (OFD) syndrome type I; affected females have cleft jaw and tongue plus other malformations of the skull, face and hands, and mental retardation.

2. Passes from father to son, so it occurs only in males; and (if fully penetrant) all the sons of an affected male should be affected.

3. (a) Autosomal dominant or X-linked dominant. (b) In addition to (a), could also be autosomal recessive or X-linked recessive if the mother were heterozygous. These ambiguities point up the importance of having extensive family data and also knowing something about the frequency of the trait.

4. X-linked recessive.

5. Autosomal recessive. Note that the parents of the affected individual are first cousins.

6. Autosomal dominant. If this were an X-linked dominant: in Family 1, male 3 should not be affected; in Family 2, males 4 and 7 would be severely affected (as is male 13); and in Family 4, female 8 would be affected moderately (like 1 and 4) rather than severely.

7. (b) In the first edition of this text we stated that the trait was probably an autosomal dominant, and that the absence of a specific enzyme caused the excess of an odoriferous intermediate product of metabolism (methanethiol), which is then excreted in the urine.

The latter part turns out to be wrong, for a very interesting reason: people were originally tested for the ability to smell this substance in their *own* urine, but not in the urine of others. When the latter approach was taken, however, it was found that *everybody excretes this substance, but not everybody can smell it!* "Thresholds for detecting the odour appeared to be bimodal in distribution, with 10% of 307 subjects tested able to smell it at high dilutions, suggesting a genetically determined specific hypersensitivity" (Lison et al. 1980). For other studies on the genetics of smell, see Beauchamp et al. (1985) and Wysocki and Beauchamp (1984).

Chapter 4 Chromosomes

Chromosomes, Genes, and DNA
Gross Structure of Chromosomes
The Components of Genes and Chromosomes
Fine Structure of Chromosomes

Human Chromosomes
Methods of Chromosome Preparation and Analysis
The Normal Human Karyotype

> *Up to now I have passed over another very remarkable phenomenon: the threads divide themselves in half lengthwise . . . Later the threads move apart from one another along their entire length . . . When I discovered this, I immediately thought . . . that one longitudinal half of each thread might move into one half of the nuclear figure, and the other half thread into the other nuclear half, in other words, each into a future daughter nucleus . . .*
>
> *I have presented this hypothesis here because the longitudinal splitting of the threads appears to me to be too remarkable not to be worthy of some attempt at explanation. At the same time I throw this out purely as a possibility without insisting upon it.*
>
> *Walther Flemming (1879)*

About the time that Mendel began his experiments with peas, the German pathologist Rudolf Virchow proposed that all cells come from pre-existing cells rather than from nonliving material. This new theory stimulated research on cell division, the process fundamental to reproduction and development in both plants and animals. Walther Flemming, a German cytologist, was among the first to devise ways of staining cells so that their internal features could be seen clearly. What he then observed in stained nuclei was "a very delicately interconnected basket-work of winding threads of uniform thickness." Flemming described cyclical changes in these threads, which were later named **chromosomes** (Greek *chroma*, "color"; *soma*, "body"), and recognized their central role in cell division.

A few years earlier, Friedrich Miescher had developed ways of separating intact nuclei from the cytoplasm that surrounds them. From these isolated nuclei he extracted an acidic substance, rich in both phosphorus and nitrogen, that was "not comparable with any other group at present known." Miescher called it **nuclein** and speculated that "a whole family of such phosphorus-containing substances, differing somewhat from each other, will emerge . . . which perhaps will deserve equal consideration with the proteins." "Here," he suggested, "lies the most essential physiological role of phosphorus in the organism," noting that in plants it is concentrated in the growing tips, where cell multiplication occurs (Miescher 1871).

These two lines of research, one on the structure and behavior of chromosomes and the other on the chemistry of the hereditary material deoxyribonucleic acid, did not actually converge until around 1950. Now they are very closely interwoven. In this chapter we present a brief overview of what is known about the physical and chemical nature of chromosomes a century after the inquiries began.

CHROMOSOMES, GENES, AND DNA

Only in dividing cells do chromosomes appear as compact bodies (Figure 1). At other times, when cells are going about their metabolic business, the chromosomes exist as ultrafine threads of chromatin dispersed throughout the nucleus. Despite these striking changes in form, which result from different degrees of compaction (due to coiling and folding), each chromosome retains its structural continuity and individuality throughout suc-

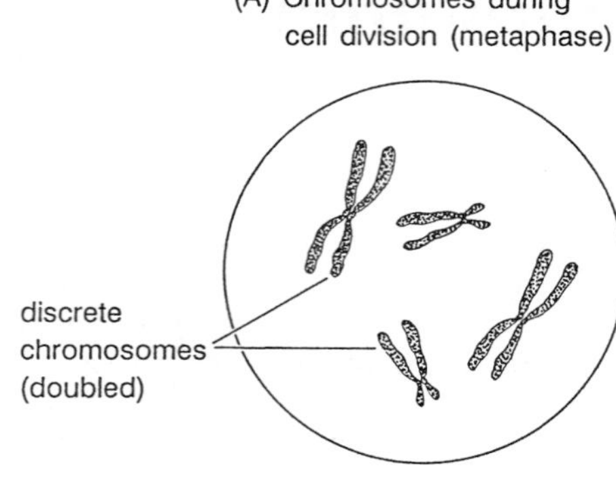

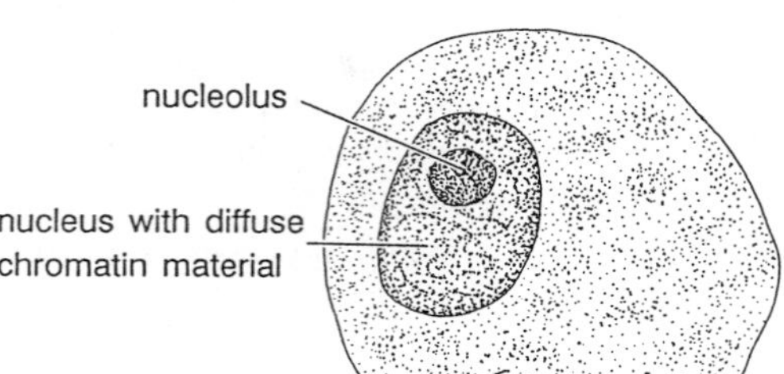

Figure 1 (A) Dividing and (B) nondividing cells. These drawings show the chromosomal material as it would appear in a light microscope; the cells are not drawn to exactly the same scale. It is assumed that this species has two pairs of chromosomes.

cessive cell cycles. With few exceptions, the *number* of chromosomes is the same for all cells in an organism and, indeed, within a given species. Humans, for example, have 46 chromosomes; dogs, 78; carp, 104; red ants, 48; and fruit flies, 8. Among plants, potatoes have 48 chromosomes; broad beans, 12; and white oaks, 24. The relation, if any, between chromosome number and size or complexity of the organism is not clear.

Gross Structure of Chromosomes

During cell division, individual chromosomes can usually be identified by size, shape, and banding patterns produced by various biological stains (Figure 2). These properties remain constant from cell generation to cell generation. Every chromosome has an indented region, the **centromere**, which is necessary for movement during cell division (Chapter 5). The position of the centromere is constant for a given chromosome, dividing it into two **arms** of specific lengths. Some staining procedures reveal unique **banding patterns** for each chromosome, whereas others show that each centromere is flanked by darkly staining blocks of highly condensed material called **heterochromatin**. Heterochromatin contains few known genes, and its function is not well understood. The more lightly staining material, called **euchromatin**, contains most of the genes.

Chromosomes duplicate themselves *before* the onset of cell division. After replication, each chromosome consists of two identical copies known as **sister chromatids**, lying side by side but still connected at the centromere region. After cell division begins, these tremendously long and threadlike chromosomes gradually shorten and thicken, becoming more rodlike. During their most condensed period (a brief stage called **metaphase**), the chromatids are partly separated; consequently, each chromosome—depending on the relative lengths of its arms—takes on a somewhat X-like or V-like shape. (See especially Figures 2 and 10).

By counting and analyzing the shapes of these metaphase figures, cytologists can describe each species by its distinctive array of chromosomes—known as the **karyotype** (Greek *karyon*, "nut" or "nucleus"). Deviations from the normal karyotype are often associated with abnormalities or even death of the organism (Chapters 7–9). In most species the chromosomes of all cells (except eggs and sperm) occur in *pairs*, the two members of a pair looking alike and carrying the same gene loci. For example, human cells have 23 pairs, carp cells 52 pairs, and white oak cells 12 pairs. The normal number of chromosomes in a nucleus is called the **diploid** (or **2n**) number. During the production of gametes, this number is halved in such a way that an egg or sperm contains only one member of each chromosome pair; such cells are said to contain the **haploid** (or **n**) chromosome number. This special reduction process prevents the doubling and redoubling of the chromosome number in successive generations. Instead, the union of two haploid (n) gametes restores the diploid (2n) chromosome number in the zygotes of each generation.

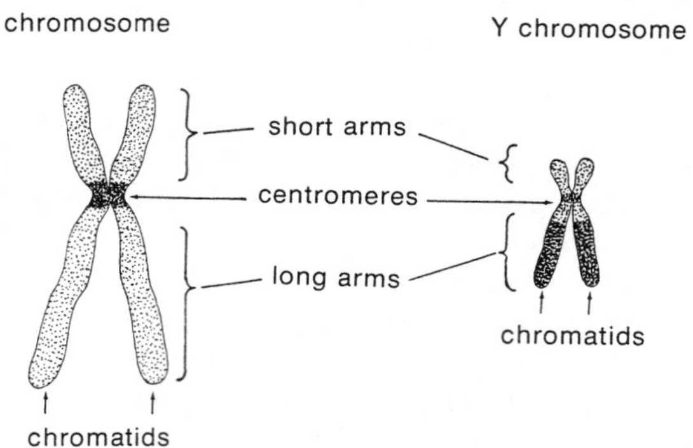

Figure 2 Drawing of human X and Y chromosomes during division to show the heterochromatin (solid) and the euchromatin (open). Most genes that control the production of proteins are located in euchromatic regions. The darkly stained areas are the C-bands. Drawn about 8,000 times actual size.

Some early cytologists suspected that chromosomes played a role in heredity, but they had no firm evidence for this until Mendel's work was uncovered in 1900. Soon thereafter, a Columbia University graduate student named Walter Sutton and a German cytologist named Theodor Boveri independently advanced the **chromosome theory of heredity**—which was based on comparisons between the behavior of genes and the behavior of chromosomes. Details will be presented after our discussion of cell division (Chapter 5), but their conclusion is undoubtedly familiar to most readers: the chromosome was visualized as a structure carrying many genes in linear order, a hypothesis that was subsequently verified by a wide variety of experiments.

The Components of Genes and Chromosomes

Half a century later at Cambridge University, James Watson and Francis Crick (1953) proposed a structural model for the stuff that genes are made of, **deoxyribonucleic acid (DNA)**. Two other workers, Rosalind Franklin and Maurice Wilkins at King's College of the University of London, had been analyzing the way that the atoms in DNA scattered a beam of X rays—and their photographs led Watson and Crick to deduce that the molecule consists of two antiparallel chains wound about each other to form a **double helix**. This structure can be crudely pictured as a twisted ladder (Figure 3), the vertical support or backbone of each chain being a monotonous series of alternating **(deoxyribose) sugar** and **phosphate** groups. Projecting inward to form the horizontal rungs are pairs of **bases**, each base being attached to a sugar group on the outside and connected by hydrogen bonds to its complementary base on the inside. (See Appendix 3 for a discussion of chemical bonds.) A single base plus its attached sugar and phosphate group together make up a **nucleotide**, the repeating unit of DNA. Thus each **nucleotide pair** forms a rung plus a small section of each backbone.

Four different bases—two **pyrimidines** (adenine and guanine) and two **purines** (thymine and cytosine)—provide variability within the molecule (Chapter 13). A unique feature of this model is the way in which the two chains are joined: The base pairs that form the rungs normally consist of a (larger) purine that is **hydrogen-bonded** to a (smaller) pyrimidine. In addition to this size constraint, the molecular configurations of the bases impose a further, very specific type of pairing: *adenine* can only bond with *thymine* (forming AT pairs), and *guanine* bonds only with *cytosine* (forming GC pairs). Any other base-pair combination leads to instability in the double helix. This finding explained a puzzling fact previously discovered by Erwin Chargaff of Columbia University—that the degradation of DNA yields 1:1 ratios of adenine to thymine and guanine to cytosine. The

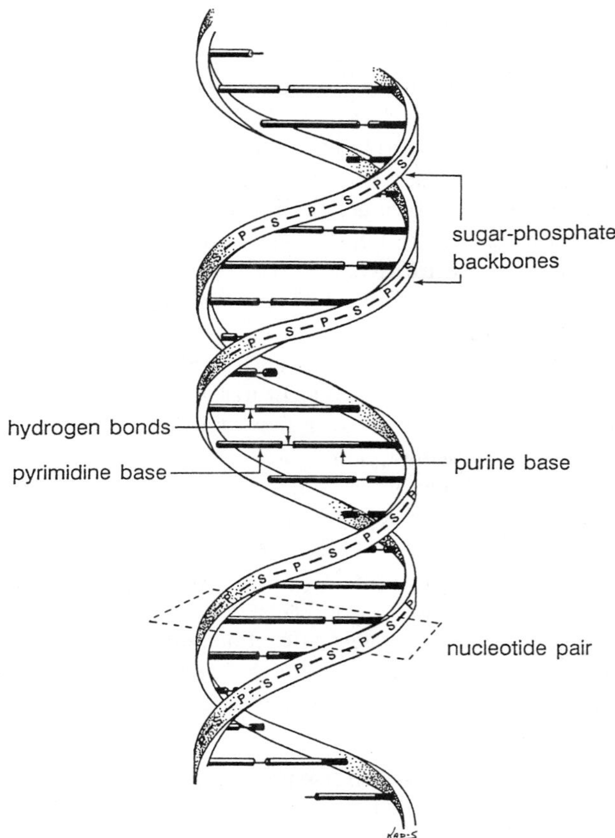

Figure 3 The structure of DNA. The nature of hydrogen bonds is explained in Appendix 3. This drawing shows the components of DNA about 15 million times actual size.

significance of these ratios now became clear. Wherever adenine occurs, it must be paired with thymine; likewise, guanine is always found opposite cytosine. The sequence of bases along the *length* of the molecule can and does vary enormously—but whatever may be the array of bases on one chain, the other chain is constrained to have the *complementary* sequence.

Almost a century after Mendel proposed the existence of discrete hereditary factors, their chemical nature was finally elucidated. The parallels in the behavior of genes inferred from breeding experiments and the physicochemical behavior of DNA were obvious. First of all, both genes and DNA scrupulously *reproduce* themselves. As Watson and Crick stated in their one-page publication (1953): "It has not escaped our notice that the specific pairing we have postulated immediately suggests a possible copying mechanism for the genetic material." A DNA double helix "unzips" when the weak hydrogen bonds break down. The two chains separate locally, and each acts as the template for the assembly of a chain of complementary base sequence. During this process, each old chain becomes hydrogen-bonded to its new complementary chain, forming two identical helices from one original (Figure 4).

Although genes and DNA usually copy themselves perfectly, both are known to exhibit rare mistakes. It has been found that during DNA replication, bases are occasionally mispaired, left out, inserted too often, or rearranged. These **mutations**, if not repaired, are perpetuated when complementary copies are made. Thus arises new material for evolution.

In addition to replicating and mutating, genes and DNA must *store the information* that is transmitted from mother to daughter cells or from parents to offspring. The stored information in DNA resides within the sequence of millions of bases along the DNA molecules—a developmental blueprint that is converted into each organism's unique set of characteristics through the process of protein synthesis. A gene is said to be *expressed* when its information is used by a cell to make a specific protein, this protein consisting of a special sequence of units called amino acids. How it happens will be described more fully in Chapters 13 and 14.

Fine Structure of Chromosomes

Chromosomes of higher organisms (*eukaryotes*) consist of more than DNA. Chemical analyses reveal three other major components: histone proteins, nonhistone proteins, and ribonucleic acid. DNA and histones occur in constant and roughly equal proportions within a cell's nucleus, but the amounts of the latter two constituents vary with the metabolic activity of the cell.

Histone proteins are small molecules (about 100–200 amino acids) that regulate the degree of chromosomal compaction of the DNA double helix. There are five classes of histones (H1, H2A, H2B, H3 and H4); all are rich in *basic* amino acids (i.e., those with a net positive charge at neutral pH), especially lysine and arginine, and all lack tryptophan. Produced in the cytoplasm, histones immediately migrate into the nucleus and become tightly bound to the negatively charged phosphate groups of newly replicated DNA.* This DNA–histone complex, the **nucleosome** (Chapter 5), forms the elementary structural unit of chromosomes. Histones (especially types H3 and H4) have scarcely changed throughout evolution; indeed, those taken from organisms as different as calves and pea plants are virtually identical! Such extreme *conservation of sequence* probably means that (1) every amino acid in the molecule is functionally important, and (2) histones have the same function in all organisms.

Nonhistone proteins (NHP) are a heterogeneous mix of several hundred kinds of chromosomal proteins that are not histones. Many of them contain the more *acidic* amino acids (i.e., those bearing a negative charge in neutral pH). Found in this potpourri are some enzyme systems, including those that direct the synthesis, modification, and degradation of nucleic acids. In addition, a sizable fraction of NHPs function as structural components of the chromosomes. Some studies suggest that certain NHPs may act as specific gene regulators, but isolating and analyzing such relatively rare molecules is a formidable task. Unlike histones, NHPs seem to vary—in amount and composition—with the degree of metabolic activity in different cell types, or in the same cell at different times, or in differentially active regions of the same chromosome. Within a nucleus, phosphate groups are added to and detached from NHPs in patterns that seem to be closely correlated with specific stages of cellular differentiation. Some NHPs are both species-specific and tissue-specific in their ability to bind DNA.

Researchers have identified two subclasses of NHP, but are not yet certain of their exact functions. *Ubiquitin*, as its name implies, occurs in all cells—from bacteria to mammals. A very short molecule, it is found linked to a

*During the formation of sperm in some animals (including humans), most histones are converted to highly basic proteins called **protamines**; these form an even tighter complex with DNA than do histones.

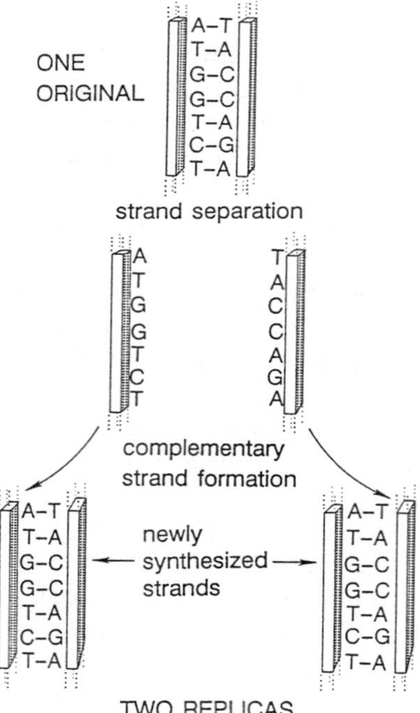

Figure 4 The replication of a portion of a DNA molecule. The helix is shown untwisted, with the sugar–phosphate backbones drawn as bars. The dashes between pairs of bases represent hydrogen bonds. The building blocks (sugars, phosphate groups, and bases) of the newly synthesized complementary strands are present in the cell.

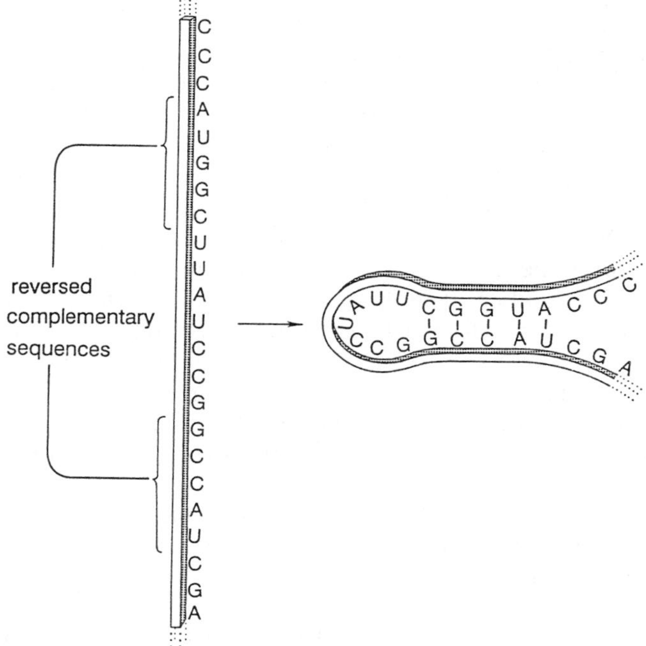

Figure 5 Reversed complementary sequences of bases (palindromes) in a single strand of nucleic acid can lead to a looped double-stranded region by hydrogen bond formation between lengths of (now) complementary bases.

2A histone subunit in 10–30% of the nucleosomes, mostly in euchromatic genes that have been expressed. Ubiquitin also occurs in the cytoplasm, often attached to proteins that are later degraded. The second NHP subclass consists of four types of *high mobility group (HMG) proteins*, so named because their small size allows them to migrate rapidly in an electrophoretic gel (see Chapter 15 for details). HMG proteins are found in all tissues. Although too common to be regulators of specific genes, they do bind directly to DNA and change its conformation—a process that seems to be a precondition for gene expression.

Also associated with chromosomes is a product of gene expression, **ribonucleic acid (RNA)**, which differs from DNA in several ways. The simple sugar in its sugar-phosphate backbone is *ribose* rather than deoxyribose. Like DNA, RNA contains the bases adenine, guanine, and cytosine; but instead of thymine it has *uracil*, which pairs by hydrogen bonding with adenine. Aside from being vastly shorter than DNA, RNA molecules are *single-stranded* rather than double-stranded. Some, however, have palindromic sequences that loop back to form limited areas of double helical structure (Figure 5). Rather than being a self-reproducing molecule, RNA is the **primary gene product**, synthesized in the nucleus from a DNA template. Three types of RNA (messenger, transfer, and ribosomal) are made and dispatched to the cytoplasm to play leading roles in protein synthesis, the process by which genes are expressed (Chapter 13).

Exactly how are all these macromolecules organized into functional chromosomes of such exquisite complexity? Nobody knows for sure, but this is an area of very active research coupled with lively speculation and debate. It is clear, however, that a chromosome contains just *one very long molecule of DNA*. Very gentle disruption of individual chromosomes yields extremely large, unbroken molecules of DNA whose length and weight matches that expected for the total DNA in those chromosomes. Thus, for example, in the largest human chromosome a DNA molecule 160 million nm* long and 2 nm wide gets packed into a metaphase chromosome 10,000 nm long and 500 nm wide. (Stretched out, the total DNA in a single human egg or sperm nucleus would measure just under 1 meter.)

The packing process involves repeated coiling and folding of the basic morphological unit of the chromosome, the DNA–histone fiber; but the process is not well understood—perhaps because, as shown in Figure 6, "spread whole chromosomes under the electron microscope look even at their best something like a bad day at a macaroni factory" (Swift 1974). Although it is impossible to follow the incredibly tortuous path of a single DNA-histone fiber through its entire length, much has been learned from electron microscopic studies† combined with biochemical and crystallographic analyses.

[Whatever the approach, all investigators must grapple with the following problem: after treatment that may include centrifuging, extracting, shearing, digesting, and staining—often with harsh agents—can they be sure that what they finally analyze in the test tube or under the microscope bears a reasonably close resemblance to what exists in functional cells? Might their experimental methods exclude or destroy some of the cell constituents they hope to study or, conversely, induce the formation of new constituents (artifacts) that simply do not exist in the native material? These are tough questions whose answers must be continually sought and reassessed.]

A substantial body of research begun in the early 1970s has established that the elementary chromatin fiber in all higher organisms is made up of repeating units called **nucleosomes** (Figure 7). Each nucleosome includes a total length of about 200 nucleotide pairs of DNA, of which 146 base pairs make 1.8 turns around a cluster of 8 histone molecules to form the **core particle**.

*For units of measurement in the metric system, see Appendix 1.

†The resolving power of a good light microscope is about 300 nm, which means two dots closer together than this will appear as one. The resolving power of a good electron microscope, with biological specimens, is about 2 nm, or 150 times higher than that of the light microscope. The maximum useful magnification obtainable with a light microcope is about 1,500 times (1,500×), and with an electron microscope, about 200,000×.

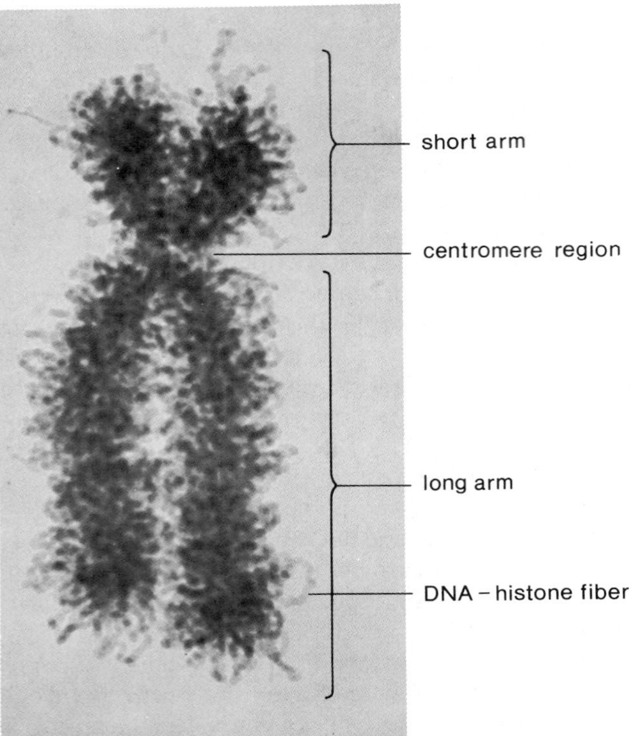

Figure 6 Human chromosome 12 during division (electron micrograph). Note the two chromatids joined at the centromere (constricted region). Each chromatid consists of a single DNA–histone fiber. Shown about 25,000 times actual size. (From DuPraw 1970.)

Most of the remaining DNA segment forms a **linker** that connects to the next nucleosome. The core particle is a disk composed of two identical halves, each containing one each of four different histone molecules. The fifth type of histone molecule lies outside the core particle, sealing off the region where the DNA enters and exits (Kornberg and Klug 1981).

When organized into nucleosomes, a DNA double helix undergoes a sevenfold shortening or compaction to form a fiber 10 nm in diameter—but how this in turn is coiled and folded to form a 30-nm-wide chromatin fiber is not entirely clear. Even less well understood are the additional levels of coiling and folding that ultimately yield a human metaphase chromatid whose DNA double helix has been compacted 8,000- to 10,000-fold (See Chapter 5).

A structural role for certain nonhistone proteins was recently proposed by a group of researchers who examined metaphase human chromosomes after the removal of all histones (Laemmli et al. 1978). Although the DNA was greatly unraveled in these histone-depleted chromosomes, they still retained their identifiable shapes. Each one had "a central scaffold surrounded by a halo of DNA. The scaffolding consists of two fibrous backbones, one for each chromatid, joined at the centromere . . . The DNA is evidently attached to the scaffold in loops" (Figure 8). When the DNA was also digested away from these histone-depleted chromosomes, the scaffolds remained intact and were shown to consist exclusively of a few nonhistone proteins. How these structures arise is not known, but Laemmli and colleagues hypothesize that "the scaffolding proteins persist in some form throughout the cell cycle and, indeed, throughout the life cycle of the organism."

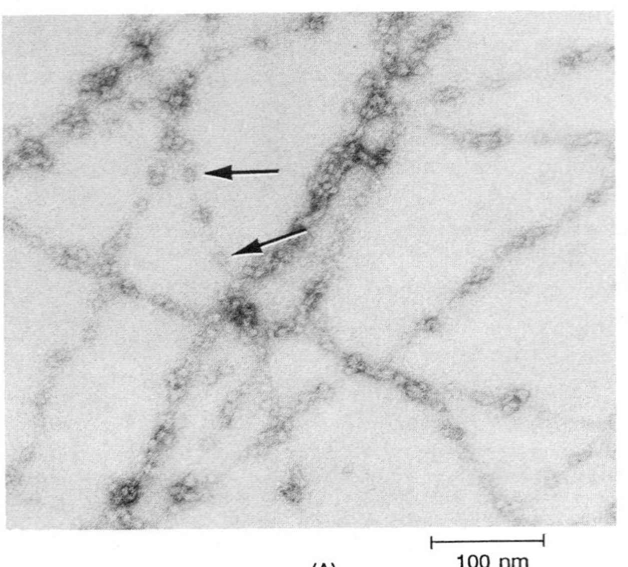

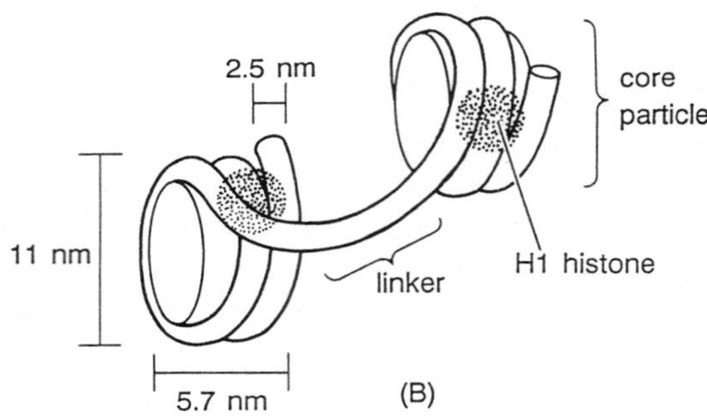

Figure 7 Nucleosome structure. (A) Chromatin from chicken blood cells. In this electron micrograph (×150,000), individual nucleosomes (arrows) look like tiny, clear beads on a string. (B) Two nucleosomes. A 146-base-pair length of DNA is wrapped 1.8 times around two identical histone tetramers (a group of four different molecules) to form each disk-shaped core particle. The remaining stretch of DNA, the linker, is associated with a fifth type of histone molecule, H1. (A courtesy of Christopher Woodcock.)

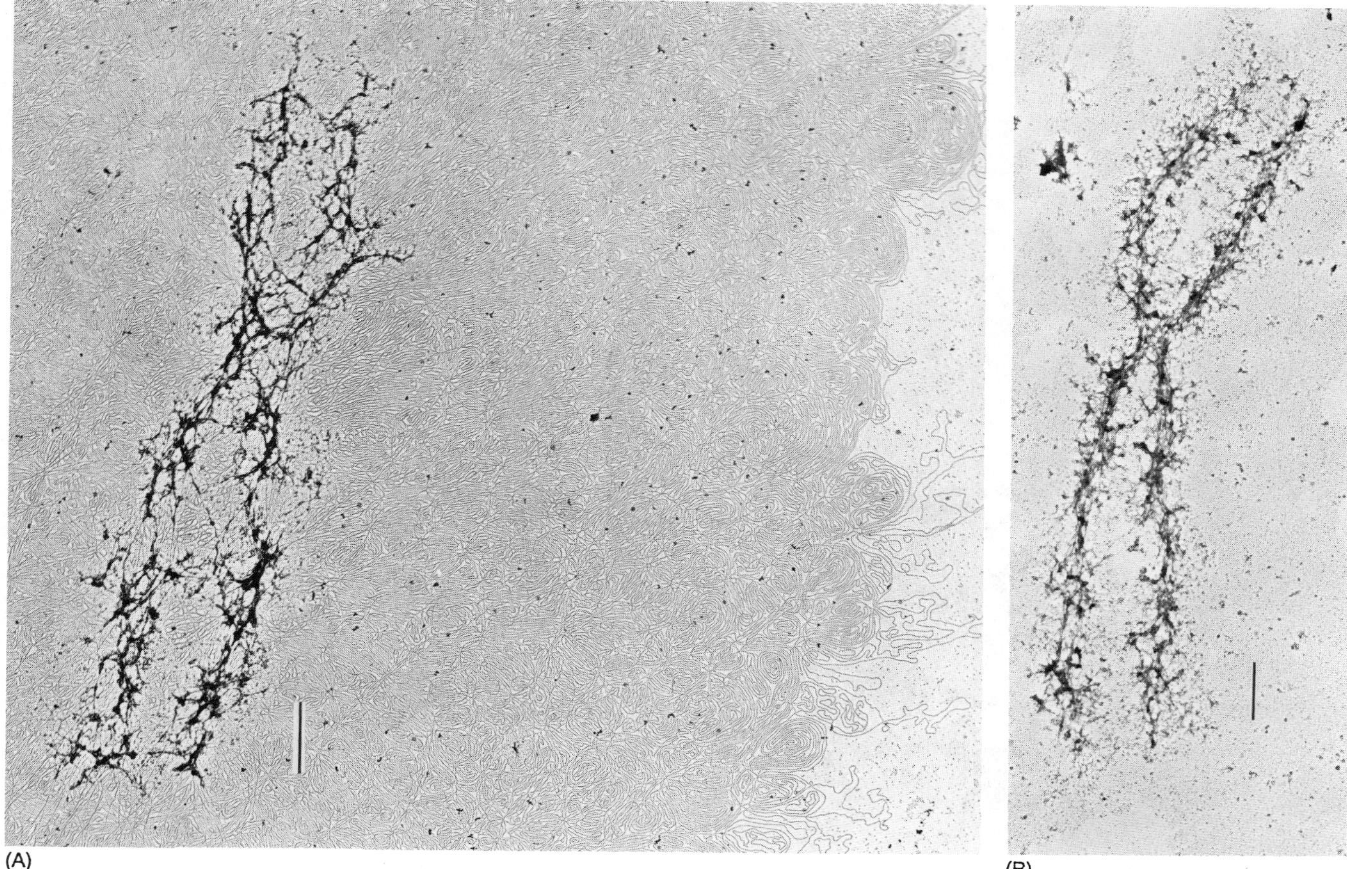

Figure 8 Human chromosomes, each with a dark-staining central scaffold of nonhistone protein outlining the shape of two chromatids joined at the centromere region. (A) After all histones are removed from this chromosome, an immense and tortuous maze of DNA loops out from the scaffold. (B) With both histones and DNA removed, all that remains is the central scaffold. The bar in each photo represents 1 μm. (Courtesy of Ulrich K. Laemmli.)

HUMAN CHROMOSOMES

It would be unfair to suggest that cytologists before the 1950s were careless in examining human chromosomes; rather, the techniques then available were simply not up to the task. The American cytologist Theophilus Painter (1923) published unequivocal drawings of 48 human chromosomes. He wrote that he was able to do this because, unlike previous workers, he used fresh material and because, in his own "modern era of cytology," he utilized better techniques of fixation, sectioning, and staining. Although he mentioned that the counting error "might well amount to one chromosome" and although he had previously written that the best preparations showed 46 chromosomes, Painter nonetheless settled on 48. The question of the number of human chromosomes thus seemed to be settled, and for more than three decades cytologists and geneticists deferred to his drawings or to those of others who also found 48. Some authors, including Painter, did not publish the photographs on which their drawings were based. Others juxtaposed a small photo with an interpretive drawing. Figure 9, one of the better examples from this period, shows that accurate counting was hampered by vague outlines and touching or overlapping chromosomes. But Painter's conclusion was accepted without question by his peers.

In the 1950s Joe-Hin Tjio and Albert Levan, working with better techniques, were surprised to find only 46 chromosomes. In 1956 they wrote: "The almost exclusive occurrence of the chromosome number 46 in one somatic tissue (lung) derived from four individual human embryos is a very unexpected finding." Cautiously they went on to say that, until another count is done on human spermatogonial mitoses, "we do not wish to generalize our present findings into a statement that the chromosome number of man is 2n = 46, but it is hard to avoid the conclusion that this would be the most natural explanation of our observations."

Within months this finding was verified in human spermatocytes (Ford and Hamerton 1956); this account makes one wonder how many other workers had previously counted 46, but failed to report the "loss" of two chromosomes. Tjio and Levan's experience teaches us

(A)

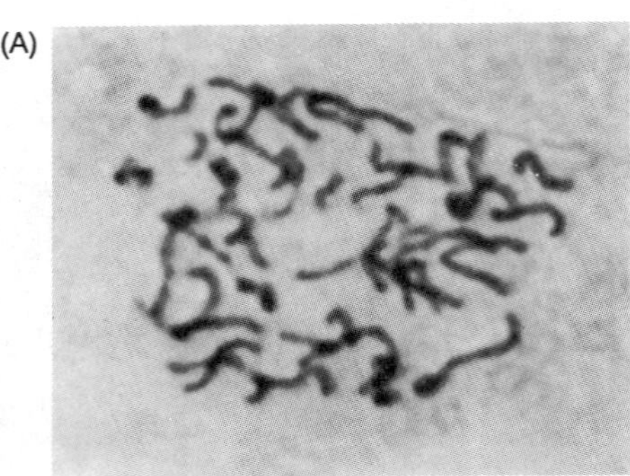

(B)

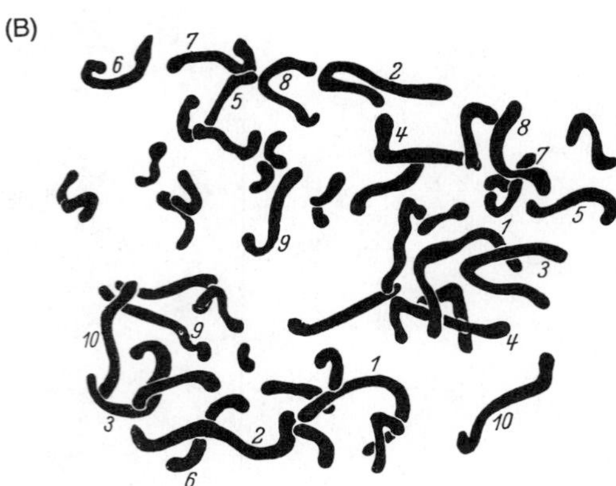

Figure 9 (A) A 1936 photograph (×2500) of the divided chromosomes from a culture of embryonic lung cells. (B) An interpretive drawing by the investigators of the chromosomes seen in the photograph. The numbers purport to show the largest ten pairs of chromosomes. (As recently as 1970, the unequivocal matching of some of these pairs was considered impossible.) Although these authors supported the human chromosome number of 48—well, maybe 47—this drawing seems to show fewer. (From Andres and Navaschin 1936.)

that objective reexamination and reevaluation of "established facts" are crucial elements of scientific progress. Figure 10, from white blood cells, provides more evidence that the chromosome number in our species is 46.

Methods of Chromosome Preparation and Analysis

Tjio and Levan utilized four technical innovations that provided highly reproducible results: the use of tissue culture, colchicine, hypotonic saline, and the squashing technique.

TISSUE CULTURE. Chromosomes are clearly visible only in rapidly dividing cells. But rapidly dividing adult cells are rare and not easily obtainable from humans. Thus the development of artificial environments in which living cells and tissues could survive and reproduce was a major breakthrough in the study of biological systems. Human cells obtained from various sites (including blood, skin, bone marrow, amniotic fluid, and cancerous tissues) are now routinely cultured in vitro (literally, "in glass") for many cell generations—and in some cases indefinitely.

COLCHICINE. This alkaloid extract of the autumn crocus (genus *Colchicum*) was used by the ancient Greeks to treat gout, a practice that continues today. After cytologists discovered in the early 1900s that it affects cell division, botanists used the drug extensively to produce *polyploids*, plants with high multiples of the haploid chromosome number. When added to cells in culture, colchicine arrests their growth at the *metaphase* stage of cell division—when chromosomes are most compact, with distinctly visible centromeres, and therefore easiest to identify. This accumulation of metaphase cells greatly increases the likelihood of finding good chromosomal spreads among the preparations of cultured cells.

HYPOTONIC SALINE. When put in hypotonic medium (Greek *hypo*, "down" or "less than," in contrast with *hyper*, "up" or "more than," or *iso*, "same as"), cells absorb water by osmosis and swell. Quite by accident, a laboratory worker discovered that within the nucleus the metaphase chromosomes also become dispersed; consequently, after being fixed (killed and preserved) and stained, they show little or no overlapping when viewed under the microscope (Hsu 1952).

Figure 10 The chromosomes at metaphase (middle of cell division) from a white blood cell of a man (×2000). The chromosomes are stained with Giemsa, which here colors all parts roughly the same. Each chromosome appears somewhat X-shaped because it has replicated, but the replicas remain attached at the centromere region. How many are there? (Courtesy of Irene Uchida, McMaster University.)

SQUASHING TECHNIQUE. The chromosomes that Painter counted were in very thin slices of human testes. This standard sectioning technique is not well suited for chromosomal analyses, however. It may exclude some chromosomes that lie just outside the plane of the slice, or it may break some chromosomes into pieces—an outcome leading to losses or gains in the true chromosome number. Tjio and Levan, on the other hand, *squashed a suspension of cells* (obtained from tissue culture) between a microscope slide and a cover slip, thereby producing flattened whole cells. Among those that broke open, they found some metaphase chromosomes well spread out and isolated from those of other cells. Nowadays, however, cells are no longer squashed. Instead, they are dropped onto a slide and air-dried.

Since 1956, there have been some other very important additions to the methods and reagents used for preparing and analyzing chromosomes: the use of phytohemagglutinin, chromosome banding procedures, bromodeoxyuridine, high-resolution banding, supplementation of culture medium with folic acid and thymidine, and silver staining techniques.

PHYTOHEMAGGLUTININ. (Greek *phyto*, "plant;" Latin *haem*, "blood;" *agglutinate*, "to glue to.") For decades, blood-group workers had used an extract of the red kidney bean to induce red blood cells (in vitro) to clump together and separate from white blood cells. In 1960, P. C. Nowell reported a chance discovery that phytohemagglutinin also stimulates white blood cells to synthesize DNA and to divide. He and others then devised a simplified method for culturing white blood cells in just 2–3 days.*

CHROMOSOME BANDING PROCEDURES. The original techniques for staining chromosomes, using **orcein** and **Giemsa** dyes, produced uniform coloration throughout each human chromosome (see Figures 9 and 10). Thus, the homologues in a metaphase spread could be only crudely matched by size and shape—a process both difficult and inexact. But in 1970, T. Caspersson and his colleagues at the Stockholm Karolinska Institute described a new staining procedure. They used a **quinacrine** dye, which (through a complex process that involves preferential attachment to AT-rich† regions of DNA) produces *bands* that fluoresce when viewed in ultraviolet light with a special microscope (Figure 11).

*Previously, researchers had to use either uncultured bone marrow cells or fibroblast cells that required several weeks to culture.

†AT-rich means rich in the DNA bases adenine and thymine; GC-rich means rich in the bases guanine and cytosine. AT-rich sequences comprise 51% of the *L1* family of *long interspersed repeated sequences* (Chapter 15), which dominates in Giemsa/quinacrine positive bands. GC-rich sequences make up 56% of the Alu family of *short interspersed repeated sequences*, which dominates in R-bands (Korenberg and Rykowski 1988).

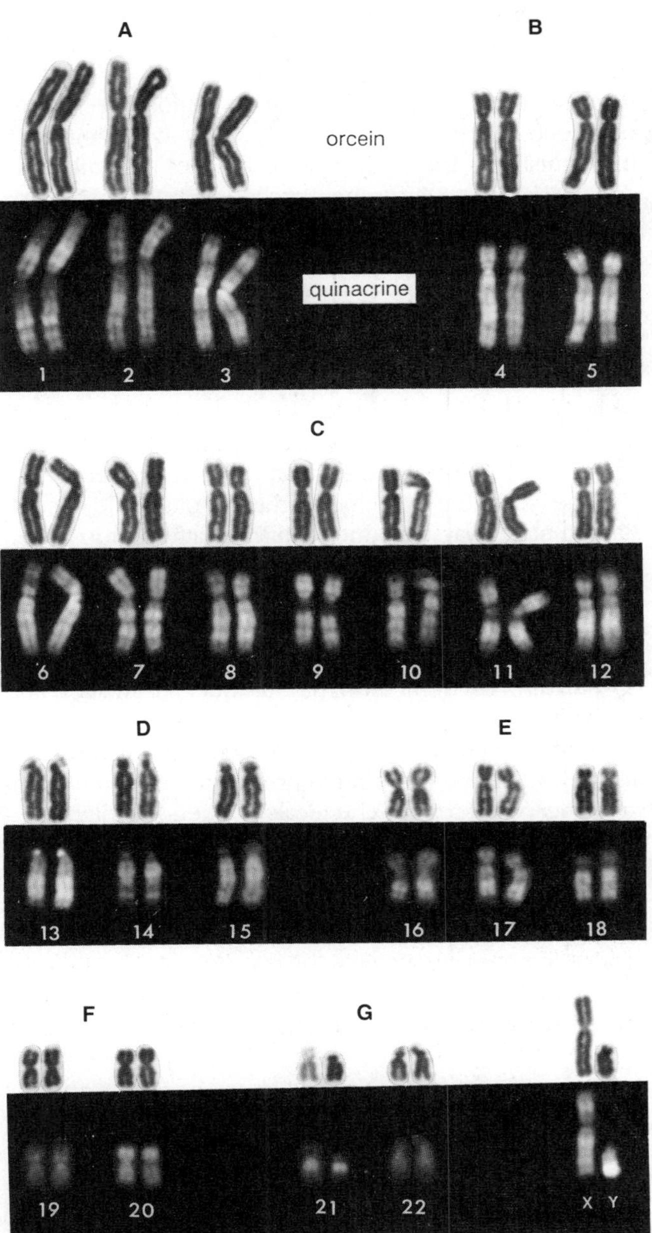

Figure 11 On the black background are quinacrine-stained human metaphase chromosomes arranged in homologous pairs according to similarities in their banding patterns. Above each pair, on the white background, are the same chromosomes stained with orcein after the quinacrine treatment. Although orcein is a good stain for giving defined chromosomal outlines, quinacrine or some other banding stain is necessary for matching up many homologous pairs. For example, it is difficult to distinguish pairs among the four orcein-stained chromosomes labeled 19 and 20; however, with quinacrine, two of the chromosomes fluoresce brighter than the other two. (*Note:* Human autosomes are arranged in groups (A–G) according to size and centromere position, and numbered (1–22) according to size. This pictorial representation of a person's chromosomes is called a **karyotype**.) (Courtesy of Irene Uchida, McMaster University.)

Because each chromosome (in a haploid set) has a unique banding pattern, every pair of homologues is distinguishable from other chromosomes of similar size and shape and thus can be matched without guesswork. (The dye Hoechst 33258 also produces fluorescent bands in AT-rich DNA.)

Because quinacrine-stained Y chromosomes fluoresce very brightly in nondividing cells as well as in metaphase spreads, the number of Y chromosomes an individual carries can be determined by staining and looking at white blood cells, or at buccal mucosa cells scraped from the inside of the mouth (Figure 12). Likewise, Y-bearing sperm can be distinguished from X-bearing sperm.

Other techniques, including the use of the old Giemsa stain in novel ways, produce banding patterns, too. When applied after pretreating chromosomes with an enzyme (trypsin) that partially digests the chromosomal proteins, Giemsa gives rise to dark bands (called **G-bands**) that are virtually identical to the fluorescent quinacrine bands (**Q-bands**). Each method has unique advantages and disadvantages.*

Another technique, called **reverse-** or **R-banding**, also uses Giemsa—but after pretreatment of the chromosomes with certain chemicals (buffers) at high temperature and low pH (6.8), so that the dye now combines with the GC-rich DNA. Consequently, the dark G-bands are light, and the light bands (called **R-bands**) look dark.

A different kind of pretreatment (acid followed by alkali) removes about 60% of the DNA from chromosomes; then the Giemsa stains only those regions (called **constitutive heterochromatin**) where DNA remains. The resultant **C-bands** mark the centromeric heterochromatin in every chromosome, plus the distal end of the Y chromosome. Figure 13 shows chromosome 1 after preparation by several different banding methods.

*Advantages of the Giemsa method over quinacrine are that (1) it does not require the expensive equipment of fluorescence microscopy, (2) the chromosomes can be analyzed directly on the slide, rather than just from photographs, and (3) the chromosome preparations can be made permanent. On the other hand, the quinacrine method (1) requires no pretreatment of the chromosomes and (2) reveals inherited variants that may be used to identify the parental origin of many chromosomes.

BROMODEOXYURIDINE (BrdU). Within a given cell, the various chromosomes and chromosome segments do not all replicate at the same time. This fact was known from older studies in which one of the four bases is in a radioactive form called *tritiated (^3H) thymidine*. Through a laborious process called *autoradiography* (Chapter 15), the replication patterns can be captured on film: wherever thymidine is incorporated into newly replicated DNA, its radioactivity develops the photographic emulsion that overlays the chromosomal preparations. The resultant arrays of silver grains indicate which chromosomal segments have replicated during a particular time period.

In the early 1970s, researchers tried out a culture medium that substituted BrdU for the nucleotide containing the base thymine. They found that BrdU alters the staining properties of chromosomes that incorporate it during their replication, and consequently enables researchers to determine much more easily whether a particular chromosome region replicates early or late. Specifically, so-called **late-replicating DNA**, which is found in all heterochromatin and also comprises one of the two X's in a normal female cell (Chapter 8), shows up very distinctly after treatment with BrdU and staining with Giemsa or a fluorescent dye (Latt 1974). Another technique, involving BrdU substitution followed by Giemsa or Hoechst staining or both, gives rise to "harlequin"-type chromosomes: alternating dark and light segments indicate the sites where sister chromatids have exchanged parts in somatic cells (Chapter 14).

HIGH-RESOLUTION BANDING. The original Q- and G-banding methods could produce up to 320 metaphase bands in a haploid set of human autosomes. This technique allowed cytogeneticists to detect many new chromosomal anomalies in humans and to study the evolutionary relations among sets of normal chromosomes

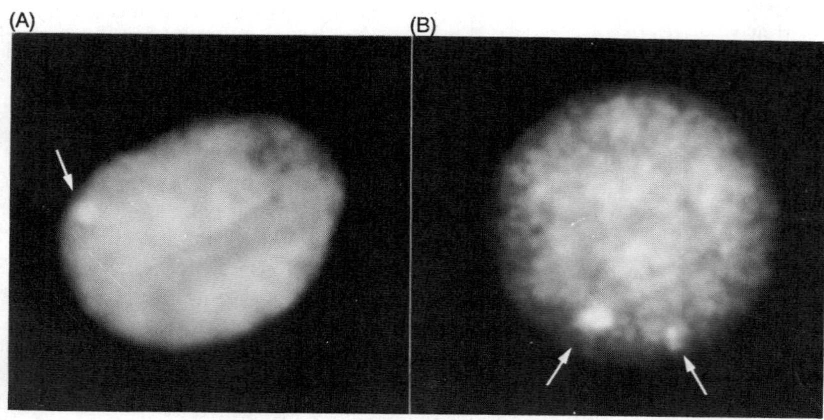

Figure 12 Fluorescent Y chromosomes in the nuclei from human male interphase cells. (A) Nucleus of a buccal mucosa cell from a 46,XY male. The arrow points to the single Y chromatin body. (B) Nucleus of a white blood cells from a 47,XYY male. The arrows point to the two Y chromatin bodies of different size. (Courtesy of Irene Uchida, McMaster University.)

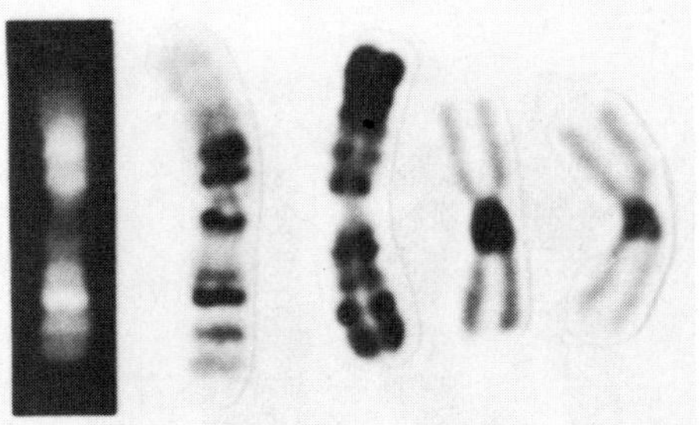

Figure 13 Human chromosome 1 prepared by four different methods. From left: Q-banding, G-banding, R-banding, and C-banding (two examples of C-banding, showing both homologues from a heterozygote). (From Therman 1986.)

from various primate species. From the late 1970s on, the power of such analyses has been expanded manyfold, to 1,000 to 2,000 bands, through some additional tricks developed by Jorge Yunis at the University of Minnesota. First he synchronized the division stages among the cells in a culture by adding a substance called *methotrexate* to the growth medium, followed by thymidine and a *brief* exposure to colchicine (in order to minimize condensation). This method gave rise to many high-quality mitoses—not only in metaphase, but also in the immediately preceding stages of *prometaphase* and *middle-to-late prophase*. Comparisons of a given chromosome in various stages of early division (Yunis 1976) showed that "the major dark and light bands of metaphase chromosomes result from the close apposition of smaller and multiple units found in the more elongated late prophase and prometaphase chromosomes" (Figure 14). With this new high-resolution technique for examining prophase chromosomes, Yunis and his co-workers were able to detect "minute chromosome defects previously unidentified" in several patients with congenital disorders. Later analyses have extended such findings to many other conditions. For example, they found chromosomal defects in the malignant cells from *over 90%* of a large group of leukemia patients—whereas only 50% of these individuals would have shown detectable abnormalities with the older banding methods (Yunis 1983). Much more on chromosomes and cancer will be presented in Chapter 19.

Although cytogeneticists do not yet understand all the biochemical complexities of the various staining reactions, they have learned quite a bit about the relations between banding patterns and chromosome structure. It is clear, for example, that chromatin exists in three main forms: **constitutive heterochromatin** (C-bands) flanking the centromeres, **intercalary heterochromatin** (G-dark and Q-bright bands), and **euchromatin** (R-bands) in the chromosome arms. For details, see Macgregor and Varley (1988), Therman (1986), Bickmore and Sumner (1989), Bennett et al. (1984), and Comings (1978).

ALTERATION OF THE STANDARD CULTURE MEDIUM TO DETECT FRAGILE SITES. In 1977 several Australian cytogeneticists described an interesting abnormality—a nonstaining constriction connected to a tiny knob at the tip of the long arm of the X chromosome—and proposed that this **fragile site** is associated with **heritable X-linked mental retardation** in males (Sutherland 1977, 1983; Harvey et al. 1977). In the 1960s there had been a few reports of such X chromosome variants in families with retarded males, but they seemed to be rare and of limited significance. We now know, however, that "the fragile X syndrome is of major importance. Next to trisomy 21, it is the most common of the causes of mental retardation that can be specifically diagnosed" (Gerald 1980; see also Sutherland and Hecht 1985).

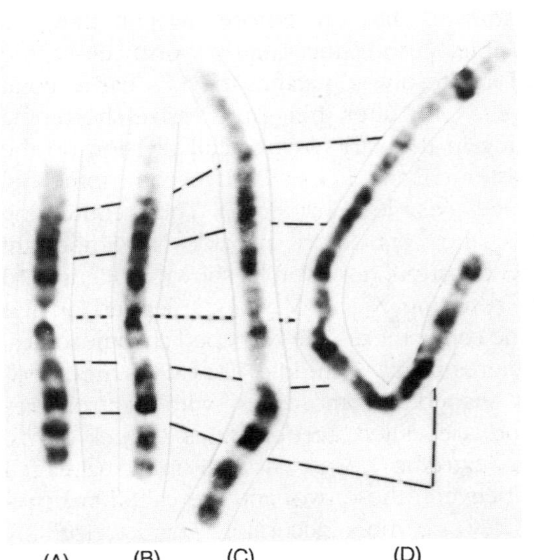

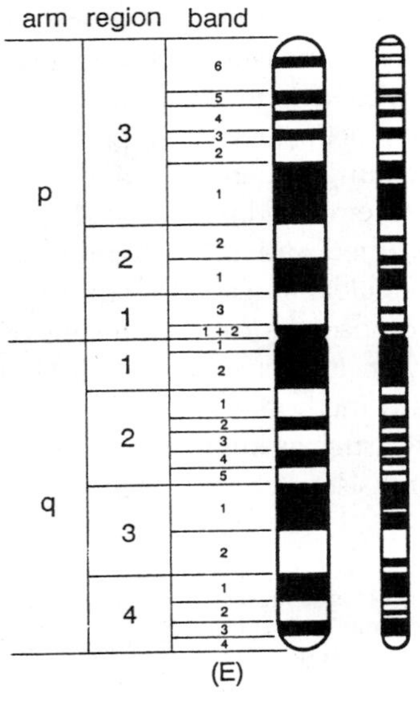

Figure 14 Human chromosome 1 as it looks at (A) mid-metaphase, (B) early metaphase, (C) just before metaphase, and (D) late prophase. In mid-metaphase the bands are fewer and thicker, while in progressively earlier stages the bands are more numerous and thinner. (E) In the diagram of chromosome 1, the chromatid on the left shows the mid-metaphase banding pattern and the chromatid on the right shows the G-bands seen in late prophase. (From Yunis 1976; diagrams after ISCN 1985.)

Why did it take so long to verify such an important discovery? The reason lay in the change to a new culture medium, which was supplemented with folic acid and thymidine—substances that were later found to inhibit the expression of fragile sites. Only by switching back to the older "deficient" recipe were cytogeneticists able to detect these sites (Figure 15) in X chromosomes, as well as about 20 heritable fragile sites in the autosomes. The latter, however, seem to be correlated with certain kinds of cancer rather than with mental retardation (LeBeau and Rowley 1984).

SILVER STAINING TECHNIQUES. Silver staining enables investigators to identify special areas, called **nucleolar-organizing regions (NORs)**, that in humans are present on five different chromosomes. These consist of multiple copies of genes giving rise to large amounts of a special kind of RNA, called ribosomal RNA, but to no proteins (Chapters 5 and 13).

Since the methods chosen for preparing chromosome spreads depend upon the type of analysis being undertaken, there is no single all-purpose protocol. Common to most methods, however, are the following steps: (1) A heparinized blood sample is drawn, then centrifuged to separate the plasma from the red blood cells.* (2) Phytohemagglutinin is added to the plasma to stimulate the white cells to divide. (3) The plasma is put into the culture medium (with or without folic acid and thymidine) to enable the cells to multiply. (4) After several days, the culture is treated with colchicine to arrest cell division in metaphase. The cells are then rinsed with hypotonic saline to enlarge the cells and disperse their chromosomes inside the cytoplasm; by now the nuclear membrane is gone. (5) The cells are treated with a fixative to kill and preserve the chromosomes, then dropped onto a microscope slide and air-dried (the latter a substitute for squashing). (6) The slides are then stained, and well spread chromosomes are photographed. (7) A karyotype is prepared by cutting out and pairing the chromosomes according to their banding patterns. (8) If desired, the slide is destained and restained with a different dye, and the same nuclear spread is rekaryotyped, so that the identical chromosomes can be compared with different stains, as shown in Figures 11 and 16. Minor variations in the standard method are introduced for special staining techniques—such as high resolution chromosomes, R-banding, and sister chromatid exchange.

*Many other types of tissues or cells—such as cells from amniotic fluid, skin fibroblasts, bone marrow cells, embryonic lung cells, testicular tubule meiocytes, and (in plants) root tips and pollen mother cells—can be used for G-banding chromosomal analyses.

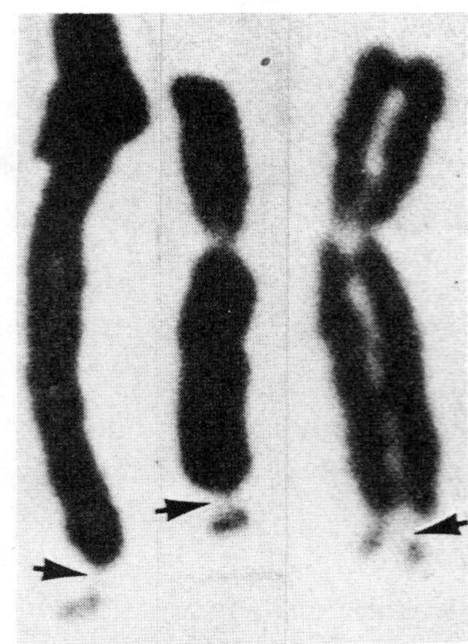

Figure 15 Three human X chromosomes from cells that were grown in the old, "deficient" type of culture medium and then stained. Each is from a retarded male with the fragile X syndrome. The fragile sites, located near the bottom of each chromosome, are the nonstaining regions (arrows) that separate the tiny, stained tips from the rest of the chromosome. Many heritable fragile sites have been identified in autosomes, too. (From *Medical World News*, August 20, 1979, p. 27.)

Another major development is automated analysis of standard slide preparations. Very promising but still fairly experimental are methods for staining cells in culture (rather than on slides) and having the chromosomes automatically sorted and karyotyped by machines with special photometric and computer capabilities (Yu et al. 1984).

The Normal Human Karyotype

Metaphase and late prophase chromosomes have the unique property of appearing visibly double. Chromosomes do *not* look this way during most of their cycle of metabolism and division. Before division they are virtually invisible, an indistinct tangle of vastly elongated threads; and in late division stages they appear as compact singletons. Only after their progressive shortening and thickening in the early part of cell division do the replicated sister chromatids of each chromosome become separately visible (Chapter 5). Their continuing attachment at the centromere provides an important means of identification, however. As shown in Figures 10 and 11, the resulting X and V shapes depend on the position of the centromere. The X-shaped chromosomes, with centromeres near the middle, are called **metacentrics**; the V-shaped chromosomes, with centromeres near one end are called **acrocentrics** (Greek *akros*, "topmost" or "extreme"). Most chromosomal shapes fall somewhere between these two, and are called **submetacentric**. They are more accurately characterized by their **centromeric index**, defined as the percentage of the total length that is spanned by the short arm. Table 1 shows that the most metacentric chromosome has an index of 48, while the most acrocentric chromosome has an index of 17.

Twenty-two of the 23 pairs of chromosomes occur in the cells of both sexes and are called the **autosomes** (Greek *autos*, "same"); note in Table 1 that they are numbered from 1 to 22 according to length. The two remaining chromosomes, because they occur in different combinations in males and females, are known as the **sex chromosomes**; the larger, metacentric one is the **X**, and the smaller, acrocentric one is the **Y**. Normally, male cells contain one X and one Y chromosome, whereas female cells contain two X chromosomes.

Although the absolute length of a given chromosome may differ somewhat from one metaphase spread to another, its **relative length**—that is, its length as a percentage of the total combined length of a haploid set of 22 autosomes—remains fairly constant from cell to cell. Thus relative rather than absolute lengths are usually given. As shown in Table 1, the longest chromosome (No. 1, with an actual metaphase length of roughly 10 μm) constitutes about 8.4% of the length of all the autosomes.

Note that the relative lengths of human chromosomes decrease with increasing chromosome number—except for the last two, 21 being shorter than 22. This misordering is due to a historical fluke involving the extra chromosome present in Down syndrome individuals

Table 1 Characteristics of human somatic cell metaphase chromosomes (Data from ISCN 1985)

Group	*Number*	*Diagrammatic representation*	*Relative length*[a]	*Centromeric index*[b]
LARGE CHROMOSOMES				
A	1		8.4	48 (M)
	2		8.0	39
	3		6.8	47 (M)
B	4		6.3	29
	5		6.1	29
MEDIUM CHROMOSOMES				
C	6		5.9	39
	7		5.4	39
	8		4.9	34
	9		4.8	35
	10		4.6	34
	11		4.6	40
	12		4.7	30
D	13		3.7	17 (A)
	14		3.6	19 (A)
	15		3.5	20 (A)
SMALL CHROMOSOMES				
E	16		3.4	41
	17		3.3	34
	18		2.9	31
F	19		2.7	47 (M)
	20		2.6	45 (M)
G	21		1.9	31
	22		2.0	30
SEX CHROMOSOMES				
	X		5.1 (group C)	40
	Y		2.2 (group G)	27 (A)

a. Percentage of the total combined length of a haploid set of 22 autosomes.
b. Percentage of a chromosome's length spanned by its short arm. The four most metacentric chromosomes are indicated by an (M); the four most acrocentric by an (A).

(see Chapter 7), which was originally named 21 in the mistaken belief that it was the *next-to-shortest* autosome. Later, with improved techniques, cytogeneticists discovered that it was really the *smallest* autosome and should be called 22. But by this time the error was so entrenched in the scientific literature that it was easier to accept the inconsistency in relative length than to correct the number.

Before human autosomes were distinguishable by banding methods, they could only be arranged according to length and centromere position into seven groups (A through G). As shown in Table 1, within each of these groups the centromeric position is quite similar. For example, Groups A and F contain the four most metacentric chromosomes; Groups D and G consist of acrocentric chromosomes; and the chromosomes in Groups B, C, and E are submetacentric.

In addition to the centromere, or **primary constriction**, some chromosomes have other pinched-in sites called **secondary constrictions**. When appropriately treated and magnified, for example, the five pairs of chromosomes in Groups D and G exhibit secondary constrictions near the tips of the short arms (Figure 16). These **nucleolar-organizing regions (NORs)** give rise to the nucleoli, which contain the special RNA and proteins destined to form ribosomes in the cytoplasm (Chapters 5 and 13). The sites and number of NORs are constant for a given species; in human diploid cells there are 10. Nucleoli tend to fuse together, however, so that fewer than 10 may be seen in an interphase nucleus. The tiny specks of chromosomal material seen at the tips of the short arms of Groups D and G chromosomes are called **satellites**. In other chromosomes there may be additional secondary constrictions, not associated with either nucleoli or satellites.

Within and among the chromosomes of *normal* individuals, the two members of a homologous pair of

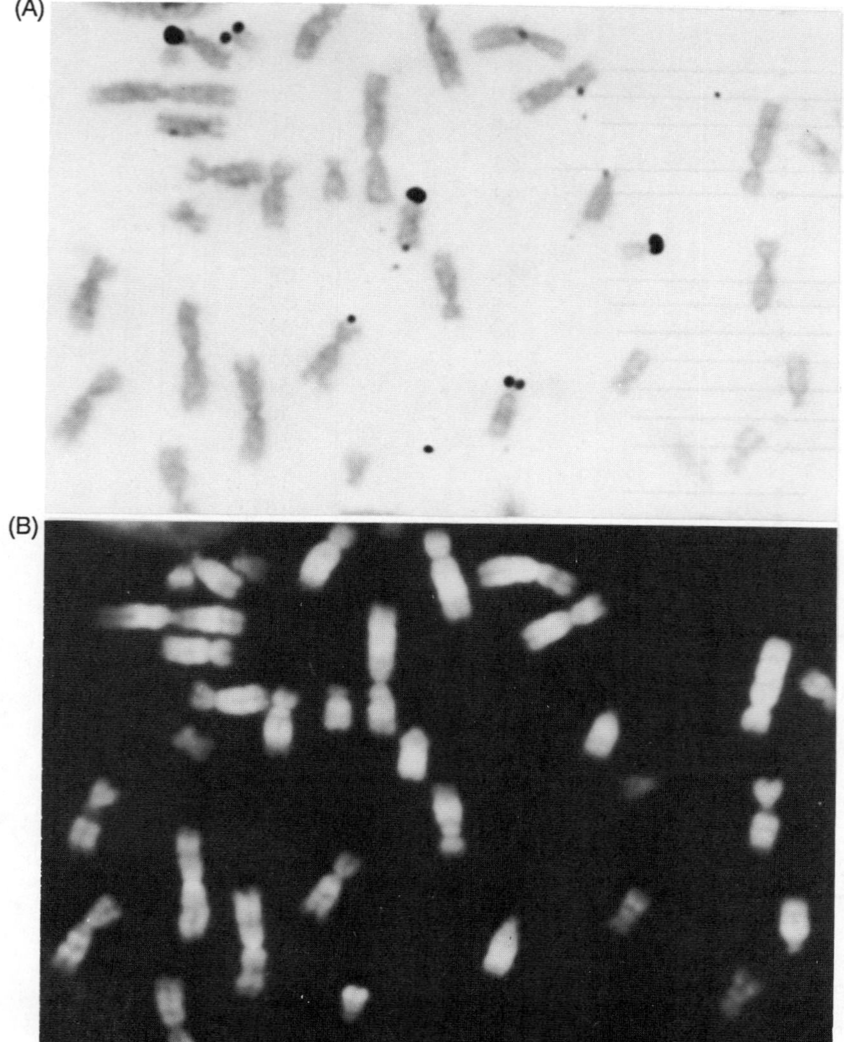

Figure 16 A sequentially stained human chromosome preparation that shows (A) the silver-stained nucleolar organizing regions (NORs), and (B) the quinacrine-stained Q-bands. (Courtesy of Irene Uchida, McMaster University.)

autosomes may often be distinguishable by slight differences in banding patterns. These observed variations were at first thought to be artifacts induced by the treatment procedures. But recent evidence indicates that some chromosome variations are inherited, unaccompanied by any phenotypic abnormalities. The best example of this is seen in the Y chromosome, whose long arm is unusually long in some males and unusually short in others. When these differences involve the C-band-positive segments, they are not correlated with any differences in maleness or other characteristics. Likewise, heritable C-band variants have been detected for chromosomes 1, 9, and 16 in normal subjects. In a study comparing the Q-banding of chromosomes 3, 4, 13, 14, 15, 21, 22, and Y among a group of 57 people (including some who were related to each other), "no two persons were found to have the same set of variants" (Olson et al. 1986). Although individual sets of banding patterns may prove to be unique, there now exists a more precise test—called DNA fingerprinting—that is very useful for excluding paternity (Chapter 15).

It is often convenient to describe an individual's karyotype with cytogenetic shorthand rather than by a picture. At the simplest level this description is merely the total number of chromosomes and the sex chromosomal complement; thus a normal female is designated 46,XX and a normal male 46,XY. To describe a site on a particular chromosome (see Figure 17), one first lists the chromosome number, then the arm (**p** designating the short arm and **q** the long arm), then the region number within an arm, and finally the specific band within that region. For example, 7p13 refers to chromosome 7, short arm, region 1, band 3—a narrow, light band. The fact that the bands appear much fuzzier in a photograph (Figure 11) than in a diagram (Figure 17) points up the need for great technical expertise in analyzing karyotypes.

To describe abnormalities in chromosome structure and number, cytogeneticists use standardized symbols. For example, a boy with typical Down syndrome (Chapter 7) is designated 47,XY,+21. He has 47 chromosomes, including normal X and Y sex chromosomes, with the forty-seventh chromosome being an *extra* chromosome 21—that is, three 21's altogether. The karyotype of a female with Turner syndrome (Chapter 8) is 45,X—which means that she lacks one sex chromosome. Other symbols, and some interspecies comparisons, will be presented in later chapters.

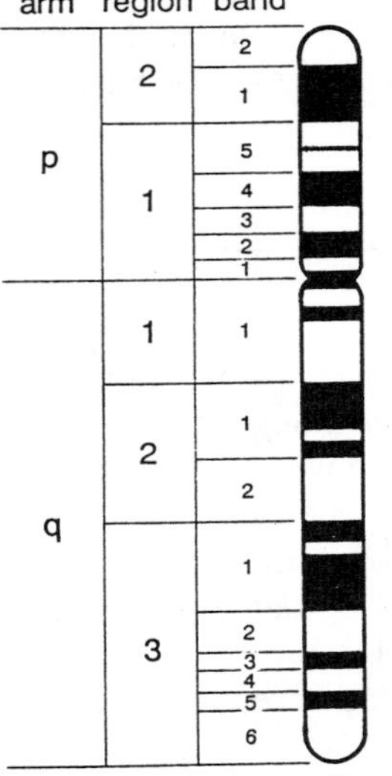

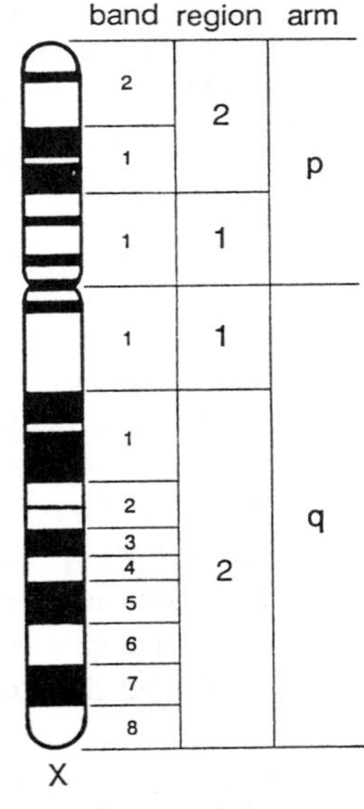

Figure 17 The quinacrine and Giemsa banding patterns (Q- and G-bands) of chromosomes 7 and X at metaphase. The dark bands are those that fluoresce brightly with quinacrine or are dark with Giemsa. These two chromosomes are almost the same length (p = short arm, q = long arm) and have almost the same centromeric index, but they can be distinguished by their different banding patterns. (From ISCN 1985.)

SUMMARY

1. Chromosomes, which carry hereditary determiners called genes, are the genetic link between generations. They occur in homologous pairs in body cells, one member of each pair derived from the female parent and the other from the male parent.

2. Chromosomes can best be seen during cell division, when they are tightly coiled. Each replicated chromosome at first consists of two identical chromatids joined by an undivided centromere.

3. Chromosomes of higher organisms contain DNA,

histone proteins, nonhistone proteins, and RNA. The basic chromosomal fiber is a DNA–histone complex, and chromosomal structure is provided by some nonhistone proteins.

4. DNA is a long molecule composed of two complementary chains wound about each other to form a double helix. Hydrogen bonds join a purine base on one chain with its complementary pyrimidine base on the other.
5. During DNA replication the two chains separate, and each makes a copy of its complement. When rare mutations occur, they are replicated, too.
6. Genetic information is encoded in the sequence of bases in DNA. With the aid of RNA molecules, this information is decoded (in the cytoplasm) into sequences of amino acids that form specific proteins.
7. Fruitful investigations of the 46 human chromosomes began with the work of Tjio and Levan published in 1956. Since then, there have been many improvements in cytogenetic methods. These include the short-term culture of white blood cells, the use of specific additives in the culture media, the development of various banding techniques (especially high-resolution banding), and the detection of fragile sites.
8. Each chromosome can be unequivocally identified in metaphase or late prophase by its size, shape, and banding pattern. The human karyotype consists of 44 autosomes (= 22 pairs) plus 2 sex chromosomes (= 1 pair). Males are designated 46,XY and females 46,XX. In addition, the karyotypes of normal individuals sometimes have special features, such as variations in Y chromosome length.

FURTHER READING

Current concepts of chromosome structure are set forth in books by Alberts et al. (1989), Watson et al. (1987), B. Lewin (1987), Brachet (1985), De Duve (1984), Bradbury et al. (1981), and Thorpe (1984); also see articles by O. J. Miller (1983), Comings (1978), Felsenfeld (1985), Haapala (1983), and Rattner and Lin (1985).

For very readable accounts of the history of human cytogenetics, see Jacobs (1982), Hsu (1979), Makino (1975), and German (1970). For details on the human karyotype and its nomenclature, see ISCN 1985. Cytogenetic techniques are described by Macgregor and Varley (1988), Bickman and Sumner (1989), and Therman (1986); chromosome evolution is discussed by Seuanez (1979).

QUESTIONS

1. Compare and contrast the following sets of terms:

acrocentric—metacentric
centromere—NOR
chromosome—chromatid
fragile site—satellite
heterochromatin—euchromatin
nucleic acid—nucleotide
purine—pyrimidine
G-bands—Q-bands—R-bands—C-bands—high resolution prometaphase bands
autosome—sex chromosome
centromeric index—relative length
diploid—haploid
genotype—karyotype
histone—nonhistone
nucleus—nucleolus—nucleosome
RNA—DNA

2. If two breaks occur in a chromosome, the intervening section may be turned around 180° before the broken ends heal, a process producing an **inversion**. Within the inversion, gene order is reversed and the location of the centromere may be altered (for example, *a b c* ● *d* changes to *a* ● *c b d*. For each of the hypothetical inversions indicated below, indicate (1) the new gene order, centromere (= ●) position, and banding pattern; (2) whether the inversion would be detectable in unbanded chromosome spreads, and (3) whether it would be detectable in banded spreads.

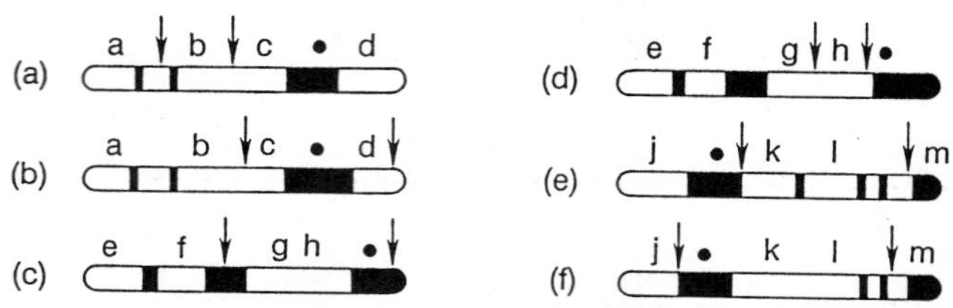

3. Review the four major attributes or functions of genes, and note briefly how each is fulfilled by the structure and/or behavior of nucleic acids.
4. Using "shorthand" symbols, give karyotypes for the

following: (a) Males with Klinefelter syndrome have an extra X chromosome; they also have small testes, sparse body hair, and some breast development. (b) Females with *Turner syndrome* are missing one X chromosome; they are short and sexually underdeveloped.

5. Although the presence of an extra autosome is usually lethal before birth, some cases survive to birth. These include: (a) Down syndrome, with an extra chromosome 21—by far the most frequent (1 in 700 births) and most likely to survive to adulthood. (b) *Trisomy 13*, with an extra chromosome 13, is rare; it leads to gross malformations and early death. (c) Also rare and characterized by serious deformities and a very short life span is trisomy 18, caused by an extra chromosome 18. Describe these karyotypes, using the standard shorthand symbols, and assuming that (a) is female, (b) is male, and (c) is female.

ANSWERS

1. See boldfaced terms in text.

2.

	New order	Detectable in unbanded spreads?	Detectable in banded spreads?
(a)	a b c • d	no	yes
(b)	a b d • c	no	no
(c)	e f • h g	yes	yes
(d)	e f g h •	no	no
(e)	j • l k m	no	yes
(f)	j l k • m	yes	yes

3. *Replication*: by strand separation and then formation of a new complementary strand on each old one. *Mutation*: by changes of one kind or another in the bases. *Information storage*: by the sequence of bases. *Information decoding*: by the mechanism leading to protein synthesis.

4. (a) 47,XXY (b) 45,X

5. (a) 47,XX,+21 (b) 47,XY,+13 (c) 47,XX,+18

Chapter 5 Gametes and Cell Division

Gametes and Fertlization
The Mature Egg and Its Cell Structures
The Mature Sperm and Fertilization

The Cell Cycle and Mitosis
Interphase
Mitosis

Meiosis and Gametogenesis
Meiosis I
Meiosis II

Cell Cycle Mutants

> *It has, I believe, been often remarked, that a hen is only an egg's way of making another egg... Why the fowl should be considered more alive than the egg, and why it should be said that the hen lays the egg, and not that the egg lays the hen, these are questions which lie beyond the power of philosophical explanation... But, perhaps, after all, the real reason is, that the egg does not cackle when it has laid the hen.*
>
> *Samuel Butler (1877)*

With the union of an egg and a sperm, another life begins. This new cell, the **zygote**, immediately divides and redivides during its journey from the upper reaches of the mother's oviduct, where (in most mammals) it was fertilized, down to the uterus. In humans, the zygote undergoes 50 cell replications in about 38 weeks, ending up as a marvelously complex newborn of more than a trillion cells. Continued division occurs throughout life, with certain cells undergoing a special process of meiosis to produce eggs or sperm; thus the cycle is complete.

GAMETES AND FERTILIZATION

The Mature Egg and Its Cell Structures

Roughly five times the diameter of an average human somatic cell, the mature **egg** measures about 0.1 mm* (or 100 μm) across (Figure 1). When ready for fertilization, the mammalian egg is actually suspended part way through cell division; thus it lacks a well-defined **nucleus**. Compare it with the nerve cell shown in Figure 2, whose nucleus is sharply separated from the surrounding **cytoplasm** by a double membrane called the **nuclear envelope**. At the stage shown in Figure 1, the egg's nuclear envelope and its associated pore complexes have disintegrated into bits and pieces that will later be reused. The chromosomes, having already undergone one meiotic division and part of the next, are lined up on a **spindle**, awaiting the signal to complete a second meiotic division (see p. 74). *Note:* All organisms whose cells contain nuclei and complex chromosomes are called **eukaryotes**. This structural complexity contrasts with that of **prokaryotes** (organisms such as bacteria and viruses), which lack nuclei but do have "chromosomes" consisting of near-naked DNA or RNA.

The cytoplasmic structures and substances present in an egg or a somatic cell, rather than just sloshing around at random, are distributed among several different compartments. One of these is the protein-rich, jelly-like **cytosol**, the matrix in which organelles and intricate membrane systems are suspended—each cell type has a somewhat different proportion and arrangement of these various components. Supporting and giving shape to the cytosol is a network of continuously changing **microfilaments** (containing the protein actin) and **microtubules** (containing the protein tubulin)—plus the more stable and more heterogeneous **intermediate filaments** (made up of various proteins). Microtubules also provide an internal transport system upon which cytoplasmic particles and vesicles can be shuttled in both directions to wherever they are needed in the cell (Allen 1987).

*See Appendix 1 on some common measurements.

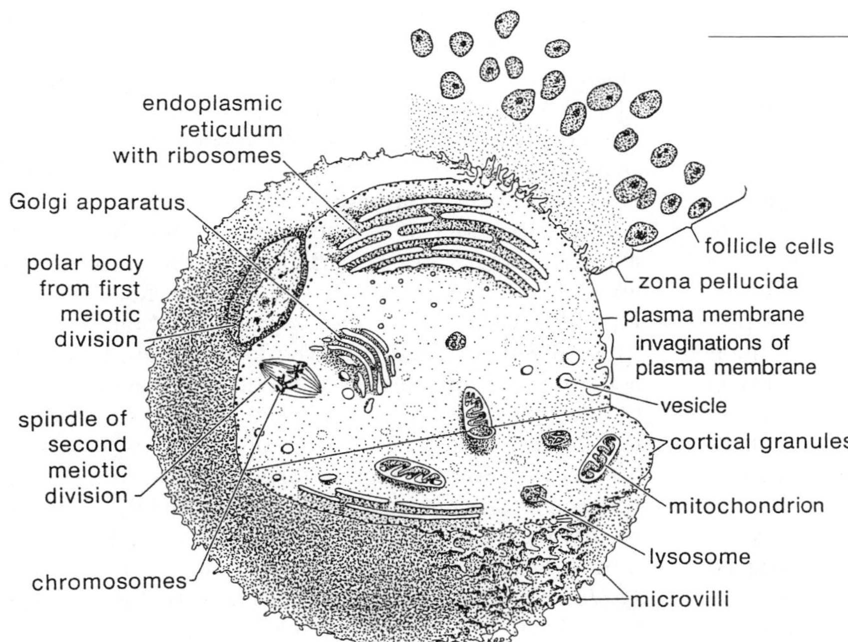

Figure 1 Egg cell (idealized and sectioned) ready for fertilization. The various inclusions shown here are suspended in the cytosol. At this stage of development, the nuclear membrane has broken down. Drawn about 500 times actual size.

Actin, a major constituent of muscle cells but also appearing in several closely related forms in nonmuscle cells, occurs in three main groups: α, β, and γ. The actin gene family in humans consists of about 20 genes. **Tubulin** consists of two kinds of proteins (α and β), whose production varies with cell type and is under the complex control of whole families of genes (Cleveland and Sullivan 1985).

A system of membrane-enclosed compartments—forming narrow channels in some areas and large cavities or small vesicles in others—provide the huge surface area needed for myriad biochemical processes. They are involved in the distribution of substances to different regions of the cell. Attached to these membranes or to the cell's plasma membrane and acting as "door openers" are numerous **receptor proteins**, each specialized to recognize and admit (or bind to) only a certain type of molecule (given the generic name **ligand**), such as a hormone or neurotransmitter. On the cell surface there may be 500–100,000 receptors for a given ligand, and they may be distributed over the entire surface or restricted to specific regions. Each **receptor–ligand complex** triggers a specific reaction inside the cell; for example, steroid hormone–receptor complexes enter the nucleus and bind to chromatin, thereby regulating the expression of specific genes.

Receptors are extremely important, for their malfunctioning or absence can have very serious consequences. One example is the disorder *hypercholesterolemia*, inherited as an autosomal dominant. It involves defects in the low-density lipoprotein (LDL) receptors that normally help remove excess cholesterol from the bloodstream (Chapter 16). Another example is *androgen insensitivity syndrome*, usually inherited as an X-linked recessive. Due to the absence or lowered activity of a cytoplasmic receptor for the male sex hormone androgen, XY individuals who would otherwise become males end up as perfectly normal-looking but infertile females (Chapter 6).

Enzyme systems are everywhere: some are embedded in or loosely bound to membranes, and others are lying free in the cytosol. Processes of destruction and

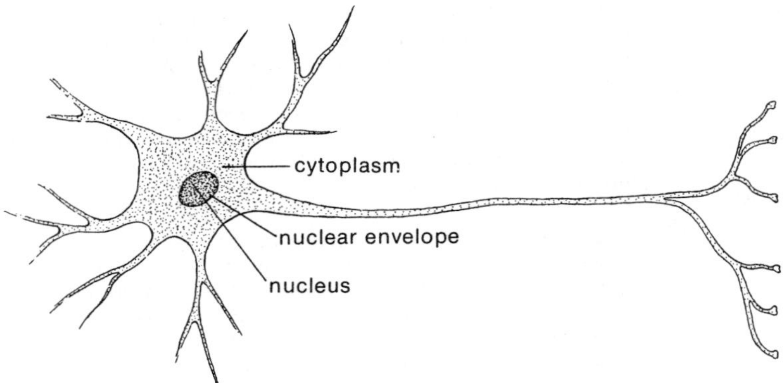

Figure 2 A human nerve cell. Except for the delineation of nucleus and cytoplasm, the subcellular components are not shown. Drawn about 1,000 times actual size.

construction are constantly going on in cells—but in a highly ordered and integrated fashion, with each cellular structure specialized to carry on a certain activity. For example, all worn-out cell parts and invading organisms plus many imported materials end up in the **lysosomes**—the membrane-bounded vacuoles that act as the cell's garbage disposal and recycling systems. Inside the lysosomes, all large molecules (nucleic acids, proteins, lipids, carbohydrates) are broken down into their smaller components by powerful enzymes called *acid hydrolases*. Were it not for the fact that they can function only in the highly acidic innards of lysosomes, these enzymes would destroy the entire cell if released into the surrounding cytoplasm!

Several dozen hereditary conditions called *lysosomal storage diseases* are known, each caused by the absence or malfunctioning of a specific acid hydrolase enzyme (Chapter 16). In each case, the substance upon which the enzyme normally acts will gradually accumulate and greatly distend the lysosomes—interfering with normal cellular activity. Most of these disorders are very serious, involving skeletal and nervous system abnormalities (the X-linked recessive Hunter syndrome, for example), and some (such as Tay-Sachs disease and Hurler syndrome, both autosomal recessives) are lethal.

Other tiny membrane-bound structures called **peroxisomes** contain enzymes that use oxygen to detoxify certain substances, including ethanol. The hydrogen peroxide generated as one of the intermediates is then converted to water by another peroxisomal enzyme called *catalase*. Among those rare humans who lack catalase (as the result of an autosomal recessive disorder called *acatalasia*), most show no ill effects whatsoever. In those who do show symptoms, the usual condition, seen only during childhood, is ulceration (or occasionally gangrene) of soft oral tissues, especially around the teeth.

Digestion products released from the lysosomes, as well as many other substances coming into the cell, are gradually degraded to extract the energy needed for doing the work of the cell—a process known as **intermediary metabolism**. During the first stage, which requires no oxygen, glycolytic enzymes in the cytosol partially break down small molecules (sugars, amino acids, fatty acids, glycerol). This process yields pyruvate and some energy-rich ATP molecules. But much more energy is released during the second stage, called **cellular respiration**, in which pyruvate is gradually degraded to carbon dioxide and water.

The latter, oxygen-requiring process takes place inside **mitochondria**, the power plants of the cell. These large, mobile structures rapidly change their shapes to adapt to cellular conditions. They are bounded by two membranes, the highly convoluted inner one containing or surrounding hundreds of the key mitochondrial enzymes.* Energy released by glycolysis and cellular respiration is used to form new molecules, both large and small, and to drive many other metabolic reactions.

Proteins, the end product of gene action, are synthesized on tiny particles called **ribosomes**, which number over a million per cell. Those lying free in the cytosol make the enzymes for intermediary metabolism, the proteins in mitochondria and peroxisomes, and the histones and many other chromosomal proteins. Other ribosomes are bound to the outer membrane of the nuclear envelope, and especially to the "rough" membranes of the **endoplasmic reticulum**. The latter system is a large and ever-changing network of tubules and compartments of various shapes and sizes. It is involved in the production of all secretory and membrane proteins, nearly all lipids, and certain complex carbohydrates. Indeed, the rough endoplasmic reticulum makes virtually all of the membranes found in a cell's various organelles.† Also, most proteins destined for export have a certain *oligosaccharide* (chain of sugars) attached to them as they travel through the rough endoplasmic reticulum on their way to the Golgi complex, another system that makes up part of the secretory pathway leading from the endoplasmic reticulum to the plasma membrane bounding the cell.

The **Golgi complex** is the cell's "shipping agent"—receiving, sorting, tagging, and directing many newly made proteins and membrane parts to their final destinations inside or outside the cell. Abutting the endoplasmic reticulum on one face (often near the nucleus) and the nuclear membrane on the other, it is made up of arrays of smooth membrane sacs, flattened and stacked like saucers. Enzymes in the Golgi complex modify or add oligosaccharide groups to the proteins and lipids passing through. The resultant *glycoproteins* and *glycolipids*, surrounded by sections of membrane, form secretory or storage *vesicles* that later fuse with the plasma membrane and release their contents to the outside of the cell. Alternatively, these molecules may become a structural part of the plasma membrane itself. The Golgi complex also packages the cell's lysosomes and recycles various components of the outer plasma membrane. In the egg cell, both the specialized cortical granules (whose contents are expelled at the time of fertilization)

*Mitochondria (and in plant cells, chloroplasts too) are actually self-replicating organelles. They contain unique forms of DNA and ribosomes, much more like those found in prokaryotes than eukaryotes, and they even produce a few of their own proteins (Grivell 1983). For these and other reasons, it seems very likely that they evolved from free-living, aerobic bacteria that invaded and were retained by primitive, anaerobic eukaryotic cells. For discussions of this, see Margulis (1982).

†The "smooth" endoplasmic reticulum has no associated ribosomes, but in specialized cells (such as liver or hormone-producing cells) it contains enzymes that catalyze other important reactions—such as the synthesis of cholesterol, steroid hormones, and bile acids, the detoxification of various harmful substances, and the regulation of blood glucose levels.

and the nutritive yolk granules are probably formed in the Golgi complex.

Enclosing the cytoplasm of the egg is the **plasma membrane**. In egg cells (but not in all cells) its entire surface is fuzzy with slender cytoplasmic extensions (*microvilli*) that protrude into the surrounding space (see Figure 1). Like all cell membranes, it is made up of lipids (including cholesterol) that form a bilayer in which many specialized membrane proteins are dissolved. It acts as the "gatekeeper" that controls which substances will pass in and out of the cell. This function is accomplished by means of specialized receptor proteins on the surface of the plasma membrane and by molecule-specific channels that pass specific ions and molecules through the membrane.

The plasma membrane "swallows" certain outside materials by engulfing them and pinching off internally to form vesicles of various sizes. During **phagocytosis**, a sizable region of plasma membrane actively expands outward and fuses around large particles, such as bacteria or parts of degenerating cells, before ingesting them. For **endocytosis**, a much more limited region of the plasma membrane invaginates to receive small molecules or droplets of fluid or to specifically bind (by means of membrane receptors) to macromolecules before fusion occurs. However they are formed, the new vesicles may fuse with other intracellular structures (e.g., lysosomes, to be degraded), or they may be stored, or they may return to the plasma membrane and release their contents elsewhere. The release of ingested materials and the secretion of cell products (e.g., hormones, neurotransmitters, enzyme precursors, serum proteins, and collagen) occurs through a reverse process called **exocytosis**. In this process ingested vesicles or secretory vesicles budded off from the Golgi complex fuse with the plasma membrane; they then open to expel their contents from the cell surface. Some vesicle membrane components return to the cytoplasm for recycling.

Now consider the exterior of an egg ready for fertilization. In the space just outside the egg membrane lies the tiny **first polar body**, a product of the first meiotic division (see p. 74). It carries the same number of chromosomes found on the egg spindle, but very little cytoplasm. Secreted by the egg and enveloping both the first polar body and the egg is the **zona pellucida**—a dense, jelly-like, glycoprotein layer whose surface is thickly peppered with species-specific sperm receptors (Barnes 1988). It protects the egg and early embryo, prevents the entry of foreign sperm, and induces the acrosome reaction (see below). Surrounding the zona are follicle cells, which by the time of fertilization have begun to decompose and to separate from each other.

The Mature Sperm and Fertilization

The **sperm**, extremely tiny compared with the egg, is a cell stripped down to bare essentials—basically a **nucleus** enclosing the chromosomes, plus **mitochondria** for energy, and a **tail** for locomotion (Figure 3). The total length of a human sperm is about 60 μm (0.06 mm), 93% of this being the tail. The head contains the nucleus with very densely packed DNA whose previously associated histones have been replaced by highly basic proteins called **protamines**. Hugging the upper part of the nucleus is a thin **acrosomal cap** filled with several kinds of enzymes. A short neck, composed of a **centriole** surrounded by segmented columns, attaches to the long tail. The core of the tail (see the cross section in Figure 3) arose from a second centriole. It is a **flagellum** with the characteristic pattern of two central microtubules surrounded by nine double microtubules, circled in turn by nine dense protein fibrils. Mitochondria are spirally wrapped around the core of the upper part of the tail, providing the energy needed for movement; and a tough, fibrous sheath covers most of the remaining length. A plasma membrane encloses the entire sperm.

For unknown reasons, a surprisingly high proportion of human sperm show obvious defects:

> Since the examination of ejaculates by light microscopy became a technically simple matter, the disconcerting conclusion has emerged that man is the species that has the greatest percentage of malformed, anomalous and immotile

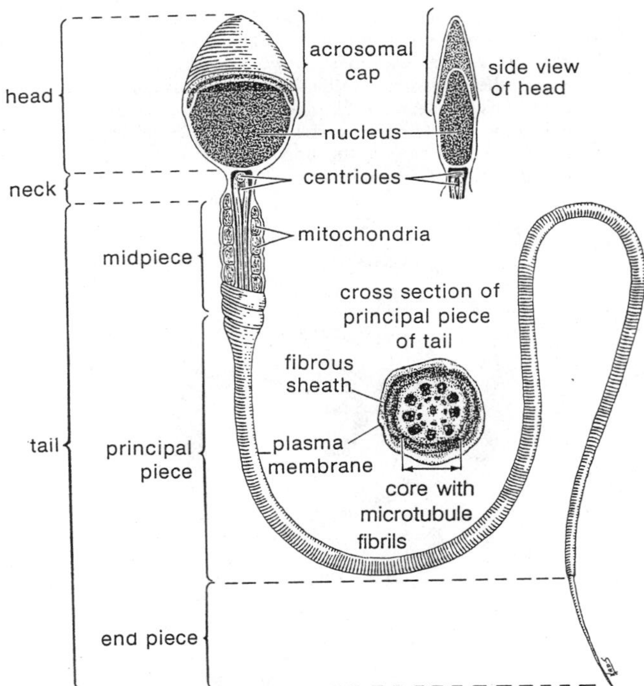

Figure 3 Sperm cell (idealized and sectioned). After fertilization all but the chromatin and the centrioles will disintegrate inside the egg cytoplasm. Drawn about 4,000 times actual size.

spermatozoa, even in samples from healthy individuals of proven fertility. A possible explanation for this is that the temperature of the scrotum in which spermatogenesis occurs is raised above its natural physiological level as [a] result of wearing clothing . . . but there are also a significant number of known genetic defects that affect the entire sperm population of an individual (Baccetti 1984).

Frequently observed abnormalities (Figure 4) include immature nucleus with uncompacted DNA (perhaps the most common), malformed acrosome, round-headed sperm without an acrosome, abnormally shaped mitochondrial helix, extra cytoplasm surrounding and immobilizing the tail flagellum, abnormal attachment of tail to head, short tail, missing or incompletely formed tail, two or more tails, and completely double sperm. Although many abnormal sperm are unable to progress through a female's reproductive tract, some not only pass through but even manage to fertilize an egg.

One ejaculate from a fertile male contains at least 250 million sperm, of which up to a hundred may find and attach to the material surrounding the egg. Sperm are swept to the fertilization site mainly by rhythmic contractions of the female reproductive tract, and they need to swim only during their final approach.* Whether they are attracted to the egg or simply collide with it is not clear. In any event, some manage to penetrate the egg's coatings to reach the plasma membrane, but normally only *one* sperm enters the egg to form a zygote.

*Some forms of hereditary male sterility (usually autosomal recessive) involve defects in various substructures of the flagellum in sperm tails. Patients with these *immotile cilia syndromes* also have defects in their respiratory cilia that give rise to chronic bronchitis and sinusitus. Those with *Kartagener syndrome* have an additional complication whereby the heart, liver, intestines, and appendix end up on the wrong side of the body—perhaps as a result of malfunctioning cilia in the embryo.

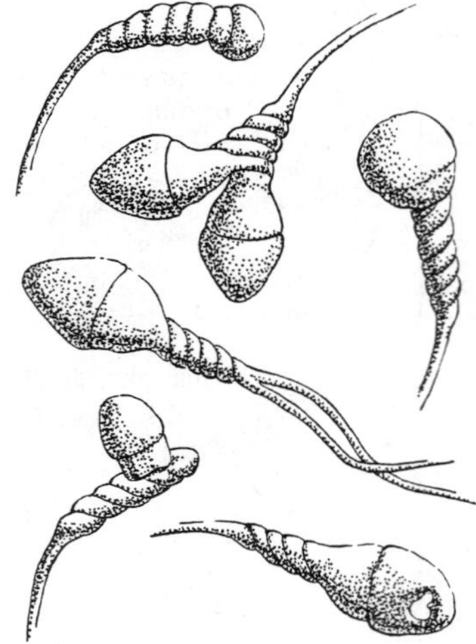

Figure 4 Some types of abnormal human spermatozoa. [After Fujita 1975, which is included (along with many scanning electron micrographs) in Hafez and Kenemans (1982).]

Before they are capable of fertilization, however, mammalian sperm must undergo physiological changes in the female reproductive tract. The process of **capacitation** involves the removal of some proteins that coat the sperm, and alteration of the plasma and acrosomal membranes. This change increases sperm motility and makes the acrosome permeable to free calcium ions near the fertilization site. As the sperm make their way through the protective layer of follicle cells, some acrosomal caps rupture and release the enzyme *hyaluronidase*, which helps to dissolve the substance cementing together these cells. Yet many sperm do not undergo this **acrosomal reaction** until after they have attached to the glycoprotein receptors in the zona pellucida (Figure 5). Additional acrosomal enzymes may help forge pathways through the zona pellucida—the end of the line for all but one of the invading sperm.

The binding of the first sperm (sideways rather than head on) to the egg's plasma membrane triggers two events. First, ionic changes (of calcium, sodium, and hydrogen) in the egg membrane and cytosol cause instantaneous but temporary depolarization (change in electric charge) of the egg membrane, rendering it impervious to further sperm for a short while. As a second line of defense, the underlying cortical granules expel their contents into the surrounding space. This **cortical reaction** alters the glycoprotein receptors in the zona pellucida and permanently blocks the adherence of more sperm to the egg.

Following fusion of its plasma membrane with the egg's plasma membrane, the entire sperm is engulfed by microvilli and pulled into the egg's cytoplasm. The mitotic apparatus is suddenly activated (perhaps by the release of calcium ions) and completes the second meiotic division. The egg chromosomes separate into two groups, one of which is pinched off to form the **second polar body**. The remaining egg chromosomes decondense, and a newly constituted nuclear envelope encloses them to form the **female pronucleus**. (As we shall see in Chapter 9, an amazingly high proportion of human eggs—perhaps around 50%—contain chromosomal abnormalities; here we concentrate on normal gametes.)

Meanwhile, the sperm's tail, midpiece, and nuclear envelope disintegrate. Its chromatin material becomes more swollen and dispersed, and the basic protamines that were present in the sperm nucleus are replaced by histones produced by the oocyte. Formation of a nuclear envelope (also from oocyte raw materials) around this chromatin completes the **male pronucleus**. The two pronuclei approach each other but do not fuse (Figures 6 and 14). Within each pronucleus, the chromosomes duplicate, condense, and (following breakdown of the two nuclear envelopes) arrange themselves on the first

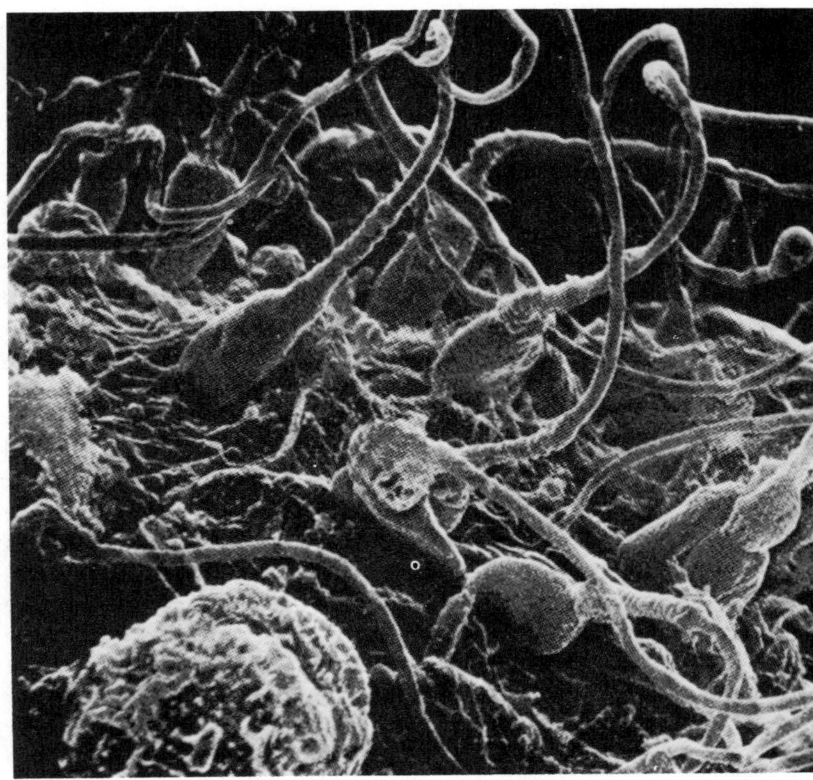

Figure 5 Many human sperm on the zona pellucida surrounding the oocyte. (From Nilsson 1975.)

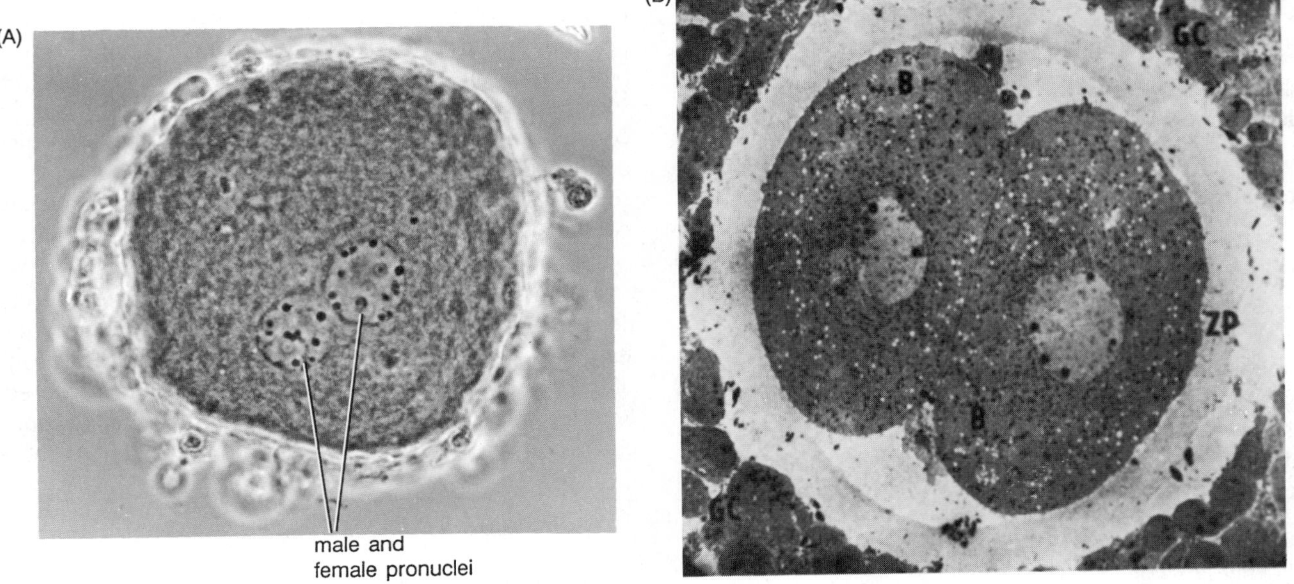

Figure 6 (A) The male and female pronuclei lying side by side, shortly after penetration of the sperm has triggered completion of the second meiotic division in the egg. The first mitotic division of the zygote will follow. Shown about 300 times actual size. (From Dickmann et al. 1965.) (B) Human two-cell embryo. ZP, zona pellucida; B, blastomere cells; GC, granulosa (follicle) cells. (From Dvořák et al. 1982.)

cleavage spindle.* Maternal and paternal chromosomes are not enclosed within the same nuclear envelope until the two-cell state. In humans, the first cleavage division occurs 24–36 hours after fertilization.

THE CELL CYCLE AND MITOSIS

Adult humans, like all multicellular organisms, are formed by the proliferation of a single cell first into 2 identical cells, then into 4, 8, 16, and so on, followed by gradual differentiation of the resultant cellular mass into tissues, organs, and organ systems. Whatever their developmental fate, all of these cells exhibit the same basic behavioral pattern: a period of growth and metabolism, known as **interphase**, alternating with a period of cell division called **mitosis**. These two delicately integrated functions constitute the **cell cycle**, whose orderly course includes four distinct phases. The first three, **G_1**, **S**, and **G_2**, make up interphase; the final phase, **M**, is the mitotic division.† Most cells, including those of mammals, complete the entire cycle in 16–24 hours, of which only 1–2 hours are spent in division. This is summarized in the following diagram:

G_1	S	G_2	M
5-9 hours	6-9 hours	2-5 hours	1-2 hours
–I–N–T–E–R–P–H–A–S–E→			DIVISION

Each cell type seems to have its own precise protocol, which is not easily altered by experimental means.

Interphase

The first phase, **G_1**, is the initial time *gap* between cell "birth" at division and the beginning of DNA synthesis. All the cellular activities described in the preceding section are carried on during this period. By far the most variable in duration among different cell types, this phase may be too brief to measure in rapidly dividing cells—or it may last for weeks, months, or years in cells that seldom or never divide. G_1 in the "average" dividing cell takes 5–9 hours. Late in this period is the **R** (resting or restriction) **point**, at which time a cell may slow down or stop dividing if conditions are unsuitable.* R is a "point of no return," too, because cells that pass it will almost always proceed unabated through S, G_2, and M—regardless of local conditions. Also at the end of G_1, the cell's centriole pair† (page 68) begins to replicate.

Then an as yet unidentified signal at the end of G_1 triggers the onset of the **S** (synthesis) period, during which nuclear DNA replicates. Every eukaryotic chromosome contains just one very long molecule of DNA; in mammals, these molecules contain from 25 million to several hundred million base pairs—one to several centimeters in length. In interphase the DNA is associated with proteins to form a chromatin fiber—the form of metaphase chromosomes (Chapter 4). As shown in Figure 7, DNA is associated with histones to form nucleosomes, and the nucleosomes are compacted into a 30-nm wide chromatin fiber which gets folded into looped *domains* that are attached to and stabilized by a network of nonhistone proteins (Laemmli et al. 1983; McGhee 1986). Adolph (1986) notes that "the domain size of interphase chromatin is from 75 to 220 kb, though values of around 85 kb are most common. This value is very similar to the average length of DNA loops in metaphase chromosomes (70–90 kb)."** Indeed, it appears that the DNA attachment sites remain stable throughout most or all of the cell cycle—with the active genes located closer to the scaffold and inactive genes farther away (Watson et al. 1987).

How does DNA in this chromatin fiber manage to replicate? Rather than beginning at just one point on each chromosome, the unwinding of nucleosomes and replication of DNA starts at one specific spot in each domain (the **origin of replication**) and continues in both directions at a rate of about 50 base pairs per second. Prescott (1987) elaborates: "On the average the origins of replication are about 90,000 bp apart, which defines the average length of the replicating units or replicons. Therefore, the mammalian diploid nucleus contains about 30,000 replicons, all of which replicate once (and only once) within an 8-hour S period."

He further points out that it takes 15 minutes to replicate the average replicon/domain, that the replication starts are staggered, and that these staggered initiations of the replicons in a mammalian cell are highly ordered. Regions that in metaphase are recognized as Q-bands each replicate during a specific time frame during the S

*In many organisms the sperm centrioles survive in the egg cytoplasm and organize the first cleavage spindle. In mammals, however, the sperm centriole disintegrates and no centrioles are seen in the early cleavage divisions.

†Some workers make a distinction between *mitosis* (nuclear division) and *cytokinesis* (cytoplasmic division), because these two processes can occur independently of each other.

*Nondividing cells are considered to be in a G_0 state.

†Exactly how centrioles replicate, and whether they contain their own nucleic acid, is still a mystery. But we know that daughter centrioles always form at right angles to the original, and first contain a circlet of 9 *single microtubules*—each of which later becomes a *triplet*.

**1 kb = 1 kilobase = 1000 bp (base pairs).

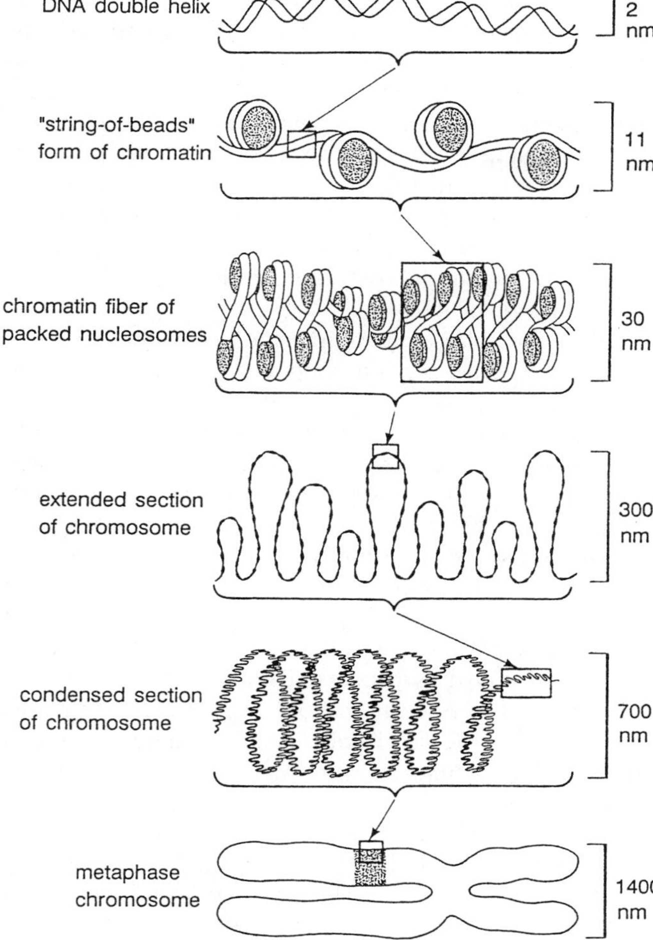

Figure 7 The different orders of chromatin packing assumed to give rise to a metaphase chromosome. (After Alberts et al. 1989.)

period (Ris and Korenberg 1979). So-called housekeeping (i.e., always active) genes, which reside in the R-bands, replicate in *early* S. Inactive tissue-specific genes present in the G-bands usually replicate *late* in S, but those few tissue-specific genes that happen to be active may replicate early (Holmquist 1987). This study confirms observations that euchromatin replicates early while heterochromatin replicates late in the S period. How all of it is controlled remains a mystery.

Meanwhile, histones are being synthesized in the cytoplasm and transported through pores of the nuclear envelope into the nucleus. There they combine with replicating DNA to form new nucleosomes and chromatin fibers. (The two copies of each chromosome, when they become visible during mitosis, are called **sister chromatids**.) The centrioles also complete their replication during the S period, after which a second gap of 2 to 5 hours, **G_2**, occurs before the nucleus is somehow stimulated to undergo mitosis—during **M**.

Much remains to be learned about chromosome condensation and nuclear division, and how the cell cycle is controlled; but studies of cultured cells have provided some clues. In some cases it appears that a certain size must be attained or a certain period of time must elapse before cells can move from G_1 to S, or from G_2 to M, but in other cases these requirements do not seem to hold.* The degree of cell spreading is very important (rounded cells having longer division cycles than flattened cells), as is the need for a solid, flat anchorage area. In its normal bodily environment, a cell's position relative to other cells and tissues is very important, perhaps because of complex cell-to-cell interactions. Striking surface changes (such as the extrusion and retraction of hairlike microvilli) and changes in electrical properties of the membrane have been correlated with different division phases.

Within the cell, the concentrations of specific enzymes vary through the cell cycle. Intracellular calcium and its binding protein *calmodulin* are also very important factors, although external changes in certain ions and molecules (including certain trace metals, amino acids, and fatty acids) are also known to affect the cell cycle. The plasma membrane, as the site of action for many growth factors, and the nuclear membrane, as an anchoring site for chromosome telomeres, play key roles in the normal progression of cell division.

Especially important for the control of cell replication in mammals are certain **hormones** and **growth factors** produced by other cells. Growth factors are regulatory chemical messengers (usually polypeptides) that cause their target cells to grow larger or to divide or both. Although the mechanism of action is not well understood, it starts with the attachment of the growth factors to specific receptors on the target cell's plasma membrane. This binding generates a signal that initiates certain cellular responses (phosphorylation of certain proteins and synthesis of DNA and proteins). Then the growth factor–receptor complexes are drawn into the cell and broken down by lysosomal enzymes (James and Bradshaw 1984).

Over 36 hormones and growth factors have been identified as necessary constituents in the serum or artificial media that bathe cultured cells. Examples include somatotropin (growth hormone), insulin, insulin-like growth factors,† epidermal growth factor, nerve growth factor, fibroblast growth factor, platelet-derived growth factor, colony-stimulating factors, and somatomedins (substances produced in the liver under the influence of

*Indeed, some cells do not have a G_1 period at all (Rothstein 1982).

†A deficiency of insulin-like growth factor in African pygmies is apparently what prevents the growth spurt that (in other human populations) occurs at puberty (Merimee et al. 1987).

somatotropin). Some other hormones that induce mitosis in specific tissues are thyroxin, estrogen, testosterone, and prolactin. As we shall see in Chapter 19, both normal and cancerous cells produce transforming growth factor, which can change normal cells into cancer cells. Indeed, it appears that the genes controlling some growth factors in normal cells— if moved to certain new chromosomal locations by various means, including some viruses—are turned into cancer-producing agents called **oncogenes**.

Growth-inhibiting factors called **chalones** are also found in many tissues. One example is epidermal chalone, which is produced by the outer layer of skin cells and regulates (inhibits) cell division in the lower (dermal) layer of skin. Overproduction of dermal cells can lead to *psoriasis*, the abnormal thickening and scaling of the epidermis.

Some cells, especially those of blood-forming and epithelial tissues, continue to divide regularly throughout the lifetime of the individual. Other cells, including those of nerve and muscle tissues, do not normally divide at all after birth. Such "noncycling" (G_0) cells, when removed from the body, can often be induced to divide in tissue culture; thus the capacity for cycling has not been lost. Conversely, cancerous cells are those whose cycling controls have run amok; consequently, they divide uncontrollably. If scientists can learn how the division cycle (especially the S phase) is regulated in normal cells, they could use this knowledge (1) to prevent the growth of cancer cells and (2) to promote wound and tissue healing. Thus, in addition to its scientific value, the study of the cell cycle has immense practical value.

Mitosis

In the 1870s, with the development of special dyes and improved microscopes, European cytologists first discovered how cells divide. Since then the process has been studied in ever more exquisite detail by focusing with more and more sophisticated microscopes on all kinds of cells and cell parts that have been treated with increasingly complicated physical and biochemical techniques. Yet, strange as it may seem to beginning students, we still do not fully understand the most basic events to be described below: exactly how chromosomes replicate and condense, how spindles form and operate, how chromosomes attach to the spindle, how daughter chromosomes separate and move to opposite poles of the spindle, how the cytoplasm divides, what limits the number of cell divisions in normal cells (as opposed to transformed or cancerous cells); and, finally, how all these processes are controlled and integrated.

Along with these more familiar problems, we know very little about a less obvious but equally important aspect of normal differentiation, called **programmed cell death**, whereby certain types of cells are eliminated at certain stages of development. [One example: around the seventh week of fetal development, the Müllerian inhibiting substance causes degeneration of the primitive female reproductive system that is normally present in all male fetuses.] This orderly interplay between cell proliferation and cell death controls the shape and functioning of organs in both the developing embryo and the adult organism.

All higher plants and animals follow, with some variations, the same general routine during nuclear replication and cell division. Although it is a continuous process, cytologists have for the sake of convenience separated it into several stages: prophase, prometaphase, metaphase, anaphase, telophase, and cytokinesis. These are shown in Figure 8.

The longest stage, **prophase**, begins as the nucleus swells and its chromatin fibers, previously stretched out to their ultimate interphase length, start to condense (Figure 8A). Heavy phosphorylation of H1 histones is associated with this process. The looped domains that are attached to nuclear scaffold proteins also gradually shorten or coalesce. Finally each gossamer thread is densely packed into a short, thick, coiled chromosome (Figure 9) whose looped domains radiate out in all directions from a central scaffold. Its two identical sister chromatids, replicated in the preceding S phase, become visible by late prophase (Figure 8B). At the same time the gradual dispersal of the *nucleoli* (containing ribosomal RNA and proteins) is completed by late prophase, while in the cytoplasm the *cytoskeletal microtubules* break down, releasing a large pool of tubulin molecules.* Two pairs of **centrioles**, replicated during the previous interphase, are located just outside the nuclear envelope; they separate and start to migrate in opposite directions around the nucleus. As they do this, a dense array of spindle microtubules is assembled from free tubulin by the diffuse material surrounding the centrioles. (A centriole pair plus this surrounding material is called a **centrosome**.) The *polar microtubules* extend from the centrosomes (poles) to the middle of the nucleus, forming two *half-spindles*. Another group of microtubules, the *kinetochore microtubules*, will connect chromatid centromeres to the polar regions. Both types of fibers together comprise the **spindle**. The remaining microtubules radiating from the centrosomes make up the two *asters*, whose role in cell division is not clear.

*The *intermediate fibers*, which seem to form a protective "cage" around the nuclear area, remain intact throughout the entire division cycle. The actin *microfilaments*, many of which are concentrated under the cell's plasma membrane, break down and re-form in response to various cellular activities; during cytokinesis, for example, actin subunits regroup and (in combination with myosin) form the *contractile ring*.

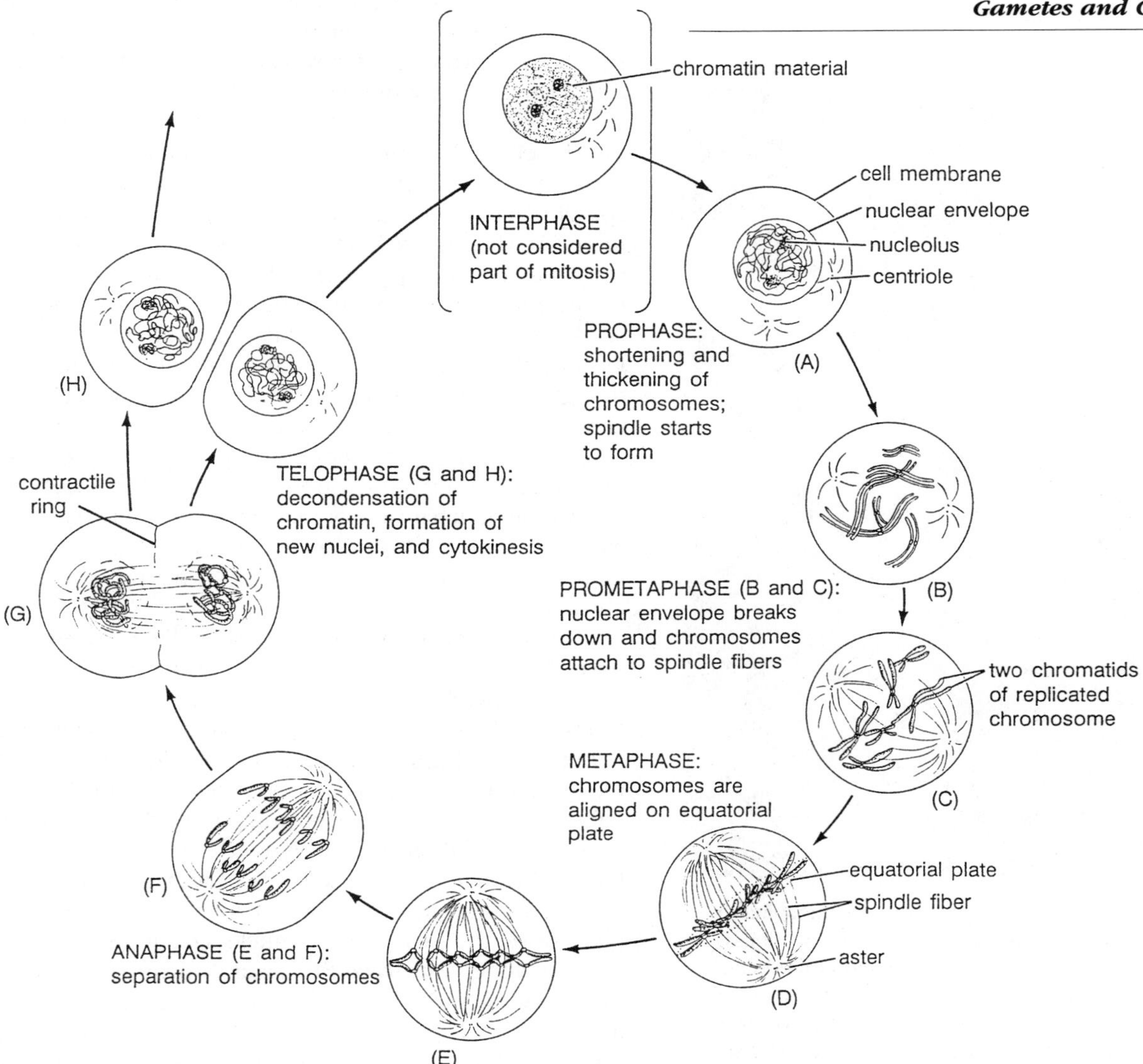

Figure 8 Chromosome behavior during mitosis. Only six chromosomes (three homologous pairs) are shown. One member of each pair came from the female parent and one came from the male parent. Two nucleoli are shown, and the centromeres are drawn as small circles rather than as constrictions. Most of the named phases of mitosis are represented by two drawings. Note that the nucleoli and the nuclear envelope disappear in prophase and prometaphase and reappear in telophase.

The next division stage, **prometaphase**, begins when the nuclear envelope starts to undulate in the polar regions. Ruptures then appear, and the whole envelope rapidly breaks down into small vesicles—now indistinguishable from pieces of endoplasmic reticulum—that remain scattered just outside the developing spindle. At the same time, the *laminar proteins* that lined the inside of the nuclear membrane break down into polypeptides and disperse throughout the cell—later to be repolymerized inside new nuclear membranes. Meanwhile, the centriole pairs have become stationed at opposite poles of the spindle, which has taken its final shape as fibers of the two half-spindles intermingle in the equatorial region. The centromeres* of replicated (mammalian) chromosomes, which have started to converge toward the midregion of the spindle, develop three-layered, platelike **kinetochores**. The kinetochores, which contain proteins and special centromeric DNA, become attached to bundles of kinetochore microtubules (Ris and Witt 1981; Rieder 1982; Aleixandre et al. 1987). The latter may originate at the kinetochore regions, or they may be polar microtubules "captured" by the kinetochores. In any event, each chromatid pair moves back and forth as one chromatid's kinetochore attaches to a bundle that goes to one pole, and its counterpart then gets connected

*Centromeres have been implicated in at least two human ailments. For unknown reasons, patients with *scleroderma* (a connective tissue disorder) make antibodies to centromeres, and patients with *Roberts syndrome* (a rare disorder characterized by cleft lip-palate and the absence or deficiency of leg and arm bones) show puffing and separation in the centromere regions of their chromosomes.

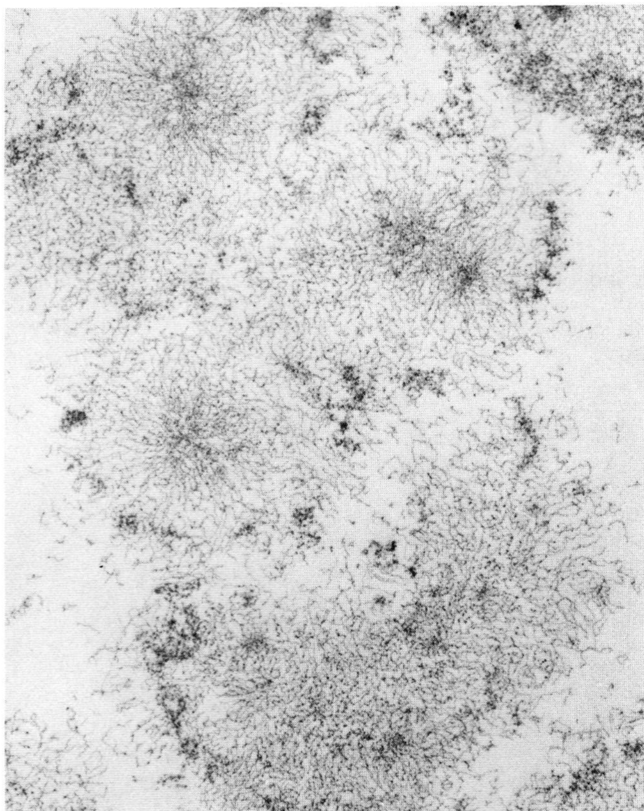

Figure 9 The radial arrangement of the chromosomal fibers. The chromosomes in this thin section are swollen as a result of EDTA treatment prior to fixation and embedding. The nucleoprotein fibers are in the 10-nm-thick ("string-of-beads") configuration. Several cross sections of chromatids appear in the middle of the micrograph; a longitudinal section is at the bottom. (From Marsden and Laemmli 1979.)

to the opposite pole. The key rule governing proper chromatid segregation is that the *two members of a chromatid pair must never be connected to the same pole at anaphase*. However, they may go through a temporarily misaligned state during prometaphase.

At **metaphase** the chromosomes, now short and thick, have all collected midway across the spindle in a region called the **equatorial plate** (Figure 8C and D). Each chromatid pair is suspended between the two poles by the equal tension exerted on each centromere through the kinetochore fibers. Keep in mind, however, that *during mitosis in somatic cells the two (replicated) members of a homologous pair of chromosomes do not interact in any way, so they may end up close together or far apart on the spindle.*

Anaphase begins when the centromeres of each sister chromatid pair suddenly jump apart. Then the kinetochore fibers appear to tow the attached centromeres, with the two chromosome arms trailing behind them, toward opposite poles (Figure 8E and F). At this point, each sister chromatid is now called a chromosome. Exactly how chromosomes get moved on the spindle is not well understood, but cytologists have identified two clearly separable components. (1) The spindle elongates, apparently as the polar microtubules of the two half-spindles lengthen and push apart in the equatorial region. (2) The kinetochore microtubules move the chromosomes to the separating poles—through shortening by the removal of tubulin at the kinetochores (Gorbsky et al. 1987; Koshland et al. 1988), or perhaps by some kind of sliding interaction with polar microtubules. Although controversial, it has even been suggested that, in certain organisms, kinetochores can propel chromosomes along the microtubules. "Indeed, as more functionally significant information is gathered, the possibility of multiple mechanisms of mitosis becomes increasingly attractive" (King 1983). The net result is that each side of the elongated dividing cell gets one copy of each chromosome.

The events of **telophase** are essentially the reverse of what happened during prophase and prometaphase. Chromosomes begin to decondense, and nuclear envelopes and their lamina and pore complexes reaggregate (from preexisting vesicles and macromolecules) around the two identical groups of chromosomes. Chromosomes continue to decondense, becoming individually indistinguishable, and their telomeres attach to the laminar proteins. Nucleoli, produced by special ribosomal DNA (rDNA) sites on nucleolar-organizing chromosomes, appear inside the two daughter nuclei (Figure 8G and H).

Meanwhile, a **contractile ring** of actin and myosin forms just under the plasma membrane, in the plane that the equatorial plate had occupied—that is, midway across the long axis of the cell. Like a tightening belt, it gradually pinches in until the cytoplasm and spindle are split into two. This process of bisection, called **cytokinesis**, is independent of nuclear division. The cytoplasm in cells whose nuclei have been experimentally removed can still divide; conversely, some nuclei can undergo mitosis without accompanying cytoplasmic division, thereby yielding multinucleate cells.

The end result of mitosis is the production, from one original cell, of two daughter cells with identical nuclei. *One interphase chromosome doubling (S period) has been followed by one division (M); so chromosome number and composition remain constant from one cell generation to the next.* The cytoplasms of daughter cells are usually about the same, too; but because there is no mechanism for the precise distribution of cytoplasmic components, differences may arise between the two cells from a mitotic division. This variation is particularly true of the early divisions of the fertilized egg (zygote) whose organelles and inclusions, such as yolk, may be present at different concentrations in various regions. In some organisms (such as frogs), such gradients result in the formation of daughter cells with greatly disparate cytoplasms.

All the cells descended from one zygote generally have the same nuclear genotype, yet they somehow manage to differentiate into many different types of cells, each with a specialized function. Among these specialized types are the **germ cells**, which will divide by a special process called meiosis to give rise to eggs and sperm.

MEIOSIS AND GAMETOGENESIS

Every egg that a human female will ever produce is already present, although immature and in a state of "suspended animation," at birth; thus any egg is as old as its host is at the time of the egg's release from the ovary. Although ovulated at the rate of one (or perhaps a few) per month for the 35–40 years between puberty and menopause, these eggs will not fully complete their meiotic divisions unless fertilized. It has been suggested that the increased risk of offspring with congenital abnormalities among older mothers may be somehow related to the longer storage time of their eggs.

Although estimates vary widely, a female produces perhaps several million immature eggs, of which only a few hundred are matured and ovulated during her lifetime—whereas a male probably makes and releases several trillion sperm in his lifetime. Sperm production does not begin until puberty, however, and usually continues for at least 50 years. From start to finish a human sperm takes only about 3 months to mature, and it is released immediately. Yet, for unknown reasons, the quality, quantity, and fertilizing ability of sperm from older men are markedly decreased (Baccetti et al. 1982; Hafez and Kenemans 1982).

Egg formation (**oogenesis**) and sperm formation (**spermatogenesis**) occur by a special pair of cell divisions known as **meiosis**, during which **diploid (2n)** cells give rise to **haploid (n)** cells. In humans, this means that cells containing 46 chromosomes (23 pairs) produce gametes with just 23 chromosomes. Unlike the mitotic cycle, in which one chromosome replication is followed by one division, meiosis involves a single replication (**premeiotic S period**) followed by a brief G_2 phase and *two* divisions, called **meiosis I** and **meiosis II**. Along the way, the maternal and paternal members of each pair of chromosomes become separated from each other as the result of a unique feature of meiosis, the pairing of homologous chromosomes and the exchange of parts during prophase of the first meiotic division. These processes, called **synapsis** and **crossing over**, usually ensure that the resultant daughter cells contain one member of each chromosome pair, either the maternal or paternal one. Both meiotic divisions include the stages observed in mitosis, but by far the longest and most complex of these—occupying up to 90% of the total time—is prophase of the first division, or **prophase I**.

First let us follow the developmental fate of some primordial germ cells in a female. These cells migrate from the embryo's yolk sac (Chapter 6) into the undifferentiated gonadal tissue of the fetus during the fifth week of development. By the twelfth week, they have undergone repeated mitotic divisions to form daughter cells called **oogonia** in the developing ovaries of the females. Some of these oogonia continue dividing mitotically to produce more oogonia, but others begin to divide by meiosis. The behavior of the latter cells is illustrated in Figure 10. The first sign of change is the enlargement of an oogonium, which is then called a **primary oocyte**. Like the chromosomes of any cell in interphase, its 46 chromosomes appear diffuse and nearly invisible in stained cells.

Meiosis I

Because prophase I is such a long and important stage of meiosis, cytologists have divided it into five substages: leptotene (Greek *leptos*, "slender"; Latin *taenia*, "thread"); zygotene (Greek *zygon*, "yoke," "pair"); pachytene (Greek *pachys*, "thick"); diplotene (Greek *diplos*, "double"); and diakinesis (Greek *dia*, "apart" or "across"; *kinesis*, "motion"). To understand meiosis, one must keep in mind that *maternal and paternal chromosomes are associated as homologous pairs during the first division*.

As a primary oocyte enters **leptotene** of prophase I, its chromosomes become visible as delicate threads (Figure 10A) that are indistinguishable from one another. Although already doubled during the previous S period of interphase, they are not visibly double when viewed through a light microscope. In some species, their **telomeres** (chromosome tips) are attached to the inside of the nuclear envelope. As in mitosis, the progressive shortening and thickening of chromosomes during prophase is due to folding and coiling of the fine DNA-histone fibers and to coalescence of the scaffold proteins.

Then, during **zygotene**, something occurs that is unique to prophase I and that controls the orderly halving of the chromosome number: by means that are still a mystery, each chromosome manages to find and pair intimately with its homologue (Figure 10B). This process is called **synapsis**. In many (but not all) organisms, it starts as telomeres migrate toward the centriolar region of the nuclear envelope, causing the chromosomes to form a *bouquet pattern*. Then, often starting in the vicin-

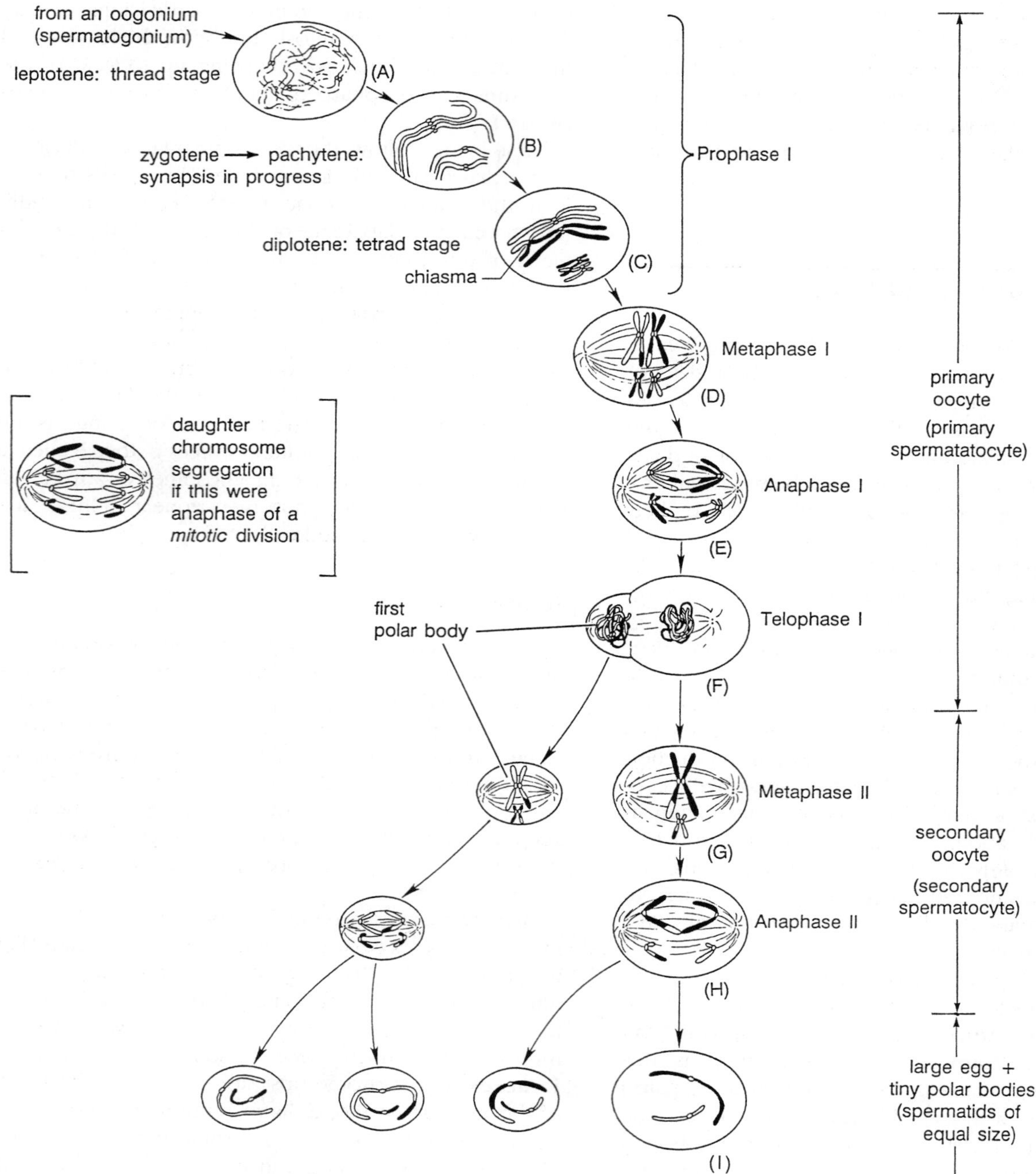

Figure 10 Chromosome behavior during the meiotic divisions of gametogenesis. In females, the smaller cells represent polar bodies; in males (nomenclature in parentheses), the four products of meiosis are the same in size and function. Only two sets of homologous chromosomes are shown, one member of each being open and the other black. It is convenient to assume that the open chromosomes were received from one parent, and their black homologues from the other. Only a few stages of the continuous process are shown and two crossovers are indicated. The nuclear membrane and nucleoli are omitted, and the centromeres are drawn as small circles rather than as constrictions. Please note that (1) the segregation of maternal and paternal chromosomes at anaphase I could equally well have been both black chromosomes to one pole and both open ones to the other, and (2) each of the four end products contains one of the four chromatids that make up each tetrad.

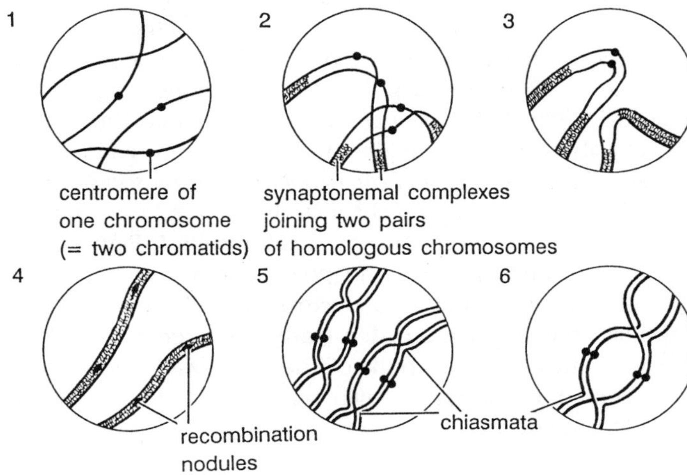

ity of the telomeric regions, homologous chromosomes become bound together by a ribbon-like proteinaceous structure called the **synaptonemal complex** that seems to zip up gradually (Figure 11). As they synapse, the homologous pairs often get tangled up with other chromosomes—but such interlocks are usually resolved by breakage and proper reconnection of synaptonemal complexes and homologues. Pairing, when complete, is so specific that if the order of genes on one chromosome should be different from that on its homologue, as sometimes occurs (Chapter 9), the two homologues will contort themselves like pretzels into a form that aligns them region by region.

Because sister chromatids are not individually recognizable until later in prophase I, the synapsed homologues appear to be a pair of threads; hence the term **bivalent** describes their association from pachytene on through to the end of metaphase I.

Pachytene, by definition, starts when synapsis is complete; chromatin condensation, which began during leptotene, continues through pachytene. Then another event that is unique to meiosis takes place. **Crossing over**, a process of breakage and exchange of parts between (at any one site) two of the four homologous chromatids, results in some shifting of genes from each chromosome to its homologue. (In Chapter 11 we discuss the genetic consequence of such swapping.) It occurs only at **recombination nodules**, spherical structures found along the central component of each synaptonemal complex, within which very low levels of DNA synthesis have been shown to occur (Carpenter 1981).

Physical evidence of these exchanges is not observed until the **diplotene** stage of prophase I, however, when the synaptonemal complexes are shed and the doubleness of individual chromosomes finally becomes visible in a light microscope (Figure 10C). Sister chromatids, however, remain together until the onset of anaphase I. Because each chromosome consists of two sister chromatids, the bivalent has four strands and is therefore called a **tetrad**. Along the length of each bivalent are found one or more X-shaped connections between nonsister chromatids; these **chiasmata** (singular, chiasma) are the places where crossovers have occurred.* When —as if mutually repelled—the pairs of homologues within each bivalent move apart during **diakinesis**, the

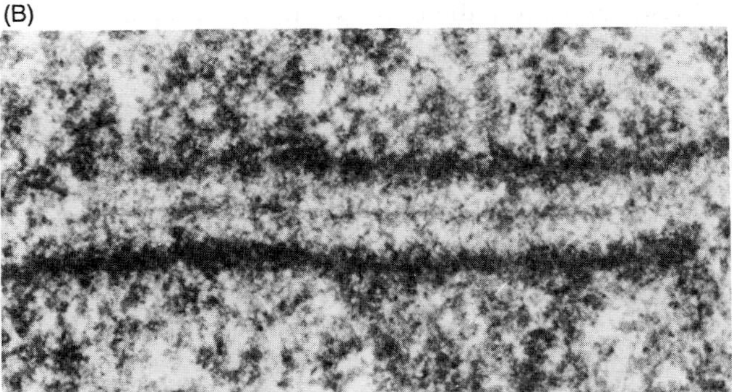

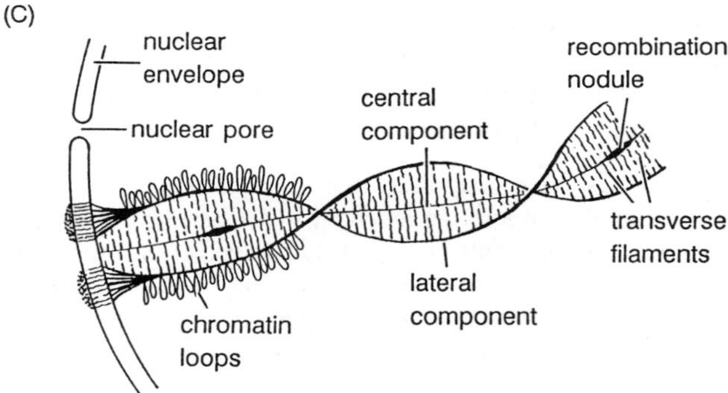

Figure 11 (A) Chromosome changes during meiotic prophase. 1, leptotene; 2, early zygotene; 3, late zygotene; 4, pachytene; 5, diplotene; 6, diakinesis, closeup of a bivalent. (After Risley 1986a.) (B) Synaptonemal complex from a human spermatocyte during the pachytene stage of meiotic prophase. LC, dense lateral components of the synaptonemal complex; each is 30–40 nm in diameter. CR, the central region (110–120 nm wide) between the lateral components. CC, the amorphous central component (10–30 nm in diameter) lies within the central region and is connected to the lateral components by irregularly spaced transverse filaments. The bar represents 0.1 μm. (From Chandley 1988; courtesy of Preben Holm.) (C) Bivalent with synaptonemal complex. Only a portion of the chromatin loops extending from each lateral component has been included. Actually, these loops radiate in all directions, except toward the central component. (After Risley 1986a.)

*In humans, the total length of synaptonemal complexes in oocytes is more than twice that observed in spermatocytes (von Wettstein et al. 1984).

chiasmata prevent the homologues from separating completely. Depending on the number and positions of these connections, the chromosome pairs may become shaped like X's, figure 8's, or more complex forms (not shown in Figure 10; but see Figure 6 in Chapter 11).

Cytogeneticists have known for decades that synapsis and crossing over are necessary for the proper disjunction (separation) of homologues. Whenever chromosomal abnormalities interfere with normal pairing or crossing over, an increase in **nondisjunction** leads to offspring with too many or two few chromosomes. Down syndrome, associated with an extra chromosome 21, is the preeminent human example of this (Chapter 7). It occurs with a much higher frequency among the offspring of older (compared with younger) mothers, although the reasons for this are not clear. Furthermore, Chandley (1988) notes:

> . . . it has been suggested that it is pairing failure that is actually the underlying mechanism for germ-cell death, not only in males but also in females . . . Speed (1988) has shown that, compared with spermatocytes, human oocytes are particularly error-prone in terms of pairing, showing many cells with partially or completely unpaired chromosomal axes, triple associations, interlocks and nonhomologous switches of pairing partners. Speed suggests that oocytes with such gross pairing abnormalities may not survive to maturity in the adult ovary. They may contribute to the normal high level of oocyte atresis (degeneration) found in the ovary during fetal development.*

By the time of a female's birth, all her oocytes have progressed to diplotene of prophase I. They remain arrested at this stage for anywhere from about one to five decades, their chromosomes still in bivalents but with the chromatin in a diffuse and greatly extended state. Then beginning at puberty, every month one (or a few) of these oocytes undergoes **maturation**; that is, it enlarges, progresses to metaphase II, and is released from the ovary. This process has the following stages.

At the end of prophase I, during meiotic prometaphase, the paired homologues move toward the equatorial plate. (At this time, whether or not their chiasmata move toward the chromosome tips is a matter of controversy.) Some events in prophase I are similar to those in prophase of mitosis: the nucleoli disappear and the nuclear envelope starts to break down while spindle microtubules arise.

During **metaphase I** (Figure 10D), the spindle is completed. Lined up on it are 23 bivalents, their homologous chromosomes now held together by chiasmata, their centromeres ready for departure in opposite directions. At the onset of **anaphase I** (Figure 10E), the cohesion of sister chromatids finally lapses and the two members of each chromosome pair finally separate (disjoin), moving to opposite poles. Unlike mitotic anaphase—during which the sister centromeres separate, thus bringing about the separation of their two chromatids—we see here the *disjunction of whole chromosomes*, each still composed of two chromatids that are held together at their centromere regions. At **telophase I** (Figure 10F), the chromosomes at one pole, together with a small amount of cytoplasm, are extruded and pinched off to form the tiny **first polar body**. The remaining chromosomes plus the bulk of the cytoplasm constitute a **secondary oocyte**, which begins to divide again immediately.

*A 5-month-old human female embryo has a peak number of 5–7 million germ cells, but at birth only about 2 million primary oocytes remain. At puberty, about 300,000 are present, and at menopause, fewer than 10,000.

Meiosis II

Like mitosis, **meiosis II** involves the separation of sister centromeres and thus the *separation of the two chromatids of each chromosome*. There are 23 rather than 46 participants, however (Figure 10G). During the very brief (in some species nonexistent) **prophase II**, a new spindle arises just under the surface of the egg. In **metaphase II**, the stage at which mature human eggs are expelled from the ovary, chromosomes are lined up in the equatorial region (Figure 12) and remain so until the penetration of a sperm stimulates them to continue to **anaphase II** (Figure 10H). At **telophase II** (not shown in Figure 10), one of the haploid chromosome groups is expelled, again with only a small glob of cytoplasm, to form the **second polar body**. By this time the first polar body might also have divided into two tiny cells. Thus, meiosis in the female leads to the formation of four haploid cells: **one egg** and three very small, nonfunctional polar bodies (Figure 13).

The whole process from oogonium to egg takes from 12 to 50 years. Its most important feature is the accurate reduction of chromosome number from diploid to haploid, but the genetic shakeup that occurs may also be important. Because the orientation of chromosome pairs on the metaphase I spindle is the result of a *random* process (i.e., it is equally likely that a maternal or a paternal centromere points toward a given pole), the chromosomes found at telophase I poles are mixtures of those originally derived from both the mother and the father. In each oocyte undergoing meiosis, different combinations of maternal and paternal chromosomes will be formed. In addition to this **independent assortment** of chromosome pairs, the process of **crossing over** causes a further reshuffling of genes: superimposed on the number of different whole chromosome combinations possible from a single individual (2^{23}, or over 8 million) is the additional variability produced by exchanges of chromosome parts between homologues.

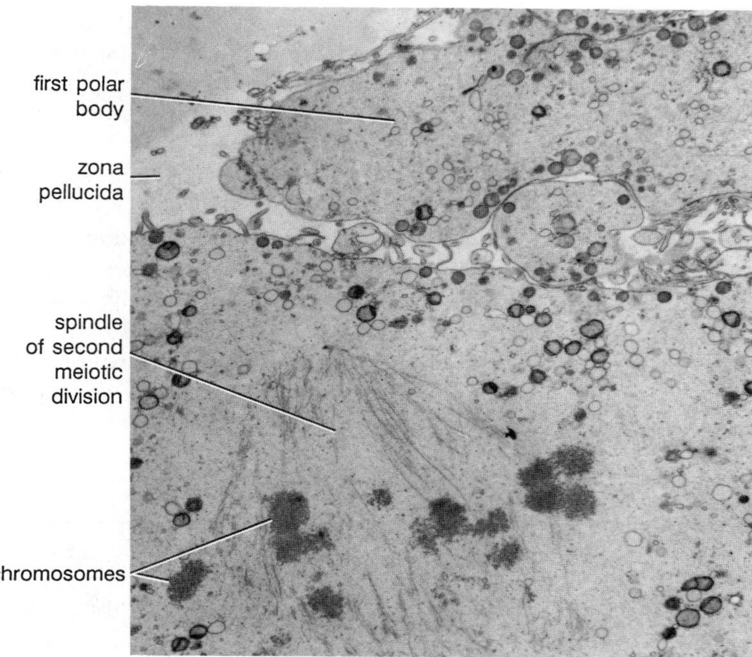

Figure 12 Part of a human egg at metaphase II. The egg has just been released from the ovary and the meiotic divisions will proceed no further unless the egg is fertilized. The first polar body is one product of the first meiotic division. (Electron micrograph courtesy of Luciano Zamboni, University of California, Los Angeles.)

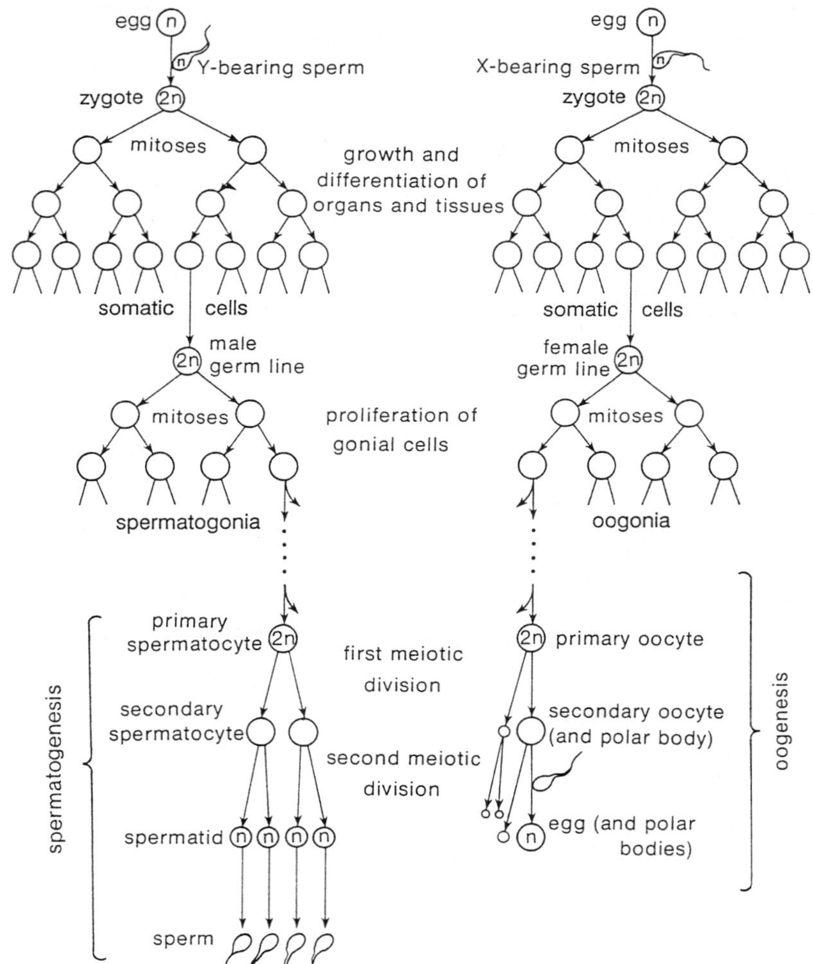

Figure 13 The overall scheme of mitotic and meiotic divisions in the two sexes, from zygotes to gametes and back to zygotes. The haploid number of chromosomes found in the gametes is designated n. In males, the spermatogonia present at puberty divide about 20 times per year to provide a continuing supply of primary spermatocytes. In females, all oogonia differentiate into primary oocytes before birth.

Hence, no two eggs (or sperm) produced by an individual are ever expected to be identical.

The events of meiosis in the male are the same as those in the female with regard to the nucleus, but they begin at puberty (rather than during fetal development) and proceed without interruption once started. After puberty, some diploid **spermatogonia** (descendants of primordial germ cells) in the walls of seminiferous tubules in the testes continually enter meiosis as **primary spermatocytes** (Figure 10A).* From one primary spermatocyte, meiosis I produces two **secondary spermatocytes** (Figure 10A–F), each of which in turn forms two **spermatids** (Figure 10G–I) after meiosis II. Division of the cytoplasm at telophase I and telophase II is *equal*, however, so that *four haploid cells of equivalent size and functional ability* are formed (Figure 13). There follows a complex developmental process, **spermiogenesis**, whereby each spermatid becomes an extremely specialized spermatozoon (**sperm**). Much cytoplasm is lost. Of the remainder, the Golgi complex produces the acrosome, mitochondria are incorporated into the midpiece of the tail, and centrioles form the neck and fibrils of the tail (Figure 3). Chromosomal proteins are replaced by protamines, which very tightly pack the chromosomes into the sperm head.

*In spermatocytes of XY males, the two sex chromosomes are paired just at one end, where they share a tiny region of homology called the pseudoautosomal region (Chapter 8).

Human spermatogonia develop into mature sperm in roughly nine weeks, of which about 24 days are spent in meiosis. Some men may continue to produce sperm throughout their lifetime—but in most males the fertilizing ability of sperm is greatly reduced by the time they are in their sixties (Holstein 1983; Honore 1978).

As depicted in Figure 14, fertilization restores the diploid chromosome number to the newly created zygote.

CELL CYCLE MUTANTS

As far as we know, the cell cycle progresses by a series of ordered steps, each of which is usually required before the next step can proceed and is presumably catalyzed by one or more specialized proteins. Hence, it seems reasonable to suppose that the cell cycle must itself be under genetic control, and that these genes must be subject to occasional mutation.

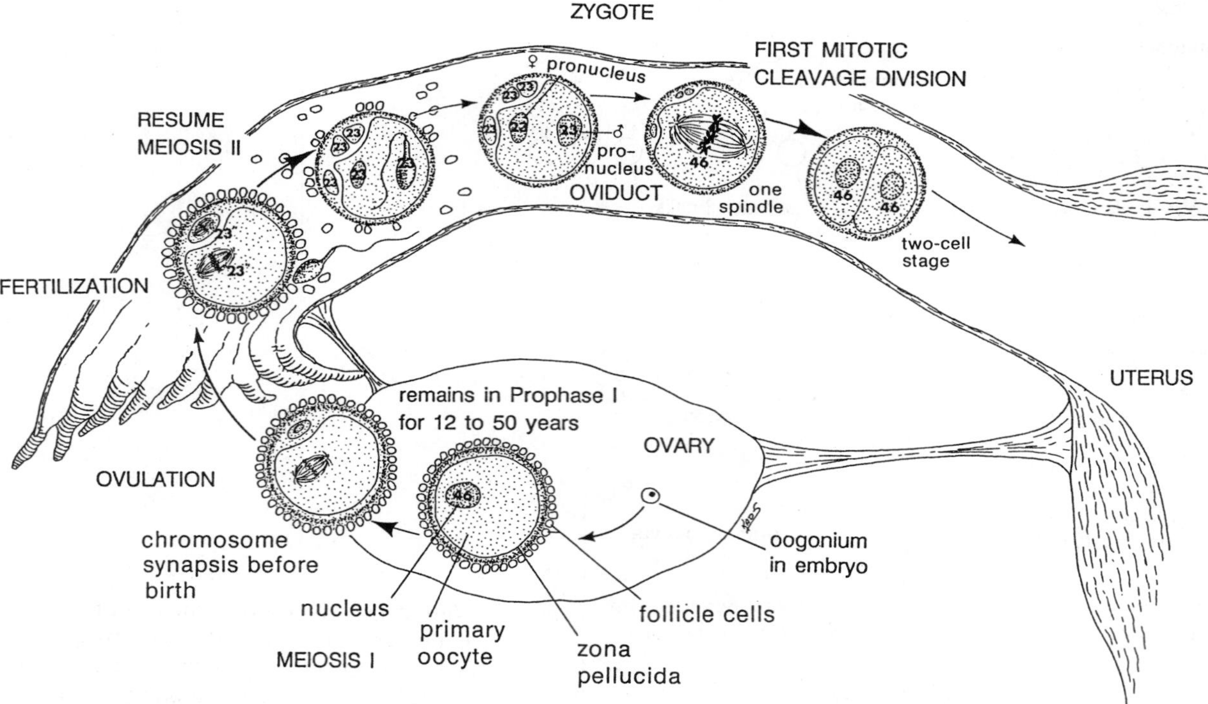

Figure 14 The location and timing of oogenesis, fertilization, and cleavage. The larger arrowheads indicate the completion of a cell division. The numbers of chromosomes per nucleus is shown, but between the two meiotic divisions the "23" refers strictly to the number of centromeres seen through the light microscope. Counting chromosomes in this "in between" cell is somewhat ambiguous, because, after meiosis I, each chromosome consists of two chromatids that become chromosomes after meiosis II.

By 1960 the search had begun for mutants that could be used to dissect the cell cycle into its many steps and to study the roles played by many of its specialized proteins. But investigators immediately faced a dilemma. Although minor alterations and errors in the cell cycle might be detected, isolated, and maintained in culture, any major error or block caused by the mutation of a cell cycle gene is likely to be lethal to the cell and thus go undetected because that cell is dead—or even if detected, be impossible to maintain in culture. But it is the major mutants, that is, those affecting the most critical steps in the cell cycle, that are of the greatest interest to geneticists and cytologists. How could they get around this problem?

Fortunately, geneticists hit upon three indirect but ingenious approaches to the problem (Hartwell 1978; Zimmerman and Forer 1981; Lloyd et al. 1982; Gatti et al. 1984) and used them on a wide variety of cells—from protozoans and yeast to Chinese hamsters and humans. One method is to isolate *mutants resistant to drugs that interfere with specific steps in the cell cycle*, with the idea that the altered drug susceptibility is due to some change in a target protein or structure. The second approach is to isolate mutants carrying so-called **temperature-sensitive *(ts)* mutations**, which are lethal at a temperature within the normal range tolerated by the wildtype organism but not lethal at a different (higher or lower) temperature. Presumably, the altered gene product is nonfunctional at the former (**restrictive**) temperature, but functional at the latter (**permissive**) one. Operationally, this means that cells are grown at the permissive temperature and treated with a substance that causes mutations; after some further growth, part of the culture is switched to the restrictive temperature to determine whether any new temperature-sensitive mutants have appeared. The third approach is to *take advantage of the recessiveness of most mutations* — maintaining these lethal mutations in heterozygotes, but generating homozygous (and nonviable) individuals for study at will via meiosis and fertilization.

Although work on cell division cycle mutants is difficult and slow, progress has been made in studying the "many hundreds of cell cycle genes" assumed to exist (Marcus et al. 1985). From yeast cells, animal cells in tissue culture, and fruit flies, researchers have isolated numerous mutants, assigned their time of action to particular stages, and in some cases even elucidated their mode of action. In budding yeast, for example, about 150 mutants have been mapped to 32 loci (Hartwell 1978) and in fission yeast, over 90 mutants have been mapped to 26 loci (Nurse and Fantes 1981). In mam-
have appeared. The third approach is to *take advantage of the recessiveness of most mutations*—maintaining
in human cells, several cell cycle genes are known. In *Drosophila*, the number of such loci is 50 and rising.

Most cell cycle mutants isolated so far have been recessive. Interestingly, among the normal cell cycle genes, there are inhibitors as well as activators, and "the initiation of mitosis is regulated by the competitive activities of positive and negative control elements" (Russell and Nurse 1987). Perhaps as a result of the techniques of selection, many G_1 mutants but relatively few S period mutants have been isolated. Among those mutants affecting the M period, cytological studies have revealed abnormalities in chromosome organization and stability, chromosome condensation, spindle formation, chromosome segregation, or cytokinesis (Smith et al. 1985; Marcus 1985). Other cell cycle mutants involve genes for growth factors and growth factor receptors.

Virtually all *mitotic* mutants exhibit similar defects in the comparable stages of meiosis. In addition, geneticists have used criteria such as reduced fertility, abnormal chromosomal segregation, altered recombination rates and patterns, and altered mutability, to isolate a large number of *meiotic* mutants in a wide range of organisms, including fruit flies. These meiotic mutants affect specialized functions such as premeiotic DNA synthesis, synapsis, recombination, nonhomologous segregation, transition from meiosis I to meiosis II, and the return to mitotic division following meiosis II (Baker et al. 1976). Not surprisingly, because meiosis II is functionally equivalent to a mitotic division, most meiotic mutants act during meiosis I rather than in meiosis II.

What can we say about cell cycle mutants in humans? Other than generalizing from what we know about studies in other mammalian systems (especially the Chinese hamster), not a lot yet. But some recent high-tech cloning experiments (summarized by Murray 1987) have opened up the very exciting possibility of using cell cycle mutants from *any* organism as functional probes for "fishing out" comparable genes in human cells.

For example, British cancer researchers Melanie Lee and Paul Nurse (1987) worked with some cell cycle genes obtained from three organisms: fission yeast, budding yeast, and humans. They started with a culture of *cdc2*, a temperature-sensitive *c*ell *d*ivision *c*ycle mutant from fission yeast that blocks the key G_1 start signal (comparable to the restriction point in mammals) and can only grow at 29°C. [The normal allele of this gene commits the cell to the mitotic cycle if cell size is right and nutrients are available; these wildtype cultures can grow at 36°C.] Then, using recombinant DNA techniques that will be described in Chapter 15, Lee and Nurse added to this culture some chopped-up DNA from nonmutant budding yeast, and looked for new "hybrids" that could grow at the higher, restrictive temperature. Further tests showed that all such hybrid colonies had incorporated from the budding yeast's DNA the normal allele of a similar gene, *CDC28*, whose ts mutant allele blocks budding yeast's G_1 start signal. Thus the budding yeast's *CDC28*$^+$ gene was able to complement (cure) the defective fission yeast *cdc2* mutation. That is, it allowed the blocked cells to undergo mitosis, presumably by supplying a very similar protein.

Thus assured that the normal allele of a cell cycle gene from budding yeast could function in another species, and guessing that homologous genes might be present in *all* species, Lee and Nurse began looking for a similar piece of *human* DNA that could likewise "cure" the *cdc2* fission yeast mutant. Using recombinant DNA techniques, they were indeed able to find a stretch of human DNA that complemented the mitotic block in mutant *(cdc2)* fission yeast! This newly isolated human gene was dubbed *CDC2Hs*, for *Homo sapiens*.

These three cell cycle genes (from budding yeast, fission yeast, and humans) each code for a protein of close to 300 amino acids. Furthermore, the protein sequences are about 60% identical—a striking conservation of homology between such distantly related organisms. Some researchers think the gene product in humans might be *maturation promotion factor (MPF)*, a substance also necessary for the transition between G_2 and M in vertebrates. In any event, it now appears that all cells probably have very similar cycling controls—a finding that is bound to stimulate renewed research efforts and new approaches to the genetic analysis of mitosis (Broach 1986).

SUMMARY

1. In the human life cycle, a haploid egg and a haploid sperm, each with 23 chromosomes, unite at fertilization to make a diploid zygote with 46 chromosomes. Repeated mitotic divisions form diploid body cells that differentiate into all tissues and organs. Diploid oocytes and spermatocytes undergo meiosis to yield haploid eggs and sperm, thus completing the life cycle.

2. The cell cycle consists of a long interphase (G_1, S, G_2), during which protein synthesis and DNA replication occur, alternating with a short mitotic division (M), during which the replicas are distributed to daughter cells. Hormones and growth factors are important cell cycle regulators.

3. The dimensions of a chromosome vary during the cell cycle, but its basic structure remains constant. One DNA double helix combines with histones to make a long, 10 nm wide, string of nucleosomes which then coils into a chromatin fiber that is 30 nm wide. This fiber folds into a series of looped domains, which attach at their "necks" to scaffold proteins. The interphase structure is diffuse and still highly extended. During cell division, the fiber condenses much further, at the metaphase stage forming a chromatid 700 nm wide. By this time the length of the DNA helix has undergone about a 10,000-fold compaction.

4. Mitosis has five main stages. Chromosomes shorten and thicken in prophase and prometaphase, and line up on the spindle in metaphase; at anaphase the centromeres separate and pull identical sister chromatids (now called chromosomes) to opposite poles; in telophase, chromosomes uncoil, nuclear envelopes reform, and the cytoplasm splits. Thus two genetically identical cells are formed from one. No pairing of homologous chromosomes occurs during mitosis.

5. The formation of haploid eggs and sperm by meiosis involves one replication and two successive cell divisions. Meiosis I results in the separation of pairs of homologous chromosomes, each individual chromosome consisting of sister chromatids held together in their centromere regions. Meiosis II results in the separation of sister chromatids.

6. A primary oocyte forms one egg and up to three nonfunctional polar bodies; a primary spermatocyte yields four functional sperm. Oocytes are in prophase I by the time of birth and in metaphase II at ovulation and will complete the second meiotic division only if fertilized. Spermatocytes do not begin to develop until after puberty.

7. Two key meiotic processes—the independent assortment of nonhomologous chromosomes, and crossing over between homologous chromosomes—result in a tremendous reshuffling of gene combinations.

8. The cell cycle consists of an ordered series of genetically controlled steps that can be studied by isolating cell cycle mutants. Many of the genes that control the cell cycle are probably the same or similar in all species.

FURTHER READING

The literature on chromosomes, cells, and cell division is vast. Much of the information in this chapter was gleaned from three superbly written and illustrated books: Alberts et al. (1989), Darnell et al. (1986), and Watson et al. (1987). De Duve (1984) gives an interesting, scaled-up cell tour, and Risley (1986b) contains excellent articles on chromosome structure and function. On gametes and fertilization, see Austin and Short (1982), Van Blerkom and Motta (1984), Hafez and Kenemans (1982), and Zamboni (1971). On fertilization, articles by Wassarman (1987, 1988), Grobstein (1979), and Epel (1977, 1980) are very readable. More technical works include Metz and Monroy (1985a,b,c), Browder (1985), Hartmann (1983), and Dale (1983).

The cell cycle and cell division are described in articles by Prescott (1987), Mazia (1974, 1987), and Inoue (1981), and in books by Moens (1987), Brachet (1985), John (1981), Baserga (1981), Prescott (1976), Mitchison (1971), and Dirksen et al. (1978). Cell cycle mutants are discussed in articles by Lee and Nurse (1988), Marcus et al. (1985), Baker et al. (1987), Carpenter (1984), Simchen (1978), and Hartwell (1978).

QUESTIONS

1. Review oogenesis in human females (a) by listing the meiotic stages that occur during each of the following periods: embryo/fetus until birth; birth until start of oocyte maturation; oocyte maturation until ovulation; fertilization until first cleavage division. (b) Give the amount of time that elapses for each period.

2. (a) Explain why a metacentric chromosome has an X shape at mitotic metaphase and a V shape at mitotic anaphase. (b) Explain why an acrocentric chromosome has a nearly V shape at mitotic metaphase and a sort of J shape (or nearly rod shape) at mitotic anaphase.

3. Consider the metaphase homologues:

(a) After one division, the two daughter cells had the chromosomes diagramed here. Was the division mitosis or meiosis?

(b) After one division, the two daughter cells had the chromosomes diagramed below. Was the division mitosis or meiosis?

(c) What would the homologous set look like (diagrammatically) before and after the second meiotic division?

4. Consider a chromosome, 1, and its homologue, 1′. If we assume that there were no crossovers between the genes we are following, two different kinds of gametes are possible for this one pair: those containing 1 and those containing 1′. Now consider a second pair of homologues, 2 and 2′, which also gives rise to two kinds of gametes when taken by itself. But what are the possible chromosomal constitutions of a gamete when both pairs are considered together? Choose one of these possibilities and reconstruct the meiotic divisions that produced it.

5. Consider a spermatogonium with three sets of homologues: 1, 1′; 2, 2′; X, Y. List, in a systematic fashion, the eight different kinds of sperm that can be formed, considering the three pairs together and assuming no crossing over within the intervals we are following. Choose one of these possibilities and reconstruct the meiotic divisions that produced it.

6. Consider two sets of homologues: 1, 1′; 2, 2′. (a) If a secondary oocyte contains the centromeres of 1 and 2, what does the first polar body contain? (b) If an egg cell contains the centromeres of 1 and 2, what does the second polar body contain?

7. During interphase and before DNA replication, a diploid human cell contains 46 chromosomes, and the term chromatid is inapplicable. How many chromosomes and chromatids should be visible in this cell at metaphase? (Recall that, at anaphase, chromatids become the chromosomes of incipient daughter cells.)

8. How many human chromosomes and chromatids (if applicable) are present at these stages of meiosis: (a) a spermatogonium before the S period? (b) a primary spermatocyte at metaphase I? (c) a secondary spermatocyte at metaphase? (d) a spermatid? (e) a sperm?

9. Among spontaneously aborted human fetuses, many are triploid (3n), and some are tetraploid (4n). How many chromosomes do these fetuses have in their somatic cells?

10. How many chromosomes has a mule whose mother (a horse) had 64, and whose father (an ass) had 62?

11. On the basis of Questions 5 and 6, plus your intuition, fill in the following table:

Number of homologous pairs being considered	Number of different kinds of gametes
1	______
2	______
3	______
4	______
5	______
n	______

12. For a human, n in Question 11 equals 23. Find the approximate value of 2^{23}. [This can be done fairly easily, if you note that 2^{10} is about 1000, and recall that $(a^x)(a^y)(a^z) = a^{x+y+z}$.]

13. Show that a human being inherits one of about 64 trillion ($= 64 \times 10^{12}$) different chromosomal combinations from his/her parents. (Recall that 10^z is the number 1 followed by z zeroes).

ANSWERS

1. *Embryo/fetus until birth*: Oogonia become primary oocytes and enter a very long meiosis I (leptotene, zygotene, pachytene). This period takes roughly 3–7 months. *Birth until beginning of oocyte maturation*: Arrest in meiosis I (diplotene). This period takes roughly 12–50 years. *Oocyte maturation until ovulation*: end of meiosis I (diakinesis), metaphase I, anaphase I, telophase I, cytokinesis (to produce secondary oocyte and 1st polar body), prophase II, metaphase II. This period takes about 1 month. *Fertilization until first cleavage division*: anaphase II, telophase II, cytokinesis (to produce egg and second polar body). This period takes 24–36 hours.

2. At metaphase the chromosome arms extend out to the sides of each centromere, but at anaphase they are pulled behind the centromere.

3. (a). Mitosis. (b) First meiotic division. (c) either

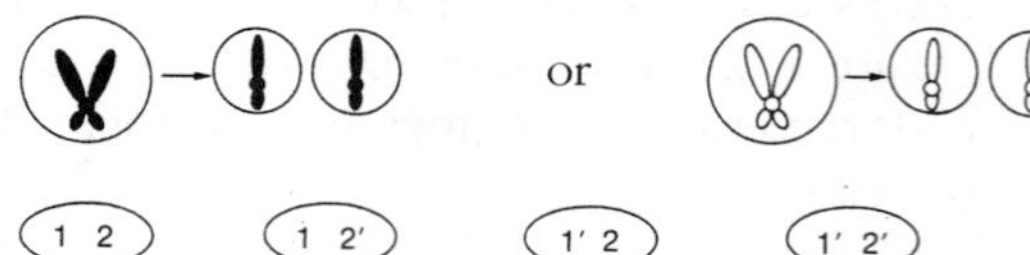

4. (1 2) (1 2′) (1′ 2) (1′ 2′)

5. One convenient way to systematize such combinations of sets of things is by setting up a branching diagram like the one below. This method works not only for homologous chromosomes but also for gene pairs or any other combinatorial data.

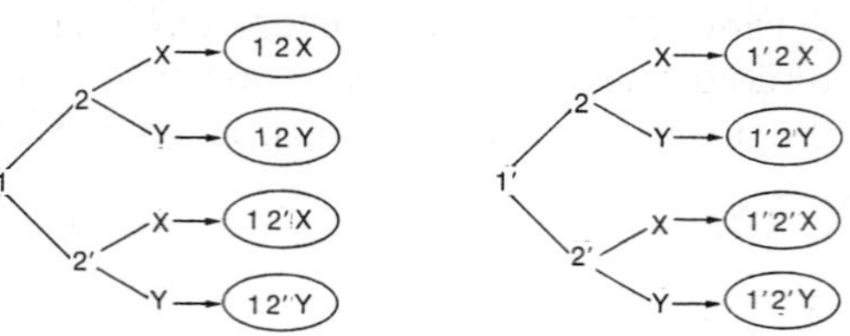

6. (a) (1′ 2′) (b) (1 2)

7. 46 chromosomes, 92 chromatids

8.

	Chromosomes	Chromatids
(a)	46	—
(b)	46	92
(c)	23	46
(d)	23	—
(e)	23	—

9. $3(23) = 69$; $4(23) = 92$

10. $(64/2) + (62/2) = 32 + 31 = 63$

11.

Number of homologous pairs	Number of kinds of gametes
1	2
2	4
3	8
4	16
5	32
n	2^n

12. $2^{23} = (2^3)(2^{10})(2^{10}) \cong 8(1000)(1000) \cong 8{,}000{,}000$

13. $(2^{23})(2^{23}) \cong (8 \times 10^6)(8 \times 10^6) \cong 64 \times 10^{12}$

Chapter 6 Genetics of Development

Embryonic Development

Genetics of Embryonic Development
Maternal Effects
Segmentation and Pattern Formation
The Need for Both Maternal and Paternal Genomes in Mammals
Future Prospects

Usual Sexual Development
Prenatal Development: Genetic and Hormonal Factors
Postnatal Development: Environmental Factors
Postnatal Development: Physiological Factors

Some Errors in Sexual Development
True Hermaphrodites
Pseudohermaphrodites
Gender Identity

> *Tom... is now a great man of science... and knows everything about everything, except why a hen's egg don't turn into a crocodile, and two or three other little things.*
>
> *Charles Kingsley (1863)*

In this chapter we discuss some aspects of how a fertilized egg develops into a complex, multicellular organism and also how this organism normally develops into a male or female.

EMBRYONIC DEVELOPMENT

Much of what we know about human embryology comes from the study of spontaneous or induced abortions at various stages of development. Improvements in optical equipment have also made it possible to get some stunning photographs of embryos and fetuses in utero (Nilsson 1986). The very earliest stages of development, before the embryo is attached to the uterus, can be studied in the laboratory (Fowler and Edwards 1973). Immature eggs are removed from ovarian follicles and grown in culture medium. After the completion of meiosis I and artificial fertilization, the resultant zygotes continue to develop in culture until about the 16-cell stage. Studies of early cell divisions can help overcome certain problems of infertility, such as blocked oviducts. These and other approaches have yielded much information about normal and abnormal human development, but a great deal remains to be learned. Following is a brief overview of the normal process.

Shortly after fertilization in the upper reaches of the oviduct, the zygote begins to divide (cleave) as it slowly migrates down the oviduct to the uterus—yielding two cells, then four, eight, and so on, without any intervening periods of cell growth. Thus the total cytoplasmic mass initially remains constant as it is partitioned among smaller and smaller cells whose metabolic needs are met by yolk and other materials originally laid down in the egg cytoplasm.* After about three days, a solid ball of about 16 cells enters the uterine cavity. Further division and the absorption of fluid convert it by the fourth day into an asymmetrically hollow **blastocyst** of not more than 100 cells. Among these, a tiny clump (the **inner cell mass**) destined to form the embryo proper can already be distinguished from the remaining cells, which will contribute to its surrounding membranes (Figure

*The biology of twinning, as well as its usefulness and limitations for genetic studies, will be discussed in Chapter 23.

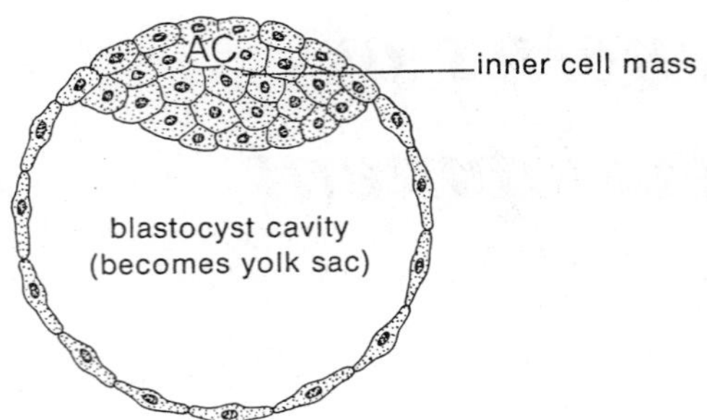

Figure 1 A human blastocyst (diagrammatic cross section) at about four and a half days. The inner cell mass will form the embryo proper, and the outer circle of cells the chorion. The amniotic cavity will begin forming in the region marked AC at about eight days. The blastocyst consists of about 100 cells and is drawn here about 400 times actual size.

1). After floating free for about two days, the blastocyst attaches to and begins to burrow into the spongy lining of the uterus, whose rich blood supply will nourish the developing embryo. By the end of the second week, **implantation** is complete. Inside the rapidly growing blastocyst, the tiny embryo begins to absorb nutrients diffusing from maternal tissue.

By the third week, the inner cell mass has become a flat pear-shaped embryo (Figure 2). Above a central rod called the **notochord**, primitive neural tissue arises, becomes grooved, and then folds together to form a **neural tube**, forerunner of the brain and spinal cord. Small cellular blocks called **somites** develop on both sides and ultimately differentiate into muscle, skeletal, and dermal tissues. Blood vessels and primitive blood cells soon make up a simple **cardiovascular system** that distributes oxygen and nutrients throughout the embryo while removing waste materials. The cardiovascular system is the first organ system to become functional. It links the embryo with specialized maternal–fetal tissue called the **placenta**, which is embedded in the uterine wall. Here the embryonic and maternal bloods flow very close to each other but remain separated by cell membranes. The placenta also produces hormones that maintain pregnancy and prevent menstruation.

No signs of sexual development appear until the fourth week, when swellings that will ultimately form the external sex organs arise and **primordial germ cells** develop in the wall of a primitive cavity called the yolk sac (Figure 2). These cells, the descendants of which are destined to become eggs or sperm, migrate during the fifth week into the newly forming internal **gonadal ridges**.

The most crucial and sensitive developmental period spans weeks 4 through 7, a time when pregnancy may still go undetected. Because by the end of this stage every major organ system has begun to take form, the presence of drugs, viruses, or ionizing radiation may lead to gross malformations. Early in this period, the flat, disk-shaped embryo bends to form a C-shaped cylinder with a disproportionately large primitive brain (Figure 3). By the seventh week, rudimentary eyes, ears, nose, jaws, heart, liver, and intestines are distinctly visible, and the development of other structures (lungs, kidneys, sex organs, and skeleton) is also well underway. By the end of the seventh week, the embryo is over 30 mm long and weighs less than 5 gm (about 1/6 ounce). Its tail has disappeared, limbs with fingers and toes are clearly formed, and the head constitutes almost half of its total length (Figure 4). It is fully encased in two transparent membranes: an inner **amnion** pressed close to an outer **chorion**. Anchored to the placenta by an umbilical cord, the embryo moves quite freely in the fluid-filled amniotic cavity. The latter acts as a protective shock absorber, prevents the embryo from forming adhesions with the surrounding membranes and promotes normal symmetry in development. Later on, the amniotic fluid produced by the mother is swallowed and absorbed by the fetus, and urine is excreted into it; but there is also rapid turnover in the amniotic fluid.

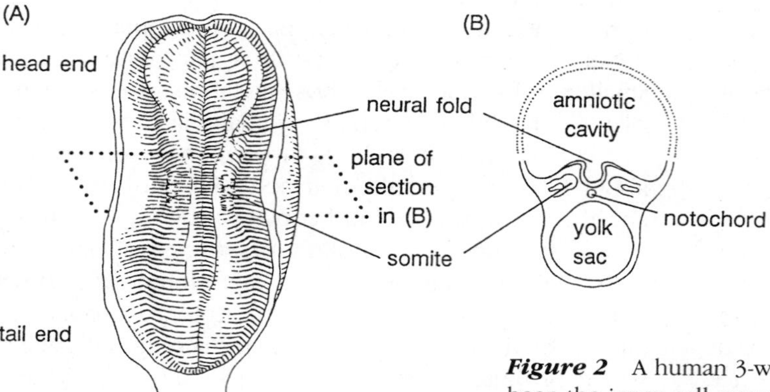

Figure 2 A human 3-week embryo. (A) Top view, looking down on what had been the inner cell mass. (B) Cross section of an embryo at the level indicated by the dotted lines in (A). The yolk sac will become much smaller as the embryo enlarges into the amniotic cavity. (Adapted from Moore 1988b.)

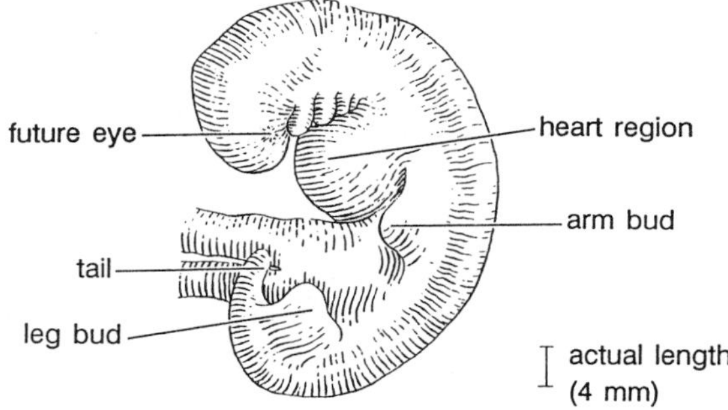

Figure 3 A human 4-week embryo. The extraembryonic membranes are not shown. (Adapted from Moore 1988b.)

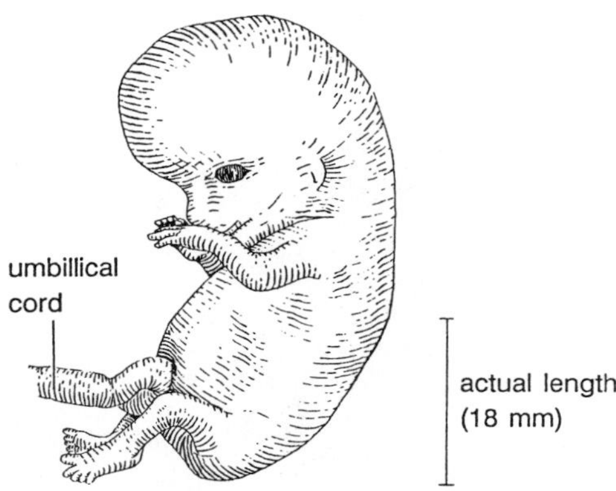

Figure 4 A human 7-week embryo. All major organ systems are developing. The toes and fingers have separated. (Adapted from Moore 1988b.)

From the eighth week of development to birth at 38 weeks the fetus will increase about 650-fold in weight and about 17-fold in length, while development of all the organ systems continues (Figure 5). Because the respiratory system is the last to develop, lung function sets the limits of viability. Actually, a 26-week-old fetus may survive if born prematurely, even though it has attained only 30% of the normal birth weight. With increased knowledge, improved technology, and extraordinary effort and expense, it is becoming possible to keep alive some babies born even earlier than this. Such heroic measures present an unresolved ethical dilemma, however, because many of these very premature survivors will suffer from serious problems for years to come.

GENETICS OF EMBRYONIC DEVELOPMENT

How does a single cell, the zygote, become transformed into an exquisitely complex organism made up of many distinctive cell and tissue types? And how do the cells and tissues get organized and distributed throughout the body in patterns and shapes that are unique to each species? These questions have fascinated and baffled scientists for more than a century. Over many decades, embryologists described in immense detail the changes and interactions of specific embryonic cells during their orderly course of differentiation. But what they found was not one grand scheme to fit all embryos; instead, they observed a bewildering array of developmental plans among the various groups of organisms studied. And although important discoveries and useful concepts helped to focus the search, clear-cut and universally applicable answers to this major biological problem continued to elude them.

In particular, neither embryologists nor geneticists could explain how such striking differences can arise among cells whose nuclei remain genetically identical! Sidney Brenner explained the situation in 1974:

> We know very little about the molecular mechanisms used to switch genes on and off in eukaryotes. We know nothing about the logic with which sets of genes might be connected to control the development of the assemblages of different cells that we find in multicellular organisms.
>
> These questions arise in a particularly acute form in elaborate structures like nervous systems .... How is this complexity represented in the genetic program? Is it the outcome of a global dynamical system with a very large number of interactions? Or are there defined subprograms that different cells can get a hold of and execute for themselves? What controls the temporal sequences that we see in development?

In the 1970s and 1980s the availability of some organisms in which many mutants of early development had been isolated and genetically analyzed, together with the use of high-powered molecular techniques (Chapter 15), brought new insights. Some of these are briefly sketched below. Much of this recent knowledge in developmental genetics comes from intensive studies of a few representative animals, including a nematode, an insect, and a mammal.

Caenorhabditis elegans is a small, unsegmented, free-living roundworm, whose haploid genome contains about 20 times the amount of DNA found in the bacterium *E. coli* and codes for at least 2,000 genes. Its 959 adult cells can all be traced back through a virtually invariant developmental lineage through the larva to the fertilized egg (Sulston et al. 1983). This newcomer to developmental genetic studies was chosen specifically because it is fairly complex but yet of a manageable size both physically and genetically. It is also easy to propagate and study in the laboratory.

The fruit fly *(Drosophila melanogaster)* and the mouse *(Mus musculus)*, although much more complex and resistant to developmental analysis, have been ex-

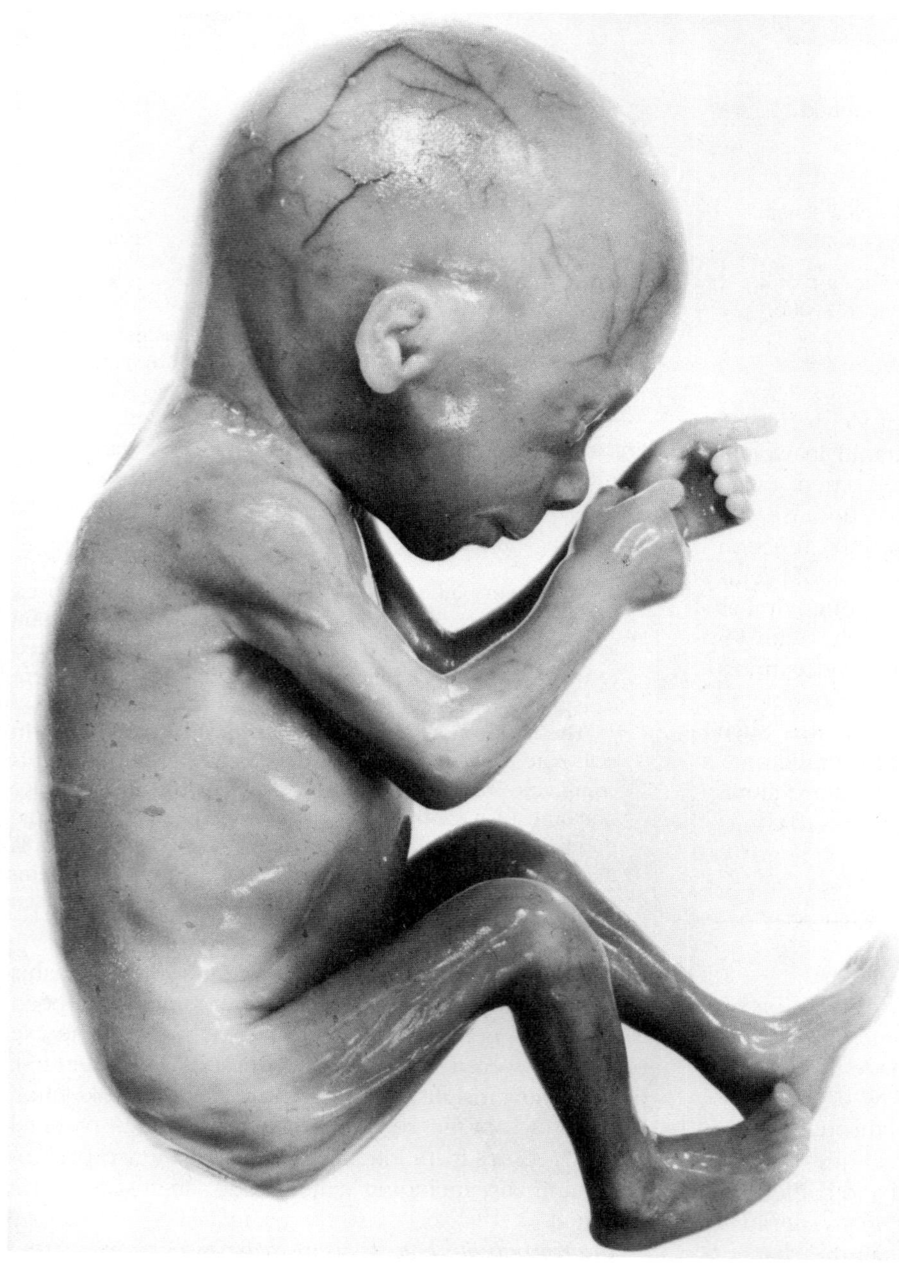

Figure 5 A human 17-week fetus, actual size. (From Moore 1988b.)

haustively studied since the early 1900s. With such a wealth of accumulated knowledge to guide them, researchers could apply the new technologies to a few particularly fascinating and promising genes. The fruit fly's haploid genome contains about 180,000 kb pairs of DNA, specifying at least 5,000 genes. Its developmental plan—which includes larval and pupal stages before metamorphosis to the adult form—is more flexible than that of a roundworm but less flexible than that of a mouse. The mouse's haploid genome of 3 million kb pairs is about 20 times that of *Drosophila*, and specifies at least 50,000 genes. Its developmental plan is quite flexible. "An early mouse-embryo cell may end up in almost any adult structure. Its fate is determined not by its place in a fixed lineage but by the spatial position it happens to take up in the early embryo" (Gehring 1985).

Maternal Effects

Long before the advent of molecular genetics, it was known that the egg contains stores of nutritive material (e.g., yolk) great enough to support the early embryo through at least its first few cleavage divisions. In addition, the egg was known to exert some genetic influence over the very early embryo. In snails, for example, the direction of shell coiling (left versus right) is controlled by the maternal genotype—which determines the spindle position and thus the planes of cleavage in the zygote and embryo—rather than by the snail's own genotype (Sturtevant 1923).

Later studies in a wide variety of organisms showed that these maternal effects involve the stockpiling in the egg cell of various materials in addition to yolk. These include everything needed for cleavage divisions: nucleotides and enzymes for DNA replication, as well as tubulin for spindle formation. Another category of reserves includes constituents needed for protein synthesis, including ribosomes and messenger RNA (mRNA) molecules that encode genetic information transcribed from maternal DNA (Davidson 1986). No wonder the egg is so much larger than other animal cells!

The distribution of these various developmentally significant inclusions within an egg cell is not uniform, however. Some types of eggs show definite *gradients* of concentration, whereas others exhibit a *mosaic* partitioning of important substances. Either pattern eventually causes an unequal apportionment of certain substances among the various cells in a developing embryo, but neither pattern can by itself account for all the observed complexities of development; interactions between cells are critical, too.

Developmental and molecular studies indicate that a three-level hierarchy of gene families controls development in *Drosophila* embryos: **maternal effect genes, segmentation genes,** and **pattern formation (homeotic) genes** (Akam 1987; Ingham 1988; Gehring 1985). The first step in this hierarchy occurs when maternal effect genes in the mother set up important spatial coordinates along both the anterior-posterior and dorsal-ventral axes in the presumptive embryo. Over a dozen mutants of maternal effect "coordinate" (pattern-shifting) genes have been analyzed. They alter the segmental fate and the germ layer (ectoderm, mesoderm, endoderm) fate of future embryonic cells, and are best explained by the existence of one or more gradient systems (Wilkins 1986). Mutants of maternal effect loci lead to embryos with extra or missing heads, tails, dorsal structures, or ventral structures (Gilbert 1988; Nüsslein-Volhard et al. 1987).

Strong evidence for the gradient hypothesis is provided by a maternal effect mutant gene called *bicoid* (*bcd*). The eggs of *bcd*/*bcd* females develop into embryos that lack both a head and a thorax, an outcome indicating that the normal gene product is needed for the development of the head and thorax. In the eggs of females with at least one normal *bicoid* allele, the protein product becomes distributed as a steep concentration gradient that is strongest at the anterior end of the egg (and early embryo) and extends to about midway down its length. Wolfgang Driever and Christiane Nüsslein-Volhard (1988a and b, 1989) of the Max Planck Institute in West Germany, and their colleagues, have shown that early in development the bicoid protein acts on two zygotic *gap* genes (see below) called *hunchback (hb)* and *Krüppel (Kr)*. Specifically, it binds to several DNA sites next to the *hb* gene, thereby activating its expression (transcription) and leading to formation of head and thorax segments. At the *Kr* locus, however, the bicoid protein has the opposite effect, repressing gene expression. Thus *Kr* is expressed only in regions where the bicoid protein is sparse or absent.

In *Caenorhabditis*, early development is also controlled mainly or totally by maternal components stored in the embryo. A few dozen temperature-sensitive mutants have been isolated, but "the molecular basis of the nematode egg's morphogenetic system still remains a mystery" (Wilkins 1986).

Segmentation and Pattern Formation

At some point, not always easily defined, the embryo's own genes become activated and take over from the maternal genes.* Detailed studies in *Drosophila* of many developmental mutants revealed that the establishment of the basic body plan is accomplished by two kinds of events. First the functioning of the segmentation genes divides the blastoderm (early embryo) into metameric units, a linear series of primitively similar segments. Then the action of homeotic genes assigns a regional identity to each resulting domain.

Studies of the regions within which the segmentation genes are active and of the consequences of their mutation (Figure 6) show that the genes interact with one another in a "developmental cascade" to subdivide the embryo into progressively smaller regions. Of the three classes of segmentation genes, the *gap* genes act first in response to the products of the maternal effect loci mentioned above; the action of the *gap* gene products establish several relatively large regions. Thereafter, genes of the *pair-rule* and *segment polarity* classes are expressed, forming a series of repeated patterns along the length of the embryo. The "stripes" within which different genes are expressed are offset somewhat along this axis, and complex interactions between the genes partition the embryo into a series of repeated morphological units (Ingham 1988).

*The time of activation of the zygotic genes is much earlier in mammals than in less complex animals. In humans, this first occurs between the 4- and 8-cell stages. In mice it occurs at the 2-cell stage, in pigs at the 4-cell stage, and in sheep at the 8-cell stage (Braude et al. 1988).

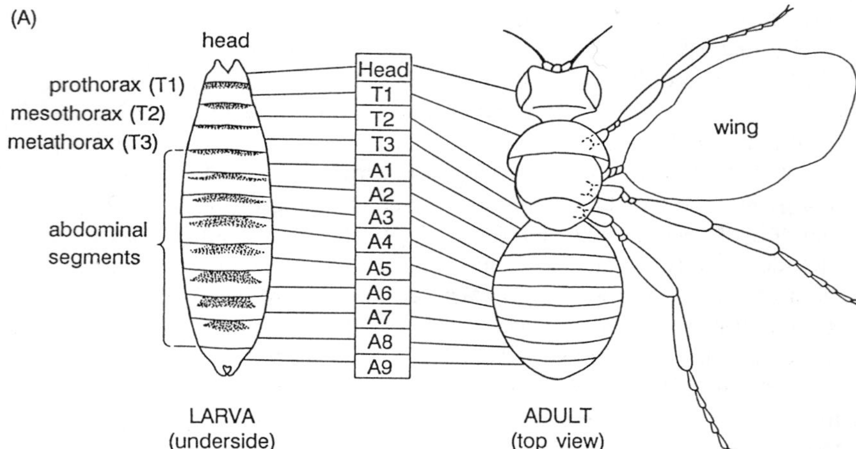

Figure 6 (A) Segmentation in *Drosophila* larva (left) and adult (right). Solid lines indicate the boundaries between individual thoracic (T) and abdominal (A) segments. The wedge-shaped, darkly stippled regions on the underside of the larva indicate bands of toothlike projections called *denticles*, which abut the anterior margins of T and A segments and provide a useful phenotypic marker. In the adult, each T segment has a unique set of appendages: T1 (prothorax) has legs only; T2 (mesothorax) has wings and legs; T3 (metathorax) has haltares (balancing organs) and legs.

Pattern formation occurs through the action of **homeotic* loci**, which act by "naming" or specifying the structures unique to each metameric unit. These genes were originally identified because flies carrying mutations in homeotic genes display bizarre effects such as legs sprouting from antennal sockets or from the mouth region, wings emerging from the region where eyes belong, and so on. Thus, when their normal function is disrupted by mutation, the fate assigned to various regions is inappropriate to their location. The important homeotic genes are located in two well-studied gene clusters. The *Bithorax* complex controls developmental fate in the posterior thoracic and abdominal segments (Lewis 1985; Duncan 1987), while the *Antennapedia* complex controls those in the head and anterior thoracic segments (Scott 1985).

Using molecular techniques, geneticists isolated and sequenced the genes and gene products of some segmental and homeotic loci. Those studied first included the pair-rule gene *fushi tarazu*† *(ftz)*, and the homeotic genes *Ultrabithorax (Ubx)* and *Antennapedia (Antp)*. The latter two are huge; all were found to be complex and interrelated. In particular, they shared a highly similar sequence of about 180 DNA base pairs, which was dubbed the **homeobox**. In the protein products of these genes, the 60 amino acids defined by the homeobox sequences collectively make up the **homeodomain** (Scott and Weiner 1984; McGinnis et al. 1984a). Further analyses involving over a dozen other maternal, segmental, and homeotic genes revealed that the homeobox was present in many of them.

*William Bateson (1894) first defined *homeosis* as the assumption by one member of a segmental series of the form or character proper to another member of that series.

†In Japanese, this means "not enough segments." The embryos have 7 rather than 14 segments and die very early in development.

Even more exciting than the elucidation of these master sequences in fruit flies, however, was the finding by McGinnis et al. (1984b) and Carrasco et al. (1984) that they also exist in many other organisms: some molluscs, annelids, amphibians, birds, and mammals, including mice and humans. [Although their segmentation patterns are not so obvious as those of insects, these organisms are segmented in one way or another— including the structure of their central nervous systems, muscles, and (where they exist) vertebral columns.] The homeobox and homeodomain sequences are highly *conserved* (unchanged) through evolution; for example, some human and mouse homeodomains differ by only one amino acid (out of 60) from that present in the *Antp* gene of *Drosophila*. This similarity demonstrates that at least some developmental genetic mechanisms are widespread among the animal kingdom (Gehring and Hiromi 1986; Gehring 1987).

A great deal of accumulated developmental genetic and molecular evidence suggests that the homeobox-containing genes function to regulate the expression of one another and of other developmentally significant genes. In addition, all known homeodomain-containing proteins are localized to the nucleus, which would be expected of molecules that regulate gene expression. Furthermore, within their homeodomain sequences is a short, nine-amino acid section that is identical in all organisms studied and thought to function by binding directly to DNA. (Similar kinds of domains are known to regulate transcription of certain genes in bacteria and to control mating type genes in yeast.) Experiments in which the effects of homeodomain binding sites from the *Drosophila* paired-rule gene *ftz* and the segment-polarity gene *engrailed (en)* were tested (alone and in combination) in cultured cells suggest that they compete with one another for DNA binding sites. Thus, in this system the *ftz* protein activated DNA transcription, while

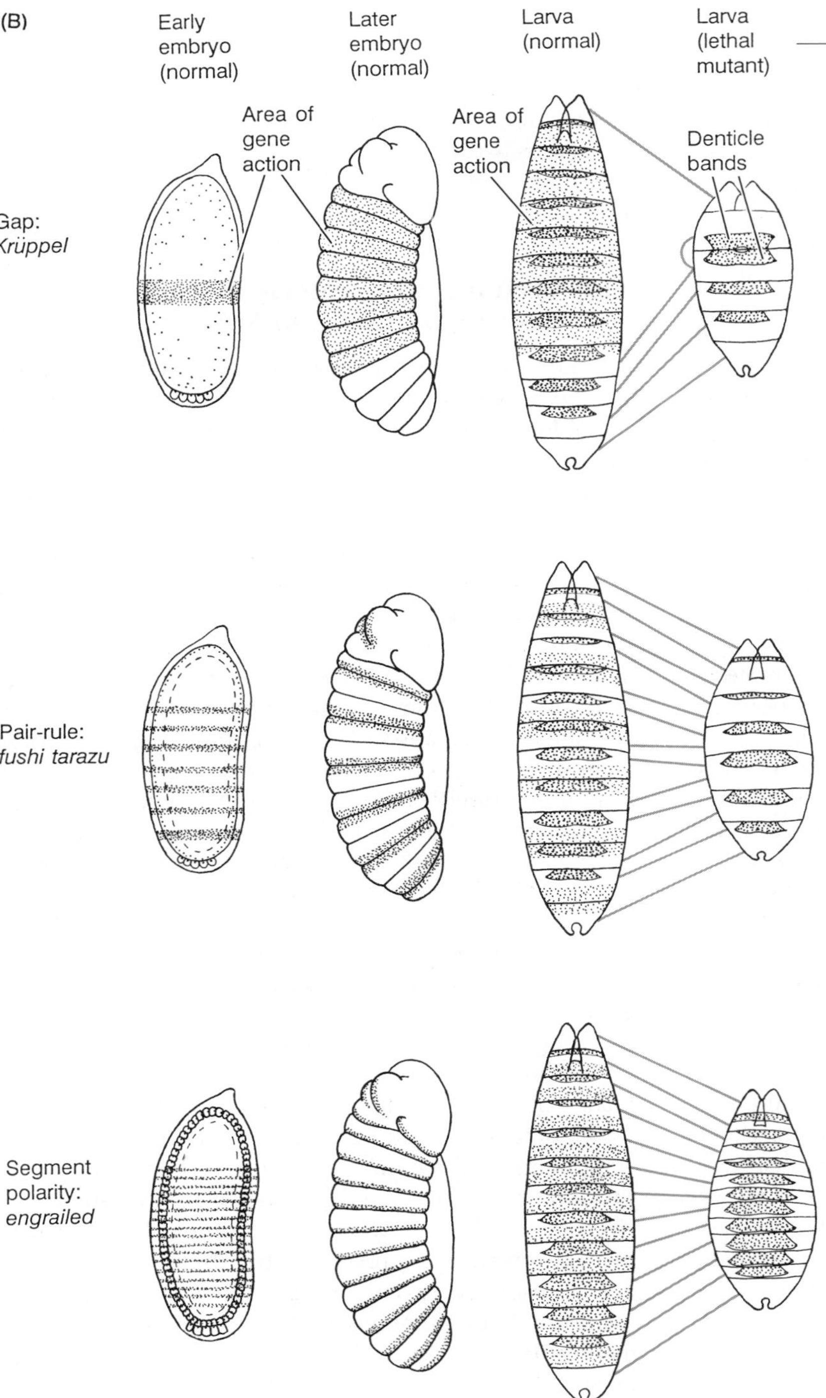

Figure 6 (B) *Drosophila* segmentation genes. Light stippling indicates the patterns of gene expression in normal embryos and larvae. Lethal alleles of these types of genes cause disruptions in the normal segmentation process, leading to deletion or malformation of certain segments and death of the embryo/larva (far right). **Gap** loci divide the early embryo into three main regions; a given gene is expressed in one of these regions. The normal gap gene *Krüppel* acts in the midregion of the embryo; thus mutant embryos lack thoracic and some abdominal segments. **Pair-rule** loci collectively subdivide the embryo into a series of 14 offset segments, including 3 thoracic (T) and 8 abdominal (A) segments, but a given gene is expressed only in alternate offset segments. The normal pair-rule gene *fushi tarazu* is expressed in 7 alternate offset segments; thus mutant embryos retain only 7 segments rather than 14. **Segment polarity** loci specify anterior and posterior compartments of each segment; thus a given gene is expressed in every segment. The normal segment polarity gene *engrailed* is expressed in the posterior compartment of every segment, but also has some attributes of pair-rule loci. Mutant *engrailed* embryos are highly variable, some showing fusion of pairs of segments (i.e., anterior regions) and others showing fusion of large groups of segments. Here we include just one of the many *engrailed* phenotypes known. (Adapted from Scott and O'Farrell 1986, and from Nüsslein-Volhard and Wieschaus 1980.)

the *en* protein repressed it by competing for binding sites (Jaynes and O'Farrell 1988). The authors point out that there are precedents for this type of gene regulation in prokaryotes, and suggest that "multiple patterns of gene expression can be generated by competition among regulators for binding to related but different . . . sites," especially depending on the relative affinities of the homeodomain proteins for each DNA binding site.

Homeodomain protein gradients associated with pattern formation have been identified in vertebrates as well. In the developing forelimbs of *Xenopus* (a frog), a gradient of the so-called **X1Hbox 1 homeodomain protein** was analyzed by the staining of antibodies specific to this protein. Investigators at the University of California in Los Angeles (Oliver et al. 1988) found that nuclei in the most anterior part of the **forelimb bud**

mesoderm (the region or "field" destined to give rise to the forelimbs) were most heavily stained, whereas nuclei in the posterior part of this field remained unstained. By contrast, no X1Hbox 1 protein was detected in any nuclei of the **hindlimb bud mesoderm**, which gives rise to the same types of tissues that differ only in their spatial arrangement. This finding provides strong evidence of an early molecular difference between presumptive arms and legs. The authors further report: "Similar results are obtained in developing mouse limbs, suggesting that X1Hbox 1 participates in forelimb development in a variety of tetrapods . . . These results suggest a molecular link between morphogenetic fields, gradients, and homeobox genes in vertebrate development." They also suggest that there may exist gradients of other homeodomain proteins that cooperate in pattern formation of limb buds. In any event, "developing and regenerating limbs provide one of the most favorable systems for the study of pattern formation in vertebrates:" Over 50 mutations in mice and at least 8 mutations in chickens are known to affect limb development.

In mammals a new class of homeobox genes has also been found. Although similar to one another, the homeobox proteins (oct-1, oct-2, and GHF-1/Pit-1) that these genes code for share at most only 21 (out of 60) amino acid residues with other known homeobox proteins (Levine and Hoey 1988). The oct proteins probably act by binding to a highly conserved 8-base DNA sequence (ATTTGCAT) called the **octamer**, found near many mammalian genes. The GHF-1 homeobox protein may bind to specific sequences of a growth hormone gene in the pituitary gland.

In addition to the homeodomain proteins found in a broad spectrum of organisms, two other types of DNA-binding protein domains have been identified that may play major roles in development (Dressler and Gruss 1988). Some gap genes of *Drosophila* code for so-called **finger proteins,** which closely resemble zinc finger proteins present in frogs, mice, and humans. Certain pair-rule genes of *Drosophila* also code for so-called **paired box** domains found in mice, where they are apparently involved in formation of the vertebral column.

Homeotic-type genes occur in the *Caenorhabditis*, too. These include several mutants of *lineage (lin)* genes, which have the ability to switch the expression of genes in several sets of vulval cells between two possible pathways (Greenwald et al. 1983). The product of one of these genes, *lin-12*, contains a series of repeated amino acid sequences that are very similar to that of the *epidermal growth factor (EGF)*. Both the lin-12 and EGF peptides are thought to stimulate mitosis. Similarly in *Drosophila*, a switching gene called *Notch* causes certain cells to become either nerve cells or skin cells. Its protein product, too, contains repeated EGF-like sequences. Although such findings are still a long way from answering all the questions about how genes control development—including the question of what regulates the regulators—they have begun to provide some important clues and approaches to this major unsolved biological problem.

The Need for Both Maternal and Paternal Genomes in Mammals

Although many invertebrates (such as bees and wasps) and some vertebrates (including an occasional turkey) are able to develop by **parthenogenesis** (i.e., from an unfertilized egg), there have never been any genetically authenticated cases of such "virgin birth" occurring naturally among mammals. Nevertheless, geneticists had always assumed that the mammalian genome inherited from a female parent and that inherited from a male parent were functionally equivalent during the course of the offspring's development. This expectation appeared at first to be borne out by a report on "microsurgically produced homozygous-diploid uniparental mice" (Hoppe and Illmensee 1977). By removing one pronucleus from fertilized mouse eggs, treating the zygote with a drug that doubles the chromosomes of the remaining haploid nucleus, allowing the "rediploidized" zygotes to develop in vitro to the blastocyst stage, and then transplanting them into foster mothers, these researchers appeared to have produced some viable baby mice that inherited *all their genes from one parent*.

Unfortunately, however, all attempts to repeat these experiments ended in failure (Figure 7). Instead, researchers were surprised to discover that the two parental genomes are *not* functionally equivalent, so mammalian offspring need both a mom and a dad for normal development to occur. Studies of early embryonic development in which mouse zygotes were manipulated to produce two egg-derived pronuclei, or two sperm-derived pronuclei, or fusion of a second polar body with the egg pronucleus, revealed that these zygotes could not develop. The two pronuclei come together and form the normal diploid number of chromosomes just as normal fertilization would. Yet female-derived embryos are lacking in the development of extraembryonic membranes, and male-derived embryos show highly retarded development of the embryo proper. Thus viable offspring are never produced (Barton et al. 1984; McGrath and Solter 1984; Surani et al. 1986; Surani 1986). In human zygotes, the same situation apparently prevails: an occasional pregnancy produces, instead of a normally developing fetus, a uterine growth called a **hydatiform mole**. Such placental masses contain extraembryonic membranes but no fetal tissue, and their chromosomal composition is entirely paternal—presumably the result of the fertilization of an egg without a pronucleus and subsequent doubling of the sperm chromosomes (Jacobs et al. 1980).

What is the nature of these genomic differences, and how do they arise? One approach to these questions involves the use of mouse strains that have different **translocations**—that is, aberrations that join together parts of nonhomologous chromosomes (Chapter 9). When crossed in various combinations, these strains produce some zygotes that have both representatives of a given chromosome arm derived from one parent and none from the other. If the genes on these arms behave the same whether inherited from the mother or the father, the zygote should develop normally. But if the maternally and paternally derived genes do not function interchangeably, the embryos containing these particular translocations will not develop properly. Solter (1988) found that:

> ... regions in which maternal duplication/paternal deficiency is lethal but maternal deficiency/paternal duplication is without effect can be found, in addition to regions in which paternal duplication/maternal deficiency is lethal but paternal deficiency/maternal duplication is normal ... We can explain these regions relatively easily by assuming the existence of a gene that is active only when contributed by one parent but not by the other ...

Among 11 mouse chromosomes tested this way, six were found to foster normal development while the other five did not (Miller 1987).

It has been hypothesized that during the processes of gamete formation, certain genes are differentially **imprinted** through various patterns of **methylation** (the attachment of methyl groups to cytosine bases in DNA). In a few cases, certain genes are found to become more highly methylated in the egg than in the sperm. Of course, this imprinting procedure would have to be erased some time during the embryo's development, so that each individual can re-imprint its genes according to its own sex during gametogenesis. Differential methylation is also thought to be involved with the *X-chromosome inactivation* that occurs in cells of female mammals (Chapter 8).

The relationship between gene activity and methylation has been explored in chickens, rabbits, mice, and humans for some time (Gilbert 1988). Studies of hemoglobin production show that increased activity (i.e., transcription into RNA) of *globin* genes is associated with undermethylation of the cytosine bases—whereas in tissues where these genes are inactive, the cytosine bases of the globin genes were more highly methylated (van der Ploeg and Flavell 1980). Activity levels of the *ovalbumin* gene in chickens show the same kinds of correlations with levels of methylation. Conversely, methyl groups can be experimentally added to or subtracted from the DNA in cloned genes or in cultured cells. When this is done, gene activity is altered: the unmethylated genes are transcribed, and the completely methylated genes are not transcribed (Taylor and Jones 1982; Busslinger et al. 1983).

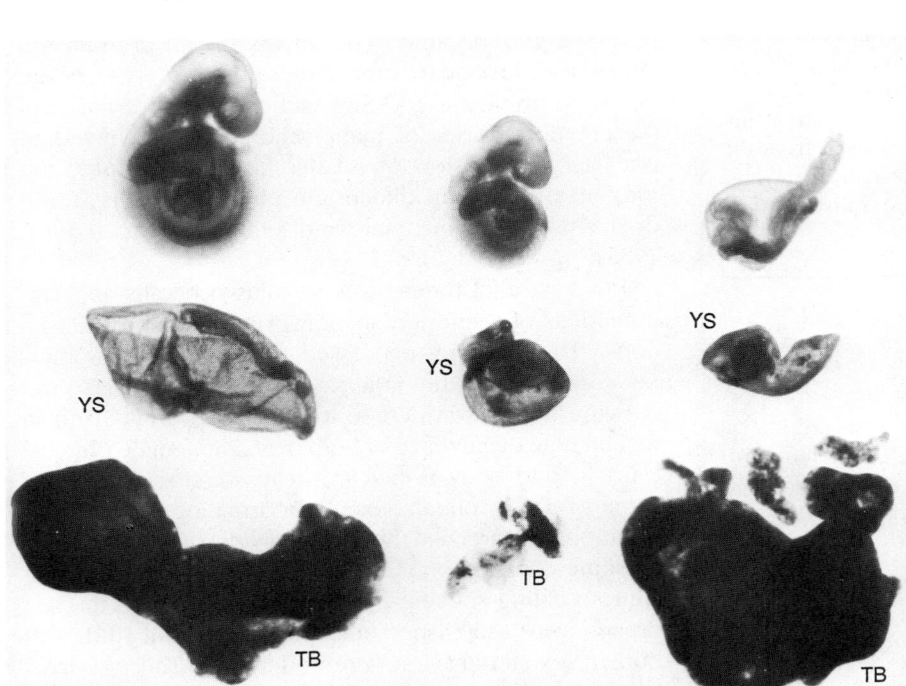

Figure 7 Results of experiments in which one pronucleus was removed from individual mouse zygotes and then replaced by a pronucleus from another zygote. In this manner the researchers produced *control* embryos containing one male and one female genome, or embryos with *two female genomes*, or embryos with *two male genomes*. Compare the three types of embryos, yolk sacs (YS), and extraembryonic membranes (TB) that resulted from these nuclear transplantations. Note that the zygote with two maternal genomes gives rise to a small, well-developed embryo but scanty extraembryonic membranes, whereas the zygote with two paternal genomes gives rise to a poorly developed embryo but extensive extraembryonic membranes. (From Surani et al. 1986.)

Although not yet proved, there are suggestions of differential imprinting in a few inherited human diseases (Solter 1988). The age of onset and severity of these disorders seem to differ according to which parent contributed the gene. *Huntington disease* (HD), an autosomal dominant neurological condition that is lethal in adulthood, provides the best example.* Its age of onset is highly variable, the average being at 30–40 years. But a "significantly greater proportion of persons with juvenile- or adolescent-onset of HD (onset before age 20) have inherited the gene from an affected father than from an affected mother. In addition, significantly more cases of HD with late-onset (initial symptoms after age 50) have inherited the HD gene from an affected mother than from an affected father" (Myers et al. 1985).

Future Prospects

Despite these exciting beginnings in describing the genetics of development, not all researchers are convinced that the techniques of molecular biology will ever reveal all the answers to all questions of how differentiation is regulated. As Wilkins (1986) cautions:

> The idea in this instance is that development is guided and accompanied by a changing pattern of gene expression. A quarter century of experimentation has amply confirmed this proposition, and another quarter century can be expected to do the same. What has singularly failed to emerge is a set of ideas that explain the underlying dynamics of the whole process. Nor is it clear how a preoccupation with the controls of gene expression will explain (1) what the biological functions of the gene products within cells are or (2) the interrelationships between gene products that determine individual cell-type properties. Finally, the phenomena of cell–cell interactions, which comprise such an important part of development, are so far removed from the processes of gene expression control that they are almost certain to be left unilluminated by the gene control studies.

USUAL SEXUAL DEVELOPMENT

Most of us have strong opinions about what constitutes maleness or femaleness, and our daily lives are permeated from birth to death by societal expectations or ideals for sexual development and behavior. Indeed, these sex differences are so much a part of our lives that we may not think to inquire as to their evolutionary usefulness.

Biologists surveying the plant and animal kingdoms have found great variability in modes of sex determination, systems of mating, and mating behavior. In most cases, sex is determined genetically, but there are also some fascinating examples—in fish, lizards, and birds, for example—in which environmental factors (e.g., temperature, local sex ratio) play a major role, even causing sex reversal under some circumstances (Chan and Wai-Sum O 1981; Crew 1965; Bacci 1965). Biologists generally agree that the evolution of sexual reproduction allows for more rapid mixing of mutations within a species, so a greater number of gene combinations can be tested by natural selection in a given period of time. For example, if mutant A occurs in one line and mutant B in another, then the double mutant AB should turn up among the offspring of matings between these two lines. The only way for a double mutant to occur in an asexually reproducing species, however, is for both mutations to take place in the same line—a much less likely event.

Some sexually reproducing plants and animals (including *Caenorhabditis*) maintain both sexes in the same body but practice either self-fertilization or cross-fertilization. For those plants and animals (including humans) in which the two sexes are normally maintained separately, however, individuals whose body structure or behavior cannot be classified as completely male or completely female have always been regarded with special interest. Indeed, the concept of male and female joined harmoniously in one body has been idealized (Figure 8), and the term **hermaphrodite** (or **intersex**) derives from Hermaphroditus, the name of the son of Hermes (the Greek god of commerce) and Aphrodite (the Greek goddess of love). According to myth, Hermaphroditus grew together with the nymph Salmacis, thus combining in one body both male and female characters. Hermaphroditic gods abound in the myths and art of India and Persia, too; Ardhanarisvara, for example, represents the joining of the male god Siva with the female Sakti. And the creation stories of many other cultures (including one Hebrew version of Adam's creation) feature "the idea of a hermaphrodite origin of man and his subsequent bisection" into male and female (Mittwoch 1981, 1986).

But a parallel theme is what Mittwoch calls an "irrational fear of seemingly inappropriate sexual manifestations." This feeling is expressed, for example, by numerous versions of the folk saying "Whistling girls and crowing hens always come to some bad ends." And in real life, the occurrence of both male and female physical traits in one body is not so ideal. We now know that some of these human sexual abnormalities result from variations in prenatal hormone levels or in sex chromosome constitution (Chapter 8). In the 1950s, the first clinics dealing specifically with birth defects of the sex organs were established, and the first careful studies of human sexual behavior were undertaken. These sources of information, augmented by anthropologists' studies

*This disease, which killed the folk singer Woody Guthrie, is described in Chapter 15; its gene was one of the first to be mapped (to chromosome 4) by the new techniques of molecular biology.

(A)

(B)

Figure 8 Hermaphrodites from two cultures. (A) From Greece (about third century B.C.). This terracotta statuette was recovered from the ancient city of Myrina on an island in the Aegean Sea. (Courtesy of the Museum of Fine Arts, Boston.) (B) From Melanesia (about 1900). This small "uli" figure was crafted of wood, fiber, shell, and plaster. It was used by secret cults on the island of New Ireland in memory of departed souls. (Courtesy of the St. Louis Art Museum; gift of Morton D. May.)

on widely divergent cultures throughout the world, have led to a deeper understanding of how maleness and femaleness normally develop in humans. The picture that emerges involves a series of steps, some occurring before birth and some after.

Prenatal Development: Genetic and Hormonal Factors

The initial step in sex determination occurs at the time of fertilization, when either an X-bearing or a Y-bearing sperm enters the egg (McLaren 1988). This establishment of **chromosomal sex** sets the stage for future changes, although the two sexes are physically indistinguishable at first. By the sixth week of development, both XX and XY embryos possess identical pairs of undifferentiated gonads called **gonadal ridges**, plus two sets of primitive duct systems: the male-type **Wolffian ducts** and the female-type **Müllerian ducts** (Figure 9A). These duct systems eventually form the internal sex organs other than the ovaries and testes. Externally, both XX and XY embryos develop a knoblike **genital tubercle**, a **urogenital groove** flanked by the **urethral folds**, and **labioscrotal swellings**.

Thus, in the beginning, the external genitalia of any embryo are equipped to form either a male or a female. Unless male hormones intervene at later stages of fetal life, it will develop along female lines. Money and Tucker (1975) put it this way:

> Nature's first choice is to make Eve. Everybody has one X-chromosome and everybody is surrounded by a mother's estrogens during prenatal life. Although not enough for full development as a fertile female, this gives enough momentum to support female development. Development as a male

requires effective propulsion in the male direction at each critical stage. Unless the required 'something more,' the Adam principle, is provided in the correct proportions and at the proper times, the individual's subsequent development follows the female pattern.

Extremists on both sides of the battle of the sexes have seized upon this developmental fact to "prove" either that females are more primitive than males, or, conversely, that females represent the basic sex and males are an anomalous offshoot. We think that such leaps of fancy from biology to ideology are unwarranted and foolish.

For presumptive males, the second critical step occurs late in the sixth week of development, when the *TDF* (testis determining factor) gene on the short arm of the Y chromosome somehow triggers the inner parts of the gonadal ridges to begin developing into male gonads, the **testes**. Although necessary for differentiation of the mammalian testis, this *TDF* gene is not sufficient; autosomal and X-linked genes are also involved (Eicher and Washburn 1986). *TDF* is now known to be completely separate from a gene that is located near the centromere on the long arm of the Y and induces production of the so-called H-Y antigen (E. Simpson et al. 1987; Kolata 1986). The function of the latter gene (or genes) is not clear, but it may be involved with sperm production at a later stage of development.

The third critical step occurs when two kinds of cells in the fetal testes begin to produce two different hormones (Josso 1981; Jost and Magre 1984). The **Müllerian-inhibiting hormone** causes regression of the Müllerian ducts, while the male sex hormone **testosterone** develops and maintains the Wolffian ducts. If the appropriate tissues are responsive to these hormones, the following changes occur by the seventh week (Wilson et al. 1981). Internally, the Müllerian ducts begin to degenerate. Under the local influence of testosterone, the Wolffian ducts begin to form the **prostate gland, seminal vesicles**, and tubes **(vas deferens)** connecting the whole system (Figure 9B). In addition, neural pathways in the hypothalamus (part of the brain) that control menstrual cycling in females are altered in male embryos. Finally, a very potent metabolite of testosterone (called **dihydrotestosterone**) organizes the shaping of external genitals. The genital tubercle and urethral folds lengthen and fuse to form a **penis**, while the labioscrotal swellings develop and fuse to form the **scrotum**.

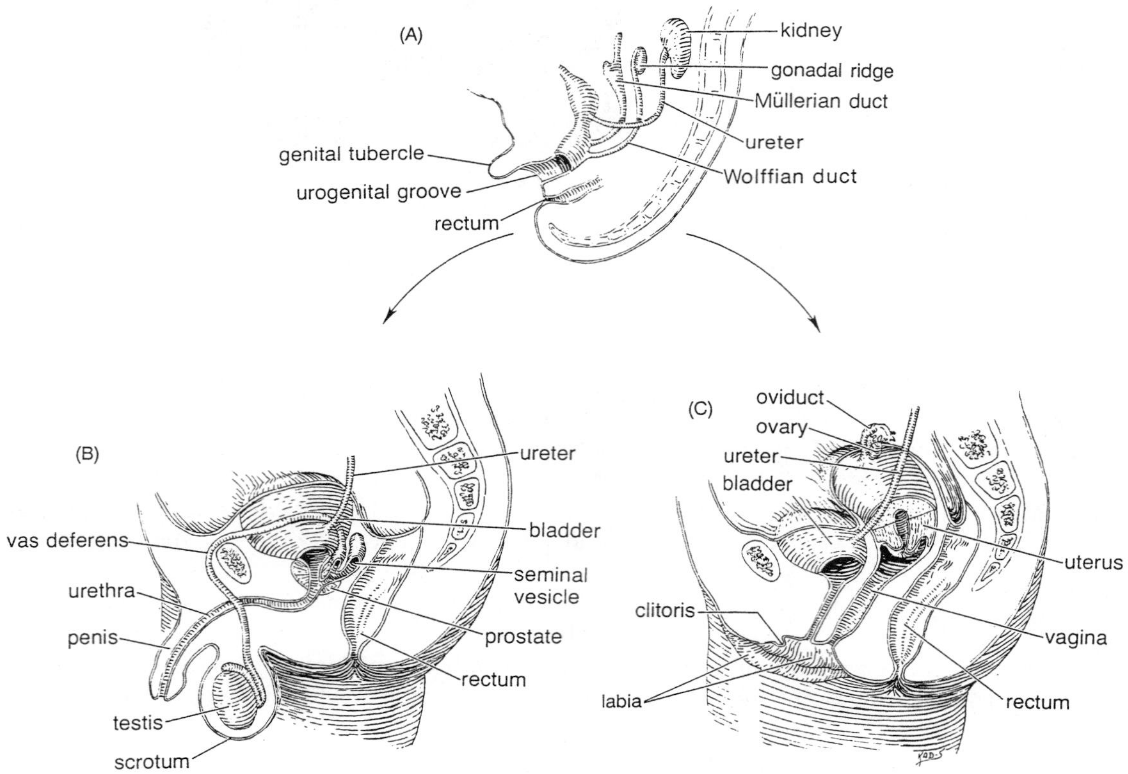

Figure 9 Development of the internal sexual structures during intrauterine life. (A) A sexually undetermined embryo at six weeks. (B) A male at birth. (C) A female at birth.

In presumptive females, the products of various genes cause the outer parts of the genital ridges to begin to develop as **ovaries** around the twelfth week. Quite independently, the Wolffian tubes regress (because of the absence of testosterone secretion), and (in the absence of Müllerian inhibiting hormone secretion at an appropriate time) the Müllerian tubes form the female ducts. The forward parts form the **oviducts** (Fallopian tubes) and the end parts fuse to form the **uterus** and most of the **vagina** (Figure 9C). The external genitals also develop: a **clitoris** from the genital tubercle, small and large **labia** (lips or folds) from the urethral folds and labioscrotal swellings, respectively, and the lower part of the vagina from the urogenital groove. Thus the sex of a fetus is clearly evident by the end of the fourth month.

Postnatal Development: Environmental Factors

On the basis of the appearance of its external genitals, a newborn's sex is assigned and recorded on the birth certificate. Sex-specific names and clothing (including color) usually follow, perhaps along with sex-specific treatment. Some (but not all) studies find that baby boys are cuddled and spoken to less frequently, but encouraged to be energetic and independent—whereas little girls are rewarded for cleanliness, passivity, and dependence. Responding to such cues, most young children begin to develop a sense of sexual identity by about 18 months of age.

Sex-specific toys and games are common, too, and by the age of 5 or 6 most children prefer companions of their own sex. Despite some recent changes, certain political and religious philosophies, the various media, and especially advertisers still project different roles and aspirations for the two sexes: Males are doers and rescuers—dependable, rational, and stoic; females are helpers, observers, or victims—unreliable, intuitive, and emotional.

How well do such assumed differences in behavior hold up under close scrutiny? Stanford University psychologists Eleanor Maccoby and Carol Jacklin (1974), after surveying and analyzing research on sex differences, concluded that girls are *not* more social or suggestible, nor are they better at rote learning or worse at analytical thinking. Whether girls are more compliant and nurturing, or less active and competitive, is not clear. But there is considerable evidence that by the age of 12 or 13 girls excel in verbal ability, while boys excel in visual–spatial ability and mathematics, and are more aggressive. Are the observed differences in these four traits related to differences in brain structure and/or function? This idea is still a matter of hot debate (Goldfoot and Neff 1985; Bleier 1984; Teitelbaum 1976). But even if it were true, the expression of these traits varies so widely *within* both sexes and shows so much overlap *between* the sexes that one could never predict, on the basis of sex alone, how any given male or female will behave.

Anthropologists find that sexual norms of behavior do indeed vary tremendously from one culture to another. As Margaret Mead (1949) commented:

> In every known society, mankind has elaborated the biological division of labor into forms often very remotely related to the original biological differences. Sometimes one quality has been assigned to one sex, sometimes to the other. Now it is the boys who are thought of as infinitely vulnerable and in need of special cherishing care, now it is the girls . . . Some peoples think of women as too weak to work out of doors, others regard women as the appropriate bearers of heavy burdens . . . In some cultures women are regarded as sieves through whom the best-guarded secrets will sift, in others it is the men who are the gossips. Whether we deal with small matters or large, . . . we find this great variety of ways, often flatly contradictory, in which the roles of the two sexes have been patterned.

These very evident cultural attributes are superimposed, layer by layer, upon a basic physiological distinction between the sexes: males inseminate; females menstruate, gestate, and lactate. The developmental events of puberty that underlie these irreducible sexual functions are described in the following section.

Postnatal Development: Physiological Factors

Between the ages of 10 and 15, the production of certain hormones sets off a chain of events that culminate in sexual maturity, the ability to reproduce. Triggered by genetic and environmental factors that are poorly understood, puberty starts when substances produced by the hypothalamus cause the tiny pituitary gland just beneath it to begin secreting **gonadotropic hormones**. These in turn stimulate certain cells in the gonads to make **sex hormones**, which bring about striking changes (secondary sex characteristics) in many sensitive tissues throughout the body. Both sexes produce the same hormones, but in much different proportions.* Under the influence of nerve pathways set up in the hypothalamus during embryonic development, female hormone levels vary considerably during the menstrual cycle. Because this brain circuitry is modified in male fetuses, the cycling of gonadotropic hormone production in adult males is greatly reduced but not eliminated (Curtis 1983; Jones 1984).

*Indeed, male and female hormones are produced from the same precursors: cholesterol → progestins → androgens → estrogens.

The two major gonadotropic hormones are named for their function in females. **Follicle-stimulating hormone (FSH)** activates a few egg follicles in the ovary to undergo maturation each month. With the increased production of **luteinizing hormone (LH)** in midcycle, one follicle is stimulated to release its egg and then to become a *corpus luteum* (yellow body). Before ovulation, the maturing follicle secretes a sex hormone, **estrogen**, that leads to development of the uterine lining (endometrium). Other effects of estrogen at puberty include increased growth of the sex organs and breasts, widening of the pelvis, and deposition of subcutaneous fat. Following ovulation, the corpus luteum secretes the hormones estrogen and **progesterone**. The latter stimulates further thickening of and increased blood supply to the endometrium in preparation for implantation of a fertilized egg. Both sex hormones also act on the hypothalamus and pituitary gland to decrease production of FSH and LH. A third gonadotropic hormone, **prolactin**, is secreted mainly after pregnancy and stimulates milk production; actually, small amounts of prolactin are secreted at all times by both sexes.

In males and females at puberty, the adrenal glands atop the kidneys make the hormone **androstenedione**, about 10% of which is transformed into **testosterone**. In females, it is responsible for the growth of pubic and underarm hair and the development of oil and sweat glands in the skin. An excess of adrenal androgens in females (say, due to adrenal tumors) can cause the abnormal development of male secondary sex characteristics, such as facial hair and deepening of the voice. It can also enlarge the clitoris.

In pubertal males, FSH and LH stimulate the maturation of germ cells into sperm in the seminiferous tubules of the testes. Small amounts of estrogen are produced by the testes and the adrenal glands, but in the presence of normal amounts of male hormone, it has no obvious effects. However, normal teenage males often show slight breast development at puberty, which usually regresses. When such tissues persist, an excess of estrogen may be the cause. In males, progesterone is produced mainly as a precursor for testosterone in the testes and the adrenal steroids in the adrenal glands. Very high concentrations of **testosterone** are necessary within the testes for complete sperm maturation. At puberty, testosterone also brings about most of the masculine changes and acts on the pituitary and hypothalamus to inhibit the production of LH and FSH.

The steps leading to the development of normal males are summarized in Figure 10. Except for the specific male characteristics mentioned, the lower part of the chart applies to females as well.

SOME ERRORS IN SEXUAL DEVELOPMENT

McKusick's catalogue of inherited phenotypes (1988) lists several dozen aberrations in sexual development. For some of these, the modes of inheritance and the basic biochemical defects are well established, but for others the evidence remains scanty or ambiguous. Polani (1981, 1985a,b) suggests that of about 48 "sex genes," 20 are X-linked, 2 (*TDF* and *H-Y*) are Y-linked, and the remainder are autosomal. Methods of classifying defects in sexual development differ somewhat among researchers; one commonly accepted way is as follows:

1. Abnormalities of sex chromosomes
   a. Abnormalities of number: 1, 3, or more than 3 sex chromosomes
   b. Abnormalities of arrangement: some XX males and XY females

2. True hermaphrodites

3. Pseudohermaphrodites
   a. Male pseudohermaphrodites
   b. Female pseudohermphrodites

Abnormalities of chromosomal sex will be considered in Chapter 8. Here we confine our discussion to a few examples of individuals who are 46,XX or 46,XY, that is, those in categories 2 and 3.

Usually, ovaries develop only in XX embryos and testes develop only in XY embryos. But when neither or both occur, or when gonadal sex does not correspond to chromosomal sex, the result is a hermaphrodite, of which there are numerous types.*

True Hermaphrodites

So-called **true hermaphrodites**, possessing *both* ovarian and testicular tissue (van Niekerk and Retief 1981; Mittwoch 1986), are rare (Figure 11). Although their external genitals are often ambiguous, roughly two-thirds of them are raised as males, irrespective of their chromosomal sex. They often produce eggs, but rarely sperm, and they are almost always sterile. Simpson (1982) reports that "most true hermaphrodites with a uterus menstruate, and at least four 46,XX true hermaphrodites have become pregnant." According to Polani (1981a), two gave birth "after a normal pregnancy . . . But only one of the children survived the neonatal period." McKusick (1988) includes a few reports of apparent autosomal recessive

*In most plants and in some animals, hermaphrodites represent the normal state of affairs and are classified differently from what follows. Regarding human hermaphrodites, different researchers may use somewhat different systems of classification; Migeon (1980) presents one commonly accepted system of categories.

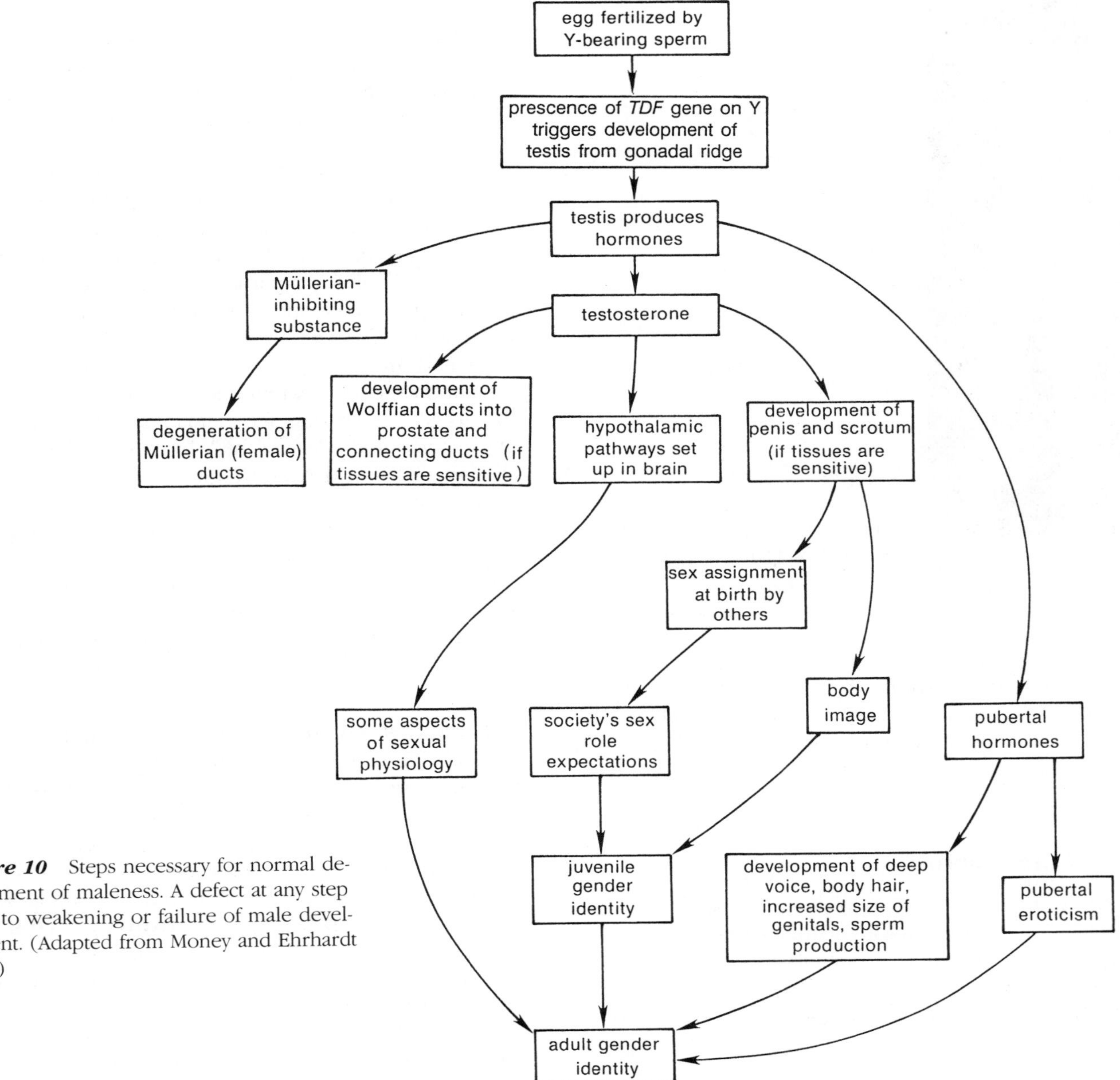

Figure 10 Steps necessary for normal development of maleness. A defect at any step leads to weakening or failure of male development. (Adapted from Money and Ehrhardt 1972.)

mutations for true hermaphroditism. He comments: "The autosomal inheritance of this disorder is one of the numerous pieces of information indicating that genes controlling sexual development and differentiation are not limited to the sex chromosomes."

Most true hermaphrodites are 46,XX, but 46,XY karyotypes and mosaics containing both XX and XY tissue are known, too (Polani 1970, 1981a). How testicular tissue manages to develop in an XX individual is not clear, but there are several possibilities: (1) During embryonic development, some XY cells were present that later disappeared or became very rare. In one case of a 46,XX/46,XY mosaic, extensive blood group analyses showed that the individual arose from double fertilization, that is, two eggs fertilized by two genetically different sperm (Gartler et al. 1962). (2) In other cases, part of the father's Y chromosome was translocated (see Chapter 8) to the X chromosome or to an autosome inherited by the hermaphrodite. (3) Among the extremely rare familial cases known, two seem to show autosomal dominant inheri-

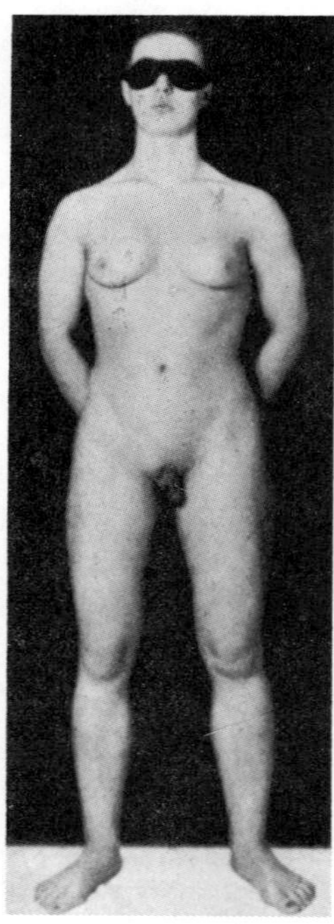

Figure 11 A patient with true hermaphroditism. (From Overzier 1963.)

tance and one looks like an autosomal recessive. The dominant *male-determining factor* "apparently causes the indifferent gonad of genetic females to differentiate partially or completely into a testis" (McKusick 1988). In some cases, however, the cause remains unknown (Ramsay et al. 1988).

Pseudohermaphrodites

In contrast to true hermaphrodites, whose gonadal sex is mixed, **pseudohermaphrodites** have either testes or ovaries—but the external genitals are opposite of the gonadal sex, or ambiguous, or abnormal in some other way. The two main pseudohermaphrodite classes are

Type	Gonads	External genitals
male	testes	female-like or abnormal
female	ovaries	male-like or abnormal

GONADAL DYSGENESIS. The failure of the gonadal ridges to differentiate into either type of gonad (**pure gonadal dysgenesis***) has been reported in XX and XY individuals. Both chromosomal types are reared as females, as they have normal female-appearing external genitals. Because their "streak gonads" are incapable of producing any sex hormones, they never undergo puberty. Development of Müllerian structures is present if the external genitals are entirely female. In subjects with an XY karyotype, the gonadal dysgenesis can be partial, and the degree of masculinization of the genitals will depend upon the amount of testicular tissue. There is evidence for inheritance of an X-linked gene in some XY "females" and an autosomal recessive gene in some XX individuals.

De la Chapelle (1972) summarized over 40 cases of apparent **sex reversal**—whereby 46,XX individuals develop testes (but no ovaries), a generally normal penis and scrotum, and secondary sex characteristics that range from typically male to partially female. They had a male psychological orientation (gender role) and some were even married, but all were infertile. In fruit flies, goats, dogs, and mice, the presence of an autosomal recessive mutant transforms XX females into males, but this is not the usual explanation for human sex reversal. Other events include the loss of the Y from an XXY embryo, or the translocation of the *TDF* gene from the Y onto an X or autosome (Chapter 8).

The cause for absence of one or both testes in otherwise normal XY males is likewise unclear. The testes must have existed in early development or these males would have had female external genitals. Because the condition (sometimes called the syndrome of vanishing testes) occasionally affects several members of a family, it is thought to have a genetic component.

ANDROGEN INSENSITIVITY SYNDROMES. Androgens produced by fetal testes normally permit and enhance the the development of the Wolffian ducts, but problems arise when the tissues upon which they act fail to respond because of receptor defects (Pinsky and Kaufman 1987; Brown and Migeon 1986; Wilson and Griffin 1985). One of the most astonishing errors in sexual development is the **complete androgen insensitivity syndrome** (formerly called the testicular feminizing syndrome), which causes 46,XY persons with the usual levels of androgens to nevertheless develop at puberty into phenotypic females with breasts (Figure 12). The disorder, with a frequency of about 1 in 65,000 genetic males, is inherited as an X-linked recessive, the androgen receptor gene being located close to the centromere on the long arm of the X (Migeon et al. 1981; Lubahn et al. 1988). Similar conditions have also been described in *Tfm* mice, rats, and cows; indeed, the mouse and human loci appear to be the same.

Although the testes are intact and functional inside the abdomen, no Wolffian (male) structures are formed in the fetus. Nor are there oviducts, uterus, and upper vagina—which means that the response to Müllerian-inhibiting hormone produced by the fetal testes is appropriate. Because the external genitals are typically female and the short, blind lower vagina looks normal

**Dysgenesis* means defective development. *Gonadal dysgenesis* refers to individuals (of any chromosomal makeup) who show defective development of the gonads. If the defect is complete (pure gonadal dysgenesis), there are streak gonads.

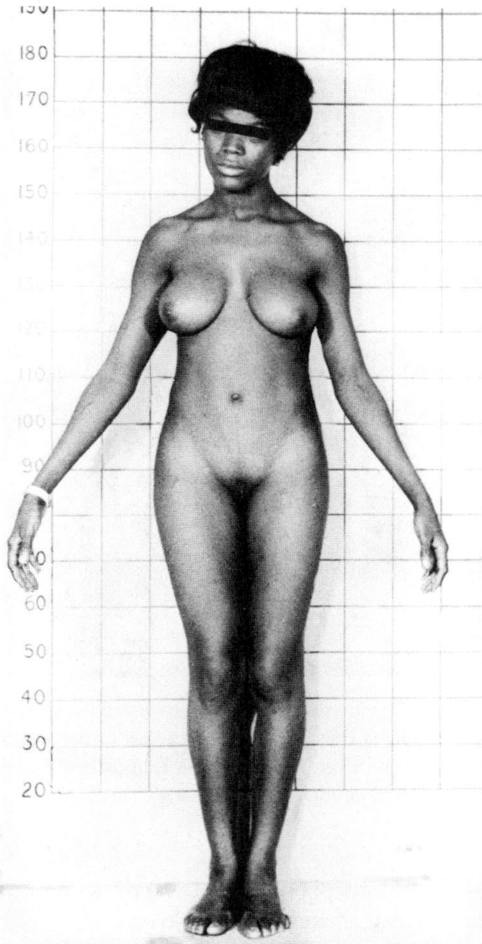

Figure 12 The androgen insensitivity syndrome in a 46,XY individual. Note the well-developed breasts, female-type body build, and absence of pubic hair. The patient is about 180 cm (5 ft 11 in) tall. (From Zourlas and Jones 1965.)

from the outside, the newborn is declared a girl and raised as such.

At the time of puberty, the estrogens arising from peripheral metabolism of testosterone stimulates breast growth and other female secondary sexual characteristics. Normal male levels of testosterone also circulate in the bloodstream of these women, but to no avail: the tissues that would normally be stimulated to maleness do not respond. Insensitivity to androgens also prevents the growth of sexual hair in the pubic and underarm areas, although hair on the head is often luxuriant and shows no male patterns of receding hairline. If affected individuals have not developed hernias during childhood (due to the attempted descent of the testes into nonexistent scrota), what finally brings them to the doctor's office at the time of puberty is the failure to menstruate. In all other ways, including sexual relations, they feel and act like women.

A similar but independent condition, called **incomplete androgen sensitivity**, is inherited as an X-linked recessive. The individual's tissues show a partial responsiveness to androgen and produce a more variable phenotype. These 46,XY individuals also have undescended testes and normal male levels of testosterone; they lack both Wolffian- and Müllerian-derived sex organs, are infertile, and at puberty develop breasts. But their external genitals, instead of being entirely feminine, are partly male: an enlarged clitoris and partially fused labioscrotal folds. At puberty they become more hairy, but their phallus remains small. Thus the preferred sex of rearing is female. (The so-called *Reifenstein syndrome* is an older term that described subjects with partial androgen insensitivity. It was used prior to our knowledge of androgen receptors and androgen insensitivity.)

CONGENITAL ADRENAL HYPERPLASIA. Some cases of male and female pseudohermaphroditism have been traced to defects in the biosynthesis of testosterone in the testes or of androstenedione in the adrenal glands, or both. In these patients the adrenal glands grow excessively large in an attempt to compensate for the deficiency, hence the name **congenital adrenal hyperplasia (CAH)** or adrenogenital syndrome. Recessive autosomal mutations affecting about half a dozen specific enzymes are known, each blocking a different step in the multibranched biochemical network. This metabolic pathway is summarized very simply as follows:

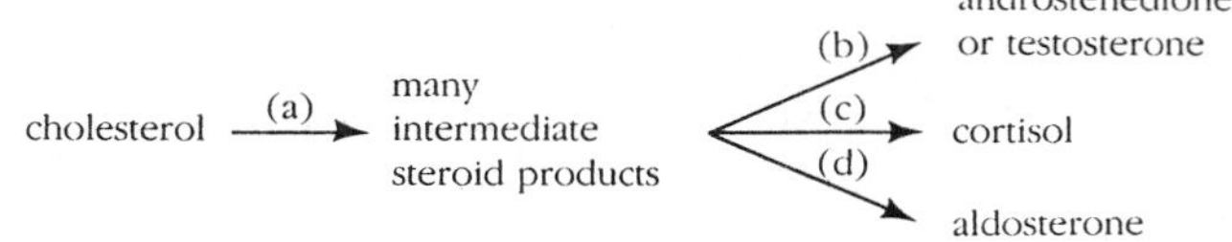

The hormones cortisol and aldosterone control sugar and salt balance in the body. Cortisol also acts as a "feedback inhibitor," regulating the amount of a substance (ACTH, or adrenocorticotropic hormone) that stimulates the activity of the adrenal cortex to produce precursor substances. Thus a deficiency of cortisol leads to an overproduction of hormone precursors and an accumulation of certain hormone products.

If step (a) is blocked in a 46,XY fetus, no testosterone is produced and a gonadal male with female genitals—a **male pseudohermaphrodite**—will result. These infants also suffer severe loss of salt and water because of the absence of cortisol and aldosterone, and they may die within a few weeks.

If the mutant block occurs somewhere in the intermediate steps of the pathway, there may be a biochemical detour around it that results in the production of *some* testosterone. Hence the genitals of these male pseudohermaphrodites may be *partly masculinized* (Figure 13). In these cases, high blood pressure rather than extreme salt and water loss is usually observed. Early diagnosis is important so that the deficient hormones can be administered; following a decision on whether to raise the infant as male or female, the genitals can be surgically corrected.

When the block occurs at (c) or (d) in a 46,XY fetus, some of the backed-up intermediates are shunted into the testosterone branch. Thus an *excess* of male hormone is produced. The male infant appears normal at birth, and unless there are problems with salt and water loss, the condition may not be immediately detected. Long before the normal age of puberty (in some cases by the age of 4), however, he begins to show secondary sexual characteristics: rapid growth and maturation of the skeleton, muscle development, voice changes, growth and frequent erections of the penis, and early acne (Figure 14).

The most commonly observed **female pseudohermaphrodites** result from autosomal recessive mutations that block steps (c) or (d) or both, or the steps immediately preceding these in a 46,XX fetus. Thus intermediate products pile up behind this block and get sidetracked into a branch leading to adrenal androgen production. (These are the same mutations that bring on early puberty in males.) The excess formation of male hormone in XX females during fetal life causes their external genitals to be masculinized, occasionally to the extent that they are erroneously raised as males. There are no testes, however, and the internal organs are female (ovaries, Fallopian tubes, and uterus). Salt and water loss can present serious problems, however, sometimes leading to early death if inadequately treated.

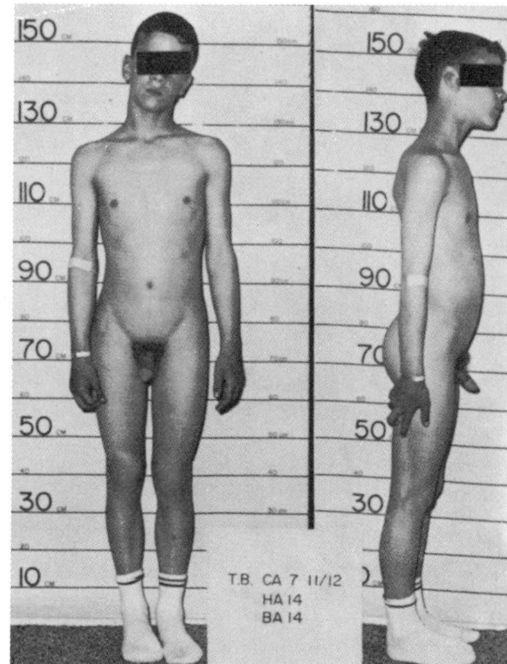

Figure 14 A 7-year-old XY male with the type of congenital adrenal hyperplasia that leads to an excess of testosterone and precocious puberty. Note that his height is about 155 cm (5 ft 1 in). (From Bongiovanni et al. 1967.)

The classic case of this type of female pseudohermaphroditism was first described in 1865 by Luigi De Crecchio, an Italian anatomist who was presented with the cadaver of an apparent male. Bongiovanni et al. (1967) summarize his findings:

(A)

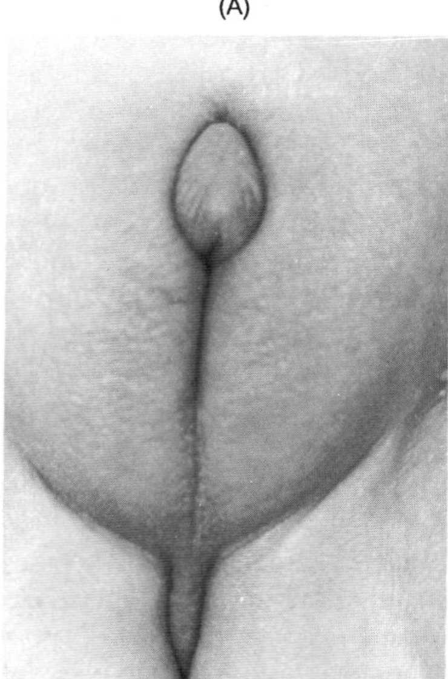

(B)

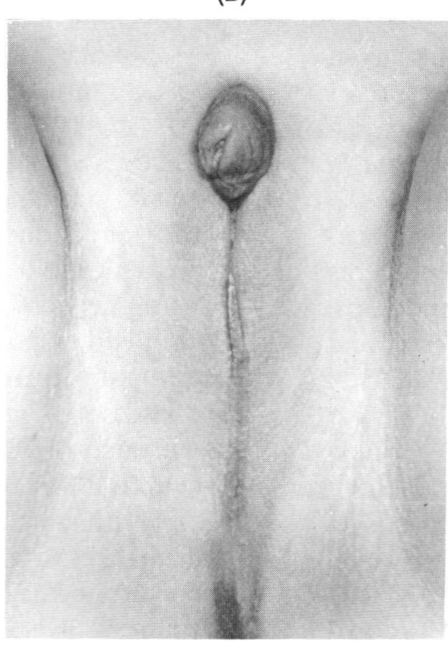

Figure 13 (A) The extremely small penis of an XY newborn with male pseudohermaphroditism. (B) The same individual after undergoing corrective surgery in preparation for rearing as a girl. (From Money and Ehrhardt 1972.)

> He noted the somewhat frail proportions of the individual. Externally the penis seemed normal enough, measuring 10 cm in length . . . On dissection, to the surprise of [De Crecchio] and his assistants, there was present a normal vagina, uterus, Fallopian tubes, and ovaries . . . and the adrenal glands were noted to be very large . . . However the anatomist did not stop here. He conducted an interrogation of those who had known 'this man' in life to discover that he comported himself in all ways, including sexual intercourse, as a male, and it was learned that he contracted the 'French disease' . . . on two occasions.

This autosomal recessively inherited deficiency of **21-hydroxylase** (the enzyme, also known as P450c21, that converts progesterone to deoxycorticosterone) is the most common type of congenital adrenal hyperplasia. The severe classic form occurs equally in males and females with a frequency of 1 in 5,000–10,000 Caucasian births, and accounts for 90–95% of all cases of CAH. (In Alaskan Eskimos, however, its frequency is about 1 in 700.) The *CA21H* gene is located on chromosome 6, very close to the HLA complex that produces cell surface substances important in transplantations (Chapter 18). Thus, for families known to be at risk, it is now possible to detect heterozygote carriers as well as affected fetuses by combining HLA typing with tests for elevated levels of hydroxyprogesterone, the substrate for 21-hydroxylase (Levine 1986).

Molecular analyses of 21-hydroxylase mutants further reveal that the *CA21H* gene exists in two side-by-side copies, although it appears that only one of them is functional. Given its duplicate nature as well as its close proximity to the HLA locus, the *CA21H* locus might be part of a recombinational "hot spot." Mutant alleles of the *CA21H* gene have been found with either deletions or point mutations; affected subjects may be homozygous for one mutant allele or heterozygous for different mutant alleles (Anonymous 1987; Miller 1988).

The severe "salt-wasting" form involves defects in aldosterone as well as androgen synthesis and accounts for over half of all cases of classic 21-hydroxylase deficiency. In addition, however, milder forms exist. Subjects with classic 21-hydroxylase deficiency but normal aldosterone synthesis have what is called simple virilizing disease (Figure 14). Patients with the nonclassic, "late-onset" form show no signs of enzyme deficiency until late childhood or puberty, and those with nonclassic "cryptic" deficiency (who are usually detected only through family studies) have no disease symptoms at all. The milder forms are thought to result from the combination of one "classical" and one "mild" allele. The nonclassic form is very common, affecting up to 1% of all Caucasians worldwide and roughly 3% of Jews of European origin; indeed, it may be one cause of infertility (White et al. 1987a,b; Mulaikal et al. 1987).

Congenital adrenal hyperplasia also results from defects in any of several other enzymes necessary for cortisol synthesis. These are: *11-hydroxylase* (also called P450c11), *17-hydroxylase* (or P450c17), *3-β-hydroxysteroid dehydrogenase*, and *cholesterol desmolase* (or P450scc). All of these deficiencies are quite rare—except for 11-hydroxylase deficiency, which occurs with a frequency of about 1 in 100,000 among Caucasians in general and accounts for about 5–8% of all cases of CAH. It is particularly common among Moroccan Jews in Israel. This deficiency, too, is quite heterogeneous, but females with the classic form are born with ambiguous genitals and tend to develop high blood pressure if untreated. The gene for the enzyme is on the short arm of chromosome 8.

In 1950 it was found that administration of the adrenal hormone cortisone inhibits the action of testosterone and is effective in the treatment of both female and male subjects. If adrenal hyperplasia is diagnosed early enough, long-term cortisone therapy will prevent further masculinization. When the genitals are surgically corrected and female hormones are also administered at the time of puberty to encourage the development of secondary sex characteristics, these females can grow up to be normal, fertile women.

Female pseudohermaphrodites have also resulted from the treatment of their mothers during pregnancy with certain hormones, such as progesterone-like steroids, once thought to prevent miscarriage. In other cases the pregnant mother may herself produce excess androgens because of a tumor on the adrenal gland or the ovary. Depending on how much hormone the fetus was exposed to and at what stage of development, masculinization of the external genitals ranges from slight to extreme. After birth, of course, the hormone levels in these infants return to normal, and there is no need for cortisone therapy. Following corrective surgery, these infants develop as normal females.

Gender Identity

The process by which an individual comes to identify himself or herself as a male or a female is immensely complex. This private **gender identity** is expressed publicly as a **sex role**, the sum of those actions that indicate a degree of maleness or femaleness to others. Analysis of this developmental process is extremely difficult. "It is easy to get trapped in circular argument as to whether boys and girls develop different patterns of preferred behavior because they are treated differently, or whether they are treated differently because they demonstrate different behavioral patterns right from the beginning" (Money and Ehrhardt 1972). Few would argue against the view that *both* physiological and environmental factors must play a role—but are the observed differences in behavior due *more* to genetic differences between the sexes, or *more* to differences in upbringing? The issue is still being hotly debated, each side trotting out case histories and comparative studies to support its point of view.

EVIDENCE OF HORMONE EFFECTS. In birds and in many mammals, including primates, the sexually stereotyped behaviors expressed in courting, mating, and caring for young are definitely controlled by hormones and neural pathways. It is also true, however, that these behavioral differences may be more of degree than kind. For example, normal females may occasionally exhibit mounting and thrusting behavior; normal males may sometimes crouch to be mounted or display other components of female mating behavior.

With human sexual behavior, hormones do control the most basic differences between the sexes: the ability of males to produce sperm and to inseminate, and the ability of females to produce eggs, to become pregnant, and to provide milk for offspring. Beyond this, much less is known about the components of mating behavior and their relationships to hormone levels or brain circuitry. Aside from the sexual act itself, there is not even general agreement on what constitutes sexual behavior.

One approach is to observe the behavior patterns of people with sexual and hormonal abnormalities. Possible (but weak) evidence for hormone-dependent behavioral differences is suggested by a study of about 25 females who were masculinized during fetal life by adrenal hyperplasia or by progesterone treatment of their mothers. All were surgically corrected, given cortisone and hormone treatment if necessary, and raised as normal girls. Compared with a control group of 25 normal females who were matched for age, IQ, race, and socioeconomic background, the fetally masculinized group included more "tomboys" and more late-marrying individuals (not very unusual traits), but no sign of homosexual tendencies (Money and Erhardt 1972). Whether or not this provides strong support for hormonal effects is still a matter of contention. In a later study, however, Money et al. (1984) report a much higher rate of bisexuality and homosexuality among 30 young women with treated congenital adrenal hyperplasia than among (1) a control group of women with androgen insensitivity syndrome or a phenotypically similar congenital sexual disorder and (2) women interviewed for Kinsey's study of sexual behavior in females. The most likely explanation, they hypothesize, is "a prenatal and/or neonatal masculinizing effect on sexual dimorphism of the brain in interaction with other developmental variables."

A recent study by Rose Mulaikal and colleagues (1987) examined fertility rates in 80 female patients with CAH due to to 21-hydroxylase deficiency. Of these, 46 were heterosexually active (with a total of 25 pregnancies and 20 normal children), 4 were homosexual or bisexual, and 30 had no sexual experience. Among these subjects, however, heterosexual activity seemed to depend much more upon the adequacy of surgical reconstruction of the vagina (i.e., whether or not it permitted comfortable intercourse) than on the degree of prenatal exposure to androgen. Commenting on this study, Federman (1987) suggests that adult psychosexual functioning of treated CAH patients could also be strongly influenced by "a direct effect of androgen on the developing fetal brain."

EVIDENCE FOR ENVIRONMENTAL EFFECTS. Especially instructive are studies of hermaphrodites with ambiguous genitals. Some were raised as males and others as females, irrespective of their chromosomal sex; several such cases have been discussed in the preceding sections.

John Money and Anke Ehrhardt (1972) of Johns Hopkins University described the development of six pseudohermaphrodites who at birth had ambiguous genitals resulting from CAH syndromes. All had the same chromosomal (46,XX), gonadal, sex duct, and external genital sex. Yet three became males and three females—following surgery, therapy, and rearing practices instituted at various stages in their lives (ranging from early childhood to adolescence) and after various degrees of emotional suffering.

Even more suggestive of environmental effects are the finding of van Niekerk (1974) on 24 true hermaphrodites whom he treated in Africa. Although 21 of them had a 46,XX chromosome constitution, 20 of the 24 had been raised as males—and their gender identity was not changed by the development of female sex characteristics at puberty. The most striking case was a 16-year-old with a tiny penis, who menstruated irregularly and had well-developed breasts. Yet he "was quite emphatic about his desire to be a male." The author summarizes, "In a vast majority of true hermaphrodites, gender identity is the same as sex of rearing; this was true for all the patients I have examined. None expressed the slightest doubt as to what sex they belonged and what sex they wanted to be."

Perhaps hermaphrodites do not develop gender identities in the same way as normal males do, and perhaps the choice of maleness among the African patients reflected some cultural preference—say, for power or freedom from restrictions rather than for maleness per se. In any event, the wide variety of outcomes from very similar beginnings suggests that gender identity does not develop automatically and unswervingly on the basis of chromosomal sex or even genital sex.

Dramatic evidence for a change in gender identity not

involving hermaphrodites occurred with a pair of identical twin males who were circumcised at the age of seven months. As the result of medical error, one boy lost his entire penis, and 10 months later it was suggested that he be raised as a girl. Following surgical feminization, he was given a girl's name, a new hairdo, new clothes, new toys, and so on—and treated quite differently from the identical twin brother. Following is the mother's account of how this child's personality changed in response to altered rearing practices (Money and Ehrhardt 1972):

> She doesn't like to be dirty, and yet my son is quite different... She seems to be daintier. Maybe it's because I encourage it... One thing that really amazes me is that she is so feminine. I've never seen a little girl so neat and tidy... She is very proud of herself, when she puts on a new dress, or I set her hair... I've tried to teach her not to be rough... to be polite and quiet... and ladylike.

OTHER VARIATIONS OF GENDER IDENTITY. Other commonly encountered problems of gender identity are poorly understood (Money 1981). **Transvestites**, nearly always males who periodically dress in female clothing, alternate between male and female identities and seem unable to permanently settle on one. They are not necessarily homosexual, however, and some prefer female sexual partners exclusively. **Transsexuals**, on the other hand, have a single gender identity that is at variance with their chromosomal and genital sex: a female identity in a male body, or vice versa. Following sex-change operations and the adoption of new sex roles, such individuals may be more satisfied with their new lives than they were with the old ones. **Homosexuals** maintain a gender identity that matches their body sex, but they are erotically attracted to individuals of the same genital sex. Anthropologists report that different cultures set up different standards of sexual behavior, some of which may accept homosexuality as normal during part or all of the life cycle. In other cultures, however, such behavior is considered taboo at all times. As for the relative roles of genetic and hormonal versus environmental differences in the development of the variations in gender identities, the results of many studies are so contradictory and hard to interpret that we do not care to speculate.

To summarize: Certainly both genetic and environmental imperatives must be expressed in human sex differentiation and behavior. After all, people are neither automatons nor blank slates. Whether or not these patterns of human development are comparable to those known to operate in other organisms—even in the most closely related primates—is hard to say.

SUMMARY

1. The human zygote undergoes repeated divisions to form an embryo and surrounding membranes. The most critical and sensitive developmental period spans weeks 4 to 7, during which all the major organ systems are forming.

2. In fruit flies and other segmented organisms, several hierarchies of gene action control early development. These include the maternal effect genes of the mother, as well as the segmentation and pattern formation genes present in the zygote.

3. Pattern formation (homeotic) genes in *Drosophila* and many other organisms (including humans) contain a highly conserved 180-base sequence called the homeobox. This codes for a 60-amino acid homeodomain that is extremely similar even between distantly related organisms. Part of the homeodomain is a 9-amino acid sequence thought to regulate groups of genes by competitive binding to their DNA.

4. Mammalian zygotes do not develop properly unless they contain both maternal and paternal genomes. Investigators have found that some genes in both sexes become differentially imprinted (probably by methylation) during gamete formation and are expressed differently during development.

5. Prenatal development of the sex organs goes through an undifferentiated period characterized internally by the presence of gonadal ridges and primitive sex ducts, and externally by a genital tubercle plus urethral and labioscrotal folds.

6. Presence of a Y chromosome (more specifically, the *TDF* gene) contributes to the development of a testis, but other genes are also necessary. The testes in turn produce testosterone and Müllerian-inhibiting hormone. Testosterone stimulates development of the Wolffian ducts into the internal male organs and external penis and scrotum. The Müllerian-inhibiting hormone prevents the formation of female internal organs.

7. Presence of two X chromosomes and the absence of a Y chromosome, testosterone, and Müllerian-inhibiting hormone leads to feminine development—but yet-uncharacterized genes probably play a role in ovarian formation. The Wolffian ducts regress spontaneously, and the Müllerian ducts

form female internal organs. Externally the clitoris, labia, and lower vagina develop.

8. Sex assignment at birth, based on appearance of the external genitals, sets the stage for differences in treatment and expectations that can strongly influence the behavior patterns of a child. Behavior varies greatly within each sex and shows considerable overlap between the sexes. Norms of behavior for the two sexes vary markedly from culture to culture.

9. The physical changes occurring at puberty are orchestrated by two major classes of hormones. Gonadotropins (follicle-stimulating hormone and luteinizing hormone) are secreted by the pituitary and act on the gonads. Sex hormones (estrogen and testosterone), are secreted by the gonads and to some extent by the adrenal glands; the sex hormones act on many parts of the body. Both sexes secrete all these hormones, but in different proportions.

10. Disorders of sex development may initially affect the differentiation of gonads, sex ducts, or external genitals. The initial defect will usually lead to further aberrations of the sexual phenotype.

11. True hermaphrodites, usually 46,XX, have both ovarian and testicular tissue. They may be raised as males or as females.

12. The androgen insensitivity syndrome occurs when 46,XY individuals, with testes that produce normal amounts of testosterone, nevertheless develop into normal-looking and normal-acting females. Because their tissues are unable to respond to the male hormone, these individuals do not develop a male phenotype. The tissues are unresponsive because they lack androgen receptors.

13. Pseudohermaphrodites, either 46,XX with ovaries or 46,XY with testes, have external genitals that are ambiguous or characteristic of the opposite sex. The most common genetic disorder (one or another form of congenital adrenal hyperplasia) is caused by an enzymatic block at some step during the syntheses of testosterone, aldosterone (a sugar-regulating hormone), and cortisol (a salt-regulating hormone). Affected individuals may be raised as males or females, irrespective of gonadal sex—but with varying degrees of success.

14. How a person privately perceives himself as a male or herself as a female is known as gender identity. The public expression of this, through word and deed, constitutes a sex role. In humans, both genetic and environmental factors operate in the development of gender identity, but the details of these processes are poorly understood.

FURTHER READING

Prenatal human development is clearly described by Moore (1988a,b) and Nilsson (1986). The genetics of development is well covered by Gilbert (1988), Watson et al. (1987), Wilkins (1986), Gehring (1985), Davidson (1986), Solter (1988), and Garcia-Bellido et al. (1979).

McLaren (1985), Ghesquiere et al. (1985), Serio et al. (1984), Austin and Edwards (1981), Naftolin and Butz (1981), Vallet and Porter (1979), and McCarrey and Abbott (1979) present fine overviews of sexual differentiation in animals, including humans. Mittwoch (1973) provides a good general background on the genetics of sex differentiation—but see E. Simpson et al. (1987), Sandberg (1985), and Goodfellow et al. (1985, 1986, 1987) for updates on the human Y chromosome.

Among the many books and articles on inherited disorders of human sexual development, we recommend (in addition to those listed above) the following: Grumbach and Conte (1985), J. Simpson (1983, 1982), and Fichman et al. (1980) on various abnormalities, New (1985) and New et al. (1983) on congenital adrenal hyperplasia, Wilson et al. (1983) on the androgen resistance syndromes, and van Niekerk (1974) on true hermaphroditism. Waber (1985), Harris (1985), Bleier (1984), Ehrhardt and Meyer-Bahlburg (1981), *Sex, Hormones and Behavior* (1979), Maccoby and Jacklin (1980, 1974), Money (1981), Money and Tucker (1975), and Money and Ehrhardt (1972) discuss research and theories of gender-related behavior in humans.

QUESTIONS

1. Female athletes at the Olympic games have sometimes been required to undergo simple tests (indicating the number of X chromosomes) to guarantee that no female teams have any male members. Can this test unequivocally accomplish its purpose? Explain.

2. In a few cases described in the literature, males possess oviducts and a uterus as well as the usual male internal organs; hernias occur as the testes descend into the scrotum. The condition is probably inherited as an autosomal recessive. What type of error in fetal development could account for this phenotype?

3. Here is a pedigree (Bowen et al. 1965) for one type of male pseudohermaphroditism with partial androgen insensitivity. Subjects have testes that show early stages of spermatogenesis but produce no functional sperm. The external genitals are highly variable; in many patients the penis is normal size, but its urethral opening is on the underside. Despite the development of breasts and a female-type distribution of body hair and fat, many of these individuals develop a strong male gender identity; some even marry. From this pedigree, what is the most likely mode of inheritance?

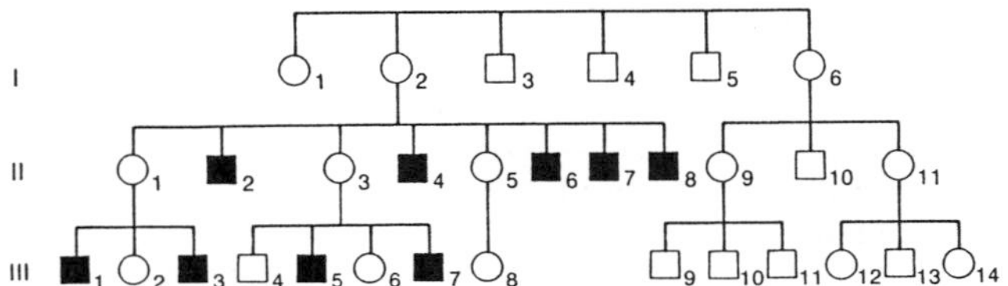

4. Complete the chart below.

Condition	Chromosomal sex	Gonadal sex	External genital sex	Assigned sex
Pure gonadal dysgenesis				
Androgen insensitivity				
CAH	XX			
CAH	XY			
True hermaphrodite				

5. Why are males generally more vulnerable to errors in sexual development than are females?

6. Imperato-McGinley and Gautier (1986) describe a large, inbred Dominican kindred with *male pseudohermaphroditism* due to deficiency of *5-α-reductase*, the enzyme which normally catalyzes the conversion of testosterone to dihydrotestosterone. (Recall that the latter hormone is necessary for the formation of external genitals in male fetuses.) The autosomal recessive mutant gene has no obvious effect in 46,XX females, who show normal sexual phenotype and fertility; but the defective steroid hormone can be detected in the urine of heterozygotes and homozygotes of both sexes.

Homozygous affected 46,XY males have normal internal male organs, but their external genitals are ambiguous at birth, and many subjects are raised as females. At puberty, male patterns of facial and body hair growth do *not* occur, and the prostate gland fails to enlarge. Yet the scrotum becomes darkly pigmented, the phallus grows, the voice deepens, and height increases to normal male levels.

Among 23 families, there are 38 known male pseudohermaphrodites. Of these, 18 "were unambiguously raised as girls. Despite this, 17 [of the 18] changed to a male gender identity (that is, they began to feel as males) and 16 [of the 17] adopted a male gender role during or following puberty."

(a) Judging from this summary, which physical changes occurring at puberty appear to be mediated by testosterone, and which ones by dihydrotestosterone? (b) Some researchers have suggested that gender identity is fixed by the age of 4 years. What is suggested by the reported switch in gender identity at puberty among some of these affected subjects?

ANSWERS

1. What criteria are the "correct" ones to use in defining sex? Although nearly all males are indeed XY and females XX, exceptions exist. Androgen-insensitive females are XY, for example, as are some male pseudohermaphrodites brought up as females (with appropriate surgery and hormone treatment) from the time of birth. Thus, it can no longer be assumed that chromosomal sex leads inexorably to a given pattern of adult sexual development and behavior. For discussions of sex testing for Olympic athletes, see de la Chapelle (1986) and Turnbull (1988).

2. Absence or inactivity of the Müllerian-inhibiting substance.

3. The simplest possibility is X-linked recessive inheritance, but it could also be an autosomal dominant gene that is expressed only in males.

4.

4. Condition	Chromosomal sex	Gonadal sex	External genital sex	Assigned sex
Pure gonadal dysgenesis	XX or YY	none	female	female
Androgen insensitivity	XY	male	female	female
CAH	XX	female	ambiguous or male-like	should be female
CAH	XY	male	ambiguous or male-like	should be male
True hermaphrodite	usually XX, some XY or XX/XY	both	ambiguous	male or female

5. All human embryos tend to develop as females unless deflected from this path by the action of testosterone at several critical steps. Prenatal estrogen is not required for normal female development, but prenatal testosterone is required for normal male development. "If any one of your sexual systems fail, you can coast down the female road, but not down the male" (Money and Ehrhardt, 1972).

6. (a) Dihydrotestosterone appears to control male patterns of facial and body hair growth as well as growth of the prostate gland. Testosterone appears to control pigmentation of the scrotum, growth of the phallus, voice change, and increased height. (b) Gender identity might not be completely fixed during the early years; perhaps it remains somewhat flexible through puberty. Societal factors should also be seriously considered, however (Money 1981). These patients, who probably all had the same mutation, lived in a remote area where the locals knew what happened to others with the 5-α-reductase defect. In this Latin society male gender is strongly preferred, even if accompanied by a short penis. In the United States, most subjects with 5-α-reductase deficiency are raised as females, usually quite successfully.

PART TWO **CYTOGENETICS**

7. Nondisjunction
8. Sex Chromosomes and Their Abnormalities
9. Other Chromosomal Abnormalities

Chapter 7 Nondisjunction

Mistakes in Cell Division
Nondisjunction and the Chromosome Theory
Meiotic Nondisjunction
Mitotic Nondisjunction

Nondisjunction of Human Autosomes: Sporadic Down Syndrome
General Features
Possible Relation to Alzheimer Disease
Frequency of Down Syndrome
Chromosomal–Genetic Basis for Down Syndrome
Search for Factors Altering Nondisjunction Rates
Down Syndrome Mosaics

Other Autosomal Aneuploids
Trisomy 18 and Trisomy 13
Very Rare Autosomal Aneuploids

> *The medical practitioner . . . may be pressed as to the question, whether the supposed defect dates from any cause subsequent to birth or not. Has the nurse dosed the child with opium? Has the little one met with any accident? Has the instrumental interference which maternal safety demanded, been the cause of what seems to the anxious parents, a vacant future? Can it be that when away from the family attendant the calomel powders were judiciously prescribed? Can, in fact, the strange anomalies which the child presents, be attributed to the numerous causes which maternal solicitude conjures to the imagination . . . rather than [to] hereditary taint or parental influence?*
>
> *John Langdon Down (1866)*

Dr. Down could not possibly have guessed that these "strange anomalies" resulted from the presence of an extra chromosome, because his paper preceded the discovery of chromosomes by about 15 years. Indeed, 90 years were to pass before geneticists determined the correct human chromosome number, and not until 1959 did they realize that Down syndrome patients carry the smallest chromosome in triplicate (Lejeune et al. 1959). By then, decades of work—mainly with fruit flies—had suggested mechanisms by which departures from the normal diploid number might come about.

Part Two of this text deals with such *changes in quantity* of otherwise normal genetic material, changes involving hundreds or even thousands of genes. This type of variation contrasts with the genetic variants discussed previously, which arose from *changes in quality* (mutations) of individual genes in organisms that had a normal chromosome number. Cells with normal chromosome sets have **euploid** karyotypes (Greek *eu*, "good"; *ploid*, "set"). For humans, haploid gametes and diploid zygotes are the euploid conditions. In this chapter we discuss **aneuploid** organisms, those with **unbalanced sets** of chromosomes due to an excess or deficiency of individual chromosomes. It is the *imbalance* among genes, rather than the nature of individual genes, that causes the observed phenotypic consequences or even death of affected organisms.

Aneuploidy can arise by a variety of mechanisms. The first such phenomenon discovered is called **nondisjunction**, the failure of two homologous chromosomes or of sister chromatids to separate (disjoin) during cell division. If it happens during *meiosis*, a gamete may end up with one too many or one too few chromosomes. Union of this aneuploid gamete with a euploid one leads to aneuploidy in the zygote and its descendant cells. During a *mitotic* division, nondisjunction occurs when the two sister chromatids of a chromosome fail to separate, and this can lead to a mixture of euploid and aneuploid cells within one individual. Experiments with a vast range of species have shown that nondisjunction is a major cause of aneuploidy, and it is now known to be the source of much grief to humans.

MISTAKES IN CELL DIVISION

We begin by describing patterns of inheritance in fruit flies that first revealed the phenomenon of nondisjunction and also provided direct physical proof that genes reside on chromosomes.

Nondisjunction and the Chromosome Theory

The key to the principle that genes are on chromosomes was a mutant male fruit fly first described in 1910 by Thomas Hunt Morgan. Starting with this fly, which had *white eyes* rather than the usual red eyes, Morgan made many crosses and found that the white-eye gene (w) was recessive to its normal allele for red eyes (+). He also showed that the locus was on the X chromosome, the first assignment of a specific gene to a specific chromosome in any organism. Thus the genotypes and phenotypes of these flies can be summarized as follows:

	Genotype	
Phenotype	**Females**	**Males**
Red eyes	+/+ or +/w	+/(Y)
White eyes	w/w	w/(Y)

Readers should verify that *matings of white-eyed females with red-eyed males yield all red-eyed daughters and white-eyed sons.*

Furthermore, Morgan discovered that in fruit flies (as opposed to humans) the Y chromosome is *not* crucial in sex determination. Although female flies are usually XX and males XY, it is the number of X's (rather than the presence or absence of a Y) that sets in motion the development of sexual characteristics. Flies with two X's are females, and flies with one X are males.*

Then around 1913 Calvin B. Bridges, an undergraduate colleague of Morgan, found a few **exceptional** progeny from matings of white-eyed females × red-eyed males. Although nearly all the sons looked like their mother and nearly all the daughters like their father (i.e., the regular "crisscross" inheritance of recessive X-linked traits), there were also *a few white-eyed daughters* among the F_1. (Later he found some *red-eyed sons*, too.) Bridges proposed that the exceptional white-eyed females developed when an *egg with two X chromosomes was fertilized by a Y-bearing sperm*, and microscopic analyses verified that these females were indeed XXY. Exceptional red-eyed males, on the other hand, had one X chromosome and no Y, so they must have come from the fertilization of a no-X egg by an X-bearing sperm (Figure 1).

How were the XX and no-X eggs formed? Bridges suggested that they arose during a meiotic cell division when the *two X chromosomes failed to disjoin* (i.e., separate) from each other, a phenomenon he called **nondisjunction**. He also noted that OY zygotes (no-X eggs fertilized by Y-bearing sperm) never survive, and XXX zygotes (from XX eggs fertilized by X-bearing sperm) rarely survive—so only *half* the products of nondisjunction were recovered. (In humans the OY karyotype is likewise lethal, but XXX females are fully viable.)

Bridges found that the frequency of exceptional offspring from diploid females was about 1 in 2,000. Non-

*Later research showed that it was actually the *ratio of X chromosomes to the "ploidy" of autosomes* (i.e., 2X:2A versus 1X:2A in diploids) that determines femaleness or maleness. And males without a Y turn out to be sterile, because the Y chromosome carries some male fertility factors.

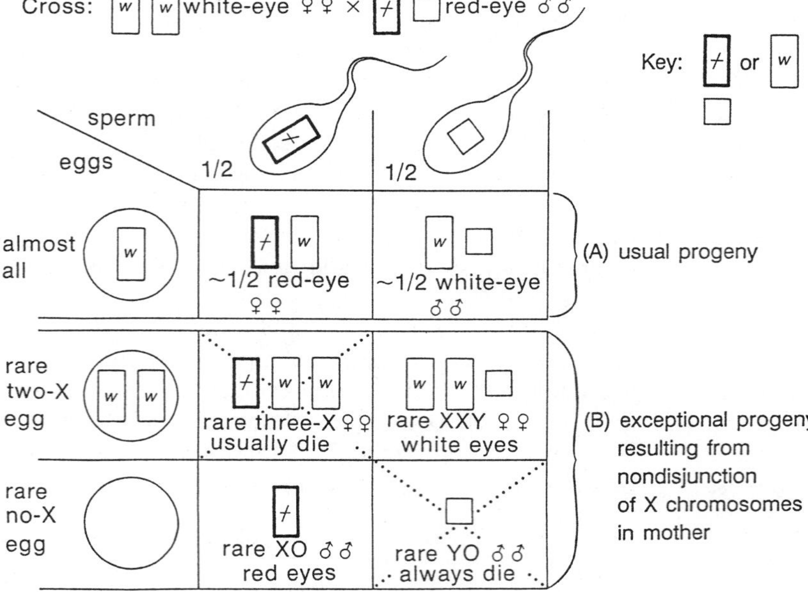

Figure 1 Results and interpretation of Bridges's experiment. (A) White-eyed females mated to red-eyed males give about half red-eyed females and about half white-eyed males. (B) In addition to the usual progeny, about 1 in 2,000 were exceptional white-eyed females (XXY) or red-eyed males (XO). The "O" in XO or OY denotes the absence of any other sex chromosome. All offspring have the correct number of autosomes. *Note:* In fruit flies, in contrast to people, XXY is female and XO is male.

disjunction also occurred in the aneuploid XXY daughters, because they produced some gametes containing two or zero rather than one sex chromosome (see Problem 7). But here it happened at a much higher frequency, yielding about 4% exceptional offspring (i.e., males receiving an X from the father, and females getting both Xs from their mother). To distinguish between the two kinds of events, he called nondisjunction in a normal diploid parent **primary nondisjunction**, and that in an aneuploid parent **secondary nondisjunction**. Both types of nondisjunction occur in humans, too.

The discovery of nondisjunction was a great genetic breakthrough, an exception that proved the rule of Mendelian inheritance and provided direct physical evidence that genes reside on chromosomes. As Bridges (1914) stated, "... the complete parallelism between the unique behavior of the chromosomes and the behavior of sex-linked genes and sex in this case means that the sex-linked genes are located in and borne by the X-chromosomes." Subsequent research has shown that nondisjunction can take place in any dividing cell and can involve any pair of chromosomes. Consequences vary widely, depending on where and when the misdivision occurs.

Meiotic Nondisjunction

Consider the most common situation, in which only one pair of chromosomes misbehaves. For simplicity's sake, we shall focus on spermatogenesis, in which all four products of meiosis are functional. Standard terminology is as follows: For a gamete, **nullisomic** (n − 1) and **disomic** (n + 1) mean, respectively, the presence of *no* and of *two* homologous chromosomes rather than one. For a zygote, **monosomic** (2n − 1) and **trisomic** (2n + 1) refer, respectively, to the presence of *one* and of *three* homologous chromosomes, rather than the normal pair. (Not included here is the possibility of adding or subtracting *whole sets* of chromosomes; this happens by a totally different process, and will be discussed in Chapter 9.)

Nondisjunction may happen during either the first or second meiotic division. As shown in Figure 2, the timing makes a difference. If nondisjunction occurs during *meiosis I*, *all* the sperm derived from that primary spermatocyte will be abnormal. Specifically, half of them will contain *neither* member of the given chromosome pair, and half will carry *both*. In humans, for example, these sperm will end up with a total of 22 or 24 chromosomes rather than the normal 23. It follows that after the fertilization of normal eggs, the resultant zygotes will have either 45 or 47, rather than 46, chromosomes.

If nondisjunction occurs in a secondary spermatocyte undergoing *meiosis II*, then only *two* of the four sperm will be abnormal. As before, these two will bear either

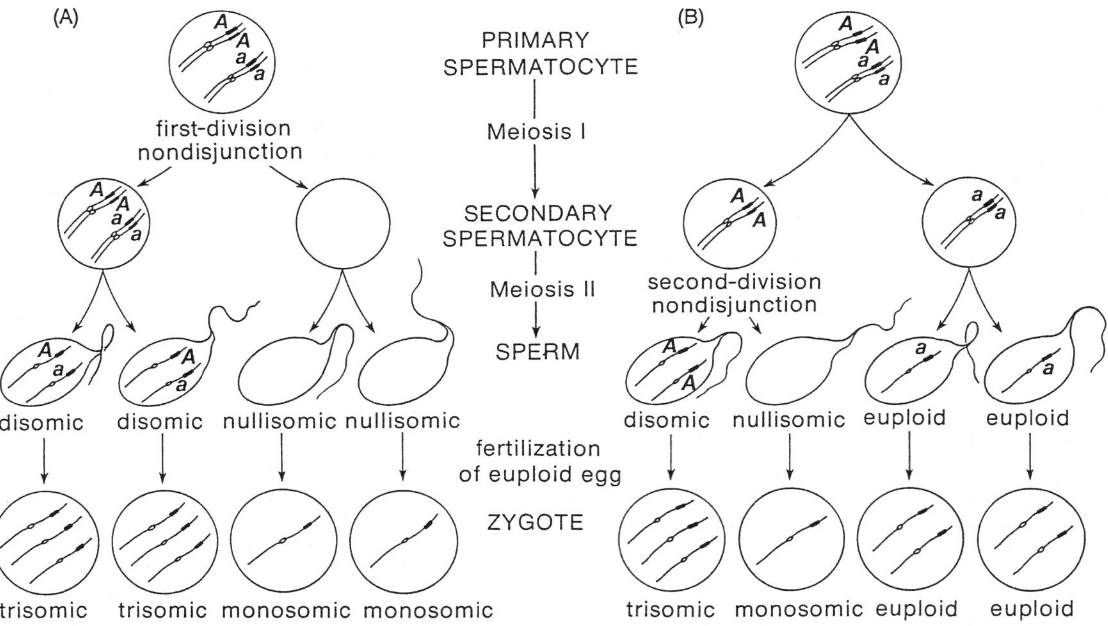

Figure 2 Results of meiotic nondisjunction involving a single pair of chromosomes with a heterozygous gene. Assume that all other chromosome pairs segregate normally. (A) When nondisjunction occurs in the first division, the eventual gametes are two disomic, two nullisomic, and no euploid. (B) When nondisjunction occurs in the second division (in just one of the two secondary spermatocytes), the gametes are one disomic, one nullisomic, and two euploid.

one too few or one too many chromosomes. Note also that the disomic sperm from the two types of nondisjunctions are different. If a male is heterozygous *A/a* (as indicated), then the abnormal disomic sperm from first division nondisjunction will have one chromosome with *A* and another with *a*. On the other hand, the disomic sperm from second division nondisjunction will have both chromosomes carrying either *A* or *a*.

Normal development of the zygote in all diploid species depends upon the presence of exactly two of each chromosome type, any deviation from this pattern causing abnormalities or death. Nondisjunction of sex chromosomes is better tolerated than that of autosomes, and an excess of chromosomes is tolerated better than a deficiency. Thus, among autosomal aneuploidies, trisomies are more viable than monosomies. But only those trisomies involving the smallest or most heterochromatic (i.e., least gene-rich) chromosomes are able to survive at all. Among these, the greater the imbalance among autosomal genes, the more abnormal the phenotype. In humans, for example, the only truly viable autosomal trisomy is Down syndrome, which involves the smallest chromosome.

In fruit flies it is quite clear that the phenotypic abnormalities associated with aneuploidy are due to the additive small effects of many genes rather than to the distribution throughout the genome of a few genes that are deleterious or lethal when present in anything but two doses. In a huge experiment, Lindsley et al. (1972) used irradiation to break chromosomes, from which they recovered 467 special chromosome *translocations* (Chapter 9) that consisted of pieces of the Y chromosome attached to pieces of *Drosophila* autosomes 2 or 3. These translocations included segments of various sizes, from small to large, collectively spanning 85% of the two autosomes. They set up thousands of carefully designed crosses of euploid (but translocation-bearing) parents that yielded progeny that were partly aneuploid, having a particular segment of chromosome 2 or 3 either duplicated or deficient. Then they observed the effects of each aneuploidy—finding many cases where small aneuploid segments were not lethal by themselves, but were lethal when added together. Altogether they detected 57 loci in which aneuploidy caused a recognizable effect on the fly; but among these, only one locus was lethal when tripled or haploid. This near-absence of aneuploid-lethal loci means that "the deleterious effects of aneuploidy are mostly the consequence of the additive effects of genes that are slightly sensitive to abnormal dosage." These researchers also verified that the effects of having a gene in triplicate were less severe than having it single.

Commenting on these results, Sandler and Hecht (1973) point out that, in *Drosophila*, triplicate segments exceeding about 10% of the haploid genome are usually lethal. Most deficiencies of up to 0.5% of the haploid genome survive, over half of the deficiencies up to 1% of the genome survive, and the largest viable deficiency comprises about 3% of the genome. In humans, too, deficiencies are less viable than excess chromosome material. Regarding the amount of excess autosomal material that still permits viability, they calculate that trisomies 21, 18, 13, and 8 (the latter due to a translocation) involve 1.8%, 2.8%, 3.6%, and about 5%, respectively, of the haploid autosomal complement. "Thus the maximal length of autosomal material that can be in excess and not be lethal (before birth) appears to be approximately 5–6% in humans compared with about 10% in Drosophila" (Sandler and Hecht 1973).

Mitotic Nondisjunction

In all the cases discussed so far, an abnormal chromosome number is established in the zygote stage, the result of fertilization involving an aneuploid gamete. Consequently, *all* the cells of the body are expected to be aneuploid. In mitotic nondisjunction, however, the zygote may be normal, with aneuploidy occurring sometime during its later development. If the resultant cells survive and continue to divide, they give rise to descendant line(s) of aneuploid cells. The individual then has a mixture of cells with different chromosome numbers and is called a **mosaic**. This anomaly can occur in somatic cells or in cells destined to give rise to gametes. The latter, called **germinal mosaicism**, can lead to aneuploidy among offspring as well.

The earlier that nondisjunction takes place, the larger the proportion of aneuploid cells that might be found in the mosaic. If nondisjunction occurs during the *first* cleavage division of the zygote, as shown in Figure 3A, then *all* the daughter cells will be aneuploid; half will be monosomic and half trisomic. If it occurs during the *second* cleavage division (Figure 3B), only *half* of the resultant cells will be aneuploid. Nondisjunction in the third cleavage division will initially make 1/4 of the cells aneuploid, and so on. If nondisjunction occurs late in development, only a tiny group of aneuploid cells will be formed, and they may escape detection.

Actually, the proportion of aneuploid cells in an individual does not necessarily reflect what happened in the embryo, because aneuploid cells have a lower probability of survival than do normal cells. Indeed, embryos with high proportions of abnormal cells may not survive at all, as evidenced by the high frequency of aneuploidy in spontaneously aborted fetuses (Chapter 9).

Mosaics can also arise through other processes. The most frequent of these is **chromosome loss**, whereby a chromosome lags so far behind during anaphase that it is not included in either daughter nucleus. Consequently, one of the daughter cells will contain one less chromosome than the other. If the original cell was euploid, one of the daughter cells will be monosomic (Fig-

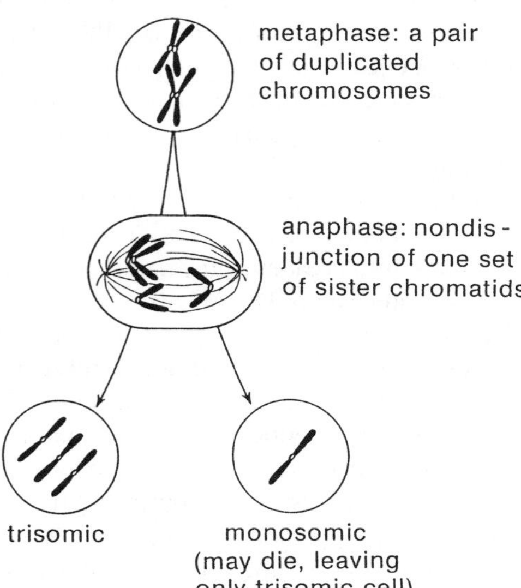

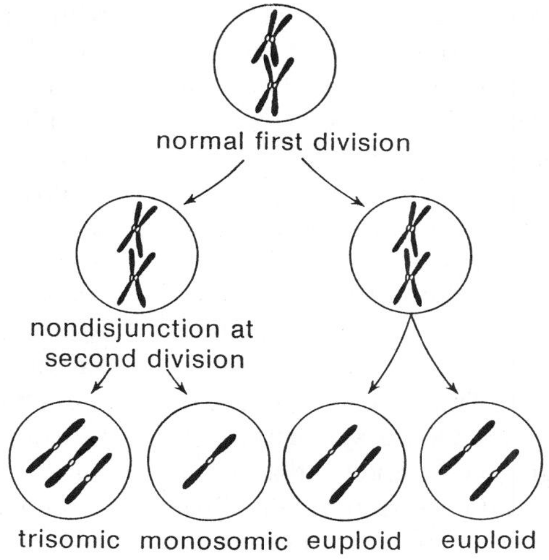

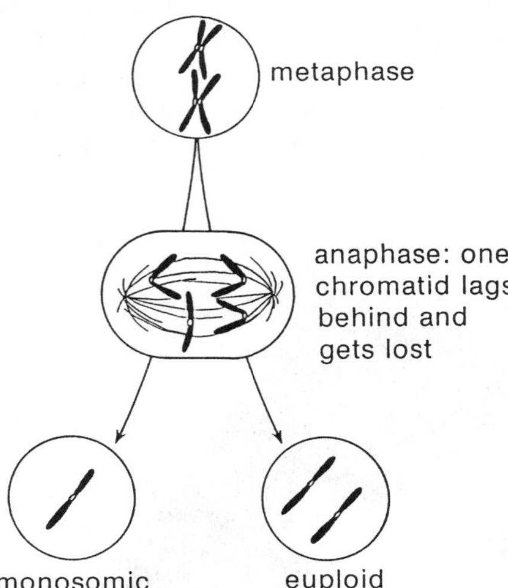

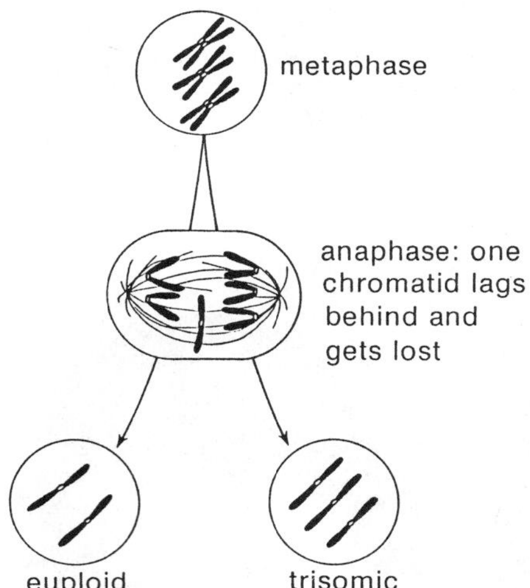

Figure 3 How mosaicism arises from mitotic nondisjunction (A and B) or from chromosome loss (C and D).

ure 3C). But if the original cell was *trisomic*, then following the loss of the extra chromosome, one daughter cell will be normal rather than trisomic (Figure 3D). Clearly, this event could lead to mosaicism in an individual who started out as a trisomic; indeed, it has been documented in several Down syndrome mosaics (Niikawa and Kajii 1984). Like nondisjunction, chromosome loss may occur during any cell division, mitotic or meiotic, and can involve autosomes or sex chromosomes.

In any given cell division, more than one chromosome may undergo nondisjunction or loss, but the result is usually lethal. Those few cases that do survive beyond birth generally involve either mosaicism or autosomal aneuploidy in conjunction with a sex chromosome aneuploidy. We shall see in Chapter 19, however, that specific single or multiple aneuploidies characterize certain types of cancer.

NONDISJUNCTION OF HUMAN AUTOSOMES: SPORADIC DOWN SYNDROME

One year after Mendel's work was published, the English physician John Langdon Down wrote a paper entitled "Observations on an Ethnic Classification of Idiots." As medical superintendent of an asylum for the severely retarded, he noted that about 10% of all the inmates resembled each other and could be readily distinguished from the rest of the patients: "So marked is this, that when placed side to side, it is difficult to believe that the specimens compared are not children of the same parents." Because they had broad faces and slanted eyelids, he assumed they were "typical Mongols," which we now know to be completely false.* Thus, what used to be called "mongolism" is now called **Down syndrome**. (A **syndrome** is a group of signs and symptoms that occur together and characterize a particular abnormality. Not all symptoms necessarily appear in every individual, however.)

*Down also suggested that "the observations which I have recorded are indications that the differences in the races are not specific but variable. These examples of the result of degeneracy among mankind . . . furnish some arguments in favour of the unity of the human species." On these grounds he vigorously opposed slavery. He also supported higher education for women, denying that it would increase the risk of their producing feeble-minded children, a popular concern of the day.

General Features

In addition to being retarded, Down syndrome patients exhibit some combination of the following traits (Figure 4), which may vary considerably from one individual to another. All parts of the body are shortened, signifying poor skeletal development. The face is broad and flat with a small nose, irregular teeth, and abnormally shaped ears. The eyes may be close-set with narrow, slanting eyelids; a large, furrowed tongue may protrude from a mouth framed by rather thick lips. Hip bones are abnormally shaped and aligned, and the feet often display a sizable gap between the first and second toes; the little finger is often short and curved inward. Some highly unusual patterns of hand creases plus hand and foot prints are also associated with Down syndrome—but with the advent of chromosome banding techniques, karyotype analysis has replaced **dermatoglyphic** (Greek *derma*, "skin"; *glyph*, "carving") analysis as a diagnostic tool. Nevertheless, attention to dermal patterns (along with other phenotypic traits) can, in the case of certain chromosomal disorders, alert physicians to the need for karyotype studies (Uchida and Soltan 1975).

Weak reflexes, loose joints, and poor muscle tone render these patients rather limp and floppy. Defects of the heart, digestive tract, kidneys, thyroid gland, and adrenal glands are also common. Males have poorly developed genitalia and are invariably sterile; in females, ovarian defects and irregular menstruation are the rule. Many Down syndrome babies die within a year, often from heart defects that occur in 40–50% of them. Susceptibility to infection (especially pneumonia) is also common, probably due to defects in the immune

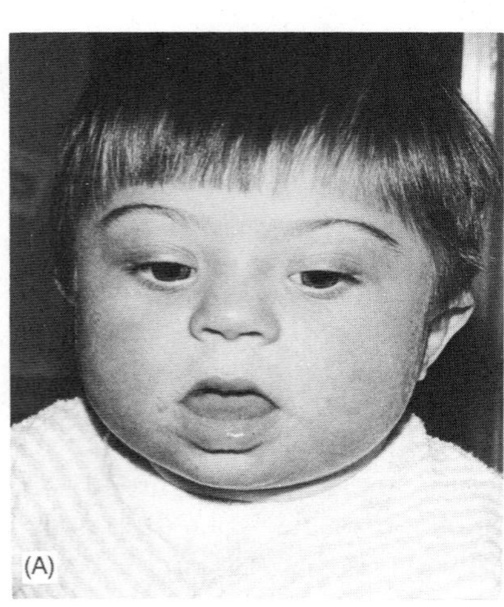

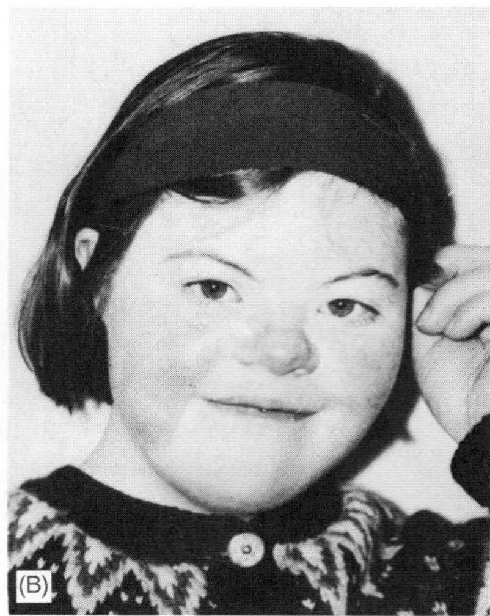

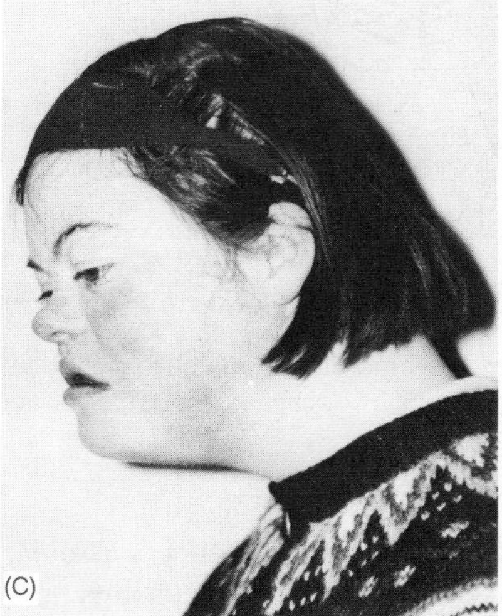

Figure 4 Down syndrome children. (A) One-year-old boy. (B,C) Sixteen-year-old girl. (Photographs courtesy of Paul Polani, University of London.)

system—but antibiotics and other medical improvements have extended the mean life expectancy to over 30 years, with 25% of affected individuals surviving to the age of 50. Leukemia is 15–20 times more frequent in Down syndrome patients than in the general population, and other types of cancer are more common too.

Motor development is slow, and bladder and bowel control may take years to develop. Many patients learn to talk, but speech is usually thick and harsh sounding, perhaps in part because of hearing defects. Lively, cheerful, and very affectionate, Down syndrome patients are noted for their impishness, flair for mimicry, and enjoyment of music and dance. Although clumsy with their hands, they respond well to early intensive stimulation and training for simple tasks; despite IQs that seldom exceed 60, some learn to read and can attend regular classes.

Roughly 1–2% of all live-born Down syndrome cases are mosaics, possessing both normal and abnormal cells. Not surprisingly, these individuals show great phenotypic variability in all respects—presumably depending upon the proportion and tissue location of the abnormal cells.

Possible Relation to Alzheimer Disease

Ironically, delayed development is usually followed by premature aging—often including, among those who survive beyond 30 years, dementia* of the type suffered by people with **Alzheimer disease**. Indeed, recent neuropathological studies have uncovered some striking similarities in the brain abnormalities (Figure 5) associated with these two conditions: **neuritic plaques** (lesions resulting from the degeneration of nerve cell endings) and **neurofibrillary tangles** (clusters of protein filaments in the cell bodies of neurons). The relations among these findings are not clear, however (Kolata 1985b). Although it appears that all Alzheimer patients

*Memory loss, confusion, anxiety, loss of ability to perform simple tasks like dressing or feeding themselves, and ultimately the loss of all functional capabilities.

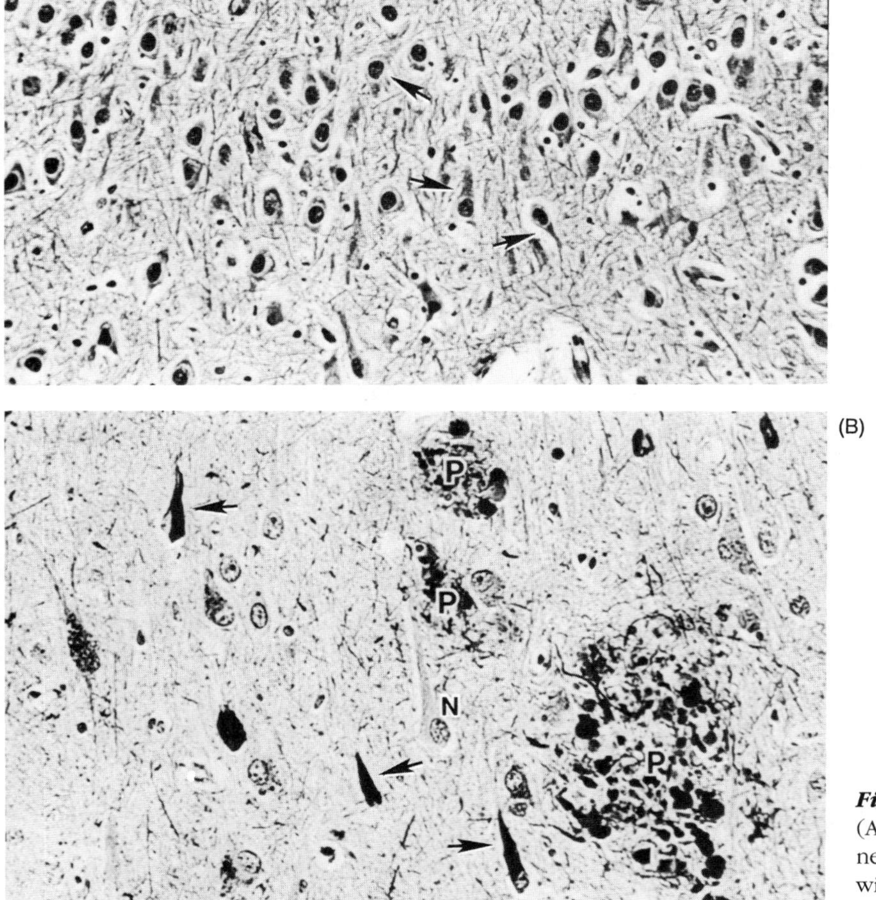

Figure 5 Plaques and tangles. (A) Normal brain tissue. Arrows point to neurons. (B) Brain tissue from an adult with Down syndrome. P, plaques; N, neurons. Arrows point to tangles.

and all Down syndrome adults over age 30 have plaques and tangles in their brains, only 25–40% of Down syndrome adults with plaques and tangles seem to have dementia. Perhaps, rather than a simple cause-and-effect connection, a certain *threshold* number of plaques and tangles must accumulate before dementia sets in; or perhaps the association is even more obscure.

There are other poorly understood similarities between the neurological defects in Down syndrome and Alzheimer disease. These include severe loss (20–50%) of nerve cells in the hippocampus (the center for receiving, storing, and sending messages to and from various brain areas) and in certain parts of the cerebral cortex; fewer and smaller contacts (synapses) between nerve cells; large numbers of vacuoles in the nerve cells; deficiencies in certain enzymes and neurotransmitters; and possible excess of nerve growth factor (Edwards 1986).

Frequency of Down Syndrome

Down syndrome occurs in approximately 1 out of every 800* live births in all ethnic groups (Hassold and Jacobs 1984). It afflicts families of wealth or intellectual achievement as well as families of the poor or uneducated; some surveys have included an excess of affected males, but this is not always observed. Nearly all cases, being single occurrences within a given family, are referred to as **sporadic Down syndrome**. (The 1–2% of cases that run in families are discussed in Chapter 9.)

Over a century ago it was noted that Down syndrome babies are often the last-born members of a sibship, an observation suggesting that birth order, maternal age, paternal age, or a combination of these, might be implicated. In the 1930s, **maternal age** was shown to be the critical factor in most cases: women over 35 years of age produced over half of all Down syndrome babies, although they accounted for only about 15% of all births. Of course, older women have older husbands and more children, but appropriate statistical procedures can often disentangle such closely correlated factors and measure their individual effects.

Where maternal age is not a factor, nondisjunction in *fathers* must be considered. Banding techniques and molecular methods using so-called RFLPs (Chapter 15) sometimes reveal variations that make it possible to distinguish the maternal 21 chromosomes from the paternal 21s. Such studies have shown unequivocally that *20–25% of sporadic Down syndrome cases are paternal in origin* (Magenis et al. 1977). Researchers agree that the great majority of these are not age-related, but they disagree on whether or not a small proportion of cases might be associated with increasing paternal age.

Figure 6 shows more specifically how the frequency of Down syndrome births varies among different maternal age groups. Note that beyond age 30 the frequency rises dramatically; consequently, a 45-year-old woman is 60 times more likely to produce a Down syndrome child than is a 20-year-old woman. Some researchers report an *increased incidence of Down syndrome among very young mothers* (Hook 1981), but others suggest that the difference may not be statistically significant. Studies of spontaneous abortions (Chapter 9) have shown that the incidence of this disorder is *at least four times higher at conception than among live births*. And as with Down syndrome live births, about 20% of these spontaneously aborted fetuses are paternal in origin (Hassold and Jacobs 1984).

With the advent of prenatal testing in the late 1960s, the routine screening of pregnant women over the age of 35 became possible. As a result of this and a lowered birth rate among this age group, fewer Down syndrome

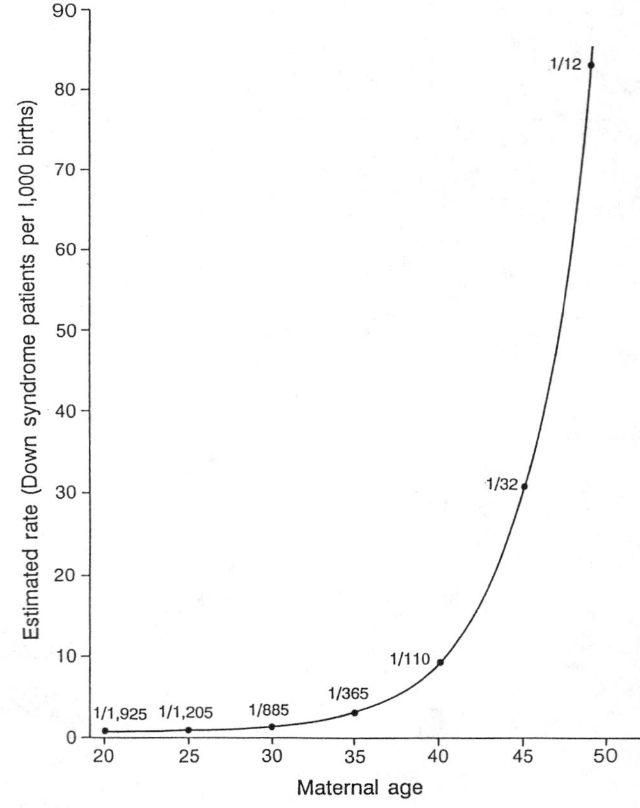

Figure 6 Estimated rate of occurrence of Down syndrome for different maternal ages. The number on the graph, 1/1,925, for 20-year-old mothers means one Down syndrome offspring per 1,925 births, and so on. (Data from Hook and Chambers 1977.)

*This is a decrease from the 1 in 700 frequency that occurred before prenatal testing was possible, but not as small as the 1 in 1,100 now being reported by some workers.

babies have been born to older women, and the frequency of Down syndrome among live births has dropped somewhat. Whereas "over-35" mothers used to produce a large proportion of Down syndrome babies, they now account for only about 20% of them. The remaining 80% of Down syndrome babies are now being born to women under 35 years of age—who, although their risk factor is very much lower, vastly outnumber older mothers.

Although it has not yet been feasible to do prenatal tests on *all* women under the age of 35, researchers have been searching for some additional clues to help them identify those younger mothers who are at increased risk of producing Down syndrome babies. In 1984 it was noted that the concentration of a substance called **α-fetoprotein (AFP)** is often *lower than average* in the blood serum of mothers who give birth to Down syndrome babies (Cuckle et al. 1984).* Using this as a preliminary test, over a two-year period Di Maio et al. (1987) screened 34,354 women under the age of 35, offered amniocenteses to 1,451 of them who showed low levels of α-fetoprotein, and among these latter women detected 9 fetuses with Down syndrome. On the other hand, 18 Down syndrome babies were born to women whose α-fetoprotein levels were *not* reduced below the arbitrary threshold (and thus not offered amniocentesis)—so this screening procedure missed 2/3 of the Down syndrome pregnancies in the study. This experience has led some people to question whether AFP is a useful screening test for Down syndrome (Pueschel 1987).

*Conversely, when present in *excess* in a pregnant mother's serum, it may signal the presence of spinal cord abnormalities in the fetus.

Chromosomal–Genetic Basis for Down Syndrome

Long before the cause of Down syndrome was known, studies of this disorder in twins suggested a genetic role. (**Identical** twins arise from a single zygote and are expected to have identical genotypes; **fraternal** twins develop from two fertilized eggs and so are genetically equivalent to sibs.) Such studies are usually done by finding one affected member of a twin pair, then determining (1) the type of twinning and (2) whether or not the other member is also affected. Among identical twins it was found that if one member had Down syndrome, then almost invariably the other one did, too.† In contrast, if one member of a pair of fraternal twins had Down syndrome, the other was in almost all cases normal.

Because the pattern of transmission did not fit the usual models for single gene inheritance, however, it was suggested in the 1930s that Down syndrome might result from a chromosomal abnormality. Cytological methods then available were too primitive to test this hypothesis, but just three years after the normal human chromosome number was established came the exciting discovery that *somatic cells of sporadic Down syndrome patients contain 47 chromosomes* (Lejeune et al. 1959). The extra chromosome was one of the G group, at first thought to be the next-to-smallest—that is, no. 21; so **trisomy 21** became another name for Down syndrome. Improved techniques later showed that the extra chromosome was actually the *smallest* member of the G group—but rather than trying to change the name, geneticists live with the inconsistency.

†It can happen that, after the separation of identical twins, one or the other loses a chromosome.

Saying that Down syndrome patients have an extra chromosome does not really explain how their abnormalities come about. Does chromosome 21—which contains only about 1.5% of the total DNA in humans—carry genes involved with the development of the brain, skeleton, and the other organ systems that are affected? More specifically, does chromosome 21 carry genes for those enzymes, hormones, blood groups, and so on that might be expressed in excess in trisomy 21 patients? Unfortunately, not enough is known about networks or cascades of biochemical events in normal human development to say how an extra chromosome interferes with these processes. Certainly, there must be complex interactions among the gene products of many chromosomes.

From cytological studies we know that all of chromosome 21 need not be present in triplicate to produce Down syndrome: in a small percentage of affected individuals, the only extra chromosomal material is the distal half of the long arm. And as geneticists use molecular techniques to dissect and characterize chromosomes, some additional clues are beginning to emerge (Patterson 1987). Of the 1,000–1,500 genes thought to be on chromosome 21, only about 20 have been mapped. Among these, about 10* are known to reside within bands 21q22.1 through 21q22.3—the region that, when triplicated, is associated with Down syndrome.

As noted before, studies that relate Down syndrome to Alzheimer disease (AD) are also being reported. For example, several research teams† have isolated and mapped to chromosome 21 the gene for the larger protein precursor of **β amyloid protein**, which is found

*These include the following genes: *SOD-1*, for soluble superoxide dismutase; *PAIS* and *PRGS*, for enzymes involved in purine synthesis; *Ets-2*, an oncogene that might be associated with leukemia; *PFKL*, for the enzyme phosphofructokinase in the liver; *CRYA1*, for the mammalian lens-specific polypeptide α-A-crystallin; *IFRC*, for an interferon receptor; and *BCEI*, for the breast cancer estrogen-inducible sequence.

†St George-Hyslop et al. (1987), Goldgaber et al. (1987), Kang et al. (1987).

in excess in the plaques and tangles of both AD and older Down syndrome patients. And after a gene for **familial Alzheimer Disease (FAD)** was also discovered to be associated with chromosome 21, some workers suggested that the two genes might be one and the same. Furthermore, one research group (Delabar et al. 1987) reported finding three copies of the β amyloid gene in three patients with Alzheimer disease, as well as in two patients with nontrisomy Down syndrome (Chapter 9)—a finding suggesting that people with Alzheimer disease possess a small duplicated region on one chromosome 21. If this is true, perhaps Alzheimer disease results from overproduction of the β amyloid protein or some mutant form of it.

But subsequent research showed that the situation is not so simple (Anderton 1987b). Combining family studies with molecular studies, one European research team (Van Broeckhoven et al. 1987) and one American team (Tanzi et al. 1987a,b) discovered that the genes for the β amyloid protein and FAD could actually be separated from each other—the latter being closer to the centromere than previously thought, perhaps near the gene for superoxide dismutase (SOD). A third gene with possible neurological effects has been mapped by means of various molecular techniques to the distal half of the long arm of chromosome 21 (Allore et al. 1988). Its product, the β subunit of the S100 protein, is part of a calcium-binding protein found throughout the nervous system but most prevalent in certain brain cells.

Molecular studies have also revealed the presence on chromosome 21 of over a dozen unique DNA sequences, numerous RFLP sites, and a few cytological markers (Korenberg et al. 1987; Stewart et al. 1988). These studies should allow researchers to start answering some key questions: In which parent did the primary nondisjunction occur? Does the phenotype of the affected child vary with parental origin of the extra chromosome? Are some parents at a high risk for nondisjunction—and if so, can they be identified in advance? What is the correlation, if any, between crossing over and nondisjunction on chromosome 21? Is the maternal age effect due to increased nondisjunction in older women, or is it due to decreased destruction of their aneuploid embryos? Is the Down syndrome phenotype associated with just a few key genes on chromosome 21, or (as with fruit flies) is it mostly due to a generalized imbalance of genes?

Another line of research on Down syndrome has been opened up by the creation of an animal model for human trisomy 21. Charles Epstein et al. (1985) at the University of California in San Francisco have produced mice that are trisomic for mouse chromosome 16, which carries several genes (including AP, ETS-2, SOD-1, and PRGS) that are homologous to human chromosome 21 genes. Although these fetuses do not survive quite until birth, their multiple defects can be studied in great detail. Many of these trisomic mice exhibit the same type of heart defect that is found in about one-third of Down syndrome patients; they also show defects in the nervous system and immunological abnormalities.

By combining trisomic mouse cells with normal diploid mouse cells, Epstein and his co-workers have also produced *mosaic mice* that survive beyond birth—and these, too, are being intensively studied. But researchers stress that, because mouse chromosome 16 is a much larger chromosome than human chromosome 21, the two trisomies cannot be entirely homologous. Thus, a *partial* trisomy of mouse chromosome 16, but carrying genes similar to those found on the distal region of human chromosome 21q, would be much better for comparative studies.

In discussing the effects of aneuploidy for any chromosome or chromosome segment, Epstein (1988) points out that there may not be a clear distinction between aneuploidy and single-gene disorders. Molecular analyses show that the genes for several "monogenic" disorders (including Duchenne muscular dystrophy) contain sizable *deletions* rather than simple point mutations. So aneuploidy may be a continuum affecting single genes as well as chromosome segments and whole chromosomes.

There is good evidence that differences in gene dosage lead to proportional differences in the amounts of enzyme produced. "For 37 different human gene products, the mean ratios of triplex (3 gene copies):duplex (2 gene copies, diploid):uniplex (1 gene copy) were 1.61 : 1.00 : 0.52" (Epstein 1988). But the effects of aneuploidy must also involve dosage differences for the numbers and binding ability of gene-determined cell receptors, for regulatory genes, for homeotic and other developmental genes, and for genes that produce subunits for important structural proteins such as collagen. Although no links have yet been proved between the imbalance of any specific genes and specific phenotypes, methods for detecting these are now becoming available.

Search for Factors Altering Nondisjunction Rates

Most cases of trisomy (for any chromosome) are due to nondisjunction during the formation of a parent's egg or sperm. As is usual in science, the solution to one problem merely raises another: What causes the observed differences in nondisjunction rates? The well-documented maternal age effect in the incidence of Down syndrome has focused the greatest attention on the mother, particularly the older mother. (Recall that oocytes may spend several decades in prophase of meiosis I, whereas sperm are newly formed.) To date, however, nobody knows for sure what factor(s) cause nondisjunction in women of any age.

Over the years, many hypotheses have been advanced

(thyroid disorders, viral infections, caffeine, effects of certain drugs or hormones, contraceptives, delayed fertilization due to reduced frequency of coitus, and so on) and then usually rejected for lack of convincing evidence. For example, clusters of Down syndrome cases sometimes occur following epidemics of infectious hepatitis or German measles. One possibility, which has by no means been proved but merits careful consideration, is that X-irradiation increases the rate of nondisjunction in aging females (Uchida 1977). This outcome is known to happen in fruit flies, and may also occur in mice. Unfortunately, the data on humans are not clear-cut. Some studies find a significant effect of irradiation on the rate of nondisjunction, but others find no such effect. Even if X rays have no effect on the incidence of aneuploidy, they are known to cause developmental defects in fetuses. Thus prospective mothers should avoid exposure to X rays, especially during early pregnancy.

Most of these studies are **retrospective** (Latin *retro*, "backward"); that is, one compiles a sample of Down syndrome patients and then inquires as to how much abdominal irradiation each mother received prior to conception (Figure 7). (Abdominal irradiation is likely to hit the ovaries.) As a *control* (comparison), investigators may compile a sample of normal children or patients with some other congenital defect, such as cleft lip, and ask the same question of their mothers. Faulty memories and inadequate records are common problems with retrospective studies. Furthermore, they give no direct information on those families in which Down syndrome did *not* occur after mothers received abdominal irradiation. Thus, by preselecting only affected offspring, retrospective studies may tend to exaggerate the effect of a given treatment. But this straightforward approach to getting a large sample of affected individuals and its relative cost effectiveness are obvious advantages.

A better way to investigate such problems of cause and effect is by a **prospective** study. Here one uses hospital records to compile a large sample of irradiated mothers, plus another sample of nonirradiated mothers for a control, and then checks out the frequency of Down syndrome among their subsequent offspring. Prospective studies are more reliable because they indicate how often a given treatment causes *no effect*, as well as how often the effect under consideration does occur following treatment. They are, however, immensely time consuming and expensive: to find just a few affected offspring, the investigator must keep track of several hundred families, often for a period of years. For example, Uchida et al. (1968) found 10 cases of autosomal aneuploidy (mostly Down syndrome) among the 972 children born to 861 irradiated mothers, compared with only 1 case among the 972 control children born to nonirradiated mothers. (It is possible, however, that a prospective study of spontaneous abortions—in which the frequency of aneuploidy is so much higher—would be much less expensive to conduct.)

Down syndrome in conjunction with sex chromosome aneuploidy has occasionally been observed. The double trisomic (for sex chromosomes and the 21s), 48,XXY,+21, for example, occurs in newborns with a frequency of about 1 in 11,000. Two other Down syndrome double trisomics, 48,XXX,+21 and 48,XYY,+21, are found less frequently. Although the data are scanty, it appears that in some families these double trisomics may occur more often than would be expected by chance alone, which has led some geneticists to postulate the existence of genes that increase the probability of nondisjunction. Such mutant genes have been identified in

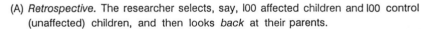

(A) *Retrospective.* The researcher selects, say, 100 affected children and 100 control (unaffected) children, and then looks *back* at their parents.

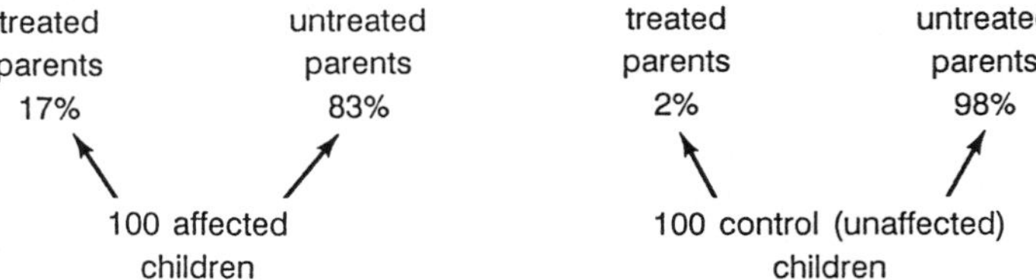

(B) *Prospective*. The researcher selects say, 100 treated parents and 100 control (untreated) parents, and then looks *forward* to their subsequent children, if any.

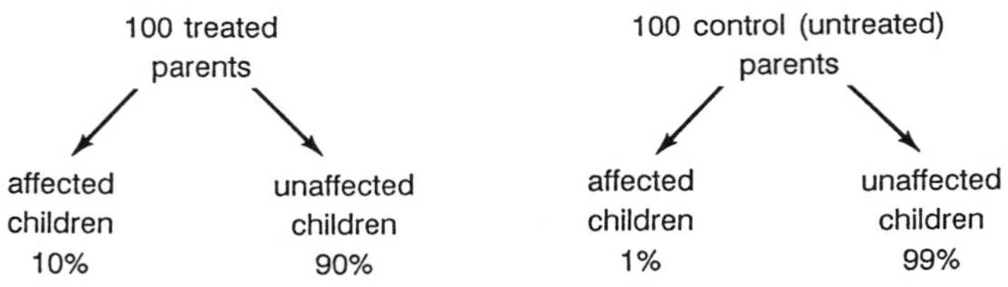

Figure 7 A comparison of retrospective and prospective studies of cause and effect. The arrows show the time course of a hypothetical investigation. The percentages at the arrowheads are predicted on the basis of two hypothetical assumptions: (1) A treated parent produces, on the average, 10% affected children; an untreated parent has 1% affected. Thus the treatment increases the effect 10 times. (2) In the population only 1 person in 50 is actually treated.

fruit flies and in corn. Additional evidence for their presence in humans are reports of a few families in which several members have different single aneuploidies—say, one trisomy 21 and one 47,XXY—again, more frequently than expected by chance alone. But even if a genetic predisposition to nondisjunction were proved in some families, it could not account for more than a fraction of all the cases that occur.

Down Syndrome Mosaics

Some small fraction of sporadic Down syndrome cases may be due to mitotic nondisjunction in the embryo rather than to meiotic nondisjunction in a parent. Recall that nondisjunction during the first cleavage division would result in one trisomic daughter cell and one monosomic daughter cell. The latter usually dies, leaving the trisomy 21 and its descendant cells to make up the entire embryo. But nondisjunction at the second cleavage division (Figure 3B) produces two normal cells, one trisomic cell, and one monosomic cell. The resultant individual will be a mosaic, exhibiting some characteristics of Down syndrome. It is estimated that about 10% of patients with doubtful diagnoses are in fact mosaics. The degree of abnormality varies depending on (1) what proportion of tissues end up being trisomic, and (2) which tissues these are. If cells of the nervous system turn out to be euploid, for example, the individual may have normal intelligence. With respect to dermatoglyphic analyses, Down syndrome mosaics often show intermediate dermal patterns. Those individuals with only a tiny proportion of aneuploid cells may have a completely normal phenotype and escape detection entirely. In some families, multiple cases of trisomy 21 can be traced to aneuploidy in a mosaic parent's gonadal tissue. These instances of germinal mosaicism are important to detect early, because, unlike the usual sporadic pattern, the occurrence of **secondary nondisjunction** in the parent's trisomic germ cells can lead to multiple Down syndrome births in a family.

OTHER AUTOSOMAL ANEUPLOIDS

Trisomy 18 and Trisomy 13

The only two other autosomal trisomies that occur with any significant frequency in newborns, **trisomy 18 (Edwards syndrome)** and **trisomy 13 (Patau syndrome)**, were first reported in 1960. Both are associated with gross abnormalities of nearly all systems, thus leading to early death. In fact, the two syndromes share many characteristics, with considerable overlap. Profound mental and developmental retardation occur without exception. Heart defects are the rule, and kidney defects are very common. Low-set and malformed ears, small eyes, and a small, receding jaw are almost invariably observed. Prominent heel bones and flexion (bending) deformities of various joints are often present. Because the numbers of cases reported are relatively few, varying both regionally and temporally, the best estimates for frequency of either of the two syndromes are not very good—ranging from 1 in 4,000 to 1 in 8,000. There is a maternal age effect, but it is not so striking as that found for Down syndrome. The mosaics that have been identified for either syndrome are often intermediate between fully affected and normal. Additional traits also are associated with these two syndromes.

The most distinctive additional features of *trisomy 18 syndrome* (Figure 8) are the occurrence of simple arch patterns on three or more fingertips (rare in normal individuals) and a long skull that bulges in the back. Because of abnormally increased muscle tone, the fingers are tightly clenched, with the index finger bent sideways across the third finger, and the limbs and the hip joints are so rigid that they can hardly be moved. About three-fourths of trisomy 18 patients have "rocker bottom" feet with big toes that are short and bent upward. Approximately half have extra folds of skin on the neck. The distinctive facial features—including a round face, small and wide-spaced eyes, and a small mouth—are also an aid to diagnosis of this syndrome. Birth weight is low, and for unknown reasons about 80% of all trisomy 18 births are female. The mean survival time is about 2.5 months, 90% of all cases dying within a year.

For *trisomy 13 syndrome*, the most distinctive additional traits, occurring in about three-fourths of patients, are a small head, cleft palate or lip (or both), port-wine-colored birth marks, and an excess number of fingers and toes (Figure 9). Unusual dermatoglyphics are the rule, too. Almost all trisomy 13 patients appear to be deaf, and some are blind. Extreme "jitteriness" and seizures are not uncommon. Genitalia may be affected; males may have undescended testes, and females may have a uterus divided into two parts. The fingers may exhibit flexion deformities. The nose is often large and triangular, and the eyes are frequently wide-spaced and defective. In red blood cells, the hemoglobin (oxygen-carrying protein) is abnormal, and certain white blood cells are irregularly shaped. Among trisomy 13 cases, there is a slight excess of females. The mean survival time is about 3 months, 80% of all cases dying within a year of birth.

Very Rare Autosomal Aneuploidies

Over two dozen cases of trisomy 22 have been reported. The phenotypes are variable but usually include mental

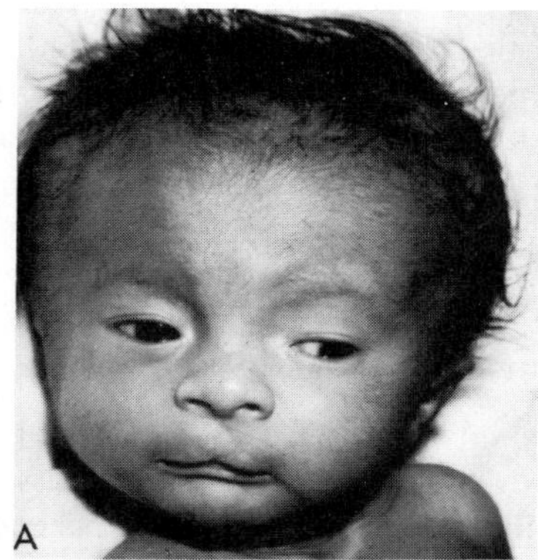

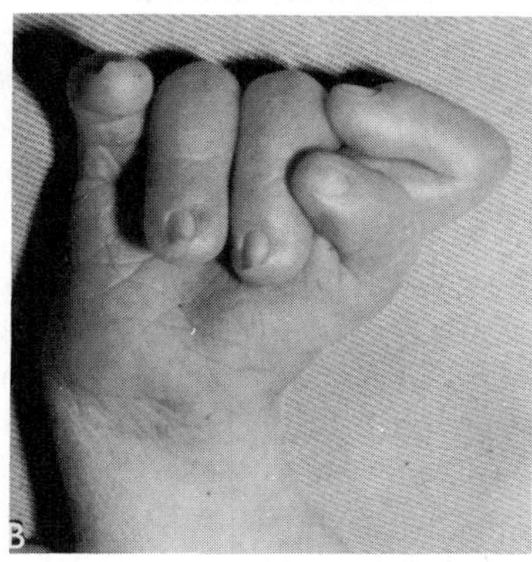

Figure 8 Trisomy 18 (Edwards syndrome). (A) Child with trisomy 18. (B) The hand of an infant with trisomy 18 demonstrates typical flexion deformities. (From Summitt 1973.)

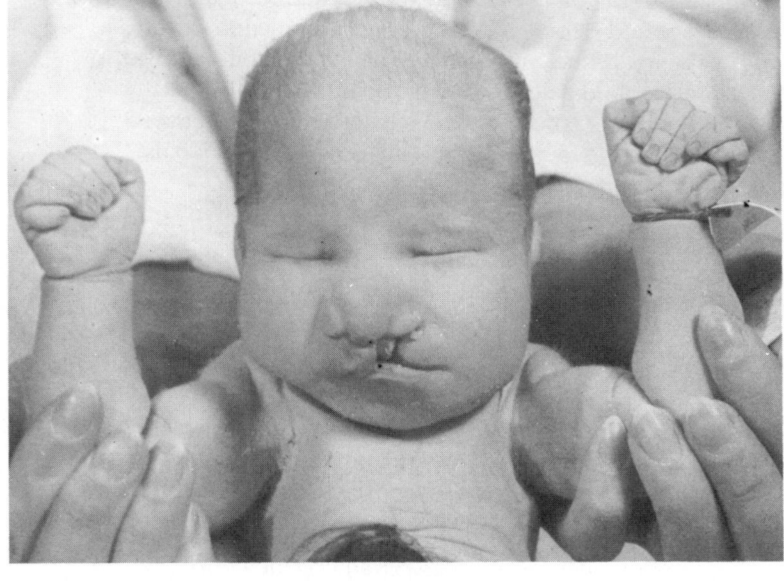

Figure 9 Trisomy 13. This infant was not photographed as a living patient but was found in the Department of Anatomy of the University of Western Ontario, where it had been preserved for 30 years. The features typical of trisomy 13 leave no doubt as to the diagnosis. Note the cleft lip, extra fingers, large triangular nose, and wide-spaced eyes. (Photograph by Murray Barr; courtesy of G. H. Valentine, 1975.)

and growth retardation, small skull and lower jaw, low-set and malformed ears, cleft palate, heart disease, thumb abnormalities, and deformed lower limbs.

Although monosomy of autosomes is generally lethal in humans, several cases of apparent *monosomy 21* have survived beyond birth, albeit with severe mental retardation and multiple abnormalities of other systems. Unlike some features of Down syndrome patients, the eyelids slant downward, the ears are large, the nose is rather beaklike, the face is narrow, and the lower jaw is small and receding. Increased muscle tone is also observed. Several mosaics for monosomy 21 have been described, as well as a few cases in which only part of chromosome 21 is missing.

In conclusion, we emphasize that these disorders result from chromosome imbalance rather than from gene mutation. That so many systems of the body are deranged when one small chromosome is present in excess suggests that normal metabolism and development are finely attuned not just to the nature of the gene products but to their concentration as well. This finding is also true of other species that have been studied. Furthermore, the relative rarity of aneuploidies among live-born children indicates that most imbalances of even small chromosomes interfere so seriously with metabolism that embryonic development is aborted in early pregnancy.

SUMMARY

1. Nondisjunction is either the failure of homologous chromosomes to separate (disjoin) during a meiotic division or the failure of sister chromatids to separate during mitosis or meiosis. In either case, the two daughter cells will have unbalanced chromosome sets; that is, they will be aneuploid.
2. Meiotic nondisjunction in a parent leads to aneuploidy of gametes, which can lead to aneuploidy in all cells of an offspring. When it occurs in euploid cells, it is called primary nondisjunction; when it occurs in aneuploid cells, it is called secondary nondisjunction.

3. Mitotic nondisjunction leads to mosaicism within an individual. The degree of abnormality expressed depends on when nondisjunction occurs, which tissues are affected, and the viability of the aneuploid cells relative to the normal cells. Usually the phenotype of mosaics is intermediate between that of seriously affected and normal individuals.

4. Aneuploidy often causes gross abnormalities or death of cells or organisms, as a result of generalized gene imbalance. Sex chromosome aneuploidy is better tolerated than autosomal aneuploidy, and aneuploidy of large autosomes is usually lethal. Trisomies are better tolerated than are monosomies.

5. Nondisjunction of X chromosomes in fruit flies provided the first physical proof of the chromosome theory of heredity.

6. In humans, only three well-defined autosomal aneuploidies are observed among newborns: trisomy 21, trisomy 18, and trisomy 13. All are retarded and exhibit deformities of most organ systems.

7. By far the most common and best understood of these is trisomy 21—Down syndrome—which accounts for about 10% of the retarded population. Down syndrome patients are also characterized by poor skeletal development, heart and intestinal defects, characteristic dermal patterns on their hands and feet, and certain facial traits. They live much longer than other trisomics, in many cases for decades. Extremely sociable, they can learn to speak and to perform simple tasks.

8. With a frequency of about 1 in 800 live births, sporadic Down syndrome usually appears only once in a family, and usually among the later-born in a sibship. Advanced maternal age is a strong determining factor, but about 25% of all trisomy 21 cases are due to nondisjunction in the father. Multiple cases in a family may be due to secondary nondisjunction in a phenotypically normal parent with germinal mosaicism for trisomy 21.

9. Many older Down syndrome patients develop Alzheimer disease (AD), which is characterized behaviorally by dementia and neurologically by degenerative changes in certain areas of the brain. Amyloid protein, which accumulates in and around abnormal nerve cells, is the product of a gene on chromosome 21. On the same chromosome, but separate, is a gene (normal function unknown) that is associated with familial AD.

FURTHER READING

A number of books cover the material in Chapters 7, 8, and 9, that is, aneuploidy in general. These include Valentine (1986), Epstein (1986), Dellarco et al. (1985), Bond and Chandley (1983), De Grouchy and Turleau (1984), and Borgaonkar (1984). Hassold and Jacobs (1984) and Hook (1985) have written good review articles on this general topic.

Many books are available on Down syndrome alone. More technically oriented are Pueschel et al. (1987), G. Smith (1985), Lane and Stratford (1985), De la Cruz and Gerald (1981), and Burgio et al. (1981). Books written for the lay public include Pueschel (1984), Cunningham (1982), and Stray-Gunderson (1986). Seagoe (1964) contains large parts of a diary kept over many years by a Down syndrome patient.

General articles on Down syndrome include Patterson (1987) and Siwolop and Mohs (1985). The relation between Down syndrome and Alzheimer disease is treated by Oliver and Holland (1986), Barnes (1987b), Anderton (1987a), Edwards (1986), Goodfellow (1987), and Sinex and Merrill (1982).

QUESTIONS

1. Diagram nondisjunction of one chromosome pair, occurring in (a) meiosis I and (b) meiosis II of oogenesis. Compare the results with those in Figure 2 to explain the following statement by Hamerton (1971): "Thus, if a nondisjunctional event occurs in the first meiotic division of spermatogenesis, only disomic and nullisomic sperm will be produced, while if it occurs in the second division then normal sperm may also be produced. *However, nondisjunction at either division during oogenesis results in a disomic or nullisomic ovum*" (italics added).

2. Diagram the loss of one member of a chromosome pair during (a) meiosis I and (b) meiosis II of spermatogenesis. Compare the results with those in Figure 2.

3. Describe the zygote formed by the fertilization of a 21-disomic egg by a 21-nullisomic sperm.

4. With reference to chromosome 21, what kind(s) of eggs could be produced by a Down syndrome female? Diagram(s), please.

5. Suppose that several offspring in one family have Down syndrome. Based on your knowledge to date, can you suggest an explanation?

6. At least one case has been reported of a pair of otherwise identical twins (same blood types, and so on), of which one member is normal and the other has Down syndrome. How might this have come about?

7. Bridges detected about 4% secondary nondisjunc-

tion in *Drosophila* XXY females. (The exceptional progeny always had chromosomes without crossovers.) Consider a cross between an XXY female and a normal XY male. Label her X chromosomes X_1 and X_2 to distinguish between them and assume that her three sex chromosome first pair and then disjoin randomly at anaphase I into "2 + 1". Assume that the XXX and YY progeny die, but that XXY and XYY progeny are viable. (a) List all possible gametes that can be formed from the XXY female. (b) If these gametes were formed in equal proportions, what kinds of viable offspring, and in what frequencies, would result?

ANSWERS

1. (a) Nondisjunction at meiosis I (b) Nondisjunction at meiosis II

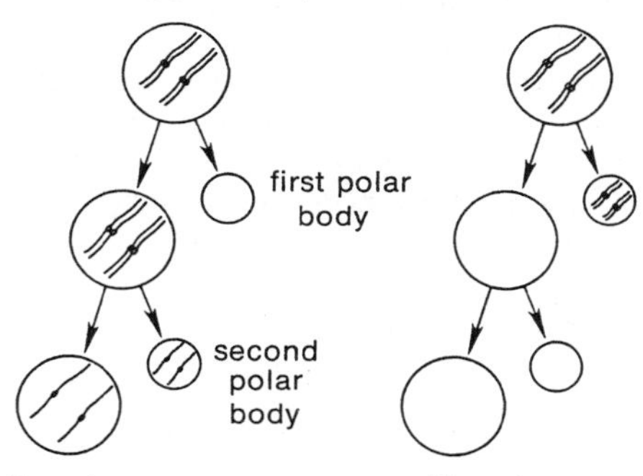

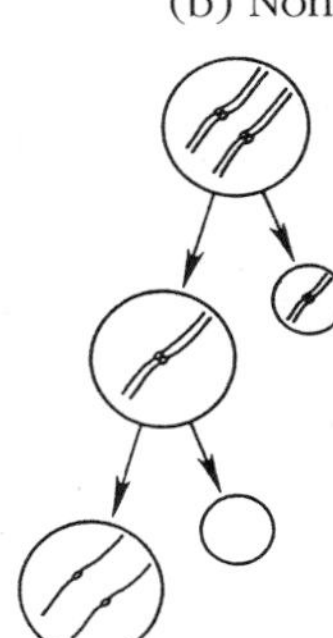

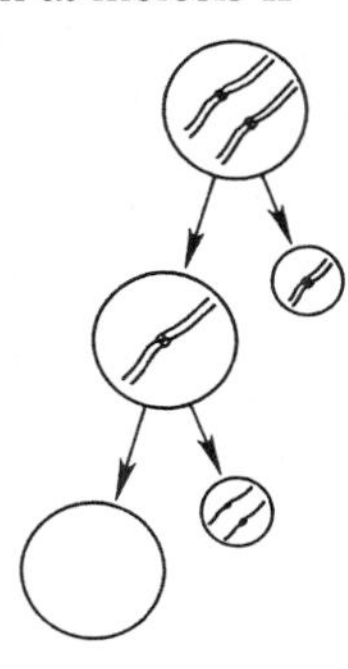

disomic egg or nullisomic egg disomic egg or nullisomic egg

Because there is only *one* secondary oocyte, nondisjunction in meiosis II must necessarily lead to aneuploidy in the egg and second polar body. (Note that Hamerton must not consider the possible division of the first polar body as part of oogenesis.) But there are *two* secondary spermatocytes (Figure 2), so if nondisjunction occurs in one of them during meiosis II, the other will still give rise to normal sperm.

2. (a) Chromosome loss at meiosis I (b) Chromosome loss at meiosis II

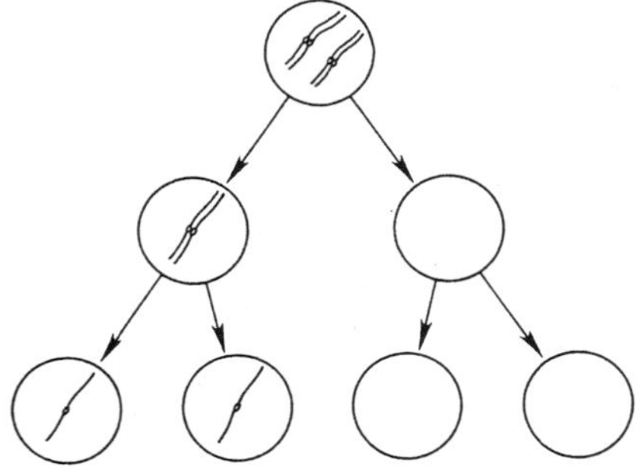

2 normal:2 nullisomic sperm

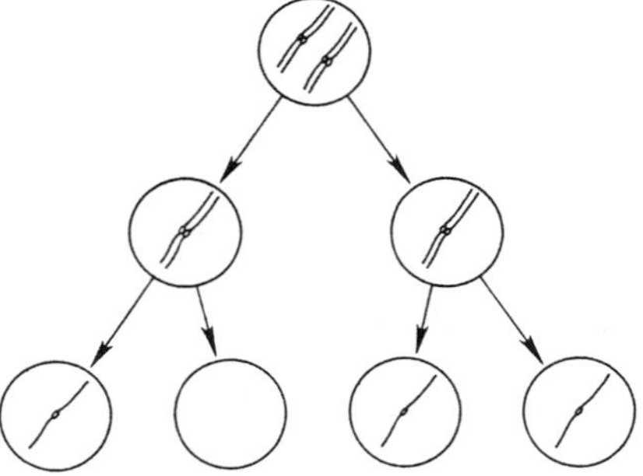

3 normal:1 nullisomic sperm

(Some cytologists see chromosome loss as a form of nondisjunction, but others distinguish between the two.)

3. The zygote will be euploid, although both chromosomes 21 came from the mother.

4.

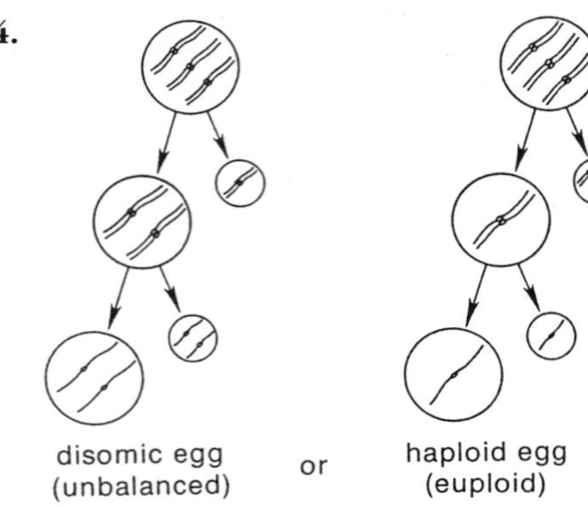

Of the dozen or so cases in which Down syndrome females have given birth, about half of their children were normal and half had Down syndrome. This phenomenon, in which an already aneuploid individual gives rise to further aneuploidy, is known as secondary nondisjunction, to distinguish it from the situation in which a euploid individual gives rise to aneuploid gametes (primary nondisjunction).

5. Undetected mosaicism, due to mitotic nondisjunction, affects one parent's gonadal tissue. Trisomic oogonia or spermatogonia would then regularly give rise (as diagramed for Question 4) to some disomic gametes.

6. If a split occurred at the two-cell stage and then mitotic nondisjunction occurred in one of these cells, the result would be one normal twin and one with Down syndrome.

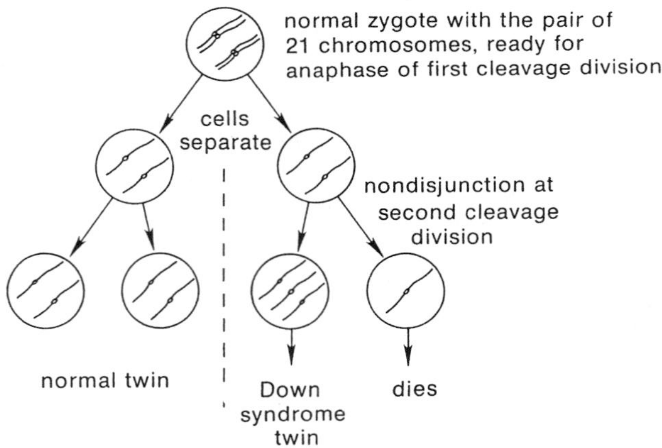

Most mosaics, however, probably begin life as a trisomic zygote from which one chromosome is lost to form a diploid cell line.

7. (a) X_1Y, X_2, X_2Y, X_1, X_1X_2, and Y. Note that the last two kinds of eggs result from nondisjunction of the two X chromosomes; thus they will give rise to *exceptional* progeny. (b) If meiotic pairing and disjunction were completely random among the female's three sex chromosomes, one would expect among the *living* offspring to find 2/10 XX females, 3/10 XY males, 3/10 XXY females, and 2/10 XYY males. This is shown by the following chart:

	Eggs					
Sperm	1/6 X_1Y	1/6 X_2	1/6 X_2Y	1/6 X_1	1/6 X_1X_2	1/6 Y
1/2 X	XX_1Y	XX_2	XX_2Y	XX_1	Dies	$\underline{XY}$
1/2 Y	X_1YY	X_2Y	X_2YY	X_1Y	$\underline{X_1X_2Y}$	Dies

Exceptional progeny

The two underlined progeny classes (on the right) are exceptional in that the XY males inherit their Y chromosome from their mother, and the X_1X_2Y females inherit both X chromosomes from their mother. These two classes together total 2/10 of the living offspring and represent the maximum frequency of exceptional offspring that could be expected, given the assumptions stated in this problem. *Note:* In this case, not all XXY and XYY offspring are exceptional. Only those that arise from nondisjunction of two Xs in the mother—that is, from two chromosomes that normally separate from each other during meiosis—are exceptional.

Chapter 8 Sex Chromosomes and Their Abnormalities

Y Chromosome Present
Klinefelter Syndrome: 47,XXY Males
The 47,XYY Karyotype

Y Chromosome Absent
Turner Syndrome: 45,X Females
47,XXX and 48,XXXX Females
Other Poly-X Karyotypes

Sex Chromatin and the Lyon Hypothesis
Detection of Sex Chromatin in Humans
The Inactive-X Hypothesis
X Chromosome Inactivation and Reactivation

Structure of the Mammalian X and Y Chromosomes
The Pseudoautosomal Region of the X and Y Chromosomes
XX Males, XY Females, and the Testis Determining Factor

The Fragile X Syndrome
Fragile X Syndrome in Males and Females
Genetics of the Fragile X Syndrome

> *[I]n our case, and in the case reported by Jacobs and Strong, the XXY individual presented an essentially male phenotype, whereas in* Drosophila *the XXY fly is a fertile female.*
>
> *Charles E. Ford et al. (1959b)*

In the late 1950s it was discovered that the human Y chromosome plays a major role in sex determination. Until then it had been thought that the critical factor was, as in the fruit fly, the number of X chromosomes—two Xs leading to femaleness and one to maleness, with the Y chromosome needed only for fertility in males.

But soon after the human chromosome number was established, the role of the human Y chromosome in sex determination was elucidated by two previously known syndromes. People with **Klinefelter syndrome** have one too many sex chromosomes: their karyotype is 47,XXY and their phenotype is male. Conversely, people with **Turner syndrome** have one too few sex chromosomes: their karyotype is 45,X and their phenotype is female. Thus two Xs do not always make a female, and one X does not always make a male. Findings on other karyotypes rounded out the picture: in humans, the presence or absence of a Y chromosome is much more important to male development than is the number of X chromosomes.

Later, some **47,XYY males** were found among the inmates of penal institutions, raising the question of whether an extra Y chromosome may somehow predispose its host to criminal behavior. In the 1970s it became clear that a different and larger institutionalized subgroup, that is, the substantial excess of males among the mentally retarded, was due to X-linked inheritance. This **X-linked mental retardation** is associated with a cytological defect called a **fragile site**.

In this chapter we describe the major sex chromosome anomalies and some aspects of their influence on behavior. We also explain how two doses of an X-linked gene in normal females produce basically the same phenotype as one dose of the same gene in normal males.

Y CHROMOSOME PRESENT

The following two karyotypes established the role of the Y chromosome as a male-determiner in humans. But attempts to study the effects of an extra Y chromosome in 47,XYY males brought problems.

Klinefelter Syndrome: 47,XXY Males

In the 1940s Harry Klinefelter and his colleagues at Massachusetts General Hospital described a syndrome occurring in males and usually not detected until after puberty (Figure 1). The signs almost always observed in these patients are very small testes (about one-third of normal size), absence of sperm due to abnormal structure of the seminiferous tubules in the testes, and androgen (male sex hormone) deficiency. The penis and scrotum are usually of normal size, however.

Additional but much more variable symptoms include poorly developed male secondary sexual characteristics (such as scanty facial hair) and the presence of some feminine traits (such as breast development). Sexual behavior is fairly normal among young Klinefelter males, who experience spontaneous erections and (usually spermless) ejaculations. A few have fathered children, however. Many marry and maintain good sexual relations, but impotence is common among older patients.

Klinefelter males often have abnormally long limbs and average about 2–4.5 inches taller than normal males; their hands and feet may be large too. Many Klinefelter males have normal intelligence and function well in society. But *some* exhibit mild retardation, which—together with an increased tendency to emotional and social problems—may account for their clear overrepresentation in mental and penal institutions. "Yet in the typical patient these disturbances are seldom severe enough to lead to conviction or admission to hospital. Rather, the patient tends to lead a quiet, passive life on a low key" (de la Chapelle 1983). Indeed, some 47,XXY males exhibit *no* symptoms of Klinefelter syndrome except infertility, and they may live their entire lives without the slightest inkling that they are in any way unusual.

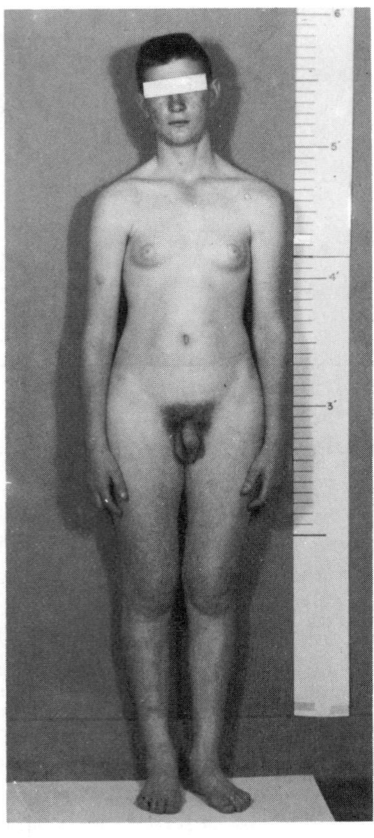

Figure 1 A 47,XXY male with Klinefelter syndrome. Note the breast development and female pattern of pubic hair growth. (Photograph by Earl Plunkett; courtesy of G. H. Valentine [1975].)

KARYOTYPE. In 1959 the Scottish cytogeneticists Patricia Jacobs and J. A. Strong showed that Klinefelter males are usually, but not always, 47,XXY. Some even have more than two X chromosomes and/or more than one Y chromosome, and some are mosaics. Unexpectedly, a few males with Klinefelter syndrome were found to have the (apparently nonmosaic) karyotype of 46,XX.*

INCIDENCE AND ORIGINS. With an overall frequency of 1 in 500—2,000 newborn males, Klinefelter syndrome is not rare in the general population (Borgaonkar 1984; Hassold and Jacobs 1984). In subpopulations of tall men (over 183 cm, or 6 ft), the frequency of Klinefelter males may be as high as 1 in 260. Among patients in mental institutions, it is even greater—about 1 in 100; and roughly the same percentage is found in penal institutions. Perhaps 1 in 20 patients seen in subfertility clinics has Klinefelter syndrome. Because of underdevelopment of secondary sexual characteristics, they also appear relatively frequently among hospital patients. These differences in incidence among various groups of males point up the importance of carefully defining each test group and of not trying to extrapolate results from any one group to the general population.

Within families, Klinefelter syndrome seems to arise randomly, as no increased incidence among close relatives of affected males has been detected. The extra chro-

*The 30% of patients who have clinical signs of Klinefelter syndrome but who are karyotypically normal 46,XY males (i.e., the so-called false Klinefelters) will not be discussed here. Some investigators consider the presence of the chromatin body (see later), which is diagnostic of the extra X chromosome, to be part of the definition of Klinefelters.

mosome in 47,XXY males is thought to arise almost entirely through nondisjunction in a parent. Pedigree studies of the inheritance of the X-linked blood group (*Xg*) suggest that about 67% of the nondisjunction occurs in mothers of XXY patients and about 33% in the fathers. The average age of mothers of XXY males is 32 years, compared with 28 years for mothers of normal males. As with Down syndrome, it appears that the chromosomes in older eggs are more prone to nondisjunction, but the reasons remain unknown. No paternal age effect has been observed for Klinefelter syndrome.

KLINEFELTER VARIANTS. About 10% of Klinefelter males exhibit karyotypes that are different from the usual 47,XXY. These include 46,XX, 48,XXXY, 48,XXYY, 49,XXXXY, and 49,XXXYY. The 46,XX males have testes that are defective in size and structure, but otherwise present a more normal phenotype. But Klinefelter males with single cell lines of 48 or 49 chromosomes seem to have more extreme problems of all sorts, including severe mental retardation. Indeed, most of them are detected through chromosome surveys in mental and penal institutions.

Sometimes classed as a separate (non-Klinefelter) phenotypic group are 49,XXXXY males. All reported cases have been profoundly retarded. Fusion near the elbow of the two bones of the lower arm is common; loose-jointedness, skeletal anomalies of the hips and feet, and curvature of the spine may also occur. Facial features include a broad, flat nose, a large mouth, and wide-set eyes. The ears may be malformed.

How do all these unusual karyotypes arise? The 46,XX Klinefelter males, who differ from 47,XXY males only in being within the normal range for height and intelligence, will be discussed later in this chapter. To account for the karyotypes with 48 or 49 chromosomes, two nondisjunctional events are required. For example, a 48,XXXY zygote could arise in several ways, as shown in Figure 2: (A) a Y-bearing sperm fertilizing an XXX egg; (B) an XY sperm fertilizing an XX egg, or (C) an XXY sperm fertilizing an X-bearing egg. Sometimes the pattern of inheritance of X-linked genes or chromosomal banding patterns or RFLPs (Chapter 15) can distinguish among the various possibilities.

Many different types of *mosaics* collectively account for about 10–15% of Klinefelter males. As with any mosaic, the phenotypes of the patients differ, the difference depending on what proportion of the cells are aberrant and also on how the aberrant cells are distributed among the various tissues. The mosaic karyotype 46,XY/47,XXY is found in about 7–10% of Klinefelter males. Not surprisingly, these males tend to exhibit more normal phenotypes: about 20% have normal-size testes, and about 30% may be fertile. The mosaic 46,XX/47,XXY accounts for about 1% of Klinefelter males, but these cases do not differ from the usual 47,XXY cases. One patient with six different cell lines suffered from multiple abnormalities as well as from profound retardation. Mitotic nondisjunction or chromosome loss, or both, must be invoked to explain the origin of these different mosaics. Perhaps some began as aneuploid zygotes following meiotic nondisjunction in the parents.

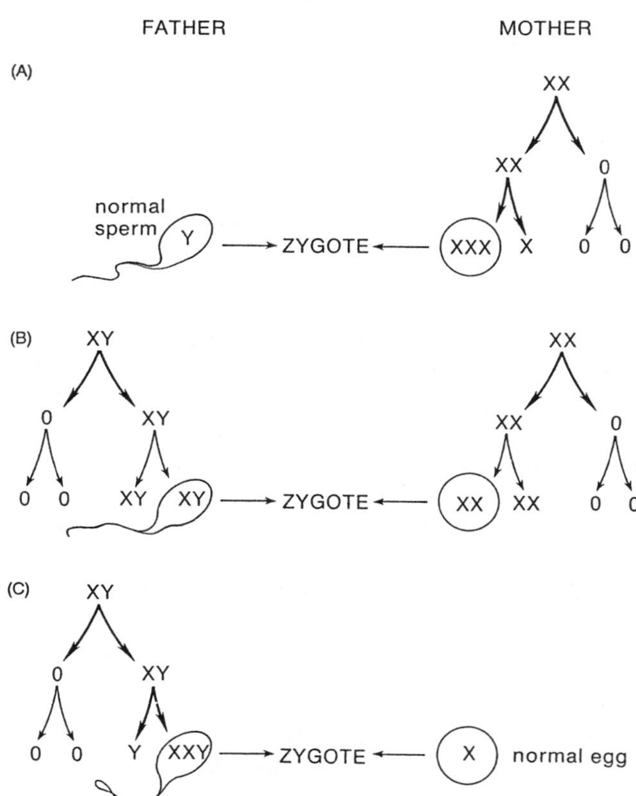

Figure 2 Three ways, each requiring two nondisjunctional events, by which a 48,XXXY zygote could be formed. Nondisjunction is indicated by heavy arrows at (A) first and second meiotic divisions in mother; (B) first divisions in both mother and father; and (C) first and second meiotic divisions in father.

TREATMENT. Surgical removal of breast tissue can relieve some psychological stress. Although the sterility is inalterable, treatment with testosterone does promote development of the sex organs, body hair, musculature, deeper voice, and so on; it may also improve the social adjustment, behavior, and learning ability of some Klinefelter males. Several follow-up studies of children whose sex chromosome abnormalities were discovered at birth (Ratcliffe and Paul 1986; Robinson et al. 1979) also stress the importance of a warm and stimulating family and school environment early in life. Indeed, some researchers suggest that optimizing all these factors "may well eliminate all risk of impaired mental or emotional development" (Nielsen et al. 1979).

The 47,XYY Karyotype

In 1961 Avery Sandberg and some colleagues at the Roswell Park Memorial Institute in Buffalo, New York, discovered the first 47,XYY male by chance—as the father of a Down syndrome child. He was tall and of average intelligence, and had no serious physical problems (Hauschka et al. 1962). Clearly, his chromosomal constitution resulted from paternal nondisjunction during meiosis II, producing a YY sperm. Several other 47,XYY karyotypes, such as the one in Figure 3, were reported later, but they did not begin to attract widespread interest until 1965. Studying the chromosomes of 197 "mentally sub-normal male patients with dangerous, violent or criminal propensities" in an institution in Scotland, Jacobs et al. (1965) found seven males with the 47,XYY karyotype—a stunning frequency of 1 in 28 that was not observed among other groups of males they tested. For example, in 266 randomly selected newborn males and in 209 randomly selected adult males, they found no 47,XYY individuals; in 2,500 other males, they found only one. Their preliminary report also noted that the average height of the seven 47,XYY inmates was 186 cm (6 ft 1 in), compared with a mean height of 170 cm (5 ft 7 in) among the 46,XY males in the same institutional sample. The authors commented:

> the finding that 3.5 per cent of the population we studied were XYY males must represent a marked increase in frequency by comparison with the frequency of such males at birth... At present it is not clear whether the increased frequency of XYY males found in this institution is related to their aggressive behavior or to their mental deficiency or to a combination of these factors.

Other geneticists began testing tall institutionalized males and finding additional 47,XYY males. Unfortunately, not all investigators were careful to define their samples or to study and describe control males from the same populations.* Based sometimes on individual cases found in preselected environments and without much knowledge of how 47,XYY males in the general population behaved, some very extravagant claims were made about the phenotypic effects of an extra Y chromosome. Soon articles linking Y chromosomes to violent or criminal behavior began appearing in the popular press as well as in journals read by psychologists, sociologists, and lawyers. The so-called XYY syndrome was also sensationalized in two 1968 murder trials (in France and Australia), where it was suggested that the defendants' 47,XYY karyotypes rendered them less responsible for their actions. The same year it was widely and *erroneously* reported that Richard Speck, an American mass murderer, was 47,XYY—a misconception that lingers on. Have post-1965 reports altered the early impressions about the phenotype(s) associated with this karyotype?

*One cannot help but find criminals when testing is restricted to prisons, but do these 47,XYY males act differently from the 46,XY males in the same institutions?

(A)

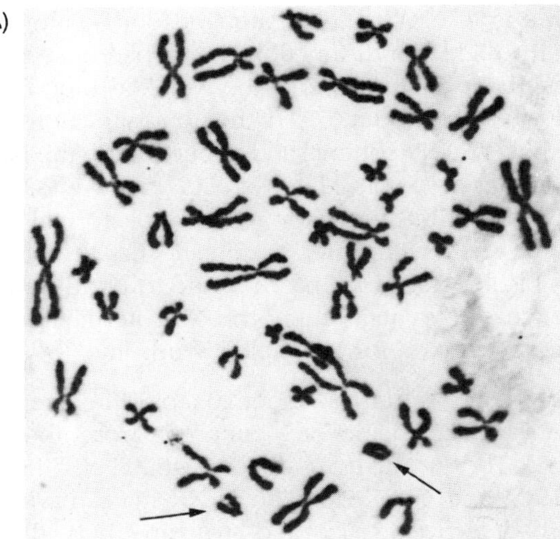

(B)

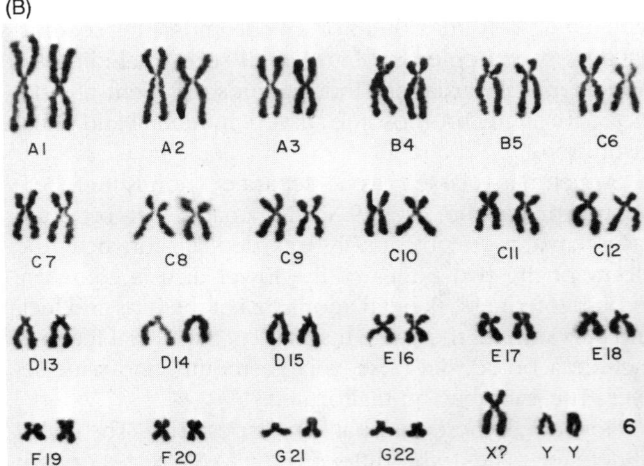

Figure 3 (A) Metaphase of a dividing white blood cell from a 47,XYY male. The two Y chromosomes lie near the bottom of the preparation (see arrows). (B) Karyotype prepared by enlarging the above photo, clipping out the chromosomes, and arranging them in order. The sex chromosomes are at the lower right. One Y chromosome is more condensed than the other. (From Hauschka et al. 1962.)

INCIDENCE. One would think that reliable estimates of 47,XYY frequencies among various groups of males could be estimated by pooling the results of several comparable (but often small) studies. But tabulated reviews by Hook (1973) and Owen (1972) show, for example, that even a simple trait like tallness is not standardized, and criteria for studying other subpopulations are even more difficult to sort out. The same reviews cite highly variable categories of mental patients, inmates in prisons, and inmates in mental–penal institutions. Clearly, any one category of pooled males (tall, mental patients, penal inmates, or mental–penal inmates) contains a great mix of observations.

Thus it is not surprising that estimated frequencies vary widely from study to study. And coupled with the tallying problem is the possibility that different geographic or ethnic groups may exhibit different frequencies of the trait in question. Considering such uncertainties—which are common in many types of research—readers should always keep in mind that estimates of a given trait in a certain population category can vary by several orders of magnitude.

With these caveats, we present some estimates of the 47,XYY frequencies. Among males in the *general population*, the frequency of 47,XYY is on the order of 1 in 1,000 males; among *tall* males in the general population it is perhaps 1 in 325. As samples narrow down to *tall mental patients* or *tall penal inmates* and then to *tall mental–penal inmates*, frequency estimates increase to about 1 in 29. Even taking into account the levels of uncertainty in these estimates, it is clear that the frequency of 47,XYY among tall males and among institutionalized males is much higher than in the general population.

WHY ARE XYY MALES OVERREPRESENTED IN INSTITUTIONALIZED POPULATIONS? Are they born more often into lower socioeconomic groups, where adverse environmental conditions may enhance behavioral problems and where lack of access to legal services prevents them from "getting off easy" in encounters with police and the courts? Are they less intelligent and thus less able to earn an honest living or to elude the authorities when they commit crimes? Are they more aggressive or violent than comparable 46,XY males?

As noted, many of the early claims about XYY males were based on small numbers of cases found in highly selected populations without adequate control studies of 46,XY males in the same selected populations. Or controls, when used, were not always well matched. The **method of ascertainment**, that is, how the cases were discovered, is an important but sometimes overlooked factor in such investigations (Chapter 12). Most early studies had another serious defect: the investigators who worked up psychological profiles on these men often knew what their karyotypes were, a condition that could have led to an unconscious bias in the evaluations. The only way to eliminate this possibility is to conduct **double-blind investigations**, in which those who do the karyotyping know nothing of the subjects' histories and those who compile the histories know nothing about their subjects' karyotypes. Only after both sets of data are independently and fully completed should they be matched up to see how the behavioral profiles of 47,XYY males compare with those of 46,XY males.

One attempt to conduct a large-scale study that would avoid many of these pitfalls is reported by Witkin et al. (1976), who made use of extensive records kept by the Danish government on every 26-year-old male reporting to his draft board. These data included the results of a physical exam and intelligence test as well as educational history, social class, and criminal convictions. For economic reasons, this search for XYY males was restricted to 4,139 males born in Copenhagen between 1944 and 1947, whose height was at least 184 cm (6 ft 1/2 in)—the latter representing the top 16% of the height distribution. Among these tall males, Witkin et al. found *12 XYY karyotypes*, a frequency of 1 in 345.* This sample left over 4,000 tall, karyotypically normal 46,XY males as the control group for the study.

How did the 12 tall 47,XYY males compare with their tall 46,XY counterparts? The main results are as follows: There was no significant difference in mean parental socioeconomic status, but the mean height of the XYY males (191 cm or 6 ft 3 in) was significantly greater than that of the controls (187 cm or 6 ft 2 in). Five of the XYY males (= 42%) had criminal records, compared with only 9% of the controls—but their crimes (petty theft, burglary, arson) were *not violent*, and their sentences were light. The mean educational level of the XYY males and their test scores on the army intelligence test were significantly lower than that of tall control males. Furthermore, all five 47,XYY males with criminal records had test scores well below the control mean, and one was retarded. But even among the control males, those with criminal records had a significantly lower mean educational level and lower test scores than those without criminal records. *This finding suggests that tall males with lower intellectual functioning are more likely to be convicted of crimes, irrespective of their karyotypes.*

In summary, the Danish study verified early reports of increased height and increased frequency of criminal activity for 47,XYY males, but there was no evidence that they are more violent or aggressive than control males. Rather, it appears that their elevated crime rate is at least partly explained by their decreased mental abilities. Regarding personality traits, there are no predictable patterns either: "the few differences which do appear tend to refute the notion that XYY males are predisposed toward aggressive and violent behavior. Indeed, in security mental hospitals and prisons, the XYY subjects

*Also identified were 16 47,XXY males and a few 46,XY males with other chromosomal aberrations.

generally show a better adjustment than do XY controls" (Borgaonkar and Shah 1974).

PROSPECTIVE STUDIES. Rather than doing retrospective studies on adult males, why not study the development of XYY males from birth onward? This has not been done to any extent, for these reasons: (1) Long-term prospective studies are extremely expensive and time-consuming. (2) As a result of all the adverse publicity surrounding the 47,XYY karyotype, it became impossible to undertake large-scale prospective studies. Indeed, one such project was forced to a halt (Roblin 1975) by opponents who feared that identifying XYY males at birth would cause them to be treated differently by their family and to develop behavioral problems as a self-fulfilling prophecy. But researchers who try to avoid this situation by not revealing babies' karyotypes may be charged with denying a patient's right to be informed. Opponents also argue that, because there is no known treatment for the XYY karyotype, studying these individuals serves no useful purpose.

Proponents had hoped that carefully designed and controlled studies could be done without jeopardizing the well-being of XYY males, if parents were told that the karyotype does not necessarily lead to deviant behavior. Indeed, most XYY males will become ordinary citizens, indistinguishable from the range of phenotypes that characterize 46,XY males. Yet proponents felt that if behavioral problems did turn up, early identification and counseling might be of help to XYY males and their parents. Finally, contrary to popular opinion, there is no reason why so-called genetic conditions should be untreatable—but, of course, treatment is impossible without prior identification and study.

Actually, several dozen 47,XYY males have been followed from birth through adolescence (Ratcliffe and Paul 1986; Stewart 1982). Some researchers are enthusiastic about the positive effects of these studies on the subjects and their parents (Nielsen et al. 1979), whereas others have expressed some misgivings (Valentine 1979).

Y CHROMOSOME ABSENT

The following karyotypes are much less common than those described above, and present none of the controversial social problems that have attended the study of some males with sex chromosome anomalies.

Turner Syndrome: 45,X Females

In 1938 an American endocrinologist, Henry Turner, described seven females, ages 15 to 23, who lacked breasts and other secondary sexual characteristics (except for some body hair), failed to menstruate, and were sterile. Although the body form was reasonably well proportioned, they were very short (adult height under 150 cm or 4 ft 11 in) because of generalized growth retardation. Their necks were short and webbed, and their forearms showed greater than normal angling away from the body when the palms faced forward (Figure 4). A few years later, females with **Turner syndrome** were found to have (instead of ovaries) primitive *streak gonads* that lack both germs cells* and hormone-producing tissue—a deficiency that explains the failure of sexual development. Their oviducts, uterus, and vagina (though anatomically normal) remain small and immature, and the ex-

*Due to degeneration of previously existing follicle cells. This degeneration occurs even in normal females: "When the fetal ovaries differentiate, they contain about 7 million oocytes. Through an unknown process, these oocytes begin immediately to disappear; by birth there are about 3 million left, by the menarche about 400,000, and the menopause occurs when under 10,000 remain" (Federman 1987).

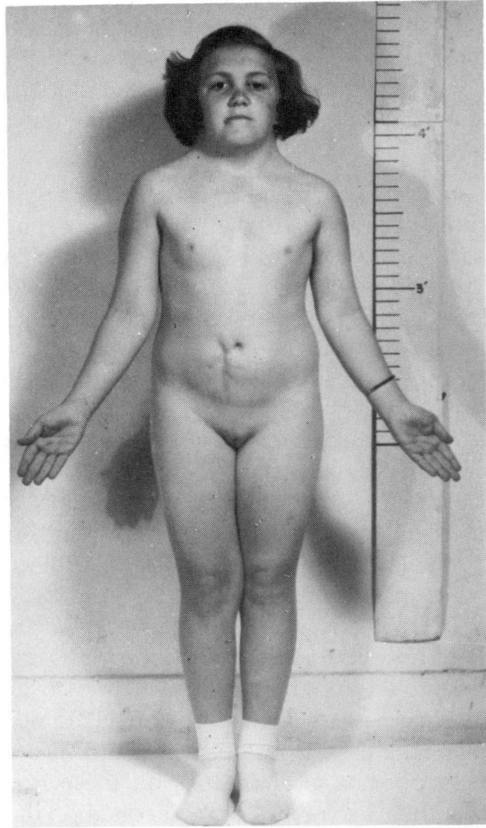

Figure 4 A 45,X female with Turner syndrome. This 14-year-old girl is 142 cm (4 ft 8 in) tall. Note the lack of sexual development, the webbed neck, broad chest with wide-spaced nipples, the old-looking face, and bent forearms. (Photograph by Earl Plunkett; courtesy of G. H. Valentine [1975].)

ternal genitals are also infantile. Up to 10% menstruate and ovulate, however, and in rare cases have given birth.

Other defects appearing in at least 25% of these females include widely spaced nipples on a broad chest, short ring fingers, poorly developed nails, many small pigmented moles on the neck and torso, a narrowed aorta (large vessel that carries blood away from the heart), horseshoe-shaped kidneys, and double ureters. Additional skeletal deformities are sometimes seen: a high arched palate, receding chin, mismatching of the upper and lower teeth, low-set ears, and abnormally low bone density. The hairline at the back of the neck may be low, and there may be an increased tendency to scarring after injury. In contrast to the childlike body form, the face often looks old. Indeed, premature aging sometimes occurs, and life expectancy may be reduced from abnormalities of the heart and other organs.

Turner females are often identified at birth by the presence of characteristic skin folds on the back of the neck, by swelling of the hands and feet, by abnormally large fingerprint patterns, and by low birth weight. But many Turner females show no phenotypic abnormalities even as adults, except for below-average height and infertility.

KARYOTYPE. The first report of a 45,X Turner female came in 1959 (Ford et al. 1959a). *Nondisjunction* in the father accounts for about 75% of all live-born cases, and there is no maternal or paternal age effect. *Chromosome loss* during early cleavage of the zygote may account for another 10% of Turner females; among the many types of mosaics that have been found, the 45,X/46,XX and 45,X/46,XY karyotypes are by far the most common. Phenotypes of mosaics range from fully affected to normal, depending on the proportion and distribution of the different cell lines. In a few cases, only *part* of an X chromosome is missing: those lacking one short arm (Xp) express the usual Turner phenotypes, but those missing one long arm (Xq) are taller and may appear normal except for the absence of sexual development.

BEHAVIOR. Although some Turner females may show a slight to moderate decrease in IQ (as low as 70), most are completely normal. It appears, however, that some Turner females have trouble with a certain type of space perception. Their general behavior is normal during childhood.* But failure to undergo puberty, coupled with small size, makes it difficult for them to keep up with their peers socially during the teenage years. It is often reported that these females tend to be timid, unambitious, impressionable, and generally immature—but this could be due to their smallness and lack of sexual development more than to their specific karyotype.

*Indeed, Money and Ehrhardt (1972) report that 15 young Turner females they studied were in some respects even more "feminine" than control females, an observation suggesting that the development of what is considered to be female behavior requires neither ovaries nor a second X chromosome.

FREQUENCY. The usual problems of determining frequencies are magnified here because the syndrome is rarer than those described above. The numbers from individual studies vary widely, but an overall frequency of about 1 in 2,500–10,000 live-born females is the usual estimate (Borgaonkar 1984; Simpson 1982).

Conversely, however, the 45,X karyotype is highly lethal in embryos—being the *most common karyotype among spontaneous abortions*, accounting for about 20% of all chromosomally abnormal aborted embryos. Indeed, it appears that at least 98% of all 45,X zygotes are lost during the first three months of pregnancy (Epstein 1986). But, for unknown reasons, the few that survive are not so severely deformed—especially with regard to mental capacities—as are the other types of live-born chromosomal aneuploidies.

TREATMENT. In younger patients, androgen treatment, although controversial, may lead to a modest increase in height. Later estrogen treatment, although it may further inhibit skeletal growth, does promote the development of breasts and other secondary sex characteristics. This, plus corrective surgery if necessary, may alleviate some social problems and also make normal sexual relations possible. Several Turner females are known to have married and become good mothers to adoptive (or, very rarely, their own) children.

47,XXX and 48,XXXX Females

Females with the **47,XXX karyotype**, first reported by Jacobs et al. (1959a), present no distinctive phenotype aside from a tendency to be tall and thin—and many of them seem to be completely normal. Their frequency among newborns in the general population is roughly 1 in 1,000–2,000 (de la Chapelle 1983; Hassold and Jacobs 1984). Some triple-X females do show a tendency to mental retardation and/or psychosis, however, because they are detected more frequently among institutionalized females (1 in 225–425) than among newborn females. Prospective studies (Ratcliffe and Paul 1986) have also detected some educational and behavioral problems among 47,XXX subjects. As with other chromosomal anomalies, mosaics are also observed.

Although some XXX females have menstrual difficulties, many menstruate regularly and are fertile. At least a dozen have delivered a total of over 30 children, most of whom were karyotypically and phenotypically normal. This outcome is surprising, because one would expect that half the eggs from a 47,XXX female would be XX, resulting in progeny that are 47,XXX or 47,XXY. Perhaps, during meiosis in XXX females, the XX nuclei are more likely to be shunted into polar bodies; aberrant segregation of this type is known to occur in other organisms.*

Studies of the inheritance and expression of X-linked genes suggest that the origin of the extra X chromosome in 47,XXX females is usually nondisjunction in meiosis I of the mother; a maternal age effect has also been noted. On trisomies, Hassold and Jacobs (1984) comment:

> While much has been learned about the origin, frequency, and clinical consequences of trisomy, virtually nothing is known about the mechanisms leading to trisomy in man . . . While studies of maternal age imply that there must be several mechanisms resulting in trisomy, we are totally ignorant of their nature. We do not even know whether the primary event in trisomy involves the chromosome itself, the spindle, or some other cellular organelle.

Other Poly-X Karyotypes

Females with more than three X chromosomes are extremely rare, perhaps fewer than 1 in 10,000 live births. Only about 30 cases of the 48,XXXX karyotype are known (Nielsen et al. 1977); although normal in sexual development, these females are severely retarded and also exhibit a wide range of physical abnormalities. Even fewer 49,XXXXX females have been found in populations of mental defectives, and none in the general population. But E. H. R. Ford (1973) sounds a note of caution:

> there is a general impression from recorded cases that the greater the number of sex chromosomes, the greater the likelihood (or the degree) of mental subnormality and of developmental anomalies generally. But it must always be remembered that . . . most cases are ascertained because they are abnormal, and that individuals who are completely or relatively normal are unlikely to be discovered. The rarer the condition, the more likely is this to be so.

SEX CHROMATIN AND THE LYON HYPOTHESIS

The lethality of an aneuploidy involving any but the smallest human autosome, and the serious mental and physical defects caused by overdosage of even that chromosome, attest to the importance of maintaining a very close balance among genes and chromosomes. But as we have just seen, aneuploidy of the much larger X chromosome is much less severe than that of the smallest autosome. And even more puzzling to consider is the *normal* chromosomal makeup of human females (who have two Xs), compared with that of males (who have one X and a small, nearly gene-less, Y chromosome). Because the X is fairly large (ranking eighth in relative length and containing about 6% of the haploid genome) and because females have twice as many X-linked alleles as males do, should not one or the other sex suffer from a relative excess or deficiency of X-linked gene products? By what mechanism(s) do the two sexes' cells compensate for this twofold difference in X chromosome dosage?

In mammalian somatic cells, *hemizygosity of the X chromosome* turns out to be the normal state, so there must be a way of silencing one X chromosome in females. We now know that the silencing mechanism involves the **sex chromatin**, whose discovery is a good example of serendipity in science— accidentally finding something new and different while looking for something else. During the 1940s, while studying cytological changes that occur in the nerve cells of cats, the Canadian scientists Murray L. Barr and Ewart G. Bertram noted differences in the position of a darkly staining mass in the nuclei of interphase cells (Figure 5A). This nuclear mass had been observed routinely in the cells of many mammals by many researchers—but its function was unknown.

Barr and Bertram noted that the spot was missing

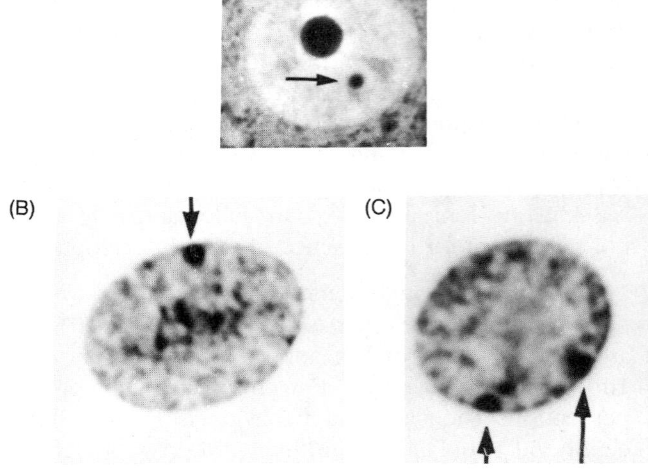

Figure 5 Sex chromatin bodies (indicated by arrows) in nuclei of female cells. (A) The light-colored nucleus in this nerve cell from a marten contains one sex chromatin body that, in this photograph, lies between the nucleolus and the nuclear envelope (×2,000). (B) Nucleus in a buccal smear from a 46,XX female. One sex chromatin body lies against the nuclear envelope (×3,000). (C) Nucleus in a buccal smear from a 47,XXX female. Two sex chromatin bodies are visible (×3,000). (A from Moore 1966. B and C courtesy of Murray L. Barr, University of Western Ontario.)

*A similar deficiency of aneuploid offspring has also been noted from 47,XYY fathers, who thus far have failed to produce any 47, XXY or 47,XYY sons.

from several of their tissue preparations and found that the only trait to show a consistent association with the nuclear mass was sex: *interphase cells from females exhibited dark-staining spots in their nuclei, but cells from males did not*. Examinations of human nerve cells showed the same sex difference. Barr and Bertram (1949) correctly speculated that each **Barr body** consists of one tightly condensed (heterochromatic) X chromosome—and we now know that it replicates later in the S phase of the cell cycle than do the other chromosomes.

Because of its correlation with sex, the spot was named the **sex chromatin**, although the term *Barr body* is also used. It can be seen in many types of somatic cells, usually lying against the inner surface of the nuclear membrane; it is particularly prominent among primates and carnivores. Cells with one or more Barr bodies are said to be **chromatin-positive**, while those lacking any are **chromatin-negative**.

Detection of Sex Chromatin in Humans

The observability of sex chromatin in an individual is *not* an all-or-none situation, because some fraction of cells from females may not show it (although they are presumed to carry it). The simplest and most commonly used method of checking for sex chromatin is the **buccal smear**: cells are gently scraped from inside the cheek, spread on a glass slide, stained, and examined with a light microscope. Usually 20–70% of the buccal cells from normal females contain a Barr body. For unknown reasons, only about 3% of newborn females' cells show it, but within a few days the percentage increases to the usual state. In female embryos it is first observed around the sixteenth day of development.

In 1959 it was reported that cells from Klinefelter (47,XXY) males are chromatin-positive, whereas those from Turner (46,X) females are chromatin-negative (Jacobs and Strong 1959; Ford et al. 1959). These exceptions to the "rule" that the sex chromatin is found only in females alerted cytogeneticists to the possibility— later verified by chromosome counts—that these two syndromes involved sex chromosome abnormalities. Further investigations showed that more than one sex chromatin occurs in persons with additional X chromosomes, but not in persons with additional Y chromosomes. The more precise rule states that *the number of Barr bodies is one less than the number of X chromosomes* in otherwise diploid organisms, as follows:

Number of Barr bodies	Karyotype(s)
0	45,X; 46,XY; 47,XYY
1	46,XX; 47,XXY; 48,XXYY
2	47,XXX; 48,XXXY; 49,XXXYY
3	48,XXXX; 49,XXXXY
4	49,XXXXX

The Inactive-X Hypothesis

What happens to the chromosome that forms the Barr body? Are its genes still functional? Is the same X chromosome (i.e., maternal *or* paternal) condensed in every cell? What determines whether or not a given chromosome will be condensed? Based primarily on studies with mice, the English geneticist Mary Lyon (1961) proposed the **inactive-X hypothesis**:

1. The condensed X chromosome found in female mammalian cells is genetically inactive.
2. Inactivation first occurs very early in embryonic development, during the blastocyst stage, at which time either the maternal or the paternal X chromosome is equally likely to be inactivated.*
3. This inactivation event occurs independently and randomly in each blastocyst cell. But once the decision is made for a given cell, the same X chromosome will be inactivated in all its descendant cells—thereby producing a clone from each of the original cells.

The net effect of this whole process—called **lyonization**—is to equalize the phenotypes in males and females. This phenomenon is known as **dosage compensation**. Although normal females possess twice as many X chromosomes as males, they generally have the same number of active X-linked genes.

Evidence for lyonization was sought in females *heterozygous* for X-linked genes; these females should have clones of cells in which one allele or the other is active, but never cells with two active alleles. If entire clones remain in place during development, good-sized phenotypic "patches" will result, each patch expressing only one X-linked allele of a gene. If cells from different clones intermingle during development, however, then an intermediate or finely variegated "salt-and-pepper" phenotype will be observed. One line of evidence given by Lyon involved the mottling or dappling on the fur of female mice heterozygous for a number of X-linked coat color genes. This effect is also seen in the familiar calico or tortoise-shell cats (Figure 6), whose patches of black or orange hair indicate which X-linked gene is active in each heterozygous cell. With rare exceptions (see Question 11), calico cats are always female.

*Actually, there are a few additional twists: during the early to middle blastocyst stages of mammalian embryonic development, the outer cell layer of the female blastula (i.e., the future extraembryonic membranes) develop Barr bodies that are *all paternal* in origin. Only in the late blastocyst stage do X chromosomes of all cells of the inner cell mass (i.e., the future embryo) become randomly inactivated. Also, in most adult cells of female *marsupials* (kangaroos, etc.) there is preferential inactivation of the paternal X.

The next requirement was to show that the genetically inactive allele in a given cell is carried on the late-replicating X chromosome that condenses to form the sex chromatin. A nice demonstration of this involved fibroblast cells from female mules (Rattazzi and Cohen 1972). The two X chromosomes of a female mule, one coming from a horse and one from a donkey, are cytologically distinguishable. The X-linked gene investigated was the one controlling production of the enzyme *G6PD (glucose-6-phosphate dehydrogenase)*. This enzyme, which occurs in nearly all plants and animals, acts in glucose metabolism. Different organisms make slightly different forms of the enzyme, and in this case the G6PD produced by the horse X chromosome could easily be distinguished (electrophoretically) from that variant made by the donkey X chromosome. Rattazzi and Cohen found that cells containing a late-replicating (i.e., inactive) *horse* chromosome produced the donkey-type G6PD enzyme exclusively and cells with a late-replicating *donkey* chromosome produced the horse-type enzyme exclusively.

Figure 6 Many autosomal genes collectively affect the color and pattern (i.e., banded versus solid) of the individual hairs on a cat. Here we only consider one X-linked locus and its role in forming the light and dark patches of fur on a calico cat. This female is heterozygous for a dominant allele (*O*) that leads to orange (or yellow) hairs, and a wildtype recessive allele (*o*) that when hemizygous leads to black (or dark speckled) hairs. In any cell, either one allele or the other is active; the inactive allele is on the X chromosome forming the Barr body. The size of each patch of fur depends on how many cells descended from each cell present at the time of the random inactivation decision.

X Chromosome Inactivation and Reactivation

The G6PD trait was also used to test the applicability of the Lyon hypothesis to humans. It was known that human 46,XX females with *two* doses of the normal X-linked G6PD allele produced no more of the enzyme in their cells than did 46,XY males with *one* dose of the normal allele, although one might expect the former to produce twice as much. Furthermore, 47,XXY Klinefelter males and 47,XXX females produce no more G6PD enzyme than do normal males—a fact suggesting that, regardless of how many X chromosomes are present in a given cell, only one X is active.

Even more to the point were these observations: (1) A recessive allele (*gd*) of this gene produces a defective enzyme with a much lower level of activity than that of the normal allele (*Gd*). When skin cells from females heterozygous for the normal and mutant G6PD alleles (i.e., *Gd/gd*) were individually cloned and then tested for enzyme activity, some clones expresses the normal allele (*Gd*) and others expressed the mutant (*gd*)—even though all clones were genotypically identical (*Gd/gd*) (Davidson et al. 1963). (2) Individual red blood cells from heterozygous females also show high or low activity but never intermediate activity of G6PD, as is found in heterozygotes for enzymes produced by autosomal genes. (3) Similar types of studies with other X-linked enzyme-producing genes yielded similar results. The inescapable conclusion is that, *in a given XX mammalian cell, only one of the two X-linked alleles remains active*.

Because humans lack fur, we cannot observe calico people. Perhaps the next best trait, however, is phenotypic expression of the recessive X-linked trait *anhidrotic ectodermal dysplasia* (Freier-Maia and Pinheiro 1984). Some affected males were described by Charles Darwin as the "toothless men of Scinde"; they lack sweat glands, teeth, and most of their body hair. Studies of females heterozygous for this gene reveal considerable variability in phenotypic expression—in the form of irregular skin patches with few or no sweat glands (Figure 7) and some regions of the jaw with missing or defective teeth. Note also in Figure 7 that the two pairs of identical twins in generations I and III show differing patterns of mosaicism. Many other human X-linked genes are also known to show mosaic expression in heterozygous females.

POSTULATED STEPS IN X CHROMOSOME INACTIVATION. For many decades cell biologists have puzzled over the behavior, structure, and evolution of sex chromosomes. Although full understanding still eludes them, researchers hope that the latest molecular genetic techniques will help solve these mysteries. Interest in X chromosome inactivation is particularly keen, because the biochemical scenarios involved in this process might be the same or similar to those that control the selective inactivation of autosomal genes. By analyzing inactivation in whole Xs as well as in various pieces of X chromosomes, some attached by translocation to autosomal pieces (Chapter 9), they have investigated several key steps (Lyon 1988; Gartler and Riggs 1983; Grant and Chapman 1988).

Initiation of X inactivation occurs at a control center located in the long arm (Xq), near the centromere. Some workers think that there are one or two additional control centers near the tip(s) of Xp and/or Xq. But nobody knows what happens at the control center(s) and what prevents *all* the X chromosomes in a cell from being inactivated. There must be some **counting mechanism** that preserves just one active X per diploid autosomal set; thus a few models propose systems of feedback inhibition by autosomally produced repressor proteins.

The third step is a virtually error-free **spreading** of inactivation to over 100 million base pairs of DNA; this chromatin condensation is perhaps akin to some kind of crystallization.* Once accomplished, it is remarkably stable, a condition implying the need for a final step of

*When autosomal genes are translocated to pieces of X chromosome that become inactivated, "the inactive state may spread into the autosome, may fail to do so, or may 'skip' proximal autosomal regions while spreading into more distal ones" (Goldman et al. 1987). These workers describe a large chicken autosomal gene that, when inserted into a mouse X chromosome, always escapes inactivation.

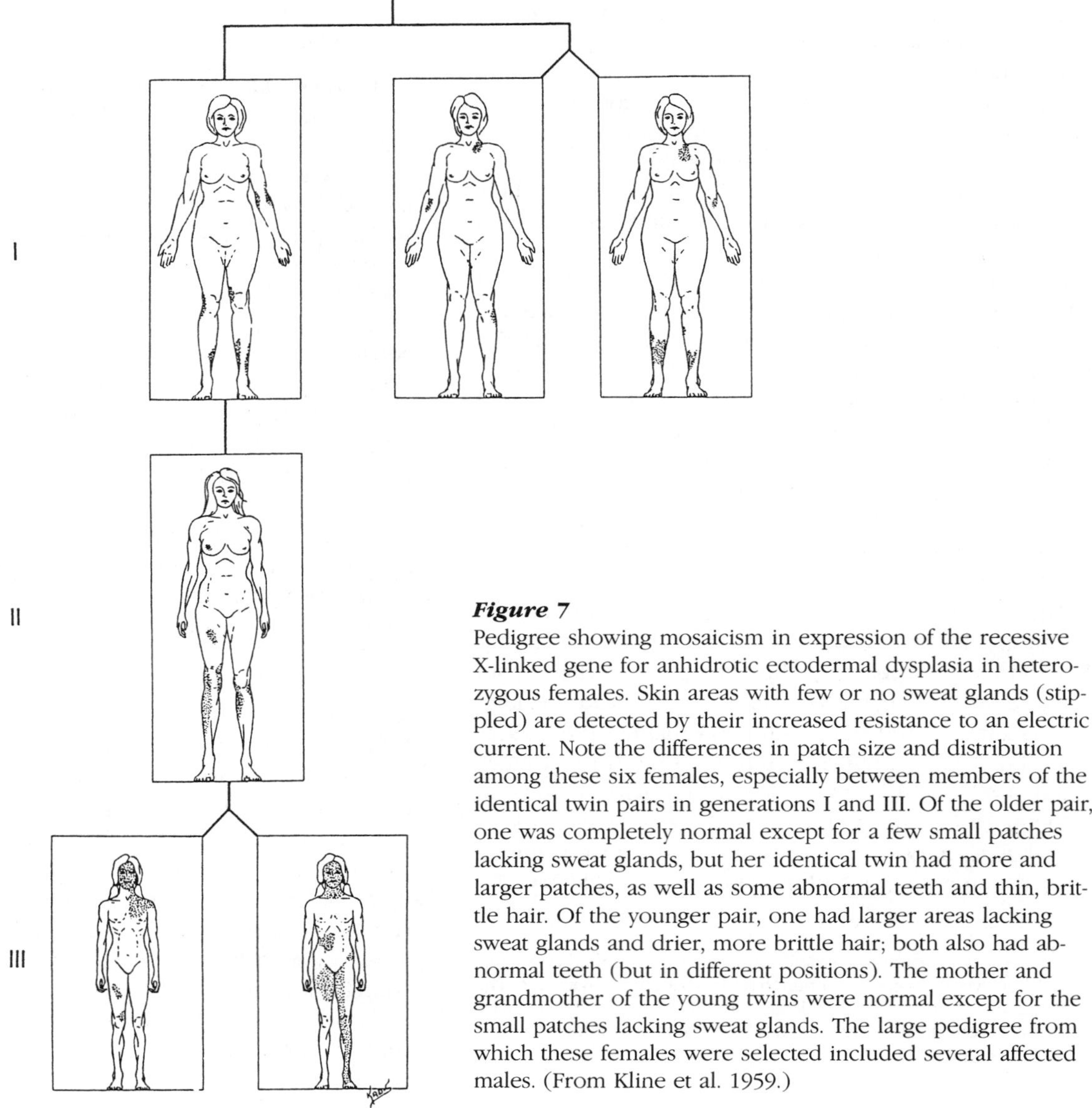

Figure 7
Pedigree showing mosaicism in expression of the recessive X-linked gene for anhidrotic ectodermal dysplasia in heterozygous females. Skin areas with few or no sweat glands (stippled) are detected by their increased resistance to an electric current. Note the differences in patch size and distribution among these six females, especially between members of the identical twin pairs in generations I and III. Of the older pair, one was completely normal except for a few small patches lacking sweat glands, but her identical twin had more and larger patches, as well as some abnormal teeth and thin, brittle hair. Of the younger pair, one had larger areas lacking sweat glands and drier, more brittle hair; both also had abnormal teeth (but in different positions). The mother and grandmother of the young twins were normal except for the small patches lacking sweat glands. The large pedigree from which these females were selected included several affected males. (From Kline et al. 1959.)

fixation and maintenance. There is strong evidence that the methylation of cytosine bases could provide this stability (Gartler et al. 1985), although not all studies agree (Wolf and Migeon 1982).

IS THE ENTIRE HUMAN X CHROMOSOME INACTIVATED? If all but one X chromosome is completely inactivated in every individual, then why do Turner females differ from normal females, and why do Klinefelter males differ from normal males? To explain these abnormal phenotypes, it was suggested that some X-linked genes must always escape inactivation. These genes would be expressed doubly in normal females and singly in normal males, but singly in Turner females and doubly in Klinefelter males. We now know that this is indeed the case: a few X-linked genes do *not* exhibit mosaicism in heterozygous females. Yet these genes do not appear to account for the phenotypic features of Turner syndrome.

One example is the *Xg* blood group gene, which is expressed as follows. Males of the genotype Xg^a/(Y) and females homozygous for this gene, have the Xg^a antigen on their red blood cells. Males of the genotype *Xg*/(Y), and homozygous *Xg*/*Xg* females, lack this antigen. But rather than showing mosaicism for this gene, Xg^a/*Xg* female heterozygotes express the Xg^a antigen on *all* their red blood cells—and all attempts to find cells without the antigen have failed.

Further studies have verified that the *Xg* gene is not inactivated. One (Buckton et al. 1971) involved a woman who inherited a normal X chromosome carrying Xg^a from her father but got from her mother an *Xg*-bearing X chromosome whose short arm was broken off and attached to an autosome. In this heterozygous woman the normal (paternal) X chromosome was late replicating (i.e., formed the sex chromatin), but she nevertheless expressed its Xg^a allele. Thus that part of the inactivated paternal X chromosome that carried Xg^a must have remained active.

The *Xg* locus resides near the tip of the short arm of the human X chromosome. Also located close to the tip of Xp and *not inactivated* in human females is the steroid sulfatase (*STS*) gene, mutant forms of which cause ichthyosis* (scaling of the skin) in males (Shapiro 1983, 1985; Yen et al. 1987). Lying even closer to the tip is a locus called *MIC2*, which codes for a cell surface antigen (Chapter 18). Closest to the tip of Xp lie several "marker" loci, that is, anonymous DNA sequences that generate restriction fragment length polymorphisms (RFLPs). The latter loci are defined, not by any particular phenotype or protein product, but by the presence (or absence) of special sites cut by restriction enzymes (Chapter 15).

*This X-linked condition is different from the autosomal dominant, Lambert-type "porcupine" phenotype described in Chapter 3. The frequency of affected STS males is about 1 in 6,000 males. Females heterozygous for the mutant STS allele show no scaling but often have dry skin.

X CHROMOSOMES IN GERM CELLS. The rules of X inactivation in somatic cells are reversed when it comes to germ cells: primary oocytes of females have *no* Barr bodies. In female fetuses, presumptive germ cells do undergo lyonization along with somatic cells. But as they develop into oocytes and enter meiosis, their inactivated X chromosomes become *reactivated*. In males, conversely, the X chromosome undergoes precocious condensation in all primary spermatocytes just as they enter meiosis—but this process is poorly understood and may not be true lyonization. The reasons for these switches are not clear. Perhaps female germ cells need two active X chromosomes for meiotic pairing and separation, whereas male meiosis might proceed normally only in the absence of an active X chromosome. This requirement might explain the infertility of Turner females (who lack a second X chromosome) and Klinefelter males (if one or two of their X chromosomes remain active in spermatocytes).

AGE-RELATED REACTIVATION OF X-LINKED GENES. Although the inactivation of X-linked genes is extremely stable, there is now good evidence that some random **gene reactivation** occurs as mammals age. Studies of mouse chromosomes have been particularly informative. When a piece of autosome is stuck into the middle of an X chromosome, the autosomal genes are usually inactivated right along with the flanking X-linked genes. One study (Cattanach 1974) involved female mice heterozygous for a wild-type tyrosinase gene whose recessive allele caused albinism. Ordinarily these autosomal genes do not undergo inactivation, but in this case the wild-type tyrosinase allele was stuck in the X chromosome. Thus the fur was mottled with colored and albino patches, similar to that of calico cats. "Cattanach made the remarkable observation that as animals aged the tyrosinase gene is often reactivated, so that animals become more pigmented with age (the reverse of one well-known effect of ageing on hair pigmentation)" (Holliday 1987).

Welsh researchers Kathryn Wareham et al. (1987) did some very clever experiments with mice, using a mutant X-linked gene called *OCT* and a special hybrid (X–autosome) chromosome called Searle's translocation. With this complex system, the stained liver cells in which the normal (inactivated) *OCT* allele had become reactivated would show up distinctly as a dark cell (or patch) on a light background. When mice of various ages were sacrificed and their livers were sliced, stained and examined, some striking differences were noted. Livers from young mice showed few dark-staining patches; but those from adult mice showed a 5-fold increase in the percentage of dark-staining area, and old mice showed a 50-

fold increase over that seen in young mice. The authors state: "We have ... clearly shown that in mice with Searle's translocation there is a progressive age-related reactivation of the *OCT* locus on the inactivated normal X chromosome and we believe that this may reflect a general mechanism of ageing."

These various changes that take place in certain cells also need to be explained: How does X chromosome reactivation occur in oocytes? How are paternal X chromosomes "imprinted" so that they can be selectively inactivated in certain extraembryonic cells in mammals, or in most embryonic cells in marsupials? What causes gene reactivation in aging cells? There are still many more questions than answers to the whole problem of chromosome or gene inactivation and reactivation.

STRUCTURE OF THE MAMMALIAN X AND Y CHROMOSOMES

In the 1930s some cytologists proposed that the short arms of mammalian X and Y chromosomes shared a small region of homology, within which there was pairing and obligatory crossing over during male meioses. In males this homologous region would be present in *duplicate* (rather than singly, as with regular X-linked loci), and any genes within it would be inherited as though they were autosomal—hence the term **pseudoautosomal region**. Recently this concept was verified—along the way explaining some unsolved cases of 46,XX males and 46,XY females.

The Pseudoautosomal Region of the X and Y Chromosomes

So far, the only known loci in the human pseudoautosomal region are *MIC2* and the RFLP markers distal to it (Figure 8). During male meiosis, *one chiasma always forms* in this tiny pairing region, resulting in a regular exchange of material between the very tips of the paired X and Y chromosomes (Goodfellow et al. 1986; Rouyer et al. 1986). Normal pairing—but *without* regular crossing over between the X and Y—also extends into the nonpseudoautosomal region nearby, past *TDF* (testis determining factor) and almost to the centromere of the Y chromosome.

XX Males, XY Females, and the Testis Determining Factor

Very rarely an "illegitimate" crossover occurs in this latter region—resulting in the *transfer of the TDF locus to the X chromosome and the loss of the TDF locus from the Y chromosome* in that particular spermatocyte. If the resulting sperm fertilizes a normal egg, a **46,XX male** or a **46,XY female** will be formed (Figure 8B; de la Chapelle 1986a). The frequency of 46,XX males is about 1 in 20,000 male births.*

Only with the new ability to identify RFLPs has it become possible for geneticists to distinguish so clearly between the two chromosomes of a homologous pair or (in this case) between the pairing regions of an XY duo. And with the ability to analyze the structures of X and Y chromosomes in various groups of vertebrates have come new insights into the evolution of mammalian sex chromosomes. Some comparisons among pseudoautosomal and noninactivated regions of the X chromosome in three subgroups of mammals (monotremes, marsupials, and placental mammals) suggest, for example, that the dissimilar XY pairs in placental mammals might have arisen gradually (by loss of material in the Y) from a fully homologous pair in a common ancestor (Graves 1987).

Additional clues come from the recent isolation of a candidate for the *TDF* gene and its protein product. An international team of researchers (Page et al. 1987b) did this by analyzing the DNA from several dozen 46,XX males and 46,XY females—each of whom was missing part (and/or had an extra bit) of a sex chromosome.† The two most extreme and useful cases were a 46,XY female who lacked only 0.2% of the Y chromosome and a 46,XX male who had only a tiny piece (0.5%) of a Y chromosome attached to one of his two Xs. But the tiny deletion in the former, and the extra piece in the latter, were both located in the same small region of the Y chromosome—that is, where *TDF* lies—thus clinching its key role in sex determination. The *TDF* product encoded by this gene—a so-called **zinc-finger protein** —has an unusual shape of a type "known to bind to genes and regulate the amount of RNA, and thus protein, that they make" (Roberts 1987).

A totally unexpected and still unexplained finding by Page and his colleagues, however, was the additional discovery of a very similar locus on the comparable region of the X chromosome! Homologous X-linked *TDF* loci were also found in all mammals tested and in other vertebrate species, even birds. The role of this highly conserved *TDF* gene in sex determination is unknown.

*Anneren et al. (1987) report on a 49,XXX male that resulted from such paternal X–Y interchange, with the X-bearing sperm fertilizing an XX egg that was the product of nondisjunction in the mother. "If maternal X–X nondisjunction occurs in approximately one meiosis in 1,000 ... and XX males have an incidence of approximately one in 20,000 newborn males ... then the coincidence of these events, resulting in an XXX male, might be expected approximately once in 20 million male births."

†Sex-reversed (46,XX) males are sterile and show some features of Klinefelter syndrome, but they may look quite normal. The 46,XY females are also sterile and often show abnormal breast development and/or fail to menstruate.

Does it function at all? Does it operate in conjunction with the "regular" *TDF* gene on the Y? In opposition? Independently? Also, will this serendipitous discovery of an X-linked *TDF* lead to a new theory of sex determination? Nobody knows for sure.

Equally surprising was the discovery that the candidate *TDF* gene, now called *ZFY* (zinc finger on the Y) by many investigators, has homologous sequences in marsupials —but on *autosomal* DNA rather than on the sex chromosomes! "This implies *ZFY* is not the primary sex-determining gene in marsupials. Either the genetic pathways of sex determination in marsupials and eutherians differ, or they are identical and *ZFY* is not the primary signal in human sex determination" (Sinclair et al. 1988). Autosomal sequences homologous to *ZFY* have also been found in reptiles and birds (Bull et al. 1988). But, as Hodgkin (1988) points out, sex determination mechanisms often differ extensively even within a single taxonomic group, so perhaps we should not be discouraged by the complexities of the TDF story. Instead, such results draw attention to *ZFX* (the homologous zinc finger sequences on the X) "as a possible major player in the process of sex determination."

THE FRAGILE X SYNDROME

Decades ago it was noted that males outnumber females by about 25% among the mentally retarded at institu-

Figure 8 (A) Structure of the human X (only half shown) and Y chromosomes. The **X–Y pairing region** comprises the distal quarter of Xp and most of Yp. The tiny **pseudoautosomal regions** (black) on Xp and Yp carry homologous *MIC2* alleles and RFLP markers; during male meioses, one obligatory crossover occurs in this region. In the nonpseudoautosomal (hatched) region, rare "illegitimate" crossovers can also occur, however. The hatched regions carry different genes, except for the recently discovered *X-linked TDF* gene, whose function is a mystery. We omit it from this diagram, but it probably lies between the *Xg* and *STS* loci, which are never inactivated. Because they carry different genes, the hatched areas plus the rest of the X and Y (clear) make up the **strictly sex-linked regions** of these chromosomes. (B) Diagram of an illegitimate crossover (line with arrows) in the (hatched) X–Y pairing region, proximal to the *STS* and *TDF* loci. The recombinant chromosome gives rise to an **XX male** (lacking *Xg* and/or *STS* but carrying *TDF* on one X) or an **XY female** (carrying *Xg* and/or *STS* but lacking *TDF* on the X). Analyses of the segregation of these loci, as well as RFLP loci and/or deletions also carried in the pairing region, have verified that this does indeed occur (Page 1986; Page et al, 1987a; Yen et al. 1987). (C) Abnormal crossover between the nonpseudoautosomal X–Y pairing region of the Y, proximal to *TDF*, and the pseudoautosomal region of the X, distal to *MIC2*. The resulting sperm with an oversized hybrid X chromosome (carrying *TDF* plus *two* doses of *MIC2*) fertilized an X-bearing egg, producing an **XX male** with three doses of *MIC2* (Rouyer et al. 1987).

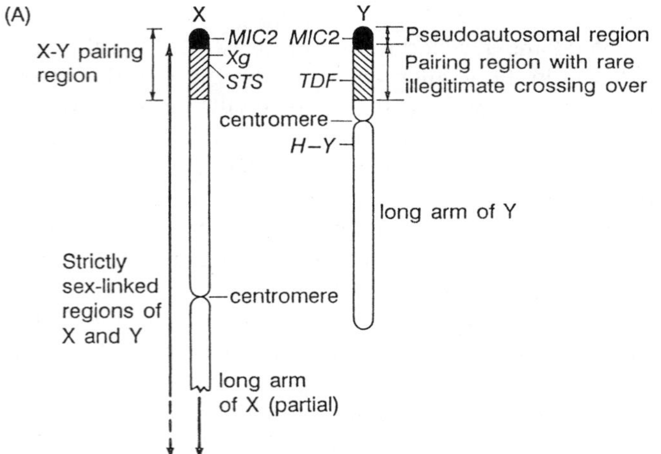

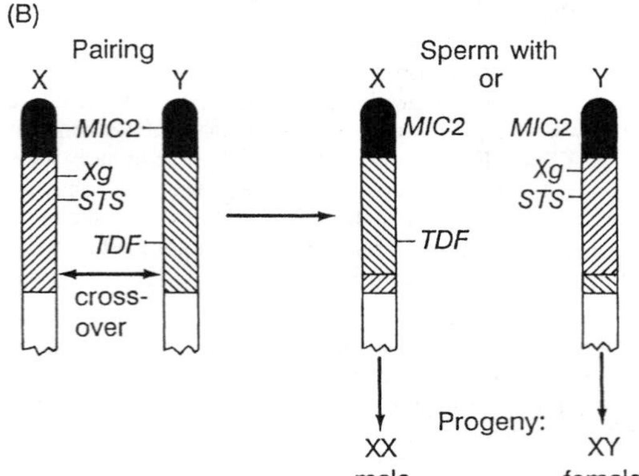

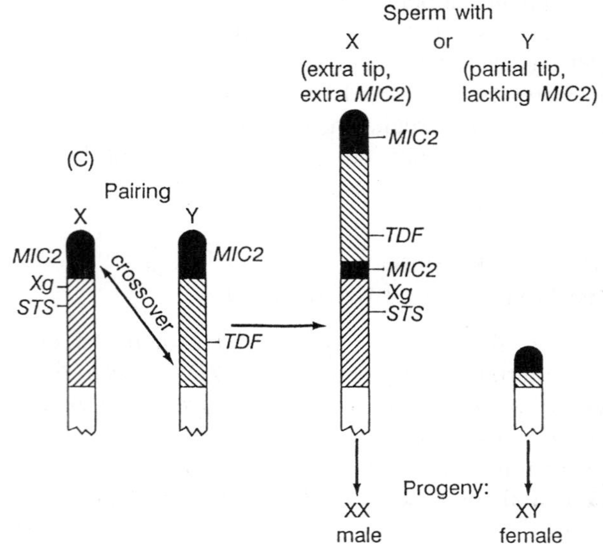

tions. Starting in 1943 with a report by the English geneticists J. P. Martin and Julia Bell, several workers suggested that at least part of the excess could be due to X-linked recessive inheritance, and they published pedigrees that supported this idea. Not much attention was paid, even when Lubs (1969) reported that a cytological abnormality—a secondary constriction—was present on the long arm of the X chromosome of four retarded males in three generations of a family. Because it was associated with chromosome breakage, the constriction was called a **fragile site**, and the related X-linked mental retardation was dubbed **fragile X syndrome**. By the late 1970s this syndrome was recognized as the most frequent X-linked disorder, indeed, the most common form of inherited mental retardation and second only to Down syndrome as a chromosomal cause of retardation. Why and how the fragile site is related to retardation remains unknown, however; about 20 different heritable fragile sites exist on other chromosomes, but none of these are associated with a particular phenotype.

About 50 X-linked recessive disorders are now known to contribute to mental retardation. Most are rare and will not be discussed here, but taken together they affect perhaps 1 in 600 males. Roughly one-fourth of these cases, however, are thought to result from the fragile X syndrome.

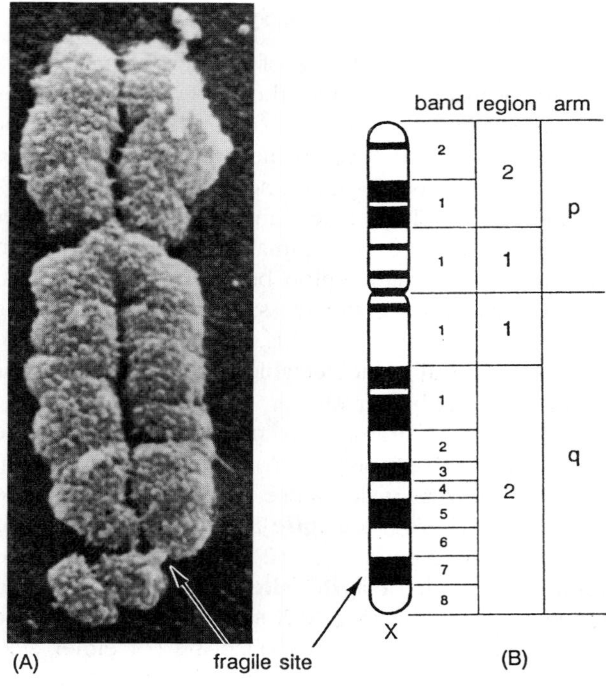

Figure 9 (A) The fragile X chromosome. This scanning electron micrograph shows a G-banded fragile X metaphase chromosome. The constricted centromere region clearly separates the short arm (above) from the long arm (below). The fragile site on each chromatid appears as a gap or pinched-in region near the tip of the long arm, at region 2, band 7 (arrows). (B) Diagram of banding patterns on the human X chromosome, showing the metaphase Q- and G-positive bands seen in a karyotype of about 550 bands total. (Micrograph from Harrison et al. 1983.)

Fragile X Syndrome in Males and Females

Roughly 1 in 2,000 males exhibits a fragile site in band q27.3 of the X chromosome (Figure 9), in up to 50% of his tested cells. (Recall from Chapter 4 that the detection of fragile sites is rather tricky and very sensitive to the amount of folic acid in the culture medium.) Most of these males are moderately to severely retarded; but some are only mildly retarded, and a few males with this fragile site may show normal intelligence (Turner and Jacobs 1983).*

Almost all affected males have large testes (after puberty, 2 to 3.5 times normal volume) and big, usually protruding, ears. A prominent jaw and large head are common, and stubby hands, lax joints, and a heart defect (mitral valve prolapse) may occur, too. Affected males tend to be tall as children but short as adults. Their speech is usually delayed in development and often high-pitched and repetitive in nature. Although some may be hyperactive or autistic as children, their behavior as teenagers and adults tends to be shy and nonassertive, but friendly. Few if any affected males are known to have produced offspring. As Sutherland and Hecht (1985) point out, "Many physical features associated with the fragile X are baffling . . . Certain hypotheses such as overgrowth and connective tissue dysplasia may serve to make us think that we understand the fragile X physical phenotype. We do not. On a basic level, we know very little."

The frequency of mentally affected females is roughly 1 in 2,400, and the percentage of their cells in which the fragile site can be seen (about 30%) is lower than that in affected males. These individuals, who are thought to represent about one-third of all carrier females,* show mild retardation or some kind of learning disability; but beyond this they exhibit no consistent physical abnormalities. There is some evidence that the degree of mental defect may be correlated with the percentage of cells in which the fragile X chromosome is active—the more severely affected females having a much greater proportion of their normal X chromosomes inactivated (Uchida and Joyce 1982).

*Conversely, there are occasional males who appear to have the clinical signs of X-linked retardation, but lack the fragile site.

*Homozygotes are rarely if ever found.

Genetics of the Fragile X Syndrome

The fragile site is near the tip of the long arm, part of a gene cluster that also includes the G6PD locus, the factor VIII hemophilia gene, and the color blindness genes. Nearby (closer to the centromere) is a second gene cluster that contains another hemophilia gene (factor IX), the HPRT gene (Chapter 16), and several RFLP marker sites. Although there is minimal crossing over within each cluster, the small region between them seems to show greater than average crossing over (Mandel et al. 1986). Some workers think this is due to the presence of a microscopically undetectable DNA amplification or rearrangement (Chapter 9).

Something very strange is going on with regard to the expression of this disorder across generations. Geneticists have tried to make sense of it by assuming the following: (1) There is a **fragile X gene (*FRAX*)** located in or very close to band Xq27. (2) This gene is expressed cytologically as the **fragile site** at band Xq27.3, and phenotypically as the **fragile X syndrome**. (3) Because carrier females produce 50% offspring (of either sex) who possess the gene and 50% who do not, segregation of *FRAX* is normal. (4) But, unlike all other X-linked genes, *FRAX* shows a fairly high degree (approximately 20%) of nonexpression (nonpenetrance) in males—who should, of course, express all their hemizygous genes.

With this latter assumption, the plot thickens (Sherman et al. 1985). Severely affected males and females, although fertile, rarely produce offspring because they rarely mate.* Rather, as shown in Figure 10, the inheritance of fragile X syndrome mainly involves normal carrier females and so-called **transmitter males**. The latter possess the *FRAX* gene but do not express it: they are phenotypically normal and do *not* show the fragile X site in any of their cells. Consider such a male (generation II, Figure 10). His brothers who possess the *FRAX* gene also tend to be transmitters (41/50 = 82%) rather than affected (9/50 = 18%). Likewise among his heterozygous sisters with *FRAX*, 90% (= 45/50) are normal and 10% (= 5/50) are affected. These unaffected carrier sisters of transmitter males show the fragile site in only 1–2% of their cells. But they produce a much higher proportion of affected sons (37–40%) and daughters (16–17%) than do their carrier mothers, who presumably had the same genotype.

Likewise, the *heterozygous daughters of transmitters* (who all inherit the same *FRAX* gene) show no retardation and few or no fragile sites—but among *their* sons who have the *FRAX* gene, four times as many (74–80%) are affected and only one-third as many (20–26%) are transmitters compared with the numbers of affected and transmitter offspring of their carrier grandmothers. And among the heterozygous daughters of the generation II and III carrier females, about 3.5 times as many (32–34%) are mentally subnormal, and significantly fewer (66–68%) are unaffected, compared with the offspring from their carrier grandmothers. Thus, within a given family's pedigree, *affected offspring tend to cluster in certain generations* (Sherman et al. 1985). The key differences are summarized as follows:

	Percentage of mother's offspring expressing *FRAX*	
Type of mother	**Among *FRAX*-bearing sons**	**Among *FRAX*-bearing daughters**
carrier mother of transmitter ♂	18%	10%
carrier daughter of transmitter ♂	74%	34%

GENETIC MODELS. Any genetic model has to explain both the occurrence of transmitter males and the above-described "Sherman paradox." Several schemes have been proposed. Currently favored models invoke a two-step process. The first step occurs before or during passage of the *FRAX*-bearing X through the *mothers* of transmitter males, and the second step occurs in the *daughters and/or sisters* of transmitter males. One such model (Nussbaum and Ledbetter 1986) proposes that the first step, a **premutation**, is some sort of *amplification (overreproduction) of the DNA in the* FRAX *region*; this change leads to the second step, abnormal pairing and unequal crossing over (Chapters 9 and 18) between the *FRAX* region and its normal homologue, in those carrier females who inherit the premutated *FRAX* chromosome.

Another two-step model (Laird 1987; Laird et al. 1987) focuses on the *process of X-chromosome reactivation in female oocytes*. First, a new mutation is hypothesized to occur at band Xq27 of an X chromosome during oogenesis in a generation I (carrier) female. This *FRAX* mutation is capable of acting (in future generations) as a local block to X chromosome reactivation in developing oocytes in a carrier female's ovary. Eggs that contain the mutated chromosome are inherited by unaffected generation II carrier daughters, or by transmitter males who then pass them on to all their (generation III) daughters. The postulated second (**imprinting**) step occurs in the oocytes of these unaffected carrier daughters and granddaughters, when a small region of Xq distal to the *FRAX* mutation is prevented from being reactivated. Sons who inherit this partly inactivated, imprinted X will express the fragile site in some of their cells and also—because of permanent inactivation (i.e., nonexpression) of certain X-linked genes—will be mentally retarded. Daughters

*Among the offspring of affected females (who are usually less severely retarded), 50% of the sons and 28% of the daughters are affected.

who inherit this imprinted X will range in phenotype from normal (if a majority of their somatic cells have the fragile X chromosome lyonized) to mildly retarded (if a majority of their somatic cells have the normal chromosome inactivated).

PRENATAL DIAGNOSIS. Because only about 50% of all known carriers show the fragile site, it is very difficult to predict which sisters of an affected male are at risk for producing affected children. Prenatal testing is not yet very reliable, but it can be done after withdrawing some blood from the fetus (a tricky procedure) or some amniotic cells from the fluid-filled sac surrounding the fetus. In some cases the fragile X will be detected. Even among these positives, however, the predictive value varies: "While fragile X males will usually be retarded, there is no way of distinguishing fragile X females who will be retarded from those who will be normal" (Sutherland and Hecht 1985). Furthermore, testing by amniocentesis has been plagued with *false negatives*, that is, the fragile X is not detected in fetuses who are later found to be affected. Because of these and other problems, counseling fragile X families is particularly complex and difficult.

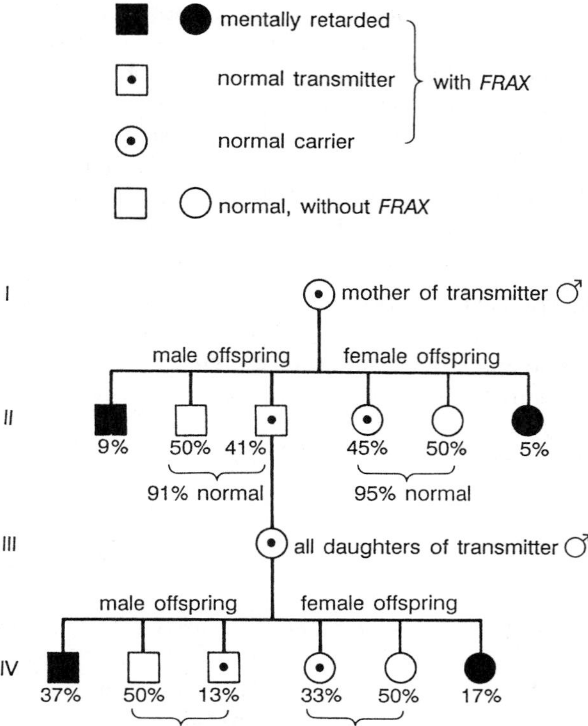

Figure 10 X-linked mental retardation inheritance patterns. Compare the frequencies of retardation in generation II with the frequencies of retardation in generation IV. The former are offspring of the generation I mother of the transmitter males, and the latter are offspring of the generation III daughters of transmitter males. These generation I and generation III carrier females should have the same fragile X genotypes—yet the percentages of affected offspring are 3.5–4 times greater from the latter than from the former. *Note:* Generation II carrier females (i.e., sisters of the transmitter male) have the same frequencies of affected offspring as do their generation III carrier nieces. (Adapted from Brown et al. 1986.)

SUMMARY

1. In humans the presence of a Y chromosome determines maleness, irrespective of how many X chromosomes are present. The absence of a Y chromosome determines femaleness.

2. Males with a 47,XXY karyotype have Klinefelter syndrome, characterized by small testes, infertility, and underdeveloped secondary sexual characteristics. They also tend to be tall and may exhibit mild mental retardation and/or some breast development.

3. Males with the XYY karyotype tend to be tall, and some may show below-average intelligence; but they have no other distinctive traits. Misconceptions about 47,XYY males arose partly because data from small and selective studies were prematurely generalized. Very little is known about 47,XYY males in the general population.

4. Females with the 45,X karyotype have Turner syndrome. They are short, have primitive streak gonads, and lack secondary sexual characteristics. They usually are of normal intelligence.

5. Females with the 47,XXX karyotype show no distinctive phenotype but may have slightly lowered intelligence. The 47,XXX karyotype is expected in half of their children, but this frequency is not observed.

6. The sex chromatin, or Barr body, is a small, darkly staining mass present in interphase somatic cells of female mammals. It consists of a late-replicating X chromosome. The number of Barr bodies seen in a given cell is one less than the number of X chromosomes in that cell.

7. Lyon's inactive-X hypothesis states that each Barr

body is an inactivated X chromosome, and every somatic cell has only one active X chromosome. The initial inactivation (during the late blastocyst stage) may affect either X chromosome—but once this selection is made, all descendants of a given cell have the same X chromosome(s) inactive. Thus females heterozygous for X-linked genes are phenotypic mosaics, and the dosage of active X-linked genes in females equals that in males.

8. The Lyon hypothesis has been verified for a number of genes on the X chromosome, and X chromosome inactivation is thought to proceed by methylation of cytosine bases in the DNA.

9. But some genes that reside near the tip of the Xp are not inactivated, and a few at the very tip reside in a "pseudoautosomal" region that (during male meiosis) regularly pairs and crosses over with a short homologous region of the Y chromosome.

10. Regular pairing between the X and Y extends more proximally, however, through a region that includes the *TDF* (testis-determining factor) locus on the Y and some noninactivated loci on the X. Rare illegitimate crossing over in this segment can give rise to "sex-reversed" 46,XX males and 46,XY females.

11. A fragile site on the X chromosome is associated with mental retardation (fragile X syndrome) in various proportions of male and female carriers. Its mode of inheritance and phenotypic expression are not well understood.

FURTHER READING

On aneuploidy in general, see Dellarco et al. (1985), Bond and Chandley (1983), De Grouchy and Turleau (1984), Epstein (1988, 1986), and Hassold and Jacobs (1984). On sex chromosome anomalies, see Roberts (1987), de la Chapelle et al. (1986), Ratcliffe and Paul (1986), Sandberg (1983b, 1985b), San Roman and McDermott (1984), Simpson (1982), Bandmann and Breit (1984), de la Chapelle (1983), and Summitt (1981). On lyonization, see Lyon (1968, 1988), Grant and Chapman (1988), Gartler and Riggs (1983), Gartler et al. (1985), and Graves (1987). On the pseudoautosomal region, see Craig and Tolley (1986), Page (1986), Rouyer et al. (1986), and Shapiro (1985). On the fragile X syndrome, see Sutherland and Hecht (1985), Nussbaum and Ledbetter (1986), Brown et al. (1986), Turner and Jacobs (1983), Laird (1987), Laird et al. (1987), Lubs (1983), and Hagerman and McBogg (1983). On the mammalian Y chromosome, see Goodfellow et al. (1987).

QUESTIONS

1. Polani et al. (1956, 1958) reported that 4 out of 20 sex chromatin-negative females showed red–green color vision defects, yet none of the 55 chromatin-positive males they tested were affected. Usually color blindness is expressed much more frequently in males than in females. What is the genetic explanation for these observations?

2. Green-shifted color perception has been noted in both Turner females and Klinefelter males having both mothers and fathers with normal color vision. Does this give any information on the nondisjunctional events that occurred in the parents? Assume no mutation. Explain.

3. Turpin et al. (1961) described a pair of twins who had identical blood groups and could successfully exchange skin grafts (another criterion for genotypic identity). One was a normal male, but the other was a female with Turner syndrome. How can this be explained?

4. Diagram all types of nondisjunction of the sex chromosomes that can occur in a 46,XX female and a 46,XY male during meiosis I alone, meiosis II alone, or in both meiosis I and II. Assume that no more than one pair of centromeres fails to disjoin in any one division.

5. Referring to the diagrams for Question 4, list the ways by which a 47,XXY zygote can be formed from karyotypically normal parents. What is the *least* number of nondisjunctional events needed for each?

6. List all the ways by which a 47,XYY zygote could be formed from karyotypically normal parents. Which is/are most likely to occur?

7. List all types of sperm that can be formed by a 47,XYY male. Assume that the three chromosomes pair as a trio and then separate as 1 + 2, and that there is no secondary nondisjunction. (To distin-

guish between the two Y chromosomes, call them Y_1 and Y_2.)

8. The most common type of mosaicism found in Klinefelter males is XXY/XY. What two types of postzygotic events could account for this?

9. Court-Brown (1967) points out that "the clinical features of chromatin-positive males show wide variations between subjects." As shown by the following data, "males identified at subfertility clinics are on the whole more masculine than, for example, those identified at an endocrinology clinic."

Sample	Poor growth of facial hair	Breast development	Feminine pubic hair
Subfertile males	32.2%	25.8%	32.2%
Endocrinology patients	87.5%	62.5%	54.2%

How can the relationship between phenotype of these Klinefelter males and their method of ascertainment be explained?

10. How many Barr bodies would you expect to find in cells from the following types of individuals: (a) karyotypically normal female; (b) karyotypically normal male; (c) female with Turner syndrome; (d) 46,XY female with pure gonadal dysgenesis; (e) 46,XY female with androgen insensitivity syndrome; (f) 46,XX male pseudohermaphrodite; (g) 46,XY female pseudohermaphrodite; (h) 46,XX sex-reversed male?

11. Although calico cats are usually female, a few rare males have been reported. What could be the genetic explanation for this?

12. Why are phenotypic differences between members

of identical female twin pairs likely to be greater than those between identical male twin pairs?

13. Linder and Gartler (1965) studied certain benign uterine tumors up to several centimeters in diameter, removed from women heterozygous *A/B* at the *G6PD* locus. In a given individual, some tumors consisted of all A cells and others of all B cells, but none were mixtures of A and B cells. On the other hand, pieces of adjacent uterine tissue as small as 1 mm in diameter contained both A cells and B cells. What can be concluded about the origin of these uterine tumors?

14. One way to estimate the embryonic stage at which X chromosome inactivation occurs is to determine what fraction of females genetically heterozygous for X-linked genes show *no* phenotypic mosaicism. Consider, for example, a group of *A/B* mouse embryos. If random inactivation first occurred at the 2-cell stage, then within a single embryo the two cells and their descendants could be phenotypically A and A, or A and B, or B and A, or B and B. Note that two of the four possible combinations lead to *absence* of mosaicism; thus, on the average, 1/2 of the resulting mice would be phenotypically all-A or all-B, and 1/2 would be mosaic. (a) Make a tree diagram to show all possible combinations of cell phenotypes that could result if X-inactivation occurred at the 4-cell stage. What proportion of mice would be expected to be nonmosaic? (b) In general terms, what is the relationship between the proportion of nonmosaic individuals and the stage at which inactivation occurs?

ANSWERS

1. Turner females will express color blindness whenever the gene occurs on their single X chromosome, as is true for 46,XY males. Klinefelter males, however, need two doses of the mutant gene to be color blind, as is the situation with 46,XX females. Thus the frequencies of X-linked genes in Turner females and Klinefelter males should parallel the frequencies usually found, respectively, among males and females in the general population.

2. The allele for color blindness must have been present in the heterozygous mother (*G/g*), and the father was hemizygous normal, *G*/(Y). The Turner females must have resulted from fertilization of a *g*-bearing egg by a sperm lacking a sex chromosome; nondisjunction in the father could have occurred at meiosis I or II. The Klinefelter males must have resulted from fertilization of a *g/g* egg by a Y-bearing sperm, with nondisjunction in the mother occurring at meiosis II.

3. The twins were derived from a single 46,XY zygote. If the presumptive embryo split in half at the 2-cell stage and one of the two cells lost a Y chromosome, then one twin would be a normal male and the other a Turner female.

4. Nondisjunction in females in division(s):

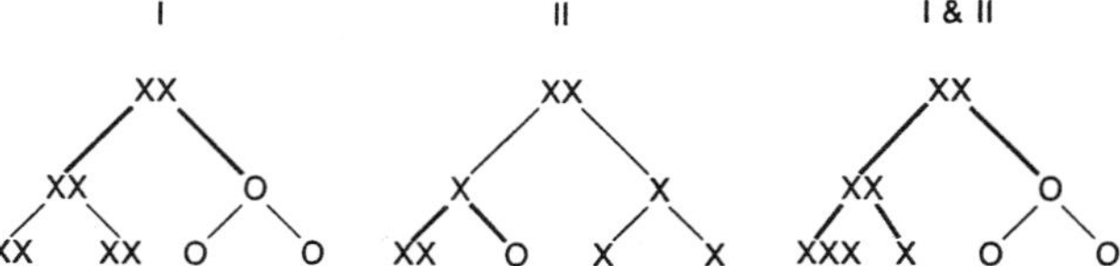

Nondisjunction in males in division(s):

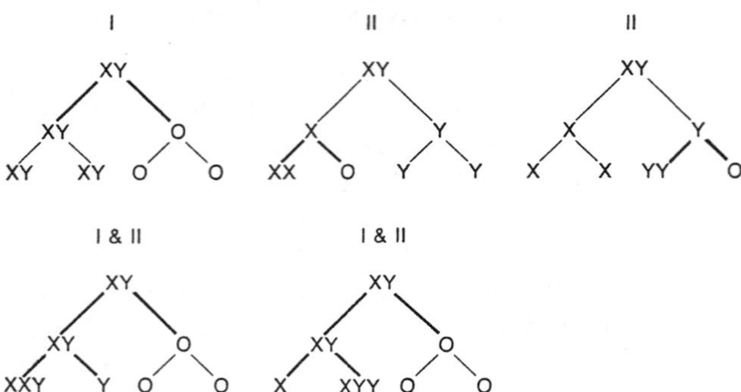

5. Of the eight possible ways, three require only one nondisjunctional event.

Egg	Sperm	Total	Nondisjunctional events: In mother	Nondisjunctional events: In father
XX	Y	1	meiosis I	
XX	Y	1	meiosis II	
XX	Y	3	meiosis I	meiosis I & II
XX	Y	3	meiosis II	meiosis I & II
X	XY	1		meiosis I
X	XY	3	meiosis I & II	meiosis I
O	XXY	3	meiosis I	meiosis I & II
O	XXY	3	meiosis II	meiosis I & II

6. Of the four possible ways, one requires only one nondisjunctional event.

Egg	Sperm	Total	Nondisjunctional events: In mother	Nondisjunctional events: In father
X	YY	1		meiosis II
X	YY	3	meiosis I & II	meiosis II
O	XYY	3	meiosis I	meiosis I & II
O	XYY	3	meiosis II	meiosis I & II

7. Six types of sperm: XY_1; Y_2; XY_2; Y_1; Y_1Y_2; X.

8. Mitotic loss of one X chromosome from a cell that was originally XXY, or nondisjunction of an X chromosome during mitotic division of an XY cell (in any division after the first cleavage) to give an XXY cell + a Y cell, of which the latter dies. These two possibilities are diagramed as follows:

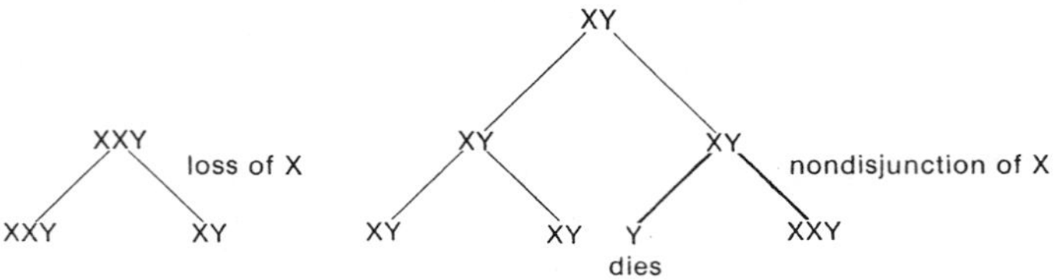

9. Endocrinology clinics would attract those Klinefelter males who seek treatment for failure to develop secondary sexual characteristics, that is, those who are more seriously affected. Subfertility clinics would attract married Klinefelter males (that is, those who are masculine enough to find mates but later discover they are infertile).

10. (a) 1; (b) 0; (c) 0; (d) 0; (e) 0; (f) 1; (g) 0; (h) 1.

11. Calico males have two or more X chromosomes (Klinefelter cats?) rather than the normal XY karyotype.

12. As illustrated in Figure 7, genotypically identical twin females may differ in the phenotypic expression of heterozygous X-linked genes. Because of the random inactivation of X chromosomes, it is unlikely that mosaic patches in twin females would be identical in size and distribution. Identical twin 46,XY males, however, must express all the same genes on their single X chromosome.

13. Each tumor must have arisen from a single cell (either *A* or *B*) rather than from a group of cells.

14. (a) Two out of 16 combinations, or 1/8, are expected to be nonmosaic.

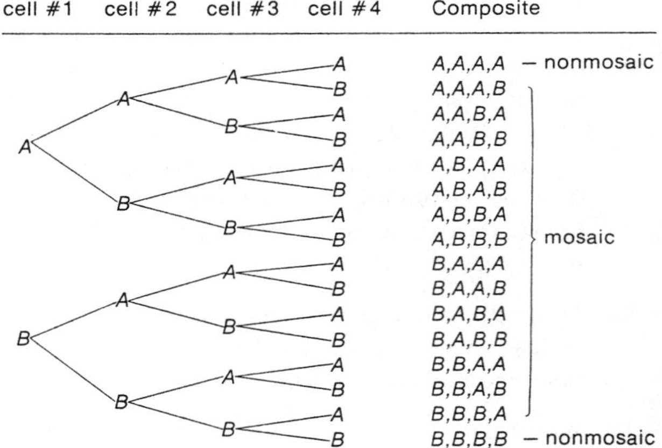

(b) The later the stage at which inactivation occurs, the smaller the expected fraction of nonmosaic individuals. Inactivation at the 8-cell stage, for example, should by chance yield $(1/2)^8 = 1/256$ all-*A* embryos and $(1/2)^8 = 1/256$ all-*B* embryos, or a total of 1/128 nonmosaic individuals. The formula for the proportion of nonmosaics is $2(1/2)^x$ or $(1/2)^{x-1}$, where x is the number of cells present at the time of inactivation. This expression assumes that all the cell types are equally viable throughout the course of development, which may not always be the case.

Chapter 9 Other Chromosomal Abnormalities

Deletions and Duplications
Origins and Examples of Deletions
Origins and Examples of Duplications

Translocations
Origins of Translocations
Translocations in Meiosis
Translocation Down Syndrome
Other Translocations

Inversions
Behavior of Inversions during Meiosis
Inversions in Humans

Other Abnormalities
Ring Chromosomes
Polyploidy

Frequencies of Chromosomal Abnormalities
Gametes
Prenatal Tests
Abortions, Stillbirths and Neonatal Deaths
Unselected Newborns and Some Adult Groups
Conclusions on Fetal Wastage

Chromosomal Abnormalities and Gene Mapping
Variant Chromosomes and Inversions
Deletions, Duplications, and Translocations

Of those of us who are born alive 1 in every 100, certainly 1 in 200, has a visible and major chromosome anomaly. . . [Down syndrome], the commonest chromosome disorder among the living, is at least twice as common as the much publicized cystic fibrosis and maybe nearly 20 times more common than muscular dystrophy or phenylketonuria (PKU). It causes untold personal tragedy and incalculable public cost. The chromosome disorders account for perhaps 20,000 new cases of mental retardation and physical defect each year in the United States alone.

G. H. Valentine (1975)

Thus far we have discussed abnormalities in chromosome number only. In this chapter we describe what happens when *parts* of chromosomes are lost, gained, or moved to new positions in the genome. Some of these changes occur when chromosomes break but then fail to reattach properly. Instead of rejoining to each other, the two ends from one break may attach to the ends from *other* breaks. Thus, depending on the number and positions of breaks, many kinds of rearrangements can be formed.

Such changes and their causes have been extensively analyzed in fruit flies, corn, mice, and other experimental organisms. It is known that chromosome breaks can be induced by radiation, by a wide variety of chemicals, and by certain viruses. But detailed analyses of structural abnormalities in humans have become possible only recently, as technical advances allowed investigators to identify both the finer banding patterns and the molecular structure of human chromosomes. These developments have led to the discovery of many new chromosomal syndromes. Also, through some spectacularly detailed analyses of rare chromosomal aberrations, geneticists have been able to find the chromosomal locations of several important human genes that had eluded them for many years.

But however they arise and whatever they may contribute to genetic analyses, these aneuploidies* inflict tremendous pain and suffering on human carriers and their families (Hamerton 1984). Virtually all of the chromosomal syndromes involve mental retardation and growth defects; seizures are very common, as are unusually low birth weight and failure to thrive. Heart and other vascular defects as well as abnormalities in the sex organs are frequently observed, and patients often exhibit unusual arrays of fingerprint patterns. Different syndromes are recognized by different combinations of these abnormalities; only rarely is there a unique phenotypic trait that distinguishes one from another. Among the 2,000+ chromosomal abnormalities that have been catalogued, about 100 are classified as bona fide syndromes—roughly half of which show quite widespread uniformity in phenotypic traits (Borgaonkar 1984; de Grouchy and Turleau 1984; Schinzel 1984; Yunis 1977).

Beginning in 1960 and reconvening every few years, cytogeneticists have held international Chromosome Conferences for the purpose of standardizing the description of human karyotypes and dealing with new technologies as they arise. In 1971, for example, an International System for Cytogenetic Nomenclature (ISCN) was set up to refine and expand an already fairly complex system of terminology to include the new banding methods. Their codes for aberrations include over two dozen categories; some of these represent subdivisions of the main types to be discussed below, but others represent rarer abnormalities that will not be presented in this text (ISCN 1985).

*Originally, the term *euploid* referred to the presence of whole chromosome sets (haploid, diploid, and even higher multiples known as polyploids), whereas *aneuploidy* referred to deviations (by loss or gain of individual chromosomes) from the whole chromosome sets. Later, however, the term *aneuploid* was broadened to include structural rearrangements that do not necessarily change the regular chromosome number (Hook 1985). We generally use the latter, broader definition.

DELETIONS AND DUPLICATIONS

Pieces of chromosomes can be lost or gained. Although human geneticists often refer to deletions as *partial monosomies* and to duplications as *partial trisomies*, we prefer the simpler terms.

Origins and Examples of Deletions

The loss of a chromosome piece, that is, a **deletion** or **deficiency**, can happen in several ways. As shown in Figure 1A, a single break could lead to a terminal deletion, which is the type usually found in humans. Less frequent in humans is the occurrence of two breaks with loss of the intervening segment (Figure 1B). If breaks occur at both chromosome ends, the acentric fragments will get lost and a ring may be formed by union of the broken ends (Figure 1C). Deletions may also occur as a by-product of rearrangements between different chromosomes.

Except in the X chromosome, deletions of more than a few genes tend to be lethal even when heterozygous (i.e., paired with a normal, nondeleted chromosome), and those deletions that do survive are often associated with severe phenotypic defects. Deletions of chromosome arms 4p, 5p, 9p, 11p, 11q, 13q, 18p, and 18q are the ones most often seen in infants, but deletions for every chromosome tip and many interstitial (internal) segments are known, too (Therman 1986). Chromosomes 4 and 5, detected in various abnormalities with more than random frequency, appear to be particularly susceptible to breakage. Or perhaps breaks in these two chromosomes are simply less lethal than those occurring in some other chromosome, and thus lead to a type of sampling bias.

The most common deletion found in humans (about 1 in 50,000 live births) is loss of the short arm of chromosome 5.* Because newborns with 5p− usually have a

*Recall that p refers to the short arm of a chromosome and q to the long arm. A minus sign following the p or q means that the arm is too short; a plus sign means it is too long.

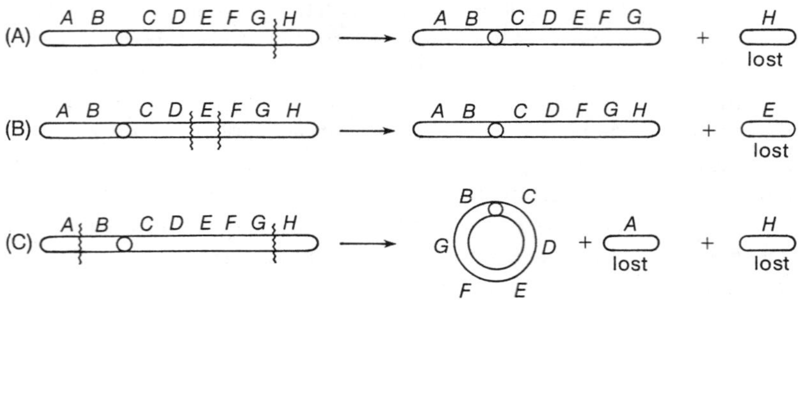

(A) A B C D E F G H → A B C D E F G + H lost

(B) A B C D E F G H → A B C D F G H + E lost

(C) A B C D E F G H → B C D E F G (ring) + A lost + H lost

Figure 1 Origin of deletions. (A) One break near a chromosome tip. A small piece is lost. (B) Two breaks followed by the loss of a small interstitial piece. (C) Two breaks followed by the loss of both tips and formation of a ring chromosome. Chromosome fragments that are acentric (i.e., lacking a centromere) will be lost from the nucleus in the subsequent cell division.

high-pitched mewing cry like that of a kitten, this aberration is called the *cat-cry syndrome* (Figure 2). Other features are variable, but they usually include a small head with a round face, wide-set eyes with oriental-type (epicanthal) folds on the upper lids, low-set ears, and slow growth. Most patients survive beyond childhood, albeit with severe mental retardation. Deletion sizes differ from case to case, but the crucial region absent in all individuals with the cat-cry syndrome is a tiny segment near the middle of 5p15.

The next most common deletion involves the short arm of chromosome 4 (i.e., 4p−). More severe than 5p deletions, they are often associated with improper fusion of the midline of the body. Affected infants usually have wide-set eyes, a broad nose, low-set and malformed ears, a very small lower jaw, and a cleft palate. Heart, lung, and skeletal abnormalities are common, and in males the penis may be improperly fused. Birth weight is very low. These babies generally fail to thrive, and most of them die young.

Large deletions of chromosome 18 are also known. Loss of the short arm (18p−) results in growth retardation and mental retardation. Loss of part of the long arm (18q−) causes these anomalies and also defects of the ears, eyes, mouth, and facial structure (Figure 3). Heart and minor skeletal defects are also frequent. In addition, both deletions are often associated with the lack of a blood serum protein called immunoglobulin A (IgA).

Small deletions in 15q, near the centromere, are associated with the *Prader-Willi Syndrome* (PWS)—a disorder characterized in infants by poor muscle tone, poor reflexes, poor feeding, and underdevelopment of the gonads and external genitalia.* PWS children are mentally retarded and extremely obese; they are also very short, with small hands and feet (Mattei et al. 1984; Ledbetter et al. 1981; Riccardi 1984).

At least two conditions are known to involve deletions of chromosome 21. Although 21p− has been noted in about 1% of the general population without any apparent ill effects, deletion of the long arm (21q−) is associated with the following traits: low birth weight, facial defects, wide-set but downward-slanting eyes—often with cataracts or other defects—abnormally shaped and low-set ears, high nose bridge, skeletal and heart defects, and sometimes very tight muscle tone. Deletions of G group chromosomes are often found as rings.

*But almost 50% of PWS patients show no detectable chromosomal defects.

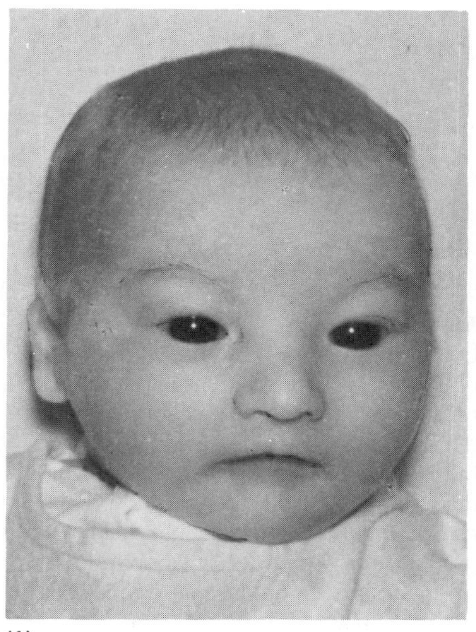

(A)

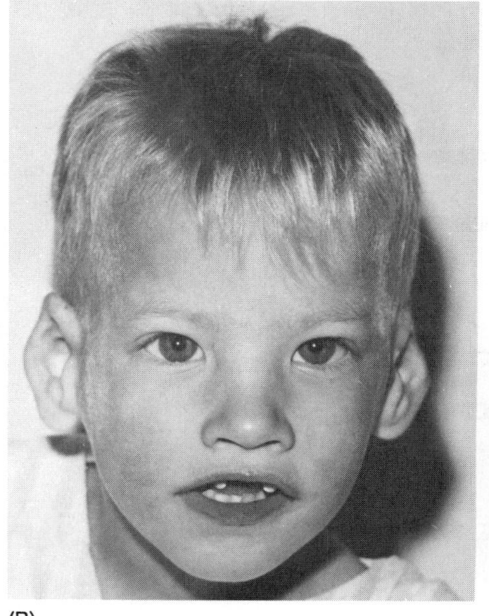

(B)

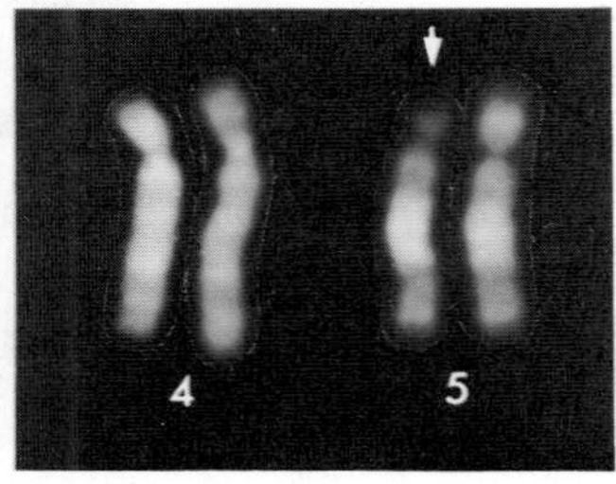

(C)

Figure 2 A patient with the cat-cry syndrome, associated with a 5p− deletion. (A) Affected newborn, showing moon-face and wide-set eyes. (B) Same child at four years; the cat-cry and moon-face have disappeared, but the mongoloid type of skin fold (extending down to the inner corners of the eyes) has persisted. The ears are somewhat misshapen and low set. Valentine (1986), from whom these photographs were obtained, states that young children with the 5p− syndrome tend to look very much alike. (C) Chromosomes 4 and 5 from an affected individual, showing a deletion of part of the short arm of one chromosome 5. (Chromosome photographs courtesy of Irene Uchida, McMaster University.)

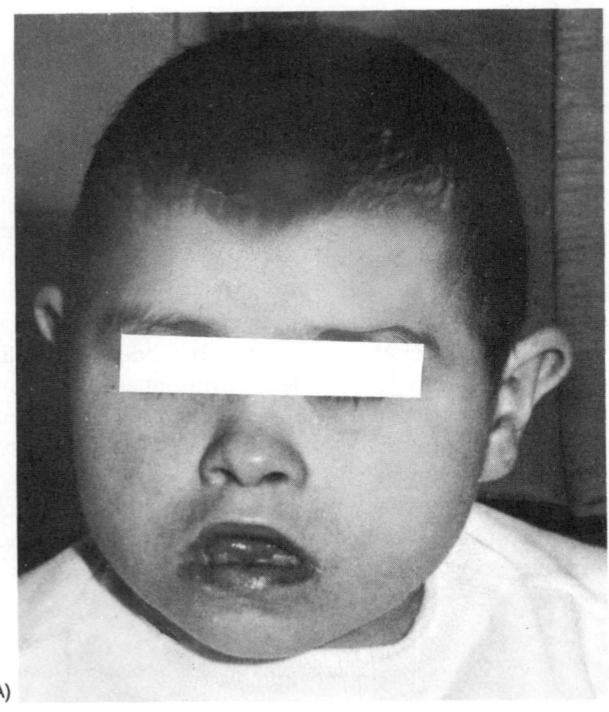

Interstitial deletions in 13q (region 14) are associated with retinoblastoma (tumor of the retina of the eye); their incidence is about 1 in 100,000 live births. A deletion in 11p (band 13) causes the WAGR syndrome, a complex that includes **W**ilms tumor (a cancer of the kidney), **a**niridia (absence of the iris of the eye), ambiguous **g**enitalia and **g**onadoblastoma (tumor of the gonad), and mental **r**etardation. These and the many other cancer-related deletions are discussed in Chapter 19.

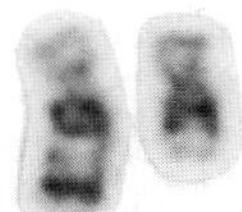

Figure 3 (A) A child with deletion of one-fourth to one-half of the long arm of chromosome 18. The downturned mouth, absence of the midline indentation that runs between the nose and the mouth, and the peculiar shape of the ears are characteristic of these patients. The carp mouth trait also occurs in certain other chromosomal syndromes. (Photograph by F. Sergovich, courtesy of G. H. Valentine [1986].) (B) Chromosomes from an affected individual. (Photographs courtesy of Irene Uchida, McMaster University.)

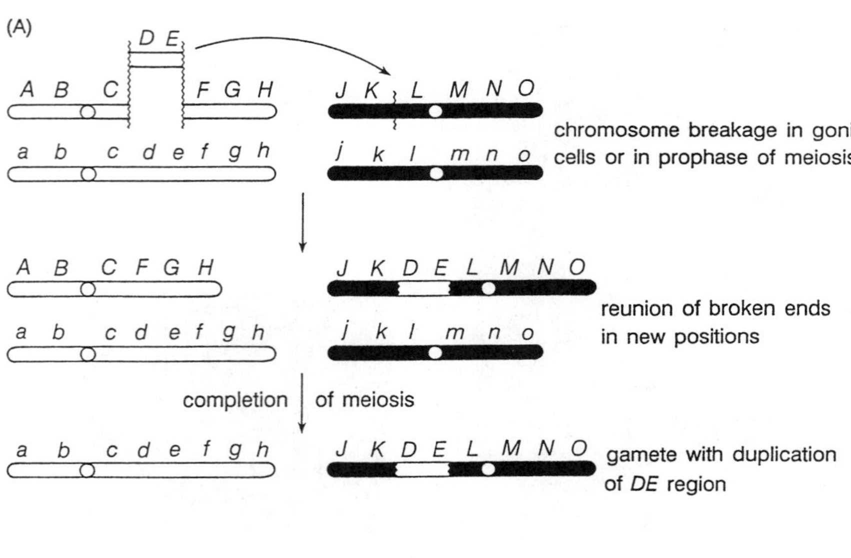

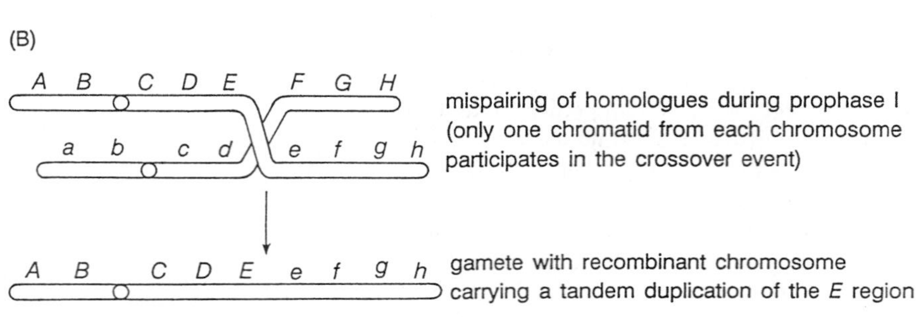

Figure 4 Origin of duplications. (A) Two breaks in one chromosome, with insertion of the intervening segment into a third break in a different chromosome. This duplication can occur in somatic cells, too, but without any imbalance of chromatid material because all the rejoined pieces remain together in the daughter cell. (B) Rare out-of-register synapsis followed by crossing over. Such small duplicated segments have been observed in fruit flies, in which very precise banding analysis is possible. In humans, this phenomenon accounts for the formation of a rare hemoglobin gene (Chapter 14). The gametes with duplicated regions, when combined with normal gametes, produce zygotes with extra genetic material. Gametes with deletions can also be produced in these processes, but these have not been indicated on the diagrams.

Origins and Examples of Duplications

Pieces of chromosomes can be gained as well as lost. These **duplications** may result when, following three chromosome breaks, a segment of one chromosome is inserted elsewhere in the homologous chromosome or into a different chromosome (Figure 4A). In fruit flies such insertions, occurring spontaneously or induced by radiation, have been analyzed in great detail. In humans, relatively few examples are known thus far, but more are likely to be found. Extra-long but normal variants of chromosomes 1, 9, and 16 have been reported; these are usually due to increased length of the heterochromatic region of the long arm near the centromere and not to insertions.

Duplications can also arise through errors in chromosome replication, through errors in crossing over (Figure 4B), or by normal crossing over between chromosomes that are heterozygous for certain other structural abnormalities. When the resultant unbalanced gamete combines with a normal gamete, the zygote will possess an extra batch of genes—ranging from just a few genes to an entire chromosome arm.

In organisms where duplications have been studied extensively, the phenotypic effects vary according to their size and position—but, generally speaking, small duplications are less harmful (even when homozygous) than are deletions of comparable size. This differential effect may explain why there seem to be few well-defined human syndromes associated with duplications alone. Geneticists believe, however, that duplications have played an important role in the evolution of many species by providing additional DNA, which is then free to mutate to different genes. There is good evidence, for example, that the independent but very similar genes controlling production of certain protein chains in human hemoglobin all arose from duplications of a single ancestral gene (Chapter 14).

TRANSLOCATIONS

Origins of Translocations

Breaks in two or more nonhomologous chromosomes, followed by reattachments in new combinations, can lead to the formation of **translocations** (Figure 5). This process is fairly common, and with the right circumstances the rearranged chromosomes may be transmitted through many successive generations. If the rearrangement of chromosome parts is complete, with no leftover pieces, the translocation is **reciprocal**. Because all of the genetic material is still present but in a different

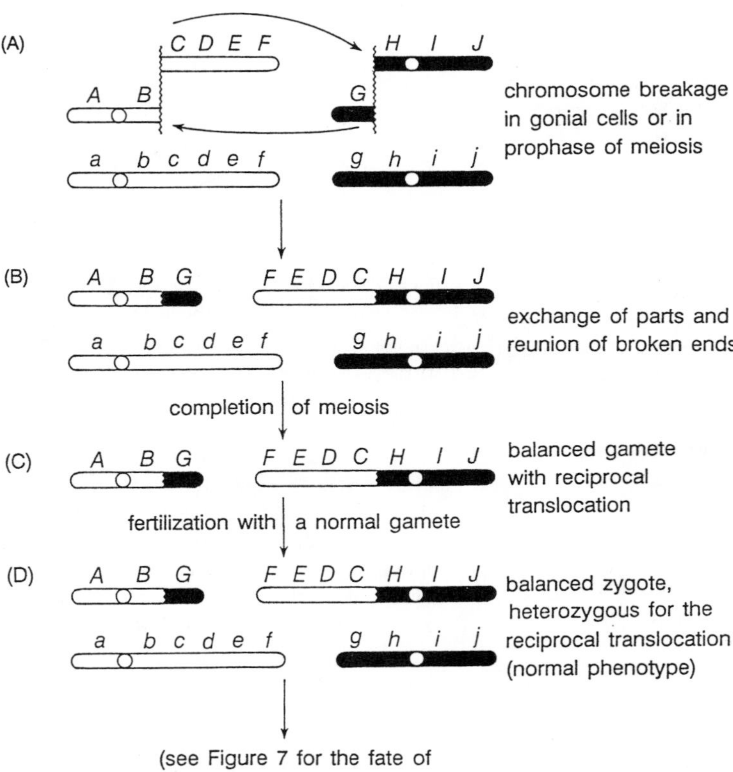

Figure 5 Origin of a reciprocal translocation. (A) Breaks occur in each of two nonhomologous chromosomes. (B) The pieces without centromeres exchange places. (C) One possible pattern of segregation puts the two translocated chromosomes into one gamete. This condition is balanced because all chromosomal segments are present; they are simply rearranged. (D) Combination of this gamete with a normal one gives a zygote that is heterozygous for the translocation, that is, two chromosomes carry translocations and their homologues do not. Note that two pairs of chromosomes are involved.

arrangement, a heterozygote for such a translocation is said to be **balanced**, and the phenotype is usually normal.

Translocations between and among the two groups of acrocentric chromosomes (Groups D and G, numbers 13–15 and 21–22) represent a special case known as a **Robertsonian translocation**, in honor of an investigator who studied such translocations in other organisms. As shown in Figure 6A, a reciprocal exchange occurs; the long arms of two acrocentric chromosomes are joined together at the centromere, as are the two heterochromatic tips. The tiny heterochromatic chromosome so formed, which apparently carries no essential genes, is usually lost. Thus the longer translocated chromosome contains the full complement of genes from two nonhomologous chromosomes. This balanced heterozygote has a total of only 45 chromosomes, yet it is phenotypically normal. Certain Robertsonian translocations are more common than would be expected on the basis of chance, probably because of the close association of these nucleolar organizing chromosomes during cell division. The acrocentric chromosomes may also be more prone to breakage. Yet "while (13q14q)* and (14q21q) translocations are found fairly frequently in the population, all other types of Robertsonian translocations are rather rare" (Jacobs et al. 1974).

*(13q14q) signifies a translocated chromosome joining together the long arms of chromosomes 13 and 14.

Another special type of aberration is a metacentric chromosome consisting of two identical arms joined to create a palindrome (Figure 6B); this aberration may or may not represent an actual translocation. Such chromosomes are called **isochromosomes** (Greek *iso*, "alike," "equal") and are thought to be formed by misdivision of sister centromeres. Instead of a normal lengthwise separation between duplicate chromatids, the centromere appears to divide transversely, thereby joining two identical chromatid arms. Union of broken sister chromatid arms also can produce an isochromosome. During the course of meiosis, the two isochromosomes separate; consequently 50% of the gametes contain a duplication of one arm and a deletion of the other.

Isochromosomes for the long arm of Group D or G chromosomes, when heterozygous with normal chromosomes, lead to effective trisomy for that chromosome, even though the total number of chromosomes is only 46. Indeed, a few cases of Down syndrome are known to have a long-arm isochromosome of 21 and a normal chromosome 21.

Translocations in Meiosis

Meiotic behavior of reciprocal translocations when heterozygous with structurally normal chromosomes is rather complicated. Because prophase I synapsis involves

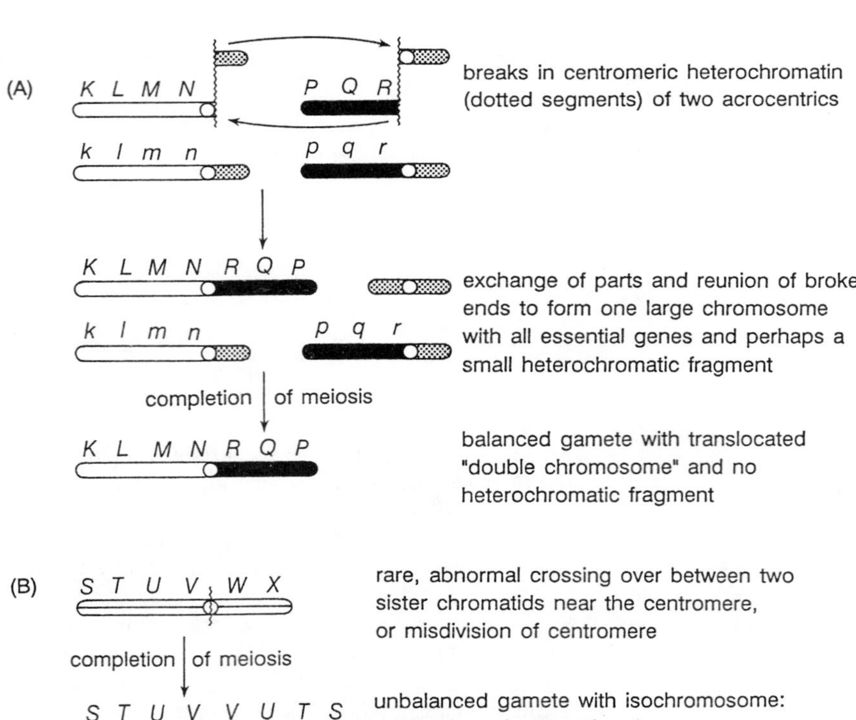

Figure 6 Special types of translocations. (A) *Robertsonian translocation*. This anomaly is similar to the reciprocal translocation in Figure 5, except that the two breaks occur very near the centromeres of acrocentric chromosomes (Groups D and G). The two very short heterochromatic arms form a small chromosome that apparently carries no essential genes. (Alternatively, they could be lost without joining.) Thus a gamete that carries only the chromosome with the joined long arms is essentially balanced. (B) *Isochromosome*. Sometimes loosely classed as a translocation, an isochromosome is formed by the union of two identical chromosome arms, usually by misdivision of the centromere. Isochromosomes for either arm of the X have been identified in humans.

precise gene-by-gene pairing, the translocated chromosomes must form some unusual configurations in order to match up with their partners. The simplest translocations will arrange themselves into a distinctive crosslike pattern (Figure 7), from which pairs of chromosomes segregate in three possible ways during Anaphase I. If segregation is alternate (i.e., if chromosomes that are diagonally opposite move to the same pole), then the resultant gametes and zygotes are balanced. Note that among these balanced gametes, half will carry the two translocated chromosomes and half will carry the two normal chromosomes. If separation is not alternate (i.e., if it occurs in the vertical plane or in the horizontal plane shown in Figure 7), then the resultant gametes will be unbalanced—having extra pieces of some chromosomes and missing pieces of other chromosomes. Any zygote formed by combination with a normal gamete will be unbalanced and abnormal.

Although the three kinds of segregation do not occur with equal frequencies, enough unbalanced gametes are produced to reduce the fertility of translocation heterozygotes. In experimental plants and animals from which large numbers of offspring can be obtained, such reduced fertility is regularly passed on to about half the offspring (i.e., to translocation carriers). This pattern of inherited semisterility is a strong indication that a translocation is present in that particular stock. Such a dramatic decline in reproductive fitness may not be so obvious in humans, because they have few offspring and because they are likely to "replace" miscarriages, stillbirths, or early infant deaths with subsequent pregnancies. Nevertheless, some of these translocation heterozygotes will show a clear reduction in fertility, and most are likely to exhibit a higher-than-normal rate of prenatal and postnatal loss of progeny.

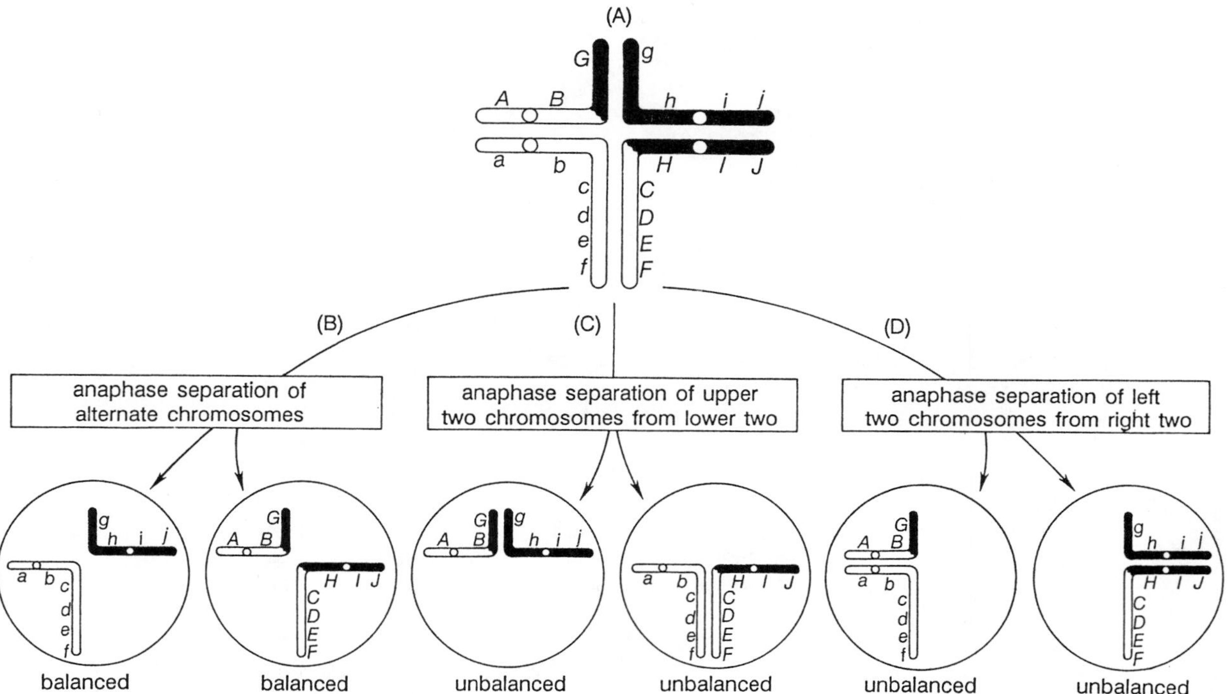

Figure 7 Meiotic synapsis and segregation from the reciprocal translocation heterozygote shown in Figure 5. (A) The only configuration that allows gene-by-gene pairing between two translocated and two nontranslocated chromosomes is a crosslike configuration. For simplicity, we have not shown the chromosomes in their actual doubled state. From this, the chromosomes usually segregate two-by-two at anaphase I, yielding six possible gamete types. (B) Of the two balanced combinations, one has two normal chromosomes and the other, two translocated chromosomes. The latter, if combined with gametes carrying normal chromosomes, will again give rise to translocation heterozygotes. (C, D) The four unbalanced combinations all have a duplication of one segment and deletion of another. (*Note:* For ease of identification, the six segregants are shown in their pairing positions; actually, these configurations do not persist beyond metaphase 1. The six gamete types do not occur in equal proportions.)

Translocation Down Syndrome

Recall that nearly all cases of Down syndrome occur sporadically in families whose other members are entirely normal. These patients have the karyotype 47,XX,+21 or 47,XY,+21, the extra chromosome coming from nondisjunction in one of the karyotypically normal parents, usually the mother. In contrast to this situation, about 4% of Down syndrome patients have *46 rather than 47 chromosomes*, but one of them is abnormally long. This long chromosome is a Robertsonian translocation, usually involving chromosomes 14 and 21 (or in some cases, two chromosomes 21). Because nondisjunction is not the cause, no maternal age effect is seen here. Although the percentage of such patients is very small, they are important to identify because this form of Down syndrome may recur in the same family.

Unlike the karyotypically normal parents of patients with sporadic Down syndrome, the carrier parents of individuals with translocation Down syndrome may have only *45* chromosomes—one of which is (14q21q). As shown in Figure 8A, three chromosomes in these translocation heterozygotes contain nearly all of the material present in the original four chromosomes. The tiny tips of chromosomes 14 and 21 are lost, but the fact that the translocation carrier is normal means that no essential genes are missing.

Consider now the kinds of gametes that a carrier might produce (Figure 8B). During meiosis, two chromosomes will usually proceed to one pole and one to the other; thus six types of gametes are possible. Of these, only two types are balanced: one bearing the normal chromosomes 14 and 21, and another carrying the single large (14q21q) translocation. When combined with structurally normal chromosomes of a gamete from the opposite sex (Figure 8C), the former will yield a normal zygote, and the latter will give rise to a translocation

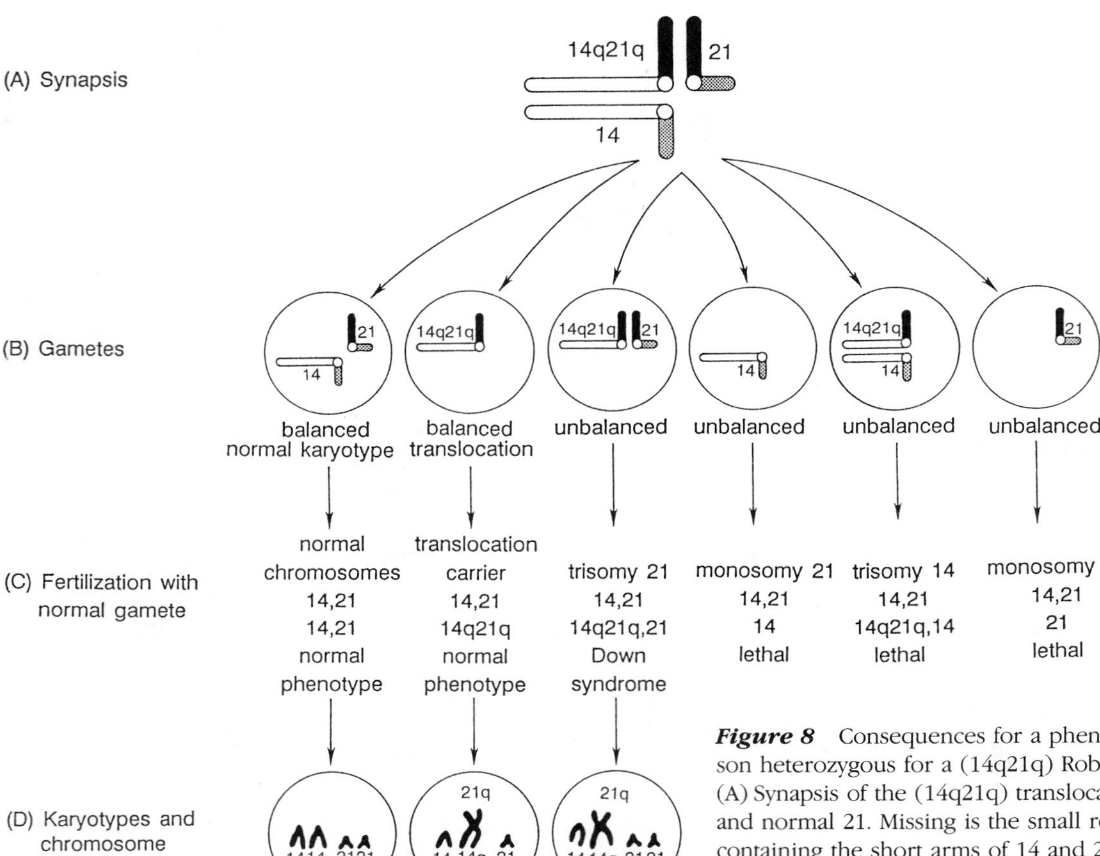

Figure 8 Consequences for a phenotypically normal person heterozygous for a (14q21q) Robertsonian translocation. (A) Synapsis of the (14q21q) translocation with the normal 14 and normal 21. Missing is the small reciprocal chromosome containing the short arms of 14 and 21. Compare with Figures 6A and 7. (B) Segregation (diagonal, up–down, left–right) gives rise to six types of gametes, of which four are unbalanced and two are balanced. (C) Fertilization with normal gametes results in three lethal zygotes plus one zygote leading to Down syndrome and two producing normal phenotypes. Of the two normal phenotypes, one is karyotypically normal and one is again heterozygous for the translocation, like the parent. (D) Karyotypes.

heterozygote with a normal phenotype. Thus the translocation may pass unnoticed through a number of generations. Among the types of unbalanced gametes produced by a carrier, only one will lead to a viable zygote after combining with a gamete with normal chromosomes. This zygote will have an extra dose of chromosome 21 and will develop Down syndrome. So a carrier parent may produce up to three kinds of offspring: phenotypically and karyotypically normal, phenotypically normal but translocation heterozygotes, and translocation Down syndrome (Figure 9).

One final comment: Although the multiple occurrence of Down syndrome in a family may suggest the presence of a (14q21q) translocation, this aberration may not be found. In fact, most cases of multiple Down syndrome are due to multiple instances of nondisjunction.

Other Translocations

Translocations are not uncommon in humans, occurring with a frequency of about 1 in 500 among newborns (Hook and Hamerton 1977) and a few percent among the mentally retarded. Breakpoints occur almost exclusively in the Q-dark regions of the chromosomes. Collectively they involve all chromosome arms, more or less at random—with the exception of a few high-affinity combinations such as 11q with 22q, 9 with 22, and 9 with 15 (Therman 1986). The unbalanced translocation t(11;22), for example, has given rise to several disorders associated with partial trisomies of 11q. But people with balanced translocations are usually phenotypically normal, although their fertility may be reduced by repeated spontaneous abortions.

X;autosome translocations are interesting for two reasons. First, their inactivation patterns: In females carrying *balanced* X;autosome translocations, the normal (i.e., untranslocated) X is inactivated. This event may lead to the expression of X-linked recessive disorders (such as Duchenne muscular dystrophy), when the translocated X also includes a deletion in that gene (Worton et al. 1986). But in females carrying *unbalanced* translocations, the translocated X chromosome is usually inactivated, although the situation may vary somewhat from cell to cell. Virtually all of these unbalanced translocation females exhibit multiple defects, including mental retardation. The second reason for special interest in X;autosome translocations is their key role in mapping the genes for Duchenne muscular dystrophy and at least eight other X-linked recessive diseases (Worton and Thompson 1988).

Autosomal translocations are sometimes involved in specific types of cancer; balanced reciprocal translocations, for example, occur in leukemias and lymphomas. The best known are the Philadelphia chromosome t(9;22) for chronic myelogenous leukemia, and t(8;14) for Burkitt's lymphoma. These and others will be discussed further in Chapter 19.

INVERSIONS

When two breaks occur in one chromosome and the intervening segment gets inverted before the broken ends rejoin, the resultant aberration is known as an **inversion** (Figure 10A). If the reversed segment includes the centromere, the inversion is *pericentric* (Greek *perí*, "around"). But if it does not include the centromere, the inversion is *paracentric* (Greek *pará*, "apart from," "beside"). An inversion does not ordinarily produce a distinctive phenotype; its presence is detected by unusual meiotic events or unusual segregation patterns in organisms heterozygous for the inversion. This type of struc-

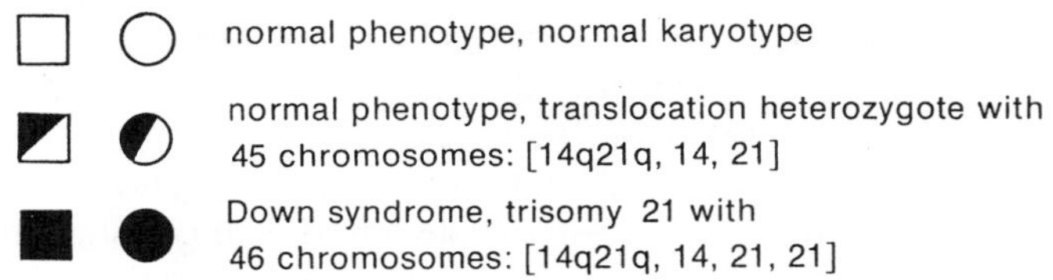

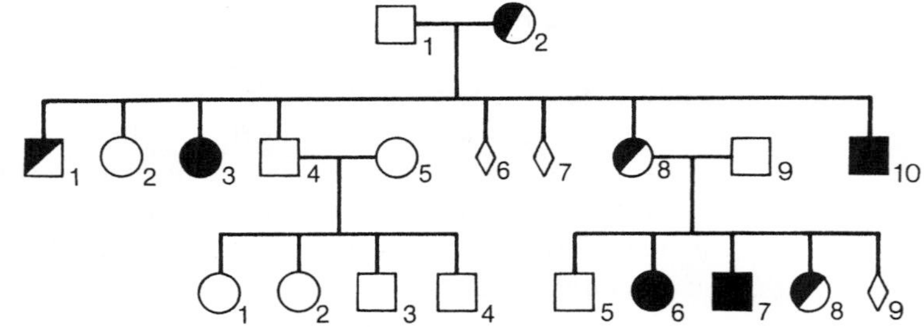

Figure 9 A hypothetical pedigree of translocation Down syndrome. Note that several affected persons may appear in the same sibship and be related to other affected persons through translocation carriers with normal phenotypes. This pattern contrasts with that of nondisjunction Down syndrome, in which the occurrence of affected persons is usually sporadic.

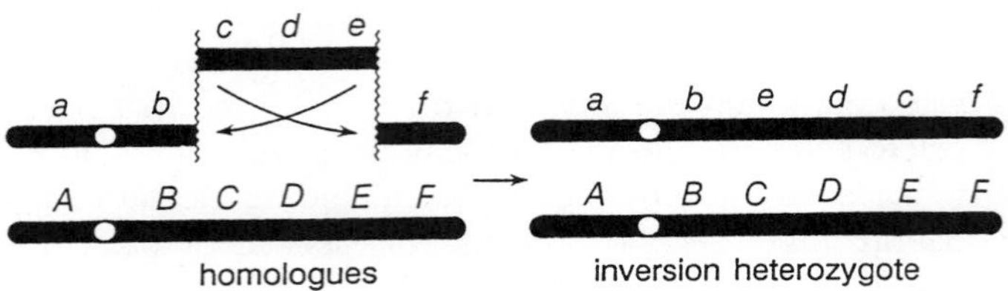

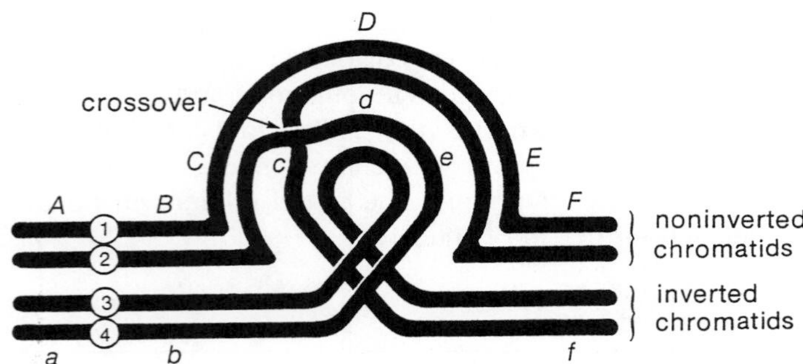

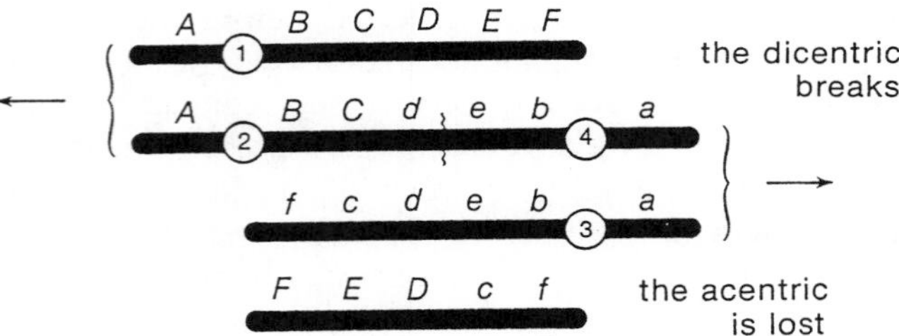

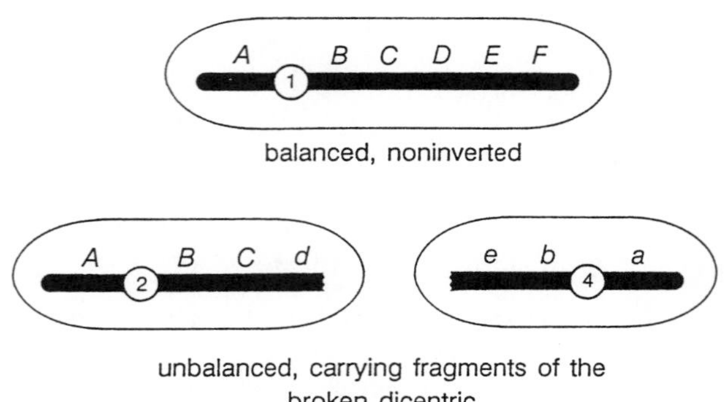

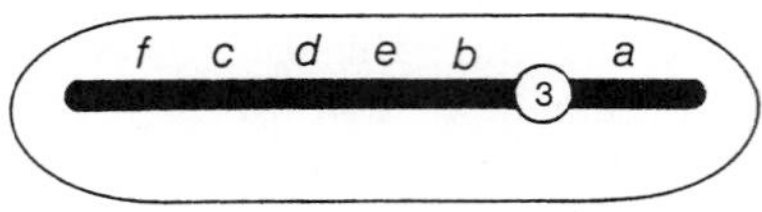

Figure 10 Origin of a paracentric inversion, and segregation of crossover and noncrossover chromatids from the inversion heterozygote. (A) Two breaks in one member of a homologous chromosome pair, followed by rejoining of the inverted segment, which does not include the centromere. (B) The two doubled chromosomes form a loop configuration during synapsis in prophase I; centromeres 1 and 2 mark identical noninverted chromatids; centromeres 3 and 4 indicate identical inverted chromatids. A single crossover between chromatids 2 and 4 within the inverted segment produces one dicentric chromatid and one acentric fragment. The former has duplications of the A and B loci and is deficient for F; the latter has a duplication at F but is deficient at the A and B loci. (C) During anaphase I separation of the two homologous chromosomes, the dicentric is broken at some point between the two centromeres that are being pulled in opposite directions. The acentric fragment will be carried to neither pole. (D) Two of the resulting gametes, one of which contains the inverted chromosome, are balanced. The other two gametes are unbalanced and will form lethal zygotes.

tural abnormality has been analyzed extensively in many species, but the routine detection of inversions in humans had to await the development of banding techniques.

Behavior of Inversions during Meiosis

Gene-by-gene pairing during synapsis requires an inverted chromosome and its normal homologue to form a loop during prophase I (Figure 10B). What happens thereafter depends upon (1) whether or not crossing over occurs within the inversion loop and (2) whether or not the centromere is included within the inversion. Figure 10B depicts the situation when there is a single crossover within the inversion loop of a paracentric inversion. By tracing out the four chromatids making up the inversion heterozygote tetrad, one can show that the two crossover chromatids emerge with both duplications and deletions. Superimposed on these aberrations is the presence of two centromeres on one chromatid (**dicentric**) and no centromeres on the reciprocal chromatid (**acentric**). Further deletions occur in the dicentric chromatid during anaphase I (Figure 10C), when it is broken into two parts by the force of its two centromeres pulling toward opposite poles. The gametes thus formed will be lethal if fertilized. Of the two balanced gamete types, one carries the inversion and the other is karyotypically normal.

In the case of a pericentric inversion, recombinant chromatids resulting from crossovers in the inversion loop will also exhibit duplications and deletions. As readers can verify for themselves, no dicentrics or acentrics will be formed. When no crossovers occur within the inversion loop, no unbalanced chromosomes are formed.

The overall effect on fertility in inversion heterozygotes is related to the size of the inversion. Because the frequency of crossing over within a given region depends on the length of that region, a large inversion may be associated with more aberrations and consequent infertility than will a small inversion.

When the reduction in crossover progeny from inversion heterozygotes was first noted in fruit flies, it was thought that the inversions actually suppressed the crossing over process in some unknown way. Researchers found, however, that crossing over proceeds normally within the inverted segment, but the crossover products were usually inviable.

This phenomenon has some interesting evolutionary ramifications that have been explored in great detail in experimental organisms. Normally, crossing over continually shuffles genes between the homologous chromosomes present in a population over a number of generations, allowing many gene combinations to be "tried out." But crossing over also has the effect of *separating* any favorable combinations of genes that may exist together on the same chromosome. If an inversion spanned a favorable gene cluster, however, the genes within it would thereafter be "protected" from the effects of crossing over because only noncrossover chromosomes could be passed on to succeeding generations. So preserved, this group of genes may increase in the population at a greater rate than would the individual genes in the absence of an inversion. Many examples of these adaptively superior inversions are known in fruit flies. (From a practical point of view, heterozygous inversions are used to preserve specific gene combinations in laboratory stocks.) Of course, the opposite situation might sometimes prevail, too: a favorable gene locked into an inversion with a lethal gene could be eliminated from the population.

Inversions in Humans

Before the advent of chromosome banding, knowledge of inversions in humans was extremely limited. Only those pericentric inversions that shifted the position of the centromere, thereby altering the relative lengths of the two chromosome arms, could be identified. Undetectable by nonbanding methods were all paracentric inversions and those pericentrics whose inverted segments on either side of the centromere were of equal length. Now, however, our knowledge of inversions is rapidly expanding.

Although estimates differ widely, the incidence of inversions in humans might be as high as 1–2%. Borgaonkar (1984) lists about 250 inversions, of which 90% are pericentric. Pericentrics have been found on all chromosomes except number 20 (Kaiser 1984), whereas paracentrics are listed for only nine chromosomes. Whether paracentrics actually occur this infrequently is not clear, however, because (1) they are more difficult to identify than are pericentrics and (2) their carriers produce fewer viable offspring, because of the generation of acentrics and dicentrics during meiosis.

Furthermore, the particular chromosomes involved and the locations of breakpoints appear to be highly nonrandom. Chromosome 9, for example, accounts for about 11% of all known inversions, and certain breakpoints recur with great frequency.

OTHER ABNORMALITIES

In humans, ring chromosomes are compatible with survival beyond the fetal stage, but most polyploid karyotypes are not.

Ring Chromosomes

Recall that when two breaks occur in one chromosome, rings can be formed (Figure 1). **Acentric rings**, lacking a centromere, usually get lost in subsequent cell divisions; but **centric rings**, because they have a centromere, can be passed on to daughter cells. During meiosis (or mitosis), if there is no prophase crossing over between the two sister strands of a centric ring (Figure 11), it will be passed on intact. But if *sister chromatid exchange* does occur, a double-size **dicentric ring** will be formed. During anaphase, there are two possible outcomes: (1) one of the daughter cells will have the dicentric and the other will have no ring, or (2) the two centromeres of the dicentric ring proceed to opposite poles. In this case the dicentric ring will break, and both daughter cells may receive centric rings of different sizes. This process can be repeated in subsequent cell divisions, too, giving rise to many different cell lines in one individual.

Rings have been reported for every human chromosome, but there is great variation in phenotypic effects, even for apparently identical rings—ranging from normal or near-normal to highly abnormal phenotypes (Therman 1986). Although syndromes associated with rings have been described for at least 14 different chromosomes, only those involving ring chromosomes 13 and 14 present any uniformity of phenotypic features (Borgaonkar 1984).

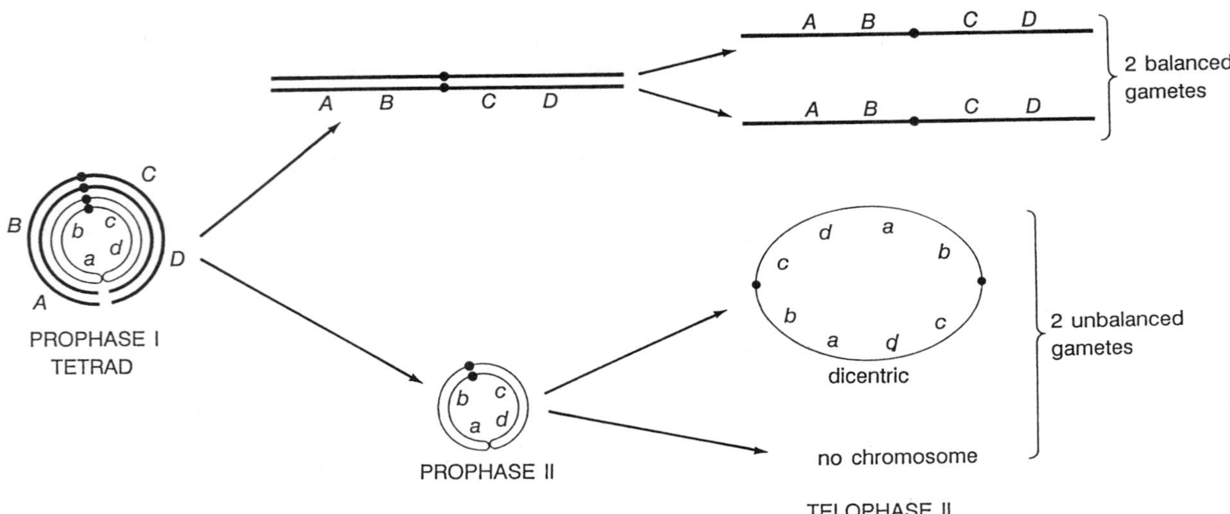

Figure 11 One possible outcome when a ring chromosome (heterozygous with a normal chromosome) undergoes sister-strand crossing over. What do you think happens to the dicentric during mitosis (following fertilization)? Would the result be the same if there were two sister-strand crossovers within the ring? If there were one crossover between the ring and the normal chromosome?

Polyploidy

In contrast to polysomy of individual chromosomes in an otherwise diploid organism, entire chromosome sets may be present in excess. An organism may be **triploid (3n)**, with three representatives of each chromosome, or **tetraploid (4n)**, with four of each chromosome, and so on. **Polyploidy** (Greek *poly*, "many"; English *ploid*, "set") can arise when chromosome replication during meiosis is not followed by cell division—yielding tetraploid rather than diploid nuclei. Triploids, on the other hand, may arise when two sperm fertilize one egg.

POLYPLOIDY IN PLANTS. Polyploids are quite common in plants. Some are derived from a single species, whereas others result from a cross of two closely related species and a subsequent chromosome doubling. Polyploidy can be artificially induced through the use of **colchicine**, a drug that prevents the formation of spindle fibers and thus inhibits chromosome separation. Because polyploid plants tend to be larger or more hardy than their diploid counterparts, they can have great commercial value. Many garden flowers, ornamental plants, fruit and forest trees, grasses, and crop plants fall into this category.

As is the case with translocations and large inversions, infertility is also a hallmark of polyploids—because of irregularities in synapsis during meiotic prophase. (This trait is characteristic of virtually all polyploids except those with even-numbered sets of chromosomes, usually 4n, that are derived from different species.) Although no more than two chromosomes can be paired for any given segment of a chromosome, different segments of the same chromosome may be paired with different partners. A group of three homologues, for example, may pair like this:

In a triploid, two chromosomes of each homologous trio will usually migrate to one pole and a single chromosome to the other. Because each homologous chromosome group segregates independently of the others, however, a gamete may contain one representative of some chromosomes and two representatives of others. The resultant zygote cannot survive. In a tetraploid whose chromosomes all came from the same species, the same problem exists: the chromosomes may segregate two by two, three by one, and so on. For this reason, many commercially useful polyploid plants must be propagated by asexual means, such as cuttings. In some cases this inability of polyploids to set seeds is a virtue, as it is in seedless varieties of fruits.

POLYPLOIDY IN ANIMALS. For a number of reasons, polyploidy is rare among animals. It occurs in only a few species (such as crustacean shrimp, sow bugs, and snout beetles) that can reproduce asexually by development of the egg without fertilization **(parthenogenesis)**. A few rare cases have been reported in bisexual vertebrates such as salamanders, frogs, fish, and a lone chicken; triploid fruit flies and salamanders have been produced in the laboratory, too. It is thought that when sex determination depends upon a certain balance between sex chromosomes and autosomes, or upon the XY karyotype, changes in ploidy generally lead to the formation of many intersexes. Thus, even tetraploids derived from crosses between different species (which in plants are often fertile) might not be able to reproduce. Such crosses between closely related species are much less common in animals than in plants (Jackson and Hauber 1983).

Two lines of evidence suggest that in humans polyploidy is not compatible with life. First of all, triploidy is the second most common aberration among embryos that are spontaneously aborted in early pregnancy (Figure 12). Second, the few cases of live-born triploids or tetraploids who have survived beyond a few days have turned out to be mosaics or chimeras with some diploid tissues; even so, they all had multiple but highly variable defects (Niebuhr 1974).

Triploidy is not associated with a distinctive syndrome, the only defect universally observed being an enlarged placenta with or without grape-shaped cysts called **moles**. Most human triploids are lost during the first three months of pregnancy. Although many triploid abortuses are partly decomposed or incomplete, others show few gross abnormalities; some even look completely normal. Growth of the embryos is stunted after four to five weeks of development, and this developmental delay is also seen in the low birth weight of the few live-born triploids. Intersexuality is also the rule. Webbing of the fingers and toes is common, and various abnormalities of the head, skeleton, heart, kidneys, and brain have also been reported. Yet, even though these defects do not seem so severe as to preclude survival, the few mosaics or chimeras who have survived for several years almost invariably show severe mental and motor retardation.

Although lethal when present in all cells of the body, polyploidy is a common situation in the cells of a few human tissues—the liver and the lining of the bronchus, for example. Why these cells are regularly polyploid is not known. Nor is it known why cancer cells from all types of tissues are often polyploid.

FREQUENCIES OF CHROMOSOMAL ABNORMALITIES

Table 1 presents some karyotype data for several categories of cells and phenotypes in humans, ranging from gametes to newborns to mentally retarded adults. Some

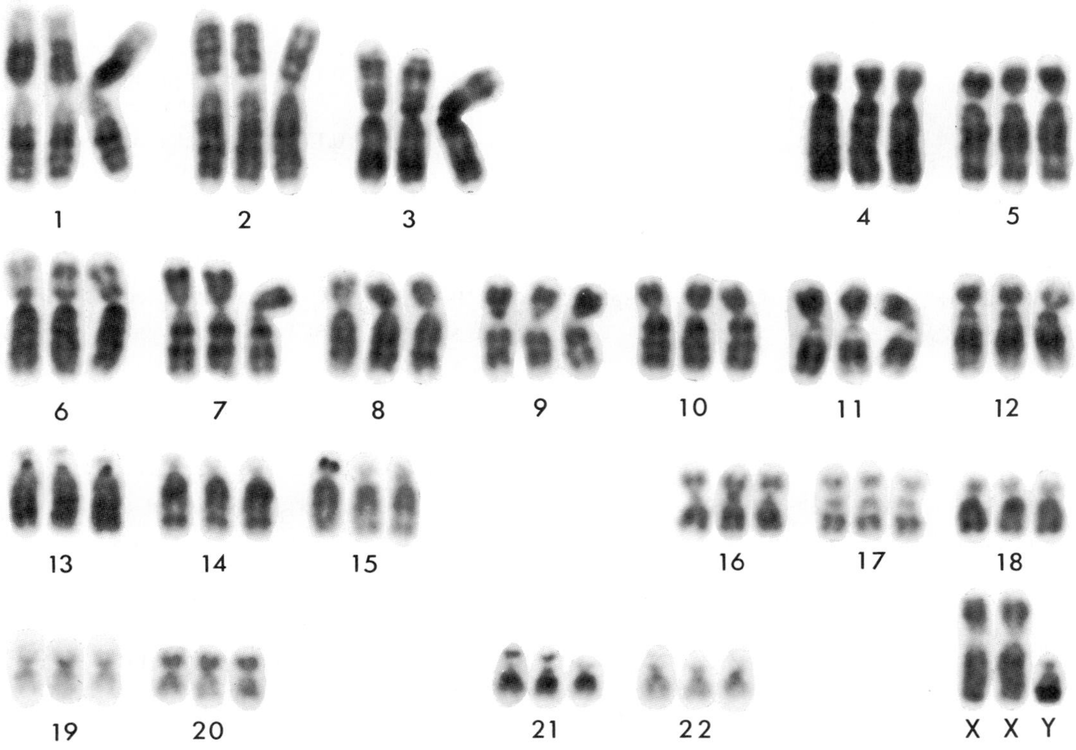

Figure 12 The triploid karyotype 69,XXY of a stillborn infant with congenital malformations. Quinacrine staining allows each set of three homologous chromosomes to be distinguished from others. (Courtesy of Irene Uchida, McMaster University.)

Table 1 Estimated frequencies of chromosomal aberrations in various samples

Sample	Sample Size	Total[b]	Tri-somic	Mono-somic	Poly-ploid	Structural anomalies	Reference
		APPROXIMATE PERCENTAGE OF ABNORMAL KARYOTYPES[a] IN SAMPLES					
Sperm from fertile males	1,572 (from 30 males)	10.4	1.2	3.6	—	6.2	Martin et al. (1987)
	2,468 (from 11 males)	9.4	0.7	0.9	—	7.7	Brandriff et al. (1985)
	1,091 (from 4 males)	13.9	0.5	0.5	—	13.0	Kamiguchi & Mikamo (1986)
Somatic cells from infertile males[c]	106 no sperm	14.1	12.3	—	—	1.9	Retief et al. (1984)
	390 low sperm	5.1	3.1	—	—	2.1	Retief et al. (1984)
	2,372 low sperm	2.2	1.4	—	—	0.8	Chandley (1979)
Oocytes from infertile females	22 (from 17 females)	45.4	9.1	36.4	—	—	Wramsby et al. (1987)
Prenatal tests: 10-wk embryos in women of age 35–49[d]	3,848	2.8	2.3	0.3	0.1	0.1	Hook et al. (1988)
Prenatal tests: 14–24 wk fetuses in women age 35–49[e]	61,000	—	—	—	—	0.2	Hook & Cross (1987)
	56,094	—	1.1 autosomal	—	—	—	Hook et al. (1984)
	52,965	2.2	1.8	0.04	—	0.4	Ferguson-Smith and Yates (1984)
	19,672	1.8	1.5	0.1	0.03	0.1	Schreinemachers et al. (1982)
Spontaneous abortions at various ages of development (from ovulation rather than from last menstrual period)	1,498 <12 wk	61.5	33.0	9.4	16.2	2.5	Boué et al. (1975)
	941 6–25 wk	30.5	16.5	7.4	5.3	1.3	Creasy et al. (1976)
	3,589 <20 wk	51.9	27.0	9.3	12.0	<3.6	Carr & Gedeon (1977)
	447 <6–28 wk	53.9	33.0	10.1	8.3	2.5	Kajii et al. (1980)
	1,000 0–18 wk	46.3	22.7	11.3	10.3	2.0	Hassold et al. (1980)
	2,517 <6–25 wk	38.5	21.1	6.9	8.0	1.5	Warburton et al. (1986)
Induced abortions	728 5–15 wk	3.2	1.8	—	1.0	0.4	Kajii et al. (1978)
	1,250 <10 wk	6.4	4.4	0.7	1.1	0.2	Yamamoto & Watanabe (1979)
Stillbirths (>28 wk) and neonatal deaths	624	5.1	4.0	0.1	0.2	0.8	Hassold (1986)
	500	5.6	3.8	0.2	0.2	1.4	Machin (1974)
	153 >20 wk	7.2	5.9	—	0.7	0.7	Bauld et al. (1974)
Live births	56,952	0.6	0.3	0.01	—	0.3	Hook & Hamerton (1977)
	68,159	0.65	0.3	0.01	—	0.3	Hsu (1986)

Table 1 (continued)

Sample	Sample Size	APPROXIMATE PERCENTAGE OF ABNORMAL KARYOTYPES[a] IN SAMPLES: Total[b]	Trisomic	Monosomic	Polyploid	Structural anomalies	Reference
Members of couples with multiple miscarriages	2,136	3.6	—	—	—	3.6	Fryns et al. (1984)
	11,332	2.4	—	—	—	2.4	Simpson (1986)
Mentally retarded patients in institutions	588 IQ <20–85	15.3	12.9 (12.4 DS[f])	—	—	2.4	Sutherland et al. (1976)
	475 IQ 0–70+	12.0	8.2 (7.6 DS[f])	—	—	3.8	Jacobs et al. (1978)
	6,163	11.8	11.2 (9.7 DS[f])	—	—	0.6	Jacobs et al. (1978) (6-study survey)
	1,062 IQ <50?	33.0	27.3 (27.0 DS[f])	0.5	—	5.2	Gripenberg et al. (1980)

a. Including mosaics and marker chromosomes.
b. Total may not be the sum of the four figures to the right, because of rounding off or because some individual cases show more than one abnormality.
c. In the study by Retief et al., 106 males produced no sperm and 390 males had low sperm counts.
d. By chorionic villus sampling.
e. By amniocentesis.
f. Down syndrome accounted for the vast majority of these trisomies.

of the frequency estimates are fairly crude, different investigators reporting quite different values. This is partly due to the different methods of sampling a given group, the different karyotyping techniques used over a period of years, and small sample size for some groups.

Gametes

Because human eggs and spermatids are very difficult for researchers to come by and because there was until recently no way to analyze the chromosomes that are tightly packed into sperm heads, studies of human chromosomes have been limited mainly to mitotic cells.

SPERM. However, some researchers at the University of Hawaii found a way to analyze the chromosomes from human sperm. It so happens that hamster eggs whose zona pellucidae have been removed will accept sperm from a few other species. Indeed, one or more sperm from fertile humans will penetrate about 75% of denuded hamster eggs in a culture dish (although for technical reasons only a fraction of the inseminated oocytes are analyzable). Within each hamster–human zygote, the sperm head swells up and becomes a pronucleus; the 23 human chromosomes decondense, replicate, condense, and form chromatid pairs that get ready to arrange themselves on the first metaphase spindle that is forming. At this point researchers treat the cells so that the male chromosomes (which are distinguishable from hamster chromosomes in the female pronucleus) can be karyotyped. "The sperm chromosomes are large, with parallel chromatids in which distinctive coils can be seen" (Rudak et al. 1978).

Quite surprisingly, even normal fertile males produce about 10% abnormal sperm—about 1–3% containing too many or too few chromosomes, and the rest showing structural abnormalities. With increasing age in males, the numerical abnormalities seem to decrease and the structural abnormalities increase (Martin et al. 1987).

Sperm from infertile males are sometimes less successful at fertilizing hamster eggs, so chromosome studies of such individuals are usually done on somatic cells rather than on germ cells. Retief et al. (1984) analyzed lymphocytes (white blood cells) from about 500 infertile males. Those males who produced no sperm had a frequency of about 15% chromosomal abnormalities, whereas those with very low sperm counts had about 7% chromosomal abnormalities. Unfortunately, because of variations in the way investigators have defined the conditions of "infertility" and "chromosomal abnormalities,"* it is difficult to compare studies of infertile males. But all surveys do seem to show an increasing frequency

*Whether minor chromosomal variants (markers) and/or mosaics are included, for example.

of chromosomal abnormalities with decreasing sperm count (Retief et al. 1984).

It is also not clear whether the rate of chromosomal abnormalities found in a male's lymphocytes is comparable to that occurring in his sperm; indeed, the former may be considerably lower than the latter. Thus studies done on somatic cells of infertile males may not be directly comparable to studies done on sperm of fertile males. Although we would expect to find higher frequencies of abnormal sperm among infertile males, verification of this will require much more research.

OOCYTES. Human eggs are still very difficult to obtain, but "extras" sometimes become available from (usually infertile) females involved in attempts at *in vitro fertilization* (Chapter 24). Wramsby's 1987 study found that almost 50% of such oocytes are chromosomally abnormal.

Prenatal Tests

Chorionic villus sampling (CVS), described in Chapter 25, provides a means of detecting certain chromosomal and metabolic disorders in fetal tissue during early pregnancy. Data accumulated from several thousand CVS tests (Hook et al. 1988) indicate that almost 3% of 10-week-old embryos harbor chromosomal abnormalities. This frequency is significantly higher than the roughly 2% of chromosomal abnormalities found with *amniocentesis* (Chapter 25), a prenatal test that is done at approximately the fifteenth or sixteenth week of pregnancy—and probably accounts for the estimated rate of loss (from spontaneous abortion) of embryos and fetuses between weeks 10 and 16+ of pregnancy.

Abortions, Stillbirths, and Neonatal Deaths

SPONTANEOUS ABORTIONS. The first inkling that a substantial fraction of human embryos are abnormal came from a morphological study done by pathologists Arthur Hertig and John Rock (1949) of Harvard Medical School. From 136 women of proven fertility who were having their uteruses removed, they recovered 28 trophoblasts or embryos of less than two weeks development. Twelve specimens (42.9%) were abnormal, "and of these seven were certainly destined to abort." Because none of the women in whom abnormal ova were found showed any evidence of uterine abnormality, Hertig and Rock concluded that "the defective fertilized ovum is due to intrinsic 'germ plasm' quality rather than to its environment and is the main factor in the production of spontaneous abortion."

At that time the human "germ plasm" was virtually unknown territory. Even now, getting reliable data on spontaneous abortions is very difficult (Leridon 1976). For example, it is not always clear whether an abortion (defined as termination of pregnancy before 20 to 22 weeks of development, or embryonic weight of less than 400 to 500 gm) is spontaneous or induced. Yet the distinction is critical because the former show a much higher frequency of abnormalities of all types than do the latter. In some hospitals the number of supposedly spontaneous abortions dropped by as much as 50% in the early 1970s following the adoption of liberalized abortion laws, while the numbers of live births remained essentially unchanged (Carr and Gedeon 1977). On the other hand, errors in the opposite direction can occur if physicians who collect the abortuses save a higher proportion of abnormal than normal embryos.

It is known that 15–20% of all conceptions end up as a *detectable* spontaneous abortion, but this estimate undoubtedly underrepresents the actual frequency of early embryonic loss, which may be as high as 50–60% (Hassold 1986).* A woman who does not realize she is pregnant, for example, is unlikely to notice a very early abortion, and other abortions occurring at home may not be reported.

Recent technical improvements have allowed researchers to detect extremely minute amounts of the hormone *human chorionic gonadotropin* (hCG) secreted by a week-old implanted embryo; their studies have verified this fact. Wilcox et al. (1988) analyzed hCG levels as a measure of early pregnancy loss in 221 healthy women who were trying to conceive. After analyzing daily urine specimens from each woman over a period of several months, they detected 198 pregnancies—among which 22% ended before the pregnancy was clinically noticeable, that is, after implantation but before the sixth week of development. An additional 9% were lost following clinical confirmation; together these data showed that 31% of pregnancies ended *after* implantation.

What is still not known is how much loss occurs *before* implantation, that is, during the first week of development. Estimates range widely, from about 15% to 55% (Kline et al. 1986; Schlesselman 1979). Researchers basically agree, however, that about 90% of spontaneous abortions occur during the first trimester of pregnancy, and that "the developmental age of the embryo† is generally less than 8 weeks" (Boué et al. 1985).

Fetal losses occur for a variety of reasons, a major one

*Indeed, Roberts and Lowe (1975) have postulated that 78% of all human conceptions are spontaneously aborted, most of them before the first missed menstrual period.

†Many abortuses are not expelled at the time of death but remain in the uterus for several weeks; thus the time of abortion does not necessarily match the developmental age of the embryo.

being the presence of chromosomal abnormalities. In one large study, about 1,500 *early* spontaneous abortions (embryos up to 12 weeks old) occurring either at home or in a Paris hospital over a 6-year period were karyotyped (Boué et al. 1975). Of these, 61% had abnormal karyotypes. Increased rates of aberrations were observed for mothers who had received ovulation-inducing therapy and for fathers whose occupations exposed them to radiation. For younger mothers who smoked and inhaled there was a *lower* rate of chromosomal aberrations among abortuses, because they lost more karyotypically normal embryos than did nonsmoking mothers. Among the 39% of abortuses with normal karyotypes, about half showed developmental abnormalities that could be due to gene mutations or structural aberrations too small to detect.

Summarizing the data from several studies, Hassold (1986), Boué et al. (1985), and Hassold and Jacobs (1984) estimate that *about one-half of all known spontaneous losses (up to 28 weeks of pregnancy) involve chromosomal abnormalities.** In the youngest abortuses (up to 4 weeks old) the frequency of abnormalities is 90–100%; among those aged 5–8 weeks, it is about 60%; and in those 9–12 weeks of age, it ranges from 12–32%. The single most common chromosomal aberration is 45,X, which accounts for about 18% of all karyotypically abnormal abortuses; indeed, it is thought that only 1% of all 45,X zygotes ever reach full term. As a group, trisomics are the most frequent class of abnormality (52%), with trisomy 16 by far the most common among these (15%). Polyploids are the next most frequent (18%) class of abnormality. Studies also suggest that in later stages (after 16–18 weeks) of pregnancy, the rate of loss of fetuses with Down syndrome or with trisomy 18 is much higher than that of fetuses with normal karyotypes (Hook 1978). Because autosomal monosomies are only rarely observed among abortuses or in any other category studied, it is assumed that virtually all are lost very early in development, probably before implantation.

INDUCED ABORTIONS. In striking contrast to spontaneous abortions, only about 5% of *induced* abortions exhibit chromosomal abnormalities. But here, too, the earlier the gestational age, the more frequent the abnormalities—ranging from 9% in 3-to-4-week-old abortuses, to about 5% in 9-to-10-week-old abortuses. As was also true of spontaneous abortuses, approximately half the abnormalities consist of autosomal trisomies. 45,X and triploidies each accounted for about 11% of the abnormalities.

STILLBIRTHS AND NEONATAL DEATHS. Roughly 1% of all recognized pregnancies end in stillbirths. Furthermore, surveys of stillbirths and early infant deaths indicate that the observed rate of chromosomal abnormalities (Table 1) is about one-tenth of that found in spontaneous abortuses, that is, about the same rate as that found in induced abortions. Trisomy 18 is by far the most common aberration in all these subjects, comprising about 35% of the abnormal karyotypes reported (Bauld et al. 1974). Because this particular trisomy is not common among spontaneous abortions, it could be that trisomy 18 is compatible with early embryonic life but exhibits some lethal effects about the time of birth. A striking maternal age effect has been noted in these studies: about 35% of stillbirths and neonatal deaths from mothers over age 40 had karyotypic aberrations, whereas only 6% of those born to mothers under 40 showed chromosomal abnormalities.

*Up to a certain arbitrary point in pregnancy, losses are called **spontaneous abortions** or **miscarriages**; after that they are called **stillbirths**. The dividing line varies somewhat—usually being designated as some point between 20 and 28 weeks.

Unselected Newborns and Some Adult Groups

LIVE BIRTHS. There is a further, tenfold decrease in the frequency of chromosomal abnormalities found in live births below that found in stillbirths (Table 1). The reported rate of abnormal karyotypes—1 per 160–200 births—slightly underestimates the true frequency, because some infants who die within 24 hours of birth are not karyotyped. Among the various abnormal karyotypes, about 35% are considered to be of clinical importance; these include the autosomal trisomics, the Klinefelter and Turner syndromes, and the unbalanced translocations that are accompanied by congenital malformations. All told, the frequency of karyotypically abnormal newborns who are likely to require treatment in later life is about 1 in 440 births.

Hook et al. (1983) estimated the frequencies of these six categories of chromosomal abnormalities (including five trisomies) detected by amniocentesis in older mothers for each year of age from 33 to 49. Then they used these rates, along with other medical data, to calculate the maternal age-specific rates of the same abnormalities in live births "if selective abortion had not occurred." The derived frequencies of all clinically significant anomalies taken together were about 0.5% at maternal age 35 years, 1.5% at age 40 years, and 5% at age 45 years.

UNSELECTED ADULTS. No comprehensive karyotypic studies of randomly chosen adults have been reported, although a small British survey by Court-Brown et al. (1966) turned up a variety of chromosomal abnormalities: translocations and inversions, variations in the length of the Y chromosome in males, and an increase in mosaicism of sex chromosome aneuploidy (for example, the number of 45,X cells) in both aging males and females.

INFERTILE INDIVIDUALS. Among couples who have experienced two or more spontaneous abortions, the rate of chromosomal abnormality is greater than in randomly selected adults. In 7% of these couples, one member has a balanced structural rearrangement (translocation or inversion)—a rate that is 20 times greater than among parents of newborns (Boué et al. 1985).

MENTALLY RETARDED INDIVIDUALS. Studies of mentally retarded individuals reveal many types of abnormalities. Recall that Down syndrome (trisomy 21) accounts for about 10% of all retarded children and that the fragile X syndrome comes in second. In addition, among subnormal males who are institutionalized, about 3% are either 47,XXY or 47,XYY. Balanced translocations and inversions also appear more frequently among mentally subnormal populations (Breg 1977; Funderburk et al. 1977) than among newborns.

Conclusions on Fetal Wastage

It is clear that chromosomal abnormalities constitute a significant medical problem. Yet the burden to families and to society would be enormously increased if a large proportion of karyotypically abnormal conceptions were not spontaneously aborted early in pregnancy. As banding techniques make it increasingly possible to detect translocations and inversions as well as small deletions and duplications, and as the data on their effects accumulate, we continue to learn more about their impact on individuals and on society.

We already know that live-born humans exhibit a vastly higher rate of chromosomal abnormalities than that found in any other species (Chandley 1981)—even when coupled with an astonishing rate of embryonic and fetal wastage that eliminates over 95% of defective conceptions. The reasons for this situation are not clear. Some biologists have suggested that it is the price we pay for human evolution. Others speculate that a high rate of mutation, chromosomal abnormalities, and spontaneous abortions may have the advantage of spacing out births in a species that requires an extremely long period for child raising. In any event, as Dorothy Warburton (1987) points out:

> In these days of efficient contraception, postponement of childbearing, and a small number of carefully planned pregnancies, low fertility and embryonic or fetal loss often cause disappointment and grief when they occur. Few couples realize how common such problems are, and many think that they, or their physicians, are somehow inadequate... Except for maternal age, almost no genetic or environmental factors are known to alter the rates of human chromosomal abnormalities. Thus, physicians and the public should be taught that some degree of reproductive loss is normal and that the best insurance against being childless may be not to allot too short a period for childbearing.

CHROMOSOMAL ABNORMALITIES AND GENE MAPPING

Geneticists have found chromosomal abnormalities to be an extremely useful tool for narrowing down the locations of specific genes. This statement is particularly true of human geneticists, who for obvious reasons cannot develop the pure lines or arrange the testcrosses that are the stock-in-trade of mapping studies in other organisms. Rather, up until the recent application of somatic cell hybridization and molecular techniques to human genetics (Chapter 15), they had to rely entirely on pedigree studies (Chapter 11) and whatever turned up in the way of chromosomal variants.

Variant Chromosomes and Inversions

Occasionally a chromosome carries a distinctive structural feature that is visible in several generations of a family. Such chromosomes are known as **variant chromosomes**. If some members of the family also display a phenotypic trait whose genetic locus is present on the unusual chromosome, it should be possible to observe their joint inheritance. This is in fact how the first assignment of an autosomal gene to a specific human chromosome came about (Donahue et al. 1968). While karyotyping members of his own family, a graduate student at Johns Hopkins University discovered a peculiar "uncoiled" region in the centromeric heterochromatin of the long arm of chromosome 1. Chromosomes with this variant are considerably longer than their normal homologues (Figure 13A), but carriers of the variant are phenotypically normal. It happened that the family also showed segregation patterns for the Duffy blood group (*Fy*) that coincided with the inheritance of the extra-long chromosome (Figure 13B). Additional studies and statistical analyses of several such pedigrees in which *uncoiler* and *Fy* were segregating now place the Duffy locus within a few map units of the heterochromatin on the long arm of chromosome 1.

In a similar manner the major histocompatibility locus, *HLA* (Chapter 18), was assigned to chromosome 6 through pedigree analyses of a family with a detectable pericentric inversion (Lamm et al. 1974). Here the correlation was perfect: a mother, four of her five children, and one grandchild carried the inverted chromosome

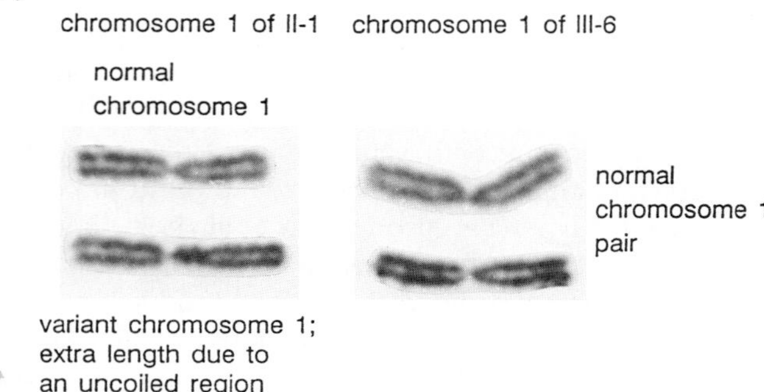

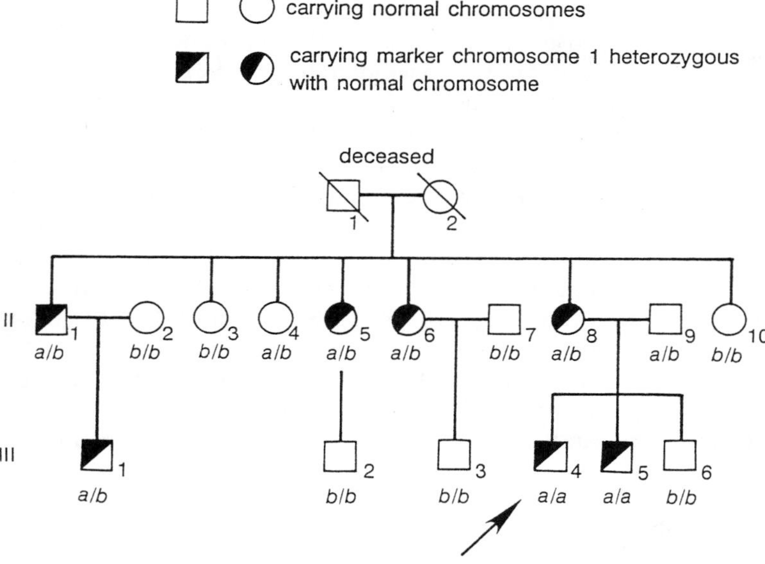

Figure 13 Linkage of a chromosome variant with a blood group. (A) Appearance of a marker chromosome. (B) Redrawn partial pedigree, showing the inheritance of the marker chromosome and the Duffy blood group alleles, *a* and *b* (more complete designations, Fy^a and Fy^b). Note that individuals II-1, II-5, II-6, and II-8 are heterozygous for both the marker chromosome and for Duffy; none of their six children is a recombinant, assuming the marker chromosome carries Fy^a. (From Donahue et al. 1968. Photographs of banded chromosomes courtesy of Irene Uchida, McMaster University.)

and expressed a particular *HLA* allele. This allele was absent in other family members, all of whom had structurally normal chromosomes. From this and other studies, the *HLA* locus (actually a complex of genes) is known to reside in the distal half of the short arm of chromosome 6.

Deletions, Duplications, and Translocations

Deletion mapping hinges on the principle that the absence of a chromosome segment should be correlated with the disappearance of the expression of genes on that segment. For example, Ferguson-Smith et al. (1973) of Glasgow, Scotland, reported that the gene for acid phosphatase is located on the short arm of chromosome 2. This conclusion came from studies of a child with severe congenital malformations due to loss of the tip of 2p and duplication of the tip of 5q (Figure 14). The large translocation chromosome, along with a normal chromosome 5, was inherited from his mother, a phenotypically normal translocation heterozygote. The mother was homozygous (P^a/P^a) for an acid phosphatase allele controlling the production of one isozyme, and the father was homozygous (P^b/P^b) for an allele that controlled a different, electrophoretically distinguishable isozyme. Thus, both electrophoretic bands should have appeared when enzyme samples from the child were tested. Instead, he expressed only the paternally inherited P^b allele. Because he also lacked the tip of the maternal chromosome 2, it followed that this is where the P^a allele must reside. The total activity of acid phosphatase in the son was only 55% of normal, confirming that he was hemizygous for the acid phosphatase gene. This was the first time that a gene assignment had been made by deletion mapping alone; the finding has since been verified by several other methods.

Deletions have in some cases been useful in *ruling out* a postulated map position for a given gene, if, in the presence of a deletion including a particular chromosome segment, the gene in question is still expressed. This is hardly an efficient means of mapping, however. Deletions on the X chromosome cannot be used at all for this method because random X inactivation is followed by survival of cells with the most balanced karyotype. Thus an X chromosome with a deletion is always found to be inactivated in surviving somatic cells. This special case of *nonrandomness* of the Lyon effect means that genes on the structurally normal X chromosome are always expressed in deletion heterozygotes.

Attempts to locate genes by *trisomy mapping*, a standard technique in corn genetics, have yielded mixed results when applied to humans. In theory, individuals with three representatives of a chromosome or chromosome segment should produce an excess of those gene products whose loci are present in triplicate rather than in duplicate. This effect has been found for some loci, but studies of Down syndrome patients and other trisomics have not always borne this out. Yet trisomy mapping has occasionally been useful in confirming a gene assignment that was initially made by other methods. For example, two loci on chromosome 21—*superoxide dismutase* and *anti-viral protein*—do show dosage effects of the expected type: trisomics produce excess protein and monosomics have a reduced amount (Feaster et al. 1977).

Partial trisomies or monosomies have been utilized to assign genes to specific regions of chromosomes. From analyses of enzyme levels in several individuals who carry independently occurring duplications or deletions for the same chromosome, it is occasionally possible to determine with some accuracy the location of the gene responsible for that enzyme. One important study of this type (Ferguson-Smith et al. 1976) involved eight patients with aberrations that collectively spanned the entire chromosome 9 (Figure 15). Production of the red blood cell enzyme adenylate kinase (AK), known from hybrid cell studies to be on chromosome 9, was measured in these patients and compared with that of karyotypically normal controls. The AK enzyme levels of those persons with deletions or duplications fell within the normal range except for one patient (Case 7), who exhibited a highly significant 43% increase in AK activity. Because hers was the only duplication that included the tip of 9q, it was concluded that the *AK* gene must reside there. What made this finding particularly noteworthy was that the *AK* locus had already been shown by pedigree studies to be closely linked to two other loci, the ABO blood group gene (*I*) and the nail–patella syndrome locus (*Np*). But until these chromosomal analyses were done, it was not known which autosome the three loci occupied. Now it was clear that they all resided near the tip of the long arm of 9. Assignment of the ABO blood group locus—the first simply inherited human trait to be discovered and one of the most intensively studied of all human genes— was particularly satisfying.

Studies of pedigrees with chromosome 16 transloca-

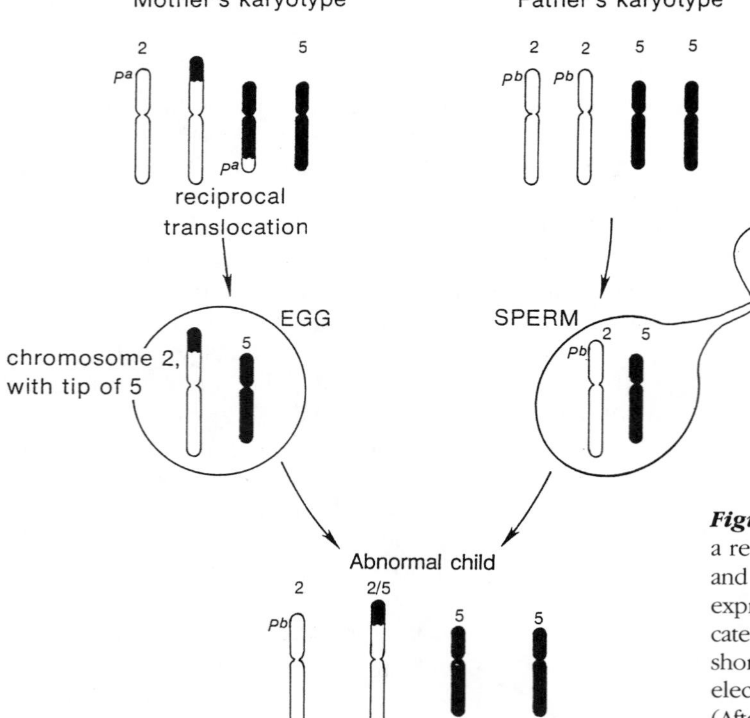

Figure 14 Results of a mating between a P^a/P^a female, with a reciprocal translocation between chromosomes 2 and 5, and a P^b/P^b male. Their child, who should have been P^a/P^b, expressed only the paternally inherited P^b locus, which indicates that a P^a gene must reside on the missing tip of the short arm of chromosome 2. The alleles P^a and P^b control electrophoretically different isozymes of acid phosphatase. (After Ferguson-Smith et al. 1973.)

tions first suggested that the gene *Hpa*, which controls the α polypeptide chain of a serum protein known as haptoglobin, resides on the long arm of that chromosome. Clinching evidence came with the discovery of a large pedigree in which many members had a fragile site on chromosome 16 (Magenis et al. 1970). Many white blood cells from these individuals showed chromosome gaps at a specific site about two-thirds of the way from the centromere to the tip of the long arm—a property that was inherited as a simple Mendelian dominant trait. Of 33 offspring from persons heterozygous for both the fragile site and the *Hpa* gene in this pedigree, only three were recombinants. This finding indicates close linkage of *Hpa* with the fragile site.

Since the early 1980s, the precision of all mapping studies, including those that involve chromosomal abnormalities, has been greatly refined and enhanced by molecular genetic technologies (Chapter 15). But the general principles remain the same. Recall that isolation of the Y-linked locus for *TDF* (testis-determining factor) was made possible by DNA analyses of several dozen "sex-reversed" individuals. These patients were XY females who had a deletion in their Y chromosome and XX males who turned out to have a tiny piece of a Y chromosome attached to one X. By relating the sex of the carrier to the original location of the deleted or added piece of Y, and also by looking for the smallest "common denominator"— just as was done for the *AK* gene in Figure 15—the researchers were able to pinpoint the critical male-determining region (Page et al. 1987a).

The power of deletion mapping when combined with the newer technologies is perhaps best illustrated by DNA analyses of a few rare Duchenne muscular dystrophy patients who suffered from additional X-linked recessive disorders. One of these affected males also had chronic granulomatous disease, retinitis pigmentosa, McLeod syndrome—and a tiny deletion in the midsection of the short arm, that is, region 21 of Xp (Francke et al. 1985). Another affected male had glycerol kinase deficiency and adrenal hypoplasia—and a similar-sized, partially overlapping deletion in the same region (Wieringa et al. 1985). An affected female carrier of an X;21 translocation had a deletion at the breakpoint of X in Xq21 on her translocated chromosome (Ray et al. 1985). Certain molecular markers (highly specific sites cut by so-called restriction enzymes) were also identified in each DNA sample, and they were compared among themselves. They were also compared to DNAs from four different (non-DMD) glycerol kinase-deficient patients, each having a small deletion in Xp21. By fitting together all the various puzzle pieces of cytogenetic, phenotypic, and molecular data, a team of Dutch, American, and Canadian researchers (van Ommen et al. 1986) was able to come up with a concise description of the chromosomal region involved in these diseases. They showed, for example, that the chronic granulomatous disease (*CGD*) locus must be *proximal* to (i.e., closer to the centromere than) the *DMD* locus, whereas the glycerol kinase (*GK*) and adrenal hypoplasia (*AH*) loci are *distal* to (i.e., further from the centromere than) *DMD*. They also constructed a map of the molecular landmarks in this region, covering at least 4 million DNA base pairs.

Series of chromosomes with overlapping deletions have also been used to determine the order of several genes on the short arm of chromosome 11 (van Heyningen and Porteus 1986). Comparable studies of other important disease loci (such as cystic fibrosis, phenylketonuria, hemophilia A, and Huntington disease) and any chromosomal abnormalities they are linked to will undoubtedly add greatly to our understanding. Indeed, the discovery of so many deletions associated with some of these loci has already muddied the distinction between simple mutations and tiny chromosomal aberrations.

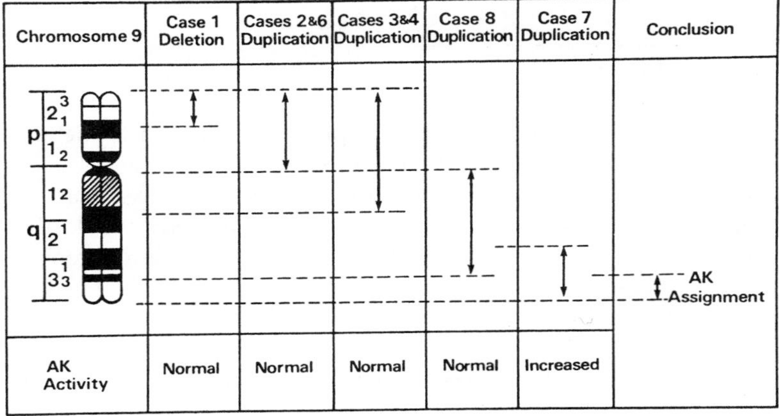

Figure 15 Regional mapping of the adenylate kinase (*AK*) locus based on data from individuals with deletions or duplications that cover different chromosome regions. Only one individual (Case 7 on the far right) showed abnormal enzyme production, an increase that was correlated with the presence of a duplicated region at the tip of 9q. Thus the *AK* gene must reside in this region. (From Ferguson-Smith et al. 1976.)

SUMMARY

1. Chromosomal rearrangements arise when segments of broken chromosomes reattach in different ways. Detached acentric pieces of chromosomes are lost.
2. Deletions of more than a few genes are usually lethal; those in group B chromosomes (especially 5p−) are the type of deletion most often seen in newborns.
3. Deletions and duplications are often the by-products of meiotic segregation from other rearrangements. Extra chromosomal material (duplications, trisomies) is phenotypically less harmful than the loss of chromosomal material (deletions, monosomies). Duplications are thought to have played an important role in evolution.
4. Meiotic segregation from heterozygous translocations leads to aberrations in some gametes. Translocation Down syndrome and Prader-Willi syndrome are associated with specific types of rearrangements.
5. Crossing over within the loop of pericentric and paracentric inversion heterozygotes usually leads to imbalance in some gametes. In paracentric inversions, dicentrics and acentrics are also formed.
6. Polyploidy is rare in live-born animals, but triploidy accounts for about 6% of early human spontaneous abortions.
7. The frequency of chromosomal abnormalities (mostly trisomies, polyploids, and 45,X) among early spontaneous abortuses is at least 50%. The observed spontaneous abortion rate of 15–20% would be much higher if very early (preimplantation) losses could be detected.
8. About 5% of stillbirths and roughly 0.5% of live births exhibit chromosome abnormalities. About half of the latter consist of structural aberrations (mostly translocations), and half are trisomies.
9. Many mentally subnormal persons are karyotypically abnormal, having either Down syndrome or the fragile X syndrome.
10. Traditional pedigree studies involving chromosome aberrations, unusual variant sites, deletions, inversions, and translocations have proved useful —especially when combined with newer molecular methods—in localizing genes to specific regions of the chromosomes.

FURTHER READING

Books by Therman (1986) and Valentine (1986) deal with human cytogenetics in general. Dellarco et al. (1985), Bond and Chandley (1983), Porter and Hook (1980), and Epstein (1986) concentrate on various aspects of aneuploidy. For reviews of frequencies of chromosomal abnormalities in abortions, live births and other categories, see Hook (1985), Hassold (1986), Boué et al. (1985), Hassold and Jacobs (1984), and Hsu (1986). For details about chromosome structure and abnormalities, consult de Grouchy and Turleau's (1984) illustrated clinical atlas of human chromosomes and Borgaonkar's (1984) catalogue of chromosomal variants.

QUESTIONS

1. Diagram the synaptic configurations (with gene-by-gene pairing) of a pair of chromosomes *heterozygous* for (a) an interstitial deletion; (b) a tandem duplication; (c) a ring; (d) a simple (i.e., nonreciprocal) translocation with one chromosome arm being attached to the end of a different chromosome.
2. Diagram the synaptic configurations that would be seen in chromosome pairs *homozygous* (a) for an inversion; (b) for a reciprocal translocation.
3. Among the parents of Down syndrome patients, a few have 45 chromosomes—including a Robertsonian translocation between the two chromosomes 21. (a) What types of gametes would they form? (b) Could they produce normal progeny?
4. How can a metacentric chromosome be changed to an acrocentric one?
5. What types of gametes would be produced if, within the inversion loop shown in Figure 10B, a

second crossover occurred and involved the same two chromatids that took part in the first crossover?

6. Diagram the gametes that would be produced if, within the inversion loop shown in Figure 10B, the centromere were located between genes *D* and *E* rather than between genes *A* and *B*.

7. List two ways by which a triploid could arise.

8. Is there any way by which fertilization between two aneuploid gametes produced by individuals with the same type of translocation heterozygote could produce a euploid zygote? (*Hint:* See Figure 7.)

9. Lucas (1969) describes a woman with 45 chromosomes, one of which was a Robertsonian translocation between the two number 15 chromosomes. She was childless, having had 12 spontaneous abortions. What was the reason for this?

10. Jacobs et al. (1974) report an apparent excess of translocations involving a terminal and centromeric breakpoint. One possible explanation is that the terminal and centromeric regions are simply more susceptible to breakage and reunion than are the intervening (medial) regions. They propose, however, that the more likely possibility is an observational bias. Suggest an explanation. Is this bias likely to occur in current studies? Explain.

11. Individuals who have an isochromosome of the long arm of the X chromosome are often mosaics, possessing a 45,X cell line as well. How did this come about?

ANSWERS

1. Both (a) and (b) will show a looping out of one chromosome:

(a)

c d
a b e f
A B E F

(b)

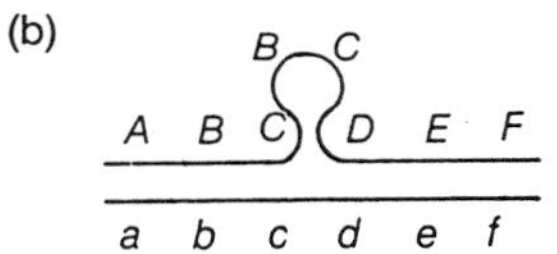

(c)

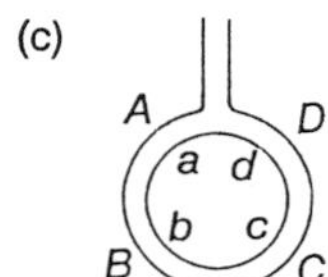

(d)

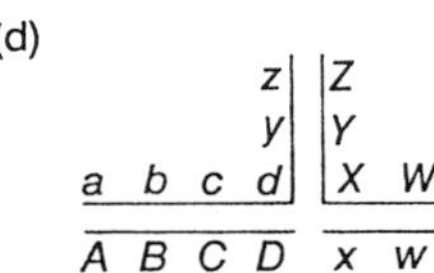

2. Neither (a) nor (b) require any unusual pairing configurations, since the homologues match:

(a) A B C (E D) F G
a b c (e d) f g

(b) A B C X Y Z U V W D E F
a b c x y z u v w d e f

3. (a) Half the gametes would contain the (21q21q) translocation chromosome (giving rise to progeny with Down syndrome), and half would contain no chromosome 21 at all (causing lethal monosomy 21). (b) No.

4. By a pericentric inversion:

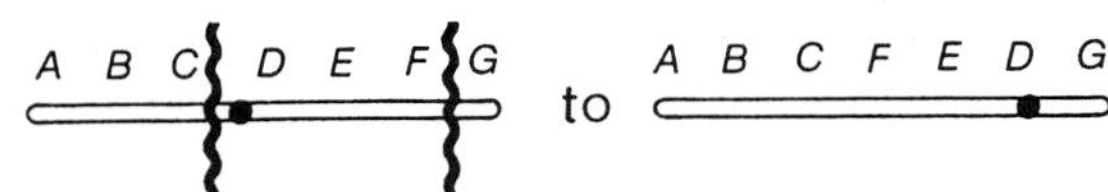

The reverse of this phenomenon is probably what accounts for the few metacentric Y chromosomes that have been reported in human males.

5. The second crossover, if it involved the same chromatids as the first one, will "undo" the aneuploidy that would otherwise have resulted. Although the two crossover chromatids will exchange some alleles, they will not possess any duplications or deficiencies, that is, they will be euploid. As a result, the more rarely occurring double crossovers from inversion heterozygotes are detected more frequently than are single crossovers—a peculiar situation indeed.

6.

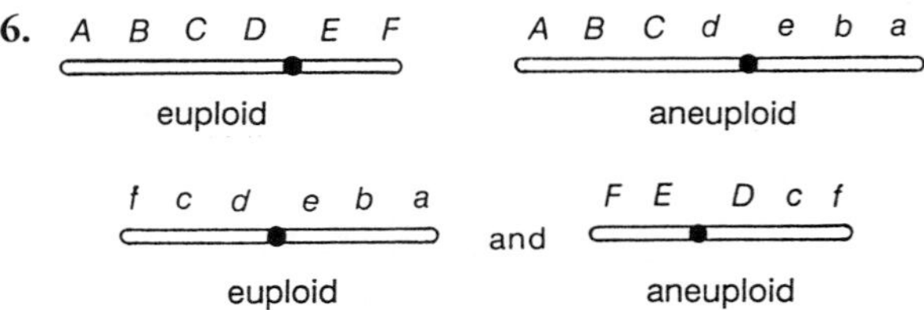

7. A triploid could arise (1) if the egg and the polar body nuclei fused to form a diploid egg nucleus that was fertilized by a haploid sperm, or (2) if two sperm fertilized a haploid egg.

8. Only if the two aneuploidies were precisely complementary, so that the duplicated chromosomes or chromosome segments present in one were balanced by deletions of the same genetic material in the other, and vice versa (i.e., mating of successive pairs of gametes in Figure 7). This event is highly unlikely, even if the two parents were related and heterozygous for the same translocation.

9. Because both chromosomes 15 were stuck together, all her eggs contained either *two* attached chromosomes 15 or *no* chromosome 15. Fertilized by normal sperm, all the resultant zygotes were therefore either trisomic or monosomic for a medium-sized chromosome and thus

inviable. One abortus was in fact recovered and found to be trisomic for chromosome 15.

10. Until 1971, human chromosome rearrangements could be detected only through changes in chromosome length or centromeric index, which selected for translocations of unequal-sized segments. The difference in length between the two exchanged parts is maximized when one piece is broken off near the tip (terminal break) and the other segment is an entire arm (centromeric break).

11. If a gamete with an X chromosome long arm resulting from misdivision of the centromere (the short arm being lost) united with a normal X-bearing gamete, events in the dividing zygote could be as follows: (1) Both chromosomes replicate during prophase of the first cleavage division. (2) The two sister chromatids of the long arm joined together by the misdivided centromere to form one large isochromosome. (3) Following mitosis, each daughter cell contains one normal X chromosome, but the isochromosome will be present in only one of the two cells. (4) Thus a mosaic results, in which one cell line has 46 chromosomes (including the isochromosome of Xq) and the other cell line is 45,X.

PART THREE **GENE TRANSMISSION**

10. Allelic Segregation and Probability
11. Recombination of Nonallelic Genes
12. Complicating Factors

Chapter 10 *Allelic Segregation and Probability*

Multiple Alleles
Origin of Multiple Allelic Series
Combining Alleles into Genotypes

The Phenotype of a Heterozygote
Levels of Observation

The Rules of Probability
Definition
Five Rules for Combining Probabilities

Problem Solving
The Binomial
Conditional Probability

I'm very well acquainted too with matters mathematical,
I understand equations, both simple and quadratical,
About binomial theorem I'm teeming with a lot o' news,
With many cheerful facts about the square of the hypotenuse.

W. S. Gilbert (1879)

In earlier chapters we discussed Mendel's first law and the chromosomal basis of inheritance. In the next several chapters we consider both of Mendel's rules of transmission. We examine traditional methods for locating several genes on a chromosome with respect to each other and discuss some complications in human genetic analysis. This material is in part mathematical (the binomial theorem, for example) and some readers may wish to review Appendix 2 first.

MULTIPLE ALLELES

Recall that Mendel's law of segregation states that paired alleles segregate during the formation of gametes, so only one of the two alleles is included in an egg or sperm. Which of the two alleles ends up in a specific gamete is a matter of chance. Thus from an *A/a* heterozygote, half of the gametes on the average contain *A* and the other half *a*. In such a heterozygote the alleles do not "contaminate" each other: for example, the phenotype of *a/a* is the same whether it results from a mating of *A/a* × *A/a*, or *A/a* × *a/a*, or *a/a* × *a/a*.

Considering a single gene with two alleles, *A* and *a*, recall also that these can be combined into three genotypes, and these genotypes can be combined into six types of matings. The expected offspring classes from each of these matings are easily predictable from a checkerboard and have ideal genotypic ratios of 1:2:1, or 1:1, or 1:0.

Origin of Multiple Allelic Series

The various forms of a gene—the alleles—are generated by the process of mutation. This is the chemical change of one allele, say, the "standard" or "wildtype" one, into another, the "variant" or "mutant" one (or vice versa). So far, we have only considered genes with two alleles, but there is nothing magical about this number. Mutations could theoretically generate as many different alleles as there are chemical changes possible in a gene. Some genes have hundreds of known alleles.

To designate the various allelic forms of a gene it is convenient to use sequential superscripts (or subscripts): $A^1, A^2, A^3, \ldots, A^N$, where N is the highest number in the allelic series.* The important thing to remember about any series of multiple alleles is that *any one individual still has only two alleles* (or *one*, for X-linked genes in a human male). These can be selected from a larger number, N, existing in the population as a whole.

*Where appropriate, we follow the international system for naming human genes and alleles (Shows et al. 1987). This terminology is designed for researchers, however, and is usually more cumbersome than needed for this book.

A good example of a multiple allelic series is the gene that codes for the beta polypeptide chain of the hemoglobin molecule (Chapter 14). Over 100 alleles of this gene have been discovered, and more will certainly be found. Four of the most common alleles of this gene (in at least some regions of the world) are often designated Hb^A, Hb^S, Hb^E, and Hb^C. The superscript *A* refers to the normal allele, and *S*, *E*, and *C* refer to allelic forms of the gene that lead to various abnormal beta polypeptide chains. Phenotypically, homozygotes for Hb^S have the disease sickle-cell anemia and are often quite ill. Homozygotes for Hb^E and homozygotes for Hb^C have only a mild or moderate anemia. Individuals who are heterozygous for the normal allele combined with either Hb^S, Hb^E, or Hb^C are normal, or nearly so.

Combining Alleles into Genotypes

Consider a gene with *three* alleles, B^1, B^2, and B^3. For each allele, there is a *homozygous genotype*. These are listed in the left column below. For each pairing of different alleles selected from the three available, there is a *heterozygous genotype*. These are on the right.

$$
\begin{array}{ll}
B^1/B^1 & B^1/B^2 \\
B^2/B^2 & B^1/B^3 \\
B^3/B^3 & B^2/B^3
\end{array}
$$

With *four* alleles of a gene, C^1, C^2, C^3, C^4, the reader can confirm that there are 10 genotypes: four homozygous and six heterozygous. One methodical way for writing down all the genotypes is to first combine C^1 with itself (C^1/C^1) and sequentially with each allele to the right of it in the list (C^1/C^2, C^1/C^3, C^1/C^4), a subtotal of four genotypes. Then combine C^2 with itself (C^2/C^2) and sequentially with each one to the right of it (C^2/C^3, C^2/C^4), a subtotal of three genotypes. Continue this process until you have run out of combinations, in this case as far as C^4/C^4, a subtotal of one genotype.

In general, for a gene with *N* alleles, A^1, A^2, A^3, . . . , A^N, this listing and combining will yield all possible genotypes, albeit in a tedious fashion. Readers proficient in algebra can derive a formula for the number of possible different genotypes for a gene with *N* alleles:

$$\frac{N(N + 1)}{2}$$

Of this number, *N* are homozygous and the rest are heterozygous. Using this formula, confirm that it gives the correct number of possible different genotypes for $N = 2$, 3, or 4 alleles (noted above). As the number of alleles of a gene increases, the number of possible different genotypes increases steeply; with 10 alleles there are 55 different genotypes, and with 20 alleles 210 different genotypes. All these genotypes may occur in some populations but some may be missing, that is, not present in any individual now living, or even once living. In any event, we emphasize again that any individual genotype is made up of only two alleles for each gene.

THE PHENOTYPE OF A HETEROZYGOTE

How a person looks depends both on that person's genes and on the environment in which the person develops and lives. It often happens, however, that a specific phenotype is consistently associated with the same genotype, even in a wide range of environments and in the presence of various alleles at other loci (the "background" genotype). The phenotypes produced by various combinations of alleles of one gene may be exactly the same, or somewhat different, or very different (as in the case of a serious inherited disorder). Geneticists use a shorthand description to indicate the relationship between the phenotypes produced by the alleles of a gene. The terms, dominant and recessive, describe a common situation we have met in previous chapters. These and other terms refer to the comparative phenotype of a heterozygote. Considering a gene with two alleles, A^1 and A^2, and therefore three genotypes, A^1/A^1, A^1/A^2, and A^2/A^2, four ways of relating genotypes with phenotypes have such shorthand names:

ALLELES DOMINANT AND RECESSIVE. If the heterozygote, A^1/A^2, looks like one of the homozygotes, we speak of **dominance** and **recessiveness**. Specifically, if the heterozygote has the same phenotype as A^1/A^1, we say that A^1 is a dominant allele and A^2 a recessive allele. Albinism is a good example: the gene A^1 is responsible for the production of an enzyme (tyrosinase); its allele A^2 does not produce a normal enzyme. Phenotypically,

$A^1/A^1 = A^1/A^2 =$ normal pigmentation, a dominant trait
$A^2/A^2 =$ albinism, a recessive trait

Usually a dominant allele is designated by an uppercase letter, and the recessive by the corresponding lowercase. Thus, rather than using superscripts, the example of albinism is usually designated

$C/C = C/c =$ normal
$c/c =$ albino (*c* for colorless)

Dominance versus recessiveness is a common situation, because an individual with only one wildtype allele often makes enough good enzyme to allow normal development; that is, half the normal amount of enzyme will suffice. In general, enzymes are needed in only small amounts to catalyze metabolic reactions. Dominance is a reflection of a safety factor built into organisms through the course of their evolution. It is not an essential part of Mendel's law of segregation, nor does it imply any kind of active struggle; dominance is just a convenient shorthand description of the way the products of alleles interact to produce a phenotype.

HETEROZYGOTE INTERMEDIATE. If the heterozygote, A^1/A^2, looks like a homogeneous blend of the characteristics of the two homozygotes, we speak of an **intermediate heterozygote.** Other terms for this situation are *semidominance*, *intermediate* or *partial dominance*, *incomplete dominance*, and *incomplete recessiveness*. Clear-cut cases of intermediate heterozygotes for easily observable phenotypic traits are rare in human genetics, but nice examples exist for flower color in plants and for fur color in mammals. For example, red flowers result from a pigment called anthocyanin, whose synthesis requires an enzyme made by *R* but not by its allele *r*. In snapdragons, *R/R* plants are red; *R/r*, pink; and *r/r*, white.

In people, the inheritance of so-called quantitative traits (Chapter 23) like height, IQ, or blood pressure is often presumed to involve many different genes, each with a small, similar, additive effect on phenotype. For each gene, the heterozygote is often considered to be intermediate in expression between the effects of the corresponding homozygotes.

ALLELES CODOMINANT. If the heterozygote, A^1/A^2, expresses both homozygous phenotypes—not a blend of the two but an unmixed combination—we speak of **codominance**. Each allele maintains its distinctive homozygous expression, largely unaltered, even when heterozygous. The blood group genes (Chapter 17) provide good examples of codominant expression. Very briefly, each allele causes a chemical, called an *antigen*, to be present on the surface of red blood cells. The several alleles of a gene lead to slightly different, but distinguishable, antigens. A person heterozygous for two alleles possesses both antigens, and each is detectable as it would be in a homozygote.

Codominance is often the rule at the level of the immediate gene product, that is, the polypeptide. The two variant polypeptides coded by the two different alleles of a gene are usually both present in roughly equal amounts in a heterozygous person. The polypeptides may interact during development to produce, say, intermediate expression at the level of the external phenotype, but at an underlying level the alleles are acting codominantly.

HETEROZYGOTE OVERDOMINANT. If the heterozygote, A^1/A^2, expresses a phenotype more extreme than either homozygote, we speak of **overdominance.** "More extreme" can mean more adapted to environmental circumstances or more fertile, and in these cases we speak of *heterozygous advantage*. That a heterozygote may be better off than either homozygote is an important idea in evolutionary theory (Chapter 22), but how widespread the phenomenon might be is unknown.

Overdominance plays a role in the production of some high-yielding food crops. Hybrid corn plants, for example, are uniform in appearance, consistently big-eared, and, because of the way they were derived, highly heterozygous. It is not entirely clear *why* hybrid corn has a superior phenotype, but heterozygous advantage for some genes is thought to provide a partial explanation.

Note that the "phenotypic value" of the heterozygote can be less than one of the homozygotes, equal to a homozygote, or greater than either homozygote. This is illustrated in Figure 1.

Levels of Observation

The label one uses to describe the way the alleles interact to produce a phenotype depends on how closely one

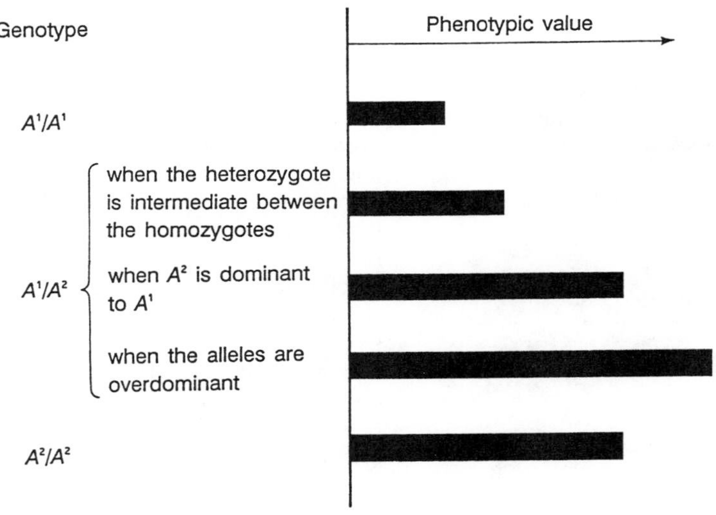

Figure 1 A graphical display of the meaning of intermediate heterozygotes, dominance, and overdominance. The phenotypic value (horizontal axis) can be given in pounds, inches, color gradations, number of offspring, IQ points, or whatever measures the trait in question.

examines the phenotype. (By analogy, an Ansel Adams photograph, however exquisite, looks like disordered specks of silver when viewed under high magnification.) The gene, Hb^A, coding for a normal beta polypeptide chain of hemoglobin, and its allele, Hb^S, coding for the abnormal sickle-cell chain, illustrate this phenomenon:

Hb^S RECESSIVE TO Hb^A. To clinicians, only homozygous individuals (Hb^S/Hb^S)—those with sickle-cell anemia—need be hospitalized. Their fragile red blood cells rupture and clog capillaries, thereby leading to excruciating pain in the abdomen or in the joints. It is a physically disabling disease, characterized by frequent infections and a failure to thrive. Heterozygotes generally do not come to clinical attention, although they may occasionally experience milder symptoms in certain stressful situations.

Hb^S DOMINANT TO Hb^A. To laboratory workers examining blood microscopically, the Hb^S allele behaves as a dominant rather than as a recessive. In the standard test for sickling, a drop of blood on a microscope slide is sealed under a cover slip. Technicians then see irregularly shaped, i.e., sickled, red blood cells from *both* homozygotes with sickle-cell anemia and heterozygotes with no clinical symptoms (Figure 2). The sickling of the red blood cells of heterozygotes occurs only in this test situation (i.e., in vitro—literally, "in glass"), not within their circulatory systems (in vivo—literally, "in life"), The heterozygotes are sometimes said to have the sickle-cell trait, but it does them no harm.

Hb^S AND Hb^A CODOMINANT. To biochemists analyzing the beta chains isolated from a heterozygote, the presence of two different polypeptides (albeit, differing by a single amino acid) indicates codominance. One of the polypeptides is coded by the Hb^S allele, and the other by Hb^A.

Hb^S/Hb^A HETEROZYGOTE INTERMEDIATE. To carriers themselves, occasional subclinical symptoms suggest that their phenotype is intermediate. During conditions of lowered oxygen supply (which could occur, for example, during high-altitude flight in nonpressurized cabins), some heterozygotes may experience abdominal pain and produce bloody urine, symptoms intermediate between normal and affected. Heterozygotes also have some sickled red blood cells in vivo, but far fewer than the 30% characteristic of those with sickle-cell anemia.

Hb^S/Hb^A HETEROZYGOTE OVERDOMINANT. To population geneticists, the sickle-cell gene has become the best known example of heterozygote advantage. Medical detective work during the 1950s showed that the heterozygote was almost completely resistant to malaria. Thus, in areas of the world where malaria is present, heterozygotes are better able to live and reproduce than either those with sickle-cell anemia or those who are homozygous normal. (See Chapter 22 for details.) Note that the advantage of being heterozygous does not extend to areas of the world where malaria is absent.

THE RULES OF PROBABILITY

Up to this point we have used checkerboards to predict the offspring of crosses involving alleles of one gene. In doing so, we have used concepts of probability without actually saying so. These procedures could be expanded to deal with several genes, but there are quicker, less tedious ways of arriving at the same predictions—using explicit rules of probability. Although some aspects of this kind of thinking might be new, much of the following material you have probably used intuitively.

Flipping a coin, for example, we can ask, "What is the *chance* or *likelihood* of a head?" Flipping the coin 100 times, we might make a bet on the question, "How often will the *proportion* or *frequency* of heads be greater than

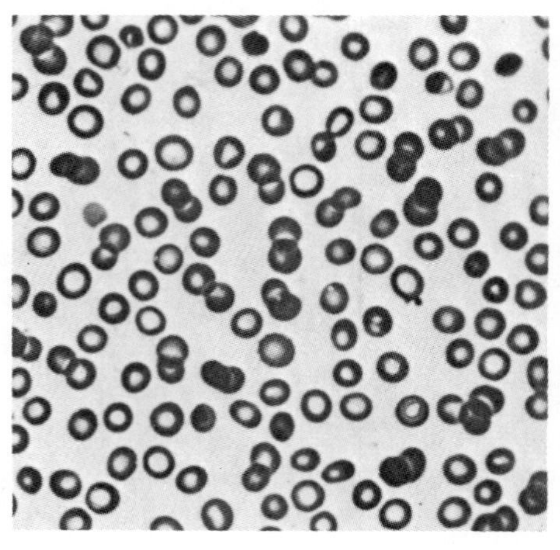

Hb^A/Hb^A

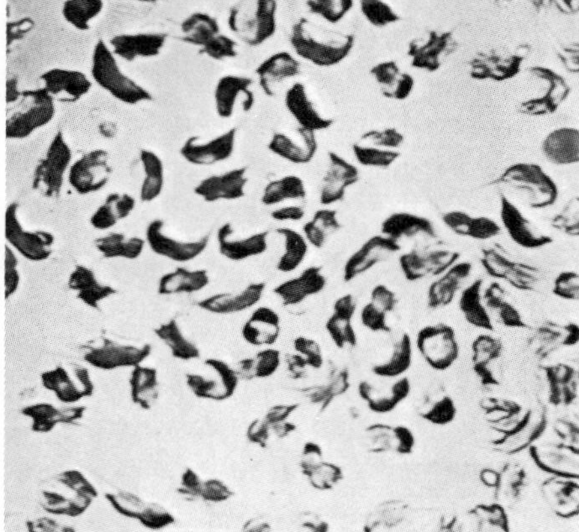

Hb^A/Hb^S or Hb^S/Hb^S

Figure 2 A laboratory test for sickling. In a thin, sealed (to reduce the amount of oxygen) film of blood, normal red blood cells retain the rounded appearance on the left. Red blood cells with some amount of sickle-cell hemoglobin assume the shapes shown on the right. Sickling is shown by the red blood cells for *both* a person homozygous and a person heterozygous for the sickle-cell allele (×500). Only the homozygote becomes ill, however. (Micrographs courtesy of C. L. Conley.)

80 (or some other number)?" These terms are roughly synonymous with *probability*. To be specific, however, probability is always written as a number between 0 and 1, expressed as a fraction (like 1/3) or a decimal fraction (like 0.333).

Definition

More rigorously, we can conceive of doing an experiment like picking a card, rolling a die (plural, dice), or selecting a gamete. In each case, there is a certain number of equally likely outcomes: *52* for a deck of cards, *6* for a die, or *2* for the case of a person heterozygous at a given locus. Before doing the experiment, we ask about the likelihood of some event, A, occurring; its probability can be written

$$\mathbf{P(A)} = \frac{\textbf{number of outcomes "favorable" to A}}{\textbf{total number of equally likely outcomes}}$$

The notation P(A) should be read "the probability of event A occurring," and its definition is a commonsensical one. In the case of drawing a card randomly from a deck, there are 52 equally likely outcomes (the denominator). Of these, 13 are in a suit "favorable," say, to the event "drawing a heart" (the numerator). Therefore, the probability of drawing a heart is

$$\text{P(heart)} = 13/52 = 1/4 = 0.250$$

Using the definition again, readers will note that for a die, fairly balanced and thrown,

$$\text{P(even number)} = 3/6 = 1/2 = 0.500$$
$$\text{P(at least a 3)} = 4/6 = 2/3 = 0.667$$

The phrase "at least a 3" means throwing a 3 or higher number: a 3 or a 4 or a 5 or a 6, a total of four favorable outcomes. For the formation of gametes in a *B/b* heterozygote

$$\text{P}(B\text{ gamete}) = 1/2 = 0.500$$
$$\text{P}(b\text{ gamete}) = 1/2 = 0.500$$

If an event is *impossible*, there will be *no* outcomes favorable to it, so the numerator is 0. On the other hand, if an event is *inevitable*, then *every* outcome is favorable to it, and the numerator is the same as the denominator. Thus,

$$\text{P(impossible event)} = 0$$
$$\text{P(inevitable event)} = 1$$

For example, the probability of rolling a 7 with one die is 0, and the probability of rolling at least a 1 is 1.

If the probability of event A occurring is some number, x, what is the probability of A *not* occurring? Something must happen; it is inevitable that either A occurs or it does not occur; so the probabilities of events A and not-A must add to 1. Therefore, if

$$\text{P(A occurring)} = x$$

then

$$\text{P(A not occurring)} = 1 - x$$

For example, if the probability of some snow tomorrow is 1/10, then the probability of no snow is 9/10.

We can generalize the foregoing a bit by saying that the probabilities of a set of events must add to 1 if the events form an exhaustive set, that is, if every possible event is included. Thus, the sum of probabilities of an exhaustive set of events equals 1. This statement often provides a useful check on your arithmetic. For example, from the mating, $B/b \times B/b$, the sum of the probability of heterozygous offspring plus the probability of homozygous offspring is 1.

Five Rules for Combining Probabilities

Sometimes the probability of interest is not a simple one that can be gotten directly from the definition. For example, the probability of a straight in poker (five cards in sequence regardless of suit) is far from obvious.* What we need are some notions for combining several simple probabilities to give us the probability of a more complex event. We present, in an intuitive fashion, five such rules.

THE ADDITION RULE (THE "OR" RULE). Consider an experiment that can result in the events A, B, C, D, . . . Our interest is not in just one of these events, but in, say, A *or* B occurring. Event A could be "picking a heart" and B, "picking a spade." In shorthand notation, we are interested in P(A *or* B). Intuitively,

$$\text{P(A or B)} = \text{P(A)} + \text{P(B)}$$

or, for more than two events,

$$\mathbf{P(A}\ \boldsymbol{or}\ \mathbf{B}\ \boldsymbol{or}\ \mathbf{C}\ \boldsymbol{or}\ \mathbf{\ldots) = P(A) + P(B) + P(C) + \ldots}$$

In words, if we want the probability of one event *or* another *or* yet another, and so on, we *add* their individual probabilities together.

Whenever the conjunction "or" is used in everyday language, the addition rule is usually appropriate. For example, for a gene with two alleles, *B* dominant to *b*, what is the probability of a person having the dominant phenotype? The numerical answer would depend upon the parental genotypes, but, in any case, note that the dominant phenotype arises if a person is either *B/B* or *B/b*. Therefore,

$$\text{P(dominant phenotype)} = \text{P}(B/-) = \text{P}(B/B) + \text{P}(B/b)$$

There is a restriction in the use of the addition rule: *the events in question must be mutually exclusive*. This means that the occurrence of one event excludes the

*The probability is 768/216,580. See Feller (1968).

other(s) from happening. The reader will discover that, in genetics, this is usually the case. If an organism has the genotype *B/B*, it does not have the genotype *B/b*, or any other genotype. If a card is a heart, it is not a spade, and so forth; thus,

P(heart *or* spade) = 1/4 + 1/4 = 1/2
P(heart *or* spade *or* diamond *or* club)
= 4(1/4) = 1 (inevitably)

To illustrate events that are not mutually exclusive, consider A = picking a heart and B = picking a queen. Because it is entirely possible for a card to be both a heart and a queen at the same time,

$$P(\text{heart or queen}) \neq 13/52 + 4/52$$

The sign ≠ means "does not equal." Since we cannot use the addition rule here, the probability, P(heart or queen), must be obtained some other way. It can be calculated directly from the definition of probability as 16/52 (that is, 13 hearts plus 3 other queens out of the 52 cards in the deck).

THE MULTIPLICATION RULE (THE "AND" RULE). Consider an experiment that has as possible results the events A, B, C, D, . . . If the experiment is repeated several times, we might be interested in the joint outcome of successive trials: for example, A on the first trial and B on the second. With the restriction given below, this joint probability is

$$P(A \textit{ and } B) = P(A) \times P(B)$$

or, for more repetitions,

P(A *and* B *and* C *and* . . .)
= P(A) × P(B) × P(C) × . . .

In words, if we want the joint probability of one event *and* another *and* another, and so on, just multiply their individual probabilities together.

With two decks of cards before you, what is the probability of drawing an ace from the first *and* another ace from the second? By the multiplication rule, we obtain

$$P(\text{ace } \textit{and} \text{ ace, two decks}) = (1/13)(1/13) = 1/169$$

Note that each factor of 1/13 is itself obtained either from the addition rule (1/52 + 1/52 + 1/52 + 1/52) or directly from the definition of probability: 4 favorable outcomes (aces) out of 52 (cards).

The restriction in the use of the multiplication rule is that *the events must be independent*. This means that the occurrence of one event does not affect in any way the probability of other events occurring. Selections from separate decks are independent events because handling one deck of cards cannot affect the other.

If we draw two cards from the same deck, however, the outcomes are *not* independent. Having drawn one card, ace or not, the probabilities applying to the second draw are changed. To compute the probability of picking two aces from the same deck *without replacing the first card* we need to take note of the altered second probability: 4/52 (ace on first draw) × 3/51 (ace on second draw) = 1/221.

Questions of independent events arise in contexts very different from those for questions of mutually exclusive events, a condition of the addition rule above. Mutually exclusive events can never be independent since excluding another event affects drastically the probability of its occurrence.

Whenever the conjunction *and* is used in everyday language, the multiplication rule is usually appropriate. What is the probability that a sibship of four will consist of all girls? This can be interpreted as

$$\begin{aligned} P(\text{4 girls}) &= P(\text{1st born a girl } \textit{and} \text{ 2nd a girl } \textit{and} \ldots) \\ &= P(\text{girl}) \times P(\text{girl}) \times P(\text{girl}) \times P(\text{girl}) \\ &= \text{approximately } (1/2)^4 \\ &= 1/16 = 0.0625 \end{aligned}$$

On the basis of actual birth records, the likelihood of a live-born girl has been found to be slightly, but significantly, less than 1/2, about 0.48 or 0.49 depending on which real population is sampled. Thus a slightly more accurate answer might be

$$P(\text{4 girls}) = (0.49)^4 = 0.0576$$

(To be even more exact, the probability of a girl birth varies a little depending upon whether she is the first born, or second, or third, and so on [Novitski and Sandler 1956]. Thus the sexes of successive births are not quite independent, but we can neglect such small effects here.)

As an example of using both the addition and multiplication rules, consider the offspring from *A/a* × *A/a*. From a checkerboard square the genotypic results are

P(*A/A*) = 1/4 P(*A/a*) = 1/2 P(*a/a*) = 1/4

The rules of probability that led to these results can be applied as follows:

P(*A/a*) = P(*A* egg *and A* sperm) = (1/2 × 1/2) = 1/4
P(*A/A*) = P(*A* egg *and a* sperm, *or a* egg *and A* sperm)
= (1/2 × 1/2) + (1/2 × 1/2) = 1/4 + 1/4 = 1/2
P(*a/a*) = P(*a* egg *and a* sperm) = (1/2 × 1/2) = 1/4
P(dominant phenotype)
= P(*A/A*) + P(*A/a*) = 1/4 + 1/2 = 3/4

THE MIXED RULE (THE BINOMIAL). For a sibship of size three, an exhaustive list of the eight different sequences of birth orders of boys and girls is given in Table 1. Only the multiplication rule is needed to calculate the probabilities in the next-to-last column. Had we been a little more approximate and let P(B) = P(G) = 1/2, all the calculations would have been $(1/2)^3$ = 1/8 = 0.125.

Note that there are three different sequences of birth orders that yield the same summed event, two boys and a girl. Each one (B-B-G or B-G-B or G-B-B) has the same probability (0.128), because the likelihood of a boy or girl does not depend on the sex of preceding births. (It is said that chance has no memory.) Thus the summed result can be written

$$P(2B, 1G) = 3\,[(0.51)^2(0.49)]$$

In the notation P(2B, 1G), it is understood that all orders producing two boys and a girl are included. The probability of one boy and two girls is written in a similar fashion:

$$P(1B, 2G) = 3\,[(0.51)\,(0.49)^2]$$

These probabilities involve both the multiplication and the addition rules. The product of 0.51 and 0.49 each raised to its appropriate power [within the square brackets] is called the **exponential**. It is the probability of one specific sequence of boys and girls in a family. We then add together the probabilities of all events involving the same totals for each sex. The number of such sequences with the same overall sex distribution—three in the example above—is called the **coefficient**. The product of the exponential and the coefficient is the probability in question.

Trying to analyze in a similar fashion situations in which the number of repetitive trials is much greater than three, one quickly gets bogged down. Hence a general formulation becomes necessary. For example, in a sibship of size nine, using the same values for the probability of a boy birth or a girl birth, we can write without difficulty

$$P(9B, 0G) = 1\,[(0.51)^9(0.49)^0] = (0.51)^9$$
$$P(8B, 1G) = 9\,[(0.51)^8(0.49)^1]$$

In the second expression above, the coefficient 9 represents the number of different sequences of birth orders yielding eight boys and a girl. The exponential portion represents the probability of any one of these nine sequences.

Continuing further with this example, What coefficient goes with the probability of 7 boys and 2 girls? The methodical reader might care to confirm that this is 36, but for obvious reasons we really need a general, easily applied method. This is supplied by the **binomial formula**, which occupies an important place in probability theory and statistics.

The binomial formula closely follows our examples above. Conceive of an experiment in which there are just two mutually exclusive outcomes, A and B (boy and girl, for example). The corresponding probabilities are represented as P(A) and P(B), and they add to one (like 0.49 and 0.51). Let the experiment be repeated independently n times. The question is this: What is the probability that, out of the n trials (9, say), the outcome A will occur r times (7, say) and B will occur s times (2, say)? We abbreviate this probability as P(r of A, and s of B), or more simply, P(rA, sB). The value of this probability is provided by the binomial formula:

$$\mathbf{P(rA, sB) = \frac{n!}{r!s!}\,[P(A)]^r[P(B)]^s}$$

The exponential part of the formula consists of the factors P(A) and P(B) raised to the powers r and s, respectively. The coefficient involves the numbers, n, r, and s, and exclamation points called *factorials*. The number "n!" is read "n factorial" and it means

Table 1 The probabilities of the distributions of sex in a sibship of size three

Distribution of sex (summed event)[a]	BIRTH ORDER *1*	*2*	*3*	*Probability of a specific birth order*		*Probability of the summed event*
3B, 0G	B	B	B	$(0.51)(0.51)(0.51) = (0.51)^3$	= 0.133	0.133
2B, 1G	B	B	G	$(0.51)(0.51)(0.49) = (0.51)^2(0.49)$	= 0.128	0.384
	B	G	B	$(0.51)(0.49)(0.51) = (0.51)^2(0.49)$	= 0.128	
	G	B	B	$(0.49)(0.51)(0.51) = (0.51)^2(0.49)$	= 0.128	
1B, 2G	G	G	B	$(0.49)(0.49)(0.51) = (0.51)(0.49)^2$	= 0.122	0.366
	G	B	G	$(0.49)(0.51)(0.49) = (0.51)(0.49)^2$	= 0.122	
	B	G	G	$(0.51)(0.49)(0.49) = (0.51)(0.49)^2$	= 0.122	
0B, 3G	G	G	G	$(0.49)(0.49)(0.49) = (0.49)^3$	= 0.118	0.118
Sum					= 1.001[b]	1.001

a. B, boy, G, girl. The probability calculation assumes that the births are independent and that P(B) = 0.51 and P(G) = 0.49. The actual values for P(B) and P(G) vary from population to population.

b. Slight rounding error.

$$n(n-1)(n-2)(n-3)(\ldots)(1)$$

that is, the product of all the numbers starting with n and going down to 1. For example, $4! = (4)(3)(2)(1) = 24$.

Returning to 9 sibs, the probability of there being 7 boys and 2 girls is, by the binomial formula,

$$P(7B, 2G) = \frac{9!}{7!2!}(0.51)^7(0.49)^2$$

The arithmetic for the binomial coefficient is

$$\frac{9!}{7!2!} = \frac{(9)(8)\cancel{(7)(6)(5)(4)(3)(2)(1)}}{\cancel{(7)(6)(5)(4)(3)(2)(1)} \times (2)(1)} = \frac{(9)(8)}{(2)(1)} = \frac{72}{2} = 36$$

Note that the 7! cancels out all of the 9! except for the factors 9 and 8. Here are a few additional "housekeeping" hints for using the binomial:

1. $r + s = n$. That is, the numbers of the two kinds of outcomes have to add up to the total trials of the experiment.
2. $P(A) + P(B) = 1$. That is, the events A and B are exhaustive.
3. The exponents are the logical ones: for example, the power of P(A) is the number of times that A occurs.
4. When one or the other outcome occurs 0 times, we have to deal with the number 0!. To make the binomial and other mathematical formulae reasonable, the value of 0! is defined as 1 (not 0).

All of the probabilities for the distributions of sex in a sibship of size nine are given in Table 2. For simplicity, we have assigned equal probabilities to the two sexes. Since the list of 10 possible outcomes is exhaustive, the probabilities add to 1. Note that no sex distribution is *very* rare. If you knew of 512 families, each with nine children, one of them, ideally, would be expected to have all boys (top line) and another one to have all girls (bottom line). Adding these together, 1 out of 256 nine-child families are expected to be unisexual, and, in fact, this statistic is seen (approximately) in populations with large families.

Consider now another simply inherited trait, albinism. We can ask, "What is the expected distribution of albinos in a family of size four when both parents are heterozygous?" The answer is summarized in Table 3. Because the probabilities of the two possible outcomes, albino birth versus normal birth, are not equal, the value of the exponential part of the binomial formula varies depending on the particular distribution. All calculations, however, are made in *exactly* the same way as they are for Table 2. Note that the ideal (theoretical or expected) outcome, one albino and three normally pigmented children, has a probability of 0.422 (second line). Thus in this case the ideal result should actually occur less than half the time; more often than not, the distribution will be something other than the expected 3:1 Mendelian ratio. Departures from expected results (which are to be expected) are discussed further in Chapter 12.

Table 2 The probabilities of the distributions of sex in a sibship of size nine

Distribution of sex[a]	*Binomial coefficient*[b]	*Exponential*[b]	*Probability of the distribution*
9B, 0G	1	$(1/2)^9$	1/512 = 0.002
8B, 1G	9	$(1/2)^9$	9/512 = 0.018
7B, 2G	36	$(1/2)^9$	36/512 = 0.070
6B, 3G	84	$(1/2)^9$	84/512 = 0.164
5B, 4G	126	$(1/2)^9$	126/512 = 0.246
4B, 5G	126	$(1/2)^9$	126/512 = 0.246
3B, 6G	84	$(1/2)^9$	84/512 = 0.164
2B, 7G	36	$(1/2)^9$	36/512 = 0.070
1B, 8G	9	$(1/2)^9$	9/512 = 0.018
0B, 9G	1	$(1/2)^9$	1/512 = 0.002
Sum			512/512 = 1.000

a. B, boy; G, girl. For simplicity we assume P(B) = P(G) = 1/2. The order of boys and girls for a distribution is not specified. Each probability is the product of the binomial coefficient and the exponential.

b. Sample calculation for 6B, 3G:

$$\text{binominal coefficient} = \frac{9!}{6!\,3!} = \frac{9{\cdot}8{\cdot}7{\cdot}6{\cdot}5{\cdot}4{\cdot}3{\cdot}2{\cdot}1}{6{\cdot}5{\cdot}4{\cdot}3{\cdot}2{\cdot}1{\cdot}3{\cdot}2{\cdot}1} = \frac{9{\cdot}8{\cdot}7}{3{\cdot}2} = 3{\cdot}4{\cdot}7 = 84$$

= number of different sequences of 6B, 3G

$$\text{exponential} = (1/2)^6(1/2)^3 = (1/2)^9 = 1/512$$

The carrier status for albinism cannot be determined from any biochemical test. Therefore the knowledge that both parents are heterozygous can come only "after the fact," that is, after the birth of one or more albino children. In the analysis of family data (Chapter 12) it is sometimes important to find the probability of at least one affected child. In the situation of Table 3, we can compute

$$\begin{aligned}\text{P(at least one albino)} &= \text{P(1 or 2 or 3 or 4 albinos)}\\ &= 0.422 + 0.211 + 0.047 + 0.004\\ &= 0.684\end{aligned}$$

Note that this probability is the sum of all the probabilities in the list except for the first one. Thus an alternative—shorter—method of calculation is

$$\begin{aligned}\text{P(at least one albino)} &= 1 - \text{P(0 albinos)}\\ &= 1 - 0.316 = 0.684 \text{ (as above)}\end{aligned}$$

Two additional remarks about the binomial formula are in order:

1. Because we are dealing with *two* possible outcomes, the probability formula is called a *bi*nomial. When *more* than two outcomes are possible, the binomial is not applicable, but a more generalized *multi*nomial expression can be used. For example, if an experiment has three outcomes, A, B, and C, with probabilities P(A), P(B), and P(C), then

$$\text{P}(r \text{ of A}, s \text{ of B}, t \text{ of C}) = \frac{n!}{r!s!t!}[\text{P(A)}]^r[\text{P(B)}]^s[\text{P(C)}]^t$$

In this expression, r, s, and t are the number of times the events A, B, and C occur, respectively, and their sum equals n. For example, in mating a roan (heterozygous R/r) cow with a roan bull, there are three phenotypic outcomes:

$$\text{P(red)} = 1/4 \qquad \text{P(roan)} = 1/2 \qquad \text{P(white)} = 1/4$$

From a roan × roan mating, what is the probability of getting one red calf, two roans, and one white? The answer is

$$\frac{4!}{1!2!1!}(1/4)^1(1/2)^2(1/4)^1 = \frac{(4)(3)}{(4)(2)(2)(4)} = 3/16 = 0.188$$

2. The arithmetic values of the binomial coefficients can be obtained by another method called Pascal's triangle. Each number in the triangular array is the sum of the two numbers above it, slightly to its left and right. The first six lines are

```
first line             1
                      / \
second line          1   1
                    / \ / \
third line         1   2   1
                  / \ / \ / \
fourth line      1   3   3   1
                / \ / \ / \ / \
fifth line     1   4   6   4   1
              / \ / \ / \ / \ / \
sixth line   1   5  10  10   5   1
```

The fifth line gives the coefficients in Table 3. Readers who care to continue the array to the tenth line can check the coefficients in Table 2. Pascal's triangle is too cumbersome to use beyond this.

CONDITIONAL PROBABILITY. So far, the probabilities we have been calculating refer to the ratio of favorable events to *all* possible events that could occur. Sometimes, however, our outlook is on *fewer* than all possible outcomes. When our viewpoint is restricted in this way, we speak of **conditional probability**.

Table 3 The probabilities of the distributions of albinos in a sibship of size four when both parents are known to be heterozygous

Number of albinos in the sibship	*Binomial coefficient*[a]	*Exponential*	*Probability of the distribution*
0	1	$(3/4)^4(1/4)^0$	81/256 = 0.316
1	4	$(3/4)^3(1/4)^1$	108/256 = 0.422
2	6	$(3/4)^2(1/4)^2$	54/256 = 0.211
3	4	$(3/4)^1(1/4)^3$	12/256 = 0.047
4	1	$(3/4)^0(1/4)^4$	1/256 = 0.004
Sum			256/256 = 1.000

a. Sample calculation for sibship with one albino (3 normal):

binominal coefficient $= \frac{4!}{3!\,1!} = \frac{4\cdot3\cdot2\cdot1}{3\cdot2\cdot1\cdot1} = 4/1 = 4$

= number of different sequences of three normals and one albino

exponential $= (3/4)^3(1/4)^1 = \frac{(3)^3}{(4)^4} = 27/256$

For any one birth, P(normal pigmentation) = 3/4, and P(albino) = 1/4.

Consider the offspring from the mating, $A/a \times A/a$, where A is dominant to a:

$$\left.\begin{matrix} P(A/A) = 1/4 \\ P(A/a) = 1/2 \end{matrix}\right\} P(A/-) = 3/4$$
$$P(a/a) = 1/4$$

If we look at all offspring, then the probability of a heterozygote is 1/2. But *given that an offspring expresses the dominant phenotype*, what then is the probability of its being a heterozygote? The answer is 1/2 out of 3/4 (the *dominant* outcomes) rather than 1/2 out of 1 (*all* outcomes). Thus we write

P(heterozygote, considering only dominant offspring)
$$= \frac{1/2}{3/4} = (1/2)(4/3) = 2/3$$

Although not completely general, we can broaden this viewpoint for our purposes as follows:

P(A, given a condition C)
$$= \frac{\textbf{P(A) as previously calculated}}{\textbf{P(C)}}$$

In words, this formula says that when our viewpoint is restricted to a portion of outcomes, the probabilities should be divided by the probability of that portion. The condition defines (it is said) a smaller sample space by ruling out certain outcomes as impossible. The *relative probabilities* within the smaller space are the same as before, however.

The previous example becomes clearer if you consider the allele a to be a recessive lethal. Assume a/a zygotes are aborted early in pregnancy. *Among the smaller group of live-born offspring* then, what is the probability of a heterozygote? The answer is the same as above: 2/3.

In slightly different form we have met this conditional probability of 2/3 twice before (Chapter 2):

1. Among the phenotypically tall plants in Mendel's F_2 generation, what fraction were heterozygous? The same 2/3.

2. We considered an X-linked gene whose lethal allele, l, was recessive to its normal allele, +. Among live-born children from the mating, $+/l$ female $\times$ $+/(Y)$ male, what is the probability of a female? Again, 2/3.

Another way of looking at conditional probabilities is to first express the results as *ratios* rather than as probabilities. Among the progeny from $A/a \times A/a$ we have

Genotype	Ratio
A/A	1
A/a	2
a/a	1

These ratios are converted to probabilities by dividing the appropriate number by the sum of whatever set of outcomes you care to specify. The prior pair of examples was

P(heterozygote among *all* offspring)
$$= 2/(1 + 2 + 1) = 2/4$$
P(heterozygote among *dominant* offspring)
$$= 2/(1 + 2) = 2/3$$

Another pair of contrasting viewpoints is

P(*A/A* among *all* offspring) = 1/(1 + 2 + 1) = 1/4
P(*A/A* among *homozygous* offspring) = 1/(1 + 1) = 1/2

Going back to Table 3, we can present the figures as ratios in a similar way:

Number of albinos	Ratio
0	81
1	108
2	54
3	12
4	1
	256 (sum)

Among all sibships, the probability of having exactly one albino is 108/256. We had noted, however, that sibships with no albinos cannot be identified as having heterozygous parents. Thus only 175 (that is, 256 minus 81) of the sibships are "available" when we start to analyze sibships. *Among the sibships with at least one albino*, the probability of having exactly one albino is 108/175 (not 108/256). Using the boldfaced formula above, this conditional probability can also be calculated as

$$\frac{108/256}{175/256} = 108/175$$

BAYES' THEOREM. Probabilities are used, consciously or unconsciously, by genetic counselors, doctors, and others to distinguish among different explanations for a "state of affairs." Is a phenotypically normal person heterozygous for cystic fibrosis, *or* is the person homozygous for the normal dominant allele? Does a patient with a rash and a fever have disease A or B or C? Are phenotypically similar twins monozygous (MZ, arising from a single fertilized egg) or dizygous (DZ, arising from separate fertilized eggs and therefore no more alike on the average than siblings)? In these and other situations, the investigator has an initial notion about the relative probabilities of the competing hypotheses: Carrier or not? Disease A or B or C? MZ or DZ? These first ideas are called **prior probabilities**; they may be based, for example, on Mendel's laws or on previous experience such as disease incidences or on twinning rates in a given group of people. Additional data are often collected to help decide between the hypotheses. For the case of twins, the geneticist then calculates P(MZ, given the additional data) or P(DZ, given the additional data). These new probabilities are called **posterior probabilities**; their values have been modified on the basis of observed facts. As more facts are accumulated, the posterior probabilities can be further refined.

This method was introduced by Thomas Bayes (1763), an English mathematician and Presbyterian minister. It involves so-called *inverse probabilities*—inverse, because the event and the condition are reversed from the usual situation. In short, we are after P(B, given A) rather than P(A, given B)(where A and B are any events). Given parental genotypes, we can easily calculate, for example, the probability that twins will both be blood type O (say) given that they are MZ, but the Bayesian method provides us with the probability that the twins are MZ given that blood tests show that they are both type O. (We should add that some concepts in probability and statistics involve philosophical subtleties about which experts disagree. Although Bayes' method is not universally accepted, geneticists find it useful.)

As an example, suppose we learn that a friend gave birth to twins last night. With no additional information, the best guess—the prior probability—that they are monozygous (or identical) is 30% because the average known frequency of MZ twinning among the various U.S. populations is about 0.3 [= P(MZ)], whereas that for dizygous (or fraternal) twinning is 0.7 [= P(DZ)]. But if we learn that the twins have the same phenotype for several heritable traits, we can make a better guess—the posterior probability—that they are MZ. (Note that if the twins differ in *any* heritable trait, including sex, they are certainly DZ, but if they are the same for all traits, they could be either MZ or DZ. The more inherited traits for which the twins agree, the more likely it is that they are MZ.)

Suppose our hypothetical case involves the recessive trait, albinism (*c/c*), and the dominant trait, brachydactyly (*B/–*), and the parents of the twins had the following genotypes:

Mother: *C/c B/b*
Father: *C/c b/b*

Both parents are heterozygous for albinism; the mother has brachydactyly and the father has normal digits. Assume that the newborn twins are female albinos with normal digits. The probabilities of these outcomes, given that the twins are MZ or DZ, are

Data about twins	P(data, given MZ)	P(data, given DZ)
female	1/2	1/4
albino	1/4	1/16
normal digits	1/2	1/4

The values above follow from the genotypes of the parents; for example, the probability of an albino child from heterozygous parents is 1/4. If the twins are MZ (middle column), it is as if they are one child. If the twins are DZ, it is as if they were independently born sibs, so the probability of both being albino is $(1/4)^2$. Bayes' theorem (presented here without derivation) allows us to write the probability of interest:

$$\text{P(MZ, given the data)} = \frac{\text{P(MZ)} \times \text{P(data, given MZ)}}{\begin{bmatrix} \text{P(MZ)} \times \text{P(data, given MZ)} \\ + \text{P(DZ)} \times \text{P(data, given DZ)} \end{bmatrix}}$$

where the denominator is given in square brackets. A similar formula expresses P(DZ, given the data) and the two probabilities must add to one. With the additional information about the sex, pigment, and fingers of the twins, we compute

P(MZ, given female and albino and normal digits)

$$= \frac{(0.3) \times (1/2)(1/4)(1/2)}{[(0.3) \times (1/2)(1/4)(1/2)] + [(0.7) \times (1/4)(1/16)(1/4)]}$$

$$= \frac{0.01875}{0.01875 + 0.00273} = 0.87$$

Thus, with more data, the likelihood that the twins are monozygous has increased from 30 to 87%. When data about ABO, Rh, and other blood characteristics are obtained, the probability that twins are MZ could approach 100% if the twins are the same for all traits.

A more general formulation of Bayes' theorem is

$$\text{P}(\text{H}_1/\text{data}) = \frac{\text{P}(\text{H}_1) \times \text{P}(\text{data}/\text{H}_1)}{\begin{bmatrix} \text{P}(\text{H}_1) \times \text{P}(\text{data}/\text{H}_1) \\ + \text{P}(\text{H}_2) \times \text{P}(\text{data}/\text{H}_2) \\ + \text{P}(\text{H}_3) \times \text{P}(\text{data}/\text{H}_3) \\ + \dots \end{bmatrix}}$$

where the denominator is given in square brackets. Here, H_1, H_2, H_3, etc. are competing hypotheses about the state of nature, and $P(H_1)$, $P(H_2)$, $P(H_3)$, . . . are their prior probabilities. The slash (/) is the usual notation for the "given" of conditional probabilities. An expression for $P(H_2/\text{data})$ has the numerator $P(H_2) \times P(\text{data}/H_2)$ and the same denominator as above, etc.

PROBLEM SOLVING

The Binomial

We mentioned that usually the probability of a boy being born is slightly but significantly greater than 0.50. Its value varies from country to country, Korea having one of the highest, about 0.54. In a Korean sibship of size five, how many times more likely is it to have three boys than it would be to have three girls? We set up the two binomial expressions:

$$P(3B, 2G) = \frac{5!}{3!2!}[0.54]^3[0.46]^2$$

$$P(2B, 3G) = \frac{5!}{2!3!}[0.54]^2[0.46]^3$$

The ratio of these numbers is 0.54/0.46 = 1.17, because all other factors in the two binomials cancel. With a calculator, you could confirm that the arithmetic values of P(3B, 2G) and P(2B, 3G) are 0.333 and 0.284, respectively.

Continuing with sibships of size five where P(B) = 0.54, what are the probabilities of all boys or all girls?

$$P(5B) = [0.54]^5 = 0.046 \qquad P(5G) = [0.46]^5 = 0.021$$

These values are quite different from the probability of a unisexual sibship of size five when boy and girl births are equally likely; under this circumstance, P(5B) = P(5G) = $(0.5)^5$ = 1/32 = 0.031.

Conditional Probability

Jean has a normal phenotype but she has an albino brother and an albino husband. What is the probability that Jean's first child will be an albino? The relevant pedigree and all known genotypes are

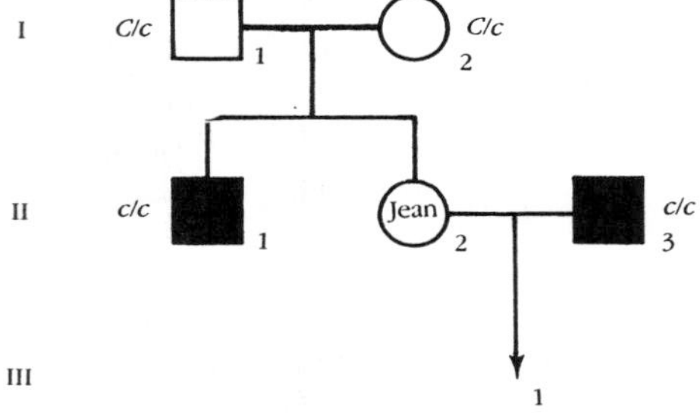

The unborn child, III-1, will receive an albino allele from its father. How likely is it that the unborn child will also receive an albino allele from Jean and therefore be affected? Jean came from the mating, *C/c* × *C/c*, but, with a normal phenotype, she can only be *C/C* or *C/c* (and not *c/c*). The 1:2 ratio applies:

P(Jean is *C/C* given dominant phenotype) = 1/3
P(Jean is *C/c* given dominant phenotype) = 2/3

If Jean is a heterozygote, she could transmit either *C* with probability 1/2 or *c* with probability 1/2. Altogether,

P(III-1 is albino)
= P(Jean is *C/c* given dominant phenotype)
× P(Jean transmits *c*)
= 2/3 × 1/2 = 1/3

If the first child turns out to be an albino in fact, there are no longer any probabilities attached to it: it is an albino. We might then ask about a second child. The albino first child, genotypically *c/c*, tells us that Jean is a heterozygote *for sure*; therefore, for a second child, P(albino) = 1/2.

If the first child, and all additional children as well, turn out to have normal phenotypes in fact, the probabilities for the next (unborn) child are more difficult to assess. Using 1/3 and 2/3 as the prior probabilities of Jean being *C/C* and *C/c*, Bayes' theorem can be used to calculate P(Jean is *C/c* given all normal children). This calculation is considered in Question 11.

SUMMARY

1. More than two—multiple—alleles may exist for one gene. The number is determined by the chemical changes—i.e., mutations—that lead to recognizable phenotypes. Although one individual possesses only two alleles for each gene, multiple alleles are the basis for a large number of possible genotypes.

2. The description of the way alleles interact is based on the relative phenotype of a heterozygote. Which description is appropriate depends on the level and technique of observation. The interaction of just two alleles may be described in more than one way.

3. The notion of a probability is similar to intuitive ideas of chance or frequency. Probabilities are expressed as numbers between 0 and 1 inclusive.

4. The way probabilities are combined depends on the kind of question and the nature of the events. When *or* is used in describing mutually exclusive events, the combined probability is obtained by addition. When *and* is used for independent events, the joint probability is obtained by multiplication.

5. The binomial formula has two factors: the exponential, which gives the probability of a specific

ordering of independent events, and the coefficient, which gives the number of different orderings that are possible.

6. Conditional probabilities are used when our viewpoint is restricted to a portion of, rather than all, the possible outcomes.

7. Bayes' theorem is used to refine prior probabilities about the state of nature on the basis of additional pertinent information.

FURTHER READING

Textbooks by Crow (1983), Strickberger (1985), and Levitan (1988) have good, short accounts of probability. Lindgren and McElrath (1978) treat conditional probability clearly, as does Maynard Smith (1968) in a delightful little book. Murphy (1979) considers probability in medicine more fully. Wolf et al. (1985) set forth an interesting test of whether doctors apply Bayesian logic in medical diagnosis. The many uses of Bayes' theorem in human genetics are found in Murphy and Chase (1975). For additional practice in problem solving, consult Stansfield (1983) and Harrison (1970). Multiple allelic series for genes that affect coat color in many mammals, and the nature of allelic interactions, are examined comparatively by Searle (1968).

QUESTIONS

1. List methodically all the genotypes possible when considering the four alleles of the hemoglobin beta chain gene: Hb^A, Hb^S, Hb^E, and Hb^C. (Your work will be easier if you omit the basic gene designation, *Hb*, and just use the superscripts.) How many genotypes are homozygous? Heterozygous?

2. Give the genotypes and probabilities for the offspring of the following matings. (The basic gene designation, *Hb*, has been omitted.) (a) *A/A* × *A/S*; (b) *A/C* × *A/S*; (c) *A/E* × *C/S*.

3. Assume that e_1 is dominant to its allele e_2 which, in turn, is dominant to a third allele e_3. We can speak of the three homozygotes as having the phenotype trait-1, trait-2, or trait-3. List all genotypes and their corresponding phenotypes.

4. Considering the gene in Question 3, what mating could yield each of the following progeny sets?
(a) 3/4 trait-1 and 1/4 trait-3
(b) 3/4 trait-2 and 1/4 trait-3
(c) 1/2 trait-1, 1/4 trait-2, and 1/2 trait-3

5. When Mendel allowed the heterozygous F_1 plants from pure-breeding tall × pure-breeding short to self-fertilize, he got these results:

Phenotype	**Genotype**
tall	787
short	277
all	1064

What is the probability of getting a result like this? Just set up the expression; do not try to evaluate it.

6. Two parents are heterozygous for albinism. Assume that they will have five children.
(a) What is the probability that the children will have the following phenotypes in the order stated: 1st born = normal, 2nd = albino, 3rd = normal, 4th = albino, 5th = normal?
(b) How many different orderings of the 3 normals and 2 albinos are there?
(c) What is the probability that, among the 5 children, 3 will be normal and 2 albino, in any order?
(d) What is the probability that all 5 will be normal?
(e) What is the probability that at least one child will be albino?
(f) Are the events in (d) and (e) an exhaustive set?

7. Assume P(boy birth) = P(girl birth) = 1/2. In a sibship of size 4 what is the probability of
(a) the expected (ideal) sex distribution?
(b) no girls?
(c) exactly one girl?
(d) at least one girl?
(e) at least two girls?

8. Repeat Question 8 for a sibship of size 7.

9. Draw the pedigree of a normal boy who has normal parents and a brother who died from the recessive Tay-Sachs disease. What is the probability

that the boy is heterozygous for Tay-Sachs disease (prior to any testing)?

10. If all you knew was that twins were like-sexed, use Bayes' theorem to calculate P(MZ/like-sexed) = 6/13. Assume the prior probabilities are P(MZ) = 0.3 and P(DZ) = 0.7.

11. Assume that Jean in the pedigree on page 180 gave birth to three normally pigmented children. Use Bayes's theorem to calculate P(Jean is *C/c* given three normal children) = 1/5. The prior probability is P(Jean is *C/c*) = 2/3. What then is the probability that the next (unborn) child will be albino?

ANSWERS

1. Four homozygous and six heterozygous:

A/A	*A/S*	*A/E*	*A/C*
	S/S	*S/E*	*S/C*
		E/E	*E/C*
			C/C

2.

(a)	(b)	(c)
1/2 *A/A*	1/4 *A/A*	1/4 *A/C*
1/2 *A/S*	1/4 *A/S*	1/4 *A/S*
	1/4 *A/C*	1/4 *E/C*
	1/4 *C/S*	1/4 *E/S*

3. e_1e_1 or e_1/e_2 or e_1/e_3 = trait-1
e_2/e_2 or e_2/e_3 = trait-2
e_3/e_3 = trait-3

4. Omitting the basic letter designation, *e*:
(a) $1/3 \times 1/3$ (b) $2/3 \times 2/3$ (c) $1/3 \times 2/3$

5. $\frac{1064!}{787!\,277!}[0.75]^{787}[0.25]^{277}$

6. (a) $(3/4)^3(1/4)^2 = 27/1024$
(b) $5!/(3!2!) = 10$
(c) 270/1024
(d) $(3/4)^5$
(e) $1 - (3/4)^5$
(f) yes

7. (a) $P(2B, 2G) = (4!/2!2!)[(1/2)^2(1/2)^2] = 6/16$
(b) 1/16
(c) 4/16
(d) $1 - 1/16 = 15/16$
(e) $1 - 1/16 - 4/16 = 11/16$

8. (a) The mathematical ideal is 3.5 of each sex, which is impossible: P = 0
(b) $(1/2)^7 = 1/128$
(c) 7/128
(d) $1 - 1/128 = 127/128$
(e) $1 - 1/128 - 7/128 = 120/128$

9. 2/3

10. P(MZ/like-sexed)

$$= \frac{P(MZ) \times P(\text{like-sexed}/MZ)}{P(MZ) \times P(\text{like-sexed}/MZ) + P(DZ) \times P(\text{like-sexed}/DZ)}$$

$$= \frac{(0.3) \times (1)}{(0.3)(1) + (0.7)(0.5)} = 0.30/0.65 = 6/13$$

11. P(Jean is *C/c* given three normal children)

$$= \frac{(2/3) \times [(1/2)(1/2)(1/2)]}{(2/3) \times [(1/2)(1/2)(1/2)] + (1/3) \times [(1)(1)(1)]}$$

$$= \frac{1/12}{1/12 + 4/12} = 1/5$$

The probability that the next child will be albino is (1/5)(1/2) = 1/10.

Chapter 11 Recombination of Nonallelic Genes

The Concept of Recombination

Genes on Different Chromosomes: Mendel's Second Law
The Proportions of the Gametes
Offspring Types for Two Genes
Gene × Gene Method
Meiosis and the Second Law

Genes on the Same Chromosome: Crossing Over
The Proportions of the Gametes
The Physical Basis of Recombination
The Limits of Recombination
The Phase Problem

Mapping the Human X Chromosome by Family Linkage Studies
Green-Shift and Red-Shift
Two Clusters of X-linked Genes
Further Considerations

Significance of Gene Mapping
Building and Using Gene Maps
Gene Mapping and Evolution

> *Sutton ('02)... [theorized that]... homologous pairs unite in synapsis, and in the reduction division are separated into groups which are neither purely paternal nor purely maternal... It is strange that so careful an observer should have overlooked the very thing that was present (the unequal tetrad was first found on Sutton's slides) offering definite chromosomic proof for his theory.*
>
> *E. Eleanor Carothers (1913)*

Sutton showed that the inheritance of alleles of one gene depends upon the behavior of one pair of chromosomes during meiosis. In this chapter, we present evidence that the assortment of alleles of several different genes is also based on chromosome choreography, as Carothers first showed in 1913. Although genes were abstractions to Mendel, his laws are valid because chromosomes act in special ways during meiosis.

THE CONCEPT OF RECOMBINATION

An individual has a pair of alleles for each of perhaps 50,000 genes that code for proteins. So far, we have considered only one at a time, but now we will follow, simultaneously, the fate of two genes.

Consider first the simplest case: What kinds of eggs can be made by a woman who is homozygous, *T/T U/U*? Mendel's first law states that every egg will contain one of the *T* alleles. Likewise, the first law states that every egg will contain one of the *U* alleles. Thus all of her eggs can be symbolized (*TU*). For a man of genotype *t/t u/u* all sperm are of the type (*tu*).

From these parents, all offspring come from the fertilization (*TU*) × (*tu*) and so have the genotype *T/t U/u*. An individual with this genotype is a **double heterozygote**. (In Mendel's terminology, an organism heterozygous for two genes is a *dihybrid*.) Of such doubly heterozygous persons, we ask:

- What kinds of gametes can be made?
- In what proportions are the various kinds of gametes formed?

The first question is the easier one to answer. According to the first law, a gamete will contain one of the alleles *T* or *t* and one of the alleles *U* or *u*. Thus, depending on how the *T* and *t* alleles are combined with the *U* and *u* alleles, four kinds of gametes are possible:

The first two gametes are the same as the **parental gametes** that formed the offspring (see above). The last two gametes, however, contain new assortments of genes, different from those in the immediately preceding gamete generation; these are called **recombinant gametes**. This process is outlined in Figure 1, which includes two ways in which a double heterozygote can be realized. Case I is that presented above, in which a fertilizing gamete has two uppercase alleles or two lowercase alleles. In Case II, by contrast, each of the gametes giving rise to the double heterozygote carries one uppercase allele and one lowercase allele. The resulting double heterozygotes and the gametes they make are identical in the two cases, but the labeling of parental and recombinant types reverses. *The designation parental gamete or recombinant gamete depends only upon how the nonallelic genes were combined in the first place.*

The rest of this chapter deals with the second question: the proportions of the four kinds of gametes made by a double heterozygote. These proportions vary, depending on the two particular genes in question. We look first at genes that are on different (nonhomologous) chromosomes.

GENES ON DIFFERENT CHROMOSOMES: MENDEL'S SECOND LAW

We illustrate the second law by following simultaneously the inheritance of a type of dwarfism and the ABO blood type. Short stature has many causations, one type stemming from a deficiency of growth hormone produced by the small pituitary gland at the base of the brain. These *pituitary dwarfs* (Type I, the most common form, Figure 2) have normal body proportions, pinched facial features with high foreheads, and high-pitched voices (Rimoin 1983).* Let *d* represent the recessive gene for pituitary dwarfism. Affected persons are *d*/*d*, and normal-height persons are *D*/*D* or *D*/*d*.

The ABO blood type, expressed as antigens on the surface of red blood cells, depends on multiple alleles,

*Therapy for pituitary dwarfism has been limited because human growth hormone has not been readily available. Recombinant DNA techniques, however, promise a much increased supply. In this method of production, the human gene for growth hormone is transferred into bacteria, which are then grown in large vats to produce the gene product. See Underwood (1984) and Chapter 15.

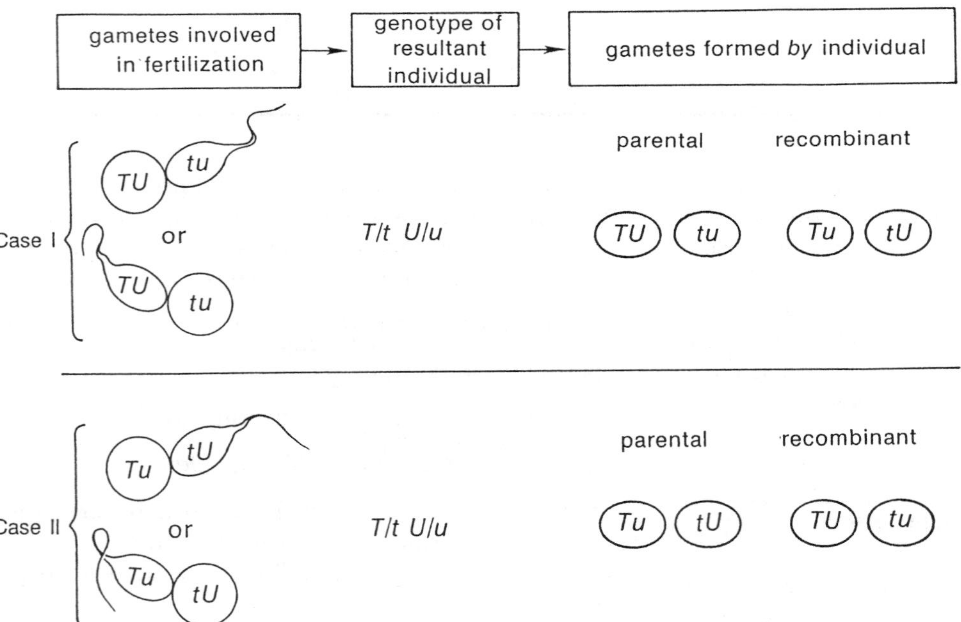

Figure 1 The meaning of parental and recombinant gametes. Notice that the same doubly heterozygous genotype can be formed in two different ways, Cases I and II. In either case, the *recombinant* gametes made by the double heterozygote are those that are *different* from the gametes at the preceding fertilization. As a consequence, the parental gametes from one case are the same as the recombinant gametes from the other.

Figure 2 Charles S. Stratton (standing 3 feet, 2 inches) and Lavinia Bump (2 feet, 8 inches), both normally healthy and intelligent, at their marriage in 1863. Their short stature apparently stemmed from homozygosity for recessive genes that interfered with the synthesis of pituitary growth hormone. Stratton's parents were first cousins, and Bump's were third cousins. Stratton (better known as General Tom Thumb), Bump, her sister, and her sister's husband (both also of short stature) were protégés of the circus showman P. T. Barnum. For more information, see McKusick and Rimoin (1967). (Photograph courtesy of Circus World Museum, Baraboo, Wisconsin.)

I^A, I^B, and I^O, of a gene. The alleles I^A and I^B are codominant to each other and cause antigens A and B, respectively, to be present. These two alleles are both dominant to I^O. (The chemistry of the antigens is presented in Chapter 17.) Summarizing the genotype/phenotype correspondences, we have:

Full genotype	Shorthand genotype	Blood type (phenotype)
I^A/I^A or I^A/I^O	*A/A* or *A/O*	A
I^B/I^B or I^B/I^O	*B/B* or *B/O*	B
I^A/I^B	*A/B*	AB
I^O/I^O	*O/O*	O

The shorthand genotypes, which we use throughout this chapter, omit the basic locus designation, *I*, leaving just the superscripts.

The Proportions of the Gametes

A doubly heterozygous male, *D/d A/B*, produces four kinds of gametes: (*D A*) (*D B*) (*d A*) (*d B*). In what proportions are these gametes formed? Mendel's law of segregation states that 1/2 the gametes are expected to carry *D* and 1/2 to carry *d*. The same law states that 1/2 the gametes are expected to carry *A* and 1/2 to carry *B*. The question is this: How are these two pairs of alleles distributed into gametes with respect to each other? Mendel was able to show that alleles of different genes are assorted into gametes independently: *The segregation of one pair of alleles in no way alters the segregation of the other pair of alleles.* Each gene pair is acting alone, paying no heed to the other. In the end, the various kinds of gametes have random frequencies; that is, the probability of any kind of gamete is calculated by multiplying together the probabilities for each allele contained therein.

A "tree diagram" is useful to determine the different kinds of gametes from a double heterozygote. Recalling the "and" rule for probabilities, the diagram for our example is:

Dwarfism alleles	ABO alleles	Gamete constitution
1/2 *D*	→ 1/2 *A* →	1/4 (*D A*)
	→ 1/2 *B* →	1/4 (*D B*)
1/2 *d*	→ 1/2 *A* →	1/4 (*d A*)
	→ 1/2 *B* →	1/4 (*d B*)

This diagram is the essence of Mendel's second law, the **law of independent assortment**.

For genes that are inherited independently of each other, 50% of the gametes from a double heterozygote are parental and 50% are recombinant. If the fertilization that produced the double heterozygote was (*D A*)(*d B*), then the first and last gametes listed above are parental and the middle two are recombinant. If the fertilization was (*D B*)(*d A*), then the first and last gametes are the recombinants and the middle two are parental. In either case, independently inherited genes like pituitary dwarfism and ABO are said to exhibit **50% recombination**. All of the pairs of genes that Mendel described had this property.

Offspring Types for Two Genes

Let us now determine the offspring phenotypes from the mating:

D/d A/B male × *D/d O/O* female

In a checkerboard, rows and columns are labeled with the various kinds of gametes and their probabilities. The four different kinds of sperm cells are given above in the tree diagram. For the eggs, the two types are derived from a similar diagram, except it does not branch because the female is homozygous for the ABO gene:

Dwarfism alleles		**ABO alleles**		**Gamete constitution**
1/2 *D*	⟶	all *O*	⟶	1/2 (*D O*)
1/2 *d*	⟶	all *O*	⟶	1/2 (*d O*)

Thus the offspring checkerboard becomes:

		Egg types	
		1/2 (*D O*)	1/2 (*d O*)
Sperm types	1/4 (*D A*)	*D/D A/O* 1/8 normal type A	*D/d A/O* 1/8 normal type A
	1/4 (*D B*)	1/8 normal type B	1/8 normal type B
	1/4 (*d A*)	1/8 normal type A	1/8 affected type A
	1/4 (*d B*)	1/8 normal type B	1/8 affected type B

Each box in the checkerboard corresponds to a probability of 1/8, the product of 1/2 and 1/4, the gamete fractions heading the rows and columns. Using the squares in the first row as examples, readers should fill in the remaining genotypes. Since pituitary dwarfism is recessive, each offspring type with a capital *D* is normal; this includes the top two rows and the first column. Summing similar boxes, the following offspring phenotypic expectations are obtained:

3/8 normal, type A
3/8 normal, type B
1/8 affected, type A
1/8 affected, type B
} sum = 1

The 3/8 normal, type A offspring come from combining probabilities according to the "or" rule:

upper row, first column: 1/8 *D/D A/O*
upper row, second column: 1/8 *D/d A/O*
third row, first column: 1/8 *D/d A/O*
} 3/8 normal, type A

Notice that among normal, type A offspring there is a 1:2 ratio of *D/D* to *D/d*. Thus 1/3 of normal, type A are homozygous normal and 2/3 are heterozygous normal for growth hormone. The normal, type B offspring are constituted in the same way.

Gene × Gene Method

The checkerboard could be called the **gamete × gamete method** for solving genetics problems, because the various types of gametes are first determined and their proportions are then multiplied together. Another procedure, called the **gene × gene method**, is especially useful if many genes are considered simultaneously and just one or a few of the possible results are of interest. We illustrate the gene × gene method using the same mating: *D/d A/B* male × *D/d O/O* female. Considering one pair of alleles at a time, we know that Mendel's first law predicts

D/d × *D/d* ⟶ genotypes: 1/4 *D/D*, 1/2 *D/d*, 1/4 *d/d*
phenotypes: 3/4 normal, 1/4 affected

A/B × *O/O* ⟶ genotypes: 1/2 *A/O*, 1/2 *B/O*
phenotypes: 1/2 type A, 1/2 type B

Now we make use of the fact that the two genes are inherited independently and so multiply the individual expectations. Considering the phenotypic results above, we obtain

$$\begin{aligned}
\text{P(normal, type A)} &= \text{P(normal)} \times \text{P(type A)} \\
&= (3/4)(1/2) = 3/8 \\
\text{P(normal, type B)} &= \text{P(normal)} \times \text{P(type B)} \\
&= (3/4)(1/2) = 3/8 \\
\text{P(affected, type A)} &= \text{P(affected)} \times \text{P(type A)} \\
&= (1/4)(1/2) = 1/8 \\
\text{P(affected, type B)} &= \text{P(affected)} \times \text{P(type B)} \\
&= (1/4)(1/2) = 1/8
\end{aligned}$$

These gene × gene results are, of course, the same as the gamete × gamete results (and in either case we might have considered genotypes rather than phenotypes).

We now apply the gene × gene method to the offspring of parents who are *both* doubly heterozygous, *D/d A/O* × *D/d A/O*, say. The growth hormone alleles, considered alone, yield 3/4 normal and 1/4 affected with pituitary dwarfism. The ABO alleles, considered alone, yield 3/4 type A and 1/4 type O. Multiplying, gene × gene, we get the joint phenotypic expectations

$$\begin{aligned}
\text{P(normal, type A)} &= (3/4)(3/4) = 9/16 \\
\text{P(normal, type O)} &= (3/4)(1/4) = 3/16 \\
\text{P(affected, type A)} &= (1/4)(3/4) = 3/16 \\
\text{P(affected, type O)} &= (1/4)(1/4) = 1/16
\end{aligned}$$

This is the familiar 9:3:3:1 dihybrid ratio that Mendel observed in pea plants when he studied the joint inheritance of two independent traits.

The two methods of analysis, gamete × gamete and gene × gene, involve about the same amount of work when considering just two genes. But consider a situation involving more than two genes: What proportion of the offspring from the following cross is expected to show the dominant phenotype for all genes? (Assume that each uppercase allele is dominant to the corresponding lowercase allele.)

$$C/c\ D/d\ E/e\ F/F \times C/c\ D/d\ e/e\ F/f$$

For this, the checkerboard is quite tedious, but the gene × gene analysis is easy:

Mating by gene	P(dominant offspring)
C/c × *C/c*	3/4
D/d × *D/d*	3/4
E/e × *e/e*	1/2
F/F × *F/f*	1

Multiplying the fractions together, we get 9/32, the proportion of offspring who have the dominant phenotype for all genes. Trying to solve this problem by means of a checkerboard would require an 8 × 8 checkerboard. Should you care to try it, the different kinds of gametes from each parent can be obtained by a tree diagram that branches for each heterozygous gene.

In summary, one can always set up a checkerboard, but the gene × gene method may be much less work.

Meiosis and the Second Law

With regard to the joint inheritance of pituitary dwarfism and ABO, it is known that the two genes are borne on different pairs of homologous chromosomes. The gene that causes pituitary dwarfism is near the middle of the long arm of chromosome 17; the ABO gene is located near the tip of the long arm of chromosome 9 (McKusick 1988). The genotype and karyotype of the doubly heterozygous father, *D/d A/B*, are roughly diagrammed as follows:

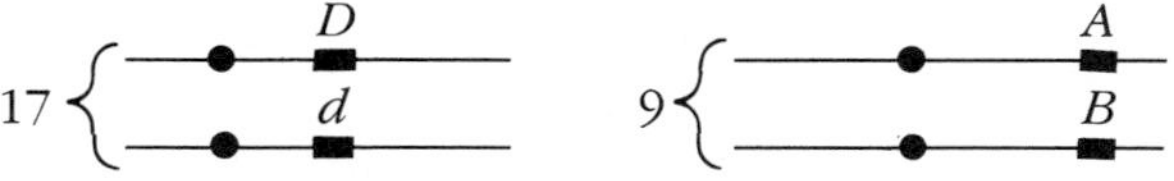

Since *D* and *d* are alleles of each other, they occupy exactly corresponding places (loci) on homologous chromosomes. The same is true for alleles *A* and *B*. This male inherited one chromosome 9 and one chromosome 17 from his mother, and one each from his father. When he makes sperm, the maternal and paternal chromosomes 9 segregate at the first meiotic division independently of the segregation of maternal and paternal chromosomes 17. As shown in Figure 3, the genetic result is equivalent to the diagram shown in an earlier section.

Can it be shown directly, that is, by microscopic examination, that one pair of chromosomes segregates independently of another? This cannot be done if homologous chromosomes look exactly alike. A direct cytological proof requires, in fact, *two pairs* of visibly different homologues. The X and Y sex chromosomes constitute one such pair in many species. Early in this century, E. Eleanor Carothers (1913) found in the grasshopper not only a sex chromosome difference but also an additional set of homologous chromosomes in which one member of the pair had an extra piece of chromosomal material stuck to it. Thus she was able to see differences between the homologues of two sets. In microscopic examinations of testes, Carothers found the four types of secondary spermatocytes (second column in Figure 3) in roughly equal numbers. This showed quite conclusively that chromosome behavior underlies both Mendelian rules of transmission.

GENES ON THE SAME CHROMOSOME: CROSSING OVER

Now consider the inheritance of genes located fairly close to each other on the same chromosome. We have chosen two loci on the X, because X-linked genes are easier to deal with than are autosomal genes. Because homologous loci are absent from the Y, all the genes, including recessives, on a male's single X chromosome are directly expressed in his phenotype. Furthermore, the precise genetic makeup of a female's two X chromosomes can usually be inferred from the phenotypes of her father and sons. Two X-linked genes (described in Chapter 2) are

G = dominant allele for normal color vision
g = recessive allele for green-shift (deuteranomaly)

H = dominant allele for normal blood clotting
h = recessive allele for hemophilia A (classical type)

First, consider the genotypes of two females, Jan and Fay, in Figure 4. Jan's father, who was both green-shifted and hemophilic, transmitted his X chromosome with *g* and *h* to his daughter. Because the pedigree shows that Jan has a normal phenotype, she must carry both a dominant *G* gene and a dominant *H* gene on her other X chromosome, the one received from her mother. Her two X chromosomes can be represented as follows:

genotype of Jan =
g *h* (paternal X)
G *H* (maternal X)

She is doubly heterozygous, with the two nonallelic re-

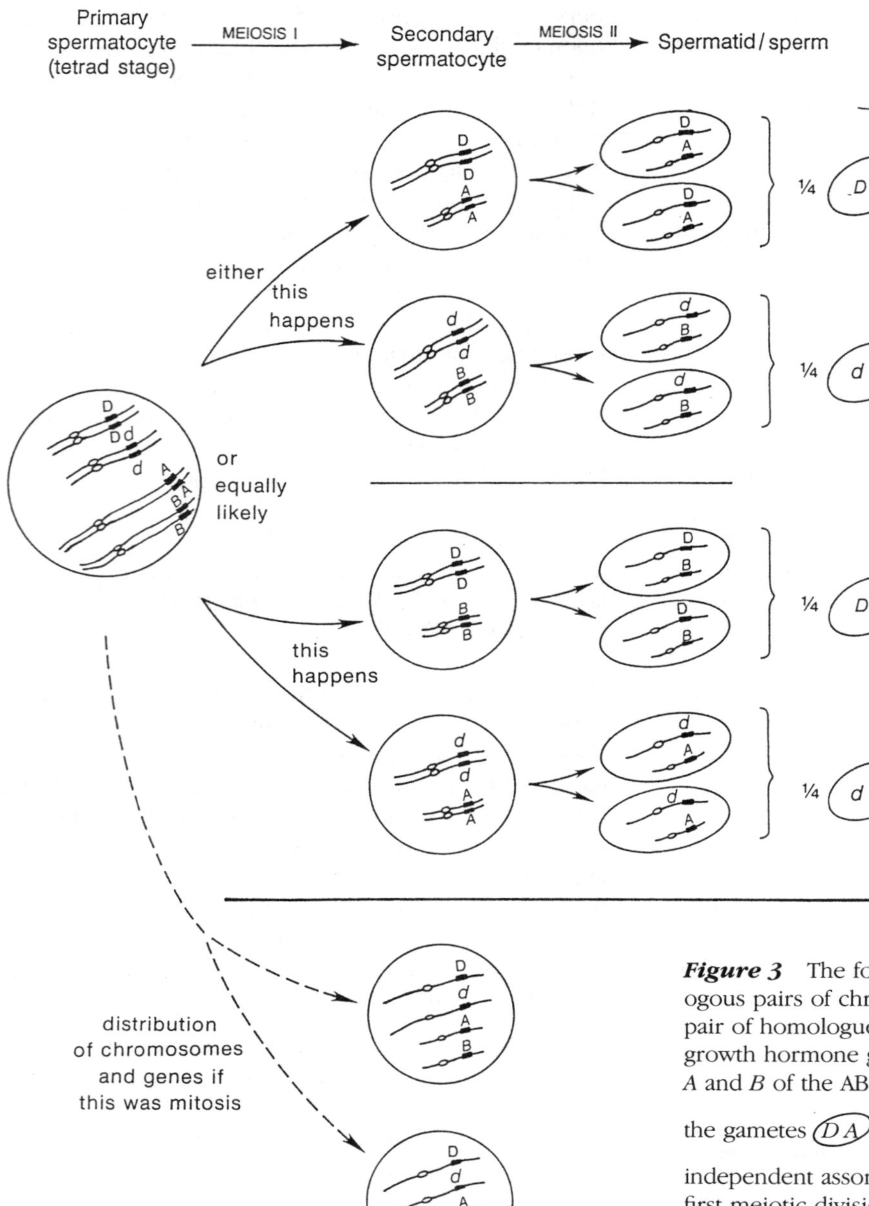

Figure 3 The four possible assortments of the two homologous pairs of chromosomes during spermatogenesis. One pair of homologues is "marked" by alleles *D* and *d* of the growth hormone gene. The other pair is "marked" by alleles *A* and *B* of the ABO gene. The genetic result—1/4 of each of the gametes *DA*, *dB*, *DB*, and *dA*—arises because of independent assortment of different homologous sets at the first meiotic division. The results of oogenesis would be the same, except that only one of the four products of any one meiosis is a functional gamete.

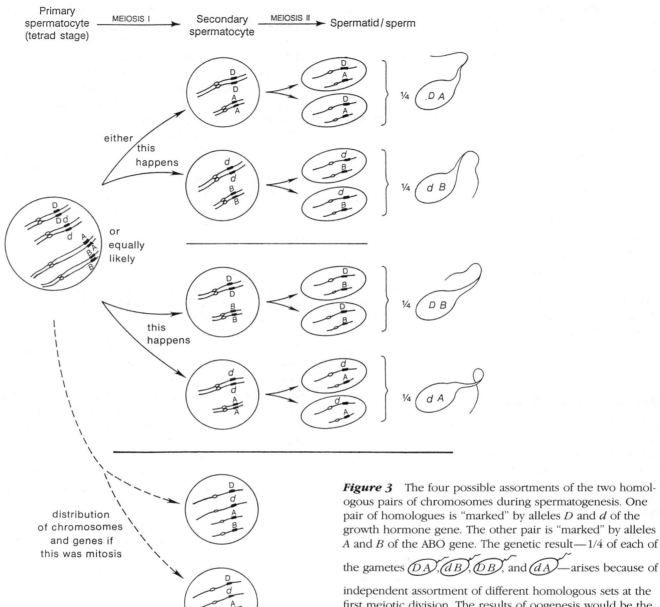

cessive genes, *g* and *h*, on the same chromosome; this is often written $\underline{gh}/\underline{GH}$.

On the other hand, Fay received a paternal X chromosome carrying *g* and *H*. Being phenotypically normal, Fay must have *G* on her other, maternal, X chromosome. This chromosome must also carry the hemophilia gene, *h*, since Fay transmitted *h* to her son. Therefore

genotype of Fay =

g	*H*	(paternal X)
G	*h*	(maternal X)

She, too, is doubly heterozygous, but the two nonallelic recessive genes, *g* and *h*, are on "opposite" homologous chromosomes ($\underline{gH}/\underline{Gh}$).

These two females differ in what is called the **phase** of the double heterozygote. The term **coupling** (or *cis*) is sometimes used to refer to the two recessive nonallelic genes when they are on the same member of a homologous pair, as seen in Jan's genotype. **Repulsion** (or *trans*) may be used when they are on opposite members, as seen in Fay's genotype. The terms derive from the work of Punnett (of checkerboard fame) and others who

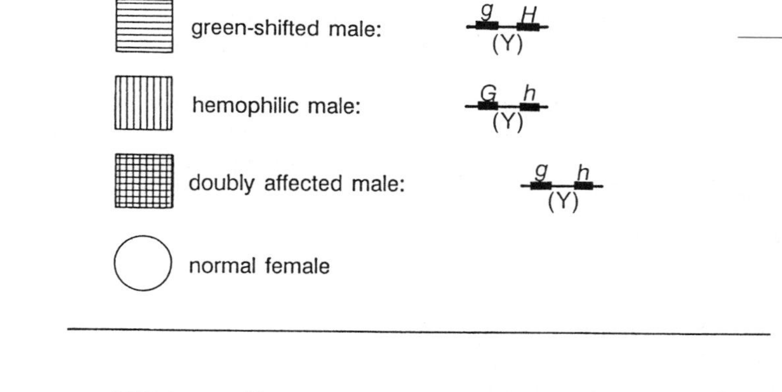

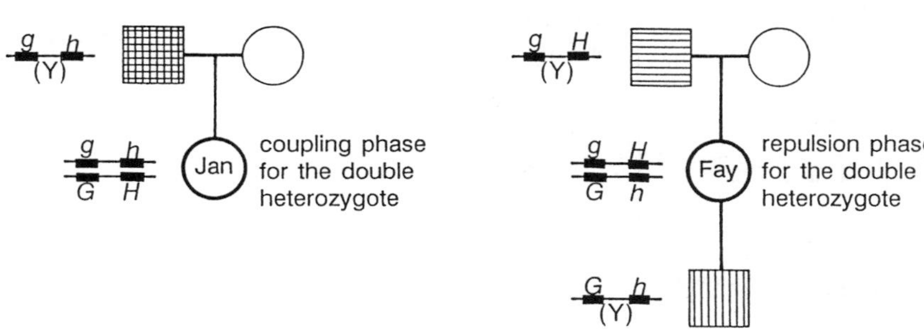

Figure 4 Pedigrees, with genotypes, showing the two ways a person can be a double heterozygote. Jan has the two recessive, nonallelic genes, *g* and *h*, on the *same* X chromosome, whereas Fay has the two recessive genes on *different* X chromosomes. A short horizontal line is used to represent an X chromosome, with dashes for the loci of green-shift and hemophilia. The symbol (Y) indicates the presence of a Y chromosome, but this has neither a locus for green-shift nor one for hemophilia.

tried unsuccessfully to explain departures from Mendel's independent assortment by elaborate systems of attractions and repulsions. (Codominant alleles do not fit this system of nomenclature.)

The Proportions of the Gametes

A number of females like those in Figure 4 have been identified and the phenotypes of their sons have been tallied. A completely normal son from a woman like Jan must have inherited his mother's X chromosome carrying the genes *G* and *H*; a doubly affected son inherited her other X chromosome, the one carrying the genes *g* and *h*. Both her $\underline{GH}$ and $\underline{gh}$ sons are derived from parental type gametes. These are not the only types of sons possible from the double heterozygote in coupling phase, however. Two other types, involving combinations of the mother's X chromosomes, are included in the left-hand column of Figure 5. The two additional types of sons are (1) those with green-shift and normal blood clotting, and (2) those with normal color vision and hemophilia. They arise from gametes carrying X chromosomes represented by $\underline{gH}$ on the one hand and $\underline{Gh}$ on the other, shown at the bottom left of the figure. These are the *recombinant-type* gametes, since the alleles of the two genes appear in a "new" assortment. The two recombinant gametes *in this case* are much less frequent (1.5% each, 3.0% total) than the two parental gametes (48.5% each, 97.0% total).

Figure 5 also shows the gametes produced by the female whose recessive genes are in repulsion. *The percentage of recombination is the same* (3.0% total), but the assortment of genes in a recombinant gamete is different. What is a recombinant gamete from a repulsion phase is a parental gamete from a coupling phase, and vice versa. The percentage of recombination is a function of the loci involved, not of the particular alleles that happen to be at those loci.

In summary, the genes for green-shift and for hemophilia A are seen to exhibit 3.0% recombination, a figure derived from data on the phenotypes of sons from doubly heterozygous females. Had we chosen other genes to investigate, the percentage recombination would generally be some other value than 3.0. It will be demonstrated in the following sections that the percentage recombination depends on the physical distance along the length of the chromosome separating the two genes. Such genes on the same chromosome fairly close to each other are said to be **linked**. Although genes physically located on the same chromosome tend to be inherited together, they can recombine with their alleles on the homologous chromosome by the remarkable process described below.

The Physical Basis of Recombination

In the early 1900s researchers discovered that certain pairs of genes did not obey Mendel's law of independent assortment, which corresponds to 50% recombination. Thomas Hunt Morgan (1911) saw the situation as "a simple mechanical result . . . of the relative location of the factors in the chromosomes," that is, the genes were physically linked. They would remain together in a parental combination so long as the chromosome on which they resided remained intact, which corresponds to 0% recombination. But percentages between 0 (so-called

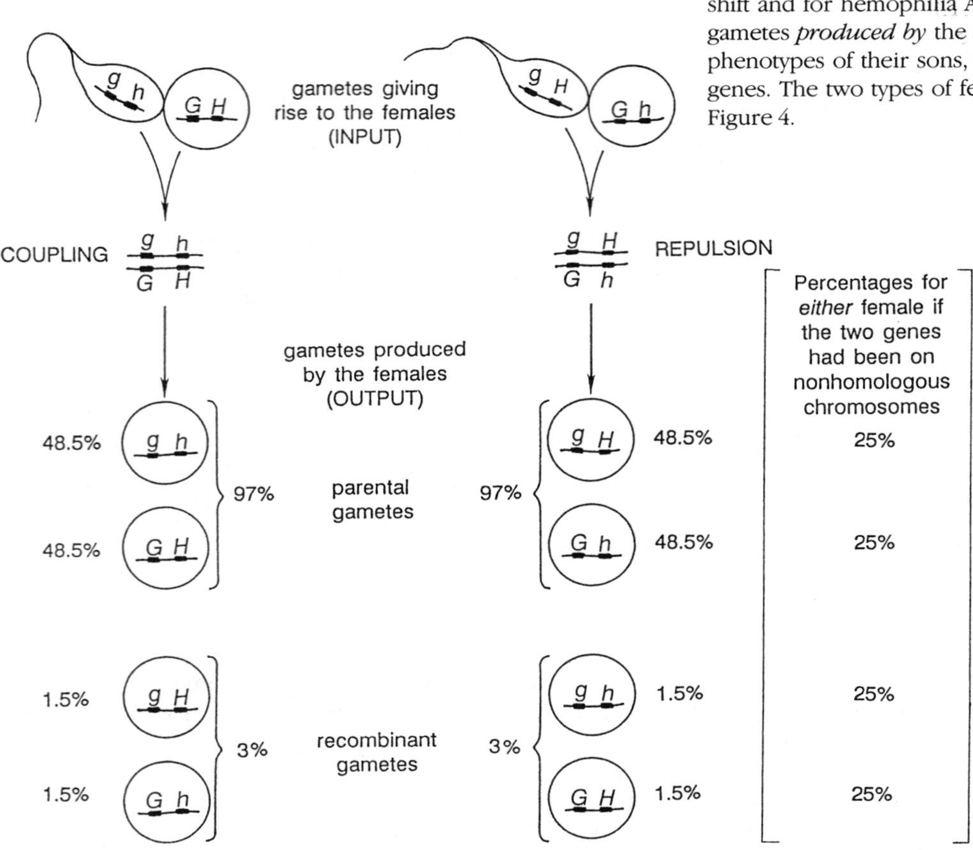

Figure 5 The approximate percentages of different gametes made by females who are doubly heterozygous for green-shift and for hemophilia A. The genetic constitutions of the gametes *produced by* the females were inferred from the phenotypes of their sons, who are hemizygous for X-linked genes. The two types of females correspond to Jan and Fay in Figure 4.

complete linkage) and 50 (independent assortment) were more difficult to explain.

The clue that helped Morgan's group solve this problem was the observation of interconnections between synapsed chromosomes of fruit flies during the tetrad stage of meiosis I. As we noted in Chapter 5, these **chiasmata** form "swaps" between homologous chromatids, resulting in **X**-shaped configurations at given points along the tetrad. As shown in Figure 6, cytologists can see chiasmata in stained meiotic cells from a variety of different organisms. Note that there may be several chiasmata per tetrad. Unfortunately, they are difficult to make out clearly in the meiotic cells of most mammals, including humans, partly because their chromosomes tend to be small. Very little information is available on human females, but in human males the tetrads of autosomes show from one to four (sometimes five, rarely six), with an average of about two chiasmata per chromosome pair. The XY pair in human males seems to have none.

Whatever the organism, each chiasma represents a place where nonsister chromatids have broken and rejoined in crisscross fashion, as shown in Figure 7. This exchange of corresponding segments of nonsister chromatids is called a *crossover*, and the resulting strands are known as *crossover chromatids* or *crossover strands*. The process itself, called **crossing over**, is a normal, relatively frequent event. It occurs early in meiosis, before the chromosomes are condensed enough for microscopists to make out individual chiasmata. In prokaryotes, research has begun to unravel the mechanism of the amazingly exact exchange of chromatid parts at the level of DNA molecules (Stahl 1987; Watson et al. 1987).

Figure 7 depicts crossing over and its genetic consequences in a woman who is doubly heterozygous for green-shift and hemophilia in the coupling phase. Strands labeled 1 and 2 are sister chromatids and exact replicas of each other; similarly, strands 3 and 4 are sister chromatids and exact replicas of each other. For the sake of clarity, the crossover is shown as occurring between nonsisters 2 and 3, but a 1–3 crossover, or a 1–4, or 2–4 (each a nonsister pairing) is equally likely and gives the same results. (In real life all four chromatids are touching one another in three dimensions rather than lying flat on a book page.) As a consequence of the crossover, the future egg may contain, on its X chromosome, the *parental* associations of genes, *gh* or *GH*, or the *recombinant* associations, *gH* or *Gh*. The choice of which of these four chromosomes is included in the functional egg is random; the other three chromosomes end up in polar bodies.

Two aspects of crossing over need to be emphasized. One is that chiasmata appear fairly randomly along the length of all chromosome tetrads. In a given meiosis, a crossover and the resultant chiasma may or may not occur in the length of chromosome between the locus for green-shift and that for hemophilia A. If a crossover does *not* occur in the region, then all products of this meiosis must necessarily be parental with respect to the *G* and *H* loci. A crossover to the left of the *G* locus or to the right of the *H* locus cannot recombine alleles of *G* with those of *H*.

The second aspect points up the difference between

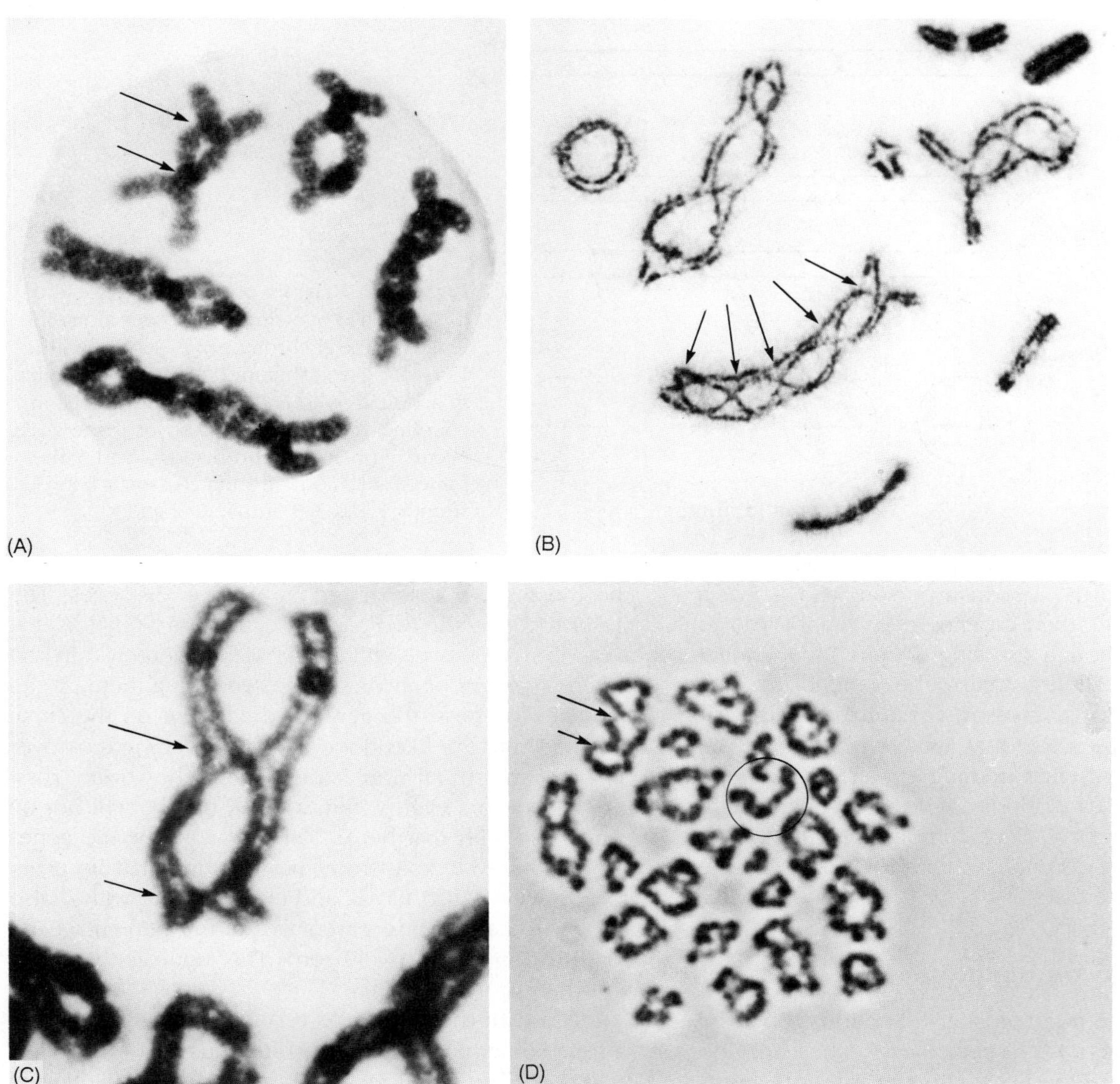

Figure 6 Photographs of chiasmata. These X-shaped configurations (some indicated by arrows) are seen in late prophase of the first meiotic division, when homologous chromosomes start to move apart from each other. (A) During pollen formation in the trillium plant. (Photograph by R. F. Smith; courtesy of A. H. Sparrow, Brookhaven National Laboratory.) (B) During sperm formation in a grasshopper. Five chiasmata can be seen in one tetrad. (Courtesy of Bernard John, The Australian National University.) (C) During sperm formation in a salamander. The centromeres and interchange points show particularly clearly. (Courtesy of James Kezer, University of Oregon.) (D) During sperm formation in a man. The circled X and Y chromosomes appear to be synapsed end to end. (Courtesy of Paul Polani, Guy's Hospital, London.)

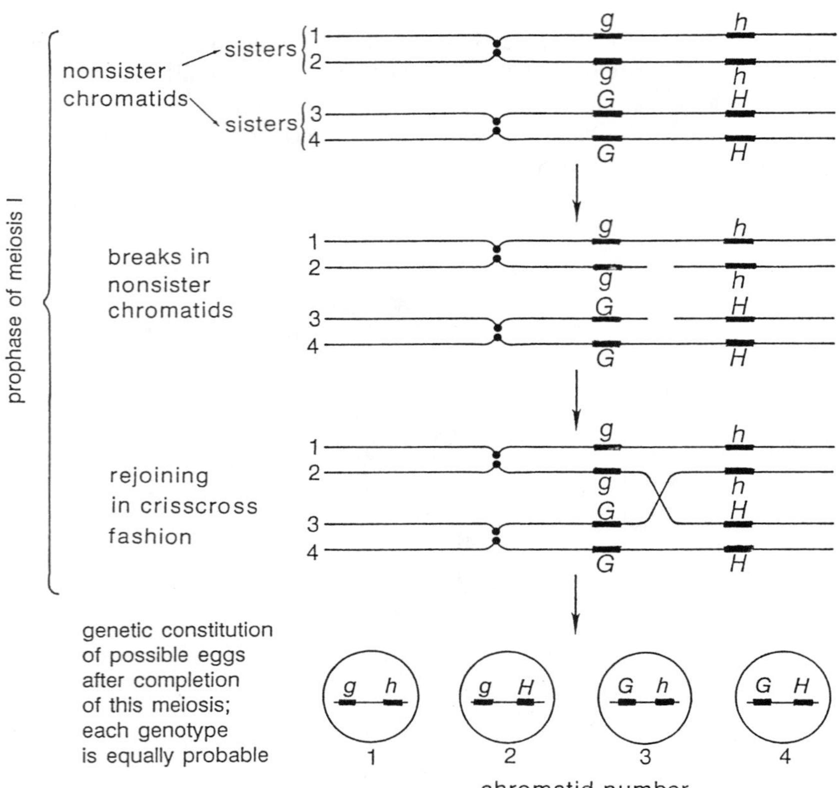

Figure 7 Diagram of the occurrence and consequence of a single crossover along the length of the X chromosome, separating the locus for green-shift and that for hemophilia in a woman who is doubly heterozygous in coupling. Although the crossover is shown as occurring between chromatids 2 and 3, the same results come from a crossover between strands 1 and 3, 1 and 4, or 2 and 4.

crossing over and recombination. If crossing over occurs as depicted in Figure 7 but in a person who is *not* a double heterozygote, then crossing over cannot lead to recombination. Although crossing over is independent of what particular alleles occupy the genetic loci, the *detection* of a particular crossover requires that the exchange occur between loci that are both heterozygous. Readers should verify that nothing genetically different from parental-type associations can result when a person is homozygous *g/g* or *G/G* or *h/h* or *H/H*; this is true whether a crossover occurs between the *G* locus and the *H* locus, or anyplace else.

The Limits of Recombination

Assume that we have no preconceived notions about the location of two genes with respect to each other. Imagine a zygote is formed by joining a gamete carrying *A* and *B* with a gamete carrying *a* and *b*. What sorts of gametes can this double heterozygote make? Table 1 gives the answer for all possible locations of the *A* locus relative to the *B* locus.

In Case I, each gene is on a *different* chromosome. The two nonhomologous chromosomes assort randomly in accordance with Mendel's second law. Independent assortment leads to 1/4 each of the four kinds of gametes, which means 50 percent recombinant gametes and 50 percent parental.

In the remaining cases, the genes are on the *same* chromosome, but separated by various distances. The closer together the two genes are, the less likely it is that a random crossover will occur somewhere within the *specific segment* of chromosome separating them. At the limit of closeness, the genes are adjacent on the chromosome, and the likelihood of an intervening crossover becomes vanishingly rare. (It is not zero, however. Crossing over occurs even within a gene, but we will not be concerned with that here.) Thus, in Case IV, the genes are absolutely linked; from that person, virtually every gamete with *A* also has *B*, and every gamete with *a* also has *b*. All gametes are parental, and the percentage recombination is effectively zero. This value is the lower limit for recombination.

In Case II, the genes are very far apart on the same chromosome, perhaps at opposite ends of a large chromosome. (The position of the centromere is irrelevant because the individual chromatids are continuous through the centromere region.) In this case, virtually all the tetrads of the first meiotic prophase would have a crossover between the genes. But we noted that a single crossover involves *just two* of the four strands of the tetrad. Thus, when a crossover occurs, there remain two chromatids that are *not* involved in an exchange at that particular spot, and these lead to *parental* gametes. Consequently, 50% of the gametes, not 100%, would be recombinants. This value is the upper limit for recombination.

When genes are very far apart, two or more crossovers may occur along the length of the chromosome between them. Remarkably enough, the effect of multiple crossovers is not different from the effect of one crossover, because the two strands involved in a second crossover are *random* with respect to a first crossover. Thus a second crossover is as likely to "undo" the recombination resulting from a first crossover as it is to make more recombination. The added effect of a second, or third, or fourth, or *n*th crossover is statistically zero.

Note that two genes exhibiting maximum recombination—50%—do not reveal to us whether they are on different chromosomes or whether they are far apart on the same chromosome. The term *linked* is reserved for those genes that show significantly less than 50% recombination: Cases III and IV in Table 1.

Case III in Table 1 is the general case for two genes that are neither very close together nor very far apart on the same chromosome. The percentage recombination depends on their separation and falls between 0 and 50. The genius of Morgan and his student, Alfred H. Sturtevant, was to equate percentage recombination with *physical distance along the chromosome*. The randomness of chiasmata along the chromosome suggested that this could be done. Three percent recombination, as in the particular case of green-shift and hemophilia A, is taken to mean that the loci are three map units apart. More generally, we simply define:

x percent recombination = x map units

That this construct is a logical interpretation of the genetic system has been amply verified by extensive mapping of genes in many species. Especially in the fruit fly and increasingly in humans, one can also assign some genes to short, microscopically distinguishable segments by cytogenetic analysis of chromosomal aberrations. The placement of genes this way correlates well with the placement of genes based on genetic recombination.

Although we omit a full analysis of the effects of multiple crossing over, one point needs to be mentioned. The system for measuring distance in map units relies on being able to detect the recombination that results from *each* crossover. Second or third crossovers between two heterozygous loci are statistically undetectable because they do not lead to more recombination, on the average, than is produced by one crossover. Therefore, we can accurately measure only those distances within which multiple crossing over is negligible. In practice, this means restricting our attention to genes that are less than about 20 or 25 map units apart.

The Phase Problem

Because human families are small, the offspring of many doubly heterozygous persons must be grouped together in order to collect enough information for linkage studies. In practice, the phases of double heterozygotes are often in doubt. When this is the case, it is difficult to determine whether the genes are independently inherited or linked.

Suppose, for example, that genes *A* and *B* are linked,

Table 1 Gametes formed by a person who developed from the joining of gametes (*AB*) and (*ab*)

Location of genes A and B with respect to each other	PERCENTAGE OF EACH GAMETE TYPE: *Parental* (*AB*)	(*ab*)	*Recombinant* (*Ab*)	(*aB*)	*Percentage of recombination*	*Description of double heterozygote*
I. On different (nonhomologous) chromosomes	25	25	25	25	50	*A* *B* / *a* *b*
II. On the same chromosome, but very far apart	25	25	25	25	50	*A* *B* / *a* *b*
III. On the same chromosome, neither very far apart nor very close[a]	$\frac{100-x}{2}$	$\frac{100-x}{2}$	$\frac{x}{2}$	$\frac{x}{2}$	x = a number between 50 and 0	*A* *B* / *a* *b*
IV. On the same chromosome, very close together	50	50	0	0	0	*AB* / *ab*

a. Only in Cases III and IV do we refer to *A* and *B* as linked genes.

showing 20% recombination—20 map units apart by the definition above. From double heterozygotes in coupling and repulsion the proportion of gametes are:

Gamete type	Percentage of gamete type: From coupling heterozygotes	From repulsion heterozygotes	Average
AB	40	10	25
ab	40	10	25
Ab	10	40	25
aB	10	40	25

A sample of double heterozygotes might well contain roughly equal numbers of the two phases of double heterozygotes, in which case the four gamete types would be found in about equal proportions (last column above). This proportion would also result if the genes were independently inherited. Thus, not knowing the phase of a double heterozygote, or not being able to make a good guess, is a serious drawback to mapping genes.

This difficulty is peculiar to human genetics. With experimental plants and animals, it is usually possible to arrange a series of crosses leading to double heterozygotes of *known* phase. In human families, three-generational data can sometimes provide knowledge of the phase of a double heterozygote, but the older ancestors may not be available for examination. Furthermore, when dealing with rare inherited human disorders, the double heterozygotes themselves are often rare, although this problem is being overcome with DNA markers spread generously throughout the genome (see Chapter 15).

MAPPING THE HUMAN X CHROMOSOME BY FAMILY LINKAGE STUDIES

All the genes on a chromosome are attached in series—like successive kilometer sections on a highway. It is a very long and narrow chromosomal highway: if the DNA double helix of a human X chromosome were magnified to stretch the 5000 km from New York to Los Angeles, it would be about 10 cm (4 inches) wide. In theory, each gene can be mapped with respect to its near neighbors, and those neighbors with additional neighbors, to yield the relative positions of the perhaps 2,000 X-linked genes. Like distances along a highway, the distances along the chromosome are also additive. Consider a chromosome with genes *a*, *b*, and *c* in that order. If genes *a* and *b* are found to be 4 map units apart, and *b* and *c* are 7 map units apart, then *a* and *c* will be 11 map units apart.

Green-Shift and Red-Shift

We have seen that the distance between *g* (green-shift) and *h* (hemophilia A) is 3.0 map units. By finding mothers who are heterozygous for one of these two genes and also heterozygous for a third nearby gene, the map can be extended left or right. The pedigree in Figure 8 is a clear-cut example involving the joint inheritance of both green-shift (*g*) and red-shift (*r*) in a family from Sardinia. Recall that these two traits are separable entities leading to somewhat different defects in color vision (Chapter 2).

Readers should write out for themselves the X-linked genotypes for all the males. For example, I-1 carries the X chromosome *G r*, and he transmitted this to his daughter, II-2. She received from her mother the X chromosome *g R*, this conclusion being confirmed by noting that two of her brothers (II-3 and II-4) also carry *g R*. Thus female II-2 is doubly heterozygous in repulsion.

Each of the three sons of II-2 has a different phenotype. III-2, the oldest, is both green-shifted and red-shifted; III-4, the middle son, is green-shifted only; III-5, the youngest, is red-shifted only. The oldest son's phenotype reflects recombination between the two color vision loci, whereas his younger brothers' phenotypes represent the two parental (noncrossover) chromosomes. Thus, in this particular sibship, we observe 33% recombination. (The phenotypes of the two daughters are uninformative.) Combined with all other pedigrees that have provided information, however, the best estimate for the distance between the two color vision genes turns out to be 6.3 map units, and, as we noted before, the distance between the green-shift gene and that for hemophilia A is 3.0 map units.

Mapping studies that combine the information from many families are complicated, and we omit the details (Conneally and Rivas 1980). The *lod score method*, invented by the mathematical geneticist Newton Morton (1955), compiles data sequentially as more and more informative families are found. The term *lod* stands for "*lo*garithm of the *od*ds favoring linkage"; the analyses are now handled largely by computer programs.

Two Clusters of X-Linked Genes

The next step in building up our map of the X chromosome is to decide the relative positions of the three genes *g*, *h*, and *r*. Knowing just the *g–h* distance and the *g–r* distance does not provide a unique map; two alternatives are possible:

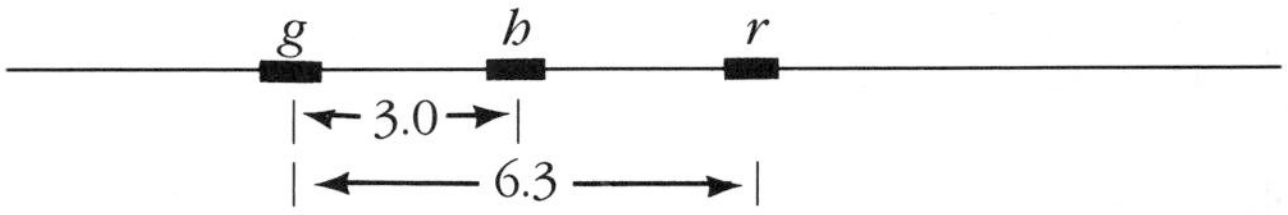

g — *h* — *r*
|← 3.0 →| (*g* to *h*)
|← 6.3 →| (*g* to *r*)

or

h — *g* — *r*
|← 3.0 →|← 6.3 →|

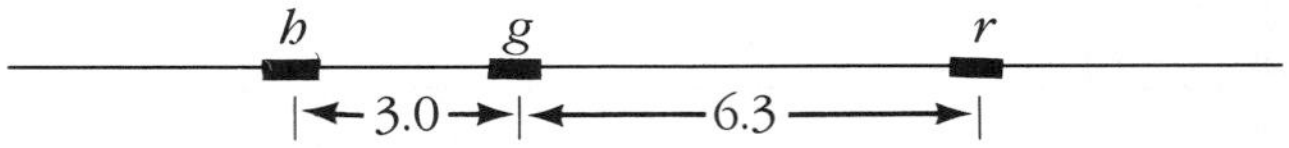

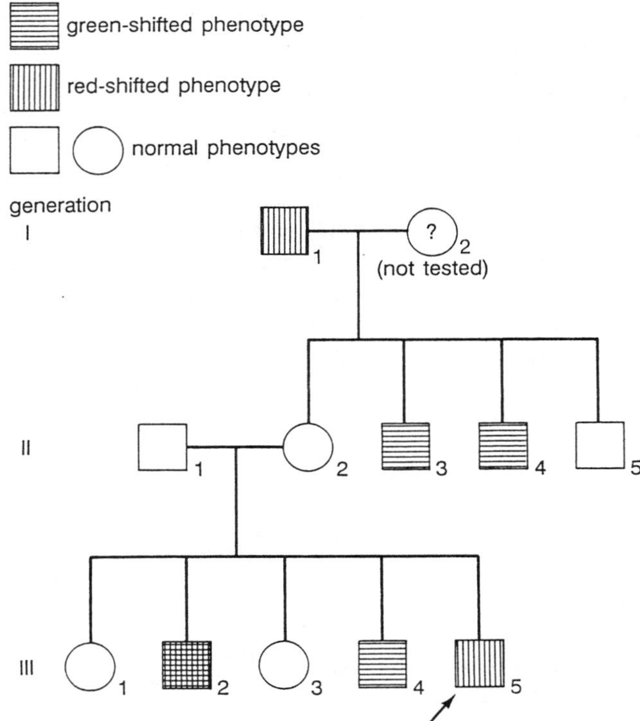

Figure 8 A pedigree from Sardinia showing the joint inheritance of green-shift and red-shift due to linked genes on the X chromosome. The members of the family were tested in several different ways and the diagnoses appear unambiguous. (Data from Siniscalco et al. 1964.)

To choose between the two possibilities, we need information about the *h–r* distance. No families have been found to provide this information directly; such pedigrees would have to include sons from women doubly heterozygous (in known phase) for hemophilia A and red-shift. Nevertheless, each of these genes has been mapped with respect to additional loci in the immediate neighborhood, so it is likely that the upper of the two alternatives is the correct one. The additional genes are *Xm* and *Gd*. The former codes for a protein found in blood serum and the latter for an enzyme, glucose-6-phosphate dehydrogenase (G6PD), found in many types of cells but especially important in red blood cells. It is also known (from families with X chromosome aberrations) that this group of five genes is near the tip of the long arm of the X chromosome, with *r* being closest to the tip. The map of this cluster of five genes is:

red-shift (near tip)
3.3 map units
hemophilia A
0.3 map units
glucose-6-phosphate dehydrogenase
2.7 map units
green-shift
9.6 map units
Xm serum protein

These map distances (Keats 1983) are not known with great accuracy, however, because they are based on limited family studies. Although derived by a sophisticated statistical method called segregation analysis, the results can be no better than the raw data on which they are based. (Mathematically minded readers interested in human linkage analysis should read Conneally and Rivas [1980].)

The discovery in 1962 of an X-linked blood group gene gave new impetus to mapping the human X chromosome. Because the gene had two frequently occurring alleles, symbolized Xg^a and *Xg*, heterozygotes were common. The Xg^a allele produces an antigen on the surface of red cells (not related to the antigens of the ABO or Rh genes). This antigen is absent in *Xg*/(Y) males and in *Xg*/*Xg* females. In most populations, between 40 and 50% of the females are heterozygous Xg^a/Xg, so the locus can be used as a convenient marker in linkage studies. For this reason "map makers" are understandably disappointed when mothers heterozygous for a rare X-linked gene turn out to be homozygous for Xg^a or *Xg*.

The English researchers R. R. Race and Ruth Sanger (1975) have summarized the results for about 50 X-borne genes that were tested for linkage with *Xg*. Definitely linked to *Xg* were three genes, those whose mutant alleles lead to the diseases ichthyosis, ocular albinism, and retinoschisis. A fourth disorder, called Xk-related chronic granulomatous disease, is now also known to be caused by a gene in this group (Table 2). All these genes turn out to be near the tip of the short arm, far removed from the color vision group located near the opposite tip. Thus any gene from the *Xg* cluster shows independent assortment with any gene in the color vision cluster. Although the distance from *Xg* of each of the genes in Table 2 has been approximately determined, their positions with respect to one another are still in doubt (see Question 12).

Further Considerations

ADDING SHORT DISTANCES. We noted that long distances along a chromosome cannot be measured directly because of inaccuracies introduced by multiple crossing over. But long distances can be calculated by adding together the shorter distances between intervening genes. For example, consider the following hypothetical map:

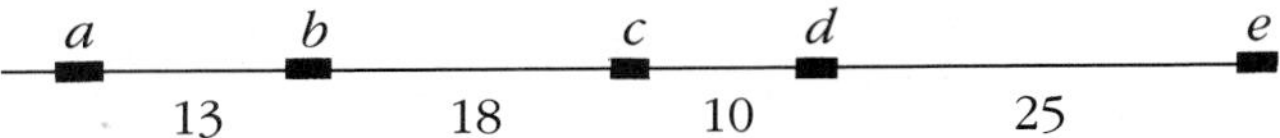

Double heterozygotes involving *adjacent* genes would lead to reasonably accurate map distances. But in double heterozygotes for, say, genes *a* and *d*, the occurrence of undetectable multiple crossovers would lead to an inaccurate *a–d* map distance. Instead of measuring 13 +

18 + 10 = 41, we would actually observe less recombination. The value of 41 map units is more accurate than what would be observed by direct measurement. At the extremes of the map above, genes *a* and *e* are 66 map units apart. By a direct recombination test, however, they would appear to be unlinked, showing the upper limit of 50% recombination.

THE TOTAL LENGTH OF THE X CHROMOSOME. The DNA double helix within the X chromosome, if stretched out, is estimated to be about 10 cm long. To measure the length, not in physical units, but in terms of recombinational map units would require genes known to be at opposite ends and other known genes at short intervals along the entire length. This array is not currently available, but a guess for the genetic length of the X chromosome can be based on the counts of chiasmata in meiotic cells, each chiasma representing 50 map units.

Unfortunately, the meiotic cells of females are difficult to observe, and males do not have two X chromosomes between which chiasmata may be observed. Investigators therefore extrapolate to the female X chromosome chiasmata frequencies based on male autosomes. This procedure is prone to error, however, because the frequency of crossing over in females is greater than that in males

Table 2 Four genes known to be linked to *Xg*

Disease, gene symbol, and distance from Xg[a]	*Description of disease*
Xk-related chronic granulomatous disease ***xk*** about 1 map unit from *Xg*	Hemolytic anemia due to abnormal projections from red blood cell membranes. Also, increased susceptibility to bacterial infections because phagocytosis of foreign cells by white blood cells is impaired. In the most severe form, affected boys die from overwhelming infection in the first years of life (Marsh 1978).
Ichthyosis ***ic*** about 11 map units from *Xg*	Scaling of the skin over much of the body starting from infancy. Scaling on limbs and trunk usually worsens with age, while that on scalp, neck, and side of face clears at bit. Dark scales, about 4 mm, slough off. Ointments give very little relief, but discomfort is less in warm weather. An occasional heterozygous female complains of skin dryness but shows no clinical signs (Wells and Jennings 1967). Patients lack the enzyme steroid sulfatase (McKusick 1988).
Ocular albinism ***oa*** about 14 map units from *Xg*	No pigmentation in the iris or in the retina and other layers at the back of the eye. The translucent iris causes poor vision and great discomfort in light. Involuntary jerking of the eyeball muscles and a peculiar head nodding in infancy also occur. Pigmentation is normal elsewhere in the body. Heterozygous females have normal vision but may show some depigmented patches in the retina (Gillespie 1961).
Retinoschisis ***rs*** about 27 map units from *Xg*	Separation of cell layers within the retina. Poor vision in childhood, due to retinal lesions, slowly worsens to blindness in middle age as a layer of the retina may completely split off from the back of the eye (Gieser and Falls 1961).

a. The recessive mutant allele of each is fairly rare, so virtually all affected persons are male. Since the positions of the four genes with respect to one another have not been established, several mapping arrangements are possible, as indicated by double-headed arrows:

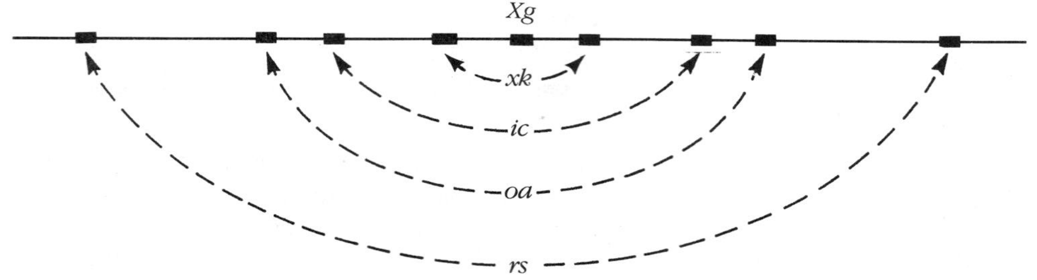

—by about 80% in some regions of chromosome 1 (Morton and Burns 1987). (Such sex-related differences in recombination rates are common, but poorly understood. In male fruit flies, for example, there is no crossing over at all!) For these reasons, a guess of about 250 map units for the recombinational length of the X chromosome is tentative.

SIGNIFICANCE OF GENE MAPPING

"There is a certain satisfaction in knowing where a piece of a puzzle fits and what its relationship is to neighboring pieces. So it is with the 'puzzle' of how the vast amount of genetic information is packaged and distributed throughout the genome" (Conneally and Rivas 1980). The construction of the human gene map can, indeed, be likened to the assembly of a jigsaw puzzle. Initially there is disorder. Then, just as border puzzle pieces are separated from inside pieces, the X-linked genes are distinguished from autosomal genes and arranged linearly wherever possible. As clusters of interconnected puzzle pieces are foci for adding other pieces, previously mapped clusters of genes acquire new members, or two previously separated clusters on the same chromosome are joined by the discovery of intervening genes.

Beyond the delight which some take in mapping for its own sake, however, there are important reasons for working on this puzzle. First of all, the functioning and evolution of the genetic machinery can only be understood by dissecting and analyzing its component parts—individually and in relation to each other. "Personally, I have little taste for formal genetics in general, let alone that of man. But formal genetics is beyond any possible doubt an absolutely essential prerequisite to the attack on any other problem involving genetic mechanisms" (Ephrussi 1972). And beyond helping to understand genetic mechanisms, gene mapping has many practical applications in the fields of genetic counseling and medicine.

Building and Using Gene Maps

Human linkage studies that involve a search for recombination in families with double heterozygotes for easily observed traits are tedious. In the 1970s entirely new mapping methods based on *somatic cell genetics* eliminated the need for any kind of family data. Instead, cells from various sources are grown and manipulated in cell cultures. Such studies yield a different type of information: they assign a gene to a particular chromosome or to a specific segment of a chromosome. If two genes are assigned to the same segment, linkage is established or at least suggested. By 1988, 771 genes had been assigned to specific autosomes by somatic cell methods, whereas only 329 had been established by more traditional family linkage studies (McKusick 1988).

The 1980s have contributed several additional mapping aids that involve manipulations of nucleic acids. It is now possible, for example, by chemically analyzing a person's DNA, to determine heterozygosity for nucleotide sequences at so-called *restriction enzyme sites*. Although no change in one's external phenotype is necessarily associated with a change in a restriction enzyme site, the heterozygosity can nonetheless be used as a Mendelian marker in family linkage studies. Thus the newer molecular and somatic cell technologies, which we discuss in detail in Chapter 15, have come to augment and complement the older mapping methods.

A complete description of the human gene map can be found in McKusick (1988). Recent data for chromosome 1, which is one of the best known autosomes, are presented in Table 3, which lists some of the approximately 50 genes whose locations there have been confirmed. (Unconfirmed observations suggest an additional 50 genes on chromosome 1.) By current convention, each gene symbol consists of one to four characters—capital letters or arabic numbers. Many of the genes in Table 3 will be unfamiliar to readers; a large group of them are genes for enzymes that can be identified by electrophoresis of proteins extracted from somatic cells grown in culture.

The best map of recombinational distances among the chromosome 1 loci whose position is known to some extent is presented in Figure 9, which also includes the numbered dark and light bands and a few regional assignments (from the last column of Table 3.) There are no inconsistencies between the order of genes determined by recombinational mapping (left column) and the regional assignments determined by somatic cell methods. Thus the genetic map is said to be *colinear* with the cytological map. Figure 9 indicates, however, that crossing over must occur *nonrandomly* along the length of this chromosome. For example, the genes *UMPK* and *PGM1* are separated by 16 map units genetically and about five bands cytologically. On the other hand, the genes *PGM1* and *AMY1,2* are genetically further apart (32 map units), but cytologically closer together (about 1 band). Thus crossing over seems to occur more frequently in the latter region than in the former. (For the standardized banding designations of all human chromosomes, see ISCN [1985].)

The human gene map is increasingly important in medicine. For example, if a mutant gene for some disease *cannot* be detected directly, its presence may be revealed if it is closely linked to an innocuous gene that *can* be detected. Molecular biologists have been able to isolate and analyze some disease-causing genes by successively

Table 3 Confirmed gene assignments to chromosome 1

Gene symbol	*Gene name*	*Mode*[a]	*Regional assignment*[b]
ACTA	Actin skeletal muscle, α chain	S	p21–qter
AK2	Adenylate kinase-2	F,S	p34
AMY1	Salivary amylase	A, F,S	p21
AMY2	Pancreatic amylase	A, F,S	p21
AT3	Antithrombin III	A,D,F,S	q23.1–q23.9
C8A	Complement component 8, α chain	F	p22
CAE	Cataract, zonular pulverulent	F	q21–q25
CMT1	Charcot-Marie tooth disease	F	p22–q23
DIZ1	Satellite DNA III	A	q11
EL1	Elliptocytosis-1 (Rh-linked)	F	p36.2–p34
ENO1	Enolase-1	F,S	pter–p36.13
FH	Fumarate hydratase	S	q42.1
FUCA	α-L-Fucosidase	F,S	p34
FY	Duffy blood group	F	q12–q21
GALE	UDP-galactose-4-epimerase	S	pter–p32
GBA	Acid β-glucosidase	A,D, S	q21
GDH	Glucose dehydrogenase	F,S	pter–p36.13
GUK1	Guanylate kinase-1	D, S	q32.1–q42
GUK2	Guanylate kinase-2	D, S	q32.1–q42
NRAS1	Oncogene NRAS1	A, S	p22
PEPC	Peptidase C	S	q42
PGD	Phosphogluconate dehydrogenase	F,S	p36.2–p36.13
PGM1	Phosphoglucomutase-1	D,F,S	p22.1
RH	Rhesus blood group	D,F	p36.2–p34
RN5S	5S ribosomal RNA genes	A	q42–q43
RNU1	U1 small nuclear RNA	A, S	p36.3
SC	Scianna blood group	F	p36.2–p22.1
UGP1	UDP-glucose pyrophosphorylase-1	S	q21–q23
UMPK	Uridine monophosphate kinase	F,S	p32

Source: Morton and Burns (1987) and McKusick (1988).

a. The methods by which the genes have been assigned to chromosome 1: A, annealing of nucleic acids (hybridization in situ); D, deletion mapping; F, family linkage studies; S, somatic cell genetics. The methods (except family linkage studies) are discussed in Chapter 15.

b. The shortest segment of chromosome 1 that encompasses the locus of the gene. The numbers represent band designations p, short arm; q, long arm; ter, terminus (tip) of short or long arm.

mapping them to closer and closer restriction enzyme sites. In addition, knowledge of linkage can be useful in the following genetic counseling situations:

1. To test fetuses in early pregnancy to determine whether or not they are affected with a genetic disorder (see Chapter 25). This could be done by analyzing amniotic fluid cells for the products of the alleles of the innocuous gene. As one example, the dominant gene for *myotonic dystrophy*, a severe muscle-wasting disease, is closely linked on chromosome 19 to the so-called *secretor* locus, which has no medical significance. Myotonic dystrophy cannot be detected prenatally, but the phenotypes associated with the secretor locus can be (Teichler-Zallen and Doherty 1980). In some pedigrees the secretor status of the fetus can reveal with high probability the presence of the gene for myotonic dystrophy. (See also Question 19.)

2. To test persons at risk for an autosomal dominant disorder before the time when symptoms of the disease usually appear, to see whether or not the person has the gene. This information—good or bad—might affect decisions regarding treatment and child bearing. *Huntington disease (HD)* is a severely debilitating condition that afflicted the folk singer Woody Guthrie. Combining extensive pedigree studies with DNA analyses, investigators have recently shown that the *HD* gene, although itself "invisible," is often linked to a particular restriction enzyme site that can be identified by mo-

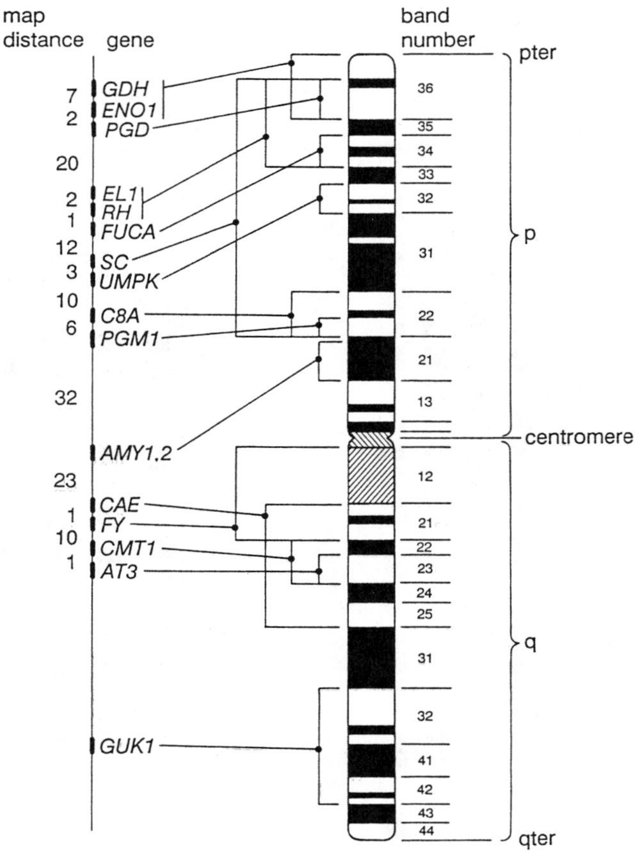

Figure 9 Recombinational map distances (for males) along chromosome 1 with regional assignments of some genes, or gene groups, to the cytological map. The total length of chromosome 1 is about 200 map units based on an average of about four chiasmata per chromosome 1 tetrad seen in spermatocytes. (Data from Hamerton et al. 1984, and Morton and Burns 1987.)

lecular techniques. This is further discussed in Chapter 15.

3. To test the normal sibs of persons affected with an autosomal recessive disorder to see whether or not they are carriers. Here, too, the test results might affect reproductive decisions.

Gene mapping can also help detect or explain multiple causes of the same genetic disease. For example, abnormal blood clotting occurs for many different reasons related to mutant genes at different X-linked and autosomal loci. Proper therapy requires knowledge of the cause of the defect. Although both the hemophilia A and Christmas disease genes are X-linked, only that for hemophilia A maps near the green-shift locus. (This was the first linkage to be established for humans [Bell and Haldane 1937]. At that time, however, the distinction between hemophilia A and Christmas disease had not been made, so their map distance was inaccurate.)

Another reason for mapping is to inquire into whether the positions of genes relate in any way to their functions. Here the answers are mixed. Several genes coding for part of the hemoglobin molecule (the α-like chains) are linked with one another on chromosome 16; another set of genes coding for a different part of the *same* molecule (the β-like chains) are linked to one another, but on chromosome 11. Several genes important in the immune system (white blood cell antigens, complement, and immune response genes) are close together on chromosome 6, but genes that code for enzymes catalyzing successive steps in the same metabolic pathway are usually located in seemingly haphazard places on the genome. Additional research will be needed to answer a major question of gene organization: Why are some genes clustered and how did they get to be that way?

Gene Mapping and Evolution

A great deal can be learned about the evolutionary history of related species by comparing their chromosomes and, when possible, their gene maps. The more closely related two species are, the more similar their chromosomes and gene maps will be. On this principle, a number of large-scale and small-scale evolutionary trees have been constructed.

The study of primate chromosomes has evoked much interest. Most similar to the human karyotype are those of the chimpanzee, gorilla, and the orangutan, all having 48 chromosomes. Comparisons of banding patterns among these four species reveal virtual identity of the X chromosomes and remarkable similarity of the autosomes as well (Figure 10). Only the Y chromosomes vary considerably from species to species. Chromosome 2 of humans seems to represent a Robertsonian fusion of two ancestral acrocentric autosomes, a conjecture that explains our having 22 rather than 23 autosomes per haploid set. Other differences among the four species can be accounted for by inversions within chromosomes and changes in heterochromatic regions. This approach suggests that humans are related most closely to the chimpanzee (13 seemingly identical chromosomes), and, in turn, to the gorilla (9 matches) and orangutan (8 matches). The gibbon is a much more distant relative; except for the X chromosome, there is very little similarity between its chromosomes and those of the other four primates. The same evolutionary relationships are obtained when comparing DNA sequences by molecular hybridizations (Lewin 1984).

Considerable progress has been made in mapping primate genes, especially those of the chimpanzee, by methods outlined in this chapter. Many homologies are found. For example, some chimpanzee structural genes that are counterparts of human genes have been mapped to comparable chromosomes. The production by humans and chimpanzees of hemoglobin chains and some other polypeptides that are identical means that the structural genes for these proteins must also be virtually identical. So similar are the karyotypes, gene maps, and proteins of humans and chimpanzees that it has been proposed that the obvious phenotypic differences between the two species must reside in differences in their genetic regulatory mechanisms rather than in the structural genes themselves (King and Wilson 1975). Even Down syndrome traits have been described in chimpanzees (McClure et al. 1969), associated with trisomy of a small acrocentric chromosome that is the counterpart of human chromosome 21.

The X chromosome is especially interesting. It is remarkably constant in size, banding patterns, and gene content, not only among primates, but also among all mammals that have been studied. For example, the *G6PD* gene is X-linked in the mouse, hamster, rabbit, kangaroo, donkey, and horse, as well as in all primates surveyed. Why the X chromosome has been preserved intact throughout mammalian evolution is an intriguing question. Recall that inactivation of one of the X chromosomes in females occurs in *all* mammalian species. This mechanism compensates for the double dose of X-linked genes in the female. Susumu Ohno (1969, 1973), of the City of Hope Medical Center in California, proposes that once X inactivation evolved early in the mammalian line, genes on the mammalian X were stuck there. Translocation of a part of the X to an autosome would remove the genes on the translocated piece from inactivation and the consequent dosage change would result in a serious disadvantage to the organism.

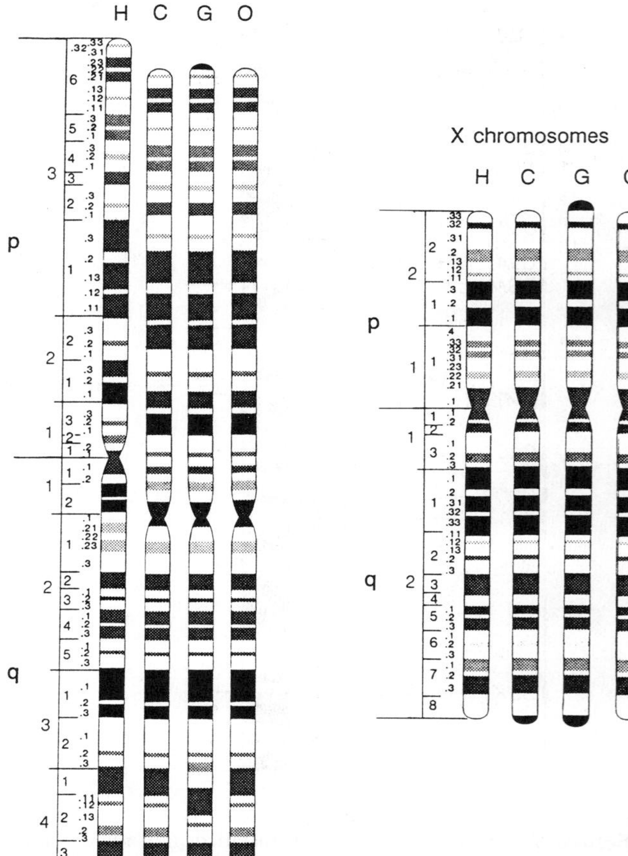

Figure 10 Comparison of corresponding chromosomes from four primates: humans (H), chimpanzees (C), gorillas (G), and orangutans (O). The diagrams of the four 1's and four X's represent Giemsa-stained chromosomes, not from metaphase, but from late prophase, when about 1,000 bands per haploid set can be resolved. More bands are resolved here than in the chromosome 1 pictured in Figure 9. The similarities in banding, especially for the X chromosome, are striking. (From Yunis and Prakash 1982.)

SUMMARY

1. A gamete carries one allele of *each* gene. Whether two loci are on different (nonhomologous) chromosomes or on the same chromosome, a double heterozygote can make four kinds of gametes, of which two are recombinants.

2. For loci on *different* chromosomes:
(a) Mendel's Law of Independent Assortment applies.
(b) 50% of the gametes from a double heterozygote will be recombinant.
(c) Offspring types can be determined by the gamete × gamete (checkerboard) method, or by the gene × gene method.
(d) The behavior of chromosomes during meiosis is the physical basis for independent assortment of genes.

3. For loci on the *same* chromosome:
(a) Independent assortment occurs only if the loci are some distance apart.
(b) The closer together the genes are, the less will be the percentage of recombination.
(c) The microscopically visible chiasmata repre-

sent the positions of crossovers, the exact exchange of pieces of nonsister chromatids.
(d) The double heterozygotes needed to detect recombination may be in either the coupling or the repulsion phase.

4. Crossing over provides a measure of physical distance along a chromosome. This equivalence is reasonably accurate as long as recombination is under about 25%, in which case x% recombination is defined as x map units.

5. Genes on the human X chromosome are mapped from data about the sons of doubly heterozygous mothers. The phenotype of each son reflects the genetic constitution of the egg from which he arose. Investigators have identified two clusters of X-linked genes, one including the color vision loci and the other including the Xg blood group locus.

6. Chromosomes can be mapped by family linkage studies (this chapter), or by somatic cell and molecular techniques (Chapter 15). Mapping information provides fundamental knowledge of genome organization and can help in genetic counseling.

7. Comparisons of mammalian gene locations and chromosome banding patterns provide information on evolutionary sequences.

FURTHER READING

The reader may wish to learn about independent assortment from Mendel himself (1865); except for some minor changes in terminology, his paper (in translation) is quite clear. Not as clear, but historically important, are the 1905–1908 papers of Bateson and Punnett. The most detailed information on human linkage is provided by Human Gene Mapping 9 (1987) and the foreword to McKusick (1988). Shows et al. (1982) have written a lengthy review that is molecularly oriented, and Conneally and Rivas (1980) have written one that is mathematically oriented. The more general articles by Francke (1983) and by McKusick and Ruddle (1977) provide a comprehensive introduction to mapping human genes.

QUESTIONS

1. What kinds of gametes, in what proportions, can be made by a person who develops from an egg carrying genes A and B and a sperm carrying a and b, if the two loci are (a) on different (nonhomologous) chromosomes? (b) on the same chromosome adjacent to each other? (c) on the same chromosome 8 map units apart? What is the percentage recombination in each case?

2. What kinds of gametes, in what proportion, can be made by a double heterozygote in coupling, $\underline{AB}/\underline{ab}$, if the A and B loci are 80 map units apart?

3. List all the kinds of eggs that can be formed by a woman who is heterozygous for the three independently inherited genes, brachydactyly, pituitary dwarfism, and sickle-cell anemia: B/b D/d Hb^A/Hb^S. What is the proportion of each kind of gamete?

4. Does the list that answers Question 3 also apply to the genes contained in the second polar body (formed as a sister to the functional egg)?

5. If the woman in Question 3 were heterozygous for a fourth independent gene, Xg^a/Xg, how many different kinds of gametes could she make?

6. Develop the algebraic formula for the number of different kinds of gametes that can be made by a person who is heterozygous for n independent genes. Refer to questions 1, 3, and 5 for instances in which $n = 2$, 3, and 4, respectively.

7. What phenotypes, in what proportions, are possible from the following mating involving the ABO and growth hormone genes:

$$B/O\ D/d \times B/O\ D/d$$

8. The genotype/phenotype correspondences for the Xg blood group system are given below. Considering together the ABO and Xg types as well as the sex of the offspring, predict the phenotypic probabilities for the offspring of the mating of an $A/O\ Xg/Xg$ female × $B/O\ Xg^a/(Y)$ male.

Genotype	Phenotype
$Xg^a/-$ ♀ $Xg^a/(Y)$ ♂	Xg(a+)
Xg/Xg ♀ $Xg/(Y)$ ♂	Xg(a−)

9. Consider these three independent genes:

B = brachydactyly (short fingers) and b = its normal recessive allele
P = polydactyly (extra fingers) and p = its normal recessive allele
S = syndactyly (joined fingers) and s = its normal recessive allele

From the mating $B/b\ P/p\ S/s \times B/b\ P/p\ S/s$, (a) What is the one most likely *genotype* among the offspring? What is its probability? (b) What is the one most likely *phenotype* among the offspring and its probability? (c) What is the probability of a child with completely *normal* fingers?

10. A certain mating gave among the offspring four different phenotypes in equal proportions. Explain this result genotypically on the basis of (a) multiple alleles at one locus and (b) two loci.

11. Why is a double heterozygote needed in order to map two genes on the same chromosome by observations on recombination?

12. Why are the positions of the genes for ichthyosis, ocular albinism, and retinoschisis known approximately with respect to *Xg* (and also to *Xk*) but not with respect to each other?

13. Crossing over seems to be less frequent in males than in females. To establish the fact that two autosomal genes are linked, would you rather find double heterozygotes who are male or female? Think about this a bit before hopping to the answer. (Assume all other pedigree parameters are the same.)

14. Assume two X-linked genes: A dominant to a, and B dominant to b. Max had both recessive traits and fathered many daughters, all with both dominant traits. These daughters, in turn, had a total of 40 sons (Max's grandsons): 15 had both recessive traits, 17 had both dominant traits, and 8 had the A dominant trait but not B, or vice versa. What is the genotype of Max's daughters? How far apart are the two genes?

15. A female with normal color vision is heterozygous in coupling for the X-linked genes for green-shift (G/g) and also for red-shift (R/r). The loci are about 6 map units apart. What percentage of her sons are expected to have normal color vision?

16. A husband and wife have the indicated genotypes with respect to green-shift and red-shift. What percent of their daughter's sons are expected to have normal color vision? (For this to happen, the daughters must have both G and R.)

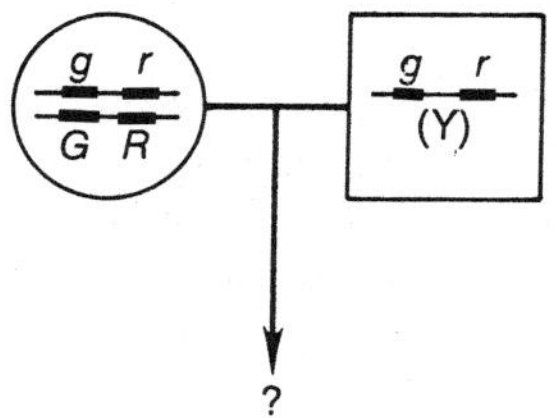

17. Repeat the previous question for the case of a wife who is a double heterozygote in repulsion.

18. (a) What is the genotype of each person in the pedigree below? Be as specific as possible, giving alternative genotypes if necessary. (b) Which person is definitely a recombinant? (c) For individual III-3, what are the probabilities of being H/H? H/h? h/h? (These numbers must add to 1.) Assume green-shift and hemophilia A are 3 map units apart. Note also that III-3 is known to be homozygous g/g.

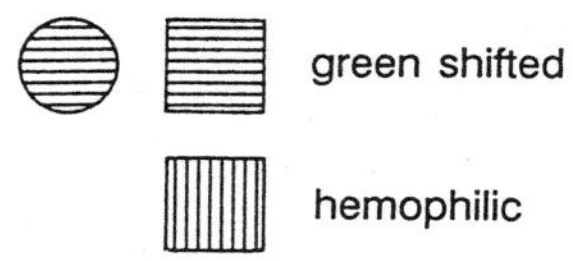

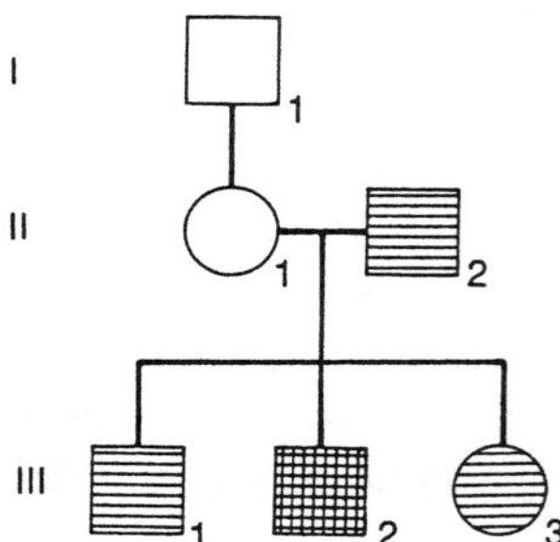

19. The nail-patella syndrome is an autosomal dominant (Np) characterized by absence or malformation of the fingernails, toenails, elbows, and kneecaps. Kidney disease may also be present. The gene locus is about 10 map units from the ABO locus on chromosome 9. A man with the nail-patella syndrome and blood type AB married a normal woman (np/np) of type O. Three children were type A and normal, one was type B and affected. (a) What is the most likely phase of the doubly heterozygous parent? (b) The woman is pregnant with her fifth child. A few red blood cells from the fetus reveal that it is type A. What is the probability that it will develop normally (without the nail-patella syndrome)? Assume the phase specified in part (a). (c) What would be the probability of normality had the fetus been type B?

ANSWERS

1.

	GAMETES (%)				percentage of recombination
	AB	*ab*	*Ab*	*aB*	
(a)	25	25	25	25	50
(b)	50	50	0	0	0
(c)	46	46	4	4	8

2. The two loci would assort independently. Thus there would be the four types of gametes in equal proportions.

3. 1/8 each of the gametes carrying the following sets of genes (only superscripts used for the *Hb* gene). Can the reader discern how the sets of three follow from a tree diagram?

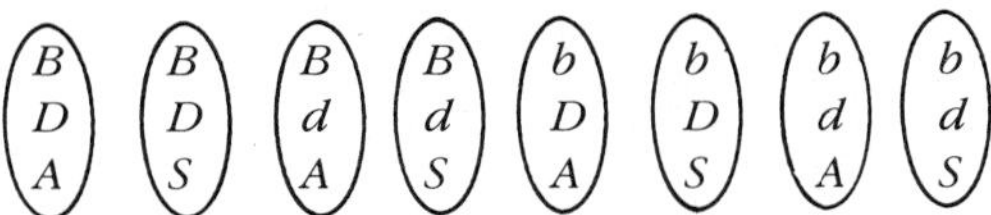

4. Yes. (But it would not apply to the first polar body.)

5. 16. The eight kinds above, with Xg^a or *Xg* added to each.

6. 2^n. For each additional heterozygous gene, the number of different kinds of gametes doubles.

7. 9/16 type B, normal
3/16 type B, affected
3/16 type O, normal
1/16 type O, affected
This is the Mendelian 9:3:3:1 dihybrid ratio for two independent genes with dominance.

8. 1/8 each of

A, Xg(a+) female	A, Xg(a−) male
AB, Xg(a+) female	AB, Xg(a−) male
B, Xg(a+) female	B, Xg(a−) male
O, Xg(a+) female	O, Xg(a−) male

9. (a) P(*B/b P/p S/s*) = $(1/2)^3$ = 1/8
(b) P(all three dominant traits) = $(3/4)^3$ = 27/64
(c) P(*b/b p/p s/s*) = $(1/4)^3$ = 1/64

10. (a) $A_1/A_2 \times A_3/A_4$ for example
(b) *A/a B/b* × *a/a b/b* for example

11. To distinguish a recombinant from a parental assortment. If either gene is homozygous, crossing over produces no new assortment.

12. The disorders are all quite rare. Females doubly heterozygous for two of the disorders are therefore also rare, and presumably never found.

13. Double heterozygotes who are *male* would make it easier to establish linkage. The less frequently genes recombine (as in males), the clearer it becomes that they are linked. The more genes recombine (as in females), the harder it becomes to distinguish linkage from independent assortment.

14. Daughters are $\underline{A\,B}$ / $\underline{a\,b}$. Among their sons, 8 out of 40 are recombinants = 20% recombination = 20 map units.

15. 47%. If this is not clear, write out the *four* types of gametes that the $\underline{g\,r}$ / $\underline{G\,R}$ female can form.

16. $(0.47)^2$ = 22%. The probability that the daughters are $\underline{g\,r}$ / $\underline{G\,R}$ is 0.47. From these daughters, the probability that a gamete carries $\underline{G\,R}$ is another 0.47.

17. (0.03)(0.47) = 1.4%. The probability that the daughters are $\underline{g\,r}$ / $\underline{G\,R}$ is 0.03. From these daughters, the probability that a gamete carries $\underline{G\,R}$ is 0.47.

18. (a) I-1 = $\underline{G\,H}$ / (Y)
II-1 = $\underline{G\,H}$ / $\underline{g\,h}$
II-2 = $\underline{g\,H}$ / (Y)
III-1 = $\underline{g\,H}$ / (Y)
III-2 = $\underline{g\,h}$ / (Y)
III-3 = $\underline{g\,H}$ / $\underline{g\,H}$ or $\underline{g\,H}$ / $\underline{g\,h}$
(b) III-1, getting $\underline{g\,H}$ from his mother and a Y chromosome from his father.
(c) P(*H/H*) = 0.06; P(*H/h*) = 0.94; P(*h/h*) = 0. III-3 got $\underline{g\,H}$ from her father. Since III-3 is green-shifted, we know that she got a chromosome with *g* from her mother. Given *g*, was this a parental chromosome with *h* (high probability) or a recombinant with *H* (low probability)?

19. (a) $\underline{A\,np}$ / $\underline{B\,Np}$
(b and c) The four types of gametes from the father are indicated below. If a gamete contains *A*, it will contain the normal (*np*) allele 90% of the time (45/50). If a gamete contains *B*, it will contain the normal allele 10% of the time (5/50).
$\underline{A\,np}$ 0.45
$\underline{A\,Np}$ 0.05
$\underline{B\,np}$ 0.05
$\underline{B\,Np}$ 0.45

Chapter 12 Complicating Factors

Variation in Gene Expression
Sex Differences
Variable Expressivity
Incomplete Penetrance
One Gene with Several Effects
One Phenotype from Different Causes

Ascertainment Bias
The Always-Missing Families
The Sometimes-Missing Families

Testing Hypotheses
The Chi-Square Method
Interpretation
Comment

Is the Trait Genetic?
Kuru

> *The plague of kuru upon the Fore people [of New Guinea] has attracted the wonder, curiosity, sympathy, and imagination of both the lay and the medical world . . . Kuru is a Fore word for the shivering and trembling associated with fear or cold, and the name is applied by the Fore people themselves to the rapidly progressive, acute, degenerative disorder of the central nervous system . . . predisposition to which may be determined by a rather simple genetic mechanism.*
>
> *D. Carleton Gajdusek (1963)*

For about eight years after kuru was characterized, its pattern of transmission remained ambiguous. The genetic hypothesis turned out *not* to stand the test of time any better than the idea of sorcery to which the Fore people attributed their fatal illness. This chapter describes some difficulties encountered in trying to establish a genetic basis for many traits, and some of the seemingly roundabout methods geneticists must use for handling data. In the last section we explain why a genetic hypothesis for kuru was first considered and then abandoned.

VARIATION IN GENE EXPRESSION

From many observations on gene transmission in plants and animals, it is known that possession of a gene does not guarantee that it will be expressed the same way in every individual. When a gene is "turned on," its RNA and polypeptide products interact with elements of the internal environment (including other gene products and metabolites) and external environment (including nutrition, medicines, climate, pollutants, exercise, and mental stress). Some genes, such as those determining the blood groups, are always expressed in essentially the same way, regardless of the "background" genotypes or outside conditions. Other genes, such as that for brachydactyly (Chapter 3), may be expressed in various ways, or not at all, depending on the presence or absence of additional factors that are often not specifically identifiable.

Sex Differences

The sex hormones can exert a powerful influence on the internal environment (Chapter 6). The product of any gene thus finds itself in one or the other of the two different chemical milieus that characterize the sexes. Although this does not seem to affect the expression of many genes, some inherited traits do differ between males and females. In general, we are considering here the effects of *autosomal* loci, whose pattern of inheritance would be distinct from sex-linked genes.

An example of such **sex-influenced inheritance** is *pattern baldness*, in which premature hair loss occurs on the front and top of the head but not on the sides (Figure 1). Its inheritance is difficult to study because hair loss may be caused by many factors—both genetic and environmental. In addition, hair loss begins at different ages and varies in extent. Pattern baldness affects mainly males; females who carry the same alleles may show only thinning of hair rather than complete loss. Male hormones, the androgens, are implicated in the expression of the trait. For example, a full head of hair, sometimes luxuriant, is seen in men—even old men—who do not mature sexually because of testicular injury or castration prior to puberty. But after androgen treatment, these men sometimes lose their head hair (Hamilton 1951). Likewise, females with tumors of the adrenal gland may secrete large amounts of androgens and become bald. Removal of the tumor in these cases again promotes growth of head hair. Thus expression of this type of baldness requires *both* the possession of a specific allele (which appears to be common) and the presence of male hormones. The actual gene product and the biochemistry of its interaction with androgens are unknown.

Father-to-son transmission (as in the Adams family) and other data tend to rule out X-linked inheritance in favor of autosomal inheritance, but the exact mode or modes of transmission are still not clear. Osborn (1916) suggested that the autosomal gene responsible for pat-

Figure 1 Pattern baldness in the Adams family. Each man fathered the next one in this list. (A) John Adams (1735–1826) at about age 65, second President. (B) John Quincy Adams (1767–1848) at age 52, sixth President. (C) Charles Francis Adams (1807–1886) at age 41, diplomat. (D) Henry Adams (1838–1918) at about age 45, historian. (A, lithograph by Nathaniel Currier, courtesy of the New York Historical Society. B, engraving by Francis Kearney, courtesy of the New York Historical Society. C, lithograph by Nathaniel Currier, courtesy of the Library of Congress. D, photograph by his wife, Marian Hooper Adams, courtesy of the Massachusetts Historical Society.)

tern baldness, B_1, acts dominantly in males and recessively in females. If so, the genotype/phenotype correspondences are as follows:

Genotype	Female phenotype	Male phenotype
B_1/B_1	pattern bald	pattern bald
B_1/B_2	nonbald	pattern bald
B_2/B_2	nonbald	nonbald

The particular pedigree in Figure 2 is consistent with this view, but it is also consistent with other genetic hypotheses (see Question 3).

Traits affecting structures or processes that exist in only one sex are called **sex-limited**. The manner of gene transmission is irrelevant here; it is the interaction of gene products with the internal environment of the body that results in the sex difference. For example, *hydrometrocolpos*, a condition characterized by the accumulation of fluid in the uterus and vagina, occurs in women homozygous for a certain autosomal recessive gene. Obviously, males can never exhibit this symptom, even if homozygous. In dairy cattle, both the quantity and quality of milk are known to be influenced by autosomal genes inherited equally through both parents. Because bulls cannot express these economically important genes, breeders must evaluate their genetic worth by assessing the milk production of their female relatives. (A prize bull can father thousands of offspring after his death via frozen sperm and artificial insemination.) Genes that influence human milk production undoubtedly exist, although little is known of their inheritance or specific effects.

Variable Expressivity

The expression of sexual traits is just one factor that interacts with gene products. More generally, the totality of both genotype and environment can modify the way a particular gene product is utilized. **Variable expressivity** refers to the differences in the observed effects of a given allele in different individuals. A sex-influenced trait could be considered a specialized case of variable expressivity.

In experimental organisms, a number of genes that have been studied affect the expression of other genes in clearly defined ways. *Suppressor mutations* reduce the effect of another mutant gene to a normal or near normal phenotype. Suppression has been studied in exquisite detail in bacteria and viruses, where specific types of base changes in DNA have been implicated (Chapter 14). *Modifiers* are genes that change the degree of expression of other genes. In mice, mink, and other mammals, for example, many genes are responsible for coloring the

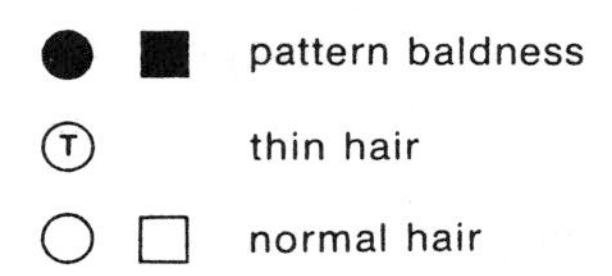

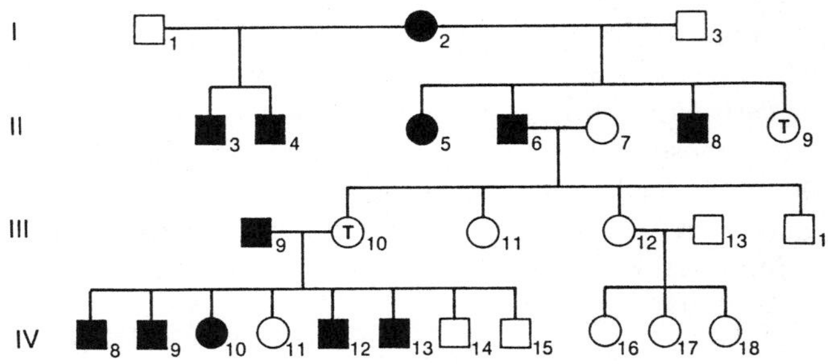

Figure 2 Part of a pedigree of pattern baldness, presumed to be dominant in males ($B_1/-$) but recessive in females (B_1/B_1). Female I-2 was reported to be "very bald on the top and front of the head" from an early age. Her sons, II-6 and II-8, and her great-grandsons, IV-8, IV-9, IV-12, and IV-13, became bald by age 25 in the same pattern. (Little was known of II-3 and II-4, her sons by a previous marriage.) If pattern baldness is recessive in females, then II-5 should not have been affected. Since she was reported to be chronically ill and did not become bald until rather late in life, other factors may have been responsible. (From Osborn 1916, with original numbering of the pedigree.)

fur various shades of black, brown, or yellow, but a modifier gene called *dilute* reduces their intensities, leading to faded grays, tans, or buffs.

Siamese cats and Himalayan rabbits present a familiar example of *environmentally caused* variation in a gene-controlled trait (Figure 3). These animals are homozygous for a mutant allele of a coat color gene that produces light or white fur all over the body except on the extremities. The paws, ears, nose area, and tail have dark fur, the particular color variation being determined by other coat color genes. The development of the dark fur in these animals is dependent on temperature; the lower the temperature, the darker the fur and the more extensive the dark areas. Apparently, the pigment-producing enzyme coded for by this mutant allele works best at the somewhat lower temperatures that occur farthest from the core of the body. If a patch of white fur is shaved from the side of a Himalayan rabbit and an ice pack is applied (or if the rabbit is simply put in a cold environment), the new hair that grows back will be dark.

Variable expressivity is the rule in humans, but it is not usually possible to pinpoint specific modifying genes or specific environmental agents. Nearly all mutant phenotypes, especially those due to dominant genes, can be shown to vary in some respect from one affected person to another. The autosomal dominant disease *neurofibro-*

(A)

Figure 3 Animals possessing a pigment-producing enzyme that works best at the somewhat lower temperatures of the extremities. (A) Siamese cats. (B) Californian rabbits, a commercial breed derived in part from the Himalayan breed. (A, copyright J.F. Glisson; courtesy of Hatcher R. Granville, Rich-Hat Cattery, Atlanta, and the Siamese Cat Society of America. B, Barry L. Runk, Grant Heilman Photography.)

(B)

matosis (more accurately, *von Recklinghausen neurofibromatosis* to distinguish it from related disorders) provides a good example (Figure 4). Mild cases involve only a few pigmented areas (so-called *café-au-lait spots*, the color of coffee with milk) and benign skin tumors (*neurofibromas*). In more serious cases, however, patients may develop thousands of neurofibromas of various sizes and a wide range of additional symptoms. Some of these are tabulated in Table 1. "The disorder has so many facets and is so variable that both hope and despair are simultaneously present" (Riccardi 1981). In some cases, cosmetic surgery may be attempted to lessen severe disfigurement and thereby help with the obstacles to social interaction. Despite the relatively high frequency of neruofibromatosis—about 1 in 3,000 births—and despite much research, investigators still cannot predict how it will be manifested in a given individual, nor can they treat the major symptoms effectively. The gene has recently been mapped to restriction enzyme sites near the centromere of chromosome 17 (Barker et al. 1987). This is the first step toward prenatal diagnosis, analysis of the gene itself, identification of the gene product, and possible therapy.

VARIABLE AGE OF ONSET. Another aspect of gene expression is the age at which the phenotypic consequences first appear. Although many genetic conditions, such as albinism or polydactyly, are present at birth or even prenatally, others only appear at a later age. For example, even though the metabolic error in phenylketonuria (Chapter 16) can be diagnosed at birth by a simple urine test, there are no overt clinical abnormalities for a few months. The change of an apparently healthy baby into a severely abnormal one (if untreated) occurs gradually over the first year of life, with variation from patient to patient in the **age of onset** of various symptoms.

Diabetes is exceedingly variable in this regard. The basic defect involves the hormone *insulin*, which is required for proper utilization of the sugar glucose (Rotter and Rimoin 1983). In diabetic patients glucose builds up in the blood and is excreted in the urine. (The full name is *diabetes mellitus*, from Latin words meaning "to pass sweetness.") There are two major forms of the disease: a juvenile, more severe form in which insulin is absent, and an adult, less severe form in which insulin is present but is not working correctly.* Each form is due in part to a different genetic system involving one or more genes. The expression of symptoms seems to require the interaction of certain mutant alleles and particular dietary and health-related factors, including obesity and viral or autoimmune diseases that damage the cells of the pan-

*The juvenile onset form is also known as Type 1 or insulin-dependent diabetes mellitus (IDDM). The adult or maturity onset form is Type 2 or non-insulin-dependent (NIDDM). About 20% of cases are Type 1, and 80% are Type 2.

creas that manufacture insulin. Variation in the age of onset exists within each form, and various clinical features begin anytime from childhood to old age. Without treatment, diabetes leads to progressive debilities: blindness, kidney failure, nerve damage, gangrene in the feet, heart attacks, and strokes. Treatments can lessen the severity of the symptoms and allow a near-normal life style in some cases. In people under 25 years of age the frequency of diabetes in many populations is about 1 in 1,000, rising to about 1 in 25 in persons over the age of 60.

Huntington disease is transmitted as an autosomal dominant trait. The primary action of this mutant gene is unknown, but it leads to progressive degeneration of brain cells, which in turn causes severe muscle spasms and personality disorders. There is no effective treatment, and death comes 10 to 15 years after the onset of symptoms. Only occasionally do the first signs of disease

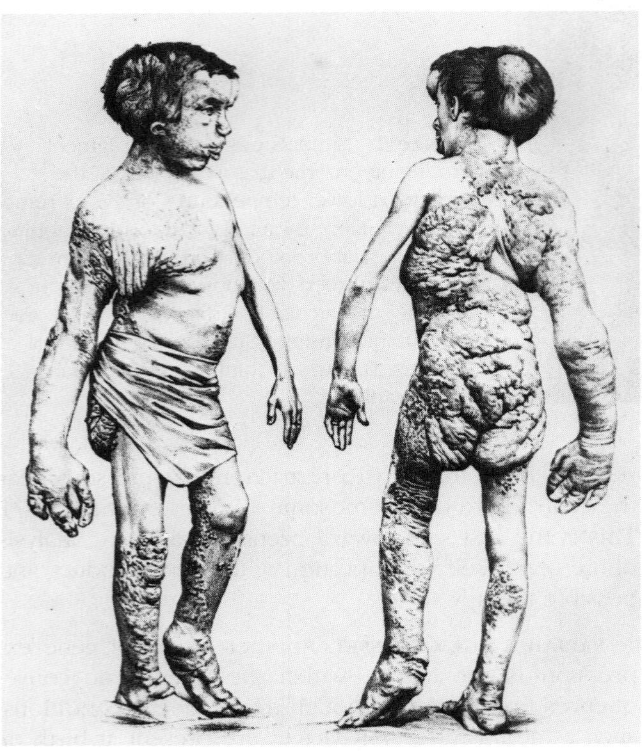

Figure 4 A possible, extreme form of neurofibromatosis characterized by skeletal deformities and masses of cauliflower-like skin tumors. This drawing is of Joseph Merrick (1862–1890), whose painful life was portrayed in a Broadway play and movie, "The Elephant Man." Reexamination of Merrick's skeleton (preserved at London Hospital) suggests, however, that he may also have suffered from one or more bone disorders that contributed to his disfigurement (Carswell 1982). (Picture from Treves 1885.)

Table 1 Symptoms that may accompany neurofibromatosis

Symptom[a]	*Frequency among neurofibromatosis patients*
Leukemia and other malignancies	Common
Headaches	Common
Intellectual handicap (usually mild)	40%
Speech impediments	30–40%
Large head (head size at or above the 97th percentile)	27%
Short stature (height at or below the 3rd percentile)	16%
Constipation	10%
Tumors of the eye, brain, or spinal cord	5–10%
Other internal tumors	few percent
Curvature of the spine	2%
Demineralization of the arm and leg bones	0.5–1%
Decreased blood supply to the brain	Uncommon
High blood pressure	Unknown frequency
Seizures	Unknown frequency
Itching associated with neurofibromas	Unknown frequency
Difficult social adjustment	Unknown frequency

Source: Riccardi (1981).

a. These symptoms occur in addition to café-au-lait spots, freckling in the armpits, skin neurofibromas, and small nodules in the iris of the eye—features that are present in most adult patients.

Figure 5 A pedigree of polydactyly. The three persons indicated by asterisks undoubtedly possessed the dominant gene for polydactyly but did not express it—examples of incomplete penetrance. (From Neel and Schull 1954.)

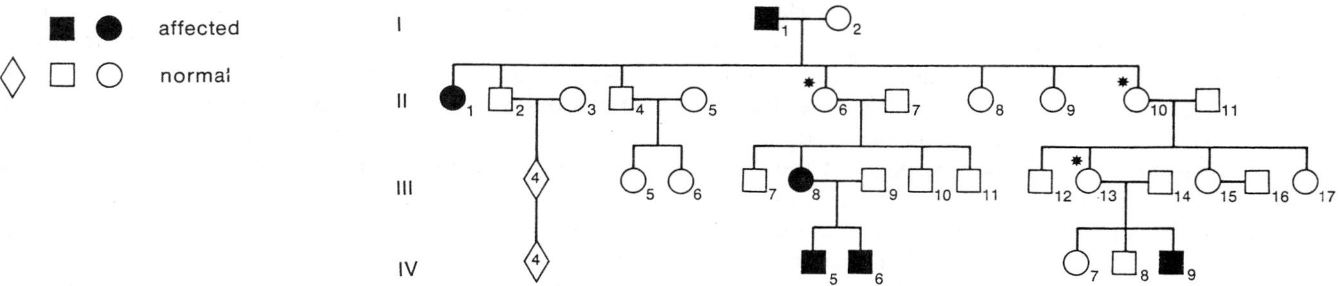

appear in children; about 60% of cases are diagnosed between the ages of 35 and 50, with the remainder occurring equally before and after this age range. Since persons with the allele may reproduce before symptoms occur, their children live with the knowledge that they too may some day be stricken. In some cases, DNA analysis can reveal the presence or absence of the causative gene before symptoms arise (Chapter 15); for persons at risk, however, this poses the agonizing problem of whether or not they should have such a test done.

Incomplete Penetrance

An extreme form of variable expressivity is the total absence of expression in persons known to carry the mutant allele. Such nonexpression of a gene that usually has a characteristic phenotype is called **incomplete penetrance**. It may occur for many reasons, both genetic and environmental. We noted an apparent example in the pedigree of Figure 3 in Chapter 3. The father, sister, and five children of female III-11 all had brachydactyly, but she is reported to have been normal. This woman probably possessed the dominant allele for shortened fingers but, for unknown reasons, did not express it.

In the case of a dominantly inherited trait, incomplete penetrance is manifested as "skipped generations." This situation in a pedigree of polydactyly (extra digits) is seen in Figure 5. Since the disease is fairly rare, it is highly likely that all six affected individuals possessed copies of precisely the same dominant allele. Individual II-6 must have received the allele from her father and transmitted it to her daughter, although she herself was not affected. A two-generation "skip" is seen on the right side of the pedigree. (Although dominant inheritance is fairly clear in this pedigree, a lower degree of penetrance might obscure the mode of inheritance.) Polydactyly also illustrates variable expressivity. Figure 6 illustrates a case in which six well-formed digits appear on both hands and both feet, although each appendage reveals slightly different bone development.*

Usually the precise cause of incomplete penetrance is unknown. But in some special cases, it has been discov-

*A reference to a polydactylous giant is found in I Chronicles 20:6: "And yet again there was war at Gath, where there was a man of great stature, whose fingers and toes were four and twenty, six on each hand, and six on each foot: and he also was the son of a giant."

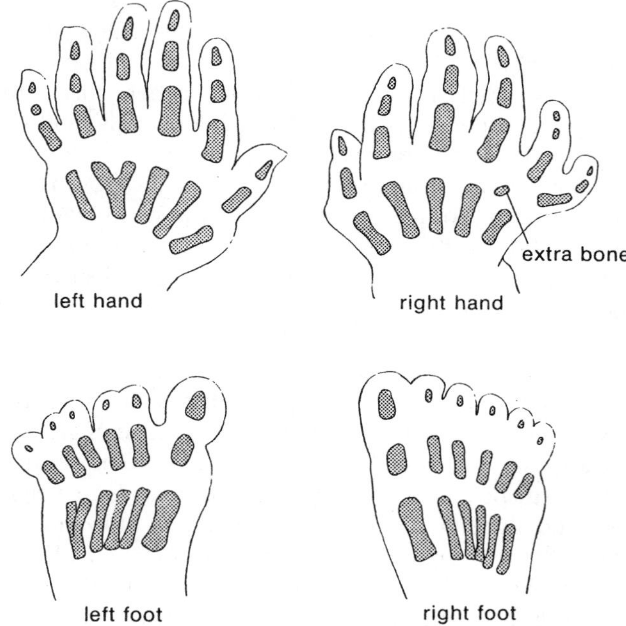

Figure 6 X rays of a 7-day-old Norwegian boy (not from the pedigree in Figure 5) with polydactyly of both hands and both feet. The sixth digit is fairly well formed on each appendage, but in different ways. A bifurcation in one of the bones of the palm of the left hand and in the left foot is clearly seen. Curiously, the extra small bone in the palm of the right hand also occurred in the right hand of the polydactylous mother. (After Sverdrup 1922.)

ered that a specific allele at one locus masks the characteristic effects of an allele at *another* locus. This phenomenon is called **epistasis**. It should not be confused with dominance, which is the masking of one allele by another allele at the *same* locus. In the mouse, for example, the genotype *b/b* is brown and *B/*− is black, but *neither* brown nor black pigment appears in the fur, if the mouse is also homozygous *c/c* at another locus. The genotype *c/c* leads to an albino mouse, which develops *no* fur pigment, so color alleles at other loci are irrelevant. We say that the genotype *c/c* is epistatic to alleles at the *B* locus and other coat color loci.

The best human example of epistasis involves the suppression of the ABO blood group in persons with the so-called Bombay phenotype, to be discussed in Chapter 17.

One Gene with Several Effects

Given the complexity of embryonic development, it is not surprising that a single gene gone wrong can have multiple, sometimes seemingly unrelated consequences. One gene product can be involved in many biochemical processes affecting different pathways of growth and maturation. The resulting cascade of effects is called **pleiotropy** (literally, "many turnings"). As with variable expressivity, pleiotropy is the rule rather than the exception, especially when a given phenotype is carefully examined at different levels.

Consider, for example, the multiple effects of insulin deficiency in juvenile onset diabetes shown in Figure 7. Diagramed on the left are some of the physiological consequences (thirst, constipation, itching, and so on) of the impaired, insulin-dependent membrane system that transports glucose inside cells. Glucose therefore builds to high levels in the blood. Even more serious are the effects (diagramed on the right) that stem from the body's attempt to compensate for the absence of glucose metabolism within cells by mobilizing proteins and fats for energy production. Protein depletion leads to weight loss and general debilitation. The excess utilization of fats for energy results in the production of acid substances (ketone bodies) in such high concentrations that the body is unable to neutralize them completely; this contributes to the possibility of coma and death.

One Phenotype from Different Causes

We have just described how one gene can have multiple effects. Conversely, any one of several different genes can result in similar effects. Polydactyly (Figures 5 and 6) is a case in point. Extra digits can be due to mutant alleles of any of about a half dozen different genes, usually showing dominant inheritance with incomplete penetrance. (We are considering here phenotypes in which polydactyly is the only observable malformation; other cases are known in which it is associated with more serious syndromes.) In one form, a reasonably functional digit appears on the little-finger side of the hand (or little-toe side of the foot). In a second form, an extra digit in the same position is much reduced in size. Other types of polydactyly involve an extra thumb, or a thumb with an extra joint, or a duplicated index finger.

The frequencies of the various types of polydactyly among newborns differ considerably, the extra little finger being the most common. Frequencies among newborns are not the same as the frequencies of polydactyly among persons who present themselves for corrective surgery, however. Such differences in frequency that depend on how cases come to light can be troublesome for human geneticists (discussed below under ascertainment bias).

An additional aspect of polydactyly is that not all cases are genetic. Errors of development brought about by slight changes in the prenatal environment can also produce bone malformations. Such environmental mimics of a characteristically genetic condition are called **phenocopies**. A person with polydactyly who has no affected relatives—a so-called sporadic case—*may* be a phenocopy, but other causes of sporadic cases are possible (see Question 8).

The term used to describe multiple genetic causes of the same, or nearly the same, phenotype is **genetic heterogeneity**. From the point of view of an affected person, the determination of a precise causation can be important. For example, hemophilia A results from a defect in so-called clotting factor VIII, whereas Christmas disease, with a similar phenotype, results from a defect in clotting factor IX. The symptoms of hemophilia A can be alleviated by using a concentrate of factor VIII, but this treatment would be of no use to a person suffering from Christmas disease, and vice versa.

An interesting and increasingly useful way to demonstrate genetic heterogeneity was first demonstrated by Morton (1956). In Figure 9 in Chapter 11 we note that the gene (*EL1*) for the rare, benign condition known as *elliptocytosis* (oval-shaped red blood cells) is closely linked to the *Rh* gene on chromosome 1. This linkage is true in some families, but in others, elliptocytosis and *Rh* are inherited independently of each other. Thus elliptocytosis must result from gene mutations at two (possibly more) different loci—*EL1* being *Rh*-linked and *EL2* being *Rh-un*linked. The phenotypes controlled by the two genes are indistinguishable by other criteria: unless linkage data are obtained, it cannot be determined which gene is being expressed. This methodology has been used recently to demonstrate heterogeneity for manic depression and schizophrenia (see Chapter 23).

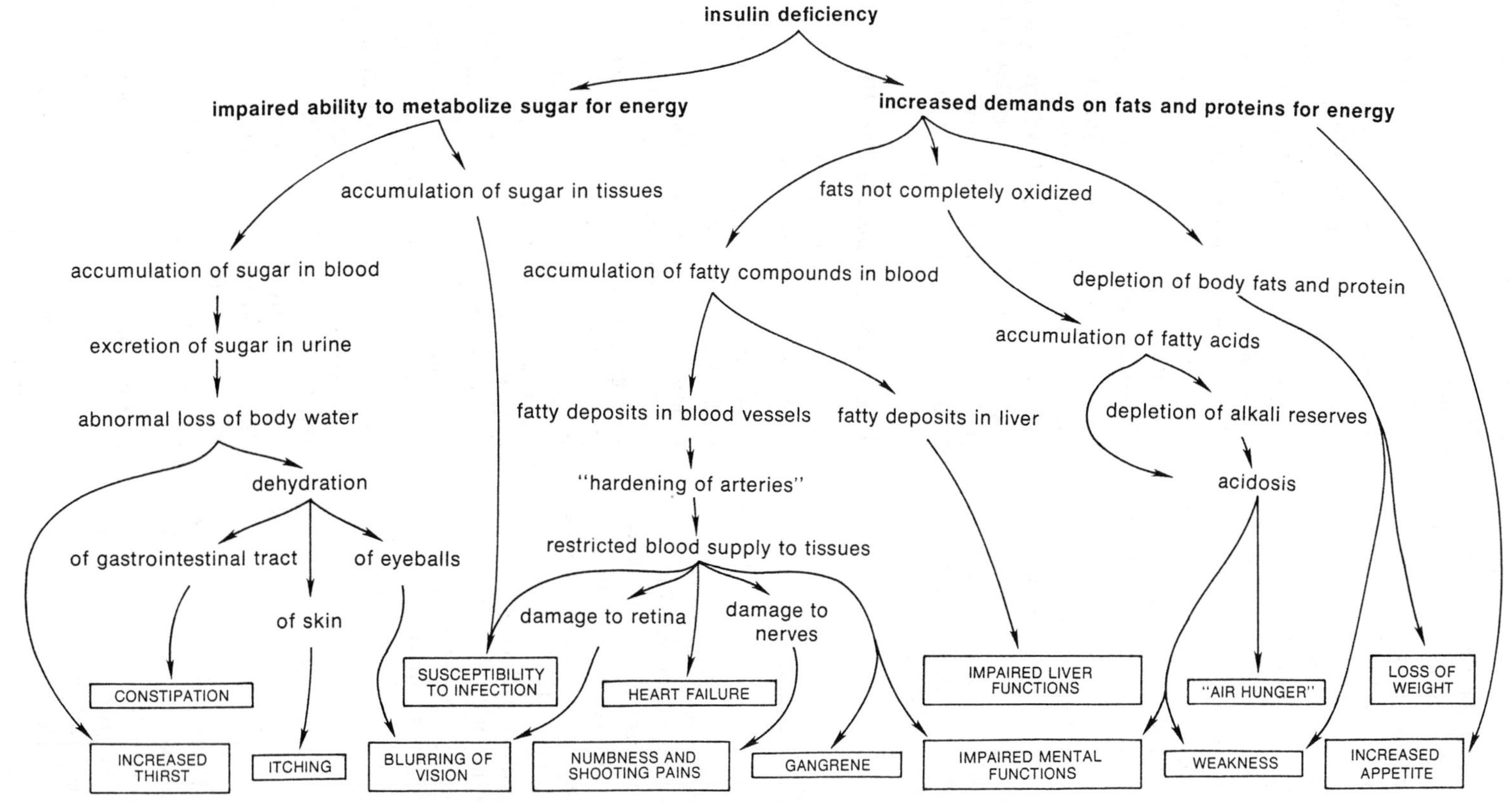

Figure 7 Pleiotropic consequences of insulin deficiency occurring in juvenile onset diabetes mellitus. (Redrawn from Neel and Schull 1954.)

Genetic heterogeneity is always expected for complex anatomical, physiological, or behavioral traits. The ability to hear, for example, depends on the proper functioning of perhaps hundreds or thousands of genes. Deafness, of a greater or lesser degree, can result when one of these genes mutates to an allele that impairs the development of the intricate structures of the middle or inner ear, or the nerve connections to the brain. It has been estimated that about half of all childhood deafness stems from such genetic factors, with the remainder—phenocopies—resulting from certain viral infections of the mother during pregnancy (measles and rubella especially), prematurity, or other types of environmentally imposed harm to the fetus or newborn. Based on audiological tests and associated pleiotropic effects, Nance and McConnell (1973) describe about 50 different genetic hearing loss syndromes. Roughly two-thirds of these genetic cases appear to be autosomal recessives, about one-third are autosomal dominants, and a few are X-linked. With refinements of testing procedures, these entities may be found to have multiple causations.

ASCERTAINMENT BIAS

Even when a single, fully penetrant gene leads to a uniform, easily recognized, congenital phenotype (and no phenocopies exist), human geneticists may still not observe a clear-cut Mendelian ratio that would confirm a specific genetic hypothesis. Departures from standard ratios may occur even in large samples, because raw data can be biased by indirect methods of collection.

An analogy illustrates the nature of the problem. Suppose a hundred doctors tabulate by sex the living members in their own sibships, including themselves. Assuming an average sibship size of three, typical raw data might be

Males	**Females**	**Total**	**Sex ratio (% male)**
189	111	300	63.0

If the ratio of 189 males to 111 females seems surprisingly male-biased, the problem is resolved when we realize that the doctors themselves—the probands—are included in the ratio, and they are likely to be predominantly male. Assume, for example, that 90 of the doctors are men and 10 are women. Subtracting these probands from the raw data, leaving only their brothers and sisters, provides an unbiased sample for calculating an adult sex ratio:

	Males	**Females**	**Total**	**Sex ratio (% male)**
raw data	189	111	300	—
probands	−90	−10	−100	—
unbiased sample	99	101	200	49.5

An uncritical person might conclude from the raw data that the sisters of doctors have terribly high death rates (supporting who-knows-what biological or sociological hypothesis). We see, however, that the data are simply biased by the method of collection: an excess of males is "forced" into the observations. This example illustrates **ascertainment bias**, the term *ascertainment* referring to the method used to discover and record data. The bias does not mean that the collected information is useless or wrong, only that it must be handled with care to avoid drawing unwarranted conclusions. We illustrate ascertainment bias in the case of autosomal recessive inheritance, for which one might reasonably expect to find 25% affected children from heterozygous parents. In fact, the reasonable expectation may *not* be 25% affected or even close to 25%.

The Always-Missing Families

There are two problems in trying to establish ratios indicative of autosomal recessive inheritance.

DOMINANCE. If the normal phenotype is truly dominant (the abnormal trait recessive), then a heterozygote cannot be distinguished from a normal homozygote by his or her phenotype alone. A heterozygote can only be identified as the child or the parent of an affected person (Figure 8). If the recessive phenotype substantially reduces survival or fertility, or is simply very obvious, then the cases that come to medical attention are largely affected children from normal heterozygous parents (Figure 8B). But, heterozygous parents who happen not to have produced any affected children are *never* found. Yet the expected 25% affected is valid only if all children from all heterozygous parents are accounted for.

In essence, geneticists locate heterozygote × heterozygote parents indirectly, through their affected child, rather than directly, by looking at the parents themselves. This method leads to a bias, because at least one child must be affected before the *A/a* × *A/a* matings can be ascertained. Because some families are cut off from observations—that is, those *A/a* × *A/a* matings with no affected offspring—this method of data collection is appropriately called **truncate ascertainment**.

SMALL FAMILIES. The problem of overlooking *A/a* × *A/a* families who happen *not* to produce any *a/a* offspring is minimal in large sibships. For example, what is the probability that *no a/a* child will show up in a sibship

affected: *a/a*

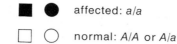

normal: *A/A* or *A/a*

(A)
Persons with a normal phenotype have an affected parent

a/a *A/–*

A/a *A/a* *A/a* known heterozygotes

(B)
Persons with a normal phenotype have an affected child

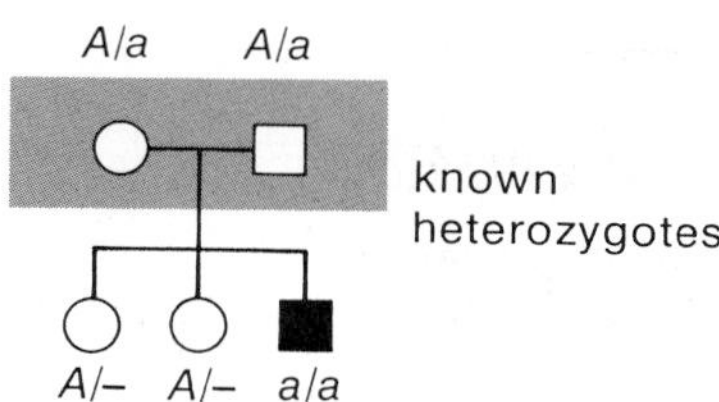

Figure 8 The two ways of establishing heterozygosity for a strictly recessive trait. We assume that no aspect of phenotype allows the normal heterozygote, *A/a*, to be distinguished from the normal homozygote, *A/A*.

of size 15? For one child, P(*a/a*) = 1/4 and P(*A/*−) = 3/4, and the binomial formula predicts

$$\text{P}(15\ A/\!-\text{ and }0\ a/a) = \frac{15!}{15!0!}(3/4)^{15}(1/4)^{0}$$
$$= (3/4)^{15} = 1.34\%$$

Thus, in the investigation of recessive inheritance among families with 15 offspring each, only about 1% of families would be missing by virtue of lacking an affected child.

On the other hand, if *A/a* × *A/a* parents produce only one child, 75% (rather than merely 1%) would be undetectable because the one and only child is normal. Table 2 lists the likelihood of missing heterozygous matings as a function of family size.

The probabilities listed in Table 2 provide the basis for accounting for ascertainment bias. Assume, for concreteness, that we identified all persons with alkaptonuria in a given area. We tabulate them and their normal brothers and sisters. This procedure is called **complete truncate ascertainment**, in the sense that *all* families with affected children are found. For the moment, we concentrate on those sibships with four children and ask: Among these, what is the expected percentage of affected children? Figure 9 shows the five types of sibships whose parents are heterozygous for alkaptonuria. If all families could be ascertained, then simple Mendelian expecta-

Table 2 The probability that an *A/a* × *A/a* family can be found through an affected (*a/a*) child (truncate ascertainment)

Number of children in sibship n	*Missing families: P(all children are A/–)* $(3/4)^n$	*Findable families:*[a] *P(at least one child is a/a)* $1 - (3/4)^n$
1	0.750	0.250
2	0.562	0.438
3	0.422	0.578
4	0.316	0.684[b]
5	0.237	0.763
6	0.178	0.822
7	0.133	0.867
8	0.100	0.900
9	0.075	0.925
10	0.056	0.944
⋮	⋮	⋮
15	0.013	0.987
⋮	⋮	⋮
20	0.003	0.997

a. This column shows that the likelihood of finding the family increases to near certainty only for very large sibships.
b. See Figure 9 for more details.

◆ affected: k/k

○ □ ◇ normal: K/K or K/k

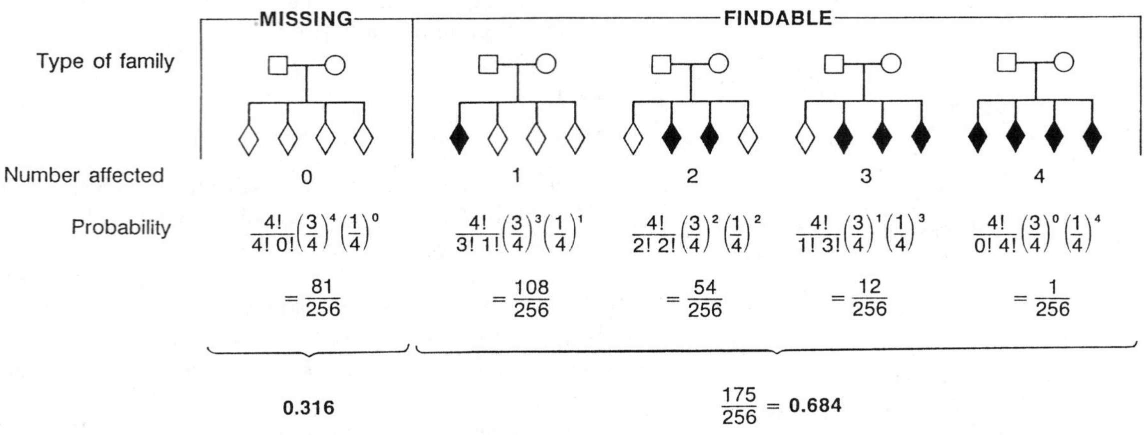

Figure 9 Segregation of a recessive trait (alkaptonuria) in sibships of four members, from heterozygote × heterozygote matings. On the basis of the number of affected children, one can list five possible family types.

tions would hold and there would be 25% affected. But because only those sibships with some affected are actually found—0.684 of all sibships—we expect our sample to be "enriched" with alkaptonuria patients. The situation is one to which *conditional probability* applies: instead of looking at all families, our viewpoint is restricted to a portion of those families (which, however, still includes all affected persons). Therefore, following the procedures of Chapter 10, we calculate

$$\begin{aligned}\text{P(being affected, given sibships of size four with } \textit{at least} \text{ one affected)} &= \frac{0.250}{0.684} \\ &= 0.366 = 36.6\%\end{aligned}$$

Thus, with complete truncate ascertainment in families with four children, we expect not 25% but 36.6% affected offspring. An alternative method for establishing this value is presented in Figure 10. This method is longer, but perhaps more straightforward.

For families with more than four children, it becomes increasingly likely that at least one child will be affected. Thus the bias becomes less, and the percentage affected gradually approaches 25.0% (Figure 11). Notice that for families with a single child, the probability of being affected is 1, because if the single child is normal, the family cannot be found.

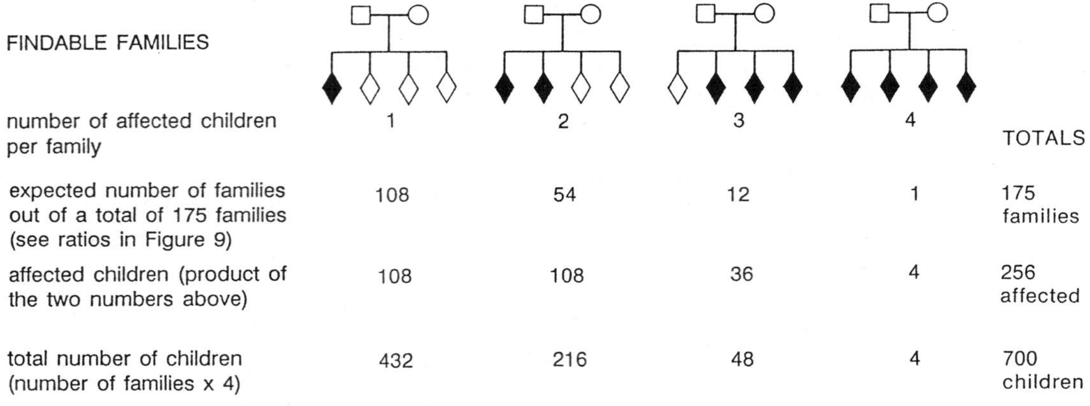

$$\text{percentage affected} = \frac{256}{700}(100) = 36.6\%$$

Figure 10 Method for establishing that 36.6% of children from four-child sibships are affected under complete truncate assessment.

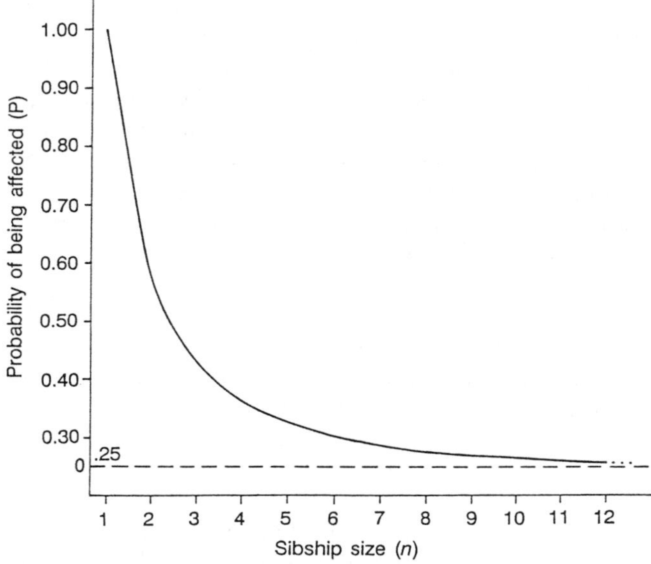

Figure 11 The probability (P) of being affected with a recessive trait when data are collected through affected children with normal parents. For small sibship sizes (n), this probability is much larger than 0.25 because of ascertainment bias. The points on the line are given by the equation $P = 0.250/[1 - (0.75)^n]$.

In summary, for recessive inheritance in humans where experimental matings are not possible and the geneticist must make use of whatever data can be made available, the simple Mendelian ratios may not be the expected values.

The Sometimes-Missing Families

Although additional mathematical analysis is beyond the scope of this book, we should point out the source of some further difficulties. Suppose that a research team decides to investigate the inheritance of, say, albinism by surveying not an entire population—too expensive—but only a portion of it over a limited time period. They enlist the aid of some third-grade teachers to report all albinos in their classes and tabulate the phenotypes of the albinos' brothers and sisters. In this way, a horizontal slice of the population is sampled.

It is still true that the method of data collection is truncated—no families will be tallied unless they have at least one albino child. But the ascertainment is no longer complete—not all families in the area with at least one affected child will be in the collected sample. An additional bias enters the data: a family with, say, four albinos is four times more likely to have a child in the third grade than a family with one albino. Thus we need to make adjustments in binomial expectations. This data collection method contrasts with complete truncate ascertainment (previous section), where these two types of families are equally likely to be found.

Some genetic surveys are neither complete nor incomplete in the specific way noted above. Geneticists may use medical or dental records, death certificates, or other data that were collected for other purposes and with various degrees of thoroughness and care over an extended period. Questionnaires directed to practicing physicians may yield different degrees of completeness depending on the phenotype in question: a serious condition that is easy to treat will bring affected persons to doctors more often than a mild condition that is difficult to treat. Reliance on cases reported in the scientific literature may also be biased: families with one or a few affected persons may be missing because they are less interesting than families with many affected.

Thus, in the analysis of patterns of inheritance, some fairly sophisticated analyses are often needed. Even the most refined mathematics, however, cannot correct for ignorance of how affected persons first came to the attention of the data collector. Poor data yield unreliable conclusions regardless of how they are manipulated.

TESTING HYPOTHESES

We have noted previously that observed data are usually not *exactly* the same as ideal expectations based on some hypothesis; there have been **deviations** between the two of a greater or lesser amount. For example, Mendel counted 787 tall and 277 short plants out of 1064 offspring from heterozygous parents (Chapter 2). But these numbers are not exactly 798 (i.e., 3/4 of 1064) and 266 (i.e., 1/4 of 1064) as predicted by the law of segregation; the comparison is

Class	Observed number	Expected number	Deviation
tall	787	798	−11
short	277	266	+11
total	1064	1064	0

The deviations of −11 in one phenotypic class and +11 in the other might be due to either (1) mere chance or (2) a wrong hypothesis. Because the deviations are "small," they are likely to be due to chance and therefore inconsequential; we pay them no heed and state with some assurance that the observations are consistent with the Mendelian hypothesis. But, had the deviations been "large," we would propose that the *expected numbers* are not appropriate to the comparison, and some other hypothesis would have given a better fit.

Coin tossing provides another example. Suppose you flip a penny 20 times, getting 15 heads and 5 tails; the corresponding table is

Class	Observed number	Expected number	Deviation
heads	15	10	+5
tails	5	10	−5
total	20	20	0

Here, the expected 10:10 ratio is calculated from the hypothesis of a true coin, fairly flipped. Are the deviations sufficiently small that we chalk them up to mere chance? Or are the deviations so large that we are suspicious of the coin or of the way it was tossed? Some rules regarding probabilities allow us to decide between these possibilities.

The Chi-Square Method

Essentially, we want to use the size of the deviations to evaluate whatever hypothesis is set forth to explain the observations. *As the deviations become larger, at what point do we begin to doubt the hypothesis?* An answer of sorts is provided by a statistical tool called the *chi-square method* (the Greek letter chi is written χ). The end result of this procedure is a single **probability, P**. Conceptually, P reflects the credibility of the hypothesis—the smaller the value of P, the less plausible the hypothesis. Operationally, P measures the likelihood of getting, by chance, deviations as large or larger than those actually found, given that the hypothesis is true. The larger the deviations are, the smaller P is, and the less likely the hypothesis becomes. We then make a decision based on the value of P.

If P is low, then we *reject* the hypothesis. We say that the observed results are not consistent with the hypothesis, that the hypothesis is not believable, or that the large deviations between observed and expected are significant—that they are not likely to be due to mere chance. The definition of "low" is arbitrary, but in many biological applications, it is taken as 0.05 or smaller. This is sometimes spoken of as the 5% level of significance.

If P is high, then we *accept* the hypothesis. We say that the observed results are consistent with the hypothesis, that the hypothesis seems like a good one, or that the deviations are not significant. "High" is defined as more than 0.05.

Consider the rationale for doing a calculation that compares the goodness of fit of observed to expected results. We want to know whether or not some idea we have is worth pursuing, and we gather some evidence to help us decide. The reasoning is similar to that in a court trial: The hypothesis is that the accused is innocent. If the evidence supporting the hypothesis is incredible—has low P—the hypothesis of innocence is *rejected*. The jurors, of course, have no mathematical means for reaching a decision but must grapple with concepts like "reasonable doubt."

To find the value of P in genetical situations, we first calculate an intermediate quantity called **chi-square, χ^2**, and, using this number, look up P in a table. The calculation of χ^2 is straightforward: For each class of observations, square the deviation and divide by the corresponding expected value. Then add these terms together. Algebraically, this is expressed as

$$\chi^2 = \Sigma(d^2/e)$$

where d is the deviation for a given class, e is the expected value for that class, and Σ is the sign for summation. For the case of penny tossing,

$$\chi^2 = \frac{(5)^2}{10} + \frac{(-5)^2}{10} = \frac{25}{10} + \frac{25}{10} = 2.5 + 2.5 = 5.0$$

It remains to look up this value of χ^2 in a table to find the value of P. For our purposes, it will suffice to note that in this situation, a χ^2 value of 3.8 corresponds to a probability of 0.05; if the χ^2 value is larger than 3.8, then the probability is less than 0.05. Thus in the present case, we would, by our 5% decision rule, reject the hypothesis of a true coin, fairly thrown.

Interpretation

Notice that the chi-square procedure does not tell us *what* is wrong with the coin experiment or even that there *is* something wrong, only that we should be suspicious. When the value of chi-square is greater than 3.8, it does not mean that the hypothesis is false, only that it is likely to be false. A mistake can be made. We might reject an hypothesis that is really true, or accept an hypothesis that is really false. The statistical procedure is designed to minimize these sorts of errors—it helps us make judgments about evidence, just as the adversary system does for jurors in a court of law. Just as a "hung jury" may cause a new trial, so a borderline decision based on the chi-square method may cause an investigator to gather new evidence. In any event, the 5% level of significance is arbitrary. If the consequences of rejecting a true hypothesis or accepting a false one are serious, we can choose another level of significance.

When the value of χ^2 is far from 3.8 (for two classes of results as in the head-or-tail and tall-or-short examples), we can have greater confidence that we have made the right decision. This situation is illustrated in Figure 12. When χ^2 is as large as 8, then the P value is 0.005. This means that if the hypothesis is true, only one time out of 200 (0.5%) would we expect to see deviations as large or larger than those actually observed; we reject the hypothesis with some confidence. It is important to understand the relationships between the various quantities: *as deviations get larger, the value of χ^2 gets larger,*

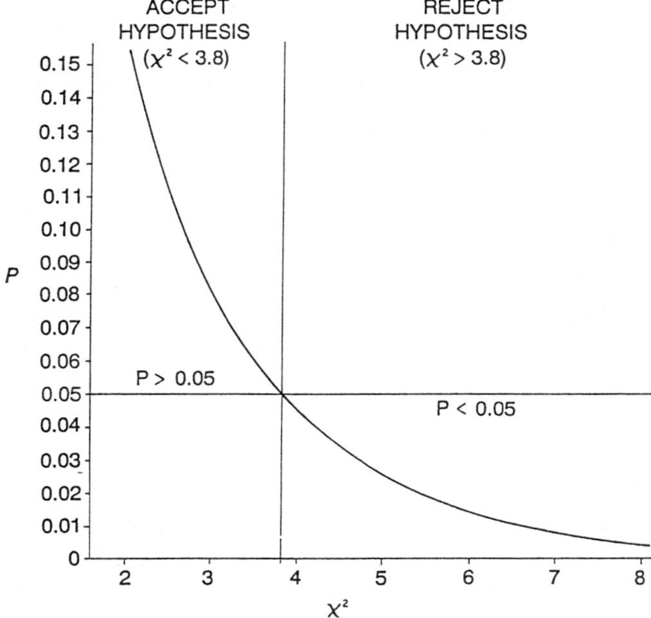

Figure 12 The distribution of χ^2 for *two classes* of observations (say, heads and tails). In testing a hypothesis, the arbitrary cutoff of $\chi^2 = 3.8$ is often used. When χ^2 exceeds 3.8 (right portion of graph), the probability (P) that chance could account for such large deviations becomes untenable, and another hypothesis is sought. (For more than two classes of observations, different values of χ^2 apply.)

and the value of the probability gets smaller—it becomes less and less likely that the deviations can be ascribed to chance. Another hypothesis that predicts smaller deviations should be sought.

Returning to the case of Mendel's peas, we find that

$$\chi^2 = \frac{(-11)^2}{798} + \frac{(11)^2}{266} = \frac{121}{798} + \frac{121}{266} = 0.15 + 0.45 = 0.60$$

On the graph in Figure 12, we try to place the 0.60 χ^2 value on the bottom axis, but it is off the scale at the far left. The corresponding probability on the vertical axis is off the scale at the top—a high probability. We have great confidence that this experiment performed by Mendel supported his ideas regarding inheritance.*

*There is a continuing controversy over Mendel's results. Largely by a series of chi-square tests similar to those outlined above, the English geneticist and statistician Sir Ronald Fisher (1936) established that, as a whole, Mendel's results are *unbelievably* good! He suggested that Mendel, or an over-eager assistant, might have fudged the data a bit. Others think that classification of some peas with intermediate phenotypes was influenced by the hypothesis, or pea counts were *un*intentionally stopped at favorable places (Wright 1966; Beadle 1967; Root-Bernstein 1983). Recall that Mendel was among the first biologists to do any substantial counting at all. (See Question 11.)

Comments

HISTORY. The chi-square method was invented about 1900 by the Englishman Karl Pearson (Hacking 1984). It provided a practical, standardized test applicable to any discipline. For better or worse, statistics now pervades our world, from medical prognoses to election-night predictions. In a sense, statistical thinking flowers when knowledge is scant, and wilts as knowledge increases. But there is little doubt that probabilities help guide the direction of many human activities. Readers who have followed the chi-square discussion above will appreciate, however, the subtleties of statistical reasoning.

APPROXIMATION TO THE BINOMIAL. The chi-square method is a way to get a close approximation to a probability value that can be obtained exactly by a more cumbersome route. For example, in coin tossing, we can compute precisely the probability of getting 15 heads and 5 tails by using the binomial formula:

$$\frac{20!}{15!5!}(1/2)^{15}(1/2)^5$$

For comparison with the chi-square probability, however, we want the probability of this outcome plus that of *any other result* with deviations as large or larger. Thus we must add up the probabilities of getting 15, 16, 17, 18, 19, or 20 heads, or 15, 16, 17, 18, 19, or 20 tails. Better to utilize the chi-square method, which approximates the sum of the binomial probabilities. The approximation is good when the total number of observations is about 50 but becomes progressively worse with smaller numbers. The chi-square method should not be used (without corrective terms) if an expected value in any class is five or less.

RAW NUMBERS. It is important to use the actual numbers of observations, not the percentages they represent. For example, getting 57 heads on 100 flips of a penny is fully consistent with a 1 : 1 hypothesis ($\chi^2 = 2.0$; $P = 0.15$). Getting 570 heads on 1,000 flips (the same percentage of heads) is, however, frankly incredible if the coin is true and fairly tossed ($\chi^2 = 20$; P = much less than 0.001). The greater amount of data provides a more discriminating test. If the coin tossing *is* unfair in some slight way, then tossing it only 100 times may not be enough to reveal the bias.

MORE THAN TWO CLASSES. The chart of χ^2 in Figure 12 is valid only for two classes of observations. But genetic results often fall into more than two classes. For example, *four* different ABO blood groups are possible among the offspring of *A/O* × *B/O*. The combination of genes at different loci can lead to even more offspring phenotypes. Although the calculation of χ^2 is similar to that above, the table or chart that relates χ^2 to probability is different. We will not consider such cases here.

IS THE TRAIT GENETIC?

We have seen that gene interactions are sometimes complex and the methods for investigating genetic traits are not always straightforward. An extension of these considerations is to ask whether a particular phenotype is genetically determined at all. At least two important criteria, taken together, are necessary to establish a genetic causation:

1. *The trait must occur more often among the relatives of probands than among the general population.* If the trait has a one-gene mode of inheritance without complicating factors, the finding of definite numerical ratios may clinch the genetic argument. If two or more genes are involved, or if complications occur, then the computation of expected ratios becomes increasingly difficult and the genetic causation becomes less and less clear.

2. *The trait must not spread to unrelated persons exposed to similar environmental situations.* This criterion helps to exclude infections, environmental poisons, nutritional deficiencies, and other conditions that may run in families or be prevalent in local populations. But some infectious agents may mimic biological inheritance. For example, syphilis and some viral diseases can be transmitted from mother to fetus across the placental barrier.

These criteria are necessary but not always sufficient. Additional, usually less forceful, observations supporting a genetic basis may be applicable in some situations:

3. *The trait appears at a characteristic age without any obvious precipitating event.* Many genetic disorders are present at birth, but so are developmental abnormalities not related to the genotype of the newborn. Other genetic disorders, Huntington disease for example, appear at later ages but with some variation in the age of onset. The absence of an obvious trigger may reflect ignorance of environmental agents rather than the operation of genes.

4. *The trait varies in frequency among separate populations.* Human populations differ in the frequencies of many alleles. Racial or ethnic characteristics *may* indicate genetic differences, but it is often difficult to disentangle genetic from environmental causations.

5. *Identical twins share the trait more often than do nonidentical twins.* Also, adopted twins resemble their biological parents more than their adopting parents. Some substantial difficulties in the interpretations of twin studies, however, are presented in Chapter 23.

6. *A human trait resembles one known to be inherited in experimental or farm animals.* Scientists can gain knowledge of a human genetic disease by studying animals that appear to have the same inherited defect.

Kuru

The difficulties of establishing a genetic hypothesis are well illustrated by the neurological disease **kuru** mentioned at the start of this chapter. The facts as they became known in the 1950s were these:

1. The disease was frequent among the Fore-language Melanesians living in remote rain forests of New Guinea and also among some near neighbors of the Fore. Villages fought with each other and practiced ritual cannibalism of dead relatives to honor them.

2. The signs of brain damage appeared suddenly and included uncoordinated movements, trembling, loss of balance, inability to talk or swallow, and emotional instability. No therapy was successful, and death came within a year after the onset of symptoms.

3. Outsiders never developed kuru, even when living with the Fore.

4. A few Fore people developed kuru while living, for up to six years, far from their villages.

5. About 75% of kuru victims were adult women, and the rest were mostly children of both sexes above age four (Figure 13). The disease was rare among adult men and very young children.

The evidence favoring a genetic hypothesis thus satisfied criteria 1 through 4 above. To account for the unusual age and sex distribution, an autosomal disease-causing allele was considered to be dominant in females and recessive in males, with the onset of symptoms occurring early in homozygotes and late in female heterozygotes. The major difficulty with this scheme lay in explaining how a highly lethal allele could ever come to have high frequency. One view was that the kuru allele had become lethal only recently through some environmental change, and, prior to this time, it had been neutral or perhaps beneficial.

Figure 13 A Fore child incapacitated by kuru. Although still alert, the child could not stand, sit, or talk, and died a few months after this photograph was taken. (From Gajducek 1977.)

The first clue to a nongenetic causation came from noting the similarity of kuru symptoms to those of *scrapie*, a fatal disease of sheep and goats. Scrapie has also been transmitted experimentally to small laboratory mammals such as mice and hamsters. The infectious agent of scrapie is a puzzle, consisting largely or entirely of protein with little or no nucleic acid. The term **prion** has been introduced to distinguish it from viruses made up of protein and DNA or RNA, and from viroids with RNA only (Prusiner 1984a). Remarkably, the prion protein itself seems to be encoded by the *host* DNA and it is also present in normal, *uninfected* cells. If the prion protein alone is responsible for disease symptoms, why is it not pathological in all cells all the time? Perhaps a small associated nucleic acid, which has escaped detection so far, is responsible for triggering the disease process. (See Robertson et al. [1985] for a review.)

The progress of the infection in scrapie, kuru, and another human illness called *Creutzfeldt-Jakob disease* is similar: an extraordinarily long period between exposure and the onset of symptoms, brain damage leading to loss of coordination and mental deterioration, and usually death. (The latent period, up to several years, originally led scientists to call the disease agent a "slow virus.") The immune system of infected animals does not respond effectively, and unusual fibrils (presumed to be aggregations of the prion protein) can be found in infected brain tissue. Similarities to some characteristics of Alzheimer's disease—mental deterioration, primarily in the elderly—are intriguing (Prusiner 1984b).

The infectious nature of kuru was confirmed by inoculating chimpanzees with the brain tissue of kuru victims. After year-long incubation periods, the animals developed a fatal disease closely resembling kuru. An explanation for the unique age and sex distribution of kuru among the Fore tribes lay in the ceremonial cannibalism of close kin. Preparing for the rites, women and young children of both sexes handled the dead body, including the brain, which was packed into bamboo cylinders and steamed. The infectious agents probably entered through breaks in the skin or through the mucous membranes of the eyes or nose. Men and older boys seldom took part in the ritual. As the practice of cannibalism has been abandoned by the Fore people, the incidence of kuru has been much reduced (Gajdusek 1977).

Nobody knows how prion infectivity survives in nature, or why the Fore people were the only ones afflicted with kuru. D. Carleton Gajdusek, a 1976 Nobel laureate, has suggested that the agents of kuru, Creutzfeldt-Jakob disease, and scrapie may all be variants of a single "unconventional" virus, and their infectivities may be influenced by genetic differences among the host animals.

SUMMARY

1. Sex-influenced genes are expressed differentially in the two sexes because of underlying anatomical or physiological differences.
2. Variable expressivity of a gene, including variable age of onset, results from variation in environments and in background genotypes.
3. Incomplete penetrance of a gene refers to the nonexpression of a characteristic phenotype. This outcome may be caused by epistasis, the masking of a phenotype by the action of a nonallelic gene.
4. A pleiotropic gene affects various aspects of an individual's phenotype.
5. Genetic heterogeneity refers to multiple genetic causes of nearly the same phenotype. A phenocopy is a nongenetic condition that mimics a genetic condition.
6. Indirect methods of data collection produce biases that require special methods of statistical analysis. The usual 3:1 ratio typical of recessive inheritance, for example, is not the proper expectation when

heterozygous parents are ascertained through affected children.

7. The deviations of observed data from those expected on the basis of some hypothesis can be used to test the credibility of that hypothesis by the chi-square method. The larger the deviations, the smaller the probability that they can be attributed to mere chance. When the probability is very small, we reject the hypothesis.

8. A trait that is genetic will run in families and will not spread to unrelated persons in the same circumstances, but these conditions may not be enough to establish a genetic hypothesis.

9. The disease, kuru, spread to close kin because of ritual cannibalism rather than being transmitted genetically. It is caused by an unusual infectious agent, a "slow virus" or prion, which has little or no nucleic acid.

FURTHER READING

A synthesis of genetic heterogeneity in humans appears in Chapter 9 of the text by Harry Harris (1980). The reference book by Stanbury et al. (1983) gives details on over 100 inherited disorders, but this and the Harris book require a background in biochemistry. Stine (1977) and Greenblatt (1974) provide short human interest accounts of many genetic conditions. Pines (1984) describes the impact of Huntington disease on families. Howell and Ford (1980) poignantly recount the life of Joseph Merrick.

Readers with a good background in calculus and statistics can pursue the topic of ascertainment first in Cavalli-Sforza and Bodmer (1971) and then in Elston (1981). The older textbook by Neel and Schull (1954) is also good. The chi-square method is treated more fully in Strickberger (1985) and some other textbooks of general genetics. The theory and practice of hypothesis testing, in general, can be found in statistics texts—Sokal and Rohlf (1987) is good.

The Nobel address on kuru by Gajdusek (1977) requires that a medical dictionary be handy, but his historically interesting 1963 paper does not. Bingham (1981) gives an interesting profile of the life and personality of Gajdusek.

QUESTIONS

1. This chapter includes a fair amount of terminology, and you may care to review the meaning of the boldfaced words.

2. If pattern baldness is indeed inherited as an autosomal trait, dominant in males ($B_1/-$) but recessive in females (B_1/B_1), what are the expected genotypes and phenotypes of the children of a bald woman and a nonbald man?

3. Is the pattern of transmission of baldness in Figure 2 consistent with the hypothesis of an X-linked recessive? Assume b/(Y) = bald man, B/(Y) = nonbald man, b/b = bald woman, $B/-$ = nonbald woman.

4. In mice, $B/-$ is black and b/b is brown. At another locus, $C/-$ is colored and c/c is colorless (albino). The albino genotype is epistatic to all genotypes at the B locus. What offspring phenotypes are expected, in what proportions, from two doubly heterozygous black mice: $B/b\ C/c \times B/b\ C/c$?

5. In mice, the recessive allele d (dilute) interacts with B genotypes in this way:

   $B/-\ D/-$ = black $\quad$ $b/b\ D/-$ = brown
   $B/-\ d/d$ = grey $\quad$ $b/b\ d/d$ = light chocolate

   What offspring are expected, in what proportions, when a triple heterozygote is crossed with a complete homozygote: $B/b\ C/c\ D/d \times b/b\ c/c\ d/d$?

6. In rabbits, the three alleles C, c^b, and c give these phenotypes:

   C/C or C/c^b or C/c = full color
   c^b/c^b or c^b/c = Himalayan
   c/c = albino

   What are the genotypes of the parents if a litter of rabbits contains all three phenotypes?

7. In humans, a few families have been reported in which both parents are albino, but all their children have normal pigmentation. A number of families are known in which both parents are deaf but all their children have normal hearing. What explanation is reasonable for these cases? Suggest genotypes for the albino parents and their offspring.

8. What pattern of occurrence in a pedigree is expected if a rare phenotype in a proband is due to (a) phenocopy; (b) newly occurring dominant lethal mutation; (c) mistake in diagnosis; (d) chromosomal nondisjunction?

9. Friedreich ataxia is a rare degenerative disorder of the brain and spinal column leading to death before age 40. Neel and Schull (1954) summarize data obtained by complete truncate ascertainment of affected persons born to normal parents: (a) In sibships of size four, 11 cases were found among 36 persons; and (b) in sibships of size three, 14 cases were found among 33 persons. What is the expected number of affected persons in (a) and (b)? Assume that the disease is due to a recessive gene and refer to Figure 11.

10. Use Table 2 to determine the sibship size necessary for 80% or more of sibships to be found by complete truncate ascertainment.

11. Mendel's "worst" monohybrid experiment (the one with an F_2 furthest from an exact 3:1 ratio) involved plant height (see Chapter 2). In his "best" experiment, he observed 6,022 yellow seeds ($Y/-$) and 2,001 green seeds (y/y) from the self-fertilization of heterozygous (Y/y) parents. (a) What are the deviations from expected in his best experiment? (Round off to whole numbers.) (b) What is the value of χ^2? (c) Is it reasonable to chalk the deviations up to chance?

12. (a) A hospital recorded 95 girls and 105 boys among 200 successive births. Are these observations consistent (at the 5% level of significance) with the hypothesis that, in humans, the sex ratio at birth is 1:1? (b) Continuing, the hospital recorded 950 girls and 1,050 boys among 2,000 successive births. Is this more extensive set of data (with the same sex ratio) consistent with the 1:1 expectation?

13. For algebra fans: The machine calculation of χ^2 is facilitated by putting the formula into a different form. If the observations in two classes are a and b, the corresponding expectations w and z, and the total t, show that χ^2 can be computed as $a^2/w + b^2/z - t$.

ANSWERS

2. All sons are bald (B_1/B_2), all daughters nonbald (B_1/B_2).

3. Yes, except for female II-5. She should have received the B allele from her father and been nonbald. As noted in the text, however, environmental factors may have contributed to her baldness.

4. Use a checkerboard or the gene × gene method to obtain

$$\begin{aligned} &9/16\ B/-\ C/- = \text{black} \\ &3/16\ b/b\ C/- = \text{brown} \\ &\left.\begin{aligned} &3/16\ B/-\ c/c \\ &1/16\ b/b\ c/c \end{aligned}\right\} = 4/16 \text{ albino} \end{aligned}$$

5. Use a gene × gene method to obtain

$$\begin{aligned} &1/2\ -/-\ c/c\ -/- = \text{albino} \\ &1/8\ B/b\ C/c\ D/d = \text{black} \\ &1/8\ B/b\ C/c\ d/d = \text{grey} \\ &1/8\ b/b\ C/c\ D/d = \text{brown} \\ &1/8\ b/b\ C/c\ d/d = \text{light chocolate} \end{aligned}$$

6. $C/c \times c^b/c$

7. The parents were homozygous for different autosomal recessives leading to albinism (or deafness). Assume that c/c is albino, and that, at another locus, w/w is albino. (The c is allelic to its normal dominant allele C; likewise w is allelic to W.) One parent could be $c/c\ W/W$, and the other parent $C/C\ w/w$. All children would then be doubly heterozygous, $C/c\ W/w$, and phenotypically normal.

8. In all cases, affected persons are likely to be sporadic, without any relatives being affected.

9. The expected fraction affected is $[0.250]/[1 - (0.75)^n]$.

Problem part	Value of n	Expected fraction affected	Expected number affected
a	4	0.366	(0.366)(36) = 13.2
b	3	0.432	(0.432)(33) = 14.3

10. 6

11. (a) deviations: +5, −5. (b) $\chi^2 = 25/6017 + 25/2006 = 0.017$. (c) The probability is fairly close to 1. That is, if you were to repeat this experiment, your results would almost always be worse than Mendel's. His deviations can certainly be due to chance.

12. (a) Yes, $\chi^2 = 0.5$; probability is quite large. (b) No, $\chi^2 = 5.0$; probability is about 0.025 from Figure 12.

13. Write the formula in the standard way, expand the squared expressions, and cancel w's and z's where possible. Note that $a + b = t$, and $w + z = t$.

$$\chi^2 = \frac{(a-w)^2}{w} + \frac{(b-z)^2}{z}$$

$$= \frac{a^2}{w} - \frac{2aw}{w} + \frac{w^2}{w} + \frac{b^2}{z} - \frac{2bz}{z} + \frac{z^2}{z}$$

and so on.

Part Four GENES, METABOLISM, AND DISEASE

---

13. Gene Structure and Function
14. Genetic Information and Misinformation
15. New Genetic Technologies
16. Inborn Errors of Metabolism
17. Genetics of Blood Groups
18. Genetics of Immunity
19. Genetics of Cancer

Chapter 13 Gene Structure and Function

DNA Structure
Base Complementarity

Replication: DNA → DNA
Bubbles and Forks

Transcription: DNA → RNA
Types of RNA

Protein Synthesis: RNA → Protein
Amino Acids
Translation

Gene Organization in Chromosomes
Interrupted Genes and mRNA Processing
Gene Clusters and Pseudogenes
Repetitive DNA
Transposable Elements

Mitochondrial Genes

> *When I got to our still empty office the following morning [February 28, 1953], I quickly cleared away the papers from my desk top so that I would have a large, flat surface on which to form pairs of [cardboard] bases... Suddenly I became aware that an adenine–thymine pair held together by two hydrogen bonds was identical in shape to a guanine–cytosine pair held together by at least two hydrogen bonds. All the hydrogen bonds seemed to form naturally; no fudging was required to make the two types of base pairs identical in shape...*
>
> *Upon his arrival Francis [Crick] did not get more than halfway through the door before I let loose that the answer to everything was in our hands.*
>
> James D. Watson (1968)

In the week that followed this simple observation, James Watson and Francis Crick excitedly reworked their wire model of the DNA double helix. Less than two months later, their historic one-page report announcing the structure of the genetic material appeared in print (Watson and Crick 1953). Then, as now, nucleic acid research stirred up ideas in biological and medical sciences—solving some long-standing problems and at the same time raising new questions about basic cellular processes. Biochemical expertise has now been so sharpened that DNA molecules can be sequenced, sized, sliced, tagged, rearranged, hybridized, and cloned in precise ways. The resultant explosion of genetic knowledge has opened up provocative medical and moral choices that scientists and lay people alike will be discussing and struggling with for decades to come.

In Chapters 13 and 14, we examine the structure and function of genetic molecules, and why they sometimes fail to work properly. Chapter 15 describes how DNA can be manipulated in various ways to yield both basic knowledge and practical results. Later chapters discuss inherited metabolic disorders, some blood group systems, the complexities of immune reactions, and the growing understanding of cancer. Before proceeding, some readers may wish to review the elementary chemistry in Appendix 3.

DNA STRUCTURE

We present here enough DNA chemistry so that readers can appreciate in a general way how genes replicate, mutate, store information, and have this information converted into protein structure. For more details, readers should consult a textbook of general genetics, molecular biology, or biochemistry (see Further Reading).

Watson, a phage geneticist, and Crick, a physicist, began their collaboration at Cambridge University in 1951. They realized that determining the structure of DNA would be a grand coup, because evidence was accumulating that DNA, rather than protein, was the genetic material. The amount of DNA was known to be constant for all cells of a species (except for gametes, which contained half as much), a pattern that was not observed for other cell components. In 1944, it had been shown that free DNA, but not protein, could enter and change the genotype of bacteria, producing phenotypic traits that appeared in subsequent, untreated generations. By 1952, even skeptics were convinced by these and other findings that heredity was controlled by the nucleic acids, but relatively few scientists were actually investigating their properties.

Watson and Crick knew the chemical components of DNA:

- the sugar (S) deoxyribose (Figure 1)
- phosphoric acid (P)
- four bases: adenine (**A**), thymine (**T**), guanine (**G**) and cytosine (**C**) (Figure 2)

Furthermore, Erwin Chargaff of Columbia University had discovered that in DNA the amount of A equals the amount of T, and the amount of G equals the amount of C. The percentage of A (or T) versus the percentage of G (or C) varies from organism to organism, however. (Human DNA has about 60% A + T and 40% G + C.)

Watson and Crick also knew about the X-ray studies of DNA fibers by Rosalind Franklin and Maurice Wilkins at King's College of the University of London. Their photographs permitted crystallographers to deduce that DNA had a helical structure, but it was unclear how many strands were present, and whether the bases pointed outward from a narrow helical core or inward from a wider helix. The fit of the DNA parts had to satisfy the laws of structural chemistry, Chargaff's rules, and some dimensions of the helix revealed by the X-ray photographs. The structure also had to make some biological sense. Before they assembled the components in the right way (amply confirmed later), they invented and discarded a long series of wrong choices. Chargaff (1974) referred to Watson and Crick as "two pitchmen in search of a helix."

The predominant form of DNA in vivo is called **B-DNA**, a *right-handed* helix with certain physical dimensions. (Right-handed helices are like screws that advance into wood blocks when twisted to the right.) B-DNA is the form modeled by Watson and Crick and sketched in Chapter 4: two smoothly spiraling sugar-phosphate backbones with bases extending toward each other like rungs of a twisted ladder. The width of the helix is about 2 nm (20 Å), and its length may be thousands to millions of times more than its width, depending on the source. (See the labyrinthine traces of DNA in Figure 8 of Chapter 4.)

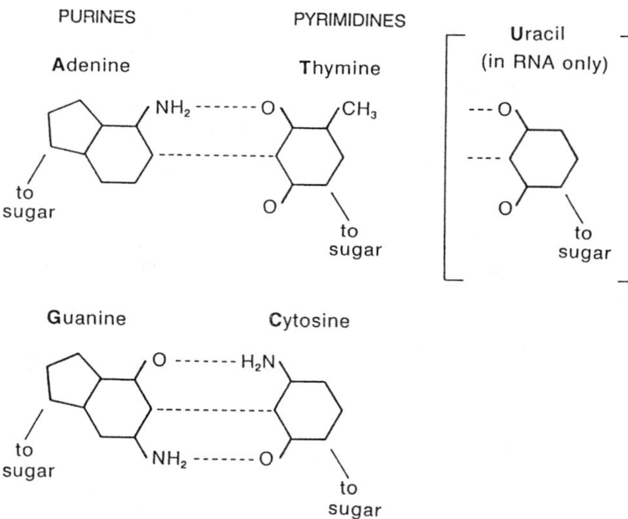

Figure 2 Partial structures of the bases present in nucleic acids. At each angle of the five- and six-membered rings is a carbon or nitrogen atom. Differences in the placement of hydrogen atoms and double bonds in the rings are omitted. The four bases in DNA are shown in their base-pairing configurations, with hydrogen bonds indicated by dashed lines. In RNA, uracil substitutes for thymine, and the four bases are not usually paired (RNA being single-stranded).

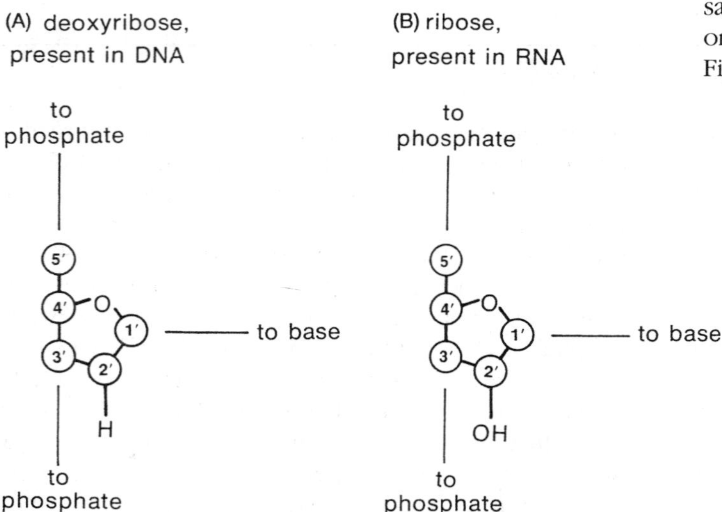

Figure 1 Partial structures of the sugars present in nucleic acids. Each sugar includes a ring of four carbon atoms and one oxygen (O). The carbon atoms are conventionally labeled with primed numerals as indicated. The 3′ and 5′ carbons make the attachments to phosphoric acid, which alternates with sugar to form a backbone structure. The attachments of hydrogens (—H) and hydroxyl groups (—OH) to the carbon atoms are not shown, except for the one difference between deoxyribose and ribose.

Structures for DNA other than the B form have been investigated in vitro. Several different right-handed helices are stable structures under certain conditions. A remarkable *left-handed* helix called **Z-DNA** can exist when a repeating –G–C–G–C–G–C– sequence occurs. The sugar–phosphate backbones of Z-DNA zigzag as they spiral to the left. The roles played by the various types of DNA in normal cells are unclear, but researchers suggest that the physical conformation of DNA might be important in its binding to regulatory proteins present in the nucleus (Dickerson 1983; Marx 1985).

Base Complementarity

The point we emphasize here is not the three-dimensional measurements of the molecule but *the pairing of the bases*. The uniform width throughout the B-form helix means that each rung of the ladder is the same size, a larger purine (A or G) being paired with a smaller pyrimidine (T or C). Chemical restraints require, further, that purine A pair only with the pyrimidine T, and purine G only with the pyrimidine C (Figure 3). Thus the sequence of bases in one strand is always complementary to the bases on the other strand according to the pairing rules: A opposite T (or vice versa) and G opposite C (or vice versa). This complementary base pairing explains Chargaff's rules. The percentage of pairs that are AT or GC depends on the sequence along the length of one strand, however, and this varies from region to region, from molecule to molecule, and from organism to organism.

The basic building block of DNA is a **nucleotide*** consisting of a deoxyribose sugar, a phosphate attached to the sugar's 5′ carbon atom, and a base attached to the sugar's 1′ carbon. (The sugar's 3′ carbon is attached to the *next* nucleotide.) The nucleotides that are incorporated into DNA are synthesized from simpler compounds in the cell. Because the four nucleotides present in DNA differ from one another only in which base they carry, the terms *nucleotide pairing* and *base pairing* express the same complementarity.

Another point of DNA structure will concern us: the two sugar–phosphate backbones of the double helix run in "opposite" directions, like two-way traffic. Specifically, the orientations of the sugars, marked by labeling the 3′ and 5′ carbons, are reversed on the two strands. Note also that at the tip of a DNA molecule (refer to Figure 3), the sugar on one strand has a 5′ carbon unattached to any more nucleotides, whereas the sugar on the other strand has its 3′ carbon so exposed.

*More precisely, the nucleotides in DNA are called *deoxyribo*nucleotides and those in RNA are called *ribo*nucleotides. Besides being components of nucleic acids, the nucleotides have other functions in cells. The adenine ribonucleotide, ATP (= adenosine triphosphate), for example, is particularly important in energy conversions (see Appendix 3).

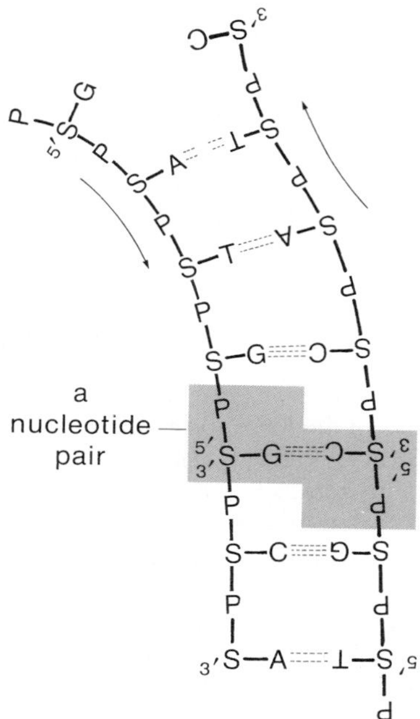

Figure 3 A portion of a double-stranded molecule of DNA (omitting the helical form). The two chains are held together by hydrogen bonds between the bases (A and T; G and C). At the top, the chains are separating as they might during DNA replication. The order of bases on one of the chains is irregular, but given the bases on one chain, those on the other are determined. The sugars (S) in one backbone are upside-down with respect to those in the other; this can be noted by designating the 3′ and 5′ attachments of the phosphates (P) to the sugars.

REPLICATION: DNA → DNA

Complementary base pairing suggested to Watson and Crick a mechanism for the replication of DNA. As shown in Figure 4 of Chapter 4, replication could occur by separation of the two strands, with each single strand acting as a **template** (that is, a mold or pattern) for the formation of a new strand. Individual nucleotides, present in the cell nucleus, are captured and strung together opposite their complements in the old template strands, a process resulting in two identical double helices where there was once just one.

The separation of the helix is itself a remarkable event because the very long strands can come apart only by unwinding. In the DNA of the bacterium *Escherichia coli*, for example, the roughly 400,000 turns of the double helix are unwound for replication within a 30-minute division time, aided by so-called unwinding proteins. Thus the helix must be spinning around its axis faster

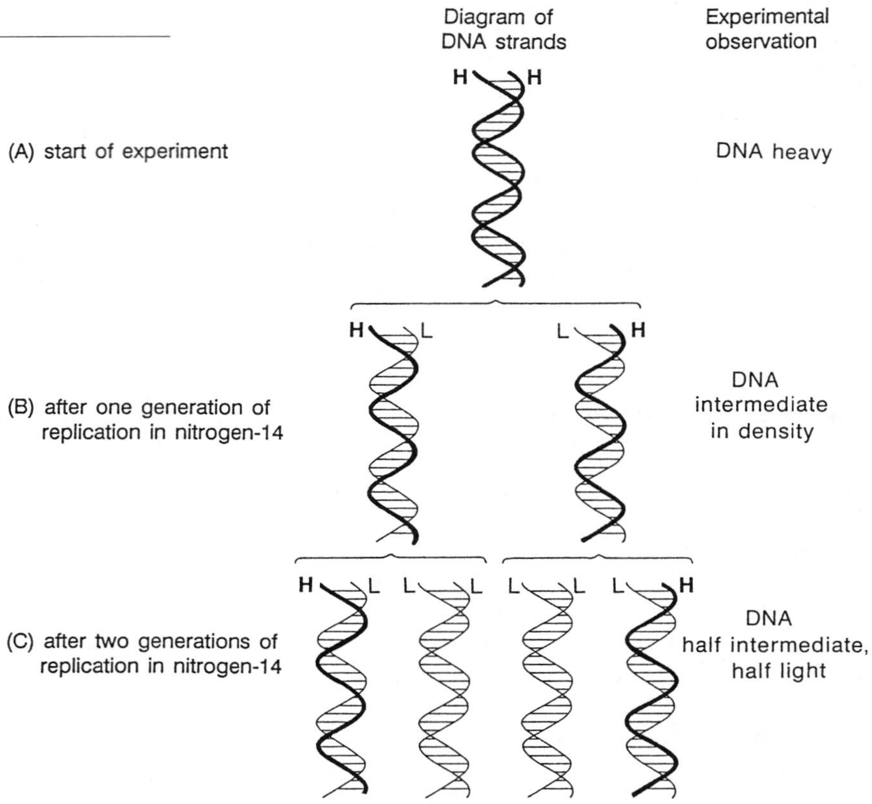

Diagram of DNA strands

Experimental observation

(A) start of experiment — H H — DNA heavy

(B) after one generation of replication in nitrogen-14 — H L, L H — DNA intermediate in density

(C) after two generations of replication in nitrogen-14 — H L, L L, L L, L H — DNA half intermediate, half light

Figure 4 Results of the Meselson-Stahl experiment. H (heavy) refers to a strand of DNA containing nitrogen-15 in the ring structures of the purines and pyrimidines. L (light) refers to a DNA strand with nitrogen-14. At the start of the experiment, *E. coli* had been grown for several generations in medium containing only the heavy isotope nitrogen-15. The actual densities of the DNA double helices were measured by centrifuging the isolated DNA solutions in salt solutions of various densities.

than 200 turns per second if, as is thought, DNA synthesis begins at just one point in the bacterial molecule. The helical thread is so fine (20 Å), however, that the actual distance traveled by a point on the surface of the helix is minuscule, and the energy needed for this trick is easily available in the cell.

That DNA actually replicates in the way suggested by Watson and Crick was shown by ingenious experiments with *E. coli* by Matthew Meselson and Franklin Stahl (1958) at the California Institute of Technology. The bacterial cells were grown for several generations in medium containing the heavy isotope nitrogen-15, rather than the normal, lighter isotope nitrogen-14. (Isotopes are discussed in Appendix 3.) The purines and pyrimidines synthesized by the bacteria incorporated nitrogen-15 into their structure, and therefore the bacterial DNA became slightly heavier than if the cells had been grown in the normal isotope of nitrogen. At the start of the experiment, then, both strands of DNA in the double helix contained nitrogen-15, and the DNA was considered *heavy* (Figure 4A).

Next, the *E. coli* cells were grown for one generation in a medium containing only nutrients with nitrogen-14. If DNA replicates as outlined here (Figure 4B), then each new double helix should contain one heavy strand and one light strand, thus being *intermediate* in density. This was indeed the observed result. As further predicted, when the *E. coli* cells were allowed to undergo *two* generations of growth in nitrogen-14 (Figure 4C), half the DNA double helices were of *intermediate* density and half were *light*. (See also Question 4.)

Bubbles and Forks

In higher organisms, the single DNA molecule that makes up each chromosome begins to replicate at hundreds or even thousands of sites, rather than at just one site as in *E. coli*. Local untwistings of the two strands at many places along the length of the helix expose sequences of bases to which enzymes can attach and orient complementary nucleotides (Figure 5). Under an electron microscope, each local separation appears as a bubble in the otherwise still closely paired double helix. The untwistings put strains on the molecule, perhaps analogous to those encountered when one tries to separate the strands within a length of rope. It is thought that an enzyme nicks one of the sugar-phosphate backbones, breaking a chemical bond that is later rejoined. This break then allows one of the strands to rotate, relieving the pressure.

The replication of DNA by complementary base pairing requires the presence of an enzyme called **DNA polymerase**, which travels along the old DNA strand and aids in the selection, placement, and bonding together of the new nucleotides. One of the interesting facts of biochemical life is that the new nucleotides are

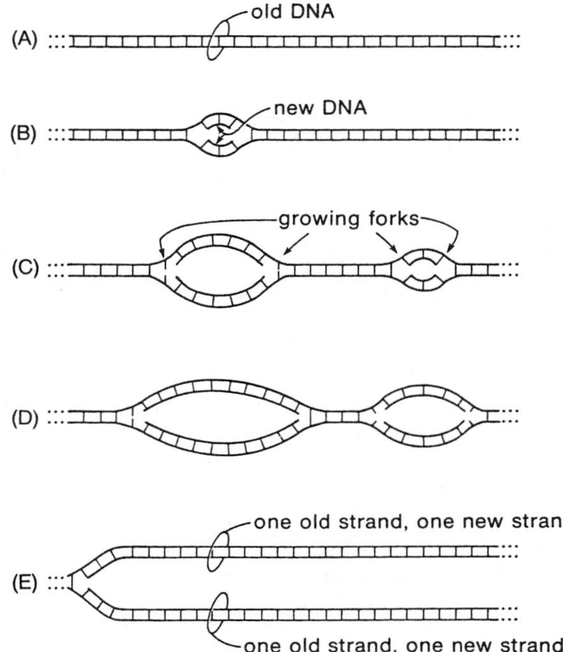

Figure 5 Replicating bubbles arising through local separations of the two strands of DNA. In (B) and (C), two bubbles are initiated; the new strands of DNA formed by complementary base pairing use the now-separated old strands as templates. (D) The bubbles increase in size at the growing points, the Y-shaped forks at the left and right ends of a bubble. (E) The bubbles have joined with each other and with a third bubble off the diagram to the right. During cell division, the newly formed double helices segregate to the two daughter cells.

always added to the 3′ end of the sugar of the nucleotide already in place in the growing, new strand of DNA. The two strands of the original double helix run in opposite directions; consequently, the two new strands forming at one of the ends of a replicating bubble cannot both be synthesized in the same direction. Consider one of the Y-shaped replication forks at the right side of a bubble (Figure 6). The upper branch of the replication fork can continually add nucleotides to the 3′ end, but replication on the lower branch must wait until a length of the template strand has been exposed and then proceed "backward." The new fragments thus formed—called *Okazaki fragments* after their discoverer— are about 150 nucleotides long in mammalian cells. They are eventually joined together by another enzyme called **DNA ligase**, thus making this new strand continuous too.

A second interesting and unexpected facet of DNA polymerase is its inability to *initiate* a new complementary nucleotide chain; it can only *add* to a preexisting sequence. A new strand gets started, not with DNA at all, but with the synthesis by an *RNA* polymerase of a short stretch of *RNA* complementary to the template DNA. This **RNA primer** is then extended as DNA by DNA polymerase. The ribonucleotides of the primer are later replaced with deoxyribonucleotides by still other enzymes, leaving a new all-DNA strand.

The discovery and characterization of many of the enzymes needed in the replication of DNA derive from the work of Nobel laureate Arthur Kornberg (1968) of Stanford University. One dramatic experiment involved a favorite tool of geneticists, the tiny phage, ϕX174, whose circular DNA molecule consists of only about 10 genes. In a test tube, Kornberg combined enzymes, nucleotide precursors, and the isolated phage DNA. This

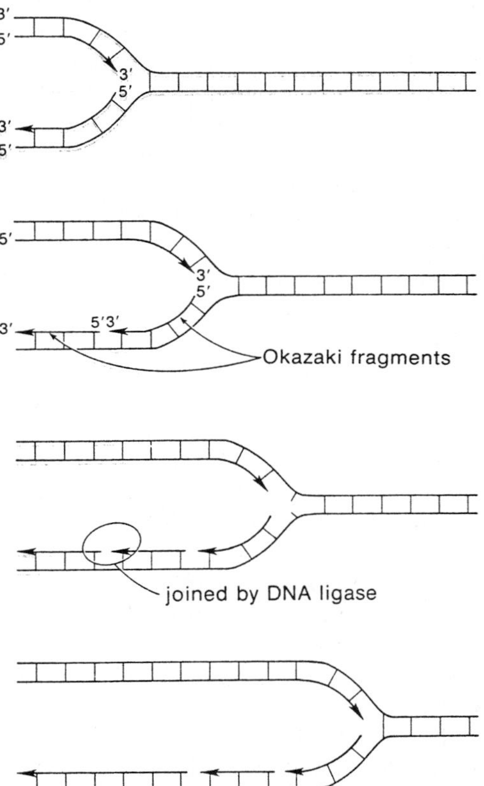

Figure 6 Events at a replicating fork (at the right end of a bubble). The DNA polymerases add new nucleotides only to the 3′ side of the sugar in the previously incorporated nucleotide. Because of this directionality and because the two strands of the original double helix run in opposite directions, replication of one strand must occur in discontinuous steps: on the lower branch the fork must open for a considerable length before synthesis of a new DNA strand can occur "backward." Thus on the lower branch, although overall replication proceeds from left to right (as indicated by the movement of the fork), individual nucleotides are added in the reverse direction in fragments of about 150 nucleotides. (It may be that synthesis on the upper branch is also discontinuous, but this would not be required by what is now known of DNA synthesis.)

reaction mixture produced new synthetic molecules of ϕX174 DNA that were just as infective as if they had been generated by the virus itself during the normal course of infection of *E. coli*.

In higher organisms, DNA is replicated in preparation for each mitotic cell division, and prior to the first meiotic division of gametogenesis. Accurate replication is the prerequisite for transmission of exact genetic information from cell to daughter cells and from one generation to the next. High-fidelity reproduction is accomplished by several mechanisms, including what is called the **proofreading** mechanism of DNA polymerase (Radman and Wagner 1988). The last-attached nucleotide at the growing tip of the new strand must "settle in," becoming firmly base-paired to its complementary nucleotide on the template strand *before* another nucleotide can be added. DNA polymerase will remove any improperly matched nucleotide and try again for an exact AT or GC fit before continuing down the line. It is estimated that only one replication mistake is allowed to remain for every 10^{10} base pairings, or an average of one mistake for each replication of a human cell (having about 6×10^9 nucleotide pairs). For comparison, one typographical error in this book of about 3×10^6 characters would constitute an error rate 2,000 times greater than that occurring in DNA replication.

TRANSCRIPTION: DNA → RNA

Within cells, information encoded in DNA helps to direct the development and maintenance of the organism. Decoding of the information proceeds in two clearly defined steps. First, base sequences in DNA are **transcribed** into complementary sequences of bases in RNA, ribonucleic acid (Figure 7). These molecules of RNA, made at chromosomal sites, move across the nuclear membrane into the cytoplasm. In the second step, occurring at ribosomes, the information now encoded in the base sequence of one type of RNA is **translated** into a sequence of amino acids, that is, a polypeptide. Many polypeptides are the enzymes (or parts of enzymes) required for cell metabolism; other polypeptides are, for example, hormones, antibodies, or structural elements.

The terms *transcription* and *translation* are well chosen. The former means to convert into a different form of the *same* language: from oral to written English, or from a sequence of deoxyribonucleotides to one of ribonucleotides. The latter describes a more profound conversion: from one language to another, say from English to French, or from nucleotides to amino acids.

The process of transcription is similar to DNA replication, except that the complementary molecule that is formed is RNA, containing the sugar ribose (rather than deoxyribose) and the base uracil (instead of thymine). Transcription is controlled by the enzyme **RNA polymerase**, which links together ribonucleotides in a sequence complementary to the sequence of bases along one of the strands of DNA.

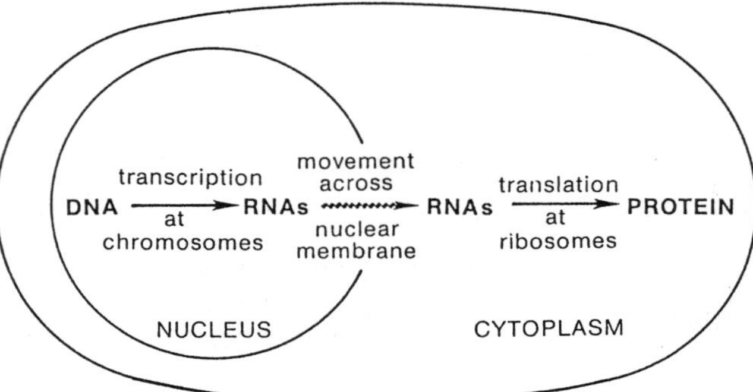

Figure 7 Decoding a gene. A particular sequence of nucleotides in DNA causes a cell to assemble a particular sequence of amino acids into a protein. It is a two-step process in which RNA molecules are intermediaries, carrying copies of genetic information across the nuclear membrane. The first step (DNA→ RNA) is called *transcription*, and the second step (RNA→ protein) is called *translation*. Some genes are transcribed into RNA but not translated (see text).

The process is illustrated in Figure 8. The place where RNA polymerase first attaches is a specific succession of bases called the *promoter*. Although promoters vary from gene to gene and from organism to organism, they all contain certain sequences in common: in eukaryotes a seven-base sequence of virtually all T's and A's (the so-called TATA box) is part of all promoters. Transcription begins beyond the promoter as new ribonucleotides are added in turn to the 3′ end of the preceding ribonucleotides, this being the same direction of synthesis used in DNA replication. The RNA polymerase moves off the DNA molecule at a certain *termination* sequence, completing the transcription of a length of DNA into a complementary RNA copy. Because a promoter is found upstream and a terminator downstream from a gene, transcription encompasses an entire gene, or in some cases, several genes together. (Upstream is before the 3′ end of the transcribed gene, and downstream is beyond the 5′ end.)

The question of which strand of a double helix acts as the template for the formation of complementary RNA is crucial, since the RNA complement of one strand is different from the RNA complement of the other. For a particular gene, the strand that is actually transcribed is the one with the specific promoter sequence and is called the **sense strand** (Figure 9); the other, the **antisense strand**, is not used in transcription. The strand of the DNA double helix that makes sense at one place,

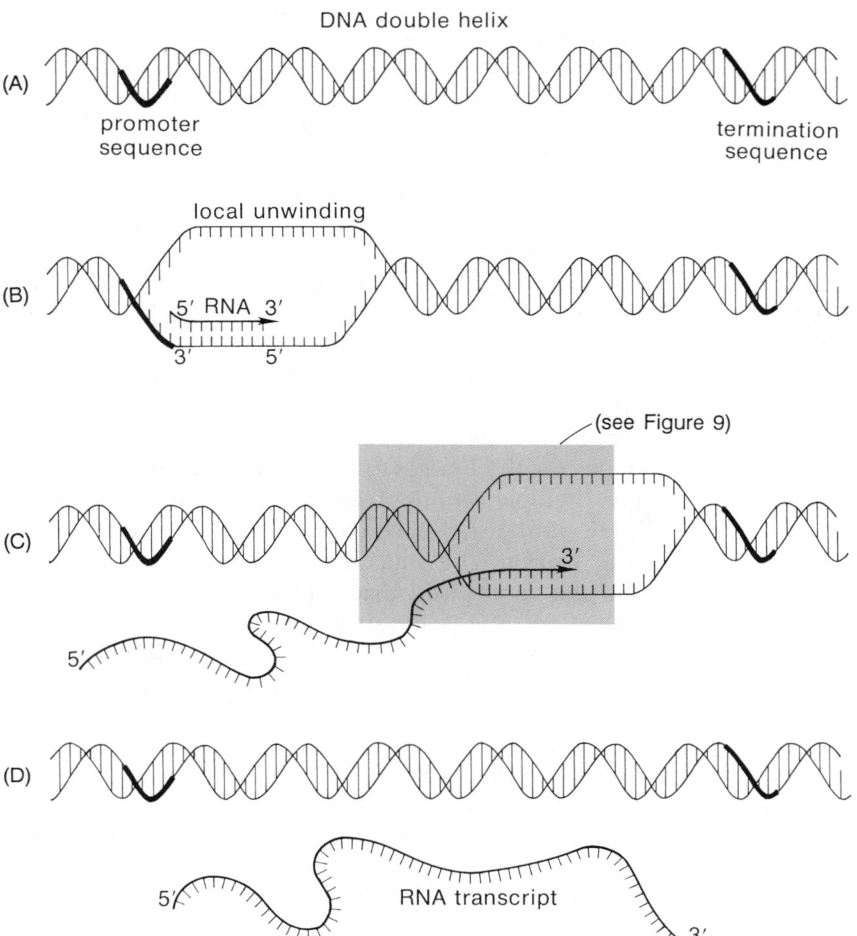

Figure 8 Transcription of a molecule of RNA. (A) Before attachment of the enzyme RNA polymerase at the promoter site, the strands of the DNA double helix are held together by hydrogen bonds. (B) Local unwinding of the DNA occurs when RNA polymerase (not shown) attaches to one strand of the DNA. Transcription proceeds downstream along this strand, ribonucleotides being added to the growing RNA transcript at its 3′ end. (C) The complementary RNA transcript does not remain bound to the DNA template but peels off as the molecule is formed. The bubble formed by the separation of the DNA double helix travels along the sense strand with the RNA polymerase molecule. (D) When the polymerase reaches the termination site, both the enzyme and the new RNA transcript separate from the DNA double helix, which assumes its original hydrogen-bonded state.

however, may not be the one that makes sense elsewhere in a chromosome; that is, the sense strand for one gene may be continuous with the antisense strand of another.

Not all genes are transcribed all the time. Intricate controlling mechanisms in the cell regulate the flow of enzymes and other proteins, thereby determining which genes should be "turned on." Some genes may be transcribed only during a brief period in the life cycle of an organism; others, such as those needed in energy production, may be operating all the time. Animals that undergo dramatic metamorphosis emphasize the important role of gene regulation: a caterpillar uses many genes not needed by the resultant butterfly, and vice versa; yet exactly the same genes are present in both.

Before describing how the cell uses RNA transcripts, we must point out that genetic information is expressed in a triplet code; that is, three adjacent bases on the sense strand of DNA stand for one of the 20 amino acids found in proteins. Consequently, during protein synthesis, three ribonucleotides in sequence along an RNA molecule are needed to position one amino acid in a polypeptide chain. But neither DNA nor the intermediary RNA are

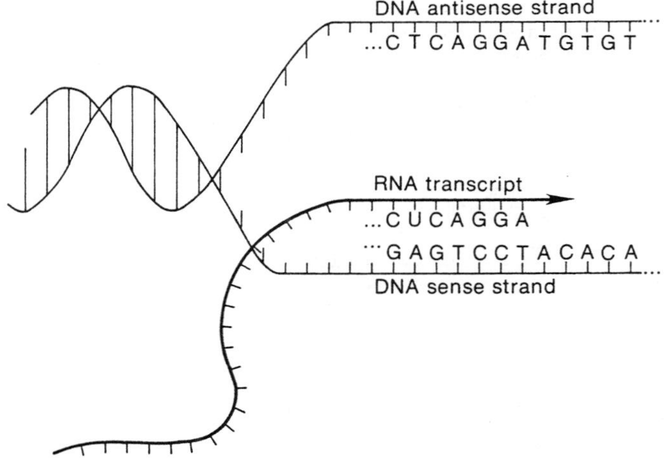

Figure 9 Base pairing during transcription. The RNA base uracil (U) is complementary to A. Note that the sequence of bases on the RNA transcript is the same as that on the unread (antisense) strand of DNA, except for the substitution of U for T.

actually marked off into threes by any special chemical structure; rather, *successive groups of three bases* are interpreted by the cellular chemistry as units of information. We present some details of the genetic code in the next chapter.

Types of RNA

There are three main types of RNA: *messenger* RNA (mRNA), *transfer* RNA (tRNA), and *ribosomal* RNA (rRNA), each transcribed from different genes. The RNAs enter the cytoplasm, where they play specific roles during translation.

Messenger RNA plays the *key informational role* during protein synthesis. Only mRNA is actually translated into polypeptide chains, so mRNA is meant in the shorthand scheme DNA→ RNA→ PROTEIN. Indeed, a gene is often defined as the sequence of nucleotides that dictates the order of amino acids in a polypeptide via an mRNA transcript. The lengths of different mRNA molecules are not the same. Because most polypeptides are at least 100 amino acids long, most mRNA molecules are at least 300 nucleotides long. Some may be much longer, containing the message for several polypeptides. And some mRNA molecules may be much longer for other reasons (see later).

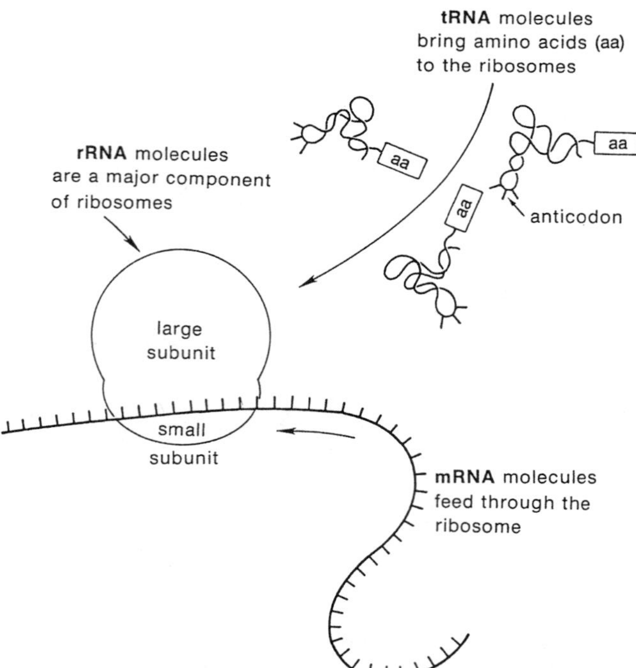

Figure 10 The roles played in translation by the three types of RNA.

In mRNA each successive group of three nucleotides represents the code for one amino acid and is called a **codon**. There are as many codons strung together in a particular mRNA as there are amino acids in the corresponding polypeptide. Because there are tens of thousands of different polypeptides, there are tens of thousands of different mRNA molecules. Any one mRNA molecule may be translated many times, although it is eventually degraded.

Transfer RNA molecules act as adapter molecules, binding to specific amino acids in the cytoplasm and fitting them to appropriate codons of mRNA. They accomplish this task through a double specificity: one end of a particular tRNA molecule is capable of binding one and only one of the 20 amino acids. The other end can recognize a specific codon of mRNA by means of a sequence of three bases called the **anticodon** (Figure 10). For example, the tRNA anticodon CCC recognizes and binds to the mRNA codon GGG by the usual rules of base pairing. The anticodon of a given tRNA molecule and the amino acid that it picks up are "matched" to each other; that is, the structure of a tRNA with a specific anticodon (CCC, say) allows it to enzymatically bind to only one specific amino acid (proline, in this case).

There are several dozen different tRNA molecules, one or a few different types for each amino acid. Each tRNA molecule is about 80 nucleotides long and is coiled and folded into an intricate three-dimensional structure. The folding is brought about by complementary base pairing between different parts of the single-stranded molecule (see Rich and Kim 1978). An interesting feature of tRNA is the presence of several unusual bases in addition to A, U, G, and C—for example, I (= inosine), which is similar to guanine but lacks the amino group (see Figure 2). The unusual bases arise by chemical modification of the usual ones after a tRNA molecule is transcribed from a tRNA gene. Because they are transcribed *but not translated*, tRNA genes have no protein product.

Ribosomal RNA, the third type of RNA, helps form the ribosomes, the "workbenches" on which protein synthesis occurs. The ribosomes in eukaryotes are composed of four types of rRNA and about 80 types of protein molecules arranged in a complex three-dimensional structure (Lake 1981). Cells that are actively translating mRNA into protein are rich in ribosomes, which line the endoplasmic reticulum. New ribosomes are synthesized as needed in what must be a well-synchronized process: the proteins for new ribosomes are translated on old ribosomes at the same time as new rRNA is being transcribed in the nucleus (Nomura 1984). We have noted before that some of the rRNA is transcribed off special chromosomal regions called *nucleolus organizers*. There are 10 of these regions in the human genome, one on each short arm of the five pairs of acrocentric chromosomes. These regions have multiple copies—

perhaps hundreds—of some of the rRNA genes. *Like the tRNA genes, the rRNA genes are transcribed but not translated.*

The ribosomes come apart into two subunits—one small and one large—in order to clamp onto a molecule of mRNA, which then feeds through the ribosome during protein synthesis. The amino acids needed for incorporation into the polypeptide chains are transported to the ribosomes by the tRNA molecules. Thus all three types of RNA are brought together at the ribosomes as indicated in Figure 10.

PROTEIN SYNTHESIS: RNA → PROTEIN

During the process of translation, anywhere from a dozen to a thousand or more amino acids are strung together to form unbranched **polypeptide chains**. One or more of these chains—coiled, folded, cross-linked, intertwined—constitute a protein. The marvelous diversity of this one class of chemical compounds is suggested by a short, rather random list of proteins: spider webs, tiger claws, snake venoms, eye lenses, blood clots, wheat gluten, and catalysts for DNA replication (Table 1). The properties of a cell are often determined by the proteins it makes. For example, red blood cells manufacture large amounts of hemoglobin, certain white blood cells make antibodies, muscle cells make myosin and actin, liver cells make an array of enzymes, and hair follicle cells make keratin (the supporting element not only of hair, nails, and skin, but also of feathers, horns, silk, and wool). The human body contains more than 10,000 different proteins, each with a characteristic structure and function and each assembled from the same set of 20 amino acids.

Table 1 Representative proteins

Name of protein[a]	*Specific function*	*Class of protein*	*Polypeptide chains*
1. DNA polymerase	replicates and repairs DNA	intracellular enzyme	one chain of about 1,000 aa's
2. Trypsin	cleaves proteins on the carboxyl side of the two basic amino acids lysine and arginine (in small intestine)	digestive enzyme	one chain of 223 aa's (bovine)
3. Histone H4	one of four DNA-binding proteins found in nucleosomes	chromosomal protein	one chain with 102 aa's (highly conserved in evolution)
4. Insulin	aids in the movement of glucose across cell membranes	hormone	two chains with a total of 51 aa's: one chain with 21, the other with 30
5. Egg albumin	the major component of egg whites; provides amino acids to nourish the developing chick	storage protein	one chain of about 400 aa's (attached to a nonprotein, carbohydrate part)
6. Hemoglobin	carries oxygen from the lungs to the tissues via the red blood cells	transport protein	four chains with a total of 574 aa's (globin part): two chains with 141, two with 146 (each chain attached to a nonprotein, heme group)
7. Gamma globulin	antibody molecules, which react with and incapacitate foreign substances	protective protein	four chains with a total of 1,320 aa's: two chains with 214, two with 446
8. Myosin	parallel arrays of myosin molecules slide past sets of actin molecules during muscle contraction	contractile protein	six chains with a total of about 4,300 aa's: two fiber-like chains of about 1,800 aa's, four small globular chains

a. The first seven are globular in shape and water soluble; myosin is part globular and part fiber-like. Intracellular proteins (class of line 1) are by far the largest class of proteins.

Amino Acids

The amino acids all have the same general structure: a central carbon atom (C) bonded to four groupings, two of which—the basic or amino group (—NH_2) and the acid or carboxyl group (—COOH)—give the class its name. A third group is a single hydrogen atom (—H). The fourth group, the **side chain** (Figure 11), accounts for the differences in the properties of the various amino acids. The simplest amino acid is *glycine*, in which the side chain is another hydrogen atom. In *alanine*, the side chain is a methyl group (—CH_3). In *phenylalanine*, a six-membered ring of carbons (a phenyl or benzene group) substitutes for one of the side chain hydrogens of alanine. A few more structures are shown in Figure 11, and all 20 amino acids are listed in Table 2 along with their common abbreviations.

Amino acids are joined to one another by covalent bonding between the carboxyl group of one amino acid and the amino group of the next (Figure 12). The successive linking of any number of amino acids into polypeptide chains always leaves a free amino group on one end of the molecule and a free carboxyl group on the other. The complete sequence of amino acids in a small two-chain protein, human insulin, is illustrated in Figure 13. Note that covalent bonding between the sulfur atoms of some of the cysteine side chains (called disulfide bonds) holds the two chains together. In addition, there is a disulfide bond between the cysteines at positions 6 and 11 of the A chains. Not all proteins with multiple chains are held together this way, however.

The three-dimensional shape of a protein is crucial to its biological activity. Most proteins are globular, in part or in whole, because the polypeptide chains are compactly folded and looped, often in complex ways. In enzyme molecules, some groups of amino acids fold in from the surface to form a precisely shaped cavity called

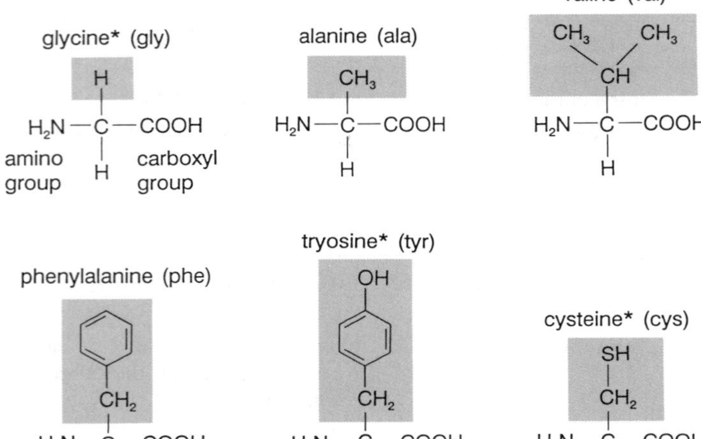

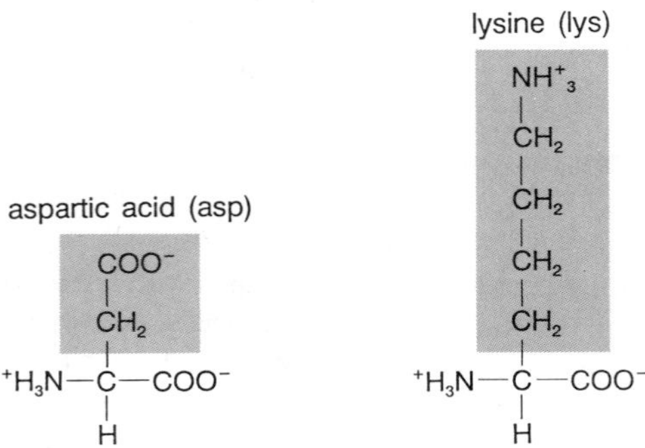

*Even though these side chains are uncharged overall, there is a separation of charge within the grouping, so that the side chain is polar.

Figure 11 Representative amino acids found in proteins. The unshaded portion of each is common to all.

Table 2 The common abbreviations for the amino acids found in proteins

Amino acid[a]	*Abbreviation*
alanine	ala
*arginine	arg
asparagine	asn
aspartic acid	asp
cysteine	cys
glutamine	gln
glutamic acid	glu
glycine	gly
*histidine	his
*isoleucine	ile
*leucine	leu
*lysine	lys
*methionine	met
*phenylalanine	phe
proline	pro
serine	ser
*threonine	thr
*tryptophan	trp
tyrosine	tyr
*valine	val

a. The so-called essential amino acids are marked with an asterisk. We cannot synthesize these particular amino acids for ourselves, but must have them premade in the proteins of the foods we eat. Because most plant proteins are very low in lysine and tryptophan, vegetarians must have a varied diet.

aa_1 aa_2

$$H_2N-\underset{H}{\overset{R_1}{C}}-C{\overset{O}{\diagup\!\!\diagup}}\!-OH \;\; H-N(H)-\underset{H}{\overset{R_2}{C}}-COOH$$

$$\downarrow$$

$$H_2N-\underset{H}{\overset{R_1}{C}}-\overset{O}{C}-\underset{\text{peptide bond}}{\rule{2em}{0.4pt}}-\overset{H}{N}-\underset{H}{\overset{R_2}{C}}-COOH \quad + \quad H_2O$$

a dipeptide

Figure 12 The formation of the peptide bond between two amino acids. Joining of additional amino acids can lead to long polypeptide chains. The symbols R_1 and R_2 stand for any side chains. The artificial sweetener aspartame is a dipeptide in which aa_1 is aspartic acid, aa_2 is phenylalanine, and the far right H on aa_2 is replaced by a methyl group.

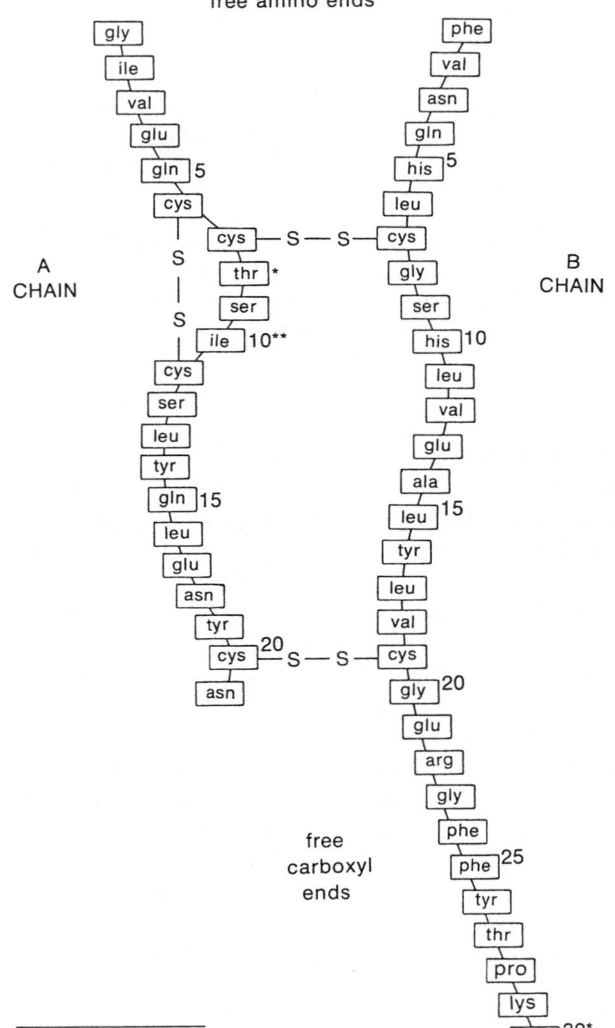

the *active site*, into which fit the substances that the enzyme acts upon. Critical sites of function are also present in other proteins, such as the oxygen-binding sites of hemoglobin or the antigen-binding sites of antibody molecules.

The folding of a protein appears to be "automatic"; that is, a particular sequence of amino acids will fold just one way. This stable configuration is determined by the angles of successive peptide bonds, hydrogen bonding and disulfide bonding, interactions between nearby side chains, and interactions between side chains and the external environment. For example, the charged amino acids, and other so-called polar ones, are often on the outside of the molecule, oriented toward the ions in solution or toward the polar water molecules. When proteins are gently unfolded (*denatured*) in vitro, some will spontaneously refold (*renature*) into the original conformation, thereby regaining their biological function (Dickerson and Geis 1969; Alberts et al. 1983).

Translation

The way a protein works ultimately comes down to its amino acid sequence, which is determined by the sequence of codons in mRNA. As noted, translation occurs on the ribosomes with the aid of tRNAs that have been loaded with specific amino acids. In Figure 14, we sketch how the sixth amino acid of a growing polypeptide chain becomes bonded to the first five, which are already linked together. The large subunit of the ribosome has pockets for *two* tRNA molecules with their attached amino acids. The right pocket will fit only the tRNA with a *particular anticodon*, because a *particular codon* of mRNA forms the lower boundary of the pocket (Figure 14A). For example, if the sixth mRNA codon were CUC, then, by the rules of base pairing, only that tRNA with the anticodon GAG will fit the pocket. The tRNA molecule that has the anticodon GAG is specific for the amino acid leucine (see the genetic code in the next chapter). In this way, whenever the codon CUC passes through the ribosome, leucine is brought into a nearby position (with its amino group to the left and its carboxyl to the right).

Figure 13 The amino acid sequence and disulfide bonds of human insulin. The amino acids are numbered from the free amino ends of the chains. This diagram does not indicate the three-dimensional structure. Human insulin lacks the amino acids aspartic acid, methionine, and tryptophan and is rich in the amino acids cysteine and leucine. The amino acid sequence of bovine insulin, which differs from human insulin in the three positions indicated, was the first protein to be so analyzed; this was accomplished by the University of Cambridge chemist Frederick Sanger, a Nobel laureate in 1958 (Sanger 1959).

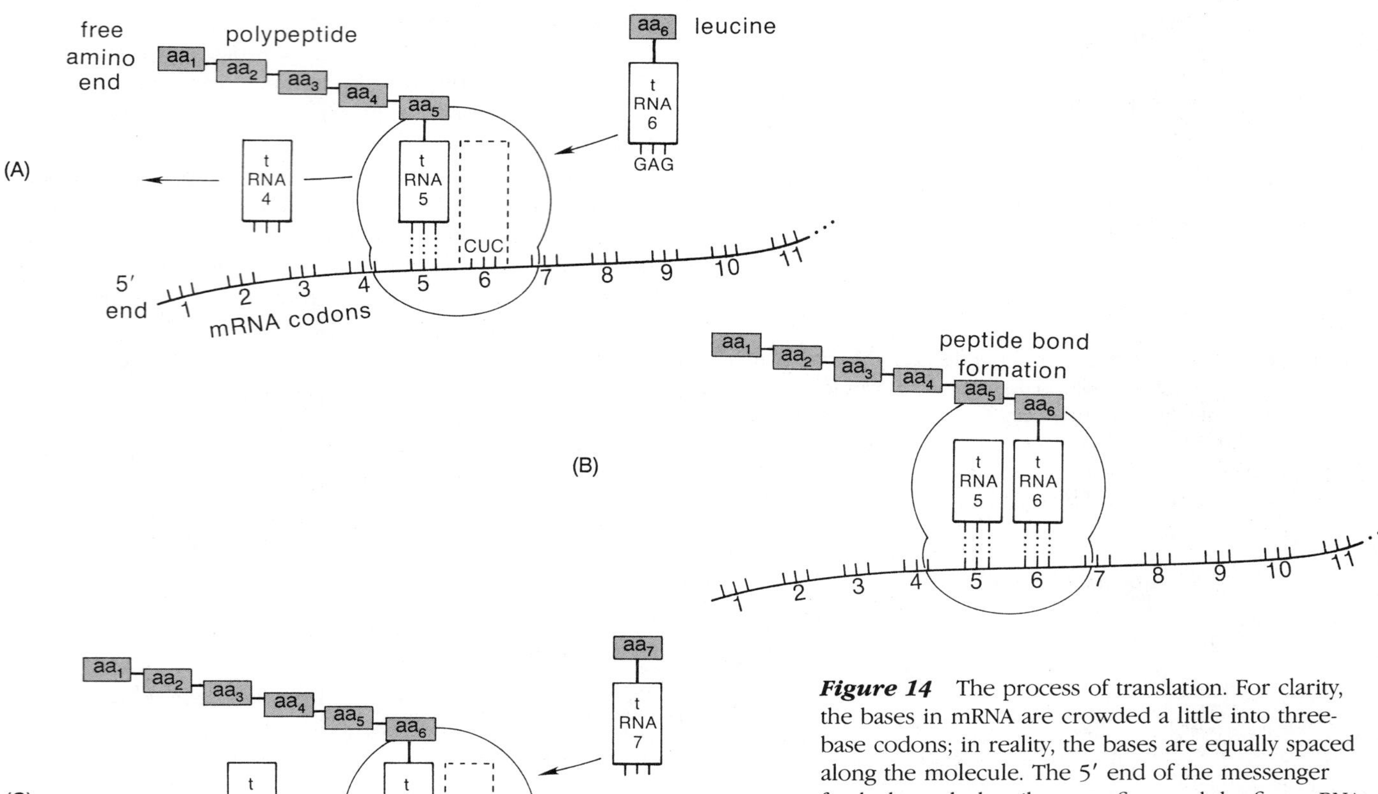

Figure 14 The process of translation. For clarity, the bases in mRNA are crowded a little into three-base codons; in reality, the bases are equally spaced along the molecule. The 5′ end of the messenger feeds through the ribosome first, and the first mRNA codon leads to the placement of the amino acid with a free amino group. Thus the first peptide bond is between the carboxyl of amino acid 1 and the amino group of amino acid 2. See text for further explanation.

In Figure 14B, the tRNA-leucine unit is stationed in the right pocket. In the left pocket is the fifth amino acid attached to its tRNA. As the fifth amino acid is released from its tRNA and becomes attached instead to leucine by a peptide bond, the length of the polypeptide is increased by one amino acid.

In Figure 14C, the mRNA molecule has moved one codon to the left, and at the same time, the leucine-tRNA has skipped to the left pocket, since the two RNAs are joined by base pairing—codon to anticodon. We are back to a diagram like that in Figure 14A, *but* now one amino acid has been added to the polypeptide. The right pocket is free for the seventh tRNA and its associated amino acid. The tRNA molecule labeled five, having contributed its amino acid to the polypeptide, is free to bind another molecule of the same amino acid to be incorporated later (if called for) as the mRNA feeds through the ribosome.

The synthesis of a *new* protein requires the participation of many *old* proteins. In addition to the 80-odd proteins that form the structure of ribosomes, special proteins are involved in attaching the mRNA to the ribosome, and several enzymes are needed to make and break covalent bonds. Initiating, lengthening, and ending the polypeptide and then releasing the mRNA from the ribosome require still other proteins.

The mRNA molecule contains not only the codons for the polypeptide but also certain head and tail sequences that are not translated. In particular, the first translated codon (some distance in from the 5′ tip of the molecule) always positions the amino acid methionine containing a modified amino group. (After protein synthesis is completed, this first amino acid may be cut off from the polypeptide.) Following the last translated codon is a so-called *stop codon*, or *termination signal*. At any given time, many ribosomes may be "working" the same mRNA molecule in single file, looking like beads on a string.

Note that the processes of transcription and translation both depend on base pairing of T with A (or U with A) and G with C (Figure 15). To say that a gene "codes for" a protein means that successive DNA triplets provide the information for the assembly of successive amino acids

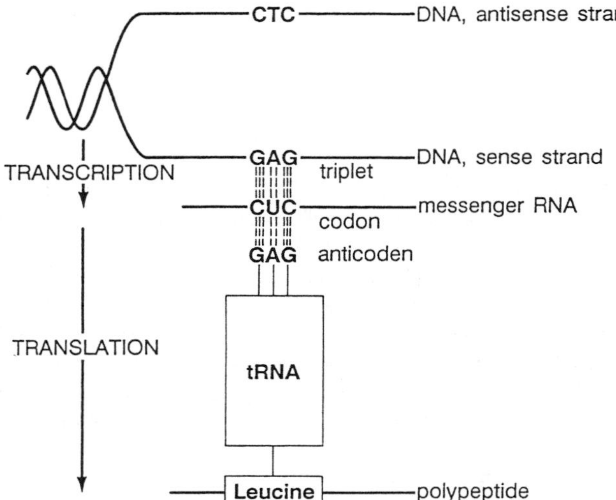

Figure 15 Base pairing during transcription and translation. The DNA triplet GAG in a polynucleotide provides the information that finally leads to the incorporation of the amino acid leucine in a polypeptide.

—through two rounds of base pairing. In bacteria, in which there is no separation of nucleus from cytoplasm, one end of the mRNA molecule (the 3′ end) is still being transcribed while the other end (5′) is already being translated. Some remarkable electron microscope pictures of a bacterial gene being multiply transcribed as each of the mRNA transcripts is being multiply translated can be seen in Miller (1973).

GENE ORGANIZATION IN CHROMOSOMES

The preceding sketch of DNA construction, replication, and expression can only suggest the true complexity of genetic processes that biochemists have partly unraveled in the last few decades. In this section we add a few details about some remarkable orders and distributions of bases, leading to yet further mysteries about the nature and operation of genes. (The way DNA is packaged into nucleosomes is covered in Chapter 4.)

Interrupted Genes and mRNA Processing

In the years after the 1953 discovery of DNA structure by Watson and Crick, geneticists developed the view that chromosomal DNA consisted of sequential genes, each an *unbroken* string of nucleotides (separated, perhaps, by "spacer" DNA that allowed polymerases to act). This traditional view turns out *not* to be true for most of the genes of higher organisms. Observations leading to a more complex view included electron micrographs of the sense strand of a gene base-paired in vitro with its complementary messenger RNA isolated from the cytoplasm. The traditional picture would be a simple double-stranded structure throughout its length (Figure 16A), because DNA, base after base, would pair exactly with the complementary mRNA, base after base. In fact, what was sometimes seen were single-stranded loops of DNA that had *no* counterpart in the mRNA found in the cytoplasm (Figure 16B). Separated from one another by double-stranded regions, the unpaired loops of DNA, called **introns** or **intervening sequences**, were often many and long, some being longer than the paired regions themselves (Chambon 1981). In the gene in Figure 16B, *seven* introns split the gene into *eight* regions (labeled 1–8) coding for successive segments of the polypeptide.

What an extraordinary surprise to discover that several nucleotide sections within the boundaries of a gene had no cytoplasmic mRNA complement, and therefore coded for no amino acids in the polypeptide product! An analogy would be for readers to confront long discourses on music or sports, or perhaps just pages of nonsense, scattered throughout this book. Investigators have discovered, however, that most genes of vertebrates are interrupted this way by one or more introns—segments that may be much longer than the coding regions of the gene they interrupt. One of many remarkable findings about introns was the discovery in the fruit fly of a gene (coding for a particular protein) *within* an intron of a second gene (coding for another protein) (Henikoff et al. 1986). The first gene is present on the strand of DNA opposite that of the second gene, and the first gene even has its own intron!

The genes that code for the polypeptides of hemoglobin provide additional well-studied examples (Collins and Weissman 1984).* The protein or globin part of the molecule (see next section for more details) includes an α (or α-like) polypeptide chain and a β (or β-like) chain. For each of the polypeptides, the appropriate gene includes two introns separating the three regions coding for amino acids. Called **exons**, or **expressed sequences**, the coding regions account for less than half the bases. The sites where the coding sequences are interrupted by the two introns are similar in all globin genes that have been studied, including those of several mammals, birds, and frogs. The greatest variation in the organization of the globin genes is in the length of the second intron, which is 150 bases long in the β gene of the mouse and 850 bases long in the β gene of humans.

The gene that codes for the β polypeptide chain of

*Collins and Weissman note that, in a recent four-year period, over 10,000 references to globin appeared in *Index Medicus*, the major international bibliographical service for medicine and human physiology. Indeed, more has been learned about the globin genes than about any other eukaryotic gene system.

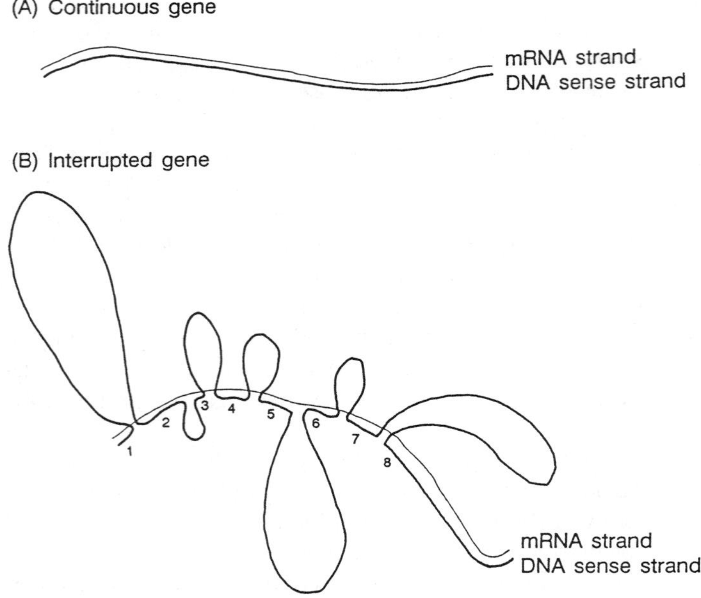

Figure 16 The results of hybridizing the DNA sense strand with its messenger RNA isolated from the cytoplasm and used in protein synthesis. In vitro DNA–RNA hybridization produces double-stranded structures by complementary base pairing. Single-stranded structures result whenever either strand has nothing complementary with which to pair. (A) The traditional concept of a gene, which still holds for bacterial and viral DNA, is an unbroken sequence of nucleotides coding for amino acids. Hybridization of the gene to its mRNA would therefore lead to an uncomplicated double-stranded structure. (B) Interrupted genes are split by regions called *introns*, or *intervening sequences*, which do not code for amino acids. Hybridization with mRNA results in the looping out of those regions of the gene that have no corresponding sequences in the mRNA. This drawing, based on the chicken gene coding for the egg-white protein ovalbumin (Chambon 1981), shows seven introns that split the gene into the eight numbered regions.

human hemoglobin contains 1,606 bases. The *initial* RNA transcript of the human β-globin gene (Figure 17) contains 1,606 nucleotides, an exact complement of the entire gene. The 130 bases complementary to intron 1 and the 850 bases complementary to intron 2 are then *cut out* of this RNA molecule before it leaves the nucleus. The splicing together of the cut ends of successive *exons* in RNA must be done with pinpoint precision to preserve the step-by-step progression of three-base codons in the different segments. How this is accomplished is not well understood, but in higher organisms it involves the base sequence GU at the 5′ beginning and AG at the 3′ end

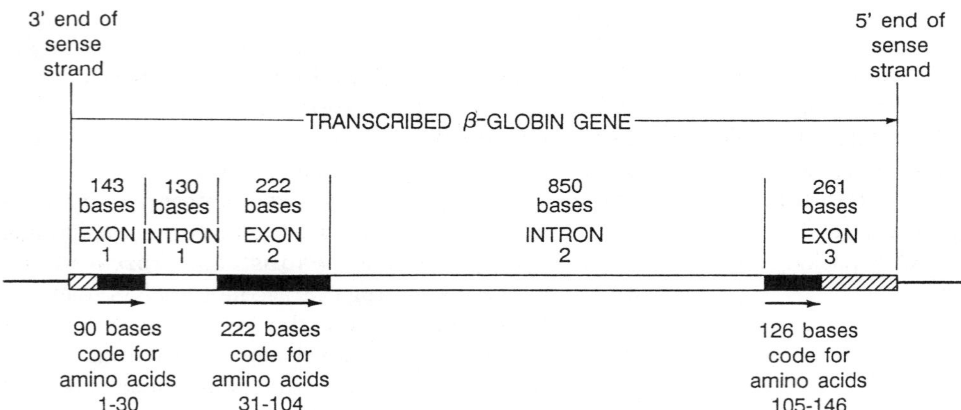

Figure 17 The structure of the gene for the β polypeptide chain of human hemoglobin. The *exons* (expressed sequences) include the regions that code for the amino acids of the β chain. The hatched leader segment of exon 1 is an untranslated 50-base sequence plus a 3-base sequence translated as methionine, the first amino acid, which is subsequently cut off the polypeptide. The hatched trailing segment of exon 3 is an untranslated sequence of 135 bases. The remaining 90 + 222 + 126 = 438 bases of the three exons specify the 146 amino acids of the β chain. The function of the two *introns* (intervening sequences) is unclear, because the complementary sequences are spliced out of the RNA molecule. Note that intron 2 is much bigger than all three coding regions put together. (Data from Poncz et al. 1983.)

of the RNA introns. Because these short sequences also occur elsewhere along the RNA, additional signaling devices must be at work.

In some genes, individual exons seem to code for key functional regions in the resultant protein, such as binding sites. Such regions, called domains, might be useful to the organism as parts of other proteins with similar functions. The *same* exon might therefore be expected to be found in *different* genes. Such is the case for the human gene (LDL receptor) coding for a protein that transports cholesterol into cells. Of the 18 exons, 13 code for polypeptide segments that are homologous to segments in other proteins: a hormone that stimulates cells to divide (epidermal growth factor, or EGF), three blood-clotting factors (IX, X, and C), and a serum protein involved in the so-called complement system (C9 component). This dramatic work on "mix-and-match" exons, summarized by Gilbert (1985), suggests a mechanism for gene evolution, an idea to which we return in the next section.

Removing intron sequences and splicing together the exons is one aspect of **mRNA processing**—the changes occurring in the molecule between the time of transcription and the time of translation (Darnell 1983; Sharp 1988). In addition to the loss of introns, RNA (which will become mRNA) is modified in two other ways before it leaves the nucleus. After transcription, some nucleotides are cut off from the 3′ end of the molecule, and a string of 50–200 adenine nucleotides is added in their place; this *poly(A) tail* aids in transporting the RNA molecule into the cytoplasm. At the 5′ beginning of the molecule, a *cap* consisting of one special nucleotide (a modified G nucleotide in reverse orientation) is added; its function is to attach the RNA to the ribosome.

The term *messenger RNA* is usually reserved for the finished product that attaches to ribosomes, that is, the sequence of codons specifying amino acids and leader and trailing sequences. RNA molecules found in the nucleus have a large range of sizes; a portion of this so-called heterogeneous nuclear RNA is that which matures by the processes described above into functional messenger RNA. Also present in the nucleus are the rRNA and tRNA molecules which will also migrate to the cytoplasm. Some RNA molecules, however, never leave the nucleus and their function is unknown.

Gene Clusters and Pseudogenes

The globin genes in different vertebrate species are all very similar, findings that suggest a common evolutionary origin of the genes and therefore of the species as well. Even within an individual, the similarity of different segments of DNA implies an evolutionary development. We present evidence here that certain ancestral genes can be duplicated within a genome so that they are present twice rather than once (or, for diploids, in four copies rather than two). Over successive generations, the duplicated loci can evolve independently of each other to produce genes that code for variants of a single polypeptide, or given enough time, perhaps even polypeptides with different functions. Further replications of similar loci can lead to a family of many related genes. The members of such a **gene family** may be clustered on a chromosome, or if translocation break points occur nearby, they may be scattered on different chromosomes. Known gene families include those that code for histones, antibodies, and ribosomal RNA components, but the best understood is the one that codes for the β and β-like polypeptides of hemoglobin.

First we present some background information about human hemoglobin (which colors red blood cells red). In the capillaries of the lungs, one molecule of hemoglobin binds four molecules of oxygen that are later released in the capillaries of other tissues. (The oxygen delivered to all body cells is essential for cell respiration, the process by which the chemical energy of glucose is converted to the high-energy compound ATP [see Appendix 3].) Each oxygen molecule carried by hemoglobin is associated with an iron atom contained within a nonprotein **heme** group. The four heme groups are pocketed in folds of four polypeptides, the **globin** part of the molecule.

Humans produce several kinds of hemoglobin. In adults, about 98% of the hemoglobin is characterized as $\alpha_2\beta_2$: two of the four polypeptides are identical **alpha (α) chains** with 141 amino acids each, and two are identical **beta (β) chains** with 146 amino acids each (Figure 18). The α and β polypeptides are coded for by the two unlinked genes, *HBA* and *HBB*, whose intron–exon organization is outlined above.

About 2% of hemoglobin in adults consists of two α chains and two chains similar to, but not the same as, the β chains. These are the **delta (δ) chains**, in which just 10 of the 146 amino acids differ from those in β chains. A very small portion of adult hemoglobin (under 1%) has two α chains and two **gamma (γ) chains**, in which 39 of 146 amino acids differ from those of β. There are two different types of γ chains, some with the amino acid glycine in position 136 (called $^G\gamma$ or G-gamma chains) and some with the amino acid alanine in position 136 (called $^A\gamma$ or A-gamma chains). During most of fetal life, hemoglobin with two α chains and two γ chains is the predominant form of hemoglobin. *Early* embryos, however, possess yet other β-like polypeptides, the **epsilon (ϵ) chains**, which differ from β chains in 35 amino acids. (Epsilon chains differ from the other β-like chains by about the same number of amino acids, but not necessarily the same ones.)

To summarize, the non-α chains of hemoglobin are of *five* types, which first appear as follows:

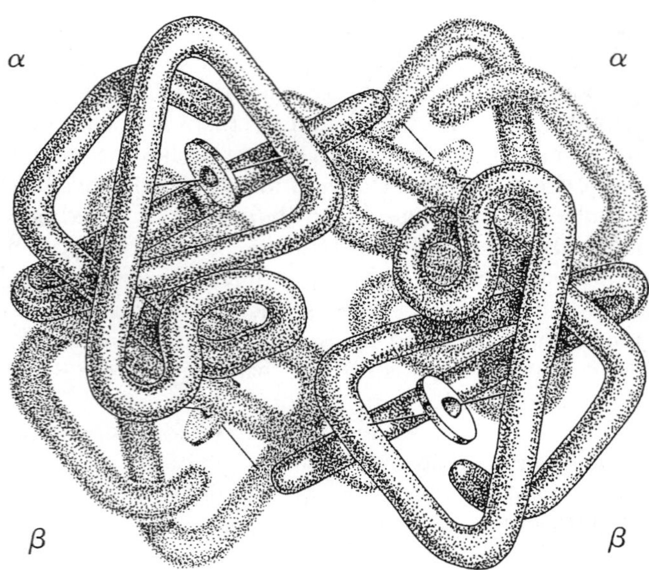

Figure 18 The folding of the four polypeptide chains in normal adult hemoglobin, $\alpha_2\beta_2$. Each chain envelopes a heme group (disk) containing an iron atom (sphere) that reversibly attaches a molecule of oxygen. Each iron atom is held in position by two bonds (lines) to the side groups of nearby histidines in the polypeptides. (After Dickerson and Geis 1969.)

Early embryo	Fetus	Newborns
	gamma with gly ($^G\gamma$)	delta (δ)
epsilon (ϵ)	or	or
	gamma with ala ($^A\gamma$)	beta (β)

Each of these polypeptides is coded for by a different gene, and the five genes form a cluster on the short arm of chromosome 11. Each has three exons and two introns similar to those of the β gene depicted in Figure 17. Spanning a total distance of about 45,000 bases (45 kilobases, or 45 kb), the five genes, at 1.6 kb each including introns, only occupy about 15% of the DNA sequence (Figure 19). What does the rest of the DNA in this region do? The question is largely unanswered. A tiny fraction is untranscribed promoter regions of the β and β-like genes, but no other genes occupy this space. The intergenic DNA, however, has many intriguing aspects, including the following two types of sequences.

One is a gene-sized length of DNA called **psi beta** (ψβ), which in structure and nucleotide sequence is similar to the β-like genes. Because it has no polypeptide product, however, it is called a **pseudogene**. Pseudogenes represent evolutionary dead ends, having accumulated one or more critical mutations that inactivate the gene: usually it cannot be transcribed, or if transcribed, its mRNA cannot be translated (Little 1982). The human β pseudogene, for example, has small insertions that muddle the reading of the triplet code and the intron–exon junctions. Globin-like pseudogenes have been found elsewhere; two are present in a clustered family of α-globin genes. It is interesting that in the several mammals (human, lemur, goat, rabbit, and mouse) whose β-globin gene families have been analyzed, a pseudogene is positioned between the genes that code for fetal and adult polypeptides. Also, in each case the order of the genes is the same as the order of their appearance in development. These observations, among others, point to a common evolutionary origin.

A second curious aspect of the intergenic DNA in the β-globin cluster are four very similar sequences of about 310 bases at the positions indicated by the vertical arrows in Figure 19. These repetitious sequences are called **Alu repeats**, because they are identified by using a restriction enzyme named *Alu*I (see Chapter 15). Other Alu

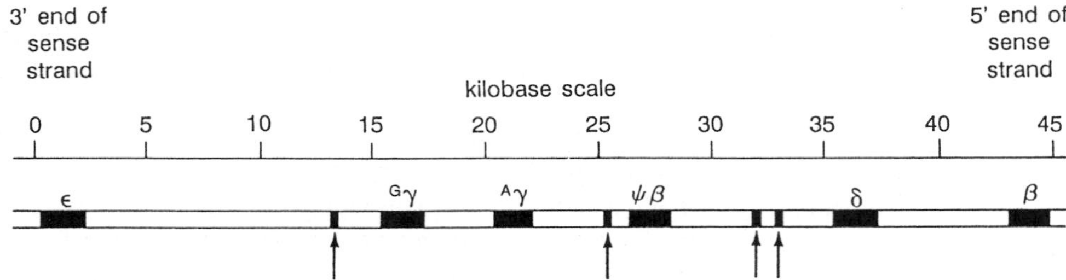

Figure 19 The human β-globin gene cluster on the short arm of chromosome 11, drawn roughly to scale. The cluster spans about 45 kb, a segment including less than one-thousandth of the DNA of the chromosome. The cluster consists of five active genes and one *pseudogene* (ψβ), each about 1.6 kb long and each including two introns and three exons. The arrows indicate similar runs of about 310 bases, called *Alu sequences*. The entire cluster plus surrounding DNA (about 50 kb total) has been sequenced. The data that form the basis for this figure are summarized and analyzed by Collins and Weissman (1984). See text for further discussion.

repeats occur nearby, upstream from the ε gene and downstream from the β gene. In fact, Alu sequences are a striking feature of our genome, and will be discussed further in the next section.

Repetitive DNA

Geneticists have traditionally studied genes that are responsible for the production of enzymes and other protein molecules. These genes are called the **structural genes**, and each is present as a pair of alleles in diploid cells. Taken collectively, the structural genes are known as **unique sequence**, or **single-copy DNA** because the bases exhibit no particular repeating patterns.

But unique sequence DNA accounts for only part of the total DNA in higher organisms. In the 1960s researchers found that a sizable proportion of the DNA of higher species exists as multiples of relatively short sequences of nucleotides. The function of most, but not all, of this **repetitive DNA** is still a mystery. Most repetitive DNA never gives rise to protein products; indeed, much of it is not even transcribed into RNA. Thus repeated sequences are generally not associated with any detectable phenotypes and cannot be mapped by the kinds of techniques described previously.

Repetitive DNA is classified as **highly repetitive** or **moderately repetitive**. The highly repetitive portion consists of short lengths of DNA, up to a few hundred bases, each of which may be repeated 10^5 to 10^6 times. The moderately repetitive portion consists of gene-sized lengths of DNA that are repeated tens to thousands of times in the genome. These distinctions were discovered by so-called *reassociation* experiments: Double-stranded DNA isolated from cells is put into a blender and broken into lengths of about 1,000 bases each. By heating the solution, the pieces are *denatured*, that is, separated into single strands. When this mixture is slowly cooled under carefully controlled conditions, single-stranded segments that possess sequences complementary to each other may join again to form duplexes. The *speed of reassociation* depends upon how many of the single strands have base sequences complementary to other single strands. Sequences that are present in numerous copies in the genome are relatively likely to collide by chance and stick together. Thus the most rapidly reassociating fraction in the experiment (Figure 20) corresponds to highly repetitive DNA. At the other extreme, among the millions of different 1,000-base segments per haploid genome, a single-copy sequence has only one or at most a few complementary segments. The reassociation is thus very slow for unique sequence DNA. Human DNA appears to be about 60% unique, about 20% highly repetitive, and 20% moderately repetitive (reassociating at intermediate speeds) (Evans 1977).

TYPES OF HIGHLY REPETITIVE DNA. Highly repetitive sequences include the Alu repeats that are *interspersed*

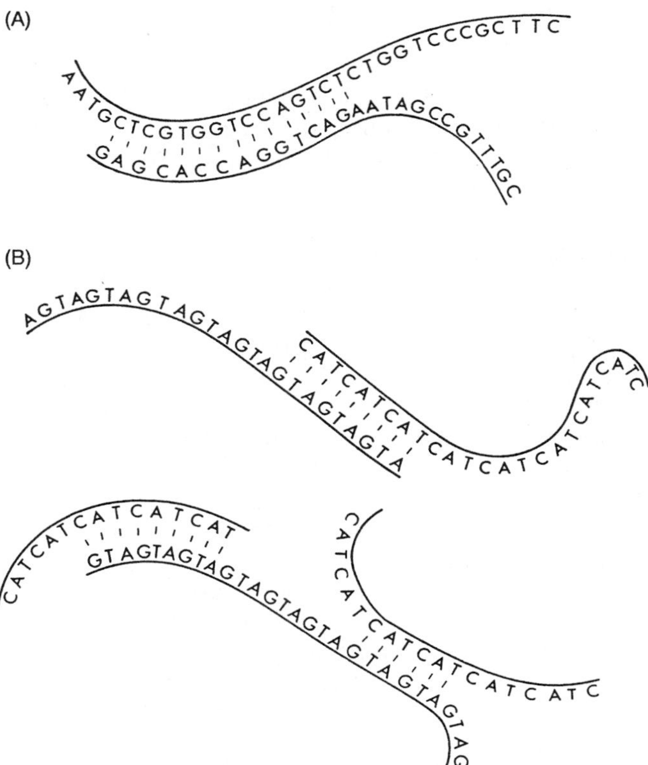

Figure 20 Representation of duplex formation during reassociation experiments. When DNA fragments are heated, the double helices disperse as single strands. When this mixture is slowly cooled, complementary strands may rejoin to form duplexes. (A) Small parts of a unique sequence (structural) gene, which has no repeating groups of nucleotides; pairing must be exactly in register. Because these exist in only two copies per genome, they are slower to "find" each other and reassociate, if they reassociate at all. (B) Short (hypothetical) sequences, $\begin{matrix}\text{CAT,}\\ |||\\ \text{GTA}\end{matrix}$ repeated up to a million times, would be the first to reassociate and need not pair exactly as in the original double helix.

among the structural genes, such as those in the globin region. There may be as many as a half-million Alu copies widely spread throughout the haploid complement of 23 chromosomes. This amount of DNA represents about 5% of human DNA, the rough equivalent of a whole average-size chromosome! (Two Alu repeats have been identified in a 3.4-kb segment of DNA extracted from the skin of the leg on a 2400-year-old Egyptian mummy [Pääbo 1985].) Alu and other families of repeats have been found in intergenic DNA, in introns, and in the untranslated 3′ trailing regions of genes. The origin and function of Alu repeats are matters of speculation (see, for example, Sharp 1983).

Much of the highly repetitive DNA consists of sequences of about a half-dozen bases tandemly repeated many times. Rather than being interspersed among the structural genes, this type of highly repetitive DNA is *concentrated* in the centromere regions of all chromosomes; it also accounts for over half of the DNA in the human Y chromosome.

Some tandemly repeated short sequences are called **satellite DNA**. When double-stranded DNA molecules in a salt solution are centrifuged at high speed for a long time, the DNA pieces can form bands at different levels in the test tube, the levels depending on their densities. The higher the density of the pieces, the further down the tube they will be moved by the centrifugal force. The denser segments of double-stranded DNA have relatively more GC base pairs than AT pairs, a condition causing the helix to be held together more tightly. (Recall that three hydrogen bonds connect a GC pair, whereas only two bonds connect an AT pair.) When human DNA is centrifuged in this way, most of the DNA forms a broad *main band* in a central position; additional distinct bands, the *satellite bands*, form above and below the main one. Four satellite bands make up a total of about 6% of human DNA. Each band represents one short sequence tandemly repeated over and over and richer than average in AT or in GC pairs. Satellite DNA, which is present in centromeric heterochromatin, might be involved in chromosome pairing and segregation.

Sometimes considered as an entirely separate class is a fraction of highly repetitive DNA that rejoins instantaneously in reassociation experiments. The segments of this fraction are repeated complementary sequences in reverse order (say, AATTGGG . . . CCCAATT) that fold back on themselves like a hairpin rather than having to form duplexes with other DNA pieces. In the human DNA genome, there are about 100,000 such *foldback* or *snapback* sequences. They appear in the controlling regions of many genes and also in the ends of so-called transposable DNA segments, which move around from site to site in the genome (see next section).

TYPES OF MODERATELY REPETITIVE DNA. Moderately repetitive sequences include some genes with well-known functions. For example, the genes that code for ribosomal RNA molecules are present in multiple copies, a redundancy reflecting the need for large amounts of the product. (About 80–90% of the total mass of RNA in both prokaryotic and eukaryotic cells is rRNA.) In humans, two types of rRNA (the so-called 18S and 28S fractions) are coded by DNA sequences localized in the nucleolar organizing regions in the short arms of all five acrocentric chromosomes (numbers 13, 14, 15, 21, and 22). The genes are tandemly repeated and present in about 280 copies per haploid genome. A third major type of rRNA (5S) is produced by genes that are present in about 2,000 copies per haploid genome, clustered not in the nucleolar organizer regions but near the tip of the long arm of chromosome 1.

The genes that code for the molecules of transfer RNA also appear to be moderately repeated. (Recall that tRNA molecules act as amino acid carriers during protein synthesis.) There are about 40 different types of tRNA, each produced by repeated sequences of DNA. Scattered throughout the genome are a total of 1,300 copies of one or another of the 40 tRNA genes.

Other sequences that have repetitive elements are the gene families that code for histones and for antibodies. The histone genes in mammals, however, are not nearly so repetitive as they are in lower organisms. The extraordinary nature of the antibody genes is discussed in Chapter 18. These and the RNA genes account for only a portion of the moderately repetitive DNA.

We end this section with an evolutionary problem that exists for all genes that are present in multiple copies: How are the copies kept free of deleterious mutations? For genes in single copies, deleterious mutations can be eliminated by the failure of their carriers to reproduce. But if there are 100 copies of the same gene, the effects of one or several mutations will be swamped by the good copies. What then prevents the accumulation of many mutations that, in time, would inevitably cause weakening of the function of the gene product? Analysis of repeated genes show that they are usually exactly the same. Although several mechanisms have been postulated that would "homogenize" them, the actual mechanism is not known (Watson et al. 1987).

Transposable Elements

Similar genetic maps in related species suggest that the organization of DNA is *stable* over long periods of time. Yet there are now known to be genetic elements that actually move around within the genome. This phenomenon was first demonstrated in maize by Barbara McClintock of Cold Spring Harbor Laboratory, New York, in the 1940s. She was able to show (by examining color patterns on kernels and abnormal behavior of meiotic chromosomes) that certain genetic elements within the maize genome modified the expression of other genes at adjacent sites (McClintock 1984). From time to time these *controlling elements* seemed to disappear from their old locations and reappear at new locations! These interpretations were considered innovative and unusual at the time, but did not lead to studies on other organisms until the 1960s when somewhat similar DNA segments called **transposable elements** or **transposons** were discovered in bacteria. Various types of "jumping genes" have since been investigated in many species including yeast, fruit flies, and mice. Transposition seems to happen, not just within nuclear DNA, but also between the genes of nuclei, mitochondria, and (in plants) chloroplasts. A transposable element that becomes inserted into the middle of a gene may represent a kind of mutation. For the most part, however, the biological significance of transposition is unclear.

Typically, a piece of DNA that can move from one position to another consists of one or several genes in the middle flanked by end sequences that are the same as each other. Within an end segment, there may be shorter repetitive sequences (the whole segment being present at the other end, too). In any event, it is believed that these *terminal repeats* help in the insertion and excision of the element. The Alu repeats, discussed above, have structures like this; although their origin and function are unknown, they may represent transposable elements.

Some animal viruses, called **retroviruses**, are similar to transposable elements. The nucleic acid of a retrovirus is a single strand of *RNA* with terminal repeats. The middle portion encodes only a few proteins, including an enzyme called *reverse transcriptase* and the viral coat that envelops the RNA core. When a retrovirus invades a cell, the reverse transcriptase catalyzes the synthesis of double stranded DNA molecules corresponding to the viral RNA (Latin *retro*, "backward"). One or several of the DNA copies integrate themselves into random positions within the host genome. An integrated copy, called a *provirus*, is then replicated and transcribed along with the host DNA. Transcription and translation of the proviral DNA provide the component parts of new retroviruses, which assemble and escape the cell to infect other cells. The presence and activity of proviruses in the cell's chromosomes may disrupt the metabolic activities of the cell. For example, scientists have isolated and characterized in detail the retrovirus causing acquired immune deficiency syndrome (AIDS) (Gallo and Montagnier 1988). Within their own genome, some retroviruses may also include additional sequences, called *oncogenes*, which can transform the host cell into a cancerous cell (discussed further in Chapter 19).

MITOCHONDRIAL GENES

Typical human cells have hundreds of mitochondria dispersed throughout the cytoplasm. They are most numerous in the metabolically active cells of muscle, heart, kidney, nerve, and secretory tissue. In living preparations, mitochondria are seen to continually alter their shapes —swelling, contracting, branching, splitting, and fusing with other mitochondria. Inside, complex enzyme systems and other specialized proteins are responsible for extracting energy from the breakdown products of the foods we eat. Mitochondria also have a genetic life of their own as first revealed in the 1940s by the French geneticist Boris Ephrussi. His slow-growing strains of yeast ("petites") had defective cell respiration because they inherited mutant mitochondria through the female parent (Chapter 3). These studies opened up a new field of mitochondrial genetics (Ephrussi 1953).

In the 1960s, scientists were able to detect the small amount of DNA in mitochondria (and in the chloroplasts of plants). In humans, each of the hundreds of mitochondria per cell contains about 10 identical circles of DNA, which therefore constitutes a type of middle repetitive DNA. Although some proteins that are used within the mitochondria are encoded by mitochondrial genes and translated on mitochondrial ribosomes, most of the needed proteins are imported from the surrounding cytoplasm, having been made under the direction of nuclear genes. A few of the mitochondrial enzymes are actually "hybrids," having some polypeptide subunits derived from mitochondrial DNA and some from nuclear DNA.

A milestone in molecular biology was the determination by biochemists at Cambridge University of the complete sequence of the 16,569 bases in human mitochondrial DNA (Anderson et al. 1981). Analysis of the arrangement of bases helped show that one circular molecule of mitochondrial DNA encodes 22 transfer RNA molecules, 2 ribosomal RNA molecules, and about 13 proteins. (Polypeptides have not been discovered for some of the potential protein-coding regions.) The tRNA and rRNA molecules are used within the mitochondria during translation of the mRNAs transcribed from the protein-encoding genes. A distinctive type of mitochondrial ribosome resembling those found in bacteria is one line of evidence suggesting that mitochondria actually evolved from primitive free-living bacteria that invaded and took up symbiotic residence in the larger cells (Margulis 1982).

The organization of information in human mitochondrial DNA, unlike that in nuclear DNA, is remarkably compact. No introns appear in any of the three dozen genes, and many of them butt up against one another so that the last base of one gene is followed by the first base of the next. There are even several instances of genes that overlap each other by having the last base of one gene be the first base of the next. Human mitochondrial genes also lack the leader and trailing regions that in nuclear genes are transcribed but not translated.

The two strands of the double helix of mitochondrial DNA can be separated and isolated from each other on the basis of their slightly different densities. The H (heavy) strand is the sense strand for most, but not all, of the genes. For further compactness, each strand has only a single promoter, so each is transcribed into one RNA molecule containing the messages for many genes. The giant transcripts are then snipped at the correct places to yield the tRNAs, rRNAs, and protein-encoding mRNAs. An additional surprise was the discovery that the mitochondrial genetic code—specifying which particular triplets code for which amino acids— is slightly different from that of nuclear DNA. (The genetic code, overlapping genes, and yet another surprising twist to mitochondrial DNA will be discussed in the next chapter.)

The mitochondrial DNA of a variety of species has been examined, but only in mammals is the genetic information so concisely organized. For example, that of yeast has about the same number of genes but five times as much DNA. An even greater amount of mitochondrial DNA in some plant species is distributed into several different circular molecules. Some of the mutations of yeast and plant mitochondrial DNA have considerable basic and practical importance, and a few rare human phenotypes involving abnormal cell respiration may result from mutated mitochondrial DNA.

SUMMARY

1. DNA, the genetic material, consists of two nucleotide chains wound in a helix. In the backbone of each chain, deoxyribose sugars alternate with phosphate groups. Attached to the sugars of both strands are the paired bases: A opposite T and G opposite C.

2. DNA is replicated by separation of the helix, each old strand then becoming a template for the enzymatic synthesis of a new complementary strand.

3. DNA polymerase, the replicating enzyme, proceeds in only one direction along the helix but starts from many sites (in eukaryotes). Because the enzyme works in only one direction, replication is continuous on one strand and discontinuous on the other.

4. DNA is transcribed by the enzyme RNA polymerase into complementary single-stranded copies of RNA, which are modified before entering the cytoplasm.

5. (a) Messenger RNA molecules are the processed transcripts of genes that code for polypeptides; in mRNA molecules, successive groups of three bases are the amino acid coding units, the codons.

   (b) Transfer RNA molecules act like adapters. One end binds a particular amino acid; the other end has a three-base sequence, the anticodon, that pairs with a particular codon on messenger RNA.

   (c) Ribosomal RNA molecules are structural elements of ribosomes.

6. As it feeds through a ribosome, the mRNA is translated into a polypeptide. Thus a protein-coding gene, via the mRNA intermediary and tRNA adapters, determines the sequence of amino acids to be strung together by peptide bonds.

7. The particular sequence of amino acids causes a protein molecule to fold in a unique three-dimensional shape that determines its enzymatic properties or other characteristics. Proteins consist of one or several polypeptide chains with a total of dozens to thousands of amino acids.

8. Eukaryotic genes are often interrupted by noncoding regions called introns. Although the introns are transcribed, they are later cut out during RNA processing.

9. The β-globin cluster of related genes is an example of a gene family that arose from duplication of genetic material. The cluster consists of the β or β-like chains of hemoglobin and a pseudogene that has no polypeptide product.

10. (a) Moderately repetitive DNA sequences are present up to thousands of times per genome. Examples include the genes that encode ribosomal and transfer RNA molecules.

    (b) Highly repetitive DNA sequences may each occur as many as a million times per genome. Some short tandem sequences are concentrated in centromeric regions; some longer sequences, like the Alu repeats, are interspersed throughout the genome.

11. Some segments of DNA with terminal repeats can move around in the genome. The nucleic acid of retroviruses resembles that of such transposable elements.

12. Mitochondria have their own genetic system. Human mitochondrial DNA is extremely compact, with no introns and very little intergenic DNA.

FURTHER READING

For more detailed information on all aspects of gene structure and function, we recommend the books by Strickberger (1985), by Lewin (1987), and by Watson et al. (1987). Alberts et al. (1989) and Darnell et al. (1986) provide broad accounts of molecular and cell biology. Among the best of several histories of the events leading to the Watson-Crick double helix is Judson (1979). See Watson (1968) and Crick (1988) for lively personal accounts, and Sayre (1975) for another view on the role of Franklin in the discovery of the DNA double helix.

Among the wealth of *Scientific American* articles on gene structure and function, we recommend Kornberg (1968) on the test-tube synthesis of DNA, Stent (1972) on "premature" discoveries, Temin (1972) on the transcription of DNA from RNA, Miller (1973) for beautiful micrographs of transcription and translation, Rich and Kim (1978) on transfer RNA, Chambon (1981) on split genes, Dickerson (1983) on Z-DNA, Nomura (1984) on ribosome structure and function, and Steitz (1988) on intron splicing. In addition, the October 1985 issue of *Scientific American* is devoted to the molecules of life and includes articles on DNA by Felsenfeld, RNA by Darnell, and proteins by Doolittle.

QUESTIONS

1. Review the meanings of the boldfaced words in this chapter.

2. The elements C, H, N, O, P, and S are the common constituents of organic matter. Which is present in DNA but not in protein? Which is present in protein but not in DNA?

3. Chargaff's rules (A = T; G = C) do *not* hold for the DNA in the virus ϕX174 and in some other small viruses, nor do comparable rules (A = U; G = C) hold for the RNA of retroviruses and most other molecules of RNA. Explain these exceptions.

4. If the *Escherichia coli* cells in a Meselson-Stahl experiment (Figure 4) were grown for *three* generations in nitrogen-14, what fraction of the DNA double helices would be intermediate in density? What fraction would be light?

5. The DNA of *E. coli* consists primarily of one double helix in the form of a closed loop. Some genes are transcribed in a clockwise fashion, and others, counterclockwise. What information does this give about the sense and antisense strands of DNA?

6. A messenger RNA molecule contains the following base composition: 21% A; 33% U; 28% G; and 18% C. What is the base composition of (a) the sense strand of DNA from which it was transcribed? (b) the antisense strand? (c) both strands of DNA considered together? Assume there are no introns in this gene.

7. A tRNA that binds the amino acid glycine has the anticodon CCC; that binding phenylalanine has AAA. Give the sequences of bases in mRNA and in the sense strand of DNA that code for the dipeptide glycine-phenylalanine. (Ignore the directionality of the molecules.)

8. The digestive enzyme trypsin specifically cleaves a protein molecule on the carboxyl side of lysines and arginines. What peptide pieces result from the trypsin digestion of the B chain (only) of human insulin (Figure 13)?

9. What fraction of the bases transcribed from the β-globin gene actually codes for amino acids in the β polypeptide chain of hemoglobin?

10. The β-globin cluster of genes extends over about 50 kilobases. It apparently includes only five genes that code for polypeptide products (β, δ, ${}^{A}\gamma$, ${}^{G}\gamma$, and ϵ chains of hemoglobin). Each of these chains is 146 amino acids long. What fraction of this segment of DNA actually codes for amino acids in polypeptides?

11. From the following data, calculate the average number of amino acids in the 13 proteins coded for within the 16.6 kilobases of human mitochondrial DNA. Of the total, about 600 bases do not code for either protein molecules or RNA molecules. The two ribosomal RNA molecules coded for in mitochondrial DNA are approximately 950 and 1,560 bases long. The average length of the 22 transfer RNA molecules coded for is close to 70 bases. The remaining bases code for the protein molecules.

ANSWERS

2. P is present in DNA but not in protein; S is present in virtually all proteins but not in DNA. These differences make possible some elegant experiments in which *either* DNA *or* protein is radioactively labeled.

3. Chargaff's rules only apply to double-stranded nucleic acids. The DNA molecules in some viruses and most RNAs are single-stranded (but with some pairing between different sections of the same molecule).

4. 1/4 intermediate, 3/4 light; that is, of the eight double helices, two are intermediate and six are light.

5. The sense strand for some genes is the antisense strand for others. Because the two DNA strands run in opposite directions and new nucleotides are only added to the growing 3′ end, transcription off one strand is opposite to transcription off the other.

6.

	(a)	(b)	(c)
A	33%	21%	27%
T	21%	33%	27%
G	18%	28%	23%
C	28%	18%	23%

7. mRNA: GGGUUU
DNA: CCCAAA

8. Three pieces having amino acids 1–22, 23–29, and 30 only.

9. The β polypeptide has 146 amino acids and so is coded for by (146)(3) = 438 bases. 438/1,606 = 27.3% (see Figure 17).

10. The number of nucleotides required to code for the five polypeptides is (5)(438) = 2,190 bases. 2,190/50,000 = 4.4%.

11.

noncoding bases	600
ribosomal RNA bases (950 + 1560)	2,510
transfer RNA bases (22 × 70)	1,540
total *nonprotein-coding bases*	4,650

Protein-coding bases = 16,600–4,650 = 11,950
Average number of bases per protein = 11,950/13 = 919
Average number of amino acids per protein = 919/3 = 306.

Chapter 14 Genetic Information and Misinformation

The Triplet Code
Deciphering Specific Codons
Properties of the Code

Mutations
Types of Base Substitutions
Spontaneous and Induced Mutations

Hemoglobin Mutations
Sickle-Cell Anemia
Other Amino Acid Substitutions
Thalassemias

Chemical Mutagenesis
Screening Systems
The Ames Test
Sister Chromatid Exchange

Radiation Mutagenesis
Human Radiation Exposures
Evaluation of Radiation Effects

> *Rather than believe that Watson and Crick made the DNA structure, I would rather stress that the structure made Watson and Crick. After all, I was almost totally unknown at the time and Watson was regarded, in most circles, as too bright to be really sound. But what is overlooked in such arguments is the intrinsic beauty of the DNA double helix. It is the molecule which has style.*
>
> *Francis Crick (1974)*

By 1960 molecular biologists had established the importance of base pairing in the reactions of DNA and RNA. But they were pessimistic about quickly cracking the code—that is, discovering *which* specific sequences of nucleotides caused the assembly of *which* specific amino acids. Yet, by 1965 several groups of investigators had ingeniously worked out the complete coding dictionary for nuclear genes. In this chapter, we describe that outstanding scientific accomplishment and examine how errors in DNA bring about changes in polypeptides and in the resultant phenotypes.

THE TRIPLET CODE

Because proteins are assembled from 20 amino acids, a sequence of three nucleotides was considered to be the most reasonable size for a coding unit: *one* nucleotide can only specify $4^1 = 4$ items (Table 1A); *two* nucleotides can specify only $4^2 = 16$ items (Table 1B); *three* nucleotides, however, carry more than enough information, $4^3 = 64$ items (Table 1C), to code for all 20 amino acids. But nature does not always act in ways that are mathematically logical, so a scheme that seems reasonable to us need not be the correct one.

Francis Crick and associates (1961) were primarily responsible for establishing the general features of the genetic code, including the first evidence that a three-nucleotide segment was indeed the correct length for the coding unit. They accomplished this with T4, a virus that parasitizes and kills the bacterium *Escherichia coli*. The virus was treated with a mutagenic agent whose mode of action is to remove or add one nucleotide from the interior of DNA. By crossing different mutant strains of T4, the experimenters were able to make a series of stocks in which one, two, or three nearby nucleotides were either missing or extra.

When one nucleotide was deleted (Figure 1B), all of the triplets beyond the point of the deletion were garbled, thus making the viral gene product nonfunctional. We now know that the deletion causes a shift in the

Table 1 The informational content of one-, two-, and three-letter words drawn from only four letters, the four bases in DNA

(A) The 4 different one-letter words: A, T, G, C

(B) The 16 different two-letter words:

First letter	ADD SECOND LETTER *A*	*T*	*G*	*C*
A	AA	AT	AG	AC
T	TA	TT	TG	TC
G	GA	GT	GG	GC
C	CA	CT	GC	CC

(C) To list the 64 different three-letter words, add to each of the two-letter words an A, T, G, or C:

Two-letter word	ADD THIRD LETTER *A*	*T*	*G*	*C*
AA	AAA	AAT	AAG	AAC
AT	ATA	ATT	ATG	ATC
AG	AGA	AGT	AGC	AGC
AC	ACA	ACT	ACG	ACC
and so on.				

reading frame, that is, each successive group of three in a string of bases. A misaligned reading frame yields all (or nearly all) wrong triplets in DNA, and therefore wrong codons in mRNA and wrong amino acids in the polypeptide product. When a second deletion occurs near the first one (Figure 1C), the virus is no better off, because the reading frame is still out of register. But the third deletion (Figure 1D) restores the proper reading frame: three wrongs make a right! Some of the three-deletion mutant viruses were able to infect and kill *E. coli* cells, an outcome showing that the polypeptide product, although wrong in a few amino acids, was capable of normal or near-normal function. Virus strains with four or five nearby deletions had the same incorrect reading frames as those with one or two deletions. Thus the evidence supported three-base units as the code words for amino acids. Supporting the triplet nature of the code, Crick and his associates also got normal function from viruses that carried *three nearby additions*, but not one or two additions, or any number of additions not a multiple of three. (Furthermore, they were able to show normal function from viruses carrying one addition and one nearby deletion, which together also maintained the proper reading frame.)

The correct reading frame for translation is established in the first place by a special methionine codon in the nucleotide sequence of mRNA. Once the starting point is determined, successive groups of three bases are automatically "counted off" without any chemical

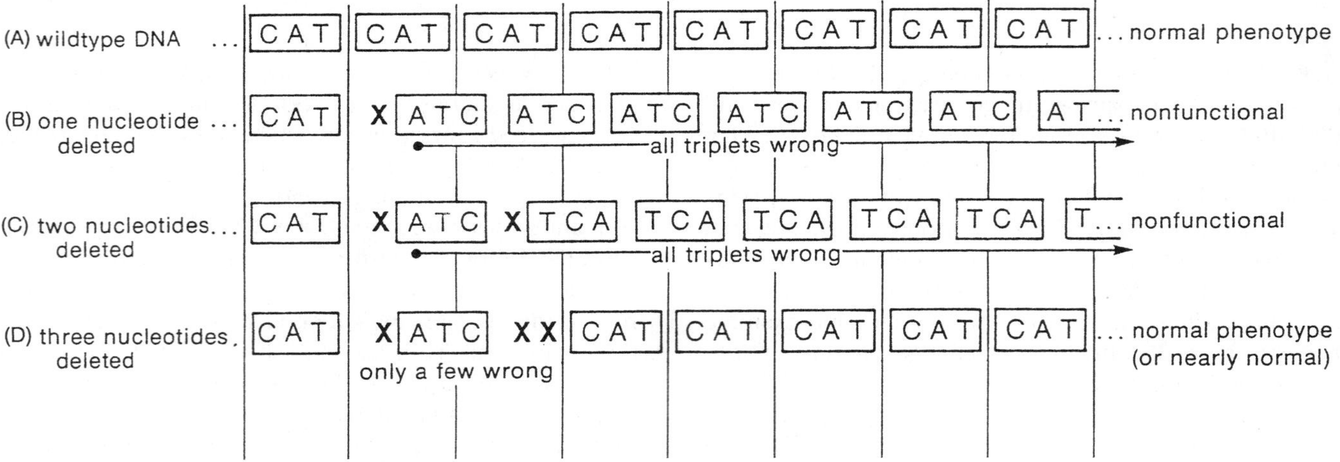

Figure 1 An elegant demonstration that the genetic code is triplet. Crick and his co-workers derived mutant viruses that possessed nucleotide deletions in an essential gene. A mutant strain with one or two deletions was nonfunctional (B, C), whereas one with three nearby deletions was normal or near-normal (D). Deletions (or additions) of nucleotides are called *frameshift mutations* because they put the reading of the code out of register. Three deletions restore the proper register beyond the point of the third deletion. (The specific triplet sequences shown here are hypothetical, chosen simply to emphasize the correct reading frame.)

punctuation marks to provide the informational units for protein synthesis.

Deciphering Specific Codons

Early experiments seeking to determine *which* specific group of three bases coded for *which* amino acid used artificial molecules of RNA added to a test tube with the following ingredients:

- ribosomes extracted from *E. coli*
- all the individual amino acids
- all the transfer RNA molecules
- ATP and other energy-rich compounds
- a batch of appropriate enzymes

Remarkably, these in vitro systems spontaneously synthesized polypeptides. For example, when the artificial RNA was polyuracil, which has a monotonous sequence of one base,

··· UUUUUUUUUUUU ···

the test tube was found to contain a polypeptide containing only phenylalanine,

··· phe−phe−phe−phe ···

The conclusion was clear: UUU in RNA (or, by inference, AAA in DNA) was the genetic word for phenylalanine.

As biochemists developed more sophisticated means for hooking together different ribonucleotides in specific orders, the nucleic acid vocabulary was extended. We give two examples (see also Question 3).

1. When synthetic mRNA with a repeating set of two nucleotides,

··· UGUGUGUGUGUG ···

was added to the protein-synthesizing system, the polypeptide

··· val−cys−val−cys ···

was made. Thus, valine is coded for by UGU and cysteine by GUG, or vice versa. Because the in vitro system lacks a proper signal for initiating the translation process, all three reading frames are utilized, but in this example, all reading frames give the same polypeptide product.

2. The synthetic mRNA

··· UUCUUCUUCUUC ···

(a repeating trinucleotide) made *three different* polypeptides:

··· phe−phe−phe−phe ···
··· ser−ser−ser−ser ···
··· leu−leu−leu−leu ···

The three reading frames for the mRNA lead to repeated codons UUC or UCU or CUU; thus these three codons stand for the three amino acids, although which stands for which is not clear from this experiment by itself. Readers should confirm that a coding unit of two bases or four bases would have generated molecules consisting of three different amino acids in one polypeptide.

On the basis of similar experiments, and especially of investigations using RNA molecules only three nucleotides long, researchers gave all 64 possible codons in mRNA the unambiguous assignments seen in Table 2. Crick (1966a, 1988) provides engaging accounts of this exciting period of biochemical sleuthing.

Properties of the Code

1. *Several codons often stand for the same amino acid.* This so-called **code degeneracy** might be expected when 64 codons are available to specify just 20 amino acids. The most striking examples are the six codons for leucine, the six for serine, and the six for arginine (Table 2). Fourfold degeneracy is seen for several other amino acids (for example, valine); here the third base of the codons can be ignored, effectively reducing the genetic dictionary to two-letter words. Twofold degeneracy is seen for many others (for example, phenylalanine); here the third base is either of the two pyrimidines (U or C) or either of the two purines (A or G). Thus code degeneracy usually involves the third base. The number of synonymous codons for an amino acid (1, 2, 3, 4, or 6) is roughly proportional to the actual frequency of that amino acid in proteins (see Question 6). For example, tryptophan, the least frequent amino acid, has a single codon; serine, the most frequent, has six codons.

A related phenomenon, called **wobble**, is the looseness of base pairing in the third position during translation; that is, pairing between the third base of a mRNA codon and the corresponding base of a tRNA anticodon does not always follow the rigid rules of base complementarity. For example, a serine-tRNA has the anticodon AGI, where I represents *inosine*, a purine somewhat similar to guanine. This single tRNA molecule is able to fit into the ribosome pocket whenever the mRNA codon, UCU or UCC or UCA, is present. Thus, during translation, I in the third *anticodon* position can pair with either U or C or A in the third *codon* position. In other cases as well, one tRNA molecule can recognize several synonymous mRNA codons.

Although the code is degenerate and the third base is subject to wobble, the genetic language is not imprecise: a specific DNA triplet, through its mRNA codon, will direct the assembly of a specific amino acid.

Table 2 The genetic code displayed as DNA triplets and mRNA codons

First base[a]	TRIPLETS[b] WHEN SECOND BASE IS							
	A *(U)*		***G*** *(C)*		***T*** *(A)*		***C*** *(G)*	
A (U)	**AAA** (UUU)	phe	**AGA** (UCU)	ser	**ATA** (UAU)	tyr	**ACA** (UGU)	cys
	AAG (UUC)		**AGG** (UCC)		**ATG** (UAC)		**ACG** (UGC)	
	AAT (UUA)	leu	**AGT** (UCA)		**ATT** (UAA)	stop	**ACT** (UGA)	stop
	AAC (UUG)		**AGC** (UCG)		**ATC** (UAG)		**ACC** (UGG)	trp
G (C)	**GAA** (CUU)	leu	**GGA** (CCU)	pro	**GTA** (CAU)	his	**GCA** (CGU)	arg
	GAG (CUC)		**GGG** (CCC)		**GTG** (CAC)		**GCG** (CGC)	
	GAT (CUA)		**GGT** (CCA)		**GTT** (CAA)	gln	**GCT** (CGA)	
	GAC (CUG		**GGC** (CCG)		**GTC** (CAG)		**GCC** (CGG)	
T (A)	**TAA** (AUU)	ile	**TGA** (ACU	thr	**TTA** (AAU)	asn	**TCA** (AGU)	ser
	TAG (AUC)		**TGG** (ACC)		**TTG** (AAC)		**TCG** (AGC)	
	TAT (AUA)		**TGT** (ACA)		**TTT** (AAA)	lys	**TCT** (AGA)	arg
	TAC (AUG	met	**TGC** (ACG)		**TTC** (AAG)		**TCC** (AGC)	
C (G)	**CAA** (GUU)	val	**CGA** (GCU)	ala	**CTA** (GAU)	asp	**CCA** (GGU)	gly
	CAG (GUC)		**CGG** (GCC)		**CTG** (GAC)		**CCG** (GGC)	
	CAT (GUA)		**CGT** (GCA)		**CTT** (GAA)	glu	**CCT** (GGA)	
	CAC (GUG)		**CGC** (GCG)		**CTC** (GAG)		**CCC** (GGG)	

a. By convention, the first base is the one on the 3′ side of the DNA triplet and on the 5′ side of the mRNA codon.

b. The DNA triplets are boldfaced and the complementary mRNA codons are in parentheses. This code is found in *nuclear* genes and their transcripts. Of the 64 mRNA codons, 61 specify amino acids and 3 (UAA, UAG, and UGA) cause the termination of translation (stop). In human *mitochondria*, four of the mRNA codons have different meanings:

UGA specifies trp instead of stop
AGA specifies stop instead of arg
AGG specifies stop instead of arg
AUA specifies met instead of ile

2. *Amino acids with similar chemical properties have related codons.* The amino acids are not distributed randomly around the code table. For example, all the amino acids in the first column (having U as the middle base of their codons) have uncharged and nonpolar side groups, which repel water molecules. These amino acids are usually found in the interior folds of a globular protein rather than on the outside in contact with the aqueous surroundings. (The only other uncharged and nonpolar amino acids are proline, alanine, and tryptophan.) The amino acids with a charged side group also tend to be clustered: the two with negatively charged R groups (aspartic and glutamic acid) have codons beginning GA, and the two positively charged amino acids (lysine and arginine) are nearby. It is not clear how this pattern evolved, but one result is to minimize the consequences of translational mistakes or of mutations that change one base: a protein is more likely to retain its essential function if a wrong amino acid is chemically similar to the right one.

3. *Three of the 64 codons do not specify any amino acid.* While deciphering the code, investigators found that the codons UAA, UAG, and UGA failed to incorporate any amino acid. These sequences are sometimes called *nonsense codons* because they do not code for amino acids; the term **chain termination codons**, however, is more accurate, because UAA, UAG, and UGA stop the translation process, releasing the now-completed polypeptide from the ribosome. The evidence for this action came from mutant strains of microorganisms in which the synthesis of a polypeptide stopped prematurely. This could happen, for example, if the third base in a tyrosine mRNA codon was changed to A or G (refer to Table 2).

4. *The code is nearly universal.* Although the code was first deciphered using synthetic RNA molecules and translation components from the bacterium *E. coli*, further tests showed that almost all organisms use the same array of codons. Biochemists can obtain specific genes or messenger RNA molecules in large enough quantities to determine the ordering of all bases; matching the codons with the amino acids in the resultant proteins confirms that the genes of T4, *E. coli*, yeast, maize, flies, humans, and most other organisms rely on the same genetic blueprint.

Furthermore, protein synthesis can run smoothly even with mixed components. Viral proteins, for example, are normally translated on the ribosomes of the host species. Toad oocytes that are injected with the messenger RNA for rabbit globin chains synthesize rabbit hemoglobin (Lane 1976). If injected instead with bee venom mRNA, the toad eggs synthesize bee venom. As described in the next chapter, *E. coli* cells can transcribe and translate human genes that are incorporated into the bacterial DNA. In this last case, however, the introns of the human genes are removed before incorporating them into the *E. coli* DNA, because prokaryotes have neither introns nor the enzymes needed to remove them from mRNA molecules after transcription.

Scientists have found several exceptions to the assignments in Table 2. For example, in all mitochondrial DNAs that have been studied, the mRNA codon UGA encodes tryptophan rather than chain termination. Specific mitochondrial DNAs have additional variations, those of human DNA being indicated in Table 2. Recently, it has also been discovered that the nuclear genes of some ciliated protozoa (*Paramecium*, for example) utilize UAA and UAG to encode the amino acid glutamine rather than chain termination (Fox 1985). A few other exceptions are known and others may be discovered.

The near universality of the code suggests that it evolved just once, through ancient life forms perhaps 3 billion years ago. Possibly the code began as a sequence of two nucleotides, because some proteins in very primitive bacterial species have fewer than 16 amino acids. It is a matter of speculation how each of the 20 amino acids in the current coding dictionary came to be specified by a particular nucleotide triplet.

MUTATIONS

A mutation is a change in a gene. Almost all changes adversely affect the way a gene functions, but harm to the organism is not inevitable. A variety of cellular processes act to repair the damage to DNA before it is transmitted to daughter cells during mitosis or meiosis. The change is often a **base substitution mutation**, a replacement of one base by another; a second major type is a **frameshift mutation**, a reading frame error due to the deletion or addition of a base. Both types are called *point mutations* because they are not visible when viewing chromosomes under a microscope. (More broadly, mutation is sometimes taken to include microscopically visible chromosomal aberrations of the types discussed in Chapters 7-9: deletions, duplications, inversions, translocations, aneuploidy and polyploidy. In this chapter, however, our discussion is restricted to point mutations.)

Usually the term *mutation* refers to the process of genic change and to the altered form of the gene. The mutation may produce in an individual a *mutant phenotype* of greater or lesser severity. Mutant alleles often

cause harm, but not always in straightforward ways (Chapter 12). Indeed, the occurrence of a mutant phenotype may be far removed in time from the mutational event itself, especially if the mutation behaves in development as a recessive, as most do. In these cases, it must be combined in a zygote with another allelic recessive mutation in order to be revealed to an observer.

A mutation may occur in any cell at any time, although many mutations occur during the replication of DNA. A **germinal mutation** occurs in a cell that is destined to become an egg or sperm. A **somatic mutation** occurs in a body cell (liver, lung, intestine, and so on) not ancestral to gametes. A somatic mutation can affect the phenotype of its carrier, but it will not be transmitted to offspring. A germinal mutation, on the other hand, does not affect in any obvious way the person in whom it occurs, but it can be transmitted to future generations. Note that if a mutation occurs very early in embryonic development, before the primordial germ cells are determined, it could be transmitted to both somatic and germinal tissues of an individual. Somatic mutations are believed to be involved in one of the early steps that lead to the development of cancer (Chapter 19). It is with germinal mutations, however, that we are primarily concerned in this chapter.

Types of Base Substitutions

As a consequence of a mutation in a gene coding for a polypeptide, one or more amino acids in the protein product may be changed. The classic example of a human mutation is the base substitution that leads to sickle-cell hemoglobin instead of normal hemoglobin. In this mutation, the middle base of a DNA triplet coding for *glutamic acid* is altered so that the triplet now codes for *valine* instead (Figure 2). It is unclear how many times this particular mutation has occurred in human populations, but mutations of a gene usually recur at a low frequency (Chapter 22).

The effect of a mutated gene on the phenotype of an organism ranges from trivial to lethal. The most innocuous mutation would change a DNA triplet to a synonym for the same amino acid. For example, a DNA base substitution that changes AAA to AAG would still lead to the incorporation of phenylalanine; such a mutation would probably be undetectable except by sequence analysis of DNA or mRNA molecules. Because one base can change to any of the other three bases, a DNA triplet can mutate in nine different ways (Figure 3).

If a base substitution does cause an amino acid change, the consequences depend on how the new side chain alters the properties of the protein and how that protein affects development and maintenance of the or-

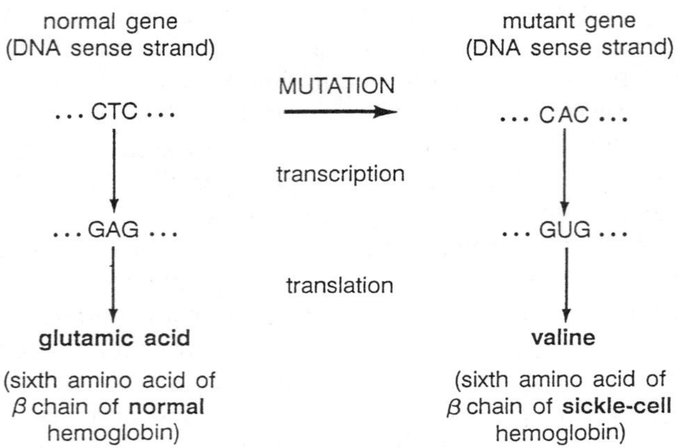

Figure 2 The sickle-cell mutation. The normal gene for the β polypeptide chain of hemoglobin incorporates glutamic acid as the sixth amino acid from the free amino end. The mutation shown here is a single base substitution in one of the two possible codons for glutamic acid. Readers should verify the coding by reference to Table 2.

ganism. Many mutations in humans produce unimportant phenotypic effects: examples include clockwise versus counterclockwise cowlick patterns on top of the head, hairy elbows, soft versus brittle ear wax, and a small pit in the margin of the ear (McKusick 1988) (Figure 4). Arrayed against these inconsequential mutant phenotypes are genes that cause death during infancy (Tay-Sachs disease), young adulthood (Duchenne muscular dystrophy, cystic fibrosis) or later (Huntington's disease). In between are genetic conditions that are inconvenient or burdensome to lesser or greater degrees (brachydactyly, pattern baldness, ichthyosis, hemophilia, sickle-cell anemia, albinism, dwarfism, diabetes, and so on).

Mutations to and from the three DNA triplets that transcribe into stop codons make a polypeptide that is too short or too long. In one such case, an Italian patient with *thalassemia* (see later) failed to produce any functional β chains of hemoglobin. Analysis of his DNA revealed that the codon for amino acid 39 had mutated from GTC, coding for glutamine, to ATC, coding for chain termination (Orkin and Goff 1981). Thus β-globin synthesis stopped after only 38 amino acids had been translated (rather than the normal 146). The abbreviated polypeptide was completely functionless. Conversely, mutations *from* a stop codon *to* an amino acid codon are also known; an α-globin mutation of this type is described later in this chapter.

Base substitutions in the genes that code for transfer RNA molecules can have a variety of interesting effects (Lewin 1987; Watson et al. 1987). For example, if a mutation changes an anticodon base of a tRNA gene, but *not* the amino acid binding property of the tRNA, then a wrong amino acid can be incorporated into *every* polypeptide made by the organism. This occurs because the amino acid carried by the tRNA arrives at the ribosome pocket under false pretenses—the mutated anticodon of the tRNA base pairs with the wrong mRNA codon.

Spontaneous and Induced Mutations

Mutations that occur haphazardly, irrespective of known environmental conditions, are called **spontaneous mu-**

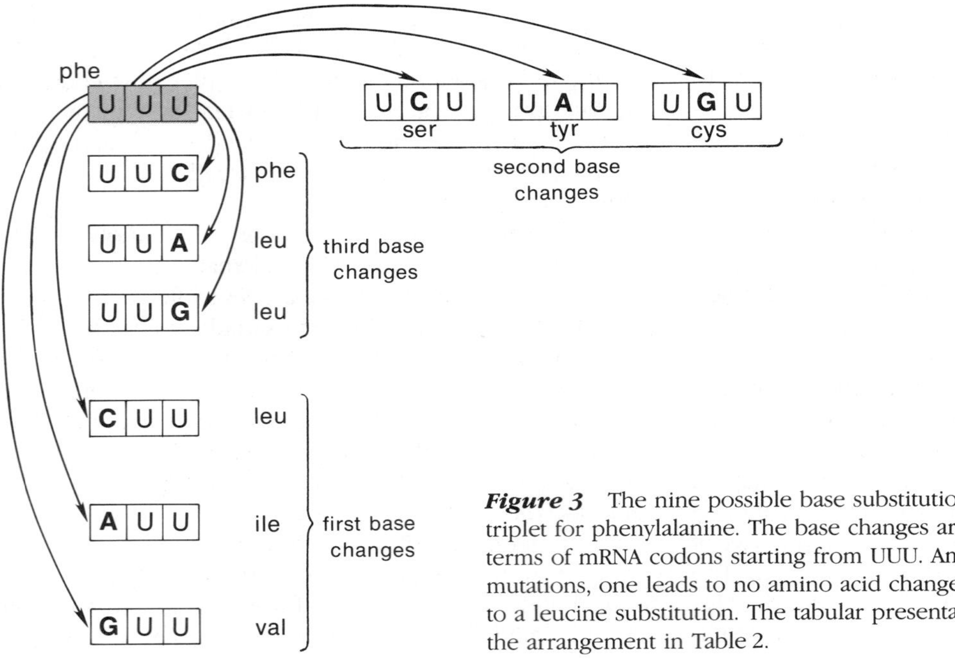

Figure 3 The nine possible base substitutions in the AAA triplet for phenylalanine. The base changes are shown in terms of mRNA codons starting from UUU. Among the nine mutations, one leads to no amino acid change and three lead to a leucine substitution. The tabular presentation parallels the arrangement in Table 2.

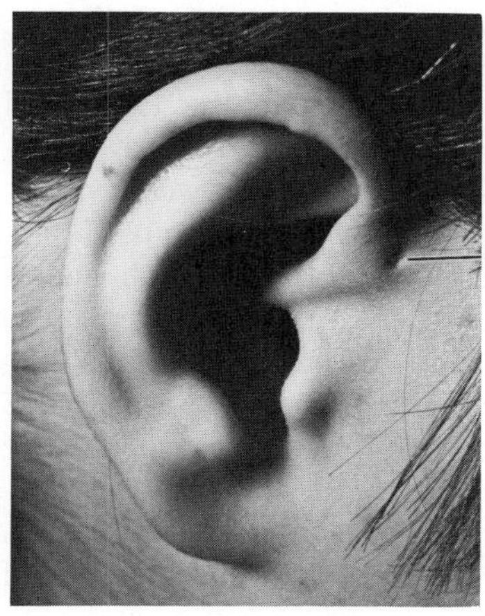

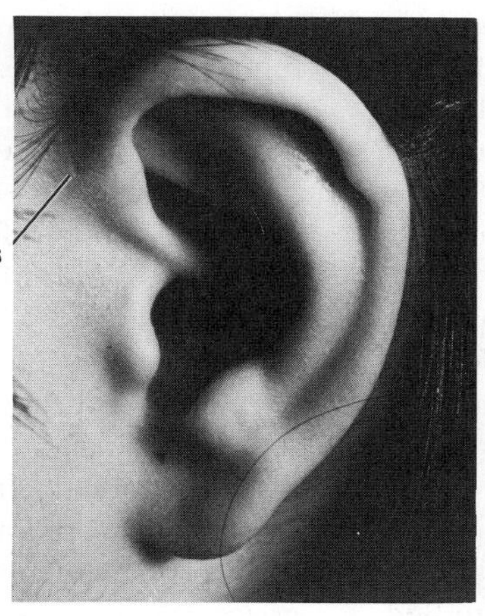

Figure 4 An inconsequential genetic phenotype, ear pits (arrows), appearing in one of the authors' students. This person, who has normal hearing, reported similar pits in many relatives (some of whom had hearing problems). For references on this condition, see McKusick (1988).

tations. To a large extent, they result from random thermal motions of atoms and molecules in and near DNA, often at the time of DNA replication, but at other times as well. This motion, due to the kinetic energy of molecules, is characteristic of all matter. In addition to this thermal effect, it is likely that a small proportion of "spontaneous" mutations are due to foreign chemicals or radiation that get inside cells and near DNA. Because we cannot observe the mutational events, it is not possible to pinpoint the precise cause (heat, chemicals, radiation) of any spontaneous mutation, especially since the observable phenotypic effects may not be evident for a long time—many generations in some instances.

Induced mutations refer to gene changes in an organism following exposure to an agent, a **mutagen**, known to produce mutations above the rate at which they occur spontaneously. Mutagens of concern to us include some industrial chemicals, pesticides, various substances in waste landfills, food additives, cigarette smoke components, drugs, medical and dental X-rays, and radiation from past atomic weapons tests, atomic warfare, and the nuclear power industry—a list that reflects our technological society and modern life styles. On the other hand, some largely inescapable *natural* products are known to be mutagens. These include radioactivity in soil, in rocks, and within our bodies. Many common raw or cooked foods also contain substantial amounts of mutagenic chemicals; some of these are manufactured by plants as natural pesticides, toxic to the species that would otherwise eat them (Ames 1983, 1984) (Figure 5).

HEMOGLOBIN MUTATIONS

Before researchers discovered the structure and function of DNA, the American Nobel chemist Linus Pauling had shown that hemoglobin from persons with sickle-cell anemia moved in an electric field at a rate different from that of normal hemoglobin (Pauling et al. 1949). Affected persons suffered from what was dubbed a **molecular disease**, one brought about by an altered protein encoded by a mutated allele. (With only a little evidence, Garrod had proposed this scheme in 1909.) The precise change in the protein was later discovered by Vernon

Figure 5 Although risk taking is part of everyday activities, risk–benefit analyses are not always easy to do. (Drawing by S. Harris; copyright © 1979, *The New Yorker Magazine*, Inc.)

Ingram (1957) who was working at Cambridge University in the same research unit as Watson and Crick. Skillfully splitting, isolating, and analyzing the parts of the hemoglobin molecule, Ingram provided the first direct evidence that a mutant allele leads to one amino acid change in the resultant polypeptide. Continuing research on the hemoglobin molecule and the genes that encode it has revealed over 350 variations in the α or β polypeptides. Because this work has helped to unravel the molecular biology of all organisms, we use it as an example of mutation research and some clinical ramifications.

Sickle-Cell Anemia

Because of its medical importance, the most studied mutation of the hemoglobin genes is the one that leads to **sickle-cell anemia** or **sickle-cell disease**. Affected persons are homozygous (Hb^S/Hb^S, or more simply *S/S*) for the sickling allele; heterozygotes (*A/S*, *A* being the normal allele) are said to have the sickle-cell *trait*; they are essentially normal and rarely come to medical attention.

In the sickle-cell mutation, valine replaces glutamic acid at the sixth position of the β chain (Figure 2). In homozygotes, this substitution slightly alters the shape of hemoglobin. In venous blood, the altered molecules tend to stack into narrow crystals, distorting the smooth contour of red blood cells (see Figure 2 in Chapter 10). Upon reoxygenation in the lungs, mildly sickled cells will resume their normal doughnut-like shapes. The irregularities of sickled cells, however, impede blood flow, which lengthens the time of return to the lungs and increases the degree of sickling (Figure 6). Irreversibly distorted cells have a life span of only a few weeks rather than the normal life span of about four months. In addition, the sickled cells become trapped in small vessels throughout the body, leading to local oxygen depletion, tissue damage, infections, and painful crises. Thus, one wrong base in the DNA can cause death from the accumulated damage to several vital organs.

Despite much clinical research, there is no good treatment for the pleiotropic complications of the disease. Pain can be relieved, and the frequency and duration of crises can be lessened by rest, transfusions, and antibiotics. Some drugs reduce sickling to various degrees, but they have major side effects and remain experimental (Orkin 1983). As of this writing, the basic defect remains largely resistant to the concerted efforts of medical researchers. Prenatal diagnosis of sickle-cell anemia has recently become possible, however, so heterozygous couples have the option of aborting affected fetuses (Chapters 15 and 25).

Sickle-cell anemia usually occurs in African Blacks and in their descendants. Among newborn Blacks in the United States, the frequency of the disease is about 1 in

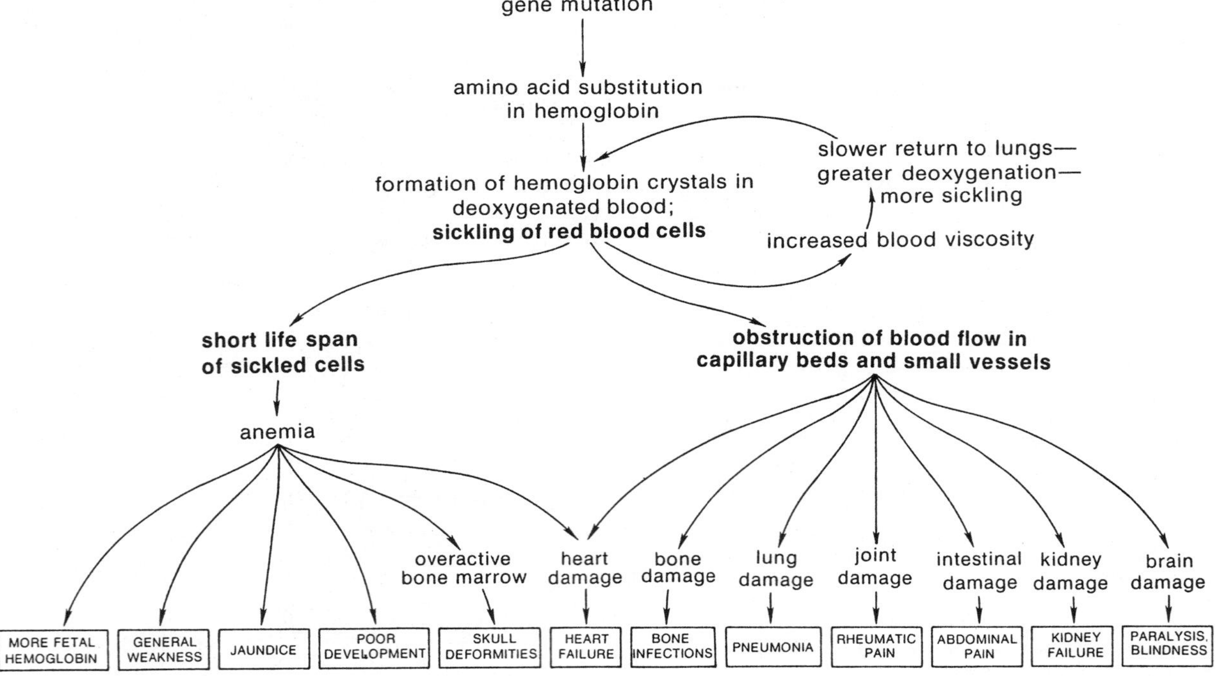

Figure 6 The pedigree of effects in sickle-cell anemia (homozygous *S/S*). The symptoms of the disease are quite variable, and not all patients experience all symptoms with equal severity. Although many sufferers die as children, survival to adulthood is not uncommon. (Modified from Neel and Schull 1954.)

600, considerably higher than that of any other inherited disorder in the American population or in large subpopulations. The reasons for the high frequency of sickle-cell anemia are detailed in Chapter 22. In essence, persons who are heterozygous, *A/S*, are resistant to malaria—a resistance *not* possessed by normal homozygotes, *A/A*. As a consequence, in regions of the world where malaria is still a major health problem, the sickle-cell allele is maintained by selection at relatively high frequencies. In nonmalarial regions, like the United States, heterozygous carriers enjoy no such selective advantage over normal homozygotes, and therefore the frequency of the sickle-cell genotypes is decreasing slowly over successive generations.

Other Amino Acid Substitutions

Like sickle-cell hemoglobin, the multitude of other globin chain variations (which continue to be discovered) are usually single amino acid changes. For the α polypeptide, McKusick (1988) lists 154 different variants, all but 11 of them being single amino acid changes at one or another of the 141 positions making up the chain. An interesting set of variants occurs at position 141: the normal arginine (arg) has changed to a different amino acid in each of six different variants—leu, pro, his, cys, ser, and gly. (Readers should note the position of these variants on the code table, Table 2, and deduce that the DNA triplet encoding amino acid 141 is either GCA or GCG.) All of the 143 simple missense mutations in the α chain can be accounted for in the code table by simple horizontal or vertical jumps of the types noted in Figure 3. A jump along a diagonal line might be expected if a mutation were composed of two nucleotide changes, or if a different code table applied. Thus the observed mutations in the α hemoglobin gene (and many others) show that a typical mutation is a base substitution in DNA and that the code table (derived largely from work with microorganisms) clearly applies to human beings.

The tabulated amino acid changes in the globin chains are not random: a large proportion of them involve a substitution of a charged for an uncharged amino acid, or vice versa. This experimental bias occurs because most of the mutant forms were discovered by electrophoretic screening of hemoglobin from normal individuals. Because electrophoresis separates proteins largely on the basis of net charge, mutations that substitute a differently charged amino acid are most easily detected. Using the codon table, one can show that only about one-fourth of randomly occurring base substitutions in DNA lead to an amino acid substitution with a charge difference. Thus, although more mutant forms are known for hemoglobin than for any other mammalian protein, most amino acid substitutions in human hemoglobin are overlooked by electrophoretic technology.

Most amino acid substitutions in human hemoglobin are very rare—often discovered in one or a few families—and not known to cause serious clinical abnormalities. Because of the rarity, however, they have been found only in heterozygotes, and the consequences of homozygosity are unknown. Some exceptions are sickle-cell hemoglobin and a few additional hemoglobins (called C, D, and E) that also cause sickling of red blood cells. People homozygous for C, D, or E show only slight sickling with very mild anemia, and lead fairly normal lives. Because the C, D, and E hemoglobins are relatively common in some malarial regions of the world, it is thought that the heterozygotes may also enjoy some resistance to the infectious disease.

Rather than sickling, some amino acid substitutions cause instability of the hemoglobin molecule, a property that also reduces the life span of the red blood cells. In other variants, the oxygen attaches too firmly to the iron atoms, so less oxygen is given up in the tissue capillaries. A number of mutations cause substitutions for the histidines that help orient the heme groups within their peptide pockets. As a consequence, the valence of the iron is kept in the ferric form (Fe^{3+}) rather than in the normal ferrous form (Fe^{2+}) and no oxygen is bound at all. Among light-skinned peoples, homozygotes for one of these so-called M hemoglobins have cyanotic (blue-gray) skin and a variable, although usually mild, anemia.

A few hemoglobin variants are known in which *two* amino acid substitutions occur in the same polypeptide. The one called *Arlington Park* (named, as almost all are, after the place of discovery*) has two substitutions in the β chain:

> position 6: glutamic acid is replaced by lysine
> position 95: lysine is replaced by glutamic acid

Each of the individual substitutions is known in its own right: the first is the common variant *hemoglobin C*, and the second is known as *hemoglobin N (Baltimore)*. The double mutant could have arisen through two mutations, or perhaps more likely, by crossing over within the β-globin gene in a person heterozygous for the two single mutations (see Question 13).

Thalassemias

In the disorders mentioned above, hemoglobin may function abnormally, but it is usually formed in roughly normal amounts. In the **thalassemias**, however, normal globin polypeptides may be synthesized in much reduced amounts, or not at all—a seriously debilitating or

*Readers might care to test their geographical expertise by identifying the locales involved in these variant hemoglobin names: Akita, Århus, Camperdown, Guantánamo, Ibadan, Icaria, Kankakee, Khartoum, Kuala Lumpur, Lufkin, Pôrto Alegre, Saskatoon, and Yakima. For more, see McKusick (1988).

lethal condition common in some Mediterranean peoples and in their worldwide descendants. (*Thalass-emia*, from Greek, means "sea-blood.") Like sickle-cell anemia, thalassemia occurs in unusually high frequencies, perhaps because of an interaction with malaria. It cannot be treated very satisfactorily in affected children, but it can be diagnosed prenatally.

Different forms are designated by the globin chain that is deficient. The **β-thalassemias** constitute one of the most serious health problems worldwide, accounting for hundreds of thousands of childhood deaths per year (Orkin and Kazazian 1984). In all its various forms, the synthesis of β chains is very much reduced. Heterozygotes have *β-thalassemia minor* and usually only a mild anemia, but homozygotes—afflicted with *β-thalassemia major*, originally called *Cooley's anemia**—often die before age 10. Individuals with β-thalassemia major partly compensate for their deficiency of β chains by producing more fetal hemoglobin ($\alpha_2\gamma_2$), and more of the minor adult component ($\alpha_2\delta_2$). But the compensation is generally incomplete, and a severe anemia results. Furthermore, the excess α chains, lacking β partners, precipitate inside red blood cells and shorten their life spans. β-thalassemia is found primarily in a belt surrounding the Mediterranean and extending eastward through the Middle East, India, and China (see Figure 2 in Chapter 22). Its frequency varies locally and is as high as 1% of all births on the island of Sardinia.

*Thomas B. Cooley (1871–1945), a Detroit pediatrician, specialized in childhood anemias.

Many kinds of mutations produce β-thalassemia (Stamatoyannopoulos et al. 1987). They affect not only the actual amino acid coding regions of the β-globin gene, but also sites before and after the gene, and even within the noncoding introns. Some persons with severe β-thalassemia may actually be heterozygous for different β-globin mutations, although the effect may be similar to being homozygous for the one mutant allele or the other. Some of these mutations are shown in Figure 7.

Mutation 1 in the figure alters the promoter region upstream from the gene, thereby reducing the efficiency of transcription. Persons homozygous for this mutation have about 25% of the normal amount of β-globin messenger RNA and are mildly anemic.

Mutation 2 changes codon 39 from a triplet specifying glutamine to a stop codon. (What was the base substitution? See Table 2.) As a consequence, no functional β chains are formed and homozygotes are severely affected. This particular mutation accounts for about 30% of β-thalassemia alleles in the Mediterranean region and for all of the disease on Sardinia.

Mutations 3, 4, and 7 affect intervening sequences. Recall from Chapter 13 that specific sequences at the beginning and end of introns provide signals for splicing together the coding regions, while cutting the intron out of the RNA molecule. A base substitution either may mutate the normal signal at one end of an intron so splicing does not occur, or alternatively, may create a sequence within an intron that makes an additional aberrant splicing site. The three mutations within introns that

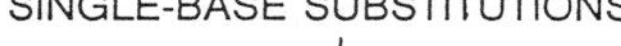
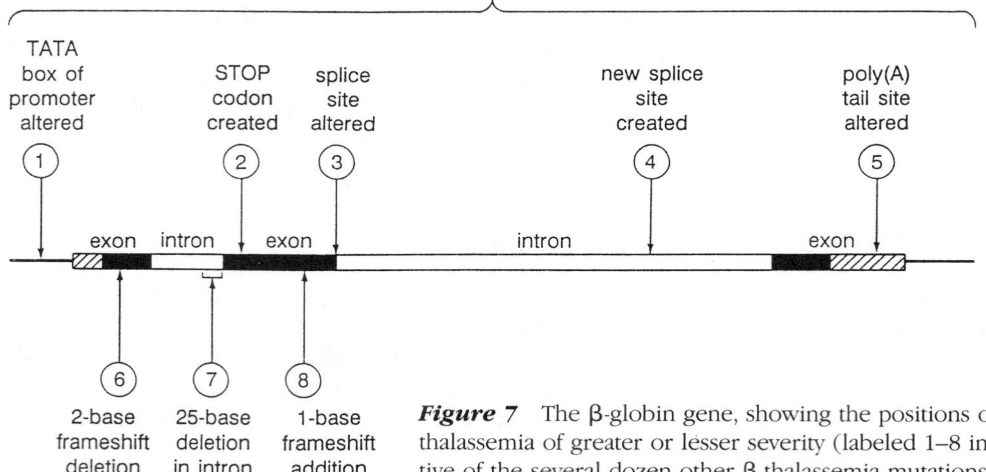

Figure 7 The β-globin gene, showing the positions of eight mutations that produce β-thalassemia of greater or lesser severity (labeled 1–8 in circles). These eight are representative of the several dozen other β-thalassemia mutations that have been studied. The exon–intron structure of the gene is taken from Figure 17 in Chapter 13. See text for further discussion. (Data from Orkin and Kazazian 1984.)

are shown in Figure 7 all result in the complete absence of functional β-globin, perhaps because the mRNA is translated into highly abnormal polypeptides.

Mutation 5, at the site where the RNA is cleaved for the addition of the poly(A) tail, causes the tail to be attached 900 bases further downstream where, by chance, the correct signal sequence again occurs. As a consequence, the mRNA is unstable and a mild thalassemia results.

Mutations 6 and 8 are frameshift mutations, one a two-base deletion and the other a single-base addition. Both cause a complete absence of functional β-globin (see Question 15).

An odd type of β-thalassemia occurs in persons with so-called *Lepore hemoglobin* (first found in a family with that name). The non-α chain has the usual 146 amino acids, but the first part is like a normal δ chain (present in the minor adult hemoglobin component), and the remainder is like a normal β chain. To understand how this could come about, recall that normal δ and β chains differ by only 10 amino acids. Because the corresponding genes have almost identical sequences and are closely linked on chromosome 11, it has been suggested that during meiosis they might sometimes mistakenly synapse with each other rather than with their own alleles (Figure 8). If a crossover should then occur along the length of the misaligned DNA sequences, the result would be mixed genes. One crossover product, the Lepore chromosome, having neither an intact δ gene nor an intact β gene, leads to a thalassemia phenotype. The other crossover product, the "anti-Lepore" chromosome, is also known and produces a near-normal phenotype.

The **α-thalassemias** usually result from large deletions that eliminate one or both of the two duplicate α-globin genes present on chromosome 16. The severity of the disease is proportional to how many of the α-globin genes (1, 2, 3, or 4) are deleted in a given individual. Some hemoglobin is made out of four γ chains (during embryonic life) or out of four β chains (postnatally). But these γ_4 or β_4 hemoglobins are poor oxygen carriers. When all four α-globin genes are deleted, chain synthesis is totally suppressed; the fetus dies in utero, and is miscarried late in pregnancy.

An odd type of α-thalassemia occurs in persons with so-called hemoglobin *Constant Spring* (a town in Jamaica). Instead of the normal 141 amino acids, the α chain has 172 amino acids, with the general sequence:

free amino end 1⋯⋯⋯⋯141 — 142 — 143⋯⋯⋯172 free carboxyl end

(1⋯141: normal α chain; 142: glutamine; 143⋯172: 30 amino acids resembling no known polypeptide)

Note that glutamine, the first extra amino acid, is coded by a triplet derived from a stop signal by a single nucleotide mutation—the reverse of mutation 2 in Figure 7. Without a proper chain termination signal, translation continues through the trailing region which is normally transcribed but not translated, until, by chance, another stop triplet is encountered. (This sequence is worked out in Question 14.)

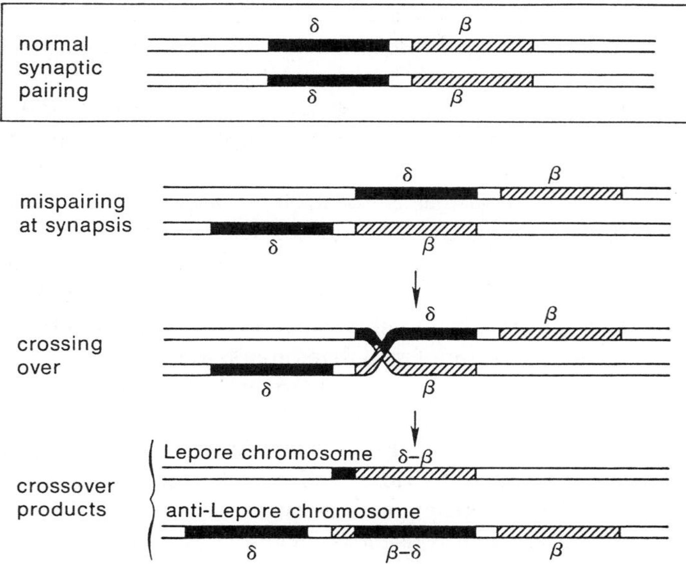

Figure 8 The origin of Lepore and anti-Lepore chromosomes. The generation of a new gene from parts of two misaligned genes is referred to as *unequal crossing over.*

CHEMICAL MUTAGENESIS

The agents that are known to induce mutations fall into two broad categories, chemicals and radiation. In this and the following section we take note of how these mutagens can be identified and their effects measured. Geneticists who work with *experimental* organisms are able to obtain some fairly precise data that may or may not apply to human populations. They can, for example, deliberately subject laboratory animals to known mutagens in order to produce some phenotypic variations. We list three laboratory chemicals that are often used for this purpose:

1. *Proflavine*, a brownish dye that is used as an antiseptic in veterinary medicine, causes the addition or deletion of single bases during the synthesis of new DNA strands. It was used by Crick to produce the frameshift mutations previously described. Proflavine's mutagenic action depends upon its shape: it is a very flat molecule that wedges itself between the base-pair rungs of the DNA helix.

2. *Nitrous acid* is a weak acid but a powerful mutagen, replacing amino groups ($—NH_2$) with keto groups ($=O$). By this action, the pyrimidine cytosine is converted to uracil (see Figure 2 in Chapter 13) and the purine adenine is converted to a base called hypoxanthine (a normal intermediate in the metabolism of purines). Because the pairing properties of changed bases are modified, DNA treated with nitrous acid produces errors during subsequent replications.

3. *5-bromouracil* is one of many known *base analogues*, which so closely resemble the natural bases that they can be mistakenly incorporated into DNA at the time of replication. A molecule of 5-bromouracil is like thymine except that the methyl group of thymine is replaced by a bromine atom. (Thymine itself is sometimes called 5-methyluracil.) Whereas thymine base pairs with adenine, 5-bromouracil can sometimes (but not always) pair with guanine and lead to mistakes during subsequent replications.

Although these three chemicals are not part of the general environment, the human population is daily exposed to a variety of substances that may be mutagenic. (Some 70,000 synthetic chemicals are used in commerce, for example, and perhaps a third of these are relatively common.) It is possible that substantial genetic damage is occurring, and may go on occurring, without our ever being able to link a particular mutagen with its eventual deleterious phenotype. We are unable to match a mutagen to its effect because (1) induced mutations are generally not different in their phenotypic effects from spontaneous mutations, (2) these effects may not appear for generations, and (3) the mutant phenotype may be mimicked by phenocopies. Therefore, as stated by the Council of the Environmental Mutagen Society (1975), "it is crucial to identify potential mutagens before they can induce genetic damage in the population at large: sensitive and reliable test systems are needed which can be used to screen large numbers of potential mutagens."

Screening Systems

Because it is neither ethical nor feasible to deliberately expose humans to a chemical in order to detect possible mutations, no reliable data have been obtained directly on human populations. Instead, the dozens of test systems that have been developed use bacteria, fungi, plants, insects, mammals, or mammalian cells in tissue culture. In principle, because DNA is DNA wherever it occurs, mutagenicity in one organism means mutagenicity in another. But in practice, differences exist between species, both in the degree of protection afforded germinal tissue and in some types of metabolic reactions by which an organism adapts to its surroundings. For example, in *E. coli* a single cell membrane and cell wall separate its DNA from the cell's environment. In humans, an ingested potential mutagen is first exposed to the harsh environments of the mouth, stomach, and intestines before it is absorbed into the bloodstream via the capillaries of the digestive organs. This blood is then shunted to the capillaries of the liver, where absorbed materials may be stored or processed by myriad liver enzymes. Only then are ingested substances or their by-products transported to the germinal cells and other sites throughout the body.

Although some mutagenic chemicals are neutralized by the enzyme systems they encounter, other chemicals may be changed *into* mutagens after entering the body. Different mammals, and even different strains of a given mammal, may metabolize particular chemicals in different ways. In addition, two chemicals may be more mutagenic together than the sum of their separate effects, or a chemical that is not itself mutagenic may enhance the mutagenic effect of another. Thus mutagenicity is not always straightforward, and the results from one species may not be directly applicable to another. Even within a species, certain members may be more susceptible to the action of a mutagen than others, so the average effect of a mutagen may not apply to an individual.

Caffeine is a case in point (Haynes and Collins 1984). It is present naturally or artificially in coffee, tea, chocolate, many soft drinks, and a wide range of common medicines. The plants that synthesize caffeine in their tissues apparently use it as a natural pesticide (Nathanson 1984). Caffeine is a purine, similar in structure to adenine and guanine, but different enough (with three methyl groups) that it does not substitute for the normal purines in DNA. When tested on most bacteria and fungi, caffeine produces mutations. It is thought that caffeine may act, not by being a mutagen itself, but by interfering with the ability of cells to repair damage to DNA from other causes. In plant, hamster, and human cell cultures, high concentrations of caffeine have been shown to cause chromosome breaks (including sister-chromatid exchanges—see below), but the relationship between breakage and point mutations is unclear. With fruit flies, the evidence is contradictory. Extensive experiments with mice yield no indication of caffeine-induced mutations. In humans, ingested caffeine is rapidly demethylated and excreted, although it *does* reach the gonads and *does* cross the placental barrier. Fetuses appear to be *unaffected* by even heavy maternal coffee consumption, however (Linn et al. 1982*). Thus, while pondering all the data over a cup of coffee, we probably need not worry too much about the caffeine in it. On the other hand, risk-taking is personal, and readers may wish to investigate whether caffeine has other detrimental or beneficial effects.

The Ames Test

Because direct in vivo testing for mutagenicity or carcinogenicity in small mammals is time consuming and costly, simple, short-term assays are used to prescreen a wide variety of chemicals. Then, those chemicals that give positive results can be investigated further by using more difficult, but perhaps more pertinent, test systems. Of the many screening tests for chemical mutagenicity, perhaps the most rapid, simple, sensitive, and economical one was developed by Ames (1979). The **Ames test**, as it is called, can be used to suggest whether a compound is mutagenic or carcinogenic (cancer-causing) in animals by determining whether it is mutagenic in bacteria. Over 90% of chemicals known (by other tests) to be carcinogenic in animals are mutagenic in the Ames bacterial test; similarly, most chemicals that are known to be *non*carcinogens in animals are *non*mutagens in bacteria.

The Ames test makes use of tester strains of *Salmonella* bacteria that are unable to make the amino acid histidine because they carry a mutant allele his^- (coding for a defective enzyme). Histidine is a vital amino acid and the bacteria need it to grow. At the start of the test, his^- bacterial cells are spread over the surface of nutrient medium (in a petri dish) that lacks histidine. Therefore, the cells can grow *only* if a mutation occurs from the allele coding for the defective enzyme to an allele coding for a functioning enzyme (a so-called back mutation, symbolized $his^- \rightarrow his^+$). Individual bacteria with back mutations will eventually be visible as discrete bacterial colonies (Figure 9).

Several different histidine-requiring strains of *Salmonella* can be used in different tests of the same potential mutagen. Some tester strains are unable to synthesize their own histidine because of a base substitution mutation, and others because of a frameshift mutation. Thus a mutagen with a particular mode of action that might be missed using one tester strain may be detected using another. For example, a frameshift mutation in a gene due to a single-base *deletion* can be most simply "corrected" by a nearby single-base *addition*. No base substitution can correct a frameshift mutation.

The tester strains of *Salmonella* also carry other heritable traits that make the test very sensitive. A modification of the cell wall allows easy entry of the test chemical into the bacteria. In addition, the bacterial enzymes that are normally used to repair DNA damage are inactivated. Sometimes added to the growth medium are rat liver enzymes that are capable of modifying some test chemicals from nonmutagenic to mutagenic form, or vice versa. In this way, the bacterial test is made to resemble, to some extent, a mammalian system in which nutrients absorbed into the bloodstream first travel to the liver.

Sister Chromatid Exchange

To further assess the possible mutagenic effect of a given chemical in humans, investigators can study human cells grown in tissue culture. After exposing the cells in vitro to the test chemical, researchers look at metaphase for chromosomal defects of various types. They do this because there is some, but not perfect, correlation between the induction of chromosomal defects and point mutations and cancer (CEM 1983).

One such cytogenetic test, **sister chromatid exchange (SCE)**, is very sensitive to low concentrations of test chemicals and is easy to perform and to score. Sister chromatid exchange is like crossing over except that *sister* chromatids rather than *non-sister* chromatids swap corresponding segments (at, or soon after, the time of DNA replication). Because sister chromatids are exact duplicates of each other, SCE has no harmful genetic consequences; for the same reason, SCE is usually impossible to detect in meiosis (where crossing over normally occurs), or in mitosis (where crossing over may also occur). Mitotic exchanges between sisters can be made visible in a microscope, however, by marking the chromatids differentially by chemical means. Cells are exposed for two replications to a form of the base analogue, 5-bromouracil. An outline of what happens is depicted in Figure 10.

*Although this study found that coffee drinking had no significant effect, it confirmed previous evidence that maternal cigarette smoking tends to produce underweight newborns.

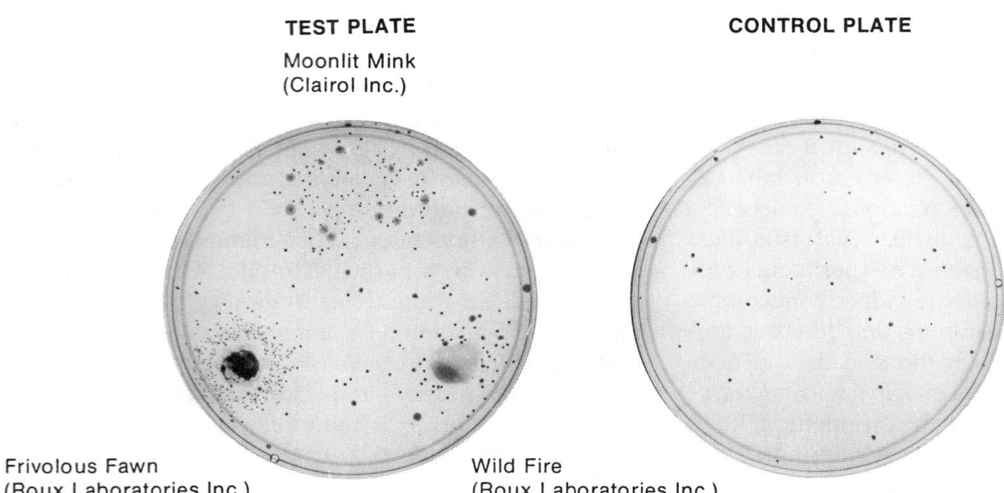

Figure 9 An Ames test of three hair dyes (Ames et al. 1975). A thin lawn of *Salmonella* bacteria is spread on the growth surface of the two plates. The cells (genetically *his*$^-$) do not reproduce, because they are unable to synthesize the vital amino acid histidine, which is also missing from the growth medium. A mutation from *his*$^-$ to *his*$^+$, however, permits a cell to make its own histidine and thus to reproduce. After incubation for a day, each mutated cell, dividing and redividing, forms a visible colony—the small dark spots. On the test plate, drops of three dyes are placed on the lawn. The halos of colonies surrounding the test drops indicate that the test compounds are mutagenic in this test. It is uncertain whether the chemicals should be a source of concern when applied to human scalps. The control plate was treated the same, except that no dyes were added; a few spontaneous mutants are seen.

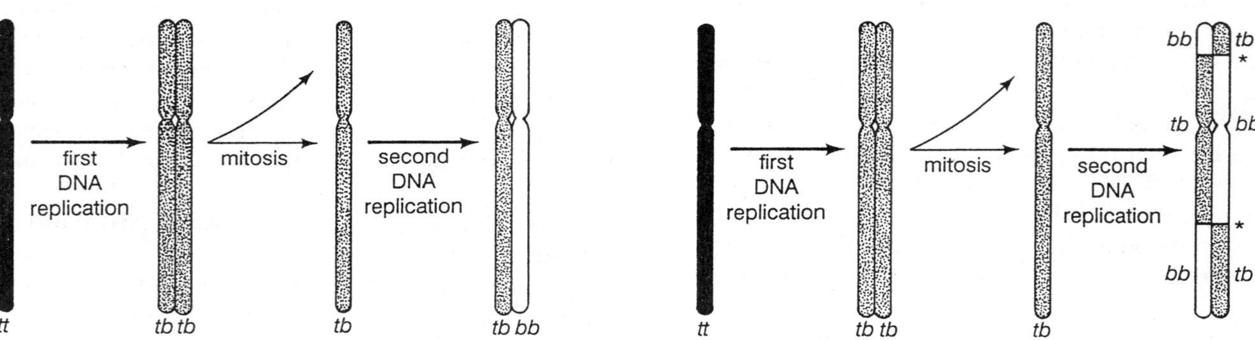

Figure 10 Detection of sister chromatid exchange (SCE). Cells were grown in culture for two generations in medium containing 5-bromouracil, an analogue of the DNA base thymine. Each chromosome or chromatid, consisting of one DNA double helix, is labeled:

tt (black) if both strands have thymine as one of the four DNA bases
tb (stippled) if one strand has thymine, and the other 5-bromouracil
bb (white) if both strands have 5-bromouracil

Note that when the original *tt* chromosome replicates, each *t* strand is a template for the formation of a complementary strand containing *b*. In Case I, the second replication produces one *tb* chromatid and one *bb* chromatid, which stain differently (see text). This diagram is similar to the Meselson-Stahl experiment depicted in Figure 5 in Chapter 13, when *t* is substituted for H, and *b* for L. In Case II, the situation is similar except that two SCEs are shown (perhaps as a consequence of exposure to a test chemical). The SCEs occurred during the second replication. Any SCE occurring during the first replication would not be detectable. Why?

In Case I we see the results after two DNA replications when there is no SCE. Consider the metaphase chromosome after the second replication. The left-hand chromatid has one strand of its double helix labeled with thymine and one strand with 5-bromouracil (*tb*), whereas the right-hand chromatid has both strands labeled with 5-bromouracil (*bb*). The chemical difference between the chromatids is made evident by a dye (Giemsa or a fluorescent chemical) that binds better to chromatids having more thymine. Thus *tb* chromatids stain darker with Giemsa than *bb* chromatids do, as indicated by the stippling. When SCE occurs (at or soon after the time of replication), as in Case II, complementary dark and light segments alternate in the two chromatids (still joined together in a metaphase chromosome). Such switching of stained and unstained segments produces *harlequin* chromosomes (Figure 11), named after the comic characters dressed in checkerboard costumes.

In one variation of the SCE system, researchers do not use tissue culture; rather, they inject the 5-bromodeoxyuridine label *directly* (a) into laboratory mammals that have been exposed to a test chemical and (b) into unexposed control animals. All aspects of mammalian physiology are thereby automatically incorporated into the test. Cells from blood, bone marrow, gonads, and other organs can then be extracted and examined for SCE. In this combination of in vivo and in vitro technologies, the action of a test chemical or radiation *within* a live experimental animal can be evaluated.

In another variation, human populations (industrial workers, hospital employees, residents near hazardous waste sites, and so on) can be monitored for effects of exposure to some long-acting environmental agents. White blood cells of exposed persons are extracted and then labeled in vitro with 5-bromouracil to reveal SCE. By this method, for example, most, but not all, studies of cigarette smokers reveal a slight increase of SCE in white blood cells (DeMarini 1983; Heath et al. 1984). In summary, because of the correlation between the occurrence of SCE (itself innocuous) and various measures of damage to DNA, SCE can be used as evidence of increased risk of mutation.

RADIATION MUTAGENESIS

Very soon after the discovery of X-rays (by Wilhelm Roentgen in 1895) and radioactivity (by Antoine Becquerel in 1896), it became clear that somatic tissues

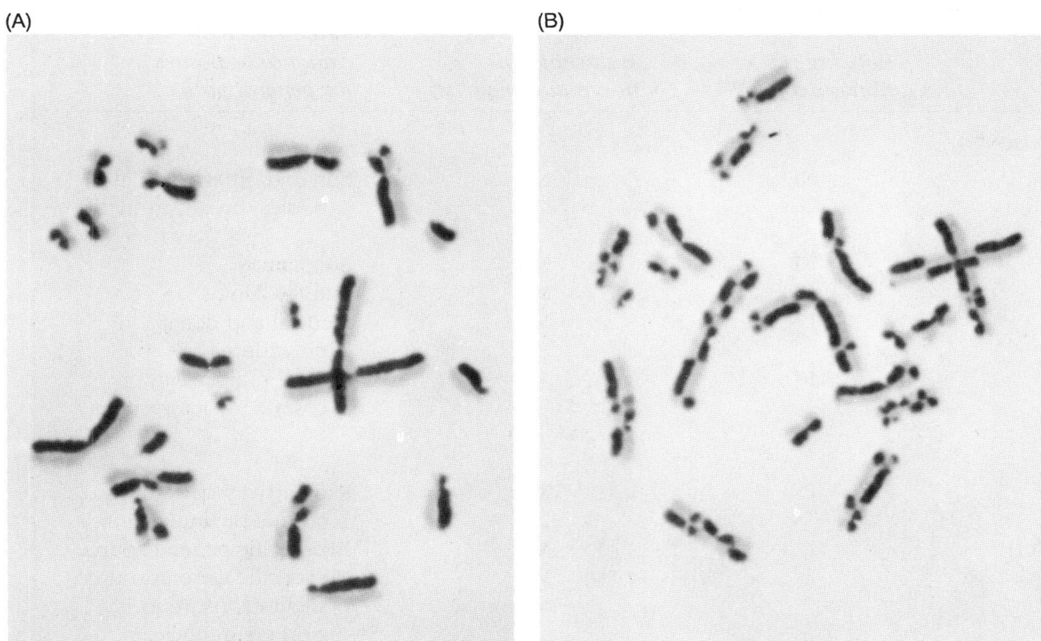

Figure 11 Sister chromatid exchanges seen in metaphase chromosomes of a hamster after differentially marking chromatids with thymine (t) or a thymine analogue (b). Compare with Figure 10. (A) Control cells not exposed to any test chemical show a few SCEs. (B) Cells treated with a chemical carcinogen show many SCEs producing so-called harlequin chromosomes. Note that the two chromatids of each chromosome are held closely together side by side. (Courtesy of Sheldon Wolff and Judy Bodycote, University of California, San Francisco.)

could be damaged by agents that could not be seen, felt, or smelled. In the 1920s, Hermann J. Muller (1927), then at the University of Texas, reported that X-rays induce mutations in the germinal cells of fruit flies; at about the same time Lewis J. Stadler of the University of Missouri got the same effects in barley plants. Public awareness of the biological consequences of radiation began in the 1950s as the result of radioactive fallout from the bombing of Japanese cities and the continued atmospheric testing of nuclear weapons. Current medical uses and the presence of about 90 nuclear power plants in the United States (about 400 worldwide) continue to focus public attention on the implications of radiation. Interest has been heightened by nuclear power accidents at Three Mile Island in 1979 and Chernobyl in 1986 (Eisenbud 1987).

All living things are exposed to natural, and largely unavoidable, radiation emanating from the sun, from cosmic rays, and from the radioactivity of uranium, radium, and other unstable elements in rocks, soils, food, and air. The types of radiation striking us from these and from artificial sources are summarized in Table 3. The effectiveness of electromagnetic radiation in inducing mutations depends primarily on the energy carried per photon, which is inversely related to wavelength. For the subatomic particles, mutagenicity depends primarily on their speed, mass, and electric charge.

To be mutagenic, radiant energy must reach DNA. Then a sufficiently energetic photon or subatomic particle can knock an electron out of an atom in DNA; this event leaves the atom as a charged ion and in a very reactive state, subject to further chemical changes. A direct hit on DNA is not necessary, however: nearby ionized chemicals can subsequently affect DNA. Highly reactive hydrogen peroxide, for example, is known to form when water—the major ingredient of all living things—is irradiated. Because ultraviolet light (see Table 3) is strongly absorbed by tissue and consequently does not penetrate below the outer layers of skin, it does not induce human *germinal* mutations. By inducing *somatic* mutations, however, high doses of ultraviolet radiation can lead to skin cancer (Chapter 19).

Human Radiation Exposures

The amount of radiation absorbed by a given amount of a substance is called the *radiation dose*. It can be calculated from physical principles and measured by various devises (e.g., Geiger counters, badges containing sensitive film) that determine the number of ions produced or the amount of energy absorbed. Common units

Table 3 Naturally occurring and artificially produced radiation impinging on humans

Radiation	*Ionizing (I) or nonionizing (NI)*	*Mutagenic (M) nonmutagenic (NM)*	*Main source of radiation exposure for general public*
ELECTROMAGNETIC RADIATION[a]			
Long wavelengths: radio, TV, microwave, and infrared radiation	NI	NM	Sun and other celestial bodies; broadcast media
Visible light	NI	NM	Sun; lamps
Ultraviolet radiation	NI	M	Sun; UV lamps
X-rays	I	M	Medical and dental procedures
γ rays (very short wavelengths)	I	M	Natural radioactivity; cosmic ray interactions
SUBATOMIC PARTICLES[b]			
Electrons	I	M	Radioactivity (β rays; cosmic ray interactions)
Protons (hydrogen nuclei)	I	M	No specific general source
Neutrons	I	M	No specific general source; slight exposure to workers in nuclear facilities
Muons	I	M	Cosmic ray interactions
α-particles (helium nuclei)	I	M	Radioactivity

a. The energy levels of electromagnetic radiation are lowest for the long wavelength types and are highest for the short wavelength types.
b. The energy levels of the subatomic particles depend on their speed and mass.

of radiation dose are roentgen, rad, rem, gray (100 rad), and the sievert (100 rem). Their definitions are beyond the scope of this book (see Shapiro 1981); but for medical X-rays, the first three units are almost the same. Radiation can be received at different rates. A dose received from a high-intensity source is called **acute radiation**, whereas a dose received at low intensity is called **chronic** or **low-level radiation**. The same total dose to tissue can be delivered quickly as an acute dose or spread out over time as a chronic dose. We will be concerned primarily with chronic doses, because these characterize many natural and artificial sources to which the general population is exposed.

A typical American receives 5–6 rem of radiation from all sources over the average reproductive cycle of 30 years (Table 4). Roughly half of this is from natural sources, including radioactive isotopes in our environment and within our bodies. Recently, increased attention has focused on radioactive radon gas, which appears to pose a risk greater than that from other radiation sources in certain geographical areas (UNEP 1985; Eisenbud 1987). (Radon seeps into houses from soil and ground water; the tighter and more energy efficient the house, the higher its concentration.)

The major artificial source of radiation is from medical and dental procedures. Although there is wide variation due to the type of equipment used and degree of resolution required, typical doses absorbed by a patient during chest or dental X-rays or mammography of the breast are in the range of 0.01 to 1 rem. Only a fraction of the radiation reaches the gonads, where it could induce germinal mutations. For abdominal or pelvic X-rays, a greater fraction is genetically significant because the gonads (especially the ovaries) cannot always be shielded effectively. Much larger doses may be needed for fluoroscopic examinations, and radiation therapy for cancer may deliver immense doses—hundreds or thousands of rem directed at the cancerous growths over a period of time. (Were these same amounts given to the body as a whole, severe radiation sickness or death could result.) Ironically, radiation can both *cause* cancer and *treat* cancer—the former by inducing somatic mutations, and the latter by killing cancer cells. Unfortunately, radiation therapy also kills other quickly growing tissues such as the intestinal linings and hair follicle cells.

Table 4 Radiation exposure of a typical person in the United States over a 30-year span

Source	*Dose (in rem) per generation (30 years)*[a]	*Percentage of total*
NATURAL SOURCES		
Cosmic radiation	0.84 rems	15%
Terrestrial radioactivity	0.78	14
Internal radioactivity[b]	0.78	14
Subtotal	2.40	43
ARTIFICIAL SOURCES		
Medical and dental radiology	2.77	50
Fallout from atomic tests	0.14	3
Consumer products[c]	0.12	2
Miscellaneous sources[d]	0.12	2
Nuclear power industry[e]	0.01	0.2
Subtotal	3.16	57
TOTAL	5.56	100

a. Data derived primarily from Table III-23 of BEIR (1980). The doses are whole body exposures (for the most part). The gonadal dose is a small proportion of these figures.

b. Includes radioactive potassium-40 from dietary sources and radon-222 gas present in indoor air.

c. Includes building materials, color television, fire detectors, luminous dials, radioactive polonium-210 in tobacco smoke.

d. Includes airline travel (excess cosmic rays), occupational exposures, effluent from coal-fired power plants, phosphate fertilizers.

e. Shapiro (1981) and Eisenbud (1987) discuss radiation accidents.

Evaluation of Radiation Effects

Direct information on the effects of radiation on humans is scanty because of (1) the absence of planned experiments, (2) the lack of precise dose measurements in accidents, atomic bomb casualties, and occupational exposures, and (3) the inherent difficulties of measuring small effects. On the other hand, a massive amount of data has been obtained from experimental organisms. How valid are the extrapolations from fruit flies and mice to humans? Nobody knows for sure, but it is clearly better to make the estimates and to be conservative in doing so than not to make them at all. Even if we had full knowledge of the rate of spontaneous and induced mutations in human populations, however, evaluating the consequences for families and society in future generations would be difficult and uncertain (Crow 1982).

Radiation has many different kinds of effects. These include chromosome breaks and point mutations in germinal and somatic tissues, the induction of cancer (particularly leukemia), localized radiation burns and tissue scarring, and at even higher doses, generalized radiation sickness (nausea, vomiting, diarrhea, internal bleeding, general weakness) and death. One important and much-debated question is whether or not there is a "safe dose" or **threshold**, an amount of radiation below which there is no risk. To answer this question, investigators irradiate experimental animals with smaller and smaller doses (necessitating greater costs and the use of more and more test organisms) and then measure a particular effect, for example, point mutations or sister chromatid exchanges. The result is a *dose-response curve* (Figure 12), which unfortunately will have no experimental points below a certain dose. The form of the curve, however, gives both theoretical information on the way radiation induces damage, and practical information to help formulate medical and public health policies.

The *somatic* effect of major concern with respect to low-level ionizing radiation is the possible induction of cancer. Using primarily data on Japanese atomic bomb survivors and on patients treated with radiation for diseases other than cancer, researchers have estimated that 1 rem of X-radiation per person (above that normally received) leads very roughly to 1–3 cancer deaths per 10,000 people (BEIR 1980). For comparison, roughly 1,500 of 10,000 births (15%) currently end in a cancer death. No single cancer fatality can be unequivocally identified as radiation-induced, because (1) the time elapsing between irradiation and the detection of the malignancy is long—years or even decades, and (2) the cancers induced by excess radiation are not different in kind from those occurring from other causes (although

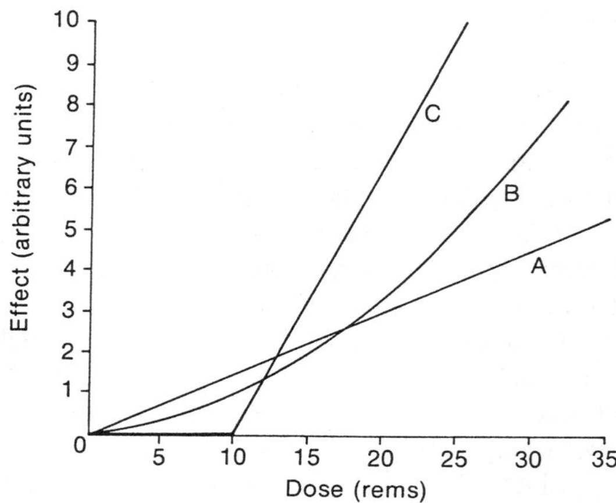

Figure 12 Types of radiation dose–response curves. Curve A is strictly linear, the effect being directly proportional to the dose, down to zero dose. Curve B is quadratic, the effect being proportional to the square of the dose. Curve C is linear, but with a threshold—below a certain dose (10 rems in the case illustrated) there is no effect at all.

leukemia seems to be induced at a greater rate by radiation than by other agents). Thus it is unclear whether or not a radiation threshold exists for the induction of cancer.

The *germinal* effect of major concern is the possible induction of point mutations or chromosomal aberrations that may cause harm to the immediate offspring or later descendants (Muller 1955; Denniston 1982). Reliable estimates of the germinal effect of radiation are especially difficult to obtain directly from humans because many new mutations are recessive; therefore the appearance of the mutant phenotype may be several generations removed from the mutational event (if it is ever expressed at all). On the basis of animal and human data, researchers have estimated that 1 rad per generation produces roughly one case of genetic disease (Mendelian or chromosomal) per 50,000 births in the children of those exposed (UNEP 1985). If the same rate of exposure continues over many generations, ultimately about one child in 7,000 will be born with a genetic disease. For comparison, approximately 3% of all newborns are currently affected with a moderate to severe Mendelian or chromosomal disorder.

The direct studies from Hiroshima and Nagasaki of the germinal effects of radiation have been plagued by great uncertainties about the doses received by the survivors. Researchers have found no significant increase in stillbirths, infant mortality, birth weight, congenital abnormalities, childhood mortality, leukemia, or sex ratio as a consequence of the parental exposures. Nevertheless, on theoretical grounds and from the experiments with flies and mice, most geneticists think that the dose–response curve is likely to be linear down to very low doses. If so, then any amount of radiation carries with it a proportional risk of mutation—*there is probably no threshold, no safe dose, for germinal effects.*

Experiments with mice have focused on radiation-induced mutations in *spermatogonial or oogonial cells* (that is, the germline prior to meiosis), because this stage is quite long in humans. To test for spermatogonial mutations, irradiated males are held for several weeks to allow the exposed spermatogonial cells to develop into mature sperm. Then the mice are mated to special homozygous recessive females, and their offspring examined to see whether they exhibit mutant phenotypes. Chronic irradiation of mice induces recessive point mutations at an average rate of about 1.5×10^{-7} mutations per locus per rem (UNSCEAR 1982); assuming a mouse has 30,000 loci, each added rem therefore induces 1 mutation in about 200 gametes.

On the basis of all information available, we conclude that the genetic hazards of radiation for the general human population are small. It has been estimated that about 50 to 250 rem of chronic radiation would be required to double the rate at which mutations occur *spontaneously* (Eisenbud 1987). This amount is called the **doubling dose.** (A higher estimate of the doubling dose, about 450 rem, is given by Schull et al. [1981] on the basis of Hiroshima and Nagasaki data, which were difficult to interpret.) Over a reproductive lifetime (Table 4), a typical person receives only 5–6 rem from both natural and artificial sources; so it appears that the mutations that occur spontaneously are not substantially due to the radiation we normally receive.

Studies of risk-taking have shown that peoples' perceptions of the hazards of life often correlate poorly with real dangers. For example, actuarial estimates of the causes of death in the United States show that cigarettes, alcohol, motor vehicles, and handguns are far more dangerous than police work, plane travel, nuclear power, and pesticides (Upton 1982). Yet we often tolerate the more serious risks while concentrating on the lesser ones—perhaps because some risks are voluntary and others are imposed upon us. With regard to radiation, we tend to ignore medical and dental X-rays as a significant source of radiation (about half a person's lifetime exposure on the average) over which we have some personal control. Whenever possible, and certainly with dental or chest X-rays, patients should insist that their gonads be protected by the lead shields that come with all X-ray machines. Technological and operational improvements can also reduce radiation exposures, and patients should not hesitate to inquire about the age and efficiency of the X-ray equipment being used. Although the benefits from diagnostic and therapeutic X-rays are often substantial, their risks should be evaluated and minimized whenever possible, and alternative methods should be considered.

For both chemical and radiation mutagens, even a tiny increase in the *rate* of germinal damage, undetectable by any feasible methodology and of little concern to an individual or a couple, could have an absolutely large effect on a population. For example, assume that certain increased environmental insults over the last 70-odd years had led to additional birth defects in just 1 out of 10,000 persons. In a population of about 250,000,000 (the current U.S. population), we would then expect 25,000 *more* affected newborns over some scores of years. If they were identifiable—like casualties from a drought or earthquake or typhoon—the tragic aspects of mutagens would be more comprehensible. Scientists' inability to demonstrate that "factor X" causes a specific human death or deaths does not mean that "factor X" is innocent. Nor does it mean that they should give up on trying to identify mutagens and carcinogens. Rather, they should continue to devise carefully controlled experiments and cautiously extrapolate their results from test organisms to humans.

SUMMARY

1. The triplet nature of the genetic code was deduced by observing the effects of base deletions and additions. Specific codons were deciphered by adding artificial RNA molecules of known composition to an in vitro protein-synthesizing system.
2. One genetic code, which includes several codons for most amino acids and signals to stop translation, applies to almost all forms of life.
3. Point mutations are usually base substitutions or frameshift mutations (small deletions or additions) occurring in somatic or germinal tissue.
4. The serious symptoms of sickle-cell anemia stem from a single base substitution. The disease has a high frequency among some Black populations and is largely untreatable at present.
5. Hundreds of different base substitutions have been found in the globin genes. The specific amino acid change that each causes is predictable from the genetic code table. The phenotypic consequences depend on how the change affects the properties of the hemoglobin molecule.
6. Thalassemias result from the underproduction of

the α or β chains of hemoglobin. Studies of the mutational sites have shown many different ways by which transcription and translation of the globin genes can be modified.

7. Spontaneous mutations occur without known cause, perhaps as the result of random thermal energies of atoms, or from stray radiation or chemicals in the environment. Induced mutations are caused by known agents (mutagens) that increase the rate at which mutations occur spontaneously.

8. Chemical mutagens in our environment can be identified by screening tests using microorganisms, cells grown in tissue culture, or small mammals. The often-used Ames *Salmonella* test can be made to partially mimic a mammalian system. The sister chromatid exchange test detects agents that damage chromosomes.

9. The human population is exposed to both natural and artificial sources of ionizing radiation, which can cause cancer via somatic mutations and hereditary defects via germinal mutations.

10. There appears to be no safe dose of radiation for inducing germinal mutations. The small risks associated with some common medical uses of radiation might be reduced further by carefully weighing the benefits, hazards, and alternative methods.

FURTHER READING

As recommended for the last chapter, Strickberger (1985), Lewin (1987), and Watson et al. (1987) give much more information on molecular genetics than we present here. CEM (1983), Heddle (1982), and McElheny and Abrahamson (1979) review chemical mutagenesis, and Searle (1987) briefly reviews radiation mutagenesis. Shapiro (1981) and Eisenbud (1987) have written detailed guides to radiation; and Brill et al. (1982) and UNEP (1985) summarize some of the wealth of radiation data in the BEIR (1980) and UNSCEAR (1982, 1986) reports. Dickerson and Geis (1983) summarize current knowledge of the hemoglobin molecule, and Harris (1980) and Stamatoyannopoulos et al. (1987) explain the molecular biology of hemoglobin variants. Emery and Rimoin (1983), Stanbury et al. (1983), and Greenblatt (1974) include chapters on sickle-cell anemia and thalassemia.

Useful *Scientific American* articles include those by Crick (1966b) on the genetic code, by Perutz (1978) on hemoglobin structure, by Howard-Flanders (1981) on DNA repair, and by Upton (1982) on low-level radiation. In addition, the September 1959 issue of *Scientific American* is devoted to ionizing radiation.

QUESTIONS

1. Let + stand for a one-nucleotide addition and − for a one-nucleotide deletion. Would the following mutant genes be likely to produce a functional product? Assume that the mutations are near each other. (a) + − (b) + − + (c) + + + (d) + − − + (e) + + + −

2. The human brain hormone somatostatin was the first product of recombinant DNA technology (Itakura et al. 1977). Bacteria synthesized the hormone after biochemists made and inserted into the cells DNA with the following base sequence (sense strand):

   3′ CGA CCA ACA TTC TTG AAG AAA
   ACC TTC TGA AAG TGA AGC ACA 5′

   What is the amino acid sequence of somatostatin?

3. What polypeptides could be made by each of the following artificial mRNA molecules? (a) poly(A) (b) ···UCAUCAUCAUCA··· (c) ···UUCCUUCC UUCC···

4. In the virus φX174, the DNA coding sequence for the E gene is, quite remarkably, contained *within* the coding sequence for the D gene. The amino acids in the E polypeptide are different from those in the corresponding stretch of the D polypeptide. Suggest how the virus might perform this elegant trick (see Fiddes 1977).

5. (a) The end of the mRNA of the D gene of φX174 has the sequence ··· GCGGAAGGAGUGAUGU AA, in which the reading frame starts ··· GCG GAA ···. What are the last several amino acids in the polypeptide product of the D gene? (b) The end

of the mRNA from the E gene is the same molecule as depicted above except that the reading frame starts ··· G CGG AAG ···. What are the last several amino acids in the polypeptide product of the E gene?

6. Tabulate how many synonymous codons there are for each amino acid in Table 2. For example, phe has 2 codons, leu 6, and ile 3. From Figure 14 in Chapter 13, tabulate how often each amino acid occurs in insulin. Find the *average* number of occurrences in insulin for the amino acids with a single codon (met and trp). Similarly, find the *average* number of occurrences for the amino acids with 2, 3, 4, and 6 codons. Plot the averages (vertical axis) against the number of synonymous codons (horizontal axis) to show that there is rough proportionality even for this very small sample of 51 amino acids.

7. (a) List the nine possible DNA triplets and corresponding amino acids that would be formed by a single base substitution mutation in the CCC triplet for glycine. (b) How many lead to no amino acid change? (c) How many lead to a different but uncharged amino acid like glycine? (d) How many would likely lead to an electrophoretic variant of the polypeptide containing the amino acid?

8. For the amino acid asp (aspartic acid) at position 75 of the α chain of hemoglobin, researchers have found *five* different substitutions: tyr, his, asn, ala, and gly. What other amino acids could theoretically substitute for asp as the result of a single base substitution mutation?

9. In what way is the Ames test for chemical mutagenicity in bacteria made to resemble a mammalian test system?

10. A man who worked in a radiation laboratory sued his company because his son was born with hemophilia, which is due to a recessive X-linked allele. He claimed that his working environment was responsible for his son's disease, inasmuch as it was not present in him, his wife, or any other relative. You are asked to testify in a pretrial hearing as an expert witness in genetics. What would you say?

11. What is the argument for the statement that most spontaneous mutations in humans are not caused by radiation?

12. Newborns homozygous for β-thalassemia are often healthy, the severe disease symptoms developing during the first several months *after* birth. On the other hand, newborns for α-thalassemia are often severely affected *at* birth or are spontaneously aborted. Explain.

13. Assume that hemoglobin Arlington Park arose by crossing over somewhere along the chromosome between codons 6 and 95 in a person heterozygous for hemoglobin C and hemoglobin N (Baltimore). The amino acids in the β chain at positions 6 and 95 are

Hemoglobin	**6**	**95**
A (normal)	glu	lys
C	lys	lys
N (Baltimore)	glu	glu
Arlington Park	lys	glu

What type of hemoglobin is coded by the complementary crossover product?

14. Below are the amino acids near the ends of the α chains of hemoglobin Constant Spring (CS) and hemoglobin Wayne (W). CS results from a substitution in the first base of the normal stop signal, which is now amino acid 142. W results from a frameshift deletion (X) of the first base coding for the normal amino acid 139. In the boxes below give the base sequence of the normal gene. (It is remarkable that the overlapping triplets provide a unique answer.)

15. The *normal* sequence of the seventh through the twenty-third codons of human β-globin mRNA is given below. Orkin and Goff (1981) describe a Turkish patient with β-thalassemia in which the two adjacent A nucleotides (underlined) coding for the eighth amino acid are missing. What happens as a consequence of this frameshift mutation?

Q. 14

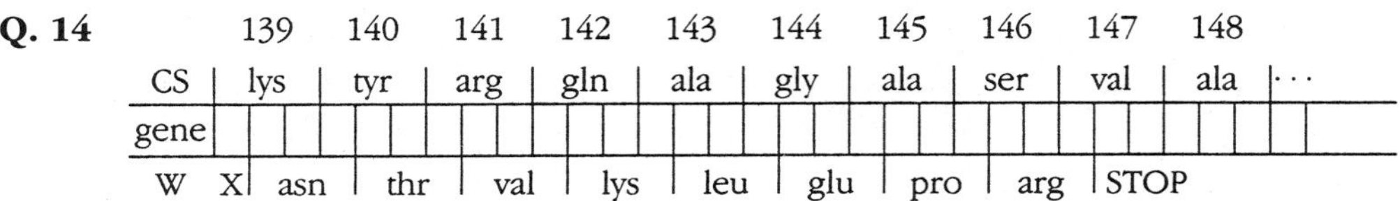

Q. 15

7	8	9	10	11	12	13	14	15	16	17	18	19	20	21	22	23
···GAG	<u>AA</u>G	UCU	GCC	GUU	ACU	GCC	CUG	UGG	GGC	AAG	GUG	AAC	GUG	GAU	GAA	GUU···
··· glu —	lys —	ser —	ala —	val —	thr —	ala —	leu —	trp —	gly —	lys —	val —	asn —	val —	asp —	glu —	val ···

ANSWERS

1. In (a), (c), and (d), the product might be functional since only a few amino acids are wrong. In (b) and (e), the product is unlikely to be functional unless the mutations are near the end of the gene.

2. NH_2: ala–gly–cys–lys–asn–phe–phe –trp–lys–thr–phe–thr–ser–cys :COOH

3. (a) ··· lys–lys–lys–lys ···
(b) ··· ser–ser–ser–ser ··· and ··· his–his–his–his ··· and ··· ile–ile–ile–ile ···
(c) ··· phe–leu–pro–ser ··· (this foursome tandemly repeated)

4. In theory, they could be read off opposite strands. In actuality, the two genes are read off the same strand in different reading frames.

5. (a) ··· ala–glu–gly–val–met (STOP)
(b) ··· arg–lys–glu (STOP)

6.

Number of codons for amino acids	Amino acid (number of occurences in insulin)	Average number of occurrences for group
1	met (0), trp (0)	0
2	asn (3), asp (0), cys (6), gln (3), glu (4), his (2), lys (1), phe (3), tyr (4)	2.9
3	ile (2)	2.0
4	ala (1), gly (4), pro (1), thr (3), val (4)	2.6
6	arg (1), leu (6), ser (3)	3.3

7. (a) ACC: trp; CAC: val; CCA: gly
GCC: arg(+); CGC: ala; CCG: gly
TCC: arg(+); CTC: glu(−); CCT: gly

(b) 3; (c) 3; (d) 3

8. val and glu

9. Enzymes from mammalian liver are sometimes added to the growth medium for whatever effect they might have in metabolizing the tested chemicals.

10. Although radiation can cause mutations, this particular man has no case. He transmitted a *Y chromosome* (largely devoid of mutable genes) to his son; the mutant hemophilia gene in question was inherited from the mother.

11. It is estimated that it would take 50–450 rem of chronic radiation to yield the observed rate of spontaneous mutations in humans. The average exposure of a human is much less than this amount.

12. α Chains are a major component of hemoglobin both pre- and postnatally, but β chains are synthesized in quantity only after birth.

13. Hemoglobin A (normal)

14.

	139	140	141	142	143	144	145	146			
CS	lys	tyr	arg	gln	ala	gly	ala	ser	val	ala	···
gene	TTT	ATG	GCA	GTT	CGA	CCT	CGG	AGC	CAT	CG	

gene	T T	TAT	GGC	AGT	TCG	ACC	TCG	GAG	CCA	TCG
W	X	asn	thr	val	lys	leu	glu	pro	arg	STOP

15.

7	8	9	10	11	12	13	14	15	16	17	18	19	20	21
···GAG	GUC	UGC	CGU	UAC	UGC	CCU	GUG	GGG	CAA	GGU	GAA	CGU	GGA	UGA···
··· glu —	val —	cys —	arg —	tyr —	cys —	pro —	val —	gly —	gln —	gly —	glu —	arg —	gly.	

Note that by chance the twenty-first triplet becomes a stop codon, so the now functionless polypeptide is only 20 amino acids long.

Chapter 15 New Genetic Technologies

Recombinant DNA
Making Recombinant DNA
Manufacturing Proteins
Public Concerns

DNA Manipulations
Electrophoresis and Restriction Maps
Sequencing
Making a Gene Library
Screening a Gene Library
Southern Blotting and the Molecular Biology of Color Vision
Restriction Fragment Length Polymorphism (RFLP)
Huntington Disease and Sickle-Cell Anemia
DNA Fingerprinting and the Polymerase Chain Reaction

Somatic Cell Genetics
Selecting and Isolating Hybrid Cells
Chromosome Assignment by a Gene's Protein Product
Methods for Regional Assignments

A Further Note on Mapping

> *Looking back over this historical development of molecular biology, one thing is clear; the future of the subject has always been unpredictable. Some advances, it is true, have been the result of a long, well-planned and sustained campaign, but right from the very beginning happenstance has played its part. That makes it extremely difficult to predict future developments with any degree of certainty. One can make a vague generalization—say, that biologists will gain a deeper understanding of the fundamental process of gene regulation and control—and be confident that the prediction will come true. But when it comes to specific forecasts, and especially when one attaches time to those predictions, one is liable to go very awry.*
>
> *Jeremy Cherfas (1982)*

Starting with almost any cell type—bacteria, white blood cells, sperm, mammalian cells grown in culture, or a bit of minced tissue or organ from an adult or fetus—scientists (or basic biology students) can isolate large molecules of DNA in nearly pure form in a few hours. First, the cells are broken open with sodium dodecyl sulfate (SDS), a detergent that dissolves the lipids (fatty molecules) in cell membranes and lets the insides spill out. Then the proteins in the resultant cellular soup are coagulated by gentle shaking with phenol, an organic solvent that does not mix with water. This process leaves a separable aqueous layer containing primarily DNA. The aqueous solution is now very viscous and elastic because the extremely long, thin, stretchy threads of DNA intertwine with each other. Adding ethyl alcohol precipitates the DNA molecules, which can be wound around the tip of a glass rod and lifted out from the liquid as a visible, glistening globule of nearly pure, high-molecular weight DNA.

Obtained this way, or by other methods, DNA is the basis of a expanding area of science called molecular biology, biotechnology, bioengineering, or genetic engineering. Even though molecular biologists may use strange terminology, expensive equipment, or lengthy laboratory protocols, the fundamental concepts that they deal with are not complex. Included in the field are techniques for handling and investigating the properties of nucleic acids, for attaching together DNA molecules from different species, and for fusing together cells and nuclei of different species.* A major feature is the speed with which the new discoveries are currently reported. In his forward to the 125 papers in "Molecular Biology of *Homo sapiens*" (Cold Spring Harbor Symposia 1986),

*Sometimes considered a part of genetic engineering are procedures such as artificial insemination and in vitro fertilization that alter the usual course of reproduction. We present these topics in Chapter 24.

James D. Watson writes, "The scientific advances... amaze, stimulate, and increasingly often overwhelm us. Facts that until recently were virtually unobtainable now flow forth almost effortlessly." The practical benefits of the new information will eventually permeate medicine, agriculture, and other human endeavors in important ways that cannot be fully foreseen.

To a large extent, the same basic chemicals and metabolic reactions occur in all cells, from bacterial to human. This biochemical unity of life includes the processes of DNA replication, transcription, translation, and mutation, as well as the common enzymatic reactions for energy transfer (using ATP) and for many other metabolic pathways. Because their intracellular environments are so similar, it is not surprising that many different species can interact readily at this level. For example, viruses do not make their own ribosomes; instead, they cause viral proteins to be synthesized on the ribosomes of their host organisms. Frog eggs, when injected with the messenger RNA from bees or rabbits, can be tricked into making honeybee venom or rabbit hemoglobin (Lane 1976). Whole human cells and mouse cells can be fused together to form hybrid cells that grow and reproduce themselves in tissue culture (Weiss and Green 1967). What is even more dramatic, a human gene coding for growth hormone, when injected into a fertilized mouse egg, can function during development to produce a mouse about twice normal size (Palmiter et al. 1983). And tobacco seedlings can be made to glow like fireflies by transferring the causative gene (for the enzyme luciferase) from the animal to the plant (Ow et al. 1986). Given this high degree of biochemical and developmental compatibility of gene products and processes among different species, it was inevitable that scientists would try to mix and match the genes themselves.

Some of these newer techniques are described here. Because it is difficult to present any one of them without making some reference to others, we bypass the strictly chronological approach that would begin with somatic cell genetics. Instead, we start with recombinant DNA, a methodology developed several years later. Along the way, we present a few of the many exciting applications to the physiology and mapping of human genes.

RECOMBINANT DNA

By 1973 the technology for gene splicing had been developed by researchers at two neighboring institutions, Stanford University and the University of California at San Francisco (Jackson et al. 1972). Using complementary base pairing as a kind of glue, they (and others) were able, for example, to insert DNA from the primitive frog *Xenopus* into the DNA of the bacterium *E. coli*, where the frog DNA subsequently reproduced itself as if it were "at home." Such a hybrid molecule (in this case a frog-bacterial molecule) is called **recombinant DNA**. Its formation does *not* involve reassortment of independent or linked genes of a species, as described in Chapter 11. Rather, recombinant DNA refers to the attachment of a piece of DNA from one species to that of a second species, followed by insertion of the hybrid molecule into a host organism (often a bacterium).

Making Recombinant DNA

The high-precision biological tool for making recombinant DNA is a special class of enzymes that cuts across a double helix in an interesting way. These special cutting enzymes, called restriction endonucleases, or **restriction enzymes**, are widespread among bacteria. Their properties were first investigated in the 1960s by Werner Arber, a Swiss biochemist at the University of Geneva and a 1978 Nobel laureate. Acting like scissors that can cut the DNA of viruses, they protect their owners from viral infection and *restrict* the range of viruses that can successfully invade. (Methylation of its own DNA protects the host from attack by its own restriction enzymes.) Hundreds of different restriction enzymes have been isolated from many species of bacteria; each recognizes a unique target sequence of usually four to seven bases (Old and Primrose 1985).

For example, one such enzyme called ***Eco*RI** recognizes the following sequence of six base pairs

··· G A A T T C ···
··· C T T A A G ···

and breaks the covalent bonds between the G and adjacent A nucleotides on *both* strands. Then the weak hydrogen bonds between the four intervening base pairs also break, and a staggered cut results:

··· G A A T T C ··· → ··· G + A A T T C ···
··· C T T A A G ··· → ··· C T T A A + G ···

Double-stranded DNA molecules with protruding single-stranded tips like these are said to have **sticky** or **cohesive ends**. This is because each broken end has four unpaired bases, AATT (A being the terminal base), which tend to spontaneously pair with any complementary end that is available—with the single-stranded tip from which it has just been separated, or with the tip of *any* other DNA molecule *cut with the same enzyme*.

A key point is that the *DNA molecules that join together need not be from the same organism, or even from the same species*. The hosts for much recombinant DNA work are special laboratory strains of *E. coli* carrying deleterious mutant genes that prevent them from surviving in the wild. Each *E. coli* cell has a circular chromosome, a DNA double helix coding for about 4,000 polypeptides. In addition to chromosomal genes, some strains of *E. coli* possess extra genes that confer attributes that the cells can usually do without. These dispensable genes are present not on the large bacterial chromosome but on small, circular pieces of DNA called **plasmids**, which often contain just a few genes. One to a few hundred copies of a plasmid may be present in a cell. Some plasmids possess genes that code for toxins against related bacterial species, and other plasmids possess genes that make the *E. coli* resistant to antibiotics of microbial origin.*

One "beauty of a plasmid" (Cherfas 1982), dubbed pBR322, is of particular interest to recombinant DNA workers. It has exactly 4,363 base pairs, within which just *one* sequence is recognized by the restriction enzyme *Eco*RI. After pBR322 plasmids are extracted, isolated in a test tube, and treated with *Eco*RI, each DNA circle is opened at only *one* point, exposing two sticky, single-stranded ends. *The crux of gene splicing is to mix the opened plasmids with foreign DNA that has been treated with the same restriction enzyme* (in this case *Eco*RI). When the two DNAs join, the plasmids become a vehicle (or **vector**) for transporting the foreign DNA into other cells.

The general process of vector preparation is diagramed in Figure 1. The human DNA is split into many different pieces determined by the positions of successive recognition sequences. Because any piece may be incorporated into the plasmid without foreknowledge by

*The latter plasmids, called R factors, are worrisome, because a given plasmid may confer resistance to several antibiotics and be transmissible to related pathogenic species of bacteria.

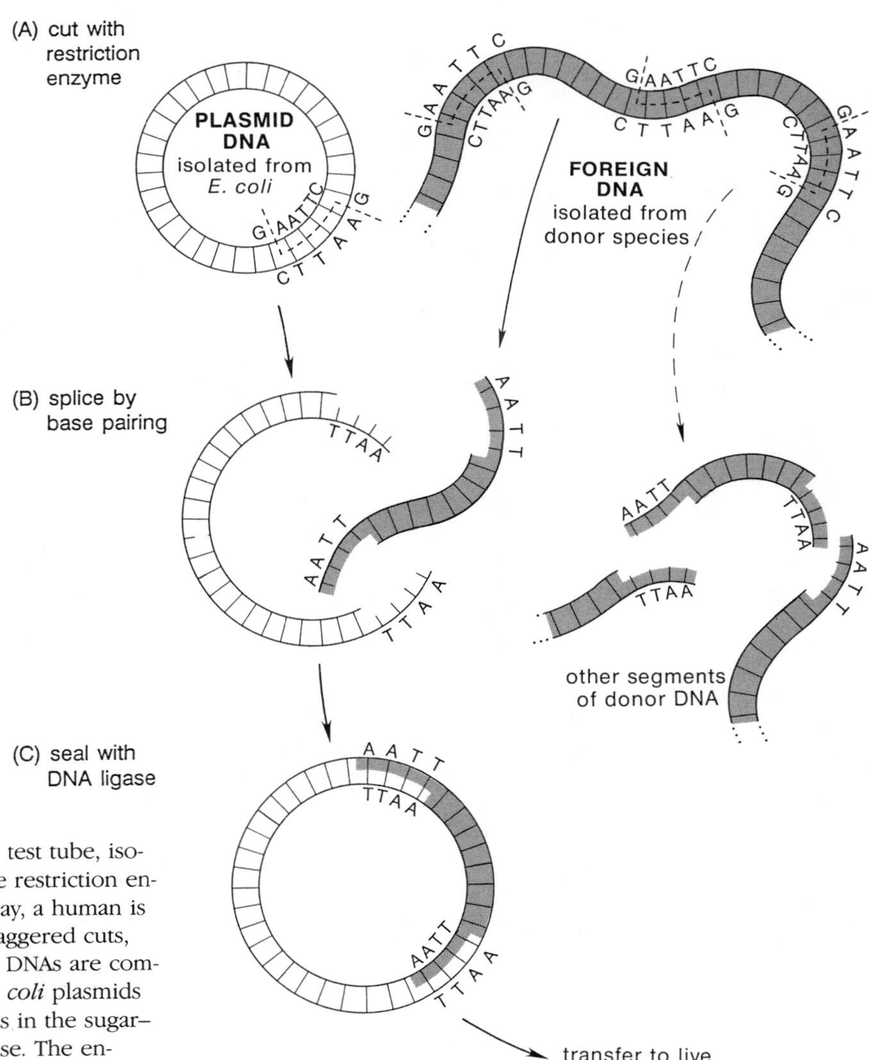

Figure 1 Preparation of a vector. (A) In one test tube, isolated plasmids from *E. coli* are treated with the restriction enzyme *Eco*RI. In another test tube, DNA from, say, a human is also treated with *Eco*RI. The enzyme makes staggered cuts, leaving sticky AATT ends. (B) When the cut-up DNAs are combined, some human DNA is spliced into the *E. coli* plasmids by complementary base pairing. (C) The bonds in the sugar–phosphate backbone are rejoined by DNA ligase. The enlarged plasmid is then transferred back into live *E. coli* cells.

the experimenter of which one it will be, this particular method is called a "shotgun" experiment. Reconstituted plasmids, enlarged by the addition of a piece of foreign DNA, can be put into other live *E. coli* cells that do not contain plasmids. The recipient bacteria are treated with calcium chloride, which alters the cell wall and permits an occasional plasmid to enter. Because any bacterium carrying the new plasmid has thousands of its own genes but only a few genes from the foreign source (equivalent, say, to adding just one word to six pages of this book), one cannot really call it a new species. The bacterium is essentially its old self—except that it harbors a bit of foreign DNA that replicates along with the plasmid DNA, and may control the synthesis of a few foreign RNA and polypeptide molecules. (Note that foreign DNA cannot function unless it is integrated into the host plasmid or chromosome; any unattached bits of foreign DNA that happen to get inside a cell are degraded by the host's nucleases.)

But only rarely does an *E. coli* cell treated with calcium chloride actually take up from the medium a plasmid with foreign DNA. How, then, can an experimenter isolate (and propagate) those *few host cells* from a background of millions of *E. coli* cells that have not picked up the plasmid? One solution is to start out with a plasmid that carries a gene for resistance to a specific drug, say, tetracycline. The plasmids carrying *both* the drug resistance gene and the foreign DNA are then added to some *E. coli* cells that are *not* resistant to the drug. Finally, tetracycline is added to the culture medium, whereupon all bacteria will be killed except for those few that have taken up the plasmid carrying both the foreign DNA and the tetracycline-resistant gene (Figure 2).

Manufacturing Proteins

A piece of foreign DNA that has been incorporated into a plasmid and later reproduces inside viable *E. coli* cells or (as we will note later) inside other host cells is said to be *cloned* (Watson et al. 1983). The plasmid is often called a cloning vector, and the process of making the recombinant DNA of the plasmid is called **molecular cloning**. The term *clone* here means multiple copies of

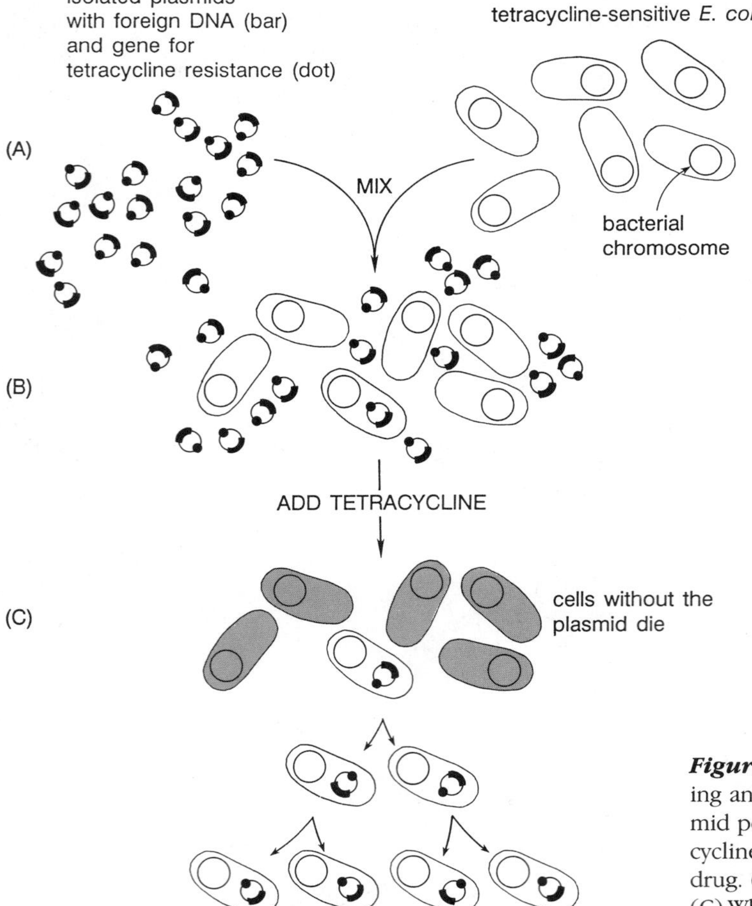

Figure 2 A selection procedure for identifying and isolating an *E. coli* cell that has taken up a plasmid. (A) The plasmid possesses a gene that makes its carrier resistant to tetracycline. *E. coli* cells not carrying the plasmid are killed by this drug. (B) An occasional bacterium takes up the plasmid. (C) When tetracycline is added to the growth medium, only cells that have taken up the plasmid survive and reproduce.

a given piece of DNA. Because the bacteria can be grown in prodigious quantities (a 10,000 gallon industrial fermenter has roughly the volume of a 12 × 14-foot living room), the foreign gene itself, its messenger RNA transcript, or its polypeptide product can sometimes be gotten in substantial amounts for investigative or practical purposes—all this from as little as one copy of one gene inserted into one plasmid that was taken up by one bacterial cell!

A California research team (Itakura et al. 1977) was the first to report on a human gene product "engineered" this way, having tricked *E. coli* cells into making a small peptide hormone called **somatostatin**. Just 14 amino acids long, somatostatin is normally made in the brain and acts to limit the production of other hormones. A deficiency of somatostatin can lead to the overproduction of growth hormone and a form of gigantism called *acromegaly*.

Knowing the genetic code and the amino acid sequence of somatostatin, these researchers first deduced a possible base sequence for the gene. Then they constructed a double-stranded helix, both sense and antisense strands, by artificially linking together the appropriate nucleotides (from bottles off their shelves). They tacked the DNA triplet for methionine onto the beginning and *two* stop triplets (just to be sure) onto the end (Figure 3A). Affixing appropriate single-stranded tips to this double-stranded structure of 17 coding units, they used two different restriction enzymes and DNA ligase to combine it with a section of an *E. coli* plasmid. Using an antibiotic selection technique, they screened for *E. coli* cells carrying the enlarged plasmids. Remarkably, the bacteria transcribed and translated the foreign insert into somatostatin as if it were a host gene.

Actually, the somatostatin molecule that the researchers isolated was attached to the end of another polypeptide, because the somatostatin gene had been deliber-

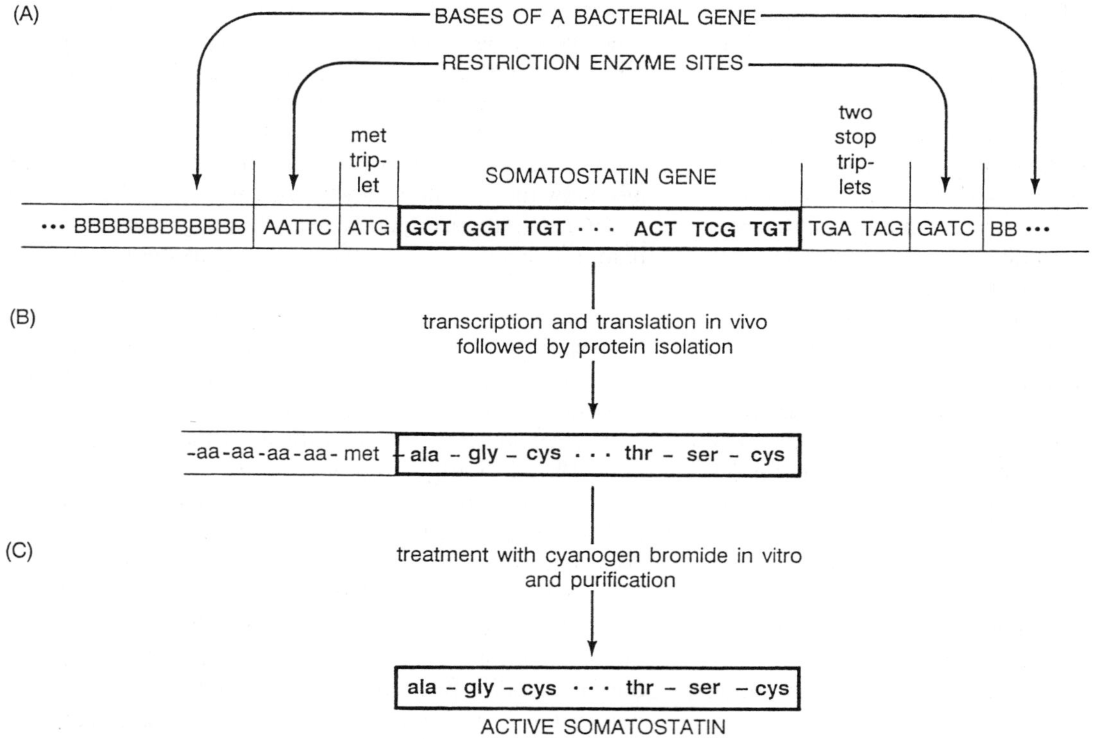

Figure 3 The experimental plan for making active human somatostatin in *E. coli*. Eight triplets are omitted from the middle of the somatostatin gene and the corresponding eight amino acids in somatostatin are also omitted. (A) The bases coding for somatostatin are inserted into a bacterial gene carried on an *E. coli* plasmid. Only one strand of the double helix is shown. By convention, this is the antisense strand, which is the same as mRNA, only with T instead of U. (B) Translation yields amino acids of the bacterial protein (—aa— · · ·) attached to methionine which, in turn, is attached to the 14 amino acids constituting somatostatin. This protein is isolated by standard techniques. (C) Treatment with cyanogen bromide frees somatostatin from methionine and fragments of the bacterial protein. (After Itakura et al. 1977.)

ately spliced into the middle of a bacterial gene, in just the right reading frame (Figure 3B). Thus the *E. coli* could use the promoter of the bacterial gene to start transcription of this section of the plasmid DNA. Fortunately, Itakura and co-workers were able to retrieve the translated somatostatin intact, by treating the hybrid protein molecule with a reagent (cyanogen bromide) that cleaves a polypeptide after any methionine residue (Figure 3C). But this clever biochemical trick worked only because somatostatin itself contains *no* methionine—so it cannot be used for typical proteins of a hundred or more amino acids, which usually contain methionine. Altogether, the in vivo bacterial synthesis of human somatostatin was a biochemical tour de force.

Many valuable pharmaceuticals for medicine and agriculture have already been produced by recombinant DNA technology, and more will certainly follow. Human gene products include insulin (for diabetics), growth hormone (for pituitary dwarfs), clotting factor VIII (for hemophiliacs), several different interferons (for possible use in viral diseases and cancers), and tissue plasminogen activator (for dissolving blood clots) (Old and Primrose 1985). At the time of this writing, the first two have gone through years of clinical testing and are available for general use, but the others are in various stages of evaluation and only available experimentally. Whether the basic discoveries, testing, and development are carried out in universities or in private industries, the general aim is the same: to have a constant and reliable source of pure, concentrated polypeptides (which may or may not be available from other sources). Biotechnology companies—about 200 of them in the United States—are, of course, interested in turning a profit, too.

The recombinant DNA production of human **insulin** was very similar to that of somatostatin (Crea et al. 1978; Johnson 1983). Researchers synthesized the gene for the A polypeptide chain from scratch (see Figure 14 in Chapter 13) and inserted it into a plasmid in one strain of *E. coli*. Likewise, they made the gene for the B chain from scratch and inserted it into a second strain of *E. coli*. Then the methionine–cyanogen bromide trick was used to obtain separate A and B chains, which were later linked together in vitro by the formation of the disulfide bonds.

The formation of disulfide bonds between and within the A and B chains was inefficient, however, so researchers tried an alternative method of making insulin—allowing the disulfide bonds to form naturally. To explain what "naturally" means, we point out that a *single* human insulin gene codes for a *single* polypeptide of 109 amino acids called *preproinsulin*, from which two sections are cut out to make the folded-and-joined A and B chains of active insulin (Figure 4). First, 23 amino acids at the free amino end (the "pre"peptide) are removed as the molecule passes through the internal membrane systems of the pancreatic B cell (Figure 4B). Then the resultant *proinsulin* molecule folds up spontaneously, the disulfide bonds form, and the "C-peptide" between the A and B chains is removed (Figure 4C). As readers may by now have guessed, in the alternative method for bacterial production of human insulin, investigators inserted into a single *E. coli* plasmid the nucleotides for the *proinsulin* polypeptide. Thus the natural folding of proinsulin brought together the appropriate cysteine residues, and the proper disulfide bonds were formed.

The recombinant DNA production of **human growth hormone (HGH)** differed from that of somatostatin and insulin in that the gene was not primarily synthesized from individual nucleotides. Instead, most of the gene was copied (by the enzyme reverse transcriptase) from the HGH messenger RNA that had been isolated from pituitary glands (Goeddel et al. 1979). The synthetic HGH differs from the normal human hormone in having an extra amino acid, methionine, ahead of the 191 amino acids of normal HGH. (This methionine was not added as a cutting site for cyanogen bromide, but as a start signal for translation. Researchers thought the methionine would be removed, but it was not.) Nonetheless, synthetic HGH seems to be safe and effective, and is currently being used for the treatment of pituitary dwarfism. Synthetic HGH was approved by the Food and Drug Administration in October 1985. This approval occurred just a few months after they banned the only other source of HGH—pituitary glands from human cadavers (Norman 1985; Sun 1985)—because several people treated with cadaver-derived HGH had died of Creutzfeldt-Jakob disease, a very rare and mysterious viral disease. If the bacterially derived HGH had not existed, several thousand American children and adolescents with HGH deficiencies would have been without effective treatment.

The recombinant DNA synthesis of **factor VIII**, a very large blood clotting protein (2,332 amino acids long), will be a boon to hemophiliacs who are unable to make it themselves (Lawn and Vehar 1986). Several years of clinical testing and scaling up of production are still required, however, before it becomes generally available. Since the 1960s the disease has been treatable with frequent injections of a protein concentrate from pooled donor blood, a program costing about $8,000 per year per patient. Furthermore, treated hemophiliacs are chronically infected with hepatitis viruses (and were at risk to acquired immune deficiency syndrome until appropriate assays became available). The production of factor VIII by recombinant DNA techniques required unprecedented technical creativity. (Some of the methods are described in more detail in the next section.) After screening a library of thousands of randomly cloned human DNA segments, investigators found the factor VIII gene; it turned out to be 186,000 bases long, with 26

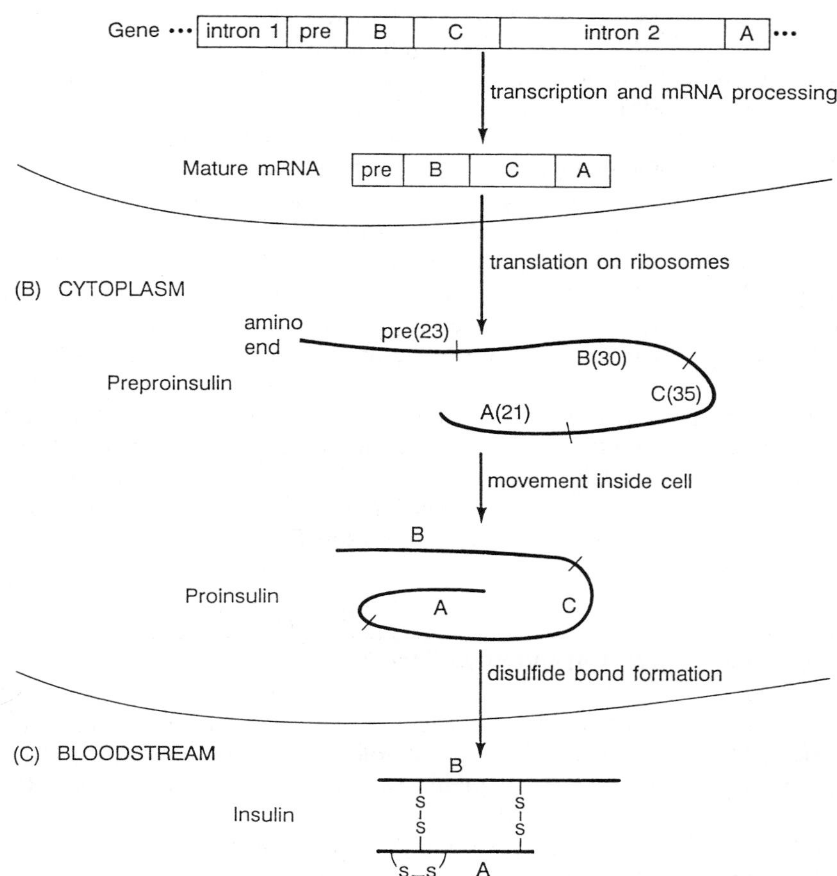

Figure 4 The biosynthesis of insulin in the pancreatic B cell. (A) *In the nucleus*. The insulin gene (on the short arm of chromosome 11) has two introns that are cut out of the primary RNA transcript to yield mature messenger RNA (whose cap and poly(A) tail are not shown). (B) *In the cytoplasm*. The numbers in parentheses represent the number of amino acids in each section of the primary polypeptide product. The "pre" peptide is needed to move the molecule within the cell's membrane systems. Before it is cut out, the C peptide helps fold the molecule for the formation of the disulfide bonds. (C) Insulin is secreted directly into the *bloodstream*. (After Robbins et al. 1984.)

separate exons that totaled about 9,000 base pairs. Eventually they managed to clone a gene sequence containing just the exons.

Another application of recombinant DNA technology is the production of **vaccines** for medical and veterinary use. Researchers try to clone the gene for the surface protein molecule responsible for the immunizing activity of a whole disease organism (Godson 1985). Vaccines consisting of just a *part* of the surface coat of a virus, for example, are inherently safer than vaccines containing killed or weakened organisms. It remains to be seen whether such synthetic vaccines will be sufficiently effective.

Organisms other than *E. coli* can also be used for cloning. The human factor VIII gene, for example, was cloned in hamster cells grown in tissue culture. An increasing favorite (for many types of genetic studies) is bakers' yeast, *Saccharomyces cerevisiae*. This one-celled organism is easily grown and manipulated, and can be maintained in either a haploid or a diploid state. Yeast is, moreover, a eukaryotic organism having a defined nucleus and a cellular organization like our own. With only about four times the DNA content of *E. coli*, the 16 discrete chromosomes range in size from about 200 to 2,000 kilobase pairs. Researchers have mapped about 600 yeast genes and can shuttle some very useful plasmids back and forth between yeast and *E. coli* (Watson et al. 1987).

Public Concerns

Many people believe that the techniques just outlined promise great benefits to human welfare; they are confident that accompanying risks can be evaluated, minimized, and adequately managed. But others object, stating that the risks to public health or environmental safety are unacceptably large, and/or it is morally wrong to tinker with nature in this manner, especially if the techniques touch on human reproduction.

In fact, some molecular biologists were the first to voice safety concerns. In the summer of 1973 a small group attending a conference on nucleic acid research requested that the National Academy of Sciences (NAS) study the potential hazards of recombinant DNA research. One year later, a distinguished NAS committee headed by Paul Berg (a 1980 Nobel laureate in chemistry) recommended that scientists throughout the world refrain from certain types of recombinant DNA experiments, pending further evaluation. This call for a moratorium on research by the researchers themselves was unprecedented in the history of science.

As one consequence, an international panel of scientists, lawyers, and others met in 1975 at the Asilomar Conference Center in Pacific Grove, California. After days of spirited and sometimes confused discussion, the conferees decided that research with recombinant DNA should be regulated by a set of safety rules, which they then helped to devise. By the time the National Institutes of Health (NIH) refined and promulgated these rules in July 1976, considerable public interest had been generated by eminent scientists who spoke out for or against such governmental regulation. (For a lively documentary history of the recombinant DNA debate, see Watson and Tooze [1981].)

The original NIH Guidelines for recombinant DNA research provided successively more rigorous standards for experiments that were perceived to be increasingly dangerous to public health and safety. (Strictly speaking, the Guidelines applied only to research funded by NIH, but it was hoped that other groups—including pharmaceutical companies—would also observe them.) At the lowest level (P_1), common sense laboratory methods for handling nonpathogenic microorganisms are sufficient. At intermediate levels (P_2 and P_3), laboratory access is limited, experimental methodology is much more meticulous, and special construction and ventilation features must be built in. For handling microorganisms that are extremely hazardous, a P_4 laboratory is needed.

In addition to the successive levels of *physical containment*, increasingly risky experiments require increasingly enfeebled strains of *E. coli*. This type of safeguard is referred to as *biological containment* and seeks to ensure that the engineered organisms do not survive if released into the environment. The Guidelines also called for an outright ban on some experiments, such as cloning in *E. coli* any DNA derived from known pathogens or cancer-causing viruses. Experience to date suggests that the original NIH Guidelines for laboratory research were too stringent; indeed, they have been relaxed several times since 1976 without untoward effects. In thousands of laboratories around the world, no known harm has yet been documented or even suspected to have occurred as a consequence of recombinant DNA research. Some biologists feel that, in complying with the NIH Guidelines, the hazards of handling recombinant organisms are no greater than the hazards of handling the same species as nonrecombinants.

But controversy continues, with recent debate focusing on the environmental safety of certain agricultural products. For example, wildtype *Pseudomonas* are bacteria commonly found on strawberry leaves, and certain protein molecules on the bacterial surface act as centers of ice crystal formation at near-freezing temperatures. Academic and industrial researchers have constructed strains of *Pseudomonas* that are missing the gene for the production of the cell-surface protein in question. When the engineered strain—called *ice-minus*—is sprayed on strawberry plants in extensive greenhouse experiments, it colonizes the leaf surfaces, excluding the wildtype populations and, to a degree, protecting the plants against frost damage. But when it came to testing these bacteria on actual crops in open fields, a tremendous row ensued. Academic and industrial personnel, environmental activists, biotechnology critics and proponents, the Environmental Protection Agency (EPA), the United States Department of Agriculture (USDA), and residents near the proposed test fields all became involved in a very spirited give-and-take of fact and opinion (Maranto 1986). At this writing, field testing is under way.

DNA MANIPULATIONS

A few additional "tricks of the trade" have increased our specific knowledge of color vision, Huntington disease, and sickle-cell anemia as well as our general understanding of human individuality. These research techniques have been applied to some other important genetic disorders as well, including cystic fibrosis, Duchenne muscular dystrophy, Alzheimer disease, and manic depressive disease.

Electrophoresis and Restriction Maps

We have seen that large molecules of DNA can be treated with restriction enzymes to give small molecules of various sizes. These can be separated from one another and visualized by **gel electrophoresis**, a major tool in biology and biochemistry. In this technique, DNA (or other) molecules migrate through a thin slab of gelatin-like material placed in an electric field. Because their regularly spaced phosphate groups are negatively charged, DNA molecules will move toward the positive pole—the smaller the piece, the faster the migration. Under some conditions (see Sequencing, below), a DNA segment with x bases will migrate *noticeably* further than a piece with $x + 1$ bases!

An apparatus for gel electrophoresis of DNA is illustrated in Figure 5. The wells at the left of the gel hold the solutions in the starting gates. As an example, let us assume we have linear (not circular) DNA molecules 20 kb long. We cut different samples with restriction enzymes—*Eco*RI, or *Bam*HI, or both enzymes—and place them in the wells of the gel as shown below:

- lane 1: DNA cut with *Eco*RI
- lane 2: DNA cut with *Bam*HI
- lane 3: DNA cut with both *Eco*RI and *Bam*HI
- lane 4: uncut DNA
- lane 5: a standard that is known to possess six different species of DNA with the following number of kilobase pairs: 2, 5, 7, 13, 18, and 20.*

*A commonly used set of control segments having lengths between 0.1 and 23 kb is the DNA of λ (a bacteriophage) that has been cut with the restriction enzyme *Hin*dIII.

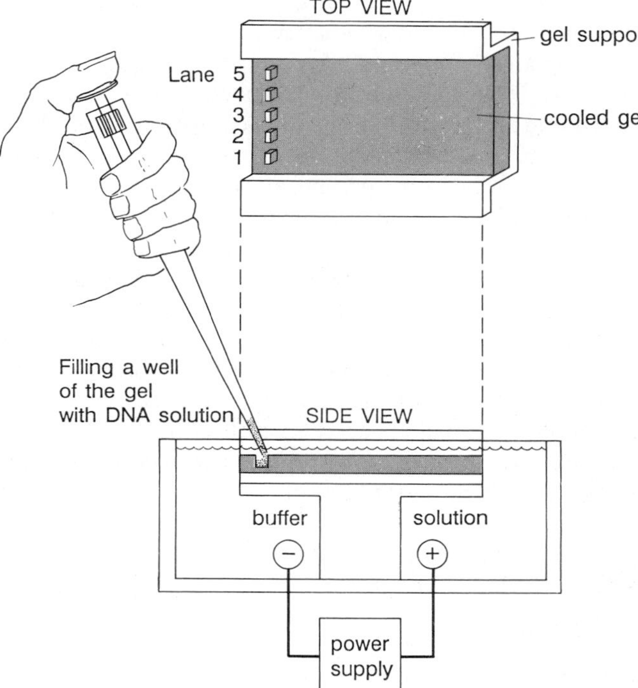

Figure 5 An electrophoresis apparatus for separating, identifying, and purifying pieces of DNA by size. The agarose gel contains ethidium bromide, a chemical that inserts between the base pairs of DNA and fluoresces bright orange under ultraviolet light. Hot agarose is poured into the plastic support, which is taped at the open ends and which holds a comb with rectangular teeth (five in the case illustrated) to make the wells. After the gel cools and sets, the tape and the comb are removed. Test solutions are pipetted into the wells of the slightly submerged gel. A heavy synthetic polymer is added to the DNA sample so that the solution sinks into the well. The samples also may contain *visible* dyes whose known migration rates mark approximately the progress of the *invisible* DNA molecules. A potential difference of 20 to 200 volts is applied across the gel for one to several hours. This potential difference causes the negatively charged DNA molecules to move toward the positive pole at a rate inversely proportional to the logarithm of their molecular weights. That is, the smallest pieces move the fastest. For details, see Maniatis et al. (1982).

Question: How many recognition sites does the original DNA have for the restriction enzymes, and how far apart are they? The hypothetical results of electrophoresis shown in Figure 6 provide a unique answer to this problem of **mapping restriction sites**. We note first that *Eco*RI (lane 1) cuts the DNA into two pieces (called **restriction fragments**) that are 2 and 18 kb long, so there must be one *Eco*RI recognition site relatively close to, say, the left end. The results of lane 2 show that the one *Bam*HI site is nearer the middle, since it cuts the DNA into two pieces 7 and 13 kb long. Whether the 7-kb or 13-kb *Bam*HI segment contains the *Eco*RI site is answered by lane 3 where the DNA is cut with both enzymes: the 7-kb piece is replaced by pieces 2 and 5 kb long, but the 13-kb piece is still intact. We conclude that the map of the DNA molecule is

```
  EcoRI        BamHI
----x------------x----------------------------------
 2  |     5      |              13                  |
```

Mapping the recognition sites of restriction enzymes in this way is a common practice in molecular biology because it provides signposts along a length of DNA, and points to where genetic manipulations may be possible. In addition, investigators can cut out the piece of gel which contains a single band, chemically remove the gel material, and thereby obtain a single, purified segment of DNA. This DNA could, for example, be the starting material for cloning or sequence determination (see below).

Conventional gel electrophoresis with a single, constant electric current (as described above) separates DNA fragments that are less than about 50 kb long, but all larger DNA molecules run at about the same rate. A variant process, called **pulsed-field gel electrophoresis**, uses electric fields that are set at angles to one another and are alternately turned on and off for a few seconds or minutes at a time (Schwartz and Cantor 1984). In one setup, 24 electrodes are arranged around the periphery of a hexagonal gel (Clark et al. 1988). Switching the field causes very long, wormlike DNA molecules to zigzag as they try to orient and reorient themselves parallel to the pores of the gel. The longer the molecules, the more their progress is hindered. Pulsed-field electrophoresis can separate and isolate DNA pieces up to several thousand kilobases, including even the DNA molecules of entire chromosomes of yeast and other organisms.

Sequencing

Sequencing DNA means determining the exact order of bases along a length of single-stranded DNA. One widely

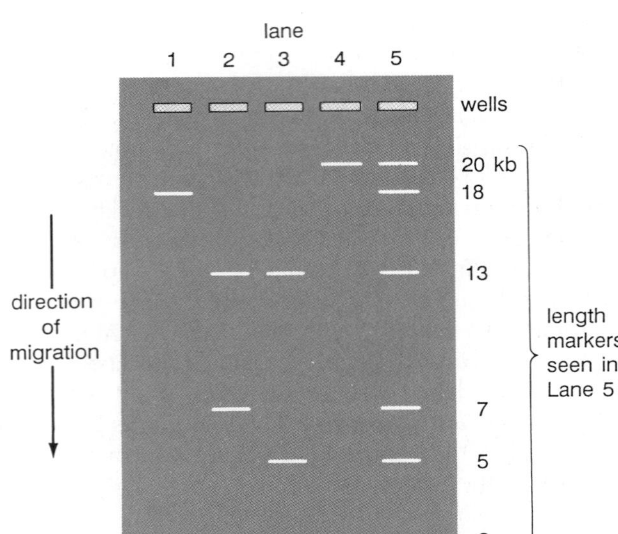

	Lanes			
	1	2	3	4
DNA cut with:	*Eco*RI	*Bam*HI	*Eco*RI & *Bam*HI	uncut
Lengths of DNA in kilobases:	18 2	13 7	13 5 2	20
Total length	20	20	20	20

Figure 6 A hypothetical gel as it would appear when linear DNA 20 kb long is cut with restriction enzymes, separated by electrophoresis, and then photographed under ultraviolet light. The bright bands contain DNA that has been rendered visible by binding the fluorescent chemical ethidium bromide. By comparing the positions of the bands in lanes 1 through 4 with the positions of the standards in lane 5, we see that they contain DNA molecules with the following number of base pairs. See text for further explanation.

used method was invented by Frederick Sanger* of the Medical Research Council Laboratory of Molecular Biology, Cambridge, England (Sanger et al. 1975). Called **dideoxy sequencing**, it relies on the incorporation into growing strands of DNA of some slightly altered nucleotides that are missing *two* hydroxyl groups on the sugar—hence these sugars are called *di*deoxy rather than just *deoxy*, like those in naturally occurring DNA. The second missing hydroxyl is from the 3′ carbon atom of the sugar (see Figure 1 in Chapter 13) where the next nucleotide in a growing chain ought to bond, but, because of the chemical change, *cannot*. As a consequence, when a strand of DNA is being synthesized in the presence of, say, the *adenine* dideoxynucleotide, this nucleotide adds normally into the growing strand, but it creates a dead end. *The chain is terminated at adenine.*

*Having invented first a method for determining the sequence of amino acids in proteins, and then a method for determining the sequence of nucleotides in DNA, Sanger is one of a very few individuals to win *two* Nobel Prizes, in 1958 and 1980. The latter was shared with Walter Gilbert of Harvard University, who helped devise another way to sequence DNA (the Maxam-Gilbert method), and Paul Berg of Stanford University, who pioneered recombinant DNA procedures. (The first person to win two Nobel Prizes was Marie Curie, in 1903 and 1911.)

The researcher who wishes to sequence a particular stretch of DNA sets up four test tubes labeled G, C, A, and T. Each of the four tubes contains DNA polymerase as well as all four normal deoxynucleotides (one or more of which are radioactively labeled). The tubes also contain identical copies of the purified, single-stranded "mystery" DNA, acting as the template against which the polymerase enzyme will zip the complementary bases into place. Critically important is the one variant ingredient: each tube contains one nucleotide present in the *dideoxy* form (in addition to its regular *deoxy* form), but in low concentration. The tube labeled A, for example, has some adenine dideoxynucleotide plus normal amounts of the A, T, G, and C deoxynucleotides. When the replication machinery calls for an adenine nucleotide (opposite T on the template strand), the deoxy form will usually be incorporated (because there is more of it) and replication will proceed normally. *But sometimes the dideoxy form will be incorporated, and then replication stops.* Thus, in the A tube, dead ends occur after any and all adenine dideoxynucleotides incorporated into the growing chains. As an example, assume a DNA template strand that looks like this:

3′ A T A G G C T A T C C C T T··· 5′

Let G, C, A, and T represent the normal deoxynucleotides, and **A** represent the dideoxy analogue. During replication in the A tube, all five molecules below will be made:

position 2

1. 3′ A T A G G C T A T C C C T T··· 5′
   5′ T **A** 3′

position 7

2. 3′ A T A G G C T A T C C C T T··· 5′
   5′ T A T C C G **A** 3′

position 9

3. 3′ A T A G G C T A T C C C T T··· 5′
   5′ T A T C C G A T **A** 3′

position 13

4. 3′ A T A G G C T A T C C C T T··· 5′
   5′ T A T C C G A T A G G G **A** 3′

position
14

5. 3′ ATAGGCTATCCCTT··· 5′
5′ TATCCGATAGGGA**A** 3′

When the DNA in this tube is denatured, the full-length template plus a population of five single-stranded segments will be present, *all of the latter ending in adenine*.

In a similar fashion, the tubes labeled G, C, and T synthesize populations of DNA segments ending, respectively, in all the G's, in all the C's, and in all the T's. Taken together (although they are physically kept separate) the four tubes contain all possible lengths of DNA: one nucleotide, two, three, four, and so on up to the total length (perhaps several hundred bases) of the template DNA. Furthermore, they can all be separated from one another and identified by gel electrophoresis.

To achieve the maximum resolution, the researcher makes the gels for DNA sequencing quite long (40 cm or more) and very thin (the gel material, polyacrylamide, is poured between glass plates spaced only a fraction of a millimeter apart). The researcher pipettes the contents of each reaction tube (G, C, A, or T) into one of four adjacent wells and turns on the current. After a suitable time, a DNA segment ending in a known nucleotide (because it is seen to be in either the G, C, A, or T lane) will have migrated discernibly further than a DNA segment just one nucleotide longer (and present in the same or a different lane). The fastest migrant is called position 1, as illustrated and further explained in Figure 7.

Thus the sequence of nucleotides complementary to a template DNA is read *directly* off an X-ray film exposed to the gel. Called an **autoradiograph**, the film shows the positions of radioactive bands in the gel. A length of several hundred nucleotides can be read off one gel (Figure 8), and confirmed by using as template the opposite (complementary) strand of the original double helix. Longer sequences can be obtained by combining the information of several cloned DNA segments that overlap one another. One very useful option is an electronic marking pen: by merely touching the bands successively from the bottom up, one automatically registers the sequence data in an attached computer.

The DNA sequencing procedure is now so good that in some cases it is easier to sequence a *protein* by first sequencing the *gene* for that protein and then inferring the amino acid sequence from knowledge of the genetic code! This approach was taken for two X-linked disorders: *hemophilia A*, for which the gene product, clotting factor VIII, was previously identified but not sequenced, and *chronic granulomatous disease (CGD)*, for which the gene product was unknown. The latter condition involves severe and recurrent bacterial infections due to unknown metabolic defects in phagocytic white blood cells. The gene, closely linked to the one for Duchenne muscular dystrophy, has now been cloned by somewhat

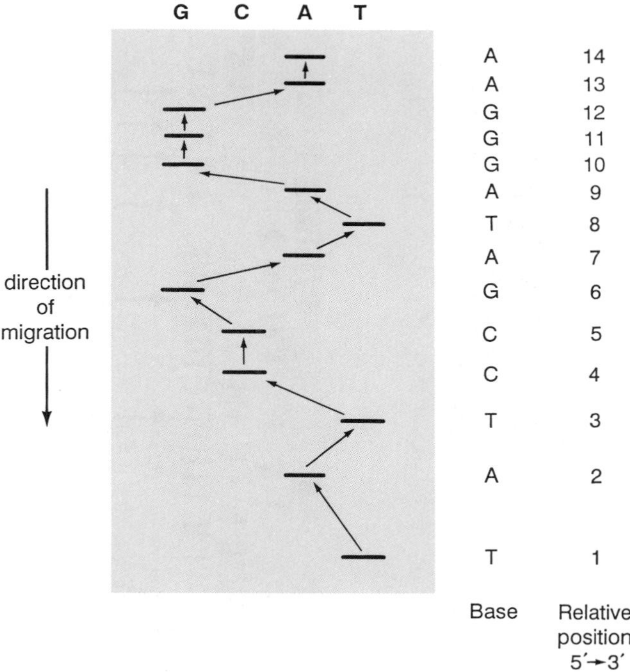

Figure 7 Drawing of an autoradiograph of a dideoxy sequencing gel. The bands in the four, parallel, vertical lanes represent the final positions of different sized segments of DNA ending in either G's, C's, A's, or T's. The smallest piece (length = 1 nucleotide: T) traveled in the T lane, the next smallest (length = 2 nucleotides: TA) in the A lane, and so on. Continuing to switch back and forth between the lanes as indicated by the arrows, the bands give the sequence of bases on the synthesized strand, the one complementary to the original DNA template. For example, the five bands seen in the A lane represent the five strands (with 2, 7, 9, 13, and 14 nucleotides) that are diagramed in the text. The bands appear dark because a radioactive element (for example, phosphorus-32) is present in the nucleotides incorporated into the DNA segments. When a sheet of X-ray film is exposed to the gel, it becomes sensitized wherever struck by radioactive emissions (high-energy electrons in the case of ^{32}P). Thus the drawing represents the developed film. In practice, a given gel may show several hundred readable bands (plus several hundred more toward the top that are too close together to be useful), as seen in Figure 8.

roundabout methods (see next section), and its protein is being investigated (Royer-Pokora et al. 1986). For some genetic diseases (such as cystic fibrosis and Huntington disease), the abnormal protein has not been positively identified by *any* technique at the time of this writing.

Some extended sections of human DNA have been sequenced—for example, the 45 kb encompassing the β-globin cluster of six genes. Included are start and stop signals, introns, repetitive DNA, and other genetic fea-

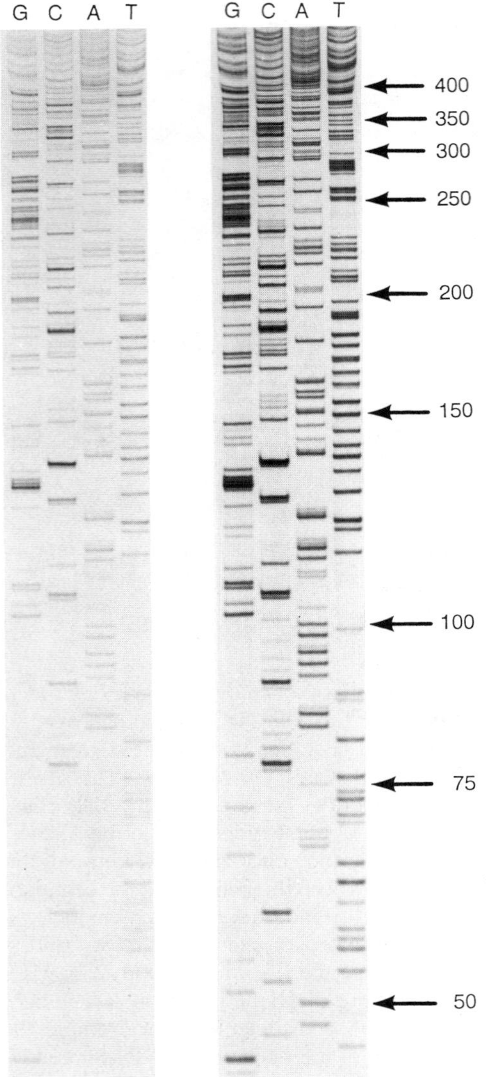

Figure 8 Autoradiograph of part of a dideoxy electrophoresis gel showing sequential nucleotides in the chromosome of a special single-stranded DNA virus of *E. coli*. The virus, named M13, is a favorite cloning vehicle for sequencing foreign DNA segments, which are inserted into one or another of the many unique M13 restriction sites that have been tailored for the purpose. This particular 40-cm long gel represents about one-half a day's laboratory work. X-ray film was exposed to the gel for 2 hours (left panel) and for 6 hours (right panel) in order to enhance the resolution in the top and bottom portions. The numbers measure the position of bands from a known start. The researchers were able to read nucleotides 1 through about 375 from this one electrophoresis run. Note that when two C's are adjacent (e.g., bases 58–59, 77–78) the band representing the first one (the bottom of the pair) is often lighter than the second—a peculiarity of the system. Confirm that the sequence from band 100 to band 110 is ACGAGCCGGAA. (Courtesy of Williams et al. 1986.)

tures (depicted in Figure 19 of Chapter 13). These nucleotide sequences, plus many others from a wide variety of organisms, have been brought together in several computerized databases, one of which, GenBank, sponsored by the U.S. National Institutes of Health (NIH), deals with both the nucleic acid and associated protein information. Thus, a small start has been made in knowing the finest details of the human genome.

Instead of accumulating such data willy-nilly, several national and international organizations have been set up for coordinating efforts to map and sequence all 3 billion bases of the human genome. It is expected that the endeavor—recently dubbed **genomics**—will take more than a decade of work by many technical teams and will cost billions of dollars (National Research Council 1988; Office of Technology Assessment 1988b). Despite early doubts about the goals of the colossal project and worry about draining funds from other desirable research, most investigators now agree that the known and possible benefits are worth the costs. "Besides the value of genomics in the understanding, diagnosis, and management of mendelian disorders, its usefulness in the category of somatic-cell genetic disease represented by cancer is becoming ever more evident" (McKusick 1989). There are likely to be economic spin-offs as well when the methods are applied to farm crops and animals. And, as is usual in research, unexpected results and insights are anticipated.

Improvements in the technologies of mapping and sequencing have been forthcoming, but more will be required to better automate the process (see, for example, Smith et al. [1986].) Further advances in computer-based methods for collecting, storing, distributing, and analyzing the data are also anticipated. Just to print 3 billion nucleotides requires about a half million pages like this one (which could contain about 6,000 characters). Variation in nucleotides among people, and between the two haploid genomes of a single person, will necessitate still greater computer-related capabilities.

Interest in the human genome project is centered in Europe, Japan, and the United States under the umbrella of the Human Genome Mapping Organization (HUGO) and includes a new international journal called *Genomics*. In this country, organizational and financial support comes from the National Institutes of Health and the Department of Energy. James Watson heads the NIH Office of Human Genome Research. In addition to coordinating research efforts, encouraging the development of new technology, and overseeing the massive job of data analyses, these organizations and others are also concerned with the possible misuse of the information. For example, assuring freedom from coercion in personal decisions and maintaining confidentiality of medical data are already familiar problems in health care, insurance, and employment; they may become even more difficult as the increased precision of personal data becomes a reality.

Making a Gene Library

Whether researchers want to sequence a gene or express it as a polypeptide, the appropriate DNA is often cloned inside an *E. coli* cell. *But where and how do researchers get the gene to put into the cloning vector?* Sometimes, as we have seen in the case of insulin, researchers can synthesize the gene from scratch if they know the amino acid sequence of its polypeptide product. There are several other methods, however, including searches through what is called a **human gene library**. In principle, this powerful technique allows investigators to find *any* gene of interest—by hunting for the right piece of DNA in a flask containing cloned pieces representing the *entire* human genome.

A library may consist of millions of bacteria, each containing a plasmid with a different insert of human DNA. If every base pair of human DNA is represented in at least one insert, the library is complete—an encyclopedia of genetic information divided into millions of volumes. For several reasons, however, bacterial plasmids are not usually suitable for housing human gene libraries. Instead, the cloning vector of choice is often a bacterial virus called a **bacteriophage**, or simply a **phage**. (The word *phage* is both singular and plural.) It is easier to maintain and manipulate the phage than the bacteria they infect. Also, some types of phage can accommodate bigger pieces of DNA than do plasmids, so the number of volumes in the library can be smaller. In practice, the physical form of the human gene library is likely to be about 1 million phage, each containing a different insert of human DNA. If desired, researchers can amplify this library by allowing the entire set of viruses to reproduce; also, it can be stored indefinitely.

A favorite virus for gene libraries is **λ bacteriophage**, which infects *E. coli*. During its life cycle, a wildtype λ attaches by the tip of its tail to the bacterial surface and then injects its DNA into the *E. coli* cell (Figure 9). The invading phage DNA takes over the machinery of the bacterial cell to manufacture more of its *own* DNA and its *own* protein components. The bacterium is lysed and releases a hundred or so exact copies of the one original infecting virus.

Roughly 20 kb of the λ DNA (out of the total of about 50 kb) is not needed for the life cycle outlined in Figure

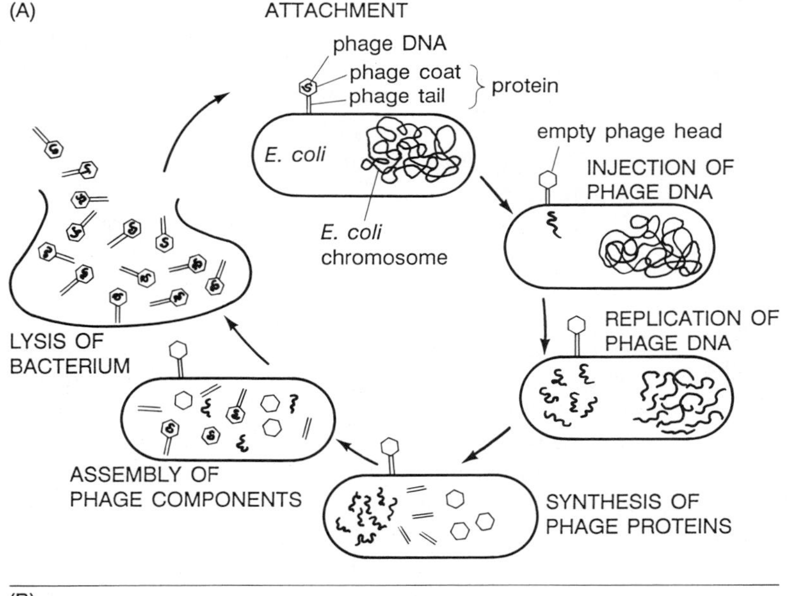

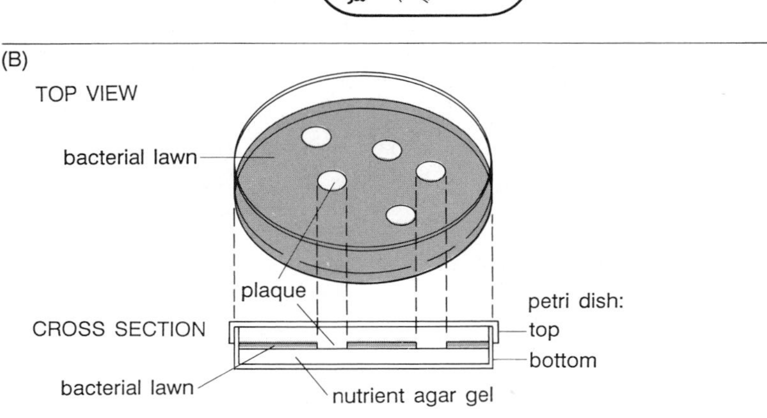

Figure 9 Aspects of λ, a virus that infects *E. coli* and is a useful cloning vehicle for making gene libraries. (A) λ life cycle. After attaching to the surface of an *E. coli* cell, the phage injects its DNA. Inside, the phage DNA replicates itself and directs the synthesis of the phage head and tail proteins. The *E. coli* ribosomes are used for the protein syntheses, while viral-encoded nucleases break up the *E. coli* chromosome. About 30 minutes after infection, lysis of the bacterium releases hundreds of progeny phage. The size of the virus particles (length = 0.1 μm) is exaggerated in relation to the bacterial cell (length approximately 2 μm). (Not illustrated here is an alternative benign pathway by which λ DNA is inserted into and replicates along with the *E. coli* chromosome, and in which the bacterium survives.) (B) A phage-counting technique. Phage particles can be counted by spreading a small number of them (suspended in a liquid) over the surface of an agar gel containing nutrients that support the growth of a solid, turbid "lawn" of *E. coli* cells. Clear spots, called *plaques*, develop wherever a *single* original phage and its many descendants kill all the bacteria in the neighborhood.

9A. In order for DNA to be efficiently packaged into λ heads, however, the total size of the genome must be approximately 50 kb. So researchers take advantage of this packaging requirement by replacing the 20 kb of nonessential λ DNA with 20 kb of random bits of human DNA. A piece of DNA of this size is the same order of magnitude as a typical human gene (including exons and introns) and perhaps some surrounding sequences.

Omitting many details, we present a typical recipe for making a human λ gene library by what is called a **shotgun** approach (Lawn et al. 1978).

1. Cut human DNA using *low* concentrations of a mixture of different restriction enzymes to produce randomized cuts in just *some* of the many recognition sites. A given recognition sequence may be cut in one copy of DNA but not in another; consequently, most DNA segments will have nucleotide sequences that overlap those of other segments. This is an important consideration for a technique called "chromosome walking" (see later).

2. Isolate DNA pieces about 20 kb long, either by electrophoresis of the population of DNA segments or by a process called *gradient centrifugation*. With the latter method, DNA molecules of different sizes settle at different density levels in a tube filled with a sucrose solution that is spun in a centrifuge at high speed.

3. Using special exonucleases, chop off the various single-stranded ends from the 20-kb DNA pieces to make them blunt. Attach artificially synthesized *linker* sequences (also blunt-ended and just 10 base pairs long) containing *Eco*RI recognition sites. Then add the restriction enzyme *Eco*RI, which cuts the linkers to produce *new* single-stranded ends —all the same.

4. Meanwhile, from special strains of λ, prepare DNA molecules with *Eco*RI ends at the sites where 20-kb pieces have been deleted from the middle and discarded. Then combine this deficient λ DNA with the *Eco*RI-cut 20-kb pieces of human DNA to get hybrid molecules that look like this:

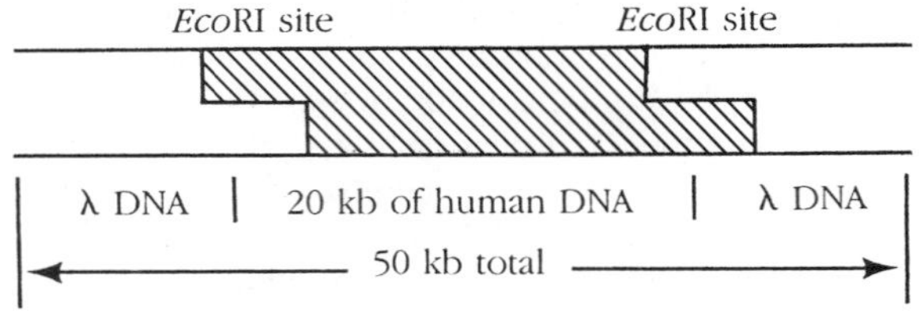

5. Using a remarkable piece of genetic technology called **in vitro packaging**, surround these hybrid DNA molecules with λ protein coats and tails (all in the absence of living *E. coli* cells). Construct roughly a million λ particles, each with a random 20-kb insert of human DNA.* The number, 10^6, is chosen so that the probability is high (about 0.99) that every piece of human DNA is present in at least one phage. Many sequences may be present more than once, a given gene or piece of a gene ending up, by chance, in a dozen or more phage.

6. As a last step, multiply the million-phage set by spreading it out on lawns of *E. coli* in the fashion of Figure 9B. (Each large petri dish may have up to 10,000 tiny plaques.) The petri dishes can be stored, or all the phage can be combined into one flask containing multiple copies of the initial library. In either case, a *specific* human gene of interest is very likely to be present within one or more phage.

Researchers make use of several sorts of human gene libraries. In one, the starting material is total human DNA isolated from any cell type, for example, a bit of liver tissue, sperm, white blood cells, or cells from a fetus. This **genomic library** contains all the DNA sequences (structural genes with introns, control sequences, pseudogenes, repetitive DNA, etc.) in proportions characteristic of the DNA in a cell nucleus.

A second type is called a **chromosome-specific library**; it contains inserts of the DNA from one human chromosome only. Researchers can isolate almost any specific human chromosome by first lysing mitotic metaphase cells and treating the isolated chromosomes with a fluorescent stain (Yu et al. 1984). Depending on its size and percentage of AT base pairs, each chromosome will absorb a characteristic amount of the stain and thus show a characteristic amount of fluorescence. The chromosomes are then passed, one at a time in tiny individual droplets, through a fluorescence detector. Every droplet showing an amount of fluorescence unique to the desired chromosome is given a small electrical charge to deflect it into a collecting tube. By this method, called *flow cytometry,* researchers achieve 80–95% chromosomal purity (as identified afterward by standard banding techniques). Droplets pass through the sorter at about 1,000 per second, and it takes about 12 hours to collect enough (i.e., a few million copies) of a specific chromosome to use for DNA isolation and cloning.

A third type of library is made of DNA that is copied

*The in vitro packaging system can be used to put *any* 50-kb piece of DNA inside a λ coat; one attaches to both ends of the 50-kb DNA the 12-base sequence found at the ends of λ DNA, the so-called *cos* sites (*cos* standing for "cohesive ends"). A bacterial plasmid packaged this way inside a λ shell is called a *cosmid*.

from messenger RNA by the enzyme reverse transcriptase. (This DNA copy is initially single-stranded, but it is later made double-stranded.) DNA derived this way is called *complementary DNA* or *cDNA*, and the library is called a **cDNA library**. The specific genes present in a cDNA library and their relative amounts depend on the distribution of the mRNA molecules that were isolated from the particular cell type used as a source. Because each tissue has specialized functions and thus produces a unique array of mRNA and proteins, the cDNA library based on, say, a liver cell would be different from a library based on a skin cell. Whatever the source, however, *genes in a cDNA library are missing their introns*, which were cut out during mRNA processing.

Screening a Gene Library

Zeroing in on a gene of interest in a library has been likened to finding a needle in a haystack. The searchers do not work haphazardly, however; they use a **radioactive probe**, which is single-stranded DNA complementary to a part of the gene being sought. The probe, which contains radioactive phosphorus-32 (usually written ^{32}P), is allowed to hybridize by base pairing with the DNA in a library (which has likewise been denatured to the single-stranded state). To bind to a gene and thus label it radioactively, the bases of the probe do not have to match perfectly, nor does the matching run have to be very long. *Thus the problem of finding the right gene in the library is essentially the problem of acquiring an adequate probe*. (Because that problem is sometimes difficult to solve, molecular biologists may begin by asking another researcher for a probe.)

One way to make a probe is to use the cDNA copied from the total mRNA of *specialized* cell types. For example, up to 95% of the mRNA in reticulocytes (immature red blood cells) is translated into the polypeptide chains of hemoglobin—so mRNA isolated from these cells is a source of globin gene probes. (In order to make a probe radioactive, researchers can replace some of its nucleotides with nucleotides that contain ^{32}P.)

Another way to get a probe is to synthesize a stretch of 15–30 nucleotides based on 5–10 known amino acids of the polypeptide product. Such a molecule is called an **oligonucleotide probe**. (The Greek prefix *oligo* means "few," in contrast with *poly*, meaning "many.") If the amino acid sequence of a given human protein is not known, researchers may guess that it matches or is similar to the known amino acids in the same or similar protein of another species. Of course, some uncertainty is introduced by degeneracy of the code, but the problems can be overcome: one makes a good guess as to which triplet is most likely to be used for coding, or one uses a mixed probe containing more than one sequence. For example, the initial probe for finding the clotting factor VIII gene in a genomic library was made this way from knowledge of 12 successive amino acids of factor VIII (Lawn and Vehar 1986). Although the researchers' 36-base probe turned out to have 6 wrong bases, it hybridized well enough anyway. Other procedures, beyond the scope of this book, have been devised for making gene probes.

Assuming that one has a probe for, say, the β-globin gene, one can screen a library as follows (Benton and Davis 1977):

1. Spread some library phage on a lawn of *E. coli* so each dish has thousands of plaques (Figure 10).

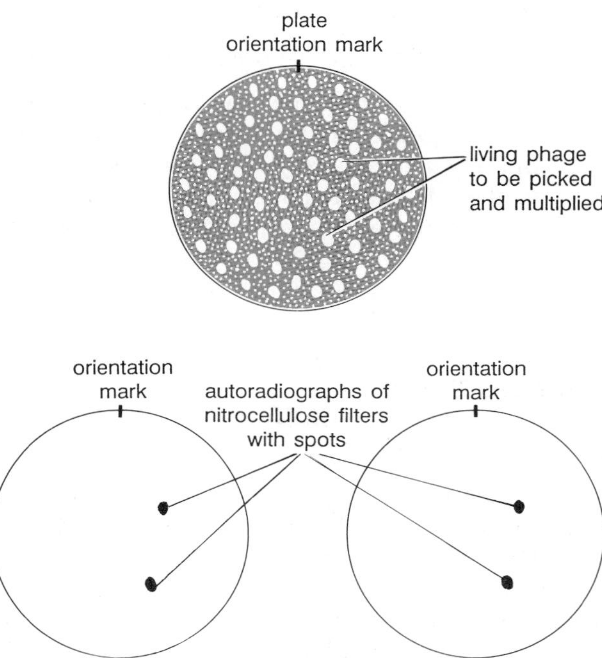

Figure 10 Screening a phage library. The petri dish at the top has a lawn of *E. coli* (the gray background) peppered with plaques (the clear spots). Each plaque arises from a single infecting phage with an insert of genomic DNA. The two circles below represent autoradiographs of nitrocellulose filters that had, in turn, been placed on top of the phage plate to pick up samples of phage from every plaque. The filters (exact replicas of the plate and of each other) were treated to break open the phage and release DNA. Then they were bathed with a radioactive probe for a specific gene—say, β-globin. After excess probe was washed off, the filters were autoradiographed to reveal where the probe had found and stuck to, by base pairing, complementary sequences. Thus the autoradiographs of the filters show the positions of phage that contain parts of the β-globin gene or closely related (say, δ-globin) genes. The plaque locations guide the investigator back to *living* phage on the original petri dish.

2. Lay a piece of nitrocellulose filter paper on the surface for a few minutes, so that a sampling of phage from each plaque is transferred to the paper. Repeat to make additional replicas. Figure 10 shows two filter papers, marked so that their relative orientations are known.

3. Treat the filter paper to break open the attached phage in situ (i.e., in place), to denature the DNA, and to attach the single-stranded DNA to the fibers of the filter.

4. Bathe the filter with the radioactive probe for the β-globin gene. During this step, the base sequence of the probe will hybridize to any complementary sequences on the filter. Wash off excess probe.

5. Cover the filter with X-ray film for autoradiography.

6. If spots appear on the developed film, it means that *the phage in the original plaques have an insert of at least part of the* β*-globin gene*. Go back to the (untreated) phage dish and pick out the plaques in the corresponding locations. Infect *E. coli* with the selected phage to amplify the β-globin segments for sequencing or other purposes.

Southern Blotting and the Molecular Biology of Color Vision

Recall (from Chapter 2) that color vision depends on three types of *cone* cells—labeled red, green, and blue—in the retina. Each type has a different light-sensitive protein pigment that absorbs a wide range of wavelengths, but with peaks of sensitivity at 560, 530, and 420 nm, respectively. (Although 560 nm corresponds to yellowish light, only these cone cells have sensitivity that extends into the red.) A fourth pigment, rhodopsin, present in *rod* cells of the retina, allows us to see in dim light, but not in color.

A remarkable series of investigations at Stanford University started with knowledge of just five adjacent amino acids of the rhodopsin molecule in cattle (Nathans and Hogness 1983). The researchers assumed there were similarities in the amino acid sequences of the several visual pigments of cows and people. From this small beginning, and after several years of work, the investigators managed to isolate and sequence not only the bovine and human rhodopsin genes, but also the genes that encode the three color vision pigments in the human retina (Nathans et al. 1986a,b).

First Nathans and his colleagues made a cDNA λ library, using mRNA from retinal cells of eyes obtained from slaughtered cows. Then they searched this library for the bovine rhodopsin gene, using a radioactive DNA probe that they made in vitro to match the five known amino acids of bovine rhodopsin. Screening the phage plaques by the nitrocellulose filter technique (see Figure 10), they identified several plaques with complementary sequences. These phage contained 1-kb segments of the bovine rhodopsin gene.

One of the bovine rhodopsin-DNA segments was then used as a probe to screen a human genomic library generated from the DNA of sperm from one of the investigators, a male with normal color vision. For this screening, the researchers varied the experimental setup:

1. Under one set of conditions, hybridization of the probe to library DNA required *long* stretches of complementary bases. This experiment identified a single human gene (now known to code for rhodopsin) showing 90% homology to the bovine rhodopsin gene.

2. Under less stringent conditions, even *short* stretches of homology led to hybridization and subsequent autoradiographic spots. A half-million plaques were screened in order to identify several weakly hybridizing clones, which proved to be parts of the human red, green, and blue pigment genes.

After sequencing these genes and then deducing the amino acids in their protein products, Nathans and his co-workers were able to compare the four human visual pigments (Table 1). Note that the red and green pigments are virtually identical to each other, 99% of their amino acids being the same or homologous; both are less similar to rhodopsin and the blue pigment, however. This similarity, and their nearby positions on the X chromosome (known from previous linkage studies), suggest that the red and green pigment loci are an example of gene duplication that occurred quite recently in evolutionary time.

Table 1 Percentage of identity and homology among the amino acids of the four human visual pigments

	Rhodopsin	*Red*	*Green*	*Blue*
Rhodopsin	100	40	41	42
Red	73	100	96	43
Green	73	99	100	44
Blue	75	79	79	100

Source: Nathans et al. (1986b)

Note: The values above the 100% diagonal are the percentages of amino acids that are *identical* in each pair identified by row and column. The larger values below the diagonal are the percentages of amino acids that are *either identical or similar* in their chemical properties. For example, the three positively charged amino acids, lysine, arginine, and histidine, are similar to one another.

One very surprising discovery was that the green gene can be present either one, two, or three times in a color-normal male. The evidence came from analyzing genomic DNA from 18 normal males—isolating each sample and digesting it with the restriction enzyme *Eco*RI, and then electrophoresing the samples in parallel lanes of a gel to separate the DNA fragments according to size. The researchers then used a favorite technique of molecular biologists, called **Southern blotting**, which is similar in concept to the phage-transfer method shown in Figure 10. After migration and denaturation, the DNA in the gel is absorbed onto a sheet of nitrocellulose filter paper, to which the single-stranded DNA sticks (Figure 11). When hybridized with a radioactive probe having a DNA sequence common to both the red and green pigment genes, the DNA on the filter produced the type of autoradiograph shown in Figure 12. In some men, the multiple copies of the green gene, represented by the darker bands, have likely arisen by **unequal crossing over** (Figure 13A and B).

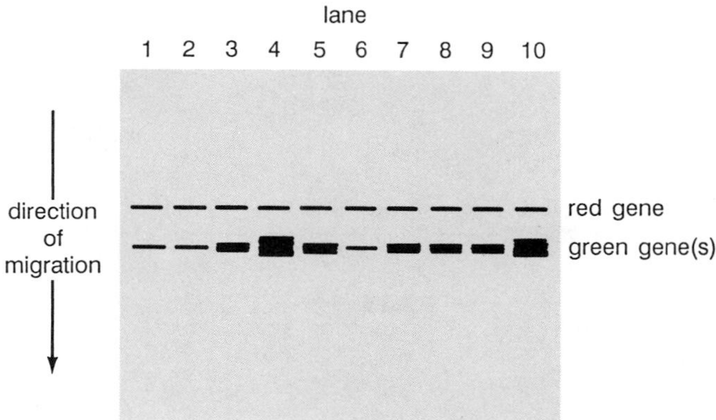

Figure 12 A drawing of an autoradiograph of a filter prepared by the Southern blotting method. The DNA of 10 phenotypically normal men was (a) digested with *Eco*RI, (b) electrophoresed individually in 10 parallel lanes, (c) denatured and transferred to a nitrocellulose filter, and (d) hybridized with a radioactive probe specific for sequences common to the red and green visual pigment genes. (e) X-ray film was then exposed to the filter to reveal where the probe bonded to complementary sequences. The upper band in each lane was identified as the red gene and the lower band as the green gene. Note that the thickness of the upper band is constant from lane to lane, but that of the lower band varies. The three levels of darkness represent, respectively, one copy (lanes 1, 2, 6), two copies (3, 5, 7, 8, and 9), or three copies (lanes 4, 10) of the green gene. (After Nathans et al. 1986b.)

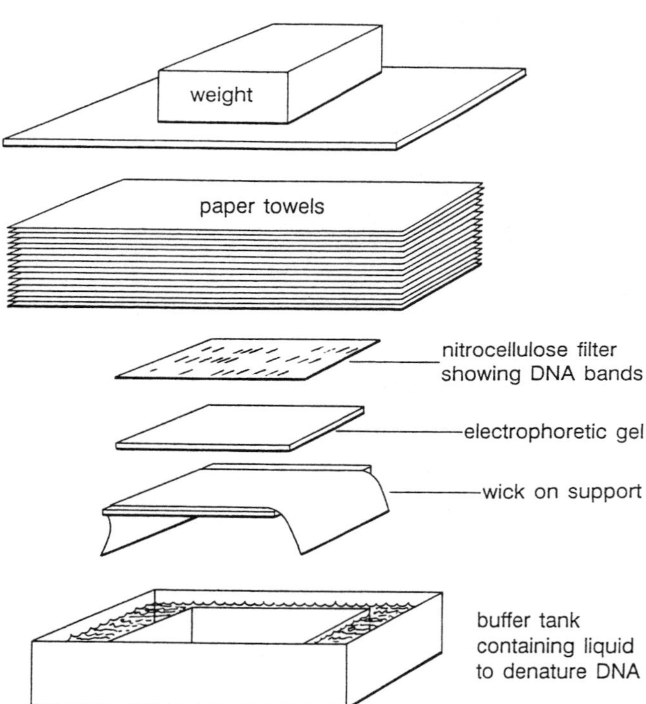

Figure 11 The setup for Southern blotting. This technique, invented by E. M. Southern of the University of Edinburgh (1975), is one of the most used procedures in molecular biology. A stack of absorbent paper draws up the liquid from the reservoir via the wick, through the gel, and through the nitrocellulose filter. DNA molecules from the gel dissolve in the upwardly moving liquid and *stick to the fibers of the filter*, where they can be easily treated. For clarity, the drawing includes extra space between the several layers. (In a play on the name Southern, a transfer technique involving RNA, rather than DNA, is called a Northern blot.)

To correlate specific *abnormal* phenotypes with specific genic defects, Nathans et al. (1986a) isolated DNA from 25 males with red–green color vision anomalies. Most of the DNA samples gave very abnormal Southern blotting patterns. This outcome suggests that mutations leading to red–green defects are *not* caused by point mutations. Rather, they seem to result from gross changes in the DNA—probably unequal crossing over of the types shown in Figure 13C. Within hybrid (i.e., partly red, partly green) genes, certain subregions seem to determine the actual spectral sensitivity of the pigments and thus the actual color vision defect. The ranges of possible breakpoints for crossovers lead to variable results that are currently under investigation.

Nathans and his colleagues also used Southern blotting to find out where the red and green pigment genes reside along the length of the X chromosome. They made use of the DNA from X chromosomes that had different segments deleted and were able to confirm that the red–green color vision genes are near the tip of the long arm of the X chromosome.

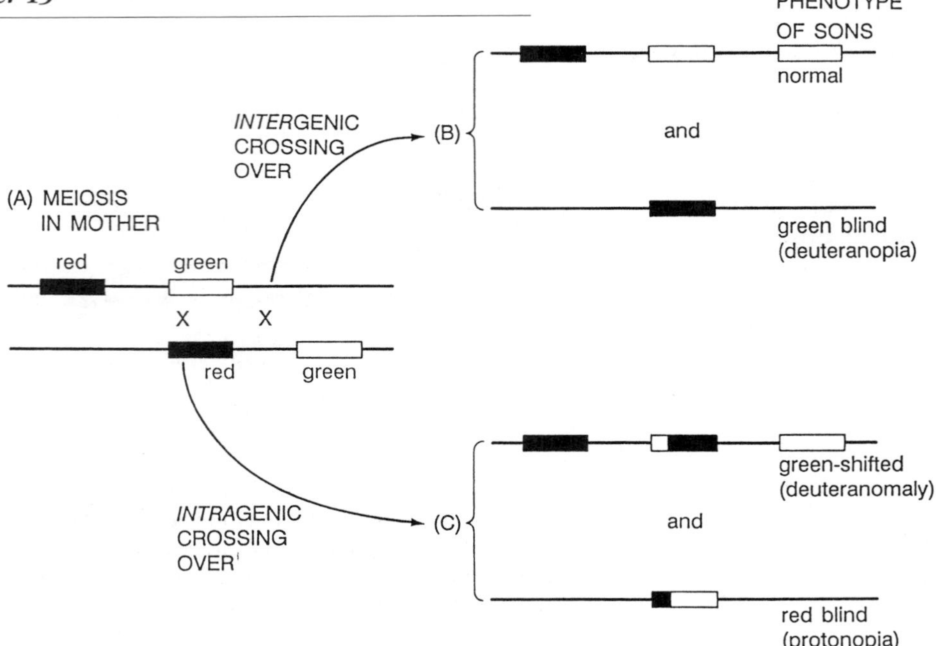

Figure 13 The results of unequal crossing over when the red and green pigment genes on the X chromosome misalign. (A) Meiosis in a female with a normal green gene (open bar) that has synapsed with a normal red gene (solid bar). Two (out of many) possible positions for a crossover are indicated, one outside the coding regions (intergenic crossover) and one within the mismatched genes (intragenic crossover). (B) The consequences of the *inter*genic crossover. A son who inherits the upper X chromosome will have normal color vision, although two copies of the green gene are present. A son who inherits the lower X chromosome is lacking the green gene and will be green blind (deuteranopia). (C) The consequences of the *intra*genic crossover. Nathans et al. (1986a) suggest that the son who inherits the upper X chromosome will be green-shifted (deuteranomaly) because some of the cone cells have the pigment coded by the hybrid gene (green → red), a "green-like" gene, but not entirely normal. Sons who inherit the lower X chromosome appear to be red blind (protonopia): no red pigment is produced because the single hybrid gene (red → green) codes for a normal green pigment.

Restriction Fragment Length Polymorphism

The use of restriction enzymes to digest the DNA of chromosomes has revolutionized the mapping and isolation of genes for inherited human disorders. In a related application, the new methodology can detect naturally occurring variation among the nucleotides of *normal* individuals. For example, the genomes of two unrelated people differ from each other in roughly one base in every several hundred bases (Botstein et al. 1980). Yet there may be no corresponding phenotypic differences, especially if the variable sites lie in the spacer regions between genes, in repetitive sequences, in introns, or in the degenerative third base of coding triplets. Nonetheless, these sites represent *allelic differences at the fundamental, DNA level.*

But how does one find such inconsequential differences? Earlier in this chapter we discussed restriction mapping. Now consider what would happen if some of the phenotypically neutral allelic changes in DNA by chance altered the recognition sites of restriction enzymes. At a particular spot, the enzyme will cut one person's DNA but not another's. For example, suppose that a certain section of John's DNA has four *Eco*RI sites (shown by **x**'s) spaced as follows:

x —— 7 kb —— **x** — 1 kb — **x** —— 4 kb —— **x**

whereas the same section of Jan's DNA has only these three *Eco*RI sites:

The difference between John and Jan could have resulted from a point mutation in any one of the bases, G A A T T C, that should make up Jan's second *Eco*RI recognition site. This mutation, perhaps occurring many generations back, produced two alleles—one allowing the restriction enzyme to cut (as it would in John) and one preventing the cut (as in Jan). Such allelic differences between John and Jan can be detected by the Southern blot technique, *if* one has available a radioactive probe whose sequence is complementary to this region. The experimenter would take the *Eco*RI digests of John and Jan's DNA, run them in an electrophoretic gel, denature the migrated DNA, then blot the single strands onto a nitrocellulose filter, hybridize with the probe, and autoradiograph. The result would look like this:

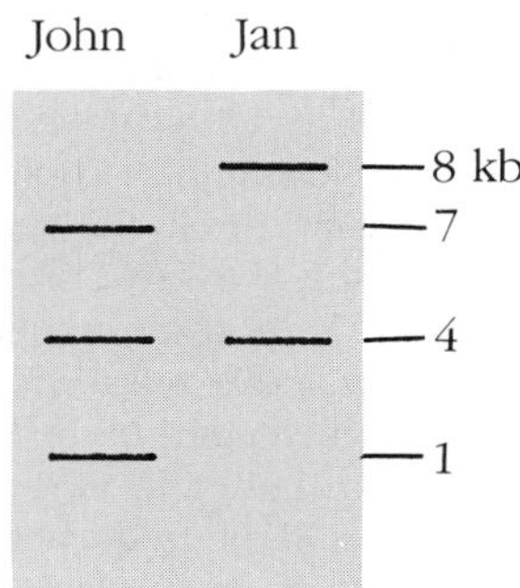

Both John and Jan have the 4-kb segment, but their other DNA restriction fragments have different lengths.

Assume that, in the general population, some people resemble John and some resemble Jan (and some may have different electrophoretic patterns, based on mutations of the other *Eco*RI sites). To describe a population in which there are several common genetic variants at one locus, the term *polymorphic* (meaning "several forms") is used. Here we are examining polymorphisms in the *lengths* of restriction fragments—thus the phrase, **restriction fragment length polymorphisms** or **RFLPs**.

A mutant site producing an RFLP represents a genetic locus with two codominant alleles that can be identified by a laboratory test (Southern blotting). The situation is similar in principle to typing blood (by agglutination) or determining the presence, absence, or variants of an enzyme (by simple electrophoresis and staining). Because they occur randomly throughout the human genome, RFLP sites are very useful for mapping.

Huntington Disease and Sickle-Cell Anemia

The pioneering work of several teams of scientists led by James Gusella of Harvard Medical School (1983, 1985) established that the dominant autosomal gene for **Huntington disease (HD)** is linked to an RFLP site located near the tip of the short arm of chromosome 4. Recall from Chapter 12 that the symptoms of HD begin in middle age, leading to progressively more severe muscular writhing, mental disturbances, and death. Many persons with HD (for example, the folk singer Woody

Guthrie) produce their offspring before any symptoms occur, so these children live with the knowledge that they have a 50% chance of being similarly affected (Conneally 1984). Huntington disease investigators hope that their work will help in the early diagnosis and treatment of the disease.

Gusella and his colleagues obtained white blood cells from the members of several extended families in which HD occurred. DNA was extracted from each cell sample and digested with the restriction enzyme *Hin*dIII. The fragments produced by *Hin*dIII were electrophoresed, transferred by the Southern blot method, and hybridized with a radioactively labeled segment from a phage library of human genomic DNA. The plan was to systematically try one random probe after another, until one of them produced a pattern of DNA bands that correlated with the inheritance of HD. That is, the researchers looked for a distinctive set of electrophoretic bands whenever the DNA sample was from a person with Huntington disease; they hoped to see a pattern different from that of normal individuals. The strength of the correlation would be a measure of the closeness of the linkage between the RFLP site and the gene for HD.

Although the investigators thought they would have to try hundreds of random probes and dozens of different restriction enzymes, they were successful within the first few runs—when a *random probe called G8 detected a* Hin*dIII RFLP site linked to HD*. As shown in Figure 14, this RFLP region actually involved *two* restriction sites located 18.7 kb apart (sites 1 and 2). The four different chromosomes or alleles (sometimes called *haplotypes*) all occur in the general population in appreciable frequencies, so the heterozygotes needed for mapping studies (see Chapter 11) are also frequently encountered. Because 1 map unit corresponds to roughly 1,000 kb, crossing over between the two *Hin*dIII sites should be quite rare. (These sites are about 18.7/1,000 or about 0.02 map unit apart).

In an initial screening of an Amish family with HD, the disease gene was always associated with the *A* allele, and no crossovers were positively identified. The constant association of *A* with HD would be expected if all affected members of the pedigree were descended from a single person in which the HD mutation happened to occur on an A chromosome, and if the linkage with the RFLP site were close. In a much larger Venezuelan pedigree of over 3,000 people,* the disease gene was always

*Nancy Wexler of the Hereditary Disease Foundation (and others) traced this extended Venezuelan family back to a single affected person living in the early 1800s (Pines 1984). P. Michael Conneally of Indiana University designed computer programs for keeping straight the complex relationships among about 100 living persons in the pedigree who are victims of HD and over 1,000 who are at risk.

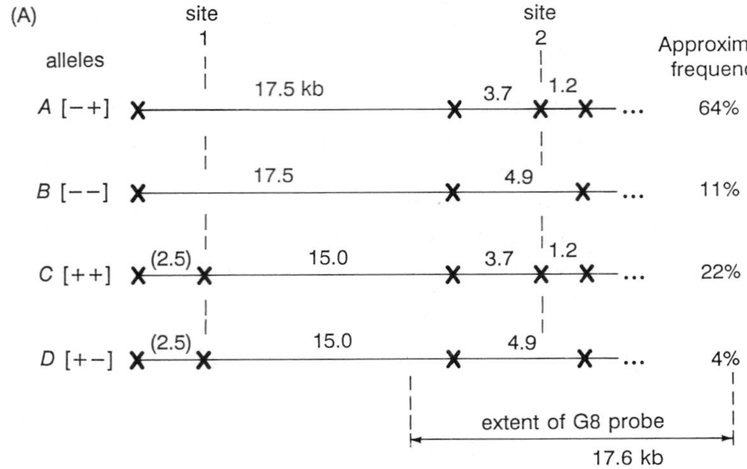

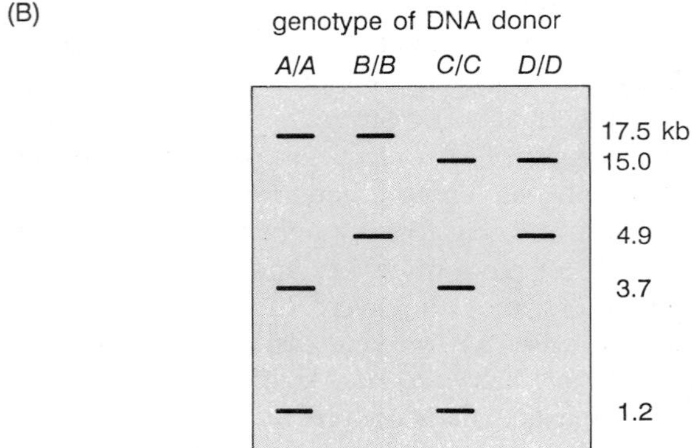

Figure 14 The *Hin*dIII RFLP sites in the region of chromosome 4 that hybridize to the G8 probe. (A) Four alleles (labeled at the left: *A*, *B*, *C*, and *D*) are produced by two variants at two nearby *Hin*dIII sites (labeled at the top: site 1 and site 2). Along each chromosome, an **x** symbolizes the position of AAGCTT, the recognition sequence for the restriction enzyme. Allele *C* has a total of five such sites in this region. The other alleles are altered at site 1 or site 2 or both; the alterations prevent cuts by the restriction enzyme. [Within brackets, + means *cut* and − means *no cut* in sites 1 and 2, respectively.] The lengths of the fragments are indicated in kilobases. These four alleles are present in the general population in the frequencies noted at the right. The G8 probe itself extends from the right part of the large segment and into a region (not shown here) that has additional *Hin*dIII sites. (B) Banding patterns in an autoradiograph of Southern blot of the DNA from the four restriction site homozygotes. Note that the 2.5-kb fragments from alleles *C* and *D*—although present on the blot—are not visible. They are invisible because they do not hybridize with the radioactive probe, whose left end fails to extend that far left. In the actual course of investigation, banding patterns like these were used to deduce the alleles shown in (A). (After Gusella et al. 1983.)

associated with the *C* allele, and again no crossovers were initially seen. At first, very close linkage was suggested, but later analyses showed that the RFLP and the HD gene are about 4 map units apart (Gusella et al. 1986). Improvement of the estimate came partly from the discovery of additional polymorphic sites within the region that hybridizes with the G8 probe, making most people RFLP heterozygotes.

Researchers hope to find an RFLP site closer to the HD gene in order to provide better presymptomatic screening for the HD gene. For the symptom-free members of HD pedigrees who choose to be tested, analysis with the current G8 probe may not be clear-cut. In the first place, the test is totally uninformative if the HD parent of the tested individual is homozygous for the RFLP site, because the HD gene cannot be associated with one particular chromosome. Or the test may be informative but not precise enough. For example, let *A*, *B*, and *C* represent the RFLP alleles, *H* represent the HD allele, and *h* represent the normal allele, and assume 4% recombination. Then one such mating would be:

HD parent	×	**non-HD parent**
$\frac{A \quad H}{B \quad h}$		$\frac{C \quad h}{C \quad h}$

which yields four genotypes among the children:

Genotype	Frequency
A H / *C h*	48%
A h / *C h*	2% (a recombinant type)
B h / *C h*	48%
B H / *C h*	2% (a recombinant type)

Among those who possess the *A* allele (first two lines), 48/50 or 96% will also carry the *H* allele; but 2/50 or 4% of the time a prediction of Huntington disease will be wrong, because the *H* allele was exchanged by crossing over. Conversely, anyone with the *B* allele will most likely carry the *h* allele as well, but 4% of the time a comforting prediction of freedom from Huntington disease will be wrong. Although not huge, a 4% error rate in such a fateful situation makes the search for a closer site imperative.

Going from G8 to a closer marker by a method called

chromosome walking is possible but tedious. First one screens a human genomic library with G8 to find a second probe (closer to the HD gene) whose base sequence overlaps a part of G8, but whose non-overlapping part extends further toward the gene. The second probe can then be used to find yet a third, and so on. Four map units, however, is about 4 million bases, and the human DNA inserts in a λ library are about 20 kilobases each. Thus about $4 \times 10^6/20 \times 10^3$ or *200 "steps" would be needed to reach the HD gene*. Rather than "walking," a method (not described here) of "jumping" in increments of perhaps 200 kb has now been undertaken. Even if a completely predictive test were available, however, in the absence of any treatment to ameliorate the disease symptoms, the decision to have the test would still be highly personal and fraught with difficulties. For example, if one family member does not want to know what the future holds, will closely related individuals be denied information to which they are rightfully entitled?

Although presymptomatic screening for relatives of Huntington's victims is not yet perfected, a completely predictive test, based on a restriction enzyme recognition site, now exists for the prenatal detection of **sickle-cell anemia** (Chang and Kan 1982, Orkin et al. 1982). Unfortunately, this condition is also one for which little can be done to treat the severe symptoms; but parents now have the option of prenatal diagnosis and possible abortion of affected fetuses. The sickle-cell test is virtually 100% reliable, because the restriction sequence in question is not just *inside* the gene coding for the β chain of hemoglobin, it *actually spans* the mutated triplet that leads to the altered polypeptide!

The restriction enzyme used in the sickle-cell test, *Mst*II, cuts DNA at the following sequence (where NN′ stands for any base pair).

··· G G A N T | C C ···
··· C C | T N′ A G G ···

In the gene for the *normal* β chain of hemoglobin, this sequence lies in the region coding for the fifth, sixth, and seventh amino acids indicated as follows (sense strand only):

*Mst*II site

··· G G A C T C C T C ···

—pro— glu— glu—

5th 6th 7th

*Mst*II recognition sites occur other places as well. In particular, the two nearest sites on the left and right of this one are positioned as follows (**x** represents the G G A N T C C recognition sequences):

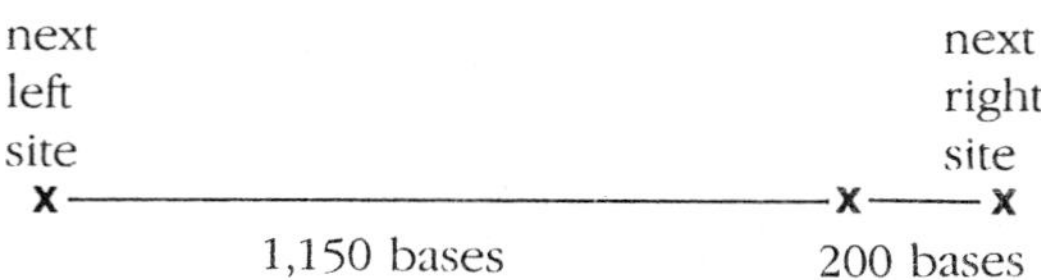

After *Mst*II digestion of DNA from cells of *normal* fetuses, two fragments (1,150 and 200 base pairs) are obtained, along with many other sequences from elsewhere in the genome.* Workers then apply the Southern blotting technique, using a radioactive probe specific for a base sequence included within the 1,150-base fragment. The 200-base fragment is not needed for the analysis of normal fetuses, so it is ignored.

Recall from Chapter 14 that sickle-cell anemia occurs when valine replaces glutamic acid as the sixth amino acid of the β chain, following a T→A mutation in the middle base of the sixth triplet of DNA. But this mutated sequence of nucleotides is not recognized by *Mst*II. Thus, after *Mst*II digestion of DNA from sickle-cell fetuses, researchers get a fragment 1,350 (i.e., 1,150 + 200) bases long that shows up at a different place on the autoradiograph (Figure 15). This test can be done with as little as a microgram (10^{-6} g) of DNA, an amount obtainable directly from the amniotic fluid surrounding a fetus. Newer techniques using about 20 nanograms (20×10^{-9} g) of DNA, have made the test even more sensitive (next section). Thus the use of restriction enzymes to produce DNA fragments of different lengths is not only useful in basic mapping studies but has a very practical value in diagnostic screening.

DNA Fingerprinting and the Polymerase Chain Reaction

The β-globin or G8 probes used to identify sickle-cell or Huntington disease genotypes work because each binds to a *unique* sequence of nucleotides. Any probe that is complementary to *repetitive* DNA, on the other hand, would bind to *many* segments—wherever the repeated sequences occurred in the genome. With this in mind, Alec Jeffreys of the University of Leicester in England used repetitive DNA probes to develop an interesting procedure called **DNA fingerprinting** (Jeffreys et al. 1985a,b). Because the test results from each person are different, this technique is the molecular equivalent of physical fingerprints for personal identification, and it promises to be very useful in genetic analysis and forensic medicine.

Recall from Chapter 13 that there are many places in

*Fetal cells are collected either from the amniotic fluid or from chorionic villi by techniques described in Chapter 25.

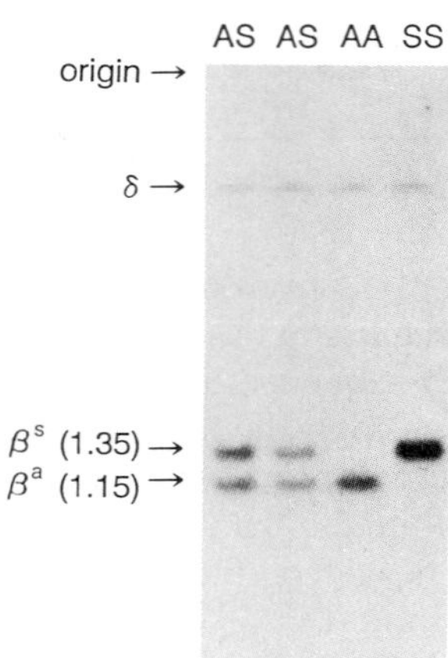

Figure 15 Autoradiograph of a Southern blot used for diagnosis of sickle-cell genotypes. DNA samples from individuals with the genotypes shown at the top of the lanes were digested with *Mst*II and probed with a radioactive β-globin gene sequence. *A*, normal allele; *S*, sickle-cell allele; AS lanes, DNA from phenotypically normal heterozygous parents; AA lane, DNA from their homozygous normal child; SS lane, DNA from amniotic fluid cells surrounding their fetus homozygous for sickle-cell anemia. The numbers 1.35 and 1.15 refer to kilobase segments explained in the text. The barely visible band in some lanes at δ is due to cross-hybridization with the δ-globin gene sequence. (After Chang and Kan 1982.)

human DNA where a particular sequence of nucleotides is repeated over and over. For example, the half-million Alu repeats (each with several hundred bases) are dispersed throughout the genome and represent 5% of the total DNA in each human cell. In other cases, a quite short sequence is tandemly repeated many times within a defined DNA region, and the same pattern with perhaps a different number of repeats occurs at other places. For example, a particular sequence includes the following 16 nucleotides:

AGAGGTGGGCAGGTGG

This sequence is tandemly repeated hundreds of times at dozens of different regions scattered over most, if not all, of the human chromosomes.

DNA fingerprinting—that is, individual-specific DNA identification—is made possible by the finding that no two people are likely to have the same number of copies of repetitive DNA sequences at all of the regions where the sequences occur. It is not known why people differ so much in the number of repetitive units, but the variation may result from unequal crossing over during many past generations. The variation in copy number within a region leads directly to variation in the lengths of the repetitive regions. Different lengths of repetitive nucleotides can be detected by treating a DNA sample with any restriction enzyme that does *not* cleave the repeat unit. The filter produced by the Southern blotting procedure contains the DNA segments resolved by size. These fragments are hybridized with a radioactive probe that binds to the repetitive DNA. Jeffreys developed one probe (called probe 33.15) that consists of 29 repetitions of the 16-base sequence shown above. Other probes for repetitive DNA detect other sequences and lead to a different set of electrophoretic bands after autoradiography.

For any individual's DNA treated with *one* restriction enzyme, run on *one* electrophoretic gel, and hybridized with *one* probe for repetitive DNA, many bands appear on the Southern blot—somewhat like a supermarket bar code. For example, the probe 33.15 detects about 15 distinct fragments that are between 4 and 20 kilobases long. (Some shorter repetitive DNA segments migrate off the gel and some bands are faint or smeared.) In general, these fragments represent nonallelic DNA segments that have been inherited from the person's parents. The degree of variation in fragment sizes from one person to another is so huge that the "probability that all resolved fragments in one individual are present in another unrelated individual is estimated at 4×10^{-11}" (1 in 25 billion) (Helminen et al. 1988). Even sibs have a low probability of identity with just a single probe that highlights, say, 15 bands, because their parents (assumed to be unrelated) are likely to possess between them four different "alleles" at any repeat-site. Thus the probability that any one fragment in sib A is also present in sib B is 1/2; the likelihood of complete identity between sibs is then $(1/2)^{15}$ or about 1 in 33,000. In essence, the various sizes of the many repetitive DNA regions are like dozens of highly polymorphic genes, all detected by a single test.

DNA fingerprinting was first used in England during a rape-murder trial in which DNA isolated from the rapist's sperm was compared with DNA from the white blood cells of suspects. It has also been used by the English immigration service to positively match children with their putative parents. More recently, the procedure has been used in a number of United States courts with increasing levels of acceptance by the legal profession (Lewis 1988).

Although the test typically requires at least microgram amounts of DNA, another important technological advance allows some types of fingerprinting to be done with nanogram amounts of DNA. This is the amount of DNA that is present in the cells at the base of a *single* human hair—perhaps all the evidence that is found at the scene of a crime. The technique, the **polymerase chain reaction**, was conceived and developed by Kary Mullis and other researchers of the Cetus Corporation of California and has wide applicability. Essentially, one nucleotide sequence (of up to about 2 kb) is singled out from all the DNA present in a sample, and *this portion alone* is multiplied about a million fold! Thus, starting with just a few copies of double-stranded DNA—or even with *one* copy—investigators can rapidly amplify it enough to do whatever analyses are called for.

The key to the polymerase chain reaction procedure is a *pair* of synthetic oligonucleotide probes whose sequences bind to the two DNA strands (sense and antisense) at opposite ends of the particular region to be replicated (Figure 16). Each probe acts as a primer to initiate DNA replication (Saiki et al. 1988a). In Figure 16A, note that the direction of replication insures that the region of interest is included in *both* new strands. (Recall from Chapter 13 that DNA polymerase only works in the 5′ to 3′ direction.) In the next cycle of replication, both old and new strands provide sites for the binding of primers (Figure 16B). After several cycles of replication, the shortest segments, *including all of the region of interest* (Figure 16C), will outnumber by far the longer segments that contain extraneous sequences to the left and right.

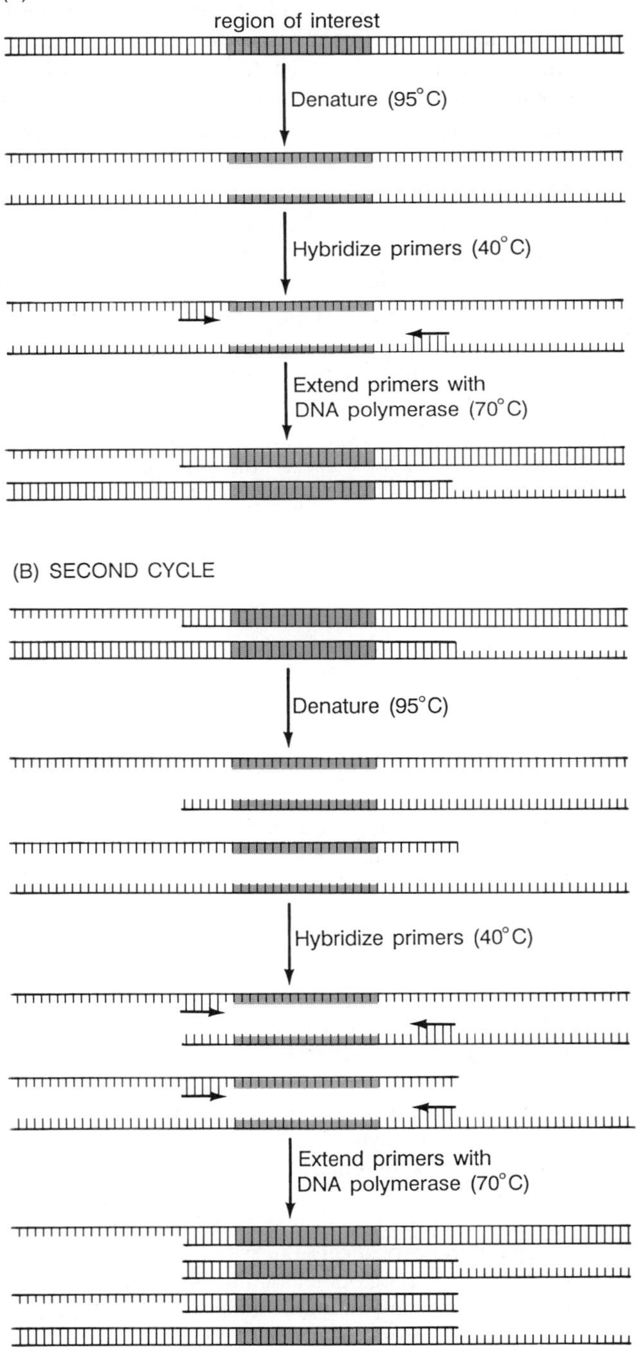

The entire process occurs continuously in a single reaction tube with the same initial set of reagents. It makes use of a special DNA polymerase that was isolated from a thermophilic ("heat loving") bacterium that lives in hot springs and survives extended incubation at 95°C—just under the boiling point of water. (Virtually all ordinary enzymes are inactivated by such a high temperature.) Changes in the temperature of the reaction mixture control the phases of successive cycles: 95°C to denature, 40°C to attach the primers, and 70°C to extend the complementary copies with polymerase. After two or three hours, the DNA region of interest—i.e., that within the sequence spanned by the two primers—is amplified a million times or more.

The generous amounts of DNA can then be used in a variety of analytical manipulations. One application is limited DNA fingerprinting from a single hair (Higuchi et al. 1988). Another is quick and sensitive prenatal diagnosis of sickle-cell anemia or β-thalassemia (Saiki et

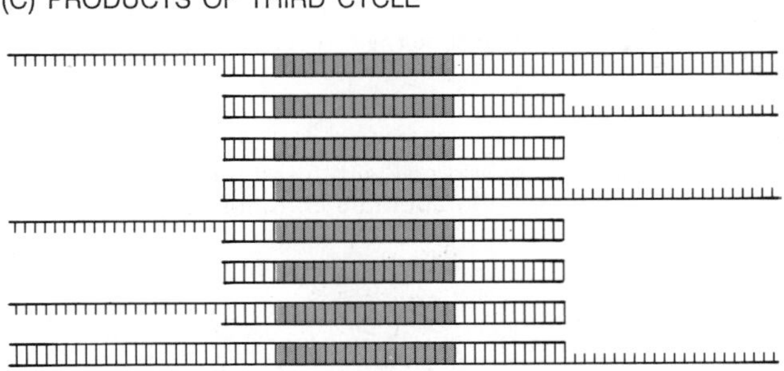

Figure 16 The polymerase chain reaction, which amplifies a DNA region of interest (shaded segment). The first two cycles are shown in detail in (A) and (B), but just the end result of the third cycle is shown in (C). In subsequent cycles, the shortest segments, spanning just the region between the two primers, multiply exponentially. Twenty cycles produce about a million (2^{20}) copies. (By contrast, the longest segments, corresponding to the original DNA double helix, remain constant at two copies, and the intermediate length segments increase by only two copies per cycle.) (After Mullis et al. 1986.)

al. 1988b). Other uses include DNA sequencing, the analysis of archeological samples of DNA, the study of cancer-causing genes, and the detection of infectious disease pathogens including the AIDS virus (references in Erlich et al. 1988).

SOMATIC CELL GENETICS

By 1966 pedigree studies had confirmed about 600 autosomally inherited and about 70 X-linked human genes, with the possibility of a genetic basis for about 800 additional traits. But not even one autosomal gene could be assigned to a specific chromosome, nor could human geneticists tell where the two known clusters of X-linked genes were localized. Now, however, dozens of human gene assignments are reported every year—and scientists foresee the day when most or all of the human genome will be mapped.

This stunning turnabout in human gene mapping started in 1967, when Mary Weiss and Howard Green (then at the New York University School of Medicine) reported on the use of human–mouse cell hybrids to assign the human gene for the enzyme thymidine kinase to chromosome number 17. This newly developed *parasexual* method, called **somatic cell fusion**, bypassed sexual reproduction altogether (Greek *para*, "beside" or "accessory to"). Used in conjunction with traditional family studies, it revitalized the field of linkage analysis; now also used in conjunction with even newer techniques described earlier in this chapter, it remains a powerful tool for human genetic analysis.

Selecting and Isolating Hybrid Cells

Somatic cells to be fused or "crossed" are first cloned—that is, generated by mitotic cell divisions from a single ancestral cell obtained from a mouse (or other rodent) or from a human.* Cells of each type are then placed together in a culture dish with nutrient medium and left for hours or days to grow and divide. In the mixed populations that result, several human–mouse hybrid cells will form spontaneously—but the frequency of fused cells can be enhanced up to 1,000-fold by adding to the culture medium either a chemical (polyethylene glycol) or an inactivated virus (the Sendai virus) that causes cells to clump together. A fused cell at first contains two separate nuclei (Figure 17), but during successive cell divisions, they coalesce and form a single hybrid

*Note that here the term *clone* refers to the replication of *whole cells*, rather than of DNA segments. The term can also apply to whole organisms that are genetically identical.

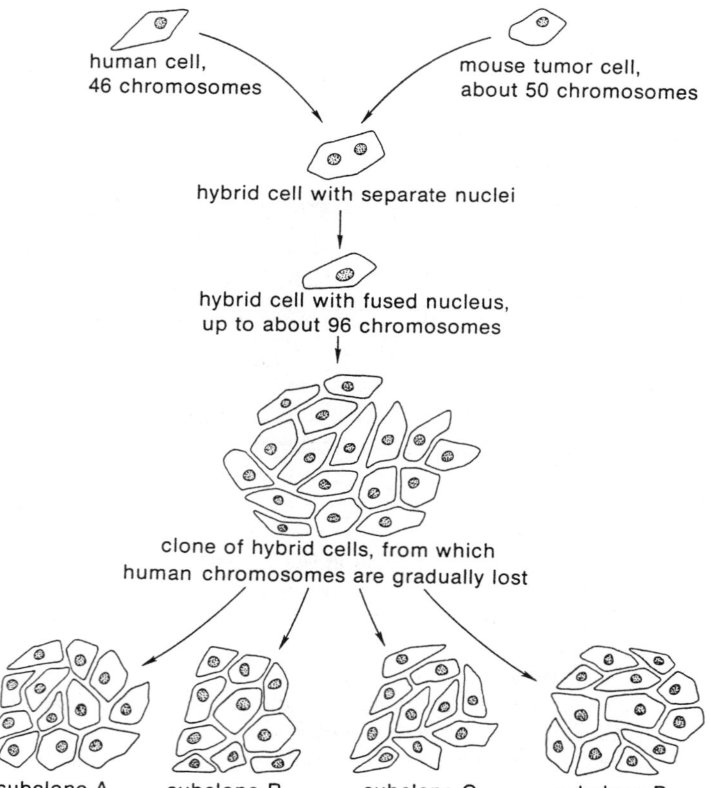

Figure 17 Human and mouse somatic cells are mixed together (on a culture plate not shown here) in the presence of Sendai virus. (Usually the mouse line consists of tumor cells with about 50 chromosomes rather than the normal 40.) Some human and mouse cells unite to form interspecies hybrids whose nuclei at first remain distinct but then join together. Each hybrid cell divides repeatedly to form a clone, from which individual hybrid cells are transferred to fresh culture dishes to form subclones. Because human chromosomes are more or less randomly lost during repeated divisions of these hybrid cells, each subclone usually ends up with a different array of chromosomes. With regard to the essential "household" functions carried out by all cells, both mouse and human genes are expressed in the hybrid cell. But specialized gene products (such as hemoglobin, albumin, and so on), which normally occur only in highly differentiated cells, may or may not be produced by cell hybrids.

nucleus. The genes of both parental types may be expressed in the hybrid cell, which divides and redivides to form a colony on an agar surface.

Colonies of hybrid cells are not necessarily any different in appearance from those formed by the parental-type unfused cells, so their detection and isolation becomes a problem. For this task, the selective techniques developed in the 1940s by microbial geneticists are useful. For example, some mutant strains of bacteria have no obvious physical phenotypes, but they can be distinguished by their nutritional requirements. Likewise, a number of recessive biochemical mutants of mammalian cells have been collected and maintained. As a hypothetical example, suppose that one mammalian line (call it L^-) lacks a functional enzyme that is necessary for the production of substance L, and that another cell line (M^-) lacks an enzyme needed to produce substance M (Figure 18). Assume that in order to grow a cell needs a source of both substances—either synthesized by the cell itself or supplied in the growth medium. Thus, if the two cell types are placed on medium containing both substance L and substance M, then *both* types can grow and divide. But if the two cell types are placed on a medium containing neither substance L nor substance M, then *neither* type can form colonies.

What happens when L^- and M^- cells fuse together? Since L^- cells carry the normal allele for the synthesis of M (they are really L^-M^+), and M^- cells carry the normal allele for the synthesis of L (L^+M^-), a combination of the two types will be mutually complementary, that is, ($L^+/L^-\ M^+/M^-$) or phenotypically L^+M^+ (Figure 18B). Thus the fused cells can actually survive and form colonies in a medium that contains neither substance L nor substance M. It is this *ability of a hybrid cell to grow where neither parental line can survive* that enables geneticists to detect and isolate even extremely rare hybrid cells from a background of mostly parental-type cells.

One commonly used selective system for mouse–human hybrids, the **HAT method** (Littlefield 1964), utilizes two mutants in a biochemical pathway leading to the synthesis of DNA. (The name for the technique comes from the first letters of two DNA precursors, hypoxanthine and thymidine, and the first letter of the drug

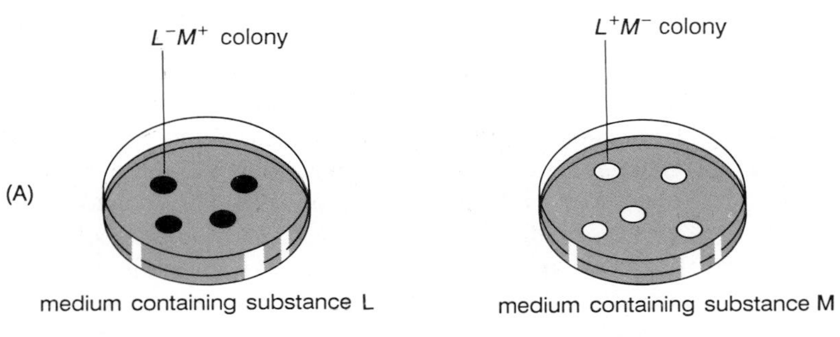

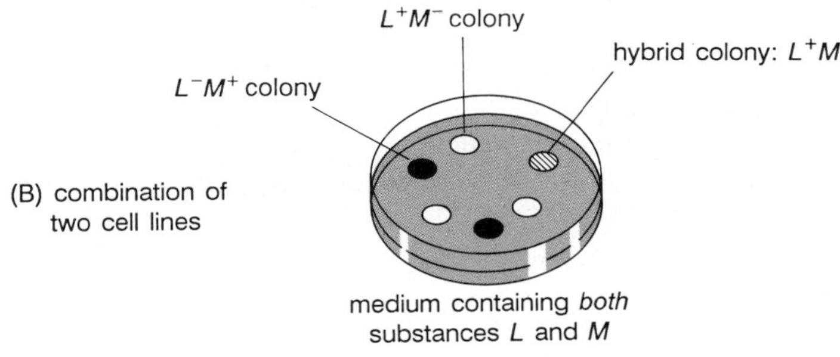

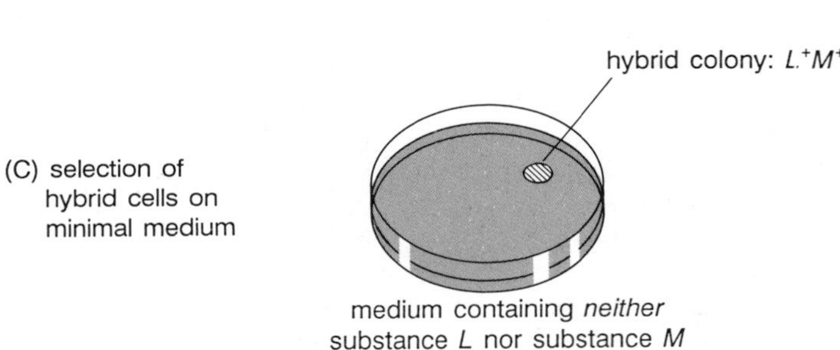

Figure 18 Selection of hybrid cells by the use of parental types with special growth requirements. (A) In the culture dish on the left, the dark spots represent colonies (clones) of L^- mutants, which can grow only if substance L is added to the growth medium. On the right, the clear spots represent colonies of M^- mutants, which can grow only if substance M is added to the growth medium. The L^- mutants carry a gene, M^+, for the synthesis of M, and the M^- mutants carry the normal L^+ gene. (B) If both L^- and M^- cells are placed on a culture medium containing both substances L and M, then both L^- and M^- colonies will grow. In addition, some hybrid clones (hatched) containing L^- and L^+, as well as M^- and M^+ genes, are also formed. (C) If a mixture of cells is transferred to a minimal medium containing neither substance L nor substance M, then only the hybrid cells will survive to form colonies.

aminopterin.) As shown in Figure 19, there are two pathways for DNA synthesis: (A) from scratch, that is, from simple sugars and amino acids and the appropriate enzymes, or (B) from other precursors (called nucleosides), if the major pathway is blocked by mutation or drugs. The second (salvage) pathway requires the presence of two functioning enzymes, HPRT (hypoxanthine guanine phosphoribosyl transferase) and TK (thymidine kinase); if either or both enzymes are defective, no DNA can be produced by this pathway.

It happens that the drug aminopterin, when added to the culture medium, blocks the synthesis of DNA from simple sugars and amino acids, thereby calling into action the enzymes HPRT and TK for the standby pathway. If *both* enzymes are functional and if the nucleosides are added to the culture medium, the cells can grow. But if *either* enzyme is nonfunctional, the cells cannot form colonies in the presence of aminopterin, even when both nucleosides are also present in the medium.

Mutants for both enzymes exist and are maintained in culture in the absence of aminopterin. Consider now what happens when the two mutant cell types (human $HPRT^-$ × mouse TK^-; or in short form, $H^- \times T^-$) are mixed and allowed to grow for a short while in nutrient medium with sugars, amino acids, but no aminopterin (Figure 20A). Three types of colonies will form: parental human cells (H^-T^+), parental mouse cells (H^+T^-), and fused cells (H^+T^+). If these cells are transferred to a new culture plate with a selective medium containing aminopterin and the two nucleosides, then only the fused cells will survive and form colonies.

Chromosome Assignment by a Gene's Protein Product

The cardinal property of fused cells that enables researchers to localize genes is the loss of chromosomes over the course of several cell generations. This loss of chromosomes is accompanied by a loss of the biochemical processes dependent upon the genes present on those chromosomes. By monitoring the loss of particular chromosomes and also the loss of specific enzymes or other proteins, researchers gather information on which functions are associated with which chromosomes.

Karyotype analyses are used to monitor the loss of individual chromosomes; interestingly, it turns out that in human–mouse hybrid cell lines, only *human* chromosomes are lost. The reason for a preferential loss is not clear, but it may hinge upon the slower replication rate of human chromosomes. Which particular human chromosomes are lost and which ones are retained seems to be fairly random, however, and the process of elimination continues until only one or a few human chromosomes are left in a hybrid nucleus. At this point, the various subclones (shown at the bottom of Figure 17) become stabilized.

The loss of particular enzymes (or other proteins) from hybrid cells can be monitored if the human and mouse forms of the same enzyme are distinguishable by *gel electrophoresis*. Here researchers utilize the fact that, although all mammalian cells perform essentially the same biochemical functions and contain the same proteins to regulate these reactions, enzymes may differ somewhat in structure from species to species. As noted before, these differences arise over the course of evolutionary separation of two populations because different base substitution mutations lead to minor amino acid changes. Thus two species may possess different molecular forms of an enzyme. (Even within an individual, or group of individuals, there may exist multiple molecular forms of an enzyme, known as *isozymes*, all of which function normally.) Other kinds of proteins, such as hemoglobin or albumin, likewise show characteristic normal variations between and within species. Such slight variations in the same enzyme that may exist between human and mouse cells can often be detected by gel electrophoresis. Like the separation of DNA segments by electrophoretic migration, the separation of protein molecules depends upon charge and size variations. In the

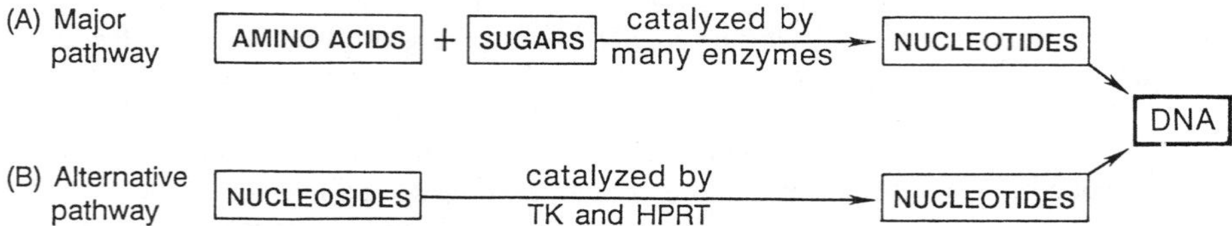

Figure 19 Two modes of DNA synthesis in cells. The major pathway (A) operates to make nucleotides "from scratch," that is, from amino acids and sugars. In the presence of aminopterin, however, the major pathway is blocked at one point, and the alternative pathway (B) converts nucleosides (hypoxanthine and thymidine) to nucleotides. But in order to do this, the cell must have both normal enzymes thymidine kinase (TK) and hypoxanthine guanine phosphoribosyl transferase (HPRT). If aminopterin is present and if either of these major enzymes is nonfunctional, the cell cannot produce DNA at all.

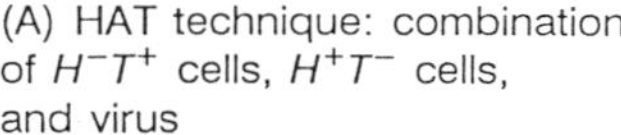

(B) Half-selective system: combination of a few normal human fibroblast cells, many mouse cells, and virus

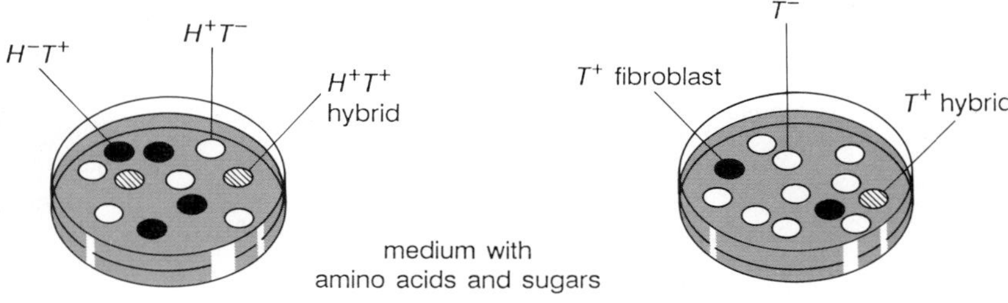

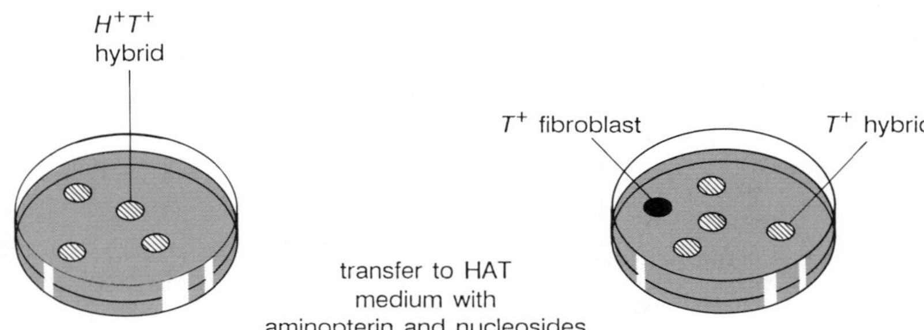

Figure 20 Two ways of selecting for hybrid cells. (A) In the HAT technique, both parental lines are mutant, one for the HPRT gene (H^-T^+) and the other for the TK gene (H^+T^-). When transferred to a HAT medium, all parental colonies die because the normal DNA pathway is blocked and they lack an enzyme for the alternative pathway. (B) In the half-selective system, only one parental type is mutant, either H^- or T^-. The other parental line, often human fibroblasts, carries both functional enzymes but is relatively slow growing. Fewer of the latter cells, in the presence of Sendai virus, are added to the culture medium. As in (A), the mutant line dies when transferred to the HAT medium. Because the other parental line is slow growing and relatively rare, the hybrid clones predominate.

case of similarly sized proteins, their differential migration rates come about primarily because of charge differences in their constituent amino acids.

When first formed by fusion, hybrid cells possess both the human and mouse forms of a protein—usually migrating at slightly different rates (Figure 21). If both of the human homologues that carry the encoding gene are subsequently lost, then the corresponding human protein, seen as a band in the hybrid lane, will also be lost.* For example, Frank H. Ruddle and his colleagues at Yale University (1972) analyzed 26 hybrid clones for the presence or absence of a human enzyme called peptidase C, which migrated a little faster than the corresponding mouse peptidase C. In 14 hybrid cell lines that retained a chromosome 1, they found the human enzyme; but in 12 lines that had lost both chromosomes 1, they failed to find the human enzyme. Thus the human gene for peptidase C must reside on human chromosome 1.

Methods for Regional Assignments

The strategy outlined above for assigning a locus to a specific chromosome is limited to genes that are *expressed*, that is, transcribed and translated into detectable amounts of the protein product. Hence, this indirect approach is suitable for only some of the genes present in the cell. A newer approach, manipulating genes directly, "has revolutionized the genetic analysis of cell hybrids. Any gene—indeed, any DNA sequence—for which a recombinant DNA probe exists can now be mapped" (D'Eustachio and Ruddle 1983).

The experimental procedure starts, as before, with the isolation of a human–mouse hybrid cell line containing a small number of human chromosomes whose identities are established by standard cytogenetic techniques. A researcher will likely have handy dozens of hybrid cell lines, each with a different array of human chromosomes. From each hybrid cell line, DNA is extracted, cut with a restriction enzyme, spread out by size on an electrophoretic gel, denatured, and transferred to a nitrocellulose filter by Southern blotting (as in Figure 11). The filter is hybridized with a radioactive probe complementary to the gene in question and autoradiographed. A band will appear in a particular lane only if the original hybrid cell contained a chromosome with the gene in question.

Gusella et al. (1983) localized the Huntington disease gene to chromosome 4 by this method. Recall that they had found (partly by luck) a probe called G8, that discriminated among four alleles of an RFLP site located about four map units from the disease gene (Figure 14). They examined 18 different human–mouse hybrid lines

*Actually, because a chromosome carries many genes, its loss from a fused cell leads to the simultaneous disappearance of many proteins from that subclone. This principle is the basis of tests for *synteny,* the joint occurrence of two or more genes on a single chromosome.

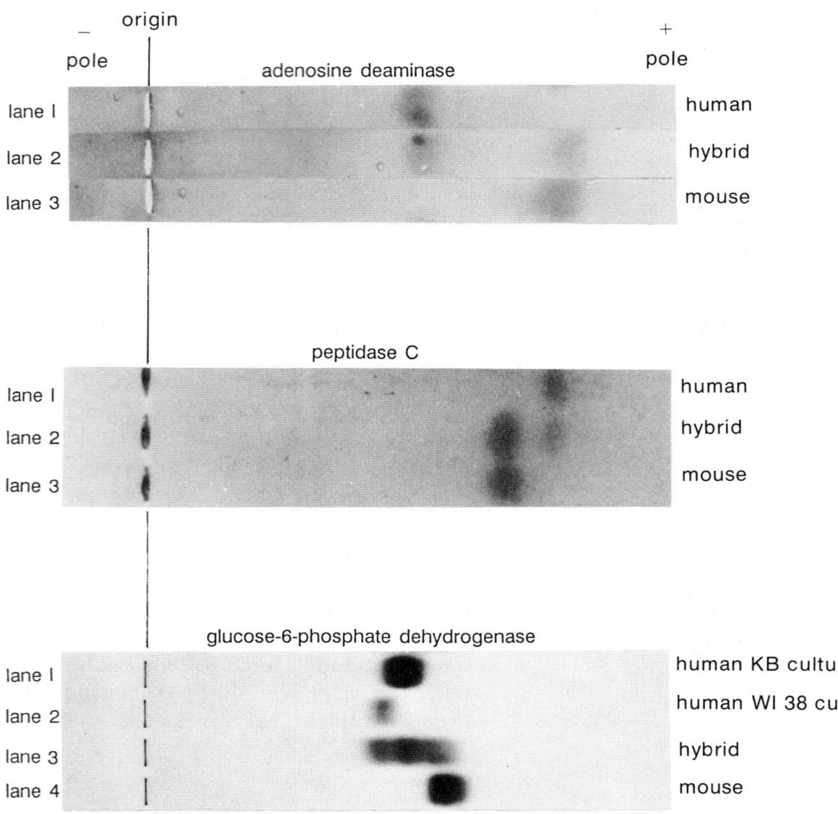

Figure 21 Three examples of enzyme variants detectable by electrophoresis. In each case the human possesses one variant, the mouse another, and the cell hybrids both. (From Nichols and Ruddle 1973.)

for the presence or absence of DNA sequences complementary to the G8 probe. The data for six of the lines are given in Table 2. The radioactive G8 probe bound only to those lanes on the Southern blot filter containing DNA from human–mouse hybrid cells containing chromosome 4. No other chromosome bears this relationship with G8. So it was this closely linked RFLP site, not the actual Huntington locus, that was localized to chromosome 4. Note that—even though the DNA from every hybrid cell line was mostly that from the full complement of mouse chromosomes, and even though only one or two copies (per cell) of the human sequences complementary to G8 can be expected—the DNA hybridization technique is still sensitive enough to detect the RFLP site wherever it is present. And, what is more important, the probe does *not* bind to any potentially homologous mouse DNA sequences.

Refinements of somatic cell techniques also let re-

Table 2 Assignment of the Huntington disease gene to chromosome 4 by analysis of human-mouse hybrid cells

Hybrid cell line	*Presence (+) or absence (−) of G8 sequences*	*Human chromosomes present*								
W-5	+	**4**	17	18	21	X				
J-22	+	**4**	6	10	11	14	17	18	20	21
N-16	+	3	**4**	5	7	12	17	18	21	X
W-2	−	8	12	17	21	X				
N-5	−	12	14	15	16	18	20			
R-11	−	11	13	16	20	21	X			

searchers identify which *part* of a chromosome carries a particular gene. One of these methods starts with mouse cells fused to human cells containing chromosomes with translocations or deletions. After the random loss of human chromosomes in successive divisions of the hybrid, what remains may include one or more chromosomes with missing parts. For example, if the original human cell carried a 4:11 translocation, the stabilized hybrid cell line might contain the *short* arm of 4 attached to the *long* arm of 11, but not the reciprocal segments. By Southern blotting with an appropriate probe, the presence or absence of the gene in question can be correlated with the presence or absence of a chromosomal part.

We illustrate this in Figure 22, which summarizes some of the cytogenetic experiments that localized the β-globin cluster of genes to band p15 of chromosome 11. (Additional genetic evidence favoring this position relies on linkage to the insulin gene and to other nearby loci.) As a first step in cytogenetic localization (Figure 22A), researchers examined mouse–human cell hybrids whose human donor had a reciprocal translocation between chromosomes 17 and 11, the breakpoint in 11 being near the centromere. A radioactive cDNA globin probe was made, using reverse transcriptase on messenger RNA isolated from immature red blood cells. In a Southern blot procedure, this probe hybridized to the filter only when the *short arm of chromosome 11* was present in hybrid cells, and never when this arm was absent. A finer localization (Figure 22B) was done with deletions that cut out all of the short arm of chromosome 11 except band p15.

A related method for localizing genes utilizes the fluorescence-detecting cell sorter described on page 282. Here, chromosome 11 by itself, or translocations involving parts of chromosome 11, are isolated directly from human mitotic metaphase cells and run through a Southern blotting protocol. This method allowed researchers to locate the β-globin genes to the region depicted in Figure 22C, a result consistent with band p15.

A rather different method, called **in situ hybridization**, also localized the β-globin genes to band p15 (Figure 22D). With this technique, researchers prepare human metaphase chromosomes on a microscope slide in the standard way, say, from white blood cells (Chapter 4). First the chromosomes are stained and photographed to reveal the banding patterns. Next the RNA within the chromosomes is digested away, and the DNA that remains in place (i.e., "in situ") is denatured to single strands. Researchers then bathe the slide with a radioactive β-globin probe, wash off excess probe, and layer a photographic emulsion on top of the slide to make an autoradiograph (Figure 23). Finally, they take microphotographs of the exposed and developed film *on top of the same chromosome spreads that had been previously stained and photographed*. Thus the developed silver grains in the film can be juxtaposed to the banded metaphase chromosomes in order to locate the position of the gene. This method was originally devised to map repetitive genes, which bind large amounts of the radioactive probe. For example, the genes encoding ribosomal RNA were assigned to the short arms of the five acrocentric chromosomes. But with improvements, the method is now sensitive enough to locate some single-copy genes.

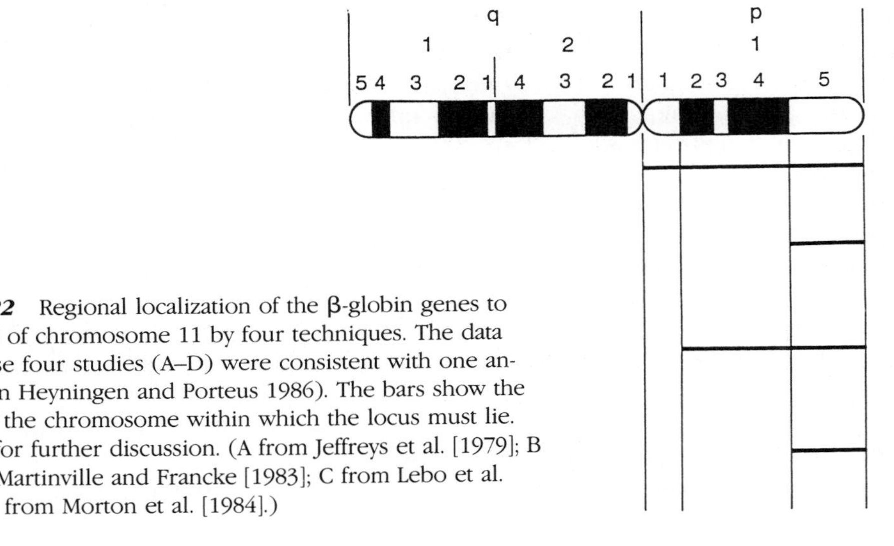

Figure 22 Regional localization of the β-globin genes to band p15 of chromosome 11 by four techniques. The data from these four studies (A–D) were consistent with one another (van Heyningen and Porteus 1986). The bars show the length of the chromosome within which the locus must lie. See text for further discussion. (A from Jeffreys et al. [1979]; B from de Martinville and Francke [1983]; C from Lebo et al. [1979]; D from Morton et al. [1984].)

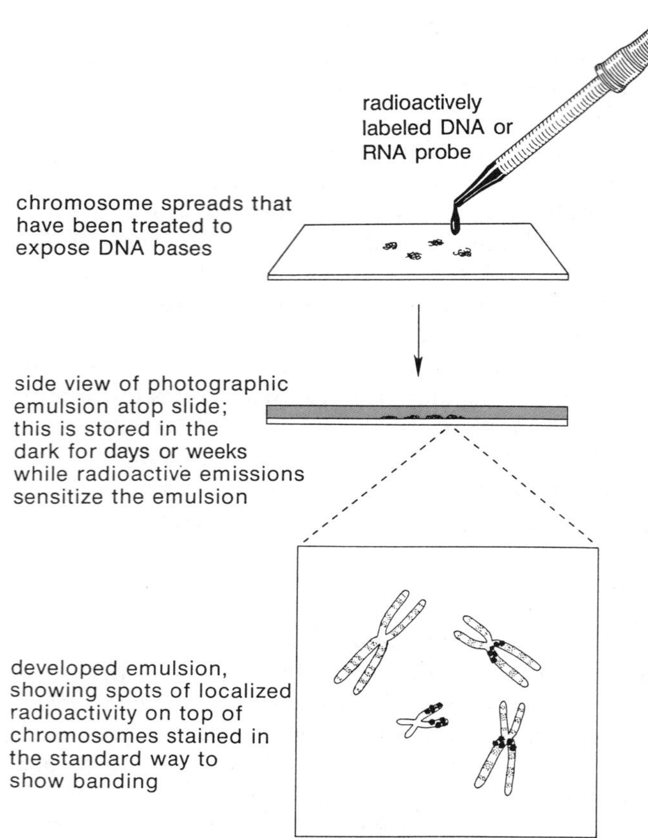
radioactively
labeled DNA or
RNA probe
chromosome spreads that
have been treated to
expose DNA bases
side view of photographic
emulsion atop slide;
this is stored in the
dark for days or weeks
while radioactive emissions
sensitize the emulsion
developed emulsion,
showing spots of localized
radioactivity on top of
chromosomes stained in
the standard way to
show banding

Figure 23 Basic methodology for in situ hybridization. See text for details.

A FURTHER NOTE ON MAPPING

Geneticists now deal with the organization of chromosomes at several levels, as reflected in the different kinds of maps they construct. The older, more traditional maps include the *cytological* and the *genetic*. At the cytological level (using a light microscope), the breakpoints of translocations and other chromosomal changes provide visible landmarks among the banding patterns. At the genetic level, map distances between two genes are measured by the percent recombination visible in the phenotypes of offspring of persons who are doubly heterozygous. The biological phenomenon underlying recombination is the frequency of crossing over during meiosis. The recombinational length of all 23 pairs of chromosomes taken together is about 3,300 map units.

At the finer *molecular* level, geneticists use the newer technologies to isolate, dissect, and analyze stretches of nucleotides that represent genes, parts of genes, or various types of intergenic DNA. Distances are measured in bases or kilobases, and the total amount of DNA in the haploid human genome is about 3 billion nucleotide pairs. Therefore, for the entire genome, about a million bases correspond to one recombinational map unit (3×10^9 total bases divided by 3.3×10^3 map units).

The various kinds of maps are roughly comparable to one another. The genes are always found in the same *order* along a chromosome, but relative *distances* are not always exactly corresponding. For example, in fruit fly chromosomes, it has long been known that sets of equidistant, microscopically located sites can represent different recombinational map distances. If there were a few "hot spots" in human chromosomes, where crossing over occurred at a rate higher than average, then the genetic maps for those particular areas would appear to be relatively longer than molecular or cytological distances would predict. There is also a sex-related difference in map distances; because crossing over does not occur in fruit fly males, recombination distances can be measured only among the offspring of appropriately marked females. In a similar phenomenon, map distances measured through genetic recombination in doubly heterozygous *human* females are routinely 40–50% greater than distances based on recombination measured through heterozygous males, although molecular distances along the chromosomes are necessarily the same in the two sexes (White and Lalouel 1987). This discrepancy arises because crossing over occurs more frequently in females than in males.

Molecular studies, themselves, involve two levels of mapping. Base sequencing gives the finest type of information, but only a tiny fraction of the 3×10^9 bases has been sequenced, and the proportion of sequenced human DNA will increase slowly. On the other hand, variations in restriction enzyme recognition sites among different individuals provide signposts throughout the genome, and the discovery of new RFLPs is proceeding rapidly. David Botstein of MIT and his colleagues (1980) have pointed out that two genes should be within about 20 genetic map units of each other in order to easily establish traditional linkage. Thus, if there were 165 (= 3,300/20) uniformly spaced RFLPs, adjacent sites could be mapped to each other, and every chromosome could be marked with known loci at regular intervals. Then *every* human gene locus could be easily mapped, since any one of them would be within 10 units of a previously mapped RFLP.

In fact, RFLPs have removed the major impediment to classical linkage studies, revitalizing a traditional genetic technique. The accurate measurement of recombination distances requires a lot of doubly heterozygous parents, but the absence of common polymorphisms had made those heterozygotes relatively rare. RFLPs, however, fill the "polymorphism gap." The current approach to studying the human gene map combines all the traditional and molecular methodologies. Gusella (1986) notes: "Improved diagnostic capability is the first benefit derived from this approach, but by far the most important result will be the new avenues it provides for isolating disease genes and exploring the mechanisms underlying their defects."

SUMMARY

1. DNA segments from different species can be spliced together to make recombinant DNA molecules. The first step in this procedure uses restriction enzymes, precision cutting tools of molecular biologists. Once inserted into a plasmid of an *E. coli* host cell, a foreign gene can be multiplied to any desired amount.

2. Genes for somatostatin, insulin, growth hormone, clotting factor VIII, and other polypeptides can be cloned in *E. coli* plasmids, permitting the bacterial synthesis of these valuable medical products.

3. DNA molecules can be separated from one another by electrophoresis, which sorts them according to size. By electrophoresing DNA that is cut with different restriction enzymes and comparing the resultant banding patterns, scientists can map the various restriction sites along the molecule.

4. The sequence of bases along a section of DNA can be determined readily. The Sanger method uses the four dideoxy nucleotides, whose random insertion in replicating DNA strands terminates synthesis, thereby generating pieces of every possible length. Electrophoresis of the four preparations allows researchers to read the base sequence directly.

5. The human genome can be cut into segments and inserted into cloning vectors such as λ bacteriophage. A collection of a million recombinant phage, each with a 20-kb segment of human DNA, constitutes a genomic library, which can be searched for a DNA sequence of interest. Smaller libraries—for individual chromosomes or for the genes expressed in a given cell—can also be constructed.

6. Any gene in a library can be detected and isolated if the researcher has available a suitable probe, that is, a radioactive segment of single-stranded, complementary DNA that can bind by base pairing to the gene of interest.

7. In the Southern blotting technique, DNA segments separated by gel electrophoresis are transferred to a nitrocellulose filter for hybridization with a radioactive probe. Using libraries and Southern blotting, researchers have analyzed the genes responsible for color vision. Multiple copies of the green pigment gene present in some people probably arose through unequal crossing over between nearly identical genes.

8. Restriction fragment length polymorphism (RFLP) occurs whenever the DNAs from individuals differ in one of the bases included in a recognition site of a restriction enzyme. Thus the cutting or not-cutting of that restriction site identifies alternative alleles that can be detected by Southern blotting.

9. An RFLP site found *near* the Huntington disease locus provides an imperfect aid for early diagnosis and genetic counseling. An RFLP site *within* the β-globin gene is now used for prenatal diagnosis of sickle-cell anemia.

10. Probes for repetitive DNA have been used for individual-specific DNA fingerprinting. Even the small amount of DNA in a single hair allows many different analyses, because it can be amplified by the polymerase chain reaction.

11. Human and mouse somatic cells, when put together in tissue culture, can fuse to make hybrid cells, which are then isolated by the HAT method. In further divisions of hybrid cells, human chromosomes are randomly lost until only one or several are left.

12. A gene that is transcribed and translated in the hybrid cell can be assigned to a particular chromosome by correlating the presence or absence of the polypeptide product with the presence or absence of a particular chromosome in hybrid subclones.

13. If a probe is available for a particular gene or other DNA sequence, Southern blotting can reveal its presence in hybrid cells, whether or not that gene is expressed. Chromosome assignment can be refined by using chromosomal translocations and deletions, plus other molecular techniques such as fluorescence-detecting cell sorting and in situ hybridization.

FURTHER READING

Broad accounts of the laboratory manipulation of DNA are given by Watson et al. (1987), Lewin (1987), Old and Primrose (1985), and in substantial but less formal fashion, by Cherfas (1982). Introductions to recombinant DNA technology are provided by Watson et al. (1983) and by Emery (1984). McKusick (1980) and Shows et al. (1982) explain the results of mapping by many methods, while Kao (1983) and Davidson (1984) review all aspects of somatic cell genetics. In a concise fashion, Weatherall (1985) and Gusella (1986) set forth applications of the new genetics to medicine. Directions of current research will be found in Messer and Porter (1983), Caskey and White (1983), Volume 51 of the Cold Spring Harbor Symposia (1986), and Childs et al. (1988).

Historical, social, and ethical aspects of molecular biology are emphasized by Watson and Tooze (1981), Krimsky (1982), the President's Commission (1982), and Nossal (1985). Articles from *Scientific American* include Anderson and Diacumakos (1981) on genetic engineering in mammalian cells, Pestka (1983) on the recombinant DNA manufacture of interferon, Godson (1985) on molecular approaches to malarial vaccines, Lawn and Vehar (1986) on the molecular genetics of hemophilia, White and Lalouel (1988) on DNA markers, and Nathans (1989) on the color vision genes.

QUESTIONS

1. A *palindrome* is a sequence of letters that reads the same forward and backward, WASITACATISAW, for example. In what way is the sequence recognized by *Eco*RI (GAATTC on one strand) palindromic?

2. The restriction enzyme *Sau*3A recognizes a 4-base sequence contained within the 6-base sequence recognized by *Bam*HI. They cut similarly, as indicated in the accompanying diagram. A researcher joins a DNA segment with a *Sau*3A sticky end (as the left piece) with a segment with a *Bam*HI sticky end (as the right piece). Will this "hybrid" site be recognized and cut by *Sau*3A? By *Bam*HI?

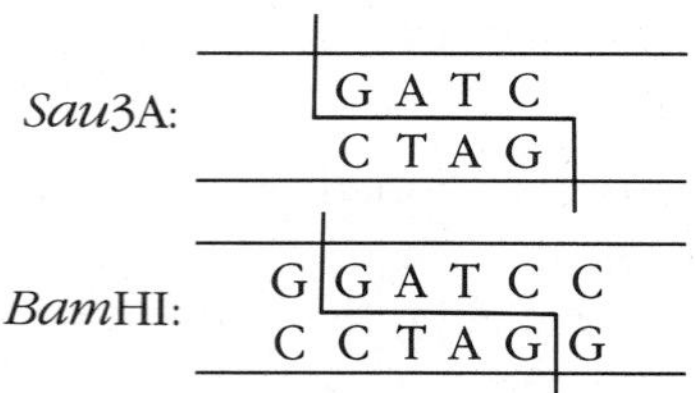

3. Assume that the four bases are in equal numbers overall and are in random order along a molecule of DNA. You cut it with a restriction enzyme that recognizes a specific 6-base sequence (say, GAATTC). How long would the pieces of DNA be, on the average?

4. Repeat Question 3 for restriction enzymes that cut at 4-base sequences (like *Sau*3A), or 8-base sequences (like *Not*I).

5. Assume restriction sites as follows, letting **E** = *Eco*RI site, and **H** = *Hin*dIII site:

3 kb —— **E** —— 6 —— **H** —— 6 —— **E** —— 7

What bands would be present in an electrophoretic gel when this DNA is digested with *Eco*RI alone, *Hin*dIII alone, or with both enzymes?

6. A researcher sets up a gel with three experimental lanes in which the same linear (not circular) DNA sample had been cut with either *Eco*RI, *Hin*dIII, or both enzymes. Comparing the resultant bands with a standard, she finds the sizes of DNA in the various bands are

*Eco*RI lane: 3 bands with lengths of 3, 7, and 12 kb
*Hin*dIII lane: 2 bands with lengths of 9 and 13 kb
"Both" lane: 4 bands with lengths of 1, 2, 7, and 12 kb

Draw the restriction map; that is, indicate on a segment of DNA where the restriction recognition sites are.

7. From Figure 8, determine the sequence of the 10 bases from number 50 to 59.

8. Assume that you have cloned a 1-kb human gene in the pBR322 plasmid using restriction enzyme *Eco*RI and that you have subsequently isolated in a test tube a high concentration of the recombinant plasmid. How do you get the gene separated from the rest of the DNA?

9. Perhaps as many as two million bases of human DNA have been sequenced at this writing. What percentage of the human genome remains to be sequenced? (Recall that the haploid number of base pairs is about 3×10^9.)

10. Lambda phage for a human gene library accepts about 20-kb lengths of DNA. To ensure that every sequence of a human genomic DNA sample is inserted into at least one phage, a researcher may make a library of 10^6 phage. Under these conditions, how many times is any given 20-kb piece of human DNA likely to be represented in the library?

11. Other than mutations *within* recognition sequences, what types of mutational changes could lead to variation in restriction fragment lengths?

12. How would you characterize the two alleles of a simple RFLP: dominant/recessive, intermediate, codominant, or overdominant?

13. From Figure 14A, predict the pattern of bands on a Southern blot from individuals *heterozygous* for any pair of *Hin*dIII RFLP alleles detected by the radioactive G8 probe. Assume that the 2.5-kb segment present in alleles *C* and *D* does not hybridize to the probe, but all other segments do. Which heterozygotes are indistinguishable from one another? Would it help if the 2.5-kb fragment could be made visible?

14. Traditional pedigree studies have always yielded a disproportionate number of X-linked genes compared to autosomal genes. Is this bias also true for studies involving somatic cell fusion, chromosome sorting, or in situ hybridization?

15. Owerbach et al. (1980) examined 15 human–mouse hybrid cell lines by Southern blotting for the presence or absence of DNA sequences complementary to the human insulin gene. The data for six of the lines are given below. Which chromosome carries the gene?

Hybrid cell line	Insulin sequences detected (+) or not (−)	Human chromosomes present
W-8	+	6 7 10 11 14 17 18 20 21 X
T-5	+	4 5 10 11 12 17 18 21
D-5	+	3 5 11 14 15 17 18 21
W-2	−	8 10 12 15 17 21 X
T-2	−	2 5 6 10 12 18 20 21 X
T-8	−	17 18 20

16. Would it be possible to assign human genes by means of human–mouse cell hybrids if mouse chromosomes were randomly lost and human chromosomes retained?

17. Recall that *Mst*II cuts DNA at the coding region for the fifth, sixth, and seventh amino acids of the β chain of normal hemoglobin:

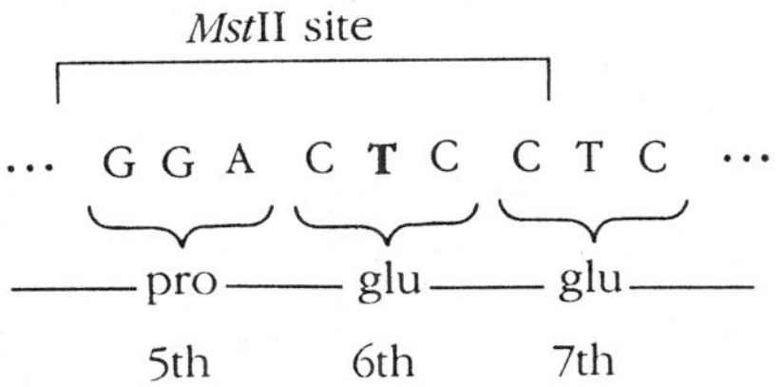

Prenatal diagnosis of sickle-cell anemia is possible because *Mst*II does not recognize the mutant sequence in which A replaces the boldfaced **T**. Would this diagnostic method work if either the normal fifth or sixth amino acids were coded by one of the several alternative triplets? (Proline is coded by four different triplets and glutamic acid by two; see Table 2 in Chapter 14.)

18. In Hemoglobin C disease, the sixth amino acid of β-globin is lysine instead of glutamic acid because the C before the boldfaced **T** (see previous question) mutated to T. (Persons homozygous for the mutated allele are less severely affected than those with sickle-cell anemia.) Why would the *Mst*II method that recognizes an RFLP in the case of the sickle-cell mutation *not* work for the Hemoglobin C mutation? (Recall that the *Mst*II recognition sequence is GGA*N*TCC, where *N* can be any base.)

19. Would you expect that genes found to be syntenic (i.e., on the same chromosome) by cell fusion methods will always be found to be linked by traditional pedigree studies? Explain.

ANSWERS

1. Reading one strand to the right yields the same sequence as reading the other strand to the left. Almost all restriction enzymes recognize palindromes. Other examples are *Sau*3A (GATC), *Bam*HI (GGATCC), *Hin*dIII (AAGCTT), and *Not*I (GCGGCCGC).

2. It will be cut by *Sau*3A. It will only be cut by *Bam*HI if the base pair to the left of the original *Sau*3A site was G–C (a 1 in 4 chance).

3. 4^6 = 4,096 bases. Each of the six consecutive, independent bases has a 1 in 4 chance of matching the corresponding base of the recognition sequence. Thus the probability of cutting any 6-base length is 1/4 raised to the sixth power. Expressed in another way, a match to the recognition site would be expected every 4,096 bases.

4. 4^4 = 256 for a 4-base recognition site, and 4^8 = 65,536 for an 8-base recognition site.

5. *Eco*RI lane: 3, 7, and 12 kb
*Hin*dIII lane: 9 and 13 kb
"Both" lane: 3, 6, and 7 kb

6. Let **E** = *Eco*RI site, **H** = *Hin*dIII site:

```
      7 kb           2     1              12
——————————————E——————H——E————————————————————
```

Reversing left for right above is the same map. All other maps that you might draw can be ruled out by trial and error. Note that the "both" lane in Question 5 is different from the "both" lane here.

7. A G C T G T T T C C

8. Treat the recombinant plasmid with *Eco*RI. This treatment yields DNA segments of two sizes: 1 kb for the gene, and about 4.3 kb for the plasmid. Run on an electrophoretic gel and cut out the gel material encompassing the 1-kb band.

9. 99.93%. (2×10^6 bases sequenced)/(3×10^9 bases total) = 0.0007 = 0.07% has been sequenced. (Actually, it is incorrect to speak of *the* human genome, since perhaps as many as 1% of the bases in unrelated persons are likely to be different.)

10. About 7 times. The total genome of 3×10^9 base pairs is cut into $(3 \times 10^9)/(20 \times 10^3)$ = 150,000 segments of DNA. These are inserted randomly into 10^6 phage, an average of $10^6/(0.15 \times 10^6)$ times.

11. A deletion or addition of nucleotides between (unmutated) recognition sites.

12. Codominant, because the banding pattern of a heterozygote would show all the bands of both homozygotes.

13.

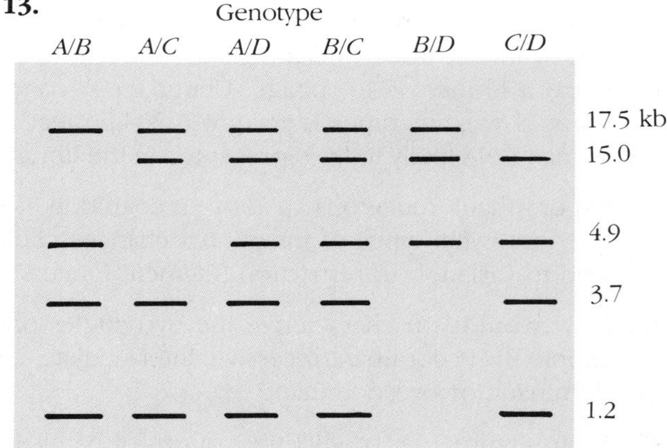

Individuals *A/D* and *B/C* are indistinguishable, and being able to detect the 2.5-kb piece would not help, since they would both have it.

14. No; human chromosomes in cell hybrids seem to be retained or lost on a random basis. Nor do chromosome sorting or in situ hybridization favor the X chromosome.

15. Chromosome 11. This chromosome is the only one present in all three "+" lines and absent in all three "−" lines.

16. No, but mouse genes could potentially be assigned to mouse chromosomes.

17. It would not work because *Mst*II would recognize *neither* the normal *nor* the mutant sequence.

18. *Mst*II's recognition sequence permits any base in the fourth position where the mutation in Hemoglobin C disease occurs.

19. No; syntenic genes that are more than 50 map units apart (without intervening marker genes) will appear to be unlinked.

Chapter 16 Inborn Errors of Metabolism

Amino Acid Metabolism
Phenylketonuria and Other Hyperphenylalaninemias
Albinism

Lipid and Lipoprotein Metabolism
Tay-Sachs Disease
Familial Hypercholesterolemia

Purine Metabolism
Lesch-Nyhan Syndrome

Medical Considerations
Detection of Metabolic Disease in Newborns
Pharmacogenetics

> *The view is daily gaining ground that each successive step in the building up and breaking down, not merely of proteins, carbohydrates, and fats in general, but even of individual fractions of proteins and of individual sugars, is the work of special enzymes set apart for each particular purpose.*
>
> *Archibald E. Garrod (1909)*

At the time that Mendel's work was being rediscovered, an English physician and biochemist, Archibald Garrod (1902) was studying an obscure condition known as **alkaptonuria**. Otherwise healthy newborns with this disorder usually had one striking feature: their urine, upon standing in contact with air, turned black. In their later years, however, most alkaptonurics also develop arthritis (La Du 1978). The dark substance in the urine was the oxidation product of alkapton, now called *homogentisic acid*, which contains a six-carbon benzene ring. Other researchers had shown that the excretion of homogentisic acid in urine was increased when alkaptonuria patients were fed excess protein or the amino acids tyrosine and phenylalanine, both of which contain benzene rings.

Although homogentisic acid was clearly seen to be involved in protein metabolism, many workers thought it was produced by a rare bacterium that infected the intestine rather than by the patient's own system. Garrod disagreed, suggesting that homogentisic acid is an ordinary product of metabolism that is broken down in normal people but is not degraded in people with alkaptonuria, thus appearing intact in the urine.

Garrod (1909) suggested that alkaptonurics are unable to properly metabolize ring-containing fractions of proteins. More specifically, he proposed that during the normal breakdown of proteins, phenylalanine is converted to tyrosine, which in turn is changed to homogentisic acid and then to simpler products (Figure 1). This idea was supported by the fact that *normal* individuals who ingested excessive amounts of homogentisic acid did not excrete any of it in their urine.

> Where the alkaptonuric differs from the normal individual is in having no power of destroying homogentisic acid when formed—in other words of breaking up the benzene ring of that compound . . . We may further conceive that the splitting of the benzene ring in normal metabolism is the work of a special enzyme, [and] that in congenital alkaptonuria this enzyme is wanting.

Garrod's analysis did not stop with the concept of metabolic blocks, however. The brilliance of his proposal lay in the genetic connection that he and population geneticist William Bateson were able to make. Garrod noted that alkaptonuria tended to occur in several sibs of a family whose parents were nevertheless normal, and also that many affected children were the offspring of first cousin marriages. This pattern of inheritance fit the requirements for a rare recessive trait, the first human application of Mendel's newly discovered laws. Garrod classified alkaptonuria and three other conditions that he had studied (albinism, cystinuria, and pentosuria) as **inborn errors of metabolism**. He regarded them as "metabolic sports, the chemical analogues of structural

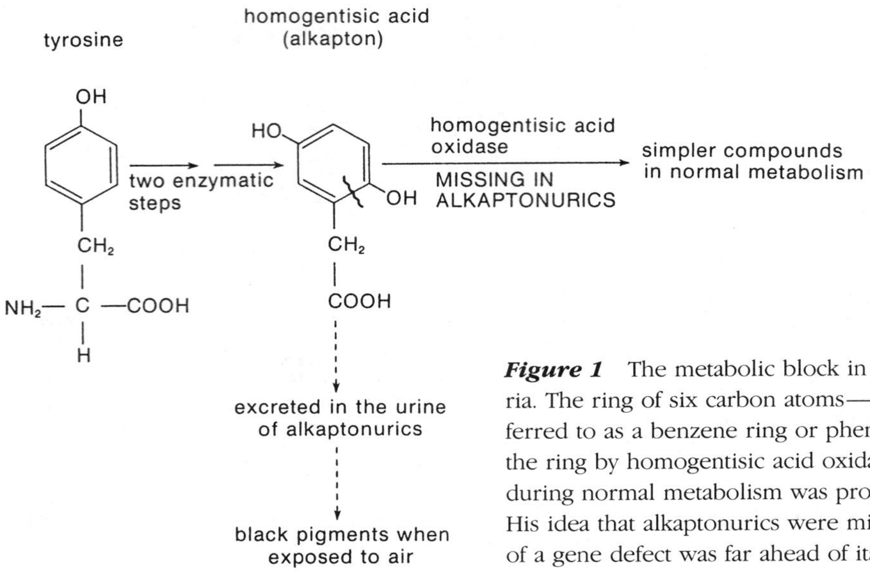

Figure 1 The metabolic block in persons with alkaptonuria. The ring of six carbon atoms—depicted ⌬ —is referred to as a benzene ring or phenyl group. The splitting of the ring by homogentisic acid oxidase—indicated by ﹏—during normal metabolism was proposed in 1902 by Garrod. His idea that alkaptonurics were missing this enzyme because of a gene defect was far ahead of its time.

malformations... Among the complex metabolic processes of which the human body is the seat there is room for an almost countless variety of such sports."

Garrod (1909) suggested that the four rare genetic conditions he studied must represent only a tiny fraction of the errors of metabolism that existed in human populations, pointing out that many of them may produce no obvious phenotypic effects. Citing known variations in hemoglobins and in muscle proteins as examples of chemical diversity within and between species, he correctly prophesied that

> it is among the highly complex proteins that such specific differences are to be looked for... [We] should expect the differences between individuals to be... subtle and difficult of detection... Even those idiosyncrasies with regard to drugs and articles of food which are summed up in the proverbial saying that what is one man's meat is another man's poison presumably have a chemical basis.

But Garrod's ideas, like Mendel's, were largely ignored for over 30 years. In the meantime, studies of pigment formation in flowers, in the fur of guinea pigs, and in the eyes of fruit flies had laid the groundwork for more extensive research on the nutritional requirements of the pink bread mold *Neurospora*. In a series of experiments done at Stanford University, George Beadle and Edward Tatum (1941) irradiated wildtype *Neurospora* to produce a class of mutants that could survive only if specific chemicals were added to the nutrient medium. By collecting and analyzing dozens of these strains, they were able to show that each one had a single metabolic block—presumably due to an enzymatic defect—that could be traced to the mutation of a single gene (Figure 2). Although Beadle and Tatum were not the first to conceive of the one gene–one enzyme hypothesis, their relatively simple testing system greatly accelerated research in this developing field of biochemical genetics and earned them a Nobel Prize as well. (Later work refined their proposal to a **one gene–one polypeptide** concept—which we now know to be an oversimplification, too, given the recent discovery of some overlapping, nested, split, and shuffled genes [Chapter 18]).

By the mid-1940s it was clear that (1) the biochemical processes that take place in an organism are genetically controlled; (2) every biochemical pathway can be resolved into a series of individual steps, each mediated by a different enzyme; and (3) each enzyme is usually coded for by one or a few genes. Indeed, by analyzing groups of mutant strains that have different blocks in the same metabolic pathway, it is possible to determine the exact order in which these metabolic steps normally occur. (See Question 2.) Clear-cut examples of biochemical mutants have now been detected in a wide variety of organisms. In humans, over 250 disorders are known to involve enzymatic defects.

Indeed, we can generalize to say that, if a condition is simply inherited, "its pathogenesis, no matter how complex, must be due to an abnormality in a single protein molecule... In many Mendelian disorders, especially those with dominant inheritance, it is not yet possible to demonstrate directly the protein that is altered by the mutation" (Stanbury et al. 1983). Not all single-gene disorders involve enzyme defects, however.

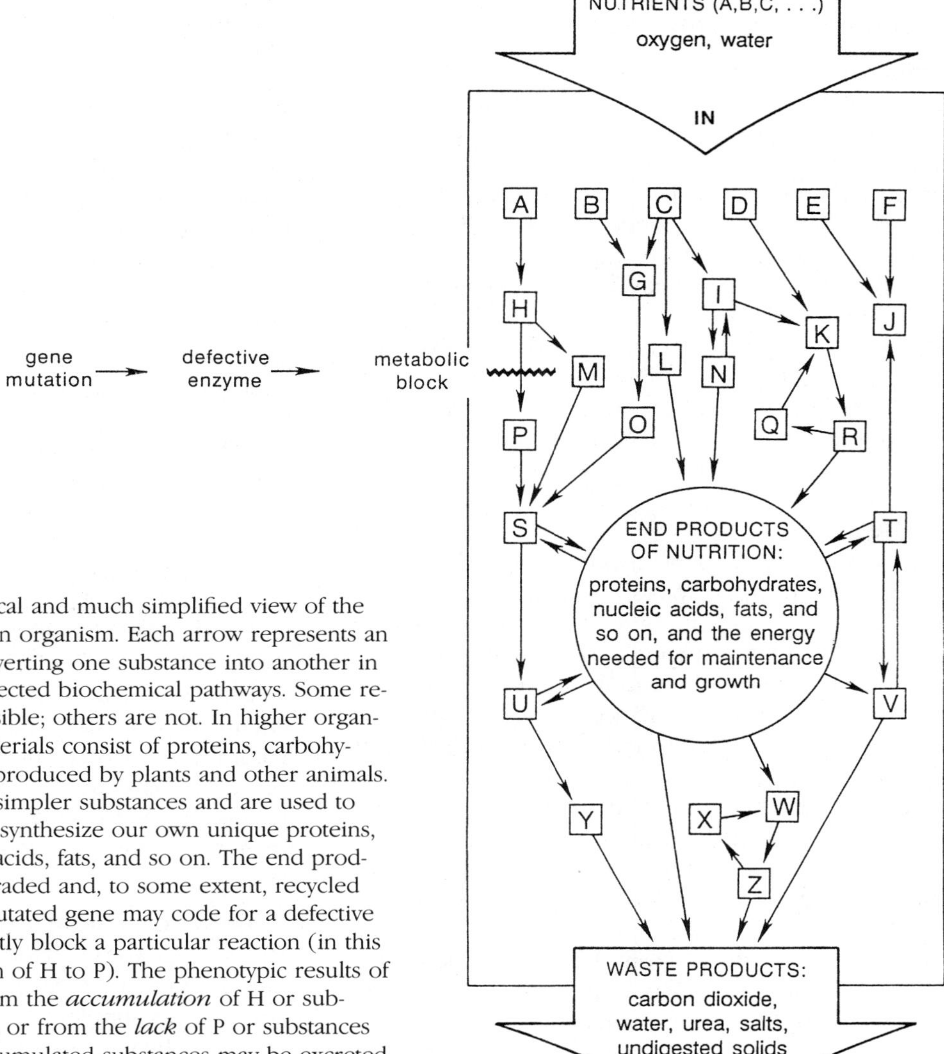

Figure 2 A hypothetical and much simplified view of the overall metabolism of an organism. Each arrow represents an enzymatic reaction converting one substance into another in complex and interconnected biochemical pathways. Some reactions are easily reversible; others are not. In higher organisms, ingested food materials consist of proteins, carbohydrates, fats, and so on, produced by plants and other animals. They are digested into simpler substances and are used to produce energy and to synthesize our own unique proteins, carbohydrates, nucleic acids, fats, and so on. The end products are eventually degraded and, to some extent, recycled within our bodies. A mutated gene may code for a defective enzyme and consequently block a particular reaction (in this diagram, the conversion of H to P). The phenotypic results of this block may arise from the *accumulation* of H or substances formed from H, or from the *lack* of P or substances synthesized from P. Accumulated substances may be excreted in the urine. The products of a blocked reaction may be available from other pathways.

Many involve receptors, hormones, transport proteins, immunoglobulins, and collagens. (Immunoglobulins include antibodies and antibody-forming proteins. Collagen is the main supportive protein of skin, tendon, bone, cartilage, and connective tissue; if boiled, it turns into gelatin.) Although nearly all are individually rare, together the known disorders account for over 5% of all pediatric hospital admissions. Their frequency in the general population is about 1%: roughly 0.65% autosomal dominants, 0.28% autosomal recessives, and 0.06% X-linked (McKusick 1988). In this chapter we present a few examples of inborn errors of metabolism.

AMINO ACID METABOLISM

Phenylketonuria, the first of the two autosomal recessive conditions described here, is one of the most common defects of amino acid metabolism and is also among the most common inherited metabolic diseases affecting brain development. If detected within a few weeks of birth, however, it is treatable. The second condition, albinism, although much less debilitating, is not yet treatable through metabolic manipulation.

Phenylketonuria and Other Hyperphenylalaninemias

Phenylketonuria (PKU) once caused hopeless mental and physical degeneration, accounting until the early 1960s for about 1% of the severely retarded patients in institutions. The first biochemically important observation about PKU was made in 1934 by the mother of two mentally retarded children. When she brought them to A. Følling, a Norwegian biochemist and physician, she pointed out that they had always exuded a peculiar "mousy" odor (Kaufman 1977).

Følling found that their urine samples turned olive green upon the addition of ferric chloride. This reaction was due to the presence of excess phenylpyruvic acid, a type of ketone (hence the name phenyl*keton*uria). He then tested the urine of hundreds of retarded children in institutions and found eight more who excreted the same substance.

It was soon established that this condition is inherited in an autosomal recessive pattern. The metabolic defect is an inability to convert the amino acid *phenylalanine* to *tyrosine* (Jervis 1947), owing to the absence of a liver enzyme called **phenylalanine hydroxylase (PAH)**. Some of the phenylalanine that accumulates in the body fluids (cerebrospinal fluid, blood plasma, and sweat) is converted to phenylpyruvic acid, which in turn is metabolized to several other derivatives (Figure 3A). Because tyrosine is deficient in these individuals, so, too, are its derivatives, including the pigment melanin. Thus, many untreated patients were fair-skinned, with blond hair and blue eyes.

Phenylketonuria affects about 1 in 10,000 newborns in the United States and Europe. It is more frequent in certain northern European populations and their descendants (1 in 4,000 among the Irish, and 1 in 8,000 among the Scotch and Scandinavians), but rare among Finns, southern and eastern Europeans, and Asians. The frequency of carriers likewise varies from group to group, but among Caucasians it averages out to about 1 in 50.

PHENOTYPE OF CLASSICAL PHENYLKETONURIA. The only clear-cut trait shown by newborn babies with PKU is the accumulation of excess phenylalanine in the blood plasma, starting several days after birth. Although some babies with PKU are underweight at birth and some suffer bouts of severe vomiting, no consistent clinical signs occur during the first month or two of life. After two or three months, however, rapid and progressive deterioration of central nervous system development sets in. Exactly how this happens at the biochemical level is not understood, but it seems to be due more to the *excess of phenylalanine* and its derivatives than to a deficiency of tyrosine and its metabolites.* From autopsies, it is known that the brains of untreated PKU patients tend to be smaller (about 2/3 the weight of normal brains), but no gross structural defects are obvious. At the cellular level there is inadequate development of *myelin*, a fatty substance that normally forms an insulating sheath around (certain) nerve fibers and thereby speeds up the transmission of nerve impulses.

Before treatment became available, the defect was manifested as profound motor and mental retardation in infants. Babies and children with untreated PKU were hyperactive but uncoordinated; some never sat, walked, or developed bowel and bladder control. The majority never learned to talk and did not progress beyond a mental age of two years. They were extremely agitated, with awkward and jerky movements caused by abnormally increased muscle tone that was sometimes so great that they assumed a characteristic "tailor's position" (Figure 4). Behavioral problems ranged from fearfulness and irritability to violent and destructive temper tantrums. Before early detection and treatment became available, PKU patients were mentally retarded and many were institutionalized. In addition to nervous system disorders, anomalies of the skin (eczema), teeth (defective enamel), and bones (small skull, growth retardation) occurred in some patients. About 75% died before the age of 30.

Unfortunately, the precise cellular actions of this biochemical defect are still unknown, and there are no naturally occurring animal models to study. Some researchers have tried to produce animal models by injecting rats and mice with excess phenylalanine, but the resulting phenotypes do not closely mimic human phenylketonuria.

TREATMENT FOR PHENYLKETONURIA. In the 1950s, physicians began to treat PKU by restricting the dietary intake of phenylalanine (Bickel et al. 1954). Synthetic mixes of amino acids supplemented with vitamins, minerals, fats, and carbohydrates were developed (Bickel 1985; Paine 1964), to be augmented by fruits, vegetables, and certain other foods. Aside from its unpalatable and extremely restrictive nature, one potential problem with this diet is that if caloric intake falls too low, tissue proteins will be metabolized, releasing some phenylalanine. Problems also arise when patients eat forbidden foods. Thus patients are continually monitored, and their diets are adjusted to maintain the optimal concentration of serum phenylalanine: either too much or too little can lead to brain damage.

Without treatment, 96–98% of babies with classical phenylketonuria would end up with IQs of 50 or less. Therapy begun after the age of three to six months also shows little or no benefit, apparently because the brain has already been irreparably damaged. But when started

*Excess phenylalanine leads to a striking decrease in the plasma concentrations of most other amino acids.

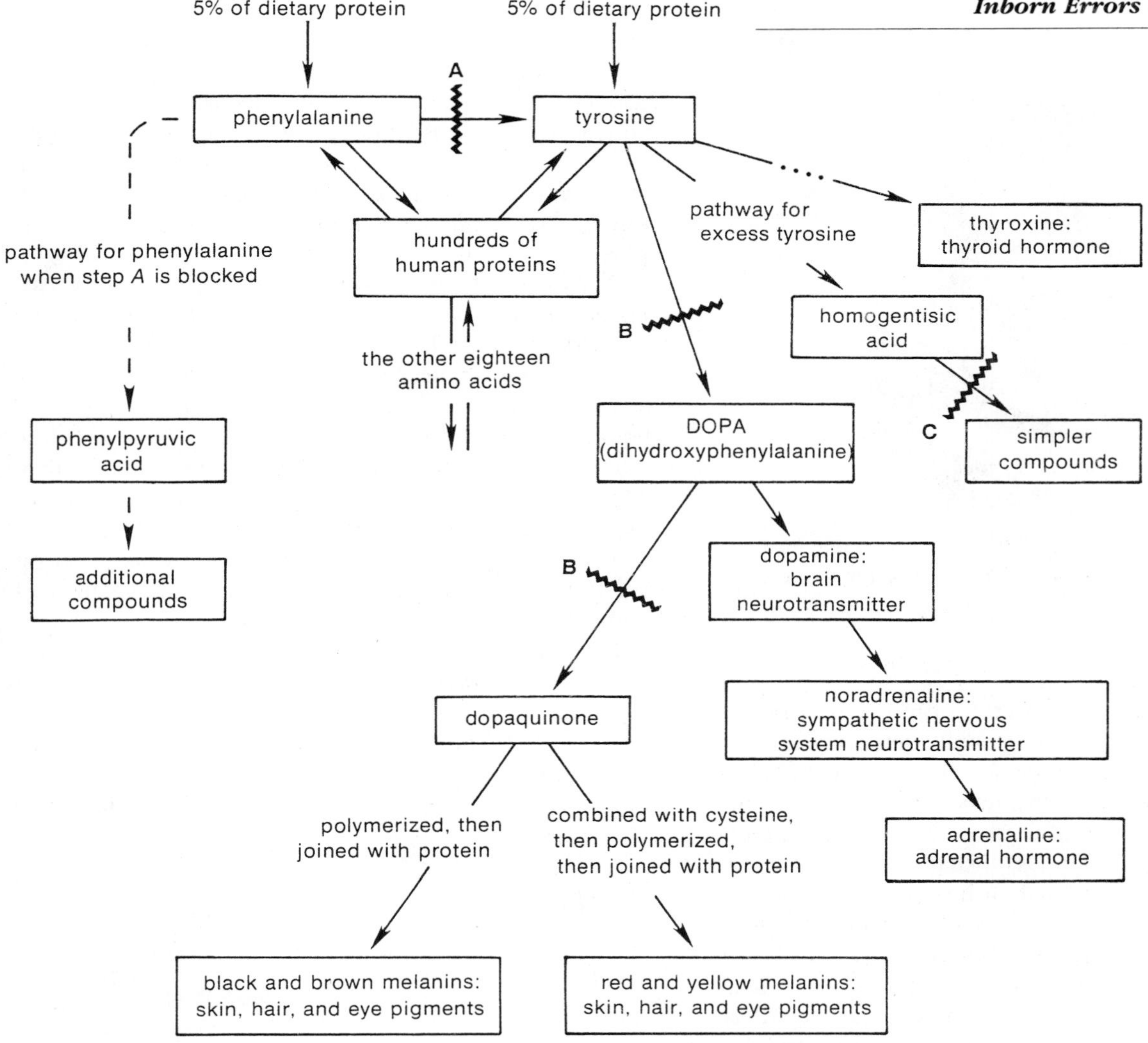

Figure 3 The metabolic blocks in three genetic disorders involving the metabolism of phenylalanine and tyrosine. (A) In phenylketonuria, phenylalanine cannot be converted to tyrosine; as a consequence, phenylalanine and its breakdown products accumulate and are excreted in the urine. The nonfunctional enzyme is phenylalanine hydroxylase. (B) In classical albinism, tyrosine cannot be converted to DOPA and dopaquinone, thereby preventing the formation of melanin pigments. The nonfunctional enzyme is tyrosinase. (C) In alkaptonuria, homogentisic acid cannot be broken down further and appears in the urine (see Figure 1). The nonfunctional enzyme is homogentisic acid oxidase.

within the first two months after birth, and properly controlled, the low phenylalanine diet is very effective. Hyperactive and difficult babies become alert and responsive; most important, they usually develop normal or near-normal intelligence, although generally not so great as that of normal sibs or parents. How long must this diet be maintained? Although it was found that some treated patients could quit by the age of five years without ill effects, others lost up to 20 IQ points. Thus the practice now is to encourage continuation of the strict diet for 12–15 years, or even indefinitely (Tourian and Sidbury 1983; Holtzman et al. 1986). This compliance is especially important for phenylketonuric females of child-bearing age (see below).

NEONATAL SCREENING AND PKU VARIANTS. In order to get affected infants on the diet before brain damage occurs, they must be identified within a few weeks after birth. This goal became possible in 1963 when Robert Guthrie and A. Susi devised a simple, reliable, and inexpensive test for excess serum phenylalanine (Chapter 25). As a result of legislation passed by most states, over

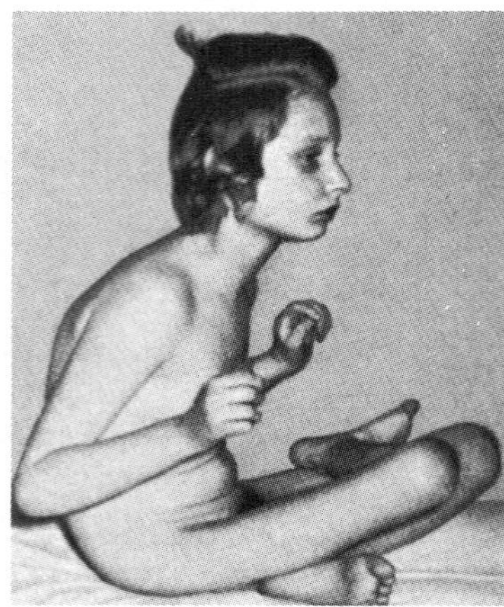

Figure 4 A severely retarded child with phenylketonuria sitting in the "tailor's position" assumed by many of these untreated patients. (From Lang 1955.)

90% of all infants are now tested for this disorder before they leave the hospital in which they were born.

Even before early screening for PKU became possible, it was known that some individuals homozygous for mutant PKU alleles developed normal or near-normal intelligence. Until the early 1960s, virtually *all* cases of PKU were ascertained because of their mental retardation, and all had high levels of phenylalanine in their blood—a condition called **hyperphenylalaninemia** (Figure 5). But was the converse situation true: Did *all* individuals with hyperphenylalaninemia become mentally retarded? Physicians suspected that this was *not* the case and that some people with hyperphenylalaninemia (perhaps 2–25%) actually developed normally.

It soon became clear that not all of these babies were destined to develop classical PKU. For one thing, the frequency of hyperphenylalaninemia determined from screening programs was considerably higher than the previously established frequency of PKU in the same populations. Second, although early studies had identified two clearly separable extremes of serum phenylalanine levels in the population (Figure 5), screening programs were now turning up new classes of individuals with *intermediate* levels (15–40 mg/dl) of serum phenylalanine. Clinicians were then faced with the vexing problems of deciding which ones should undergo dietary restriction and what concentrations of phenylalanine are safe at what ages.

Further studies revealed that the metabolism of phenylalanine involves several enzymes and coenzymes (Figure 6), each complete or partial deficiency leading to a different type of hyperphenylalaninemia (Kaufman 1983; Tourian and Sidbury 1983). Even phenylketonuria itself "is not a single genetic entity but a spectrum of conditions in which conversion of phenylalanine to tyrosine is impaired to varying degrees. The severity of the fetal damage is also variable and the mechanisms are uncertain" (Drogari et al. 1987). Here we consider the four types of hyperphenylalaninemia that represent defects in the various components of phenylalanine metabolism, omitting rarer variants that involve yet other steps of the pathway.

Type I phenylketonuria (**classic phenylketonuria** or **phenylalanine hydroxylase deficiency**) is due to a virtually complete absence (less than 1%) of activity of the liver enzyme *phenylalanine hydroxylase (PAH)*, and accounts for roughly 60% of all hyperphenylalaninemia cases (number 1 in Figure 6). Phenotypic variability and molecular studies (see below) suggest that, although there is only one *PAH* locus, found on chromosome 12, several mutations have occurred within the gene at different times and places. Despite its effect on the brain, the *PAH* gene is active only in the liver of humans.

Persistent mild hyperphenylalaninemia results from a *partial deficiency* (2%–35% activity) of the PAH enzyme, whose structural alterations are caused by other (i.e., non-PKU) mutant alleles at the *PAH* locus (number 2 in Figure 6). A transient form of mild hyperphenylalaninemia results from a maturational delay rather than from

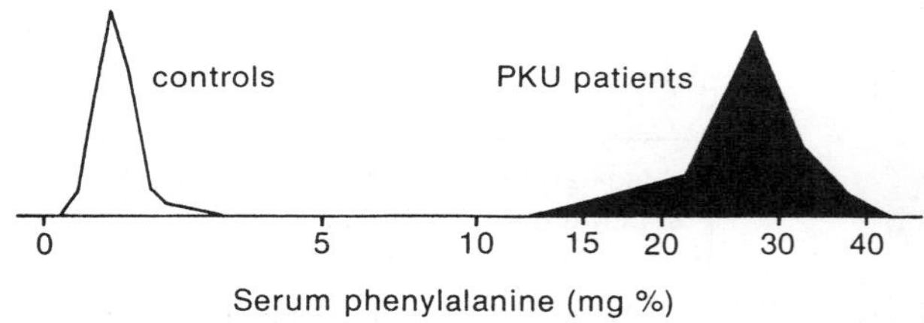

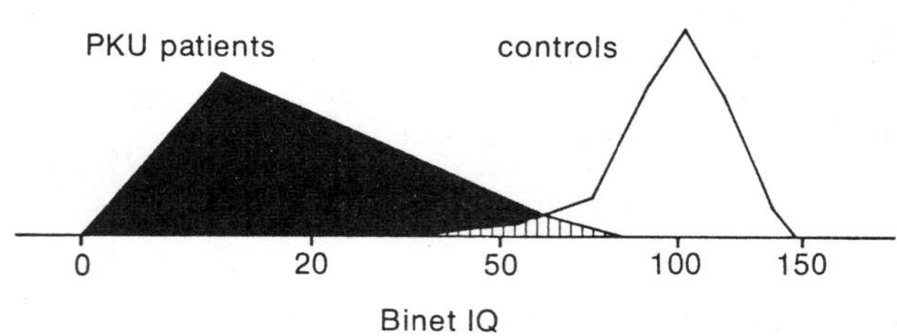

Figure 5 Distribution of phenylalanine in blood serum and distribution of IQ scores in patients with PKU and in normal individuals. This study was done before screening programs were instituted, so that most patients with PKU were in institutions. Note a complete absence of overlap in serum phenylalanine levels and only a slight overlap in IQ scores. Following the adoption of PKU screening programs in the 1960s, a few newborns with intermediate levels of serum phenylalanine were discovered, leading to problems of classification and treatment. (Redrawn from Penrose 1951.)

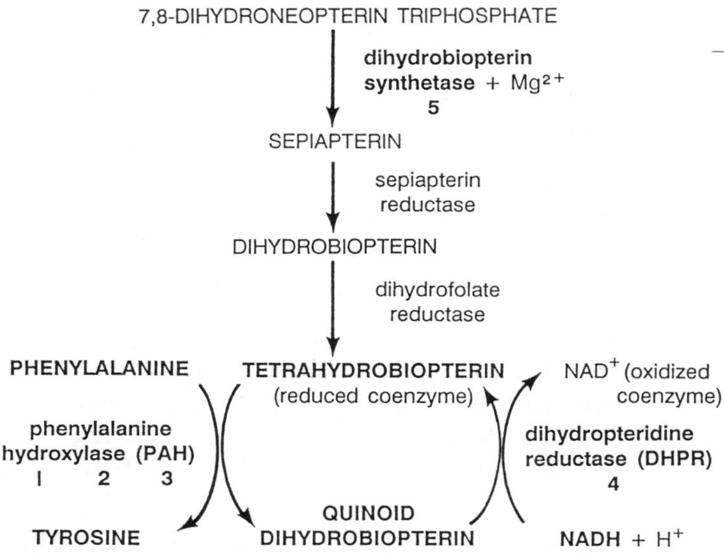

Figure 6 Partial scheme of phenylalanine metabolism, showing the key enzymes and coenzymes. Locations of the five metabolic blocks known to cause the various types of hyperphenylalaninemia are indicated by numbers 1–5, and are described in the text. (Adapted from Tourian and Sidbury 1983.)

a defect in the PAH enzyme (number 3 in Figure 6). These two mild conditions together may account for about 35% of all cases of hyperphenylalaninemia, and they require either no treatment or only early dietary restriction. These various *PAH*$^-$ phenotypes may be homozygotes, or they may be compound heterozygotes for different *PAH* alleles (Scriver and Clow 1980a,b).

Type II phenylketonuria (atypical hyperphenylalanemia, or **dihydropteridine reductase [DHPR] deficiency)** comprises about 3% of all hyperphenylalaninemia cases. Patients have a normal PAH enzyme, but their DHPR enzyme (number 4 in Figure 6) is nonfunctional. There is inadequate synthesis of the cofactor BH_4 (tetrahydrobiopterin), which in turn leads to a deficiency of BH_2 (dihydrobiopterin). The major clinical finding is an abnormal response to dietary treatment. Despite the control of hyperphenylalaninemia, there is progressive deterioration of brain function, often ending with death (Scriver and Clow 1980a,b).

The *DHPR* gene is active in many tissues, including brain cells. Besides its role in phenylalanine metabolism, the enzyme is also an essential component of the tyrosine and tryptophan pathways, which are involved in the synthesis of *neurotransmitters* (dopamine, noradrenaline, adrenaline, and serotonin). Thus its absence or deficiency has additional serious effects on nervous system functioning. In some cases, treatment with BH_4 and precursors of the missing neurotransmitters appears to slow neurological deterioration; in other cases, however, the outcome is unclear. Inherited as an autosomal recessive, the gene has been assigned to the short arm of chromosome 4 by cell hybridization and molecular techniques (Kuhl et al. 1979; Lockyer et al. 1987).

Type III phenylketonuria (dihydrobiopterin synthetase deficiency) phenotypes involve a defect of the dihydrobiopterin synthetase enzyme (number 5 in Figure 6). They comprise 1–3% of hyperphenylalaninemia cases. These patients, too, show progressive neurological deterioration even after treatment for classical PKU, and some respond to treatment with BH_4 and neurotransmitter precursors. The condition is inherited as an autosomal recessive; the gene location is unknown.

MOLECULAR BIOLOGY OF THE *PAH* GENE. Using somatic cell hybrids, in situ hybridization and various recombinant DNA techniques, researchers at Baylor University in Houston have mapped the *PAH* locus to the tip of the long arm of chromosome 12, in region 12q22-24.1 (Lidsky et al. 1984, 1985). It is a large gene of about 90,000 nucleotide base pairs and 13 exons, with the normal allele coding for a protein of 451 amino acids (Kwok et al. 1985; DiLella et al. 1986; Ledley et al. 1985). It is highly conserved, with the human and rat *PAH* genes showing about 98% homology; the amino acid sequences of the two PAH enzymes are likewise very similar.

Eight polymorphic RFLP sites have been identified within the human *PAH* locus. By scoring the action (+ or −) of each restriction enzyme on each chromosome from 33 informative Danish kindreds that had produced at least one child with PKU, RFLP profiles of the 66 carrier parents could be determined. From this, 12 unique arrays, or **haplotypes**, were defined—the RFLP haplotype of an allele being "the composite profile of the presence or absence of each of the eight different polymorphic sites within the allele" (DiLella et al. 1986). Haplotype 1, for example, contains DNA sequences that are cut by restriction enzymes *Pvu*II$_a$, and *Msp*I—but not by enzymes *Bgl*II, *Pvu*II$_b$, *Eco*RI, *Xmn*I, *Hin*dIII or *Eco*RV. Table 1 shows the frequencies of these haplotypes within the population of 66 normal and 66 PKU chromosomes, and provides some hints about the origin of the PKU mutation(s).

Table 1 Haplotype distribution of normal and mutant *PAH* alleles[a]

	NORMAL ALLELES		MUTANT ALLELES	
Haplotypes	*Number*	%	*Number*	%
1	23	34.8	12	18.2
2	3	4.6	13	19.7
3	2	3.0	25	37.9
4	21	31.8	9	13.6
5	7	10.6	0	0
6	0	0	2	3.0
7	7	10.6	1	1.5
8	1	1.5	0	0
9	0	0	1	1.5
10	1	1.5	0	0
11	1	1.5	1	1.5
12	0	0	2	3.0
Total	66	99.9	66	99.9

Source: DiLella et al. (1987).
a. The table summarizes the data gathered from 132 alleles in 66 Danish people heterozygous for PKU; 49/66 (= 74.1%) of the normal alleles and 59/66 (89.4%) of the mutant alleles were associated with haplotypes 1–4.

Note that no one haplotype accounts for all PKU chromosomes; in fact, 6 of the 12 haplotypes are associated with both normal and PKU chromosomes. Haplotypes 1 through 4 together account for almost 90% of the PKU chromosomes, and 2 and 3 together account for over 50% of them. Yet haplotypes 2 and 3 differ from each other at 5 of the 8 RFLP sites and thus are not closely related. The two haplotypes (numbers 1 and 4) most frequently associated with the *normal* chromosomes also differ at 5 of the 8 sites. Statistical analyses of linkage disequilibrium (Chapter 18) show that no one RFLP polymorphism is strongly correlated with PKU in this population. These observations, taken together, suggest that there are multiple mutations causing PKU (Chakraborty et al. 1987). How this squares with the proposed Celtic origin and subsequent spread of the *PKU* mutation through northern Europe will be interesting to see. (In Chapters 21 and 22 we discuss ways by which gene frequencies can change.)

In addition to being the more common PKU chromosomes, haplotypes 2 and 3 may be linked to the more deleterious *PKU* alleles as well (Table 2). Among 33 Danish phenylketonuric children, all those homozygous for haplotype 2 or haplotype 3, or heterozygous for haplotypes 2 and 3 (i.e., those with genotypes 2/2, 3/3, or 2/3), exhibited the "classic PKU" phenotype. On the other hand, 13 out of 16 children who were heterozygous for haplotype 1 (or 4) and any other of the first four haplotypes had a mild clinical course (Güttler et al. 1987).

Molecular analyses of the haplotype 3 chromosome carrying the most common *PKU* allele (38% in Denmark) turned up a G-to-A base substitution at the 5′ splice donor site of intron 12. This mutation leads to defective processing of PAH mRNA, in which exon 11 is skipped, with loss of PAH enzyme activity resulting from protein instability (DiLella et al. 1986). Haplotype 2, the second most prevalent *PKU* allele (20% in Denmark), carries a different mutation: a C-to-T change in exon 12, causing an arginine-to-tryptophan amino acid substitution at residue 408 of the PAH enzyme. The mutations associated with other haplotypes shown in Table 1 have not yet been characterized. But a totally different haplotype and associated *PKU* mutation—deletion of an entire exon—has been found among Yemenite Jews in Israel (Kidd 1987).

PRENATAL DIAGNOSIS AND DETECTION OF PKU CARRIERS. Because the PAH enzyme is expressed only in the liver, its activity or inactivity cannot be detected prenatally in fetal amniotic cells. But molecular techniques make it possible to detect the mutant DNA itself—not only in fetuses known to be at risk, but in the general population as well.

Because heterozygosity for the RFLP sites is about 90% among Caucasians, most families already known to be at risk for PKU can be effectively screened and diagnosed by using haplotype analyses (Woo 1986). Yet for those families with uninformative haplotype segregation patterns, and for carriers without a family history of PKU, haplotype analysis alone will not work—because about 8% of the time a normal *PAH* allele is associated with haplotype 2 or 3. Thus direct detection of the mutant alleles in the DNA of some carriers or homozygotes is

Table 2 Correlation between genotypes and phenotypes for haplotypes 1 to 4[a]

	NUMBER OF CHILDREN WITH PHENOTYPE	
Genotype	*Classic PKU*	*Mild PKU*
1/1	0	1
1/2	1	3
1/3	0	4
1/4	0	2
2/2	2	0
2/3	2	0
2/4	0	0
3/3	4	0
3/4	2	4
4/4	0	0

Source: Adapted from Güttler et al. (1987).
a. Data on 25 Danish children (of 33 probands) with classic or mild phenylketonuria whose genotypes included haplotypes 1 through 4. The remaining 8 children (omitted here) had other, less frequent haplotypes. *Note:* No children with genotypes 2/4 or 4/4 were observed in this sample.

necessary. The more tedious and expensive method involves gel electrophoresis and Southern blot analyses of DNA restriction fragments with probes for the mutant and normal *PAH* alleles. But by using the polymerase chain reaction to amplify the DNA fragment (i.e., exon 12 and flanking intronic sequences) containing the mutant sites, researchers have constructed oligonucleotide probes that do not require gel electrophoresis. The alternative "dot-blot" analyses are just as accurate but much simpler, quicker, and cheaper than older hybridization techniques. They can also be automated, an advantage that opens the door to carrier screening in the general population (DiLella et al. 1988). Perhaps far in the future, gene replacement techniques (described below and in Chapter 25) will enable investigators "to use the mouse model to investigate somatic gene therapy for PKU" (Woo et al. 1987)

MATERNAL PHENYLKETONURIA. Recall that before the advent of mass screening for PKU very few affected females had babies; instead, they were usually profoundly retarded and often institutionalized. Ironically, the successful treatment of PKU has now created a different medical problem. In the late 1950s physicians noted a few cases in which treated women with PKU gave birth to babies who did not exhibit hyperphenylalaninemia, but were nonetheless retarded. Then, as dietary treatment rescued more and more females with PKU from retardation, thereby allowing them to lead normal lives that included motherhood, the incidence of retarded non-PKU offspring increased. When mothers with PKU go untreated during pregnancy, over 92% of their babies are mentally retarded and about 75% have a small head (microcephaly). In addition, these babies often have low birth weight and heart defects. Despite the fact that they are heterozygotes for recessively inherited PKU, many also exhibit hyperphenylalaninemia (Lenke and Levy 1980).

The origin of **maternal PKU** soon became clear, however. Most of these mothers with PKU had never been identified and diagnosed, because screening programs began long after they were born. But some mothers with PKU had been treated and then allowed to abandon the diet therapy—after which their serum phenylalanine levels increased greatly. This dietary change seemed to have limited ill effects on them (especially if they remained on the diet beyond age 10), but it caused irreparable harm to their developing offspring by transferring high phenylalanine levels across the placenta. The metabolic cause of these abnormalities is not understood, although it is more likely due to the excess of phenylalanine than to the deficiency of tyrosine (Levy 1987). Indeed, it is now clear that, if these women reproduce at average rates, the incidence of PKU-related mental retardation could return to its original level within just one generation (Lowitzer 1987)!

> To families who have successfully endured the rigors of dietary treatment . . . it seems unfair that they must still be troubled by the specter of mental retardation . . . To physicians and others who have devoted their lives to the prevention of mental retardation from PKU . . . it must seem devilishly unfair that among the fruits of these efforts is the possibility of mental retardation in the children of these children. (Levy 1985)

It is hoped that this problem can be prevented if women with PKU return to diet therapy before (or perhaps within a few weeks after) they become pregnant. But many females of child-bearing age are unaware of their PKU until after they have produced an abnormal baby. Oddly enough, many have no recollection of ever being on a special diet, and their parents see no reason to tell them. "Some families put off vacations and restaurant meals until the child was off the diet. Nothing pleased them more than the feeling they could forget the whole thing" (Caldwell 1981).

The treatment of maternal PKU raises many medical, legal, and ethical questions (Robertson and Schulman 1987; Johnsen 1987). Is there a clear relationship between a mother's blood phenylalanine concentration and the degree of damage to her offspring? Could very *low* levels of phenylalanine (i.e., overtreatment of the mother) also damage the fetus? Can dietary therapy begun *after* conception prevent damage to the fetus? How can females with PKU be identified before they become pregnant? Should a mandatory registry of females with PKU be established? What if some women with PKU consciously choose not to resume diet therapy before or during pregnancy; can they be forced to do this against their will? As Scriver et al. (1988) point out: "Genetics, medicine and society meet in the hyperphenylalaninemias."

Albinism

Although not common in humans, the absence of coloration is such a striking trait that most of us have seen it at some time. **Albinism** has been noted by writers as far back as the first century A.D.; indeed, from the description of his birth in the Book of Enoch,* it appears that Noah was an albino (Sorsby 1958). European explorers were fascinated by the "white Negroes" they saw in Africa and by the "moon-eyed people" found in some Indian tribes of Central and North America.

Albinism occurs in a wide variety of animals, from insects to mammals. Usually the basic defect is the absence of melanin pigments, inherited as an autosomal

*The Book of Enoch is part of ancient Jewish Old Testament writings known as the Apocrypha.

recessive. Garrod (1909) regarded albinism as an inborn error of metabolism whose clinical manifestations are due to the lack of an end product rather than to the excess of some biochemical intermediate:

> The essential phenomenon of albinism is the absence of the pigments of the melanin group, which play the chief part in the colouration of man and lower animals, and which serve the important function of rendering the eye a dark chamber.
>
> Three possible explanations . . . suggest themselves. We might suppose that the cells which usually contain pigment fail to take up melanins formed elsewhere; or that the albino has an unusual power of destroying these pigments; or again that he fails to form them.

Garrod favored the third hypothesis, suggesting that "an intracellular enzyme is probably wanting." Again he was correct. Complete or classical albinism is now known to be caused by the absence of a functional **tyrosinase** enzyme (Figure 3B). But the biochemistry of melanin formation is very complex, and other types of albinism are also known, including a few X-linked or dominant types. Witkop et al. (1983) note that 62 different genes affect the skin and fur colors of mice, and suggest that genetic influences on human pigmentation may be equally complex.

HISTOLOGY. The tyrosinase enzyme is active only in specialized cells known as **melanocytes**. These arise in the neural crest (blocks of tissue on either side of the neural fold) during very early embryonic development and later migrate to their final destinations. Melanocytes occur in the lower layer of the skin epidermis and the hair bulbs as well as in certain parts of the eye, of the ear, of the nervous system and of mucous membranes. As Figure 7 shows, each melanocyte has many thin, finger-like dendritic processes that connect with several dozen surrounding epidermal cells called *keratinocytes*. Inside the melanocytes, the tyrosinase enzyme is restricted to small specialized structures known as *melanosomes*. During normal pigment formation, melanin is deposited in these melanosomes, which travel up the dendritic processes and are transferred to the adjacent keratinocytes; there they come to lie over the nucleus of each cell. By capping the epidermal nuclei and absorbing much of the ultraviolet radiation that enters the cell, melanin shields the DNA from the mutagenic effects of sunlight and other sources of ultraviolet radiation.

Albinos have normal melanocytes with structurally normal melanosomes that lack only pigment. Among normal individuals, skin color is determined not by the number of melanocytes—which appears to be about the same in all races—but rather by the number and distribution of melanosomes produced by these cells, as well as by the type of melanin. In Caucasians and Asians, for example, melanosomes aggregate into complexes of three or more; in Blacks and Australian aborigines, the melanosomes remain separate. In all races, the rate of production of melanosomes varies with amount of sunlight, hormonal changes, and genetic constitution.

BIOCHEMISTRY AND MOLECULAR BIOLOGY. As shown in Figure 3, tyrosine is normally converted to *dihydroxyphenylalanine (DOPA)*, a key biological compound. In melanocytes, DOPA is converted to *dopaquinone*, which finally is polymerized with proteins in the melanosomes to form black and brown *eumelanins*. Alternatively, dopaquinone may combine with cysteine before being conjugated with melanosome protein to form red and yellow *pheomelanins*. The first two steps in this scheme (tyrosine → DOPA → dopaquinone) require the enzyme **tyrosinase**, whose 548-amino-acid sequence has been deduced from the nucleotide sequence of the recently cloned human tyrosinase gene (Kwon et al. 1987). Some evidence suggests that this gene resides on chromosome 11.

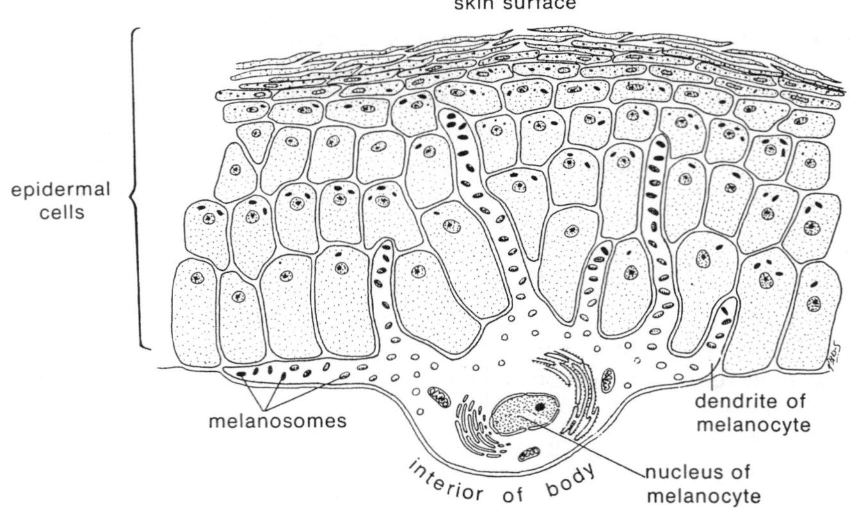

Figure 7 Association of one melanocyte with several dozen overlying epidermal cells. Pigment-containing melanosomes migrate up through the dendrites and are transferred to the overlying keratinocyte cells.

THE ALBINO PHENOTYPE. In humans, many types of albinism are known, including even a few with normally pigmented skin. Although the severity of the clinical features varies, all types of albinos show reduced or absent eye pigmentation and structural abnormalities of the eye. Severe squinting, poor vision, and nystagmus (rapid, jerky, involuntary movements of the eyeballs) also occur.

All albinos lack binocular vision, as a result of misrouting of optic nerve fiber pathways in the brain (Drager 1986). There is some evidence that even heterozygotes may show some slight defects in binocular vision (Leventhal et al. 1985). Studies on albino cats suggest that this impairment is due to a lack of pigment in certain nerve cells during early development (Guillery 1974). Researchers have found that albinism affects hearing as well as sight. In the inner ear, too, nerve fibers are normally routed to one or the other side of the brain, and melanin is apparently necessary for proper development of these neural pathways. Melanin also may act as a sound absorber, damping extreme peaks. The amount of melanin found in the inner ear correlates closely with the amount present in the iris of the eye, albinos lacking pigment in both (Creel et al. 1980).

HUMAN VARIANTS. Witkop et al. (1983) describe 14 types of albinism, all but one of which are poorly understood. They fall into two main groups. The 10 types of **oculocutaneous albinism** or **OCA** (*oculo*, "eye"; *cutaneous*, "skin") involve the absence or partial deficiency of melanin in the eyes, skin, and hair, as well as the visual defects described above. In the four types of **ocular albinism (OA)**, defects are limited to the eyes; the skin is normally pigmented or only slightly underpigmented.

The most extreme clinical form, the classic one studied by Garrod, is **tyrosinase-negative OCA**. It is the only type for which the metabolic defect is known: absence of the tyrosinase enzyme leads to the total lack of melanin in the eyes and in the skin and hair—"as witness the white hair, pink eyes, and unpigmented skin which characterize such individuals. Pigments of other kinds are not wanting, such as the lipochromes which impart their yellow tints to fats and blood serum, and haemoglobin and its derivatives" (Garrod 1909). This enzymatic defect is readily demonstrated by the **hair bulb incubation test**, in which individual hair bulbs are plucked out and placed in a solution containing a high concentration of tyrosine (or DOPA) for about 12 hours. After this time, hair bulbs from normal individuals will have become heavily blackened as a result of the conversion of tyrosine to melanin. But hair bulbs from tyrosinase-negative albinos show no melanin formation at all. The serum of tyrosinase-negative albinos contains no excess of tyrosine, however, which means that it must enter other metabolic pathways rather than accumulate in body fluids.

Extreme sensitivity to sunlight causes excessive roughness, wrinkling, and folding of the skin as well as a high frequency of skin cancer. (This is especially true in tropical regions. In Nigeria, half of all Ibo albinos develop cancer by the age of 26 and die by 40 [Brody 1981].) The irises of the eye are gray to blue, containing no visible pigment. These albinos are usually cross-eyed and very nearsighted, often to the point of legal blindness.

The overall frequency of this disorder varies considerably among ethnic subgroups, as do estimates for each group. Among the Irish, for example, the frequency is about 1 in 10,000–15,000; in the United States the frequency is about 1 in 28,000 among Blacks and about 1 in 39,000 among Whites; but in British Columbia, Canada, it is about 1 in 68,000. Heterozygotes for the various types of albinism are fairly common, however, totaling 1–2% of the general population.

The other major type of oculocutaneous albinism is called **tyrosinase-positive OCA**, because a positive reaction to the hair bulb incubation test (i.e., darkening of the bulb) reveals the *presence* of tyrosinase. Here, too, serum tyrosine levels are normal. The nature of the metabolic error is unknown, but geneticists speculate that it might involve a defect in the transport of tyrosine into the melanosomes rather than in its subsequent conversion to melanin. This defect could result from an abnormal permease enzyme in the melanosome membrane, although none has yet been identified. Tyrosinase-positive OCA is the most common type, but it is also less frequent in Causasians (about 1 in 37,000–60,000) than in blacks (1 in 14,000–15,000). Among Nigerian Ibos, the frequency is about 1 in 1,100. It is also quite prevalent among certain Native American tribes (about 1 in 140–240 for the Tele Cuna, the Hopi, the Jemez, and the Zuni), perhaps owing to the special place of albinos in these cultures.

There is considerable overlap in phenotypes of the two disorders, but some distinctions have been noted. Tyrosinase-positive albinos have skin and eye problems very similar to those of tyrosinase-negative albinos, but often to a lesser degree. For example, nystagmus and squinting may be mild to severe rather than always severe, and vision may improve somewhat with age. Some pigmentation may develop over the years in tyrosinase-positive individuals: the hair may turn from white or cream-colored in infancy to blonde or red in adulthood; the skin may change from pink to cream-colored (or light tan in Blacks), and freckles may appear; eye color may change from blue to brown, and red reflectance of the retina usually diminishes. All of these transformations are more pronounced in Black tyrosinase-positive albinos than in Caucasians of this type.

Because both of these types of albinism are autosomal recessives, matings between two tyrosinase-negative or two tyrosinase-positive albinos are expected to produce all albino progeny. But, because the two autosomal loci are independent, matings between tyrosinase-positive and tyrosinase-negative albinos produce all normal offspring.

A third category of OCA, the tyrosinase-variable **yellow mutant**, is probably allelic to tyrosine-negative albinism. Hu et al. (1980) report on a family in which both types of albinism were segregating, and they suggest that individuals with the yellow mutant phenotype were actually heterozygous for this allele and the tyrosine-negative allele. These individuals have blonde or reddish-blond hair, and their hair bulbs (after incubation with tyrosine and cysteine) show yellow rather than black pigmentation.

Two rare syndromes of complete or partial tyrosinase-positive OCA have also been studied in considerable detail. Patients with *Hermansky-Pudlak syndrome* have bleeding problems due to platelet defects, and various other problems associated with defective storage of lipids and waxy substances. The *Chediak-Higashi syndrome* is characterized by a much-increased susceptibility to infections and a certain kind of lymphatic cancer, usually leading to death during childhood.

Among the five other rare types of OCA described by Witkop et al. (1983), all but one are clearly tyrosine-positive; four are autosomal recessives and one is inherited as an autosomal dominant. Among the four types of ocular albinism, three have normal skin color and one shows normal-to-mottled skin. Two of these four types are X-linked recessives, one is autosomal dominant, and one is autosomal recessive.

LIPID AND LIPOPROTEIN METABOLISM

Here we discuss two well-known disorders that involve lipid metabolism. The first is an autosomal recessive lysosomal storage disease that is lethal in early childhood. The second is a fairly common autosomal dominant condition, a defect of membrane receptors that leads to coronary heart disease.

Tay-Sachs Disease

In 1887 an American neurologist, Bernard Sachs, described the case history of an infant who had a strange condition and died at the age of two years:

> The little girl . . . was born at full term, and appeared to be a healthy child in every respect; its body and head were well proportioned, its features beautifully regular. Nothing abnormal was noticed until the age of two to three months, when the parents observed that the child was much more listless than children of that age are apt to be . . . and that its eyes rolled about curiously . . . The child would ordinarily lie upon its back, and was never able to change its position; muscles of head, neck, and back so weak that it was not able either to hold its head straight or to sit upright. It never attempted any voluntary movements . . . It could not be made to play with any toy, did not recognize people's voices, and showed no preference for persons around it. During the first year of its life, the child was attracted by the light, and would move its eyes, following objects drawn across its field of vision; but later on absolute blindness set in . . . Hearing seemed to be very acute . . . the slightest touch and every sound were apt to startle the child . . . The child never learned to utter a single sound . . . During last summer (1886), the child grew steadily weaker, it ceased to take its food properly, its bronchial troubles increased, and finally, pneumonia setting in, it died.

Sachs had referred the sick child to an ophthalmologist, who discovered a peculiar **cherry-red spot** on the retina of each eye (Figure 8A). In the next few years, Sachs encountered several more children with the same malady, including a sib of the first case and four sibs in another family. All had red spots on the retina, and all died very young. A search of the medical literature for previous descriptions of this condition uncovered two reports, published in 1881 and 1884, by an English ophthalmologist, Warren Tay. He had described a family in which the cherry spots occurred in several members, all of whom died by the age of three years. Within a decade, about two dozen cases were recorded, "so far almost exclusively observed among the Hebrews" (Sachs 1896). Although the condition was originally called amaurotic family idiocy (Greek *amaurosis*, "loss of sight"), then infantile amaurotic familial idiocy, it is now usually called **Tay-Sachs disease (TSD)**.

FREQUENCY OF TAY-SACHS DISEASE. The condition occurs predominantly among Ashkenazi Jews, i.e., the descendants of Jews who settled in eastern and central Europe, with only about 15% of all cases being born to non-Jewish parents. The highest rates are found in the descendants of Jews from northeastern Poland, southern Lithuania, and an adjoining area of White Russia. Interestingly, the condition is rare among Sephardic and Asian Jews. Among American and Canadian Jews, the incidence is roughly 1 in 3,000 to 6,000 births, much higher than the incidence of about 1 in 550,000 births found among the non-Jewish American and Canadian populations. Because of unusual founder effects, however, high rates of TSD are also found among French Canadians in Quebec and among some Pennsylvania Dutch of non-Amish ancestry. Roughly 1 in 30–40 Jews is heterozygous for the gene, whereas only 1 in about 300–400 non-Jews is heterozygous. Why this harmful allele has been maintained at such a high frequency is not known.

MORPHOLOGICAL AND BIOCHEMICAL ABNORMALITIES. The progression of this untreatable illness is very much

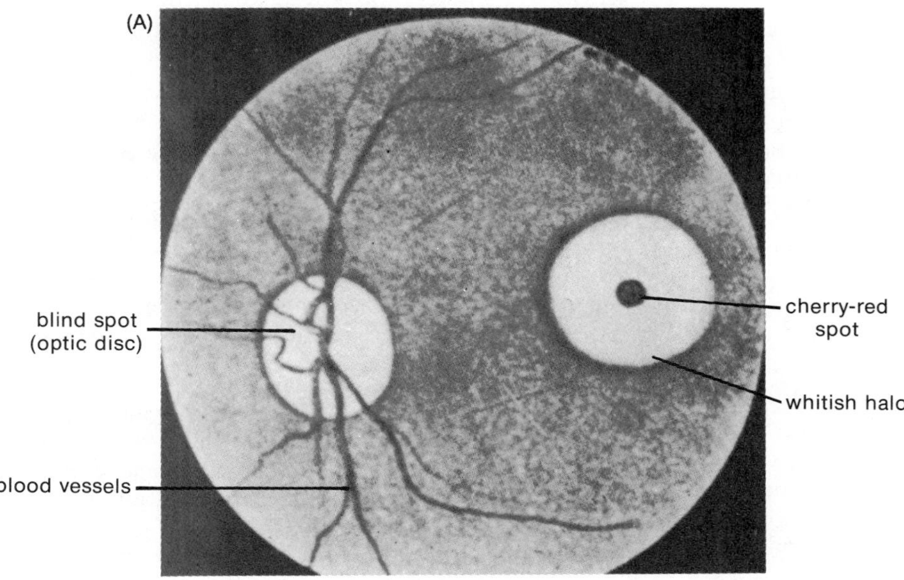

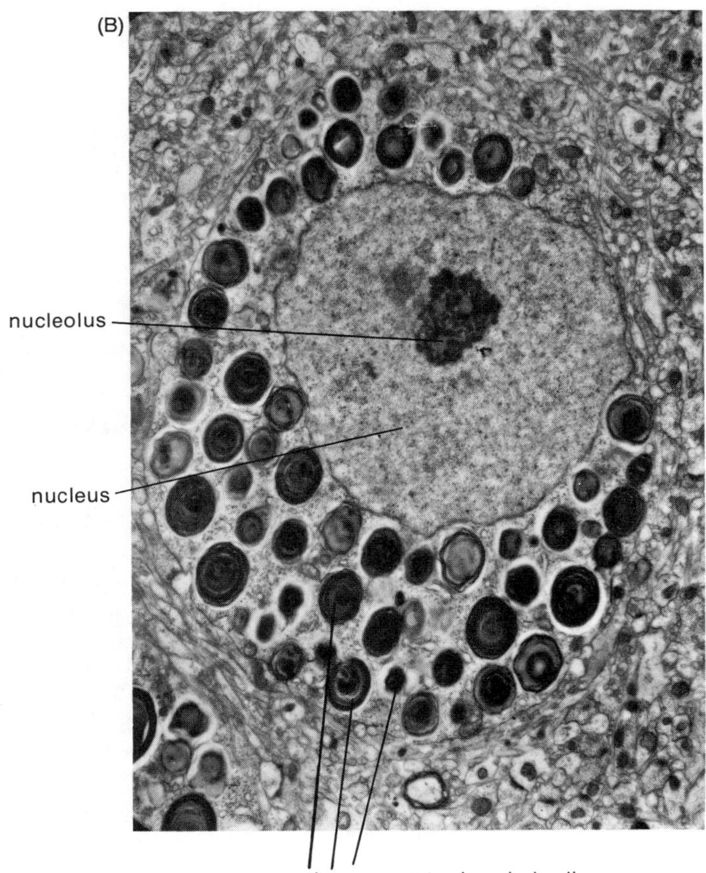

Figure 8 Manifestations of Tay-Sachs disease in the retina of the eye and in nerve cells of the brain. (A) The cherry-red spot on the retina can be seen through the pupil of the eye. The spot is caused by the absence of cells that usually overlay the fovea centralis, the region of sharpest vision. Neither the red spot nor the pronounced whitish halo is seen in normal retinas. (B) Darkly staining membranous cytoplasmic bodies in a brain cell of a Tay-Sachs patient contain excess lipid. The structures are absent in normal nerve cells. (A from Sloan and Frederickson 1972; B from Terry and Weiss 1963.)

as described above. Usually the first symptom noticed is the "startle reaction" to sharp noises. After one year of age, the child's condition degenerates rapidly; generalized paralysis, blindness, gradual loss of hearing, severe feeding difficulties, and some enlargement of the head are the rule by 18 months. By two years the patient is completely immobile and may need to be institutionalized. Most patients die from respiratory infections by the age of three or four.

In his original report Sachs (1887) included the results of the autopsy on his first patient, in which there were both gross and microscopically observable malformations of the brain. Later studies have verified and greatly expanded these findings. Mainly, the cerebrum becomes greatly enlarged by the swelling or ballooning of individual nerve cells. This abnormality is due to the accumulation of a specialized type of lipid known as a **ganglioside**. Within individual nerve cells of Tay-Sachs patients, this material collects in the form of *membranous cytoplasmic bodies* (Figure 8B), which fill up and distend the cytoplasm.

Gangliosides are important components of normal membranes. Found mostly in the brain, they are thought to act as surface membrane receptors that can bind to certain substances outside the cell, thereby transmitting information to the cytoplasm. These large molecules are synthesized by the orderly addition of several simple sugars to a branched long-chain lipid called ceramide and degraded by the removal of these sugars—one by one—from the lipid part of the molecule. It is this latter process that was thought to be defective in Tay-Sachs patients, the defect resulting in the massive accumulation (up to 300 times the normal amount in the cerebrum) of a ganglioside called G_{M2}, which is normally found almost exclusively in nervous tissue. Researchers suspected that the basic defect was a missing hexosaminidase enzyme that normally splits off the terminal sugar group (hexosamine) from the rest of the G_{M2} molecule. But they were surprised to find that Tay-Sachs patients showed nearly normal levels of total hexosaminidase activity. The mystery was solved when investigators at the University of California in San Diego distinguished *two* forms of the enzyme in the tissues of normal children but could find only *one* form in the tissues of Tay-Sachs patients (Okada and O'Brien 1969). The form missing in Tay-Sachs children was **hexosaminidase A** or **Hex-A** (Figure 9), whereas the form present in both normal and Tay-Sachs children was called *hexosaminidase B* or **Hex-B**. Heterozygous carriers of the Tay-Sachs gene have Hex-A levels that are intermediate between those of normal controls and affected patients. Okada and O'Brien concluded their report with the following observation:

> The immediate practical importance of our discovery is that hexosaminidase assay provides a means for the diagnosis of homozygotes. We have found that both hexosaminidase components are present in normal fetal amniotic fluid cells obtained by amniocentesis early in pregnancy. If component A is absent in fetal amniotic cells derived from the individuals homozygous for Tay-Sachs disease, as appears likely, the intrauterine diagnosis of this fatal human disease will be possible.

CARRIER DETECTION AND PRENATAL DIAGNOSIS. These predictions were soon borne out. Hex-A does indeed perform the proposed function in normal individuals, and heterozygotes can be detected by measuring the level of this enzyme in a small sample of their blood. In 1971 the first of a series of voluntary screening programs was set up to identify carriers in a number of Jewish populations (Chapter 25). In 1970 the first prenatal identification of a Tay-Sachs fetus occurred, when a pregnant woman whose previous child had died from Tay-Sachs disease underwent amniocentesis and was found to be carrying a second affected child. Since then, tragedies have been averted in several hundred families—by detecting carrier couples *before* they produce an affected offspring (Kaback 1977a,b) and by diagnosing Tay-Sachs disease during early pregnancy.

MOLECULAR BIOLOGY AND GENETIC VARIANTS. The

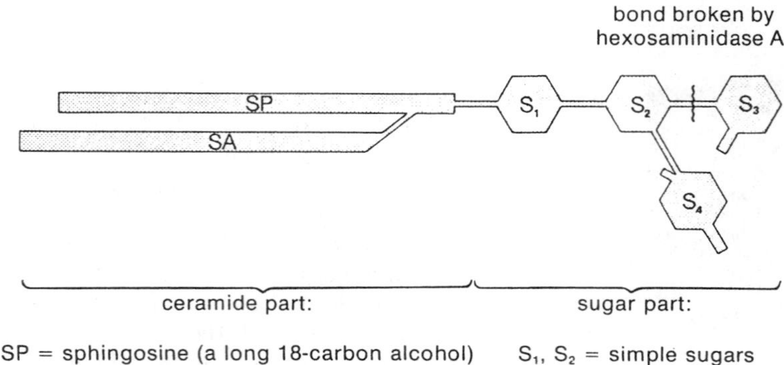

Figure 9 Ganglioside G_{M2}, the lipid that accumulates in the cells of Tay-Sachs patients. This large, complex molecule is a component of the membranes of brain cells. In normal persons, the enzyme hexosaminidase-A cuts off the terminal sugar group (S_3) in one step of a pathway that degrades the molecule. Infants with Tay-Sachs disease are homozygous for an autosomal recessive allele and lack a functional enzyme.

Hex-A enzyme actually consists of two subunits, *α and β chains*, which are coded for by genes on chromosome 15 (short arm) and chromosome 5, respectively. *TSD results from a deficiency of α chains* due to mutations in the α locus. This gene is about 35 kb long and has 14 exons (Proia and Soravia 1987). Several variants of both classic (infantile) and nonclassic TSD are known, and some patients have turned out to be genetic compounds (i.e., heterozygous for two different mutant alleles) rather than homozygous for a single allele.

1. Ashkenazi Jewish patients usually make no α chains and no complete mRNA for this locus. Yet their DNA at this locus is apparently intact, an observation suggesting that there may be a defect in mRNA processing or transport. Arpaia et al. (1988) found a *base substitution* (G to C) in the first base of intron 12, which probably results in defective splicing of the mRNA. This mutant may account for over 20% of the *TSD* alleles in Ashkenazi Jews.
2. French Canadian patients also make no α chains or mRNA for this locus, but they have a 7.6-kb *deletion* at the 5′ end of the gene that could have arisen by unequal crossing over between Alu sequences (Myerowitz and Hogikyan 1987).
3. Some patients who belong to neither of the above two groups make altered α chains that form defective Hex-A with low activity or stability. Other unusual variants have also been described (Zokaeem et al. 1987).

Upon discovering a few rare Tay-Sachs patients who possessed *normal* Hex-A and Hex-B, K. Sandhoff and colleagues at the University of Bonn (1971) correctly postulated that there must be a third constituent of the G_{M2} system. This **activator protein**, by binding to and solubilizing the lipid ganglioside substrate, allows the water-soluble Hex-A enzyme to do its job. The gene for this protein has been mapped to chromosome 5 (Burg et al. 1985).

Other types of TSD (*juvenile, adult, chronic*) also exist; in these conditions the symptoms have a later onset or are less severe, and their enzymes show weak activity. Some of these patients make α chain precursors that do not associate with β chains and are not converted to the mature form. Some may carry the classic *TSD* allele plus a milder mutant variant.

The Hex-B enzyme, consisting of beta subunits only, is produced by the chromosome 5 β chain locus (Gilbert et al. 1975). Mutation at this locus causes absence of both Hex-A and Hex-B enzyme activity and is associated with another fatal but very rare lipid storage disease that is quite similar to Tay-Sachs disease. First described in 1968, **Sandhoff disease** is characterized by the massive accumulation of a different form of G_{M2} ganglioside in the cerebrum and in other organs, including the liver, spleen, and kidney. These and several other hereditary ganglioside storage disorders are discussed at length by O'Brien (1983).

Familial Hypercholesterolemia

> Cholesterol is a Janus-faced molecule. The very property that makes it useful in cell membranes, namely its absolute insolubility in water, also makes it lethal. For when cholesterol accumulates in the wrong place, for example within the wall of an artery, . . . its presence eventually leads to the development of an atherosclerotic plaque . . . If cholesterol is to be transported safely in blood, its concentration must be kept low, and its tendency to escape from the bloodstream must be controlled (Brown and Goldstein 1986).

Since its discovery in 1784, the functions, malfunctions, and metabolism of this substance have fascinated and challenged many chemists, biologists, and physicians. Cholesterol is an essential compound in our bodies: a major component of plasma membranes as well as raw material for the production of bile acids in the liver and steroid hormones in the gonads and adrenal glands.

In 1938 a Norwegian physician, Carl Müller, described a condition characterized by hypercholesterolemia (excess cholesterol in the blood), xanthomas (yellow fatty deposits in the skin and tendons), and coronary heart disease. He recognized it as an inborn error of fat metabolism that was inherited as an autosomal dominant. **Familial hypercholesterolemia (FH)** is now known to be the most common dominant Mendelian disorder in humans: the worldwide heterozygote frequency is 1 in 500, accounting for 5% of all heart attacks in patients under the age of 60, and the homozygote frequency is about 1 in a million. In certain self-contained populations—among Lebanese, South African Afrikaners and French Canadians, for example—it is much more frequent due to founder gene effects (Chapter 21).

PHENOTYPES OF FAMILIAL HYPERCHOLESTEROLEMIA. Normal newborn babies have serum cholesterol levels of about 30 mg/dl, and in adults raised on a low-fat diet, serum cholesterol levels are between 50–80. But in adults raised on the high-fat diets characteristic of Western societies, "ideal" serum cholesterol levels range from 130 to 190 (Grundy 1986). *FH heterozygotes*, however, are born with cholesterol levels that are at least twice the adult levels, ranging from 270–550, and averaging 340. This excess is limited to the **low-density lipoprotein (LDL)** fraction of cholesterol (see below), which also tends to be deposited as fatty pads or nodules in tendons and under the skin and—with dire consequences—as components of atherosclerotic plaques in the walls of arteries. Heart attacks, which are 25 times more frequent in FH heterozygotes than in normal relatives, begin to occur at around age 35; 50% of males and 15% of females die by age 60. In both heterozygotes and homozygotes, recurrent arthritis of the ankles, knees, wrists, and digits is common. Unlike many candidates for heart disease, however, FH patients usually have a slim body build and normal blood pressure.

FH homozygotes are born with serum cholesterol values ranging from 600–1200 mg/dl, averaging about 4–6 times the normal levels. Xanthomas are more frequent and develop at a much earlier age in homozygotes; they also occur under the skin as well as in tendons. Heart attacks are common from age 5 on, and few patients survive to age 30. Because of early death or incapacitation, homozygotes rarely reproduce.

TRANSPORT AND METABOLISM OF CHOLESTEROL. Cells get their cholesterol in two ways: externally from our diet, and internally through metabolic synthesis (mostly in liver cells). These two pathways are interconnected, however, in a manner that was worked out mostly between 1972 and 1984 by Michael S. Brown and Joseph L. Goldstein of the University of Texas Health Science Center at Dallas—research that won them Nobel Prizes in 1985 (Goldstein and Brown 1983; Brown and Goldstein 1984, 1986; Goldstein et al. 1985).

Within our intestines, fats are broken down into cholesterol and triglycerides, which then move into the bloodstream and are distributed throughout the body. But because these fatty substances are insoluble in water, they cannot be transported "as is." Instead, they are packaged into **lipoprotein particles** that have a large, oily core protected by an outer detergent coat attached to one or more protein molecules. Figure 10 summarizes the interrelationships among the several types of lipoproteins: chylomicrons, chylomicron remnants, very low density lipoproteins (VLDL), VLDL remnants (= intermediate density lipoproteins, or IDL), low-density lipoproteins (LDL), and high-density lipoproteins (HDL).

The major carrier of cholesterol is the LDL fraction. Each LDL particle consists of a sizable core of cholesteryl esters surrounded by a detergent coat of phospholipid molecules and unesterified cholesterol molecules. Atop this sphere sits one large protein molecule called **apoprotein B-100**. Exactly how LDL particles get into cells and how cholesterol is metabolized inside cells were the mysteries that Goldstein and Brown set out to solve in the early 1970s. Using cultured fibroblast cells as a model system, they soon discovered that apoprotein B-100 attaches to specific regions on the cell membrane—tiny invaginations called **coated pits**. There, embedded in the plasma membrane, reside clusters of binding sites

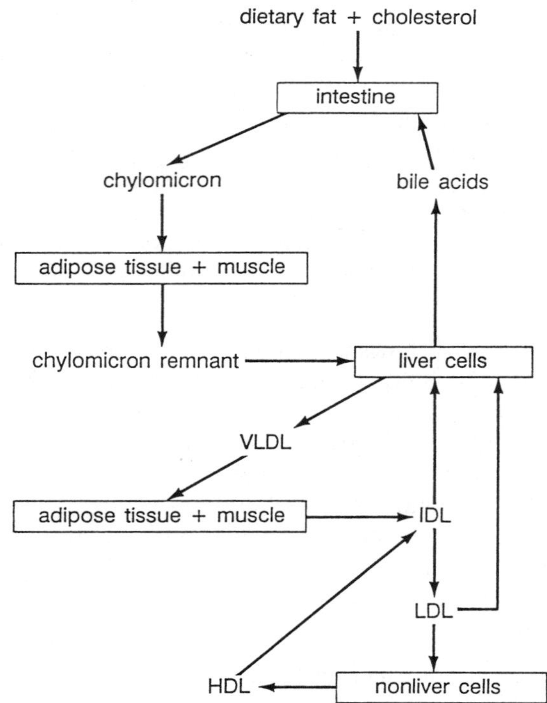

Figure 10 Partial transport pathways for both externally and internally derived fats. VLDL, very low density lipoprotein (cholesterol + digested fats); IDL, intermediate-density lipoproteins (mostly cholesterol); LDL, low-density lipoproteins (esterified cholesterol); and HDL, high-density lipoproteins (nonesterified cholesterol). (Adapted from Brown and Goldstein 1984; Grundy 1986.)

for apoprotein B-100. These **LDL receptors** and the gene that codes for them are the focus of our discussion.

The receptors are made of glycoprotein, and their number increases or decreases according to a cell's need for cholesterol. Once attached to the LDL receptors, LDL particles are engulfed by the coated pits and drawn into the cytoplasm (Figure 11). Then the coated vesicles transfer their LDL cargo to lysosomes and return to active duty at the cell surface. Each LDL receptor can make a round-trip in 10 minutes (Brown and Goldstein 1986). Lysosomes digest the LDL particles, releasing unesterified cholesterol. In all cells, this cholesterol will be used to make plasma membranes; in certain specialized cells, some will be converted to other products: estradiol in the ovary, cortisol in the adrenal gland, and bile acids in the liver. Production of bile acids makes the liver by far the heaviest user of cholesterol. But after the bile acids have done their job (aiding in the digestion of fats) in the intestine, they are returned through the bloodstream for recycling in the liver.

What happens next depends upon a cell's circumstances. If it contains *too little cholesterol*, more will be made *intracellularly* through the action of HMG CoA

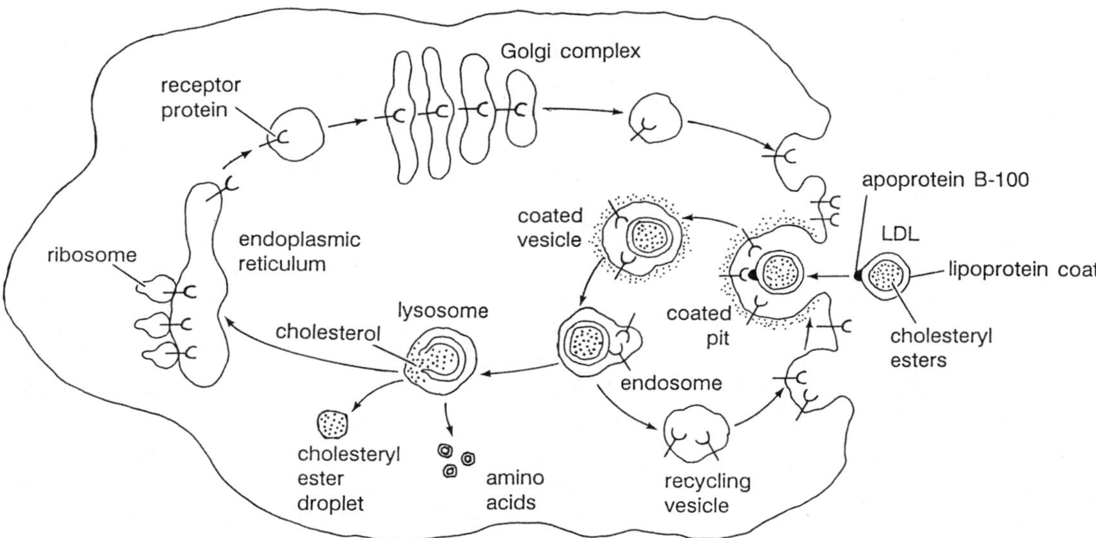

Figure 11 Itinerary of the LDL receptor in mammalian cells. The receptor is produced in the endoplasmic reticulum and then travels to the Golgi complex, the cell surface, the coated pit, a coated vesicle, an endosome, and back to the surface. The regulatory effects of cholesterol released from lysosomes include: stimulation of the ACAT enzyme to esterify excess cholesterol for storage, inhibition of the production of new LDL receptors, and inhibition of the production of new cholesterol. (Adapted from Brown and Goldstein 1986.)

reductase enzyme, and more LDL receptors will also be synthesized by an active **LDL receptor (LDLR) gene**. The new receptors migrate from the endoplasmic reticulum through the Golgi apparatus to the plasma membrane and finally into coated pits. *Too much cholesterol* has three metabolic consequences which serve to reduce cholesterol levels: (1) the HMG CoA reductase enzyme is inhibited, thereby shutting down synthesis of cholesterol; (2) transcription of the LDL receptor gene is inhibited, thereby shutting down receptor production, and (3) the ACAT enzyme is activated to esterify some excess cholesterol to be stored as cytoplasmic lipid droplets.

The latter type of scenario is called **feedback inhibition**, which works very well to control cholesterol levels inside cells. (For example, an actively dividing fibroblast that is continually making new plasma membrane maintains about 40,000 LDL receptors on its cell surface, whereas a nondividing cell might cut back to about 4,000.) But this regulatory mechanism cannot adequately control cholesterol levels outside cells when large amounts of fats and cholesterol are continually being released from the intestines into the bloodstream. Blocked from entering the liver by the decrease in LDL receptors, the *excess cholesterol remains trapped in the bloodstream*—and ultimately gets deposited inside artery walls, causing atherosclerotic plaque formation and heart disease.

MOLECULAR BIOLOGY OF THE LDL RECEPTOR PROTEIN AND GENE. By 1982 Wolfgang Schneider had isolated and purified the LDL receptor glycoprotein from cows' adrenal glands. Other colleagues of Goldstein and Brown determined its partial amino acid sequence and used its mRNA to isolate a complete cDNA for the human LDL receptor gene. From this cDNA, the 839-amino-acid sequence of the mature receptor's protein backbone was deduced, and further studies showed it to consist of five distinct domains (Russell et al. 1986).

Domain 1 (292 amino acids) is a highly convoluted and very stable binding domain that contains a cysteine-rich sequence repeated seven times. This domain, which sticks out from the plasma membrane, binds to apoprotein B. **Domain 2** (about 400 amino acids) is 35% homologous to part of the extracellular domain for the precursor of *epidermal growth factor (EGF)*. Its function in either protein is unknown. **Domain 3** (58 amino acids) is attached to some sugar chains and lies just outside the plasma membrane. **Domain 4**, which spans the plasma membrane, consists of 22 hydrophobic amino acids that are not highly conserved among species. **Domain 5**, the LDL receptor's 50-amino-acid tail, lies within the cytoplasm and is highly conserved among species. It is important for the proper clustering of receptors in coated pits.

The LDL receptor gene was mapped by somatic cell

hybridization and in situ hybridization studies to the short arm of chromosome 19, bands p13.1–13.3 (Francke et al. 1984; Lindgren et al. 1985). It is a good-sized locus, spanning about 45 kb and containing 18 exons and 17 introns, whose order correlates with the order of the LDL receptor's protein domains (Südhof et al. 1985a,b). This relation is shown in Figure 12. Some striking homologies with the exons of other proteins have been noted: exons 2–6 in domain 1 with part of *a component of complement*; and exons 7–14 in domain 2 with the *EGF precursors* as well as with several proteins in the *blood clotting system*. Indeed, the gene for EGF precursor contains eight exons that are identical to those of domain 2:

> These exons form a cassette that has been lifted out of some ancestral gene during evolution and placed in the middle of the EGF precursor gene and the LDL receptor gene. Three of these exons have also been used . . . in several proteins of the blood clotting system, including factor IX, factor X, and protein C. Thus, these exons have been used by members of at least three different gene families . . .
>
> The sharing of exons between the LDL receptor gene and other genes provides strong evidence to support Gilbert's hypothesis concerning the nature and function of introns. As originally proposed by Gilbert, introns permit functional domains encoded by discrete exons to shuffle between different proteins, thus allowing proteins to evolve as mosaic combinations of preexisting functional units. The LDL receptor is a vivid example of such a mosaic protein (Brown and Goldstein 1986).

CAUSES OF HYPERCHOLESTEROLEMIA AND VARIANTS OF THE LDL RECEPTOR GENE. In genetically normal individuals with low-fat diets the LDL pathway works very effectively, and ideal serum cholesterol levels of under 190 mg/dl can be maintained. But the high-fat diets prevalent in industrialized Western societies can swamp the LDL transport–metabolism pathways, thus causing the accumulation of cholesterol in liver cells, suppressing the number of LDL receptors on the cell surfaces, and leaving dangerous levels of cholesterol circulating in the bloodstream.

Those who carry mutant LDL receptor genes in single or double dose (especially the latter) are at greatest risk to these health problems. Researchers have found at least 12 mutations of the LDL receptor gene. They are grouped into four classes, each mutant class having a different defect in receptor production or function and involving a different gene domain.

Class I mutants have no receptors because little or no receptor protein is synthesized. This class comprises about half of the known mutants, one of which is a large deletion extending from exon 13 into exon 15. *Class II mutants* synthesize receptors that never make it to the cell surface. Instead, the receptors remain in the endoplasmic reticulum, or move toward the surface at an extremely slow rate and are ultimately degraded. In these mutants, which comprise the second most common class, the sugar moieties are not processed at all or they are processed incompletely. *Class III mutants* synthesize, process, and transport receptors to the cell surface, but they do not properly bind to the LDL particles. This defect may be due to amino acid abnormalities in the ligand binding or the EGF precursor domains. *Class IV mutants*, which are rare, produce receptors that reach the cell surface and bind LDL—but they do not cluster in coated pits and thus fail to internalize the LDL particles. All of

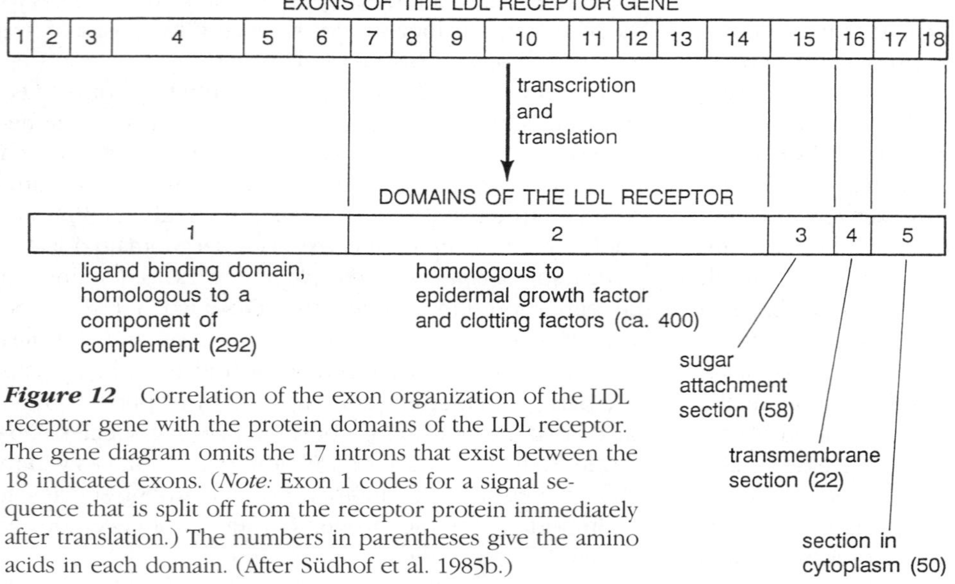

Figure 12 Correlation of the exon organization of the LDL receptor gene with the protein domains of the LDL receptor. The gene diagram omits the 17 introns that exist between the 18 indicated exons. (*Note:* Exon 1 codes for a signal sequence that is split off from the receptor protein immediately after translation.) The numbers in parentheses give the amino acids in each domain. (After Südhof et al. 1985b.)

these mutations occur in the cytoplasmic domain and involve base substitutions, duplications, or deletions that lead to defective tail regions. In a few cases, deletions remove the membrane-spanning region as well, and unanchored receptors are secreted into the extracellular medium.

Molecular analyses of LDL receptor variants in families with familial hypercholesterolemia have turned up deletions and duplications plus one or more nonsense and missense mutations. Several deletions and duplications have arisen from unequal crossing over between homologous repeated Alu sequences that occur in some of the LDL introns. As was the case with phenylketonuria, a sizable fraction of individuals previously thought to be homozygotes are actually heterozygous for two different LDL receptor mutant alleles.

At least two useful RFLP polymorphisms have been reported. About 30% of individuals are heterozygous for two alleles at a *Pvu*II site in intron 15. Thus, within genetically informative families, it should be possible to identify FH heterozygotes during their infancy, so that appropriate therapy can be instituted before the symptoms of the disease develop (Humphries et al. 1985).

FOUNDER EFFECTS. Specific LDL receptor gene defects have been observed in different populations in high frequency. For example, haplotypes for two RFLP sites within exon 8 characterize *South African Afrikaners* with FH (Brink et al. 1987). The two sites are recognized by the restriction enzymes *Stu*I and *Pvu*II, and the four haplotypes are (depending on whether or not each site is cut) + +, + −, − +, and − −. Only + − and − + were found in this group, with + − being by far the most common. These data support the idea that the extraordinarily high incidence of FH among Afrikaners is due to a founder effect—the rapid multiplication in a genetic isolate of just one or two mutant *FH* alleles introduced by the 1,526 Dutch immigrant founders who in just three centuries gave rise to the present population of over 2.5 million Afrikaners. Among Afrikaners, FH heterozygotes are 1 in 100 and homozygotes are 1 in 30,000 (Brink et al. 1987).

Similarly, there is good evidence for a founder effect among the 8,000 ancestors who gave rise to present-day *French Canadians*. Here the LDL receptor mutant present in 63% of FH heterozygotes (and a high proportion of homozygotes) is a large deletion that eliminates the promoter and exon 1, thereby preventing transcription to LDL receptor mRNA. The deletion has not been observed in any other ethnic group (Hobbs et al. 1987).

Another unique defect characterizes the LDL receptor mutant that is very common in *Lebanon*. This allele has a single base substitution leading to a nonsense mutation in domain 2. Premature termination at the codon for amino acid 660 shortens the receptor protein by 180 amino acids, a defect that prevents the receptor from ever reaching the cell surface. The mutation also creates a new restriction site for the enzyme *Hin*fI, thus permitting early diagnosis of affected individuals (Lehrman et al. 1987).

TREATMENT AND PRENATAL DIAGNOSIS OF HYPERCHOLESTEROLEMIA. Low-fat diets should be the cornerstone of cholesterol-lowering treatment for both genetically normal individuals and FH heterozygotes. Because drugs have side effects, they should be used only for high-risk cases (cholesterol levels over 240 mg/dl) or for borderline cases (200–239 mg/dl) with other risk factors—such as being male, or being a smoker, or being diabetic, or having high blood pressure (Roberts 1987a; NIH 1987). The drugs are of two types: (1) *Bile-binding resins* (such as cholestyramine) attach to bile acids and prevent their normal recycling; this defect enhances the liver's need for cholesterol, increasing the number of LDL receptors and decreasing the amount of LDL circulating in the blood. (2) *Inhibitors of HMG CoA reductase* (such as lovastatin) prevent liver cells from synthesizing their own cholesterol, further increasing LDL receptors and decreasing serum LDL. Taken together, these two kinds of drugs can double the receptor output of FH heterozygotes' one normal allele and reduce serum LDL levels by 50%, that is, into the normal range. Bile salt reabsorption can also be prevented by surgical means.

But because FH homozygotes lack a normal allele and thus cannot make receptors under any conditions, they cannot be effectively treated by either diet or drugs. Consequently, the only possibilities for treatment are repeated plasma exchanges and/or surgery—including liver–heart transplants from individuals with normally functioning LDL receptor genes.

In families or populations where FH is prevalent, the detection of FH homozygotes by prenatal diagnosis may be desirable. Class I receptor-negative homozygotes, being the most severely affected, are the main focus of concern; they can be identified around the sixteenth week of pregnancy by measuring LDL receptor activity in cultured fetal amniotic cells. The detection of heterozygotes (50% receptor activity) by this method, however, is not accurate enough for prenatal diagnosis.

PURINE METABOLISM

Defects in purine metabolism are present in an X-linked lethal disorder characterized by some highly unusual behavioral traits.

Lesch-Nyhan Syndrome

In 1962 a seriously ill boy was referred to Johns Hopkins School of Medicine. There he was examined by a physician, William Nyhan, and by a medical student, Michael Lesch. The child suffered from various problems: blood in the urine, extremely high concentrations of uric acid in the urine and blood, mental retardation, spastic cerebral palsy (a motor disorder that includes uncontrollable spasms of the legs and arms), and self-mutilation. After ruling out gout and other conditions known to be associated with excess production of uric acid, Lesch and Nyhan (1964) concluded that they were seeing a new metabolic disease.

CLINICAL SYMPTOMS. Babies with **Lesch-Nyhan syndrome** look healthy at birth and seem to develop normally for several months. Nevertheless, early signs of anomaly are sometimes noted during the first weeks of life, such as the presence of orange "sand" (uric acid crystals) in the diapers or the occurrence of colic or uncontrolled vomiting. The latter may cause dehydration, which is followed by further concentration of uric acid, kidney stones, obstruction of the urinary tract, and other problems.

Delays in motor development set in between three and eight months; between eight months and one year, the muscles develop excess tone and the child begins to exhibit uncontrollably repeated writhing movements of the hands and feet. Involuntary spasms of the arms, legs, ankles, neck, and trunk prevent the child from walking unassisted and often lead to dislocated hips or club feet. Feeding problems contribute to patients' small size and frail condition. Poor muscle control also prohibits the child from speaking clearly or from being toilet trained. Seizures occur in about half of these children.

After two or three years of life, some patients begin to exhibit the most unusual trait associated with this disorder: compulsive biting of the fingers, lips, tongue, and inside of the mouth. Also typical is aggressive behavior toward others (hitting, pinching, swearing, and spitting) while apologizing for this behavior (Kelley and Wyngaarden 1983). With IQ scores ranging between 30 and 65, Lesch-Nyhan patients are classed as severely retarded, but some researchers suspect that their poor scores are partly due to difficulties in communication. These children often seem alert, cheerful, responsive, and genuinely anxious to control their destructive tendencies. Devoted care from those around them may improve their compulsive behavior, but any stressful situation will trigger another episode of self-mutilation. (Seegmiller [1976] suggests that chewing of the finger tips could represent an extreme exaggeration of the fingernail-chewing habit practiced by many normal people when under duress.) The degree of self-mutilation varies considerably among Lesch-Nyhan patients and also in the same patient under different circumstances. Although survival into the twenties or thirties is known, most affected individuals die before then.

BIOCHEMISTRY. Lesch and Nyhan (1964) determined that the basic error was a block in purine metabolism. As shown in Figure 13, the purine nucleotides needed for DNA and RNA production can either be synthesized from scratch or be rescued from degraded nucleic acids and recycled by means of a salvage pathway. In normal humans this latter pathway works very efficiently, recycling 90% of free purines; but in Lesch-Nyhan patients it fails to function, and excess purines are converted to uric acid.

A team of scientists at the National Institute of Arthritis and Metabolic Diseases (Seegmiller et al. 1967) identified **hypoxanthine-guanine phosphoribosyl transferase (HPRT)** as the enzyme whose deficiency shuts down the salvage pathway, and found that Lesch-Nyhan patients exhibit less than 1% of normal enzyme activity. They found that in normal individuals this enzyme functions in all cells, but is most active in the brain, especially in *basal ganglia*—paired masses of gray matter that are embedded in the middle of the cerebrum and control certain movements of skeletal muscles.

The team also reported that a partial deficiency of HPRT (1–20% activity) leads to a severe form of gout, with high uric acid production. These patients, who comprise about 1 in 200 males with gout, have a normal life span and can reproduce. Although about 20% show some neurological symptoms (e.g., retardation, spasticity, seizures), none of them self-mutilate. Later studies have revealed "that there is a virtually continuous spectrum of enzyme activity in mutant hemizygous subjects with HPRT deficiency ranging from undetectable to about 50 percent of normal" (Kelley and Wyngaarden 1983).

How does a deficiency of HPRT bring about such highly stereotyped behavior? Excess uric acid alone cannot be to blame, because (1) males with partial HPRT deficiency produce excess uric acid but do not self-mutilate, (2) excess uric acid occurs in the liver but not the brain of Lesch-Nyhan patients, and (3) self-mutilation persists in Lesch-Nyhan patients even after uric acid levels are greatly reduced by medication. No obvious physical bases for these various neurological problems have been found. The cerebrospinal fluid is normal, as are results of standard tests for brain and nerve function. Autopsies reveal no structural defects of the brain (including the basal ganglia) that are unique to this syndrome, but only those changes found in all victims of severe uremia. This evidence strongly suggests that the neurological defect is biochemical rather than anatomical.

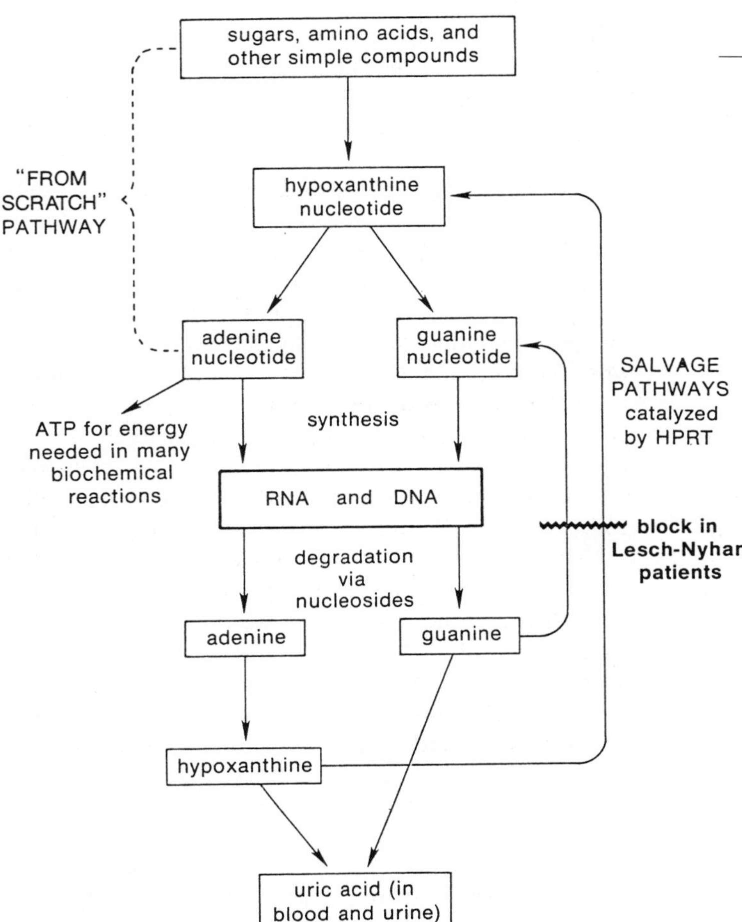

Figure 13 The metabolism of purines, showing the block in Lesch-Nyhan patients. The purine nucleotides that are incorporated into RNA and DNA are mostly synthesized from simple compounds (top center). The free purines (adenine and guanine) resulting from the degradation of old nucleic acids can be recycled, however (middle right). These salvage pathways use the enzyme hypoxanthine-guanine phosphoribosyl transferase (HPRT), which transfers a phosphate-plus-ribose group to either hypoxanthine or guanine. Lesch-Nyhan infants lack HPRT and excrete large amounts of uric acid. In this metabolic scheme, most arrows represent several individual chemical reactions.

GENETICS AND MOLECULAR BIOLOGY. Total HPRT deficiency is inherited as an X-linked recessive, expressed in males receiving the mutant allele from a carrier female. No females with Lesch-Nyhan syndrome are known, and affected males do not reproduce. The disorder is relatively rare (1 in 100,000–380,000 births) and is not associated with any particular ethnic groups. Over 15 X-linked recessive variants with partial deficiencies of HPRT enzyme activity have also been identified.

Pedigree analyses and various studies of human–rodent cell hybrids established that the *HPRT* gene is located near the tip of Xq, closely linked to *G6PD*, fragile X, and the deutan loci; specifically, it lies in region 27 of Xq (Pai et al. 1980). The full-length human gene has been isolated and sequenced by scientists at the University of California at San Diego (Jolly et al. 1983). It is large, containing about 44 kb of DNA, organized into nine exons and eight introns; its coding region of only 654 base pairs gives rise to a 218-amino-acid protein product (Patel et al. 1986). Sequence comparisons with mouse and hamster cDNAs reveal that homology between these three species is over 95% in the coding regions of these genes. Sequence differences between the mouse and human result in seven amino acid substitutions (Stout and Caskey 1986).

Now armed with the complete sequences for both the *HPRT* gene and the protein, researchers can analyze specific mutations found in Lesch-Nyhan or gout patients. The alterations turn out to be very heterogeneous and appear to consist mainly of point mutations or very tiny chromosomal aberrations. For example, Wilson et al. (1986) found that the cell lines from at least 16 out of 24 HPRT-deficient patients contained unique mutations. So far, about 35% of analyzed mutations are known to result from single base pair changes, whereas only 15% have been major deletions or rearrangements (Gibbs and Caskey 1987; Silverman et al. 1987).

CARRIER DETECTION AND PRENATAL DIAGNOSIS. Heterozygotes, although clinically normal, may show some subtle defects in purine metabolism— including increased levels of uric acid in their urine. As a result of X chromosome inactivation, heterozygous females show mosaicism for the *HPRT* gene. Fibroblast cells taken from their skin and grown in culture give rise to two types of clones, one with full activity of the HPRT enzyme and the other completely deficient. By analyzing the enzyme activity of individual follicles of hairs plucked from various places on the scalp, Gartler et al. (1971) devised another test for heterozygosity. Because each hair follicle arises from just a few cells in the embryo, one would expect a heterozygote to have three types of follicles: (1) those showing a normal level of HPRT activity because all the precursor cells happened to express the normal allele; (2) those that show no enzyme activity because all precursor cells expressed the mutant allele; and (3) those showing an intermediate range of enzyme activity because the precursor cells were a mixed population in which some expressed the normal and others expressed the mutant allele. All three types of follicles are indeed found in heterozygous females.

Because the HPRT enzyme is detectable in all cells of the body, including amnion and chorion cells, prenatal diagnosis by amniocentesis or chorionic villus sampling is an option for women known to carry the defective allele (Gibbs et al. 1984). The occurrence of mild forms of HPRT deficiency complicates the process of prenatal diagnosis, however. Great care must be taken to distinguish fetuses affected with Lesch-Nyhan disease from those with partial HPRT deficiency, because the abortion of the latter type is unwarranted. The two clinical types have never been found in the same extended family, however, so the best guide to decision-making is to look at the particular phenotype that occurs in each pedigree.

Work by some British researchers on *HPRT*$^-$ mice (see below) has raised the possibility of detecting HPRT-deficient embryos within a few days of conception, that is, before implantation. They were able to detect the absence of HPRT activity in single mouse cells removed from eight-cell preimplantation embryos derived from crosses of carrier female mice by normal males. The procedures could conceivably be extended to humans. As explained in Chapter 24, the preimplantation embryos to be diagnosed could be obtained by a number of procedures (Monk et al. 1987).

ANIMAL MODELS. No animals with phenotypic symptoms corresponding to the Lesch-Nyhan syndrome are known, but researchers can now study genetically engineered *HPRT*-deficient mice. Several methods have been used to produce them, but the first step was taken by researchers at the Fox Chase Cancer Center in Philadelphia (Dewey et al. 1977). They treated mouse *embryonic carcinoma* cells with a chemical mutagen and (using various techniques outlined in Chapter 15) selected those few that mutated to HPRT deficiency. They injected these mutant cells into normal 3-day-old mouse embryos, which were then implanted into foster mothers. Finally, they noted the distribution of HPRT$^-$ cells in the resultant newborn *mosaics*.

But embryonic carcinoma cells do not colonize the germ line of these mosaic mice, so it was not possible to produce offspring with the *HPRT*$^-$ allele. Two groups of British researchers have overcome this problem and raised some lines of *HPRT*$^-$ mice that are not mosaics. They started with cultures of *embryonic stem (ES)* cells (obtained from mouse blastocysts), which can colonize both germline and somatic tissues (Evans and Kaufman 1981). And instead of using chemicals to produce mutant embryonic stem cells, Kuehn et al. (1987) tried a more efficient technique: infecting cell cultures with *retroviruses* that randomly integrate into mouse genes, thereby causing **insertional mutations**. (The viral insertion also introduces a *Kpn*I restriction site marker into the *HPRT* gene.) Then they selected *HPRT*$^-$ stem cells from the tissue culture, tested for the presence of the new restriction site, and injected them into blastocysts of normal mice (marked with an X-linked coat color gene), which were then transferred into foster mothers. Any mosaic female mice that result are recognized by a distinctive coat color. Among these, some *HPRT*$^-$ cells are incorporated into their gamete-forming tissues and give rise to *HPRT*$^-$ eggs. Thus their offspring include some carrier females and genetically "Lesch-Nyhan" males—strains that can be bred and intensively studied, with increased likelihood of unraveling the events that link the mutant allele to its devastating effects.

University of Edinburgh scientists used the same general protocol for raising a pure line of *HPRT*$^-$ mice (Hooper et al. 1987). But rather than employing chemical or retroviral means of producing mutant *HPRT*$^-$ genes, they developed a method for selecting preexisting spontaneous *HPRT*$^-$ mutants among XY embryonic stem cells. Because these methods allow researchers to produce and isolate almost *any* biochemical mutant and incorporate it into laboratory mice, it should now become possible to study in the laboratory many other inborn errors of metabolism known to afflict humans.

GENE TARGETING AND POSSIBLE GENE THERAPY. The devastating nature of Lesch-Nyhan syndrome, the lack of effective treatment for its neurological symptoms, and the special characteristics of the *HPRT* gene—that is, being expressed in all cells, being a reasonable size, and being readily selectable—make this disorder a prime candidate for gene therapy. The technical and ethical problems associated with this issue will be discussed in Chapter 25. Here we shall only point out that some important first steps have been taken by scientists who have already substituted normal *HPRT* alleles for mutant *HPRT*$^-$ alleles (or vice versa) in mouse embryonic stem cells.

Oliver Smithies and colleagues at the University of Wisconsin and the University of Edinburgh showed that homologous recombination can occur between a chosen gene in target cells and foreign DNA that has sequences in common with the target. In this case the target gene was an *HPRT*$^-$ deletion mutant in mouse embryonic stem cells, and the foreign DNA was a plasmid containing human *HPRT*$^+$ sequences. Results were encouraging enough that gene targeting in embryonic stem cells "shows considerable promise as a practical means of obtaining specific, predetermined germline changes in experimental animals" (Doetschman et al. 1987).

At the University of Utah, Kirk Thomas and Mario Capecchi (1987) used targeting techniques to go in the opposite direction, replacing *HPRT*$^+$ alleles with *HPRT*$^-$ mutations in (hemizygous) male blastocysts. First they inserted a plasmid neomycin-resistant gene (*neo*^{r}) into the cloned fragment of a mouse *HPRT* gene, to use as the selectable marker and as a disrupter/mutator. After infecting *HPRT*$^+$ embryonic stem cells with this fragment, and allowing growth to occur, they added neomycin to the culture to kill off all colonies except those which had absorbed the neomycin gene. Then they tested these resistant survivor colonies for the presence of the *HPRT* fragment, which—if recombined with the *HPRT*$^+$ genes of the host cells— would have mutated to *HPRT*$^-$. This recombination had indeed happened with the relatively high frequency of 1 in 1,000 colonies. These *HPRT*$^-$ mutations were of two types: *replacement* of the homologous fragment, or *insertion* (and thus duplication) of the fragment into the blastocyst's *HPRT* allele. The researchers hope that these methods "will provide the means for generating mice of any desired genotype... If successful, this technology will be used in the future to dissect the developmental pathway of the mouse as well as to generate mouse models for human genetic diseases."

MEDICAL CONSIDERATIONS

Since 1950, the discovery of metabolic diseases has increased greatly, due in large measure to the development of new analytical methods for identifying metabolites of various kinds. At least 250 different disorders are known, distributed among many more categories than we have presented here. In recent years our understanding of these disorders and their genetic bases have been vastly enriched by molecular techniques. And from the blending of old and new knowledge, some useful generalizations about metabolic disorders as well as some new fields of research—such as pharmacogenetics—have emerged.

Detection of Metabolic Disease in Newborns

Physicians have noted that certain symptoms tend to be common to a wide variety of metabolic disorders. Because some syndromes are amenable to treatment if recognized early enough, it is important that physicians and parents be alert to these diagnostic signals.

Many affected babies appear normal at birth but very soon develop severe problems. Some of the most telling signs of a metabolic disorder are (1) severe vomiting, beginning in the first few days of life; (2) abnormal odor to the urine, skin, or breath; (3) abnormal concentrations of ketones or acids in the body fluids; (4) abnormalities in pigment development; (5) seizures, especially when accompanied by muscle spasms in a particular area of the body; (6) failure of muscle coordination; (7) ambiguous genitalia; (8) failure to thrive; (9) overwhelming, unexplained illness; and (10) a family history characterized by the death of sibs in early life from unknown causes. Another important sign, which may not be detected for several months, is severe mental retardation. By this time, the damage is usually irreversible. Nevertheless, once recognized in a family, a metabolic disorder can be anticipated in subsequent pregnancies or births, carefully monitored, and perhaps treated.

The mode of inheritance is a good clue to the basic defect in genetic disorders. In those cases where the basic defect is known, recessive traits usually involve enzymes (especially catabolic enzymes, which break down substances) or peptide hormones. They also tend to exhibit a fairly uniform phenotype and an early age of onset. Although far fewer basic defects are yet known, dominant traits, on the other hand, usually involve nonenzymatic or structural proteins, or (in special cases) enzymes that are rate-limiting in metabolic pathways under feedback control. They also tend to be much more variable in phenotype and often have a later (i.e., adult) age of onset. Thus, biochemical analyses of dominant disorders are more difficult (McKusick 1988; Stanbury et al. 1983).

Pharmacogenetics

Not every inborn error of metabolism causes death, mental retardation, or even serious illness. In fact, an individual can be born with a metabolic defect and live for years, perhaps a lifetime, without any health problems. If the affected biochemical pathway is a relatively unimportant one, or needed only under special circumstances, no ill effects may ever be noted.

As technology introduces new substances into our environment, however, it is inevitable that some people will be unable to metabolize some of them because of the presence of previously unrecognized enzyme variants. In recent years, the discovery of more and more cases of this type has spawned a new field of study called **pharmacogenetics**. It deals with the genetic variation leading to differences in the ways that individuals metabolize drugs, foods, and other substances in their environment. Pharmacogenetics promises to play an increasingly important role in medicine (Motulsky 1964; Szorady 1973; Gerrick 1978).

GLUCOSE-6-PHOSPHATE DEHYDROGENASE DEFICIENCY. One example is a common inborn error of sugar metabolism involving abnormalities of an enzyme called **glucose-6-phosphate dehydrogenase (G6PD)**. The *G6PD* locus lies near the tip of Xq, in region 28, next to the *HPRT* gene. The human *G6PD* gene is about 18 kb long and has 13 exons (Martini et al. 1986; Takizawa et al. 1986). About 325 variants of this X-linked recessive gene are known, many of which are rare (McKusick 1988). A few kinds of G6PD deficiency are of great medical interest—not only because of their striking effects, such as *acute hemolytic anemia* caused by breakdown of red blood cells, but also because they are surprisingly common, affecting about 100 million people (mostly males) worldwide. But virtually all the affected people live in (or come from) tropical or subtropical countries where malaria is endemic. Here the mutant genes have some selective advantage because heterozygotes are more resistant to malaria than are $G6PD^+$ homozygotes.

The G6PD enzyme functions only under special circumstances in a minor biochemical pathway of red blood cells, where it oxidizes glucose-6-phosphate to 6-phosphogluconate. Coupled to this reaction is another one that converts glutathione to reduced glutathione (GSH). Reduced glutathione is necessary for maintaining the integrity of the cell membrane. But because this pathway is scarcely utilized, under usual conditions even a low activity of G6PD is sufficient to keep red blood cell membranes intact. Only in "emergency" situations, when GSH is rapidly oxidized to glutathione, does the need arise for a normal full-strength, quick-acting G6PD enzyme to reverse this reaction (Yoshida 1973; Yoshida and Beutler 1986).

The rapid oxidation of GSH can occur when certain foods or drugs, such as *fava beans* or the antimalarial drug *primaquine*, are ingested. In persons with mutant *G6PD* genes, the low activity of the G6PD enzyme is inadequate to the challenge: the concentration of GSH falls below required levels and the red blood cells rupture. Depending on genetic and environmental circumstances, these attacks of hemolytic anemia vary from mild to acute (they can be fatal if blood transfusions are not given). But this sensitivity to oxidation is also what protects carriers against the malaria parasite, which generates hydrogen peroxide in the red blood cells it infects; subsequent loss of potassium kills both the cells and their parasites (Friedman and Trager 1981).

There are two major variants of *normal* G6PD: B (100% activity), which is common throughout the world, and A (88% activity), which is confined to Black Africans and their descendants. The amino acid sequences of these two variants are identical except at one position, where B has asparagine and A has aspartic acid (Yoshida 1967). **Favism** is associated with a mutation of the *B* allele known as the *Mediterranean* or B^- allele. Despite its severe enzyme deficiency (0–7% activity) and clinical effects,* this mutant is extraordinarily common in Greece, Sardinia, northwest India, and southern Italy, and among the Sephardic and Oriental Jews of Israel. **Primaquine sensitivity** among Blacks of African descent is associated with a mutation of the *A* allele known as A^-. Partly because the enzyme's activity (8–20%) is somewhat higher than that of the B^- mutant, it can maintain GSH in young blood cells but not in old ones. Allele frequency is over 20% among males in many parts of Africa and about 10–15% among Black American males. Two other common *G6PD* alleles are the *Canton variant* (4–24% enzyme activity) in southern Chinese and the *Constantine variant* (16% enzyme activity) among Arabs. Two rare alleles (*Hartford* and *Hektoen*) are unusual in that they exhibit 200–400% of normal enzyme activity. For more information, see Beutler (1983) or Yoshida and Beutler (1986).

PORPHYRIA. Sometimes a drug reaction unmasks the presence of a previously unsuspected metabolic disease, as happens in people who carry the dominant allele for **acute intermittent porphyria**. These individuals may live for decades without any inkling of the potentially fatal condition they harbor. When given barbiturates and certain other sedatives or anesthetics, however, they suffer attacks characterized by severe pains in the abdomen and limbs, muscular weakness, and mental instability. Sometimes abdominal surgery is performed, but when nothing is found amiss, the physician may suggest that the problem is entirely psychosomatic, especially in view of the patient's seemingly abnormal behavior. Ironically, barbiturate-type anesthetics may be used during the operation and more sedatives prescribed thereafter; life-threatening symptoms that include delerium, excruciating pain, and paralysis may follow. Other factors known to trigger these episodes include infections, increased levels of steroid hormones (especially estrogens), and severe dieting or starvation.

Because its symptoms mimic so many other known conditions, from appendicitis and gallbladder attacks to lead poisoning and schizophrenia, porphyria may be misdiagnosed. The best clue to this disorder is the striking change in the color of the urine—to a brilliant red hue like that of port wine—when it is exposed to light and air. This is due to the excessive excretion of two

*In people with favism, hemolytic reactions may also be triggered by infections (especially typhoid fever, influenza, or viral hepatitis) as well as by sulfa drugs, chloramphenical, vitamin K, and aspirin. Other oxidizing substances, such as mothballs, can cause acute reactions too.

metabolic precursors of *heme*, the red-colored part of a hemoglobin molecule. (Heme belongs to the *porphyrin* class of molecules, members of which form the base of all respiratory pigments in animals and plants.) The basic metabolic error is a deficiency of the enzyme **porphobilinogen deaminase**, formerly called uroporphyrinogen synthetase (Sassa and Kappas 1981; Mustajoki and Desnick 1985). About six variants are known, most of which exhibit only about half the enzymatic activity shown by normal controls, as would be expected for heterozygotes. (Because no homozygotes have been detected, a double dose of this gene may be lethal.) Porphyria is an exception to the general finding that enzyme deficiencies are inherited as recessives. The gene has been mapped to the long arm of chromosome 11 (Wang et al. 1981). It was once assumed to be quite rare, but the vastly increased use of barbiturates in recent years has led to its detection in many populations—notably in Sweden and Lapland, where the frequency is about 1 in 1,000. Elsewhere in the world its frequency may be about 1 in 50,000.

Another autosomal dominant disorder, **porphyria variegata**, is especially common among Whites in South Africa and in Finland. Affected adults, in addition to suffering acute drug-induced attacks like those described above, have a variety of skin problems. These include blistering and fragility of those areas exposed to sunlight, with frequent infection and scarring of the lesions, as well as excess hairiness and overpigmentation. The feces always contain certain heme metabolites in excess, and the urine may include excess porphyrins. Here the deficient enzyme is **protoporphyrinogen oxidase** (Brenner and Bloomer 1980). A few rare cases of severely affected homozygous infants have also been reported (Mustajoki et al. 1987). Two cases of genetic sleuthing on porphyria variegata are particularly interesting. Dean (1957) traces its inheritance in a huge South African kindred back to one very prolific early Dutch settler. Macalpine and Hunter (1969) review the medical history of King George III, the "mad" British monarch who reigned in the late 1700s, and suggest that he had porphyria.* For information on the many types of porphyria, see reviews by Kappas et al. (1983) and Meyer and Schmid (1978).

OTHER DRUGS. Dozens of examples are now known in which a drug that benefits most people can cause severe reactions or even death in some unlucky patients. One involves the muscle relaxant *suxamethonium* (succinylcholine chloride), which is used before the administration of general anesthesia to allow the easy insertion of a tube into the windpipe. It is also used by psychiatrists in patients undergoing electric shock treatment. Normally, the relaxing effect lasts only 2–3 minutes, because an enzyme known as *pseudocholinesterase* quickly inactivates the drug by splitting it apart. But about 1 in 2,000 Whites of European extraction and about 1 in 50 Alaskan Eskimos is homozygous for autosomal recessive mutations of the *CHE1* pseudocholinesterase gene. Because the variant enzymes are much less effective at inactivating the drug, paralysis lasts much longer—hours or even days—and may be fatal if the patient's breathing is not maintained by a respirator during that time (Lehmann and Liddell 1964). Soreq et al. (1987) have mapped this gene to the long arm of chromosome 3.

One of the most common enzyme deficiencies of humans occurs in so-called **slow metabolizers**. These people are the 5–10% of North American and European Whites and 30% of Hong Kong Chinese who show adverse side effects to ordinary doses of about 25 commonly prescribed medications—including some β-blocker blood pressure drugs, drugs for angina and abnormal heart rhythms, asthma medicines, cough medicines, and antidepressants. When given the β-blocker drug *debrisoquine*, for example, they metabolize only 0.5–10% as much of it as do normal individuals. A team of American and Swiss researchers found that the deficiency is in the *cytochrome P450 enzyme 450db1*. The P-450 cytochromes are important liver enzymes that play a major role in the detoxification of drugs and environmental pollutants. P-450 genetics is complex, involving a superfamily made up of eight gene families, each coding for 2 to 20 discrete cytochromes (Nebert and Gonzalez 1987). By cloning and sequencing the human P450db1 cDNAs from both normal and slow metabolizers, Gonzalez et al. (1988a) discovered several variants of the mutant gene. But rather than involving any of the nine exons of *450db1*, all mutations occurred within introns—giving rise to mRNAs with splicing errors and to defective enzyme products. This great variability and high total frequency (35–43%) of mutant alleles promise to create major problems for drug companies and regulatory agencies as well as for patients and physicians (Idle 1988). But the good news is that slow metabolizers also seem to show a much (20–40 times) reduced risk to cancers of the liver, gastrointestinal tract, and lung (McKusick 1988). The *P-450db1* mutations, which seem to be inherited as recessives, have been mapped to chromosome 22 (Gonzalez et al. 1988b).

In the 1950s an extremely common enzyme variant was discovered among tuberculosis patients being treated with the drug *isoniazid*. Many of them inactivated this drug so slowly that they required only a fraction of the normal dose; in fact, the standard dose produced

*In addition to the mental and physical symptoms of the type described previously, they note references to the king's dark, discolored urine. Other descendants of Mary, Queen of Scots, also exhibited intermittent "colics," progressive paralysis, and madness. Among these was her son, King James VI, who claimed to have urine the color of Alicante wine.

toxic side effects. Here the difference in response is caused by a variant of the *N-acetyl-transferase* enzyme, which is inherited as an autosomal recessive trait. Homozygous recessive slow inactivators are widespread (40-70%) in populations of European, African and Jewish origin. They are much less common (10–20%) among Far Eastern populations and are least often found (5%) among Eskimos (Motulsky 1964). In cases like this, where two forms of a gene are well established, the mutant allele is presumed to have some selective value in a population—such as the ability to better metabolize certain types of food or other substances in the environment. For details, see Iselius and Evans (1983).

Not all drug reactions are so dramatic as the ones just discussed, but there is no longer any doubt that metabolic variability must be taken into account in the dispensation of drugs. Any serious side effects should be investigated for possible genetic bases, and, if the situation warrants it, other members of the family should be tested for similar metabolic responses. Garrod's prophesy of a chemical basis for "idiosyncrasies with regard to drugs and articles of food" is amply borne out, as large numbers of people are being hospitalized—not for their original medical problems—but rather for unexpected reactions to their treatment.

SUMMARY

1. All metabolic events are mediated by enzymes under genetic control. Thus a mutation in an enzyme-producing gene is likely to induce a metabolic block in one step of a biochemical pathway. The consequences stemming from an excess of precursors or lack of end products depend on the pathway involved.

2. By 1909 Archibald Garrod had proposed the occurrence of gene-determined inborn errors of metabolism and described several examples in humans.

3. Enzymatic disorders are usually inherited as recessive traits, although molecular studies now show that many so-called homozygotes are actually compound heterozygotes, carrying two different mutant alleles of the same gene.

4. Classic phenylketonuria (PKU), a defect of amino acid metabolism (the enzyme phenylalanine hydroxylase, or PAH), leads to profound mental retardation unless detected near the time of birth and treated with a low-phenylalanine diet. Properly treated, these children have intelligence in the normal range.

5. Women with PKU who become pregnant while not on this restrictive diet give birth to babies who, although heterozygous and expected to be normal, suffer brain damage from the high levels of phenylalanine in the mother's blood. Unless women with PKU return to diet therapy throughout pregnancy, the success in treating homozygotes will be totally offset by maternal PKU occurring in their children.

6. Many PKU homozygotes and heterozygotes are identifiable by haplotype analyses made possible by the existence of several polymorphic restriction sites linked to the *PAH* locus. In some populations, certain haplotypes are associated in highly nonrandom frequencies with PKU mutations; this association can greatly aid in the screening process.

7. All albinos lack binocular vision and have visual problems due to deficiencies of melanin pigment in the eyes; most have pigment deficiencies in the skin and inner ear, too. The classic (tyrosine-negative oculocutaneous) albinism is an autosomal recessive defect of amino acid metabolism caused by the deficiency of an enzyme that normally converts tyrosine to DOPA, the precursor for all melanin pigments. About a dozen other types of albinism exist, but their genetic defects are poorly understood.

8. Tay-Sachs disease is an autosomal recessive error of lipid metabolism caused by deficiency of the Hex-A enzyme, which normally removes a terminal sugar group from G_{M2} ganglioside. Accumulation of this lipid in nerve cells leads to progressive mental deterioration, blindness, paralysis, and death in early childhood. Tay-Sachs disease occurs mostly among Ashkenazi Jews and French Canadians; several mutant alleles are known.

9. Familial hypercholesterolemia is a common autosomal dominant disorder that leads to increased risk of heart attacks. It is associated with defects in the gene coding for the low density lipoprotein (LDL) receptor protein. Various mutations (mostly deletions and duplications) in the large *LDL receptor* gene are known. Some groups of exons in this gene are identical to exons in several different gene families, an identity suggesting that some proteins evolve by the shuffling of exons to form mosaic combinations of preexisting functional units.

10. Boys with Lesch-Nyhan syndrome (an X-linked recessive disorder of purine metabolism) are retarded, lack motor control, and exhibit stereotyped biting behavior. Partial HPRT deficiency leads to a milder condition characterized by severe gout but no retardation or behavioral problems. Mutations of the *HPRT* gene are very heterogeneous and seem to consist mostly of point mutations or very tiny chromosomal aberrations. This disorder is a candidate for gene replacement therapy, which has been experimentally accomplished with the *HPRT* gene in mice.

11. Pharmacogenetics deals with inherited enzyme variants that lead to individual differences in the metabolism of drugs, diet, or other environmental substances. In particular, poor inactivation of a drug can lead to serious illness. Examples are G6PD deficiency, porphyria, and sensitivity to drugs such as suxamethonium, debrisoquine, and isoniazid.

12. G6PD deficiency is an X-linked recessive defect of sugar metabolism that causes hemolytic anemia only when affected individuals eat fava beans or take certain oxidizing drugs such as primaquine. G6PD mutant alleles are most common in populations where malaria is endemic, because heterozygotes are more resistant to malaria than are homozygotes. Over 300 variants of this gene are known.

13. The porphyrias, errors of porphyrin metabolism, are inherited as autosomal dominants—an exception to the rule that the mutations causing enzyme disorders are usually recessive.

FURTHER READING

Harris (1980) writes very lucidly on the principles of human biochemical genetics. Scriver et al. (1989), Lloyd and Scriver (1985), Stanbury et al. (1983), and Emery and Rimoin (1983) are gold mines of information, containing dozens of detailed chapters written by experts on all types of metabolic disorders. Benson and Fensom (1985) is less encyclopedic but still broad-ranging and informative. For other references on specific disorders, see citations listed within chapter sections. Motulsky (1960) provides an interesting review of metabolic variants and the role of infectious diseases in human evolution. On pharmacogenetics, see Kalow et al. (1986), Gerrick (1978), and a report by the World Health Organization (1973). For more information on general and human biochemistry, respectively, see Lehninger (1982) and Stryer (1988).

QUESTIONS

1. In humans, what is the difference between essential and nonessential amino acids?

2. The occurrence of several biochemical mutants involving the same metabolic pathway allows researchers to determine the relative positions of individual steps in a pathway. The general principles are: (1) the substance just prior to a block accumulates, and (2) the mutant organisms cannot grow on any substances prior to the block but will grow if provided with any substance after the block.

For example, the following table summarizes the growth responses of several different mutant strains of the bacterium *Salmonella typhimurium* in several different media. Wild-type *Salmonella* can grow on minimal medium, but none of the four mutant strains can. In this table, a "+" means growth and a "0" means no growth.

Diagram the correct order of all the steps in this biosynthetic pathway, indicating the location of the metabolic block imposed by each mutation. *Hint:* The fewer substances a mutant can grow on, the further along in the pathway is its metabolic block.

Mutant strain	Accumulated product	GROWTH RESPONSE Minimal medium	MINIMAL MEDIUM PLUS Indole	Anthranilic acid (ant)	Tryptophan (trp)	Indole glycerol phosphatase (ICP)
2	Ant	0	+	0	+	+
6	Indole	0	0	0	+	0
3	IGP	0	+	0	+	0
8	None of the 4 compounds	0	+	+	+	+

3. Sometimes the intermediate substance just preceding a metabolic block is *not* excreted in excess. Suggest an explanation.

4. MacCready and Hussey (1963), in a progress report on a newborn screening program in Massachusetts, state that 27 cases of phenylketonuria were detected out of 217,752 babies tested. This frequency of roughly 1 in 8,000 is considerably higher than that estimated for the country as a whole. Suggest an explanation.

5. Witkop (1971) describes a yellow variant of albinism that phenotypically resembles tyrosinase-positive albinism, but affected individuals have a distinctly yellowish cast to the hair and the ability to tan slightly when exposed to sunlight. These individuals give equivocal results with the hair bulb test, however. After incubation with tyrosine, some hair bulbs show a small amount of pigmentation, but others do not. After incubation with cysteine (in the presence of tyrosine and DOPA), some yellow pigment is formed, but no black pigment develops. Refer to Figure 3 and suggest where the biochemical block may occur in this mutant.

6. Would you expect that all DNA base substitutions (mutations) leading to single amino acid changes in an enzyme will lead to detectable electrophoretic variants? Why?

7. Childs et al. (1958) found that females who are heterozygous at the *G6PD* locus nevertheless show levels of enzyme activity that range from near-deficiency to within the normal range. Why?

8. Fibroblast cells and hair follicles taken from females who are heterozygous for HPRT deficiency show mosaicism in the expression of this enzyme's activity, being either deficient or normal. When red or white blood cells are sampled from the same females, however, all show normal HPRT activity. Suggest one or more reasons why the Lyon effect may not be randomly expressed in all tissues.

9. An article by Kretchmer (1972) discusses milk intolerance in humans. With rare exceptions, all children up to the age of two years are able to digest the milk sugar lactose; they can do so because they have large amounts of the enzyme lactase, which splits lactose into glucose and galactose. After the age of two, however, most of the world's people no longer produce much of this enzyme, and thus become lactose intolerant. Drinking more than a moderate amount of milk causes bloatedness, stomach cramps, gassiness, and severe diarrhea. Lactose-intolerant children who continue to drink milk may become malnourished and perhaps even die. (Eating cheese and yogurt presents no problem, however.) The vast majority of northern Europeans, White American ethnic groups, and a few dairying tribes in Africa possess a dominant mutation that causes the continued production of lactase through adulthood. But 70% of U.S. Blacks and nearly all Africans and Asians are lactose-intolerant.

Comment on the usefulness of aid programs that involve the shipping of large amounts of dried milk to Africa and Asia.

10. In "modern" society, milk is not the only source of lactose. According to Koch et al. (1963), either dried milk or lactose is added as a filler to most processed and prepackaged foods. How does this affect the ability of lactose-intolerant individuals and galactosemia patients to restrict their dietary intake of lactose?

11. Fratantoni et al. (1968) set up independent fibroblast cultures from patients with Hunter syndrome and Hurler syndrome, both of which are recessive lysosomal storage disorders of mucopolysaccharide metabolism. When they put Hunter-syndrome cells in medium from the Hurler-syndrome cell cultures, or Hurler-syndrome cells in medium from the Hunter-syndrome cell cultures, both metabolic defects were corrected. What does this suggest about possible allelism of the two defective mutants?

12. Ledley et al. (1986) published this pedigree of a family that is segregating both phenylketonuria (PKU) and mild hyperphenylalaninemia (MHP). The carrier parents are heterozygous for four different haplotypes (defined by reactions with an array of seven restriction enzymes); these are designated a–d. Individual phenotypes are indicated by PKU, or MHP, or nothing (for normal carriers). (a) Analyze this pedigree and indicate whether each haplotype is mutant or normal; also (under the appropriate headings of the table below) indicate which allele combinations give rise to each phenotype. (b) In this family, which haplotype combination is not observed among the children? If this haplotype did occur, what phenotype would it most likely express?

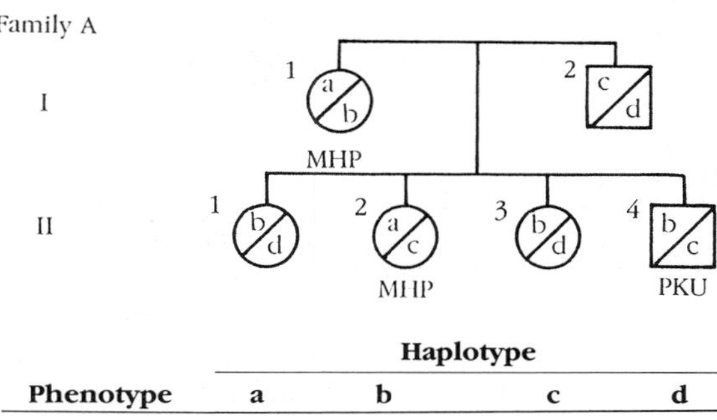

Phenotype	Haplotype a	b	c	d
PKU				
MHP				
Normal				

ANSWERS

1. Essential amino acids are those that our bodies cannot make "from scratch" (i.e., from simple sugars), because we lack the appropriate enzymes; thus, they must be obtained directly from the diet. Nonessential amino acids are those that our bodies can produce from other substances.

2. metabolites in minimal medium $\xrightarrow{8}$ ant $\xrightarrow{2}$ IGP $\xrightarrow{3}$ indole $\xrightarrow{6}$ trp

3. The intermediate may be metabolized via a different pathway.

4. This allele is more common among the Irish and their descendants, many of whom live in Massachusetts. In contrast to this, so few PKU cases were found in Washington, D.C. (most of whose citizens are Black) that the screening program there was discontinued.

5. The error might occur during the process of polymerization of dopaquinone to form black and brown pigments (left fork at bottom of Figure 3).

6. No. Only those changes that lead to a difference in the electric charge or overall shape of the protein molecule would be detected by electrophoresis.

7. Random inactivation of one X chromosome in each cell during early development would lead to a variety of phenotypes—some with a preponderance of cells in which the mutant X-linked gene is expressed and others with a majority of cells expressing the normal allele.

8. Nobody is sure of the answer to this question, but there are two main hypotheses: (1) random inactivation occurred in the precursors of the blood cells, but those expressing the mutant allele were at a disadvantage and became overgrown by their normal counterparts; (2) X chromosome inactivation was nonrandom (favoring the chromosome with the normal allele) in the blood cell precursors but random in other tissues.

9. It makes little sense. Milk can induce serious illness among the children and great discomfort in the adults; such experiences may lead recipients of this aid to reject other, more useful, food products.

10. It makes the task of restricting offending foods much more difficult, considering that even products such as processed meats, soups, fruits, salad dressings, and instant coffee are likely to contain lactose.

11. The "cross-correction" studies showed that the two mutants are nonallelic, because each excreted a normal substance that was deficient in the other (in this case, the products of two different acid hydrolase enzymes). If they had been allelic, neither could have provided the normal substance missing in the other. Complementation also occurs in cell hybrids formed from two phenotypically abnormal parental strains. If the hybrid cells show a normal phenotype, the two mutants must involve different genes (Gravel et al. 1979).

12. (a) Family A has 3 mutant alleles (a, b, and c) and 1 normal allele (d).

Phenotype	Haplotype a	b	c	d
PKU	—	with c	with b	—
MHP	with b or c	with a	with a	—
Normal	—	with d	with d	with b or c

(b) a/d. Normal phenotype.

Chapter 17 Genetics of Blood Groups

Blood Grouping
Red Blood Cells
Polymorphisms
Techniques
How Some Systems Were Discovered

The ABO Blood Group System
Transfusions
Biosynthesis of the Antigens
The Curious Case of the Bombay Bloods

Rh and the Prevention of Hemolytic Disease
Rh Genes, Antigens, and Antibodies
Incompatibility
Protection Against Rh-Caused Hemolytic Disease

> *You have doubtless heard of a* Mad-man, *that hath been lately cured and restored to his Wits by the means of the Transfusion [with the blood of a Calf]... As this second Transfusion was larger, so were the effects of it quicker and more considerable... [W]e observ'd a plentiful sweat over all his face. His pulse varied extreamly at this instant, and he complain'd of great pains in his Kidneys, and that he was not well in his Stomach, and that he was ready to choak unless they gave him his liberty.*
>
> *Presently the Pipe was taken out that convey'd the blood into his Veins, and whilst we were closing the wound, he vomited store of Bacon and Fat he had eaten half an hour before... He made a great glass full of Urine, of a colour as black as if it had been mixed with the soot of Chimneys.*
>
> *Jean Denis (1667)*

Prior to the 1900 discovery of the ABO blood group system by the Austrian physician Karl Landsteiner, interest in transfusion waxed and waned because patients exhibited an unfortunate tendency to die from the treatment.* An understanding of the significance of the ABO differences between people led again to the use of transfusions during World War I. Even then, the procedure sometimes produced unexpected reactions, especially on repeat transfusions. Only after the 1940 discovery of the Rh blood group system did multiple transfusions become generally safe, as long as donor blood was carefully matched to a recipient on a one-to-one basis. Although ABO and Rh are the most important blood groups from a medical point of view, about 20 additional systems have been discovered. The major applications of blood groups have been in patient care (transfusion and transplantation, maternal–fetal incompatibilities), but they are also important in mapping studies, paternity testing, forensic medicine, and anthropology.

*The man described above died a day after he was prepared for a *third* transfusion from a calf, though Denis, a Paris physician and mathematician, claimed that no foreign blood actually entered the man. Implicating the man's wife, Denis (1668) wrote "it was Arsenick she mingled in her Husbands Broths," that killed him, as well as a cat that shared the broth. The consequences of the second transfusion are typical of the hemolytic shock that may accompany an incompatible blood transfusion.

BLOOD GROUPING

Blood transports substances to and from various parts of the body and acts as a main line of defense against infectious agents. It makes up about 8% of body weight (5 quarts in a 125-pound person) and consists of roughly 55% liquid, or plasma. Suspended in the plasma are red blood cells, the major focus of this chapter, as well as white blood cells (which have manifold immune functions) and platelets (which initiate the clotting process). Each milliliter of blood (about a thimbleful) contains about 4×10^9 (4 billion) red blood cells and lesser numbers of the other particles.

The **plasma** is a clear, straw-colored fluid containing hundreds of dissolved substances. Nutrients (e.g., sugars, amino acids, and vitamins) move from the capillaries of the digestive tract to other tissues, and hormones travel from organs of manufacture to target cells. Intermediate metabolites, end products, and waste materials circulate between different body parts. Plasma also contains a variety of specialized proteins, among which are the clotting factors, carrier proteins, and immunoglobulins.

Clotting factors, comprise about a dozen proteins including Factor VIII (which is abnormal in hemophilia A). They occur in a complex cascade of reactions, beginning with the disruption of platelets and ending with the conversion of a soluble protein, *fibrinogen*, to an insoluble network of threadlike molecules. The mesh sponges up the liquid portion and entraps the blood cells to form a clot, thereby preventing further loss of blood. Clots gradually contract, extruding the liquid, which is usually called **serum**, that is, the plasma without its fibrinogen.

Carrier proteins specifically bind and transport substances from blood plasma into various cells. In order to cross the cell membrane, the carrier protein must join with receptor molecules on the cell surface. One carrier, transferrin, binds to iron molecules in blood plasma and delivers them to bone marrow, where iron is incorporated into new red blood cells (Dautry-Varsat and Lodish 1984). Another carrier, apoprotein B-100, binds to globules of cholesterol in the plasma to form LDL (low-density lipoprotein) particles (Brown and Goldstein 1986; see Chapter 16). Receptors for apoprotein B-100 are in high concentration in the membranes of gonadal and liver cells, allowing cholesterol to enter, where it is converted to steroid hormones and bile acids. Cholesterol also forms structural elements of all cell membranes.

Immunoglobulins are large protein molecules that are manufactured by certain white blood cells and can act as **antibodies** that recognize and incapacitate foreign intruders such as viruses and bacteria. The action of antibodies underlies vaccinations, allergies, and rejections of skin grafts and organ transplants. Antibody molecules normally distinguish "self" from "nonself," but they can turn against an individual's own tissues in so-called autoimmune diseases. A great deal is known about the structures and functions of immunoglobulins, of which the major type is called IgG (immunoglobulin G), or gamma globulin (Chapter 18).

Red Blood Cells

Red blood cells, or **erythrocytes**, are shaped like doughnuts with partly filled holes (Figure 1). Having a diameter of about 7 μm, they pass in single file through

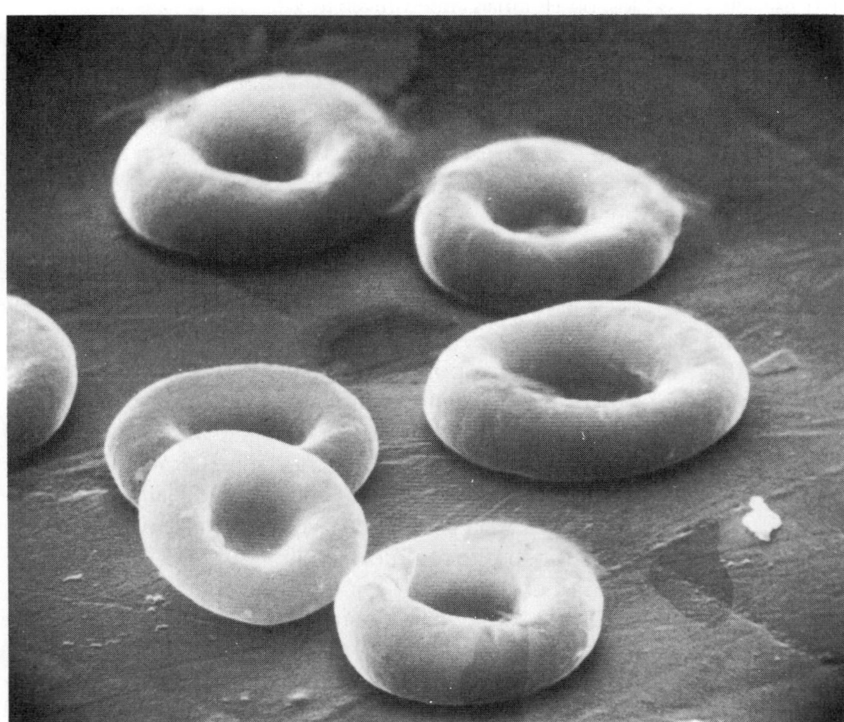

Figure 1 Red blood cells seen in a scanning electron microscope, which produces a three-dimensional appearance (×5,000). The doughnut shape (with the hole partly filled) is clearly evident. (From Morel et al. 1971.)

the finest capillaries. Mature red blood cells contain no nuclei or other organelles, although they are derived from precursor cells (*erythroblasts*) that do. The primary function of red blood cells is to carry oxygen to the tissues, where it participates in energy production, and to return carbon dioxide, the waste product of these reactions, to the lungs. The transport of both O_2 and CO_2 involves hemoglobin, which is packed inside red blood cells to the exclusion of almost everything else. The average red blood cell circulates for about four months, traveling 200 miles before it is eventually phagocytized (engulfed) by large white blood cells in the spleen. (These white blood cells are different from those that make antibodies.) It is not known exactly why red blood cells wear out, but turnover of molecules and cells within the body is the usual course of events. In one person, about 2×10^6 red blood cells are made and destroyed each second (a total of 2×10^{13} cells divided by an average life span of 1×10^7 seconds).

The hemoglobin inside red blood cells has nothing to do with the blood groups, which are properties of the red cell membrane. This extremely thin film (Figure 2) acts as a gatekeeper, allowing some substances to pass in or out while denying passage to others. It consists of two layers of lipids (of which 30% is cholesterol) in which are embedded various protein molecules. Some of the proteins are structural elements of the membrane, some are enzymes involved in energy production, some aid in the transport of substances across the membrane, and others are called *cell recognition proteins*. The latter are exposed on the outside of the cell membrane and may be combined with glucose, galactose, or other sugars into molecules called *glycoproteins*. Similar sugar groupings may be combined with the lipid molecules in the membrane to form *glycolipids*. These molecules mediate cell-to-cell recognition and communication, regulate tissue growth and differentiation, and appear to be altered in cells that become cancerous (Hakomori 1986). Each biological species has some cell surface proteins, glycoproteins, or glycolipids—called **antigens**—that are foreign to other species. Within a given species, individuals may also differ from one another in the antigens they possess, thus providing the basis for human (and animal) blood grouping.

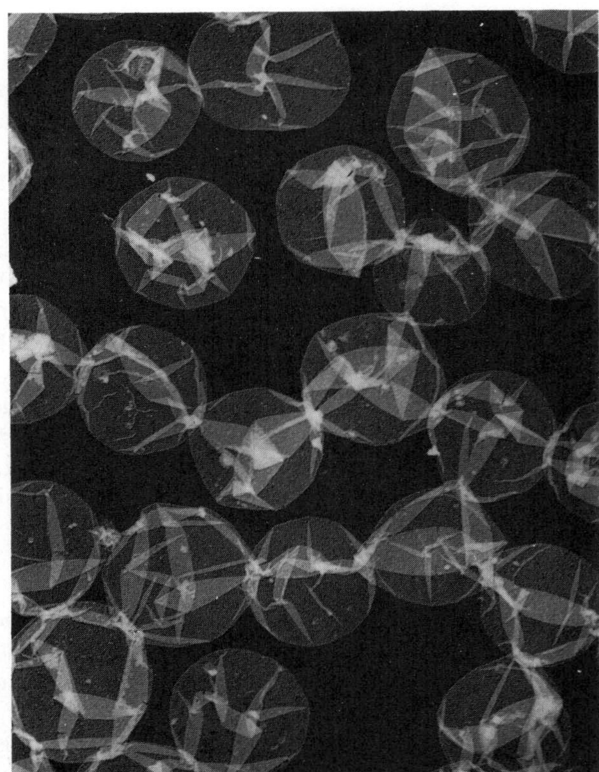

Figure 2 The membranes of red blood cells (×3,000). The so-called ghosts of red blood cells are obtained by gently breaking open the cells and washing away the released hemoglobin. The membranes are only about 75 Å thick; the whitish areas represent overlapping folds. (Courtesy of Joseph Hoffman, Yale University.)

Polymorphisms

The membranes of red blood cells are loaded with antigens. One set of antigens determines the human ABO blood group; another set determines Rh; another, Kell; and so on. It is important to recognize that these antigens do not, in themselves, produce disease, although serious interactions can occur *between* bloods (as in transfusion reactions), and a few blood types are weakly correlated with certain disorders, which we note later in the chapter. In general, blood group phenotypes are neutral attributes.

Each set of antigens is controlled by a different gene, which often has multiple alleles. When an antigen is present on red blood cells, a corresponding allele is present in a person's genotype. The alleles of one set and the corresponding antigens are referred to as a **blood group system**, or simply a **blood group**. Some blood group genes have one very common allele and one very rare allele. For such a gene, almost all people would be homozygous for the common allele; only infrequently would a heterozygote appear, and homozygotes for the uncommon allele would be very rare. We will not consider such blood group systems.

On the other hand, some blood group loci have appreciable proportions of *two or more* alleles in this or that human population. Such genes are said to be **polymorphic** (*poly*, "many"; *morph*, "form"). For example, if we tallied all the ABO alleles (designated *A*, *B*, and *O* using the shorthand form) among the English, we would find that about 28% were *A*, 6% were *B*, and 66% were *O*. The definition of polymorphism is usually taken at 1%; that is, the existence of *two or more alleles, each at a frequency of 1% or greater in a certain population, defines a polymorphic locus*. For a polymorphic gene, two people picked at random are likely to have different genotypes, and a relatively high proportion of them will be heterozygous. The 11 blood group systems listed in Table 1 are polymorphic in English populations (Race and Sanger 1975). Note that the Rh gene has the most known alleles with frequencies of 1% or greater. The last to be discovered, Xg, is the only blood group controlled by an X-linked gene. A comparable list of polymorphic genes for some other country would be similar but not exactly the same as this one. (Because the husband-wife team of R. R. Race and Ruth Sanger work in England, their compatriots have probably bled the most for science.)

Techniques

The antigens present on a person's red blood cells determine his or her blood types. The presence of any antigen is revealed by a chemical reaction with a corresponding antibody. The antigen–antibody reaction is a very specific one, but it is somewhat oversimplified to state that for any antigen, say Z, there is one and only one antibody, called anti-Z, that will react with it. Nevertheless, the basic knowledge of blood groups stems from this concept, and actual tests are sometimes so simple that "it is surprising that they have contributed so much to biological knowledge" (Race and Sanger 1975). On a microscope slide or in a test tube, a drop of red blood cells is mixed with a drop of serum containing a known antibody, say anti-Z. Either (a) the red blood cells clump, this being called **agglutination**, or (b) nothing happens. Agglutination means that the corresponding antigen, Z, is present on the red blood cells; no agglutination means that the antigen Z is absent.

Usually no microscope is needed to see the clumps of red blood cells resulting from agglutination (although individual red blood cells cannot be seen with the naked eye). The phenomenon is illustrated in Figure 3, using as an example red blood cells with an antigen called A. Two tests are indicated, one with anti-A and another with anti-B. Because antigen A and the antibody—anti-A—have complementary three-dimensional shapes and chemical properties, they are able to combine. Because most antibodies have two sites that can react with the antigen, adjacent red blood cells can be linked into a network of cells large enough to be seen easily. Anti-B is unable to recognize and react with the A antigen because it has a different shape and chemistry.

Some of the antibodies that laboratories use for blood typing are obtained by injecting rabbits with human red blood cells over a period of several weeks. The animals respond to the foreign antigens by manufacturing antibodies, which can be obtained by simply withdrawing blood from the animals. The serum, when separated from its suspended particles, is called **immune serum** (plural, **sera**); it will usually contain, in significant concentration, many kinds of antibodies against the many human red cell antigens. The immune serum sample with multiple antibodies can often be purified so that it contains just one. Several dozen such typing sera, each

Table 1 Polymorphic blood group systems in English populations

Blood group[a]	*Year of discovery*	*Frequency (%) of most common allele*	*Frequencies (%) of other alleles*[b]
ABO	1900	66	21,[c] 7,[c] 6
MNSs	1927	39	29, 24, 8
P	1927	55	45
Rh	1939/1940	41[d]	39, 14,[d] 3,[d] 1,[d] 1, 1
Lutheran	1945	96	4
Kell	1946	95	5
Lewis	1946	82	18
Duffy	1950	56	42, 2
Kidd	1951	51	49
Auberger	1961	58	42
Xg	1962	66	34

Source: Race and Sanger (1975).

a. Ordered by year of discovery, each of the 11 blood group systems is controlled by a locus with two or more alleles with frequencies of 1% or greater.

b. The table excludes some alleles with smaller frequencies among the English.

c. Different forms (A_1 and A_2) of the *A* allele.

d. Different forms of the Rh positive allele; other Rh alleles in the table are Rh negative.

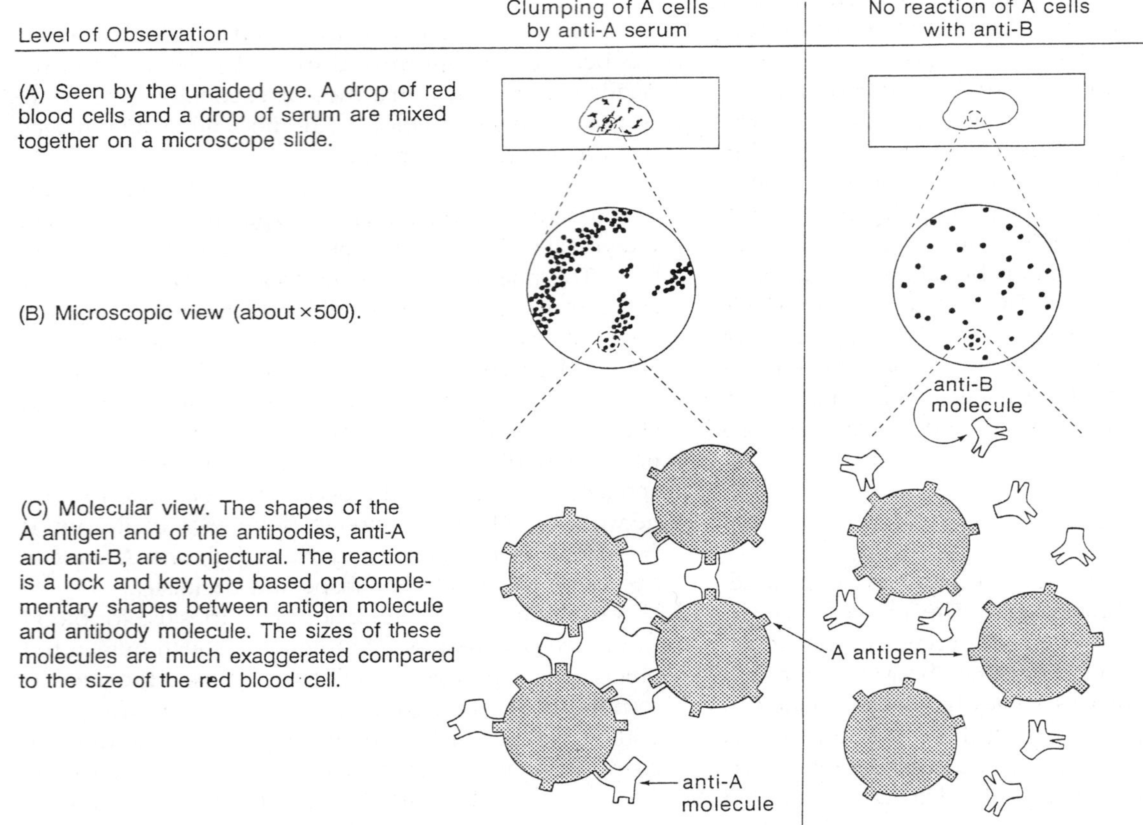

Level of Observation
Clumping of A cells by anti-A serum
No reaction of A cells with anti-B
(A) Seen by the unaided eye. A drop of red blood cells and a drop of serum are mixed together on a microscope slide.
(B) Microscopic view (about ×500).
anti-B molecule
(C) Molecular view. The shapes of the A antigen and of the antibodies, anti-A and anti-B, are conjectural. The reaction is a lock and key type based on complementary shapes between antigen molecule and antibody molecule. The sizes of these molecules are much exaggerated compared to the size of the red blood cell.
A antigen
anti-A molecule

Figure 3 The appearance of the red blood cells of a person with antigen A after reaction with sera containing anti-A or anti-B. Three levels of observation are shown: the unaided eye, a medium-power light microcope, and the molecular level.

with an antibody specific for one antigen, may be used when determining human blood groups.

This process in rabbits is similar to the immunization of people against a variety of disease organisms. Viruses and bacteria possess antigens foreign to us, and we defend ourselves by making antibodies (Chapter 18). For example, shots for tetanus contain an antigen (toxin) extracted from tetanus bacteria; a series of shots consisting of an initial injection followed by several boosters causes us to manufacture and maintain protective levels of antibodies (antitoxin).

Some antigens cause an immunized animal to respond strongly by making copious amounts of antibodies, whereas other antigens elicit little antibody response. Furthermore, not all animals are equally good at making an antibody when exposed to the same antigen. For these reasons, antibody preparations are variable. Thus, although the principles of typing are simple, some of the tests require the careful attention of skilled technicians.

In addition to immunizing animals with human red blood cells to obtain typing antibodies, blood workers rely on several other sources. One commonly used starting point is human plasma. For example, the antibodies that detect antigens of the ABO system occur naturally in certain people in a pattern to be described later. Antibodies for other systems may also occur in the plasma of people who have had multiple transfusions or pregnancies. In addition, researchers have recently discovered an in vitro method that produces so-called monoclonal antibodies (Moore et al. 1984). In this technique, discussed further in the next chapter, mouse white blood cells are first stimulated to produce specific antibodies. The cells are then fused with mouse cancer cells so that the cultured hybrids, called hybridomas, produce their programmed antibodies indefinitely.

In summary, the blood groups are defined in terms of *antigens* present on the surface of red blood cells. An antigen

1. elicits the production of the corresponding antibody when injected into an animal to which the antigen is foreign; and

2. combines with the specific antibody (produced this or another way) to cause agglutination (or other evidence of reaction) on a microscope slide or in a test tube.

How Some Systems Were Discovered

New blood groups are usually discovered by finding a serum that agglutinates the red blood cells of some persons but not all. Such a result suggests that the new serum possesses an antibody that recognizes a gene-controlled antigen present on the surface of the agglutinated cells but absent on the unagglutinated cells. Some variations on this theme are indicated in Figure 4.

ABO. Landsteiner knew that the mixing of bloods of different species produced severe reactions (as described in the introductory quote to this chapter). This knowledge prompted him to inquire into whether lesser reactions might occur between the bloods of individuals of the *same* species. At about the same time that another Viennese was rediscovering Mendel's laws, Landsteiner (1901) did the following experiment: He took blood from himself and from five of his colleagues, and, by

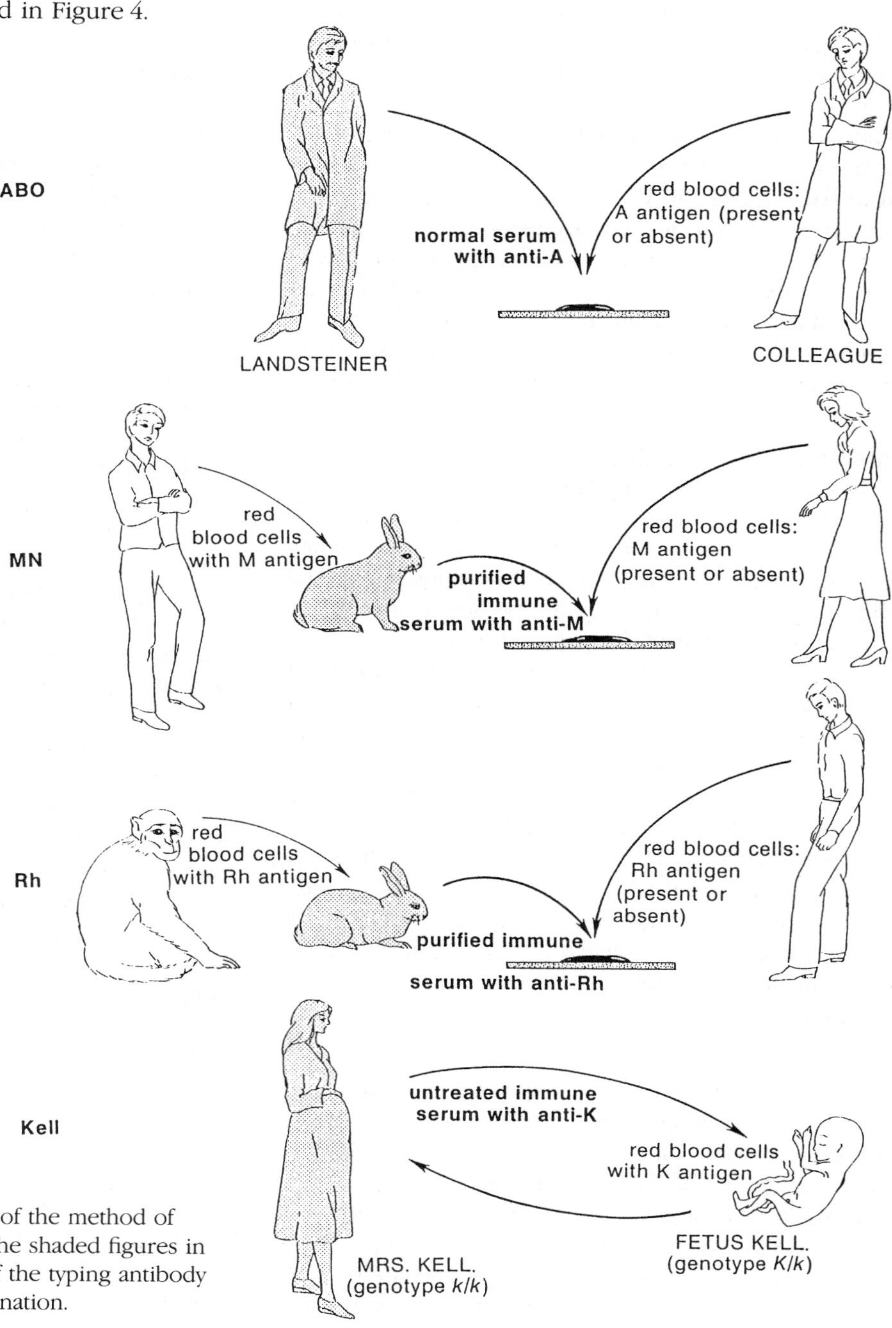

Figure 4 Diagrammatic representation of the method of discovery of four blood group systems. The shaded figures in the second column indicate the source of the typing antibody (or antibodies). See text for further explanation.

centrifugation, separated the sera from the red cells. He then recombined the sera and cells in all possible pairings. He noted, for example, that the serum of Dr. Pletschnig strongly agglutinated the cells of Dr. Sturli, but did not affect his own. (See also Question 1.) Thus the serum of Dr. Pletschnig seemed to have an antibody directed against an antigen present on one person's red blood cells but not on another's. The first letters of the alphabet were available to label the different sera and cells. For his investigations of antigen–antibody reactions over a period of several decades, Landsteiner received a Nobel Prize in 1930. He was the first of many subsequent winners whose research was related to genetics (Dixon 1984).

Only the ABO blood groups could be discovered this way, because only ABO typing antibodies occur in **normal serum**, that is, in the serum of a person who has not been exposed to foreign antigens. Also for this reason, the ABO antigens and antibodies of persons involved in transfusions must be very carefully checked in order to prevent a serious antigen–antibody reaction within the recipient's bloodstream. The typing antibodies for *all* other blood groups are derived from an explicit immunization process.

MN. In the 1920s Landsteiner (who had emigrated to the United States) and Philip Levine, a colleague at Rockefeller Institute for Medical Research, started looking for other blood groups by immunizing rabbits with human red blood cells from various donors (Figure 4). The rabbits responded by making antibodies against many foreign antigens. One of the purified immune sera contained an antibody that came to be known as anti-M. It agglutinated the red blood cells of certain people (those with the M antigen) in a pattern that showed *no correlation with the individuals' ABO blood types.* Thus Landsteiner and Levine discovered a second, independent blood group system to which still other antigens (e.g., N, S, s) were later shown to belong. Although genetically interesting, this so-called MN blood group is of little medical concern because people do not usually react strongly to the various antigens (foreign to themselves) that belong to the system. A third blood group, the P system, was found at the same time as MN by Landsteiner and Levine.

Rh. In the late 1930s Landsteiner and another colleague, Alexander Wiener, modified the search protocol by injecting rabbits with the red blood cells of Rhesus monkeys (Figure 4). In a pattern independent of all previously known blood groups, one particular purified immune serum, which they called anti-Rh, agglutinated the red blood cells of Rhesus monkeys and about 85% of a sample of New Yorkers (those positive for the Rh antigen).

The Rh antigen was immediately recognized as a strong antigen, one that could be responsible for severe transfusion reactions in persons with either prior transfusions or prior pregnancies (Figure 5). If blood with the Rh antigen is transfused to a person who lacks it, the recipient responds by making the anti-Rh antibody (Figure 5A, top). Although this antibody will not usually produce problems after a first transfusion—and it disappears after a time—the recipient's immune system is now sensitized to the Rh antigen. Upon additional transfusions of blood containing the Rh antigen, increasingly severe reactions can be expected as more and more anti-Rh antibodies are produced by the recipient (Figure 5A, bottom). Within a person's circulatory system, the reaction between the Rh antigen on the donor's cells and the anti-Rh antibodies manufactured in the recipient destroys the transfused cells rather than just agglutinating them. The destruction is called **hemolysis** (Latin *haem*, "blood"; English *lysis*, "releasing"). Some of the unhappy consequences of in vivo hemolysis are due to high concentrations in the blood of the released hemoglobin and breakdown products of hemoglobin. The resulting disturbances in blood pressure and kidney function can be fatal.

The year before Rh was named Rh by Landsteiner and Wiener, what were subsequently shown to be the same antigen and antibody were reported in the literature by Levine and Stetson (1939), but *without a name*. (Levine had been a codiscoverer of the MN and P blood groups.) They had investigated a woman, M.S., who had never previously received a transfusion. Because of blood loss during the delivery of a macerated, stillborn child (that had apparently been dead for several months), M.S. received from her husband a transfusion that quickly resulted in chills, pains, and later, dark urine. The ABO system was definitely ruled out as a cause of the transfusion reaction. Levine and Stetson concluded that the red blood cells of the husband carried a previously unknown antigen to which M.S. had *already* been exposed by the dead fetus's red blood cells that leaked into her circulation. The reaction was later shown to be due to the system that was to be named Rh: Rh antigen was on the husband's red blood cells and anti-Rh antibody was in the woman's plasma. The gene for the Rh antigen present in the fetus was inherited from its father, so they were *both* Rh positive. M.S. evidently lacked the antigen and so was Rh negative. These circumstances are summarized in Figure 5B.

KELL. Many of the blood groups discovered subsequent to Rh came to light because of a *transfusion reaction due to previously unknown antigens.* This was true for the Lutheran, Duffy, and Auberger systems listed in Table 1. The Kell and Kidd groups, however, were discovered because a child was born with a disease called **hemolytic disease of the newborn** (or **HDN**). This disease results from an antigen–antibody reaction that

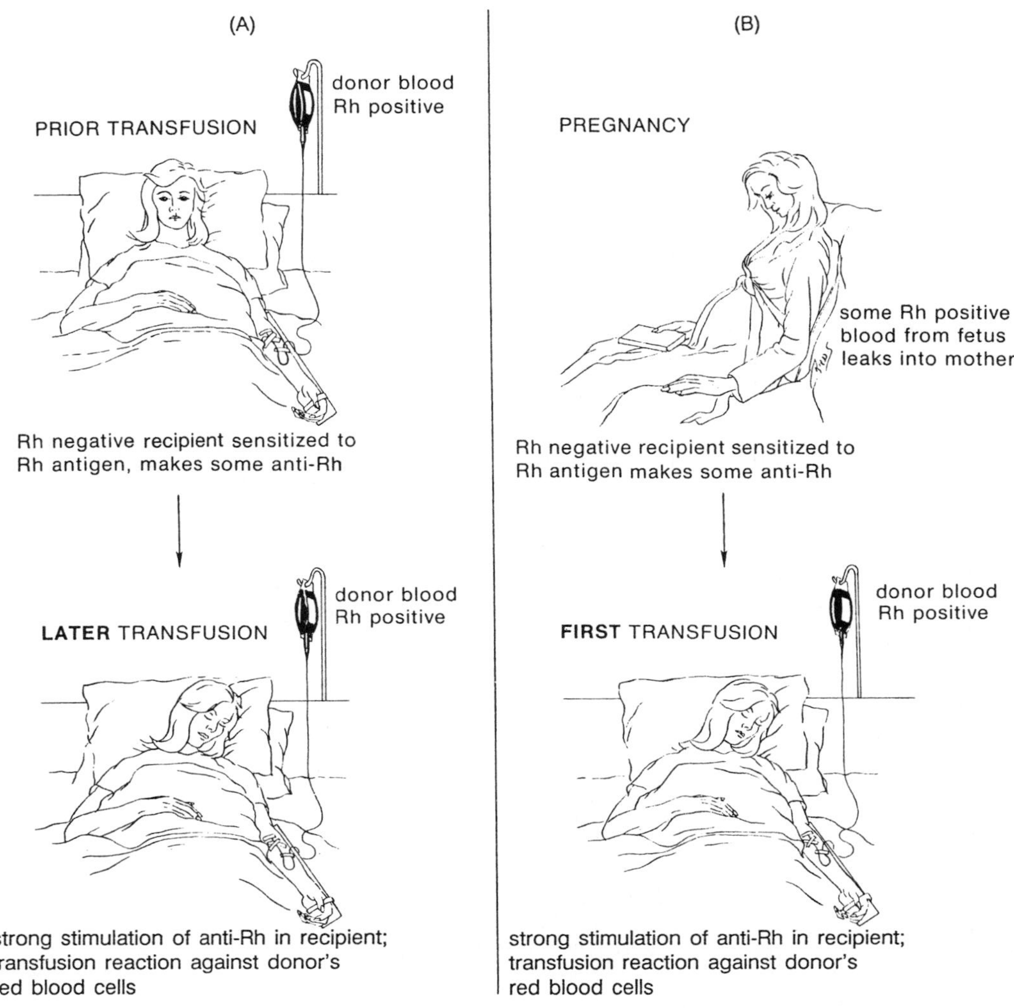

Figure 5 Transfusion reactions. In 1940 it was shown that such reactions could be due to Rh antigens. See text for further explanation.

occurs within the fetal blood circulation prior to and at the time of birth.

A woman, Mrs. Kell., (Figure 4) had lost two of her first three children soon after their births, and a fourth was born with HDN. (The child was successfully treated.) Because the woman, her husband, and the fourth child were all Rh positive, she could not have made anti-Rh. Nevertheless, present in her plasma was an antibody that reacted in vitro with the red blood cells of her newborn son, her other living child, and her husband. It was called anti-K, and the antigen with which it reacted was called K. This antigen was later shown to be controlled by an allele, *K*, that the children and the father possessed (as *K/k* heterozygotes) and that the mother lacked (being homozygous *k/k*).

The probable sequence of events is outlined in Figure 6. During one or more of her prior pregnancies, red blood cells with the K antigen (or Z in the generalized diagram) had leaked into her circulation (Figure 6A). Although the bloodstreams of mother and fetus do not generally merge, part of the placental membrane separating the two circulatory systems is just two or three cells thick. A few tiny breaks may have allowed some fetal red blood cells to enter the mother's system. At the time of birth, when the placenta separates from the uterine wall, it is possible for larger amounts of fetal blood to enter the maternal circulation. In response to the K antigen on these fetal red blood cells, Mrs. Kell. was sensitized to K (Figure 6B). When the fourth child was in utero, the leakage of just a small amount of fetal blood (Figure 6C) was sufficient to stimulate a significant production of anti-K (Figure 6D). Then this antibody was able to pass back across the placenta into the fetus (Figure 6E). Here it destroyed some red blood cells of the fetus, who at the time of birth was observed to have HDN. Jaundice, one of the distinguishing symptoms, is

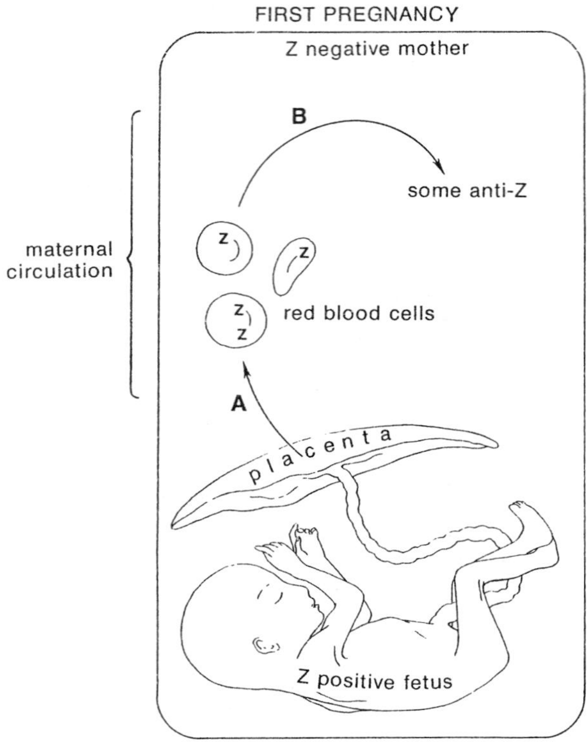

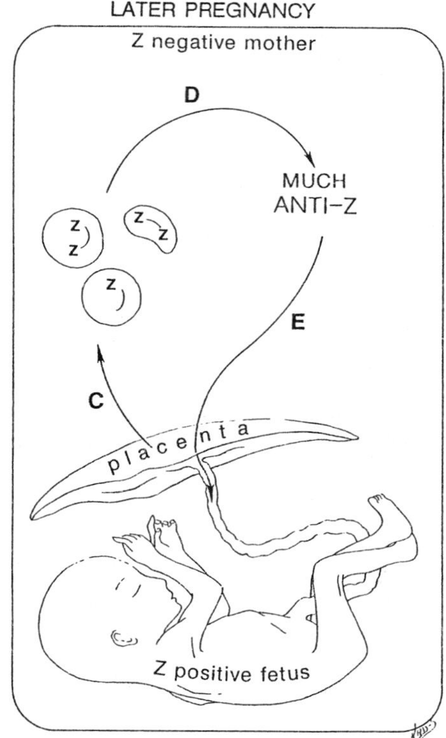

A Some fetal red blood cells with Z antigen leak through the placenta from fetus to mother, perhaps during pregnancy, but mostly at the time of birth

B Mother is sensitized to Z and makes some anti-Z, but usually the baby has been born before sufficient quantities of the antibody have built up

C Leakage of some fetal red blood cells during a later pregnancy

D Mother responds strongly with anti-Z

E Anti-Z molecules enter fetus, causing destruction of fetal red blood cells by reacting with Z antigens

Figure 6 The development of hemolytic disease of the newborn. The diagrams are generalized to indicate that any strong antigen, Z, present on fetal red blood cells but lacking in the mother, can produce hemolytic symptoms.

due to the yellow color of some of the breakdown products of the released hemoglobin. Another consequence is indicated by the medical term for HDN: *erythroblastosis fetalis*. When mature red blood cells are destroyed in large number, the body tries to compensate by releasing into the circulation immature ones, erythroblasts (*osis*, "proliferation of").

The events described above also occur sometimes with respect to the Rh antigens and much less often with other blood group antigens. (The macerated fetus in the 1939 report of Levine and Stetson was almost certainly due to Rh-caused hemolytic disease.) Medical developments in the 1970s, however, drastically reduced the occurrence of HDN due to Rh (see page 347).

THE ABO BLOOD GROUP SYSTEM

The alleles, *A*, *B*, and *O*, control the cell surface antigens of the ABO blood group system. Although primarily on red blood cells, the ABO antigens can also be detected on other cell types. Already outlined in Chapter 11 to illustrate Mendel's second law, the system is described in more detail in Table 2. The typing antibodies, anti-A and anti-B, can be obtained from the normal serum of persons lacking the corresponding antigens. Thus anti-A is found only in the serum of persons of blood type B or O, and anti-B is found only in the serum of persons of blood type A or O. It is not known why these anti-

Table 2 The ABO blood group system[a]

IN VITRO REACTION OF RED BLOOD CELLS WITH				
anti-A	*anti-B*	*Antigens on red blood cells*	*Blood type (phenotype)*	*Genotype*
+	0	A only	A	*A/A* or *A/O*
0	+	B only	B	*B/B* or *B/O*
+	+	Both A and B	AB	*A/B*
0	0	Neither A nor B	O	*O/O*

a. Three alleles, A. B, and O, are considered, although the system is somewhat more complex. For example, the A antigen is actually two closely related antigens controlled by different *A* alleles, A_1 and A_2. The anti-A normally used for typing will agglutinate cells with either antigen (or both). We have ignored this complication. Note that the symbol A can be an antigen, an antibody, a blood type, or an allele. The context in which the symbol is used should make clear which is meant. +, agglutination; 0, no reaction.

bodies are present, although their occurrence is so critically important in transfusions that there has been a good deal of speculation (see Question 8). For no other red cell antigens are the corresponding antibodies normally present in human plasma.

Note that blood types AB and O (Table 2) correspond to single genotypes, *A/B* and *O/O*, respectively. Blood types A and B can result from the corresponding homozygote or from the heterozygous combination with the *O* allele, however. Sometimes, but not always, the specific genotype can be inferred from the blood types of related individuals. For example, consider the mother, father, and daughter with the blood types shown in the accompanying diagram. The mother can not be genotypically *A/A*, for she would then have to transmit an *A* allele to the child (but, as seen, the child does not have an *A* allele). The child cannot be *B/B*, because the mother does not have a *B* allele to transmit. The father, however, can be either *B/B* or *B/O*.

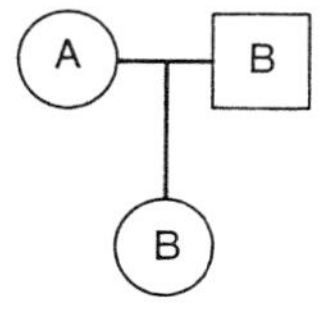

Transfusions

The presence of anti-A and anti-B in some normal human plasma was the main stumbling block to successful transfusion. For example, type A red blood cells inadvertently given to a type B person will be hemolyzed by the recipient's anti-A antibodies, the combination leading to a possible severe reaction even on a first transfusion. Such reactions involving the *donor's red cell antigens* (A or B or both) and the *recipient's plasma antibodies* (anti-A or anti-B or both) are the most important ones when whole blood is transfused. The prohibited combinations are indicated by **X**'s in Table 3. The absence of an **X**, however, does not guarantee a safe combination. There are several considerations in the choice of a donor:

1. The donor red blood cells should not, as indicated above, have the A or B antigen if the recipient has the corresponding antibody. Type A can always be given to type A, B to B, and so on—the four checked combinations in Table 3. In the reverse situation (the donor has antibodies against recipient antigens), the transfusion is also considered safe—the five dotted combinations in the table. The donor antibodies usually do no harm because they are diluted out when one or a few pints of blood are added to the several quarts of blood in the recipient. Subtle effects of low concentrations of inappropriate antibodies may be a bit worrisome, however. Moreover, they may remain concentrated enough to cause damage if a small patient needs a lot of new blood.

2. The donor is matched for Rh, positive for positive, negative for negative. It is especially important that Rh negative premenopausal females not be given Rh positive blood: A first transfusion may be uneventful, but the recipient will become more sensitized to the Rh antigen. Upon further transfusion, or when pregnant, unwanted consequences may occur to the woman or to the developing fetus.

3. After selection by the ABO and Rh label on donor blood, a final precaution is taken prior to transfusion: the donor and recipient bloods are tested for compatibility in vitro by **antibody screening**, by **cross matching**, or often by both procedures (Greendyke 1980). In antibody screening, the patient's serum is tested against specially chosen red blood cells (from several sources) that contain

Table 3 ABO blood types and transfusions

Donor *blood type (= antigens on red blood cells)*	**RECIPIENT** BLOOD TYPE (AND PLASMA ANTIBODIES)			
	A (anti-B)	*B (anti-A)*	*AB (none)*	*O (anti-A and anti-B)*
A	✓	X	•	X
B	X	✓	•	X
AB	X	X	✓	X
O	•	•	•	✓

Symbol key:
X, a transfusion that is prohibited by the potential reaction between the donor's red cell antigens and the recipient's plasma antibodies.
•, a transfusion that is generally safe (and is done), although the donor has some antibodies against the recipient's red cell antigens.
✓, a transfusion that is preferred (when considering only ABO antigens and antibodies).

most of the common antigens that cause transfusion problems. In cross matching, the patient's serum is combined with the donor cells to test for any significant antibody that is present for any reason. If positive reactions occur, another donor is chosen.

The patient antibodies that are most frequently detected by these tests are those directed against antigens of the Kell, Duffy, Kidd, and Rh blood groups systems. (In the case of Rh, they are antigens other than the main Rh positive antigen.) The antibodies may be present in the recipient's serum because of a previous transfusion or pregnancy. (The reverse test with the donor's serum and the recipient's cells, called the minor cross match, is not usually done.) Because these laboratory tests are important and are completed so quickly, they are virtually always done, even in emergency situations.

The safest transfusions of all, called **autologous blood transfusions**, occur when people donate blood to themselves (Council on Scientific Affairs 1986). About two-thirds of all transfusions accompany surgery. When it is planned sufficiently far in advance, some persons can have blood withdrawn ahead of time and get it back when needed. Autologous blood transfusions are particularly valuable for patients with rare blood types or for those who may have built up many antibodies as the result of prior transfusions or pregnancies. Furthermore, such transfusions carry no risk of blood-borne infections (hepatitis or AIDS, for example), and they reduce the demand on blood banks.

Note that the use of the term "universal donor" for persons of type O—because they have neither the A nor the B antigen on their red blood cells—is inappropriate. As indicated above, there are many reasons why a particular type O person may not be a safe donor. Similarly, the term "universal recipient" for persons of type AB—because they have neither anti-A nor anti-B—may be a dangerous label, for they may have other antibodies which could react with red blood cells transfused to them.

Biosynthesis of the Antigens

The chemical structures of the A and B antigens are now known (Watkins 1980; Hakomori 1986).* Interestingly, the first analyses were not done on red blood cells at all, but on body fluids such as saliva, gastric juice, and especially fluid from pathological ovarian cysts. This was possible because about 80% of people secrete a soluble form of the ABO antigens into many body fluids (including also bile, tears, semen, and breast milk). Secretion of the antigens is under genetic control and is due to the dominant allele, *Se*:

Genotype	**Phenotype**
Se/–	secretion of ABO antigens
se/se	nonsecretion (i.e., on red blood cells only)

Only those ABO antigens present on red blood cells are secreted; for example, a type B secretor will have the B antigen, but not A, in body fluids. The protein product of the secretor gene and its mechanism of action are not known.

Whether on red blood cells, on other cell types, or in solution, the A or B blood group antigen molecule is a short carbohydrate chain of six sugars that is attached to a carrier molecule (Figure 7). In the red blood cell membrane, the carrier can be either protein or lipid, the

*The chemistry of the antigens of the closely related Lewis and P blood groups has also been worked out, but little is known about the nature of the antigens of other red cell systems.

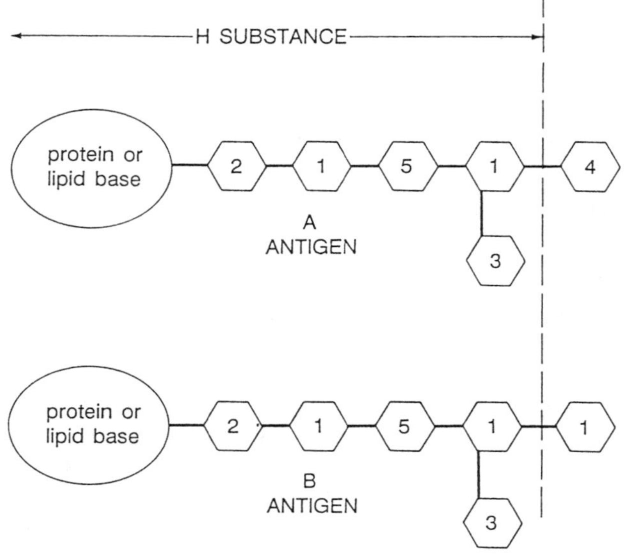

Figure 7 Chemical outlines of the A and B blood group antigens. The only difference between the two is the terminal (right-most) sugar of the six-unit carbohydrate portion of the molecules. Gene-controlled enzymes (glycosyltransferases) add the terminal sugars to a common precursor called the H substance (left of dashed line).

whole molecule being called a **glycoprotein** or a **glycolipid**, respectively. (The secreted form of the A and B substances is a glycoprotein.) On the red cell surface, the carrier part is embedded in the membrane and the carbohydrate chain sticks out to form the business end of the molecule; that is, the antigenicity—what makes A react differently from B—is a property of the carbohydrate chain. Whereas the terminal sugar in antigen B (galactose) has a hydroxyl group, —OH, the sugar in antigen A (*N*-acetylgalactosamine) has an aminoacetyl group, —NH—CO—CH_3. Yet this small difference can so alter the immunological specificity of the antigens that severe consequences ensue from a wrong transfusion. Such fine tuning of antigen–antibody reactions was first studied by Landsteiner, who showed that small chemical groups attached to larger molecules were the fundamental antigenic units.

The differences between the ABO antigens are associated with differences in susceptibility to a number of diseases. For example, persons of type A are more prone to stomach cancer than are persons of type O (who possess only the H substance shown in Figure 7). On the other hand, persons who are type O and nonsecretors are more likely to develop peptic ulcers than are persons of other blood/secretor types. Mourant et al. (1978) describe other correlations and speculate on the reasons for their occurrence.

Figure 8 outlines the steps in the biosynthesis of the A and B antigens. (We have omitted the Lewis and P blood group antigens, which also involve some of the

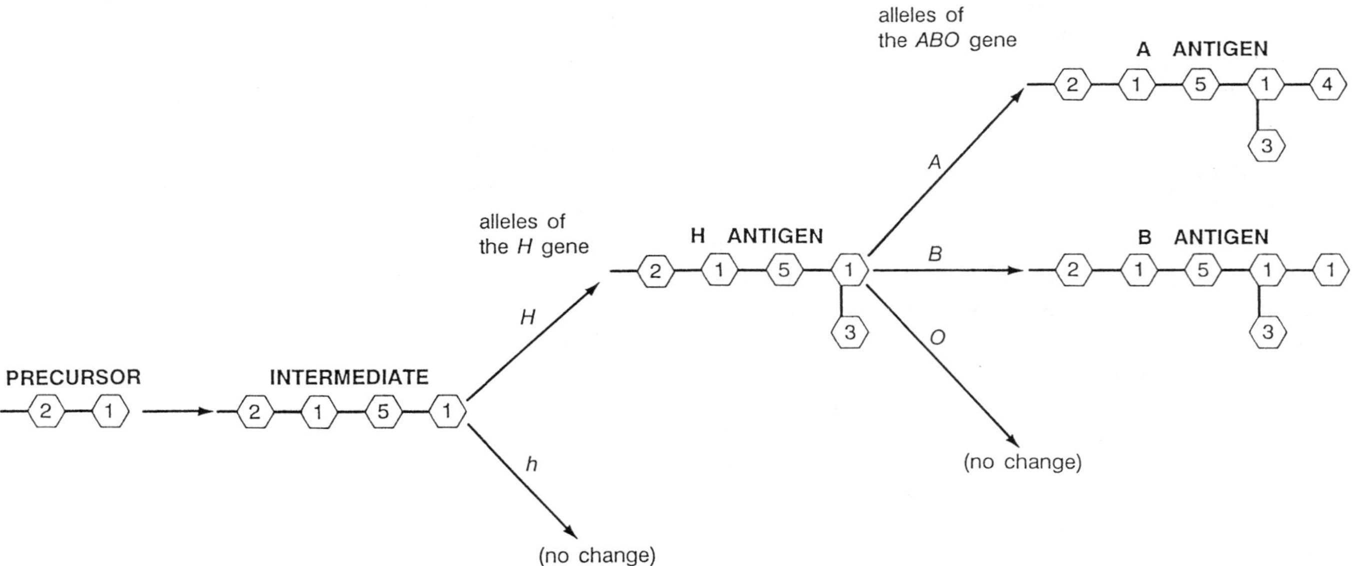

Figure 8 Biosynthesis of the ABH antigens. Enzymes coded by the unlinked genes, *H* and *ABO*, control the last two steps in the sequence of chemical reactions. The names of the sugars in the indicated carbohydrate structures are given in Figure 7. (Data from Hakomori 1986.)

same molecules.) Working backward, the *A* and *B* alleles of the ABO gene code for transferase enzymes that specifically add the one or the other sugar to the ends of the carbohydrate chains. The common substrate on which these two enzymes act is called H (also shown in Figure 7). The enzyme coded by the *O* allele (if *O* makes any effective enzyme at all) apparently adds no sugar, thus leaving H antigen unaltered on the red cell surface.

The H substance is an antigen in its own right, and it can be revealed by agglutination of O cells with the antibody anti-H, in the usual way. The sources of anti-H are, however, unusual. Human serum normally lacks this antibody, because all persons have at least a little H antigen on their red blood cells. (Even in persons with the genotype *A/A* or *B/B*, the A and B transferases do not convert all the H substance into A or B antigens.) The rule is that the antibodies of the ABO system—including anti-H—are *not present* in plasma when the corresponding antigen *is present* on red cells. Nor has it been possible to obtain anti-H by injecting laboratory animals with human red blood cells and purifying the immune serum.

One source of anti-H turned up in deliberate surveys of the plant kingdom. Recall from Chapter 4 that phytohemagglutinins are plant substances that agglutinate red blood cells. In the 1940s, investigators sought plant extracts that would distinguish one blood group from another rather than agglutinating all types nonspecifically. A useful reagent was extracted from the seeds of *Ulex europaeus*, a shrub in the bean family. This extract reacted most strongly with type O red blood cells, which have the most H antigen, and less strongly with other blood types. It is probably a fluke of chemical structure that *Ulex* possesses anti-H activity, but many plant proteins, called *lectins*, have been shown to combine specifically with sugars on cell surfaces (Sharon 1977).

The Curious Case of the Bombay Bloods

Another source of anti-H is in the sera of a few rare people. The first cases came to light when two unrelated men from Bombay needed transfusions upon admittance to a hospital (Bhende et al. 1952). (One had been hurt in a railway accident and the other has been stabbed in the abdomen.) Since both men were type O and neither had been previously transfused, most type O donor bloods should have been acceptable. But routine in vitro cross matching showed that the patients' sera agglutinated the red blood cells of every one of 115 potential type O donors! (Plasma alone was therefore transfused and the men recovered.) Subsequent investigation revealed that the two men—with the suitably named **Bombay phenotype**—possessed in their sera a high concentration of anti-H, which agglutinated *all* red blood cell samples. Thus, they differed from usual type O persons in the following way:

Phenotype	Antigens on red blood cells	Antibodies in plasma
Usual type O	H	anti-A, anti-B
Bombay phenotype	no H	anti-A, anti-B, anti-H

If only typed with anti-A and anti-B (the customary procedure), persons with the Bombay phenotype appear to be regular type O persons, because their red blood cells are not agglutinated by either antibody. But their red blood cells are not agglutinated by anti-H (from *Ulex* or from their own sera), as usual type O cells would be.

Here is the genetic explanation for the Bombay phenotype: In most people, a dominant gene, *H*, codes for a transferase that adds the sugar fucose (labeled 3 in Figures 7 and 8) to an intermediate substance to yield H. But people with the Bombay phenotype are homozygous for a rare recessive allele, *h*, that codes for a defective transferase enzyme. This defect causes a block in the metabolic pathway leading to the formation of H, and without the H substance, the enzymes controlled by the ABO alleles have no substrate on which to work. Thus, the red blood cells of people with the genotype *h/h* will not be agglutinated by anti-A, anti-B, or anti-H, *regardless of their ABO genotype*.

This interpretation was first suggested by the pedigree shown in Figure 9. Like the Bombay men, the propositus, II-5, required a transfusion, and no suitable donor could be found. Blood typing studies revealed that she and her older twin sisters had the Bombay phenotype. In addition, the children of II-5 possessed the genes *B* and *Se*, an apparently confusing situation, since *neither* of their parents (II-5 and II-6) tested positive for these two genes. Mutation was ruled out: one mutation is unlikely, and the probability of the two mutations required here would be vanishingly small. Knowledge of the family also suggested that the children were not illegitimate. It was therefore surmised that II-5 really possessed the *B* and *Se* genes but did not express them because she had no H substance.

The fact that I-1 and I-2 were actually first cousins increased the chance that a rare allele—*h* in this case—was present in both of them, having been inherited from a common ancestor. Levine et al. (1955) suggested these genotypes for II-5 and her parents:

I-1	×	I-2	→	II-5
H/h		*H/h*		*h/h*
O/O		*B/–*		*B/O*
Se/–		*se/se*		*Se/se*

The genotype and phenotype of II-5 offer a good example of **epistasis**, the masking of the effect of one gene by an allele at another locus. Neither the ABO nor the secretor genes are expressed in II-5 because of homo-

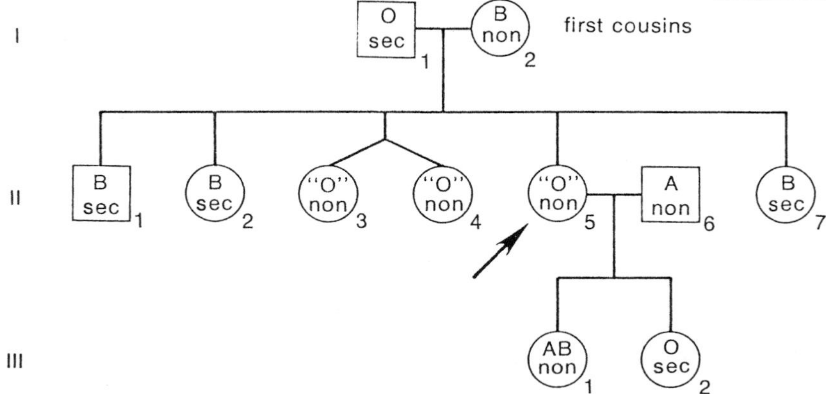

Figure 9 The ABO and secretor phenotypes of a Rhode Island family of Italian ancestry. sec, secretor; non, nonsecretor. Since the red blood cells of II-3, II-4, and II-5 failed to agglutinate with anti-A and anti-B *and* anti-H (and their sera possessed all three antibodies), they have the Bombay phenotype, indicated by "O." Note that the children in Generation III are seemingly inconsistent with their parents. In III-1, where did the *B* allele come from? In III-2, where did the *Se* allele come from? These genes were present, but masked, in their mother, II-5, and may have been present in the mother's older twin sisters as well (see text). (Adapted from Levine et al. 1955.)

zygosity for *h*. The MN, Rh, Kell, Duffy, and Kidd blood groups of the family were not unusual in any way. Therefore, the genotype that blocks the formation of H substance must not affect the formation of other red cell antigens, which are controlled by other genes and synthesized from other precursors.

Rh AND THE PREVENTION OF HEMOLYTIC DISEASE

With regard to the Rh blood groups, Race and Sanger (1975) have written, "There seems to be no limit to the marvellous complexity of this system of antigens and antibodies." Here we will indicate briefly what kinds of puzzles are involved, but then concentrate on just the two Rh phenotypes, positive versus negative, the most important distinction for clinical medicine. Unlike the antigens of the ABO system, the antigens of the Rh system are not secreted but are found only on red blood cell surfaces.

Rh Genes, Antigens, and Antibodies

Researchers have discovered a few dozen antigens belonging to the Rh system. The struggle to encompass these findings in a rational genetic model has resulted in two main notational systems, the *Wiener* scheme and the *Fisher-Race* scheme. Each is based on a different conception of the nature of the antigens and their controlling genes. (A third "genetically neutral" nomenclature for Rh is also in use.) Until information becomes available on the chemistry of the Rh antigens and on the enzymatic pathway responsible for their synthesis (if they are not direct gene products), no choice between the competing views is possible.

At the first level of complexity, Wiener envisions eight alleles of the Rh gene (Table 4, first column). Each allele is somehow responsible for the presence of several antigens on the red blood cells. The third column shows that the alleles are not equally common; in fact, the R^1, R^2, and r alleles together account for 95% of the variations of the Rh gene in England. In contrast to this, the theory of Fisher* and Race envisions three very closely linked genes called *C*, *D*, and *E*. Each gene has two alleles, designated by upper and lower case letters. There are exactly eight ways of combining *C* or *c*, with *D* or *d*, and with *E* or *e* (second column), and these gene complexes correspond to the eight alleles of Wiener. (No practical difference would emerge in the Fisher-Race view if the closely linked genes were actually three sites within the boundary of one gene. Knowledge of the polypeptide products of the gene(s) are needed to resolve the problem.)

*Sir Ronald A. Fisher, one of the leading statisticians and population geneticists of this century (see also Chapter 22).

Table 4 The genetic basis of the Rh system: Wiener and Fisher-Race

Eight alleles of Wiener	*Eight linked gene complexes of Fisher-Race*	*English frequencies (%)*	*Rh phenotype*
R^0	*cDe*	3	Rh positive: genetic symbols contain *R* or *D*
R^1	*CDe*	42	
R^2	*cDE*	14	
R^z	*CDE*	<1	
r	*cde*	39	Rh negative: genetic symbols contain *r* or *d*
r'	*Cde*	1	
r''	*cdE*	1	
r^y	*CdE*	<1	

Note: The table is simplified, because researchers have discovered additional (mostly rare) alleles. These alleles involve other sites within the gene not predicted by the *CDE* nomenclature, or quantitative differences such as a "weak *D*" variant, or other complexities.

With the Fisher-Race nomenclature, the Rh system is specified in a straightforward way: the letters included in each gene complex denote the red cell antigens that are revealed by positive reactions with the corresponding antibody. The most common gene complex, *CDe* (Wiener R^1), for example, gives rise to red blood cell antigens C, D, and e, which are detected by antibodies anti-C, anti-D, and anti-e. The genotype of a diploid individual includes two Rh gene complexes, which may specify more than three Rh antigens on the red blood cells. For example, the genotype, *CDe*/*cDE* (Wiener R^1/R^2), causes five different antigens to be formed on red blood cells: C, c, D, E, e. The other possible antigen, d, is apparently very weak (or perhaps nonexistent), because an antibody that specifically agglutinates cells that are not agglutinated by anti-D has yet to be found. Although clumsy to use when speaking, most geneticists use C, D, and E symbols, because they are easier to learn than the Wiener terminology. The context in which the C, D, and E symbols are used makes clear whether a given letter means the gene, the antigen, or the antibody. In any event, usage of either model does not necessarily mean that it is more valid than the other.

The single typing antibody anti-D (previously called anti-Rh) is sufficient by itself to distinguish the Rh positive from the Rh negative phenotype:

Reaction of red cells with anti-D	**Phenotype**	**Genotype**	
Agglutination	Rh positive	*D*/*D* or *D*/*d*	(*R*/*R* or *R*/*r*)
No reaction	Rh negative	*d*/*d*	(*r*/*r*)

Thus the *C* and *E* genes are not relevant to the positive–negative dichotomy. In either the Wiener or the Fisher-Race scheme, the upper case allele, *R* or *D*, is the Rh positive one (see Table 4). It is a dominant allele, because red blood cells from both *D*/*D* and *D*/*d* carry the Rh positive antigen and are agglutinated equally well by anti-D.

Incompatibility

We have noted that hemolytic disease of the newborn (HDN) was a factor in the initial 1939 observation of a transfusion reaction involving the Rh system and that the Kell blood group system was discovered through a case of HDN involving a previously unknown antigen. HDN is a potential problem whenever a fetus possesses a strong red cell antigen that is not present in the mother, a situation called **maternal–fetal incompatibility**. In these cases, the allele for the antigen is absent in the mother but present in the father, who transmits it to the fetus. Mothers who are incompatible with their fetuses for the D antigen of the Rh system account for most cases of HDN. Compared with cases caused by other antigens, D-caused cases are generally more severe, over 10% of affected fetuses being stillborn prior to full term. The immune processes in the case of Rh can be understood by substituting D for Z in Figure 6. Fortunately, this chain of events can now be avoided (see next section).

A mating is said to be immunologically incompatible if the male possesses an allele for a strong red cell antigen that the female lacks. Because the male may be heterozygous, however, not every fetus conceived from an incompatible mating is actually incompatible with its mother. For the two possible matings of Rh positive males to Rh negative females, we have the following:

Incompatible mating	**Proportion of fetuses incompatible with mother**
D/*D* ♂ × *d*/*d* ♀	all
D/*d* ♂ × *d*/*d* ♀	1/2

Even when the male is homozygous and all fetuses destined to be incompatible, the complications of HDN only arose in less than 10% of pregnancies in Rh negative women prior to about 1970. Levine (1958) reported an overall incidence in Caucasian populations of 1 case of Rh-caused HDN out of 150 full-term pregnancies.* Apparently, the leakage of fetal red blood cells into the maternal circulation (Figure 6C), or the antibody response of the mother (Figure 6D), or the susceptibility of the fetus (Figure 6E) is not always significant. Consistent with the immune nature of the disease, first pregnancies of Rh negative women with Rh positive fetuses rarely lead to hemolytic disease. The probability of being affected increases with subsequent Rh positive pregnancies. No Rh negative fetus (from a heterozygous father) is ever at risk as a result of the D antigen.

Other than D, the antigens implicated in HDN are usually A or B of the ABO system, c of the Rh system, and, to a much lesser extent, antigens of the Kell, Duffy, Kidd, and other blood groups. Note that designating an Rh antigen with a lower case letter (using Fisher-Race symbols) does not mean that the antigen is a weak one.

Hemolytic disease of the newborn due to ABO antigens occurs perhaps once in a thousand births. The symptoms are usually mild. One puzzle is why it occurs so seldom, because the antibodies are normally present in the mother's plasma. For example, an O mother carrying an A fetus need not be stimulated by this or by any prior A-pregnancy to produce anti-A, since she already has it! These normally occurring anti-A antibodies, however, are mostly of a class called IgM (immunoglobulin M). They *do not cross* the placental membrane, being much larger molecules than the IgG (immunoglobulin G) antibodies, which do cross the placenta. This antibody class difference may explain why HDN due to ABO incompatibility does not occur very frequently. When it does occur, however, antibodies of the IgG type can be demonstrated in the mother's plasma. Some investigators speculate that the mother may make IgG anti-A or anti-B in response to materials in the environment that resemble the A and B antigens. It is known that substances chemically similar to A, B, and H are present in various animals and plants, and especially in the bacterial flora of our digestive system. This fact would explain why HDN due to ABO incompatibility, when it occurs, is almost as likely to be the first ABO-incompatible pregnancy as any other.

Protection Against Rh-Caused Hemolytic Disease

Since the late 1960s, couples starting their families have not needed to worry much about Rh-caused HDN, because the incidence of the disease has been reduced about 90%. This achievement is remarkable, for within just a few decades investigators described the clinical disease, discovered its mechanism of causation, and developed a preventive treatment. It is also remarkable that the action required to prevent the disease is very simple: within a few days of the birth of *every* Rh positive child to an Rh negative mother, the mother is given a dose of the anti-D antibody (Figure 10A). It is important that the Rh negative woman be injected starting with the *first* Rh positive fetus—whether it is live-born, stillborn, aborted, or miscarried.

It might seem strange to administer the very antibody that is *un*wanted. But by injecting anti-D, the mother is prevented from making her *own* anti-D. To understand why this works, bear in mind that Rh-caused HDN occurs after the mother has been sensitized by the D antigen as a consequence of a prior pregnancy. Upon each exposure to the D antigen, her own immune mechanism responds more strongly to it. Most of the leakage of fetal red blood cells seems to occur at the time of birth, when the placenta separates from the uterine wall. Thus, the preventive shot of anti-D soon thereafter *destroys the invading fetal cells before their D antigens have an opportunity to stimulate the mother's immune system*. (Or the injected anti-D simply inactivates the D antigen—the exact mechanism of action is not known.) The D antigen is never "registered" in the mother's immune memory, and the injected anti-D disappears after a few months. "Immunologically speaking, the treated mothers enter their second pregnancy as if it were their first" (Clarke 1968). Similar treatments after successive Rh positive births continue to prevent sensitization. One untreated pregnancy could eliminate such protection, although this does not necessarily happen.

Most instances in which this treatment does *not* work are caused by immunization of the mother by fetal red blood cells that cross the placenta into her circulation *during the course* of pregnancy. It has been proposed that almost all cases of HDN due to Rh from this cause can also be prevented by injecting the mother with anti-D at about the thirtieth week of pregnancy. Because the Rh status of the fetus is not normally known, however, the treatment would have to be given *routinely* to every pregnant Rh negative mother. This plan would result in approximately 100 needless injections of anti-D for every case in which immunization of the mother was a real hazard. (Most Rh positive fetuses are not at risk because of the postpartum injections after prior births and insignificant leakage of fetal cells during the current pregnancy; and Rh negative fetuses are never at risk, leakage or not.) Arguments for and against anti-D injection during pregnancy as a desirable routine procedure are summarized by Nichols (1984) and Nusbacher (1984).

*HDN due to Rh is primarily a disease of Caucasian populations, because the proportion of Caucasians who are Rh negative is higher than in other groups (Chapter 22).

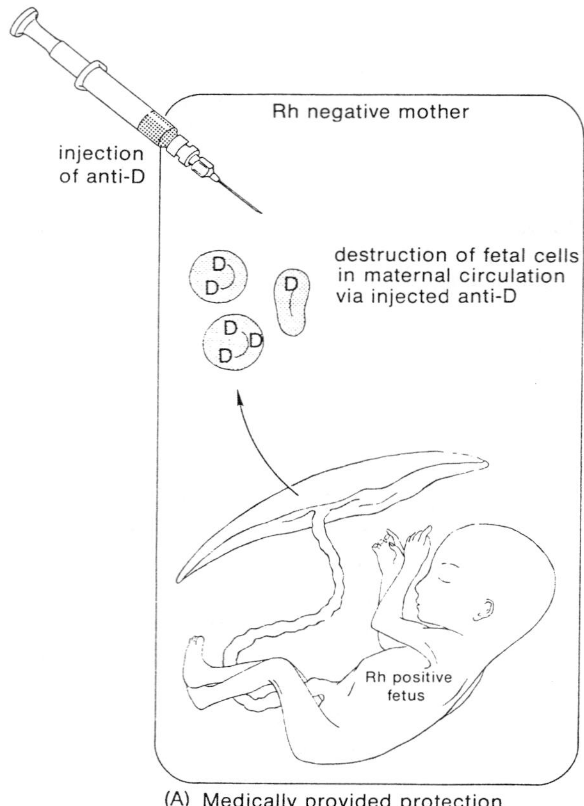

(A) Medically provided protection

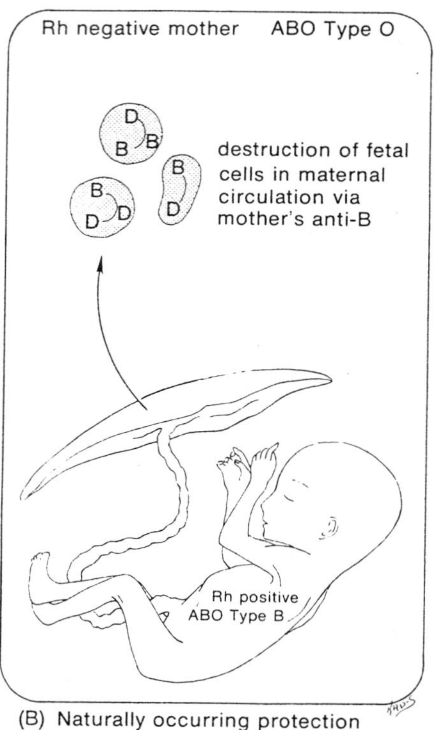

(B) Naturally occurring protection

Figure 10 Protection against hemolytic disease of the newborn due to Rh. (A) Following each Rh positive birth (live birth, stillbirth, or abortion), an injected dose of anti-D destroys fetal red blood cells that leak into the maternal circulation. The mother is therefore not sensitized to the D antigen. The injected anti-D disappears after a few months. (B) Anti-B (in the case illustrated) in the mother destroys fetal red blood cells that leak into her circulation. The mother is therefore not sensitized to the D antigen also carried on the red blood cells. Natural protection, due to either anti-A or anti-B, can occur only when the mother and fetus are ABO *in*compatible.

As early as the 1940s, researchers noted a surprising association between the occurrence of HDN due to Rh and the *ABO blood types* of family members. Summarizing 3,000 matings in which some of the children were born with Rh-caused HDN, Levine (1958) concluded that fetuses that were *ABO incompatible* with their mothers were almost never born with HDN due to Rh. (Of course, ABO-caused HDN might develop in the ABO-incompatible cases, but as noted before, this is only rarely a medical problem.) *In other words, maternal–fetal ABO incompatibility protects a fetus against Rh-caused HDN.* The reason for the ABO protection seems to be that antibodies of the ABO system do naturally what the anti-D medical treatment described above does artificially. Consider, for example, a type O mother and a type B fetus — an incompatible combination because the fetus has a strong antigen the mother lacks. The mother has anti-B normally in her plasma (Figure 10B); this antibody will destroy any fetal cells with the B antigen that leaks into her circulation. If these fetal cells also possess the D antigen, it too disappears along with the cells.

The history of the development of the anti-D treatment has been well documented (Zimmerman 1973). The idea occurred independently to two groups of physicians, first at the University of Liverpool and later at Columbia-Presbyterian Medical Center in New York. (Although Ronald Finn [1960] of the Liverpool group seems to be the first to note the idea at a scientific meeting, Sir Cyril Clarke writes that his wife was the first to think of it.) None of the principal investigators were trained immunologists, and the New York group had some difficulty getting financial support. The cooperation and competition, the good will and the frictions, that prevailed within and between the groups of investigators probably typify the way much scientific research unfolds.

SUMMARY

1. Blood consists of liquid plasma that contains many dissolved substances and suspended particles, primarily red blood cells. The membranes of the red blood cells are rich in the antigens that determine the various blood group systems.

2. Each set of antigens is controlled by a different blood group gene. When an antigen is present on red blood cells, a corresponding allele is present in the person's genotype. About a dozen blood group genes are polymorphic, having two or more common alleles in some population.

3. Red blood cells carrying a given antigen are detected by agglutination with the corresponding antibody. For some systems, the antibodies are obtained by immunizing an animal with human red blood cells of the appropriate type.

4. The blood groups were discovered by finding antibodies that agglutinated the red blood cells of some persons, but not others. In the case of Rh, Kell, and some other groups, the defining antibodies led to transfusion reactions or maternal–fetal problems.

5. The ABO antigens are the most important in transfusions because the antibodies, anti-A and anti-B, occur naturally in the plasma of persons lacking the corresponding antigen on their red blood cells. Prior to transfusion, donor red blood cells are preselected for ABO and Rh types, and then tested in vitro against a patient's serum for compatibility.

6. The A and B antigens are similar glycoproteins (or glycolipids) differing by a single sugar residue. They are synthesized from the H substance via transferase enzymes coded for by the *A* and *B* alleles, respectively. The *O* allele does not code for an active enzyme, thereby leaving a high concentration of the precursor substance H unaltered on the cell surface.

7. In Bombay phenotype bloods, a recessive mutation blocks the synthesis of the H substance, which prevents the synthesis of the A and B antigens as well, regardless of the person's ABO genotype. The serum of persons with the Bombay phenotype is a good source of the antibody, anti-H.

8. Hemolytic disease of the newborn (HDN) can be due to maternal–fetal incompatibility for any strong antigen, but most cases are due to the D antigen of the Rh system. A fetus is incompatible if it possesses a strong red cell antigen lacking in its mother.

9. Nearly complete protection against Rh-caused HDN is afforded by injections of anti-D following every Rh positive pregnancy of Rh negative women. Natural protection can occur whenever mother and fetus are ABO incompatible.

FURTHER READING

The book by Race and Sanger (1975) is the definitive work on human blood groups, but books by Mollison (1983) and by Salmon et al. (1984) cover the field in equally detailed fashion. Greendyke (1980) discusses all aspects of blood banking. Fudenberg et al. (1984) include a chapter on human blood groups, and Watkins (1980) presents the chemistry of the A, B, H, and related antigens. All these readings are quite technical.

Zimmerman's history (1973) of the prevention of hemolytic disease is very long, whereas an account by Clarke (1968) is very short. About 50 significant papers relating to Rh have been collected by Clarke (1975) and are accompanied by his useful commentary. A few of the original (and easy to read) papers relating to ABO are included in Boyer (1963). *Scientific American* articles include Hakomori (1986) and Sharon (1980) on the glycoproteins and glycolipids in the cell membrane, Sharon (1977) on lectins that combine with sugars on cell surfaces, and the collection by Burnet (1976) on immunology generally. Mourant (1983) explains how blood groups are applied to the study of anthropology. Nisonoff (1984) provides an introduction to modern immunology for biology students, while Mizel and Jaret (1985) present much of the same material in delightful form for general readers. The latter two books are applicable to both this and the next chapter.

QUESTIONS

1. These data are culled from Landsteiner's paper (1901). What was Landsteiner's ABO blood type? (0, no reaction; +, agglutination.)

	Red blood cells from		
Serum from	**Lands**	**Plecn**	**Sturl**
Lands	0	+	+
Plecn	0	0	+
Sturl	0	+	0

2. In a 1945 trial, a woman accused Charlie Chaplin of fathering her child. The ABO blood types were as follows: woman, A; Chaplin, O; child, B. When the California jury declared that Chaplin was the father of the child, the *Boston Herald* commented, "California has in effect decided that black is white, two and two are five, and up is down." (Quoted in Gradwohl 1954.) What do you think?

3. Wiener (1943) records these happenings in Chicago in 1930. When Mrs. W returned home from the hospital, she discovered the name label Y on her newborn child; the newborn given to Mrs. Y bore the label W. The ABO blood types of the six involved individuals were:

Mrs. W: O Mr. W: O Newborn with W label: O
Mrs. Y: O Mr. Y: AB Newborn with Y label: A

Can you decide whether the hospital switched the labels or the babies?

4. Give the ABO *genotypes* of a set of parents such that (a) three different phenotypes (only) are possible among the offspring; (b) four different phenotypes are possible among the offspring.

5. Among four children from the mating $A/O \times A/O$, show that the probability of getting exactly the ideal ratio (3 type A to 1 type O) is less than 1/2.

6. In Figure 9, what are the genotypes of II-5 and II-6 (parents) and III-1 and III-2 (children) for ABO and secretor. Note that II-5 is h/h for the H gene; assume II-6 is H/H.

7. Would persons with the Bombay phenotype (a) make good blood donors? (b) be able to receive blood transfusions easily?

8. Substances that resemble the A, B, and H antigens are widespread in nature, and we undoubtedly eat some. Does this suggest where normally occurring anti-A and anti-B in human plasma might come from?

9. Within the pedigree symbols is recorded the reaction of each person's red blood cells with anti-A, anti-B, and anti-D, *in that order*. Give the ABO and Rh *genotypes* of each person. 0, negative reaction; +, positive reaction.

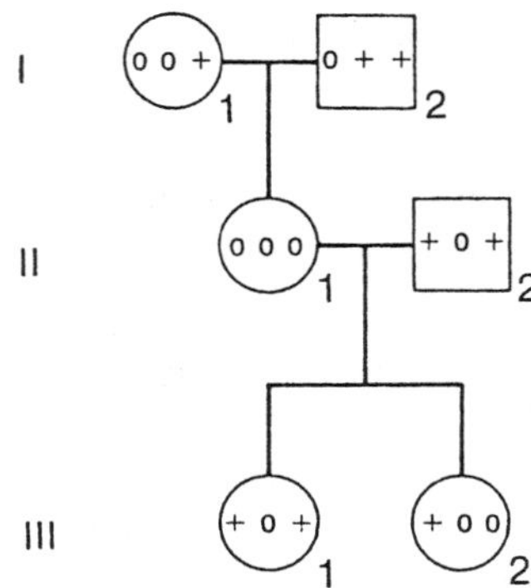

10. Considering the ABO and Rh loci together, what *phenotypes* would you expect among the offspring of the mating $A/B\ D/d \times A/O\ D/d$? What is the expected proportion of each phenotype?

11. Both the Rh and Duffy genes are on Chromosome 1, but they are more than 50 map units apart. Considering the two alleles of the Duffy gene, Fy^a and Fy^b, what kinds of gametes can be made by the double heterozygote, $\underline{D\,Fy^a}/\underline{d\,Fy^b}$? What is the expected proportion of each?

12. Using either the Wiener or the Fisher-Race symbols, list the six genotypes that can be formed from the three most common alleles listed in Table 4. For each, indicate whether the person is Rh positive or Rh negative.

13. The X's in Table 3 mean that transfusion reactions will occur between the donor's red blood cells and the recipient's antibodies. Relabel the table, substituting *fetus* for *donor*, and *mother* for *recipient*. Now what is the meaning of the X'ed combinations?

14. Reviewing the literature, Levine (1958) compiled records on 61 babies born with Rh-caused HDN in which one parent was type O and the other was type AB. In only two cases was the mother type O and the father type AB. Explain.

15. A baby is born with HDN, but both the baby and its mother are Rh positive. Give a reasonable explanation for this occurrence.

ANSWERS

1. Landsteiner must have been type O, his serum containing both anti-A and anti-B. The other two were type A and type B, although these data do not tell us which was which. (No type AB was among Landsteiner's original group.)

2. The *Boston Herald* has a point. Note that blood grouping can only *exclude* paternity: any man of type B or AB may have been the father.

3. The babies had the right labels but were given to the wrong mothers. The W's could not have had the baby with the Y label, nor could the Y's have had the W baby. (You may care to list systematically the possible offspring versus the impossible offspring from each of the ten matings based on ABO phenotypes.)

4. (a) $A/B \times (A/O$ or B/O or $A/B)$
(b) $A/O \times B/O$ only

5. $\left(\frac{4!}{3!1!}\right)\left(\frac{3}{4}\right)^3\left(\frac{1}{4}\right) = \frac{27}{64}$

6. II-5: $B/O\ Se/se$ III-1: $A/B\ se/se$
II-6: $A/O\ se/se$ III-2: $O/O\ Se/se$

7. (a) Yes; their red blood cells have neither the A, B, nor H antigens. (b) No; their anti-A, anti-B, and especially anti-H would agglutinate any donor cells except those from another person with the Bombay phenotype.

8. Anti-A and anti-B may be immune antibodies like any others, produced in response to foreign antigens. If a person is type A by inheritance, then B, a foreign material, when ingested and taken up by the circulatory system, would stimulate the production of anti-B. Substance A, not being foreign, would not lead to anti-A. (See references in Race and Sanger 1975.)

9. I-1: $O/O\ D/d$ II-1: $O/O\ d/d$ III-1: $A/O\ D/d$
I-2: $B/O\ D/d$ II-2: $A/\text{–}\ D/d$ III-2: $A/O\ d/d$

10. A, Rh pos: 3/8 B, Rh pos: 3/16 AB, Rh pos: 3/16
A, Rh neg: 1/8 B, Rh neg: 1/16 AB, Rh neg: 1/16

11. 1/4 each of $\underline{D\,Fy^a}$, $\underline{D\,Fy^b}$, $\underline{d\,Fy^a}$, $\underline{d\,Fy^b}$.

12.

Rh positive	**Rh negative**
$R^1/R^1 = CDe/CDe$	$r/r = cde/cde$
$R^2/R^2 = cDE/cDE$	
$R^1/R^2 = CDe/cDE$	
$R^1/r = CDe/cde$	
$R^2/r = cDE/cde$	

13. The X'ed combinations are ABO-*in*compatible; that is, the fetus has an antigen lacking in the mother.

14. When the mother is type O and the father type AB, all fetuses are ABO-*in*compatible with their mothers and therefore protected against Rh-caused HDN. (The two cases of Rh-caused HDN in this situation were equivocal.) In the 59 cases of Rh-caused HDN in which the mother was type AB and the father type O, no protection was afforded the fetuses.

15. Maternal–fetal incompatibility for some antigen other than D.

Chapter 18 Genetics of Immunity

The Immune Response
Main Components of the System
The B Cell Response
The T Cell Response

Antibody Structure and Genetics
Antibody Structures
Antibody Diversity
Monoclonal Antibodies

Transplantation Antigens
Tissue Typing: The HLA System
Antigens and Alleles of the *BCA* Loci
Transplantation

The Immune System and Disease
Associations with HLA
Allergies
Autoimmune Disorders
Immune Deficiency Disorders

> *It may sound presumptuous to date modern immunology from 1955 or thereabouts when every reader will be well aware that most of the great practical achievements of immunizations came well before that date . . . Effective immunization against smallpox, diphtheria, whooping cough, yellow fever, and tetanus had been in regular use for many years. Less comprehensive means of immunization against cholera, plague, typhoid fever, and influenza had also been used.*
>
> *MacFarlane Burnet (1976)*

A first golden age of immunology began toward the end of the last century. Through the work of Louis Pasteur, Robert Koch, and many others, major infectious diseases in Western countries were brought under control. For example, the smallpox virus had ravaged human populations for centuries. Around 1800, Edward Jenner, an English physician, demonstrated that immunity to smallpox could be achieved without the suffering and disfiguring pockmarks. He scratched material from the lesions caused by the related, but very mild disease, cowpox or vaccinia (from the Latin word for cow, *vacca*), into an uninfected person's skin. This process, later dubbed *vaccination* by Pasteur, was slowly accepted by medical personnel worldwide. Because the vaccines (materials containing the cowpox virus) were unstable in heat, however, smallpox remained a common, often fatal, disease in tropical countries until recent times. In 1967 the World Health Organization began a determined— and ultimately successful— global smallpox eradication program. Because the smallpox virus cannot exist for long outside the human body (except in pampered laboratory cultures), the absence of new cases anywhere in the world since 1978 means that the disease has been completely wiped out—one of many milestones in human health services.

The development of molecular biology beginning in the 1950s ushered in a second golden age of immunology, which has reshaped our understanding of the immune system. Researchers have found the mechanisms of immunity in humans and in other vertebrates to be based on a marvelously intricate web of interactions among and between cells and molecules. The processes are of practical importance, not just in the treatment and prevention of infectious diseases, but also in organ transplantation and in cancer. In addition, investigations of both normal and abnormal immune responses have revealed unusual elements of gene structure and novel patterns of gene expression. These several aspects of immune processes are discussed in this chapter.

THE IMMUNE RESPONSE

In discussing blood groups, we noted that immunological processes are characterized by three attributes: (1) the ability to recognize substances as either native to one's own body or foreign (i.e., self or nonself); (2) the ability to react specifically against any foreign antigen; and (3) the ability to remember a particular foreign antigen and to respond more strongly to it upon a later exposure.

Main Components of the System

The vital immunological properties—recognition, activation, and memory—depend on several types of **white blood cells**, which are produced in the bone marrow and enter the bloodstream. From there they are able to slip through capillary walls and wander into the spaces between tissue cells, which are filled with fluid called **lymph**, derived largely from blood plasma. The lymph drains into inconspicuous lymphatic capillaries that join into progressively larger vessels, the whole network running more or less parallel to, but separate from, the network of veins. The largest lymph vessel empties into a main vein near the heart, completing the joint blood–lymph circulation. Associated with this *lymphatic system* are several small organs: the *lymph nodes*, the *spleen*, and the *thymus*.

Lymph nodes (several dozen are spaced along major lymph vessels) and the spleen (lying behind the stomach) are rich in small, roundish white blood cells called **lymphocytes**, which respond specifically to foreign antigens. In addition, as lymph seeps through the finely divided spaces within the nodes, large, irregularly shaped white blood cells called **macrophages** engulf and digest foreign substances and interact with the lymphocytes in intricate ways. Lymph nodes in the neck may become swollen and painful during colds and other infections. The tonsils are somewhat similar to lymph nodes in structure and function.

The thymus is particularly important in the immune response. Lying in the chest cavity between the lungs, it consists largely of a mass of lymphocytes. The thymus is fully developed at birth but gradually regresses during adolescence, remaining very small throughout adult life. Experiments have shown that newborn mice whose thymus glands are removed grow fairly normally but cannot respond immunologically to some types of foreign material. Mice that are homozygous for an autosomal recessive allele (*nu/nu*, called nude because they are hairless) develop no thymus at all; they must be protected from infection in germ-free environments in order to survive for long.

Two major types of small lymphocytes with large nuclei and very few cytoplasmic organelles are involved in the immune response: the **B cells** (or **B lymphocytes**) and the **T cells** (**T lymphocytes**) (Figure 1). The functional distinction between them was first observed in chickens after removal of one or the other of two organs: the bursa of Fabricius (attached to the intestine) or the thymus. Chicks with no bursa lack B cells and fail to make antibodies, but they can still mount an immunological attack on skin grafted from a different chick. Conversely, chicks with no thymus lack T cells and fail to respond to grafted skin, but they can still produce some antibodies.

The dual nature of the immune system is seen in mammals as well, even though they have no bursa of Fabricius. The precursors of human B cells are made and mature in the bone marrow. Human T cells are also made in the bone marrow but migrate to the thymus, where they divide rapidly; finally they colonize the lymph nodes and spleen. Both B and T lymphocytes are found throughout the circulatory and lymphatic systems, some sites being particularly rich in the one type or the other.

The B and T cells can be distinguished by certain properties of their membranes (different antigens and receptor molecules, for example) and by the strengths of their responses to various foreign antigens. Human B cells respond particularly strongly to foreign antigens on the surface of invading bacteria by differentiating into antibody-producing **plasma cells** (Figure 1 and next section). In contrast, T cells of several subtypes respond particularly strongly to viral infections, to cancerous cells, and to tissue grafts. Unlike B cells, T cells do not make antibodies that circulate in blood plasma, but they do respond specifically to antigens and interact with them by direct cell-to-cell contact. In this way, T cells can bring about the death of (1) *host* cells that have been infected with a virus (thereby checking the spread of the virus); (2) *host* cells that have been transformed into cancer cells; and (3) *foreign* cells present in a tissue graft (e.g., skin or other organ, bone marrow). In cases (1) and (2), the T cells are reacting to *newly formed* (therefore, "foreign") antigens on the surface of host cells, arising by virtue of the viral infection or the malignant transformation.

Both B cells and T cells are regulated by a variety of chemical messengers called **lymphokines** (for "lymphocyte movement"), and the whole interactive system is finely tuned to produce an integrated immune response throughout the body. For example, B cells that mature to produce antibodies do so only after contact with one subtype of T cells. In addition, the phagocytic

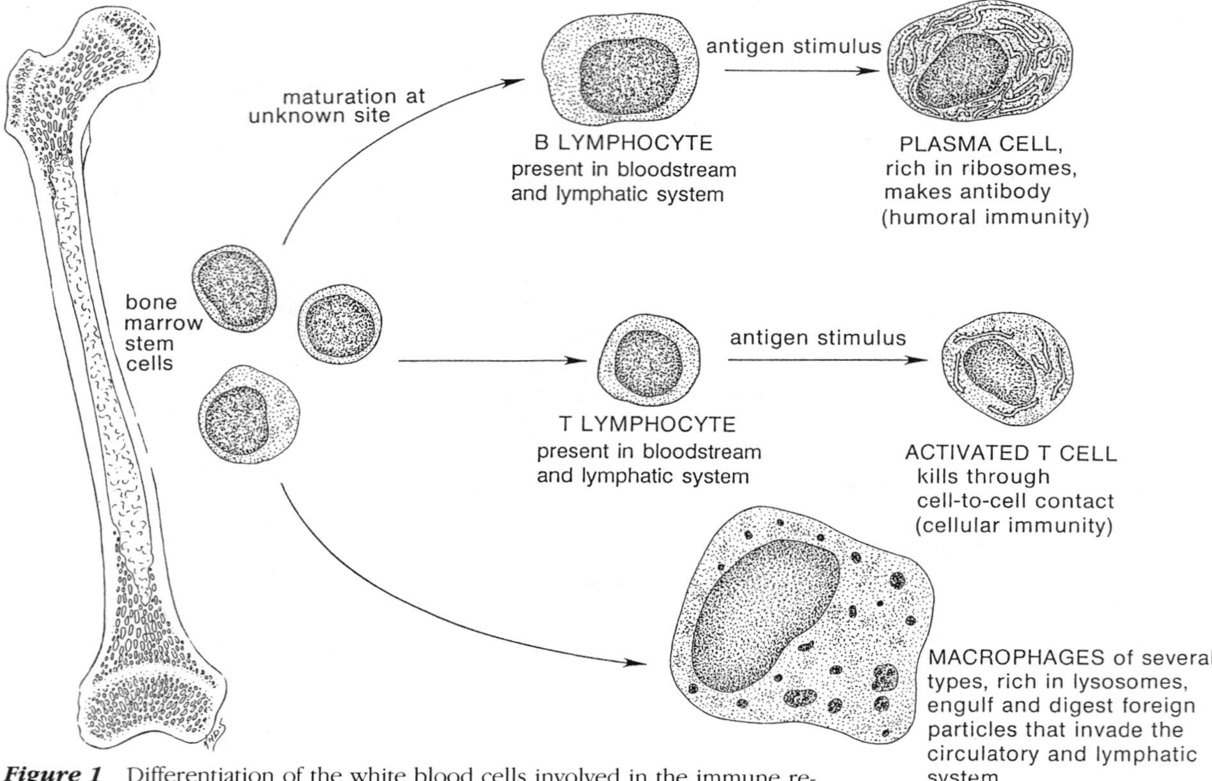

Figure 1 Differentiation of the white blood cells involved in the immune response. Although the plasma cell actually secretes antibody molecules, the T cells and macrophages are also required for initiation of antibody production. The T cells and macrophages have additional immune functions (see text). The bone marrow is also the site of manufacture of red blood cells, platelets, and other kinds of white blood cells. The cells are magnified about 4,000 times.

macrophages (which engulf foreign particles) interact with both B and T cells. One type of macrophage is stimulated to phagocytize bacteria after specific antibodies combine with the bacterial surface antigens. Another type of macrophage is needed to present antigen in a form that T cells can recognize.

The presence of antibodies circulating in solution in the blood plasma and lymph is often called *humoral immunity* (after an archaic term for a body fluid); the T cell response is called *cellular* or *cell-mediated immunity*. A foreign antigen will usually stimulate both systems of immunity, but one more strongly than the other.

The B Cell Response

By the 1950s, it was known that plasma cells congregate in large numbers at the sites of bacterial infection. These cells, rich in the organelles of protein manufacture, are the source of antibodies directed against the many different antigenic sites on the surface of the bacteria. By culturing individual plasma cells, it was possible to show that *one* plasma cell makes antibody against only *one* specific antigen (Mäkelä 1968). This result was itself surprising, but the most intriguing question about the immune response was how it could manage to identify so *many* different foreign antigens. Over the course of a lifetime, a human being must be exposed to thousands or hundreds of thousands of different nonself materials, and the immune system responds specifically to each one. Indeed, researchers have shown that plasma cells are even capable of making antibodies to *synthetic* chemicals that no vertebrate species has ever encountered over the course of evolution! The seemingly unlimited individual challenges that the immune system meets and overcomes is its most extraordinary property and is the focus of much of this chapter.

One model of the immune response that has been supported by much evidence was suggested by Niels Jerne, director of the Basel Institute of Immunology, and was developed in detail by MacFarlane Burnet, director of Melbourne's Institute of Medical Research. Both men won Nobel Prizes, Burnet in 1960 for early experimental work and Jerne in 1984 for decades of innovative contributions to modern immunology (Jerne 1985). According to Burnet's **clonal selection theory**, a lymphocyte is *preprogrammed* to recognize and respond to just one antigen, but without advance "knowledge" of whether it will ever be called on (Burnet 1962, 1968). The idea is illustrated in Figure 2, which shows a small sample of B cells. The preprogrammed antigen specificity of each cell is indicated by a (hypothetical) number. When antigen 17 is present, for example, it interacts with the surface molecules on one or a few lymphocytes specific for 17, stimulating them to divide and differentiate over a period of several days to form a clone of several hundred identical plasma cells. These mature plasma cells secrete large amounts of antibodies, **Y**-shaped molecules with two pockets at the upper tips of the **Y**. The pockets, called the **combining sites**, have shapes complementary to the shape of the initiating antigen.

Recognition of antigen 17 by the set of B cells that are preprogrammed to manufacture anti-17 antibody begins when antigen 17 joins to receptor molecules embedded in the surface of the B cells (Step 1 of Figure 3). On any given B cell, all the receptors are the same, being the **Y**-

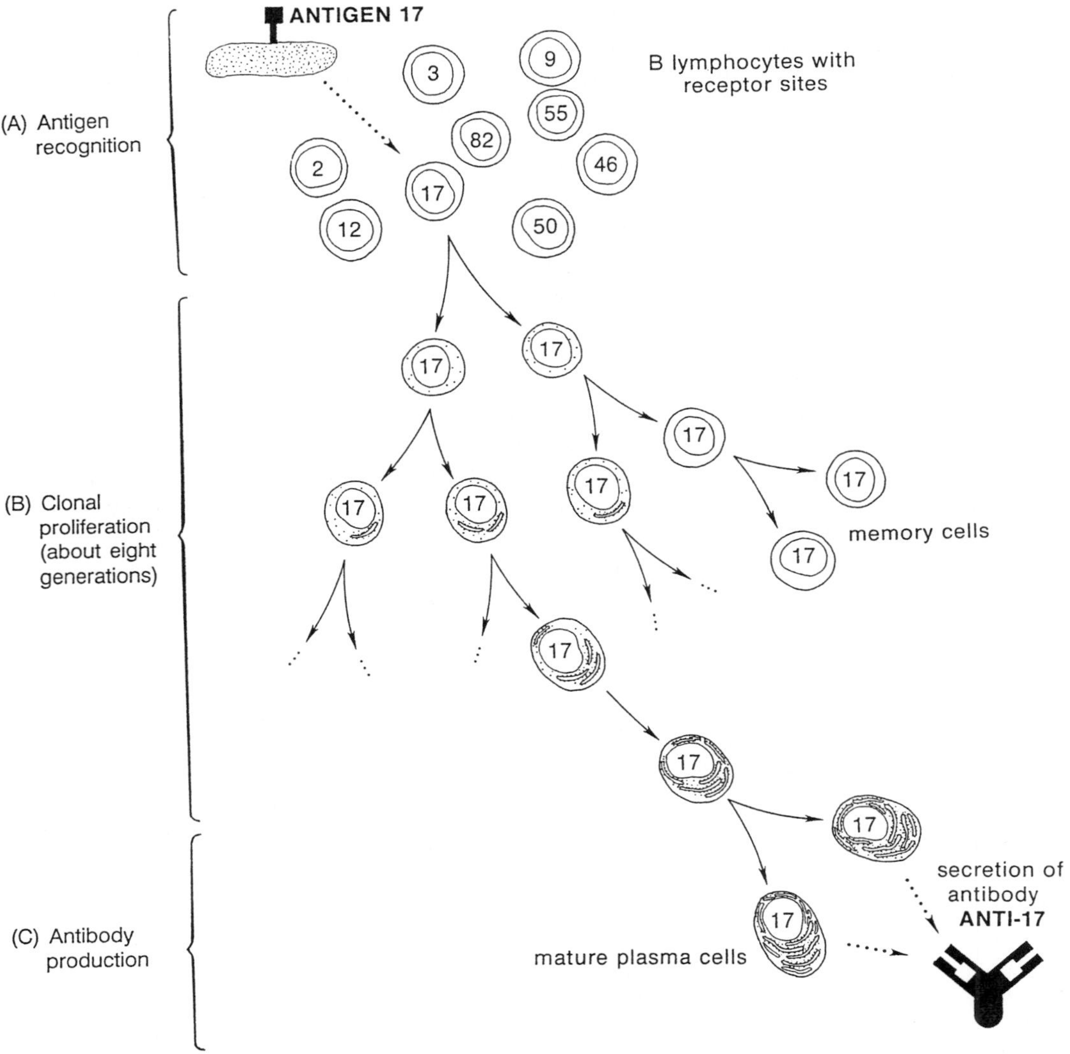

Figure 2 Burnet's clonal selection hypothesis. A particular B lymphocyte recognizes a foreign antigen (A), and proliferates (B), differentiating into antibody-producing plasma cells (C). The sizes of the antigenic determinant [i.e., antigen 17 represented by the solid, square shape on the surface of the stippled, rod-shaped bacterium in (A)] and of the antibody molecule [in (C), with the matching square holes] are much exaggerated compared with the cell sizes. Greater detail within antigen recognition (A) is shown in Figure 3.

shaped antibody molecules (with the combining sites protruding outward) that the mature plasma cell will eventually secrete in large amounts. At this stage, however, the B cell is not secreting any antibody into the extracellular space; it is just displaying its name tag so that it can be greeted accurately.

Another controlling event is needed before the B cell is able to proliferate and mature: antigen 17 must be processed, moved, and displayed on the surface of the B cell in conjunction with another protein (Step 2 in Figure 3). This membrane protein is synthesized by the B cell and is called MHC II, an abbreviation for the Class II proteins encoded by the Major Histocompatibility Complex on chromosome 6. (MHC proteins are discussed more fully in later sections.) This association of antigen plus MHC protein is then recognized by a specific type of T cell called a *helper T cell* (which had been previously stimulated by the same antigen). The helper T cell then secretes a hormone-like chemical (lymphokine interleukin-2) that stimulates the B cell to multiply and differentiate (Step 3). The complexity of this recognition system apparently helps the immune system discriminate between foreign material and the body's own substances.

A week or so after initial contact with bacterial antigen 17, the responding plasma cells begin to secrete antibodies into the fluid portion of blood and lymph (Figure 2C). The combining sites of the antibodies recognize and join with antigen 17 on the surface of an invading bacterium. When the tips of antibody 17 are bound to antigen 17, the stem of the Y-shaped antibody molecule sticks out from the surface and can react with receptor sites on the surface of macrophages. This reaction stimulates the phagocytic action of the macrophages and the bacterium is engulfed and destroyed.

Alternatively, the antigen–antibody complex that forms on some cell surfaces invites other plasma proteins, called *complement*, to join up. Complement proteins make holes in the cell membranes of the bacteria, thereby causing them to lyse. (This destructive chain of events also occurs during hemolytic diseases of the newborn discussed in the last chapter. The cells destroyed in HDN, however, are the red blood cells, not bacterial cells.) Thus we see that antibody does not itself destroy the invading bacteria. Rather, antibody molecules mark cells for destruction by other agents.

The antibody-producing plasma cells of a selected clone do not live long after the removal of the stimulating antigen. Other members of the same clone that have not fully differentiated, however—called **memory cells**—can survive for several years. Once primed by a given antigen, they will initiate the so-called *secondary immune response* when they meet the same antigen again. This response is characterized by a quicker, stronger production of antibody, primarily because more plasma cells are produced. The booster shots given in an immunization program have the effect of renewing the supply of memory cells.

Although most of the information about the B cell response comes from research over the last few decades, the conceptual father of the clonal selection idea was the German medical researcher, Paul Ehrlich, a Nobel laureate in 1908. Ehrlich (1900) proposed that antigen molecules united with specific receptors attached to a cell surface, stimulating the cell to produce more and more receptors—so many that the receptors were secreted into the blood as free antibodies (Figure 4). At the time, Ehrlich's ideas appeared to be incompatible with findings that vertebrates could respond specifically to many thousands of different foreign antigens.

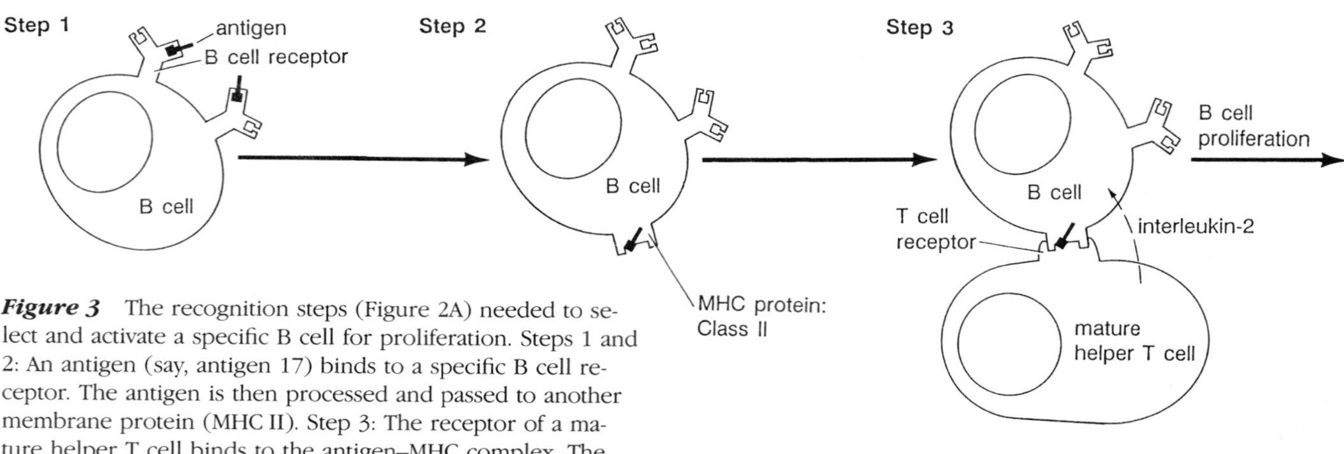

Figure 3 The recognition steps (Figure 2A) needed to select and activate a specific B cell for proliferation. Steps 1 and 2: An antigen (say, antigen 17) binds to a specific B cell receptor. The antigen is then processed and passed to another membrane protein (MHC II). Step 3: The receptor of a mature helper T cell binds to the antigen–MHC complex. The helper T cell then secretes a lymphokine (interleukin-2) that stimulates the B cell to divide and differentiate into a plasma cell.

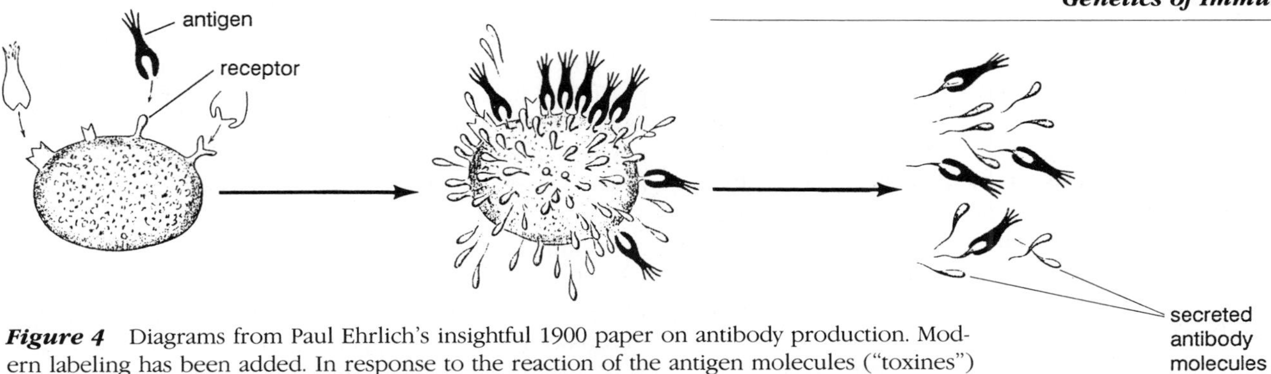

Figure 4 Diagrams from Paul Ehrlich's insightful 1900 paper on antibody production. Modern labeling has been added. In response to the reaction of the antigen molecules ("toxines") and cell surface receptors ("side-chains"), the cell manufactures the receptors in excess and secretes them as antibodies ("antitoxines"). Note that the spatial configurations of antigen and antibody are the reverse of those depicting the modern conception (Figure 3); that is, the antigen here is depicted more like a "lock" than a "key."

The T Cell Response

Like B cells, T cells also respond by clonal selection, multiplying and differentiating after they are specifically stimulated. Showing the same high degree of specificity, T lymphocytes have on their surfaces recognition molecules, called **T cell receptors**, able to make the necessary fine distinctions between foreign antigens (Marrack and Kappler 1986; Tonegawa 1988). Although the molecules that are called T cell receptors are *different* from the B cell receptor molecules, the two systems share some amino acid sequences and are presumed to be evolutionary products of a common primordial molecule. As we have seen, the B cell receptors are membrane-bound forms of the antibody molecules that the B cells will eventually secrete when they are mature. The T cell receptor molecules, by contrast, always remain embedded in the cell membrane; the T cells *never secrete* any antigen-specific molecule, performing their functions instead by cell-to-cell contact.

The antigens that stimulate immature T cells to proliferate into a clone do not join directly to the T cell receptors for which they are specific. Rather, the antigens are first phagocytized by macrophages, processed in poorly understood ways, and displayed on the macrophage surface in conjunction with a protein encoded by the major histocompatibility complex (MHC I or MHC II; see later). Only then is the antigen in a form that the T cell receptor can recognize (Figure 5). This complicated system of recognition is similar in some ways to that described above for the stimulation of B cells by mature helper T cells (compare Figures 3 and 5).

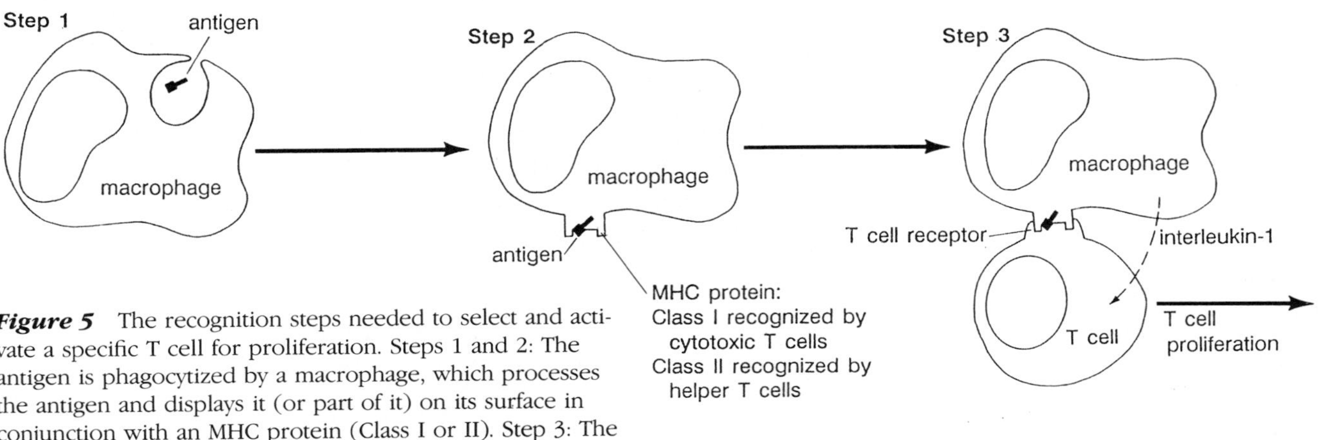

Figure 5 The recognition steps needed to select and activate a specific T cell for proliferation. Steps 1 and 2: The antigen is phagocytized by a macrophage, which processes the antigen and displays it (or part of it) on its surface in conjunction with an MHC protein (Class I or II). Step 3: The receptor of an immature T cell binds to the antigen–MHC complex. The macrophage then secretes a lymphokine (interleukin-1) that stimulates the T cell to divide and differentiate.

An additional complexity of the T cell system is the existence of several different types of T cells: **cytotoxic T cells** do the actual killing achieved by the T cell system; **helper T cells** and **suppressor T cells** modulate the activities of both B cells and other T cells. For example, lymphokines secreted by T cells induce both B cells and T cells to divide and differentiate.* Let us illustrate one mode of action of cytotoxic T cells. Assume that the immune system is stimulated by antigens of the influenza virus. After the virus enters susceptible host cells, antigens characteristic of the virus are displayed in the membrane of the infected cells. The clone of mature cytotoxic T cells that is formed against influenza antigens by the processes described above will then bind to the membrane of infected cells at the sites of the viral antigens. This reaction leads to the death of the infected cells and any influenza viruses they contain.

ANTIBODY STRUCTURE AND GENETICS

Collectively, antibodies are called **immunoglobulins** (abbreviated **Ig**), a term describing many similar types of immune molecules constituting about 20% of all plasma proteins. Because the antibodies secreted by B cells are so varied in their antigen-combining abilities, protein chemists were long intrigued by the differences in their structures. But the chemists were hampered by the large size of the molecules and by an inability to purify enough of any one antibody to analyze it chemically. This problem was solved by the discovery that persons suffering from *multiple myeloma* (a cancer of bone marrow) had in their plasma large amounts of a *single* antibody species—comprising up to 95% of all their immunoglobulins. Because cancers usually begin with the uncontrolled proliferation of one cell, scientists thought that the peculiar symptom— production of immunoglobulin with essentially one specificity—began with a single malignant antibody-producing plasma cell. Thereafter, the analysis of the structures of various immunoglobulins was accomplished by several groups of researchers including Gerald M. Edelman (1973) at Rockefeller University and Rodney R. Porter at the University of Oxford, Nobel laureates in 1972.

*Because helper T cells occupy such a central regulatory role in the immune system, their preferential destruction by the AIDS (acquired immune deficiency syndrome) virus leaves AIDS victims susceptible to a variety of infections and malignancies (Laurence 1985).

Antibody Structures

The antibody researchers were able to show that the major form of antibody, **IgG** (= **immunoglobulin G** = **gamma globulin**), consisted of four polypeptide chains connected by covalent bonds between the sulfur atoms of cysteine molecules (Figure 6). Each of the two identical **heavy (H) chains** has about 440 amino acids divided into four regions of about 110 amino acids each. The regions, called **domains**, have similar, but not exactly the same, amino acids. (They are depicted by the elliptical outlines in Figure 6.) Domains similar to those in the heavy chains also occur twice in each of the two identical **light (L) chains**, which are present in the arms of the Y-shaped molecule. The attachment of the arms to the trunk of the Y is somewhat flexible.

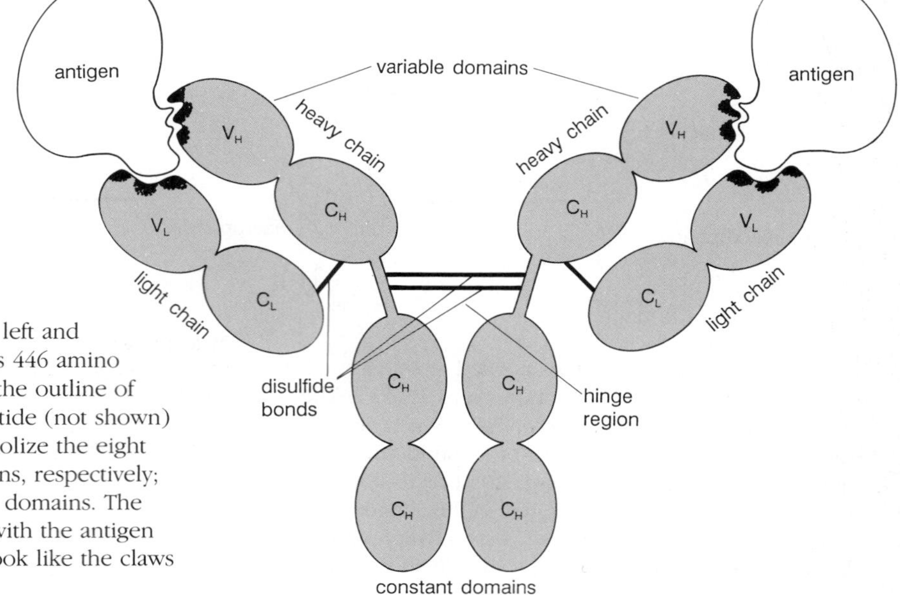

Figure 6 The Y-shaped IgG antibody. The left and right halves are identical. Each heavy chain is 446 amino acids long and each light chain, 214. Within the outline of each chain, the actual folding of the polypeptide (not shown) is quite complicated. The C_H's and C_L's symbolize the eight constant domains of the heavy and light chains, respectively; the V_H's and V_L's symbolize the four variable domains. The parts of the variable domains that combine with the antigen are shown in black; diagrammatically, they look like the claws of a lobster.

Comparison of the amino acid sequences of the *light* chains from *different* myeloma patients revealed a remarkable pattern: within the domain at the upper tips of the Y, about 25% of the amino acids in one patient differed from those in another patient; thus this region was called the **variable (V) domain** (Wu and Kabat 1970). In the other half of the light chain, the amino acid sequences were exactly the same from patient to patient (with one minor exception that we ignore); thus this region was called the **constant (C) domain** (Figure 7). When the *heavy* chains from different myeloma patients were sequenced, the domains at the tips of the Y's were also found to be variable; the remaining three domains were constant. Researchers reasoned that the amino acid differences in variable domains must relate to the antigen specificity of the clone of plasma cells from different myeloma patients. Interestingly, most of the variation in amino acids within variable domains is restricted to just three limited subregions totaling 20–30 amino acids, rather than being spread out over all 110 amino acids. These so-called **hypervariable regions**—three in each light chain and three in each heavy chain—form the lining of the antigen-binding cavity (Figure 8). Antibody specificity arises from the three-dimensional shape of the cavity and from the particular chemistry of the hypervariable amino acids. The surfaces of the antigen and the heavy and light chains conform like parts of a jigsaw puzzle, but the fit may be either snug or loose. Thus the affinity of a given antibody for a given antigen may vary.

IgG, whose structure is described above, constitutes about 80% of the immunoglobulin molecules. Altogether, five immunoglobulin classes (G, A, M, D, E, in the order of their concentration in plasma) have been analyzed: each has different amino acids in the *stem* region of the heavy chains, giving the five classes somewhat different functions. For example, IgG readily crosses the placenta and provides protection to newborns before they develop their own immunological competency. The IgA class is found not only in plasma but also in body secretions, such as saliva, nasal mucus, sweat, and breast milk. IgM molecules have somewhat larger heavy chains than the other classes do, and five of the Y-shaped molecules may be joined together in a circle. Because each IgM antibody then has ten combining sites, IgM molecules can have a property described as high *avidity*; that is, they bind more firmly to antigens. The function of IgD is not well known, and IgE is involved in allergic reactions.

It is important to note that antibody of any of the classes G, A, M, D, or E may have the same combining site, directed against, say, antigen 17. That is, the five functional classes of antibodies defined by the *stems* of the Y's can have exactly the same antigen specificity at the *tips* of the Y's. During an immune response, a B cell changes the class of antibody that it synthesizes, but each different class will have the same antigenic specificity (see *class switching* later in this chapter).

Antibody Diversity

As a group, the B cells of a person can make a million or more different IgG antibodies, each characterized by the distinctive array of amino acids lining the pockets where the antigens bind. Like any protein, antibodies are coded for by genes. Yet we could not possibly have 10^6 genes, each one coding for the amino acid sequence of a different antibody, because that is more genes than we possess! The geometry of each combining site is formed, however, by *two* polypeptide chains, one heavy and one

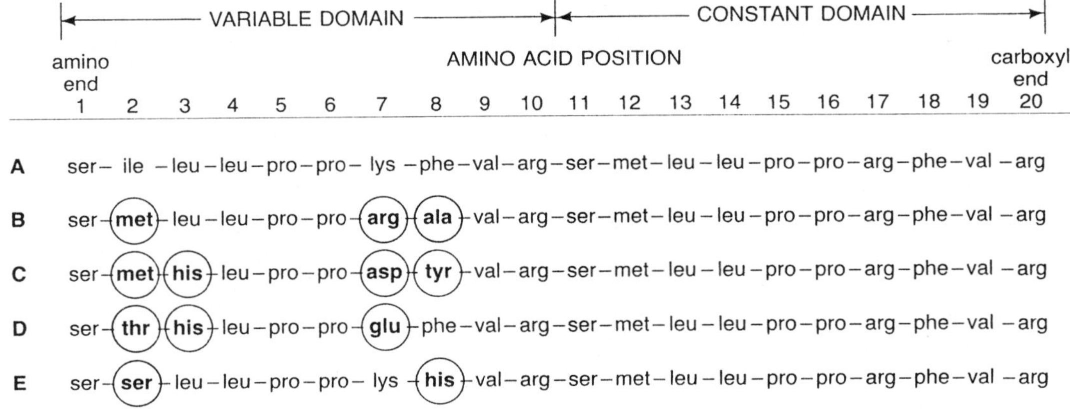

Figure 7 The meaning of variable and constant domains. Five hypothetical molecules (A, B, C, D, and E) are pictured, each with 20 amino acids. In the left half of the molecules, the variable domain, amino acid differences exist at positions 2, 3, 7, and 8. Molecule A is taken as standard; the amino acids that differ from A are boldfaced and circled. In the right half of the molecules, the constant domain, the amino acid sequences are identical to each other in all positions, and somewhat similar to the variable domain sequences.

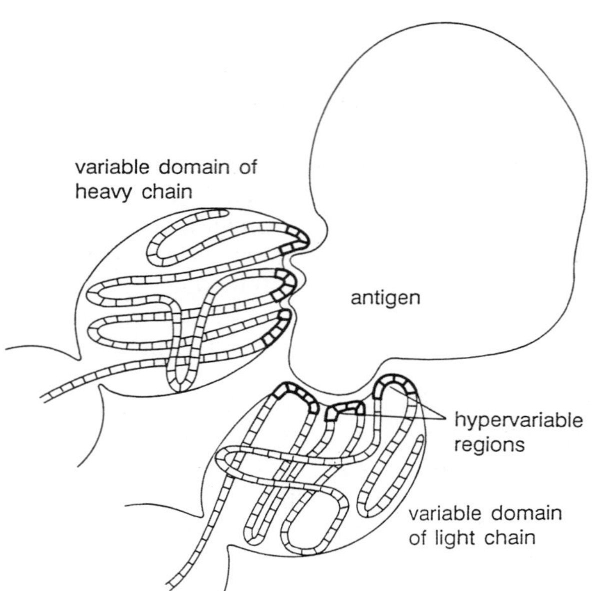

Figure 8 One combining site of an antibody. The specific looping pattern of the polypeptide chain within the variable domains of the heavy and light chains causes certain amino acids (indicated by heavy outlines) to line the pocket binding the antigen. These amino acids are in hypervariable regions, which contain most of the differences between antibodies of different specificities. The actual folding of the polypeptides is more tortuous than is indicated here.

light. Thus, if there were 10^3 different H chains (from 10^3 genes) and 10^3 different L chains (from another 10^3 genes), then all possible combinations of the two proteins would generate $10^3 \times 10^3 = 10^6$ different spatial patterns. Thus 2,000 antibody genes might suffice.

Actually, recent research shows that the enormous repertoire of antibody molecules is encoded by many fewer genetic elements: in the range of several hundred. The primary solution to this number puzzle is simply stated: we inherit and pass on in germ cells several *separated* DNA sequences that together code for the combining sites. To generate diversity, the pieces are shuffled around and combined in different ways during the formation of individual B cells. By analogy, imagine outfitting a doll with a hat, shirt, and slacks. If the doll has 10 different hats, 10 different shirts, and 10 different slacks (30 elements), you will be able to arrange a thousand (10 × 10 × 10) different styles (each befitting a particular occasion, perhaps). With a *second* doll and an *additional* 10 hats, 10 shirts, and 10 slacks, you can generate another thousand styles. All together, from just 60 unique elements, you will be able to assemble a million different pairs of outfits! In a somewhat similar multiple choice situation, the various DNA segments constituting the parts of antibody genes are selected and brought together as B cells develop from undifferentiated stem cells.

DIVERSITY THROUGH SOMATIC RECOMBINATION. A conceptually simple experiment with a restriction enzyme first showed that genetic elements move around during B cell maturation (Hozumi and Tonegawa 1976). The experimenters compared DNA from early embryonic mouse cells with DNA from mature plasma mouse cells that were making a specific antibody. They used *Bam*HI to cut up the DNA samples (cuts were made wherever the base sequence GGATCC occurred), electrophoresis to spread out the restriction fragments according to size, and radioactive probes to identify either the *V*-encoding part or the *C*-encoding part of the L chain gene. They found that in mouse embryos, one restriction fragment contained the *V* sequences, and a *different* fragment contained the *C* sequences; thus a *Bam*HI site must have occurred between them. On the other hand, in DNA from mature plasma cells, the *V* and *C* elements were on the *same* restriction fragment, one that was smaller than either of the embryonic fragments. The experimenters concluded that a substantial piece of DNA, including a *Bam*HI site, had been cut out during cell differentiation.

The shuffling is called **somatic recombination**, a process that has been studied by several researchers, including Susumu Tonegawa (a Nobel Prize winner in 1987) then at the Basel Institute of Immunology (Tonegawa 1988). He identified and mapped the DNA elements that undergo somatic recombination by cloning the antibody genes in λ phage and determining the base sequences of cloned segments (Chapter 15). Most of his work was done with mouse DNA, and only recently have the details of human antibody genes become known. Researchers have found that an extended region near the tip of the long arm of human chromosome 14 codes for the heavy chain. The top line of Figure 9 shows the general arrangement of heavy chain DNA elements as they occur in the germline and early embryonic cells. Note that there are four families of elements: ***V*** (variable), ***D*** (diversity), ***J*** (joining), and ***C*** (constant). During the maturation of a particular B cell, intervening DNA regions are cut out in order to splice together by somatic recombination *one* member of *each* family to yield a unique ***V-D-J-C* gene** (second line of Figure 9). The gene is then transcribed and translated into a heavy chain that joins with a light chain (from a somewhat similarly processed light chain gene) to make an antibody.

The combined *V-D-J* portion of the heavy chain gene codes for the variable domain (including the hypervariable regions that characterize the antigen specificity), and the *C* portion codes for all of the constant domains (that determine the class of the antibody: IgG, IgM, IgA, IgD, or IgE). The number of different *V-D-J* regions that a cell can construct is the product of the number of *V*, *D*, and

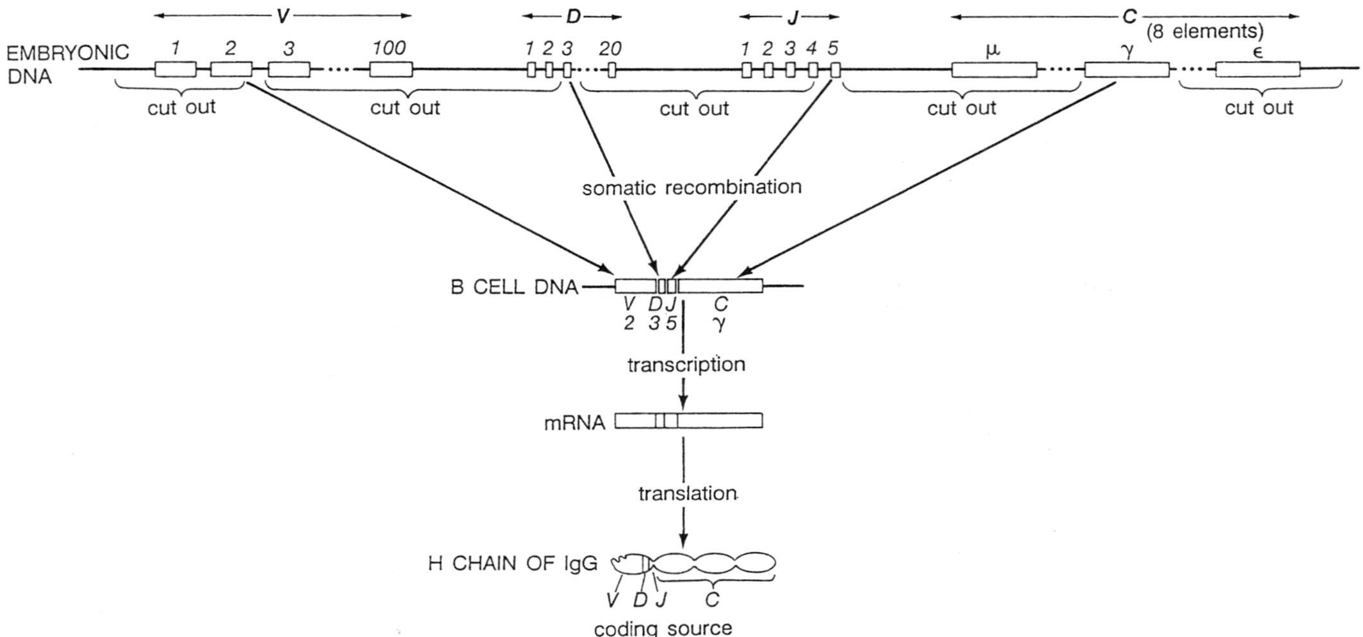

Figure 9 DNA elements on human chromosome 14 code for the immunoglobulin heavy chain. Embryonic DNA (top line) contains four families of genetic elements: *V*, *D*, *J*, and *C* (not drawn to scale). By somatic recombination, one element from each family joins together into a single *V-D-J-C* gene (second line) that is transcribed and translated into a specific polypeptide (third and fourth lines). The *V-D-J* portion codes for the variable domain, and *C* codes for the constant domains. The *V* nucleotides code for about 90 amino acids, including two hypervariable regions. *D* and *J* each code for about 10 amino acids, including (together) the third hypervariable region. *C* codes for the 330–440 amino acids of the constant domains. The recombination events that are shown join together the second *V* element, the third *D* element, and the fifth *J* element. These are depicted as joining with a γ constant region to form the heavy chain of an IgG molecule. Eight (or possibly more) constant elements include four slightly different ones (not shown) coding for the heavy chain of IgG. The many introns within the genetic elements are also not shown, nor are control regions (such as the signal sequences governing recombination sites).

J elements. For humans, the numbers are not known precisely, but they are approximately 100 *V*, 20 *D*, and 5 *J* elements. On the basis of these estimates, 100 × 20 × 5 = 10,000 different heavy chain variable domains can be made.

The process of splicing is apparently aided by complementary nucleotide sequences just beyond the ends of the DNA elements that are joined. The unwanted sequences probably loop out of the way and are deleted by currently unknown enzymes (Figure 10). Keep in mind that this process, like that of intron excision, requires exact joining in order to maintain the proper reading frame for protein synthesis.

STILL MORE DIVERSITY. The number of different heavy chain variable domain polypeptides is certainly more than the 10,000 predicted by the combinatorial lottery. The greater possibilities result from two other unusual phenomena: **imprecise *D-J* joining**, and **somatic mutation** (Golub 1987a). In the former process, the last few bases (near the 3′ end) of a selected *D* element may replace the first few bases (near the 5′ end) of the following *J* element, depending on the specific site of recombination (Figure 11). The base sequence in the completed gene is therefore variable near the joining site and may thus code for several different amino acids.

In the third diversifying process, base substitution mutations occur in one or more of the *V*, *D*, or *J* elements during the differentiation of a B cell into a mature plasma cell, that is, after stimulation by a particular antigen. The rate of these mutations during clonal proliferation is several orders of magnitude greater than the rate in the

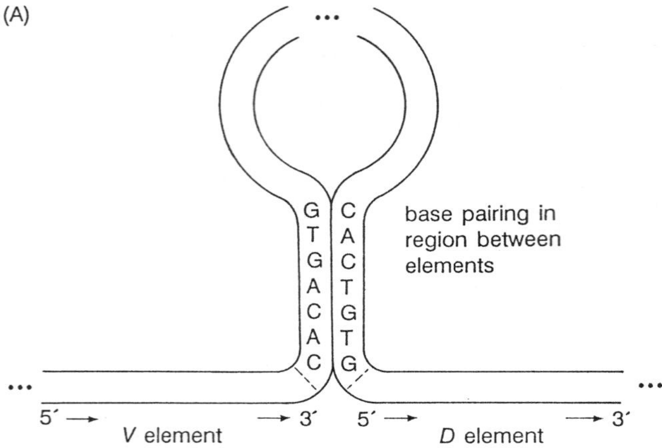

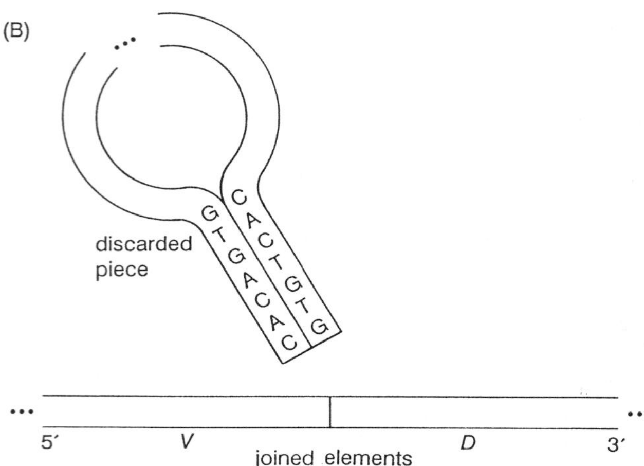

Figure 10 The probable mechanism for joining together *V* and *D* elements. (A) Just beyond the 3′ end (right end) of a *V* element, bases pair with those before the 5′ end (left end) of a *D* element. Pairing between the reverse complementary sequences brings the ends of the *V* and *D* elements next to each other. (B) The *V* and *D* elements are enzymatically joined and the loop is cut out. The particular base sequence shown is based on mouse data (From Tonegawa 1983.)

germline itself. Although their mechanism of occurrence is unknown, some base substitutions may lead to variant antibodies that fit better with the antigen. Tonegawa (1983) has suggested that such B cells may multiply more quickly. Thus the several hundred plasma cells derived from a single antigen-selected B lymphocyte may represent many different clones that make antibodies with slightly different antigen-combining ability.

In summary, the extraordinary diversity of antibodies that can be assembled by the immune system arises from

1. multiple recombinations of a moderate number of germline elements; this process occurs as bone marrow stem cells develop into differently programmed B cells
2. imprecision at the sites of recombination
3. somatic mutations occurring as the B cells, stimulated by an antigen, develop into antibody-secreting plasma cells

The germline elements consist of families of related base sequences that have arisen over evolutionary time by repeated duplication events (due perhaps to unequal crossing over) followed by independent mutations in the different units. The discovery that the variable DNA elements are physically separated from the constant DNA elements resolves the evolutionary paradox of having an immunoglobulin chain with many amino acid substitutions toward one end, but few toward the other. The general answer, as we have noted, is that several genes can cooperate to code for one polypeptide. One gene (or "minigene" or "genelet") is selected over time for variation, and another for conservation. This idea was

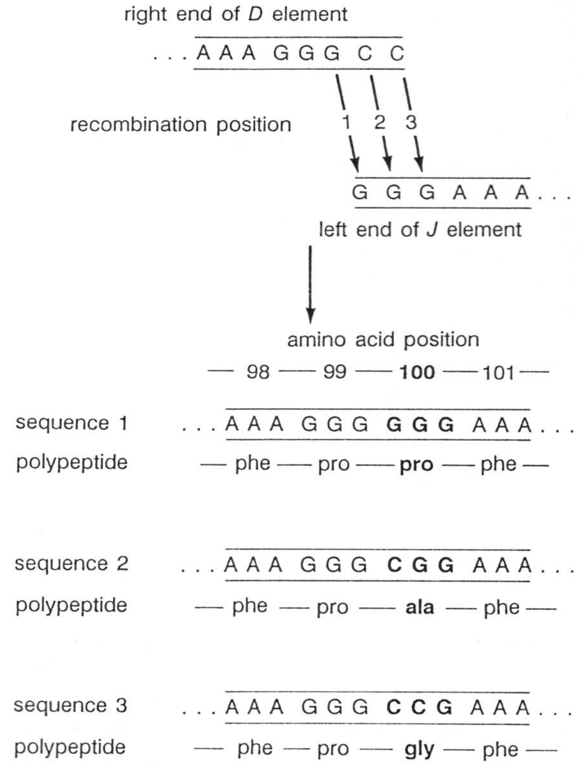

Figure 11 Imprecise joining of a *D* and *J* element at the triplet coding for amino acid number 100 (within the third hypervariable region). Depending on the exact placement of the recombinational event at position 1 or 2 or 3, the 100th triplet will code for proline, alanine, or glycine. The particular sequence shown (in the sense strand of DNA) is hypothetical.

suggested decades ago by W. J. Dreyer and J. Claude Bennett (1965) of the California Institute of Technology. Although their novel *two genes-one polypeptide* theory was not well received when proposed, it has withstood the test of time.

OTHER UNUSUAL PHENOMENA. We noted in Figure 9 that the DNA for the heavy chain includes eight elements encoding the *constant* domains. The B lymphocyte at first connects a specific *V-D-J* genetic element with the μ constant region element to make IgM. This is the antibody whose stem is anchored in the membrane of the cell with the arms of the Y outstretched to act as an antigen receptor. Later, after the B cell differentiates into an antibody-producing plasma cell, the IgM that is synthesized *and secreted* lacks the terminal amino acids that had previously anchored it in the membrane. The two forms of IgM depend in part on mRNA processing that includes or excludes a stop codon (Leder 1982). Still later, the plasma cell switches production to IgG or another class of antibody—a phenomenon appropriately called **class switching**. Note that the same *V-D-J* element is joined with one or another of the constant region elements; thus antibodies of any class produced by one plasma cell have the same antigen specificity.

We note one additional puzzle of antibody formation that involves the fact of diploidy. Although we possess two chromosomes 14, and therefore two extended heavy chain encoding regions, only one is expressed in any given plasma cell. The process is called **allelic exclusion**.

Although we have described the generation of variability only for the H chains of antibodies, similar processes operate in the production of the L chains. (The degree of light chain variation is less, since a *V* element joins directly with a *J* element—no family of *D* elements exists.) We also omit a description of the antigen receptor present on T cells, a two-chain molecule that bears some homology to antibody molecules and is synthesized by somewhat similar DNA juggling (see Steinmetz 1986).

Monoclonal Antibodies

A single antigen injected into a mouse will generally raise a mixture of antibodies, each recognizing a slightly different aspect of the antigenic surface. If the antigen is complex, like a whole virus or a cell with myriad proteins, lipids, and carbohydrates on its surface, then dozens or hundreds of antibodies will be made, each by a different clone of plasma cells. In an in vivo situation, all the antibodies mix together in the blood plasma and other fluids, where they are chemically inseparable. A special in vitro method of making very homogeneous antibodies with predetermined combining abilities has generated much interest in recent years because of its demonstrated and potential value in the diagnosis and treatment of disease (Diamond et al. 1981). Each pure and uniform antibody, called a **monoclonal antibody**, is secreted by a clone of cells that can be grown in virtually unlimited numbers. The special clone is derived by fusing together an antibody-secreting *plasma* cell with a *cancer* cell. Because plasma cells by themselves do not continue to live in culture but cancer cells do, the hybridization is required to render the plasma cell immortal. The plasma cell-cancer cell combination is called a **hybridoma**. The process of producing hybridomas is a variant of the somatic cell fusion technique described in Chapter 15.

Monoclonal antibodies were developed at Cambridge University by a West German, Georges Köhler, and an Argentine, Cesar Milstein (1975), who were awarded Nobel Prizes for this work (Milstein 1986, Köhler 1986). The basic procedure for producing monoclonal antibodies is outlined in Figure 12. A mouse is immunized with the antigen of interest in order to stimulate the proliferation of plasma cells that make antibodies against the antigen (Figure 12A). Repeating the injection produces a heightened secondary immune response. The mouse's spleen, containing a mixture of lymphocytes, is removed, and a suspension of spleen cells is combined with cells derived from a mouse myeloma (Figure 12B). The myeloma cells were previously isolated from a tumorous bone marrow cell that, initially, produced its own antibodies. During continued transfer in culture, however, the myeloma cells were selected to be deficient in both the production of antibodies and in one of the enzymes of the nucleotide salvage pathway (*HPRT*$^-$).

As in the procedure described in Chapter 15, cell fusion occurs in HAT medium, which stops the growth of unfused *HPRT*$^-$ myeloma cells. The unfused spleen cells also do not grow well in culture medium. But any hybrid cells that form—acquiring the *HPRT*$^+$ allele from the spleen cells and cancerous growth traits from the myeloma cells—rapidly outgrow both parental cell lines (as shown in Figure 19B in Chapter 15).

Several hundred fused cells may be formed from the spleen of one mouse. The resultant clone of cells from each hybrid is isolated and tested for the presence of antibody against the original immunizing antigen (Figure 12C). Any positive lines—there may be none, or at best a small number—can be frozen for later use, or propagated in culture medium or inside living mice as myeloma tumors (Figure 12D). Note that the monoclonal protocol yields a homogenous crop of antibodies, however complex the stimulating antigenic substance. Each isolated hybridoma clone arises from a single plasma cell from the mouse's spleen, and each plasma cell secretes only a single specificity of antibody molecules.

Hybridomas have been put to many practical uses. For

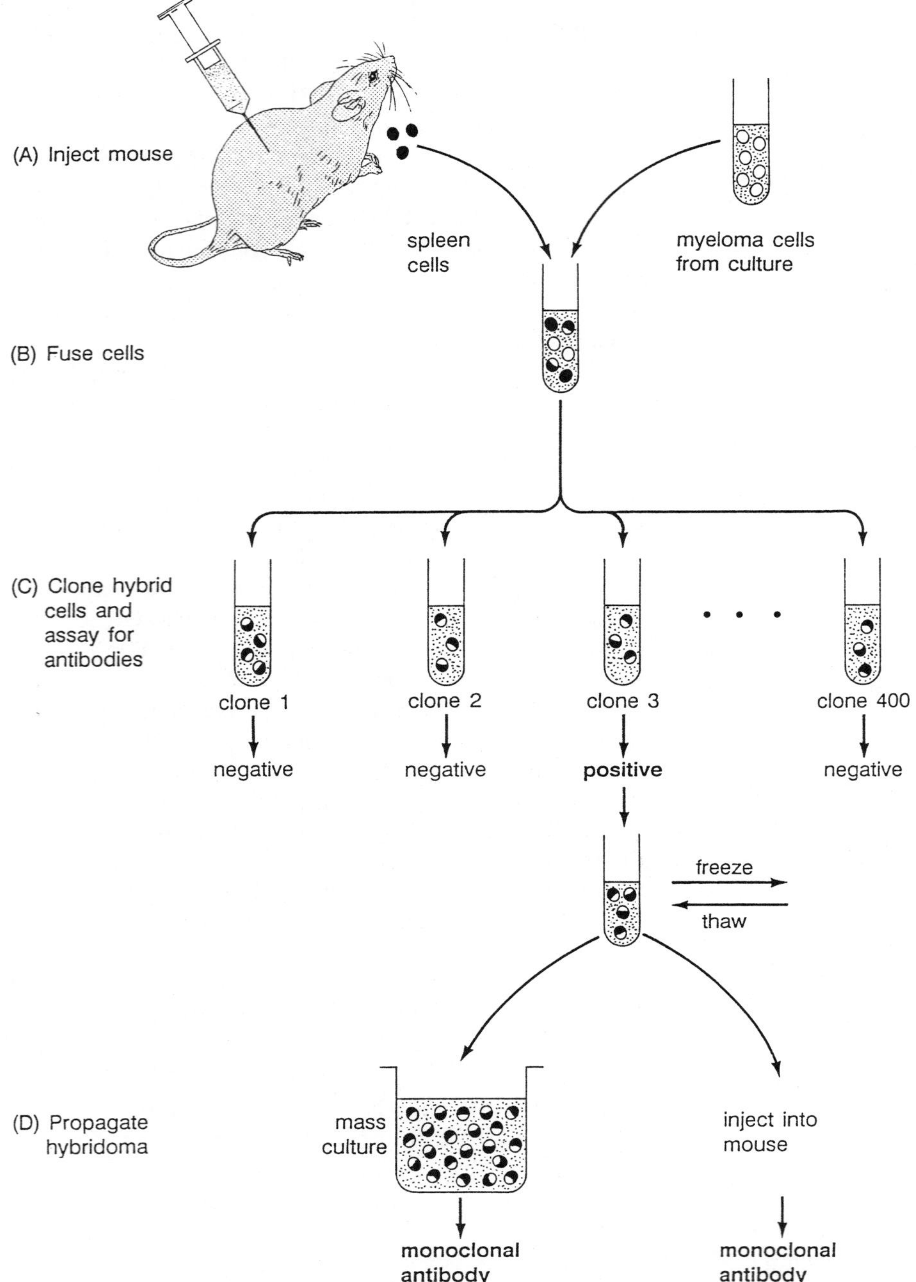

Figure 12 The standard method for deriving long-lived cells that make monoclonal antibodies having a predetermined specificity. See text for further details. (A) Inject and reinject a mouse with an antigen against which an antibody is desired; later, remove its spleen. (B) Fuse the spleen cells (including, it is hoped, specific antibody-producing plasma cells) with mouse myeloma cells derived from a culture. Only the hybrid cells will grow because unfused (parental) spleen cells are short-lived in culture and unfused (parental) myeloma cells die in HAT medium. (C) Clone the hybrid cells individually, assaying the medium in which each clone is growing for the presence of antibody directed against the original immunizing antigen. (D) Grow the hybridomas (antibody-producing hybrid cells) in tissue culture for long periods of time, or inject them into mice to form myelomas that continue to produce pure monoclonals.

example, researchers routinely use monoclonal antibodies to type red blood cells (Chapter 17), to distinguish between the various subtypes of T cells (cytotoxic, helper, suppressor), and to type white blood cells for the HLA antigens (tissue typing; see next section). The clinical use of monoclonal antibodies for diagnosis and therapy is developing but is still largely experimental. Researchers hope that monoclonal antibodies can be used to identify the cancerous cells in children with leukemia, for example. Because lymphocytes acquire a number of tumor-associated antigens in the process of becoming malignant, they are vulnerable to immunological attacks. Recently, researchers have been trying to couple the monoclonal antibodies directed against a tumor antigen with a toxic chemical to generate a so-called magic bullet, an **immunotoxin** that will kill the malignant cells but no other cells (Collier and Kaplan 1984). Cancer therapy of this type, although still hampered by technical difficulties, would be a great improvement over current chemotherapies or radiation treatments, which kill all rapidly dividing cells, including normal ones.

A problem with the immunotoxin procedure, however, is that the best monoclonal antibodies are made by *mouse* cells, not human cells. As such, they are foreign to humans and can cause a recipient to make antibodies against the monoclonal antibodies, thus destroying their intrinsic value and producing undesirable side effects. For this reason, investigators have been trying to produce monoclonal antibodies derived from a human plasma cell fused with a human myeloma cell (Pinsky 1986). A major limitation is that one cannot arbitrarily inject people with an antigen of interest and later harvest a sample of spleen cells (Figure 12). (Sometimes spleens are surgically removed for unrelated reasons; or useful lymphocyte cell lines can be grown in culture.) Nonetheless, some progress has been made, including the genetic engineering of mice so that they produce antibody sequences characteristic of *other* species (Strelkauskas 1987).

TRANSPLANTATION ANTIGENS

The Australian immunologist G. J. V. Nossal (1978) has written,

> There is perhaps something ghoulish in the thought that immediately after a person's death, a whole tribe of surgeons may descend on the body ready to remove kidneys, heart, liver, lungs, pancreas, and any other bits and pieces that they can get hold of. This block is largely in our minds. Most people in good health would agree that their own organs would be better employed, after their death, in helping to keep another person alive than in being burnt in a crematorium or buried in a coffin.

It is, of course, not the role of medicine to confer immortality by continually replacing worn-out parts. Rather, each of us hopes to live with a minimum of ill health and, after some scores of years, die with dignity. In achieving these goals, the transplantation of organs has a limited but increasing role. For an individual who might lead a productive life except for damage to one vital organ, a transplant operation is sometimes a last but logical recourse. The most common organ to be transplanted, and the one with the most hopeful prognosis over a long period of time, is a kidney. But this operation requires more than a skilled surgeon working in a sophisticated medical milieu. The body's "search and destroy" immune system must also be taken into account. If the new kidney's degree of foreignness is too great, reaction by the recipient's T lymphocytes will lead to the destruction of the intrusive tissue.

To improve the chances of a "take," two avenues are open. One is to repress the normal activity of the recipient's immune system. Drugs are available that interfere with the normal functions of T lymphocytes, but side effects may be serious. The patient's ability to ward off infection is compromised, a deficiency that may lead to a fatal infection shortly after a transplant operation, or to cancer later in the recovery period. The second method for thwarting the immune system is to choose a donor organ with a minimum degree of foreignness. This matching is done by tissue typing the recipient and potential donors for so-called transplantation antigens. By a combination of antigenic matching, surgical expertise, and immune suppression, a transplanted kidney today has a good chance of functioning for many years.

Tissue Typing: The HLA System

Knowledge of transplantation antigens was first obtained in the 1940s from grafts of skin or tumors between different inbred strains of mice. The fate of transplanted tissue depended on so-called **histocompatibility antigens** (*histo*, "tissue") on cell surfaces. Rejection occurred if tissue from the donor strain possessed histocompatibility antigens foreign to the recipient. Although researchers found that many different genes were involved, one series of closely linked genes coded for particularly strong cell surface antigens. Subsequent work has involved many challenges, partly because tissue transplants do not occur in nature and therefore cannot directly represent a real function of the antigens. Transplant rejection can only be an artificial by-product of other, essential business of the immune system.

Beginning in the 1950s, work on human transplantation genes and antigens was done by several researchers, including Jean Dausset, a Nobel Prize winner at the University of Paris (Dausset 1981). He found that the immune sera of a few patients who had received multiple transfusions agglutinated the *white* blood cells (leukocytes) of some, but not all, French people. (This methodology is analogous to the discovery of red blood cell groups.) Continuing worldwide research has revealed dozens of antigens on white blood cells and other nucleated cells. The strongest antigens are controlled by a cluster of genes on human chromosome 6 that includes at least seven loci (Figure 13). The genes and the cell surface antigens that they encode are known collectively as either the **MHC** (major histocompatibility complex, a general name applying to any vertebrate species) or the **HLA system** (for human leukocyte antigens). The gene products are polypeptides that combine with carbohydrates to form cell surface *glycoproteins.* (The carbohydrate portions are not shown on Figure 13.) Two classes of cell surface molecules are coded for by the HLA genes: Class I antigens by genes *B*, *C*, and *A*; Class II antigens by the several genes within the *D* region (called *DP*, *DQ*, *DR*). Class I molecules are found on the surface of all nucleated cells in the body, while Class II molecules are found only on the cells of the immune system.

The two classes of histocompatibility antigens were noted in the first section of this chapter in reference to the interactions of B cells, T cells, and macrophages (Figures 3 and 5). In order to be identified as foreign by the cells of one's immune system, an antigen must be

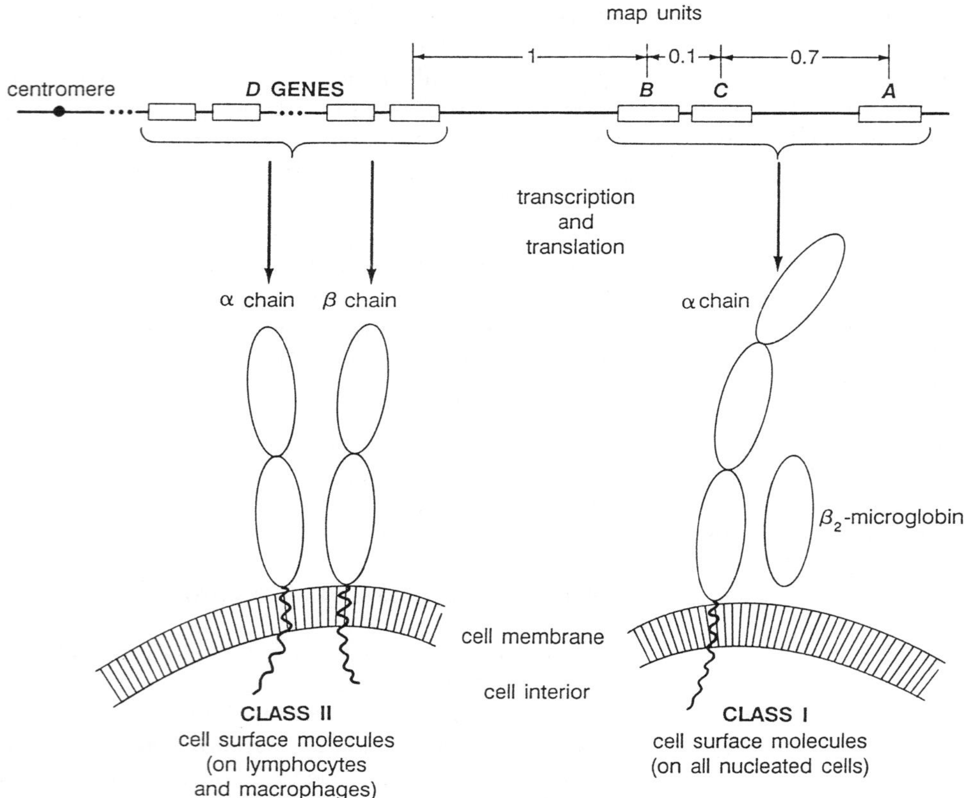

Figure 13 The genes and polypeptides of the HLA system. The genes are linked on the short arm of chromosome 6 with the map distances shown at the top. The two chains of Class I molecules (α and β_2-microglobulin) and the two chains of Class II molecules (α and β) have domain structures (depicted by the ovals), each domain having 90–95 amino acids. The domains bear some amino acid sequence similarities to one another and to the domains of the immunoglobulins and the T cell receptors, reflecting a common evolutionary ancestry. The two chains of each class are *not* held together covalently by disulfide bonds. The polypeptide chains transcribed and translated from the HLA genes also have about 25 amino acids that anchor them to the cell membrane, and another 30 amino acids that extend into the cell (the free carboxyl ends). The smaller chain of the Class I molecules, β_2-microglobulin, is not anchored in the membrane and is coded for by a gene on chromosome 15 rather than by the HLA system.

presented on cell surfaces in conjunction with a MHC protein expressed by one's own genes. For example, through its receptor, a helper T cell is activated only when it recognizes on the cell surface of a macrophage a foreign antigen associated there with a Class II MHC molecule. The recognition event is said to be **MHC-restricted**. MHC restriction appears to give T cells a finely tuned ability to perceive foreignness by "seeing" both nonself and self at once. The complex recognition apparatus on cell surfaces was derived through evolution and can detect not only the molecules of a different species but also the more subtle variations between individuals of the same species (Schwartz 1985). These properties of T cells are thought to be established while they are in the thymus.

Although only one molecule of each class is shown in Figure 13, it is important to note that genes *B*, *C*, and *A* each code for separate Class I α chains. By cloning and sequencing the DNA from different individuals, researchers have deduced that the distal domains of the polypeptides contain polymorphic regions coded for by the *many different alleles* of each of the three genes. (The shorter chain of Class I molecules is constant in amino acid sequence, however, and is coded for by a gene outside the MHC.) Also, several genes within the *D* region code for the Class II α chains (fairly constant in amino acid composition) and several code for β chains (variable in amino acids). In addition to the loci depicted, genes coding for some components of serum complement have been mapped to the DNA segment between the *D* and *B* genes. Thus this region of chromosome 6 is rich in immunological functions. We concentrate here on the polymorphisms of the Class I molecules, which are coded by the HLA genes *B*, *C*, and *A*.

THE TEST PROCEDURE. The Class I HLA antigens on the surface of white blood cells are detected by using specific antibodies. These occur in the sera of some individuals immunized to foreign HLA antigens as a result of prior transplants, multiple transfusions, or successive pregnancies. One difficulty is that the anti-HLA sera obtained this way often contain antibodies to more than one antigen. But increasingly, pure preparations of specific antibodies for detecting HLA antigens are being produced by hybridoma techniques. Because monoclonal antibodies can be produced in large amounts, samples of specific typing reagents are now shared by many laboratories. This practice has led to a much-needed worldwide standardization.

For technical reasons, the in vitro agglutination of white cells by specific antibodies is difficult to demonstrate consistently. Therefore, another method had to be developed to visualize the occurrence of an antigen–antibody reaction. This involves the addition of complement to the mixture. As noted before, complement consists of several plasma proteins that (1) join only to the *combined* antigen and antibody and (2) injure the cell membrane at the site of the antigen–antibody–complement complex. In the laboratory, the harm to a white cell membrane is then detected by a biological stain that can enter and color a damaged cell but not an undamaged cell. This so-called *complement fixation* test for, say, HLA antigen 7 on white blood cells (wbc's) can have two results:

wbc's + anti-7 + complement + dark stain (eosin) →

dark wbc's when
antigen 7 is present

or

colorless wbc's when
antigen 7 is absent

The presence of a particular antigen on wbc's means that a corresponding allele is present in the genotypic makeup of the individual. The individual's genotype may be either homozygous or heterozygous.

Antigens and Alleles of the *BCA* Loci

With large numbers of fairly common alleles, the HLA genes are the most polymorphic known in humans. About 40 alleles of the *B* gene, about 8 of the *C* gene, and about 20 of the *A* gene control the presence of a corresponding 68 white cell antigens (approximately), although only a portion of these antigens can be detected with high reliability. In any given population, the frequencies of many of the alleles are substantial (Table 1). (The frequency of a particular allele represents the proportion of that allele among all the alleles of the gene.) Among American Whites, for example, 23 of the alleles of the *B* gene have frequencies of 1% or greater. Differences between populations have been well documented by anthropologists and others. For example, in Table 1, note that the most frequent *B* allele among American Whites is the one designated *44* (at 14%), among American Blacks allele *58* (at 10%), and among Mexicans allele *35* (at 24%).

A single chromosome 6 will, of course, have all three genes, *B*, *C*, and *A*. The alleles of these genes will almost always be inherited together, because crossing over is uncommon within an interval of less than one map unit. The number of possible combinations of *B*, *C*, and *A* alleles on one chromosome is about $40 \times 8 \times 20 = 6{,}400$, since (in theory) any *B* allele can be associated with any *C* allele and with any *A* allele. A particular combination is called a **haplotype**; for example, a chromosome 6 carrying *B 7*, *C 4*, and *A 3* has the haplotype

7 4 3. (It is understood that the alleles of the three genes are written in the left-to-right order *B*, *C*, *A*, as they occur on the chromosome.) For simplicity, we omit the *D* loci, although polymorphisms for these genes are also very important in transplants.

The HLA genotype of a person consists of two haplotypes—one for each chromosome 6. Considering just the *B*, *C*, and *A* genes, we calculate that there are more than 20×10^6 [(6,400)(6,401)/2] different HLA genotypes—and orders of magnitude more if we were to consider the *D* loci as well. Because no one genotype is particularly common, no two unrelated people picked at random are likely to have exactly the same set of white cell antigens. This great diversity of haplotypes is one reason why organ transplants fail when the donor tissue has not been specifically chosen to match the recipient.*

LINKAGE DISEQUILIBRIUM. A characteristic of the set of HLA genes is that many of the haplotypes do *not* occur randomly. Were the alleles at the *B*, *C*, and *A* loci associated with one another independently, then the expected frequency of any given haplotype should be the product of the individual frequencies. For example, considering just the *B* and *A* genes for American Whites (from Table 1),

frequency of *B 8* = 0.09
frequency of *A 1* = 0.14

expected frequency of the (partial) haplotype 8 1 = 0.09 × 0.14 = 0.013 = 1.3%

The actual frequency of this haplotype among American Whites is found to be 6.1%; it is almost five times greater than random association would predict. In other words, the two alleles are not in equilibrium with each other. Rather, they are said to be in a state of **linkage disequilibrium.**

No one knows why these two alleles and many other combinations seem to be preferentially associated, or in some cases, preferentially *un*associated—when the combination is *less* frequent than predicted. It may be that the 8 1 combination, for example, is advantageous to health in subtle ways and so has increased in frequency over generations through natural selection. On the other hand, the haplotype may be selectively neutral and the high frequency is only transient. According to this view, the alleles are on their way from an initial positive association (at the time when mutation first created them) to becoming randomly assorted. Because the *B* and *A* loci are close together, the attainment of equilibrium by recombination along the intervening chromosomal region would take thousands of generations.

*Note that the great genetic variability for HLA types among a group of people has a completely different causation (multiple alleles among a set of linked genes) than the great diversity of antibodies that a given person can form (somatic recombination of linked genes).

Table 1 The more common alleles of the *HLA-B*, *-C*, and *-A* genes

	FREQUENCY (%)		
Allele[a]	*American Whites* (n = 1,028)[b]	*American Blacks* (n = 365)	*Mexicans* (n = 84)
B 7	10	9	3
B 8	9	3	3
B 35	8	6	24
B 44	14	7	19
B 58	1	10	1
C 2	5	12	4
C 3	12	9	10
C 4	10	16	22
A 1	14	3	10
A 2	27	15	25
A 3	14	7	3
A 23	3	10	1
A 24	7	3	14
A 25	2	0	11
A 30	3	15	2

a. Alleles are listed if they have a frequency of 10% or more in at least one of three groups.
b. n, The number of individuals typed in each population. Frequencies are derived from data in Tiwari and Terasaki (1985).

INHERITANCE. Figure 14 illustrates how the HLA haplotypes for the genes *B*, *C*, and *A* can be inferred from the tissue typing reactions. Note that any individual will type positive for either one or two antigens encoded by each of the three genes—three to six antigens altogether. The complete specification of a person's genotype often requires additional information from close relatives. Once the haplotypes are inferred, however, the inheritance patterns are exactly like that for codominant alleles. In other words, a given *B-C-A* haplotype is inherited as if it were a single allele of a simple gene, because crossing over occurs rarely within this region. Note that each parent can have, at most, a total of two haplotypes. Calling them, for simplicity, *f/a* in the father and *m/o* in the mother, only four different genotypes are possible among their children (excluding the possibility of recombination): *f/m*, *f/o*, *a/m*, and *a/o*.

Transplantation

The fewer foreign HLA-encoded antigens present on a donor organ, the more likely it is to escape rejection by the host's T lymphocytes. By foreign, we mean HLA an-

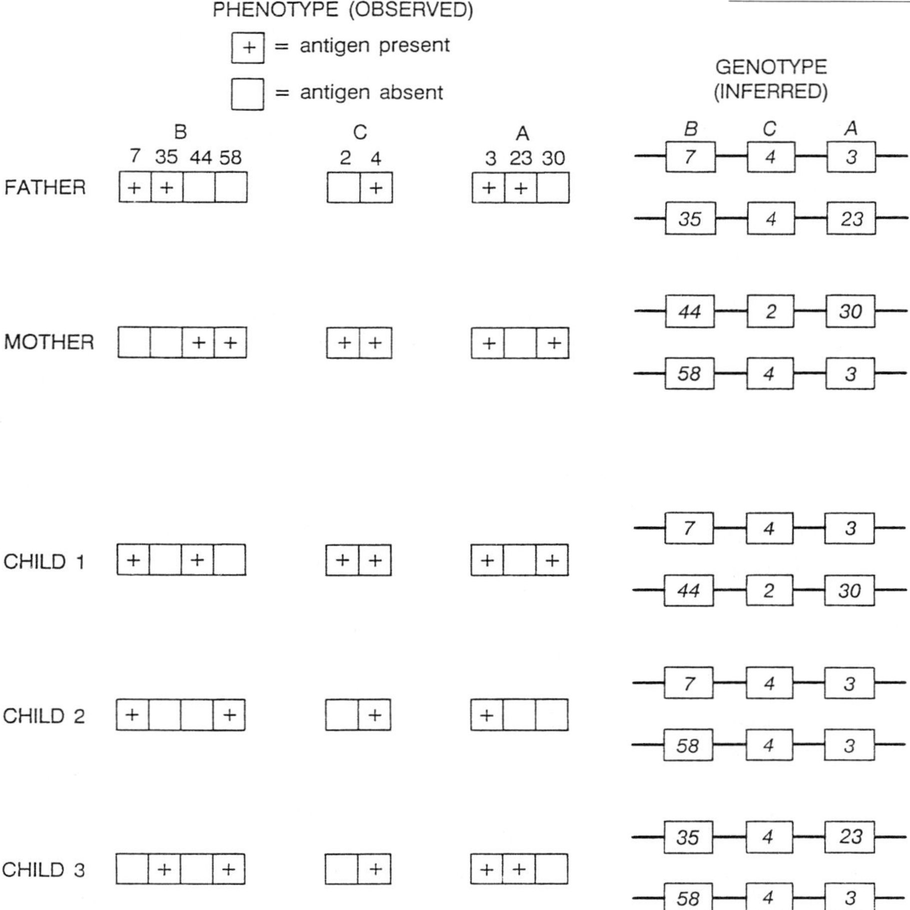

Figure 14 The inheritance of histocompatibility haplotypes in a hypothetical family. The white cell antigens that are present (i.e., the phenotypes) are determined using HLA antibodies and complement fixation test (see text). All positive reactions are indicated on the figure; the father, for example, has no C antigen other than 4. One then infers the genotypes, each consisting of two *B-C-A* haplotypes. Note that child 2 is homozygous at both the *C* and *A* loci. If we assume that no recombination occurred in this region (in this family), child 2 establishes the phases for the parental haplotypes. The father, for example, *cannot* have the *B* alleles interchanged with respect to the *A* alleles (from that shown), and still be the father of child 2. Note that the fewer the +'s for an individual, the more easily one can infer the genotype.

tigens on the donor tissue that are different from the host's own HLA antigens. The foreign HLA antigens present on grafted tissue are thought to be recognized directly by the host's T cells without MHC restriction, that is, without presentation to the host's T cells in conjunction with the host's own HLA molecules (Figure 5) (Marrack and Kappler 1986).

In addition to the HLA system, transplants need to be matched for the ABO blood group system, because the strong A and B antigens appear on many cell types besides red blood cells. There also exist a number of *minor histocompatibility systems*. That these minor antigens can be important is shown by skin grafts between siblings other than identical twins. Even when completely matched for HLA and ABO, sib-to-sib skin grafts survive an average of three weeks, an outcome indicating the existence of unanalyzed antigen systems. When there is a known mismatch for one or more HLA antigens, the survival time of sib-to-sib skin grafts is reduced to two weeks or less (Davies 1984).

KIDNEY TRANSPLANTATION. The best donor for a kidney transplant, after an identical twin, is a sib with the same HLA and ABO antigens as the recipient. (Rejection episodes due to the minor histocompatibility antigens can usually be suppressed by drug therapy.) In more than

90% of these cases, the transplanted kidney is still functioning a year later; in more than 80% of cases, the kidney survives 5 years (Joysey 1984). The survival time of a grafted kidney is reduced when a sib donor has one or more HLA antigens foreign to the recipient. In any event, the donor must be exceptionally well motivated, since the removal of an organ and subsequent existence with only one kidney entail some risk.

Kidneys from unrelated people can also be successfully transplanted, and tens of thousands of such operations have been performed since the 1960s. The donor is often an auto accident victim who is certified brain dead (the usual criterion for death), but whose other body systems can continue to function with external life-support machinery until organs are excised. Usually, the long-term survival of the graft is related to the closeness of the HLA matching. For example, one large study showed that a well-matched kidney is about 2.5 times more likely to be functioning after 10 years than a poorly matched one (Festenstein et al. 1986). Not all studies show this positive effect, however, perhaps because of variation in the strength of the many HLA and non-HLA antigens and their different distribution patterns in various populations. The minor antigens are generally more important when the donor is unrelated to the recipient, because there is a greater likelihood of chance mismatches. One problem with any kidney transplant is that the disease that destroyed the patient's own kidneys may affect the grafted one as well.

BONE MARROW TRANSPLANTATION. Unlike the solid organs, the cells of the bone marrow continually renew themselves from stem cells, and a portion may be removed from a living donor without long-term loss of function. The physician repeatedly inserts a long needle into the upper part of the hip bone of the donor (who is under general anesthesia) in order to remove about 500 ml of the spongy marrow contained within. The cell suspension is filtered to remove chips of bone and clumps of fat, and slowly injected into a vein of the recipient. Marrow transplants are increasingly successful in treating patients with certain marrow defects: leukemia, immunodeficiency diseases, and aplastic anemia (Storb and Thomas 1983). Marrow transplants have been tried experimentally on a number of patients suffering from inborn errors of metabolism (such as thalassemia and sickle-cell anemia) that are clinically expressed in the cells of bone marrow. Even more generalized genetic diseases (for example, the Lesch-Nyhan syndrome), may be treatable if the critical enzyme is transported from the grafted marrow to the tissues where it is needed.

The **leukemias** are a group of diseases due to the malignant proliferation of an immature lymphocyte, myelocyte, or other type of white blood cell. In the marrow, the massive buildup of cancer cells crowds out both red blood cells (thereby producing anemia) and platelets (producing internal bleeding). Also, vital organs are damaged by huge numbers of infiltrating leukemic cells. Advance of the disease can be either slow (chronic) or rapid (acute), the latter leading to death in a few months if untreated. *Acute lymphoblastic leukemia* is the most frequent type of childhood cancer (a lymphoblast is an immature lymphocyte).

Depending on the course of the disease and a patient's response to other forms of treatment, bone marrow transplantation is now an accepted procedure for some types of leukemia (Gale 1986). Before the transplant, physicians try to kill 100% of the leukemic cells with drugs and radiation, forms of treatment that destroy all rapidly dividing cells (and that also cause nausea and hair loss). Ideally, injected marrow cells, generally from an HLA-identical sib, will then seed the patient's bone marrow cavities and eventually restore immune functions.

But these procedures may have several adverse consequences: (1) recurrence of leukemia if not all malignant cells were destroyed, (2) serious microbial, viral, or fungal infections while the patient's immune system remains ineffective, or (3) **graft-versus-host disease**. The latter reaction occurs when the lymphocytes in the *donor* marrow mount a pervasive and painful immunological attack against the non-HLA antigens on the cells of the recipient—the reverse of the usual transplant complications. It is fatal in about 25% of transplant patients with acute leukemia. But in persons who survive graft-versus-host disease, leukemia is much less likely to recur: presumably, the T cells of the graft can kill not only the host's normal cells but also the host's leukemic cells.

When a person with leukemia does not have an HLA-identical sib, the physician may instead remove bone marrow from the patient. The marrow is returned, but first technicians try to purge it of the leukemic cells by monoclonal antibodies directed against cancer-specific surface antigens. In the United States, where a person averages little more than one sib, about 70% of the population do not have an HLA-identical sib; consequently this form of therapy is being actively investigated (Kersey et al. 1987).

Several genetic diseases can also be treated by bone marrow transplants. The first such disease to be completely corrected in this way was *severe combined immune deficiency (SCID)* in which babies lack both functional B and T cells. It is due in some cases to an X-linked recessive and in others to an autosomal recessive. The most publicized SCID patient was a Texas boy known as David, who spent virtually all his life (1971–1984) in sterile plastic bubbles (Lawrence 1985).* He died four months after receiving a bone marrow transplant from his older sister, who was the most closely (but still not perfectly) HLA-matched donor available. Prior to infusion into David, her marrow cells were treated in vitro with a monoclonal antibody against an antigen (called T12) present only on mature T cells. The object was to kill donor cells that might otherwise initiate graft-versus-host disease. Although most likely due to transplant complications, the exact cause of David's death is not clear (Simmons 1984).

The role of bone marrow transplantation in the correction of genetic diseases whose clinical symptoms are expressed *outside* bone marrow-derived tissues (e.g., liver, muscle, heart) is complicated (Parkman 1986). The critical assumptions are that the missing enzyme (1) would be produced and secreted by the transplanted bone marrow cells, and (2) would circulate to, enter, and function in the affected tissues. The bone marrow would be taken from an HLA-matched donor free of the disease. Alternatively, the bone marrow cells could be taken from the affected patient, treated in vitro with cloned DNA to (ideally) replace the defective allele with the normal allele, and returned. This contemplated form of treatment is called *gene therapy* (Chapter 25). Use of the patient's *own* bone marrow avoids the immunological consequences of transplantation discussed here.

OTHER CONSIDERATIONS. Far fewer transplant operations have involved other organs (e.g., heart and liver) and the importance of the HLA antigens on these grafts has generally not been fully evaluated (Calne 1984). A liver transplant in particular is technically very difficult and expensive ($300,000 or more), requiring great surgical skill and complex support systems. The problem of sharply limited supply hampers all organ transplants, and many patients die while waiting for their last chance to live. Questions of who gets and who pays for an organ, and whether medical resources might better be directed to entirely different endeavors involve political, social, and moral puzzles that continue to be pondered and discussed (Engelhardt 1984, for example).

Finally we ask the following transplantation-related question: Why does a pregnant female not reject every fetus "grafted" to her uterine lining? The developing fetus is likely to have several foreign histocompatibility alleles inherited from the father. Presented to the mother as an organ transplant, the encoded antigens would elicit antibody responses and graft rejection by the mother. The significant role of HLA antigens in pregnancies is underscored by some studies that show, somewhat surprisingly, that HLA *mismatches* between mother and fetus are advantageous for a successful pregnancy (Rodger and Drake 1987). Rather than being a potential threat, anti-fetal-HLA responses made by the mother may lead to successful protective measures mounted by the fetus or mother, which would not otherwise occur. In any event, to avoid rejection of the fetus and subsequent damage or spontaneous abortion, the maternal–fetal union seems to have interesting modifications of immune mechanisms.

*Because David's older brother had died of SCID, his family and physicians were prepared. David was delivered by cesarean section to avoid vaginal contamination and immediately placed in a sterile chamber.

THE IMMUNE SYSTEM AND DISEASE

Here we briefly look at four types of disorders that involve the immune system.

Associations with HLA

An intriguing aspect of the HLA genes is the statistical association of particular alleles with particular diseases. In an extensive review, Tiwari and Terasaki (1985) list scores of HLA antigens that are correlated with hundreds of diseases (Table 2). The prime example is *ankylosing spondylitis*, an inflammation of the areas where tendons and ligaments attach to bones, especially the bones of the hip and spine. In 40 studies involving Caucasian groups, 89% of patients having ankylosing spondylitis possessed the HLA antigen B 27; only 9% of healthy controls possessed the same antigen. In the absence of B 27, the disease hardly ever develops. On the other hand, ankylosing spondylitis is fairly uncommon (the overall frequency is about 1 in 2,000 persons), and most people with B 27 remain healthy.

Two different theories have been advanced to explain this association:

1. *The linkage hypothesis*. Susceptibility to ankylosing spondylitis is due, at least in part, to an allele—call it *Ank*—of a gene that is closely linked to the *HLA-B* locus. This hypothesis supposes that the particular chromosome 6 haplotype $\underline{Ank\ B27}$ is in linkage disequilibrium; that is, it is more common than would be expected on the basis of random combination.

2. *The antigen-effect hypothesis*. The polypeptide coded by the *B 27* allele leads to susceptibility to the disease. Perhaps the B 27 antigen is a cell surface receptor for a virus or other pathogen. Alternatively, the B 27 molecule may mimic antigens on a pathogen so that antibodies directed against the

Table 2 Significant associations between disease and HLA antigens in Caucasian populations

Disease[a]	HLA antigen	Percentage with antigen: Affected	Percentage with antigen: Normal	Relative risk
JOINTS				
Ankylosing spondylitis	B 27	89	9	69.1
Reiter disease	B 27	80	9	37.1
GASTROINTESTINAL TRACT				
Celiac disease	B 8	68	22	7.6
SKIN				
Psoriasis vulgaris	B 17	19	7	5.3
	B 13	19	5	4.1
	B 37	7	2	3.9
CONNECTIVE TISSUE				
Juvenile rheumatoid arthritis	B 27	25	9	3.9
Rheumatoid arthritis	DR 4	68	25	3.8
Systemic lupus erythematosus	B 8	40	20	2.7
ENDOCRINE SYSTEM				
Juvenile diabetes mellitus	DR 4	51	25	3.6
	DR 3	46	22	3.3
NERVOUS SYSTEM				
Myasthenia gravis	B 8	44	19	3.3
Multiple sclerosis	DR 2	51	27	2.7

Source: Tiwari and Terasaki (1985).
a. Included are all disease–HLA associations that are based on at least 600 patients from a dozen different studies and having a combined relative risk of at least 2.7. Relative risk can be interpreted as the risk of developing the disease when one *possesses* the antigen compared with that for persons *lacking* the antigen.

pathogen also attack host tissues carrying B 27. No firm decision is yet possible between competing hypotheses for ankylosing spondylitis and B 27, or for any of the disease–HLA correlations.

Although they affect many different body systems, the HLA-associated diseases share some general features. Typically the disorders are chronic and involve immune processes in one way or another. They also have a genetic component but do not show simple Mendelian inheritance, a finding suggesting multiple causative genes and substantial environmental influences.

For example, *celiac disease* (*celiac* means "abdomen"), which is strongly associated with antigen B 8, is associated with foul-smelling diarrhea, skeletal disorders, internal bleeding, muscle wasting, and edema. These symptoms result from injury to the intestinal lining, which is infiltrated with plasma cells that produce antibodies to gluten, a major protein in wheat (and in lesser amounts in oats, rye, and barley.)* Oldstone (1987) suggests that the anti-gluten antibodies were originally raised against a particular sequence of eight amino acids of an intestinal virus (called Ad-12), which is present in most affected people. Gluten (by chance) has exactly the same sequence of eight amino acids, and its consequent reaction with the *same* antibody (called *cross reactivity*) may play a role in celiac disease pathogenesis. In any event, most patients respond favorably and quickly to a

*Being somewhat elastic, gluten is the component in bread dough that enables it to to trap bubbles and rise during baking.

diet free of gluten. Although uncommon in the general population (perhaps 1 in 2,000), between 10 and 15% of the close relatives of people with celiac disease have some abnormal abdominal signs (Sleisenger 1982).

Allergies

The discomforts associated with allergies appear to be unfortunate by-products of certain antigen–antibody reactions. In some people, the antigens, or **allergens** (in recognition of their effects), cause immunological reactions when *inhaled* (ragweed pollen and animal danders, for example), or *eaten* (chocolate, nuts, bananas), or *touched* (wool, poison ivy), or *injected* (bee venom, penicillin)—often in very small amounts. One of the most common allergies, hay fever, appears to have a genetic component, although the mode of inheritance is not simple.

Allergic reactions that follow quickly upon exposure to the allergen are provoked by a B cell response leading to circulating antibodies of a special class, IgE, which otherwise has a limited immunological role (Buisseret 1982). Other allergies are slow to develop (taking hours or days) and are largely due to a T cell response. The antigen-antibody reaction stimulates certain tissue cells to release *histamine* (the amino acid histidine without the carboxyl group). The physiological effects of histamine include contraction of the muscles of the lungs and dilation of capillaries, which then leak fluid into the surrounding tissues. Although responsible for the discomforts of allergy, it is unclear how these responses help the body of a susceptible person deal with the allergens. *Anaphylaxis*, a severe and rapidly developing form of allergy (to a bee sting, for example), can sometimes be fatal if not quickly treated with antihistamines.

Autoimmune Disorders

One way to escape allergic symptoms is to avoid the allergen. Unfortunately this is not possible for the victims of the poorly understood *autoimmune disorders*, in which so-called **autoantibodies** are produced against one's own tissues. In *systemic lupus erythematosis*, for example, autoantibodies are directed against cell surface components, nuclear substances (double-stranded and single-stranded DNA, RNA, histones), and cytoplasmic elements (mitochondria, lysosomes, ribosomes). With periods of greater or lesser suffering, the illness affects many organs, most frequently the brain, joints, kidneys, and skin.* Serious damage to the kidneys results from deposits of antigen–antibody complexes that block fluid filtration. Like several other autoimmune diseases, lupus erythematosis is chronic and is associated with a particular HLA antigen (Table 3).

In *juvenile diabetes* (= Type I = insulin dependent diabetes mellitus = IDDM), which affects at least a half-million people in the United States, the onset of clinical symptoms may be sudden. Yet the disease develops only after many of the insulin-producing β cells in the pancreatic islets of Langerhans have been destroyed by autoantibodies over a course of time. Histocompatibility antigens DR 3 and DR 4 are important contributors to the onset of disease (but in ways that are not understood). In fact, 95% of diabetic individuals—but only 40% of the general population—have one or the other antigen or both. The risk of developing juvenile diabetes increases with the number of HLA haplotypes shared with a known diabetic sib (Eisenbarth 1986):

Number of shared haplotypes	**Risk of diabetes (%)**
0	1
1	5
2	15

*A red rash (erythema) in some patients apparently reminded a nineteenth-century name-giver of the bite of a wolf (lupus) (Rosenthal 1989).

Table 3 HLA-associated autoimmune disorders

Disorder	*Associated HLA antigen*	*Target of autoantibodies*
Rheumatoid arthritis	DR 4	Gamma globulin
Systemic lupus erythematosus	B 8	DNA and other nuclear and cytoplasmic constituents
Juvenile diabetes	DR 4, DR 3	Islet cells of pancreas
Myasthenia gravis	B 8	Acetylcholine receptors in nerve-muscle junctions
Multiple sclerosis	DR 2	Myelin sheaths around nerve cells

But since only about 1 in 20 children of a diabetic parent gets diabetes (versus 1 in 500 for the general population), several genes are probably involved. Furthermore, no one knows what genetic or environmental factors trigger the autoimmunity; a number of different viruses have been suggested, especially the rubella virus. In addition to the standard insulin-injection therapy for diabetes, experimental treatments involve drugs that suppress the immune system, and transplants of either a section of pancreas or just isolated islet cells (Wechsler 1986).

Immune Deficiency Disorders

In another group of about a dozen diseases, the immune system functions poorly or not at all. These immune deficiency disorders may involve just the B cells, just the T cells, or both (Hirschhorn and Hirschhorn 1983).

Agammaglobulinemia, as an example, is a rare, X-linked, recessive disease (Figure 15). Affected male infants have no detectable plasma cells and virtually no immunoglobulin in any of the five classes, except for their initial charge of maternally transmitted antibodies. (Recall that IgG, the major antibody class, readily crosses the placenta. A normal baby's ability to manufacture its *own* antibodies develops gradually after birth.) Infants remain well for about a year, but thereafter suffer from recurrent bacterial infections. Survival depends on periodic injections of gamma globulin that contain antibodies to common bacterial diseases. Because they possess normal T cells, the patients are no more susceptible to common viral diseases than other infants.

Studies of the sisters III-3 and III-4 in Figure 15 showed that the mutant gene (a) for agammaglobulinemia somehow blocks stem cells in the B-lineage from differentiating into B lymphocytes; that is, for a B cell to develop, it must carry and express the normal allele (A). Because the sisters—normal mothers of affected males—are carriers (A/a), each of their stem cells should express either A or a, but not both, because of X chromosome inactivation (the Lyon hypothesis: Chapter 8). Although the gene product of A is unknown, the two sisters were found to be heterozygous (G^1/G^2) for the X-linked gene for the enzyme G6PD (glucose-6-phosphate dehydrogenase). The randomness of X chromosome inactivation predicts that about half of their cells should express the allele G^1, and half the allele G^2. Investigators found, however, that *all* of their mature B cells expressed G^1; they interpreted this finding to mean that the sisters' X chromosome genotypes were $\underline{A\,G^1}/\underline{a\,G^2}$. When the $\underline{A\,G^1}$ X chromosome was active in a B stem cell, it developed normally into a B lymphocyte. When the $\underline{a\,G^2}$ chromosome was active, the stem cell failed to develop into a B lymphocyte, as did all the stem cells in their affected, a/(Y) male offspring.

Severe combined immunodeficiency disease (SCID) was noted above in regard to bone marrow transplants. SCID infants possess neither B nor T cells. They fail to thrive because of overwhelming bacterial, fungal, and viral infections, and they invariably die in their first years if untreated. A distinctive variant of SCID called *adenosine deaminase (ADA) deficiency* accounts for about half of the SCID cases that are inherited as autosomal recessives (Rosen et al. 1984). The enzyme ADA participates in the degradation of nucleic acids, allowing purines to be recycled, or excreted as uric acid (see Figure 14 in Chapter 16). But in ADA deficiency the substrates, adenosine and deoxyadenosine, accumulate in large quantities and interfere with the normal synthesis of DNA. Although ADA is normally produced in all cells of the body, it is not clear why the major phenotypic effect of its deficiency occurs just in lymphocytes.

Acquired immune deficiency syndrome (AIDS) is due to the human immunodeficiency virus (HIV), which in-

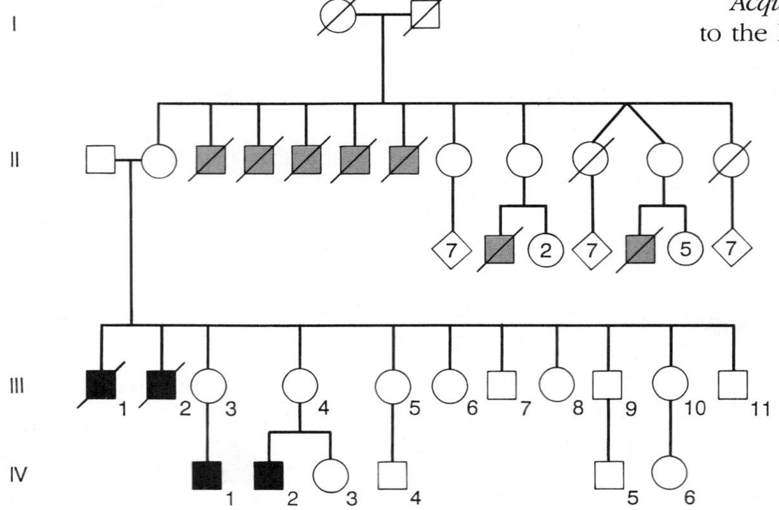

Figure 15 Pedigree illustrating X-linked inheritance of agammaglobulinemia. a, mutant allele; A, normal allele. Affected males are hemizygous for the mutant allele, a/(Y), and their normal mothers are heterozygous, A/a. Diagonal lines indicate persons who were dead at the time of investigation. (From Conley et al. 1986.)

vades several cell types, especially helper T cells (Gallo and Montagnier 1988). As a consequence, immune functions related to both T cells and B cells are compromised, leading to death in up to half of those infected with the virus. Containing RNA rather than DNA, HIV is a so-called *retrovirus*. The name is based on the fact that the virus uses its own reverse transcriptase enzyme to copy the viral RNA into DNA, which is then inserted into the genome of the host cell. The integrated form, known as a *provirus*, replicates along with the cell's own DNA. Later, the provirus can be activated, an event leading to the synthesis and release of large numbers of intact viral particles and the death of the cell. Transmitted almost entirely by sexual acts or blood exchange, and from mother to fetus, the disease is prevalent among, but not limited to, homosexual men, intravenous drug users and their sexual partners, and recipients of blood products. Because the disease involves vital immune cells in complex ways, both the development of protective vaccines and effective treatments have proved difficult. Education, testing, and modification of transmission-related habits appear to be the best current strategy for containment (Francis and Chin 1987).

SUMMARY

1. The immune system can identify an antigenic molecule as foreign, respond specifically to its shape and chemistry, and react more vigorously to it upon a subsequent exposure. Vaccines possess the antigens present on disease-causing organisms but lack the organism's harmful attributes.

2. Immune responsiveness depends on various types of white blood cells: B cells manufacture antibodies, T cells (of several subtypes) kill by cell-to-cell contact or regulate the activities of other elements of the immune system, and macrophages present antigen to T cells and engulf foreign particles.

3. As B and T cells mature from stem cells, they express receptors that respond to specific antigens. When stimulated by that antigen (or a similar one), they proliferate into a clone of immunologically active cells and a reserve of memory cells.

4. Antibodies are Y-shaped molecules composed of two heavy and two light polypeptide chains. The specificity of the antigen-binding pockets at the tips of the Y are determined by the particular amino acids of the hypervariable regions.

5. The great diversity of antibodies that plasma cells can make results largely from the multitude of ways that DNA elements recombine with each other during somatic differentiation of lymphocytes.

6. Additional diversity results from imprecision at the sites of recombination and from mutation of the variable elements during clonal proliferation.

7. While the many antibodies that are present in any person's plasma are a mixed lot, a monoclonal antibody preparation is pure and homogeneous. Monoclonals are produced in vitro by hybridomas, made by fusing a specifically activated plasma cell with a cancer cell.

8. The HLA antigens on the surface of cells are encoded by several highly polymorphic, closely linked genes—the major histocompatibility complex. For an antigen to be recognized as foreign by a T cell, it must be presented on cell surfaces in association with one of these HLA molecules (MHC restriction).

9. The particular alleles of the linked HLA genes on one chromosome are inherited as a group called a haplotype. A given haplotype may be more frequent than random combinations would predict; this phenomenon is called linkage disequilibrium.

10. Matching the HLA antigens of donor and recipient helps make transplants successful, by avoiding rejection of the cells of the graft by the host's T cells. Rejection of the foreign tissue depends also on a number of minor histocompatibility systems.

11. Bone marrow transplantation between HLA-identical siblings is useful for some types of leukemia and other diseases. Besides the usual complications of transplants, patients are subject to graft-versus-host disease caused by the immune cells of the donor.

12. A number of diseases are associated with particular HLA alleles. Many of them are due partly to genetic influences that affect the functioning of the immune system. In some people, the body makes autoantibodies that are directed against one's own tissues.

13. Another group of usually severe diseases results from poor functioning of the immune response. These may involve B cells, T cells, or both.

FURTHER READING

Bernal (1986) is a well-written overview of this chapter from a medical point of view. General books on the immunogenetics (of mice and humans) are by Fudenberg et al. (1984), Nisonoff (1984), and Golub (1987b). The cell biology books by Alberts et al. (1989) and by Darnell et al. (1986) include excellent and detailed chapters on the immune system, while Nossal's summary (1987) is very concise. Bibel (1988) has collected important papers in the history of immunology.

A review article by Ellison and Hood (1983) looks at human antibody genes, and one by Auffray and Strominger (1986) looks at the HLA system. The monograph by Tiwari and Terasaki (1985) gives a more detailed account of HLA. Articles on immune diseases are included in Stanbury et al. (1983), in Emery and Rimoin (1983), and in the November 27, 1987 issue of *JAMA* (entirely on immune diseases). Calne (1984) covers experimental and clinical aspects of transplantation while Lax (1984) gives a reporter's account of a hospital unit devoted to bone marrow transplantation.

About 25 *Scientific American* articles have been collected and updated by Burnet (1976). Many more recent *Scientific American* articles on immunology include Milstein (1980) on monoclonal antibodies, Rose (1981) on autoimmune diseases, Leder (1982) and Tonegawa (1985) on antibody diversity, Collier and Kaplan (1984) on immunotoxins, Laurence (1985) on the immune system in AIDS, Marrack and Kappler (1986) on the T cells, Ada and Nossal (1987) on the clonal selection hypothesis, and several articles in the October 1988 issue, which is devoted to AIDS.

QUESTIONS

1. What is the basis of immunological memory, the heightened antibody reaction to a second exposure to an antigen?

2. Explain how patients with multiple myeloma contributed to the determination of antibody structure.

3. An IgG molecule has a total of 12 domains. How are they spatially related to one another?

4. Porter found that the enzyme papain (from the papaya tree) split an antibody molecule into exactly three pieces. Two were identical and could still bind antigen; the third piece could not bind antigen. Where does papain split the antibody molecule?

5. The constant portions of antibody chains exhibit some differences that are unrelated to antigen binding. For example, Fudenberg et al. (1984) noted that some persons have phenylalanine at amino acid position 296 of the IgG heavy chain while other people have tyrosine. Is this difference due to a single base substitution? (See Table 2 in Chapter 14.)

6. In the mouse, there are approximately 150 *V* elements, 12 *D* elements, and 4 *J* elements coding the variable domain of the H chain of IgG (Steinmetz 1986). How many different heavy chain variable domains can a mouse make (ignoring imprecise *D-J* joining and somatic mutation)?

7. Do the somatic mutations that occur during plasma cell differentiation affect the evolutionary history of immune genes?

8. When they made the first hybridomas, Köhler and Milstein fused a normal plasma cell taken from an immunized mouse with a cultured myeloma cell *that synthesized its own antibodies* (because it was derived from an immature B cell). Why were these first hybridomas not very useful? What change was made to improve the production of monoclonals?

9. What are the HLA genotypes (for just the *B* and *A* genes) of the family members below? Indicate properly as haplotypes. +, antigen present; blank, antigen absent.

	B antigens			**A antigens**			
	5	**8**	**40**	**2**	**3**	**9**	**11**
father	+		+	+		+	
mother	+	+			+		+
first child	+			+	+		
second child	+	+		+			+
third child		+	+			+	+

10. For the family in Question 9, would the first or third child make a better kidney donor to the second child? Assume each of the HLA antigens is equally strong.

11. For the family in Question 9, how could a child of the given parents come to have antigens B 5, A 2, and A 11 only?

12. For the family in Figure 14, assuming no crossing over, what additional genotype is possible among children of the given parents?

13. Assuming no crossing over, what is the probability that *you* will have an HLA-identical sib (one or more), if you have a total of (a) 1 sib? (b) 2 sibs? (c) 3 sibs? Assume four different haplotypes among your parents.

14. Why is a sib who is matched for all HLA antigens a better kidney donor than an unrelated person *also* matched for all HLA antigens?

15. Why are the HLA white cell antigens more useful than the ABO or Rh red cell antigens in cases of disputed paternity?

16. In Caucasian populations, the haplotype *B 7 A 3* has an observed frequency of about 5%. Using the data in Table 1, show that these genes are in linkage disequilibrium.

17. The cornea (the transparent front surface of the eyeball) has no blood or lymphatic vessels. Does this suggest why corneal transplants from unrelated donors are usually successful?

18. Fill in the approximate numbers n_1–n_8 in the following table, knowing that (a) ankylosing spondylitis affects one person in 2,000 overall; (b) 89% of *affected* persons possess the HLA antigen B 27; and (c) 9% of *unaffected* persons possess B 27.

	Phenotype		
Antigen B 27	**Affected**	**Unaffected**	**Total**
present	n_1	n_2	n_3
absent	n_4	n_5	n_6
total	n_7	n_8	20,000

19. In conjunction with other modern medical practices, organ transplants have led to a reexamination of the definition of the time of death. Discuss.

ANSWERS

1. During proliferation of a specific, stimulated B cell, some cells of the clone do not fully mature to produce antibody, but remain available for further proliferation even after long time periods.

2. The large amount of a single antibody species in each patient provided a nearly pure substance for amino acid sequencing. Antibodies from different patients led to the distinction between variable and constant domains.

3. See Figure 6.

4. Somewhere on the heavy chains between the disufide bonds holding the heavy chains together and the disulfide bonds connecting the heavy and light chains.

5. Possible DNA triplets are AAA = phe and ATA = tyr, so the amino acid change could arise from a single base substitution.

6. $150 \times 12 \times 4 = 7{,}200$.

7. No; only germinal mutations affect the course of evolution.

8. The antibodies included both parental types as well as mixed molecules in which the H and L chains were from different parents. The production of true monoclonals required a mutant of the myeloma line that failed to make immunoglobulins.

9.

Person	**Haplotype**
father	5 2 / 40 9
mother	5 3 / 8 11
first child	5 2 / 5 3
second child	5 2 / 8 11
third child	40 9 / 8 11

10. The first child would be better, possessing only one antigen (A 3) foreign to the recipient.

11. The child received the chromosome *5 2* from the father and the crossover chromosome *5 11* from the mother.

12. *35 4 23* / *44 2 30*

13. (a) 0.25; (b) $1 - (0.75)^2 = 0.44$; (c) $1 - (0.75)^3 = 0.58$.

14. A sib is generally better matched for the minor histocompatibility antigens.

15. The HLA system is more polymorphic, so it is more likely to reveal differences between unrelated persons.

16. If the genes were to associate at random, the frequency of the *B7 A3* haplotype would be $(0.10)(0.14) = 0.014 =$ 1.4%.

17. T cells, the agents of immunological rejection, do not reach the site of the graft.

18. $n_1 = 9$ $n_2 = 1{,}799$ $n_3 = 1{,}808$
$n_4 = 1$ $n_5 = 18{,}191$ $n_6 = 18{,}192$
$n_7 = 10$ $n_8 = 19{,}990$ $n_9 = 20{,}000$

19. The sooner a vital organ can be removed from a dead person, the more likely its success as a transplant. Therefore, it is important to define the moment of death.

Chapter 19 Genetics of Cancer

Cancerous Cells

Oncogenes and Proto-Oncogenes
Oncogenes Carried by Retroviruses
Finding Human Oncogenes by Gene Transfer
The Proteins of Oncogenes
Transgenic Mouse Models

Cancer and Cytogenetics
Proto-Oncogenes at Translocation Breakpoints
Other Specific Chromosomal Abnormalities

Cancer and Mendelian Inheritance
Dominants: Retinoblastoma and Wilms Tumor
Recessives: Chromosome Breakage Syndromes

Genetic and Environmental Factors

Cancer research has turned into something like a running hunt. The fox is not yet within sight, but it is at least known that there is indeed a fox, and this is a great change from the sense of things twenty years ago. At that time it was generally believed that cancer was not one disease but a hundred, all fundamentally different and each requiring its own unique penetration. Today it seems much more likely that a single mechanism, or a set of mechanisms, lurks at the deep center of every form of cancer.

Lewis Thomas (1984)

Just as knowledge of nucleic acids and proteins opened up new routes for studying the immune system, so, too, has molecular biology begun to reveal the central processes whereby cells become malignant. Although cancer seems to have a multiplicity of determinants and takes many clinical forms depending on the specific tissue involved, there may be only a few precipitating causes within cells. Recent research has emphasized stepwise genetic mechanisms that alter the growth of somatic cells and the pathways by which the relevant genes and their gene products go awry. Sparking optimism among cancer researchers is the thought that new insights into basic cell biology may eventually provide much better methods for preventing, diagnosing, and treating the disease.

So far, however, most of the returns on the research investment lie in the future, for the toll from all types of cancer combined has remained high and substantially unchanged for several decades. True, the mortality rates from some types of cancer have been decreasing due to earlier detection (uterine cancer), or changing diets (stomach cancer), or more effective radiation and chemical therapies (leukemia). Yet the unrelenting increase in the incidence of lung cancer, first noticed among men in the 1930s and then among women in the 1960s, has overshadowed the decreases. The tally of new cases of all forms of cancer among Americans is now about 900,000 annually, and deaths number about 450,000 per year—16% of all deaths—more than any other cause except heart disease. (Part of the high cancer incidence stems from the conquest of infectious diseases, so our population includes proportionately more and more members in older age groups—those in which cancers are more likely to appear.) Several researchers have suggested that the much heralded "war on cancer," launched as the National Cancer Act of 1971, has advanced only on very narrow fronts (Cairns 1985; Bailar and Smith 1986). They suggest a switch in primary emphasis from cancer therapy to cancer prevention. This program involves several approaches: epidemiological studies of human populations to help identify cancer-causing agents, public health measures to reduce their prevalence, and biological studies to understand the origins of the disease at the molecular and cellular levels.

CANCEROUS CELLS

Unlike normal cells injured by infectious diseases, cancerous cells are quite healthy by the usual standards—so much so that they interfere with the well-being of neighboring tissue. A mass of cells that multiplies when it should not is called a *tumor* or *neoplasm* (the medical term meaning "new form"). A so-called *benign tumor* is not a cancer, because it remains localized in the place of origin, often separated from the surrounding cells by a layer of connective tissue. Common warts are benign skin tumors occurring primarily in children and young adults (the causative agent being a virus). Although compact and slow growing, benign tumors may eventually contain billions of cells, become quite large, and cause damage by pressing against a vital organ. They do not, however, spread to other sites nor recur following surgical removal. The karyotypes of benign tumor cells are usually normal.

On the other hand, the cells of a **malignant tumor**, or **cancer**,* have the following characteristics:

- Their growth is unresponsive to the mechanisms that normally restrict cell numbers to those required for growth or replacement. Like the growth of benign tumors, the exponential proliferation of cancer cells may be slow, but it continues unabated.

- Cancer cells become disorganized and may revert to a more primitive form as they lose their specialized structures and functions. Pathologists look for these characteristic changes under the microscope in order to distinguish—with partial, but not complete, success—between benign and malignant tumors (Franks 1986).

- Cancer cells are invasive, damaging adjacent tissues and often leading to internal bleeding. They may also slough off from the primary site and circulate through the blood or lymph to other locations in the body, where they initiate secondary cancers. The spreading to distant sites, called **metastasis**, poses a severe problem in treatment. Deaths of cancer patients may result from the destruction of vital organs at secondary sites, or from overwhelming infections due to general wasting and impaired immune functions.

- Cancer cells almost always have abnormal karyotypes, including translocations, inversions, deletions, isochromosomes, monosomies, and extra chromosomes—the last sometimes far beyond the normal diploid number. Specific chromosomal aberrations are sometimes associated with particular cancers (see later).

Cancers often pass through stages as they progress from relatively benign to highly malignant. This *tumor progression* may follow the order of characteristics given above; that is, cells first divide without control, then become disordered and less distinctive, and then invade surrounding tissue and metastasize. Each step is probably accompanied by additional mutations as well as by increasing chromosomal aneuploidy. As variant populations compete with one another, the result is evolution at the cellular level. Such a race favors the emergence of cell lines with greater growth potentials, with tragic consequences for the patient.

The naming of benign and malignant tumors is complicated. Briefly, of the more than 200 different kinds of human cancer, about 85% are **carcinomas**, solid tumors derived from epithelial tissues—skin, linings of the respiratory, digestive, urinary, and genital systems, plus various glands, the breasts, and nervous tissue. Carcinomas of the lung, large intestine, and breast account for about half of all human cancers in Western societies. **Sarcomas** (2% of all cancers) are composed of closely packed cells derived from bone, cartilage, muscle, and fat. Two types of malignancies affect white blood cells: **leukemias** (3% of all malignancies) are cancers of the white blood cells manufactured in the bone marrow; **lymphomas** (5% of all cancers) affect the white blood cells present in the spleen and lymph nodes. The remaining cancers are of mixed origin. Although cancers are primarily associated with processes of aging, they can occur in younger persons as well. Leukemia, for example, is a major cause of death in children.

A tumor is usually a *clone* of cells that enlarges by successive divisions from *one* original aberrant cell. Three approaches have provided supporting evidence for the single-cell origin of cancers (Nowell 1976):

1. *X chromosome inactivation*. Recall that in any particular cell of a female, the genes on one X chromosome are active while those on the other X are

**Cancer* is the Latin word meaning "crab." The Greek/Roman physician, Galen (A.D. 130–200), observed of a breast cancer: "Just as a crab's feet extend from every part of the body, so in this disease the veins are distended, forming a similar figure." The Greek word for crab, *karkinos* (from which *carcinogen* and *carcinoma* are derived), had been used for cancer 600 years earlier by the "father of medicine," Hippocrates (B.C. 460–370); he may have thought that invasive growths of the cancer seemed to grasp surrounding tissues (Haubrich 1984).

inactive (the Lyon hypothesis, Chapter 8). Once the active-versus-inactive decision is made in a given embryonic cell, all descendant cells reproduce true to that type. Thus women who are heterozygous for an X-linked gene have two cell populations, each expressing one allele or the other. For example, Linder and Gartler (1965) studied electrophoretic enzyme variants in females heterozygous (Gd^A/Gd^B) for the gene G6PD. They showed that a *single* cell had only a single isozyme (encoded by either Gd^A or Gd^B), and each control sample of normal uterine tissue consisting of *many* cells had both A and B isozymes. On the other hand, they discovered that all the cells of a uterine tumor were the same—type A or type B (Figure 1). Some women with several separated tumors had some of each type, but no tumor contained both A and B enzymes. That entire tumors had only one X-linked phenotype (despite the millions or billions of cells in an individual tumor) supports the idea that each arose from a single cell.

2. *Tumors of antibody-producing cells.* In Chapter 18, we noted that people with malignancies of antibody-secreting plasma cells provided the first material for determining the amino acid sequence of specific antibodies. These patients, afflicted with *multiple myeloma*, have in their blood high levels of particular molecular species of the polypeptide chains of antibodies. Different patients, however, have different antibody chains in high concentration. Because a person's blood normally contains thousands of variant antibodies, all in low concentration, the finding of one antibody species in huge excess suggests that the malignancy arises in a single B cell—but a differently programmed B cell in different people.

3. *Karyotype studies.* A typical feature of the aneuploid karyotypes of cancer cells is the presence of one or several unusual *marker chromosomes* that are seen in all of the cells of a specific tumor. The marker chromosome has a particular aberration that identifies the cells as having a common origin. Aside from marker chromosomes, however, the cells of a tumor may not always be chromosomally identical. Successive chromosome changes can appear as the tumor progresses, and the cell types that proliferate more rapidly will become more frequent.

ONCOGENES AND PROTO-ONCOGENES

A major advance in understanding cancer came with the recognition of a class of genes that regulate the division

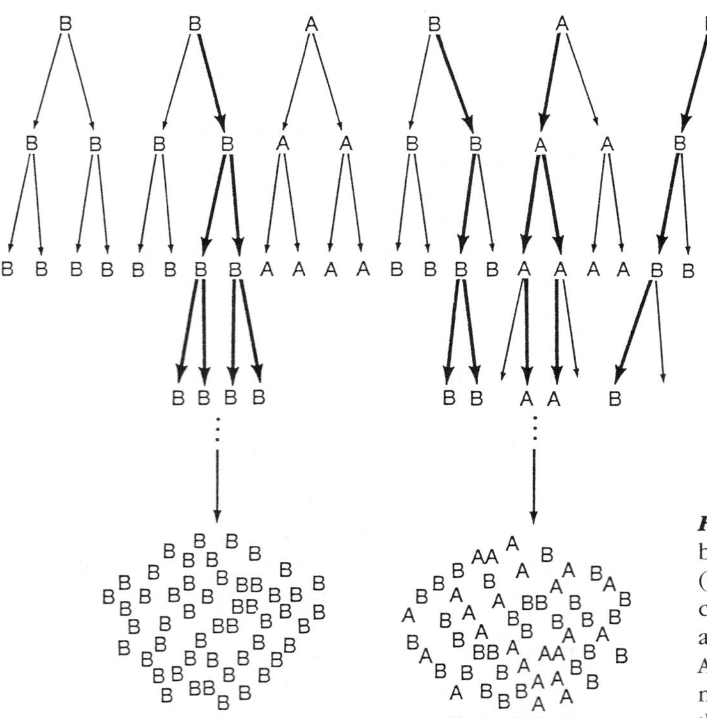

Figure 1 Evidence of the clonal origin of a tumor revealed by X chromosome inactivation. In a woman heterozygous (Gd^A/Gd^B) for the X-linked gene *G6PD*, A and B represent cells showing either the A isozyme or the B isozyme. We assume that the cells in the top row are already determined. An entire tumor always consists of a single type—A or B—never mixed. This outcome can be explained by assuming that the tumor has a single-cell origin, whereas comparably sized bits of normal tissue are of mixed origin.

of cells. These genes were first identified in chickens and mice, where they were picked up by *viruses* and carried to other cells. Some of the same genes—not generally associated with viruses—were later shown to be present in human cells.

Viruses have been pivotal agents in understanding some genetic aspects of cancer. Many viruses consist of little more than a molecule of nucleic acid—DNA or RNA, single-stranded or double-stranded—surrounded by a protein coat. The coat proteins are able to recognize specific receptors on the cells of their hosts. The specificity of the recognition determines what species, and what cell types within a species, a virus infects. By reproducing themselves inside the living cells of a host organism, the viruses may disrupt the cell's normal metabolism and lead in some cases to cell death. Viruses have been shown to be present in only a few human cancers: hepatitis B virus (a DNA virus) in a form of liver cancer, papilloma virus (DNA) in a cervical cancer, and the Epstein-Barr virus (DNA) in a lymphoma (see later). An RNA virus, HTLV-I, has been implicated in a form of human leukemia. These cancers are not common in the United States but are important worldwide.

Although not seen until the advent of electron microscopes, the presence of viruses was signaled early in this century by so-called filterable agents that passed through the pores of filters developed by Pasteur to stop the passage of bacteria. Mosaic disease in tobacco and foot-and-mouth disease in cattle were among the first that were shown to be transmissible by filterable agents. In humans, viruses are responsible for a wide range of diseases: AIDS, yellow fever, mumps, chicken pox, measles, rubella, poliomyelitis, hepatitis, shingles, warts, mononucleosis, influenza, and common colds. Other viruses infect bacteria (the bacteriophage described in Chapter 15), insects, and plants (where they are of great economic importance).

That some viruses can also be *carcinogenic* was shown early in this century by Peyton Rous (1911) while he was a young research assistant at Rockefeller Institute. He found that a cell-free filtrate from a ground-up chicken sarcoma produced the same kind of tumor when injected into other chickens. Despite careful, well-designed experimentation, his work was generally disregarded because it contradicted the then current, noninfectious views about the causation of cancer (Dulbecco 1976). (*Because* it could be transmitted by a cell-free extract, others denied that the highly malignant chicken sarcoma was a cancer!) Not until the 1950s was it generally recognized that viruses could transmit some forms of cancer. (In 1966 Rous, at age 86, was awarded a Nobel Prize for his pioneering work.)

The *Rous sarcoma virus* (*RSV*) was later found to be one of a family of viruses called **retroviruses**, which have been implicated in many types of animal cancers, but in only a few types of human cancers.

Oncogenes Carried by Retroviruses

A retroviral genome consists of single-stranded RNA, 7–10 kb long, that codes for three to six proteins (Watson et al. 1987; Varmus 1988). One of the genes codes for an enzyme called **reverse transcriptase**, which is already present in the mature virus particle. Immediately after entry of the retrovirus into a cell, this enzyme is used to reverse transcribe the RNA genome into a DNA copy. In double-stranded form, the DNA is inserted as a stable **provirus*** at a nonspecific site of a host chromosome. Like the integrated form of a λ bacteriophage (Chapter 15), the provirus is replicated along with the host DNA in successive cell divisions. The proviral DNA is transcribed (in the forward direction this time) into viral *messenger RNAs* to make viral proteins. It is also transcribed into the viral *genomic RNA*, which is packaged with the viral proteins to make complete particles. These can escape the cell and infect others. If the proviral DNA gets incorporated into the eggs or sperm of the host, it can be transmitted to successive generations, indistinguishable in its chemical nature from other genes of the host.

In addition to, or in place of, its usual genes, some retroviruses (including RSV) carry a gene that is unneeded by the virus but is able to convert normal host cells growing in culture into cancer cells. Such a gene is called an **oncogene** (Greek *onco*, "mass"). The integration of a single provirus of RSV is sufficient to accomplish this **transformation**, the change of cultured cells from normal to malignant. The cancer-inducing gene of RSV is called *src* (for sarcoma); when present in the virus, it is designated **v-*src***. The questions that naturally arise are, Where do viral oncogenes come from, and how do they induce cancer?

On the question of origin, the oncogene in the Rous sarcoma virus came from a chicken cell some time in the past. One line of supporting evidence was the demonstration that DNA complementary to v-*src* hybridizes to DNA present in *normal* chicken cells, as well as in the normal cells of turkeys, quail, ducks, and emus (Stehelin et al. 1976). The chicken gene that is the presumed progenitor of v-*src* is called **c-*src***, for **cellular oncogene**, or **proto-oncogene**. It cannot be exactly the same as the viral oncogene, however. During the acquisition by RSV of the cellular proto-oncogene sequences (by processes that are fairly well understood), mutational

*Many years before it could be verified, Howard Temin (1964) of the University of Wisconsin uncovered evidence of a DNA provirus complementary to the RNA of Rous sarcoma virus. Although the new idea that genetic information could be transcribed *from* RNA *to* DNA was resisted, he and David Baltimore of the Massachusetts Institute of Technology independently discovered the enzyme, reverse transcriptase, that does the job—and were awarded Nobel Prizes in 1975.

events lead to alterations in the protein encoded by c-*src*. The fact that the proto-oncogene is conserved among several bird species suggests that it is essential to the metabolism of all avian cells.

Researchers have now defined several dozen different proto-oncogenes that were picked up by some retrovirus to become viral oncogenes (Table 1). Remarkably, almost all of the oncogenes have homologous DNA sequences in the *human* genome. The chromosomal sites of the human genes have been determined by analyses of human-mouse somatic cell hybrids or by in situ hybridization (Marshall 1985; see also Chapter 15). In the first method, researchers make use of a group of hybrid cells, each carrying a different set of human chromosomes. DNA from each line is extracted and run on Southern blots with radioactive probes specific for the oncogene in question. In the second method, radioactive probes are hybridized to denatured metaphase chromosome spreads.

We shall return to the possible roles played by the encoded proteins after describing the direct demonstration of oncogenes and proto-oncogenes in human cells.

Finding Human Oncogenes by Gene Transfer

Because the descendants of cancer cells are themselves cancer cells, it was long ago suggested that the initial malignant change is genetic and is passed from one cell generation to the next. The work with retroviral oncogenes described above also argues that some cancerous changes in the somatic cells of animals are due to the action of single genes. Furthermore, animal carcinogens are usually found to be mutagens in the Ames test. Investigators have now shown *directly* that specifically

Table 1 Some oncogenes carried by retroviruses and the location of homologous human genes

Oncogene	*Retrovirus carrier*	*Animal source*	*Type of tumor in animal*	*Human chromosomes with a homologous proto-oncogene*
abl	Ableson murine leukemia virus	Mouse	Pre-B cell leukemia	9
	Hardy-Zuckerman feline sarcoma virus	Cat	Sarcoma	
*erb*A	Avian erythroblastosis virus	Chicken	Erythroblastosis, sarcoma (when with *erb*B)	17
*erb*B	Avian erythroblastosis virus	Chicken	Erythroblastosis, sarcoma	7
fos	Finkel-Biskis-Jinkins murine sarcoma virus	Mouse	Osteosarcoma	14
H-*ras*	Harvey murine sarcoma virus	Rat	Sarcoma, erythroleukemia	11, X
K-*ras*	Kirsten murine sarcoma virus	Rat	Sarcoma, erythroleukemia	6, 12
jun	Avian sarcoma virus 17	Chicken	Fibrosarcoma	?
mos	Moloney mouse sarcoma virus	Mouse	Sarcoma	8
myc	MC29	Chicken	Sarcoma, carcinoma, myelocytoma	2, 8
sis	Simian sarcoma virus	Monkey	Sarcoma	22
	Parodi-Irgens feline sarcoma virus	Cat	Sarcoma	
src	Rous sarcoma virus	Chicken	Sarcoma	1, 20

Source: Mostly from Teich (1986), who lists 24 oncogenes carried by 50 different retroviruses.

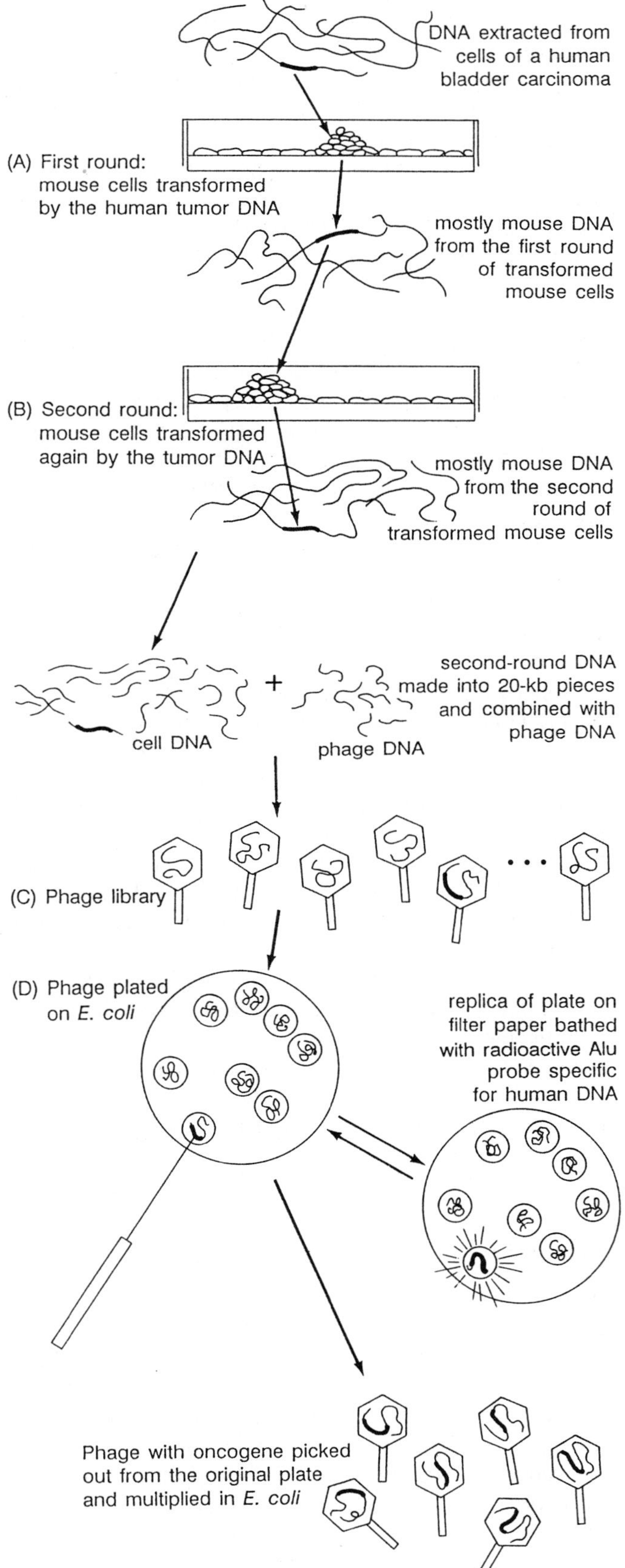

identified, dominantly acting, mutant genes can indeed influence the course of human cancer.

One line of experimentation was carried out by Robert Weinberg and Chiaho Shih at the Massachusetts Institute of Technology and by others (summarized by Weinberg 1983). These researchers began with DNA extracted from a human bladder carcinoma cell line. They used this DNA to transform a special line of noncancerous mouse fibroblast cells growing in tissue culture. During the transformation process, short sections of the human DNA enter the mouse cells and integrate with the mouse DNA, although with low frequency. (The entry of DNA into cells is facilitated by first precipitating DNA with calcium phosphate.) The transformed mouse cells are identified by their abnormal growth behavior: rather than remaining a flat layer—one cell thick—on the surface of the nutrient medium, they grow into a mound of cells during a two- to three-week period (Figure 2A). Thus the information for cancer induction had been transferred from the bladder carcinoma cells to the mouse cell line by means of DNA extracted from the carcinoma. Such entry and incorporation of foreign DNA into animal cells is called **transfection**.

The transformed mouse cells were isolated and grown further; DNA extracted from these cells was used in a second round of transfection (Figure 2B). The occurrence of new foci of abnormal growth on the fresh cells confirmed that cancer-causing DNA not only had entered the mouse cells during the first round, but also had been copied and multiplied in the descendant tumor cells. Note that only short sections of DNA survive extraction, purification, and calcium phosphate precipitation, and that only one or a few sections are likely to be taken up and stably incorporated into the DNA of a given host cell. Thus serial transfections show that *a single short segment of human DNA was causing the cancerous change in a dominantly inherited fashion*. The researchers confirmed this remarkable finding by showing that

Figure 2 Cloning and identification of a human oncogene — shown as the thick segment of DNA (▬). DNA extracted from human bladder carcinoma cells was used to transform mouse cells (A), and the process was serially continued (B). DNA from the tumorous mouse cells was cloned in bacteriophage λ (C). The tiny fraction of phage harboring human DNA (among the overwhelming number of phage with mouse DNA) was identified by use of a radioactive DNA probe (D). The probe was complementary to repetitive Alu sequences present throughout human DNA but absent in mouse DNA. See also Figures 9 and 10 in Chapter 15. (Modified from Weinberg 1983.)

the transforming DNA was contained entirely within one fragment generated by the restriction enzyme *Eco*RI.

The next step was to isolate the piece of human DNA that caused the transformation of the mouse cells (Shih and Weinberg 1982). This isolation was accomplished by cloning the DNA from the tumorous mouse cells in a library of bacteriophage λ (Figure 2C) (see the cloning procedures described in detail in Chapter 15). The total genomic DNA is broken into many thousands of short segments, which are inserted randomly into phage DNA. Each recombinant phage particle carries some of its own genes (so that it can still infect *E. coli*) and one or another of the segments of DNA from the transformed mouse cells. In the library of phage, all DNA segments from the transformed mouse cells are likely to be present, including the human oncogene—the specific piece of human DNA that was responsible for the transformation. Each of the phages in the library can be amplified by plating on a lawn of *E. coli*, which leads to the formation of phage plaques (see Figures 9 and 10 in Chapter 15).

The final step in isolating the human oncogene was to screen the phage plaques with a radioactive probe complementary to the nucleotide sequence of the human DNA. This task looks impossible: Would not the experimenters need to have prior knowledge of what they were searching for—namely, the nucleotide sequence of part of the oncogene? Recall, however, that human DNA is characterized by certain repetitive sequences, called Alu, that are present more or less randomly throughout the genome in hundreds of thousand copies and that constitute 5% of the DNA in a human cell (Chapter 13). Alu sequences are present in human DNA but *not* in mouse DNA. Thus Weinberg and his associates used an Alu probe to find those phage that contained human DNA, and presumably the bladder carcinoma oncogene as well (Figure 2D). The assumption was that an Alu sequence was close to the transforming DNA.

Evidence for the successful isolation of the bladder oncogene was clear: the cloned piece of human DNA had potent biological activity. "Whereas two micrograms of the original bladder-carcinoma DNA had induced, on the average, one colony of transformed cells, a comparable mass of the cloned transforming gene induced as many as 50,000 foci" (Weinberg 1983). Furthermore, the cloned oncogene hybridized readily and strongly with a DNA sequence in *normal* human cells—the presumed proto-oncogene.

Further research has shown that the nucleotide sequence of the bladder carcinoma oncogene is similar to that of a rodent oncogene called *ras* (for rat sarcoma) carried by a retrovirus. There are several members of the *ras* family of oncogenes, some known to be present in retroviruses and some not (Levinson 1986). Thus it appears that the *same* proto-oncogenes can be changed to oncogenic forms either by passage through a retrovirus or by some sort of intracellular mutational event not associated with viruses. Genes with homology to *ras* are widely distributed throughout the animal kingdom, including fruit flies, slime molds, sea slugs, and yeast. In humans, oncogenes from a number of tumors, including carcinomas of the lung, pancreas, and colon, are also in the *ras* family. Thus similar proto-oncogenes can give rise to different types of human cancers, depending on the specific tissue that harbored the oncogenic change.

By cloning and sequencing both the oncogene and the corresponding proto-oncogene, Weinberg and his colleagues (and other investigative teams) were able to pinpoint the mutational event that converted the one into the other. The DNA molecules differed by a single base: G in the proto-oncogene is substituted by T in the oncogene. As a consequence, the DNA triplet codes for glycine in the proto-oncogene protein and valine in the oncogenic product. Studies on the consequences of this change are discussed in the next section.

About 40 oncogenes and their corresponding proto-oncogenes have been located in the genomes of humans and other animals by procedures involving DNA transfection, by studies of oncogenic retroviruses, or by other means (Bishop 1987). Some discoveries, for example, have involved *chromosomal rearrangements* in which the breaks alter the activity of proto-oncogenes, and some have involved *amplification of genes* in which many extra copies of a particular sequence are present in a cell. It has been found that a single change by itself is probably never sufficient to cause a cell to become cancerous. Several cellular events are required for malignancy. Each misstep in the progression may foster additional abnormalities characteristic of cancer: cell proliferation, invasion of adjacent tissue, metastasis, and establishment of new tumorous sites. Although not all of these events may be directly precipitated by *genetic* damage, recent knowledge of oncogenes has provided a basic unifying mechanism to view the myriad aspects of carcinogenesis.

The Proteins of Oncogenes

Researchers are beginning to delve into the physiological functions of the protein products of oncogenes and proto-oncogenes (Hunter 1984, Weinberg 1985, Teich 1986). They hope to find important—perhaps essential—keys to the success of normal cells in controlling their growth and differentiation and to the failure of malignant derivatives to do the same. The dozen or so oncogenic proteins that have been studied appear to act in diverse ways. Although general conclusions remain elusive, we give a few examples of the patterns of action displayed by oncogenes.

PROTEIN KINASES. Currently, the largest number of products of oncogenes appear to be **protein kinases**, enzymes that modify other proteins after their translation. (Kinase comes from a Greek word meaning "to move.") The kinases act by *phosphorylation*, transferring the terminal phosphate group of ATP (adenosine triphosphate) to the hydroxyl groups on the side chains of serine, threonine, or tyrosine (leaving behind ADP, adenosine diphosphate). The transfer alters the three-dimensional shape of the phosphorylated protein, and either increases or decreases its activity (Alberts et al. 1989).

The protein kinases are themselves affected by fluctuations in cell metabolites, such as shifts in the concentrations of so-called second messengers that, in *their* turn, can be activated by the presence of hormones, growth factors, and neurotransmitters (the so-called first messengers). For example, the binding of the hormone-like protein, epidermal growth factor, to its receptor on a cell surface induces phosphorylation of the receptor and other proteins. It is thought that this kinase reaction to epidermal growth factor is what triggers division of previously nondividing cells in vitro. Thus, along with other molecules that act as chemical signals within cells, the protein kinases are important regulators of events in the cell cycle (Berridge 1985).

An example is the enzyme encoded by the *src* gene carried by the Rous sarcoma virus and homologous to human genes on chromosomes 1 and 20 (Table 1). The enzyme was the first protein kinase found to phosphorylate *tyrosine*, rather than the serine and threonine residues phosphorylated by most protein kinases (Hunter 1984). Bound to the inner surface of the cell membrane, the *src*-encoded kinase phosphorylates many different proteins. One of them, *vinculin*, is involved in a cell's cytoskeleton, especially in the structure of adhesion plaques that help bind cells to surfaces and to each other. Researchers speculate that increased phosphorylation of vinculin in cancer cells could produce their rounded shapes and decreased adhesion.

Another example of a tyrosine-specific protein kinase is the *abl* gene product. The proto-oncogene was discovered first in mouse leukemic cells, but it is also located near the tip of the long arm of human chromosome 9. The c-*abl* locus is a translocation breakpoint that produces the human Philadelphia chromosome present in some forms of leukemia (see next section). The protein product is an enzyme that has a molecular weight of 145,000 and is embedded in the plasma membrane.

OTHER PROTEIN PRODUCTS. Less is known about the action of other oncogenic proteins. It would be particularly interesting to learn the physiological function of the *ras* family of oncogenes, because they are directly involved in several human cancers. Recall from the last section that the mutation in the *ras* proto-oncogene leading to human bladder carcinoma changes one amino acid (out of 189) in the protein product. The normal *ras* product binds to the nucleotide GTP (guanosine triphosphate) at the inner surface of the cell membrane and hydrolyzes it to the less reactive form, GDP (releasing a phosphate group). This reaction in turn may modulate the activity of second messengers within the cell. Just such activity is known to occur when certain other GTP-binding proteins are stimulated after the hormone adrenaline attaches to the cell surface. The protein encoded by the *ras* oncogene hydrolyzes GTP less well, but little is known of the specific significance of this for cancer induction.

A number of oncogenic proteins have been shown to act in cell nuclei, where they might be expected to bind with DNA. One example is the protein encoded by *jun*, an oncogene first discovered in a chicken virus. The human proto-oncogene product of c-*jun* interacts with specific target DNA sequences (eight nucleotides long) to regulate gene expression (Bohmann et al. 1987). (The product of another proto-oncogene, c-*fos*, binds with the c-*jun* product to enhance its activity.) Another protein with nuclear activity is the one encoded by *myc*, also discovered originally in a chicken virus. The homologous human gene is located near the tip of the long arm of chromosome 8, at one of the breakpoints involved in translocations that are found in patients with Burkitt lymphoma (see next section). The new neighbors for c-*myc* cause it to be overexpressed; that is, the same protein of 439 amino acids is produced, but in increased amounts.

Finally we note the *sis* oncogene carried by the simian sarcoma virus. It seems to be derived from a monkey gene that codes for one of the polypeptide chains of platelet-derived growth factor (PDGF), a protein that aids cell division at the site of a wound. The similarity of PDGF and the oncogenic product was unexpectedly revealed in a computer-aided search for *any* protein with amino acid sequences like those of the few then-known fragments of PDGF. Thus (to paraphrase Watson et al. 1987), the function of the oncogene was discovered in the fastest possible way: it was identified as one whose function was already known. Researchers do not yet know how the oncogenic change causes the sarcoma. Whereas the normal PDGF is delivered to receptors at cell surfaces in a regulated manner, the oncogene-encoded protein may be produced continually or at inappropriate times or places.

In summary, carcinogenesis is at least partly precipitated by stepwise genetic damage occurring in somatic cells. The subversions from recessive proto-oncogenes to dominant oncogenes alter the quantity or quality of the encoded proteins, which in turn disturb elaborate, balanced signaling systems extending from the cell surface to the nucleus. It may be, however, that most or all oncogenes affect only a few key controlling functions. New insights are needed: they will add to our understanding of both abnormal malignant processes and normal cell growth and differentiation.

Transgenic Mouse Models

Laboratory strains of mice that have a foreign gene stably incorporated into their germline DNA are called **transgenic mice**. One example is the strain of mice carrying the human *HPRT* gene described in Chapter 16. An important new approach to the study of cancer derives from the corresponding ability to insert an oncogene into the germinal cells of a mouse, allowing copies of the oncogene to be transmitted from generation to generation (Hanahan 1988). The study of such mice has been useful in exploring how a particular oncogene affects the multistep process leading to cancer.

To produce a transgenic mouse carrying an oncogene, experimenters first clone in a bacterial plasmid the DNA sequences constituting the oncogene. A solution with this DNA is then injected via a thin glass needle (and skillful hands) into one of the pronuclei of a fertilized mouse egg (Figure 3). The eggs that survive the microinjection are then surgically implanted in a foster mother (Gordon and Ruddle 1983). Some of the embryos that come to term will have incorporated the oncogene into their own DNA (at random sites) in some or all of the cells of their bodies. In about 10% of births, the oncogene is also incorporated into the germline, and can thus be propagated by appropriate matings with standard strains. The numerous mice descended from a single transgenic progenitor will all carry the same oncogene and will be genetically similar.

The "transplanted" oncogene will often elicit tumors in a reproducible fashion in all members of a transgenic mouse strain. Different tissues appear to be particularly susceptible to transformation by specific oncogenes. For example, the *fos* oncogene expresses itself in transgenic mice by pathological changes in bone and thymus tissues that are similar in nature to those that arise via infection of mice by retroviruses carrying *fos*. The proto-oncogene *myc* (when joined with the immunoglobulin heavy chain gene) transforms the B lymphocytes of transgenic mice, but not other cell types. This expression is similar to that in human Burkitt lymphoma, where the c-*myc* proto-oncogene is juxtaposed by reciprocal translocation to genetic elements coding for immunoglobulin chains (see later).

The development of cancers in the transgenic mice is usually not direct, but is preceded in almost every case by the appearance of proliferating cell lines that are not themselves cancerous. Thus the study of transgenic mice confirms that the presence of an oncogene is a necessary but not sufficient condition for manifestation of malignancy. The primary action of the oncogenes seems to be the loss of restraints on cell division; secondarily, one or more of these cells may become cancerous as the result of further cellular changes.

CANCER AND CYTOGENETICS

For many years, researchers have seen abnormalities of chromosome structure or number in cancer cells. As cytogenetic techniques continue to improve, more and

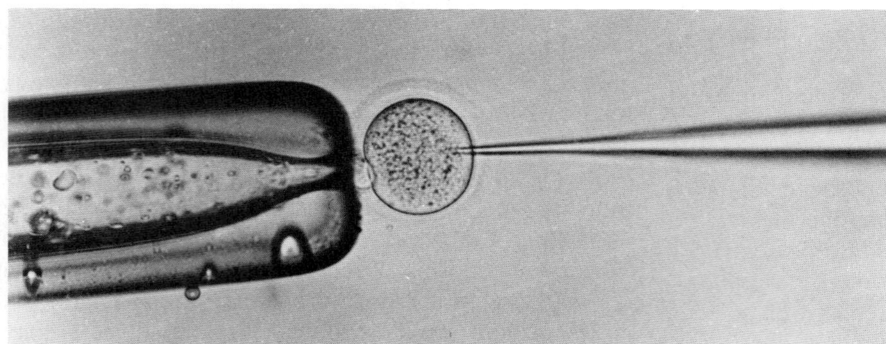

Figure 3 Injection of foreign DNA into a fertilized mouse egg. The egg is held on the blunted tip of a glass micropipette by slight suction (left). The injecting microneedle (right) is inserted into one of the two pronuclei of the egg, delivering about 1 picoliter of DNA solution. If the injection is successful, the pronucleus will swell noticeably. The eggs are in a petri dish placed on the stage of a low-power microscope (about 150×), and the movement of the needle is controlled precisely through the knobs of a micromanipulator. Successfully injected eggs are then surgically transferred to the oviducts of a foster mother for development. (Courtesy of Gordon and Ruddle 1983.)

more aberrations—sometimes subtle ones—are being detected (Yunis 1983; Sheer 1986). The karyotypes of the cancerous tissues are often observed to change over time as the cells evolve; this trend signals a poorer prognosis for patients. It is believed, however, that from the time of inception of the malignancy, some *specific* aberrant chromosomes are consistently present in some *specific* cancers. Thus, although it was once thought that chromosomal aberrations were secondary to perturbed cellular growth, it now appears that the genesis of structural change itself can contribute to cancer induction. The first such discovery—of the so-called *Philadelphia chromosome*—was made in 1960 at the University of Pennsylvania in Philadelphia (see below). More recently, this and other cancer-causing aberrations have been shown to affect the activity of proto-oncogenes (Haluska et al. 1987). We start with two well-studied malignancies, one a leukemia and one a lymphoma, each associated with a translocation and involving an oncogene at or near one of the breakpoints.

Proto-Oncogenes at Translocation Breakpoints

As already noted, the various leukemias are cancers of immature white blood cells. Induced in part by ionizing radiation, leukemia occurs with increased incidence among radiologists exposed to excessive amounts of X-irradiation, among patients receiving large doses of therapeutic radiation, and among the people exposed to the atomic bomb explosions (Clarkson 1982). In *myelogenous leukemia*, the affected cells are the stem cells (myeloblasts) that normally mature into the so-called granular white blood cells having a number of different protective functions, including phagocytosis. In *lymphocytic leukemia*, the affected cells are the stem cells (lymphoblasts) that normally mature into B and T cells. Additional leukemias arise as a result of the abnormal proliferation of other types of white blood cells.

In the bone marrow, the massive buildup of leukemic cells crowds out both red blood cells (thereby producing anemia) and platelets (producing internal bleeding). In addition, the liver, spleen, and other vital organs are damaged by large numbers of infiltrating cancer cells. Advance of the disease can either be slow (*chronic*) or rapid (*acute*), the latter type leading to death in a few weeks or months if untreated. In the acute forms, which mainly affect children, the primitive nature of the stem cells is retained in the cancer cells, whereas in the chronic forms, substantial differentiation occurs. A frequent cause of death in acute leukemias is uncontrolled infection due to the deficiency of mature white blood cells.

Chronic myelogenous leukemia (CML) affects mainly people of middle and older ages and is invariably fatal. Occurring in about 1 per 160,000 persons per year, it accounts for about 15% of all cases of leukemia. It is easily recognized because, in over 90% of patients the leukemic cells have a distinctive chromosomal marker—the so-called **Philadelphia chromosome** (**Ph^1**), which is a shortened chromosome 22 originally thought to result from a simple deletion (Nowell and Hungerford 1960). The introduction of banding techniques in the early 1970s, however, allowed investigators to show that terminal pieces of the long arms of chromosomes 9 and 22—*unequal* in size—had switched places by a reciprocal translocation (Figure 4) (Rowley 1973).

Blood examinations of the atomic bomb survivors who developed CML showed that an average of about eight years elapses between the original translocational event that produces the Ph^1 chromosome and the development of clinical symptoms. Patients whose leukemic cells later develop additional chromosomal abnormalities often die within a few months. The most commonly observed secondary alterations include the following changes, alone or in combination (Le Beau and Rowley 1986):

- gain of an additional Ph^1 chromosome
- gain of an isochromosome of the long arm of chromosome 17
- gain of other chromosomes, especially chromosome 8

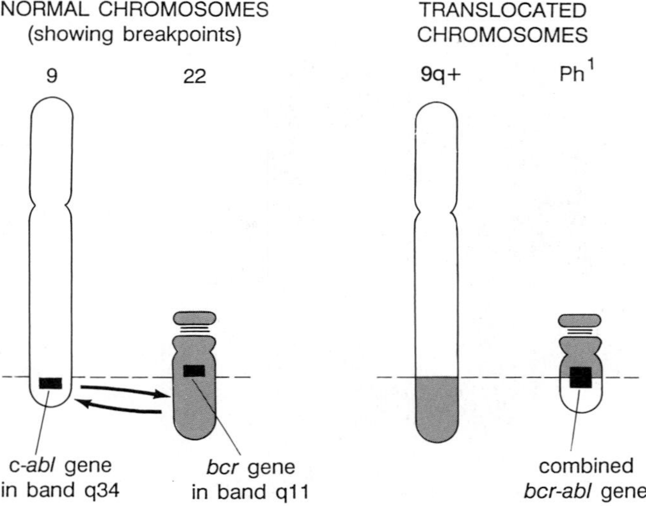

Figure 4 The reciprocal translocation between chromosomes 9 and 22 that produces the Philadelphia chromosome (Ph^1) characteristic of chronic myelogenous leukemia. Breakage and reunion joins together the c-*abl* proto-oncogene on chromosome 9 with a so-called *breakage cluster region* (*bcr*) of chromosome 22. Transcription and translation of the fusion gene yields an abnormally large *bcr-abl* protein.

It is suggested that genes on these chromosomes—as yet unidentified—confer a proliferative advantage to the leukemic cells.

The regions around the breakpoints of chromosomes 9 and 22 have been cloned and shown to involve the cellular proto-oncogene c-*abl* (Heisterkamp et al. 1983; Holt et al. 1987). This work used Southern blots to identify sequences near the fusion area in DNA samples taken from CML patients. The complementary DNA probes were homologous to v-*abl* or c-*abl* genes. The breakpoint in chromosome 9 appears to be upstream of the *abl* gene, although the exact breakpoint occurs over a range of 50 kilobases or more in different patients. In contrast, the breakpoint in chromosome 22 appears to be restricted to a 5.8-kb segment in a gene that is called the *breakage cluster region* (*bcr*), but whose function is not known. The fused gene is transcribed into an abnormally long mRNA, which (after processing) is translated into a protein in which about 25 amino acids normally present at the amino end of the *abl* protein are replaced by 600-700 amino acids coded by the *bcr* gene. Different CML patients have about the same size hybrid protein despite the variations in the position of the breakpoint. This uniformity results from the fact that the breakpoints are in introns that are spliced out of the mRNA transcript. The protein products of *both* the fusion gene and the intact cellular *abl* gene appear to have tyrosine kinase activity, so the malfunction of *bcr-abl* remains a matter of speculation.

Another translocation with a breakpoint near another proto-oncogene, c-*myc* on chromosome 8, is a consistent feature of **Burkitt lymphoma (BL).** Occurring commonly in parts of central Africa (about 1 case per 10,000 children per year), and rarely in other regions of the world, BL is a malignancy of B lymphocytes that mature into the antibody-producing plasma cells (Ziegler 1981; Burkitt 1985). In a wide belt of equatorial Africa, it is the most frequent cancer among children, the average age at onset being about seven years. First described in 1958 by Denis Burkitt, a British surgeon working at Makerere University in Uganda, the solid tumors typically affect two areas: the bones of the jaws and the organs in the abdomen. (Unlike leukemia, the bone marrow is not usually involved.) Extremely fast-growing and aggressive, the tumors can reach prodigious size in a matter of days and obstruct neighboring organs (Figure 5). The kidneys may also be damaged by the excessive elimination of uric acid, a by-product of the breakdown of DNA, which results from the rapid growth and death of cancer cells. Fortunately, treatment with anti-cancer drugs often gives dramatic improvement and leads to long-term survival in about half the cases.

The malignant B cells of BL patients have a translocation involving the c-*myc* proto-oncogene at band q24 on chromosome 8. The other participant in the translocation in most cases is chromosome 14, or less commonly chromosome 22 or 2 (Figure 6). Strikingly, the

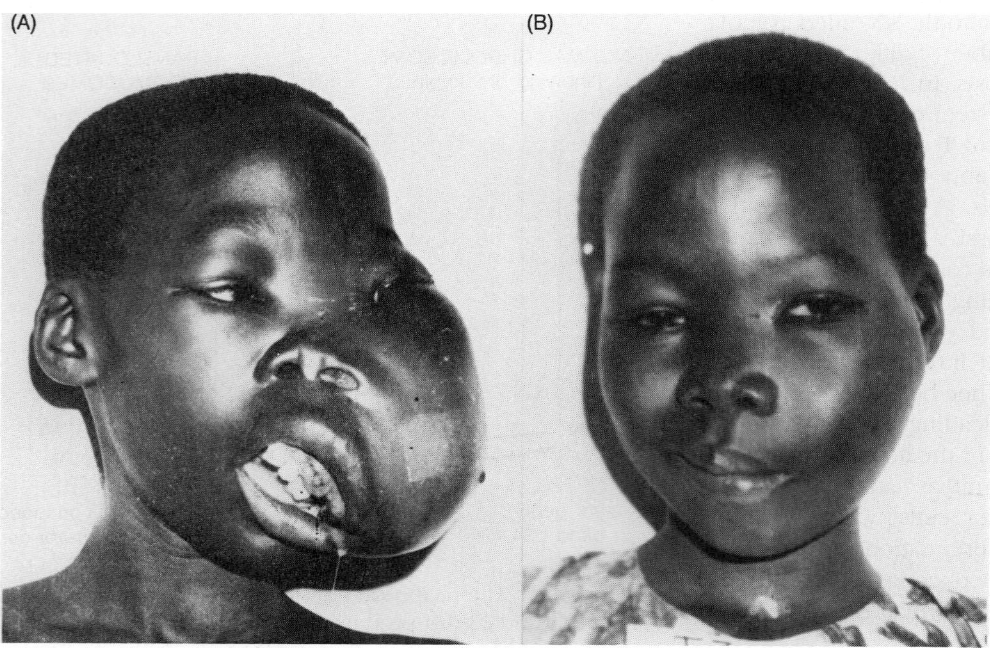

Figure 5 (A) A massive jaw tumor characteristic of Burkitt lymphoma in a 9-year-old Ugandan girl. (B) The same child 3.5 weeks after two injections of an anti-cancer drug, cyclophosphamide. (From R. Owor and C. Olweny 1978.)

Chromosome:	8	14	22	2
Breakpoint band:	q24	q32	q11	p11
Breakpoint gene:	c-*myc*	*IgH*	*Igλ*	*Igκ*
Approximate percentage of cases:	all	80-90	5-15	5

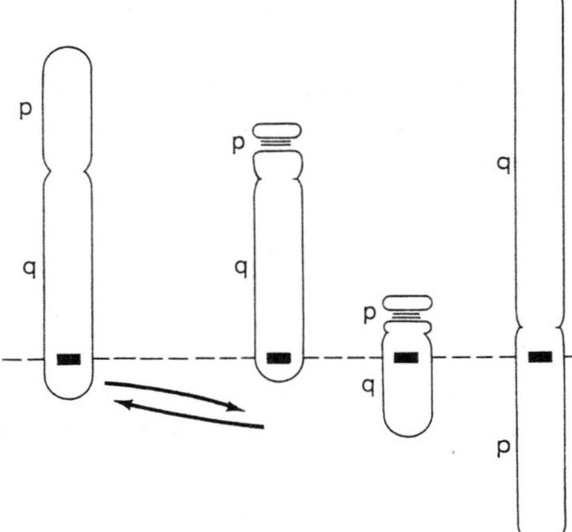

Figure 6 The breakpoints and genes involved in the translocations producing Burkitt lymphoma. In all cases, a breakpoint occurs in chromosome 8 at the c-*myc* proto-oncogene. The other breakpoint occurs as indicated in either chromosome 14, 22, or 2. *IgH* is the gene coding for the heavy chains of immunoglobulins; *Igλ* and *Igκ* are genes coding for two slightly different versions of the immunoglobulin light chains (see also Figure 6 in Chapter 18).

breakpoints in these three chromosomes are all within genes that code for the heavy or light chains of immunoglobulins (Croce and Klein 1985; Haluska et al. 1987). The 8;14 translocation shifts the c-*myc* locus from chromosome 8 to chromosome 14, whereas the less frequent translocations are in the reverse direction: they shift the light chain immunoglobulin genes from 22 or 2 to chromosome 8.

Recall from Chapter 18 that the immunoglobulin chains, which produce millions of different antibody species, are coded by adjacent groups of genetic elements that recombine with each other during the maturation of B cells. The heavy chains, for example, are coded by an extended length of DNA including four families of genetic elements, *V*, *D*, *J*, and *C* (see Figure 9 in Chapter 18). Because the antibody genes in B cells are so active in such complex ways, it is perhaps not surprising that a proto-oncogene repositioned nearby might have *its* activity altered.

As with the *bcr-abl* fusion gene, which leads to chronic myelogenous leukemia, the juxtaposed genes of c-*myc* on chromosome 8 and one of the immunoglobulin chain genes (*IgH* or two variants of *IgL*) have been analyzed at the nucleotide level. For example, the breakpoint within *IgH* may lie anywhere from the *D* elements to the *C* elements, but is often just upstream of C_μ in a region involved in *class switching*. (As noted in Chapter 18, class switching means that the immunoglobulin product of a B cell changes production from one antibody class to another.) The 8;14 translocation occurs prior to B cell maturation, so the rearrangements of the genetic elements that normally lead to antibody diversity do not occur. Although the malignant cells may produce antibodies, these are coded for by the allele on the homologous chromosome—the one not involved in the translocation.

Unlike the *bcr-abl* protein product, the normal c-*myc* protein with 439 amino acids remains intact and unattached to any other polypeptide. As a result of its new neighbor, however, c-*myc* becomes *deregulated*, being transcribed and translated at inappropriate times or at increased rates. This deregulation is thought to be one of the steps required to produce the malignant transformation, but it is probably not the only step. Information on this point is speculative, in part because the function of the c-*myc* protein is unknown, other than its being present in the nucleus and binding to DNA.

The African (or *endemic*) form of Burkitt lymphoma differs somewhat from the non-African (or *sporadic*) form. Outside of Africa, for example, BL usually presents itself in older children, and more often involves abdominal tumors and less often jaw tumors. The breakpoints of the translocations also differ. For example, in the endemic cases, the whole c-*myc* gene is intact, including an exon that does not code for protein but contains promoters and other regulatory elements; in the sporadic cases, the breakpoint in c-*myc* occurs between this exon and the protein-encoding region (Holt et al. 1987).

In addition, the role of the **Epstein-Barr virus (EBV)** differs in the endemic and sporadic forms. A herpes virus containing double-stranded DNA, EBV was first seen in electron micrographs of an African Burkitt lymphoma cell line in 1964, prompting intensive research into the underlying causal relationships (Henle et al. 1979). Epstein-Barr virus was found to be present in over 90% of African patients with BL, but in only 15–20% of cases outside of Africa. Surprisingly, researchers also discovered that EBV regularly infects almost all African children, but only a small percentage are thereby stricken with Burkitt lymphoma—or any other clinical disease. Thus EBV is not a sufficient cause of BL. Nor is it a necessary cause, since some cases—especially the sporadic ones—have no detectable EBV nor antibody to EBV antigens. That the virus selectively infects B cells, however, suggests an etiological association, but the nature of the line of causation from the virus to the cancer is not clear, and several hypotheses have been suggested (Ziegler 1981). It is unlikely that EBV is a casual passenger in the cells of patients with Burkitt lymphoma. As readers may know, EBV also causes *infectious mononucleosis*, which is characterized by fever, sore throat, swollen lymph nodes, and general malaise—often in young adults, and not uncommonly among college students. Epstein-Barr virus is also associated with another human cancer, *nasopharyngeal carcinoma*, which is very rare except among Chinese adults.

Other Specific Chromosomal Abnormalities

In addition to chronic myelogenous leukemia and Burkitt lymphoma discussed above, many other cancers of *hematopoietic* tissue (pertaining to blood cells) are associated with chromosomal abnormalities, although usually with less consistency (Yunis 1983; Sheer 1986). The most common rearrangements are translocations, and less often inversions, deletions, trisomies, and so on. The total number of chromosomes in any of the leukemias and lymphomas is generally in the diploid range—near 46. Interestingly, the *same aberrations* are sometimes seen in several *different cancers*. For example, the Ph^1 chromosome characteristic of chronic myelogenous leukemia is also seen in the cancer cells of some other leukemias (but not in the high frequency characteristic of CML). The breakpoint in chromosome 14 at the band (q32) containing the immunoglobulin heavy chain gene, in addition to its occurrence in Burkitt lymphoma, is also seen in a number of other leukemias and lymphomas.

On the other hand, several *different aberrations* may lead to the *same cancer*. In these cases, the prognosis for patients and their response to treatment may vary depending on the specific chromosomal defects. As one example, Yunis et al. (1984) found 17 different types of aberrations among 99 randomly selected adult patients with *acute nonlymphocytic leukemia*. (Seven patients showed no detectable chromosomal aberrations even with high-resolution banding techniques.) The main categories included five translocations, an inversion, a monosomy, a trisomy, and a deletion. Those carrying an inversion of chromosome 16 had the best prognosis, many of them surviving many years. Those with trisomy 8 had intermediate survival—about one year. A number of patients with complex chromosomal defects usually lived only two or three months. The authors hope that knowledge of the specific aberrations in a given patient with acute nonlympocytic leukemia can lead to more individualized therapy.

The leukemias and lymphomas whose chromosomal aspects have been so intensively studied represent only about 8% of all human cancers. Less is known about the cytogenetic aspects of the other 92%, including the common solid tumors of the breast, lung, colon, prostate, and pancreas. Dividing much more slowly than the cells of hematopoietic cancers, the cells of most solid tumors are more difficult to grow in culture and less likely to be seen during the mitotic divisions needed to obtain karyotypes (Sandberg et al. 1988). Another problem is that solid tumors, at the time they are detected, have a much greater range of chromosomal defects among cancers of the same type and even within the same tumorous growth (Figure 7). Dozens of different secondary changes, often with a great many extra chromosomes, may obscure a primary karyotypic event that led directly to the cancerous transformation.

The idea that secondary chromosomal changes in cancer cells can lead to greater malignancy has had dramatic confirmation in the discovery of repetitive chromosomal elements that contain an oncogene. In some lung cancers (*small-cell lung carcinoma*, constituting about 25% of lung cancers), and in some cancers of neural tissue (*neuroblastomas*, rare tumors of embryonic origin), researchers have found two different visual indicators of repetition:

- extra unbanded chromosomal segments called *homogeneously staining regions (HSRs)*
- numerous tiny, acentric chromosomes that look like double dots—so-called *double minute* chromosomes (*DMs*) (Figure 8)

In situ hybridization and Southern blotting procedures using a probe for the *myc* oncogene suggest that HSRs and DMs contain one or another variant of *myc* repeated many times— representing a phenomenon called **gene amplification** (Seeger et al. 1985, for example). Several HSRs may occur in the same cell line, and their positions are not necessarily near the locus of the proto-oncogene. Because they are never seen in the same cell, HSRs and DMs probably represent alternative microscopic forms of gene amplification (Sheer 1986). The increase in copy number of the *myc* proto-oncogenes often occurs in later stages of the cancers, probably contributing to the progression to more highly malignant forms. Recall also that the *myc* proto-oncogene is at a breakpoint in the translocations that lead to Burkitt lymphoma and is presumed to be deregulated by its juxtaposition to immunoglobulin genes.

We note finally that all of the chromosomal changes discussed in this section occur in the *cancerous* cells but *not* in the normal cells of the cancer patient, and *not* in the germline where they might be included in eggs or sperm. In the next section, we discuss two cancers, retinoblastoma and Wilms tumor, in which a small deletion may be present in all the cells of the body and may be passed from one generation to the next.

(A)

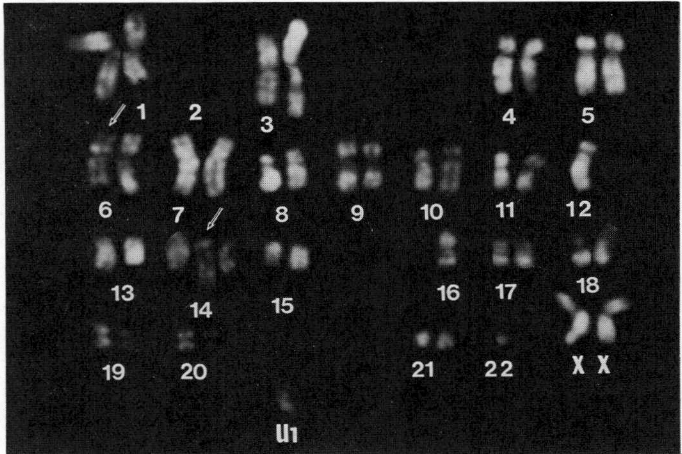

(B)

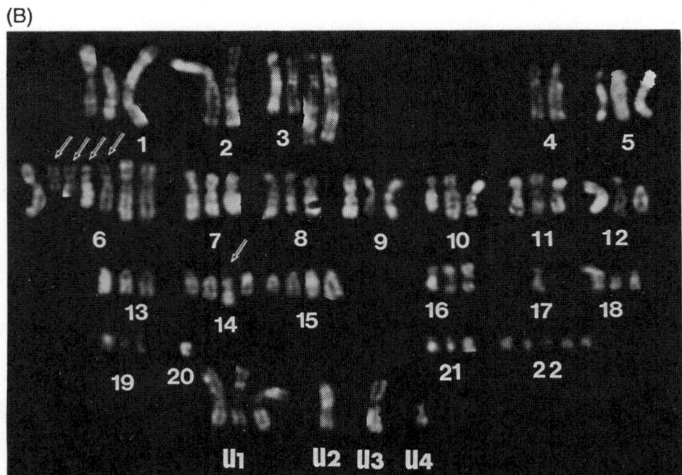

Figure 7 Two cells that are derived from the *same* ovarian cancer but show different karyotypes. (A) A cell showing a 6;14 translocation (arrows), a chromosome of unknown origin (U1), and a total of 45 chromosomes. The translocation is thought to be the primary cytogenetic change. (B) A cell showing the same 6;14 translocation, several chromosomes of unknown origin (U1–U4), and a total of 77 chromosomes. (From Sandberg 1988.)

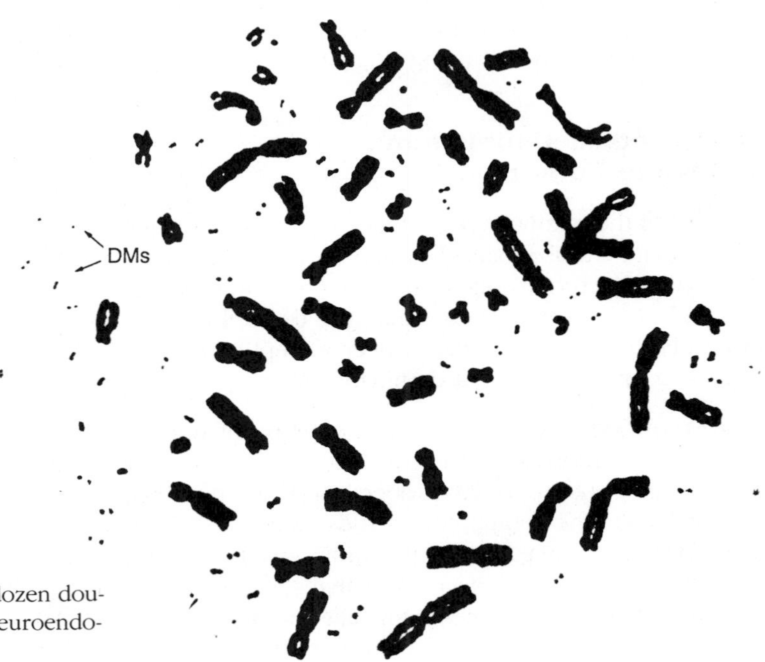

Figure 8 A metaphase spread showing several dozen double minute chromosomes (labeled DMs) from a neuroendocrine tumor cell line. (From Sheer 1986.)

CANCER AND MENDELIAN INHERITANCE

Although the development of cancer is primarily related to the behavior of *somatic* cells, malignancies can also be influenced by *germinal* mutations. Some pedigrees of families with cancer show ratios characteristic of dominant or recessive Mendelian transmission, while other pedigrees reveal only a very small increased risk among close family members (compared with the general population). Some types of cancer that have a low risk of occurrence are part of more generalized syndromes. *Each of the cancers with a familial component is uncommon, and collectively they constitute only a small percentage of all malignancies.* What is inherited in "inherited cancer" is a cellular defect that increases the probability of cancerous changes in one or more somatic cells. In most cases, researchers have not tracked down the basic gene defect or the mechanism whereby the inherited defect promotes the somatic mutations characteristic of cancer progression. Hereditary cancers tend to have an earlier age of onset than do similar cancers that have no known heritable component. They may originate in several places in the body and affect both members of paired organs.

We shall first discuss retinoblastoma, a cancer of the retina of the eye. Although susceptibility to retinoblastoma is inherited germinally as a dominant trait, the clinical expression of the cancer is not so simply stated. Nor are the effects of the causative gene limited to the retina. Included in a more general category are disorders in which cells of the body are defective in DNA replication or repair. Such mutations tend to be recessive and are discussed second.

Dominants: Retinoblastoma and Wilms Tumor

Because of the relative ease in following the inheritance of dominant traits, especially when they are associated with a visible chromosomal deletion, the following two cancers are among the best studied at the cellular and molecular level. Several other cancer susceptibilties also appear to be inherited in a dominant fashion.

RETINOBLASTOMA. A cancer of the retina of one or both eyes, retinoblastoma occurs with a frequency of about 1 in 20,000 persons (Roberts and Aherne 1983; Angier 1987). The retinal cells—rods, cones, and neurons making connections to the brain—stop dividing early in life. Because cancerous changes almost always start in *dividing* cells, the malignant tumors develop primarily in young children. (The term *blast* in the name of the disease means "immature cells.") Studies of gene expression in cultured retinoblastoma cells suggest that the cancers develop specifically in the color-perceiving cone cells. Bogenmann et al. (1988) observed that the in vitro tumor cells transcribe the red or green photopigment genes but *not* the rhodopsin gene characteristic of rod cells.

A key symptom of the disease is the glassy appearance of the pupils of the eyes in which the tumors are growing. If untreated, the cancer expands along the optic nerve to the brain; it may also metastasize to the skeletal system, liver, or other organs, thereby causing death. If the eye tumors are detected early enough, however, and the affected eye or eyes are removed by surgery, or treated by cryotherapy, chemotherapy, and radiation, patients may live to adulthood and parenthood.

About 65% of cases are not inherited, while 35% are inherited as an autosomal dominant with about 90% penetrance of the retinoblastoma gene (*RB*). Because the phenotypic expressions are the same, however, the distinction is not always apparent. Furthermore, most cases appear *sporadically*, that is, in the absence of any affected relatives. This pattern is that expected either for phenocopies or for a dominantly inherited disease that seriously affects reproduction. Thus sporadic cases may either be noninherited or be due to a *new* germinal mutation in the mother or father of the patient. When inherited and noninherited retinoblastoma can be identified, however, researchers usually find that inherited cases have tumors in *both* eyes, whereas noninherited cases involve just *one* eye. Why this might be expected is discussed below.

About 3% of the heritable cases are accompanied by a small visible deletion in the long arm of one chromosome 13. The sizes of the deletions are variable, but they all include the band q14 (Figure 9). Persons who inherit the q14 deletion or the mutant *RB* allele have it on just *one* chromosome 13 in all cells. These cells are not cancerous, although the normal function of one *RB* locus is presumed to be lost. Knudson (1971, 1986) suggested that a *second* mutation occurring at the corresponding locus on the *other* chromosome 13 in any one of the millions of retinal cells is the precipitating factor in the retinoblastoma malignancies. Thus the cancers develop only when a particular retinal cell becomes *homozygous* for the mutant *RB* allele or the deletion. Strong evidence favoring this interpretation comes from karyotypes of tumorous tissue, from biochemical analyses of proteins coded by nearby genes, and from RFLP studies (Hansen and Cavenee 1988). In a real sense, then, the expression of retinoblastoma at the cellular level is a *recessive* trait, requiring abnormal states at both of the relevant q14 sites.

The likelihood of the second somatic mutation in *inherited* retinoblastoma is sufficiently high that several tumors develop, thus explaining why these cases usually

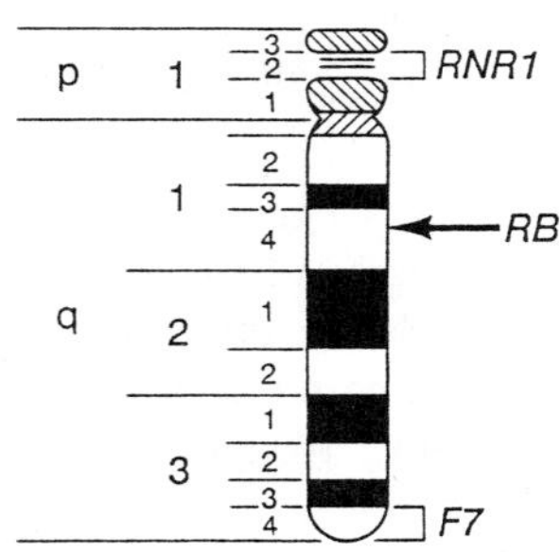

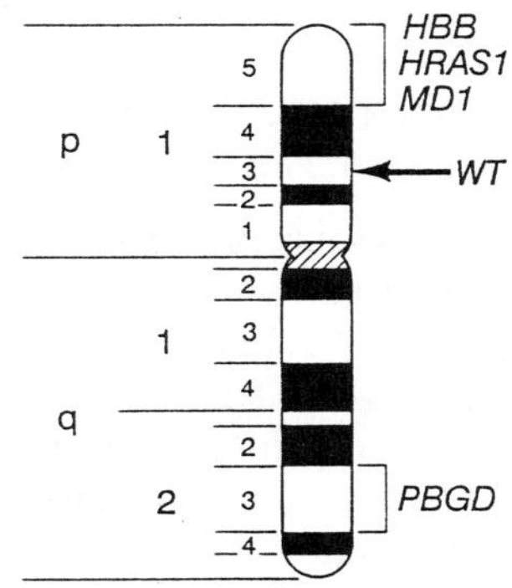

Figure 9 The chromosomal locations (arrows) of deletions that can produce autosomal dominant inheritance of retinoblastoma (A) and Wilms tumor (B). (The genetic defect in a small number of inherited Wilms tumor cases lies elsewhere.) A few other loci are also indicated: *RNR1*, ribosomal RNA; *F7*, clotting factor 7 deficiency; *HBB*, hemoglobin β (sickle-cell anemia, β-thalassemia); *HRAS1*, Harvey rat sarcoma oncogene; *MD1*, manic-depressive illness; *PBGD*, porphobilinogen deaminase (acute intermittent porphyria).

affect both eyes. Note that loss of function of both q14 sites on chromosome 13 can occur via either two mutations at the nucleotide level, or two visible deletions, or one of each. In addition, somatic nondisjunction resulting in the loss of an entire chromosome 13, or other cytological events (such as mitotic crossing over), can produce the second "mutation" needed for malignancy.

In retinoblastoma that is *not* inherited, both chromosomes 13 in the zygote are normal at the relevant q14 locus. The mutational or cytological changes of the *RB* loci must occur on *both* chromosomes 13 in the *same* retinal cell. Such persons typically have a tumor in only one eye—a tumor that develops later in childhood, as expected by the very low probability of two independent rare events occurring in a single cell. The idea that several steps are needed to produce the tumors of retinoblastoma has in fact been a model for explaining the general multistage development of cancer.

The DNA region including the *RB* gene has been identified, cloned, and sequenced (Lee et al. 1987; Weinberg 1988). The protein product, which is 928 amino acids long, is found in cell nuclei. In cultured cells that are transformed by an oncogene (*E1A*) of an adenovirus, the protein product of *RB* associates with the protein product of *E1A*. (Adenoviruses are DNA viruses and their oncogenes do not appear to have cellular homologues.) By itself, the *E1A*-encoded protein is a potent regulator of gene expression, activating transcription of both viral and cellular genes (Whyte et al. 1988).

Although the *RB* gene appears to be transcribed in cells throughout the body, it is not clear why the mutant allele affects the retinal cone cells in particular. That *other* cells can also be targeted is suggested by the increased risk of developing *bone* cancers among the survivors of retinoblastoma. In addition, some persons who do not get retinoblastoma but develop certain forms of *breast* cancer or *lung* cancer have structural abnormalities of the retinoblastoma gene (Harbour et al. 1988; Lee et al. 1988). Thus the *RB* gene may play a fundamental role in susceptibility to several forms of cancer.

As an aid to counseling families at risk, molecular analyses of RFLP polymorphisms that lie within the *RB* gene allow presymptomatic diagnosis in most cases (Wiggs et al. 1988). Benefits include very early detection of tumors and freedom from anxiety for those who are found not to possess the mutant *RB* allele.

WILMS TUMOR. Cancers that develop in the embryonic cells of one or both kidneys characterize the disease called Wilms tumor, which occurs in about 1 in 10,000 young children. Fortunately, it is often curable by surgery, radiotherapy, and chemotherapy. Its pattern of occurrence and the types of genetic changes closely resemble those in retinoblastoma patients. For example, a noninherited form usually affects one kidney; a less common inherited form due to a dominant germline gene *WT* may affect both kidneys (Knudson 1986). In a very few patients with inherited Wilms tumor, a visible deletion is seen heterozygously in all somatic cells in band p13 of chromosome 11 (Figure 9). When the deletion is present, additional effects are often seen, such as absence of the irises of the eyes (aniridia) and mental retardation. It is presumed that the p13 deletion includes genetic material affecting these other aspects of the phenotype. Some families with Wilms tumor, however, carry genetic defects that are *not* linked at all to the p13 site on chromosome 11 (Grundy et al. 1988, for example). The interpretation of these cases is unclear.

As noted for retinoblastoma, the development of kidney tumors in those cases with defects at the p13 site is thought to be contingent on the occurrence of an abnormality at both *WT* loci on homologous chromosomes. In the hereditary form, one lesion occurs in all cells because of a germline mutation or deletion. In support of these ideas, Dao et al. (1987) examined the normal and malignant tissue of 13 Wilms tumor patients by cytological and molecular (RFLP) analyses. In eight cases the researchers were able to show that, compared with *nontumor* cells, the *tumor* cells had lost DNA sequences corresponding to chromosome 11 band p13 or nearby bands. In most of these instances, the loss was not a typical deletion of chromosomal material but was based on abnormal mitotic events that resulted in homozygosity for the abnormal germline gene.

Weissman et al. (1987) have confirmed that the mutant *WT* allele is recessive at the *cellular* level by showing that *one* normal allele protects against tumor formation. The experiments involved a strain of mice called **nude**—mutant animals without thymus glands, or fur—that are prone to malignancies because they lack the T cells that are important in attacking cancer cells. *When cells from a Wilms tumor were injected into nude mice, they developed tumors.* But when the cells from a Wilms tumor were first modified to contain an extra, *normal* chromosome 11, the tumors did *not* develop. As controls, nude mice were injected with Wilms tumor cells that were augmented with a chromosome *other* than 11; they, however, still developed tumors.

In the experimental protocol above, the extra chromosomes are inserted into Wilms tumor cells by **microcell transfer.** In this technique, drugs are applied to human cells in culture to induce the formation of tiny nuclei called *microcells*, which contain one or a few chromosomes (McNeill and Brown 1980). The microcells can be isolated by centrifugation and then fused with the Wilms tumor cells by the methods described in Chapter 15 for making human–mouse hybrid cells.

Similar but more finely tuned experiments have been done with the *RB* gene transferred to a retinoblastoma cell line (Huang et al. 1988). The normal *RB* gene *by itself* (cloned in a retrovirus) suppresses the tumor-forming capability of cultured retinoblastoma cells. As a result of these experiments, the normal alleles at both the *WT* and *RB* loci are sometimes said to be acting as **tumor-suppressor genes**, or **anti-oncogenes** (Weinberg 1988). The implication is that the normal alleles (dominant at the cellular level) are acting to regulate cell growth, and that the abnormal alleles or deletions (recessive at the cellular level) release cells from the usual controls on their growth. Some researchers refer to these cancer-causing alleles as recessive oncogenes. Regardless of the nomenclatural viewpoint, however, the normal alleles are simply doing their genetically programmed jobs—coding for particular polypeptides. Knowledge of the functions of relevant polypeptides is needed to clear up the terminology.

Recessives: Chromosome Breakage Syndromes

All nucleated cells possess polymerases and ligases that participate in DNA replication. Also present are versatile enzyme systems that recognize and correct DNA damage such as breaks in the sugar–phosphate backbone or abnormal bonding of adjacent bases. (These injuries may be caused by naturally occurring radiation or chemical mutagens.) About a half-dozen autosomal recessive (germline) disorders are referred to as *chromosome breakage syndromes*. Although clinically dissimilar, they are due at least in part to defects in replication or repair (Hanawalt and Sarasin 1986). In these cases, DNA functions are compromised, a deficiency resulting in complex symptoms that include a greatly increased susceptibility to cancer. Here we briefly discuss three such syndromes.

XERODERMA PIGMENTOSUM (XP). This is a group of very rare diseases with a total frequency of about 1 in 250,000 persons.* Patients have very heavy freckling, open sores, and several types of cancers in skin areas that are exposed to sunlight (Cleaver 1983) (Figure 10). Some variant forms of XP involve not only skin lesions but also neurological abnormalities, including microcephaly and progressive mental deterioration. Death usually occurs before adulthood, often after metastases of the skin cancers.

The several different recessive genes leading to XP phenotypes code for abnormal enzymes that are unable to repair those DNA defects that are induced by ultraviolet (UV) rays (or by a few chemical mutagens). A typical UV-induced abnormality in DNA is the **pyrimidine dimer**. For example, when adjacent *thymine* bases (on one strand) bond to each other they form a double base (or dimer), distorting the DNA molecule and interrupting both transcription and replication (Howard-Flanders 1981). (In addition, each base bonds normally to the sugar residue of its respective nucleotide.) One important system for rectifying the UV-induced damage is called **excision repair,** a cut-and-patch process that has been well studied in bacteria (Figure 11). The many pyrimidine dimers that might be induced daily by sunlight in skin cells are usually temporary in normal individuals, but they become permanent in persons with xeroderma pigmentosum.

The genetic heterogeneity among XP patients has been analyzed by somatic cell hybridization. Fusing together cultured cells from different patients produces binucleate cells called *heterokaryons*. In some cases, the heterokaryons from two XP patients *are* able to carry out excision repair, a finding indicating that the recessive defects are *not* the same in the two people—being, say, *a/a* in the one and *b/b* in the other. The full genotypes would be *a/a B/B* in the first patient and *A/A b/b* in the second, so each supplies a normal dominant gene that the other lacks (see also Chapter 12). Two recessive genes that individually act abnormally but together restore normal function are said to **complement** each other. Altogether, XP patients fall into nine different complementation groups (A through I). Heterokaryons formed from cells taken from *different* complementation

*Research on xeroderma pigmentosum has been extensive. "Indeed cynics point out that the number of investigators studying the disorder considerably exceeds the number of patients afflicted with the disease" (Lehmann 1983).

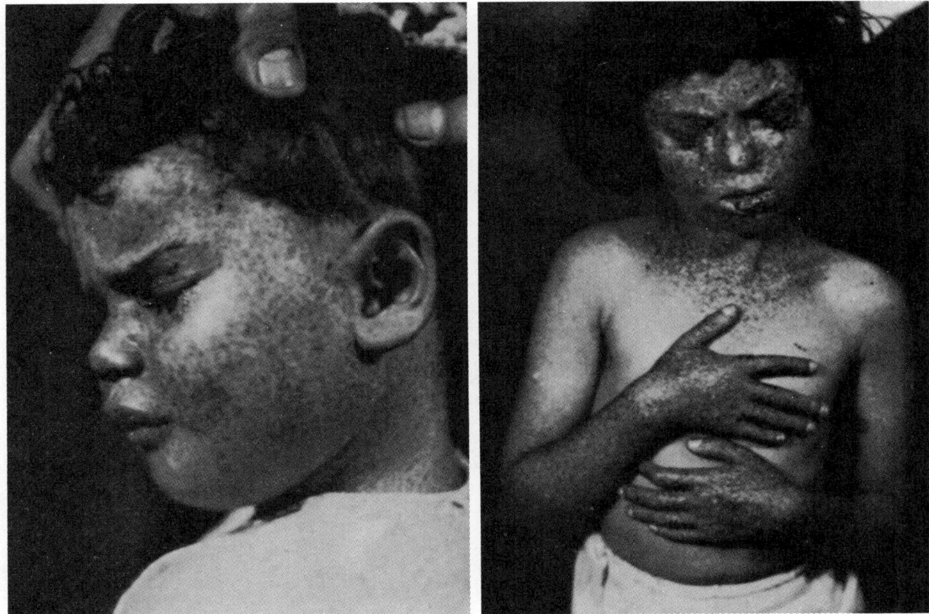

Figure 10 Two cases of xeroderma pigmentosum. The freckles and lesions are apparent. (From El-Hefnawi et al. 1965.)

groups can carry out excision repair; *within* a group, however, the heterokaryons, like the parental cells, fail to carry out excision repair, because the defect is the same (Cleaver 1983). The implication is that the first step in excision repair (the step that is unique to correcting UV-induced damage) requires the normal functioning of nine distinct polypeptides. Although the details of the enzyme systems are still unknown, Chu and Chang (1988) have identified a protein that specifically binds to DNA when it is damaged by UV; this protein is absent in the cells of patients belonging to XP complementation group E.

Some of the UV-induced DNA damage that remains *unrepaired* in XP patients because of faulty enzymes is presumed to occur in genes that control the growth of cells, perhaps in some of the proto-oncogenes that we have previously described. Interestingly, XP cells in culture *will* repair damage due to both X-rays and some chemical mutagens, so the XP defect is specific for *UV-type* damage, rather than being a generalized defect. Unfortunately, the most frequent complementation group (group A) is also the most serious form of XP, with both skin and neurological abnormalities. Prenatal diagnosis of all XP types is possible, based on the ability of fetal cells to repair UV-induced damage.

BLOOM SYNDROME (BS). This disorder is another very rare autosomal recessive disease; a total of about 100 cases have been reported worldwide since its recognition as a clinical entity in the 1950s (Passarge 1983). Patients are much smaller than normal from the time of birth; they have a narrow face, a prominent nose, and a receding chin. Mild to extremely disfiguring telangiectases (tiny dilations of capillaries) appear on the face in early childhood—worse in summer than in winter. Males are more severely affected than females. Protection from sunlight helps reduce the severity of the skin lesions. Various immunological deficiencies have been reported, and severe, sometimes life-threatening, infections of the respiratory and digestive tracts are common. Many BS patients develop malignant tumors, especially leukemias, lymphomas, and a number of different carcinomas.

Cultured cells from BS patients grow slowly, and about 10% of metaphase spreads have one or more abnormal chromosomes, including chromatid breaks, acentric fragments, and dicentric chromosomes (Ray and German 1983). Researchers find many X-shaped metaphase figures, representing crossover-like exchanges between nonsister chromatids of homologous chromosomes. Although the pairing of homologous chromosomes that must precede such exchange is a normal event during *meiosis*, it appears to occur as well for some chromosome segments in BS cells—perhaps all cells—during mitosis. In addition, a distinctive cytological feature of BS is a high frequency of sister chromatid exchanges (see Chapter 14) in some, but not all, cell populations.

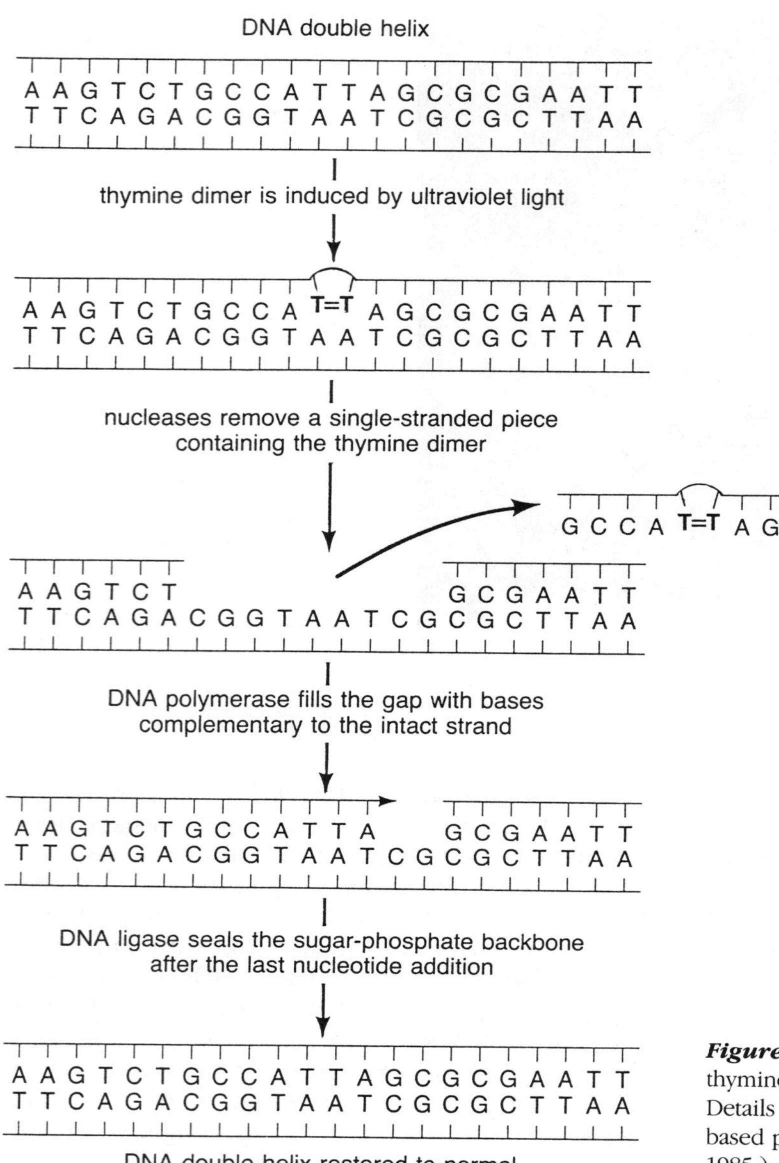

Figure 11 Induction by ultraviolet light of a DNA-distorting thymine dimer and its return to normal by excision repair. Details of this and several other DNA repair systems are based primarily on bacterial experiments. (After Strickberger 1985.)

These chromatid-exchange events are characteristic of cells from individuals with Bloom syndrome.

A clue to causation came from the discovery that cultured BS cells have slower rates of DNA replication, judged by delays in the movement of replication forks and in the linking of short fragments of replicating DNA into longer molecules (Chapter 13). These events seem to be due to a defective DNA ligase, called *ligase I* (Willis and Lindahl 1987; Chan et al. 1987). (In mammals, a second ligase, normal in Bloom syndrome cells, functions during repair processes.) The defective ligase I apparently fails to mend gaps in the sugar–phosphate backbone of DNA during replication. This lack of repair is thought to *increase* the likelihood of sister chromatid exchanges by complicated molecular processes in which the undamaged chromatid provides the base-sequence information to repair its sister (Watson et al. 1987). (Remember that the two DNA double helices formed at a replication fork *are* the sister chromatids.) It is not yet clear how some of the other symptoms of Bloom syndrome come about.

ATAXIA-TELANGIECTASIA (AT). Occurring with a frequency of about 1 in 40,000 children, this autosomal recessive disease (or set of diseases) is pleiotropic and variable in expression (Gatti and Hall 1983; McKinnon

1987). Although normal at birth, affected infants soon begin to display the diverse symptoms. *Ataxia* refers to loss of muscular control (often beginning with an unsteady gait) due to progressive damage in the cerebellum, the portion of the brain concerned with posture, balance, and coordination. Patients may later be confined to wheelchairs and have difficulty speaking. *Telangiectasia* refers to skin redness especially on the eyes, ears, and neck resulting from tiny dilations of capillaries. Immunological deficiencies associated with both B and T cells are common, contributing to a high frequency of lung infections, a major cause of early death. About 15% of patients develop malignancies, especially lymphomas and leukemias, but also carcinomas of the stomach, ovary, breast, mouth, and other organs. Additional traits include increased levels of a serum protein called α-fetoprotein, premature aging, and endocrine disorders.

The distinctive feature of AT is extreme sensitivity to ionizing radiation (but not to ultraviolet light). For example, X-rays used for therapeutic treatment of the cancers in AT patients often produce death of normal tissues. X-irradiation and some drugs also easily kill cultured cells from AT patients. In the cells that survive, X-ray-induced DNA breaks in one or both strands of the double helix may go unrepaired (in some but not all AT cell lines). Resulting chromosomal breaks, acentrics, dicentrics, and rearrangements are common, and in some cell lines certain bands on chromosomes 7 and 14 are specifically involved (Figure 12). Anomalies are also seen in the replication of DNA. In normal cells, replication is *delayed* following X-irradiation, perhaps to allow time for damaged DNA to be repaired. In AT cells, however, replication is *not* delayed, so some of the damaged DNA is permitted to replicate. Observations such as these point to abnormalities in DNA processing or repair systems among AT patients, although the precise biochemical defects are not yet known.

Different families with ataxia-telangiectasia may be variable in their responses to ionizing radiation. Part of this variation apparently stems from different genetic mutations. Like the experiments with cells from persons with xeroderma pigmentosum, genetic heterogeneity can be demonstrated in cell *hybrids* formed from different AT patients. Restoration of normal function following X-irradiation in such cell fusions indicates that whatever is wrong in the one cell line is corrected by the other, and vice versa. At least four complementation groups have been delineated, a finding suggesting that four different genes can produce somewhat similar AT phenotypes. One locus (complementation group A) has been linked to RFLP sites on the long arm of chromosome 11 (Gatti et al. 1988).

Although *heterozygotes* for AT are generally healthy, their cultured cells are sensitive to ionizing radiation at levels intermediate between normal persons and AT ho-

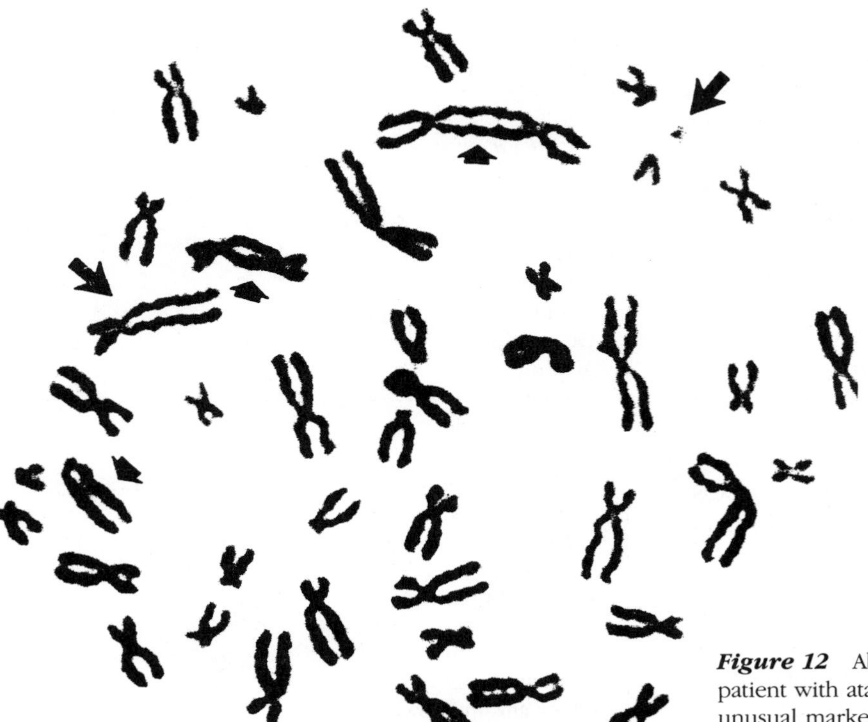

Figure 12 Abnormal karyotype of a white blood cell in a patient with ataxia-telangiectasia. The karyotype includes two unusual marker chromosomes, two dicentrics, and an abnormal acrocentric. (From Harnden 1974.)

mozygotes. Furthermore, Swift et al. (1987) have assembled evidence that heterozygotes are susceptible to cancer, especially breast cancer. In their study group, women who were heterozygous for AT with high probability (1 or 0.5) developed breast cancer about seven times more frequently than women in a matched control group consisting of normal homozygotes. Assuming that about 1 in 70 randomly chosen persons is likely to be heterozygous for AT, the authors calculate that 8.8% of all patients with breast cancer within the White population in the United States would be expected to be heterozygous for AT.

GENETIC AND ENVIRONMENTAL FACTORS

Although this chapter has been devoted primarily to the *genetics* of cancer, the disease is *not* usually transmitted from parents to offspring in the manner of typical heritable traits due to single genes, or to the combined effects of several genes (Chapter 23). Indeed, inherited susceptibility to cancer probably contributes less than 20% to overall cancer incidence (Bodmer 1986b).

All cancer is probably genetic, however, in the sense of something being transmitted from one *cell* generation to the next during the development and maintenance of a person's organs and tissues. *Thus the genetics of cancer is played out primarily in somatic rather than in germinal cells.* The environmental agents that initiate cancer in somatic cells are called **carcinogens**, and most of these apparently act by causing mutations—perhaps mutations in proto-oncogenes. In fact, about 90% of the *carcinogens* revealed by animal testing are *mutagens* (demonstrated by the Ames test) or *chromosome breaking agents* (demonstrated by sister chromatid exchange; Chapter 14). (Other substances, called *promoters*, which are not carcinogenic in themselves, are often necessary to establish the cancerous state [Franks 1986]).

Identifying environmental causes of cancer in humans is usually difficult, because years, or even decades, may pass between the action of carcinogens and the clinical detection of tumors. A clear illustration of this time lag involves the sex hormone diethylstilbestrol (DES). First administered to women during the 1940s, when it was thought to help maintain pregnancy, DES was later found to be ineffective, even detrimental, for this purpose. Among the teenage daughters resulting from these pregnancies—perhaps a million of them—about 1 in 1,000 has developed vaginal cancer, a type of malignancy that is otherwise extremely rare in young women (Prescott and Flexer 1986). The clinical effect was delayed for nearly two decades, because these cancers must have been initiated prior to birth. Note that if DES had instead induced at this rate a common type of cancer, such as lung or breast cancer, chances are its carcinogenic effect would never have been detected.

Extensive testing of substances for their carcinogenic potential stems in part from the 1958 Delaney amendment to the Food, Drug, and Cosmetic Act, which bars from the market any additive found to cause cancer in experimental animals, *whatever* the dosages tested.* Because the animal tests must be conducted within a reasonable time on a finite budget and because the cancerous transformation of any given cell is a rare event, the experimental animals, primarily rodents, are usually exposed to very high concentrations of the test agent. Only in this way is the testing procedure practical. Although the extrapolation from rodent bioassays based on high doses to humans exposed to low doses is far from direct, some reliance on animal tests is unavoidable (Ames et al. 1987). A prudent person will not ignore the results.

Most investigators—epidemiologists and others—think that 80% or more of all cancers are caused by various environmental factors, including life style (Table 2). Chemicals and radiation in the workplace, and to a

*Some critics claim that trying to legislate a 100% risk-free environment is unrealistic and that the Delaney amendment targets a tiny part of the carcinogenic burden while ignoring more important carcinogens and toxins (some of which are naturally occurring). In 1986 the Food and Drug Administration tried to permit in cosmetics certain chemical dyes whose cancer risk was conservatively estimated to be 1 in millions to 1 in billions. When the ruling was challenged, however, a Federal Court of Appeals stated that the Delaney clause had to be strictly interpreted—disallowing the use of the dyes (Curran 1988).

Table 2 Percentages of cancer deaths attributable to different environmental factors

Factor or class of factors	*Best estimate of percentage of cancer deaths*	*Range of acceptable estimates*
Tobacco	30	25 to 35
Diet	30	10 to 70
Reproductive and sexual behavior	7	1 to 13
Occupation	4	2 to 8
Alcohol	3	2 to 4
Pollution		
Atmospheric	1	<1 to 5
Water	<1	
Ionizing radiation:		
Background	1	1 to 2
Medical procedures	0.5	
Industry	<0.1	
Industrial products	<1	1 to 2
Medical drugs	<1	<1 to 2
Ultraviolet light	0.5	<0.2 to 1
Food additives	<1	−5 to 2
Other and unknown	>10(?)	?

Source: After Doll and Peto (1987).

lesser extent in the general environment, are responsible for several percent of all cancers. Some of the demonstrated carcinogenic agents in these categories include asbestos, metals, and organic compounds used in mining, and in the manufacture of pesticides, plastics, dyes, and paints. Pollution, food additives, alcohol, and medical drugs also appear to be responsible for a few percent of cancers. At the top of the table of causes, however, we find tobacco and diet. Cigarette smoke contains about 50 known carcinogens and by itself accounts for perhaps 30% of all cancer deaths (Prescott and Flexer 1986). Smoking is the greatest single *preventable* cause of cancer.

Dietary factors are thought to be responsible for another 30%, although the uncertainty in this figure is greater than that for tobacco smoke. Many food plants manufacture their own toxic chemicals to ward off insects and other attackers (Ames 1983). A number of these substances are animal carcinogens, or they are converted to carcinogens by storage or preparation of foods or by metabolic events in the body. Examples include *hydrazines* in mushrooms, *piperines* in black pepper, and *pyrrolizidines* in herbal teas. Ames (1984) states that we "are eating more than 10,000 times more of nature's pesticides than of man-made pesticides." On the other hand, some dietary components (including *vitamin E*, *β-carotene*, and *selenium*) have been proposed as *protective* agents against cancer. The consequences for cancer etiology of eating this or that type of food—for example, the fats and fiber that one can hardly escape hearing about from food companies and government agencies—are continuing areas of research (Cohen 1987).

By way of summary, we need to know more about *both* the genetic and environmental factors that contribute to the formation of cancer cells. Although the common cancers—those of the colon, breast, and lung—have no discernible heritable basis generally, there are subsets of each type of cancer that appear in increased frequency within families. The relationships between substantially environmental types of cancer and substantially heritable types, and the roles of oncogenes and their protein products, require further work. Many initially unconnected lines of investigation are now converging toward a fuller understanding. It is hoped that the new knowledge—especially at the molecular level—will eventually open up ways to alleviate the suffering that accompanies cancer.

SUMMARY

1. Cancerous cells reproduce without end; they lose their form and function, invade nearby tissue, and spread to remote sites. The cells of benign tumors also reproduce without end, but they remain localized. Tumors usually stem from single cells and progress through several stages as they become more malignant.

2. Dominantly acting oncogenes were first discovered in some retroviruses. These agents are RNA viruses that use the self-coded enzyme reverse transcriptase to copy their own genetic information from RNA to DNA and then insert it into the DNA of their host, thereby initiating a cancerous change. The viral oncogenes were picked up from normal proto-oncogenes of host species sometime in the past.

3. Human oncogenes have been detected directly by using DNA taken from human cancer cells to transfect mouse cells. Some dominant human oncogenes differ from their normal recessive alleles (the proto-oncogenes) by single base substitutions, and most of them are homologous to retroviral oncogenes.

4. The protein products of proto-oncogenes and oncogenes are involved in different aspects of cell growth and differentiation. A number of these proteins are tyrosine-specific kinases that alter the activation of other proteins by phosphorylation. Others are involved with cellular growth factors.

5. Transgenic mice that carry oncogenes in all their cells (germinal and somatic) are being used to study the development of cancer.

6. In chronic myelogenous leukemia, the proto-oncogene c-*abl* is located at a breakpoint involved in a translocation. Although the protein product of *abl* is thereby altered, its role in the genesis of cancer is unknown.

7. In Burkitt lymphoma, the proto-oncogene c-*myc* on chromosome 8 is located at a breakpoint involved in one of three different translocations. The other breakpoint lies within an immunoglobulin gene that is active in the B cells whose transformation produces the cancer. The protein product of c-*myc* is not altered, but it is synthesized at the wrong times or in the wrong amounts. The Epstein-Barr virus is associated with Burkitt lymphoma.

8. Researchers sometimes find the same chromosomal aberrations in different cancers and different

aberrations in the same cancer. Chromosomal changes are more difficult to document in solid tumors than in the hematopoietic malignancies. The progression of some cancers is accompanied by secondary chromosomal aberrations or by the occurrence of gene amplification.

9. Although cancer genetics relates largely to changes in somatic cells, a few forms of cancer involve germline mutations that influence the occurrence of the somatic changes. These cancers may involve dominant, recessive, or multifactorial modes of inheritance.

10. Retinoblastoma and Wilms tumor are examples of cancers that can either be dominantly inherited through the germline or not inherited at all. In the inherited form, all somatic cells have one lesion; a second lesion at the homologous locus is required to initiate the malignancy. Thus, the diseases are transmitted through the germline in dominant patterns, but expressed at the cellular level as recessive traits.

11. Xeroderma pigmentosum, Bloom syndrome, and ataxia-telangiectasia are examples of diseases related to cancer susceptibility due to defects in DNA replication or repair. Each susceptibility is recessively inherited and involves chromosome instability. Xeroderma pigmentosum involves enzymes that repair ultraviolet-induced damage, ataxia-telangiectasia probably involves enzymes that repair X-ray-induced damage, and Bloom syndrome involves enzymes in DNA replication.

12. Carcinogens are environmental agents that initiate and promote the malignant changes, and they often act by causing a series of mutations. Agents in tobacco smoke and diet are major factors in cancer causation, but the clinical effects of a carcinogen may be delayed for years or decades.

13. It is hoped that new information on the molecular nature of malignant changes will produce better treatments for cancer patients.

FURTHER READING

Excellent general discussions of cancer for nonscience students and lay persons include Roberts (1984), Prescott and Flexer (1986), and LaFond (1988). More details on the molecular and cellular aspects are given by Marshall (1985), Franks and Teich (1986), Watson et al. (1987), and Burck et al. (1988), and in a brief review by Bishop (1987). Overall chromosomal aspects are reviewed by Yunis (1983) and Sheer (1986). Volume 51 of the *Cold Spring Harbor Symposia* (1986) has a fairly technical section on human cancer genes. Books devoted to the genetics of cancer include those edited by Bodmer (1983) and by German (1983).

Scientific American articles about cancer have been collected by Friedberg (1986); we particularly recommend those by Weinberg (1983) and Hunter (1984), and others (not in the collection) by Croce and Klein (1985), Cairns (1985), Cohen (1987), and Weinberg (1988). For browsing on the topic of cancer, readers are directed to the Library of Congress classifications, RC 254–282, where a variety of books, journals, and review series can be found.

QUESTIONS

1. A typical cell has a volume of about 10^{-9} cc. How many cell generations are required to produce a tumor the size of a marble, about 1 cc in volume?

2. In women heterozygous for an X-linked gene, Linder and Gartler (1965) discovered that each uterine tumor had a single phenotype for the X-linked gene. This meant, first of all, that each tumor stemmed from a single cell (see text). They argue *further* that the single-cell origin provided strong evidence that the active-versus-inactive decision for the X chromosomes, once made, is *permanent*, as postulated by the Lyon hypothesis. Explain.

3. Although cancer researchers generally agree that the path to malignancy is a *multi*step process, Weinberg and his colleagues were able to permanently transform mouse cells in culture in *one* step (with DNA containing a single base substitution mutation). Can you suggest an explanation for the apparent discrepancy?

4. What experiment would show that transformation of mouse cells by DNA taken from the bladder carcinoma cell line was not also a property of normal DNA?

5. What is a tyrosine-specific protein kinase? How could a mutation in a gene encoding such a protein make a cell cancerous?

6. The proto-oncogene *erb*B (first found in a chicken retrovirus) codes for the *receptor* for epidermal growth factor, a hormone-like protein that stimulates division of cells when *joined* to the receptor. Speculate on how a mutation in the *erb*B proto-oncogene might lead to malignancy.

7. Newly discovered oncogenic retroviruses tend to have oncogenes the same as or similar to those previously known. Oncogenes discovered through transfection experiments follow the same pattern. What do these facts suggest about the total number of oncogenes?

8. Several researchers have pointed out that translocations having breakpoints *within* immunoglobulin genes are not necessarily unexpected (Haluska et al. 1987; Rabbitts et al. 1988). Why?

9. Although seen in leukemias and lymphomas, highly specific chromosomal defects are not usually identifiable in the more common solid tumors of the lung, breast, and colon. Yet investigators suggest that this lack is just an artifact of working with solid tumors (Sandberg 1988). What is the problem in finding the *specific* chromosomal defects in solid tumors that are presumed to exist?

10. In hereditary retinoblastoma, the average number of retinal tumors is about three. Assuming that a tumor is equally likely to appear in either eye, what is the probability that at least one tumor will appear in each eye?

11. Distinguish between *dominant* inheritance and *recessive* expression (for retinoblastoma or Wilms tumor).

12. In the table below, the rows and columns designate eight different patients with xeroderma pigmentosum (1 through 8). The body of the table shows whether the heterokaryons formed from their cells will (+) or will not (○) repair UV damage to DNA. Determine the complementation groups. Which patients belong to the same complementation group?

	Patient number						
Patient number	**2**	**3**	**4**	**5**	**6**	**7**	**8**
1	+	○	+	+	+	○	○
2		+	+	○	○	+	+
3			+	+	+	○	○
4				+	+	+	+
5					○	+	+
6						+	+
7							○

13. Swift et al. (1987) note that breast cancer is uncommon among *homozygotes* affected with ataxia-telangiectasia (AT), but they found that it was the most frequent type of cancer among AT *heterozygotes*. The authors think that this may reflect the age distribution of the two groups. Explain.

14. What is the Delaney clause? Although designed to protect us against cancer, some have argued that its strict application could increase the *overall* risk of medical harm. Can you suggest a way in which this might come about?

ANSWERS

1. About 30. Each generation doubles the number of cells, so we need to know the power of 2 that gives 10^9. (10^9 cells—each equal to 10^{-9} cc—have a total volume of 1 cc.) Use the to-the-power key on a hand-held calculator, or the approximation, $2^{10} \approx 10^3$.

2. Some of the uterine tumors were quite large, consisting of as many as 10^{11} cells. That all 100 billion members of the clone had the same active X chromosome means that the type of the initial cell was permanently retained.

3. The cultured mouse cells may have already undergone some of the changes on the pathway to malignancy, so only one step remained. In fact, the cultured mouse cell line commonly used for transfection experiments (called NIH/3T3) is known to be unusual in several respects.

4. Repeat the transfection experiments with DNA from *normal* cells. Transformation does not occur (unless a change occurs in the DNA during the transfection process itself). When DNA is taken from various *cancer* cell lines, transformation is not inevitable; it occurs only about 20% of the time.

5. The protein is an enzyme that phosphorylates other proteins on the —OH groups of tyrosine side chains. Errors in phosphorylation due to an altered enzyme (due to a mutant gene) could affect the activity of many proteins; one or more of these could be critical in regulating cell division.

6. The oncogenic form of the receptor might behave as if it were *always* joined to epidermal growth factor. "The cell is thus deluded by its malfunctioning receptor, being informed of high g owth factor concentration when little or none is in fact present" (Weinberg 1985). This aberrant message might provide a *continuous* stimulus for division in the absence of EGF.

7. There are not many more oncogenes to be discovered.

8. The genetic elements comprising the immunoglobulin genes *naturally* break and rejoin during lymphocyte maturation. If nucleotide sequences homologous to those involved in the joining segments (see Figure 10 in Chapter 18) occur on other chromosomes, "illegitimate" recombinational events—that is, translocations—might be expected.

9. The cells of solid tumors are more difficult to culture and are slow to divide. Furthermore, the accumulation of secondary chromosomal defects (at the time solid tumors are analyzed) can obscure a specific aberration common to all cells of a given cancer type.

10. 3/4. One way to arrive at this answer is to ask for the probability that all three tumors will be in the left eye —$(1/2)^3$—or all three in the right—$(1/2)^3$. In all other cases, there will be tumors in both eyes.

11. Pedigrees of the diseases show the standard characteristics of dominance inheritance. But in persons carrying the dominant gene, cells become cancerous only if they become homozygous for the genetic defect.

12. Patients 1, 3, 7, and 8 are in one group because their cells do not complement each other, but do complement cells outside this group. Patients 2, 5, and 6 are in another complementation group. Patient 4 is in a separate group, complementing everyone else.

13. Usually dying before age 20, patients with ataxia-telangiectasia (homozygotes) do not usually survive to ages characteristic of the onset of breast cancer.

14. A company might substitute a chemical that was *relatively* safe with respect to cancer (but failed the Delaney test) with one that was *absolutely* safe in cancer tests but more worrisome in other respects. For example, with low probability it might cause another disease.

Part Five Population and Quantitative Genetics

20. Population Concepts
21. Inbreeding and Isolates
22. Processes of Evolution
23. Quantitative and Behavioral Traits

Chapter 20 ***Population Concepts***

Genotype, Phenotype and Allele Frequencies
Relation between Allele and Genotype Frequencies

Mating Frequencies
Random Mating
Offspring from Random Mating

The Hardy-Weinberg Law
A Proof
The Meaning of the Hardy-Weinberg Law
The Array of Genotype Frequencies
Another Proof of the Hardy-Weinberg Law

Some Applications
The Sickle-Cell Allele among American Blacks
Alkaptonuria and the Problem of Dominance
ABO: A Three-Allele System
X-Linked Genes

TO THE EDITOR OF SCIENCE: I am reluctant to intrude in a discussion concerning matters of which I have no expert knowledge, and I should have expected the very simple point which I wish to make to have been familiar to biologists. However, some remarks of Mr. Udny Yule . . . suggest that it may still be worth making.

In the Proceedings of the Royal Society of Medicine *. . . Mr. Yule is reported to have suggested, as a criticism of the Mendelian position, that if brachydactyly is dominant "in the course of time one would expect, in the absence of counteracting factors, to get three brachydactylous persons to one normal."*

It is not difficult to prove, however, that such an expectation would be quite groundless . . . [using] a little mathematics of the multiplication-table type.

G. H. Hardy (1908)

Mendel's laws predict the offspring of *particular* crosses. Can they also yield statistics on the genetic makeup of *whole* populations? Shortly after Mendel's work was rediscovered, several theorists thought so; they suggested, for example, that any dominant characteristic (such as the stubby fingers seen in brachydactyly) should appear in 75% of people. Since this is obviously not seen, somebody must be wrong. That it was *not* Mendel was elegantly shown by a Cambridge University mathematician, G. H. Hardy. His solution, which was also worked out independently by the German clinical physician, Wilhelm Weinberg, is presented in this chapter. The Hardy-Weinberg concept formed the basis for the development of a new field, called **population genetics**. Based in part on logical extensions of Mendel's ideas, human population genetics seeks to understand the distribution of inherited traits in diverse groups of people, and why genetic statistics may change through time.

GENOTYPE, PHENOTYPE, AND ALLELE FREQUENCIES

A population is simply an assemblage of individuals. To describe the proportion of a certain type within a population, we use the term *frequency*, which is similar in meaning to the term *probability*. Both words express a group concept; a single individual is not usually described by a frequency or probability (except 0 or 1).

Suppose that a group of parents is tested genotypically, giving these results:

Genotype	Number	Frequency
B/*B*	114	0.57
B/*b*	56	0.28
b/*b*	30	0.15
total	200	1

The **genotype frequencies** above are the proportions of the total represented by each genotype. For example, freq(B/B), the frequency of the B/B genotype, is 114/200 = 0.57. Note that the sum of all genotype frequencies must necessarily be one. This rule for frequencies (as for exhaustive probabilities) provides a quick check on arithmetic accuracy.

If each genotype in a population corresponds to a different phenotype, then genotype frequencies and **phenotype frequencies** are the same. If B is dominant to b, however, then the genotypes B/B and B/b are phenotypically indistinguishable; thus

Phenotype	Number	Frequency
dominant	170	0.85
recessive	30	0.15
total	200	1

A little more arithmetic is needed to obtain what are called the **allele frequencies**, the proportions of the total number of alleles represented by each of the alleles. We note that each individual possesses two alleles per locus, so that among the 200 parents there are 400 alleles to be tallied as either B or b. Assuming that B/B's and B/b's are distinguishable from each other (no dominance), the 114 B/B homozygotes together possess 228 B alleles. The 56 heterozygotes add to the tally of both allelic forms, 56 of B and 56 of b. The total number of B alleles is therefore 228 + 56, and the total number of b alleles is 60 + 56. Allele frequencies can then be calculated:

$$\text{freq}(B) = (228 + 56)/400 = 284/400 = 0.71$$
$$\text{freq}(b) = (60 + 56)/400 = 116/400 = 0.29$$

$$\text{sum of freq} = 1$$

In order to make this calculation, however, homozygous and heterozygous genotypes must be distinguishable from each other. We will see later in this chapter what can be done to estimate allele frequencies when this condition is not met. The allele frequencies are often symbolized more briefly as $\boldsymbol{p} =$ **freq($\boldsymbol{B}$)** and $\boldsymbol{q} =$ **freq($\boldsymbol{b}$)**. Using this notation, $p + q = 1$.

Relation between Allele and Genotype Frequencies

An easily remembered relationship exists between allele frequencies and genotype frequencies. To find this expression in *general* form, assume

Genotype	Number	Frequency
B/B	G	G/T
B/b	H	H/T
b/b	J	J/T
total	T	1

The numbers G, H, and J can be any biologically meaningful numbers (for example, 114, 56, and 30). Calculating the allele frequencies as before, we have

$$\begin{aligned}\text{freq}(B) &= (2G + H)/2T \\ \text{freq}(b) &= (2J + H)/2T\end{aligned} \qquad (1)$$

Rearranging a little, we obtain:

$$\text{freq}(B) = 2G/2T + H/2T = G/T + (1/2)(H/T)$$
$$\text{freq}(b) = 2J/2T + H/2T = J/T + (1/2)(H/T)$$

The fractions G/T, H/T, and J/T are the genotype frequencies. Thus

$$\begin{aligned}\text{freq}(B) &= \text{freq}(B/B) + (1/2)\text{freq}(B/b) \\ \text{freq}(b) &= \text{freq}(b/b) + (1/2)\text{freq}(B/b)\end{aligned} \qquad (2)$$

In words, Equations (2) say that *the frequency of an allele equals the frequency of the corresponding homozygote plus half the frequency of the heterozygote*. This is a commonsense, weighted average.

Calculating allele frequencies in a population in which there are *three* alleles of a gene requires the recognition of more than one kind of heterozygote. For example, three alleles for the gene that encodes the β chain of hemoglobin may be present in some regions of Africa: Hb^A is the normal allele, Hb^S is the sickle-cell allele, and Hb^C is the so-called hemoglobin C allele. All six genotypes are distinguishable when the hemoglobins are compared in an electrophoretic apparatus. An intuitive extension of Equations (2) predicts that the frequency of, say, the C allele equals the frequency of the homozygote (C/C) plus half the frequencies of all genotypes heterozygous for the C allele (C/A and C/S). Note also that freq(A) + freq(C) + freq(S) = 1.

MATING FREQUENCIES

One can predict the proportions of the various genotypes in an entire offspring generation if one first considers the **mating frequencies** among all their parents. To show how this is done, we return to the group of 200 persons set forth above:

Genotype	Frequency
B/B	0.57
B/b	0.28
b/b	0.15
total	1

We seek the relative proportions of the *six* possible matings:

1. $B/B \times B/B$	4. $B/B \times B/b$
2. $B/b \times B/b$	5. $b/b \times B/b$
3. $b/b \times b/b$	6. $B/B \times b/b$

The frequency of mating 1 (for example) is the proportion this cross is of all crosses in the population. But to calculate its value, we need to know the rules of the mating game. For example, if "like attracts like," it may be that the only matings that occur are those in the left-hand column, in which the mates have the same genotype. But many other mating schemes can be envisioned.

Random Mating

One mathematically convenient and sometimes realistic assumption, called **random mating**, is that mating occurs without regard to a person's genotype or phenotype. This assumption is true (or very nearly true) for some human traits—blood groups or HLA types, for example. It means that an individual is equally likely to mate with any member of the opposite sex. Random mating frequencies depend only on the frequencies of the two pertinent genotypes (which are multiplied together). For example, for mating 1 above, we calculate

$$\begin{aligned} \text{freq}(B/B \times B/B) &= \text{freq}(B/B\ ♂) \times \text{freq}(B/B\ ♀) \\ &= (0.57)(0.57) = (0.57)^2 \end{aligned}$$

The symbol freq(B/B ♂) means the frequency of B/B among males and freq(B/B ♀) means the frequency of B/B among females. We are assuming that these are equal; that is, genotypes are distributed similarly in the sexes.

Mating 4 is a bit different from 1 in that the two mates are genotypically different from each other. As a consequence, one has to add together the frequencies of *reciprocal* crosses. For example, the mating $B/B \times b/b$ (mating 6 above) could mean either B/B ♂ $\times$ b/b ♀ or B/B ♀ $\times$ b/b ♂. Thus,

$$\begin{aligned} \text{freq}(B/B \times b/b) &= \text{freq}(B/B ♂) \times \text{freq}(b/b\ ♀) \\ &\quad + \text{freq}(B/B\ ♀) \times \text{freq}(b/b\ ♂) \\ &= (0.57)(0.15) + (0.57)(0.15) \\ &= 2(0.57)(0.15) \end{aligned}$$

The factor 2 is present in the random mating frequencies of all three matings in the right-hand column. It is not present in the matings on the left because the spouses have the same genotype—reversing the sex is not a different situation.

For the specific numbers assumed, the random mating frequencies for the three genotypes are summarized as follows:

$$\begin{aligned} &1.\ \text{freq}(B/B \times B/B) = (0.57)^2 \\ &2.\ \text{freq}(B/b \times B/b) = (0.28)^2 \\ &3.\ \text{freq}(b/b \times b/b) = (0.15)^2 \\ &4.\ \text{freq}(B/B \times B/b) = 2(0.57)(0.28) \\ &5.\ \text{freq}(b/b \times B/b) = 2(0.15)(0.28) \\ &6.\ \text{freq}(B/B \times b/b) = 2(0.57)(0.15) \end{aligned}$$

A *general* representation of random mating frequencies can be obtained from the expansion of the algebraic expression $(x + y + z)^2$, where x, y, and z are the frequencies of the three genotypes. Recall that squaring a multinomial expression yields the square of each individual term and twice the cross product of each possible pair of terms (Appendix 2). The expansion is thus equal to:

$$x^2 + y^2 + z^2 + 2xy + 2zy + 2xz$$

These terms are precisely in the form of the arithmetic example above where $x = 0.57$, $y = 0.28$, and $z = 0.15$. The random mating frequencies are summarized algebraically as follows:

$$\begin{aligned} &1.\ \text{freq}(B/B \times B/B) = x^2 \\ &2.\ \text{freq}(B/b \times B/b) = y^2 \\ &3.\ \text{freq}(b/b \times b/b) = z^2 \\ &4.\ \text{freq}(B/B \times B/b) = 2xy \\ &5.\ \text{freq}(b/b \times B/b) = 2zy \\ &6.\ \text{freq}(B/B \times b/b) = 2xz \end{aligned} \qquad (3)$$

where $x = \text{freq}(B/B)$, $y = \text{freq}(B/b)$, and $z = \text{freq}(b/b)$.

Just as the two possible *alleles* in a population have frequencies that add to 1, and the three possible *genotypes* have frequencies that add to 1, so, too, do the six *matings* have frequencies that add to 1.

Offspring from Random Mating

Using Equations (3), one can find the frequency of any particular genotype among the offspring of all parents in a randomly mating population. Note, for example, that the genotype b/b could be present in the children of matings 2, 3, or 5. From these parents, the expected fractions of b/b children based on Mendel's first law are

Mating	Random mating frequency	Probability of b/b offspring given the indicated mating
2. $B/b \times B/b$	y^2	1/4
3. $b/b \times b/b$	z^2	1
5. $b/b \times B/b$	$2zy$	1/2

Summarizing the information in this table, we obtain for the offspring generation:

$$\begin{aligned}\text{freq}(b/b) &= (1/4)y^2 + (1)z^2 + (1/2)2zy \\ &= z^2 + zy + (1/4)y^2 \quad \text{[a perfect square]} \\ &= (z + y/2)^2\end{aligned}$$

As an arithmetic example, assume a parental population of 25% *b/b* and 30% *B/b*. The derivation above then predicts that 16 percent of the children will be *b/b*: $(0.25 + 0.30/2)^2$. Note that 16% does not correspond to any simple Mendelian ratio, because the *b/b* offspring come from a *mixture* of different matings, each one having a particular Mendelian expectation.

THE HARDY-WEINBERG LAW

Hardy and Weinberg applied the rules of Mendel to randomly mating populations in a manner similar to that used in our example. As above, it is assumed that the allele frequencies are the same in the two sexes. For two alleles of an autosomal gene, *B* and *b*, the **Hardy-Weinberg law** can be stated in two parts:

1. Under so-called Hardy-Weinberg conditions (see below), the genotype frequencies in a population become, in *one* generation:

$$\begin{aligned}\textbf{freq}(\boldsymbol{B/B}) &= \boldsymbol{p^2} \\ \textbf{freq}(\boldsymbol{B/b}) &= \boldsymbol{2pq} \\ \textbf{freq}(\boldsymbol{b/b}) &= \boldsymbol{q^2}\end{aligned}$$

where p and q are allele frequencies: $p =$ freq(B) and $q =$ (b). These equations mean that the genotype frequencies can be predicted from the allele frequencies.

2. As long as the Hardy-Weinberg conditions prevail, the genotype frequencies do not change. That is, generation after generation the frequencies of the three genotypes remain constant at p^2, $2pq$, and q^2. The constancy over time is called an **equilibrium**, meaning no net change in frequencies. Of course, the individuals in succeeding generations are not the same; it is the expected frequencies of their genotypes that stay the same. The conditions under which the Hardy-Weinberg law is applicable are:

1. approximately random mating
2. a fairly large population (several hundred or more)
3. a negligible amount of mutation between the *B* and *b* alleles
4. a negligible amount of migration into and out of the population
5. a negligible amount of selection; that is, all genotypes are about equally viable and equally fertile

Strictly speaking, the Hardy-Weinberg conditions are more stringent: they include an *infinitely* large population, *no* mutation, *no* migration, and *no* selection. But for practical purposes, the law can be applied where these conditions are only approximated. There is no living population that exactly meets the mathematical model. But there are many genes in many populations that meet the conditions sufficiently well that the Hardy-Weinberg law becomes usable and useful.

A Proof

A partial proof of the Hardy-Weinberg law was set forth above under the guise of merely predicting the *b/b* children under random mating. Although random mating was the only explicitly stated condition, it was implied that the population met the other Hardy-Weinberg conditions as well. We had assumed that in a parental generation the frequencies of the three genotypes were (in the most general terms) x, y, and z. The frequency of *b/b* among the children (obtained by considering the matings that could produce such offspring) simplified to:

$$\text{freq}(b/b) \text{ among children } = (z + y/2)^2$$

The expression on the right in parentheses, $z + y/2$, is shorthand for freq(*b/b*) + (1/2)freq(*B/b*). Looking back at Equations (2), we see that this is the frequency of the *b* allele among the parents. Writing freq(b) as q, we have thus shown, for a population meeting the conditions of the Hardy-Weinberg law, that the frequency of *b/b* becomes q^2 in one generation, that is:

$$\text{freq}(b/b) = q^2$$

To complete the proof of the Hardy-Weinberg law requires that we repeat this procedure for the other two genotypes. This is done in Figure 1. Note that the demonstration that freq(B/B) $= p^2$ is symmetrical to the work above showing that freq(b/b) $= q^2$. The demonstration that freq(B/b) $= 2pq$ has one algebraic factoring step that is not too obvious, but otherwise proceeds in a similar manner.

The second part of the Hardy-Weinberg law says that the $p^2{:}2pq{:}q^2$ distribution of genotypes is stable over time. Once the population has achieved these frequencies and as long as it continues to meet the Hardy-Weinberg conditions, no further changes in frequencies will occur in successive generations. We check this by starting with a parental generation having frequencies

Genotype	Frequency
B/B	p^2
B/b	$2pq$
b/b	q^2
total	1

We ask about the frequencies of the three genotypes in the *next* generation. We could repeat the algebra in

Figure 1, letting $x = p^2$, and so on, but there is no need to do so. The derivation in Figure 1 is entirely general and leads to the p^2:$2pq$:q^2 distribution in the offspring generation regardless of what genotype frequencies existed in the parental generation.

In summary, assuming equal allele frequencies in the two sexes, one generation of mating under the Hardy-Weinberg conditions adjusts the genotype frequencies to p^2:$2pq$:q^2, and there they remain for as long as the conditions persist.

The Meaning of the Hardy-Weinberg Law

The Hardy-Weinberg equilibrium is a statement of no change—of no evolution. Over long periods of time, it is likely to be unrealistic, because we know that populations do evolve. But over short periods, the Hardy-Weinberg stability may apply. The equilibrium is said to be *neutral*. This statement means that if the allele frequencies do change, say, from p to p' and from q to q' (due to some temporary departure from the Hardy-Weinberg conditions), then the frequencies of the genotypes quickly become p'^2:$2p'q'$:q'^2. The genotypes remain at these new frequencies in successive generations without tending to return to the old values. Thus we need not know anything about the past history of a population in order to make current predictions about the frequencies of genotypes and phenotypes. The maintenance of alleles at nonzero frequencies also means that genetic variability within a population is not blended away over successive generations. This Mendelian notion of discrete hereditary units was an important contribution to evolutionary thought.

Modern ideas on evolution are rooted in concepts of population genetics. An individual does not evolve, a population does. And it does so by those mechanisms that are ignored in the derivation of the Hardy-Weinberg law: by departures from random mating, by unpredictable changes that occur because a population is small, by mutation, by migration, and especially (in many cases) by natural selection. These mechanisms lead to allele frequency changes, with corresponding changes in the average phenotype of a population. The accumulation of modifications over time brings about the evolutionary patterns that have been documented by paleontologists and others. Although the grand sweep of phylogenetic changes over millions of bygone years cannot be correlated with specific genes, more recent evolutionary events can sometimes be studied by analyzing allelic changes. The emergence of high frequencies of melanistic (darkly pigmented) forms of moths in industrially polluted regions is a well-known example. In the following chapters we look at some of the forces of evolution as they pertain to human populations. The Hardy-Weinberg law provides a basis for discussion, but it will be modified as needed.

When the mechanisms of evolution are negligible, the frequency of an allele remains constant whether it is dominant or recessive. We emphasize that dominance is *not* an evolutionary force; it is merely a label for how alleles express themselves in a heterozygote. And this brings us back to Hardy's remarks at the beginning of this chapter. He gently chides biologists who believe that

We start with a parental population in which

frequency of genotype $B/B = x$
frequency of genotype $B/b = y$
frequency of genotype $b/b = z$

$$p = \text{freq}(B) = x + (1/2)y$$
$$q = \text{freq}(b) = z + (1/2)y$$

Mating	*Random frequency of mating*	*Probability of offspring genotype* *B/B*	*B/b*	*b/b*
1. *B/B* × *B/B*	x^2	1	0	0
2. *B/b* × *B/b*	y^2	1/4	1/2	1/4
3. *b/b* × *b/b*	z^2	0	0	1
4. *B/B* × *B/b*	$2xy$	1/2	1/2	0
5. *b/b* × *B/b*	$2zy$	0	1/2	1/2
6. *B/B* × *b/b*	$2xz$	0	1	0

The frequencies of genotypes among all the offspring are

$$\text{freq}(B/B) = (1)x^2 + (1/4)y^2 + (1/2)2xy = x^2 + xy + (1/4)y^2 = \left(x + \frac{1}{2}y\right)^2 = p^2$$

$$\text{freq}(B/b) = (1/2)y^2 + (1/2)2xy + (1/2)2zy + (1)2xz$$
$$= 2\left(xz + \frac{1}{2}xy + \frac{1}{2}zy + \frac{1}{4}y^2\right) = 2\left(x + \frac{1}{2}y\right)\left(z + \frac{1}{2}y\right) = 2pq$$

$$\text{freq}(b/b) = (1/4)y^2 + (1)z^2 + (1/2)2zy = z^2 + zy + (1/4)y^2 = \left(z + \frac{1}{2}y\right)^2 = q^2$$

Figure 1 One derivation of the equilibrium frequencies for a population meeting the Hardy-Weinberg conditions.

an allele must *automatically* come to have a frequency of 1/2, so that 3/4 of the population has the dominant phenotype. But the equilibrium frequency of an allele depends primarily upon how it affects reproductive success or failure. Some dominant alleles impair the ability of persons to leave offspring, as do some recessive alleles. The frequency of an allele then changes because it affects reproduction, not because it is dominant or recessive.

The actual frequency of brachydactyly in human populations, far from being three in four, is about one in a million. The low frequency persists because the Hardy-Weinberg conditions are approximately true, although individuals with brachydactyly likely produce fewer children than others (tending to a lower frequency), and the normal allele mutates occasionally to the abnormal allele (tending to a higher frequency). Brachydactyly would have a low frequency whether it was inherited in a dominant or a recessive pattern. As Hardy says in his 1908 paper, "There is not the slightest foundation for the idea that a dominant character should show a tendency to spread over a whole population, or that a recessive should tend to die out."*

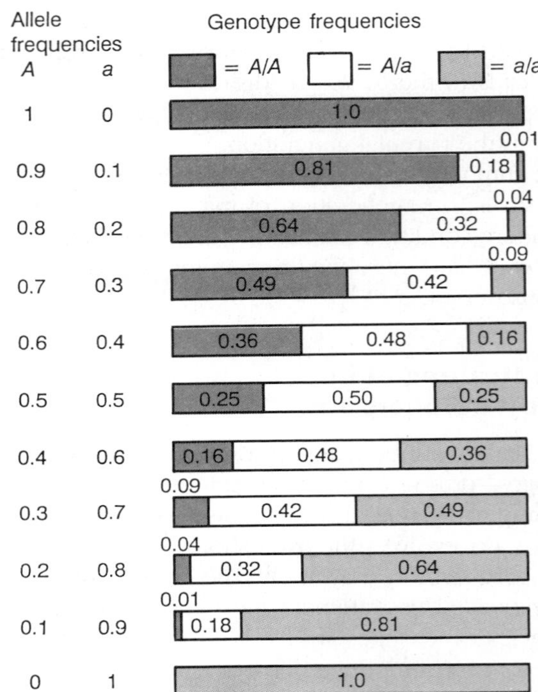

Figure 2 The expected genotype frequencies in a Hardy-Weinberg population for different values of the allele frequencies. A fuller geometrical interpretation for three of the lines is given in Figure 3.

The Array of Genotype Frequencies

The $p^2{:}2pq{:}q^2$ distribution of genotypes can be obtained by squaring the binomial expression, $p + q$:

$$(p + q)^2 = p^2 + 2pq + q^2$$

In words, this expression says that squaring the allele frequencies yields the genotype frequencies. In a stepwise extension, squaring the genotype frequencies yields the mating frequencies (as we noted before). Perhaps Hardy had these operations in mind when he referred to a little mathematics of the multiplication-table type.

If we are dealing with a gene with *three* alleles, A_1, A_2, and A_3, with corresponding frequencies, p, q, and r, the Hardy-Weinberg equilibrium frequencies can be obtained by squaring the trinomial expression $p + q + r$:

$$(p + q + r)^2 = p^2 + q^2 + r^2 + 2pq + 2pr + 2qr$$

Each of the six terms on the right is the frequency of the corresponding genotype. For example, the term $2pr$ refers to the frequency of the genotype A_1/A_3, because A_1 and A_3 are the two alleles whose frequencies are p and r.

*Hardy learned of this problem through his friend, Reginald Punnett (of Punnett squares, or checkerboards), with whom he played cricket. Punnett felt that Yule must be wrong about brachydactyly (in the quote heading this chapter), but couldn't prove it. Yule was asking the right questions, however, and he strongly supported Mendelism (Provine 1971).

The frequency of an allele can theoretically be any number from 0 to 1 inclusive. As noted, its actual frequency will be the result of a long evolutionary history. The genotypic frequencies predicted from the Hardy-Weinberg law will vary accordingly. A graphical display of some representative values is presented in Figure 2. As freq(A) varies from 1 to 0 (reading from top to bottom), the frequency of the A/A homozygote does likewise; but since the genotypic proportion is a squared function, the decrease in freq(A/A) is rapid. For example, as p decreases from 1 to 0.5, p^2 decreases from 1 to 0.25. If the frequency of an allele is quite low (say, 1 in 100), then the corresponding homozygote is rare indeed (1 in 10,000). The frequency of heterozygotes is largest when the two alleles are equal in frequency and decreases to zero at both extremes. In the middle third of the range of allele frequencies, heterozygotes are the largest genotypic class. The main features of the Hardy-Weinberg equilibrium are also nicely shown by a geometrical scheme based on checkerboards (Figure 3).

Another Proof of the Hardy-Weinberg Law

Random mating of genotypes and random combination of gametes are equivalent concepts: if it is a matter of chance which genotypes mate, it must also be true that

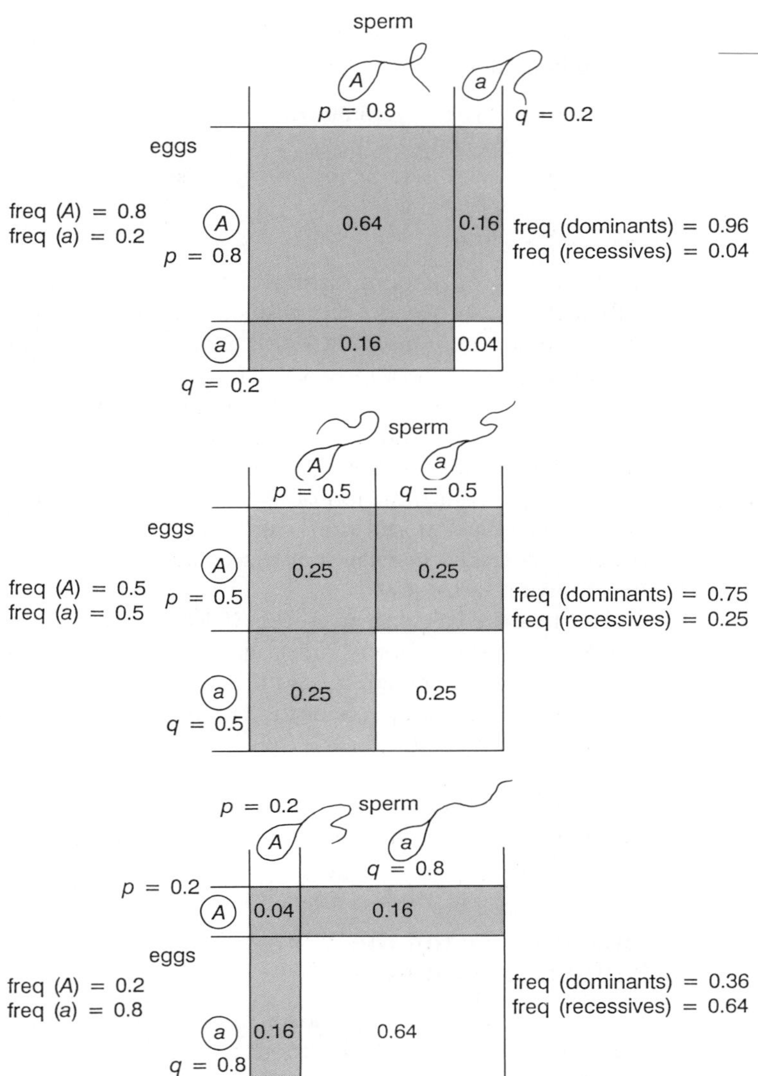

Figure 3 Checkerboards used to represent three Hardy-Weinberg equilibria. The frequencies of the parental gamete types, *A* or *a*, correspond to allele frequencies, since each gamete carries one allele. The linear dimensions of the row and column headings are proportional to the allele frequencies in the parents, and the areas within the checkerboards are proportional to the genotype frequencies in the offspring. Assume that *A* is dominant to *a*; then the shaded areas correspond to the dominant phenotype and the unshaded, to the recessive phenotype.

it is a matter of chance which allele from an egg combines with which allele from a sperm. This idea provides a simple, alternative proof of the Hardy-Weinberg law based on a checkerboard for the entire mixed population (rather than a particular mating). We set up the checkerboard for a pair of alleles, *B* and *b*, by labeling the rows and columns with the parental gametes—two sorts of eggs and two sorts of sperm with frequencies *p* and *q*:

sperm

eggs	p (B)	q (b)
p (B)		
q (b)		

The offspring genotypes are written in the four squares of the checkerboard by combining the alleles and multiplying the frequencies heading the rows and columns. Thus, among the children,

$$\text{freq}(B/B) = p^2$$
$$\text{freq}(B/b) = pq + pq = 2pq$$
$$\text{freq}(b/b) = q^2$$

The second part of the Hardy-Weinberg law, which states that the frequencies are constant over time, is shown by repeating the checkerboard through another generation. Since the allele frequencies are unchanged from *p* and *q*, the headings and the body of the checkerboard are unchanged. This proof can easily be extended to more than two alleles of a gene.

SOME APPLICATIONS

We give a few examples of how Hardy-Weinberg arithmetic and algebra can give some interesting results.

The Sickle-Cell Allele among American Blacks

We noted in Chapter 14 that the sickle-cell allele (Hb^S or more simply S) has a relatively high frequency in some populations that were exposed to the malarial parasite over many generations. The Black population of the United States, for example, although not now generally at risk for malaria, derives in large part from African populations of the seventeenth and eighteenth centuries that were at risk. In some African groups the frequency of the sickle-cell allele is as high as 20% today, and it may have had a similar frequency in centuries past. A random sample of 2,000 newborn American Blacks (successive births in a hospital, say) would typically consist of the following genotypes:

Shorthand genotype	Number	Frequency
A/A	1,804	0.902
A/S	192	0.096
S/S	4	0.002
total	2,000	1

Thus, about 1 in 500 newborns (4/2,000) has sickle-cell anemia (*S/S*) and almost 1 in 10 (192/2,000) is a carrier (*A/S*) with a normal phenotype. We calculate the frequency of the sickle-cell allele, *S*, in this sample as [(see Equations (2)]:

$$\text{freq}(S) = q = 0.002 + (1/2)(0.096) = 0.050$$

In words, about 1 in every 20 *Hb* alleles present in United States Blacks codes for sickle-cell hemoglobin.

The random frequency of matings between two carriers is

$$\text{freq}(A/S \times A/S) = (0.096)^2 = 0.0092$$

or a little less than 1% of matings among United States Blacks. It is primarily these matings that produce offspring at risk to sickle-cell anemia, since most *S/S* individuals do not become parents: they either die before reproductive age or, as adults, do not have children. To be exact about the expected fraction of heterozygote × heterozygote matings, we should modify the numbers in the table above to reflect the *mating* population rather than the *newborn* population. If we assume, for simplicity, that no *S/S* genotypes reproduce but that all others are equally viable and fertile, the effective *mating* population becomes:

Genotype	Number	Frequency
A/A	1,804	0.9038
A/S	192	0.0962
S/S	0	0
total	1,996	1

The frequency of random matings between two carriers is thus $(0.0962)^2$. We do not wish to belabor the tiny difference between this and $(0.096)^2$, except to note that the nonreproduction of most *S/S* persons over the many generations since the seventeenth century accounts in part for the decline in the frequency of the *S* allele from approximately 20% to about 5%. As will be explained in Chapter 22, the introduction of alleles from White populations into Black populations (as culturally perceived) through interracial marriages also accounts for some of the reduction in freq(S).

Note that the calculation of freq(S) from the random sample of 2,000 newborns tallies up the actual number of *S* alleles. The calculation is valid whether or not the group meets the conditions of the Hardy-Weinberg law. In fact, the sickle-cell allele is certainly *not* at equilibrium, for its frequency is constantly declining for the reasons stated above. But the rate of change is slow, and the three genotypes are approximated by Hardy-Weinberg arithmetic.

Alkaptonuria and the Problem of Dominance

Among 250,000 randomly chosen infants, perhaps just one will be affected with alkaptonuria, *k/k*. The diagnosis is simple—the darkening of urine in the diapers—but there is no test to tell which of the remaining 249,999 are normal carriers, *K/k*, and which are normal homozygotes, *K/K*. These observations are summarized as follows:

Genotype	Number	Frequency	
K/–	249,999	0.999 996	≈ 1
k/k	1	0.000 004	$= 4 \times 10^{-6}$
total	250,000	1	

When we try to calculate the frequency of the alkaptonuria allele, $q = \text{freq}(k)$, a problem arises: we do not know the number of heterozygotes. Therefore, Equations (1) or (2), which require this information, cannot be used. Perhaps some assumptions about the population might aid in calculating q. Because the disease is very rare and relatively benign, a reasonable supposition is that the distribution of genotypes approximates the Hardy-Weinberg formulation. *Making this assumption*, we would then expect

$$q^2 = \text{freq}(k/k) = 4 \times 10^{-6}$$

Taking the square root of both sides, we get an estimate of the frequency of the recessive allele for alkaptonuria

$$q = \sqrt{4 \times 10^{-6}} = 2 \times 10^{-3} = 0.002$$

Thus, among all the alleles that code for homogentisic acid oxidase, 1 in 500 gives rise to a defective enzyme, if the population in question meets the Hardy-Weinberg conditions. What we are doing is equating an *observed* frequency (4×10^{-6}) with an *expected* frequency (q^2), given some hypothesis about the population. If the hypothesis does not apply, we err in proceeding with the calculation.

Although the Hardy-Weinberg law seems to be acceptable, the frequency estimate of 0.002 is not very good for another reason. If, by chance, the sample of 250,000 had included just a few more affected individuals, our estimate might have been perhaps twice as big. On the other hand, if the sample did not happen to include an alkaptonuric, we would have an allele frequency of zero. The wide range of possible estimates arises because we are looking at very small numbers of affected individuals.* If we could identify and tally all of the normal *heterozygotes* (of which there are many more), our potential error would be much less. But we have no choice in this case, and we accept $q = 0.002$ as the best estimate of the allele frequency possible under the circumstances.

Having estimated the allele frequency, however roughly, we can now estimate the proportion of heterozygotes in the population:

$$\text{freq}(K/k) = 2pq = 2(0.998)(0.002) \approx 0.004$$

Thus about 1 person in 250 is a carrier, a somewhat surprising result when we started with the information that only 1 in 250,000 has the metabolic disorder. For every affected person, there are about 1,000 carriers. The expected frequency of matings between two carriers is $(2pq)^2 = (4 \times 10^{-3})^2 = 16 \times 10^{-6}$. For these matings, heterozygosity of the two individuals will be revealed if they produce an alkaptonuric infant. Otherwise, most heterozygotes will never know they possess the *k* allele.

As noted in the next chapter, infants affected with alkaptonuria, or any rare recessive disease, often have parents who are related to each other (first cousins, for example). The relatedness increases the chance that both parents carry precisely the same alkaptonuria allele inherited from a common ancestor.

ABO: A Three-Allele System

For a gene with three alleles, we noted that the expected Hardy-Weinberg frequency of a *homozygous* genotype is the square of the relevant allele frequency, and the frequency of a *heterozygous* genotype is twice the product of the two relevant allele frequencies. The ABO system provides a good example. We define the allele frequencies as

$$p = \text{freq}(A) \quad q = \text{freq}(B) \quad r = \text{freq}(O)$$

Using Hardy-Weinberg algebra, we find that the phenotype frequencies are:

$$\begin{aligned}
\text{freq(type A)} &= \text{freq}(A/A) + \text{freq}(A/O) &&= p^2 + 2pr \\
\text{freq(type B)} &= \text{freq}(B/B) + \text{freq}(B/O) &&= q^2 + 2qr \\
\text{freq(type AB)} &= \text{freq}(A/B) &&= 2pq \\
\text{freq(type O)} &= \text{freq}(O/O) &&= r^2
\end{aligned} \qquad (4)$$

As an arithmetic example, among Greeks in Athens (Boyd 1950), the allele frequencies are $p = 0.25$, $q = 0.10$, and $r = 0.65$. The corresponding phenotype frequencies are

$$\begin{aligned}
\text{freq(type A)} &= (0.25)^2 + 2(0.25)(0.65) = 0.39 \\
\text{freq(type B)} &= (0.10)^2 + 2(0.10)(0.65) = 0.14 \\
\text{freq(type AB)} &= 2(0.25)(0.10) = 0.05 \\
\text{freq(type O)} &= (0.65)^2 = 0.42
\end{aligned}$$

Thus blood types A and O are about equally frequent in Athens, and blood type AB is relatively rare.

In practice, the data are obtained the other way around: one gets the frequencies of the four blood types by testing blood samples and then calculates the allele frequencies. Because of the particular dominance relationships in the ABO system, these calculations are difficult to do in an efficient way. However, by substituting for the indicated phenotype frequencies using Equations (4) readers can confirm that the following estimates are reasonable ones (assuming the Hardy-Weinberg law is applicable):

$$\begin{aligned}
p &= 1 - \sqrt{\text{freq(type B)} + \text{freq(type O)}} \\
q &= 1 - \sqrt{\text{freq(type A)} + \text{freq(type O)}} \\
r &= 1 - p - q
\end{aligned}$$

X-Linked Genes

Genes on the X chromosome will be present twice in females, but only once in males. For this reason, the equilibrium frequencies under Hardy-Weinberg assumptions must be stated separately for the two sexes:

	Genotype	Hardy-Weinberg equilibrium frequency
among females:	H/H	p^2
	H/h	$2pq$
	h/h	q^2
among males:	$H/(Y)$	p
	$h/(Y)$	q

*Some political, social, economic, and medical decisions are based on estimates of this type; lacking extensive knowledge about uncommon events, one relies on whatever imperfect information is at hand.

H and h are used here to represent the two alleles of any X-linked gene (hemophilia, for example); p and q represent the frequencies of H and h, respectively, in either sex. Among females, the Hardy-Weinberg frequencies are just as they would be for an autosomal gene. A male's *one* allele, however, either H or h, determines his X-linked phenotype. Therefore, the frequencies of male genotypes or phenotypes are the same as the allele frequencies. For example, approximately 1 in 10,000 males is afflicted with classic hemophilia and therefore has the genotype h/(Y). This tells us that q = freq(h) = 1/10,000 = 10^{-4}. The Hardy-Weinberg formulation suggests why the trait is exceedingly rare (but not absent) among females. An affected female must have the genotype h/h and the expected frequency $q^2 = (10^{-4})^2 = 10^{-8}$ = one in a hundred million. If the total population of the United States (more than 200 million) were one big Hardy-Weinberg population, we would expect to find just a few female hemophiliacs. Indeed, a few cases of homozygous female bleeders, with bleeder fathers and carrier mothers, have been reported.

SUMMARY

1. Population genetics extends the Mendelian parent–offspring rules to larger groups that can be described by the frequencies of alleles, genotypes, phenotypes, and matings.
2. Allele frequencies can be calculated from genotype frequencies: the frequency of an allele is equal to the frequency of homozygotes plus *half* the frequency of heterozygotes.
3. Under random mating, the frequency of a given mating is the product of the frequencies of the genotypes involved (multiplied by 2 when the mates have different genotypes).
4. The Hardy-Weinberg law states that population statistics remain constant when evolutionary forces are negligible. For p = freq(A) and q = freq(a), the equilibrium genotype frequencies of A/A, A/a, and a/a become p^2, $2pq$, and q^2, respectively, in one generation.
5. Proofs of the Hardy-Weinberg law involve random pairings of genotypes or alleles from an arbitrarily constructed base population.
6. Quantitative considerations of evolution begin with modifications of the Hardy-Weinberg law. The frequency of a particular gene today depends primarily on evolutionary forces in the past and not on whether it acts in a dominant or recessive fashion.
7. Hardy-Weinberg arithmetic can be used for specific human traits to obtain statistical information on allele frequencies, the chance that a person may be a carrier of a detrimental recessive, and other quantities.
8. The Hardy-Weinberg law can be extended to situations involving multiple alleles or X-linked genes.

FURTHER READING

Introductions to population genetics are clearly presented in the short books by Ayala (1982) and by Hartl (1987), and in the general genetics books by Strickberger (1985) and Crow (1983). Greater depth is provided by Crow (1986), Hartl and Clark (1989), and Spiess (1977). Provine (1971) provides a history of the ideas of population genetics, and Jameson (1977) reprints important papers, including ones by Yule, Hardy, and Weinberg. Accounts of human population genetics include Bodmer and Cavalli-Sforza (1976) and Cavalli-Sforza and Bodmer (1971).

QUESTIONS

We suggest that the following problems be done in order.

1. For each of the populations below, calculate the expected frequencies of the phenotypes for the MN blood group, assuming the conditions of the Hardy-Weinberg law are met. Type M and type N are homozygotes, and type MN is heterozygous. [The data in this and some other problems are taken from Mourant et al. (1976).]

Population	Allele frequency	
	M	*N*
Pima Indians of Arizona	0.70	0.30
Bushmen of Botswana	0.60	0.40
Amish of Indiana	0.50	0.50
Polynesians of Easter Island	0.40	0.60

2. Assume that the frequency of the Rh negative allele, *r*, in a particular city is 0.1. (a) What genotypic frequencies, *R/R*, *R/r*, *r/r*, are predicted by the Hardy-Weinberg law? (b) There are 40 children in one of the city's kindergartens. What is the probability that all 40 are Rh positive, *R/*−? (Just set up the proper expression.)

3. Among native Hawaiians, about 90% of alleles at the Rh locus are *R*; thus freq(*R*) = 0.9 and freq(*r*) = 0.1. (a) In what percentage of all matings is one person Rh positive and the other Rh negative? (b) In what percentage of all matings is the woman Rh positive and the man Rh negative? (c) What fraction of Rh positive people are heterozygous?

4. A group of 100 people splits away from a larger population and establishes a separate society. With respect to the MN blood types, the emigrants number: type M = 41, type MN = 38, type N = 21. (a) What are the allele frequencies? Do you need to assume the Hardy-Weinberg Law to do these calculations? (b) If this group and their descendants meet the conditions of the Hardy-Weinberg law, what are the expected frequencies of the MN phenotypes in subsequent generations? (Because the group is fairly small, assume that genetic drift, discussed in following chapters, is negligible.)

5. Consider the sickle-cell gene, *S*, and its normal allele, *A*. In an adult African population, an investigator finds the following numbers of the three genotypes:

Genotype	Number
A/A	605
A/S	390
S/S	5
total	1,000

What numbers (out of 1,000) should the investigator have expected if this population met the conditions of the Hardy-Weinberg law?

6. Do the observed numbers in the population below conform reasonably well to the predictions of the Hardy-Weinberg law? Use a chi-square test to find out, recalling that a chi-square value of 3.8 corresponds to a probability of 0.05 for 1 degree of freedom. (Although there are three phenotypic classes here, there is only 1 degree of freedom, since both the total number and the allele frequencies are determined by the data.)

Genotype	Observed numbers
A/A	20
A/a	40
a/a	40
total	100

7. The ability to taste phenylthiocarbamide is controlled by a dominant allele, *T*. Thus tasters are *T/*− and nontasters are *t/t*. (a) For Russians and Malays, estimate q = freq(*t*). (b) For each population, what fraction of tasters is expected to be homozygous?

Phenotype	Tasters	Nontasters	Total
Russians in Kharkov and Moscow	412	235	647
Malays in Singapore	199	38	237

8. In a certain population, 1 person in 10,000 is albino (*c/c*). What fraction of this population is expected to be heterozygous? Do you need to assume the Hardy-Weinberg law?

9. Among human males, about 10% are color blind because they carry an allele of an X-linked gene. What percentage of women is expected to be color blind? Assume the Hardy-Weinberg law.

10. If 60% of men showed a *dominant* X-linked trait [having genotype *A*/(Y)], what percentage of women would be expected to show the trait? Assume the Hardy-Weinberg law.

11. Consider an X-linked gene with two alleles, *B* and *b*, with frequencies p and q, respectively. (a) List the six possible matings and their frequencies if mating is at random. (b) Assuming $p = 2/5$ and $q = 3/5$, which mating is expected to be the most frequent?

12. Among French Canadians in Quebec, the frequencies of the ABO alleles are roughly: freq(*A*) = 0.3, freq(*B*) = 0.1, freq(*O*) = 0.6. (a) What are the expected Hardy-Weinberg frequencies of the ABO blood types? (b) What fraction with blood type A is expected to be homozygous?

13. The American geneticist, William Castle, and the English mathematicians, Karl Pearson and Udny Yule, demonstrated stable genetic equilibria in certain cases before the general algebraic proofs by Hardy and Weinberg (Provine 1971). For example, Yule assumed Hardy-Weinberg conditions starting with Mendel's F_1 generation (all A/a individuals). What were the proportions of dominant and recessive phenotypes in the equilibrium population constructed by Yule?

14. Incompatible matings for Rh are those in which the woman is Rh negative, r/r, and the man is Rh positive, $R/-$. Demonstrate that the expected frequency of incompatible matings is $q^2(1-q^2)$, where $q = \text{freq}(r)$.

15. An investigator finds that a group of people meets the conditions of the Hardy-Weinberg law. She tells you that there are twice as many persons with the genotype B/B as there are heterozygotes. What is the frequency of the B allele?

16. Optional for algebra fans: Show that the sum of the random frequencies of the three matings in which both parents have the dominant phenotype (A dominant to a) simplifies to $(1-q^2)^2$, where $q = \text{freq}(a)$.

ANSWERS

1.

Population	M	MN	N
Pima	0.49	0.42	0.09
Bushmen	0.36	0.48	0.16
Amish	0.25	0.50	0.25
Polynesians	0.16	0.48	0.36

(Columns M, MN, N under the heading **Phenotype**.)

2. (a) 0.81, 0.18, 0.01. (b) $(0.99)^{40} = 0.67$ (using logarithms or the to-the-power key of a hand-held calculator).

3. The genotype frequencies are the same as those given in Question 2. (a) $2(0.99)(0.01) = 0.0198 = 1.98\% \approx 2\%$. (b) $(0.99)(0.01) = 0.0099 = 0.99\% \approx 1\%$. (c) $18/99 \approx 18\%$.

4. (a) $\text{freq}(M) = p = (82 + 38)/200 = 0.6$; $\text{freq}(N) = q = (42 + 38)/200 = 0.4$ (Hardy-Weinberg law not assumed). (b) freq(type M) $= p^2 = 0.36$; freq(type MN) $= 2pq = 0.48$; freq(type N) $= q^2 = 0.16$.

5. $p = \text{freq}(A) = (1210 + 390)/2000 = 0.80$; $q = 1 - p = 0.20$.

Genotype	Expected frequency	Expected number
A/A	$p^2 = 0.64$	640
A/S	$2pq = 0.32$	320
S/S	$q^2 = 0.04$	40
total	1	1,000

6. $p = (40 + 40)/200 = 0.4$; $q = 1 - p = 0.6$.

Genotype	Observed number	Expected number	Deviation
A/A	20	16	4
A/a	40	48	−8
a/a	40	36	4
total	100	100	—

$\chi^2 = 16/16 + 64/48 + 16/36 = 2.8$, which is less than 3.8, so the observed data conform well enough to the Hardy-Weinberg expectations.

7. (a) Assuming the Hardy-Weinberg law, for Russians, $q = \sqrt{235/647} = 0.60$; for Malays, $q = \sqrt{38/237} = 0.40$.
(b)

Population	freq (T/T)	freq (T/t)	All tasters	Fraction homozygous
Russians	0.16	0.48	0.64	16/64 = 0.25
Malays	0.36	0.48	0.84	36/84 = 0.43

8. Assuming the Hardy-Weinberg law, $q = \text{freq}(c) = \sqrt{1/10{,}000} = 1/100$; $p = \text{freq}(C) = 1 - q = 99/100$. Fraction of population that is heterozygous $= 2pq = 198/10{,}000$.

9. $(1/10)^2 = 1/100 = 1\%$.

10. $p = \text{freq}(A) =$ frequency of men with trait $= 0.6$. Among women, $\text{freq}(A/-) = p^2 + 2pq = 1 - q^2 = 84\%$.

11. (a)

Male genotype	Female genotype	Frequency
$B/(Y)$	B/B	p^3
$B/(Y)$	B/b	$2p^2q$
$B/(Y)$	b/b	pq^2
$b/(Y)$	B/B	p^2q
$b/(Y)$	B/b	$2pq^2$
$b/(Y)$	b/b	q^3

(Male and female genotype columns under the heading **Mating**.)

(b) $\text{freq}(b/Y \times B/b) = 2pq^2 = 2(2/5)(3/5)^2 = 36/125$.

12. (a) freq(type A) $= (0.3)^2 + 2(0.3)(0.6) = 0.45$
freq(type B) $= (0.1)^2 + 2(0.1)(.6) = 0.13$
freq(type AB) $= 2(0.3)(0.1) = 0.06$
freq(type O) $= (0.6)^2 = 0.36$
Sum $= 1.00$
(b) $(0.3)^2/0.45 = 9/45 = 0.20$.

13. 3/4 dominant, 1/4 recessive, because $p = q = 0.5$

14. freq(Rh negative among women) $= q^2$ and freq(Rh positive among men) $= p^2 + 2pq = 1 - q^2$. The product equals the given expression.

15. Algebraically she says that $p^2 = 2(2pq) = 4pq$. Dividing both sides by p gives $p = 4q$. Substituting $1 - p$ for q and rearranging gives $p = 4/5$.

16.

Mating	Frequency
$A/A \times A/A$	$(p^2)(p^2)$
$A/A \times A/a$	$(2)(p^2)(2pq)$
$A/a \times A/a$	$(2pq)(2pq)$

Sum of frequencies $= p^2(p^2 + 4pq + 4q^2) = p^2(p + 2q)^2$. Substituting $1 - q$ for p yields

$$\begin{aligned}\text{Sum of frequencies} &= (1-q)^2(1+q)^2\\ &= [(1-q)(1+q)]^2\\ &= (1-q^2)^2\end{aligned}$$

Chapter 21 Inbreeding and Isolates

Relatives and Their Offspring
Cousins
Inbreeding

Measuring Inbreeding from Pedigrees
Alleles Identical by Descent
Calculating the Coefficient of Inbreeding
More Complex Pedigrees

The Effects of Inbreeding
Increase in Specific Recessive Traits
Health and Mortality Data
The Prognosis for a First Cousin Marriage

Isolates
Recessive Disease in the Old Order Amish
Inbreeding in the Hutterites

> *None of you shall approach to any that is near of kin to him, to uncover their nakedness.*
>
> *Leviticus 18:6*

Speaking to Moses, the Lord issued this ban on sex between closely related persons. The Biblical injunction probably reflected awareness of both adverse biological and social consequences. Although some species successfully reproduce by close matings—even self-fertilization—other species fare less well. In human populations, the offspring of closely related mates are, on the average, not as healthy as others.

RELATIVES AND THEIR OFFSPRING

Two people are related to each other if they have one or more shared ancestors. Siblings, of course, have the same mother and father, but most other relatives have common ancestry more remote than the parental generation.

Cousins

In Western civilizations the system of naming related persons, beyond close relatives, is based on the term *cousin* and a few modifiers (Figure 1). First cousins (G and H) have a common set of grandparents (A and B), and one each of their parents (D and E) are sibs. Second cousins (I and J) have a common set of great-grandparents (A and B), and one each of their parents are first cousins. The term *removed* pertains to a difference in generations: thus G and J are first cousins once removed, and G and L are first cousins twice removed.

If D and E had the same father but different mothers (or vice versa), they would be *half sibs*, and their children would then be *half* first cousins, and so on. If G and H had two common sets of grandparents, they would then be *double* first cousins; this relationship would arise, for example, if their fathers were brothers and their mothers were sisters. Figure 2 illustrates these relationships in two ways: as standard pedigrees and as **path diagrams**, the latter method being particularly useful for this chapter. The single straight arrow connecting each parent with each child in a path diagram represents the passage of an egg or a sperm.

In some isolated human populations, one can identify many more common ancestors within the half dozen or so generations ancestral to two given people (Figure 3). Extending far enough back in time, of course, we are all related one way or another. The maximum number of ancestors n generations back is 2^n. Thus, you have 2 parents, 4 grandparents, 8 great-grandparents, and so on. This number rapidly exceeds the entire population of the world at some time in the not too distant past, and shows that all of us must have many ancestors in common.

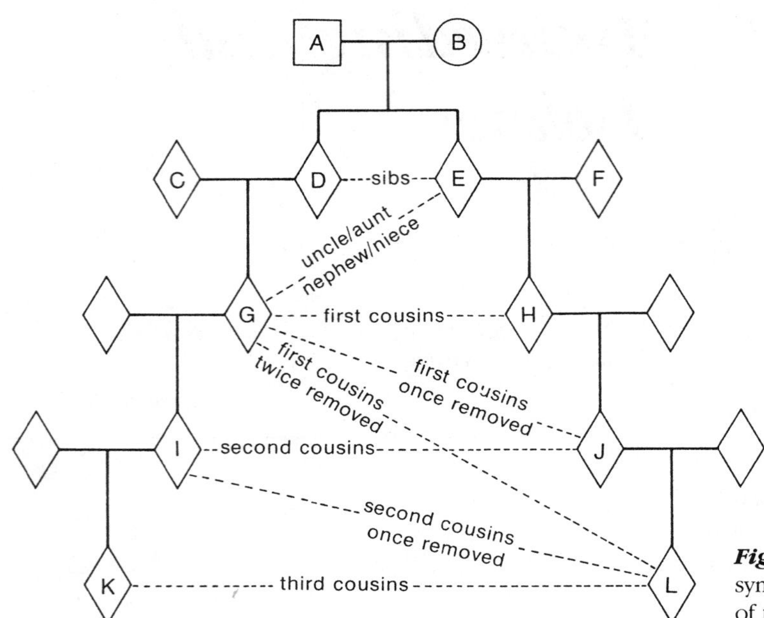

Figure 1 The terminology of relatedness. The diamond symbol is used here, because the term *cousin* leaves the sex of the individual unspecified.

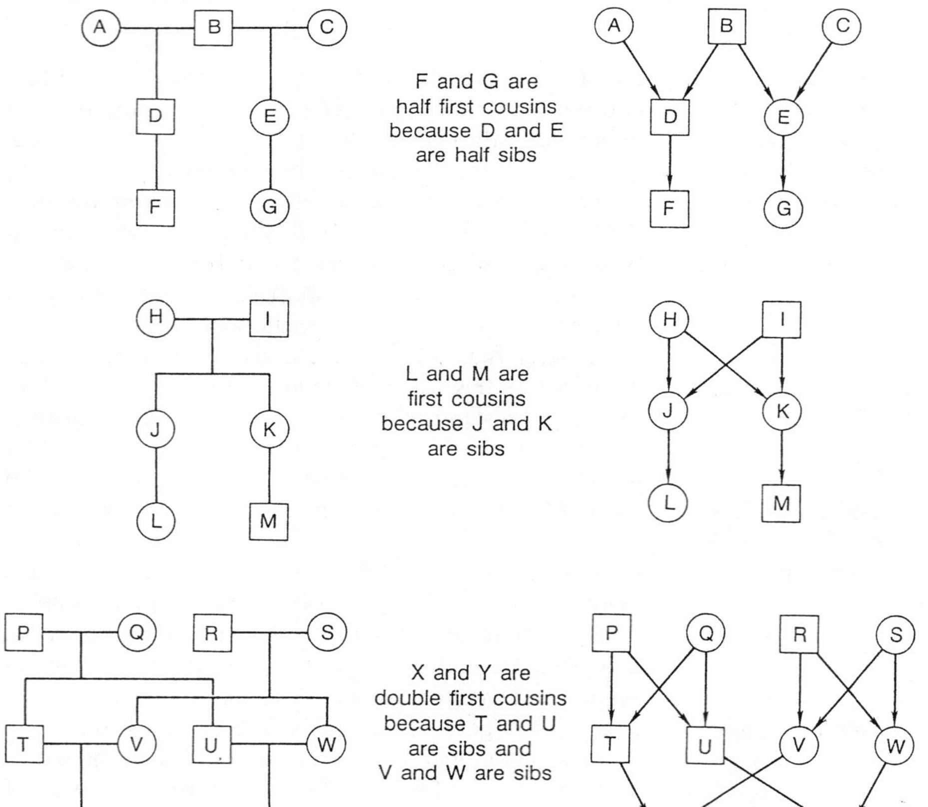

Figure 2 Half, regular, and double first cousins. In a path diagram, a single unbranched line connects each parent with each child. The form of the relationship is not dependent on the sexes of the individuals.

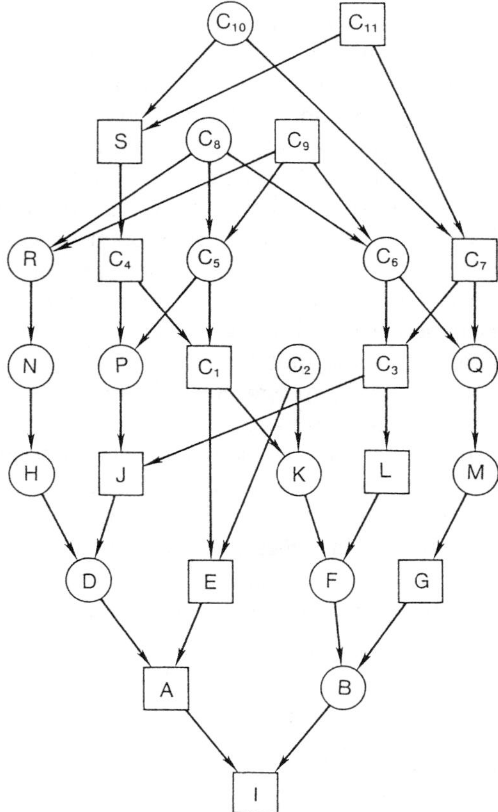

Figure 3 A path diagram from a religious isolate, the Hutterites, living on the western plains of the United States and Canada. The diagram shows the many relationships between individuals A and B, who had child I. The common ancestors of A and B are labeled C_1 through C_{11}. Other than the common ancestors, the path diagram includes only those in the direct lines of descent to A and B. The closest relationship between A and B is first cousins once removed (via common ancestors C_1 and C_2). At the same time, A and B are half second cousins (C_3), third cousins (C_4 and C_5), third cousins again (C_6 and C_7), and mutiple looser relationships. Surprisingly, the total relationship between A and B is not as great as it would be if A and B were just first cousins.

Inbreeding

A mating between relatives is called **consanguineous** ("with blood"*), and their offspring are said to be **inbred**. The degree of inbreeding depends directly on the closeness of the relationship. Overall, the human species is not very inbred, and the frequency of consanguinity seems to be declining in some places of the world (Khoury et al. 1987b). For matings that are consanguineous, however, we show how to measure the level of inbreeding and the effects on the offspring. Many of the methods for doing so were invented in the 1920s by the much honored and long-lived American geneticist Sewall Wright (Wright 1922; Crow 1988; Provine 1986). As we note, consanguinity has less and less significance for progressively looser relationships; the consequences are very small for the offspring of second cousins and become negligible for more remote consanguinity.

Inbreeding is maximized in species in which self-fertilization occurs. Mendel himself (1866) analyzed the consequences of successive generations of self-fertilization in garden peas. He considered that in generation 1, the numbers of plants of genotypes *A/A*, *A/a*, and *a/a* were 1, 2, and 1, respectively, and that each plant contributed 4 seeds to the next generation. Mendel's table is very similar to this one:

Generation	Numbers of			Fraction heterozygous
	A/A	*A/a*	*a/a*	
1	1	2	1	2/4 = 1/2
2	6	4	6	4/16 = 1/4
3	28	8	28	8/64 = 1/8
4	120	16	120	16/256 = 1/16
⋮	⋮	⋮	⋮	⋮
n				$= (1/2)^n$
⋮				⋮
limit for large n				0

These figures can be verified by noting that each self-fertilized *homozygote* produces four offspring like itself. Each self-fertilized *heterozygote* produces only two offspring like itself, and one offspring in each homozygous class. The effect of the inbreeding is the continuous decrease in the fraction of heterozygotes and the corresponding increase in the fraction of homozygotes. *This relation is the basic result for all inbreeding, but the rate of increase in homozygosis depends on the degree of the inbreeding.*

In humans, and in any species with separate sexes, the closest form of inbreeding results from parent–child or brother–sister mating. Usually prohibited by law, such matings are called *incestuous*. They are almost universally proscribed by religious scriptures as well. The incest taboo probably evolved very early in prehistory to minimize adverse effects on offspring and to help maintain the stability of the family unit (van den Berghe 1983). Because incest involves only family members and since they almost always feel guilt and shame, it is seldom a court matter, and its actual frequency is unknown.* On

*Although consanguinity depends on the shared *alleles* rather than shared *blood*, the use of the latter term persists in some everyday expressions: "bloodline," "full-blooded," "blue-blood," and so on.

*Here we define incest *narrowly* as heterosexual intercourse between pairs of postpubertal sibs, half sibs, parent–child, grandparent–grandchild, or uncle–niece (aunt–nephew)—that is, close consanguineous matings that could produce inbred offspring (see Question 14). Social scientists usually define incest more *broadly* to include abuse of young children within the home by older persons, especially adults, including step-relatives. These investigators are usually concerned with psychological harm to the victim and approaches to effective therapy (Gelinas 1983, for example). Recent investigations suggest that incest in the broad sense is more common that once thought.

the other hand, in some cultures at some periods of history, brother-sister marriages were encouraged among royalty, but not among the general population. For example, Cleopatra (died 30 B.C.) may have been a child of a brother–sister mating (Figure 4).

Geneticists use continued programs of brother–sister matings in order to produce so-called *inbred strains* of corn, flies, mice, and so on. The path diagrams for continued self-fertilization and for continued brother–sister matings are drawn in Figure 5. The probabilities that the organisms are homozygous increase quite rapidly with self-fertilization and somewhat more slowly with sib matings. In either case, however, the likelihood of any individual being homozygous approaches certainty after a moderate number of generations.

Uncle–niece and aunt–nephew marriages are very rare in human groups except where social custom dictates this form of consanguinity, if it can be arranged. In southern India, for example, Hindu men favor marriage with a sister's daughter but may not marry a brother's daughter (Dronamraju 1964). In one study of several caste groups representing very different socioeconomic levels (from well-to-do to so-called untouchables), about 19% of all marriages were between uncle and niece, and another 22% were between first cousins (Reddy 1987).

Marriage between first cousins is the most frequent type of human consanguineous mating, in part because cousins may be close in age and geography and their families are usually known to each other (Imaizumi 1987). Usually the fraction of first cousin marriages is well below 1%, but it may be higher in some cultures, as noted above for southern India. In small Japanese towns, first cousin marriages are currently 2 to 3%, but they were much more frequent (5 to 15%) a generation ago (Imaizumi 1986; Schull and Neel 1965).

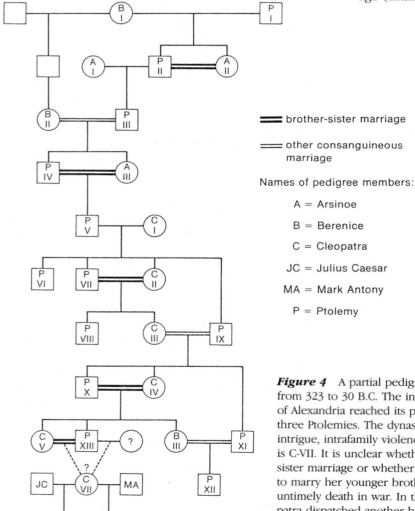

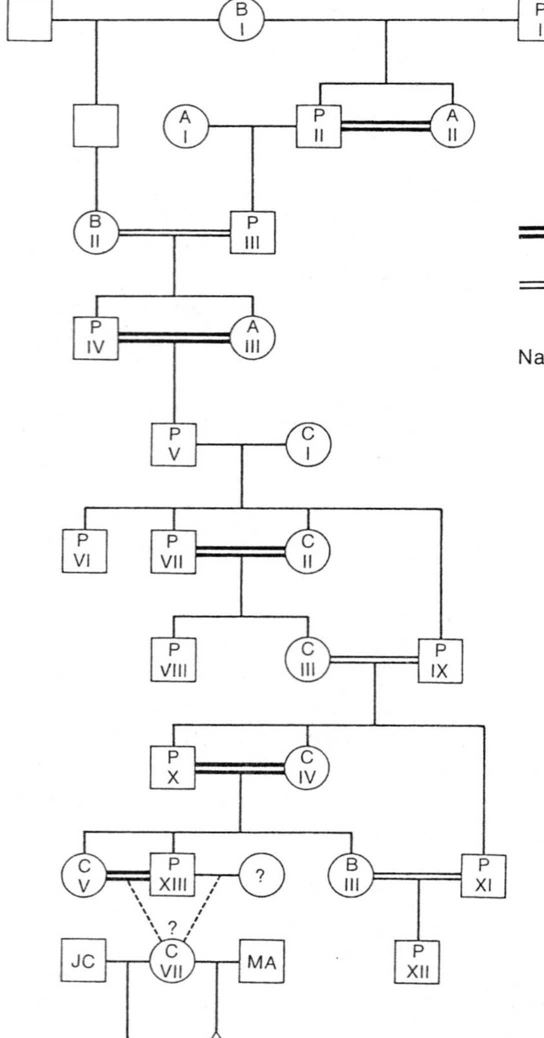

Figure 4 A partial pedigree of the Ptolemies, kings of Egypt from 323 to 30 B.C. The intellectual and commercial vitality of Alexandria reached its peak during the reigns of the first three Ptolemies. The dynasty was often turbulent, marked by intrigue, intrafamily violence, and war. The famous Cleopatra is C-VII. It is unclear whether she was a child of a brother–sister marriage or whether she was not inbred at all. She was to marry her younger brother (half-brother?), but he met an untimely death in war. In the manner of her forebears, Cleopatra dispatched another brother (or half-brother) by poison. She and her son by Caesar, P-XVI, were the last of the Ptolemaic line. (Pedigree mostly after Weigall 1924. The numbering of the Ptolemies after P-V varies from source to source.)

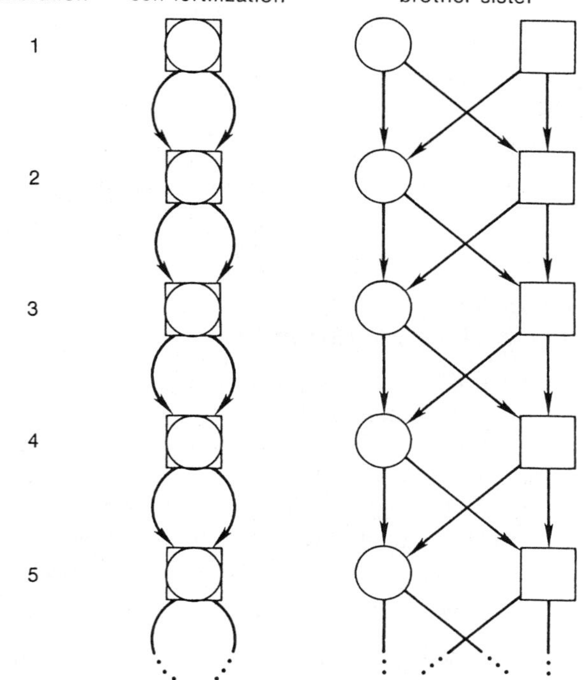

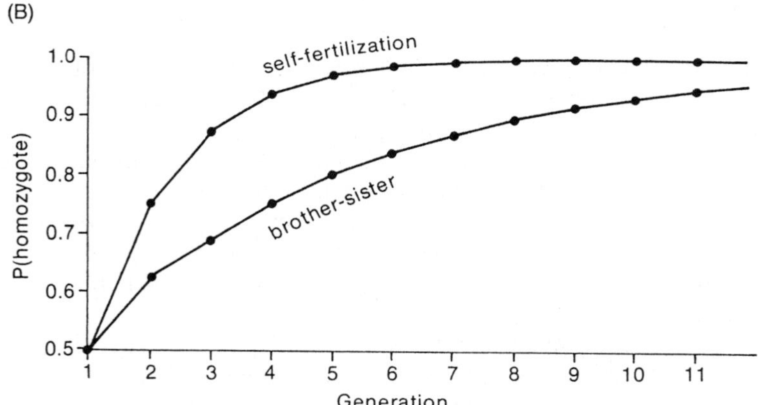

Figure 5 Path diagrams for continued self-fertilization and brother–sister mating. Note that an individual in a continued brother–sister scheme has two parents, *two* grandparents, *two* great-grandparents, and so on. (B) The corresponding increase in the probability of homozygotes over the course of generations. It is assumed that the self-fertilized individual and the mated sibs in generation 1 are taken from randomly mating populations with two alleles in equal frequency.

MEASURING INBREEDING FROM PEDIGREES

Whether as a regular system of mating, as a preferred social custom, or as an irregular happening, consanguinity increases the likelihood of homozygosity among the inbred progeny. The increase can be calculated from conventional pedigrees or, more easily, from path diagrams. A person can be homozygous without being inbred, of course, as is clear from the Hardy-Weinberg formulation: the probability of homozygosity in a randomly mating population is equal to $p^2 + q^2$. Inbreeding does not alter allele frequencies, but it pushes the probability of homozygosity above the value, $p^2 + q^2$. The *excess homozygosity* (above the random amount) is what produces the greater likelihood of adverse effects in inbred persons. The increase in homozygosity comes about because of shared ancestry; a specific deleterious allele from a joint forebear can be successively replicated and transmitted through *both* lines of descent to mates who are related, and thence to their offspring.

Alleles Identical by Descent

To make matters more concrete, consider two alleles at the alkaptonuria locus, K and k, with frequencies p and q.

Those alkaptonurics, k/k, who are *not* inbred—whose parents are *not* related—are said to be homozygous for **independent alleles**. That is, the two k alleles have independent histories of descent from separate ancestors (as far as can be demonstrated).

Alkaptonurics who are *inbred*—whose parents are related—may also be homozygous for independent alleles. But, in addition, because their parents have one or more common ancestors, inbred alkaptonurics may possess alleles that are replicas of precisely the same allele present in a joint forebear. Such alleles, which not only code for the same polypeptide but also trace back to exactly the same segment of DNA, are said to be **identical by descent**.

Thus there are two ways for an inbred person to be homozygous: for alleles with independent histories, or for alleles with a shared history—that is, descended from precisely the same allele in a common ancestor. In either case, the resultant phenotype is the same. The probability that the two alleles in a person are identical by descent is that person's **coefficient of inbreeding**.

Figure 6 illustrates these concepts. Female Q is the child of M and N, who are half first cousins via common ancestor D. It is assumed that the seven individuals in generation I are not related to each other and are not themselves inbred. Each person has a particular genotype, of course (K/K or K/k or k/k), but we do not need to know what that genotype is. We want to be able to trace the descent of individual alleles from generation I through generation IV, however, and to do so, we have simply given each allele a tag: K_1 through K_{14}.

There are many possible genotypes for Q, such as K_3/K_9, or K_7/K_8, or K_7/K_{11}, and so on. (In each case, Q could be homozygous or heterozygous, depending on the state of the two alleles.) The allele pairs in these particular examples are of independent origin. (Even though 7 and 8 happen to be in the same individual in generation I, they have no common history as far as we know, and therefore are independent.)

On the other hand, Q could have genotype K_7/K_7 or K_8/K_8. In these two cases, Q is homozygous for alleles identical by descent. This would come about if all the gametes labeled r through w on the path diagram carried replicas of K_7, or alternatively, replicas of K_8. The coefficient of inbreeding, symbolized f, is the probability of these events. Thus,

$$\begin{aligned} f_Q &= \text{coefficient of inbreeding of person Q} \\ &= P(\text{Q possesses alleles identical by descent}) \\ &= P(\text{Q is } K_7/K_7) + P(\text{Q is } K_8/K_8) \end{aligned}$$

Calculating the Coefficient of Inbreeding

First we calculate the probability that Q is K_7/K_7. (The probability of identity for K_8 has the same value.) Because of the randomness of segregation and fertilization, gamete t is equally likely to contain any one of the eight alleles K_1 through K_8. Thus P(t carries K_7) = 1/8. Also, gamete w is equally likely to carry any of the eight alleles K_7 through K_{14}, so P(w carries K_7) = 1/8. Thus,

$$P(\text{Q is } K_7/K_7) = (1/8) \times (1/8) = 1/64$$

The probability of identity for K_8 is obtained the same way. Since identity by descent can be for either K_7 or K_8,

$$f_Q = 1/64 + 1/64 = 1/32$$

This means that Q has about a 3% chance of being homozygous for alleles identical by descent. She could be identically homozygous for either K or k, but the former is much more likely, because the normal dominant allele has a much higher frequency than the mutant recessive allele has. Still, as we will see in the next section, the excess homozygosity for the rare recessive is many times greater in an *inbred* person than in a *noninbred* person.

A different way of calculating Q's coefficient of inbreeding can be generalized to any pedigree, however complex. We define a **loop** as a continuous path through a common ancestor, starting with one parent and ending

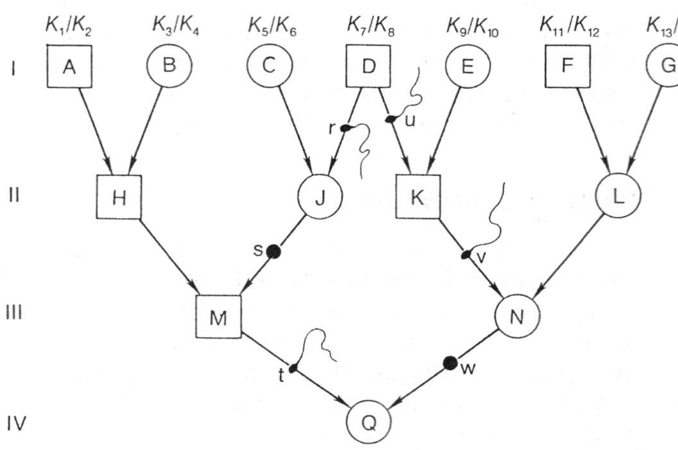

Figure 6 A path diagram showing the ancestry of a child, Q, of a half first cousin marriage. The labels K_1 through K_{14} identify the alleles in the seven ancestors in generation I. The state of the alleles, K (dominant) or k (recessive), is not specified, nor is the phenotype of any person. Note that Q has *seven* rather than the maximum of *eight* great-grandparents. The coefficient of inbreeding of individual Q is f_Q = 1/32 (see text).

with the other. The loop goes *directly* back to the common ancestor and only then reverses direction to come *directly* forward. It must not go through any person twice. For Q in Figure 6, the loop is **M–J–D–K–N**. When there is just one loop, as in this case, the inbreeding coefficient can be calculated as

$$f = (1/2)^n$$

where n is the number of ancestors in the loop, including the parents (but not the individual whose f is being calculated). For Q,

$$f_Q = (1/2)^5 = 1/32$$

The justification for this simple multiplication of powers of 1/2 is as follows. Consider the gametes leading to and from any ancestor in the loop other than the common ancestor, say M (see Diagram A). The particular allele of the alkaptonuria gene present in gamete t is a replica of either the one in s or the one in the other gamete. These are equally likely events. Using the symbol ≡ to mean "is a replica of the one in," we have $P(t \equiv s) = 1/2$.

To complete all the steps in the loop, we need to evaluate the probability that the alleles in r and u are replicas of each other (see Diagram B). These gametes are formed by the person with allelic tags K_7 and K_8. Each gamete is equally likely to carry the one allele or the other, so there are four equally likely outcomes considering both gametes together: 7-7, 8-8, 7-8, or 8-7. In the first two cases, the alleles are replicas; otherwise they are not. Thus, $P(r \equiv u) = 1/2$. (This step through a common ancestor is considered further below.)

Diagram A

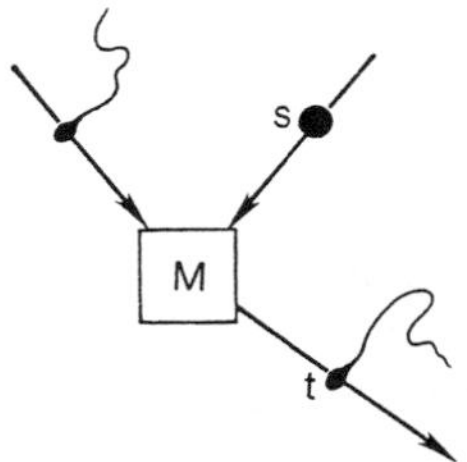

Diagram B

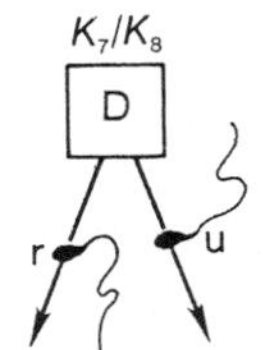

In summary, in this simple case, each and every step in the loop corresponds to a probability of 1/2 that the designated gametes carry replicas of each other. In order for Q to possess alleles both tagged K_7 or both tagged K_8, each gamete in turn must carry a replicated allele. Thus

$$f_Q = P(t \equiv s) \times P(s \equiv r) \times P(r \equiv u) \times P(u \equiv v) \times P(v \equiv w) = (1/2)^5$$

More Complex Pedigrees

Three kinds of complexities that modify the formula $f = (1/2)^n$ are sometimes found in human pedigrees. We briefly take them into account.

MORE THAN ONE COMMON ANCESTOR. Full first cousins have two common grandparents (C_1 and C_2 in Figure 7). Their offspring can have alleles identical by descent from either the one or the other common ancestor, but not from both at once. For this reason, the contributions toward the coefficient of inbreeding from several common ancestors are simply added together. Generalizing, we have

$$f = \sum(1/2)^n$$

where Σ (Greek sigma) means to add the quantities $(1/2)^n$ for each common ancestor. The number n may be different for the loops through different common ancestors; in Figure 7, the six loops through the six common ancestors of Sewall Wright's parents have n values of 5, 5, 17, 17, 18, and 18. In the same path diagram, note that two other people (and only two) are inbred: Philip Green Wright (four common ancestors) and Elizur Wright I (two common ancestors).

MORE THAN ONE LOOP THROUGH A COMMON ANCESTOR. In some pedigrees *more than one loop* connecting the parents of an inbred child can be traced through a *single* common ancestor. Loops are considered different if they differ from each other by one or more ancestors, and each separate loop makes an additive contribution to inbreeding. Thus, the summation above indicated by Σ includes not only all common ancestors but also all loops through each one of them. Because it often becomes tricky to visually find all the loops through a common ancestor, complicated pedigrees are sometimes analyzed by computers, which do not balk at tedium (Boyce 1983). For example, in the path diagram of Figure 3, there are five loops each through C_{10} and C_{11}, and eight loops each through C_8 and C_9. (Readers should not feel obliged to find them.)

AN INBRED COMMON ANCESTOR. A further modification is needed whenever a common ancestor is inbred. An example is found in Figure 4: in computing the inbreeding coefficient of Cleopatra III (or her brother), we must take into account that her parents' common father, Ptolemy V, is *himself* inbred—the product of a brother–sister marriage. (Readers may wish to redraw this portion of the pedigree as a path diagram to better illustrate the

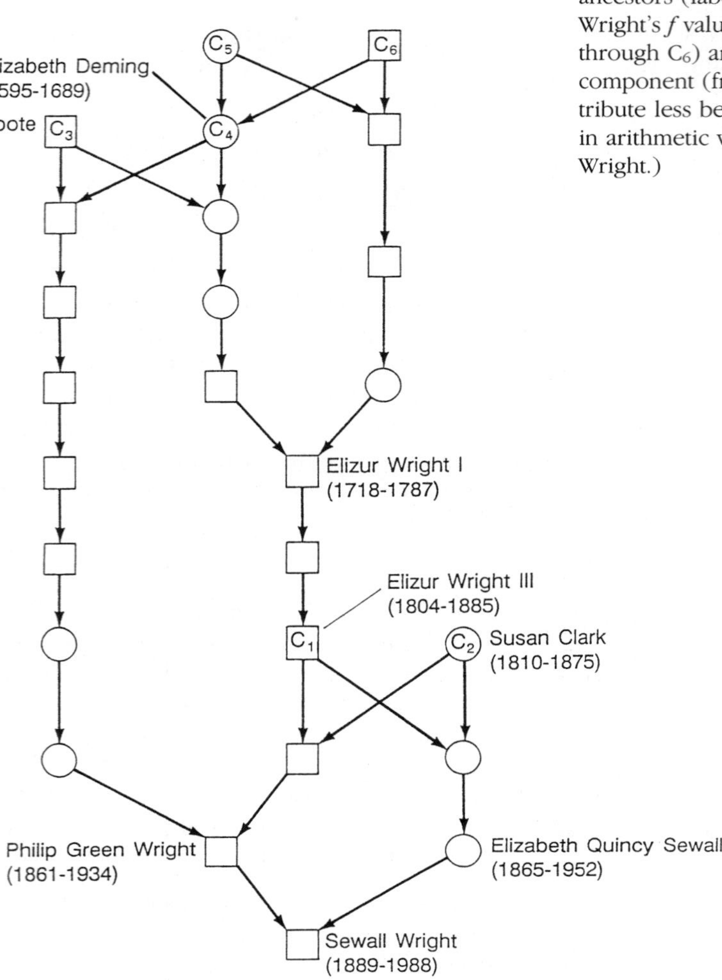

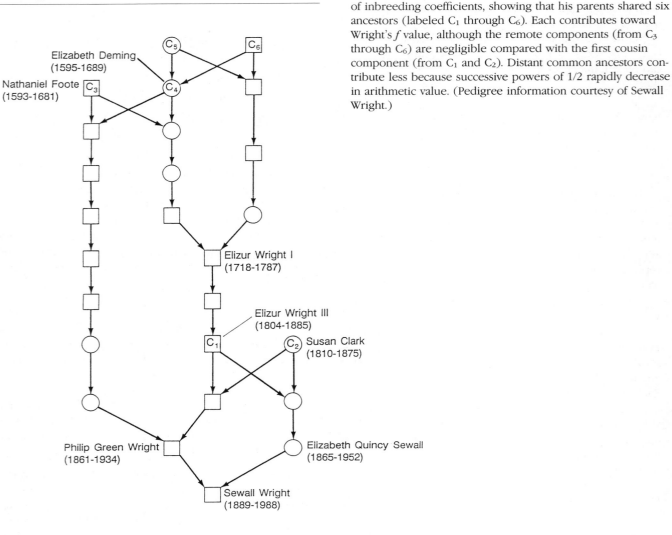

Figure 7 The path diagram for Sewall Wright, the inventor of inbreeding coefficients, showing that his parents shared six ancestors (labeled C_1 through C_6). Each contributes toward Wright's f value, although the remote components (from C_3 through C_6) are negligible compared with the first cousin component (from C_1 and C_2). Distant common ancestors contribute less because successive powers of 1/2 rapidly decrease in arithmetic value. (Pedigree information courtesy of Sewall Wright.)

relationships.) Two alleles in an *inbred* common ancestor may already be identical by descent (rather than independent as we assumed before). If so, the probability that two of a common ancestor's gametes carry alleles identical by descent is greater than 1/2. It is $1/2 + (1/2)f_C$, where f_C is the coefficient of inbreeding of the common ancestor in question. The second term recognizes that when the alleles in two gametes are *not* replicas (probability = 1/2) they may *still* be identical by descent (probability $= f_C$). Putting all the complicating factors together, the coefficient of inbreeding of an individual is calculated as follows:

$$f = \sum(1/2)^n \, (1 + f_C)$$

The exponent n is the number of ancestors in a loop. The summation is over all loops through all common ancestors, one or another of whom may be inbred with value f_C. If no common ancestor is inbred, the formula reduces to $\Sigma(1/2)^n$. In breeding programs for experimental farm animals and plants, inbreeding of common ancestors may add substantially to f values, but for most human pedigrees the increase is very small.

In summary, the inbreeding coefficient measures a particular way of being homozygous. The value for the offspring of first cousins, 1/16, means that one-sixteenth of the time a pair of alleles—for any gene being considered—is expected to be identical by descent. Alternatively, it means that, on the average, one-sixteenth of all genes in the offspring of first cousins have alleles identical by descent.

THE EFFECTS OF INBREEDING

Whether inbreeding is good or bad depends on whether increased homozygosity is good or bad. For the human species, in which matings are usually between unrelated or only remotely related persons, inbreeding often has adverse effects. We can measure these effects either in terms of single genotypes with identifiable phenotypes, such as alkaptonuria, or in terms of general health characteristics.

Increase in Specific Recessive Traits

For a population mating at random and conforming to the other conditions of the Hardy-Weinberg law, we have shown that the frequency of a recessive phenotype is q^2, where q is the frequency of the recessive allele. For alkaptonuria, where freq(k) = 1/500, the random frequency of alkaptonuria is freq(k/k) = $(1/500)^2$ = 4.0×10^{-6}.

INBRED FREQUENCY OF ALKAPTONURIA. For the half first cousin mating in Figure 6, we noted that individual Q could possess either alleles identical by descent or independent alleles. The probability of alleles identical by descent is $f_Q = 1/32$. We distinguish the two cases:

Case I: Q has alleles identical by descent; probability = 1/32
Case II: Q has independent alleles; probability = 31/32

In Case I, Q can be homozygous k/k or homozygous K/K. (Q could also be heterozygous if a mutation had occurred in the lines of descent, but we ignore this rare event.) Because both alleles are necessarily in the same state—being replicas of just one allele in the common ancestor—the probability of k homozygosity versus K homozygosity depends only on the frequencies of k versus K. Thus, given that Q has alleles *identical by descent*, freq(k/k) = q = 1/500.

In Case II, Q can be of any genotype: K/K, K/k, or k/k. Since the alleles are of independent origin, the frequencies of the three genotypes are given by the Hardy-Weinberg law. Thus, given that Q has *independent* alleles, freq(k/k) = q^2 = $(1/500)^2$.

Altogether, we can write that the frequency of alkaptonuria among the offspring of half first cousins is the sum of the two values of freq(k/k) above, each weighted by the probability of its occurrence; that is

$$\begin{aligned}\text{freq}(k/k \text{ overall}) &= (q \times \text{probability of Case I}) \\ &\quad + (q^2 \times \text{probability of Case II}) \\ &= (1/500)(1/32) + (1/500)^2(31/32) \\ &= 66.4 \times 10^{-6}\end{aligned}$$

INBRED VERSUS RANDOM FREQUENCIES. Note that the probability of alkaptonuria from this particular level of inbreeding is increased many fold over its probability from random mating: 66.4×10^{-6} versus 4.0×10^{-6}, or about 17 times. Alkaptonuria is still a rare disease among the children of half first cousins, but the increase over random occurrence applies to *any* deleterious recessive allele with the same frequency.

There are two variables in the calculation of the inbred probability of homozygosity: the frequency of the allele, q, and the coefficient of inbreeding of the offspring, f. The more general form of the formula for any recessive allele, a, is

$$\textbf{freq}(a/a) = qf + q^2(1 - f)$$

which corresponds term by term to the specific calculation for alkaptonuria. This expression extends the Hardy-Weinberg frequency to account for a departure from random mating. It applies to the inbred offspring of a particular consanguineous mating in an otherwise randomly mating population, or to the population as a whole when the average inbreeding coefficient is f. Notice that freq(a/a) reduces to q^2 when $f = 0$ (as it must).

Using the formula above, Table 1 lists the increase in the likelihood of homozygosis among the offspring of various consanguineous matings and for various allele frequencies. Reading down a column, note that the effect of inbreeding increases as the parental consanguinity increases, as expected. Reading across a row, note that the rarer the allele, the greater is the likelihood of homozygosity from consanguineous mating relative to random mating. This is because it is extremely unlikely that very rare alleles would come together in an individual independently of each other. But related persons can inherit the identical allele from a common ancestor. Mathematically, the increase across a row occurs because q is much bigger than q^2 when q is very small.

There is a useful corollary, which we mentioned briefly in Chapter 3 on pedigrees: If the parents of sibships that include a child with a particular disease are related to each other more often than the parents of all-normal sibships, then the disease may be due to homozygosity for a recessive allele. Here one looks at consanguinity from the reverse standpoint. Instead of predicting the frequency of recessive children from consanguineous matings, this corollary says that finding consanguinity among the parents of affected children gives evidence that the disease results from recessive inheritance. For example, in a study of Japanese families in the 1940s, the overall frequency of first cousin marriages was about 6%. Yet about 50% of children with albinism and about 70% of children with Tay-Sachs disease had parents who were first cousins (Neel et al. 1949).

Inbreeding also increases homozygosity for *dominant* alleles. This fact is not important, however, because the resultant phenotype is usually the same as that of the heterozygote. Note that inbreeding is not of itself harmful; if, by chance, no detrimental recessives were present in an initial population, then inbreeding would have no

Table 1 The increase in *a/a* children from consanguineous matings compared to random matings

Coefficient of inbreeding (f) of offspring (relationship between parents)	RATIO OF THE INBRED FREQ(a/a) TO THE RANDOM FREQ(a/a) FOR:[a]				
	q = 1/10	q = 1/30	q = 1/100	q = 1/300	q = 1/1,000
1/64 (second cousins)	1.1	1.5	2.5	5.7	16.6
1/32 (first cousins once removed)	1.3	1.9	4.1	10.3	32.2
1/16 (first cousins)	1.6	2.8	7.2	19.7	63.4
1/8 (uncle-niece)	2.1	4.6	13.4	38.4	125.9
1/4 (sibs)	3.2	8.2	25.8	75.8	250.8

a. q = freq(a). Each number in the body of the table is the ratio of the inbred freq(a/a) = $qf + q^2(1 - f)$ to the random freq(a/a) = q^2. For example, for an allele with q = 1/30, there are 8.2 times as many recessive homozygotes among the offspring of sibs as among the offspring of unrelated persons. Many recessively inherited diseases have q's in the neighborhood of 1/100 to 1/300.

untoward effect. Exactly the opposite is possible: inbreeding could make homozygous some *beneficial* recessives if any were present in a base population.

Another form of nonrandom mating that increases homozygosity at the expense of heterozygosity is **assortative mating**, in which mates have the same phenotype more often than would occur by chance. To the extent that similar phenotypes reflect similar genotypes, assortative mating affects genotype frequencies in the same way as inbreeding does, but not nearly as strongly. Human populations mate assortatively with regard to skin color, stature, intelligence, and other quantitative traits that are determined by both many environmental and many genetic factors (Chapter 23).

In summary, the reason why inbreeding is usually bad for people is that it tends to reveal detrimental recessive alleles (for alkaptonuria, phenylketonuria, albinism, cystic fibrosis, Tay-Sachs, and so on) that are usually hidden in heterozygotes. Recessive mutations that do not result in specific, recognizable diseases may also cause harm by their small cumulative effects when made homozygous (next section).

Health and Mortality Data

Many studies have compared aspects of health and survival of individuals with different coefficients of inbreeding. The traits that have been tallied include congenital malformations, mental illnesses, IQ, fertility, and mortality within given time periods. Examples of the latter include miscarriages, stillbirths, neonatal deaths (that is, within a month of birth), and deaths during childhood and the teens. All of these traits are influenced in many ways by both genetic and environmental factors—some large and some small.

The most dramatic data deal with incest. The largest investigation involved 161 Czechoslovakian children from 88 father–daughter matings, 72 brother–sister matings, and 1 mother–son mating (Seemanová 1971). The research was *retrospective* in that the children of incest were referred (by maternity and child care agencies) to the investigator, who then studied the parents. For comparison, Seemanová used a control group of 95 noninbred children born to some (but not all) of the same mothers. About 9% of the controls either died before one year of age or lived (for at least a year) with significant mental or physical defects. The corresponding value for the inbred children was 48%, about five times greater. This study has been criticized, however, on the grounds that more often the children of incest were born to parents with substantial mental or physical problems than were the control children (Bittles 1983). A more recent study from Canada had a better design because it was largely *prospective*, including 21 mothers who, at the time of contact, were still pregnant from the incestuous relationship (Baird and McGillivray 1982). Of the ensuing 21 children (7 from father–daughter and 14 from brother–sister matings), 9 (or 43%) had severe malformations or retardation without known cause, or (in one case) a specific autosomal recessive disorder. The authors note that, even though the parents of these children may represent a biased sample, "it remains evident that the empiric risk for abnormalities in children of incest is high."

An extensive and detailed study in postwar Japan of the effects of less severe (and more usual) inbreeding is that of William J. Schull and James V. Neel (1965) of the University of Michigan. They took great care to minimize sources of bias in both the study populations and in the investigators. Table 2 gives a small part of the results. Reading down a column, infant mortality is seen to increase a small (but significant) amount among the more inbred Japanese infants.

In a comprehensive analysis of 31 studies of inbreeding in different countries, Khoury et al. (1987b) report that practically all data show small but measurable increases in mortality (through age 20) among the offspring of cousins, relative to the mortality of the offspring of unrelated parents. The survey also finds that inbreeding has less consequential effects in places with prolonged and relatively high frequencies of consanguinity (Japan, India, and Brazil) than in places with historically lower consanguinity levels. This finding is consistent with the theory that continued inbreeding over many generations should eliminate some of the detrimental recessive genes through the nonreproduction of recessive homozygotes.

Data like those in Table 2 can be used to estimate how many harmful recessive alleles an average person carries in a *heterozygous* state (thereby causing no damage). This estimation is done by extrapolating the amount of mortality or disease among children with known f values to what it would be if $f = 1$ (completely homozygous). The straight-line extrapolation is not very precise, because the *observed* range of f values (generally 0 to 4/64) provides only a short start to the required *extension* to $f = 64/64$. Nevertheless, the results of such calculations show that the average person carries several recessives that would lead to a premature death *if homozygous*. Because there is no way to distinguish between one gene that kills 100% of the time, or two genes each causing death 50% of the time, and so on, the term **lethal equivalent** is used. The average person is probably heterozygous for several recessive lethal equivalents (Crow 1986).

The Prognosis for a First Cousin Marriage

What should be said to first cousins who contemplate marriage and are anxious about possible harm to their offspring? Popular opinion holds that such children are likely to suffer malformations or be less intelligent than their peers. But the data presented here suggest that first cousins carry only a small additional risk of having children with genetic defects of greater or lesser severity. All couples, whether related or not, face the likelihood (perhaps 2 to 3%) of having a child with a serious defect, depending on the criteria used to define "serious." Another few percent of children may fall outside a commonly accepted definition of "normal" but not have a serious defect.

A genetic counselor will certainly obtain a detailed family history from first cousins. If there is a suggestion of a deleterious recessive allele in one partner or in an ancestor, a specific probability calculation can be made. The couple can then consider both the risk and the severity of the defect. More often than not, however, there will be nothing in the family history on which to base a specific calculation, because, as we have seen, rare recessive detrimental alleles will usually remain hidden in heterozygous condition generation after generation. Only vague statements of the small but real increased

Table 2 Effect of parental consanguinity on infant mortality in Japan

Coefficient of inbreeding (f) of offspring (relationship between parents)	INFANT DEATHS/LIVE BIRTHS (= % DEATHS)[a]		
	Hiroshima	*Nagasaki*	*Total*
0 (unrelated)	145/4,089 = 3.5%	273/7,988 = 3.4%	418/12,077 = 3.46%
1/64 (second cousins)	32/722 = 4.4%	50/1,312 = 3.8%	82/2,034 = 4.03%
1/32 (first cousins once removed)	42/585 = 7.2%	53/1,073 = 4.9%	95/1,658 = 5.73%
1/16 (first cousins)	101/1,651 = 6.1%	173/3,296 = 5.2%	274/4,947 = 5.54%

Source: Schull and Neel (1965).
a. Infant mortality is defined as death under one year of age. Overall, the tabulation includes 869 deaths from 20,716 live births (4.19%). Data are from the 1950s and are based on interviews of mothers and on direct observations of some of their children. This inbreeding investigation was separate from studies on the effects of the atomic bomb radiation gathered earlier by the same investigators.

risk from first cousin consanguinity can then be made (Fuhrmann and Vogel 1983). The couple will have to evaluate this not very satisfactory information and come to a decision primarily on the basis of their own feelings.

ISOLATES

An isolate is a community whose members marry within the group for generations, thereby separating themselves genetically from the general population. Actually, semi-isolation of human populations is the rule, because marriage choices are always more or less restricted by socioeconomic, religious, cultural, and racial traits. (See, for example, Klat and Khudr 1986.) The fewer the immigrants a group receives, the more clannish it becomes. People who leave the group do not disturb the genetic isolation, although they may alter allele frequencies within it if they are genotypically unrepresentative of those they leave behind.

Isolates, intriguing in themselves, have several properties that also make them genetically useful: (1) Small, closed populations allow investigation of some aspects of human evolution, especially *genetic drift* (random changes in allele frequencies; see below). (2) Excellent genealogical records, kept because the members of isolates are often proud of their culture and ancestry, provide very thorough analyses of inbreeding. (3) Uniform living conditions, which often prevail in isolates, can aid in the study of the relative effects of genetic variation versus environmental differences.

The most isolated groups are characterized by geographical remoteness or strongly held religious beliefs. Examples that have been studied include the island populations of Pitcairn in the Pacific and Tristan da Cunha in the Atlantic, the practically inaccessible villages of the Jicaque Indians in Honduras, and the Xavante Indians in the interior of Brazil (Jacquard 1974). Religious isolates include the Old Order Amish of Pennsylvania, Ohio, and Indiana and the Hutterites of the western prairies.

Recessive Disease in the Old Order Amish

The Amish sect originated in the seventeenth century in Switzerland as an offshoot of the Mennonites. Both these groups, as well as the Hutterites, were forced to migrate when they incurred the wrath of European Catholic and Protestant churches for deviations from accepted religious practices. About 200 Amish people moved to Pennsylvania between 1720 and 1770, and their descendants in Lancaster County currently number about 14,000 (Khoury et al. 1987a). Other Amish people settled in Ohio and Indiana. Both the Amish and the Hutterites follow a Bible-centered life in which religion and economics are strongly interdependent. Remaining aloof from the outside world within highly ordered farming communities, they are characterized by high morals, devout pacifism, adult baptism, hard work, thrift, and mutual aid. The familiar horse and buggy seen on country roads in Pennsylvania symbolize Amish conservatism, which rejects most modern technology as well as formal education beyond the legal minimum.

In a survey of the Amish by Victor McKusick and his colleagues at Johns Hopkins University (1964), several ordinarily rare diseases were found in increased frequency, each one in a different subgroup of the population.* One of these diseases, the Ellis-van Creveld syndrome (Figure 8) is a form of dwarfism in which the forearms and lower legs are disproportionately short. Many patients die soon after birth, but a few milder cases reach adulthood and may rarely have children. Among the approximately 8,000 Amish then living in Lancaster County, the investigators found 43 persons with the syndrome (genotypically e/e, say). The frequency of the recessive allele (e) in this subgroup was thus estimated as the square root of 43/8,000, or about 1/14. Elsewhere, even among other Amish groups, the allele frequency is very low, less than 1/1,000. Two factors, both related to unpredictable changes in allele frequencies that may occur when a population is small, account for its exceptionally high frequency in the Lancaster County Amish: the founder effect and random genetic drift.

THE FOUNDER EFFECT. In general terms, **founder effect** refers to a newly formed population in which an allele has a frequency different from its frequency in the general population from which the founders came. The smaller the number of founders, the more likely they are to be unrepresentative of the larger population. This concept is illustrated by random sampling from a bag containing, say, 900 black marbles and 100 yellow ones. If you draw a large number, close to 10% of them are likely to be yellow. But if you randomly picked just 10 marbles, it would not be at all surprising to find as little as 0% ($P = 0.35$) or as great as 20% yellows ($P = 0.19$). (Verify these probabilities using the binomial formula.)

The ancestry of all the Lancaster County families with the Ellis-van Creveld syndrome traces back to a Mr. and Mrs. Samuel King who immigrated in 1744. The recessive abnormal allele was undoubtedly present in the one or

*The Amish people have provided a continuing source of information for geneticists. A recent example is the discovery of a heritable form of manic-depressive disease. In the affected Amish families, it is due to a dominant gene that exhibits 60–70% penetrance and is linked to RFLP sites on chromosome 11 (Egeland et al. 1987).

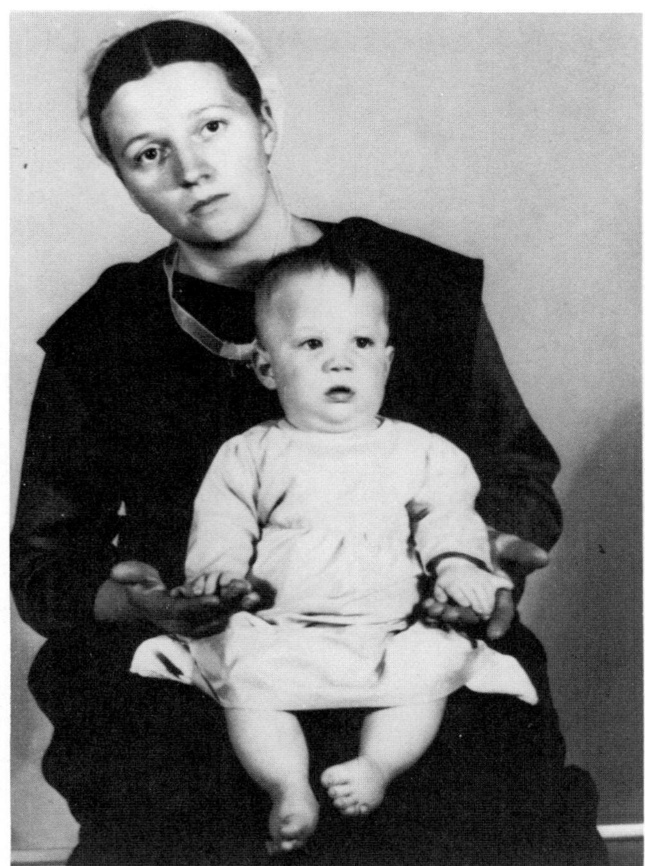

Figure 8 An Amish child with the Ellis-van Creveld syndrome. Note the shortened arms and the extra finger on the left hand. (Courtesy of Victor A. McKusick, Johns Hopkins University.)

the other in heterozygous form. Thus the frequency of the allele among the founders of the Lancaster County Amish was perhaps 1/400 if present only in Mr. or Mrs. King. But its frequency was not likely to have been as high as 1/14, its value in 1964. Another factor must have operated to increase its frequency further.

RANDOM GENETIC DRIFT. A generation of persons derived from just a small number of parents may not possess genotypes representative of the parental generation. One parental couple with an unusual genotype may be prolific while another remains childless, or Mendelian segregation of alleles within families may depart significantly from expected ratios. Thus, allele frequencies may change capriciously, irrespective of whether they are "good" alleles or "bad" ones. Such changes, which are increasingly likely the smaller the population, are called **random genetic drift**, or simply *drift*. The founder effect is really a special case of drift occurring in an initial generation. Another special case is the *bottleneck effect*, in which a community's sample of alleles is changed when the population is reduced in numbers by natural catastrophe, infectious disease, or other factors.* Broadly speaking, drift refers to haphazard changes in allele frequencies due to sampling from small numbers.

In the case of the Lancaster County Amish, the Kings and their descendants had larger families than others in their community, despite the presence of the semilethal recessive allele for a form of dwarfism. As a consequence, the frequency of the deleterious allele "drifted" higher in succeeding generations, especially in the early generations when the population was smaller.

Inbreeding in the Hutterites

The Hutterian way differs from that of the Amish, because the Hutterites pursue a communal life style based on Acts 2:44–45: "And all that believed were together, and had all things common; And sold their possessions and goods, and parted them to all men, as every man had need" (Figure 9.) Although personal possessions of the people are minimal, the Hutterites, unlike the Amish, utilize the most modern farming technologies. Religious persecution in Europe from the sixteenth to the nineteenth centuries forced the Hutterites to move repeatedly, and about 400 of them established three separate agricultural colonies in South Dakota in the 1870s. They prospered, and when one of the communities outgrew what the land could support, about half of the residents established a new colony. The branching process has produced approximately 300 colonies totaling about 23,000 persons (Hostetler 1985). Birth control measures are frowned upon and large families, averaging about 11 children, are the rule (Robinson 1986).

Since the early part of this century, there has been very little intermarriage between the descendants of the three original colonies. One group, the Schmiedenleut ("blacksmith's people") had by 1960 branched into 51 colonies mostly in South Dakota and Manitoba (Mange 1964). Only 15 surnames were present in the original colony and, for lack of male heirs, two surnames died out by 1960. The descendants of the other two colonies are geographically distant from the Schmiedenleut.

MARRIAGE PATTERNS. The average age of marriage among the Schmiedenleut in the 1950s was about 22 for women and 24 for men, a small increase over prior

*For example, the world's population of cheetahs (totaling 20,000 animals in two African locations) is now threatened by a lack of genetic variability because of severe and prolonged bottlenecks about 10,000 years ago. O'Brien et al. (1986) write that the populations "must have dropped to a very few individuals, escaping extinction by a whisker."

Figure 9 A Hutterite group. (Courtesy of Kryn Taconis, Toronto.)

generations. This change may be due to the desire of some "modern" women to have fewer children; by delaying marriage, one or two pregnancies are avoided. About 17% of marriages involve persons with the same surname, an incidence that does not differ from chance expectations. For example, the most common surname, Waldner, constitutes about 30% of the population, and approximately $(0.30)^2 = 9\%$ of marriages are between Waldner men and Waldner women.

A common practice is for several marriages to occur between two sibships. In about 28% of marriages, the spouses of two sibs (or sometimes three or four) were themselves sibs—say, two brothers marrying two sisters (Figure 10). This is *not* inbreeding: the children of these marriages are double first cousins of one another, but inbreeding would not occur unless they or their descendants marry and have children.

INBREEDING. It is possible to trace the complete ancestry of the 5,450 Schmiedenleut alive in 1960 back to exactly 68 forebears. That is, (barring mutation) all alleles present in the 1960 population are replicas of only the alleles in the set of 68. Although each person in the ancestral set has had, therefore, an average of 80 descendants, some ancestors have had many times that number because of differential fertility (Martin 1970). The continued closed nature of the small population has led to the situation in which everyone is demonstrably related to everyone else. Yet the Schmiedenleut are not inbred to the extent that it has obvious practical consequences on health or mortality. In general, they do not marry first cousins or closer. Although the relationships between spouses are often manifold (see Figure 3), all the relationships in a pedigree rarely add up to the equivalent of a first cousin marriage. For 667 marriages occurring mostly between 1900 and 1960, the average inbreeding coefficient of offspring was only $f = 0.022$, a value greater than that for offspring of second cousins, but not as great as that for first cousins once removed.

Knowledge of Schmiedenleut ancestry usually extends back in time, so fourth cousin relationships are detectable. Would the inbreeding coefficients be even higher if more pedigree information were available to investigators? There is no way to know for certain, although there are suggestions that this is true. First, inbreeding coefficients can be separated into two components: one involving possible parental relationships of second cousin or closer, and one involving possible relationships of third and fourth cousins. The two f values are as follows:

average f from second cousin relationships or closer
$= 0.010$

average f from third and fourth cousin relationships
$= 0.012$

Both near and less near relatives make substantial contributions to inbreeding, a fact suggesting that more remote ancestry might make further contributions to the average f value.

Another type of inbreeding analysis is based on **parental isonymy**, or the occurrence of like surnames among marriage partners (Crow and Mange 1965, Lasker 1985). This method takes *all* remote ancestry into account and relies on a simple relationship: the inbreeding coefficient of offspring is one-fourth the random probability of parental isonymy. For example, one-fourth of *first* cousins will have the same surname: that is, whenever their fathers are brothers, but not when their mothers are sisters, nor when they are related through both sexes. The children of first cousins have $f = 1/16$. *Second* cousins will have the same surname one-sixteenth of the time, and, commensurately, their children have $f = 1/64$. Based on the isonymy method, the average inbreeding coefficient among the Schmiedenleut is about twice as great as that based on direct pedigree information. Several types of errors are inherent in the method, and it may overestimate inbreeding levels in some populations. Nevertheless, isonymy is now much used for analyses of the structure of human populations; for example, 20 papers on isonymy were presented at a conference arranged by Gottlieb (1983).

RANDOM AND NONRANDOM INBREEDING. *In itself*, the small size of a population accounts for a certain amount of inbreeding. For example, the Schmiedenleut may well be mating *at random* among themselves with respect to relatedness and still be inbred! This seems to be the case:

we noted that the observed frequency of Waldner × Waldner marriages does not differ from random expectations, and this is true for the other surnames as well. Generally, inbreeding in a *small* population is considered to have two separate components:

1. *Inbreeding that would exist if mating were purely random within the isolate.* Formulae for making this calculation are beyond the scope of this book, but the size of the population is the critical variable: the smaller the group, the greater the inbreeding. This component is referred to as the **random component**, and it is the major factor in Schmiedenleut inbreeding. (In a large population, the random component is near zero.)

2. *Inbreeding due to choosing mates more (or less) related to each other than are average pairs within the group.* This is called the **nonrandom component**, and it is superimposed on the first component. It may work to either increase or decrease the inbreeding in a small population. (The nonrandom component would be zero in a population mating at random.) In the Schmiedenleut, there is a slight increase in inbreeding due to nonrandom mating; that is, there is a slight tendency for marriage partners to be more closely related to each other than they would be if they were randomly selected pairs. In some small populations in which consanguinity is consciously *avoided*, however, the level of inbreeding is less than would occur if mating were at random. This statement may appear contradictory, but unusual events occur in small populations.

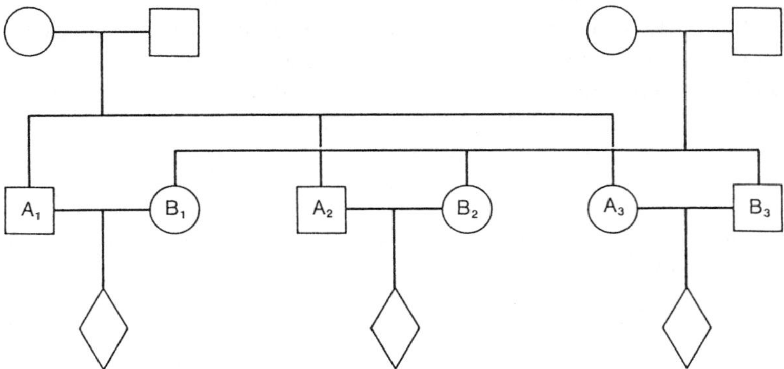

Figure 10 A set of three marriages between two sibships. Although two brothers and their sister, the A's, married two sisters and their brother, the B's, nobody depicted in this pedigree is inbred.

SUMMARY

1. People who are related to each other have one or more ancestors in common. A mating between relatives is said to be consanguineous, and the resulting children are inbred.
2. An inbred person may be homozygous not only for independently inherited alleles (as in a noninbred person), but also for alleles identical by descent, that is, replicas of a single allele present in a common ancestor. The basic effect of inbreeding is to increase homozygosity for alleles identical by descent.
3. The probability of possessing alleles identical by descent is a person's coefficient of inbreeding. It is approximately $\Sigma(1/2)^n$, where n is the number of persons in each of the loops connecting the person's parents. Tracing loops is facilitated by redrawing pedigrees as path diagrams.
4. Whether inbreeding is beneficial or detrimental depends on the consequences of increased homozygosity. On the average, inbreeding in humans has small, adverse effects on the offspring of first cousins, the most common consanguineous mating.
5. The rarer a recessive allele, the greater is the risk of its appearing homozygously in inbred persons as compared to noninbred persons. Thus, for rare recessive diseases, a large proportion of affected persons come from consanguineous matings.
6. Mortality statistics for increasingly inbred persons suggest that each of us possesses several genes that, if made homozygous, would kill us before maturity.
7. Isolates may have characteristic sets of alleles different from those of the general population. The smaller the community and the longer it maintains its isolation, the more inbred its members are likely to be.
8. The high frequency of certain alleles in isolated groups like the Amish or the Hutterites may be due to the founder effect and random genetic drift. The smaller the population, the greater is the likelihood of these chance effects.
9. In small isolates, a certain amount of inbreeding occurs even if mating is random within the group. The actual level of inbreeding may be higher (or lower) than this amount if mates tend to be more (or less) related than they would be in average random pairings.

FURTHER READING

Comprehensive coverage of human population structure, including inbreeding, is given by Cavalli-Sforza and Bodmer (1971) and by Bodmer and Cavalli-Sforza (1976). Other good presentations of inbreeding principles, with many computational examples and problems, are found in the books on population genetics by Crow (1983), Hartl (1987), and Hartl and Clark (1989). Summaries of inbreeding effects are by Schull and Neel (1965, 1972), and Khoury et al. (1987b). Hostetler (1974, 1980), Huntington (1981), and Peter (1987) provide interesting background on the Amish and the Hutterites, and McKusick (1978) has edited a large collection of medical genetic studies of the Amish. Lasker (1985) reviews the application of isonymy to several aspects of population structure.

QUESTIONS

1. Some coefficients of inbreeding are given below. Confirm these values by tracing loops on path diagrams.

For offspring of	$f =$	For offspring of	$f =$
sibs	1/4	first cousins, once removed	1/32
uncle–niece	1/8	second cousins	1/64
first cousins	1/16	second cousins, once removed	1/128

2. Draw the path diagram for P-IV in Figure 4, including just the one common ancestor and the individuals in line from her to P-IV. What is the relationship between the parents of P-IV? What is the f value of P-IV?

3. Draw the path diagram for a child of double second cousins. (There are several ways to do this.) What is the child's f value?

4. Trace just the ancestors of individual J in Figure 3. What are the relationships between J's parents? What is f_J?

5. Two first cousins, B and C, have a common grandfather who was an albino (see accompanying path diagram). What is the probability that their first child, D, as yet unborn, will be an albino? Assume that the albino allele entered this pedigree only through the affected grandfather.

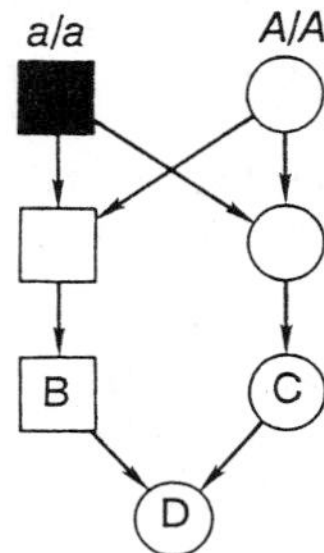

6. In Question 5, let the albino alleles in the common grandfather be labeled a_1 and a_2. What is the probability that D will be identically homozygous for the a_1 allele? For the a_2 allele? Why do these probabilities not add to 1/16, the answer to Question 5?

7. In Question 5, let the nonalbino alleles in the common grandmother be labeled A_1 and A_2. Express the f value of D in terms of probabilities involving all four grandparental alleles.

8. We noted that the frequency of the allele for the Ellis-van Creveld syndrome among the Lancaster County Amish is freq(e) = 0.07. (a) What is the expected frequency of the syndrome among the offspring of first cousins in this population? (Assume no prior knowledge of phenotypes of the pedigree members.) (b) What is the expected frequency of the syndrome among the offspring of unrelated parents? (c) How many times more frequent is the e/e genotype in Case (a) than in Case (b)?

9. Assume that outside Lancaster County the frequency of the e allele = 0.001 (although it is probably less than this). (a) What is the expected frequency of the syndrome in a randomly mating population? (b) How many times more frequent would it be among the offspring of first cousins? Find the answer in Table 1. Contrast this value with the corresponding factor for the Lancaster County Amish.

10. The ratio computed in 8(c) and 9(b) is $[qf + q^2(1-f)]/q^2$. Show, algebraically, that this expression reduces to $1 + pf/q$. Use this simplified formula to check your answers.

11. The complete generalizations of the Hardy-Weinberg law for a population in which the average coefficient of inbreeding is f are given below. Note that the zero term for the heterozygote occurs because the two alleles cannot be identical. Show that these formulae can be written in the modified forms to the right, which make more evident the Hardy-Weinberg frequencies plus and minus the changes due to inbreeding.

$$\text{freq}(a/a) = qf + q^2(1-f) = q^2 + pqf$$
$$\text{freq}(A/A) = pf + p^2(1-f) = p^2 + pqf$$
$$\text{freq}(A/a) = 0 + 2pq(1-f) = 2pq - 2pqf$$

12. Among the Samaritans of Israel and Jordan (Bonné 1963), the average f is about 0.04, this being among the highest inbreeding values for isolated human populations. What would be the expected frequency of the three genotypes for freq(A) = 0.9 and freq(a) = 0.1?

13. The "excess homozygosity" above the random amount ($p^2 + q^2$) brought about by inbreeding is $2pqf$ according to the modified formula in Question 11. Show that for the Samaritan population, excess homozygosity is less than 1% of the random amount.

14. A child of incest (in the narrow sense) is often taken as a person whose f value is x or greater. Use the relationships in the footnote on page 419 to determine the value of x.

15. In Figure 7, calculate the f value for Elizur Wright I (EWI). Is his son (EWII) or his grandson (EWIII) inbred (as far as is shown)?

16. The rules for calculating inbreeding coefficients for *X-linked* genes are different from those for autosomal gene calculations. For one thing, the concept of identity cannot apply to males, because they have a single X chromosome. Consider the accompanying diagrams, which show parts of a path diagram having (Case I) a male between two females, and (Case II) two males in succession. For each case, what is the probability that the gametes labeled c and d carry alleles identical by descent for an X-linked gene? For more details, see Crow (1986).

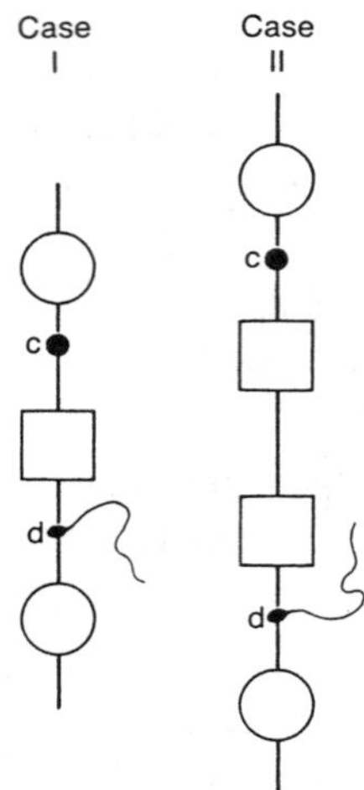

ANSWERS

1. For sibs, there are two loops, each with three persons, so that
$f = 2(1/2)^3 = 1/4$, and so on.

2. The parents of P-IV are half first cousins. $f_{\text{P-IV}} = 1/32$.

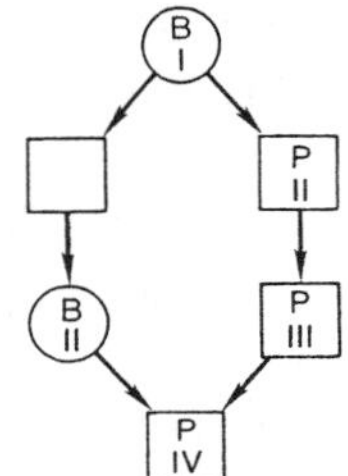

3. $f_A = 4(1/2)^7 = 1/32$

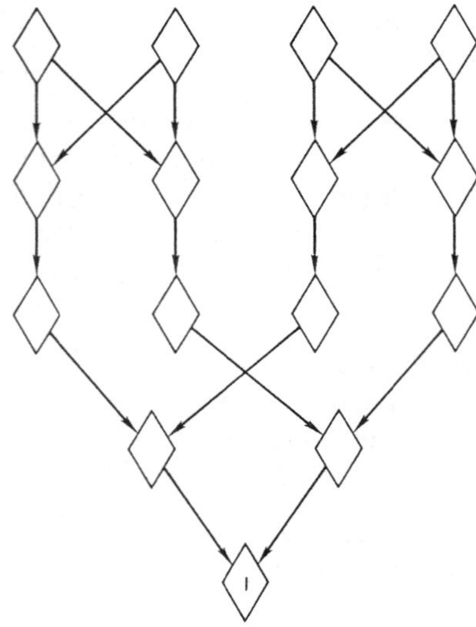

4. First cousins via C_8 and C_9. First cousins once removed via C_{10} and C_{11}. $f_J = 2(1/2)^5 + 2(1/2)^6 = 3/32$.

5. The children of the affected grandfather must have been heterozygous *A/a*. Each of their gametes in line *to* B and C had *a* with probability 1/2, and each of the gametes *from* B and C (to D) had *a* with another factor of 1/2. Thus P(D is *a/a*) $= (1/2)^4 = 1/16$.

6. $P(a_1/a_1) + P(a_2/a_2) = 1/64 + 1/64$ (see Figure 6). These do not add up to P(D is *a/a*) = 1/16 because D would also be albino if she were a_1/a_2 (and this has probability 1/32).

7. $f_D = P(a_1/a_1) + P(a_2/a_2) + P(A_1/A_1) + P(A_2/A_2) = 4(1/64)$.

8. (a) freq(e/e) $= qf + q^2(1 - f) = (0.07)(1/16) + (0.0049)(15/16) = 0.00897$. (b) $q^2 = 0.0049$. (c) $897/490 = 1.8$.

9. (a) 10^{-6}. (b) 63.4. When the allele is rare, the disease is relatively more frequent among the children of first cousins.

10. $[qf + q^2 - q^2 f)]/q^2 = [qf(1 - q) + q^2]/q^2$
$= pqf/q^2 + q^2/q^2 = pf/q + 1$

11. Expand and factor as in Question 10.

12. freq(a/a) = 0.0136; freq(A/A) = 0.8136; freq(A/a) = 0.1728.

13. $2pqf/(p^2 + q^2) = 0.0072/0.82 = 0.88\%$.

14. $x = 1/8$.

15. $f_{\text{EWI}} = 2(1/2)^8 = 1/128$. Neither EWII nor EWIII is inbred (their $f = 0$).

16. Case I: $P(c \equiv d) = 1$ because a male must transmit to any daughter the same X chromosome he got from his mother. Case II: $P(c \equiv d) = 0$ because the X chromosome in c cannot be transmitted from the first male to the second male.

Chapter 22 Processes of Evolution

Mutation
Measuring Spontaneous Mutation Rates
The Average Mutation Rate for Human Genes

Selection
Selection Against a Recessive Lethal Phenotype
Selection Against Both Homozygotes

Migration and Drift
Migration and Gene Flow
Random Genetic Drift

Differing Views on the Mechanisms of Evolution
The Role of Drift
The Nature of Variability

A Genetic Concept of Race
The Reality but Arbitrariness of Races
Human Variability
Individuality

> *Is evolution good or bad, beautiful or ugly, directed or undirected? These are largely value-judgments, and are thus not scientific. But it is the answer to them which makes evolution interesting to the ordinary educated man and woman. In making them I can, of course, claim no special standing. I can write of natural selection with authority because I am one of the three people who know most about its mathematical theory. But many of my readers know enough about evolution to justify them in passing value judgements upon it which may be different from, and even wholly opposed to, my own.*
>
> *J. B. S. Haldane (1932)*

The Englishman Haldane was indeed one of the three authorities who integrated the ideas of Mendel and Darwin into a mathematics of evolutionary processes. His countryman Sir Ronald Fisher and the American Sewall Wright were the other members of this famous triad. With most other scientists (since about 1800), they agreed on the **principle of evolution**, that populations of organisms have changed and diverged over eons to produce the untold number of current species. Each man, however, put his special stamp on the theories behind the facts; that is, they differed among themselves, sometimes vigorously, on the mechanisms that might explain how evolution occurred—and is still occurring (Crow 1987).

We shall not broach here the question of biological evolution versus special creation.* Rather, we start with the knowledge that gradual, cumulative, and nonmiraculous changes over many millions of generations account for the wonderful complexity of adaptations and the astonishing diversity of species, including our own. Evolution—the fundamental thread tying together all

*Regarding this issue, the United States Supreme Court has ruled that it is unconstitutional to *require* public school teachers to discuss special creation whenever they discuss evolution (*Edwards v. Aguillard* 1987). The justices argued that special creation reflects a particular religious doctrine (the Book of Genesis); teaching it would therefore violate the separateness of church and state mandated by the First Amendment. A prior court case emphasized that so-called "creation science" is not science in any sense of the word (*McLean v. Arkansas Board of Education* 1982). The issue is also examined in Committee on Science and Creationism (1984).

fields of biology—has been repeatedly investigated and thoroughly confirmed. Even in 1932, Haldane could write that it is "quite as well proven as most other historical facts." The documentation is presented especially well by Futuyma (1983) and Dawkins (1986), and more fully in textbooks of evolutionary biology (for example, Futuyma 1986). The many interlocking and coherent lines of evidence that support evolution include

1. The fossil record of ages past, showing temporal sequences of more or less continuous forms
2. The geographical distributions of past and present species over wide (continental) or narrow (insular) ranges
3. The use of artificial selection by plant and animal breeders to generate a prodigious variety of pets, flowers, livestock, crops, and so forth
4. The retention of developmental stages of remote ancestors during the embryology of current species
5. The similar underlying structures of some functionally dissimilar body parts as revealed by studies of comparative anatomy
6. The greater similarity of cellular and molecular structures (chromosomes, nucleotide sequences, amino acid sequences, and so on) among more closely related species
7. The nearly universal genetic code, as well as other consistent similarities and differences at the biochemical level among all living creatures.

The idea of evolution predated the work of Charles Darwin by many years. In the early 1800s, for example, the Frenchman Jean-Baptiste de Lamarck argued strongly for evolution, and reasoned quite logically (for his time) that it occurred by the inheritance of so-called acquired characteristics. But convincing scientific support for *this* particular mechanism for evolution was not brought forward by Lamarck, nor by anyone else. On the other hand, Darwin (1859) marshaled considerable evidence that evolution occurred by natural selection, which is the greater rate of reproduction by better adapted individuals. *This* mechanism for evolution, coupled with the workings of heredity (discovered in 1865 by Mendel, but not known to Darwin nor generally appreciated until after 1900), can account for the step-by-step, nonrandom, adaptive changes characteristic of the evolution of populations over space and time.

There are additional evolutionary mechanisms—mutation, migration, and drift—but these factors by themselves lack the "management skills" of selection. Their interplay with each other and with selection are what occupied Haldane, Fisher, Wright, and others. In an influential book entitled *Genetics and the Origin of Species,* the Columbia University biologist Theodosius Dobzhansky (1937) synthesized early thought and experimentation connecting genetics with evolutionary processes (Powell 1987). More recent work has emphasized variation at the molecular level. Here, we will look at evolutionary factors in a very simplified fashion—in terms of changes each factor can make in the frequencies of alleles of one gene.

In what follows, we discuss **Darwinian fitness**, or simply **fitness**, as a measure of the average number of offspring left by an individual or class of individuals. Differences in fitness depend upon differences in both *survival* and *fertility*. For example, Tay-Sachs disease represents zero fitness due to premature death, whereas Turner syndrome represents zero fitness due to infertility. Zero Darwinian fitness also applies to members of religious orders who practice celibacy. On the other hand, persons with Huntington disease, a crippling condition characterized by progressive brain destruction, may have near-average Darwinian fitness, because the dominant allele often manifests itself only *after* reproductive age. To be fruitful and multiply, one must be both viable and fertile; physical prowess is in some cases irrelevant.

The greater the fitness of individuals with a particular phenotype, the more their genes tend to be represented in future generations. This idea is the salient point of evolution by natural selection. We emphasize that fitness is not usually all-or-none. A difference of percentage points in health or fertility, rather than life versus death, may distinguish different categories of people. The catchwords *survival of the fittest** often mean "somewhat greater procreation by somewhat fitter individuals." In addition, fitness values relate to prevailing environments. If surroundings change, the fitness of individuals may change. This relation is particularly striking when severe genetic diseases are treatable: the phenylketonuria genotype is nearly *lethal* in one dietary situation but nearly *normal* in another.

MUTATION

Mutations are the source of the biochemical, physiological, anatomical, and behavioral variability upon which evolution depends (Vogel 1983). Recombination or migration can bring about reshuffling of alleles, yielding perhaps beneficial combinations, but all current genetically based differences between organisms derive from

*This rather simplistic phrase does not appear in the first edition of "The Origin." It was coined by the laissez-faire philosopher Herbert Spencer, and the accommodating Darwin used it in later editions.

germinal mutations at some time in the recent or remote past. In Chapter 14 we examined the nature of mutations in terms of DNA structure. We noted that naturally occurring mutations may be due to the inherent thermal energy of atoms, or to stray chemicals or radiation, but there is usually no way of knowing the source of a particular mutation. Most mutations are changes to *detrimental* and *recessive* alleles. They are generally detrimental because they alter a functioning system already attuned to its environment. (By analogy, how likely is it that a *random* change in the innards of an adequate TV set will improve the picture?) They tend to be recessive because one good, unmutated copy of a gene is often sufficient to produce a normal phenotype. Occasionally a mutation might be beneficial initially (especially if it produces only a small change in the phenotype), or it might become beneficial in a later generation as the result of an environmental change. For example, mutations resulting in penicillin resistance are of no use to bacteria unless they encounter penicillin in their environment.

The time required for a germinal mutation in DNA to be expressed in a phenotype varies. For example, an autosomal mutation to a *dominant* allele will be immediately evident if transmitted to an offspring. On the other hand, the number of generations that elapse between an autosomal *recessive* mutation and its eventual expression in a homozygote depends on (1) the rate at which the particular mutation recurs, (2) the pattern of matings, and (3) chance events. A high mutation rate and inbreeding will hasten the formation of homozygotes. Random genetic drift may cause the rare mutated alleles to either increase or decrease in frequency.

In addition to creating allelic differences that other forces of evolution act upon, recurrent mutation *by itself* may lead to very slow changes in allele frequencies. Assume that the gene A_1 mutates to its allele A_2 at a certain rate ($A_1 \rightarrow A_2$), but that no other evolutionary forces are operating. If the new A_2 alleles do not mutate back to A_1, or to any other allele, then eventually A_2 is *fixed*, that is, its frequency becomes 1 (and the frequencies of all other alleles of A_2 become 0). This scenario is highly unlikely for many reasons, not the least of which is the immense period of time involved. For a given human gene, the mutation rate is typically in the neighborhood of *one mutation per million gametes per generation*. In other words, if you could follow a gene in a particular gamete through all cell divisions to the next generation of gametes, the probability that it will mutate (by base substitution, deletion, and so on) is one in a million, or 10^{-6}. This rate is so low that the replacement of even half of A_1 by A_2 via mutation alone would take hundreds of thousands of generations, more time than *Homo sapiens* and its several antecedent species have been around.

Typically, the A_2 allele will also mutate, and it may change back to the original state ($A_1 \rightleftharpoons A_2$). Then neither allele is lost from the population. Given enough time, a balance point may be reached at which the frequencies of the alleles A_1 and A_2 are constant (although usually different from each other) over succeeding generations. Such a situation is called a **genetic equilibrium** and is brought about by opposing forces. No alleles of any actual gene are known to be maintained at stable frequencies by counteracting mutations, but other mechanisms that bring about genetic equilibria are discussed later in this chapter.

Measuring Spontaneous Mutation Rates

Determining a spontaneous mutation rate is inherently difficult because the events to be counted are very rare, and, in the case of recessive mutations, are generations removed from their actual expression in an animal or plant. These problems can often be overcome by using experimental organisms, especially haploid microorganisms, in which the problem of dominance and recessiveness vanishes. Bacteria and viruses can be grown in prodigious numbers, and their environments can be manipulated to allow mutants, no matter how rare, to stand out from a background of millions of unmutated cells. For example, if a drug-sensitive population is exposed to that drug, only the preexisting mutants to drug resistance will survive. But mutation can be studied in diploid organisms, too. In the 1920s, Muller invented methods to measure collectively the rate of mutation of the few thousand genes on one fruit fly chromosome, although the genes were not individually identified. Muller showed that it was especially easy to tally X-linked lethals as a class of mutations, but he also studied semilethals (having a bit less than 100% chance of death) and mutations leading to visible alterations of a fly's phenotype.

In mice, researchers have developed tester stocks that are homozygous recessive for several different genes affecting fur color or other easily scored phenotypes (Russell 1951). For one locus, the test goes as follows: Wildtype mice, homozygous for the dominant allele (B/B = black fur), are mated to the tester stock (b/b = brown fur). In the absence of mutation in the wildtype animals, all the offspring are black (B/b). But a mutation to a recessive in the germline of a wildtype parent ($B \rightarrow b$) leads to an offspring that is brown (b/b). By using a tester stock that is homozygous recessive for seven different genes (a = nonagouti, b = brown, c^{ch} = chinchilla, p = pink eye, d = dilute, s = piebald, and se = short ear), it becomes feasible—barely—to carry out mutation rate research in mice (studying, for example, the effects of a mutagen such as X irradiation).

The time, space, and money needed to determine mutation rates increase by orders of magnitude in moving from microorganisms to fruit flies to mice (Council of the Environmental Mutagen Society, 1975). In a few days, one geneticist at a small bench on a modest budget can determine fairly accurately some specific mutation rates in a bacterium or virus. To collect even some general mutation rate data in fruit flies takes many months and requires technical help to examine and count offspring. With mice, years of work, extensive animal-rearing facilities needing careful supervision and care, and a sizable investment of money can produce the approximate value of the mutation rate for relatively few genes.

For humans, experimental methods are not possible; instead investigators must ferret out informative families from among worldwide populations. Even though mutations are infrequent, there are a lot of us to use as a base population, and large collections of hospital records may be available. It is possible to systematically tally at least small numbers of mutant phenotypes and obtain, thereby, approximate values for some mutation rates. Two methods have been employed: the direct method for dominant mutations, described below, and the indirect method for either dominant or recessive mutations, which we mention only briefly.

THE DIRECT METHOD. Because *one* dose of a dominant allele is revealed in the developing phenotype, we can identify new mutations from recessive (normal) to dominant (abnormal) by noting *persons with the dominant trait who have normal parents*. The genotypes involved are:

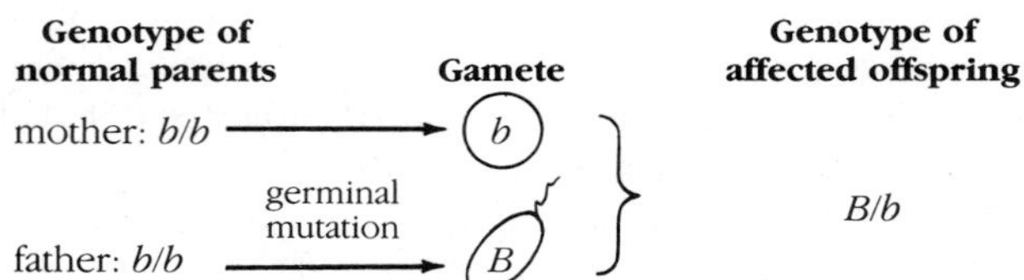

In this example, the sperm is assumed to contain the changed allele that is descended from a mutational event ($b \rightarrow B$) occurring sometime previously in his germline. (Alternatively, we could have assumed a mutation that affects the egg.) In practice, an investigator examines all the births in a given region during a given time period. Let

T = total number of newborns

M = number of affected newborns with normal parents

μ = probability of mutation of the gene per gamete per generation

Then,

$$\mu = (1/2)(M/T)$$

The factor 1/2 appears because an affected newborn arises from two gametes, only one of which carries the new mutant allele. The expression for the mutation rate can also be written $\mu = M/2T$ which more directly represents a probability: the number of mutant alleles (M) divided by the total number of alleles ($2T$) in the newborn population.

In an early study using this method, Mørch (1941) tallied 8 achondroplastic dwarfs (Greek *a*, "lack of"; *chondros*, "cartilage"; *plasia*, "molding") with normal parents among 94,075 births in a Copenhagen hospital over a 20-year period. In a recent study, Nelson and Holmes (1989) tallied 2 achondroplastic dwarfs with normal parents among 69,277 infants born at a Boston hospital over a 10-year period. This dominant disease (or set of diseases—see later) interferes with normal bone development and leads to abnormally short extremities (especially thighs and upper arms) set on a normal-sized but swayback trunk and to an overlarge head (Figure 1). The specific biochemical pathway altered in achondroplasia is unknown.* Severe cases are stillborn or die in infancy, but patients who survive the first year of life often become healthy adults of normal intelligence and may reproduce (Scott 1981, Rimoin and Lachman 1983).

For the two studies, the mutation rates are computed as follows:

for Copenhagen: $\mu = (1/2)(8/94{,}073) = 4.2 \times 10^{-5}$
for Boston: $\mu = (1/2)(2/69{,}227) = 1.4 \times 10^{-5}$

or about 1 to 4 mutations per 100,000 genes per generation.

In addition to the small number of mutations on which these calculations are based, several other reservations must be noted (Crow and Denniston 1985).

1. *The same phenotype may result from more than one cause.* In the Danish study (at least) recessive-to-dominant mutations for more than one gene were responsible for the phenotype that was tallied as affected (genetic heterogeneity). Thus the calculated mutation rate was the sum of the rates for several genes. In other cases, the phenotype may result from both dominant and recessive mutations of different genes, or from nongenetic developmental phenocopies.

2. *Penetrance may be less than 100%.* This factor

*Achondroplastic dwarfs are quite different from normally proportioned pituitary dwarfs, who are deficient in growth hormone (Chapter 15). Dwarfs of several distinct types were retained at royal courts from the days of the Pharaohs and have been exhibited at shows for centuries. An organization called The Little People of America was formed in 1960 to help meet some of the social difficulties faced by dwarfs.

Figure 1 An apparent achondroplastic dwarf, Giacomo Favorchi, painted by the Dutch artist Karel van Mander around 1600. (Courtesy of Statens Museum for Kunst, Copenhagen.)

does not seem to be important with achondroplasia. A direct mutation rate calculation would be erroneously too high, however, if a dominant allele is not expressed in a parent but is expressed in a child to whom it was transmitted intact.

3. *Ascertainment of cases may be less than 100%.* This factor also does not seem to be important for achondroplasia or other diseases that are clearly evident at birth, but in other situations some affected individuals may be overlooked. In very severe genetic diseases, death may occur prenatally or early in life, thereby leading to ascertainment errors.

THE INDIRECT METHOD. For rare *recessive* traits, the direct method will not work because affected individuals with normal parents almost never signify newly occurring mutations. Rather, the affected person is a homozygote usually arising by Mendelian segregation from parents both of whom are heterozygous: $B/b \times B/b \rightarrow b/b$. The mutational events ($B \rightarrow b$) occurred some generations in the past. A rate calculation can nevertheless be made by assuming that the frequency of affected persons is constant over time. The postulated *genetic equilibrium* means that the gain in alleles by mutation is offset by the loss of alleles through the lowered fitness of affected persons, a mathematical procedure first applied by Haldane (for details, see Crow and Denniston 1985). The assumed genetic equilibrium between mutation and selection may be approximately true for some traits, and the indirect method has been applied to both recessive and dominant conditions, autosomal and X-linked.

Even if the equilibrium assumption is true, however, the method in its simplest form is not very accurate for many, if not most genes classified as recessive. This inaccuracy arises because a (so-called) recessive allele may have a slight, usually unknown, detrimental effect on the phenotype when it it present in heterozygotes. The *heterozygous effect*, even if very small, has a large consequence on the mathematical calculation (on the loss side of the equilibrium equation), because so many more individuals in a population are heterozygous than homozygous recessive.

The Average Mutation Rate for Human Genes

Table 1 gives mutation rate data for a few autosomal dominants and X-linked recessives. The rates are heterogeneous, ranging from a high of about 73 to a low of 2 mutations per million gametes per generation. These values are likely to be overestimates of the true mutability of human genes, simply because the traits have been documented in many families. In contrast, some of the hereditary phenotypes in McKusick's catalog (1988) are known in just one or a few families. These latter, truly rare, diseases undoubtedly have much lower mutation rates. Furthermore, those genes that have never been known to mutate cannot be investigated at all. Thus, an average rate of *1 mutation per million gametes per generation* (for genes that lead to visible phenotypic effects) may be nearer the truth than the higher rates in Table 1.

The tabulated traits are also atypical in that they lead to easily identifiable, usually severe phenotypes. An "ordinary" mutation may be more benign, leading to an amino acid substitution that has little, if any, effect on the functioning of its protein. Some researchers, especially the Japanese mathematical geneticist Motoo Kimura (1979, 1983), believe that *most* base substitution mutations occurring over evolutionary time may be neutral or nearly neutral in their effect on phenotype (see later).

Analysis of rates of mutation at the molecular level in humans is just beginning. The work involves sequence data for proteins or cloned DNA (e.g., the globin polypeptide chains or the corresponding regions of DNA)

Table 1 Fairly reliable mutation rates for some human genes

Trait and description[a]	*Country*	*Mutations per million gametes per generation*[b]	
Neurofibromatosis (AD): multiple fibrous tumors of the skin and nervous tissue; scattered areas of brownish pigment in skin (known as café-au-lait spots). Biochemical defect unknown.	Russia	46	avg. = 73
	USA	100	
Duchenne-type muscular dystrophy (XR): progressive degeneration of muscular tissue from infancy; usually death before adulthood. Due to absence of a protein in the internal membrane systems of muscle cells (see Chapter 3). The many reliable estimates derive from the relatively easy diagnosis and zero fitness of affected persons.	England	43	avg. = 67
	Poland	46	
	England	47	
	Germany	48	
	Northern Ireland	60	
	Japan	65	
	Switzerland	73	
	USA	92	
	USA	95	
	England	105	
Hemophilia A (classical) (XR): internal bleeding, especially at joints, may lead to crippling deformities. The delayed clotting results from a lack of factor VIII in plasma.	Finland	32	avg. = 44
	Germany	57	
Achondroplasia (AD): dwarfism with disproportionate shortening of arms and legs. The nasal bridge is set back, and the forehead protrudes. A defect of cartilage growth.	Germany	8	avg. = 10
	Denmark	10	
	Northern Ireland	13	
Retinoblastoma (AD): a cancer of the retina of one or both eyes, appearing in children who may live to adulthood if the affected eyes are removed; otherwise fatal in childhood. Phenocopies are difficult to exclude. (See Chapter 19.)	France	5	avg. = 8
	Hungary	6	
	Japan	8	
	Netherlands	12	
Marfan syndrome (AD): unusually long limbs and digits, accompanied often by defects of the heart and eyes. An abnormality of connective tissue formation.	Northern Ireland	5	avg. = 5
Hemophilia B (Christmas disease) (XR): symptoms similar to classical hemophilia but generally milder. Defect results from the lack of factor IX in plasma.	Finland	2	avg. = 2
	Germany	3	

Source: The data are selected from a more extensive table in Vogel and Rathenberg (1975), which should be consulted for references.

a. AD, autosomal dominant; XR, X-linked recessive.

b. Multiple determinations for the same gene represent independent studies. The direct method was used for most of the autosomal dominants and the indirect method for the X-linked recessives.

and also in vitro tissue culture methods (DuBridge and Calos 1987). Researchers hope to obtain reliable information on mutation rates *per nucleotide*—either for DNA that encodes various proteins or for DNA segments that do not code for amino acids (e.g., introns, spacer regions, or repetitive DNA).

An interesting question is whether mutation rates differ between the sexes. A sex difference is predicted if, as is thought, many mutations occur at the time of DNA replication, and fewer at other times in the cell cycle. Vogel and Motulsky (1986) estimate that about 24 cell divisions ensue between an XX zygote and *any* egg that the resulting woman produces (all eggs are arrested in the first meiotic division at the time of a female's birth). On the other hand, sperm cells from a 25-year-old male have passed through about 300 divisions starting from the XY zygote. Thus division-dependent mutation rates would be expected to be greater in males—perhaps much greater—but experimental data on this point is scanty and contradictory.

Because spermatogonial stem cells *continue* to proliferate at the rate of about 23 divisions per year, the mature sperm from, say, a 38-year-old man may have passed through 300 more divisions than the mature sperm of a 25-year-old man (13 more years × 23 divisions per year). Thus, a further prediction of the division-dependent hypothesis is that older fathers should have a greater probability of passing a mutated gene on to children than younger fathers do. This outcome does seem to occur for some diseases (for example, achondroplasia), but not for all diseases for which data are available (Vogel and Motulsky 1986). This age effect would *not* be expected to occur in females.

All in all, a gene is a remarkably stable chemical entity, because a probability of 10^{-6} can be interpreted to mean that a single gene followed from gamete to gamete will change (to a form having a visible effect on phenotype) only once in a million generations—about 30 million years. Nevertheless, the total number of mutations among all of the 50,000 or so genes per individual is substantial, and quite adequate to provide the raw material for evolution.

SELECTION

Natural selection, or simply **selection**, is the differential reproduction of the various *genotypes* in a population over a period of time. The different numbers of progeny left by different genotypes may lead to gradual changes in the totality of genes—the so-called **gene pool**—of a population. The expressed *phenotypic* changes that result from these genetic changes constitute the observed record of evolution. In this section we shall examine the simplest aspects of selection: changes in allele frequencies of a *single* gene in a *constant* environment. In reality, the reproductive fitnesses of various organisms depend on a complex interaction of their total genotypes with varied and changing environments.

That natural selection occurs in nonhuman species is well documented. When species are exposed to new environmental conditions, the differential reproduction of the more fit genotypes has led to new genetic strains: penicillin-resistant bacteria, DDT-resistant mosquitoes, and poison-resistant rats. Rabbits in Australia have become resistant to a lethal virus introduced to control their numbers. The viruses have also changed over time, becoming less virulent. This latter change occurred because the viruses can only replicate in a *live* host; when the rabbit is not killed, the conditions for multiplication and spread of the viruses are enhanced. In industrial areas with soot-covered tree trunks, pale-colored moths have evolved to dark strains that are apparently better able to survive predation by birds. And more recently, in areas where tree trunks have lightened as a result of pollution control, the dark moths are evolving back to the pale forms from which they came (Bishop and Cook 1975).

New strains of animals and plants have been bred by **artificial selection**. Through this process, breeders have enhanced certain economical or esthetic traits for the benefit of people. The target species itself may or may not benefit, but if overall fitness is reduced, as often happens, the breeding stock can be coddled. Chickens laying more eggs? Mink with unusual colors? Leaner pork? Higher protein corn? Fancier goldfish? Some progress toward any goal is practically always possible— in either direction: cow's milk can be selected for either higher or lower butterfat; dogs, for long legs or short, for short fur or long. The implication of these examples is that mutation has supplied plenty of genetic variation to be exploited by artificial (or natural) selection.

It has been suggested that natural selection in our species is no longer operating—that technology preserves the more fit and the less fit alike. According to this view, the unrelenting forces of nature still operate in other species, but selection has now been *relaxed* for humans. Medical science has surely been successful in preserving a few once-harmful genotypes. Patients with phenylketonuria eat foods low in phenylalanine, hemophiliacs receive clotting factor VIII, diabetics take insulin. Surgery can correct some faults of anatomy and physiology (whether heritable or not). Myopic individuals wear eyeglasses, increasing their fitness above what it might have been in a hunting culture. To the extent that the treatments preserve previously (but no longer) detrimental alleles, selection has indeed been relaxed.

On the other hand, Dobzhansky (1962) suggests that technological civilizations may also discriminate against some genetic constitutions. Is genetic resistance to pulmonary and circulatory diseases increasingly important when individuals are challenged by cigarette smoke, by more air, water, and soil pollution, and by high-fat diets? Are psychological disorders that have some genetic basis intensified in complex social situations and crowded urban environments? How much do individuals differ in the relevant genetic factors? The answers are few, but there are enough questions to indicate that a suspension of natural selection cannot yet be proclaimed.

We should also note that medical science treats relatively few genetic diseases. People still die childless from a long list of formidable hereditary conditions for which medicine can currently do little. In addition, natural selection evidently still operates quite unchanged in that part of our life cycle from gametogenesis through the first few weeks of fetal life. It is estimated that perhaps 15% of fertilizations end in *recovered* spontaneous abortions and an unknown percentage in still earlier *undetected* abortions. Boué et al. (1975) suggest that as many as 50% of fertilizations do not come to term. Thus some weeding out of unfit alleles continues as usual during the relatively unseen stages of gametogenesis and early embryonic development.

Three specific types of simple selection are outlined in Table 2. In Case I, the dominant allele *A* is deleterious or perhaps lethal. If *A* is lethal, it will be reduced to near zero frequency in one generation. Note that "lethal" need not mean that the allele kills its carrier, only that it prevents reproduction. Each generation, all new cases of the dominant trait must arise from recurrent mutation ($a \rightarrow A$), since only *a/a*'s have progeny.

Selection Against a Recessive Lethal Phenotype

Case II considers a *recessive detrimental* allele. We will examine the extreme case, a **recessive lethal**, in which the fitness values of the genotypes can be assigned as follows:

	Genotype		
	A/A	***A/a***	***a/a***
fitness	1	1	0

The fitness values are relative measures of reproduction that span the range from 1 (assigned to the most productive genotype) to 0 (assigned to a genotype that leaves no offspring). In this case *A/A* and *A/a* are equally fit. Notice that the situation is very different from that of a dominant lethal, because the potentially harmful recessive *a* alleles that are present in normal heterozygotes are temporarily protected from selection.

We ask how much is the frequency of the *a* allele reduced in one generation? To make things concrete, assume that *a/a* individuals are live born but die prior to reproduction. (Any other assumption producing zero fitness would yield the same results.) We also assume that the population meets all the Hardy-Weinberg conditions except for the indicated selection. Table 3 outlines the steps needed to show that under these circumstances, the frequency of a recessive lethal, freq(a) or simply q, is reduced in one generation from $\boldsymbol{q}$ to $\boldsymbol{q' = q/(1 + q)}$ (q' is read "q prime.") In arriving at this change, we multiplied the frequencies of the newborn genotypes (p^2, $2pq$, and q^2) by the fitness of each (1, 1, and 0) to get the *relative* proportions in the mating population (p^2, $2pq$, 0). The *absolute* genotype frequencies are obtained by dividing each of these terms by their sum. Freq(a) in the mating population—call it q' since it will have changed in value from q—is obtained by adding the frequency of the *a/a* homozygotes, 0, to half the frequency of *A/a* heterozygotes, $2pq/(1 - q^2)$ [see Equation 2 in Chapter 20, page 406). The mating population under Hardy-Weinberg conditions then produces the genotypes of the next newborn generation without any further change in the frequency of the alleles.

The change from q to $q/(1 + q)$ is a small one when freq(a) is small to begin with (as would be the case for any detrimental or lethal allele). For example, if q = 0.02, then the frequency of a recessive lethal will be reduced in one generation to q' = 0.0196. The difference, $q' - q$, is called Δq (delta q). Here

$$\Delta q = 0.0196 - 0.0200 = -0.0004$$

That Δq is negative means that freq(a) is decreasing. The values of the percentage change, $100(\Delta q/q)$, for a recessive lethal over a wide range of frequencies are as follows:

q = freq(a)		$q' = q/(1 + q)$		Percentage change = $100(\Delta q/q)$
0.200	= 1/5	0.167	= 1/6	−16.7
0.0200	= 1/50	0.0196	= 1/51	−1.96
0.002 000	= 1/500	0.001 996	= 1/501	−0.200

Table 2 Simple models of different types of natural selection

Model	*A/A*	*A/a*	*a/a*	*Type of selection*
Case I	x	x		Against a dominant phenotype
Case II			x	Against a recessive phenotype
Case III	x		x	Against both homozygotes; heterozygote has highest fitness

Note: An "x" indicates a genotype that leaves, on the average, fewer offspring than the other genotypes; that is, the x'ed genotypes have lower Darwinian fitness.

Table 3 The change in the frequency of a recessive lethal allele over one generation[a]

	Algebraically, using genotype frequencies				*Arithmetically, using genotype numbers*			
GENERATION 0								
Uniting gametes	freq(a) = q				freq(a) = 1/10			
	A/A	*A/a*	*a/a*	*total*	*A/A*	*A/a*	*a/a*	*total*
Newborn population	p^2	$2pq$	q^2	1	810	180	10	1,000
Fitness	1	1	0		1	1	0	
Mating population	p^2	$2pq$	0	$p^2 + 2pq = 1 - q^2$	810	180	0	990
GENERATION 1								
Uniting gametes	$\text{freq}(a) = q' = \dfrac{0 + \frac{1}{2}(2pq)}{1 - q^2} = \dfrac{pq}{(1-q)(1+q)} = q/(1+q)$				$\text{freq}(a) = \dfrac{0 + 180}{2(990)} = \dfrac{2(90)}{2(990)} = 1/11$			
	A/A	*A/a*	*a/a*	*total*	*A/A*	*A/a*	*a/a*	*total*
Newborn population	p'^2	$2p'q'$	q'^2	1	827	165	8	1,000
Fitness	1	1	0		1	1	0	
Mating population	p'^2	$2p'q'$	0	$p'^2 + 2p'q'$	827	165	0	992

a. The assumptions are (a) The initial freq(a) = q = 1/10. The changed freq(a) is designated q'.
(b) a/a persons are live-born but do not reproduce.
(c) The populations meet the conditions of the Hardy-Weinberg law except for the indicated selection, that is, the fitness of a/a = 0.
(d) In the arithmetic example, the population size at birth is constant at 1,000.

The hands emphasize that, over one generation, the frequency of a recessive lethal allele changes from q to $q' = q/(1 + q)$. Confirm that for q = 1/10, q' = 1/11. See text for further explanation.

By contrast, for a dominant lethal allele, the change would have been −100% regardless of its initial frequency. Furthermore, *the rarer the recessive lethal becomes, the less effective selection is in reducing its frequency further* (reading down the last column). For example, it is possible to show, by an extension of the type of algebra above) that it takes *45 generations* to reduce the frequency of a recessive lethal from 0.2 to 0.02, but *450 generations* to go the next factor-of-ten decrement, from 0.02 to 0.002 (see Question 7). In human terms, 450 generations is a long time: 450 × 30 = 13,500 years. This many years ago, humankind was just learning primitive agriculture.

The reason why selection against a recessive lethal is progressively less effective becomes clear when we examine the ratio of heterozygotes to homozygous recessives. In a population of one million, these ratios are as follows (using Hardy-Weinberg arithmetic):

q = freq(a)	**Number of *A/a***	**Number of *a/a***	**Ratio *A/a*:*a/a***
0.2	320,000	40,000	8:1
0.02	39,200	400	98:1
0.002	3,992	4	998:1

Thus, as the allele gets rarer, heterozygotes are not reduced in numbers nearly as quickly as homozygotes are. In the last line there are nearly a thousand times more phenotypically normal heterozygotes, which *mask* the recessive lethal from selection, than homozygous recessives, which *expose* the lethal to selection.

Eugenic proposals in the early part of this century suggested that our genetic endowment could be much improved if we got rid of heritable diseases by sterilization of affected persons. As far as recessive traits are concerned, the reader can show that the plan could not succeed. For example, if albinos were considered undesirable, their nonreproduction would yield allele frequency changes over three generations of

$$0.0100 \xrightarrow{1} 0.0099 \xrightarrow{2} 0.0098 \xrightarrow{3} 0.0097$$

This *tiny* reduction would represent the progress of 100 years of sterilization of all albinos!

As a corollary, one can show that treating rare recessive lethal genotypes such as phenylketonuria so that these individuals can lead productive and fertile lives will increase the frequency of the causative alleles only very slowly (Crow 1966; Vogel and Motulsky 1986). Thousands of years may be involved for appreciable frequency changes, which, in the absence of selective pressure, still respond to other mechanisms of evolution such as mutation and random genetic drift.

The snail's pace of evolutionary change characteristic of recessive *lethals* is slower yet for recessive *detrimental* alleles, which have fitness values greater than zero but less than one. The formulae for treating these cases are developed in Strickberger (1985). He also considers the more general case in which a "recessive" allele has some effect in the heterozygote. When this condition is met, the tempo of change is much speeded up because the heterozygote is relatively much more frequent. Substantial reductions in the frequency of severe recessive diseases might be achieved by identifying heterozygous persons (by biochemical tests, for example) and counseling them as to risks. Voluntary modification of reproduction by heterozygotes could achieve what sterilization of affected persons could not.

Selection Against Both Homozygotes

For Case III in Table 2, there is a well worked out human example—that of the sickle-cell polymorphism in malarial areas. We look first at the theoretical model for **heterozygote advantage**:

	Genotype		
	A/A	***A/a***	***a/a***
fitness	$1-s$	1	$1-t$

Here, the fitness values of the homozygotes are expressed as numbers less than one by the amount s for A/A's and by the amount t for a/a's. The numbers s and t, called **selective disadvantages**, measure the mortality or infertility of a genotype relative to the best genotype, the heterozygote. The larger s and t are, the smaller the respective fitness values; a selective disadvantage of 1 (the upper limit) corresponds to zero fitness. Using s and t, rather than some other measures, leads in the end to easily stated mathematical results.

Table 4 outlines the process of getting Δp occurring over one generation. Its value is

$$\Delta p = \frac{pq(-sp + tq)}{1 - sp^2 - tq^2}$$

We ask whether Δp is negative (meaning that freq(A) is decreasing) or Δp is positive (freq(A) is increasing). Each number in the above expression (p, q, s, t) is itself zero or positive, and the denominator as a whole must be positive (because subtracting a portion of the homozygous types from 1 always leaves some heterozygotes). Therefore, looking at the numerator, Δp will be negative whenever sp is *more* than tq, and will be positive whenever sp is *less* than tq. Thus, freq(A) may either increase or decrease depending on the particular allele frequencies and the magnitude of the selective disadvantages.

Examine now the interesting consequences of Δp being *neither* negative nor positive, but *zero*. This situation will occur whenever

$$sp = tq$$

We solve this equation for p:

$$\begin{aligned} sp &= t(1-p) = t - tp \\ sp + tp &= t \\ p(s+t) &= t \\ \boldsymbol{\hat{p}} &= \boldsymbol{t/(s+t)} \text{ and} \\ \boldsymbol{\hat{q}} &= \boldsymbol{s/(s+t)} \end{aligned}$$

The caret (^) over p and q is used to indicate that these are not run-of-the-mill values of freq(A) and freq(a); they are the **equilibrium values** (read "p hat" and "q hat"). The constancy is achieved because in each generation the population loses the *same* proportion of A alleles (sp) as a alleles (tq). Freq(A) and freq(a) will remain at the equilibrium values for as long as the assumed conditions remain unchanged. Note that the equilibrium values will be 0.5 and 0.5 only if the selective disadvantages are equal. Otherwise, $\hat{p}$ is different from $\hat{q}$; the higher frequency belongs to the allele whose homozygote is the more fit.

The sickle-cell polymorphism illustrates this model. In some regions of Africa where malaria is prevalent, homozygotes for the normal allele, Hb^A/Hb^A, are more easily infected by the malaria parasite than are heterozygotes, Hb^A/Hb^S. Although persons with sickle-cell anemia, Hb^S/Hb^S, are also resistant to malaria, they often die of sickle-cell complications before reproduction, so their malaria resistance is of small consequence. Although the selective disadvantages of the two homozygotes are difficult to estimate precisely from available population data, about 15% of Hb^A/Hb^A's and 90% of Hb^S/Hb^S's die between birth and reproduction, when measured against the heterozygotes. Therefore (omitting the basic gene designation Hb), we have

	Genotype		
	A/A	***A/S***	***S/S***
selective disadvantage	$s = 0.15$	0	$t = 0.90$
fitness	0.85	1	0.10

and the equilibrium values are calculated as follows:

$$\hat{p} = \text{equilibrium value of freq}(A)$$
$$= \frac{t}{s+t} = \frac{0.90}{0.15 + 0.90} = 90/105 = 0.86$$

Table 4 The change in the frequency of an allele for a gene with a superior heterozygote[a]

United gametes for initial generation	freq(A) = p ☞			
	A/A	A/a	a/a	*total*
Newborn population	p^2	$2pq$	q^2	1
Fitness	$1 - s$	1	$1 - t$	
Mating population	$(1 - s)p^2$	$2pq$	$(1 - t)q^2$	$1 - sp^2 - tq^2$

United gametes for next generation

$$\text{freq}(A) = \frac{(1-s)p^2 + \frac{1}{2}(2pq)}{1 - sp^2 - tq^2} = \frac{p^2 - sp^2 + pq}{1 - sp^2 - tq^2}$$

$$= \frac{p(p+q) - sp^2}{1 - sp^2 - tq^2} = \frac{p - sp^2}{1 - sp^2 - tq^2} \text{ ☞}$$

$$\Delta p = p' - p = \frac{p - sp^2}{1 - sp^2 - tq^2} - p = \frac{p - sp^2 - p(1 - sp^2 - tq^2)}{1 - sp^2 - tq^2}$$

$$\begin{aligned}\text{numerator of } \Delta p &= p - sp^2 - p + sp^3 + tpq^2 \\ &= -sp^2(1-p) + tpq^2 \\ &= -sp^2q + tpq^2 \\ &= pq(-sp + tq)\end{aligned}$$

a. We assume the Hardy-Weinberg conditions except that the fraction s of A/A's and the fraction t of a/a's die before reproduction. We work out the algebra for Δp under the line. The reader may care to parallel this by finding Δq. Since, in a two-allele system, an increase in one allele must exactly balance a decrease in the other, $\Delta q = -\Delta p$.

$$\hat{q} = \text{equilibrium value of freq}(S)$$

$$= \frac{s}{s+t} = \frac{0.15}{0.15 + 0.90} = 15/105 = 0.14$$

These allele frequencies are indeed characteristic of A and S in some African populations exposed to malaria.

The distribution maps of malaria and sickle-cell anemia (Figure 2) are very similar. On the basis of the geographical correlation between malaria and another hemoglobin defect, thalassemia, Haldane first suggested an interaction between the infectious disease and the genetic one. Evidence that the sickle-cell heterozygote indeed has higher fitness than the normal homozygote in malarial environments comes from several sources:

1. In areas where malaria is prevalent, the ratio of heterozygotes to homozygous normals increases with age. This shift would be expected if survival among *A/S* persons was greater than among *A/A* individuals.
2. Population surveys and experiments with volunteers show that the red blood cells of *A/S* persons are harder to infect and, when infected, have lower counts of parasites than the red blood cells of *A/A* individuals.
3. Hospital records reveal that the death rate from malaria is much lower among heterozygotes than among homozygous normals. Some of the data supporting these lines of evidence are summarized in Allison (1964) and Harris (1980).

It appears that the malarial parasite, which spends part of its life cycle inside red blood cells, flourishes less well when these cells possess some sickle-cell hemoglobin. Heterozygotes are less susceptible to the disease, and when they do contract it, the symptoms are milder than those of homozygotes.

The equilibrium resulting from heterozygote advantage leads to what is called a **stable polymorphism**. Stability means that if the allele frequencies depart from the equilibrium values, selection will tend to bring them back. In the sickle-cell example, $\hat{p} = 0.86$. If the actual frequency of the A allele were more than this amount for some reason, the reader can confirm that Δp would be negative and selection would tend to bring the frequency back down; conversely, if freq(A) were less than 0.86, Δp would be positive.

Heterozygote advantage is one way in which alleles of a gene become *balanced* against each other at intermediate frequencies due to opposing selection forces.

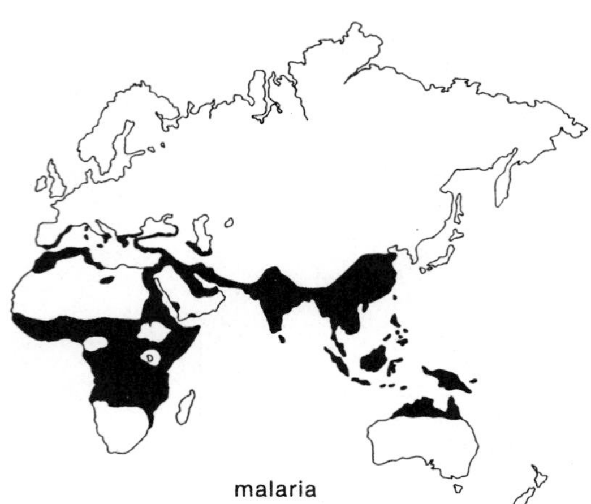

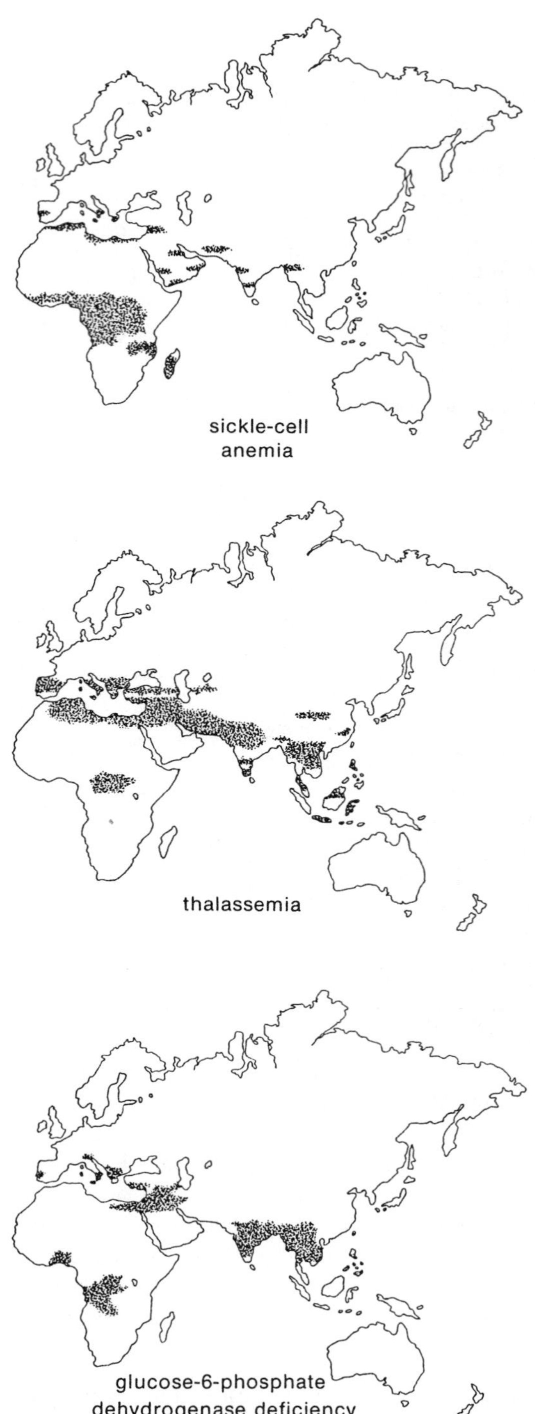

Figure 2 The geographical distribution of malaria in the Old World before 1930, compared with the distributions of three hereditary diseases: sickle-cell anemia, thalassemia, and glucose-6-phosphate dehydrogenase deficiency. The precise nature of the interaction of malaria with the last two diseases is not well understood. (After Motulsky 1960.)

Such a balance can occur for other reasons as well. For example, a stable polymorphism results if a genotype is favored when it is rare but disfavored when it is common (frequency-dependent selection). Another situation involves different genotypes being better adapted to separate ecological niches within the environment. These latter mechanisms for balanced polymorphisms (and others) are probably of greater importance than heterozygote advantage (Hartl 1988).

In regions of the world where malaria is not a problem, the fitness of *A/A* homozygotes is as high as that of *A/S* heterozygotes, because the absence of malaria erases the selective disadvantage of being homozygous *A/A*. Thus only the *S* alleles in persons with sickle-cell anemia are lost through selection. In the United States, for example, sickle-cell anemia acts like a simple recessive detrimental, and freq(*S*) continues to decrease at a slow rate in successive generations. It remains a serious health problem among U.S. Black populations, however.

MIGRATION AND DRIFT

We have described how mutation and selection can disturb the constancy predicted by the Hardy-Weinberg law. Now we consider allele frequency changes due to migration between populations and to the small size of populations.

Migration and Gene Flow

Migration is the movement of people. Prehistoric migrations (for example, Orientals into the New World) often involved movement into previously uninhabited regions where new environments may have imposed new selective pressures. More recent migrations have been into already occupied territory (for example, Europeans and Africans into the New World). Although the immigrants may remain genetically isolated from their neighbors, more often some interbreeding occurs and leads to mixed (hybrid) populations in subsequent generations. When the parental groups differ in allele frequencies, it may be possible to document the amount of **gene flow**, that is, the introduction of new alleles into a population. To the extent that a population (that retains its identity) receives new alleles, the genetic variability *within* it is increased. On the other hand, because they now share some ancestry, gene flow between populations causes them to be more similar to each other than they would otherwise be.

Quantitative analysis of gene flow has been used to complement historical knowledge of migration (Crawford and Workman 1973). Some of the hybrid populations that have been examined include

- *American Blacks*: dihybrids of Africans (of mixed cultures) and Europeans (of mixed cultures, but primarily English)
- *Brazilian subgroups*: trihybrids of native American Indians, Sudanese and Bantu Africans, and Europeans from Portugal, Italy, and Spain
- *Hawaiian subgroups*: multiple-way hybrids of native Polynesians, Whites, Japanese, Chinese, and Filipinos

To obtain information on the extent of the mixing, we can concentrate on one allele of a gene that is selectively neutral, or nearly so, for example, the R^0 allele of the Rh system. (Some selection occurs at the Rh locus because of incompatibility, but the resulting changes in allele frequencies from that cause are slight.) We define

p = frequency of R^0 in the hybrid population

$p_1, p_2, p_3, \ldots$ = frequencies of R^0 in the parental populations

$m_1, m_2, m_3, \ldots$ = the fractions of the respective parental populations making up the hybrid population (the sum of the m's = 1)

These quantities are related in the following way:

$$p = m_1p_1 + m_2p_2 + m_3p_3 + \ldots$$

An analogy is the mixing of dye solutions: the final color (p) results from combining so much (m_1) of one dye (p_1), some (m_2) of a second (p_2), and so on. The value of p will be intermediate between the extreme values of the individual frequencies, or dye colors.

In practice this expression is subject to many errors, so the estimates of component parts (the m's) are very rough. We illustrate the difficulties by using American Blacks as an example (Reed 1969). About 400,000 slaves were brought into the United States, mostly during the eighteenth century. They came from western regions of Africa, extending from Senegal in the north (13%), through central (Benin) areas (70%) to Bantu-speaking regions in the south (17%), a coastal distance of 3,000 miles. (As noted in Figure 3, the percentages are based on restriction enzyme analysis of chromosomes containing the sickle-cell mutation—see later.) In the United States, the offspring of slaves and Europeans, and their subsequent descendants were, and are, almost always considered Black. Only rarely is a person with hybrid ancestry thought of as White. Thus, by definition, the mixing has been virtually one way.

For contemporary U.S. Blacks, let m_1 be the fraction of African ancestry, and m_2 be the fraction of European ancestry (where $m_1 + m_2 = 1$). These values can be estimated from

$$p = m_1p_1 + m_2p_2$$

after substituting for p, p_1, and p_2. Although we can measure the value of p, the frequency of R^0 among contemporary U.S. Blacks, it is unclear what values to use for p_1 and p_2, the frequencies of R^0 among the parental populations that were intermixed *many years ago* with no geneticists around to test their Rh types. The best that can be done is to test contemporary samples of West Africans and Europeans and assume that the current population samples are representative of the parental populations. Another difficulty is that U.S. Blacks are not homogeneous. For example, freq(R^0) varies as follows:

freq(R^0)	Sampled in
0.56	Charleston, South Carolina
0.49	Oakland, California
0.36	New York, New York

Recognizing the uncertainties, typical figures might be as follows:

p = freq(R^0) among U.S. Blacks = 0.5

p_1 = freq(R^0) among African ancestors = 0.6

p_2 = freq(R^0) among European ancestors = 0.025

from which

$$m_1 = \frac{p - p_2}{p_1 - p_2} = \frac{0.5 - 0.025}{0.6 - 0.025} = 475/575 = 0.83$$

$$m_2 = 1 - m_1 = 0.17$$

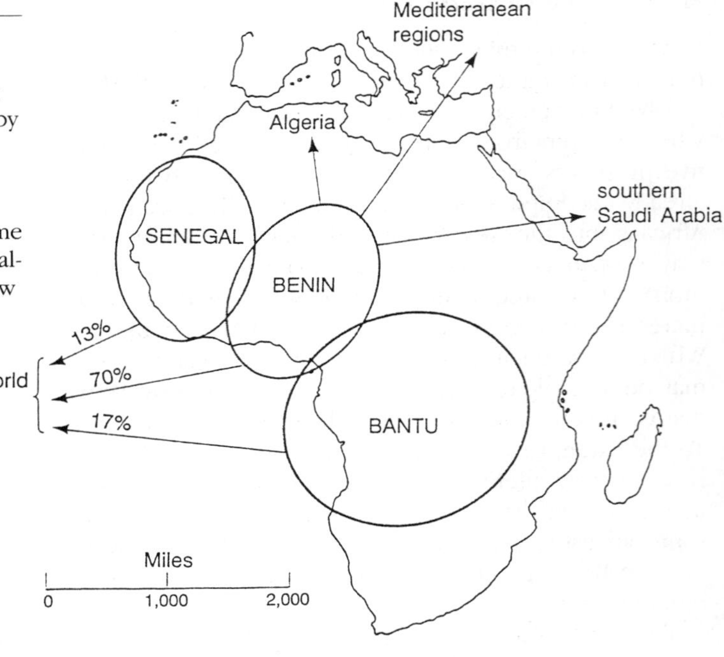

Figure 3 Origin and migration of African groups carrying three different Hb^S chromosomes (area of origin indicated by ellipses). Each chromosome, or haplotype, is defined by a distinctive pattern of RFLPs in the 60-kb length of DNA surrounding the β-globin locus. A separate mutation of $Hb^A \rightarrow Hb^S$ occurred in each background haplotype sometime in the past and is detectable today by restriction enzyme analysis. (Map after Labie et al. 1986 and Kulozik et al. 1986. New World data from Nagel 1984.)

Although this general calculation indicates that about 17% of the ancestry of U.S. Blacks is European in origin, specific calculations for Black populations in the southern United States yield lower figures (4–10%) and for northern populations higher figures (20–30%).

The sickle-cell allele of the β-globin gene has been used to identify migrations of peoples (Pagnier et al. 1984). Mutations from Hb^A to Hb^S arose and proliferated three times in malarial environments of Africa some 2,000 to 3,000 years ago: in Senegal, in Benin, and in the Bantu-speaking region (Figure 3). The different mutations have been identified by the patterns of cutting or not cutting by several restriction enzymes in the 60-kb section of DNA that surrounds the β-globin gene. Each resultant RFLP pattern constitutes a distinctive chromosomal region or *haplotype*, which provided the "background" DNA sequence within which a sickle-cell mutation occurred. Most persons with sickle-cell anemia living in each of the three defined regions are homozygous for one of the three haplotypes. Interestingly, Blacks in northern African (Algeria), in other Mediterranean regions, and in southern Saudi Arabia are exclusively of the Benin Hb^S type, this particular allele having migrated along well-documented trans-Saharan caravan routes.

Judging from a quite different RFLP haplotype, researchers have concluded that a fourth Hb^S mutation occurred in an area extending from eastern Saudi Arabia to India (not shown on Figure 3; Kulozik et al. 1986). For reasons that are not fully understood, persons homozygous for this Asian Hb^S haplotype are affected with a somewhat milder form of sickle-cell anemia.

Random Genetic Drift

The concept of **drift** was introduced in the last chapter to explain the relatively high frequency of the Ellis-van Creveld syndrome among the Pennsylvania Amish. By chance, the initial frequency of the causative allele among the group of founders was higher than among the source population. This founder effect is one aspect of drift and is illustrated further in Table 5 for a very small population. Here a generation of just two persons is picked from a base population with equal allele frequencies. Such a sample is likely to be unrepresentative of the preceding generation. In fact, only three-eighths of the time will the allele frequencies be the same as before: when the founding parents are both heterozygotes or when they are different homozygotes (middle bar of the graph). Using the same base population, readers may care to find the probabilities for samples of more than two individuals. One conclusion from such calculations will be that the sample frequencies are more likely to approach those in the source population as the sample becomes larger.

In more general terms, random genetic drift refers to unrepresentative sampling that occurs in successive generations of a population. This situation occurs in part because different matings may not leave equal numbers of offspring. For example, the Amish families that harbored the allele for the Ellis-van Creveld syndrome appeared to have been more prolific over several generations than average Amish families. Drift can also occur because the successful gametes from heterozygotes may differ from 50:50 expectations.

Table 5 A model of the founder effect for a population of size two

	GENOTYPES AND FREQUENCIES[b]			NUMBER OF ALLELES IN SAMPLE		$P = \text{FREQ}(R)$	
Sampling possibilities[a]	*R/R* 1/4	*R/r* 1/2	*r/r* 1/4	*R*	*r*	0 0.25 0.50 0.75 1	*Probability of sample*
1	x,x			4	0		$(1/4)^2 = 1/16$
2	x	x		3	1		$2(1/4)(1/2) = 4/16$
3		x,x		2	2		$(1/2)^2$ } = 6/16
4	x		x	2	2		$+ 2(1/4)^2$ } = 6/16
5		x	x	1	3		$2(1/2)(1/4) = 4/16$
6			x,x	0	4		$(1/4)^2 = 1/16$

a. Envision a new population started by a couple picked randomly from a Hardy-Weinberg population in which $p = q = 1/2$. Allele frequencies among the two might be the same as among the population from whence they came, but more likely the new frequencies will differ from the old. Note that 2/16ths of the time the parents possess only R alleles or only r alleles (top and bottom lines). If the population continues to be small in subsequent generations, fixation of the one allele or the other continues at a substantial rate. Eventually, the population will be all R or r, and no further changes can occur without mutation or migration.

b. x, male or female parent.

DIFFERING VIEWS ON THE MECHANISMS OF EVOLUTION

Presented here are some competing ideas on the relative importance of the several evolutionary forces. We emphasize that the arguments are not *whether* evolution occurred, but *how* evolution occurred. In order to explain the marvelous complexities of the entire living world, virtually all professional biologists see the many lines of evidence as favoring gradual evolutionary changes over many millions of years. Some others may try to misuse the (sometimes spirited and very public) argumentation among biologists about mechanisms of evolution to discredit evolution itself. But the arguments only attest to the vigor of the science and, to some extent, the clash of personalities.

The Role of Drift

The importance of drift in evolutionary history, for example, is disputed. Wright was an advocate of drift, whereas Fisher downplayed its role. Although the conflicting arguments are often couched in sophisticated statistics, the differences are laid out, mostly in nonmathematical language in Wright's review (1930) of Fisher's major book (1930), in a more recent article by Wright (1982), and in historical assessments by Provine (1971, 1986) and by Crow (1987). Fisher conceived of evolution as occurring gradually in large populations by the kinds of selective systems we have previously outlined. Such evolutionary changes are very slow, to be sure, but long periods of time have been available. Wright emphasized that evolution could occur more rapidly when a species is broken up into groups of moderate size between which there is a limited amount of migration. By his *shifting-balance theory*, it would be possible for a large (but subdivided) population to evolve from one well-adapted gene combination to a better one even when intermediate combinations are less fit. When populations are too small, however, drift leads inevitably to fixation, to the consequent loss of variability, and to the inability of populations to respond to environmental changes. Wright (1930) wrote that at intermediate sizes "there will be continuous kaleidoscopic shifting of the prevailing gene combinations, not adaptive itself, but providing an opportunity for the occasional appearance of new adaptive combinations of types which would never be reached by a direct selection process."

The fortuitous combinations of alleles at different loci, achieved by chance, are then in Wright's view still subject to the same kinds of selection pressures as those championed by Haldane and Fisher. In different subpopulations, the joint results of mutation, drift, and selection could yield, for example, different, well-adapted strains of the same species. By increased reproduction and outmigration of the most fit strains, the rate of evolution could be much faster than in a single large interbreeding population. With regard to human evolution, drift was possibly more important in prehistory when our species was broken up into hunting camps, tribes, and clans. Some human isolates still exist, of course, but the world's population is becoming more "Fisherian."

The Nature of Variability

How forces of evolution maintain the variability that is created by mutation is a related area of vigorous argument. Without the persistence of allelic variants, evolution would have nothing to work on. Nobody disputes that species are genetically quite variable. For example, analyses of the human globin chains reveal that a sizable fraction of single amino acid changes that *could* occur *have* occurred, although most of the variants are very rare (see listings in Dickerson and Geis 1983; McKusick 1988). In addition to rare alleles, species have many *polymorphisms*, that is, genes with two or more *common* alleles. (Recall that a polymorphic locus is defined as having two or more alleles with frequencies of 1% or greater.) In humans, the blood group genes and the HLA loci are particularly polymorphic. In addition, many individuals from many populations have been screened over many years for enzymes that may show variants when subject to electrophoresis. Some of these more or less *randomly* chosen genes from *randomly* chosen, healthy individuals are also polymorphic. The debate is over *how* the large amount of variation that is observed today in human populations has been brought about during the course of our past evolution.

Electrophoretic surveys were pioneered by the American geneticists Richard Lewontin and John Hubby on enzyme variations in fruit flies (summarized in Lewontin 1974). The work on human enzymes has been summarized by the English geneticist Harry Harris (1980), who has tabulated the results on 104 different loci. Of these, 33 loci exhibit an electrophoretically detectable polymorphism in at least one major ethnic group. For several reasons, the 32% (33/104) of loci represents a *minimum* estimate of polymorphic loci:

1. Electrophoresis separates proteins only by *charge*. But about two-thirds of possible amino acid differences brought about by single base substitution mutations do *not* lead to charge differences.

2. Two distinct amino acid charge variants of the same protein may have the same electrophoretic mobility. Altering the conditions of electrophoresis may reveal previously hidden differences (Lewontin 1985).

3. Some mutations, especially those in the third position of coding triplets, do not change an amino acid at all. Similarly, a base substitution mutation in an intron of a gene would not generally affect the protein product. These types of mutations have been increasingly investigated by restriction enzymes and by DNA sequencing.

Thus, geneticists have documented (in many species) a large amount of persistent genetic variation and, in particular, a significant number of polymorphic genes. Explanations of these observed patterns emanate from two opposing camps, the selectionists and the neutralists.

The **selectionist position** is the long-standing one begun by Darwin.* It suggests that variations are maintained by natural selection operating on randomly occurring mutational events. Most of the mutations are detrimental and are held at low frequencies because they are selected against (the classic view). Occasionally, a mutation may be advantageous when it occurs, or it may become advantageous because of an environmental change. It this case, the newly arising mutation will supplant the old one given enough time. Selectionists note that a polymorphism will exist when the frequencies of the old and new alleles are intermediate (a so-called *transient* polymorphism). Polymorphisms can also exist because selection operates against both homozygotes (a *balanced* polymorphism). Selection favoring sickle-cell heterozygotes is the best and almost the only known human example of this balancing act. Several other ways of maintaining a polymorphism have been proposed (see Harris 1980; Hartl 1988).

The **neutralist position** is quite different. It suggests that a large fraction of variations, especially polymorphisms, are selectively neutral (Kimura 1979, 1983). The occurrence of allelic variation in current populations is explained by past mutations followed by random frequency changes, without any fitness differences upon which selection could act. Of course, random drift could quickly eliminate newly occurring mutations, and it will do so with predictable and high probability. But the polymorphic variation we now see (in enzymes, for example) is due to random drift that happened to *increase* the frequencies of the initial mutations. Neutralists agree with selectionists that some mutations are deleterious, being kept at very low frequencies by selection. But neutralists hold that some rare allelic variants and most of the polymorphic ones result from neutral alleles whose appreciable frequencies are due to drift.

Neutralists suggest that over long time periods the drift of neutral genes accounts for some of the differences in the same gene in different species. Thus evidence for the neutral viewpoint comes from examining the same protein molecule (or DNA segment) in different species. For example, the normal β chain of hemoglobin in humans differs from that in other species:

*Within the selectionist camp, *punctuationists* emphasize that evolution consists of relatively brief periods of change and long periods of no change, or stasis (Gould and Eldredge 1977). Punctuationists suggest that the so-called gaps in the fossil record are not gaps at all but represent the real "jerkiness" of evolutionary changes. They also emphasize the role of small populations and random genetic drift. Others (for example, Dawkins 1986) emphasize the *gradualism* of evolutionary change. The gradualist and punctuationist positions seem to be merging.

Species	Number of amino acids that differ from those of humans
chimpanzees	0
gorillas	1
rhesus monkeys	8
other vertebrates	more than 8

Despite the differences in structure, all these hemoglobin molecules function well within their respective bodies.

Paleontologists can determine the approximate time of divergence of one species group from another. Using these dates (measured in hundreds of millions of years), Kimura and other neutralists find that the observed amino acid changes occur at one rate for protein A, say, and at another rate for protein B. Histone molecules, for example, are nearly identical in widely divergent species, and therefore the whole molecule seems to be functionally important—practically no change is tolerated. *But for any one molecule, the rate of change over evolutionary time is more or less constant.* DNA sequencing data show that nucleotide changes in introns and in the third position of coding triplets are less constrained than in more critical regions. Neutralists argue that these observations support the idea that most molecular evolution is due to random fixation of neutral mutations. Because all the molecules in different species still function well, the amino acid changes and the mutations that caused them (as well as mutations in noncoding regions) must have no phenotypic significance. Neutralists extrapolate from changes between species to changes within species.

Despite their differences, both camps are right. There can be no doubt that a fraction of polymorphic genes is maintained by selection pressure, and another fraction is due to random drift of neutral alleles. We just do not know what the fractions are.

A GENETIC CONCEPT OF RACE

The terms *species* and *race* are familiar; both suggest the idea that members of a group have more in common with one another than with outsiders. It is often difficult to delimit the groups, however, and to measure the differences between them.

We begin with the easier term: a **species** is a group whose members are capable of interbreeding with one another but incapable of producing viable and fertile offspring with members of other species. The divergence of separate species from common ancestral stocks is the main theme of evolution, proceeding in part by the mechanisms described previously. During the branching process over long time periods, geographically isolated groups come to have different allele frequencies, and perhaps different amounts of chromosomal material, but one way or another, they evolve the intersterility that distinguishes them as separate species.

The Reality but Arbitrariness of Races

Just as separate species can differentiate from each other, given isolation and enough time, so too can subgroups within a species. Extensive wanderings of *Homo sapiens* over the earth led to dispersed settlements 40,000 years ago (Davis 1974). Group differences developed, partly by natural selection operating in widely differing environments. Some differences relate to easily visible traits, such as skin color, body form, and facial features. Such morphological characters are the ones that are traditionally used to distinguish three main categories of humankind: African, Caucasian, and Oriental. Other differences, such as blood groups, HLA types, enzyme variants, and RFLPs are usually observed only by laboratory workers.

Races can be defined as the subdivisions of a species that have come to differ, to a greater or lesser degree, in the frequencies of the alleles they possess. But for two reasons, human races have not achieved anything like separate species status: (1) The time available has been relatively short. (2) The continual migrations of people throughout history have prevented the degree of isolation necessary for speciation. It is likely that few, if any, human groups have remained completely isolated genetically for more than a few generations. Although races involve finer distinctions than species, the differences are sufficient to call attention to themselves—real enough to invite some sort of classification.

Although there is general agreement on what constitutes separate species (with quibbling in a few cases), there is no single handy criterion by which to define races. Consequently, there are many opinions as to how the job should be done. Among anthropologists, the "lumpers" recognize relatively few different races—commonly the three categories mentioned above—whereas the "splitters" come up with many more divisions, especially among the Oriental group. The problems of classification relate to the number and placement of lines between more or less merging categories. Perhaps no practical purpose is served by drawing lines at all. But the groups are of scientific interest, primarily for investigating how they evolved. That is, the measurement of variability within and between divisions of humankind provides insight into the processes of biological and cultural evolution.

Because of racial prejudices, some prefer to replace the term *race* with a more neutral one, such as *ethnic group* or *population*. By whatever name, human variability is real, although the manner of categorizing it into larger or smaller groupings, shifted this way or that, is arbitrary. Here we examine the kinds of traits that are studied and comment briefly on the extent and possible causes of the differences.

Human Variability

The traditional measurements of racial differences relate to the size, shape, and color of various parts of the body. Cultural characteristics, especially language and behavior, have also been studied. These traits are substantially influenced by environmental variables and depend on many genes whose precise functions are unknown and that interact during development in unknown ways. Thus attempts to delineate racial differences by overall morphologies are made difficult by the inability to isolate environmental and genetic contributions. That these traits are measured along a continuous scale further compounds the problem of where to draw division lines.

To avoid some, but not all, of these problems, anthropologists have increasingly turned to simply inherited traits that are polymorphic in one or more populations. As early as 1918, before ABO inheritance was understood, anti-A and anti-B were used by two army physicians on the Balkan front to survey different groups of soldiers. Since then, blood samples have been channeled into anthropologists' and geneticists' test tubes from literally millions of people living in virtually every habitat around the globe. Using the Hardy-Weinberg law to calculate allele frequencies within different populations, Boyd (1950) was among the first to summarize masses of data in a way that pointed up simple differences between human races.

As an example, current knowledge of the worldwide distribution of the *B* allele of the *ABO* gene is presented in Figure 4. The maximum frequencies of about 30% appear in central Asia. Frequencies decrease westward into Europe, diminishing to about 5% and even less in the Basque populations of France and Spain. The gradual changes in the distribution of the *B* allele across Europe seem to be due to the Mongolian invasions of the fifth through the fifteenth centuries. The allele is absent among Indian populations of the New World and among the Aboriginal populations of Australia.

Other blood groups, especially MN, Rh, and Duffy, exhibit even larger allele frequency differences. (The specifics of variation differ from gene to gene.) Table 6 shows some typical Rh allele frequencies selected from a huge collection of data (Mourant et al. 1976). The Rh negative allele (*r*) has its highest frequency among Caucasians and is least common in Orientals. Note that racial origin is more important than current geographical location: compare American Whites, American Blacks, and American Indians with one another.

Table 6 Typical Rh allele frequencies in different world populations

Population	ALLELE FREQUENCIES (%) *R*	*r*
Caucasians		
Spanish Basques	50	50
American Whites, Australian Whites, Spaniards	60	40
Africans		
American Blacks	70	30
Angolan Bantus, Nigerians	75	25
Orientals		
American Indians, Australian Aborigines, Japanese, Chinese	95+	near 0

Source: Selected from the extensive compilation of Mourant et al. (1976). In these studies, only the antibody anti-D was used.

More recently, other kinds of simply inherited traits have been added to the anthropologist's repertoire of tools. These include various plasma proteins (immunoglobulins, haptoglobins, transferrins), red blood cell enzymes (dehydrogenases, phosphatases), hemoglobin types, histocompatibility types, and mitochondrial DNA variants. We can look forward to more documentation of racial differences based on restriction enzyme analysis and DNA sequencing (see also Cavalli-Sforza et al. 1986).

The general conclusion that emerges from these studies is that human populations vary considerably in the frequencies of alleles. It is usually not an all-or-none phenomenon, although in some populations the frequency of a particular allele may be 1 or 0. Nor is any particular gene sufficient to delineate racial groupings. Rather, the combinations of frequencies for many different genes, the total gene pool, can be utilized as a fairly objective criterion for determining similarities and differences between human populations. Several mathematical methods have been used to combine the information on frequencies of several genes into a single measure of genetic difference between two populations (Cavalli-Sforza and Bodmer 1971). These data complement and generally confirm the groupings based on morphological traits.

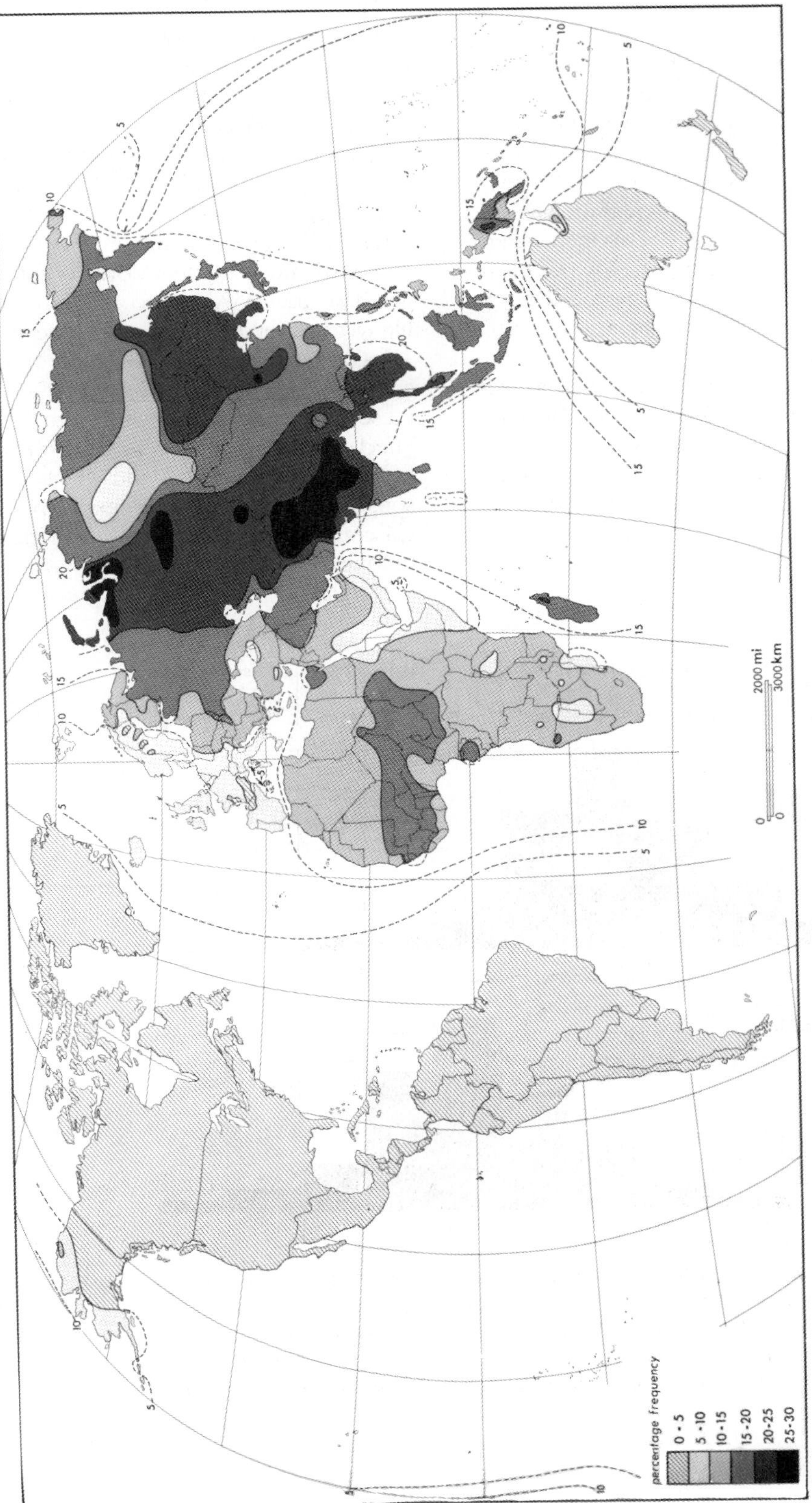

Figure 4 The distribution of the *B* allele of the *ABO* gene among indigenous populations of the world. (From Mourant et al. 1976.)

Individuality

The existence of so many polymorphisms means that the idea of a pure race is fallacious. Purity implies constancy of type, which implies homozygosity, as found in the pure-breeding strains of Mendel's peas, alike in their pods. The absence of pure races is also due to the intermixture of populations arising from migration. Gene flow is not just a recent phenomenon. Fossils increasingly indicate that several different forms of *Homo sapiens* coexisted tens of thousands of years ago. Successive separations and fusions of human groups suggest that racial classifications are not static.

It is easy to see the conspicuous traits, such as skin color or eyelid shape, which are determined by just a few of the tens of thousands of human genes. Even though these particular characteristics allow pigeonholing, we know that the degree of variability within any one grouping is very large. The genetic variability among the individual members of any race, however defined, is greater than the average difference between races. The concept of *within* and *between* variability is illustrated in Figure 5.

On the basis of electrophoretic variation, we noted that about one-third of human genes are polymorphic. Using the observed frequencies for the alleles of each of the polymorphic genes, Harris (1980) calculated that the average person is heterozygous for about 6% of those genes that code for enzymes with electrophoretic variants. Six percent of, say, 50,000 genes is 3,000. But because electrophoresis only detects a minority of enzyme variants, the actual number of heterozygous genes per person is much higher. More recent data on RFLPs suggest that the DNA of two randomly chosen persons is likely to differ by about one base in several hundred. Thus, at the DNA level, two random individuals differ in about 15,000,000 bases ($3 \times 10^9/200$), although many of these differences are not likely to have any phenotypic effect (Botstein et al. 1980). However one views our biological endowments, the chance of finding two persons with the same total genotype is effectively zero (discounting identical twins). Genetically, each person is unique.

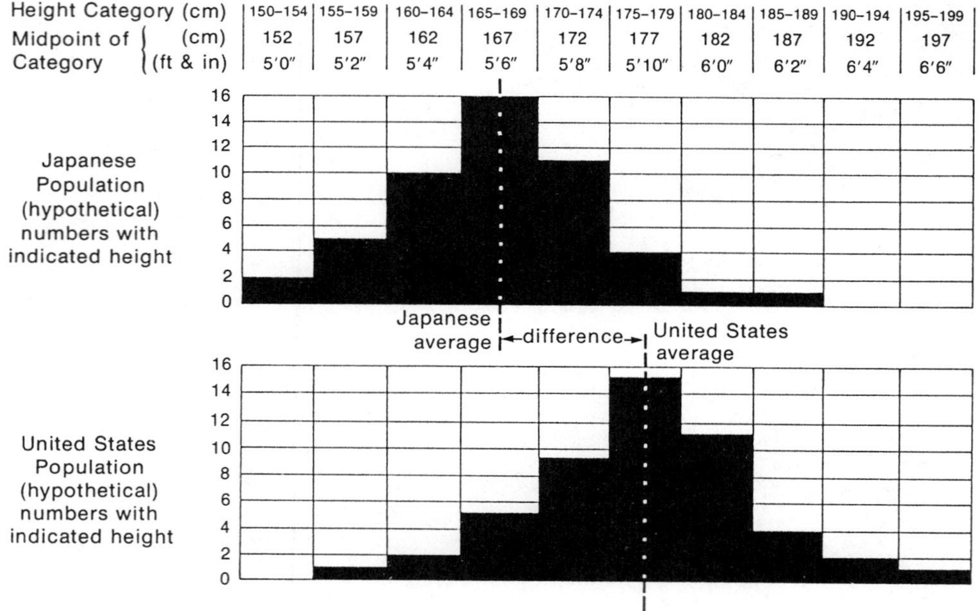

Figure 5 The meaning of variability within and between groups. The histograms give the numbers of individuals in various height categories for a Japanese population and for a United States population, each of size 50. (Although the numbers are hypothetical, the average heights are approximately correct for young adult males in 1970.) *Within* each group the variability in height is over 40 centimeters, but the difference in the group average is just 10 centimeters.

SUMMARY

1. Evolution of the vast array of living forms has occurred by mutation, random genetic drift, migration, and especially by natural selection. In a simplified sense, evolution occurs by gradual changes in allele frequencies.

2. Mutations provide the raw material for evolution. Dominant mutations are revealed by noting the abnormal dominant trait in persons with normal parents. Recessive mutations may not be phenotypically expressed for many generations.

3. Measuring the rates of mutations in humans is imprecise because of the rarity of mutational events for a given gene. A typical human gene mutates at the rate of about one mutation per million gametes per generation.

4. Selection occurs when various genotypes in a population differ in fitness. Changes in environmental conditions lead to changes in the way selection acts.

5. Selection against a rare recessive phenotype decreases the frequency of the recessive allele only very slowly, because most rare recessive alleles are hidden in heterozygotes.

6. Selection in favor of the heterozygote stabilizes the allele frequencies at intermediate levels. The sickle-cell allele in malarial environments is the best understood example in humans.

7. Migration may lead to gene flow, the introduction of new alleles into a population. Estimates of the amount of mixture can sometimes be obtained by allele frequency analysis.

8. Drift refers to haphazard changes in allele frequencies that become more likely the smaller the population. The importance of drift in evolution has been disputed since the 1930s.

9. The human species is genetically quite variable for both rare and polymorphic alleles. Some of this variability is maintained by selection pressures, and some is due to random drift of selectively neutral mutations.

10. Human races have arisen by evolutionary processes. The aggregate of allele frequencies in one race differs, to varying degrees, from the frequencies in others. The boundaries between races are fuzzy.

11. The genetic differences within human populations are large; each individual has a unique genotype.

FURTHER READING

Accounts of evolution include the informative and masterful books by Dawkins (1986) and Futuyma (1983), and the more traditional textbook by Futuyma (1986). Two collections of articles on modern evolutionary topics, including the selectionist/neutralist controversy, are edited by Bendall (1983) and by Milkman (1982). Crow (1986) and Hartl (1988) have written short accounts of population genetics. Harris (1980), Bodmer and Cavalli-Sforza (1976), and Cavalli-Sforza and Bodmer (1971) treat clearly the population genetics of our species. Volume 51 of the Cold Spring Harbor Symposia (1986) includes a section on using molecular techniques for studying human evolution. Haldane's delightful book (1932) remains useful for modern readers, and Darwin's autobiography (1969 reprint) is fascinating.

Lewontin (1985) reviews some recent topics in population genetics, and Crow and Denniston (1985) summarize material on mutation in human populations. Historical viewpoints include Provine's (1986) analysis of the work of Sewall Wright and Jameson's (1977) collection of important papers in evolutionary theory. A balanced textbook on human races is that of Molnar (1983), and a nice account of genetics and racial differences is that of Sunderland in a collection edited by Ebling (1975). Of many *Scientific American* articles on evolutionary topics, those by Stebbins and Ayala (1985) and by Wilson (1985) are of current interest.

QUESTIONS

1. If most mutations occur during DNA replication, then sperm would be expected to carry more new mutations than eggs. Why?

2. On the average, fathers are older at the time of birth of achondroplastic dwarfs than at the time of birth of their normal babies. Give a possible explanation.

3. Assume that a rare dominant allele has incomplete penetrance. Thus, some affected children might have normal-appearing parents even when a parent possesses the dominant allele. Would the calculated mutation rate by the direct method then be too high or too low?

4. Assume that all human genes stopped mutating completely. How would you demonstrate that this unusual phenomenon occurred?

5. What is meant by the relaxation of natural selection? Is it happening? With what consequences?

6. Explain why selection against a recessive lethal becomes less and less effective as the allele becomes rarer.

7. The formula for a recessive lethal, $q' = q/(1 + q)$, is particularly easy to work with when the q's are expressed as fractions. It predicts that, over successive generations, the frequencies of a recessive lethal will be 1/5, 1/6, 1/7, . . . Therefore, in g generations, the frequency will go from $1/x$ to $1/(x + g)$. How many generations does it take to go from 0.5 to 0.05? From 0.05 to 0.005? From 0.005 to 0.0005?

8. Assume these fitness values: 1 for *J/J*; 1/2 for *J/j*; 0 for *j/j*. Except for this selection, assume the Hardy-Weinberg conditions. Show that freq(j) changes from q to $(1/2)q$ in one generation. Follow the procedures in Tables 3 and 4. The fairly rapid decrease is due to the heterozygous effect.

9. An investigator notes that the frequency of a is 10% despite *a/a* having zero fitness. Using the model of a superior heterozygote, show that the expected fitness of the *A/A* homozygote is 8/9.

10. Identify the three types of genetic equilibrium situations mentioned in this chapter involving mutation or selection, or both.

11. Give two reasons why the frequency of the sickle-cell allele among American Blacks is less than among their African ancestors.

12. Select a random sample of size three from the offspring of $R/r \times R/r$. (a) Show that among the three offspring, the probability of allele r being lost or fixed is 1/32. (b) If the sample were larger, would the probabilities of loss or fixation be greater or less?

13. (a) The formulas for the fractions of African and European ancestry among American Blacks are most accurate when the difference in frequency between p_1 and p_2 is as great as possible. Why? (b) For the Fy^a allele, Reed (1969) reports

    $p_1 = 0$ = frequency among West Africans
    $p_2 = 0.4296$ = frequency among Oakland, CA Whites
    $p = 0.0941$ = frequency among Oakland, CA Blacks

    Calculate m_1 and m_2.

14. Kimura (1979) compared the amino acid differences between species for the A, B, and C segments of the proinsulin polypeptide. He found many more differences for the C segment than for A and B segments. Does this make evolutionary sense? Why? (Refer to Figure 4 in Chapter 15.)

15. (a) Give a genetic definition of race. (b) What are the advantages of using blood groups to delineate races?

ANSWERS

1. In males, the spermatogonial stem cells have many more divisions than the oogonial stem cells.

2. Sperm cells from older fathers are more likely to carry a mutation because the stem cells from which they are derived continue to divide mitotically about 23 times per year.

3. The calculated rate would be too high because some affected children would be tallied as new mutations when, really, an "old" mutant gene, unexpressed in a parent, was transmitted to them.

4. The first obvious consequence would be a sharp reduction in the appearance of new cases of genetically

straightforward, dominant diseases. Except for incomplete penetrance, all patients would have an affected parent.

5. Relaxation refers to the increase in reproductive fitness of certain genotypes, as a result of modern, largely medical, technology. It is happening for a limited number of genetic diseases. As a consequence, the frequency of these diseases, which then become treatable, will increase very slowly.

6. As the allele becomes rarer, the proportion of alleles present in heterozygotes, where they are masked, becomes greater.

7. 18; 180; 1,800.

8. uniting gametes: freq(j) = q

	J/J	***J/j***	***j/j***	**total**
newborns	p^2	$2pq$	q^2	1
fitness	1	1/2	0	
mating group	p^2	pq	0	$p^2 + pq = p$

uniting gametes forming the next generation:

$$\text{freq}(j) = \frac{0 + (1/2)pq}{p^2 + pq} = (1/2)q$$

9. $\hat{q} = s/(s + t)$ where $\hat{q} = 0.1$ and $t = 1$. Substituting and solving for s gives $s = 1/9$, the selective disadvantage of A/A.

10. (i) forward versus back mutation; (ii) selection against a recessive allele versus mutation to it; (iii) selection against both homozygotes.

11. (i) selection against the detrimental homozygote without a superior heterozygote; (ii) some admixture with White populations.

12. (a) for any one offspring, P(R/R) = P(r/r) = 1/4. Thus,

$$\text{P(all three } R/R) = (1/4)^3 = 1/64$$
$$\text{P(all three } r/r) = (1/4)^3 = 1/64$$
$$\text{Sum} = 1/32$$

(b) the effects of drift are lessened; the probability of extreme allele frequencies is reduced.

13. (a) If p_1 and p_2 are just a little different, small sampling errors lead to large errors in the denominator, $p_1 - p_2$.

(b) $m_1 = \dfrac{p - p_2}{p_1 - p_2} = \dfrac{0.0941 - 0.4286}{0 - 0.4286} = 0.78$; $m_2 = 0.22$

14. When a molecule of insulin is formed, the C segment is cut out; thus, its amino acids might be expected to be less constrained—more free to change without affecting critical functions. Neutral mutations would therefore be more likely to occur here than in the A or B sections.

15. (a) A subdivision of humankind that has formed by evolutionary processes and that differs from others in the aggregate of its allele frequencies. (b) The correlation between phenotype and genotype is clear; the allele frequency differences are sometimes large, thus allowing for a cleaner analysis of data; the blood groups are sociologically neutral, signifying neither better nor worse.

Chapter 23 Quantitative and Behavioral Traits

Genetic and Environmental Variation
Heritability

The Genetic Component
An Additive Model
Skin Color

Twins: Their Usefulness and Limitations
The Biology of Twinning
Twins in Genetic Research

Behavioral Traits
A Genetic Component to Human Behavior
Intelligence
Alcoholism

The similarities and differences that we see among our relatives often intrigue us, for we cannot help but notice when someone shares a brother's unusual mannerism, a mother's perfect complexion, a grandparent's longevity. Children whose parents have special talents are often expected to be correspondingly more athletic, or more musical, or more brainy than their schoolmates. Personality traits, too, may be attributed to common heredity, and a "bad seed" in a family may cause some apprehension. In this chapter we discuss the genetics of such composite physical and behavioral characteristics.

The juxtaposition of the words *behavior* and *genetics* may strike some readers as odd. Certainly, behavior genetics is a relatively new field of study. Early behavior scientists were uninterested in genetic influences, and early geneticists could not easily study complicated, ill-defined behavioral traits. Early in this century, studies of behavior split into either ethology or animal psychology. Ethologists—largely Europeans—studied birds, insects, and fish in natural habitats, emphasizing the role of evolution in the development of innate and learned behavior patterns. This group included the Nobel Prize winners Konrad Lorenz and Niko Tinbergen. Animal psychologists, on the other hand, were largely Americans who studied the laboratory rat almost exclusively.* They tested theories of learning, and minimized—or denied altogether—any influence of genes on behavioral traits. (The two groups of investigators also sneered at each other a lot.)

In the 1930s, Robert Tryon at the University of California, Berkeley, and others began genetic studies of behavior. Starting from a *single group* of rats, Tryon (1940) was able to breed one line that made relatively few errors in running a maze and another line that

*The spell that rats seemed to have cast over experimental psychologists was pointed out by Beach (1950), who invoked the image of the Pied Piper of Hamelin. "Now the tables are turned," he wrote. "The rat plays the tune and a large group of human beings follow."

performed poorly. The ability to separate out maze-bright or maze-dull rats by selective mating over several generations demonstrated the existence of genetic variation for a complex behavioral trait. It also showed that animal psychologists should pay attention to rats as individuals, rather than "the rat" as an unvarying member of a species.

In ensuing decades, behavior geneticists experimented with many other animals, including fruit flies, bees, birds, mice, cats, dogs, and monkeys. Publication of the first textbook of behavior genetics (Fuller and Thompson in 1960) and the first journal dealing exclusively with this topic (*Behavior Genetics* in 1970) marked the establishment of a distinct field of investigation. Although studies of human intelligence have sparked considerable scholastic, social, and political controversy (for example, Lewontin et al. 1984; Davis 1986), recent molecular genetic investigations of mental conditions such as manic depression and Alzheimer disease leave no doubt that genes can influence human behavior. This effect is accomplished not by magic, but through gene-controlled biochemical pathways that affect the anatomy and physiology of sense organs, hormones, nerves, and muscles—pathways that also affect many physical characteristics in simple or complicated ways. Of course, environmental factors also influence the same behavioral traits. We especially want to emphasize that a demonstration of genetic influence never rules out environmental effects. Any disease, such as phenylketonuria, that is clearly genetic but is treatable to some degree shows the relevance of both genes and environment.

Usually, complex physical and behavioral traits are not simply present or absent like many of the Mendelian characters that we have previously considered. Rather, they vary *continuously* and are measured by length, weight, time, color gradations, activity levels, test scores, or some other suitable scale. And when it comes to intelligence or personality, the definitions of the traits themselves may be imprecise and the resulting measurements correspondingly uncertain. Yet the difficulties imposed by such **quantitative variation** have not stopped geneticists, anthropologists, psychologists, educators, and others from attempting analyses. They have tried to measure how much of the variation is due to the fact that people are conceived with different genotypes and how much is due to the fact that people live in different environments. This approach raises the old nature–nurture question, which seeks to partition the underlying causes of observed differences. The question is asked because it deals with important or interesting human traits—including matters of health (Williams 1988).

The analysis of quantitative traits uses **biometry**, which is statistics applied to biological variation. Although the computations are largely beyond the scope of this book, we describe why the nature–nurture problem is a statistical concept. The question was first seriously addressed by the inventive and many-sided Sir Francis Galton* (1822–1911), a contemporary of Mendel (1822–1884) and a half first cousin of Darwin (1809–1882). Galton measured many physical and behavioral characteristics in humans and inquired into their inheritance. He suggested ways to improve the genetic endowment of humans and coined the term **eugenics** to describe this. He was also the first to utilize twins in his studies of stature and intelligence.

Galton's views on inheritance were seen at first as contradictory to those of Mendel, because quantitative traits that interested Galton often appeared to conform to a blending type of inheritance, whereas Mendel studied the segregation of the discrete traits. Later, however, geneticists reconciled the two views by showing that the genetic component of quantitative traits depends on the Mendelian inheritance of several genes, each with a small effect.

GENETIC AND ENVIRONMENTAL VARIATION

A person's phenotype unfolds during development and maturation when genes and gene products interact with one another and with life's circumstances, comprising "all the peculiarities of nurture both before and after birth, and every influence that may conduce to make the characteristics of one brother differ from those of another" (Galton 1889). This concept is sometimes summarized in a cause-and-effect diagram:

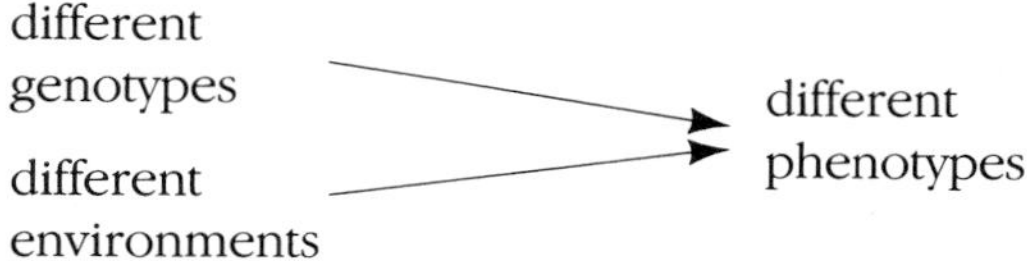

It is not appropriate to ask whether a trait is "due to genes" or "due to the environment," since a person cannot exist without both. The nature–nurture question is one of degree rather than one of kind. We can ask, for example, how much of the observable phenotypic variation in hair color is due to people having different hair-color alleles and how much is due to hair being put into

*Galton read at an early age, studied medicine and mathematics, explored Southwest Africa for the Royal Geographical Society, established a widespread network of self-recording weather stations, and began the classification of fingerprints. Measuring was his passion. He appreciated the same temperament in Mendel and felt a sentimental bond because of their common birth year, but Galton never recognized the fundamental importance of Mendel's work.

different environments. In a sense, we are asking how boldly, relative to each other, the two arrows in the diagram above should be drawn.

Because the genes provide the initial guidelines for the development of a new person, this diagram is sometimes described in the following way: the genotype of an individual determines a *range* of possible phenotypes, and within that predetermined range, a specific phenotype is molded by environmental influences. The range determined by genes may be narrow or broad, depending on the trait under consideration. For example, one's ABO blood type is gene-determined within a very narrow range. In fact, one usually does not consider any environmental influence at all. But Race and Sanger (1975) note that the strength of the A antigen may decrease with age, and very rarely the B antigen may be acquired by a person without the B allele.

At the other extreme, infectious diseases are primarily caused by contact with external agents and are seemingly independent of a person's genotype. Still, we all know people who, through all adversity, never get the colds we do. Studies of twins and adoptees reinforce the suggestion that disease resistance has a genetic component, expressed through the manifold cells and reactions of the immune system. Sørensen et al. (1988), for example, studied the medical histories of Danish children born in the 1920s and adopted early in life by unrelated persons. The researchers showed that the likelihood of death from infections was substantially increased among adoptees having a *biological* parent who died of infection. No such mortality correlation existed between the same children and their *adoptive* parents.

Thus for no characteristic of interest are differences between people entirely of genetic or entirely of environmental origin.

Heritability

A useful way to describe the origins of variability is by the algebraic expression

$$V_P = V_G + V_E$$

This equation can be understood on two levels. Most simply, it expresses the fact that the total phenotypic variation (V_P) that is observed for a given trait in a given population has two components: variation in the genotypes (V_G) and variation in the environments (V_E). The *V*'s can also be taken to represent a more precise measure called *variance*. This estimate of variability is the usual one that statisticians compute in order to express how diverse a group is. The more widespread or extreme the values, the greater the variance, as illustrated by the histograms in Figure 1. The variance that is computed here is V_P, the observed phenotypic variance in seed weight. There is no information in these histograms that would allow separate estimates of the genetic and environmental components of variation. The beans possessed a variety of genotypes and grew under somewhat variable conditions of sunlight, temperature, wind, water, and nutrients.

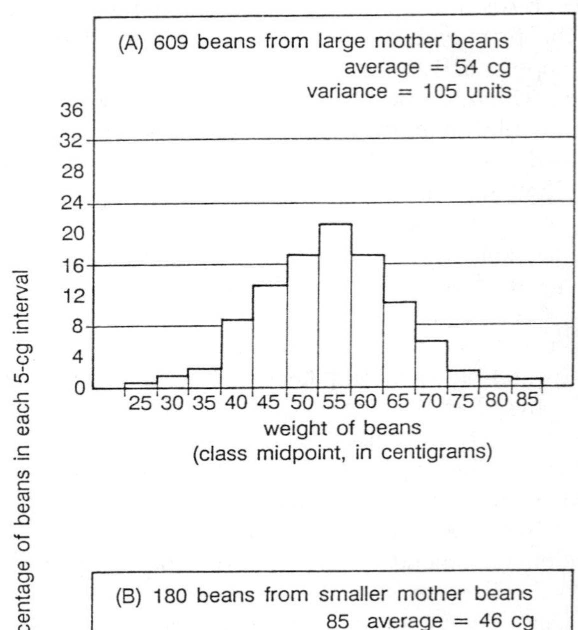

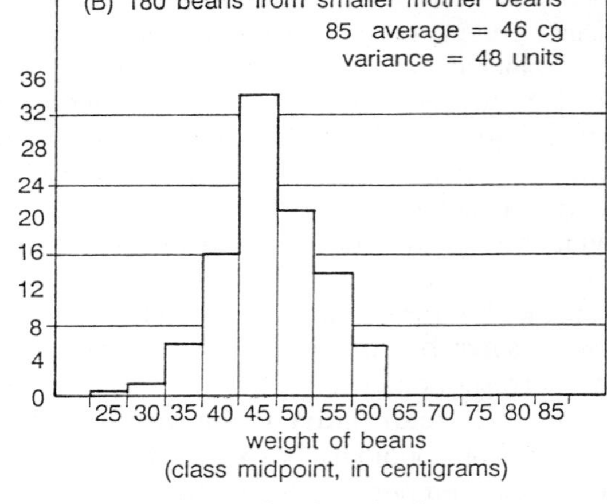

Figure 1 Variation in weight of garden beans, expressed as histograms. The data are taken from the classical work of Johannsen (1903). The horizonal axis marks off increasing weight classes in hundredths of a gram (cg, centigrams). The midpoint of each class is specified; for example, 35 cg includes all beans weighing between 32.5 and 37.5 cg. The vertical axis gives the percentage of beans within each weight class. For example, in (A), about 2% of the beans were in the 35-cg class. (A) The offspring from mother beans that weighed between 55 and 65 cg. (B) The offspring from mother beans that weighed between 15 and 25 cg. Note that the offspring beans in (B) not only are lighter in weight on the average but also are less variable. This can be judged visually by the shape of the histogram (taller and narrower) or by calculating the statistical quantity called *variance*.

The formulation $V_P = V_G + V_E$ assumes that different environments act similarly on various genotypes and that genetic and environmental sources of variation are independent of each other. These assumptions are not always met, however. The effect of smoking on the incidence of cardiovascular mortality, for example, shows *genetic–environmental interactions*. The observed cardiovascular death rate in smokers with a family history of heart attack is much greater than expected when considering the separate risks of smoking, on the one hand, or family history of heart attack, on the other (Khaw and Barrett-Connor 1986). As another example, consider the consequences of severely restricting the intake of the amino acid phenylalanine. This diet greatly benefits homozygotes for the PKU allele, but it provides no benefits—or may even bring harm—to others. To illustrate a *covariance* between genotype and environment, Crow (1986) uses the example of sending a musically talented youngster to a school of music. In all these cases, corrective terms must be included on the right side of the equation to take into account the associations between genes and environments, which are often difficult to identify and evaluate.

Heritability in the broad sense, H_b, expresses the proportion of phenotypic variation that is due to the genotypic differences among members of a population. Mathematically, it is the genetic variance divided by the total phenotypic variance, and it is expressed as either a fraction or a percentage:

$$H_b = V_G/V_P$$

Suppose that all members of a population have the same genotype, as is approximately true in an inbred line of mice. Then $V_G = 0$, and any observed variability in a trait must be attributed to environmental variation; thus the heritability in this restricted case would be zero. Conversely, suppose that all members of a population grew up under precisely controlled environmental conditions, in which case $V_E = 0$. Here any variability is attributed to genotypic differences, that is, $V_P = V_G$, so the heritability would be 100%.

In human populations it is not usually possible to control either the genotypes or the environments, and values for V_G and V_E must be estimated indirectly. This estimation can be done very roughly by measuring the correlations between the phenotypes of relatives. The correlations measure the similarities between, say, parents and offspring or between pairs of siblings. Twins, either identical or fraternal, can also be used. But keep in mind that any estimate of heritability applies *only* to a particular population in a particular set of environmental conditions and may not hold in other situations or other times. These matters are discussed further in later sections.

Another quantity, **heritability in the narrow sense, H_n,** is harder to understand, but it is crucial to plant and animal breeders. Just as *environmental* variation can be separated into various components (differences in sunlight, water, soil nutrients, etc. for growing plants), breeders identify three major components of *genetic* variability:

$$V_G = V_A + V_D + V_I$$

V_A is called the **additive** component of the genetic variance, V_D is the **dominance** component, and V_I the **interaction** component. These components have no easy interpretation but are defined in terms of the frequencies and phenotypic values of the genotypes in a population. The additive component of variance is the main determinant of the resemblance between relatives and is the major factor in the response of agricultural or other species to selection. The dominance component (for a single locus) is zero if the heterozygote is midway in phenotype between the two homozygotes—when, for example, the heights of genotypes of *t/t*, *T/t*, and *T/T* are, say, 160, 162, and 164 cm, respectively. To the extent that alleles of a gene do not behave in the additive sense, there is an increasing dominance component of genotypic variability. The interaction component measures epistatic effects and other types of interactions between different genes. If this component is zero, then genes are said to act additively between loci.

Heritability in the narrow sense is defined as

$$H_n = V_A/V_P$$

H_n is the proportion of variation in a population that is due to just the additive portion of the genetic differences among the members. H_n is less than or equal to H_b, and it provides breeders with the best prediction of offspring phenotypes from knowledge of parental phenotypes. For example, breeders can more easily alter through selective mating the butterfat percentage of cow's milk (H_n for *this* trait = 60%) than they can increase the total yield of milk (H_n = 30%). To the extent that the best individuals achieve that position by a favorable environment, or to the extent that the effect of a desirable or undesirable allele is masked by dominance or epistasis, artificial selection can make no predictable headway. Further considerations of the concept of heritability are explained in Loehlin et al. (1975), Bodmer and Cavalli-Sforza (1976), and Falconer (1981).

THE GENETIC COMPONENT

During the first decade of this century, the Danish botanist Wilhelm Johannsen determined the separate contributions of genetic and environmental variables to seed weight in garden beans. (In doing so, he coined the terms *gene*, *genotype*, and *phenotype*.) Somewhat later, investigators in Sweden and in the United States showed that Mendelian rules were adequate to explain continuous variation in the seed color of wheat and in the flower dimensions of the tobacco plant. The explanations supposed that several genes affected the trait in question, each by a small amount. This type of heredity was called **multiple gene inheritance**, or **polygenic inheritance**. (Note that the idea of multiple *genes* is quite different from that of multiple *alleles*. The former refers to many genes affecting one phenotypic characteristic; the latter to many alleles of one gene.) Below, we present a polygenic model for the inheritance of any quantitative trait, ignoring for the moment environmental influence.

An Additive Model

In an attempt to analyze genetic variability in a quantitative trait, let us assume that

1. The trait is controlled by three unlinked genes, each with two alleles: *G*, *g*; *H*, *h*; and *I*, *i*.
2. Each gene affects the trait in the same way with no interaction between loci.
3. The alleles of each gene act additively.

Suppose, also, that the genotype *g/g h/h i/i* has the phenotype 42 units (in combination with a standard environment and a constant genetic background at other loci). Assume that each uppercase allele adds 3 units beyond this. Thus the phenotype is completely specified by noting the number of uppercase alleles (Table 1). For example, a genotype with two uppercase alleles has the phenotype 48 units, regardless of whether the uppercase alleles are both at the same locus or at different loci. Altogether, there are 27 possible genotypes generated by combining any genotype at *G* (3) with any at *H* (3) with any at *I* (3), but only seven different phenotypes.

Using either the gamete-by-gamete method (checkerboard) or the gene-by-gene method for solving genetics problems, we can predict offspring phenotypes from any set of parental genotypes with the information in Table 1. For example, we ask what phenotypes, in what proportions, are expected among the progeny of two triple heterozygotes, *G/g H/h I/i*. Because the same phenotype may result from several different genotypes, some simplifications are possible. For example, a triple heterozygote can make eight types of gametes (see Question 3 in Chapter 11), but we need not resort to an 8×8 checkerboard. The gametes have 0, 1, 2, or 3 uppercase alleles, so a 4×4 checkerboard will do. The fractions of the gametes are worked out in Figure 2, and the relevant checkerboard is set up in Figure 3.

Another way to determine the fractions of offspring with each phenotype in this case makes use of the binomial formula (Chapter 10). The mating is such that any of the six genes (actually three genes with two alleles each) in an offspring is equally likely to be upper- or lowercase. The probability that an offspring will receive, say, two uppercase alleles is then,

$$\text{P(two uppercase alleles)} = \frac{6!}{2!4!}(1/2)^2(1/2)^4 = 15/64$$

The number six is the total number of alleles, of which we ask that two be uppercase and four be lowercase. The 1/2 fractions are the probabilities that any one allele will be uppercase or lowercase. By similar calculations, readers can confirm the probabilities for the other outcomes of this mating.

Table 1 An additive model of polygenic inheritance involving three genes

Representative genotype	*Number of uppercase alleles*	*Phenotype*[a]	*Number of different genotypes (same phenotype)*
g/g h/h i/i	0	42	1
G/g h/h i/i	1	45	3
G/G h/h i/i	2[b]	48	6
G/g H/h I/i	3	51	7
G/G H/H i/i	4	54	6
G/G H/H I/i	5	57	3
G/G H/H I/I	6	60	1
			sum = 27

a. We presume that each uppercase allele, *G* or *H* or *I*, contributes 3 units to the phenotypic measurement of 42 for *g/g h/h i/i*. The units may be length, weight, color grades, or whatever units the trait is measured in.

b. The 2 uppercase alleles can be at the same locus (*G/G* or *H/H* or *I/I*) or at different loci (*G* and *H*, or *G* and *I*, or *H* and *I*).

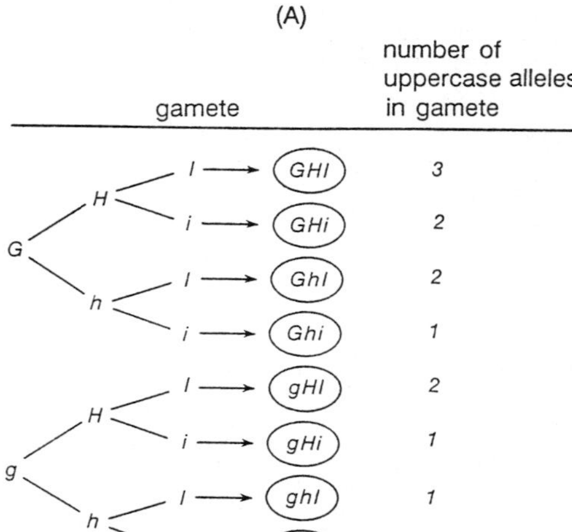

(B)

number of uppercase alleles in gamete	fraction
3	1/8
2	3/8
1	3/8
0	1/8

Figure 2 (A) The gametes made by the triple heterozygote *G/g H/h I/i*, and the number of uppercase alleles in each gamete. (B) The fractions of gametes having 0, 1, 2, or 3 uppercase alleles.

Readers may also wish to work out phenotypic distributions in offspring for some of the other matings involving three genes with additive alleles. You will discover that *the average of the offspring phenotypes is always the same as the average of the two parents*. This similarity is characteristic of genetic situations based on additive alleles, and provides important information to breeders. To the extent that V_P is due to V_A, they can be assured of producing, on the average, progeny just as superior as their hand-picked parents.

The additive model of gene action underlying quantitative traits may involve any number of genes. A more general formulation would also involve more than two alleles per locus and, perhaps, genes with one allele dominant to others at that locus. There are further complications that could be added to the model: linkage, epistasis, and the existence of one major gene and several modifiers.

If we assumed that a trait is also influenced by environmental factors, then the histogram of Figure 3 would become smoothed out and intervening phenotypic values (such as 43 and 44 units) would be filled up. The distribution would be broadened at the left and right by individuals with all lowercase or all uppercase alleles

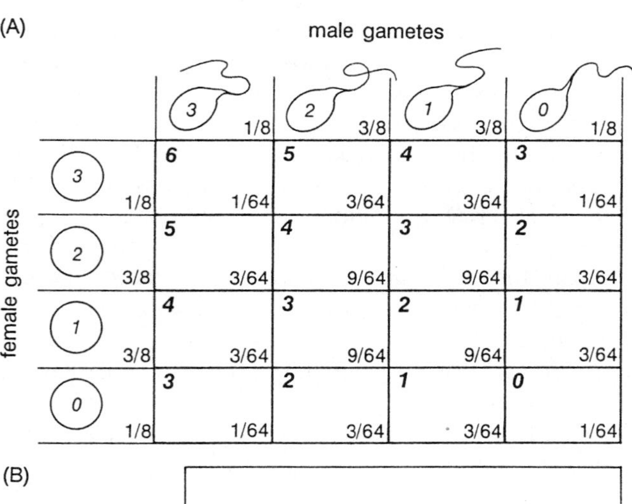

(A)
male gametes
female gametes
3 1/8
2 3/8
1 3/8
0 1/8
6 1/64
5 3/64
4 3/64
3 1/64
5 3/64
4 9/64
3 9/64
2 3/64
4 3/64
3 9/64
2 9/64
1 3/64
3 1/64
2 3/64
1 3/64
0 1/64

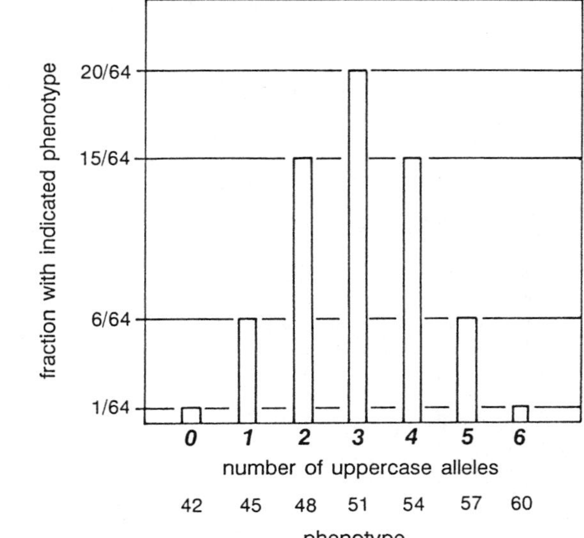

Figure 3 The offspring from *G/g H/h I/i* × *G/g H/h I/i* under the additive model of polygenic inheritance. (A) Checkerboard. The number of uppercase alleles are given within the gamete symbols. The total number of uppercase alleles in the zygotes (bold figures) and the corresponding fractions (products of row and column fractions) are given within the body of the checkerboard. (B) Summary of expected phenotypes. Zygotes with the same number of uppercase alleles are combined; the corresponding phenotypic value is obtained from Table 1. Note that the average offspring phenotype is 51, the same as that of the parents.

living in correspondingly extreme environments. Thus, as few as three additive genes, accompanied by environmental variations, can account for quite a range of phenotypes among the children in one family. Even more variation would be expected among the members of a large population.

Skin Color

The degree of skin darkness in people seems to have evolved in response to varying environmental conditions, thereby producing populations that are fairly homozygous for different genes controlling the rate of production of melanins. The mechanism of action of the control genes is not well understood, nor is the nature of the environmental agents responsible for selection. One suggestion, however, involves the ability of ultraviolet light to convert precursor substances into vitamin D under the surface layers of the skin. Lighter surface layers are conducive to greater vitamin D synthesis, compensating peoples living in far northern or far southern latitudes for weaker sunlight. Because vitamin D is needed for bone formation and its deficiency leads to rickets, strong selective pressures could exist if dietary vitamin D were restricted (Bodmer and Cavalli-Sforza 1976).

The American geneticist Charles B. Davenport (1913) studied skin color inheritance in the families of Black–White matings in Louisiana, Bermuda, and Jamaica. He suggested that *two* genes with additive alleles were involved (Table 2). His measurements of skin darkness were obtained by matching a skin patch to a continuous color scale, but his cutoff points between phenotypic classes were necessarily arbitrary. (The reader will also recognize that "white" skin is not the white of this page, nor is "black" the color of this ink.) Nevertheless, Davenport was able to show generally good agreement between his observations and the predictions of a simple two-gene model. Table 3 summarizes a small part of the data.

Table 3 Davenport's observations on the children of two doubly heterozygous parents

Phenotype of children[a]	NUMBER OF CHILDREN *Observed*	*Expected*[b]
White	3	2 (1/16)
Light	10	8 (4/16)
Medium	13	12 (6/16)
Dark	5	8 (4/16)
Black	1	2 (1/16)
Total	32	32 (16/16)

a. The parents were both products of White × Black matings; thus the children were the product of a medium × medium mating.

b. Following the methods of Figures 2 and 3 (or the binomial method based on the distribution of four alleles), the reader can confirm the expected results.

More recent studies have used a spectrophotometer to measure the percentage of light reflected from skin —the darker the skin, the less light reflected. Harrison and Owen (1964) measured skin reflectance of Whites and Blacks living in Liverpool, their first generation offspring (F_1), and the children from F_1 × Blacks, F_1 × Whites, or $F_1 \times F_1$. In agreement with other work, their analysis of the averages and the variances of the various groups suggested that more than two genes were involved, probably three or four (Cavalli-Sforza and Bodmer 1971). The investigators were able to calculate roughly the component of variance: V_E was about 35% of the total variance in skin color, V_A was about 65%, and other components of variance were essentially zero. Thus, in this population, in either the broad or narrow sense, the heritability of skin color was about 65%.

In summary, in the simplest cases the genetic contribution to variation in quantitative traits may be explained on the basis of several genes inherited according to Mendelian rules, with each gene having a small, additive, phenotypic effect.

Table 2 Davenport's hypothesis for skin color variations in Black–White crosses

Representative genotype	*Number of uppercase alleles*	PHENOTYPE[a] *Percentage black*	*Shorthand description*
a/a b/b	0	0–11	White
A/a b/b	1	12–25	Light
A/A b/b	2	26–40	Medium
A/A B/b	3	41–55	Dark
A/A B/B	4	56–78	Black

a. The percentage of black (third column) was measured by spinning a wheel that had movable sectors of black and white paper (and also red and yellow). The spun wheel was matched to the color of each person's upper arm; since this area was generally clothed, the variable effects of tanning by sunlight were minimized.

TWINS: THEIR USEFULNESS AND LIMITATIONS

Occurring in somewhat more than 1% of pregnancies, twins are always a source of family interest and general curiosity. But more often than single births, twin births cause pregnancy complications for the mothers. The twins themselves are subject to health-related problems as newborns (often premature) and as infants. Because identical twins share the same genotype, they provide human geneticists a little of what a highly inbred line gives to animal or plant geneticists. Twin data are the starting point for a more detailed analysis of the nature–nurture question.

The Biology of Twinning

There are two types of twins: identical or **monozygotic (MZ)**, and fraternal or **dizygotic (DZ)**.

Monozygotic twins derive from a single zygote (one egg fertilized by one sperm) that divides into two separate cell masses within the first two weeks of development (Figure 4). MZ twins are the same sex, their genotypes being identical except for possible somatic mutation (as is thought to happen with some regularity in the development of the cells of the immune system—Chapter 18). *Conjoined* twins may arise in those very rare instances when the cell masses remain partially joined.*

Dizygotic twins result from two zygotes (two eggs separately fertilized). DZ twins are like-sexed about half the time, and they have only half their alleles identical, on the average—the same as for sibs born at different times. Although DZ twins or sib-pairs could share *all* their genes, or *none* of their genes, depending on the vagaries of chromosomal segregation and crossing over, these extreme events are virtually impossible. We all know sib-pairs, nonetheless, who are very similar or very different in some aspects of their phenotype.

The diagnosis of MZ versus like-sexed DZ twins on the basis of physical appearances and mannerisms is usually, but not always, reliable. Visual appearances can be supplemented with information on blood groups and

*Such twins are often called *Siamese* after a famous exhibition pair, Chang and Eng, who were joined in the lower chest region by a tough, flexible ligament 3–4 inches thick and extending 5–6 inches between them. Born in Siam (Thailand) in 1811, the twins were bright, resourceful, and wry. At one New York performance, they refunded half the admission fee to a one-eyed man, because they said he could see only half what other viewers could. They eventually settled in North Carolina as farmers, married sisters, fathered a total of 21 children, and died within hours of each other at age 62 (Wallace and Wallace 1978).

Figure 4 An example of twins thought to be identical on the basis of physical traits observed by family and friends.

other traits for which one or both parents are heterozygous. Just *one* genetic difference between twins is enough to establish that they are DZ. When no genetic difference is found, the twins may be either MZ or DZ. The more traits examined *without finding a difference*, however, the more likely it is that the twins are MZ (see Bayes' theorem in Chapter 10). Because there are so many polymorphisms that can be examined, including restriction fragment length polymorphisms (RFLPs), monozygosity can be established with little doubt (assuming that the tests are performed accurately).

If, at birth, both twins are enclosed in a single chorion (outer embryonic membrane), they are always MZ. But the presence of two chorions is uninformative, because all DZ twin pairs *and* about a third of MZ twin pairs develop their own separate chorions. The number of placentas is also not informative, because placental fusion occurs whenever the twin embryos, MZ or DZ, implant near each other. In virtually all cases, the placental fusion is superficial, the blood vessel systems of the twins being completely separate.

The pattern of occurrence of MZ and DZ twinning can be examined in at least four ways (Bulmer 1970): by race; by maternal age; by family; and over time.

RACIAL VARIATION. The frequency of MZ twinning is remarkably constant throughout the world: about 4 MZ twin pairs per 1,000 pregnancies. On the other hand, the frequency of DZ twinning varies widely (Table 4), being greatest among African groups and least among Orientals. This pattern suggests that MZ twinning results from a random accident of early development equally likely to occur in any embryo. DZ twinning, however, is influenced by whatever environmental and genetic differences exist between different racial groupings.

MATERNAL AGE VARIATION. Nearly constant MZ rates but quite variable DZ rates are also observed as a function of maternal age (Figure 5). Mothers in this Italian population between 30 and 40 years have a higher rate of DZ twinning that those who are older or younger. It is evident that to produce DZ twins, mothers must ovulate two eggs in the same menstrual cycle. Ovulation is under the control of gonadotropic hormones secreted by the pituitary gland. Gonadotropin levels increase from adolescence through the entire reproductive period, paralleling the increase in DZ twin rates up to about age 37. The sharp drop after this age seems to be due to a general failing of ovarian function as menopause approaches.

A further indication of the pivotal role of gonadotropins on reproduction comes from women with certain infertility problems. Treatment with gonadotropins (or some other drugs) increases the number of eggs that are ovulated and that could be fertilized during any one menstrual cycle. Especially striking is the high frequency (10 to 40% in different studies) of twins, triplets, quadruplets, quintuplets, and even higher multiplicities that result from the use of various ovulation stimulants (Scialli 1986).

FAMILIAL PATTERNS. Monozygotic twinning shows no tendency to recur in the same mother or to run in

Table 4 Approximate twinning rates in different populations ranked by DZ rate

Population	TWIN PAIRS PER 1,000 PREGNANCIES[a] MZ	DZ
AFRICAN		
Nigerians	5	40
South African Blacks	5	22
United States Blacks	4	12
CAUCASIAN		
Italians	4	9
Swedes	3	9
United States Whites	4	7
ORIENTAL		
Koreans	5	6
Chinese	5	3
Japanese	5	3

a. Data are average values primarily from tables in Bulmer (1970).

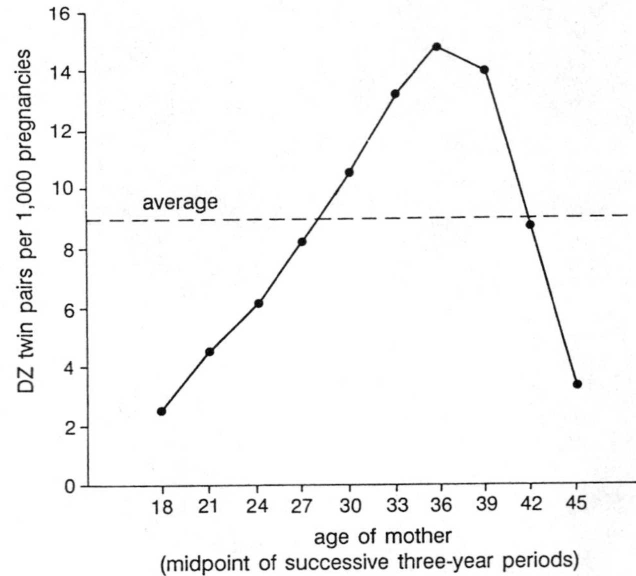

Figure 5 The dizygotic twinning rate by age of mother for Italian births, 1949 to 1965. The monozygotic twinning rate was almost constant at about 4 per 1,000 for all maternal age groups. (Adapted from Bulmer 1970.)

families. A woman with one set of DZ twins, however, is somewhat more likely to have another set of DZ twins than a "nontwinning" mother. Furthermore, the twinning rate among the close female relatives of a DZ-producing mother is slightly increased, a statistic suggesting a small genetic component for DZ twinning, but not for MZ twinning. The nature of the genes involved in DZ twinning is unknown, but they may act by increasing the level of gonadotropins. The genes are expressed only in females but can be transmitted by either sex.

TEMPORAL PATTERNS. Worldwide, dizygous twinning rates decreased significantly from the 1950s through the 1970s—as much as 40% in England and Wales. The reasons for this decline are unknown, but several hypotheses are given by Elwood (1985) and by James (1986b). An apparent leveling off of the decrease in recent years may be partly due to the increased use of drugs to treat infertility.

Twins in Genetic Research

If geneticists assembled inbred lines of mice, each line consisting of just *two* animals (and noninbred controls also in units of two), they would not have particularly good material for quantitative genetic analysis. Yet this is the type of situation that human geneticists must deal with in twin studies. Some information of interest can be obtained, but the relatively small numbers limit its reliability. Further difficulties arise because human geneticists have no control over the environmental variables they might wish to evaluate. Indeed, it is not always clear what environmental factors are relevant to variation in complex traits such as intelligence.

The analysis of twin data for a quantitative trait, such as adult height, can proceed in the following way: For each twin pair, the difference in height is obtained. Then the data from all MZ twins (or from all DZ twins, *like-sexed* for a proper comparison) can be combined into a single figure that is the average of these differences (Table 5). Alternatively, one can compute the variance statistic: V_{MZ} (or V_{DZ}). Both types of calculation express how much two MZ (or DZ) twins differ from each other, on the average. The application of twin data to the nature–nurture question then involves these quantities and how they are related to the components of variability V_E and V_G in a typical human population.

Because identical twins have the same genotype, whatever variability exists between them must be due to environmental differences. Therefore, the quantity V_{MZ} is related to V_E. These variances are not equal, however, since environmental influences on identical twins are quite different from those affecting *random, unrelated* individuals, the basis of the V_E statistic. The same family environment and similar experiences based on being the same age suggest that V_{MZ} is less, to an unknowable degree, than V_E.

These difficulties are minimized but not eliminated by examining identical twins who have been raised apart. Four studies have each included a dozen or more such twin pairs: an ongoing study of personality by Thomas J. Bouchard, Jr. at the University of Minnesota with about 44 pairs (see Holden 1987), an older study by Horatio Newman and colleagues at the University of Chicago with 19 pairs, a British study by James Shields with 44 pairs, and a Danish study by Niels Juel-Nielsen with 12 pairs (see references in Rose 1982).* But even here, V_E is not the same as the variance between MZ twins raised apart. For example, over half of the separated MZ twins studied by Shields were raised in families related to each other. In addition, adopting families generally tend to be at least middle class in socioeconomic level. Thus, although

*Throughout this chapter, we exclude the publications of the English psychologist, Sir Cyril Burt (1883–1971), who was once much admired for his intellect and expertise. Although he could be charming and generous, Burt was also cantankerous and jealous of his successful colleagues. After personal setbacks late in life, he spoiled a distinguished career by perpetrating several scientific frauds, including fabricated research on MZ twins raised apart. The data were supposedly collected by two shadowy assistants—apparently real people, but certainly not involved in Burt's alleged twin studies during his later years (Hearnshaw 1979).

Table 5 A comparison of twin pairs, or sib pairs, for three quantitative traits

	MZ TWINS			
	Raised together (50 pairs)	*Raised apart (9 pairs)*	*DZ twins (50 pairs)*	*Sibs (50 pairs)*
Height	1.7 cm	1.8 cm	4.4 cm	4.5 cm
Weight	4.1 lb	9.9 lb	10.0 lb	10.4 lb
Stanford-Binet IQ	5.9	8.2	9.9	9.8

Source: Newman et al. (1937).
Note: The figures are simply the average *differences* within the pairs.

the individual case histories of twins raised apart make fascinating reading (Rosen 1987), the collected data are not extensive and, consequently, do not provide very precise conclusions.

The difference in variances, $V_{DZ} - V_{MZ}$, is somewhat related to V_G. The relationship is based on the fact that DZ twins differ in both genotype and environment, whereas MZ twins differ in environment only. At first glance the subtraction of one from the other should therefore cancel out the environmental variation, leaving just the genetic component. But there are good reasons why this difference is not equal to V_G, the genetic variance between *random* persons. One is that the genetic difference existing between fraternal twins, or sibs, is clearly not as great as that existing between unrelated persons. In addition, the environmental influences on the two types of twins may not be equal, so the subtraction would not cancel that source of variability. For example, parents and friends sometimes treat MZ twins more uniformly, exposing them to more similar experiences, than they treat like-sexed DZ twins. Scarr and Carter-Saltzman (1983) argue that this difference is not important, whereas Lewontin et al. (1984) argue the opposite viewpoint.

Recognizing the difficulties outlined above, a quantity similar to heritability in the broad sense can be obtained from twin data by substituting $(V_{DZ} - V_{MZ})$ for V_G, and substituting V_{MZ} for V_E:

$$\frac{V_G}{V_G + V_E} \approx \frac{(V_{DZ} - V_{MZ})}{(V_{DZ} - V_{MZ}) + V_{MZ}} = \frac{V_{DZ} - V_{MZ}}{V_{DZ}}$$

The quantity on the right is called simply the H statistic, and it is a measure of the genetic portion of the variation between DZ twins (Cavalli-Sforza and Bodmer 1971). The value of the H statistic for height, for example, is about 84% (using data from Newman [1937]).

Twin data applied to many quantitative traits confirm that most of them are influenced by variation in both genotype and environment. For the traits in Table 5, it appears that the genetic component of height is greater than that of weight, which is greater than that of IQ, and this conclusion is borne out by many other sets of data as well. In summary, one can, at the least, state the following: To whatever extent MZ twins differ, environmental variation is a factor; to whatever extent DZ twins differ more than MZ twins, genetic variation is a factor.

THRESHOLD TRAITS. Some phenotypes have the following attributes. They

1. are discontinuous, being either present or absent
2. appear to have a genetic component
3. do not show simple Mendelian inheritance

Examples of such traits include a number of birth defects, such as *cleft lip/cleft palate* (separately or together); behavioral abnormalities, such as *schizophrenia*; and metabolic disorders, such as *diabetes* (Table 6). As in the case of quantitative traits, it is likely that multiple genes as well as many environmental influences are involved. The discontinuity in expression—the trait being present or absent rather than being continuously variable—is explained by the concept of a **threshold**. This explanation implies that an essentially normal phenotype results from many different gene–environment combinations; beyond a critical accumulation of (usually unknown) genetic and environmental stresses, however, the phenotype is abnormal.

The genetic component in threshold traits is indicated, in part, by twin studies. Twins are said to be **concordant** if they both have the trait in question, and **discordant** if one has the trait and the other does not. For the traits in Table 6, the concordance among MZ twins is 35–95%, whereas the concordance among DZ twins is 5–25%. Although some investigators have disputed the role of genes in the occurrence of schizophrenia, this position seems no longer to be tenable with the discovery of a mapped locus predisposing to schizophrenia in some famillies (see next section). Data on adopted children also support a genetic component in schizophrenia. For example, adopted children are more likely to develop mental disorders if their biological mother was schizophrenic than if their biological mother was not (Diederen 1983). For the two types of diabetes noted in Table 6, different MZ twin concordances have confirmed that the diseases' underlying genetic systems are quite different (Hrubec and Robinette 1984).

BEHAVIORAL TRAITS

Trying to sort out genetic and environmental influences on human behavioral traits, such as schizophrenia or intelligence, is especially difficult. One problem, noted previously, is the lack of control over important sources of environmental variation. In addition, some behavioral traits lack objective standards of measurements that are agreed upon by all investigators.

A further difficulty with regard to human behavior is the ideological frame in which the nature–nurture question is often put. Studies of IQ, for example, are construed to support either a *hereditarian* or an *environmental* outlook on society, and then perhaps used to influence political policy or the structure of social institutions. Hereditarians emphasize the role of genes in fixing the limits of behavioral characteristics, whereas environmentalists see the newborn as much more pliable, to be shaped in large degree by societal forces.

Because the debate is sometimes acrimonious, both

Table 6 Three threshold traits

Trait	CONCORDANCE (%) *DZ*	*MZ*	*Phenotype*
Cleft lip/cleft palate	5	35	A variable abnormality affecting the formation of the upper lip and the roof of the mouth during gestation. As a consequence, there is a verticle fissure in the midline of the lip (25% of cases), palate (25%), or both (50%). Modern surgery can usually correct the defect functionally and aesthetically. Although this condition often accompanies other syndromes, its frequency by itself is about 1 per 1,000 births. (Cohen 1983.)
Schizophrenia	15	45	A variable psychiatric disorder sometimes accompanied by delusions and hallucinations. Patients may lose all interest in their surroundings or may make illogical and inappropriate responses to persons or happenings. The diagnostic criteria vary, and frequency estimates range from about 1/80 to 1/200 among Americans. Biochemical and structural differences between the brains of schizophrenics and normal persons are being actively investigated. (Nicol and Gottesman 1983; Barnes 1987a.)
Diabetes mellitus			Both types result in high levels of glucose in the blood (see also Chapter 12). In Type 1, affecting young people, an insulin deficiency prevents glucose from entering cells. The deficiency stems from destruction of the insulin source, the beta cells of the pancreas. The frequency is about 1/400. In Type 2, affecting older people, insulin receptors at the cell surface do not seem to function properly. Obesity is an important predisposing factor in Type 2, and the frequency is about 1/150. Both types may be characterized by constant thirst, frequent urination, vision difficulties, slowly healing sores, and coma. (Rotter and Rimoin 1983.)
Type 1	5	50	
Type 2	25	95	

inside and outside the academic community, it is important to understand that hereditarians are not necessarily, or even often, racists, nor are environmentalists necessarily, or even often, utopian dreamers. Common sense suggests that neither dogma can be 100% true. It is very simple to find instances of critical environmental influences on behavior. For example, whether one is a smoker, or a nonsmoker, or an ex-smoker depends in part on social pressures. But it is also easy to identify at least some behavioral phenotypes that are clearly controlled by different genotypes. Before looking into the IQ problem per se, we note some of these gene-influenced behaviors.

A Genetic Component to Human Behavior

Behavior is what people do and how they do it. The mental or muscular responses by which such actions are expressed depend on environmental inputs to our sense organs. The various stimuli are coordinated and controlled by immensely complicated networks of nerve impulses and by hormonal messengers. Our activities are therefore based on aspects of anatomy and physiology that we know can be affected by gene-controlled structural proteins or enzymes. It should therefore be expected that genetic variation can lead to behavioral variation (Koshland 1987).

We have already noted several simply inherited traits that have a behavioral component as part of the phenotype, and Ehrman and Parsons (1976) list several dozen others. One of the most distressing is the Lesch-Nyhan syndrome, which is caused by an X-linked recessive gene (Chapter 16). The normal allele controls the production of an enzyme, HPRT, needed in the metabolism of purines. Boys hemizygous for a mutant allele of this gene compulsively bite and mutilate their lips and fingers. How the enzymatic abnormality brings about the bizarre behavioral component is still a mystery.

Another example of a single-gene behavioral phenotype involves *porphyria*, which may have affected King George III of England (1738–1820). The unusual abnormalities of this rare autosomal disorder stem from a defect in the synthesis of heme, the iron-containing part of the hemoglobin molecule (Chapter 16). The primary *physical* symptoms are abdominal pain, constipation, vomiting, and wine-red urine. (Porphyria derives from a Greek word meaning "purple.") In the case of George III, *neurological* symptoms were first noted when he was age 50: The king was afflicted with visual disturbances and restlessness. Then after three weeks he became delirious, with convulsions and a prolonged stupor (Macalpine and Hunter 1969). The king recovered from this bout but suffered increasingly from several subsequent attacks of what was then diagnosed as insanity. The illness of George III provided the early impetus for research in psychiatry.

More recently, a dominant gene that confers a predisposition to the psychiatric illness, *manic depression*, was mapped near the tip of the short arm of chromosome 11 by linking it to an RFLP site (Egeland et al. 1987). Affecting as many as 1% of Western peoples at some time during their lives, manic depression is characterized by large mood swings. On the one hand, patients may have racing thoughts, inflated self-esteem, and delusions of grandeur. On the other hand, there may be lethargy and feelings of worthlessness and guilt, extending to suicide. The inheritance of a causative gene, having about 60% penetrance by age 30, was traced in a group of Old Order Amish. (Their large families, clearly established paternity, and prohibitions against alcohol and drugs aided in the establishment of pedigrees and diagnoses—see Chapter 21.) Manic depression is, however, clearly heterogeneous in its biological and environmental components. In an Israeli population, for example, a different predisposing gene is not on chromosome 11 at all, but maps near the color blindness locus on the X chromosome. In yet other populations, the disease is not linked to either of these chromosomal positions (references in Robertson 1987).

In a similar fashion, a dominant gene that confers a predisposition to *schizophrenia* has been mapped near RFLPs on the long arm of chromosome 5 (Sherrington et al. 1988). The illness often develops in the late teens, and affects about 1% of people over the course of their lifetimes. Symptoms include hallucinations and delusions, inappropriate emotional responses, and inability to concentrate. The predisposing chromosome 5 allele was present in several British and Icelandic families, and may also be involved in other psychiatric conditions within these pedigrees. On the other hand, in several Swedish families with schizophrenia, the disease was not associated with any of seven RFLP sites on chromosome 5 (Kennedy et al. 1988). These results again illustrate heterogeneity in the causation of complex mental characteristics.

The peculiar language problems known as *dyslexia* may also have genetic components. Dyslexia constitutes a group of disorders in which otherwise normal children have extraordinary difficulties learning to read and write (Vellutino 1987). One form of dyslexia that runs in families may be caused by a dominant gene near the centromere of chromosome 15 (Tasset et al. 1988).

If single-gene mutations can affect behavior in obvious ways, there is no reason to doubt that multiple genes with subtle effects can affect various mental processes, even if specific loci and their products cannot be identified. For example, it is thought that extraordinary combinations of (unknown) genes may have contributed to the abilities of the great computational and mathematical marvels who received little or no formal education. The first of the incredibly fast human calculators who performed on stage was Zerah Colburn (1804–1840), the son of a poor Vermont farmer. With only a few weeks of schooling and not yet able to read, the five-year-old boy (playing on the floor near his father's workbench) was heard to quickly multiply numbers (Colburn 1833). Among the calculating feats demonstrated on European tours, the boy almost instantly multiplied two four-digit numbers, but hesitated a bit on five-digit numbers. Colburn was ordinary in other ways except that—like a brother, his father, and a great-grandmother—he was polydactylous, with six fingers on each hand and six toes on each foot (Gardner 1975).

More incredible yet was Srinivasa Ramanujan (1887–1920), the son of a poor accountant in southern India, whose inborn creative ability continues to astound even today's mathematicians (Kolata 1987, Borwein and Borwein 1988). Ramanujan left behind thousands of sophisticated formulas and theorems, without proofs, many of which have come to be important in modern number theory, physics, and computer science. Although he completed high school, his obsession with mathematics twice caused him to fail in college, and indeed led him to ignore all other activities, including eating and sleeping. His genius was recognized by G. H. Hardy (of Hardy-Weinberg fame) on the basis of an unsolicited letter containing complex, unproved results. After several years at Cambridge University with Hardy, he returned to India and died of what appears to have been severe vitamin deficiency.

In addition to gene mutations, gene imbalance (as occurs in chromosomal aberrations) can also have important effects on behavior. Recall that Down syndrome patients have distinguishing personality features, such as friendliness, joviality, love of music, and ability as mimics. Patients with Turner syndrome also exhibit an interesting behavioral trait: although the distribution of their IQ scores is not unusual, they tend to do very poorly on tests that require them to visualize forms in space (Money 1973). Furthermore, this peculiarity is unrelated to their test scores for other mental tasks. In school they often do poorly in arithmetic and mathematics despite otherwise normal or superior performance.

In summary, there are real genetic contributions to variations in behavioral traits, expressed on a continuing basis throughout life. The specific relationships between genotypes, environments, and the resulting phenotypes are often obscure, but possible neurological pathways for gene action are increasingly active areas of investigation.

Intelligence

All of us have a general idea of the meaning of intelligence (a dictionary synonym is "mental acuteness") by which we sometimes judge ourselves and others. Most psychologists emphasize abstract reasoning ability, which includes thinking rationally, solving problems, understanding the basics of a complex situation, and responding effectively to new environments more consistently than luck would allow. These meanings are sufficiently broad and applicable to such a wide variety of human endeavors that no obvious yardstick for measuring intelligence immediately presents itself. Apart from methodologies for trying to measure intelligence and ascertain the relative importance of nature and nurture, there is much interest in the relevant neural and biochemical networks, as well as in the evolution of intelligence in human and other primate species.

Galton, with his passion for measuring, appears to have made the first attempts to test mental abilities. At the 1884 International Exposition in London, he set up a booth where visitors paid threepence to be scored in several ways. Some of Galton's tests dealt with sensory perception (for example, accuracy in discriminating different weights) or with quickness of reaction, which he felt correlated with a person's intellect. Galton invented statistical tools, such as correlation analysis and percentiles, that are now much used in psychometrics, the measurement of mental abilities.

In Paris, beginning in 1904, Alfred Binet and his colleagues at the Sorbonne were given the task of identifying subnormal school children who might need special attention. Binet's easily administered tests proved to be fairly successful in predicting how well a child performed in subsequent schooling. The test items were verbal, numerical, or pictorial problems of increasing difficulty for older children. It is important to note that the selection of test items was determined empirically: questions that were poor predictors of future success were eliminated. Problems selected for, say, six-year-olds were those that many of them could answer correctly, but were too easy, on the average, for older children and too hard for younger children. Although Binet hoped to measure "general intelligence" rather than knowledge accumulated through schooling and other experiences, he recognized that test construction depended on trial-and-error standardization rather than on a theory of intelligence. His aim was to identify children who were not up to grade, so they could be provided with the special education needed to improve their position.

Psychologists at Stanford University modified the Binet procedure to evaluate White middle-class American children. Because the tests are commensurate with the children's ages, the results—IQ scores—depend on the rate of development of mental performance. The Stanford-Binet tests for six-year-olds, for example, contain vocabulary items, verbal problems of several types, numerical concepts, identification of missing parts of pictures, and maze tracing. Test items are so selected that, overall, boys and girls perform equally well. The scores are scaled to average 100 for each age group, and further adjusted to make the distribution approximately symmetrical about 100, with about two-thirds of the scores falling in the central range from 85 to 115. Gould (1981) gives a detailed history of IQ testing, including accounts of their misuse by the American eugenics movement.

Do IQ tests measure intelligence in any of its various meanings? No one doubts that the scores do predict reasonably well the future scholastic success of the children against whom the tests are standardized. That, of course, is what they were originally designed to do. In addition to the generalized Stanford-Binet, Wechsler, and Scholastic Aptitude Tests (SATs),* other quite different types of tests purport to measure one or another narrower aspect of mental ability. Educational psychologists point to the considerable degree of correlation between many heterogeneous tests and suggest that some common capability is being examined. The factor (called "g") common to these various forms of measurements is what some psychologists call intelligence. This conception of intelligence is derived from the statistical analysis of tests. It might be more satisfying if the flow of ideas were reversed so that test construction could be derived from an analysis of intelligence (Hunt 1983). Intelligence may be too complex a trait, however, to be defined by one or a few test scores.

We emphasize that intelligence is not a genotype; it is a *complex phenotype* that develops under the influence of genes and the experiences of a lifetime. An IQ test provides a simplified assessment of that phenotype at a particular time, and the results may change with age. Although IQ measurements are related to what it means to be intelligent, the results are not comprehensive measures.

*The various Wechsler scales, in addition to verbal tasks, include performance components (such as assembling an object) in order to judge aspects of mechanical ability. The Wechsler Adult Intelligence Scale (WAIS) is the most widely used IQ test for adults in the United States, having been standardized against age groups from 16 to 75 (Wechsler 1981). Many readers will be familiar with the SATs, which are taken by more than a million college-bound students every year (Jensen 1981).

HERITABILITY OF IQ. In this section we ask: To what degree do IQ variations *within* middle-class Whites (the group on whom the tests were standardized) arise from genetic differences? It is, of course, the measured IQ scores themselves that are at issue, not intelligence in its broader context. A quite different question is raised in the section following: To what degree does the difference in average IQ *between* Black and White Americans arise from genetic differences? The answers to the within and between questions need not be the same.

Many populations have yielded values for the heritability of IQ, although some individual studies are based on small numbers. A 1981 summary of 111 studies that met certain criteria is given in Figure 6. The data are presented as the correspondences (correlation coefficients) in scores between pairs of individuals on a variety of tests that purport to measure an aspect of intelligence. The pairs of individuals were raised either together or separately; they shared none of their genes (top line), or an increasing proportion of their genes (middle lines), or all of their genes (bottom two lines).

The statistical significance of these data is hard to assess, partly because the results in any one line are so heterogeneous. Overall, it appears that neither genetic nor environmental influences can be neglected. For example, if environmental influences were absent, then zero correlation should be found between unrelated persons reared together (first line); but the 23 studies summarized show that there is such an association. The last two lines in Figure 6 lead to a similar conclusion: the MZ twins raised together or apart do differ, and that difference must necessarily arise from *environmental* factors. On the other hand, the influence of *genotypic* factors on IQ is strongly suggested by the higher correlation between MZ twins reared together (about 0.84) than between either DZ twins reared together (0.57) or sibs reared together (0.45).

Estimating heritability values from correlations is complicated and involves assumptions about the study pop-

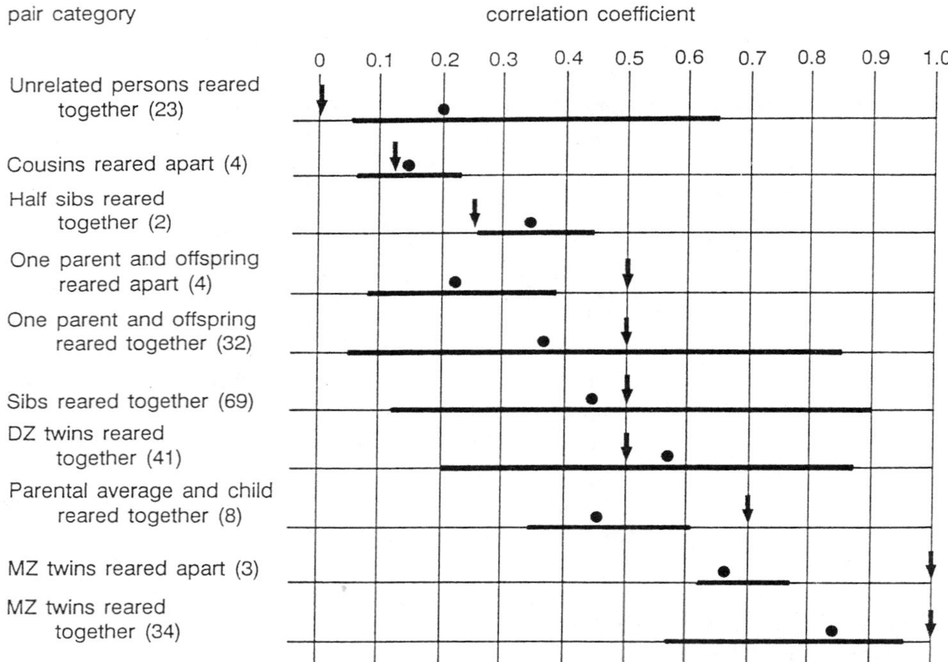

Figure 6 Correlation coefficients between pairs of individuals given a variety of IQ tests. The number of studies of each type of pairing is given in parentheses, and the range of correlation values is given by the extent of the the bar. The dot above a bar denotes the median correlation observed, and the arrow denotes the correlation that would be expected on the basis of a simple polygenic model with additive alleles. In rough terms, the similar trend of dots and arrows suggests the existence of genetic factors. That the dots and arrows are not superimposed suggests the existence of environmental factors. The first pair category, unrelated persons reared together, includes the following pairs: two adopted children, adopted and natural child, one parent and adopted child, and parental average and adopted child. (Redrawn and somewhat abridged from Bouchard and McGue 1981.)

ulation that may be only approximately true. Using an "oversimplified formula," Jencks et al. (1972) calculate from some of the data in Figure 6 that IQ heritability in the broad sense for the studied populations (H_b) is about 45%. They ascribe 35% of the IQ variability to purely environmental variations, and the remaining 20% to associations between genotypes and environments. This latter source of variation arises, for example, when parents with "favorable" genotypes not only transmit favorable alleles to their children, but also (because of the same genotypes) provide their children with an enriched environment conducive to achieving high IQ scores. Loehlin and DeFries (1987) suggest a higher value, about 30%, for this kind of genotype–environment correlation.

The 45% H_b figure is toward the low end of values reported for IQ heritability, and its derivation has been criticized in the comprehensive review of the IQ problem by Loehlin et al. (1975). They note that the shortcomings of both raw data and interpretations are such that a "sufficiently determined critic can always discover some grounds for rejecting any item of it that he happens not to care for." Most investigators wind up with values of H_b that average about 50%. For example, reviews of recent data by Plomin and DeFries (1980) and by Scarr and Carter-Saltzman (1983) conclude that about half of the differences in measured intelligence result from genetic differences among the particular American and European, White, middle-class populations that have been studied. These authors also note that this and related numerical values, by themselves, are only starting points in assessing how social and educational changes can help improve intellectual development.

We also wish to reiterate the limits in the meaning of heritability. These cautions have been set forth by many persons, including the University of California psychologist Arthur Jensen in the 1969 paper that sparked heated arguments over the race-IQ issue (next section).

1. *Population versus individual*. Heritability is not defined for an individual. It would make no sense to say, for example, that 60 points of a person's IQ score are due to his genes and 40 points to his environment. Rather, heritability is a measure of the genetic variability of persons within a population, valid only at the time of measurement. In groups with relatively homogeneous environments (low V_E), the heritability of a trait will naturally tend to be larger than for groups in a heterogeneous environment. Perhaps this tendency is one reason why heritability estimates among English populations are often higher than those estimated for the more diverse cultural and educational settings of the United States.

2. *Known versus unknown genes*. Heritability values do not depend on knowing the metabolic actions of whatever genes and alleles are involved. Without knowing about polypeptide products or neurological processes, one can state that IQ variations depend partly on genetic variations. Although it would be nice to understand the relevant biochemistry, the heritability values do not depend on this knowledge. Undoubtedly many genes with individually small effects influence the type of mental processes involved in IQ measurements. It has been calculated, for example, that rare recessive alleles at more than 300 loci may contribute to the decrease in IQ scores with an increase in inbreeding coefficients (Morton 1978). This IQ depression has been found in four separate studies.

3. *Constant versus changeable IQ*. Many times we have noted that an environmental change can modify the expression of a gene-influenced phenotype. As an additional example, consider the quantitative trait, stature. The gradual increase in human height in diverse ethnic groups over the last several generations (about 10 cm) is undoubtedly due to an improving environment (e.g, better nutrition, sanitation, and medical care), even though stature has very high heritability (Meredith 1976). Likewise, the significant heritability of IQ does not rule out environmental modifications of IQ. Although there is a high correlation between IQ scores for the same person taken at different ages (0.87; see Plomin and DeFries 1980), the correlation is not perfect. Furthermore, the correlation applies to existing circumstances of schooling and middle-class values, and novel circumstances might affect the IQ phenotype more. No one is predestined to have an assigned IQ value.

RACIAL DIFFERENCES IN IQ. In this area of investigation, the facts do not always speak for themselves. The same set of data may be interpreted by different scientists to mean different things, and these viewpoints may be oversimplified or exaggerated by journalists seeking eye-catching copy on scientific matters that impinge on public policy. But even *responsible* scientists can make things difficult for *responsible* writers. For example, the Stanford psychologist L. J. Cronbach (1975) has noted that the scientist "needs writing skills of an entirely unaccustomed order if he is to make sure that no unwanted implications will be drawn from a buried sentence, . . . and that no colorful aside will be remembered instead of his main message. We may rail against the journalist for relaying what we said instead of what we meant to say, but mindreading is not his job."

The major undisputed fact about the IQ scores among White and Black children is that the average racial difference is about 15 points (Jensen 1969). Nevertheless, there is considerable overlap, so the IQ score of a particular individual is no guide to his or her race, and vice versa. The reasons for the average IQ difference between Blacks and Whites are difficult to assess, and most investigators think that the underlying causes are unknowable on the basis of currently available data. The many interrelated environmental variables that could affect test scores include the following:

1. *Standardization*. Because Blacks were not included in the development or standardization of the Stanford-Binet tests, cultural differences could lead to lower scores for Blacks.
2. *Discrimination*. Social and educational opportunities that are available to Whites of any socioeconomic level have often been denied to Blacks. Predominantly Black schools, for example, tend to be inferior to predominantly White schools. Also difficult to measure are the effects of hundreds of years of racial prejudice, both blatant and subtle, that are rooted in slavery but continue to exist.
3. *Socioeconomic status*. One measure of socioeconomic status (SES) combines, into a single index, data on father's and mother's education, father's occupation, number of siblings, and reading materials and appliances in the home. Studies show that Blacks in general have lower SES ratings than Whites, and people with low SES ratings typically get lower IQ scores.
4. *Language*. This particular cultural difference between Blacks and Whites is especially important, because IQ tests are predominantly verbal. Designing tests to eliminate language and other cultural differences is difficult, and most investigators believe it has not yet been accomplished.
5. *Motivation*. What are the factors that allow a test-taker to do as well as possible? How are performances affected by subjects' self-esteem and by their perception of how the results will be used? How important are the attitudes and expectations of teachers, friends, and parents? These factors have not been fully evaluated.

The variables above suggest that some or all of the average IQ difference between Blacks and Whites could be due to *environmental* differences between the two ethnic groups.

The argument that at least some of the average IQ difference could be due to *genetic* differences flows from our concept of races: subdivisions of humankind that have come to differ, to a greater or lesser degree, in the frequencies of the alleles they possess. Blacks and Whites do differ in traits controlled by single genes—for example, hemoglobin variants and blood types. These differences have occurred through the processes of evolution—mutation, selection, drift, and migration. There is no reason to suspect that genes that affect behavioral processes could not also come to have different frequencies. But the idea that genetic factors for IQ performance *could* differ between races, certainly does not mean that they *do* differ.

Jensen (1980), suggests that most of the Black–White difference in IQ is due to genetic factors. His argument is that even when Black and White children are matched for environmental factors, most of the IQ difference remains. Others strongly dispute that the matching can ever be fairly done (Loehlin et al. 1975, Bodmer and Cavalli-Sforza 1970, Gould 1981, among others). Scarr and Carter-Saltzman (1983) summarize three studies that suggest that genetic differences are unimportant. For example, when Black children are adopted into socioeconomically advantaged White families, they do very well on standard IQ tests—as well as White children adopted into similar families. In short, the arguments for a genetic contribution to the average Black–White difference in IQ are weak.

Of course, the possibility or reality of an average genetic difference—or any other kind of difference—between groups should be inconsequential in a democratic society. Although many of the conclusions of Jensen (1969) have been divisive, he also notes that the full range of human talents is represented in all races, and that it is "unjust to allow the mere fact of an individual's racial or social background to affect the treatment accorded to him. All persons rightfully must be regarded on the basis of their individual qualities and merits."

Alcoholism

We conclude this chapter with a brief look at alcoholism, which people view variously as a biological problem, a psychosocial abnormality, or a moral weakness. Occurring in about 10% of American men and 4% of American women at some time during their lives, alcohol addiction can have disastrous consequences. The annual toll in the United States includes more than 50,000 deaths and $100 billion in costs (Desmond 1987, Reich 1988). The development of an alcoholic is certainly influenced by interacting personal, social, and cultural factors. The existence of some genetic variation as well is supported by finding substantial risks for children of alcoholics even when adopted into nonalcoholic families, and by finding higher concordance in MZ than in DZ twins in some, but not all studies (Goedde and Agarwal 1987).

The major psychological effect of alcohol intoxication is depression of the central nervous system, producing uninhibited behavior, poor judgment, and mood changes, as well as incoordination, slurred speech, blackouts, and coma (Kissin 1982). In advanced stages of addiction, alcoholics are obsessed with alcohol to the exclusion of everything else—family, friends, work, food—all the while denying they have a problem. Long-term alcohol abuse is associated with many degenerative diseases, especially of the liver, but also of the stomach, esophagus, pancreas, heart, and brain. In affected women who drink heavily during pregnancy, alcohol diffusing across the placenta can produce the *fetal alcohol syndrome*, the symptoms of which range from mild to profound physical and mental abnormalities (Streissguth 1986).

Three separate adoption studies were undertaken in the late 1970s in Denmark, Sweden, and the United States (references in Goodwin 1985). The results were generally consistent: the adopted sons of alcoholics were several times more likely to be alcoholic than were the adopted sons of nonalcoholics. The Swedish study (Table 7), for example, investigated all persons who were born out of wedlock within a 20-year period in Stockholm and who were adopted at a very early age by nonrelatives. The table shows that when adopted males had an alcoholic (biological) parent, they were 2.3 times more likely to develop alcoholism than when they had nonalcoholic (biological) parents. Later reevaluation of this study suggested that genetic factors play a more important role in those cases where the parent was severely alcoholic than when mildly alcoholic (Cloninger et al. 1981). Alcohol abuse thus appears to have heterogeneous causations and expressions, both genetic and environmental. The lower rate of alcoholism among females is not well understood but may be due in part to societal roles, to a lower rate of identification, or to biological factors related to the metabolism of alcohol.

Researchers have tried to find physiological or neurological characteristics that would identify those at risk for alcohol abuse, but positive results have not been confirmed. For example, Begleiter et al. (1984) discovered certain brain wave abnormalities in alcoholics who were presented with visual stimuli. These same abnormalities (in the so-called P3 component) were present in the young, nondrinking sons of alcoholic fathers (but not present in a matched control group). Other investigators have found different brain wave changes, but not generally in the P3 component (references in Peele 1986).

Enzymes that have been examined include two in the liver, alcohol dehydrogenase and aldehyde dehydrogenase, that break down alcohol via the pathway:

$$\text{alcohol} \xrightarrow{\text{alcohol dehydrogenase}} \text{acetaldehyde} \xrightarrow{\text{aldehyde dehydrogenase}} \text{acetate}$$

(The acetate is then metabolized to carbon dioxide and water, with the generation of considerable energy in the form of ATP.) No consistent differences have been demonstrated between alcoholics and nonalcoholic controls with regard to these enzymes, but an interesting racial trait has come to light. More than 80% of Orientals possess a variant, slow-acting form of the second enzyme, aldehyde dehydrogenase. It is thought that this difference (leading to a buildup of acetaldehyde) accounts for the marked facial flushing, accelerated heart rate, and other symptoms of distress that many Orientals experience upon drinking even small amounts of alcoholic beverages (Agarwal and Goedde 1987).

More recently, Tabakoff et al. (1988) investigated two enzyme systems present in blood platelets. (It is believed that the same enzymes operate similarly in brain cells.) In one set of experiments, the in vitro stimulation of the enzyme adenylate cyclase* was significantly less effective in platelets taken from alcoholics than in platelets from nonalcoholics. The lessened activity persisted even among alcoholics who had abstained from one to four years. It is not yet known, however, whether the abnormalities in platelet enzymes represent an underlying genetic predisposition to alcohol abuse or a consequence of many years of heavy consumption.

*Called a *second messenger*, adenylate cyclase is present in cell membranes and helps to transmit the effects of a *first messenger* (a hormone or neurotransmitter) from outside to inside the cell. The first messenger binds to a cell-surface receptor and thereby activates the second messenger, which, in its turn, acts upon ATP, the molecule that controls energy transformations within the cell.

Table 7 Alcoholism among male adoptees

Biological parent	*Number in sample*	*Percentage of adopted sons[a] who were alcoholic*	
Alcoholic father	89	39.4	avg = 34.0%
Alcoholic mother	42	28.6	
Nonalcoholic father	723	13.6	avg = 14.6%
Nonalcoholic mother	1,029	15.5	

Source: Bohman (1978).

a. Swedish males born illegitimately between 1930 and 1949, adopted as infants by nonrelatives, and followed through 1972. Alcoholism was determined by official records of alcohol-related fines, arrests, clinic visits, hospitalizations, and so on.

SUMMARY

1. Many physical and behavioral traits vary continuously over a wide range. The observed phenotypic variability for these and other traits result from both genetic and environmental influences.

2. In the broad sense, heritability of a trait is the proportion of phenotypic variation that is due to all genetic differences. In the narrow sense, heritability measures just the additive portion of the genetic variability. Narrow sense heritability is the best gauge of how effective selection can be in altering the average phenotype of a population over the course of generations.

3. A value for the heritability of a trait applies to the particular population in which it was measured and may not hold for the same population at other times, or for other populations. A high heritability estimate in a specific population does not rule out the possibility of altering phenotypes by novel environmental changes.

4. The genetic component of quantitative variation involves the cumulative effects of several genes that are inherited according to Mendelian rules. Analysis of Black–White crosses, for example, suggests that skin color differences are due in large part to 2–4 genes with additive alleles.

5. Monozygotic twins are genetically identical, whereas dizygotic twins are equivalent to sibs born at the same time. Apparently unaffected by genetic factors, MZ twinning seems to result from random events of early development. On the other hand, DZ twinning tends to run in families, vary between racial groups, and be affected by levels of maternal sex hormones.

6. Twin data provide only rough estimates of genetic and environmental contributions to quantitative and threshold traits. Environmental influences are suggested whenever MZ twins differ. Genetic influences are suggested whenever (like-sexed) DZ twins differ among themselves more than do MZ twins.

7. Variations in many human behavioral traits are influenced by genetic differences, although specific genes cannot always be identified and the pathways between the gene products and the behavioral phenotypes are usually obscure.

8. IQ tests were originally developed in France in order to identify school children who needed help, and IQ testing continues to be useful for that purpose. American researchers in this century have developed a large variety of tests to gauge different mental factors. The IQ tests have been standardized by trial-and-error methods rather than by theories of intelligence. Mental abilities are probably too complex to be adequately represented by one or a few scores.

9. Within White, middle-class populations (on whom most tests have been standardized), heritability of IQ, in the broad sense, is about 50%. These calculations are based on correlations in IQ between pairs of individuals of various degrees of relatedness and involve assumptions that may be only partly true. The meaning of heritability is limited —see item 3 above.

10. The average IQ difference—as judged by the standardized tests—between American Blacks and Whites is about 15 points. Some or all of this difference could be due to environmental and cultural differences between races. Some could reflect genetic differences, but the evidence for this is weak.

11. Like other behavioral traits, alcoholism develops under the varied influences of both environmental and genetic factors. The nature of the genetic factors so far remains obscure.

FURTHER READING

Broad accounts of quantitative inheritance are found in Crow (1986), Bodmer and Cavalli-Sforza (1976), Cavalli-Sforza and Bodmer (1971), and Falconer (1981). General behavior genetics is covered in the textbook by Plomin et al. (1980), and in a more popularized summary by the Staff of Research and Education Association (1982). A review of many aspects of human behavior genetics is by Henderson (1982), and some specialized topics, including schizophrenia, IQ, and the evolution of behaviors, are in the collection edited by Fuller and Simmel (1983). The textbooks by Gleitman (1986) and by Anastasi (1988) provide background information on psychology and psychological testing.

Material on intelligence testing and its interpretation

is voluminous, varied, and sometimes acrimonious. The reader should sample both Jensen (1980, 1981) and his detractors (Gould 1981, Lewontin et al. 1984). We also recommend the thorough and balanced book by Loehlin et al. (1975). McKean (1985) has written an excellent popular article on some recent theories of intelligence, and Scarr (1987) includes a personal account of being involved in a controversial field. We have not considered in this chapter genetic aspects of personality traits, but this area of investigation is likely to develop (Holden 1987).

QUESTIONS

1. Which population would tend to show the higher heritability for trait Z: (a) One that is relatively homozygous for genes affecting Z or one that is more heterozygous? (b) One that is in a relatively uniform environment for factors affecting Z or one in a more heterogeneous environment?

2. Many of Galton's views on inheritance were vague or incorrect, but his law of filial regression is often observed for quantitative traits. Using weight as an example, this law states that offspring of very heavy parents are, on the average, not so heavy, and the offspring of very light persons are not so light; that is, a portion of the filial generation that might be expected to be extreme is closer to the population average. Assuming that weight is affected by both genetic and environmental factors, how would you explain filial regression?

3. In Table 1, what six genotypes have phenotype 54 (four uppercase alleles)?

4. Although the inheritance of eye color has not been worked out, it certainly involves more than one gene. Galton (1889) distinguished eight eye colors; by combining his two darkest categories, we can fit his scheme to the model of polygenic inheritance in Table 1:

Galton's classification	Assumed number of uppercase alleles
light blue	0
blue	1
blue-green	2
hazel	3
light brown	4
brown	5
dark brown and black	6

(a) What are the genotypes of the lightest-eyed parents who could produce a child with brown eyes (5 uppercase alleles)? (b) Assume that anyone with 0–2 uppercase alleles is said to be blue-eyed and anyone with 3–6 uppercase alleles is said to be dark-eyed. (Eye color thus becomes a threshold trait.) What is the darkest-eyed child possible from two "blue-eyed" parents?

5. Under Davenport's model of skin color inheritance (Table 2), are these statements true or false? (a) When one parent is White (0 uppercase alleles), the progeny can be no darker than the other parent. (b) When one parent is light (1 uppercase allele), the progeny can be no darker than the other parent.

6. Using Davenport's model of skin color inheritance, list the phenotypic expectations for offspring from medium × medium matings, when (a) both parents are homozygous for both genes? (b) one parent is homozygous for both genes, and the other is heterozygous for both genes? (c) both parents are heterozygous for both genes?

7. Use Figure 5 to judge roughly how much more likely an Italian mother is to have DZ twins than MZ twins (a) at age 20, (b) at age 39.

8. Race and Sanger (1975) summarize 20 twin pairs in which red blood cells of two different types were present within each twin. Data on Case 10 is given below. Were the twins MZ or DZ? What were the parental genotypes? Suggest how the bloods became mixed.

Twin	O, Rh negative	AB, Rh positive
female	90%	10%
male	35%	65%

9. It has been postulated that an egg and a polar body may (very rarely) be separately fertilized by two sperm to yield an unusual type of twin pair (Bieber et al. 1981). Comment on the degree of similarity of these twins compared with regular DZ or MZ pairs. Assume that one sperm fertilized the mature egg, and the second sperm fertilized the second polar body (sister to the egg). Also assume, for simplicity, no crossing over during meiosis.

10. Below are the Stanford-Binet IQ scores of the 19 pairs of MZ twins reared apart (Newman et al. 1937). The average of the IQ differences is 8.2 points. What conclusions are warranted?

106	95	91	85	101	90	93	127	102	116
105	94	90	84	99	88	89	122	96	109

102 88 115 97 78 92 106 96 116
94 79 105 85 66 77 89 77 92

11. Very low IQ can be a consequence of homozygosity for a single recessive allele (as in untreated PKU) or the result of a developmental birth defect. On the other hand, moderately low IQ reflects, to some extent, polygenic inheritance. Generally, the sibs of the severely retarded have higher IQ scores than the sibs of the moderately retarded. Explain.

12. Assume that in country A, all schools are equally bad, learning is never encouraged, and all other environmental factors relevant to taking IQ tests are constantly unfavorable. Thus, $V_E = 0$ with regard to IQ.

Assume that in country B, all schools are equally good, children are impartially encouraged in mental tasks, and all other environmental factors relevant to IQ are constantly good. Here, too, $V_E = 0$.

What is the heritability of IQ within A and B? Comment on the cause of the likely difference in IQ scores between the two countries.

ANSWERS

1. (a) Higher heritability when genetically heterozygous. (b) Higher heritability when environmentally uniform.

2. The full explanation depends on the effects of dominance, epistasis, and environmental factors. With regard to the environment, we can note that heavy persons are often that way because of a "weighty" life style, in addition to "weighty" alleles. Whatever genes they transmit, however, the environmental contribution to weight gain among their children is not likely to be repeated to the same extreme degree. (A similar argument applies to lightweights.)

3. *G/G H/H i/i* *G/G h/h I/I* *g/g H/H I/I* *G/G H/h I/i*
G/g H/H I/i *G/g H/h I/I*

4. (a) hazel × blue-green: *G/g H/h I/i* × *G/g H/h i/i*, for example. (b) light brown.

5. a) True. In addition, no progeny can be darker than medium. b) False. The child can be medium when the other parent is light, and the child can be dark when the other parent is medium.

6.

Skin color	(a)	(b)	(c)
White	0	0	1/16
Light	0	1/4	4/16
Medium	1	2/4	6/16
Dark	0	1/4	4/16
Black	0	0	1/16

7. (a) DZ and MZ rates about the same. (b) About 3.5 times more DZ than MZ.

8. The twins were DZ. For ABO, the parents were *A/O* × *B/O*. For Rh, the parents were *R/r* × *r/*−. In a fused placenta, a common circulation allowed a two-way, but unequal, exchange of some primordial red cells that established themselves in the other twin and continued to produce mature red blood cells.

9. The paternal contribution to the unusual pair is similar to that for DZ twins, but the maternal contribution is similar to that for MZ twins. Note that in the absence of crossing over, the second polar body is genetically identical to the egg.

10. The range of differences—up to 24 points—shows that environmental factors are important in IQ variation. In addition, the average IQ score is 95.7, a figure significantly lower than 100. Other studies also show that being a twin reduces IQ an average of 5 points below that for single births.

11. The sibs of the severely retarded would generally have IQ scores in the normal range: they have only a 25% chance of being homozygous for the postulated abnormal recessive, and a much smaller chance of abnormality through repetition of a postulated developmental error. On the other hand, the sibs of the moderately retarded would tend to have relatively low IQ scores because of polygenic inheritance of the same additive alleles, or perhaps because of similar environmental factors.

12. Within both countries, heritability of IQ is 100%. There would almost certainly be an average difference in IQ scores between persons from A and B, caused in part by the obvious environmental factors. There is nothing in this situation as stated that rules out genetic factors contributing to the difference, however.

PART SIX APPLICATIONS OF GENETIC TECHNIQUES

24. Beginnings of Life
25. Genetic Practices and Prospects

Chapter 24 Beginnings of Life

Negative Eugenics
Involuntary Sterilization
Voluntary Sterilization

Positive Eugenics
Germinal Choice
Artificial Insemination

Other Reproductive Technologies
In Vitro Fertilization and Embryo Transfer
Variations on a Theme
Ethical and Legal Issues
Reproductive Technologies in Farm Animals

Sex Selection
Available Techniques (Postconception)
Contemplated Techniques (Preconception)

> *What Nature does blindly, slowly, and ruthlessly, man may do providently, quickly, and kindly. As it lies within his power, so it becomes his duty to work in that direction; just as it is his duty to succour neighbors who suffer misfortune. The improvement of our stock seems to me one of the highest objects that we can reasonably attempt.*
>
> *Francis Galton (1905)*

An idealistic, humane, and childless Victorian gentleman, Galton coined the word **eugenics** (literally, "well born") to mean the genetic improvement of the human species over generations. Although based entirely on voluntary action, his eugenic aspirations and plans were heavily tinged with nationalism. He noted (1901) that a "high human breed" is especially important for the English because "we plant our stock all over the world and lay the foundation of the dispositions and capacities of future millions of the human race." Adolf Hitler carried these notions to cruel and bizarre extremes with involuntary eugenic programs that championed Aryan elitism. From the early 1930s, the Nazis forced sterilization upon hundreds of thousands of "inferior" persons and established spa-like homes for the married or unmarried women of Hitler's special SS troops (Kevles 1985). The horror of the Holocaust was to follow. Thus high hopes but terrible consequences can accompany attempts to alter customary patterns of procreation.

Besides overestimating the role of hereditary factors in human behavioral traits, Galton naively believed that eugenic proposals would be easily and voluntarily accepted. But today, as in Galton's time, few persons give more than passing thoughts to the overall quality of the genes that are committed to future generations—and some are offended by the very idea that methods used daily to improve agricultural plants and animals could be applied to humans. Yet the means of eugenic improvement are in hand, based partly on old ideas and partly on new reproductive technologies and genetic engineering. Some techniques ought not be applied to the human species under any circumstances, but others are simply extensions of traditional medical practices that help individuals and families improve their prospects for health and happiness.

In this chapter we look first at schemes for preventing the spread of "bad" alleles (negative eugenics) or for encouraging the transmission of "good" ones (positive eugenics) by manipulating gametes in one way or another. Traditionally, this can be accomplished within the framework of the usual sexual unions and, less traditionally, by intervening in the reproductive process. The newer techniques (with or without eugenic aims) include artificial insemination of a female, and fertilization of an egg outside the body with subsequent transfer of the very early embryo back into the egg donor's uterus, or into the uterus of another female.

In the final chapter we look at additional genetic and

medical skills that can be applied before or after birth. Such procedures include attempts to detect genetic disorders prenatally (or at least before the appearance of overt symptoms), screening for heterozygous carriers with a view toward counseling parents about future children, and treatments for genetic diseases. These topics summarize and extend material that has been presented previously, especially in Chapters 15 and 16 on new genetic technologies and inborn errors of metabolism.

The problems inherent in some of these technologies are out of the ordinary, and publicity on the technical, ethical, legal, and social issues has been widespread. (See, for example, American Fertility Society 1988.) For each technique, readers might consider the following questions:

1. *Is it feasible?* Is a newly announced achievement now applicable only to microorganisms? To laboratory animals? How will it be tested for use in humans? Is the technique so complex, the cost so high, or the morality so debatable that it would only be utilized by a handful of persons? Is *my* family likely to benefit? Sensational pronouncements to the contrary, for quite a while no human babies will be decanted from flasks as in Aldous Huxley's *Brave New World* (1932).

2. *Who decides?* This question arises in several contexts. Who decides what type of research shall be supported by public funds and who selects the investigators that receive support? Some decisions on the treatment of malformed newborns or on the choices for recipients of transplantable organs are painfully difficult. Who shall be the primary decision makers? Doctors? Parents? Clergy? Judges? Administrators? Legislators? "Experts" of one sort or another?

3. *Who pays?* In a world of limited resources, we must decide not only who receives an expensive treatment but also who pays the bill. Should scarce medical processes or facilities be available on a random basis, or on the basis of one's ability to pay, or on the basis of one's ability to mount a television appeal for funds? If society is to pay for some costly individual treatment, does society have a right to prevent the birth of persons who might need that treatment?

4. *Is it morally right?* Is a particular technology good or bad? Does it increase or diminish human freedom? Does it enhance our sense of human dignity or does it dehumanize us? Paul Ramsey (1970), who was a theologian at Princeton University, discussed two types of ethical considerations: the morality of the perfected procedure and the morality of the experimentation in humans that would certainly be needed to perfect the procedure. How do we obtain informed consent to allow for human experimentation? Fetuses cannot give informed consent, nor can mentally retarded persons, nor, it has been argued, can prison inmates whose privileges may depend on cooperation (Sun 1981). A cloned frog that turns out badly can be casually discarded, but what will be done with a defective baby arising from experimentation? Opinion varies on the sanctity of life versus the quality of life. Is life worth preserving, nurturing, and extending under any circumstances?

These questions prompt the additional question, "When does a specific human life begin?" Life itself is continuous: an egg, sperm, zygote, embryo, and adult (as well as the parts of an adult such as a kidney or a blood sample) are all alive, but the question relates to the life of a particular individual. The *potential* for a given human life begins at conception, for, at this point, the joining of two gametes determines a unique genetic heritage that is not further altered in any substantive way during its specific time span. The *humanness* of the potential human life appears to develop gradually; it does not seem possible to pinpoint any embryonic state at which the developing protoplasmic blob of a zygote or blastocyst with no human features suddenly becomes human. No one developmental happening—implantation, the onset of nervous feeling, fetal movement (quickening), the potential for independent existence— marks the advent of a new human being (Ayala 1977). The American Fertility Society (1988) thinks that the "degree and nature of respect and moral value accorded to the human pre-embryo or fetus rises continuously until birth." Many people have argued that defining the beginning of a human life is not a scientific issue at all but belongs to the realm of philosophy and religion (Rosenberg 1981). We return to this question later—see Ethical and Legal Issues.

NEGATIVE EUGENICS

Lessening the incidence of hereditary disorders through prevention of child bearing, i.e., **negative eugenics**, has an unsavory history that is primarily related to the enforced sterilization of persons considered "unfit" or likely to have "socially inadequate" offspring. Most infamous was Hitler's policy of "racial hygiene," but even in the United States, tens of thousands of involuntary operations were performed between 1930 and 1960. Although 19 states still retain statutes allowing for the eugenic sterilization of institutionalized retarded persons after due process of law, its implementation is minimal in this country today (Reilly 1985).

Involuntary Sterilization

The first compulsory sterilization law in the United States was enacted by Indiana in 1907. Although this and other early statutes were declared unconstitutional, a more carefully drawn law in Virginia was upheld by the U.S. Supreme Court in the 1927 decision *Buck v. Bell* (reprinted in Bajema 1976). The case involved an institutionalized patient, Carrie Buck, who was declared feebleminded, as were her mother and Carrie's seven-month-old daughter. The trio was found to be defective due to hereditary factors by a eugenics "expert," Harry Laughlin, who never examined them and who misrepresented the daughter. (She died at age eight of infection, but her school teachers considered her to be very bright.) Furthermore, Buck's ineffective defense lawyer was in collusion with judicial and legislative proponents of the new Virginia sterilization law, so the case was a sham (Lombardo 1985). Yet, the patriotic rhetoric of Justice Oliver Wendell Holmes rings out:

> We have seen more than once that the public welfare may call upon the best citizens for their lives. It would be strange if it could not call upon those who already sap the strength of the State for these lesser sacrifices, often not felt to be such by those concerned, in order to prevent our being swamped with incompetence. It is better for all the world, if instead of waiting to execute degenerate offspring for crime, or to let them starve for the imbecility, society can prevent those who are manifestly unfit from continuing their kind. The principle that sustains compulsory vaccination is broad enough to cover cutting the Fallopian tubes. Three generations of imbeciles are enough.

Not only was the case factually incorrect, but the decision has been questioned on legal grounds (are vaccinations and sterilizations equivalent?), and its genetic suppositions are groundless. Mental retardation is not a single entity. Some cases are due to the presence of a major gene with differing modes of inheritance, some to the chance combinations arising from polygenic inheritance, and some to various chromosomal aberrations. Other types of mental deficiency are due to environmental damage occurring prenatally, to birth trauma, or to infectious diseases. To the extent that mental deficiency is traceable to recessive alleles, calculations in Chapter 22 show that even the nonreproduction of *all* recessive homozygotes produces only very slow reductions in frequency. Thus it is silly to suggest that Carrie Buck and her "kind," whatever that is, will populate the world with criminals and imbeciles.

Such a Supreme Court decision today is unthinkable, but Holmes undoubtedly reflected public opinion of the time. This opinion was formed in part by eugenic literature that supported the fallacy that all sorts of social ills resulted from single-gene inheritance. One model for eugenic legislation (Laughlin 1922) proposed the following socially inadequate classes: the feebleminded, insane, criminalistic, epileptic, inebriate, syphilitic, as well as orphans, ne'er-do-wells, and paupers. *The Kallikak Family*, a much publicized book that went through five printings, purported to be a scientific study of heritable mental defects among the descendants of one New Jersey man (Goddard 1912). We are told that Martin Kallikak, a young Minuteman, strayed from the path of virtue with a "nameless feeble-minded girl." Martin neglected mother and child, but through this line of descent he became the progenitor of 480 persons among whom were 143 feebleminded souls, 36 illegitimate children, 33 sexually immoral persons, 24 alcoholics, 8 madams, 3 epileptics, and 3 criminals.

Martin later "straightened up and married a respectable girl of good family." On this "control" side, he became the proud ancestor of 496 persons all with good credentials: "doctors, lawyers, judges, educators, traders, and landholders." From this dual family history, it was concluded that bad alleles were perpetuated on one side and good alleles on the other. The book, however, is simply a moralistic tract that almost completely ignores the role of environmental factors or, apparently, anything to do with Martin himself. Far from proving anything about heredity, the Kallikak* study could just as well constitute a plea for upgrading the social conditions that perpetuate poverty, if the data are to be believed at all. Gould (1981), for example, made the interesting discovery that the facial features in photographs of persons on the bad side of the family were crudely altered to make them look sinister or stupid. Curiously, the Kallikak author criticizes an earlier analysis of crime, pauperism, and disease in the *Jukes* family as being inconclusive because of unknowable interactions of heredity and environment.

Voluntary Sterilization

However misguided past practices have been, sterilization can be of value in specific families to reduce or prevent unhappiness and suffering. For example, high courts in several states have ruled that parents can seek to obtain sterilization of their mentally retarded children that live at home (Reilly 1985). This might allow them to live outside an institutional setting without fear of becoming a parent. Also, many normal adults have chosen voluntary sterilization as the surest contraceptive. Although these cases are primarily employed to prevent the birth of a child of *any* phenotype, rather than as eugenic measures, sterilization would be appropriate for persons carrying alleles for genetic diseases. For example, Fraser (1973) tells of a Montreal woman who gave birth to two hemophilic sons despite the use of contra-

*A pseudonym from Greek words meaning "beauty" (*kallos*) and "bad" (*kakos*).

ception (rhythm method). Although she then requested sterilization, she became pregnant once more before her application was approved by a hospital committee. "By that time . . . the baby was moving, and she just couldn't go through with [an abortion], quite understandably. The result was that she now has three hemophilic sons. She loves the little boy, and she is a good mother, but she is quite sure that she would have preferred that he had not been born."

A common method of female sterilization is *tubal ligation*, in which both Fallopian tubes (oviducts) are cut and tied so that eggs cannot pass from the ovaries to the uterus. Although the operation is relatively routine, it requires hospitalization, and the effects of surgery are felt for several weeks. More recently, the procedure has been improved by the use of a *laparoscope* (literally, "looking into the abdomen"), a pencil-like instrument with a a glass fiber optical system. This tool is inserted through a small incision near the navel, along with an instrument for delivering an electric current; the Fallopian tubes are viewed, severed, and seared shut. In males, sterilization is more easily achieved by *vasectomy*: cutting and tying off both sperm ducts leading from the testes. The operation does not require hospitalization, and scrotal soreness lasts only a few days. In neither sex do these sterilizations involve the gonads themselves. Except for blocking the passage of gametes, sexual functions are unimpaired; indeed, the sense of freedom from fear of pregnancy may increase sexual motivation and pleasure. Although these operations are generally considered permanent, the surgery can occasionally be undone. Millions of voluntary sterilizations have been performed in the United States for the purpose of individual birth control, about three-fourths of them on men. With informed consent, the operation is legal in all states.

POSITIVE EUGENICS

Rather than advocating the sterilization of the unfit (however defined), Galton stressed the need for increased propagation of men of talent and genius, those with superior health, moral strength, and high "civic worth." **Positive eugenics** could be accomplished, he felt, by educational programs that would influence popular opinion and by "the wholesome practice during all ages of wealthy persons interesting themselves in and befriending poor but promising lads" (Galton 1901). He specifically recommended that exceptionally worthy young couples be provided convenient housing at low rentals.

Germinal Choice

Although most American eugenicists were obsessed with negative eugenics, the intellectual heir to the positive idealism of Galton was Hermann J. Muller (a 1946 Nobel laureate for mutation studies). Through his 1935 book, *Out of the Night: A Biologist's View of the Future*, Muller hoped to stimulate interest in both positive eugenics and social reform. "He advocated freedom from the bondage of pregnancy, including legalized abortion, public endorsement of birth control, and public programs and facilities for child care, especially for women desiring work or higher education" (Carlson 1973). For over 30 years, Muller proselytized for positive eugenics, primarily through "germinal choice," that is, artificial insemination of women with semen freely selected from sperm banks to which eminent men of known identity made contributions. Muller (1961) recommended that the sperm be deep frozen for decades before use in order to better view the individual worth of donors and to "reduce the danger that choices will be based on hasty judgments, swayed by the fads and fashions of the moment." It was hoped that these methods would enhance basic values distinctive to humans, among which Muller included intelligence, curiosity, creativity, "genuineness and warmth of fellow feeling," and "joy in life and in achievement."

Although there has been no stampede to follow Muller's proposals, about 70 sperm banks have been established in the United States (many of which are listed in telephone Yellow Pages). As we note below, however, most of these sperm banks are not intended to fulfill the types of eugenic purposes envisioned by Muller. One exception is the Repository for Germinal Choice, which was established by Robert K. Graham in southern California and which accepts sperm only from a few men, including Nobel Prize winners, other eminent scientists, and, more recently, exceptional athletes. Many people (including Muller's heirs) have ridiculed or castigated these blatantly elitist and narrow criteria. Nevertheless, this bank has "fathered" 37 children (Chase 1987).

Artificial Insemination

Although not universally acceptable on religious or moral grounds, **AID**—Artificial Insemination Donor—is a simple, inexpensive (about $100) medical procedure dating back to the nineteenth century. Hidden to some degree by secrecy and poor record keeping, and characterized by various degrees of care in execution, it is thought that 6,000–10,000 children per year in the United States are now conceived through AID, using fresh or frozen donor semen (Curie-Cohen et al. 1979). Employed primarily when a husband is infertile because of a low sperm count or poor sperm motility, it has also been used by couples when a fertile husband could transmit a deleterious allele, or if he were exposed to mutagenic agents such as radiation or chemicals. It has also been used by unmarried women.

In the past, infertile couples desiring children often chose to adopt. But the number of adoptable Caucasian infants has decreased drastically because effective contraception and elective abortion are easily available, and because unwed mothers often choose to keep their children. The situation has provided the impetus for AID in those cases in which the male is the cause of childlessness. The choice of donor usually resides with the doctor, who may select a medical student matched as closely as possible to the husband for obvious physical characteristics, and perhaps for blood groups as well.* (The donor is usually paid about $50 per usable sample.) With a syringe, the doctor squirts the donor's semen close to the woman's cervix perhaps several times near the estimated day of ovulation and over as many months as may be necessary to attain fertilization—an average of 6–7 inseminations. Spontaneous abortions and birth anomalies among the resulting AID pregnancies appear to be no more frequent than among pregnancies achieved through sexual intercourse (Richardson 1975). Although some earlier data suggested that AID yielded an excess of male births, especially using fresh semen, the matter now seems to be unresolved (discussed again in the section on sex selection).

The donor and the "preadopting" couple remain unknown to each other (and a few doctors deliberately muddle the matter further by mixing semen from several donors). Signed forms absolve the donor of responsibility for paternity and the doctor of responsibility for birth defects. The five persons intimately involved—the couple and their child, the doctor, and the donor—bear legal relationships that are to some extent unresolved. Thirty states have laws stating that the AID child is legitimate and entitled to the same rights as a child conceived the old-fashioned way (Office of Technology Assessment 1988a). But no state has considered the status of frozen sperm left after a man's death, which complicates the meaning of the phrase "I bequeath to my children" that is often used in wills.

The American Fertility Society (1986), a professional association of infertility specialists, suggests that the number of births per donor be restricted to ten or fewer to reduce the probability of an inadvertent consanguineous mating. In a large, mixed, urban population this chance is remote, but the probability would be higher in a small, close-knit community.

The introduction of medical technology into traditional family life has, in many cases, fulfilled a valid human desire to bear children. Some people feel, however, that the procedure is repugnant in its methods and morally equivalent to adultery. Although most AID couples probably keep its origin secret from the resulting child, one informed offspring has written, "Knowing my A.I.D. origin did nothing to alter my feelings for my family. Instead, I felt grateful for the trouble they had taken to give me life. And they had given me a strong set of roots, a rich and colorful cultural heritage, a sense of being loved" (Atallah 1976).

The use of frozen sperm has made the practice of AID safer and easier (Ansbacher 1978; Kremer et al. 1987). A semen sample to be frozen is mixed with a solution of glycerol and other cryoprotectants and stored at the temperature of liquid nitrogen, −196°C, for long periods of time. It is thawed just before use. The greater safety of frozen sperm results from the greater time available for testing the donor and his semen for the pathogens that cause sexually transmitted diseases. These include the bacteria that cause syphilis, gonorrhea, and chlamydial infections, and several viruses: hepatitis B virus, cytomegalovirus, and the virus (HIV) that causes AIDS, Acquired Immune Deficiency Syndrome (Mascola and Guinan 1986). Although only three states *require* any sort of screening of the donor for venereal or genetic diseases, the Office of Technology Assessment (1988a) reports that most sperm banks do quarantine sperm while conducting follow-up tests, especially the test for HIV antibodies that may show up only months after the donor is capable of transmitting the AIDS virus through his semen. For this reason, current guidelines by several fertility associations recommend the use of only frozen semen for AID, although pregnancy rates are generally acknowledged to be somewhat lower using frozen samples (Peterson et al. 1988).

The existence of banks of frozen sperm also enlarges the inventory of available types for better husband-matching. The stored specimens, however, do not necessarily reflect the human values that would provide for eugenic programs. Indeed, people disagree on what these values are; they clearly differ between cultures, change over time, and are only partly influenced by genetics. Furthermore, because of the anonymity of donors in most current practice, the couple could not, in any case, make informed eugenic choices. Readers will also recognize that, because of meiotic segregation and polygenic inheritance (not to mention large environmental effects), the selection of, say, Babe Ruth as a sperm donor would only slightly increase the probability of producing a slugger. (Indeed, none of Ruth's 15 descendants has shown any particular talent for baseball [Thomas 1988l.) Still, the eugenic potential of a currently feasible technology, having been shown to be safe and

*Elias and Annas (1987) note that in choosing medical students as donors, AID physicians appear to be making eugenic decisions based on the belief that "society needs more individuals with the attributes of physicians."

to some extent effective (depending on the heritabilities of the traits in question), and voluntarily undertaken with only minimal governmental regulation, would undoubtedly have given hope to Galton and Muller.

OTHER REPRODUCTIVE TECHNOLOGIES

Approximately 8.5% of married couples who want children, or who want more children, are unsuccessful in producing them because of a variety of physical or emotional problems. For about 30% of infertile couples, the difficulty is attributable to the husband; about 50% of cases are due to some factor in the wife, and 20% to problems with both parents (Smolev and Forrest 1984). A variety of "assisted reproduction" techniques are available, depending on the cause of the infertility. Blocked oviducts (Fallopian tubes), which prevent the egg from traveling down to the uterus or sperm from traveling up to the egg, constitute a major source of female infertility. Surgery to open the oviducts can help some of these women, and the technique of in vitro fertilization and embryo transfer (**IVF-ET**) can help others.

In Vitro Fertilization and Embryo Transfer

The successive steps involved in IVF-ET require several outpatient hospital visits for the woman, numerous other examinations and tests, skilled medical and technical personnel, and a specialized laboratory. The general procedure is to obtain an egg directly from a female's ovary, fertilize it with the husband's sperm in nutrient medium, allow the zygote to undergo a few cleavage divisions, and transfer the developing embryo to the woman's uterus. There it might be expected to implant in the uterine wall as if it had been normally fertilized while passing through an unblocked oviduct. The first liveborn child to be conceived in this way, Louise Brown, developed under the close scrutiny of the English obstetrician Patrick Steptoe, reproductive physiologist Robert Edwards, and a swarm of eager reporters from all over the world* (Steptoe and Edwards 1976, 1978). For several years, the technical details were adjusted and improved largely in research laboratories in England and Australia, but over 100 centers in the United States now perform IVF-ET (not all successfully, and none with federal funding). The number of babies worldwide conceived this way is now in the thousands, but the pool of patients who might benefit from the procedure is thought to be in the millions.

*The zygote that gave rise to Louise Brown divided in laboratory glassware three times over 2½ days to a size smaller than the dot ending this sentence. To label her a "test tube baby" for that reason seems to overlook the 38 weeks she spent in her mother's uterus. In some other animal species, fertilization and development are normally external; this includes most fish and amphibians and many invertebrates.

The prospective mother is treated with gonadotropic hormones (or other fertility drugs) to induce her oocytes, arrested in the first meiotic division, to begin to mature (Seibel 1988). The doses and timing are carefully controlled so that the eggs, about to be released from the enlarging ovarian follicles, can be recovered at a predetermined time. Then, using a laparoscope inserted into the abdomen, the ovaries are viewed and the contents of the mature follicles on their surface are literally vacuumed into a collecting chamber. Alternatively, ultrasound imaging can be used to visualize the ovarian surface. Three or more eggs can be obtained during one operation. Easily viewed through a low-power microscope, the eggs are put into a special culture medium and bathed with fresh sperm from the husband. Before use, the washed sperm may be sedimented by light centrifugation; in this technique, the 50,000 to 100,000 sperm (about 1 in 5,000 of the total sperm in an average ejaculate) that swim up from the pellet are collected and mixed with the eggs. Fertilization is indicated by the penetration of sperm through the zona pellucida, then by the extrusion of the second polar body, by the detection of a sperm midpiece inside the egg cytoplasm, and by the formation of the male and female pronuclei (see Figure 6 in Chapter 5). The zygote cleaves to form two, four, and eight cells over several days (Figure 1). The developing embryo is transferred back to the female's uterus at the four- or eight-cell stage by means of a fine tube inserted through the cervical canal. If conception had occurred following normal intercourse, the egg, fertilized in the oviduct, would have entered the uterus on the third or fourth day after being released from the ovary, having grown to 8 or 16 cells.

IVF-ET has also been used when the woman has *normal* oviducts, but the couple needs an assist to bolster a very low probability of conception (due to one or another adverse circumstance). In one technique, the fertilized egg is transported by a catheter through the cervix and uterus to the oviduct, from which point it travels back into the uterus as it would in the normal course of events (Jansen et al. 1988).

Despite 10 years of research at dozens of centers worldwide, the success rate for IVF-ET remains disappointingly low. For each month that an attempt is made, only about 10% of starts eventually end in a live birth (although there is much variation among clinics and among different subpopulations of infertile women). This figure is called the "take-home baby rate" and is the statistic of concern to the treated couple. One can calculate that 50-50 odds of producing a take-home baby

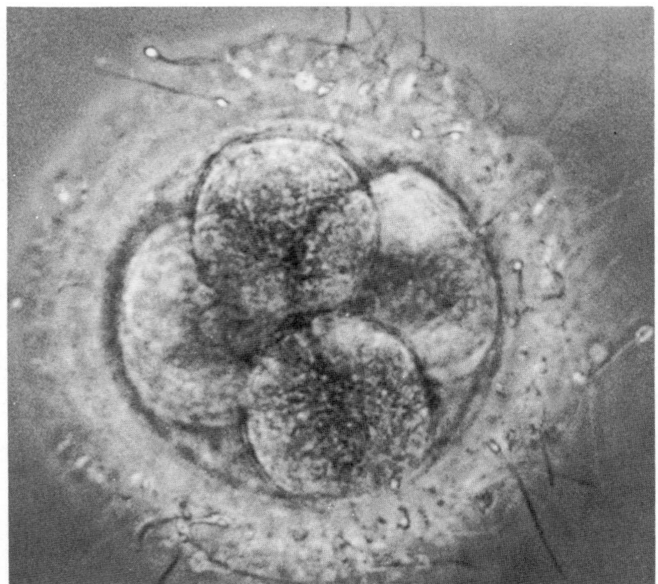

Figure 1 A four-cell human embryo derived from an egg recovered directly from a woman's ovary and fertilized in vitro. A few of the husband's sperm can be seen in the zona pellucida surrounding the embryo. (Courtesy of R. G. Edwards, Cambridge University.)

are achieved after 6–7 cycles. Because the cost of the multistepped technological assist is about $5,000 per cycle (for as many monthly cycles as the process is attempted), the financial as well as emotional costs are high. Oocytes are successfully retrieved over 90% of the time, and fertilization and cleavage occur at the same high rate. Thus the major stumbling block seems to be the failure of the embryo to implant when it is transferred to the recipient's uterus. Once pregnancy is established by IVF-ET, however, the risks of miscarriage, chromosomal aberrations, and other birth defects have generally been held to be no higher than that in the general population (although these observations have recently been questioned [Raymond 1988]). In addition, the overall low rate of success should be compared with the natural situation. For normally fertile couples, it is estimated that only 30% of eggs exposed to sperm through sexual intercourse produce a viable offspring (Seibel 1988).

To improve the chances of success, several fertilized eggs are usually transferred to the recipient's uterus. This practice has resulted in a higher than usual occurrence of multiple births, despite the low rate of establishing any pregnancy. Edwards (1985) notes that the incidence of multiple pregnancies is, in fact, far higher than predictions based on binomial probabilities. This outcome means that the implantation of one embryo somehow *helps* another embryo to implant. In one study, 18% of pregnancies achieved with the in vitro assist were twins and 3.8% were triplets (see Seibel 1988, and Question 10).

Australian scientists have developed the techniques for freezing and storing four- to eight-cell embryos (Trounson 1986). This practice reduces the risk of multiple births, because the physician can transfer just one or two fresh embryos and save the remainder for successive attempts. This technique therefore sidesteps the ethical problem of discarding four- and eight-celled human embryos, which some in vitro centers do in order to avoid multiple pregnancies. Of course, freezing also reduces the need to repeat ovulation stimulation and laparoscopy in order to retrieve additional eggs. Subsequent menstrual cycles occur without drug treatment (because no more eggs need be collected), and this may improve the chances of achieving pregnancy. The Office of Technology Assessment (1988a) reports that 60 children have been born in Australia and Europe from the transfer of thawed frozen embryos.

Variations on a Theme

Many other technological answers to various causes of infertility have been developed. These are discussed briefly below and summarized in Table 1.

A spin-off of in vitro fertilization and embryo transfer is a technique called **gamete intrafallopian transfer** (or **GIFT**) first performed in 1984 by researchers at the University of Texas Health Science Center at San Antonio. The technological assist might be indicated when the husband has defects of sperm concentration, motility, or morphology, when the wife builds up antibodies against the husband's sperm, or for a number of other causes of infertility. Unlike the woman using IVF-ET, however, the woman using GIFT must have at least one functional Fallopian tube. The induction of ovulation and egg recovery with a laparoscope are similar to the operations used for in vitro fertilization. But instead of allowing fertilization to take place externally, eggs and sperm are introduced via a thin tube (threaded through the laparoscope immediately after egg retrieval) into the upper end of the oviduct. There fertilization occurs as it would normally. The rate of success seems to be better than for IVF-ET.

When a wife does not produce viable eggs or when she chooses not to risk transmitting a dominant or X-linked recessive disorder to a child, two other techniques are sometimes used. In **egg transfer**, a donor is hormonally stimulated to produce one or more eggs, which are then obtained by laparoscopy. (The donor may be a woman undergoing IVF-ET who has "extra" eggs.) The eggs may then be fertilized within the wife with the husband's sperm using GIFT procedures. **Embryo**

Table 1 Reproductive technologies[a]

Name of Technology	*Sperm source*	*Egg source*	*Place of fertilization*	*Place of pregnancy*
Traditional situation	●	●	●	●
Gamete intrafallopian transfer[b]	●	●	●	●
In vitro fertilization and embryo transfer[c]	●	●	in vitro	●
Artificial insemination donation[d]	○	●	●	●
Egg transfer[e]	●	○	●	●
Embryo transfer[f]	○	○	○	●
Surrogate motherhood[g]	●	○	○	○

a. Solid symbols represent husband or wife; open symbols represent a third party.
b. Abbreviated GIFT. The egg, obtained via laproscopy, and sperm are injected separately into a healthy oviduct. GIFT is similar to IVF-ET, except that fertilization occurs in vivo.
c. Abbreviated IVF-ET. Either the egg or the sperm could come from a third party if needed.
d. Abbreviated AID. The husband could supply the sperm, as is done in certain cases of lowered male fertility.
e. Egg from a donor is transfered to the wife using GIFT or IVF-ET procedures.
f. Donor female is inseminated with donor sperm (or sperm from the husband) and fertilization occurs in her. The early embryo is flushed out of her uterus before implantation and transferred to the wife. Embryo transfer is also available with the fresh or frozen embryos obtained through IVF.
g. Using the techniques of AID.

transfer may be used when both husband and wife are infertile. In one scenario, an egg donor is artificially inseminated by a sperm donor. The resulting embryo is flushed out of the donor's uterus for transfer to the wife whose menstrual cycle must be synchronous. An obvious complication of this process is that failure to find and aspirate the developing embryo could result in unwanted pregnancy for the egg donor. The success rate of embryo transfer is usually low.

There is also **surrogate motherhood**, by which a husband and wife contract for the services of another woman, the surrogate. Typically, the surrogate is artificially inseminated with sperm from the husband and is legally obligated to turn over the resulting newborn to the contracting couple. About 15 surrogate recruiting and matching services are active in the United States and have been responsible for about 600 births. Surrogacy differs from the other technologies in that the fetus does not develop in the woman who intends to be the mother of rearing. This unusual social situation provoked strong public reactions in the case of so-called Baby M. The surrogate (biological) mother changed her mind and decided to keep the baby rather than accept the fee of $10,000 and relinquish her to the contracting couple (the biological father and his wife). After a series of emotional confrontations, escapes, and raids, New Jersey courts ruled that the contracting couple was to have permanent custody of the child, although the surrogate was to have considerable visitation rights (Annas 1987).

Ethical and Legal Issues

The morality and legality of the noncoital reproductive technologies that are presented above have been debated by many individuals and groups. The forcefulness of the arguments reflects clearly perceived intrusions of the new and unfamiliar techniques into intimate reproductive customs and some uncertainty about consequences and future developments. Ethical conclusions and legal responses have been highly varied, as might be expected in a pluralistic society. To sample a range of opinion, readers are directed to American Fertility Society (1988), Congregation for the Doctrine of the Faith (1987), Elias and Annas (1987), and Schneider (1985). We cannot hope to give here more than a few salient points of the debates.

One basic question is whether every couple has a right to reproduce. For persons who are fertile, the answer seems obvious. The desire to have children and create a family results from natural urges and universal social values, and the right to do so is based on concepts of individual freedom. The responsibilities for bearing and raising children rest largely with individual couples, although families also make use of community services and social institutions. (In addition, problems of regional and global overpopulation are real and increasingly urgent.) The right of fertile couples to procreate seems to extend even to those who may knowingly impose a substantial risk of disease upon their fetuses—25% chance for the offspring of parents heterozygous for, say, Tay-Sachs disease, phenylketonuria, or sickle-cell anemia. This burden is borne by fetuses without their consent and without much public discussion of the moral issues involved.

When a couple is involuntarily sterile, the right to overcome the problem and thereby have a child of their own is not always so easily appraised. To diagnose and treat infertility uses, to various degrees, social services and medical resources. The prevailing view seems to be that procreation by initially infertile couples, using whatever medical technology is appropriate, is acceptable if they can pay their own way. The cost may be minimal if, for example, only donor sperm is required to substitute for a husband's, or if drug treatment is needed to stimulate a wife's egg production. The costs may be high for surgery or for in vitro fertilization and embryo transfer. Although infertility is a widespread and often embarrassing problem and may even be a source of despair and desperation, the use of tax dollars for reproductive assistance continues to be debated.

Another basic question is what manipulations, if any, are permissible on an embryo. We have noted that one cannot pinpoint any developmental stage at which an embryo suddenly becomes a human being, or at least there is no agreement on what that stage might be. What value should we place on, say, a four-cell human embryo? Is it all right to transfer it to a third party, to discard it, to experiment on it, or to freeze and store it? Some would give consent to these and other aspects of reproductive technology because of the joy the techniques may bring to otherwise infertile couples. Supporters have pointed out that many embryos fail to implant and therefore die in the natural course of events, and that intrauterine devices used as contraceptives are specifically designed to thwart implantation. Others think that the embryo has inherent rights. The Roman Catholic Church, for example, holds that a human being exists from the moment of conception, so that even a zygote or a few-celled embryo is entitled to the same respect and protection as a mature adult.

Legally, however, neither an early embryo nor a later fetus has any standing, except for rights contingent upon an eventual live birth. The fetus is not generally considered a "person" in the legal sense. This interpretation is reinforced by the United States Supreme Court's controversial 1973 decision in *Roe v. Wade*. The Court ruled that the Fourteenth Amendment's concept of privacy includes a woman's right to abort her fetus, although this right is qualified. States may *regulate* the conditions of abortion during the second trimester when the abortion operation may be less safe for the woman than continuing to term. States may *prohibit* abortion during the third trimester—after the point of viability, that is, when the fetus is able to live outside its mother's womb. Some legal scholars have objected to the *Roe v. Wade* decision, noting that a right to privacy does not explicitly appear in the Constitution; others have objected to the use of viability as a deciding point since viability is both difficult to determine and dependent upon whatever medical technology is available to support the lives of premature infants. As a result, a number of states have enacted legislation that seeks to limit the availability of abortion. In Missouri, for example, statutes prohibit any abortion that uses public funds or public facilities, except to save the life of the mother. The restrictive statutes have been upheld as constitutional by the Supreme Court in *Webster v. Reproductive Health Services* (*New York Times*, July 4, 1989). The preamble to Missouri's law says that "the life of each human being begins at conception," but it is unclear how this legislative opinion will affect future litigation.

With regard to the morality of specific reproductive technologies, some writers distinguish between those that involve only the gametes of a husband and wife (okay) and those that introduce a third party into the marriage covenant (not morally acceptable). The reproductive techniques are arranged in Table 1 by this criterion. Note that the use of a third party is not related to the degree of artificiality. The two techniques toward the top (GIFT and IVF-ET) are technically complex but involve (or may involve) only the gametes of a husband and wife. (Much of medical science, of course, involves artificial interventions, and almost everyone agrees that artificial does not mean immoral.) AID and surrogacy, on the other hand, are simple to arrange biologically, but are morally wrong in the opinion of some because of the third party. Surrogacy, in particular, has been denounced not only because traditional motherhood is altered in such an obvious way, but also because poorer women may be exploited by the richer. Many states are now wrestling with legislation to regulate surrogate motherhood. Payments to the surrogate and establishment of a time period for her to change her mind are some of the areas of deliberation (Andrews 1987).

We see that a child may have up to five different parents—planned prior to conception: its father of rearing may differ from its genetic father, and its mother of rearing may differ from its gestational mother who may differ from its genetic mother. Thus have science and technology upset time-honored arrangements for nurturing and being nurtured. We can but hope that people are wise enough to use the new procreation techniques to further the happiness, dignity, and well-being of both parents and children.

Reproductive Technologies in Farm Animals

Development of the new reproductive manipulations for humans has been paralleled by technical advances in breeding better animals for food. Many millions of people worldwide suffer because they do not have enough to eat or because their diet lacks one or more essential nutrients. The development of higher yielding cereal crops—the so-called Green Revolution in wheat, rice, and maize—has been the greatest agricultural achievement over the past 50 years, but increased production of meat, milk, and other animal products has also been significant. One drawback to animal husbandry is that it is relatively inefficient, requiring more energy inputs than the production of a caloric equivalent of plant food. Animal products, however, are nutritious and tasty and are preferred diet items even in developing countries. A second disadvantage is that animals may compete with people for the same types of food. Livestock, however, can graze on land unsuitable for growing crops, and the ruminants (cattle, goats, and sheep) can thrive on grasses and other cellulose materials that we cannot digest. In a world that is hungry, we need to advance scientific frontiers in all fields of agriculture, even while national and international leaders are grappling with seemingly intractable political, social, and economic problems that thwart the production and distribution of food (Hulse 1982).

The most important technological advance in animal breeding has been the worldwide use of **artificial insemination (AI)**, especially in the breeding of dairy cattle, but increasingly for beef cattle, sheep, pigs, horses, and many other species (Campbell and Lasley 1985). In some countries of the world, almost all dairy cattle are produced through AI. The practice began in the 1940s, largely through research efforts at a half dozen American land-grant universities in the Northeast and Midwest. The major advantage of AI is that the sperm of superior sires is made easily available, at low cost, even to small farmers. Because an average bull in commercial AI service is mated annually to about 5,000 cows (and the best bulls to 50,000 or more), the rate of improvement of milk quantity, milk quality, or any other selected trait can be very rapid.* A bull is tricked into ejaculating into an artificial vagina, and the semen sample is diluted with extenders and cryoprotectants, colored with food dyes (to help safeguard against errors), distributed into 300–500 plastic straws, and frozen in liquid nitrogen for up to two years (or more) with little or no loss of viability. To achieve fertilization, the semen in one straw is transferred to the cervical area of one cow in estrus.

A more recent technological advance increases the reproductive potential of superior females, although not to the same extent, nor as inexpensively, as AI can increase the reproductive potential of superior males. Usually a cow produces one calf a year, but with **embryo transfer (ET)**, this rate can be increased to 10 or more calves per year (Seidel 1981; Church et al. 1985). The process is similar to that for humans (see Table 1, next to last line). It begins when a cow is treated with gonadotropins to cause her to superovulate, producing as many as 10 eggs per estrous cycle instead of the usual one. The eggs are fertilized by artificial insemination, and the developing preimplantation embryos are flushed out of the cow's uterus about a week later. Easily isolated from the flushing saline, the embryos may be kept for up to a day at 37°C (body temperature) or frozen at −196°C. They are then transferred to the uteruses of less valuable surrogate mothers who are in the same stage of estrus as that of the genetic mother. In practice, the embryos may be transported continents away in either the fresh or frozen state. Exchange of valuable strains between countries by means of frozen embryos has been a particularly useful breeding technique. Newborns arising from transferred embryos receive appropriate passive immunity through the colostrum (first milk) of their surrogate mothers and easily become attuned to their new environment. Embryo transfer has also been used by zoos to help in the preservation of rare and endangered species (Figure 2; Summers 1986).

Spin-offs of research in ET include determining the *sex of embryos*, splitting early embryos to produce *identical (MZ) twins*, and combining embryonic cells to produce *chimeras*, animals with cells of different origin. During the process of being transferred, the sex of the embryos can by determined by teasing off a few cells of the developing mass (usually at the morula or blastocyst stage) and examining the chromosomes in subsequent mitotic divisions. Sexing embryos could be commercially valuable in the production of both dairy and beef herds. (It might be even more useful to separate X- and Y-bearing bovine sperm for use in artificial insemination,

*Dairy bulls, of course, produce no milk; their breeding value is gauged by the performance of their sisters and daughters. Artificial insemination also speeds up the evaluation of bulls, and therefore increases the time of their service once they have been proven.

Figure 2 Born June 1985 following embryo transfer, a Grant's zebra foal is being successfully reared by its surrogate mother, a domestic pony. Although the Grant's zebra is not an endangered species, several other types of zebra are. (Courtesy of The Zoological Society of London.)

but this is not yet possible. Separating human X- and Y-bearing sperm is discussed in the next section.)

The production of MZ cattle twins is accomplished by surgically separating an embryo at the morula stage (about 64 cells) into two (sometimes more) masses (Figure 3) and transferring both halves into recipient cows. Researchers have noted that the coat color patterns of the resultant MZ twins are similar, but *not* identical, confirming the existence of indeterminate events during the migration of the cells that become melanocytes (Chapter 16). Cattle MZ twins will be particularly valuable in basic studies of heritability and other nature–nurture questions.

The production of chimeras is achieved by combining the cells of one early embryo with those of another. The most unusual example is the "geep," an animal with cells from both a goat and a sheep (Fehilly et al. 1984). It is interesting to note that neither a goat embryo in a sheep nor a sheep embryo in a goat survives, but the chimeric embryos come to term in mothers of either species.

In one further manipulative technique, specific genes produced through recombinant DNA technology and molecular cloning are injected into fertilized eggs (see Chapters 15 and 19). Incorporation of foreign DNA into the genome of a developing fetus produces a so-called **transgenic animal**. Preliminary work is being conducted along these lines with farm animals, and some dramatic results have already been accomplished with laboratory animals. For example, researchers have produced transgenic mice that possess and express the growth hormone gene of a rat or of a human; the mice grow to large size and transmit the foreign gene to subsequent generations (Palmiter et al. 1983). These investigations are related to concepts of gene therapy which are presented in the next chapter.

SEX SELECTION

Suppose that samples of pure X-bearing sperm or pure Y-bearing sperm were available for artificial insemination of women. Suppose further that less objectionable techniques—safe, effective, inexpensive—were devised for predetermining a child's sex. Do you think that using such technology would be morally right? What proportion of people might avail themselves of the opportunity? Would the ability to choose a son or daughter at each birth alter the composition of families? The population sex ratio? Overall fertility? What societal changes might ensue?

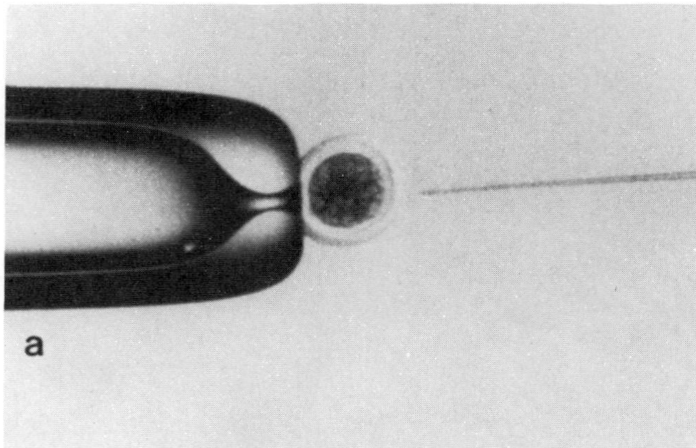

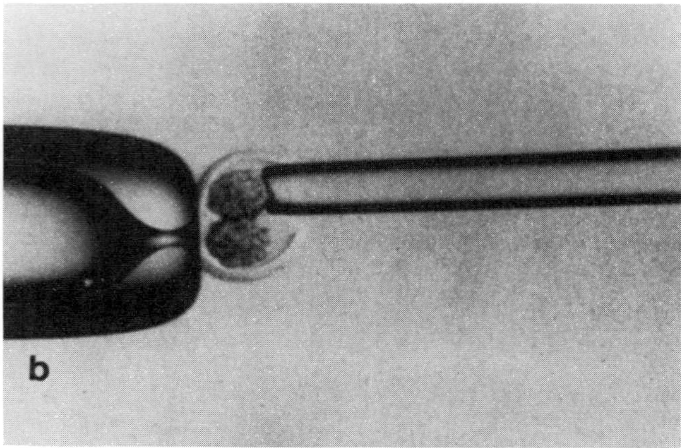

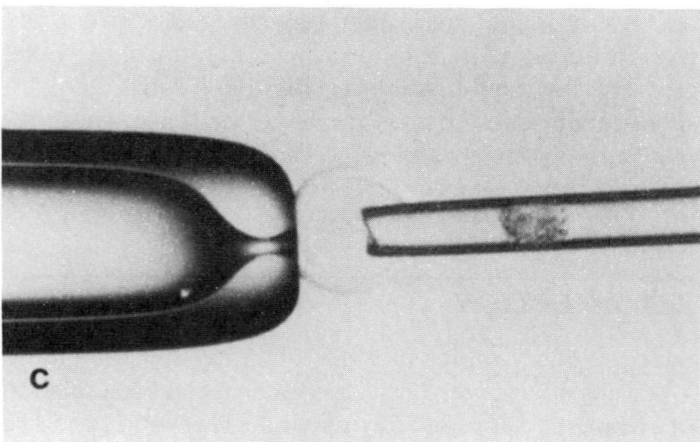

Figure 3 The production of identical cattle twins by embryo splitting. Not apparent in these pictures is the apparatus mounted on the microscope stage that is needed for micromanipulation. Approximate magnification is 100 times. (a) Recovered seven days after fertilization, a morula with about 64 cells is held in position on the mouth of a micropipette by light suction. A microscalpel is used to bisect the embryo inside its zona pellucida. (b) Another micropipette is introduced through the hole in the zona pellucida and one half of the embryo is gently sucked up, leaving behind a half-embryo in its own zona pellucida. (c) The half-embryo in the pipette is injected into an empty zona pellucida that had been previously prepared. Transfer of both embryos to a synchronized recipient cow may result in monozygotic twins. (Photographs by Brian Shea. From Church et al. 1985.)

Reliable sex selection schemes of the types envisioned above are not now available, although they may become available someday. Tantalizing reports of partial success with small numbers of births remain largely unconfirmed, and past predictions of the time needed to develop effective techniques have been wrong. For these reasons (and in contrast to other reproductive technologies), discussions of the ethics and consequences of sex preselection are preceding rather than following the advent of the techniques themselves (Bennett 1983). As one example of many sociological studies, Pebley and Westoff (1982) surveyed over 7,000 married American women regarding preferences for the sex of their next child. Although many more women wanted the firstborn to be a boy rather than a girl, the desire for a balanced family was very strong (Figure 4). Thus, if the firstborn was male, over 80% of women who expressed a preference wanted the *second* to be female, and vice versa. For a *third* child following one child of each sex, 40.4% of women had no preference, but among those who did, boys and girls were about equally desired. Thus the major consequences of sex selection in the United States would derive from more firstborn children being male. Holmes (1985) points out the many studies that show the greater achievements of the eldest child compared with those of later siblings and notes the benefits that would increasingly accrue to the male sex.

Depending upon assumptions about family sizes and the proportion of women using the techniques, one can expect the overall sex ratios in a population having the opinions expressed in Figure 4 to be 52 to 56% male.* These values are not too much higher than the actual value of about 51 to 52% male births. The role of fathers in implementing sex preferences, however, was not sampled in this study and would undoubtedly favor somewhat more males. A still larger male excess might be expected in other cultures, especially in some Asian and African countries where there exist social and religious traditions of male inheritance, or where a girl is perceived as a burden to her family because of poor employment opportunities and the need for large marriage dowries.

*Following a widespread practice, we express the sex ratio as percent males. Another convention is to use the number of males per 100 females. As an example, 55% male = 122 males per 100 females.

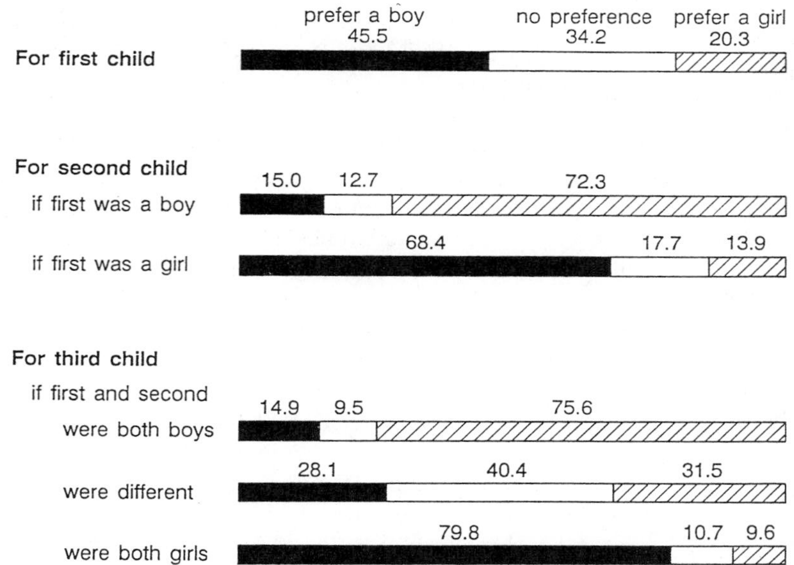

Figure 4 The preference of American married women for the sex of their next child. The figures above the bars express the percentages of women preferring a boy (■), a girl (▨), or expressing no preference (□). (Data from Pebley and Westoff 1982, based on a 1975 survey.)

Available Techniques (Postconception)

It is now technically possible for a determined couple to choose the sex of children with nearly 100% success. But the required methodology carries some small risk and is morally offensive to many persons because it involves terminating unwanted pregnancies. Two methods are available for sampling living cells of the growing fetus (discussed further in the next chapter). In the older technique, called **amniocentesis**, a physician removes a small amount of the amniotic fluid surrounding the fetus at about the sixteenth week of gestation; this procedure is carried out with a syringe and a long needle inserted through the abdominal and uterine walls. The fluid contains cells sloughed from the fetus. In the newer technique, **chorionic villus sampling (CVS)**, the physician aspirates a tiny piece of fetal tissue from the edge of the placenta at about the tenth week of gestation by means of a catheter inserted through the cervical canal, or through the abdominal and uterine walls. Fetal cells obtained by these methods can be reliably karyotyped for their sex chromosome constitution (Chapter 4). Alternatively, the cells can be reliably sexed with a DNA probe that specifically binds to base sequences from the Y chromosome. In this technique, only a small amount of DNA—purified, dotted onto filter paper, fixed and denatured, bathed with the radioactively labeled Y chromosome probe, and autoradiographed—is sufficient to determine the presence or absence of the Y (Gosden et al. 1984).

Although these procedures are useful in the prenatal diagnosis of many genetic diseases (Chapter 25), most American physicians would not agree to perform amniocentesis or chorionic villus sampling for the sole purpose of sex control. Nevertheless they might agree to abort all male fetuses that are carried by women heterozygous for hemophilia or other serious, recessive, X-linked disorders, even though half the aborted fetuses are expected to be normal. Consider, however, that in the United States, abortion before about 24 weeks is *legal* for any reason. One needs to ask whether it is *ethical* to do so in order to fulfill parental desires for a boy or a girl. Even though being the "right sex" might further the happiness of both parents and child, it is argued that being the "wrong sex" is not a disease and does not generally merit the use of medical resources. Furthermore, psychologists find that successful parenting requires the acceptance of a child's individuality—including particular strengths and weaknesses. Choosing a child's sex to the point of abortion seems at odds with these values (President's Commission 1983).

Contemplated Techniques (Preconception)

In an area of abundant scientific and popular publications, one fact stands out about planning a child's sex *prior* to conception: even a useless technique will succeed about half the time! The ancient who prescribed tying off the left testicle to produce a son could easily become a seer on the basis of small samples. Over the centuries, hundreds of theories of sex determination were constructed, all of which worked in about 50% of the trials. More modern methods have also been set forth either for altering the behavior of the two sperm types in vivo or for separating X-bearing from Y-bearing sperm in vitro, thereby yielding samples for artificial insemination (Levin 1987). Although each different methodology has its adherents, we limit our discussion to two examples.

AN IN VIVO TECHNIQUE. Many people have investigated the timing of intercourse relative to ovulation. One theory from the 1960s holds that Y-bearing sperm swim faster but are shorter lived than X-bearing sperm (Shettles and Rorvik 1984).* If this conjecture were true, then *intercourse at the time of ovulation would favor boys* because the faster Y-bearing sperm would find a waiting egg. Intercourse that stopped two to three days before ovulation, on the other hand, would favor the birth of girls because the X-bearing sperm, according to the theory, have greater staying power. Only three scientific journal reports, all from the 1970s and totaling about 400 births, provide any direct supporting data. (Negative results tend not to be published.) Also, the claimed phenotypic differences between X sperm (larger, hardier, but slower) and Y sperm (smaller, weaker, but faster) are unsubstantiated (Beatty 1970).

In fact, other ideas about sex selection predict just the opposite result! For example, James (1986a) hypothesizes that high levels of maternal gonadotropins at conception are associated with female offspring, although he proposes no mechanism for this action. Since these hormones peak as the egg matures, *intercourse at the time of ovulation would favor girls*, while intercourse either earlier or later in the cycle would favor boys. James summarizes many lines of evidence that support, and a few that fail to support, his hormone theory.

Both Shettles and James note that the practice of artificial insemination ought to provide relevant information, because the transfer of sperm is arranged to be as close as possible to the time of ovulation. A summary of 22 published reports of AID programs in Europe and the United States between 1959 and 1982 (compiled to see whether freezing of sperm changed the sex ratio) found the following (Alfredsson 1984):

Number of reports	Total number of births	Fresh or frozen sperm	Sex ratio (% male)
13	3,086	fresh	50.4
9	3,950	frozen	49.8

Compared to a general sex ratio of 51–52% male, these figures suggest very little, if any, change due to insemination at the time of ovulation (using either fresh or frozen sperm). Thus we are suspicious of all theories claiming that timing of intercourse shifts the sex ratio—up *or* down.

*In 1978 Rorvik wrote an amazing "nonfiction" story about the cloning of a man. The publisher, J. B. Lippincott, later admitted that the book was a hoax—after sales of $730,000 (Broad 1982).

AN IN VITRO TECHNIQUE. Over the years, many researchers have tried to enrich sperm samples from humans or farm animals for X-bearing or Y-bearing sperm (Amann and Seidel 1982). Basic work was slow until recent times, because the only sure way of knowing whether the separation had been achieved was to use the putatively X- or Y-enriched samples in artificial insemination and then wait nine months (for both cows and people). Several methods that are now available, however, claim to identify quickly whether a given sperm cell carries an X or a Y chromosome.

In one method, quinacrine is used to stain human sperm (which are killed in the process). When this is done, 40–45% of cells show a bright fluorescent spot called the **F-body** (Barlow and Vosa 1970). The spot is presumed to be the long arm of the Y chromosome, which is known to fluoresce brightly with quinacrine in mitotic cells. Because the F-body is seen significantly less than 50% of the time, one presumes that the Y chromosome is not always positioned inside the sperm head where it can be seen by the microscopist. Sperm cells with an F-body contain, on the average, a few percent less DNA, which suggests that their second sex chromosome is the smaller Y rather than the larger X (Sumner and Robinson 1976). The use of quinacrine to distinguish the sperm types has been criticized, however, because 5–6% of human sperm display *two* F-bodies (Gledhill 1983). No cytologist suggests that so large a fraction of sperm actually has two Y chromosomes, which casts doubt on what the spots really mean. (Sperm cells with two Y's do arise rarely by nondisjunction—see Chapter 8.)

In a newer method for sexing sperm cells, the eggs of the golden hamster (with the zona pellucida removed) are exposed to human sperm, which penetrate the egg membrane without apparent bias toward X- or Y-bearing sperm (Kamiguchi and Mikamo 1986). Once inside the egg, the human chromosomes replicate and begin to undergo division. The Y chromosome, if present, can be identified by standard Giemsa or quinacrine banding of the condensing mitotic chromosomes. This method is more complicated and time consuming than F-body staining of a sperm sample, but most workers believe it to be more reliable. There is room for debate, however (McDonough 1987).

Once these methods were available, many investigators tried to separate X-bearing from Y-bearing sperm using centrifugation, swimming tubes, or electrophoresis. The most remarkable claims of success are those of Ericsson et al. (1973), who place sperm at the top of a column containing progressively more viscous solutions of albumin. According to Ericsson and his colleagues, sperm cells that swim downward and accumulate in the most viscous layer at the bottom are enriched in Y-bearing sperm, *at least as judged by their F-body staining*. The Y-enriched samples are then used for artificial insemination. Ericsson holds several U.S. patents on the albumin gradient column technique, and operates through a firm called Gametrics Limited, which has several dozen licensed centers worldwide for sex preselection. Does the method work? Although no details are given, the latest report we have found notes the birth of 271 males and 87 females—76% male (Ericsson and Beernink 1987). After about 10 years of use, however, these numbers are relatively skimpy, and we are unsure what to make of them (but see Question 9). Complementary to methods for enriching samples for Y-bearing sperm, methods to select for X-bearing sperm have also been used by Ericsson's centers (Corson et al. 1984).

Several recent studies that sexed about a thousand sperm from two dozen men have *failed to confirm Y-enrichment* by the Ericsson albumin-layering technique. Using the hamster egg/human sperm method for determining sex chromosomes, Brandriff et al. (1986) found about 50% Y-bearing sperm *before* processing and about 43% Y-bearing sperm *after* processing—a slight but nonsignificant change in the direction of X-enrichment, rather than Y-enrichment! The sperm samples in this study had been put through the albumin-layering technique either by Ericsson himself or by a licensed center. Using *both* F-body and hamster egg/human sperm analysis, Ueda and Yanagimachi (1987) also found very small changes as a result of processing by the albumin-layering technique: before processing, about 43% Y-bearing sperm; after processing about 46% Y-bearing sperm. In most cases, the results of sexing by the F-body method paralleled those of the hamster egg method. The authors note that, at best, the albumin separation column for Y-enrichment "appears to be inefficient." Thus, if the Ericsson method of producing boys works 76% of the time, it seems unclear why it does so. At the moment, the matter is unresolved.

SUMMARY

1. New reproductive technologies are designed to further human health and happiness, but they may have eugenic overtones as well. Questions of feasibility, decision making, cost, and morality need to be aired.

2. Sterilization to prevent the transmission of presumed undesirable traits—negative eugenics—has had an unsavory history. Voluntary sterilization to prevent the transmission of specific harmful alleles, however, is fairly easy and is often used as a means of birth control.

3. Muller's proposals for positive eugenics were based on the use of artificial insemination with banked sperm of selected men. Artificial insemination of a wife with donor semen is an accepted medical practice when a husband is sterile. The use of frozen rather than fresh sperm provides extra time to test the donor for sexually transmitted disease organisms such as the AIDS virus.

4. Clinics for in vitro fertilization and embryo transfer (IVF-ET) try to help women with blocked oviducts by using eggs obtained directly from the woman's ovary. The IVF-ET process is complex, expensive, and not always successful. Multiple births may occur, but storing frozen embryos can lessen this complication by transferring fewer embryos over more cycles.

5. Other technological answers to infertility include gamete intrafallopian transfer of a wife's egg and her husband's sperm (GIFT), similar manipulations with donated eggs or donated embryos, and surrogate motherhood.

6. Ethical aspects of reproductive techniques center on whether couples have an inherent right to reproduce, on whether it is permissible to manipulate, transfer, or freeze an early embryo, and on whether a third party should be used to help a couple have a child.

7. In animal breeding, artificial insemination using frozen sperm has long been used to markedly increase the availability of superior males. More recently, embryo transfer has been used to increase the use of superior females. Additional technological advances include sexing embryos and making identical twins or chimeric animals.

8. It is not now possible to select the sex of a child using a simple, trouble-free technology. Americans who might use such a technique often want a balanced family overall but with a male as firstborn. Using abortion to try to achieve a desired sex is not generally seen as morally right, however.

9. Many investigators have sought a method for determining the sex of the fetus before conception. Methods based on the timing of intercourse do not seem to be effective. Albumin density-gradient columns for increasing Y-bearing sperm in samples to be used in artificial insemination have been commercially promoted. It is unclear whether this method works for preselecting a male or, if so, why it works.

FURTHER READING

A sampling of subjects apropos of this chapter will be found in many issues of the *Hastings Center Report*, in the *Encyclopedia of Bioethics* edited by Reich (1978), or in the three volumes on *Genetics and the Law* edited by Milunsky and Annas (1976, 1980, 1985). The Office of Technology Assessment (1988a) is a fine summary of most topics in this chapter and provides a wealth of primary references. Ramsey (1970) on the one hand, and Fletcher (1974) and Edwards (1974) on the other, provide contrasting viewpoints on the ethics of genetic manipulations.

Particularly good review articles are written by Allen (1983) on the American eugenics movement, Seibel (1988) on human reproductive technology, Mapletoft (1984) on embryo transfer in farm animals, and Levin (1987) and Holmes (1985) on human sex preselection. Books for a wide readership include Kevles (1985) on the history of eugenics, Edwards and Steptoe (1980) on the story of the first IVF-ET baby, and Andrews (1984) on reproductive technologies generally.

QUESTIONS

(Questions 1–7 are meant primarily to stimulate discussions and we do not supply answers.)

1. Do you approve or disapprove of the following processes that alter human reproduction? For yourself? For others? What conditions would you impose on their use? Which processes are the most feasible? The least feasible?

- Sterilization of sexually mature Down syndrome patients
- Artificial insemination using a husband's sperm
- IVF-ET using a wife's egg and a husband's sperm
- Banking of sperm from distinguished and identified men
- Banking of eggs from distinguished and identified women
- Banking of embryos from distinguished and identified couples
- Reproductive technology to aid homosexual couples or single women
- Amniocentesis or chorionic villus sampling for sex identification
- Abortion to avoid the birth of a boy or a girl
- Timing of intercourse to conceive a child of a given sex
- Artificial insemination with samples of purely X or Y sperm

2. If a democratic government decided that some line of research should be stopped, what could the government do? Should any line of research be prohibited? (As an example, Office of Technology Assessment [1988a] discusses the lack of federal funding for IVF research.)

3. Are Muller's criteria for selecting sperm samples for a bank of frozen samples (page 484) the same as yours? Comment on the heritability of these characteristics.

4. Curie-Cohen et al. (1979) mentions that a doctor using AID arranged 50 pregnancies from one sperm donor. Is there anything wrong with this?

5. When they were killed in a plane crash, Mario and Elsa Rios, a wealthy couple from Los Angeles, left behind two frozen embryos (from IVF using the wife's egg and donor sperm). The couple provided no instructions for the disposition of the embryos. Common law generally holds that those who have been conceived have the right to inherit if born alive. What would you do with the embryos? (See Ozar 1985.)

6. Andrews and Tiefel (1985) offer opposite opinions on the following fictitious scenario: Although now infertile, Sally has a 25-year-old daughter from a prior marriage. Sally and her new husband (himself childless) very much want to start a family. The daughter wants to donate an egg to be fertilized by the husband's sperm in vitro and to become the surrogate mother for her mother by embryo transfer. In this way, it is planned for the baby to be raised by its genetic father and genetic grandmother. Do you approve of this arrangement? Is there additional information you would like to have before deciding?

7. Is Figure 1 a photograph of a human being? If so, where are the characteristic features? If not, when would it become human?

8. Suppose that a sex preselection scheme resulted in the desired sex in 9 cases out of 12 (75% success). By the chi-square test, show that this result could easily be due to chance (at the 5% level of significance).

9. Ericsson and Beernink (1987) report 271 boys and 87 girls from artificial insemination using Y-enriched sperm samples (76% success). By chi-square, test these data against a 179:179 expectation.

10. Assume that three embryos are routinely transferred in an IVF-ET program. Assume that the probability of any one embryo coming to term is 0.1 (10%), and that the embryos implant and develop independently. Using the binomial formula (Chapter 10, page 174), what is the probability of no births, singletons, twins, and triplets on any cycle? (Note that these four probabilities must add to 1).

ANSWERS

8. Chi-square value = $(3)^2/6 + (-3)^2/6 = 3.0$. Because this value is less than 3.84 (see Figure 12 in Chapter 12), the observed results are consistent with random (6:6) expectations.

9. Chi-square value = $(92)^2/179 + (-92)^2/179 = 95$. Because this value is much more than 3.84, the observed results deviate very much from random expectations.

10. P(no births) = $(0.9)^3$ = 72.9%
P(singleton) = $3!/1!2!\ (0.1)^1\ (0.9)^2$ = 24.3%
P(twins) = $3!/2!1!\ (0.1)^2\ (0.9)^1$ = 2.7%
P(triplets) = $(0.1)^3$ = 0.1%
Note that the probabilities of twins and triplets are fairly low, much lower than the actual occurrence of multiple births in IVF-ET.

Chapter 25 Genetic Practices and Prospects

Genetic Counseling
The Procedure
Ethical Concerns

Prenatal Diagnosis
Procedures for Obtaining Fetal Cells
Indications and Results
Sex Prediction

Genetic Screening
Screening For Early Detection of Genetic Disease
Screening For Heterozygotes

Treatments for Genetic Disease
Current Practices
Gene Therapy

Some Thoughts on Science and Society
Public Influence on Science Policy
The Lysenko Experience
Conclusion

> *Increased knowledge of heredity means increased power of control over the living thing, and as we come to understand more and more the architecture of the plant or animal we realize what can and what cannot be done towards modification or improvement.*
>
> *Reginald C. Punnett (1911)*

> *I am angry at those well-meaning theologians and politicians who wish to curtail genetic engineering research. To them I would like to say: "Until you have had intimate contact with those who have suffered or have yourself experienced the pain, strokes, seizures, and leg ulcers; the ridicule from peers, low self-esteem, desire to die, and diminishing hope for the future associated with sickle cell anemia, do not deny me the right to decide for myself and my children whether to try the procedures." So long as the proper guidelines are in place, the research should proceed.*
>
> *Ola Mae Huntley (1984)*

Both Punnett (referring to Mendel's work) and Huntley (referring to advances in molecular technology) consider the consequences of the genetic knowledge of their day. Because it comes so much closer to the material basis of life, modern genetic engineering has evoked quite different points of view. Pessimists predict a host of problems: damage to human health or to the environment, unwarranted intrusions into private affairs, unethical or irreligious experimentation on animals or human embryos, or perversions of eugenic principles. Optimists, on the other hand, expect that genetics will contribute a great deal more to alleviate hunger and disease throughout the world. Especially in the 1980s, researchers have developed powerful molecular methods for probing disease, and they feel that these will equip medical personnel with new tools to fix genetic parts that are broken (Fraser 1988).

In this chapter we examine the implications of advances in human genetics for individual families and for society. We begin by surveying the growing field of genetic counseling, including the required fact-finding techniques of prenatal diagnosis and genetic screening, areas of both promise and controversy. Although very few genetic diseases can now be treated effectively, we discuss the several methods that do exist and stress that there is nothing inherent in genetic disorders that makes them untreatable. Indeed, as infectious diseases are brought under control, the relative importance of disorders that are wholly or partly genetic has been steadily increasing. We note the prospects for cures by what is called gene therapy—the ultimate application of molecular genetic technology. Finally, we describe the incredible Lysenko experience as an example of the triumph of politics over science, and conclude with some further remarks about the role of human genetics in our society.

GENETIC COUNSELING

Every year, thousands of families are affected by the birth of an abnormal child. About 0.5% of newborns have a chromosomal abnormality with moderate to severe phenotypic effects. Another 0.5–1% suffer the consequences of single-gene defects (dominant, recessive, or X-linked), and about 2% have a malformation that may be due in part to heritable factors (Porter 1982). Altogether, about 4% of newborns have a serious defect that is recognized at birth or within their first year. It has been estimated that about a third of all children in pediatric hospitals are being treated for conditions that have a genetic component. Thus there can be no doubt about the need for increased genetic information and counseling services for the many families that require it.

The aim of genetic counseling is to convey medical and genetic facts to an affected or potentially affected family and to explain the options that may be available. It involves much more than simply reeling off statements of probability, however. Counselors must make certain that the information has been understood, and they must try to see that the families are provided with a full range of medical and social services while they adjust to their genetic situation emotionally and intellectually (Erbe 1988).

Counseling requires professionals who are thoroughly grounded in both genetics and medicine. To this basic education must be added patience, sensitivity, respect, and the ability to talk easily with people who are likely to be deeply troubled. Illness and disability inevitably bring tremendous stresses to family life, and many broken marriages are found among the parents of children with genetic defects. Counselors must be prepared to deal with people who are experiencing denial, shock, anger, despair, and guilt, and who may turn against the partner supposedly "at fault." All this may require the joint efforts of a team of counselors: a clinician with genetic training, a geneticist with a mathematical background, laboratory personnel, plus a public health nurse, social worker, or genetic associate* skilled in working with families in their home environments. The first heredity clinic employing a team approach to genetic counseling was established at the University of Michigan in 1940. Today, several hundred centers worldwide—usually associated with medical schools—provide advice and guidance on genetic matters (Lynch et al. 1986; National Center for Education in Maternal and Child Health 1985).

*About a dozen training programs in the United States offer a master's degree in genetic counseling. The first one was established at Sarah Lawrence College in 1969 (Scott et al. 1988).

The Procedure

The need for counseling often arises when a child is born with a possibly hereditary disorder and the parents or other family members are concerned with the well-being of future children. In addition, women may be troubled by repeated miscarriages, or they may have heard about the greater risk of certain birth defects with advanced maternal age. People without any history of genetic disorder may also wonder whether they harbor a potentially harmful gene. This situation might occur if a husband and wife are related to each other, or if they belong to an ethnic group in which certain genetic diseases are more frequent than usual. In any event, the counseling process will often include the following steps.

MEDICAL DIAGNOSIS. Genetic counseling requires precise diagnosis, because a number of conditions have multiple causes (see genetic heterogeneity in Chapter 12). Sometimes these variants may be phenotypically indistinguishable from one another, but in other cases an experienced clinician may detect slight differences (such as hemophilia A versus hemophilia B). In addition to examining the propositus, the counselors may want to see other members of the family, or possibly family photographs. The clerical work involved in corresponding with relatives and searching for pertinent medical records and autopsy reports can be very time consuming. Specialists in pediatrics, radiology, cardiology, and so on may be consulted. Laboratory work often includes karyotyping, biochemical analyses of blood, urine, or cultured cells, or molecular analysis with restriction enzymes and DNA probes. But a correct diagnosis is the cornerstone of useful genetic counseling, for a wrong diagnosis may have devastating consequences.

PEDIGREE ANALYSIS. Not only must a complete three to four generation family pedigree be obtained, but the reliability of the collected information must also be assessed. As pointed out in Chapter 3, some important clues—miscarriages, stillbirths, or mildly affected relatives—may be entirely missing. But even if the data are complete, more often than not there will be complicating factors of the types discussed in Chapter 12. Thus deciding among various possible modes of inheritance may not be a simple matter.

ESTIMATING RECURRENCE RISKS. How likely is it that a given condition will recur in a subsequent birth? The answer is straightforward in the case of simple Mendelian traits: 50%, for example, if one parent is known to be heterozygous for a completely penetrant, dominant allele. The calculations of risk may involve complex conditional probabilities, however, if only ancestors or collateral relatives of a couple being counseled exhibit a Mendelian trait.

With the advent of recombinant DNA and other molecular techniques (Chapter 15), more and more disease-causing Mendelian genes can be detected prenatally by analyzing fetal cells obtained through amniocentesis or chorionic villus sampling (see later). Predictions of risk may depend on genetic linkage to a marker RFLP or on direct detection of the causative allele by using a DNA probe complementary to the gene in question (Fraser 1988). Thus it may be possible to refine a risk probability from 50:50 (without mapping information) to 95:5 (if an RFLP marker is, say, five map units away from the gene in question), or even to 100:0 (if there exists a DNA probe for the mutation itself). The number of hereditary diseases that can be analyzed in these ways is growing rapidly, and any list that is given will soon be out of date (Table 1).

If the trait in question is multifactorial or due to a chromosomal aberration, counselors cannot establish a risk figure based on definite rules of transmission (although chromosomal aberrations can be detected by prenatal testing in future pregnancies). In these cases they may make use of empirical risk figures that rely on the statistics of prior experience with the particular phenotype in question. For many congenital malformations, such as spina bifida, cleft lip alone or with cleft palate, or clubfoot, the risk of recurrence of the particular malformation after the birth of one affected child to normal parents is about 2–5%. For certain traits, the geographical

Table 1 Some disease-causing genes that can be directly detected with complementary DNA probes or mapped to nearby RFLPs[a]

Disease	*Chromosomal location of gene*[b]	*Directly detected or approximate map units from RFLP*[c]
Agammaglobulinemia	X long	very close
Alzheimer disease, familial	21 long	very close
Anhidrotic ectodermal dysplasia	X long	very close
Chronic granulomatous disease	X short	very close
Cystic fibrosis	7 long	1–5 map units
Fragile X mental retardation	X long	10 map units
Hemochromatosis	6 short	directly detected
Hemophilia A	X long	directly detected
Hemophilia B	X long	directly detected
Hunter syndrome	X long	10 map units
Huntington's disease	4 short	very close
Hypercholesterolemia, familial	19 short	directly detected
Lesch-Nyhan syndrome	X long	directly detected
Manic depression, autosomal form	11 short	close
Muscular dystrophy, Duchenne/Becker	X short	directly detected
Myotonic dystrophy	19 long	very close
Phenylketonuria	12 long	directly detected
Polycystic kidney disease, adult	16 short	5 map units
Porphyria	1 short	directly detected
Retinitis pigmentosa	X short	8 map units
Retinoblastoma	13 long	directly detected
Sickle-cell anemia	11 short	directly detected
Thalassemia, α	16 short	directly detected
Thalassemia, β	11 short	directly detected
Wilson disease	13 long	3–7 map units

Source: Data primarily from Cooper and Schmidtke (1986, 1987) with updates from a variety of sources.

a. Additional genes, not included here, have been mapped by these and other methods, but DNA probes and RFLPs are increasingly useful.

b. We indicate only the chromosome number and arm, although finer localization has been made in most cases.

c. For many genes, a closer RFLP or a direct probe will have been developed since this table was compiled. (Recall from Chapter 15 that one map unit is approximately a million base pairs.)

or racial background of a couple may be an important element of risk estimation. For trisomic conditions such as Down syndrome, it is important to distinguish between the presence of a translocation in a parent (relatively high risk of recurrence) and a nondisjunctional cause (relatively low risk).

Whether clear-cut or not, the risk figures must be conveyed in such a way that they are sure to be understood. In particular, it should be emphasized that chance has no memory. The risk of recurrence for a simple Mendelian trait is the same for each birth, regardless of prior outcomes. For example, if normal parents produce one child with a recessive disorder, this does *not* mean that the next three children will be normal. Rather, the risk of an affected child on each successive birth continues to be one-fourth for any given recessive trait.

OPTIONS. When the genetic prognosis is unfavorable, a couple may consider only two options for further action: to refrain from childbearing or to risk a defective birth. If abortion is an acceptable alternative, however, prenatal diagnosis followed by possible termination of pregnancy is now available in a growing number of cases (next section). Alternatively, a couple may choose adoption or an appropriate reproductive technology using either a donor egg or donor sperm (Chapter 24).

One's perception of risk and willingness to accept it are highly subjective matters, depending on one's personality, experiences, moral convictions, and especially on the burden of care imposed by the condition in question. Some newborns with serious birth defects (whether genetic or not) may die within a few months. Others may require constant attention from their family throughout childhood and beyond, or may require expensive medical treatment. We do not mean to minimize the grief of parents upon the loss of a infant, but we want to emphasize here the difference between short-term and long-term care. A couple may decide to take a chance on a relatively mild disorder, or on a more serious disorder of short duration, or on one that can be ameliorated by treatment. But they may not be willing to accept the same percentage of risk on a disorder that imposes greater physical, financial, and emotional burdens on themselves and their family.

The humanistic aspects of counseling are extremely important (Applebaum and Firestein 1983; Emery and Pullen 1984). Because a couple may need time to talk it out among themselves, their friends, and their relatives, the process should not be hurried. For example, comprehension and decision-making by the couple may be especially difficult if counseling follows too closely the birth of a child with a severe defect. The counselor must be prepared to understand and help a mother who blames herself for any one of many different actions taken during pregnancy (e.g., smoking or drinking) or a father who may at first refuse to believe that the child is his biological offspring. The setting of counseling sessions must also be considered; a hospital environment may be intimidating, even if the counselor is skillful in personal interactions.

FOLLOW-UP AND SUPPORTIVE SERVICES. Because a counselor's spoken words may be misinterpreted or forgotten, the relevant medical and genetic information is often put in writing. Such a letter usually includes the names and addresses of specialists and appropriate social services, as well as written answers to the major questions raised by the clients. Families should be informed of useful new research results and allowed to decide on the value to them of participating in experimental trials. Services can also be extended to relatives who may unknowingly be at risk, but giving unsolicited information raises a new set of ethical questions (see below). Overall, the counselor must provide expert and caring assistance to persons whose self-esteem has been hurt and whose lives may have suddenly been made complicated by the birth of a child with a genetic disorder.

Ethical Concerns

The goal of counseling is to enable couples under stress to plan their families according to their own values and with the maximum amount of information and support, but it is not always clear how this may best be done. Opinion varies among counselors as to the degree of advice that should be given. At one extreme are those who specifically suggest a course of action based on the counselor's own values or experiences (as is common in a doctor–patient relationship). Others may provide advice, if asked, but with the proviso that it is impossible for the counselor to fully appreciate all the variables that enter into an intensely personal situation. (Counselors could not, in any case, avoid all the nuances of speech and expression that might reveal a strongly held belief.) Most counselors, however, are nondirective—providing accurate information, much empathy, and complete support—but no specific advice about a course of action (Harris 1988).

Another difference in outlook involves a couple's responsibilities to the human species as a whole. Medical personnel are trained to act primarily for the well-being of their patients, but individual reproductive decisions affect the average welfare of future generations. Whatever our collective obligations to descendants are, however, the views of counseled couples, not those of the counselors, ought to prevail.

The question of withholding certain information also arises (Minogue et al. 1988). For example, in one case told to the authors by a genetic counselor, amniocentesis revealed a twin pregnancy with 46,XY and 47,XYY fetuses. The prognosis for double-Y males is uncertain, although most develop within the range of normal variation of XY males (Chapter 8). Although the existence of the two karyotypes was revealed to the prospective parents, they preferred not to be told which twin had which karyotype. Even if a counselor believes that withholding knowledge is in the best interests of the person being counseled, telling a lie might erode the trust necessary to the counseling relationship. (Also, counselors who withhold information may be subjected to law suits if the clients later learn of the secret information.) Generally, counselors tell only the truth—all of it, or virtually all of it.

The conflict between telling or withholding information is especially acute with regard to relatives who have not specifically sought counseling. Although many persons want to know such genetic information, some might not, especially if the "bad news" involves a serious late-onset condition such as Huntington disease. When no treatment is currently available, is a humanitarian purpose served by informing a person that a progressively debilitating neurological disease might strike with such and such a probability? If this person is still in child-bearing years, does the counselor bear any obligation to help prevent the conception of additional affected persons? There are no pat answers to such questions.

PRENATAL DIAGNOSIS

The ability to detect prenatally virtually all major chromosomal aberrations and many gene-controlled biochemical defects has been a tremendous aid to genetic counseling. If a disorder is diagnosed early in pregnancy, the couple has the option of aborting the fetus and beginning again. Fortunately, in about 97% of prenatal analyses the fetus is found *not* to have the disorder in question. This finding relieves anxiety for the parents, who can then anticipate a newborn no more at risk to abnormality than is a random birth. Many couples at substantial risk of having a child with a serious defect (for any reason) refrain from further pregnancies unless they are assured that the risk can be reduced. Thus, *because* of the availability of prenatal diagnosis, life is given to children that would otherwise never be conceived.

Procedures for Obtaining Fetal Cells

There are two major techniques for obtaining a sample of fetal cells for disease diagnosis; both are done on an outpatient basis with minimal discomfort for the woman. In **amniocentesis**, which has been performed since the 1960s, a physician removes a small sample of the amniotic fluid that surrounds the fetus (Fuchs 1980; Elias and Simpson 1986) (Figure 1). A long, thin needle is inserted through the abdominal wall, uterus, and the fetal membranes. The position of the needle with respect to the fetus can be accurately seen by ultrasonic scanning, which makes major surface features of the fetus visible through reflected sound waves. The needle must not, of course, penetrate the fetus, the umbilical cord, or the placenta. Suspended in the amniotic fluid that is withdrawn (about 25 ml = 5 teaspoons) are living cells that have sloughed off from the fetal skin, the linings of the respiratory or urinary systems, or the amnion. Often amniocentesis is done at about the sixteenth week of pregnancy when the fetus, about 13 cm (5 in) from crown to rump, floats in about 200 ml (about 7 oz) of amniotic fluid. After this period, there may not be enough time for a safe abortion (if chosen), because some of the tests for abnormalities require several weeks to complete. Indeed, the main disadvantage of amniocentesis is that the fetal cells obtained are not dividing. Thus, they must be put in culture medium and allowed to multiply for two to four weeks before cytogenetic analyses can be done.

Amniocentesis is more than 99% safe when done by an experienced obstetrician. The risk of abortion due to amniocentesis (that is, above the naturally occurring rate) is no greater than 0.5%, and the risk of serious injury to the mother (bleeding, infection) or to the fetus (needle damage) is very small. Although some women leak amniotic fluid through the vagina for various lengths of time following amniocentesis, this problem does not seem to compromise the delivery of a normal baby (Elias and Simpson 1986). The continuing but uneventful loss of the amniotic fluid in some women suggests that the removal of about 12% of it (25 ml/200 ml) at the time of amniocentesis is probably not harmful to the fetus.

Another method of obtaining cells for analysis, **chorionic villus sampling (CVS)**, has been performed in the United States only since about 1983, although it was used previously in other countries (Blakemore and Mahoney 1986; Rodeck 1987). The major advantage of CVS over amniocentesis is that it is done much earlier in pregnancy—from the ninth to twelfth week of gestation. Thus weeks of anxiety are avoided, the pregnancy can remain a private affair, and abortion (if chosen) is performed during the first trimester when it is medically safer for the mother. In one variation of this process, the physician inserts a catheter guided by ultrasound imaging through the cervical canal (Figure 2). A few bits of tissue, consisting of branched projections from the chorion (the chorionic villi), are sucked off from the developing placenta. In another technique for obtaining chorionic villi, a needle is inserted through the pregnant woman's abdominal and uterine walls. (This procedure is similar to

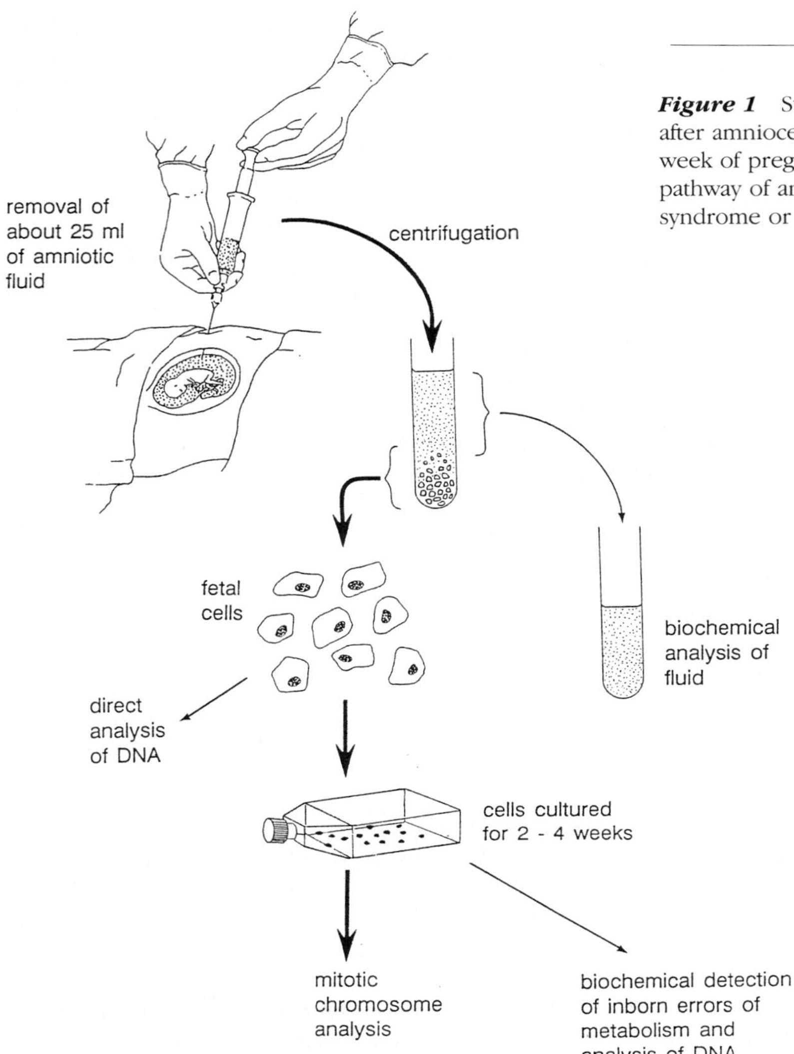

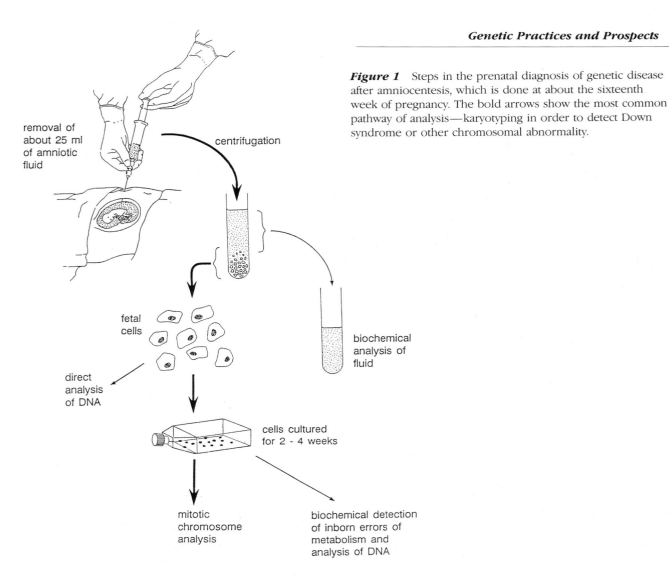

Figure 1 Steps in the prenatal diagnosis of genetic disease after amniocentesis, which is done at about the sixteenth week of pregnancy. The bold arrows show the most common pathway of analysis—karyotyping in order to detect Down syndrome or other chromosomal abnormality.

amniocentesis except that the needle does not penetrate the fetal membranes). After removal, the fetal villi are separated from any attached maternal tissue with the aid of a low power microscope.

The safety of chorionic villus sampling continues to be evaluated, but the procedure seems to be comparable to amniocentesis, or perhaps slightly more risky (Rhoads et al. 1989; Canadian Collaborative CVS-Amniocentesis Clinical Trial Group 1989). In the *natural* course of events, about 2% of pregnancies that are viable at 10 weeks subsequently miscarry, so it has been difficult to evaluate the cause of some abortions following CVS. A special problem of CVS is the risk of Rh immunization of the mother because of possible disruptions in the developing placenta, where maternal and fetal circulations are separated by only a few cell layers (Chapter 17). But overall, CVS has many advantages over amniocentesis from the woman's point of view, and promises to be a valuable procedure.

Indications and Results

Prenatal diagnosis using fetal cells or amniotic fluid is usually offered in three types of situations.

INCREASED RISK OF A CHROMOSOMAL ABNORMALITY. The major use of prenatal diagnosis is for the detection of Down syndrome or other chromosomal aberration (see bold arrows in Figure 1). The risks increase gradually, but more steeply when a pregnant woman is 35 years or older, and a large percentage of all prenatal diagnoses—as much as 85% at some centers—is done in this situation (Verp and Simpson 1985). It is estimated that about half of New York City women 35 or older actually have amniocentesis (Hsu 1986). Chromosomal analysis of fetal cells is also indicated when a couple (of any age) already has one child with a chromosomal abnormality, or when one of the parents is a carrier of a translocation, or in some other situations. Because the

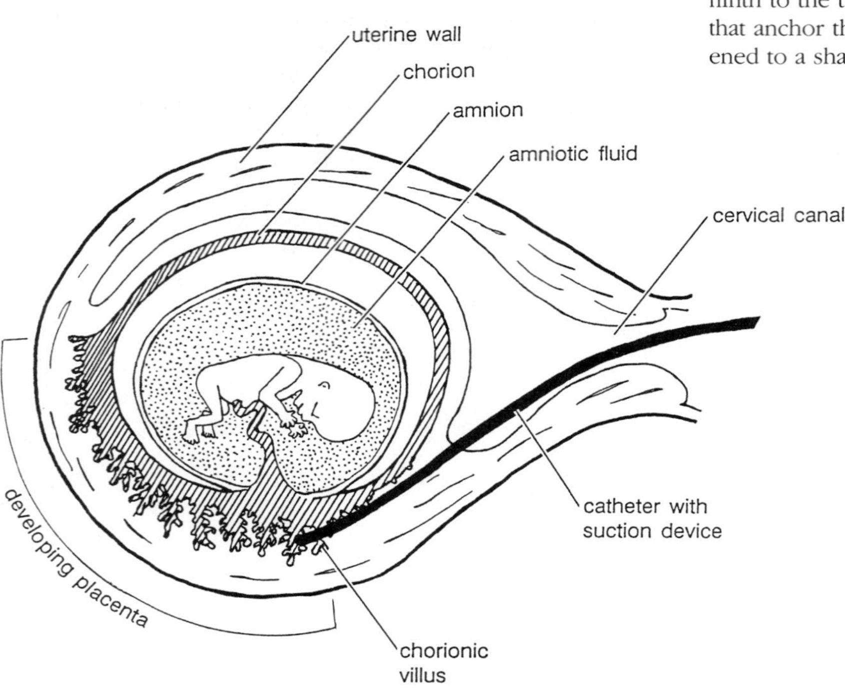

Figure 2 The transcervical method for obtaining fetal cells by chorionic villus sampling (CVS), which is done at the ninth to the twelfth week of pregnancy. The projecting villi that anchor the fetal sac to the uterine wall have been likened to a shaggy carpet.

cells are actively dividing, karyotyping can be done directly with the villous tissue obtained by CVS.

Unfortunately, prenatal diagnoses performed for the indications noted above have only minimal effects on the total burden of abnormal karyotypes among the population of newborns. The reasons are that (1) a large proportion of children (including most Down syndrome cases) are born to women under 35, and (2) in the other situations for which karyotyping is offered, at least one abnormal child will already have been born into the family. Although selecting age 35 for offering amniocentesis has become the standard of practice, many younger women also have the test. (At age 35, the risks associated with amniocentesis are more or less the same as the risks of a major chromosomal abnormality in the fetus.) Offering prenatal diagnosis to *all* pregnant women might be desirable if there were no risks involved, but this is not the case. Nevertheless, in the interest of fairness, the President's Commission (1983) recommends that amniocentesis be more generally available to younger women.

There is still no satisfactory explanation why advancing maternal age increases the likelihood of aneuploid conditions in fetuses and subsequent newborns. We have previously noted that the frequency of Down syndrome increases gradually to about one in a hundred births for 40-year-old women and becomes more frequent for yet older women (Chapter 7). Other chromosomal abnormalities that increase with maternal age include Klinefelter syndrome, trisomy-13, trisomy-18, and triple-X, so that the total risk of chromosomal abnormality in the liveborn offspring of a 40-year-old woman is about 1 in 65.

INCREASED RISK FOR A SINGLE-GENE DISEASE. Although most defects of metabolism are individually rare, over 300 Mendelian disorders are recognized on the basis of an altered (or absent) enzyme or other protein (McKusick 1988). Of these, about 140 have been detected prenatally, or could be using technology developed in other contexts (Desnick et al. 1985). A sampling is presented in Table 2.

The prenatal diagnosis is often made by determining the level of activity of a specific enzyme, *provided* that the encoding gene is expressed in the cells obtained by amniocentesis or CVS. Genes that are transcribed and translated include those responsible for the basic housekeeping functions that occur in virtually all cell types. On the other hand, phenylalanine hydroxylase (the liver enzyme whose absence results in phenylketonuria) is *not* expressed in the cells obtained by amniocentesis or CVS, so it cannot be diagnosed prenatally by this method. But phenylketonuria, and a growing list of other diseases, can be determined by directly analyzing fetal DNA. These powerful molecular methods allow prenatal diagnosis *whether or not* the gene is expressed within the extracted cells and *whether or not* the protein product of the gene is even known.

Table 2 Some inherited biochemical disorders that can be diagnosed prenatally

Error in the metabolism of	*Disorder*[a]	*Metabolic defect*	*Brief description of phenotype*
Amino acids	Maple sugar urine disease (AR)	Deficiency of enzymes needed in breakdown of some amino acids (leading to large excesses of leucine, isoleucine, and valine)	Poor development, convulsions, and early death; urine has maple sugar odor; diet therapy seems promising
	Cystinosis (AR)	Primary defect unknown (but leading to accumulation of cystine in cells)	Several forms; in severe form, kidney function is impaired, leading to poor development, rickets, and childhood death
Sugars	Galactosemia (AR)	Deficiency of enzyme needed in the metabolism of galactose (derived primarily from milk)	Liver and eye defects, mental retardation, and early death if untreated; restrictive diet can control adverse symptoms
Lipids	Tay-Sachs disease (AR)	Deficiency of enzyme needed in the breakdown of a complex lipid (allowing its accumulation in nervous tissue)	Progressive physical and mental degeneration, paralysis, blindness, and death in infancy
	Congenital adrenal hyperplasia (AR)	Deficiency of enzymes needed in the synthesis of sex hormones and of steroids controlling salt and water balance	Mild and severe forms; in the most common form, dehydration and early death; in females, virilization is often seen
Purines	Lesch-Nyhan syndrome (XR)	Deficiency of enzyme involved in the metabolism of purines (resulting in excessive production of uric acid)	Mental retardation, muscular spasms, compulsive self-mutilation; patients may survive into adulthood
Complex polysaccharides	Hurler syndrome (AR) Hunter syndrome (XR)	Defects of connective tissue (allowing accumulation of mucopolysaccharides in cells)	Dwarfism, grotesque facial features, mental retardation; Hunter syndrome less severe
Heme	Porphyria (AD)	Deficiency of enzyme needed in the synthesis of heme	Attacks of abdominal pain accompanied sometimes by impairment of mental functions

a. AR, autosomal recessive; AD, autosomal dominant; XR, X-linked recessive.

By whatever means a disease is diagnosed prenatally, it is important for clients to recognize that these tests are often quite complex and costly and are performed only in families already known to be at risk. Unlike karyotyping, in which mitotic analysis will reveal many different chromosomal abnormalities, a test for alterations in one protein does not reveal alterations in any other.

Tay-Sachs disease is an example of an inborn error of metabolism that can be detected prenatally by an enzyme assay. Recall (from Chapter 16) that Tay-Sachs disease is inherited as an autosomal recessive. A lipid (G_{M2} ganglioside) accumulates in brain cells because an enzyme, hexosaminidase A, that participates in its normal degradation, is missing. The result is progressive degeneration of physical and mental functions to a vegetative state that ends in death by age 3–4. Prenatal diagnosis is offered to couples when a prior child is affected or when both parents are otherwise determined to be carriers (see section on genetic screening). The determination of levels of hexosaminidase A (usually zero in affected fetuses and approximately half the normal amount in heterozygotes) has generally been made on cultured cells after amniocentesis, but it was one of the first assays that was also feasible using chorionic villus tissue (directly or after culturing the cells).

There are many different experimental protocols for

prenatal diagnosis based on fetal DNA (Antonarakis 1989). In Chapter 15, we presented two methods that depend on restriction fragment length polymorphisms (RFLPs). Recall that the Huntington disease gene (*HD*) has been linked to a nearby RFLP (page 287). Diagnosis of Huntington disease prior to the onset of symptoms is possible for some at-risk persons, but with a margin of error determined by the crossover distance between the *HD* gene and the RFLP site. These molecular mapping techniques for *HD* and for several dozen other disease-causing genes can be used on DNA from CVS or amniocentesis (see Table 1). Because there are so many RFLPs scattered throughout the entire genome, soon virtually no gene that is investigated should be without an RFLP to which it can be linked more or less closely. For example, there are about a dozen known RFLP sites in the 60-kb segment of the β-globin gene cluster, which makes possible the prenatal diagnosis of many forms of thalassemia (Weatherall 1985). Even if the linkage is not close, finding RFLP sites on both sides of a disease-causing gene (called *flanking markers*) allows for almost error-free diagnosis (if the necessary heterozygosity exists). For example, many RFLP sites have now been found on either side of the chromosome 7 gene for cystic fibrosis (see Chapter 3).

Also in Chapter 15 we described the 100%-accurate method for distinguishing among DNA samples from homozygotes (including fetuses) for sickle-cell anemia, heterozygotes for the sickle-cell trait, and normal homozygotes. The method is based on an unusual RFLP sequence that includes the site of the sickle-cell mutation. This case is fairly special, however; it would not be expected that many diseases would be caused by a base substitution mutation that destroyed (or created) a recognition sequence of a restriction enzyme within the causative gene and for which a complementary probe was available.

A third method of molecular prenatal diagnosis, which is generally applicable for diseases caused by known base substitutions, depends on so-called **oligonucleotide probes**—synthetic segments of about 20 nucleotides (Chapter 15). These short sequences are made complementary to the portion of a gene that includes the mutation; near the middle is the one base that corresponds to either the normal allele or the mutant allele. Thus the *normal* probe exactly matches the *normal* allele but has a one-base mismatch with the mutant allele; the *mutant* probe exactly matches the *mutant* allele but has a one-base mismatch with the normal allele. Under carefully controlled conditions of hybridization, each probe will bind only to its corresponding allele.

As an example, oligonucleotide probes have been used in the prenatal diagnosis of β-thalassemia in Sardinia (Rosatelli et al. 1985) (Figure 3). In this study of 94 couples and their resulting pregnancies, both husband and wife were heterozygous for the G-to-A mutation (on the sense strand of DNA) that changes the thirty-ninth codon of the β-globin gene from a triplet specifying the amino acid glutamine to a stop codon. The fetal DNA for analysis was obtained directly from cells obtained by CVS at 10 weeks gestation, or from cultured cells derived from amniocentesis at 16–17 weeks. (Analysis of uncultured amniotic cells was not very successful.) The DNA was digested with the restriction enzyme *Bam*HI, which generated a 1.8-kb segment that included codon 39 (Figure 3A). The digested DNA was electrophoresed to separate the fragments, then denatured and transferred to a filter (with the Southern blotting technique). Bathed with the radioactive probes, the 1.8-kb band of fetal DNA was pinpointed by either the one probe, the other, or both probes in the case of heterozygotes, as made evident by autoradiography.

INCREASED RISK FOR A NEURAL TUBE DEFECT. In addition to chromosomal abnormalities or Mendelian disorders, the possibility of **neural tube defects (NTD**s) constitutes another reason for prenatal diagnosis. By the thirtieth day of embryonic development, the furrow of neural tissue that forms along the back closes upon itself to make a tubelike structure, the forerunner of the brain and spinal cord (see Figure 2 in Chapter 6). When the sides of the neural fold fail to close near the head of the embryo, the brain tissue remains exposed and becomes disorganized, resulting in *anencephaly* (literally, no brain). Such fetuses are stillborn or survive only a few days. When the failure to close occurs lower down, the result is one or another form of *spina bifida*, usually characterized by paralysis or weakness below the level of the spinal opening, incontinence of bowel and bladder, and hydrocephalus (abnormal accumulation of fluid in the brain, producing head enlargement). Some patients with less severe types of spina bifida may live many years with greater or lesser disabilities. In the United States, NTDs occur at a frequency of 1–2 per 1,000 births, but they are more frequent (up to 7 cases per 1,000) in some other countries, including Ireland, Scotland, and Wales (Laurence 1983).

Because the recurrence rate for couples with one child with an NTD is about 2% in the United States and about 5% in England (Porter 1982), and the prospect is usually so bleak for affected children and their families, prenatal diagnosis is very useful. The determination of an NTD depends upon the concentration in the amniotic fluid of **α-fetoprotein**, a substance of unknown function that circulates in the fetal blood up to the eighteenth week of development. Small amounts are present in the amniotic fluid of normal fetuses, but its concentration in the amniotic fluid of fetuses with NTDs is usually increased manyfold (especially in the more serious cases).

(A) The β-globin gene

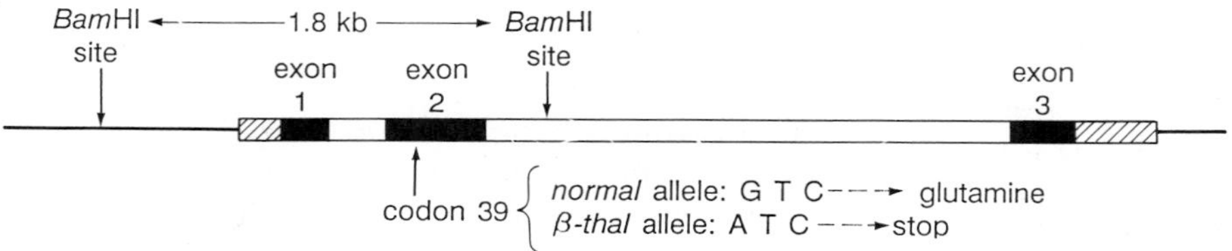

(B) Autoradiographs of Southern blots

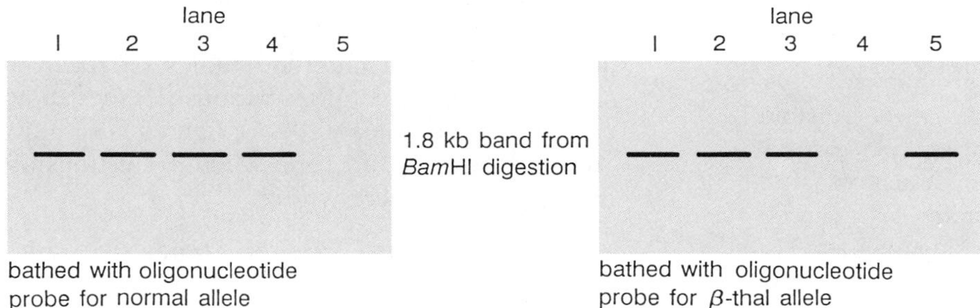

Figure 3 The use of oligonucleotide probes for prenatal diagnosis of β-thalassemia. (A) Diagram of the β-globin gene showing the position of the G-to-A mutation that accounts for most β-thalassemia in Sardinia. (This is the mutation labeled 2 in Figure 7 in Chapter 14.) The recognition sites of the restriction enzyme *Bam*HI to the left and right of codon 39 are also indicated. Two 19-base oligonucleotide probes for the region around codon 39 were identical except for the G or A base corresponding to the mutational site. (B) Sketch of autoradiographs of the Southern blot filters labeled with the oligonucleotide probes. Digestion of parental or fetal DNA was with *Bam*HI. Lane 1, father, heterozygous; Lane 2, mother, heterozygous; Lane 3, fetus, heterozygous; Lane 4, fetus, homozygous normal; Lane 5, fetus, homozygous affected. Note that heterozygotes yield DNA that binds both probes, while the two homozygote types bind either the one probe or the other but not both.

It is thought that in anencephaly and spina bifida, α-fetoprotein occurs in normal amounts in the blood, but the exposed neural tissue allows excess amounts of serum to leak into the amniotic cavity. Ultrasound scanning is also done at the time of amniocentesis to confirm the fetal age and to exclude twins (both of which affect α-fetoprotein concentrations). Ultrasound can also detect directly most of the deformities associated with neural tube defects.

It is disheartening, of course, that it usually takes the birth of one child with an NTD to initiate the prenatal diagnosis that could prevent a second one. Screening of *all* fetuses might be desirable if amniocentesis carried no risk. Recently it has been found that most (but not all) women carrying a fetus with an NTD have elevated levels of α-fetoprotein in their *own* circulation. Thus it would seem worthwhile to do amniocentesis and ultrasonography on the subpopulation of pregnant women with elevated levels of serum α-fetoprotein. But this level of testing has presented some problems that are discussed in the section on screening.

Sex Prediction

Specific prenatal diagnosis is not yet possible for many X-linked recessive disorders that lead to serious disabilities in hemizygous males. Although individually rare, they collectively constitute a significant problem. Examples are the *Wiskott-Aldrich syndrome* and the *lymphoproliferative syndrome* which impair the developing immune system so that affected boys cannot cope with recurrent and often fatal infections. Another example is *anhidrotic ectodermal dysplasia*, which is severe in males, harming especially the skin and mouth, and less severe in heterozygous females, who exhibit only patches of affected skin because of X chromosome inactivation (Figure 7 in Chapter 8).

When a woman is heterozygous for a disease-causing X-linked gene (and her mate is hemizygous normal), one way to prevent the birth of affected boys is to abort *all* male fetuses, even though half of them are expected to be phenotypically normal. The dilemma posed by the abortion of some normal fetuses to avert the birth of possibly abnormal children is not easy to resolve. However, at least six methods for the requisite sex determination of the fetus are available. (See Rogers and Shapiro 1986 for discussion and references.) These methods are arranged here roughly in order of accuracy.

1. *Karyotyping.* This method is the most reliable one—virtually error-free when proper care is taken to avoid maternal cell contamination—but is also the most trouble (see Chapter 4). It can be done after CVS, with or without culturing the cells, or after culturing amniotic fluid cells for a few weeks.

2. *Y-chromosome probes.* In Chapter 24 we noted the development of DNA probes that hybridize to repetitive sequences on the Y chromosome. The method is easy to apply to DNA from cells obtained directly by CVS or by amniocentesis, and it yields accurate results in a few days. In a recent series, the sex of all 148 fetuses was correctly predicted using a Y chromosome probe (Boehm and Drahovsky 1986). Some rare rearrangements of the Y chromosome could produce spurious results, however, because the probes are not specific for (nor close to) the gene that triggers male development (the TDF or testis determining factor—see Chapter 8).

3. *Y-chromosome fluorescence.* We also noted in Chapter 24 that the long arm of the Y chromosome (the heterochromatic region) fluoresces brightly with quinacrine staining. The resulting bright spot can be seen in 20–70% of amniotic fluid cells from male fetuses. Some autosomal sequences also fluoresce brightly, however, so the interpretation is sometimes difficult. As with the region detected by Y chromosome DNA probes, the fluorescent region is not the part that determines maleness; in fact, some apparently normal males are missing all of the fluorescent region (Chapter 8).

4. *Barr bodies in amniotic fluid cells.* We noted in Chapter 8 that Barr bodies (or sex chromatin bodies) in interphase cells represent condensed X chromosomes. Because the number of Barr bodies is one less than the number of X chromosomes, normal females should have one and normal males should have none. Unfortunately, the appearance of Barr bodies is not an all-or-none phenomenon, and their detection in uncultured amniotic fluid cells is erratic, giving a small percentage of both false positives and false negatives.

5. *Hormone levels in amniotic fluid.* Fetal sex can be determined by the levels of testosterone or other sex hormones in amniotic fluid during the second trimester. There is, however, a small area of overlap in values for male and female fetuses (in most studies), leading to a small percentage of errors.

6. *Ultrasonography.* The external genitals may be evident in sonographic determinations at the time of amniocentesis, but different studies on relatively small numbers have given various degrees of accuracy. At best, at least 10% of fetuses fail to give an informative picture, and when predictions are made, at least 5% are wrong.

GENETIC SCREENING

Genetic screening refers to the systematic search of currently healthy populations for persons having a particular genotype (or karyotype) that might cause later trouble for them or for their children (Rowley 1984). Screening has two general purposes:

- To detect a serious genetic disease before the onset of debilitating symptoms (Erbe and Boss 1983). This form of screening, typically on a population of newborns, is consistent with an increasing emphasis on preventive medicine and is similar in intent to *non*genetic screening—for example, periodic mammography or serum cholesterol determinations.
- To identify unaffected carriers in order to counsel them about the risk of producing affected children (Kaback 1983). Screening for such reproductive reasons has its primary effect not on the phenotypes of the individuals that are screened, but rather on the health of future generations.

Although few would disagree with these general goals, screening programs initiated with the best of intentions have sometimes caused ill will among the people who were supposed to be helped (see later). Planning and execution of screening requires that careful attention be paid to the following concerns (President's Commission 1983).

1. *Testing.* To be useful for screening large populations, a diagnostic test should be convenient, inexpensive, safe, and reliable. Identifications can often be made by biochemical tests involving the protein product of a gene or by molecular meth-

ods based directly on DNA. False negatives (persons who have the genotype but are not detected) and false positives (persons who are said to have the genotype but do not) must be minimized. State legislatures that provide for screening programs are also concerned with cost-effectiveness: the money spent in testing and subsequent services should be less than the money that state agencies would spend on dealing with affected persons (Holtzman 1983). Except for very rare conditions, this is likely to be true: the cost of long-term care for institutionalized patients with burdensome genetic conditions, for example, is often very great in measurable resources and beyond calculation in terms of human suffering.

2. *Treatment*. Unless the disease in question can be effectively treated, there may be little point in identifying it before the beginning of actual symptoms, although reproductive decisions might be altered in the case of diseases with late onset. In order to treat all affected persons, financial aid for some families from government or private sources will generally be required.

3. *Counseling*. In order to be effective, the benefits of a specific screening program must be made clear to the persons involved. Unless expert counseling services are made available to guide affected families through difficult times, there is no value —and there can be considerable harm—in identifying carriers of genetic disease. The people responsible for screening programs must also seek to remove social and economic stigmata that could be associated with an "adverse" genotype.

4. *Safeguards*. A difficult question is whether screening should be mandatory or voluntary. Although the genetic health of its citizens can be considered a proper governmental concern, genetic diseases do not pose a danger to others in the same way that contagious diseases such as polio or AIDS do. Although most newborn phenylketonuria screening laws require participation (see below), the constitutionality of this has never been tested in courts. In any event, most laws provide for a right of refusal on religious grounds. Strict confidentiality of medical records is another area of concern, especially with the rapid expansion and centralization of computerized records.

Screening for Early Detection of Genetic Disease

The diagnosis of phenylketonuria (PKU) in newborns was the first large-scale screening program undertaken by governmental agencies and is still the most widely used (Reilly 1975). By the late 1950s it was clear that the most severe abnormalities of PKU could be controlled by diet, but only if treatment was begun in the first few weeks of life. With the development of the Guthrie test (Guthrie and Whitney 1965), the routine screening of newborns prior to their leaving the hospital became practical. In this procedure, a technician pricks a newborn's heel and prepares a small filter paper disk of dried blood. It is placed on top of an agar surface inoculated with bacteria that do not reproduce *unless phenylalanine or closely related chemicals are added* to the nutrient medium. After incubation (along with control disks having known amounts of phenylalanine), a halo of bacterial growth around a test disk means that excess phenylalanine is present, and this is usually diagnostic of PKU (Figure 4). Confirmatory tests using more refined techniques are performed to rule out false positives.

In 1962 Massachusetts became the first state to require that newborn children be screened for PKU; today, all states and several dozen foreign countries have PKU legislation. In the United States, testing is optional in two states and the District of Columbia and mandatory in 48 states (Holtzman 1983). Tens of millions of newborns have been screened and thousands of affected infants have been detected, an incidence ranging from 1 in 8,000 births in Oregon to almost none in Washington, D.C. Unfortunately, few of the state laws provide for financial aid for counseling services or for treatment of children found to have PKU. Although PKU it is not a major public health concern, the screening program is cost effective: Tens of thousands of dollars are spent to detect one case of PKU, but considerably more would be spent on institutionalization of affected individuals in the absence of testing.

The screening programs for PKU are good but not perfect. The Guthrie test does not detect all cases of PKU: false negatives occur because the level of serum phenylalanine may not rise until the newborn has digested substantial amounts of protein. By this time—perhaps several days after birth—the baby may have been discharged from the hospital. Thus rescreening is recommended for all infants who were initially tested within 24 hours of birth (Committee on Genetics 1982). Conversely, because higher than normal serum phenylalanine levels at birth may occur for a variety of reasons, not all Guthrie-positive infants have PKU (Chapter 16). Subsequent testing can usually distinguish between true PKU and mimicking conditions, however. Oligonucleotide probes for the several base substitution mutations within the gene that causes PKU can detect homozygotes as well as heterozygotes, but these methods are currently too expensive for a newborn screening program (DiLella et al. 1988).

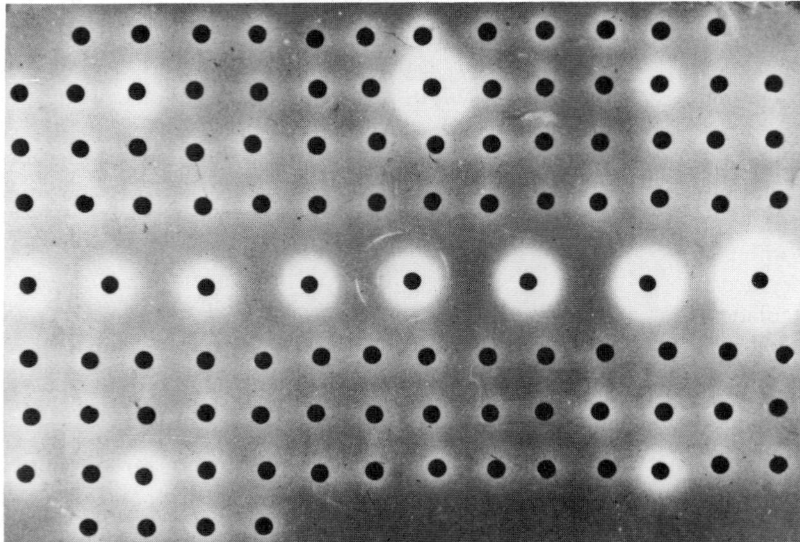

Figure 4 A Guthrie test plate for phenylketonuria screening. Filter paper disks of dried blood from 100 babies have been placed on top of bacteria that only grow if phenylalanine is present. The disks in the middle row are controls that contain increasing concentrations of phenylalanine, from 2 mg/dl (milligrams/deciliter) at the far left to 50 mg/dl on the far right; they support increasingly larger halos of bacterial growth. The test disk near the top center having over 20 mg/dl phenylalanine is from an infant with phenylketonuria. (From Levy 1973.)

OTHER SCREENING TESTS. Some states make multiple use of the blood sample obtained from newborns to test for PKU. For example, the very rare (1 in 200,000 births) autosomal recessive disorder *maple syrup urine disease* can be diagnosed by a Guthrie-type test. (The urine of infants with this disease has the odor of maple syrup.) The otherwise severe symptoms (lethargy, vomiting, poor growth, early death) can sometimes be alleviated by a diet low in the three amino acids leucine, isoleucine, and valine (Holtzman et al. 1981). Another autosomal recessive disorder *galactosemia*, with an incidence of about 1 in 75,000 births, can be diagnosed from dried newborn blood by a direct assay for the missing enzyme (galactose-1-phosphate uridyl transferase). Without treatment, babies (like those with maple syrup urine disease) suffer from severe vomiting, diarrhea, and malnutrition — symptoms that may be evident even without screening. The progress of the disease can usually be halted if the infant is quickly identified and placed on a diet that includes specially formulated milk substitutes and eliminates all other sources of the sugar galactose, which the babies cannot metabolize. For both maple syrup urine disease and galactosemia, unfortunately, infants may die before a definitive diagnosis is made and therapy begun.

Most states also screen newborns for *congenital hypothyroidism*, which has several causes and leads to general physical and mental retardation if untreated. The screening tests measure the levels in cord blood (also dried on filter paper) of the thyroid hormone thyroxine and the thyroid-stimulating hormone. Screening is very useful because the frequency of congenital hypothyroidism is moderate—about 1 in 4,500 births—and early treatment usually prevents gross retardation. Because excessive thyroid medication may be hazardous, care must be taken to avoid treating infants who would develop normally without intervention (Holtzman 1983).

More recently, general prenatal screening for open *neural tube defects* (NTDs) has been initiated in Great Britain and elsewhere by testing the blood of pregnant women for the presence of α-fetoprotein (Brock et al. 1974). Recall from the last section that higher than normal levels of α-fetoprotein are present in the amniotic fluid of fetuses with spina bifida and anencephaly. Alpha-fetoprotein in the amniotic fluid may diffuse into the maternal plasma, so about 80% of fetuses with open NTDs cause elevated levels in the blood of their mothers. Testing programs screen *all* pregnant women in a defined population rather than just those mothers who have had a prior child with an NTD. The justification for general screening is that 95% of newborns with NTDs occur in families with no prior history of abnormalities. This amounts to about 4,000 NTDs per year in the United States.

The screening programs must be carefully managed in order to minimize uncertainties for the pregnant women. Their blood is tested for α-fetoprotein at the sixteenth to eighteenth week of pregnancy, but high levels do not necessarily—or even usually—mean a defective fetus. An α-fetoprotein retest, plus ultrasonography and amniocentesis are needed to yield a more definitive diagnosis, but these additional procedures take time. Out of 21,000 American women in one pilot study (Milunsky and Alpert 1984), the initial α-fetoprotein test detected 249 women with elevated levels. Diagnostic ultrasound on these women led to the recommendation of amniocentesis in just 56 cases. Overall, the panel of tests eventually found 33 abnormal fetuses: 20 with open neural tube defects and 13 with other major congenital abnormalities. In most cases the parents chose to abort the abnormal fetuses. None of the original 21,000 women aborted a normal fetus as a consequence of the testing program.

As the numbers above show, most women with initially elevated serum α-fetoprotein deliver perfectly normal fetuses. They must be provided with emotional support in the face of a preliminary, dreadful diagnosis, which cannot be confirmed or discredited until two to four weeks later. "Many clinical geneticists, however, will bear testimony to the dismay, anger, and resultant chaos in most families who have a child with spina bifida and who were not offered maternal serum α-fetoprotein screening" (Milunsky and Alpert 1984). The question of whether maternal serum α-fetoprotein testing should be restricted to experienced centers or be generally available to all obstetricians continues to be debated (President's Commission 1983; Nightingale 1985). The discussion contrasts the need for expert counseling to allay parental anxieties and to avoid unnecessary abortions with the inequities that exist when up-to-date health care is unnecessarily restricted. In an ambitious program that includes information brochures in several languages, many testing laboratories, and 18 counseling centers, obstetricians in the state of California are now required to offer maternal serum α-fetoprotein screening to all women (Steinbrook 1986).

Screening for Heterozygotes

It is important to detect the carriers of recessive diseases, so that counseling information can be provided to marriage partners when both are heterozygous. For many autosomally inherited inborn errors of metabolism, heterozygotes show only half the level of enzyme activity exhibited by normal homozygotes. Unfortunately, variability in enzyme levels sometimes blurs the distinction between carriers and normals, making mass screening inappropriate. Of the two programs initiated in the 1970s, one was fairly successful by most criteria and one was marked, at least in the beginning, by confusion and frustration. We examine these projects briefly and note the reasons for their divergent results.

TAY-SACHS DISEASE. The opportunity for screening young adults for the recessive allele causing Tay-Sachs disease (TSD) was unusually favorable. The disease is extremely rare in almost all population groups, but it has a moderate frequency (about 1 in 4,000 births) in Ashkenazi Jews of eastern and central European descent. Furthermore, the initial screening test for heterozygosity is convenient, reliable, and discriminating (an assay for hexosaminidase A in serum). When marriage partners are both confirmed as carriers, the option of amniocentesis for all pregnancies and selective abortion of affected fetuses can be offered. Thus couples who are tested can assure themselves of a family without fear of the emotional trauma imposed by the slow, inevitable death of a TSD child.

After a year of careful planning and organization led by Michael Kaback, a geneticist then at Johns Hopkins School of Medicine, a pilot program was undertaken in the Jewish communities of Baltimore and Washington, D.C. With funds provided by a private foundation, information about genetics in general and Tay-Sachs in particular was widely disseminated through newspapers, radio, television, and special pamphlets. About 7,000 persons volunteered for testing in the initial 1971–72 period. Although some couples at first expressed shock, anger, or anxiety on learning that one or the other of them was a carrier, these reactions usually abated with further counseling. Nearly everyone was glad to have been tested, the implications were freely discussed with friends and relatives, and, in the end, the carrier state was not generally regarded as a stigma (Kaback 1977a).

Nevertheless, the feelings of those couples identified as heterozygotes, or their (perhaps untested) relatives, cannot be fully known. It could be "that forces somewhat below the surface of public assertion are at work in the couple's interpersonal dynamics when it is established that one or the other member is a TSD carrier" (Massarik and Kaback 1981). The problem is particularly acute in some Orthodox Jewish communities, where the revelation of a TSD allele in one person may adversely affect the marriageability of a whole kindred, and where abortion is prohibited. Some ultra-Orthodox New York Jews participate in an interesting program in which confidentiality is assured by the traditional matchmaker. This person suggests a meeting between a particular boy and girl, who may eventually marry if they and their parents agree. An individual's Tay-Sachs test result is known only to the screening center (by number) and to the matchmaker. *A person does not learn of his or her carrier status unless matched with another carrier*; in this case the families can say that the pair broke up for other reasons (Merz 1987).

Over 100 major Jewish communities throughout North and South America, Europe, Israel, and South Africa have started TSD screening programs. Through 1984, about 490,000 young Jewish adults have been voluntarily tested, with a carrier detection rate of about 1 in 25 (Ludman et al. 1986). About 480 couples without a prior history of affected children have been identified as *both* being carriers and therefore their pregnancies have been monitored. Recent data suggests that these efforts have contributed to about a 70% decline of TSD among Jewish populations. By 1984 the majority of infants born with Tay-Sachs disease had shifted from Jewish to non-Jewish parents.

SICKLE-CELL ANEMIA. Recall that homozygotes for the sickle-cell allele (about 1 in 500 Black Americans) suffer from severe anemia and painful crises. The symptoms are variable, however, and although many affected persons die young, other homozygotes live long, useful lives. Heterozygotes (about 1 in 10 Black Americans) have normal phenotypes. A simple slide or solubility test can clearly separate persons with at least one sickle-cell allele from homozygous normals (Figure 2 in Chapter 10), and electrophoretic analysis can easily and reliably distinguish all three genotypes.

The history of screening for sickle-cell phenotypes is intertwined with politics. The civil rights movement of the 1960s, combined with the appreciable frequency of the sickle-cell allele among Blacks and the availability of good screening tests, led to a wave of state legislation to try to stem the tide of a "neglected disease." These laws were conceived in good faith and were often passed without debate, having been sponsored by Black politicians and supported enthusiastically by community leaders. Unfortunately, later experience showed that they were too hastily drawn and poorly planned. For example, rather than being *voluntary*, some of the statutes *required* the testing of either prospective marriage partners who could use the information, or school children who could not (Reilly 1975).

In the first place, the benefits of sickle-cell testing were never very clear. Most homozygotes (those with sickle-cell *anemia*) already know the state of their ill health from early age; for some milder cases without prior symptoms, the new-found knowledge of homozygosity is not usually medically useful (but see below), because effective therapy is not generally possible. Heterozygotes (those with sickle-cell *trait*) might use this information for reproductive decisions, but unlike the situation with Tay-Sachs disease, prenatal detection of sickle-cell anemia through amniocentesis or chorionic villus sampling was not then possible. (It is now feasible by DNA analysis.) Thus the options available to a couple who were both heterozygotes were either a 25% risk of affected children or no children at all. Neither choice was easily accepted. "As the screening programs were being promoted at a time of rising racial tension in the United States, the implication that certain Blacks should not have children was seen by some members of the community as having hidden racist motivations" (President's Commission 1983). Making matters worse, the early legislation seldom mandated either medical confidentiality or counseling services. Furthermore, community-based educational programs were inadequate, causing needless fear because of confusion of the burdensome *anemia* with the harmless *trait*. As a consequence, some employers and life insurance companies discriminated unfairly against heterozygotes.

The National Sickle Cell Anemia Control Act of 1972 addressed the inequities of the then-existing state laws by providing federal funds to states and private groups that planned voluntary programs incorporating much more community representation, effective education, counseling services, and strict safeguards. Thus the initial screening difficulties have gradually been replaced with well-planned, comprehensive testing programs that are available to those who request it.

The value of screening *newborn* populations for sickle-cell genotypes has received attention recently, with the recognition that early identification of sickle-cell anemia can reduce mortality by about 15% (Gaston et al. 1986). Infants with sickle-cell anemia may die quickly from overwhelming *Pneumococcus* infections, which often progress from the onset of fever to death in less than 12 hours. If diagnosed early, however, the frequency and severity of infections can be much reduced by daily administration of oral penicillin. Newborn screening is easily and accurately done by electrophoresis of hemoglobin, using a portion of the same dried filter paper sample as is used for PKU. Such tests will also detect unaffected carriers. Because it is not uncommon to confuse the implications of being homozygous or heterozygous, these families need information and counseling to remove anxiety and to emphasize that there is little or no clinical impact on a heterozygous child (Wethers and Grover 1986; Consensus Conference 1987).

TREATMENTS FOR GENETIC DISEASE

In principle, genetically influenced conditions *can* be treated. Like the treatments for infectious diseases, effective therapy of genetic diseases usually requires an understanding of the cascade of causes and effects. In only a very small percentage of the thousands of genetic diseases, however, can anything like a normal phenotype be restored so that treated individuals can pursue life styles unhampered by their handicaps. Therapies are often restricted to reducing the severity of overt symptoms. The limitations arise, in part, because the products of mutant genes often interfere with very fundamental processes; therefore it may be difficult to provide substantial help without significantly harming other aspects of metabolism. In addition, most genetic diseases are individually uncommon; lacking private foundation or government support, they may be slighted in the pursuit of research funding or in the manufacture of therapeutic reagents. Yet even some of the better-known genetic or chromosomal disorders with a long history of well-funded research (muscular dystrophy, cystic fibrosis, albinism, Huntington disease, sickle-cell anemia, Lesch-Nyhan syndrome, Down syndrome, for example) remain largely intractable. We can hope that recent developments in molecular genetics will open up new avenues for intervention.

Actually, the most effective treatment is prevention. For genetic diseases, this means first of all eliminating some of the mutations to abnormal alleles by reducing exposure to unnecessary ionizing radiation and mutagenic chemicals (Chapter 14). Once these new mutations have been transmitted to later generations, prevention becomes a matter of genetic screening and counseling, including prenatal diagnosis, selective abortion, and alternative reproductive technologies. "It is so difficult to do only good in such matters that we are better off putting our strongest efforts into the prevention of mutations, so as to minimize the heavy moral and other burdens of decision once the gene pool has been seeded with them" (Lederberg 1970).

Current Practices

To improve everyday functions, many of us use corrective devices that may be required for nongenetic reasons or whose purpose is far removed from the primary effect of any mutant gene. The list would include eyeglasses, hearing aids, artificial limbs, wheelchairs, and so on. In extreme cases, a few children with immune deficiencies of genetic origin have been confined to environments that seal out infectious agents (Chapter 18). Such external management of the physical environment may involve ingenious biomedical engineering, but it is often cumbersome.

Therapeutic measures that act closer to the genetic defect are extremely varied (Desnick and Grabowski 1981). They may consist of diet modifications to restrict the intake of a toxic substance or to supplement a deficient metabolite. They may involve drugs, surgery, transplantation, or replacement of a polypeptide that the mutant allele is unable to supply. Research at the level of the gene itself, which theoretically could bring about a complete cure, is in experimental stages. These kinds of tactics that help persons with genetically influenced diseases are outlined in Table 3 and discussed below.

DIETARY RESTRICTION OF SUBSTRATE. The diet therapy for phenylketonuria detailed in Chapter 16 remains the prototype for this method of treatment. For phenylketonuria, as well as for galactosemia and maple syrup urine disease (see prior section on screening), the diet reduces the cellular concentration of substances that accumulate when missing enzymes block metabolic pathways. Several dozen genetic diseases are treated with various degrees of success by diet restrictions or diet exclusions of these types (Hsia 1975). In these cases, the end products that are cut off because of the blockage may be available from food or from alternative metabolic pathways.

ADDITION OF A DEFICIENT END PRODUCT. In contrast, the body can sometimes tolerate a high concentration of a precursor but cannot supply the end product. One example is the very rare disorder *orotic aciduria*, which is caused by an autosomal recessive allele and characterized by severe anemia plus mental and physical retardation. In affected persons, the biochemical steps from orotic acid to uridine are blocked. The accumulation of orotic acid causes no difficulties, but the deficiency of uridine (uracil plus ribose) is harmful, because it is utilized not only in the synthesis of RNA but also in that of DNA and some coenzymes.* In the dozen or so cases of orotic aciduria that have been studied, a daily dose of uridine alleviates the severe symptoms, especially if the diet therapy is begun early in life.

DEPLETION OF EXCESSIVE SUBSTANCE. A rare autosomal recessive condition known as *Wilson disease* results from an inborn error in the metabolism of copper, an essential trace element required for the functioning of several important enzymes (Danks 1983). In affected persons, the abnormal deposition of copper in many tissues causes liver and brain damage; if untreated, it leads to death after years of suffering. In some persons, the symptoms are primarily psychological: grossly inappropriate social behavior and bizarre personality changes, which can be mistaken for schizophrenia or manic depression. The age of onset varies from 6 to 50. Although the chain of causation of the symptoms is obscure, drugs that bind copper (so-called chelating agents) and lead to its excretion in the urine can provide dramatic improvement. One such drug is penicillamine (which is derived from penicillin but has no antibiotic activity). Early diagnosis and treatment can completely avert the serious consequences of Wilson disease.

Recall from Chapter 16 that heterozygotes for *familial hypercholesterolemia* (having a frequency of about 1 in 500 persons) develop atherosclerosis and may have heart attacks beginning in their 30s (Goldstein and Brown 1983). The symptoms are due to excessive LDL (low density lipoprotein), which is the major cholesterol-transport system in plasma. Elevated cholesterol levels result from a deficiency of cell receptors (especially in the liver) that take up LDL from the blood. Useful re-

*A coenzyme is a complex molecule that combines with a number of different enzymes in order to activate them. Some coenzymes are synthesized from vitamin precursors.

Table 3 Some genetic diseases that are treatable

Method of treatment[a]	*Disease*
DIETARY RESTRICTION OF SUBSTRATE	
Phenylalanine	Phenylketonuria
Galactose	Galactosemia
Leucine, isoleucine, valine	Maple syrup urine disease
Lactose	Lactase deficiency
Fructose	Fructose intolerance
Neutral fats	Lipoprotein lipase deficiency
Phytanic acid	Refsum syndrome
Fava beans	Favism (G6PD deficiency)
ADDITION OF A DEFICIENT END PRODUCT	
Uridine	Orotic aciduria
Vitamin D and phosphate	Hypophosphatemic rickets
Cortisol	Congenital adrenal hyperplasia
Copper	Menkes syndrome
Thyroxine	Familial goiters
DEPLETION OF EXCESSIVE SUBSTANCE	
Copper (by penicillamine)	Wilson disease
Sterol (by bile binding resins)	Familial hypercholesterolemia
Uric acid (by several drugs)	Gout
SURGICAL REPAIR AND REMOVAL	
Surgical repair	Cleft lip and cleft palate
Splenectomy	Hereditary spherocytosis
Colectomy	Familial polyposis of colon
Thyroidectomy	Medullary thyroid carcinoma syndrome
REPLACEMENT OF A MISSING GENE PRODUCT	
Insulin	Juvenile onset diabetes
Growth hormone	Pituitary dwarfism
Factor VIII	Hemophilia
Various enzymes	Lysosomal storage diseases
Gamma globulin	Agammaglobulinemia
ORGAN AND TISSUE TRANSPLANTATION	
Kidney	Fabry disease
Bone marrow	Severe combined immunodeficiency
Bone marrow	Thalassemia
Bone marrow	Lysosomal storage diseases
Liver	α_1-antitrypsin deficiency
GENE THERAPY	
Addition of normal allele; replacement or correction of abnormal allele	(Experimental)

Source: Data culled from Table 1-12 of Stanbury et al. (1983), with a few additions.
a. The degree of amelioration of symptoms is variable.

ductions in serum cholesterol levels are possible by an unusual route involving the bile acids. These substances, rich in cholesterol products, are made in the liver and transported to the intestines (via the gallbladder), where they disperse ingested fats into small droplets. Normally the bile acids are reabsorbed in the intestines and used again, but drugs are available that cause them to be excreted in the stool instead. This action forces the liver cells to take up more cholesterol from the serum in order to manufacture the missing bile acids. Blood cholesterol levels are thereby reduced about 15 to 30%.

SURGICAL REPAIR AND REMOVAL. Alteration of the phenotype by surgery can improve appearances or basic health. The disfigurement, speech difficulties, and swallowing problems of *cleft lip* and *cleft palate* (occurring by themselves or accompanying other genetic and chromosomal syndromes) can be restored to near normal by skillful surgery. The extra digits of *polydactyly* can also be easily removed. In the autosomal dominant disorder, *spherocytosis*, the red blood cells assume rounded shapes because of increased permeability of the cell membrane. Although the rounded cells carry oxygen reasonably well, their fragility leads to hemolysis and chronic anemia. The fragile cells are entrapped in the spleen, whose phagocytic cells normally remove aging red blood cells from the circulation. Removal of the spleen in patients with spherocytosis apparently allows the rounded red blood cells to survive longer and relieves most of the disease symptoms.

REPLACEMENT OF A MISSING GENE PRODUCT. *Diabetes*, *pituitary dwarfism*, and *hemophilia* can be fairly well treated with insulin, human growth hormone, and clotting factor VIII, respectively. In the first two cases, recombinant DNA techniques are currently used to make the therapeutic proteins. This manufacture involves inserting the cloned human gene into the DNA of nonhuman cells growing in culture and isolating the gene product (Chapter 15). Factor VIII of recombinant DNA origin is expected to be available soon for general medical use, although the protein can be obtained now from pooled human blood (at high cost and risk of acquiring bloodborne viral diseases like hepatitis or AIDS). Because proteins are broken down by digestive enzymes, oral administration is not usually possible. Therefore intravenous injections, often annoying or painful, are required, often over a lifetime. Still, many hemophilic and diabetic patients, using home treatments, may lead nearly normal lives.

Attempts to replace defective enzymes (rather than the nonenzymatic proteins mentioned above) have generally *not* reached the level of clinical practice. Two major difficulties are the short time that the injected enzyme persists in the body, and the inability to deliver the enzyme to the site where it is needed (while protecting it from degradation). Research is active on the *lysosomal storage diseases*. Recall that the lysosomes are subcellular globules containing many hydrolytic enzymes that cleave large molecules into smaller pieces. Defective lysosomal enzymes that allow the large molecules to accumulate cause many genetic disorders, including Fabry, Tay-Sachs, and Gaucher diseases, and the Hunter and Hurler syndromes. One way of getting the enzyme to the lysosomal organelles is by combining the enzyme with molecules that recognize specific receptors on the cell surface. After binding to the surface, the complex enters the cells by endocytosis, as described for LDL in Chapter 16. An alternative route for treating lysosomal storage diseases is by transplantation procedures (described in the next section).

ORGAN AND TISSUE TRANSPLANTATION. An interesting procedure for providing the correct genetic information to patients with an hereditary disease is to transplant to them an organ or tissue from a normal individual (Groth and Ringdén 1984; Parkman 1986). Such a graft may provide a missing enzyme within the patient's body *on a continuing basis*. The enzyme might operate in situ on circulating substrates or be secreted by the cells of the transplant to work elsewhere in the body. Because surgeons have had the greatest experience and success with kidneys (see Chapter 18), those genetic diseases that affect renal function are good candidates for this method of treatment. For example, a few dozen patients with *Fabry disease* have been given a new kidney. Fabry disease, an X-linked recessive trait, results from a missing enzyme (α-galactosidase A) that allows the abnormal accumulation of a lipid in cell lysosomes and in body fluids. (Compare this disorder with Tay-Sachs disease, which was discussed in Chapter 16.) Symptoms include painful crises on a background of chronic pain and cell damage, especially to the skin, kidneys, heart, and eyes. The transplant treatment remains controversial and has been limited to patients whose kidneys were failing anyway. Although the hope that the missing enzyme would be effectively supplied by the new kidney has not usually been realized, some transplant patients have survived for 10 years or more.

We noted in Chapter 18 that David, the "bubble boy" with *severe combined immunodeficiency disease* (*SCID*), received a bone marrow transplant from his sister. Although he did not long survive the transplant, over 100 other patients with SCID have received bone marrow and have done reasonably well (Buckley 1987). Matching for all HLA antigens is not absolutely required in transplants to infants with SCID since they lack both functioning B cells and T cells that mount immunological attacks on the transplanted tissue. On the other hand, the reverse reaction, graft-versus-host disease, remains a problem in mismatched transplants. Recent techniques to rid the donor marrow of mature T cells have sometimes prevented this problem.

The use of bone marrow transplants to treat thalassemia and other hemoglobinopathies is more difficult, because immunosuppressive agents must be used to prevent the rejection of the transplanted bone marrow cells. Nevertheless, in one study, 40 patients aged 8 to 15 years with β-thalassemia were treated in this way (Lucarelli et al. 1987). Each child received HLA-identical bone marrow from a sibling, and about three-fourths survived free of the disease for at least one to three years. About one-fourth died from graft-versus-host disease or other complications of marrow transplantation. Alternative management of thalassemia is also available; it is based on frequent blood transfusions and chelation of the excess iron that results from the treatment. Because these more conservative techniques can provide almost all affected children with 15–20 years of life, it is not clear what therapy is to be preferred.

The impressive variety of therapeutic approaches outlined above should remove any pessimism over the *possibility* of treatment for genetic diseases. On the other hand, many of the maneuvers require sophisticated medical or technical expertise and the accompanying expense. Furthermore, some are half-way measures that have significant side effects or discomfort for users and must be continued for a lifetime.

Gene Therapy

Investigators as well as patients with genetic diseases look forward to the day when it will be possible to correct the effects of abnormal alleles by supplying the normal alleles. The hope is that the transferred genes will cure the basic defect rather than merely alleviate the resulting damage. We use the word *cure* to mean a "single course of treatment that restores health," as penicillin cures pneumonia and appendectomy cures appendicitis (Rosenberg 1985). In such procedures, treatment is discontinued once normal functions are regained.

For single-gene defects, transplantation from a donor may sometimes effect a cure in this sense, but continuing treatment is usually needed to control the immunological side effects of other genes. Another procedure, **gene therapy**, cures a hereditary disease by transferring just single genes rather than whole genomes. We consider here only *somatic* gene therapy, in which the body cells, but not the germ cells, of the patient receive the corrected gene. Of course, if the germ cells were also changed from mutant to normal, the cure could extend through future generations. However, there are objections to this *germinal* gene therapy because of unsavory overtones—the possibility of unwarranted or unwise tampering with another generation's genes. Therefore, investigations of human gene therapy have been limited to the correction of disease within individuals. This is, of course, *the traditional aim of medicine* and, as we see below, is similar in concept to bone marrow transplantation. Although many different avenues are being pursued, we sketch here the steps that may achieve initial success (Anderson 1984; Ledley 1987; Nichols 1988).

CHOOSING A DISEASE. Gene therapy techniques will probably always be limited to single-gene defects, rather than being used to treat chromosomal imbalance (like Down syndrome) or polygenic disorders (like spina bifida and schizophrenia). Because of the inherent risks of the procedure, at least initially, the diseases to be treated would have to be serious ones that are not easily treated by other techniques. The best chance for success will occur when the genetic damage is restricted to a single organ or tissue rather than involving multiple sites (as in cystic fibrosis, Huntington disease, and muscular dystrophy). Diseases expressed in bone marrow cells are good candidates for gene therapy because the tissue is accessible. Possibilities include thalassemia, sickle-cell anemia, and especially severe combined immunodeficiency disease due to a deficiency of the enzyme adenosine deaminase. This form of gene therapy calls for patients to donate their *own* bone marrow and receive it back after the gene defect is corrected.

GETTING THE NORMAL DNA. Hundreds of human genes have already been cloned by the recombinant DNA techniques described in Chapter 15. Getting the nucleotide sequence of the gene itself is not enough, however. It must be combined with whatever promoters or enhancers are required to achieve transcription and translation in the proper cells (and probably not in other cells). Because the length of the DNA construct may be limited, messenger RNA or its complementary DNA copy (rather than the native gene with its lengthy introns) may be combined with promoters and enhancers from other sources. A lot of molecular manipulation is required (Gilboa et al. 1986).

PUTTING THE NORMAL DNA INTO PATIENTS' CELLS. Many techniques for getting the potentially therapeutic gene into host cells have been investigated. The most direct is to simply inject the cloned DNA into recipient cells, where it may integrate into chromosomal DNA. Although microinjection techniques are tedious, they have been effective in the germinal gene therapy of mice that have diseases mimicking those in humans. For example, researchers have injected the normal β-globin allele into zygotes of a strain of mice with the equivalent of β-thalassemia (Constantini et al. 1986). A few of the resulting **transgenic mice** have been cured of the disease and have transmitted the correct allele to their offspring and subsequent generations. Other mice that carry a transferred, specific, correct gene sequence have been cured of the equivalent of the Lesch-Nyhan syndrome (Stout et al. 1985). We must emphasize, however, that the production of transgenic mice by microinjection of zygotes is not applicable to human gene therapy. Even if germinal gene therapy were acceptable (which it is not), and safety were assured (which it is not), there is simply no way to identify which human zygotes might need a new gene.

The injection technique, in any event, is not practical

for bone marrow cells. The target is a stem cell, which is not individually recognizable and is present only at the frequency of one in a million bone marrow cells. (Stem cells are those which replicate into other stem cells while also providing the precursors of the many differentiated marrow cells: B and T lymphocytes, macrophages, other white blood cells, platelets, and red blood cells.) Rather, one technique is to package the correct genetic material into an RNA virus, a so-called retrovirus. The major advantage is that retroviruses infect cells with very high efficiency. Then, in the normal course of their life cycle, they cause a DNA copy of their genetic material to be inserted into the chromosomal DNA of the host cell. One common retrovirus available for potential human gene therapy is the Moloney murine (mouse) leukemia virus. Before using it, researchers excise some of the genetic material of the virus to render it incapable of further reproduction after it invades the target cell. The empty space in the viral RNA is then replaced with suitable human RNA (Figure 5).

It is contemplated that bone marrow cells would be taken from the patient and incubated with the genetically engineered retroviruses. Some stem cells would be infected by the retrovirus, and the DNA copy of the correct gene would be incorporated into the cell's DNA. Insertion appears to be at random places in the genome rather than at the locus of the abnormal gene; this outcome means that the inserted DNA may not be regulated and expressed correctly. There is also the risk of interruption of some normal gene activities or death of the cell. Researchers also worry that a cellular oncogene may be activated, or that infectious viruses may be generated (by recombinational processes) and spread to other cells. Addressing these possible dangers is a major effort of current research.

A different procedure for inserting a normal gene into a patient's cells may eventually solve some of the problems noted above. Called **gene targeting**, the technique was first demonstrated in yeast cells and subsequently in cultured mammalian cells (Smithies et al. 1985; Gregg and Smithies 1986). In the experiments, the mutant gene itself, *at its usual chromosomal position*, was excised and replaced by the normal allele. To do this, researchers incorporated the normal human β-globin gene into a plasmid that also contained a gene for antibiotic resistance and other markers. Cells were transformed by the plasmid, selected for antibiotic resistance, and subsequently shown to possess the new β-globin gene. Incorporation of the new allele is thought to involve *homologous recombination* of complementary DNA sequences, perhaps similar to DNA repair processes. Although the incorporation is so far inefficient (only 1 in 1,000 transformed cells carries the new β-globin gene), it is hoped that gene targeting will be useful in treating bone marrow cells taken from patients with sickle-cell anemia or thalassemia.

EXPRESSING THE GENE WITHIN THE PATIENT. The treated bone marrow would be returned to the patient. If all goes well, the inserted DNA in infected or transformed cells replicates as part of the host cell's DNA. Thus the many thousands of descendant differentiated cells of a given stem cell also carry the corrected gene. But for

LTR = long terminal repeat including regulatory signals

gag = gene for core proteins of virus

pol = gene for reverse transcriptase (a polymerase) and enzyme for integration

env = gene for outer envelope proteins of virus

neo^R = gene for resistance to neomycin (allowing selection of cells that have been infected by the vector)

pro = primate promoter sequence

hADA = gene for human adenosine deaminase

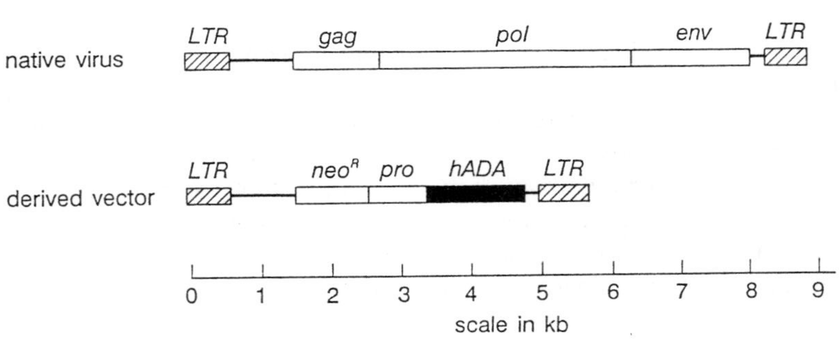

Figure 5 Comparison of the RNA of the native Moloney murine leukemia virus with the RNA of the derived vector for gene therapy of adenosine deaminase deficiency. Because the vector is missing three important viral genes, it is defective as a virus. Therefore, a so-called helper virus system is needed to package the vector RNA in a virus that is capable of infection and integration. (After Gilboa et al. 1986.)

every corrected stem cell, there could be many that have only the abnormal gene (either because they were not removed from the patient or because they were removed but not changed). Methods must be used to insure that the cells with the "good" allele have a competitive edge over the larger number of cells with the "bad" allele. This outcome may be accomplished, for example, by simultaneously incorporating into the vector a drug-resistance gene and then treating the patient with that drug. Another possibility involves limited skeletal radiation of the patient prior to receiving the treated cells, to provide empty bone marrow sites.

Finally, the gene must be expressed in the proper amounts in the patient. Because too much of a gene product may be as bad as too little, the problem is not simple. For example, α and β chains of hemoglobin must be produced in equal amounts within red blood cell precursors. Otherwise, the red cells accumulate abnormal depositions of the excess polypeptide. For this reason, severe combined immunodeficiency disease may be an easier target for successful gene therapy: even if overproduced, excess adenosine deaminase is not likely to be harmful, and even a little additional enzyme seems to be useful. Getting sufficient expression of genes transferred by retroviruses has been a problem in experimental situations.

Many of the separate procedures above have been accomplished in both cultured cells and experimental animals (Nichols 1988). Although the pace of research is unpredictable, success in human somatic gene therapy for a small number of patients is expected in the coming years. Whether the ultimate balance will favor risk or benefit is unclear, but any success will be heralded both as a "celebration of human creativity" and as a "reminder of human obligations to act responsibly" (President's Commission 1982).

SOME THOUGHTS ON SCIENCE AND SOCIETY

The impact of genetics upon the fields of medicine, public health, and agriculture has been impressive. Although increased knowledge of genes and increased skill in genetic technology promise to contribute further to humankind, progress will cause some difficulties as well. A precise accounting of future benefits and risks from genetic or any other scientific research, however, is not possible. Scientists can predict neither the outcome of their experiments nor the many ways their findings might be used. For example, the biochemists who discovered restriction enzymes could hardly have imagined all the applications to gene mapping, prenatal diagnosis, genetic screening, and who-knows-what to come.

Scientists should, of course, make every effort to consider the ramifications of their own work and to convey to the public what it is they hope to accomplish. They could perform a valuable service by also explaining what is *not* known in so many important areas of inquiry. Among some scientists, the popularizing role is looked upon with disdain as a distraction from "real" work in the laboratory. But those who take time and effort to communicate the wonders and limitations of science and who have a knack for putting facts and ideas into language that can be understood by lay people should be applauded. By confronting antiintellectualism in general, and hostility toward science in particular, they make it easier for all scientists to pursue their objectives.

Many commentators have pointed out the serious lack of scientific and technological knowledge among the general public and particularly among the makers of social policy (for example, Westheimer 1988). There is a special problem, however, in explaining scientific ideas to nonscientists. This communication problem arises because of the vertical nature of learning in science—the necessity of knowing one concept before one can understand another. For example, important new evidence for evolution over millions of years comes from recent advances in molecular biology. But this evidence cannot be readily explained to those who do not have a basic knowledge of genetics and biochemistry, which in turn depends upon more facts and ideas from organic and general chemistry. Some mathematics is needed, too—especially a clear notion of probability. (For just this reason, science courses often have prerequisites.) Similar layers of understanding are needed to fully appreciate other discoveries of the last 50 years: atomic energy, transistors, lasers, and superconductivity (to name a few that deal with the structure of matter). Scientists do not collect facts disconnected from each other, nor do they rely on anecdotal accounts. Rather, they seek a unified, reasoned, and satisfying view of the world. The most recent ideas of science—after being debated, modified, and corroborated—are always folded into the accumulated knowledge of prior years.

Public Influence on Science Policy

Many scientists are keenly aware of their responsibilities as educators and are concerned with the welfare of people and their environment. Recall that the continuing debate on recombinant DNA research was precipitated by a group of researchers who voiced their concerns to other members of the scientific establishment and to the public (Chapter 15). In some communities (including our own), the level of open discussion generated by the recombinant DNA controversy was most impressive. In public forums, citizens listened to both scientists and nonscientists, weighed the various points of view carefully, and reacted in responsible ways, neither outlawing the research nor allowing themselves to be intimidated by exuberant proponents.

Another way in which the public (or more usually their governmental representatives) can have a say in the course of scientific research is through control of funding. The views of the general citizenry should be considered when governmental agencies set broad goals and priorities for future areas of investigations. Health and energy research are often emphasized. The generous support for cancer research, or more recently for AIDS research, for example, reflects a strong sense of public urgency about these problems. Once these broad goals have been set, however, the merits of specific research proposals within a field of study can in most cases only be judged by scientific peer review, a process that appraises technical matters and scientific competence.

Public influence in the operation of science and technology has not always led to exclusively beneficial results. For example, some doubt that the much heralded war on cancer has made progress (Bailar and Smith 1986). Although certain cancers have yielded to new treatment methods, the overall mortality from cancer remains almost unchanged over the last 30 years. (Bailar and Smith suggest a shift in emphasis from research on treatments to research on prevention.) Problems associated with population growth and environmental pollution are at least partly due to social policies that fostered biomedical research and industrial development as short-term blessings without fully considering their long-term effects.

Other aspects of public influence have been controversial—for example, the intrusion of specific religious views into public schools, as occurred in the 1925 trial of John Scopes for teaching evolution. More recently, fundamentalists have tried to force Biblical creationism into public school curricula by claiming it is science. Yet courts have consistently found that creationism is a religious concept and not science in any sense of the word (Chapter 22). The potential harm in confusing the two is incalculable. "In a nation whose people depend on scientific progress for their health, economic gains, and national security, it is of utmost importance that our students understand science as a system of study, so that by building on past achievements they can maintain the pace of scientific progress and ensure the continued emergence of results that can benefit mankind" (Committee on Science and Creationism 1984).

From time to time throughout the course of history, an entire science is subverted. One of the most astonishing episodes occurred in Russia, where Communist leaders outlawed all Mendelian genetics for over a quarter of a century, supporting in its stead a brand of Lamarckian thought that they found more ideologically appealing. Because readers may not be familiar with these events, and because they continued into the period of the emergence of molecular biology, we present a short account.

The Lysenko Experience

By 1937 a hard-working agronomist, Trofim D. Lysenko, managed to convince Joseph Stalin that there was a quick solution to the critical shortages in farm production. Rather than using the tedious methods of plant breeding that were based upon Mendelian principles, he claimed to be able to permanently alter the phenotypes of plants and animals by more direct means. From a fruit horticulturist, I. V. Michurin, Lysenko borrowed the idea that environmental shocks, graftings, and the crossbreeding of widely different varieties could "shatter" the hereditary constitution, making it receptive to change by the "assimilation" of external conditions. One could convert winter rye into spring rye by planting it in a different season, he suggested, or make tomato plants bear bigger fruits by grafting them onto a strain that produces large tomatoes. Convinced that mathematics could never be applied to living things, Lysenko eliminated both control groups and statistical analyses from all his experiments (Huxley 1949). Bolstered by Stalin's favoritism toward persons of worker or peasant ancestry, Lysenko's views prevailed over the reactionary ideas of "bourgeois" geneticists, and Lysenko was given authority to begin his projects.

Nikolai I. Vavilov was a world-renowned geneticist and head of the All-Union Institute of Plant Breeding in Leningrad with a research staff of thousands. He was initially open-minded about some of the novel claims of Lysenko and Michurin, and supported their careers. But Vavilov soon recognized that their methods and conclusions were completely unscientific. Having spent his life studying the evolution of cereal crops, Vavilov hoped to use his vast collection of stocks from all over the world to develop new strains suited to Russian environments. Unfortunately, such a project requires many years of intensive work, and Russian leaders were desperately anxious for immediate results. Although the Lysenkoists were unable to deliver on *their* promises either, they managed for many years to appease the government by exaggerating or fabricating experimental results and presenting them in ways that discouraged rational discourse.

Vavilov was politically naive. Lysenko, on the other hand, was skillful in political intrigue, which more than offset his scientific illiteracy. He used his authority to promote those who agreed with him and discredit those who did not. Most traditional geneticists simply lost their posts and had to switch to other lines of work, but the more outspoken among them lost their lives as well.

Vavilov himself was slandered and vilified; arrested in 1940 and convicted on trumped-up charges of spying and sabotage, he was sentenced to be shot, although the sentence was not carried out. Unable to survive harsh and inhumane prison conditions, he died in 1943 from the effects of severe malnutrition (Dobzhansky 1947; Popovsky 1984).

In 1948 Lysenko convened a special meeting of the prestigious Lenin Academy of Agricultural Sciences, ostensibly to discuss scientific issues. His real purpose, however, was to rout the growing opposition that was threatening his programs and authority. To accomplish this, he had Stalin appoint 35 of his partisans to the Academy (including an officer of the KGB), ignoring normal election procedures (Soyfer 1989). At the meeting, the Mendelians were at first encouraged to speak their minds but then roundly attacked by the Lysenkoists. With his packed majority, Lysenko carried the day, and hundreds of the best Russian biologists were dismissed or demoted "at the whim of a group of ignoramuses" (Medvedev 1969). Scientific facts played no part during this extraordinary period, and by 1950 Mendelian genetics in Russia was dead.*

During Premier Khrushchev's tenure, Lysenko continued to thrive. Even at a time when Western biochemists were deciphering the genetic code, he persisted in denying the reality of the gene as the hereditary unit. But the worthlessness of his farm policies became increasingly evident, and his opponents began to speak up again. With the downfall of Khrushchev in 1964, due partly to unproductive agricultural policies, Lysenko also began to lose power. In 1965, the centennial anniversary of Mendel's paper, he was dismissed from all his posts; nine years later he died in obscurity. Although traditional genetics has been restored to the biological sciences in Russia, it took decades to recover from the loss of a whole generation of geneticists.

*Lysenko wrote, "Formal genetics . . . not only retards the advance of theory, it actually hinders such an important matter for farming practice as the improvement of plant varieties and animal strains" (quoted in Popovsky 1984). Lysenko ridiculed the development of hybrid corn when Vavilov endorsed it in the 1930s; in order to approach the corn yields of other countries, Russia finally had to buy hybrid seed stocks from American farmers in the 1950s.

Conclusion

One hopes that such huge aberrations cannot occur in democratic societies, where there are many opportunities for public debate on important issues. Certainly, it is much less likely to happen if both society at large and their leaders know something about the nature of evidence and the concepts of science.

As we learn more about our genetic architecture, it is also to be hoped that people will come to view themselves and their fellow creatures in a clearer and more sympathetic light. Human genetics, a science of differences, emphasizes the uniqueness of each individual.

> It is astonishing that man, a species which displays such a range of variety and which lives in a world populated by tens of thousands of other species, should be so conformist and so intolerant of diversity. If knowledge can help to counter this limitation, the study of the origins of human individual differences is very important. Some understanding of the genetic determinants of behavior—their biological qualities, their extent, and their distribution—might . . . give people an enhanced sense of their uniqueness, as well as acceptance, perhaps even tolerance, of their kinship with others. It might lead, above all, to a more charitable view of those foibles, frailties, peculiarities, and eccentricities which cause many people, their virtues notwithstanding, to be set apart. The headlines of the newspapers on any day proclaim the urgency of the need for this understanding.
>
> Barton Childs et al. 1976

SUMMARY

1. Genetic counselors provide medical and genetic diagnosis and information on the options for further action. While sympathetic to the emotional needs of affected families, most counselors are nondirective, providing accurate information but not advise.

2. Prenatal diagnosis usually relies on analyses of fetal cells obtained by chorionic villus sampling at about the tenth week of gestation or by amniocentesis at about the sixteenth week. If the fetus is found to be abnormal, abortion is an option.

3. Prenatal diagnosis is usually offered if there is an increased risk of a chromosomal abnormality or an inherited disease. Technicians may determine the concentration of a fetal enzyme or analyze fetal DNA. The latter may involve linkage with an RFLP or binding with oligonucleotide probes.

4. Diagnosis of neural tube defects can be made through α-fetoprotein in amniotic fluid; preliminary information can be obtained by α-fetoprotein levels in maternal serum. Fetal sex can be accurately determined by karyotyping or by Y chromosome probes.

5. Genetic screening is the systematic search of populations to detect a genetic abnormality prior to overt symptoms or to determine heterozygosity for recessive disorders. Programs should be carefully planned to assure acceptability in the screened community.

6. The most widely used screening programs test newborns for phenylketonuria and several other diseases from a single blood sample. Other programs screen adult Jewish populations for Tay-Sachs disease and Black populations for sickle-cell anemia.

7. Genetic diseases are treatable in principle. Some treatments rely on diet modifications to restrict a potentially toxic substance or to supply an essential metabolite. Injecting a missing gene-encoded protein or transplanting an organ or bone marrow can be effective in some cases.

8. Gene therapy would cure a genetic disease by supplying a normal allele to correct the consequences of an abnormal one. The normal allele could be provided by splicing it into a defective retrovirus; the patient's bone marrow cells would be removed, infected with the retrovirus, and returned.

9. Scientists need to make a greater effort to explain the nature of science and their own work to the general public. Lay people and their governmental leaders need to become more knowledgable about science and should have some say about the overall pace and type of scientific endeavors.

10. The study of human genetics leads to a greater appreciation of the uniqueness of each individual.

FURTHER READING

Several hefty collections cover much of the material in this chapter. Descriptions of genetic diseases—features, biochemistry, diagnosis, genetics, and management—are in Stanbury et al. (1983) and in Emery and Rimoin (1983); the latter volume has a long section on applied genetics. A work focused on the fetus, with many fine chapters, is edited by Milunsky (1986). Three volumes emphasizing legal aspects of genetics are edited by Milunsky and Annas (1976, 1980, 1985).

Porter (1982) provides a review article on counseling, prenatal diagnosis, and screening; Fuhrmann and Vogel (1983) cover the same material in a book format. The increasing role of molecular techniques in clinical practice is discussed briefly by Fraser (1988) and at length by Weatherall (1985). Similarly, readers should refer to the well-written article by Rosenberg (1985) for a short version of gene therapy and to Nichols (1988) for more detail and background. The President's Commission books on human genetic engineering (1982) and screening (1983) are both worthwhile summaries of basic information and ethical issues. Two informative articles are by Holtzman (1983), on the pitfalls of newborn screening, and by Parkman (1986), on bone marrow transplantation for the treatment of genetic diseases.

QUESTIONS

1. In parts of Africa, Asia, and Latin America, infectious diseases—malaria, schistosomiasis, and a half dozen others—afflict perhaps a billion people. Are the concerns raised in this chapter equally applicable to these populations? What funding should be given to research on tropical diseases versus genetic engineering?

2. Following is a brief resumé of one of the factual "Case Studies in Bioethics" presented in the *Hastings Center Report* (Pauli and Cassell 1978). A baby was born with trisomy-18, a serious syndrome with multiple defects including breathing difficulties, abnormal features, and an expected life span of perhaps a year (Chapter 7). Several problems had to be dealt with quickly: Should a respirator be used if necessary to maintain the baby's life? Should the parents be encouraged or discouraged from seeing or holding the baby? Two writers offer different answers to these questions. What is your opinion?

3. In a so-called wrongful life case, an infant with Tay-Sachs disease sought damages from a medical laboratory for negligence, because the lab failed to find that both her parents were carriers (Annas 1981). The child's lawsuit was based on her suffering. The laboratory's mistake, however, did not

make the child worse off than she otherwise would be. Had the laboratory made a correct diagnosis there would have been no child at all, because it would have been aborted or not conceived. If you had sat on the case, how would you have judged? (The high courts in some states have recognized wrongful life action while others have rejected it —see Botkin 1988.)

4. Can there be any point in amniocentesis or chorionic villus sampling if the parents do not agree beforehand to abort a seriously defective fetus? How do you feel about using these techniques for the sole purpose of determining sex in advance of birth?

5. Why is chorionic villus sampling not useful for the prenatal diagnosis of neural tube defects?

6. One aspect of genetic screening that has received much attention is whether it should be mandatory or voluntary. Do you think that compulsory screening for phenylketonuria represents an unwarranted governmental intrusion into personal lives, or is it a legitimate concern of a state's public health service? What about other screening programs?

7. Gene therapy is not likely to be applicable to raising the intelligence of persons already within a normal range, but it might someday be used to overcome some causes of severe mental retardation, such as that due to phenylketonuria or the fragile-X syndrome. Why is this so?

8. The initial in vivo attempts at gene therapy will be at least somewhat experimental rather than assuredly therapeutic. Thus informed consent (freely given permission from patients after learning of the risks involved) will be a minimum safeguard. Should such experimentation be done in newborn children who cannot give informed consent? Should parents give informed consent for their children?

9. With regard to the label "playing God" for genetic engineering research, the President's Commission (1982) wrote,

> At its heart, the term represents a reaction to the realization that human beings are on the threshold of understanding how the fundamental machinery of life works . . . In this view, playing God is not actually an objection to the research but an expression of a sense of awe—and concern.

Is this your interpretation of "playing God"? Should we be doing this?

APPENDICES

1. *Some Common Units of Measurement*
2. *Basic Mathematics*
3. *Basic Chemistry*
4. *The Human Gene Map*

Appendix 1 Some Common Units of Measurement

QUANTITY	UNIT	ABBREVIATION	EQUIVALENTS
Length	kilometer	km	10^3 m, 0.621 mile
	meter	m	39.4 in, 3.28 ft, 1.09 yd
	centimeter	cm	10^{-2} m (10^2 cm = 1 m), 0.394 in
	millimeter	mm	10^{-3} m (10^3 mm = 1 m)
	micrometer	μm	10^{-6} m (10^6 μm = 1 m)
	nanometer	nm	10^{-9} m (10^9 nm = 1 m)
	Angstrom unit	Å	10^{-10} m (10^{10} Å = 1 m)
	mile	mi	1.61 km
	yard	yd	0.914 m
	foot	ft	0.305 m, 30.5 cm
	inch	in	2.54 cm
Mass	kilogram	kg	10^3 g, 2.20 lb
	gram	g	0.0353 oz
	centigram	cg	10^{-2} g (10^2 cg = 1 g)
	milligram	mg	10^{-3} g (10^3 mg = 1 g)
	microgram	μg	10^{-6} g (10^6 μg = 1 g)
	nanogram	ng	10^{-9} g (10^9 ng = 1 g)
	picogram	pg	10^{-12} g (10^{12} pg = 1 g)
	ton[a]	ton	2,000 lb, 907 kg
	pound[a]	lb	454 g, 0.454 kg
	ounce[a]	oz	28.3 g
Volume (liquid)	liter	l	1.06 qt
	deciliter	dl	10^{-1} liter (10 dl = 1 liter)
	milliliter	ml	10^{-3} liter (10^3 ml = 1 liter)
	microliter	μl	10^{-6} liter (10^6 μl = 1 liter)
	quart	qt	0.946 liter
	pint	pt	0.473 liter
	tablespoon	tbsp	15 ml
	teaspoon	tsp	5 ml
Temperature	Fahrenheit	°F	0°C = 32°F
	centigrade	°C	20°C = 68°F
	(Celsius)		40°C = 104°F
			60°C = 140°F
			80°C = 176°F
			100°C = 212°F

a. Avoirdupois measurements.

Appendix 2 Basic Mathematics

Fractions
Dealing with Fractions
Fractions and Ratios

Exponents
Negative and Zero Exponents
Dealing with Exponential Numbers

Factoring

> *Minus times minus equals plus.*
> *The reason for this we need not discuss.*
> *W. H. Auden; quoted in Martin Gardner (1977)*

This brief review is meant to jog the memory of any reader whose math skills are rusty. (Others can ignore it.) The three sections—fractions, exponents, factoring—should clarify the operations used in this book. The time-honored methods we discuss are merely the rules for working with numbers and handling simple algebraic statements. The hand-held calculator can relieve any computational distress and provide more time for comprehension. We stress methodology more than computation in trying to put you on somewhat friendlier terms with mathematics.

We offer a practical hint applicable to whatever mathematical operation you perform: Ask yourself, "Is the answer plausible?" For example, in some genetics problems, an answer must lie between zero and one. In algebra, too, errors can sometimes be spotted by substituting sensible numbers for the symbols to see if the answer is reasonable.

An aspect of algebra that is difficult to teach is how to set up a mathematical statement from a verbal one, or, conversely, how to understand the meaning of an equation that peers up at you from the page. It is uncertain how people get the "mathematical intuition" to handle this, but it is clear that practice helps. We encourage readers who may once have said "I can't do math" to lay aside their anxieties and give it another try.

FRACTIONS

The analysis of Mendelian genetics in Chapters 2, 3, 10, and 11 involves *fractions* with relatively small denominators: 1/2, 1/3, 3/4, and so on. One could convert these fractions to decimals (0.5, 0.33 . . ., 0.75) but using the fractions as such is often less tedious and, in many cases, is a bit more accurate; for example, 0.33 is not quite the same as 1/3. We encourage you to operate in fractions whenever possible and to *let a fractional answer stand* without further change in form.

Dealing with Fractions

Fractions are added or subtracted by first putting each fraction over a *common denominator* and then *adding or subtracting the numerators*.

$$\frac{1}{2}+\frac{1}{3}=\frac{3}{6}+\frac{2}{6}=\frac{3+2}{6}=\frac{5}{6}$$

$$\frac{2}{3}-\frac{3}{7}=\frac{14}{21}-\frac{9}{21}=\frac{5}{21}$$

Note that a common denominator can always be given by the product of the individual denominators, as shown in the examples above. Although the smallest common denominator is the easiest one to use, any common denominator will do.

To multiply fractions, multiply separately the numerators and denominators:

$$\frac{1}{2}\times\frac{1}{2}=\frac{1\times 1}{2\times 2}=\frac{1}{4}$$

$$\left(\frac{2}{3}\right)\left(\frac{3}{7}\right)=\frac{2\times \overset{1}{\cancel{3}}}{\underset{1}{\cancel{3}}\times 7}=\frac{2}{7}$$

$$\left(\frac{1}{6}\right)\left(\frac{2}{7}\right)\left(\frac{3}{8}\right)(4)(5)=\frac{1\times \overset{1}{\cancel{2}}\times \overset{1}{\cancel{3}}\times \overset{1}{\cancel{4}}\times 5}{\underset{1}{\cancel{6}}\times 7\times \underset{2}{\cancel{8}}\times 1\times 1}=\frac{5}{14}$$

Note that sets of parentheses without any sign between them represent multiplication. Often "canceling" can be done by dividing out factors common to both numerator and denominator. Note also that an integer is equivalent to a fraction whose denominator is 1 (as in the last example), that is, 4 = 4/1.

To divide one fraction by another, first invert the denominator and then multiply:

$$\frac{\frac{1}{4}}{\frac{3}{4}} = \frac{1}{4} \div \frac{3}{4} = \left(\frac{1}{\cancel{4}_1}\right)\left(\frac{\cancel{4}^1}{3}\right) = \frac{1}{3}$$

$$\frac{1}{2} \div 5 = \left(\frac{1}{2}\right)\left(\frac{1}{5}\right) = \frac{1}{10}$$

The following problem incorporates all the rules above. The reader should verify each step.

$$\frac{\frac{3}{4}}{1 - \left(\frac{1}{4}\right)\left(\frac{1}{4}\right)} = \frac{\frac{3}{4}}{1 - \frac{1}{16}} = \frac{\frac{3}{4}}{\frac{15}{16}} = \left(\frac{\cancel{3}^1}{\cancel{4}_1}\right)\left(\frac{\cancel{16}^4}{\cancel{15}_5}\right) = \frac{4}{5}$$

Fractions and Ratios

Consider families with two children. The sequences of sex could be any of the following four types:

	first born	second born
1.	girl	girl
2.	girl	boy
3.	boy	girl
4.	boy	boy

Assuming that a girl is just as likely as a boy, each sequence is equally likely—1/4 for each. Among the four kinds of families, two include both sexes; thus, if we do not consider the order of births, the following *fractions* apply:

families with two girls: $\frac{1}{4}$

families with one of each: $\frac{2}{4} = \frac{1}{2}$

families with two boys: $\frac{1}{4}$

The **ratio** of these numbers is 1:2:1, which expresses, in a shorthand way, their relative magnitudes. *The ratios are simply the numerators when all the fractions have a common denominator.* To reverse this operation, that is, if we had the 1:2:1 ratio in hand and wanted fractions that add to 1, we would add the ratios (1 + 2 + 1 = 4) to get the denominators for each number in the ratio.

For practice, the reader should verify the following equalities:

(1) $\frac{14}{63} = \frac{2}{9}$

(2) $\frac{1}{3} + \frac{2}{5} - \frac{3}{7} = \frac{32}{105}$

(3) $\dfrac{3 - \frac{1}{2}}{5 + \frac{2}{3}} = \frac{15}{34}$

(4) Show that the ratio of the fractions $\frac{1}{2}, \frac{1}{3}, \frac{1}{4}$ is 6:4:3.

EXPONENTS

In Chapters 10, 12, and 20 to 23 it is convenient to use a shorthand notation to express how many times a number is multiplied by itself. Examples are 2^4, $(1/3)^5$, and b^6. Here, the numbers 4, 5, and 6 are the *exponents* of the *base numbers* 2, 1/3, and *b*, respectively. The evaluations of these exponential expressions are

$$2^4 = (2)(2)(2)(2) = 16$$

$$\left(\frac{1}{3}\right)^5 = \left(\frac{1}{3}\right)\left(\frac{1}{3}\right)\left(\frac{1}{3}\right)\left(\frac{1}{3}\right)\left(\frac{1}{3}\right) = \frac{1}{243}$$

$$b^6 = (b)(b)(b)(b)(b)(b)$$

Exponential numbers are sometimes very large. For example, the number of stars in the Milky Way is about 10^{11} (one hundred billion); the number of cells in your body is larger—perhaps $2^{50} \cong 10^{15}$ (one quadrillion).

Negative and Zero Exponents

A base with a *negative* exponent, for example, 2^{-3}, represents the reciprocal of the base with the corresponding positive exponent.

$$2^{-3} = \left(\frac{1}{2}\right)^3 = \frac{(1)(1)(1)}{(2)(2)(2)} = \frac{1}{2^3} = \frac{1}{8}$$

$$16 \times 20^{-4} = \frac{16}{20^4} = \frac{16}{(20)(20)(20)(20)} = \frac{16}{160{,}000} = \frac{1}{10{,}000}$$

$$6\left(\frac{3}{2}\right)^{-2} = 6\left(\frac{2}{3}\right)^2 = \left(\frac{\cancel{6}^2}{1}\right)\left(\frac{2}{\cancel{3}_1}\right)\left(\frac{2}{3}\right) = \frac{8}{3}$$

Any base with a *zero* exponent is defined as the number 1 (not zero, as you might expect). Thus $3^0 = 8^0 = b^0 = 1$. The values of some powers of the base 2 are

m	Value of 2^m
-3	1/8
-2	1/4
-1	1/2
0	1
1	2
2	4
3	8
4	16

Successive values of 2^m are bigger by the factor 2 (but this would not be so if $2^0 = 0$).

Probably because we have 10 fingers, our number system is decimal, that is, geared into base 10. Because of this, exponents of base 10 neatly place the decimal point in a string of digits (often zeroes)—a convenience for very large and very small numbers. We note the important rule: *$n \times 10^m$ means the number n with the decimal point moved m places to the right.*

$$5 \times 10^4 = (5.0)(10)(10)(10)(10) = 50{,}000$$
$$0.731 \times 10^3 = (0.731)(10)(10)(10) = 731$$

In the opposite direction, *$n \times 10^{-m}$ means the number n with the decimal point moved m places to the left.*

$$5 \times 10^{-4} = \left(\frac{5}{1}\right)\left(\frac{1}{10}\right)\left(\frac{1}{10}\right)\left(\frac{1}{10}\right)\left(\frac{1}{10}\right) = \frac{5}{10{,}000} = 0.0005$$

$$0.731 \times 10^{-3} = 0.000731$$

Dealing with Exponential Numbers

If two exponential numbers have the *same base*, they can be multiplied or divided to yield a simpler expression. The rule is that *the exponents are added together when multiplying* and *subtracted when dividing.*

$$3^2 \times 3^5 = (3)(3) \times (3)(3)(3)(3)(3) = 3^{(2+5)} = 3^7$$

$$4^4 \div 4^2 = \frac{(4)(4)(4)(4)}{(4)(4)} = 4^{(4-2)} = 4^2$$

$$b^3 \div b^5 = \frac{(b)(b)(b)}{(b)(b)(b)(b)(b)} = \frac{1}{(b)(b)} = b^{-2}$$

In the following examples, powers of 10 make the placement of the decimal point easier:

The number of seconds in a year is (60)(60)(24)(365)

$$= (6 \times 10^1)(6 \times 10^1)(2.4 \times 10^1)(3.65 \times 10^2)$$
$$= 315 \times 10^5$$
$$= 3.15 \times 10^7$$

The Earth is 15×10^7 kilometers from the sun, whose light travels 3×10^5 kilometers in a second. How long does it take sunlight to get to us? Since distance divided by rate will give us time—

$$\text{time} = (15 \times 10^7)/(3 \times 10^5) = 5 \times 10^2 = 500 \text{ seconds}$$

Thus we see the sun where it was about eight minutes before.

In the following examples, pay attention to the placement of the parentheses. Keep in mind that here the exponent 3 applies to what is *inside* the parentheses:

$$(p^2)^3 = p^2 \times p^2 \times p^2 = p^6$$
$$2(p^2)^3 = 2 \times p^2 \times p^2 \times p^2 = 2p^6$$
$$(2p^2)^3 = 2p^2 \times 2p^2 \times 2p^2 = 8p^6$$

Recall also that the product of two negative numbers, or of two positive numbers, is positive. Negative times positive is negative, however. Thus

$$(-1)^2 = (-1)(-1) = +1$$
$$(-1)^3 = (-1)(-1)(-1) = -1$$

(For a delightful commentary on how something positive can arise as the product of two things less than nothing, see Gardner [1977].)

Here are a few more problems for practice.

(1) How many seconds does it take light to travel the 4800 kilometers from New York to Los Angeles? The answer is 16×10^{-3} second.

(2) show that $9a^2b^3c^4 \div 6a^2b^{-2}c^5 = \dfrac{3b^5}{2c}$

(3) show that $\dfrac{10^3 \times 10^0 \times 10^{-5}}{10^7} = 10^{-9}$

(4) show that $\dfrac{3 \times 10^{23}}{12 \times 10^8} \times \dfrac{16 \times 10^{10}}{4 \times 10^4} = 10^{21}$

(5) What is the volume of the Earth in cubic *feet*, assuming it is a sphere with radius = 4,000 *miles*?

$$V = \frac{4}{3}\pi r^3 = \left(\frac{4}{3}\right)(3.14)(4{,}000 \times 5{,}280)^3$$
$$= 3.95 \times 10^{22} \text{ cubic feet}$$

FACTORING

We review here just a few algebraic operations, the kinds that are used in Chapters 20 to 23. The basic rule in working with equations is that whatever you do to the symbols on one side of an equal sign, you must do to the other side. For example, by subtracting 2 from both sides of

$$a + 2 = b + 5$$

we obtain $a = b + 3$. By subtracting 5 from both sides, we get $b = a - 3$. If both sides are thus treated the same, the equality remains. Simplifying and solving equations for an unknown quantity also involves adding and subtracting like terms, substituting, factoring and expanding, and remembering relationships between symbols that may not be explicit in the equation at hand. These procedures, as well as the operations outlined in the preceding sections, are fairly mechanical—no wizardry is involved. You *do* have to know the rules, however, and intuition, whatever that may be, grows with honest effort.

Parentheses must be used with care to indicate what terms are to be multiplied by what other terms. An expression of the form $(a + b)(c + d)$ indicates that each term within the first pair of parentheses must be multiplied by each term within the second pair. Expanding this expression yields

$$\begin{aligned}(a+b)(c+d) &= a(c+d) + b(c+d) \\ &= ac + ad + bc + bd\end{aligned}$$

Reversing this procedure, that is, factoring the expression on the right to produce the form on the left is not so obvious; with practice comes the ability to "see" that $ac + ad + bc + bd$ factors into $(a + b)(c + d)$. When $a = c$ and $b = d$, the expression above reduces to some common "algebraic formulae." *Three of these are used so often that they should be committed to memory:*

$$\begin{aligned}(a+b)(a+b) &= (a+b)^2 = a^2 + 2ab + b^2 \\ (a-b)(a-b) &= (a-b)^2 = a^2 - 2ab + b^2 \\ (a-b)(a+b) &= a^2 - b^2\end{aligned}$$

These equalities are not difficult to expand (going from left to right) even though the factorings (going from right to left) may not be obvious. Two other expansions and factorings are often useful:

$$(a+b+c)^2 = a^2 + b^2 + c^2 + 2ab + 2ac + 2bc$$

$$\begin{aligned}(d+g)^4 &= [(d+g)^2]^2 \\ &= [d^2 + 2dg + g^2]^2 \\ &= d^4 + 4d^2g^2 + g^4 + 4d^3g + 2d^2g^2 + 4dg^3 \\ &= d^4 + 4d^3g + 6d^2g^2 + 4dg^3 + g^4\end{aligned}$$

The first of these equations is squaring a *trinomial* (an expression with three terms). More generally, one can consider a *multinomial* (many terms); the square of a multinomial is the sum of the squares of each term ($a^2 + b^2 + \ldots$) plus twice the product of each possible pairing of different terms ($2ab + 2ac + \ldots$). The second expansion is one of a series of the form $(d + g)^m$ corresponding to various powers of the *binomial*. An expression for the general term in the expansion of $(d + g)^m$ is called the **binomial formula,** which is discussed further in Chapter 10. Note that the particular case of $(d + g)^4$ can be considered the squaring of a trinomial, that is $[d^2 + 2dg + g^2]^2$.

Here are three worked problems. Assume that $r + z = 1$ (or the equivalent statement, $z = 1 - r$). A piece of paper will hide the solution until you have tried it yourself.

(1) Show that $r^2 + rz - r = 0$

Solution: $r^2 + rz - r = r(r + z) - r$
$= r(1) - r = 0$

(2) Show that $\dfrac{2r^2z^2 + 4rz^3}{2rz^3} = \dfrac{1+z}{z}$

Solution: factor out $2rz^2$ in the numerator and cancel it:

$$\frac{2rz^2(r+2z)}{2rz^3} = \frac{(r+z)+z}{z} = \frac{1+z}{z}$$

Note that canceling means dividing both numerator and denominator by the same thing; since this amounts to multiplying the fraction by 1, it does not change anything. Also we used a bit of trickery by writing $r + 2z$ as $r + z + z$ and reassociating the three terms.

(3) Show that $\dfrac{1-r^2}{z} = 1 + r$

Solution:

$$\frac{(1-r)(1+r)}{z} = \frac{z(1+r)}{z} = 1 + r$$

Appendix 3 Basic Chemistry

Atoms and Molecules
Isotopes
Bonds

Organic Molecules
Functional Groups
Large Molecules

Biological Reactions
Enzymes
Biochemical Pathways

> *Today's brains are yesterday's mashed potatoes.*
>
> *Richard Feynman; quoted in I. S. Shklovskii and Carl Sagan (1966)*

A knowledge of some basic chemical principles can help us better understand many issues that regularly make the news: nutrition, pollution, energy, drugs, and recombinant DNA. Here, we summarize principles that underlie our discussions of molecules and chemical reactions in many chapters. Readers who are familiar with this material can skip it, and those who would like more detail should consult a biochemistry text (for example, Stryer 1988).

ATOMS AND MOLECULES

The materials around us are all composed of tiny particles, *atoms*, each one somewhat like an ultraminiature solar system. The positively charged nucleus of each atom corresponds to the sun, and the negatively charged electrons correspond to the planets. And just as our solar system is mostly space, so, too, are atoms mostly space—even those in hard materials like bricks and steel. This conception of matter is based on experiments with the passage of electric sparks through gases, with X-rays, with radioactive materials, and especially on the way beams of charged particles are scattered about by thin sheets of metal. Principles that explain the behavior of atoms were deduced by the Danish physicist Niels Bohr early in this century and account for the kinds of chemical events we will discuss.

The simplest atom is that of the element hydrogen, H, with one *electron* spinning about an atomic nucleus with one *proton*. These two particles carry exactly the same *electric charge* but the electron is negative and the proton positive. The *mass* of the proton, however, is over 1,800 times greater than that of the electron. Each succeeding atom in the periodic chart of the elements is characterized by the addition of one more proton and one more electron; thus, helium has 2 of each; lithium, 3; beryllium, 4; boron, 5; carbon, 6; nitrogen, 7; oxygen, 8; and so on. The number of protons (or electrons) is called the *atomic number*, which is the same for all atoms of an element. The arrangement of the electrons about the atomic nucleus accounts, in large measure, for the chemical reactions of an element.

Isotopes

The nucleus of any atom (except for the major form of hydrogen) also contains *neutrons*, particles that have about the same mass as a proton but no charge. The sum of *protons plus neutrons* in an atomic nucleus is its *mass number*. For example helium generally has two neutrons and always two protons, for a mass number of 4, which is written helium-4.

An element can vary in the number of neutrons without altering its chemical properties. There is, however, a preferred number of neutrons in naturally occurring elements. For example, almost all carbon atoms have six neutrons and, therefore, a mass number of 12. In nature, however, about 1% of carbon is carbon-13 (with seven neutrons), and a trace amount is carbon-14 (with eight neutrons). These various forms of carbon are called **iso-**

topes: they all have similar chemical characteristics based on an atomic number of 6 but differ in mass numbers and in some physical properties.

Isotopes may be stable or unstable. If stable, the atoms have no tendency to change into anything else without adding to the nucleus energy from outside. If unstable, or *radioactive*, the atoms spontaneously decay—bits of the atomic nucleus are ejected (often with great energy) and the remaining parts rearrange themselves to become some other atom. Carbon-12 and carbon-13 are stable, but atoms of carbon-14 periodically decay. In the process, carbon-14 changes into the stable atom nitrogen-14. Carbon-14 has a half-life of 5,730 years, which means that after this period, half of a given quantity of carbon-14 will be transmuted to nitrogen-14. Isotopes are very useful in studying the mechanism of chemical reactions and biological processes. For example, a radioactive atom can act as a tag that allows its position in a molecule to be monitored or its amount to be measured, often very accurately. Table 1 lists some isotopes that are useful in biological experiments.

Bonds

The four most frequent atoms in living material are, in order, hydrogen (H), oxygen (O), carbon (C), and nitrogen (N). These and other atoms join together to form *molecules*. The simplest molecules consist of just a few atoms of the same kind (molecular oxygen, O_2) or of different kinds (water, H_2O; carbon dioxide, CO_2; ethyl alcohol, C_2H_6O). The subscripts give the number of atoms of each element that are joined together into the molecule.

The atoms are held together by *chemical bonds*. Energy input is often required in bond formation, so a molecule (the sugar glucose, for example) contains more energy than the sum of its separated atoms. For this reason, the bonds within a molecule represent *stored chemical energy*. It is this energy in food substances (measured in calories) that enables organisms to grow, to breathe, to move about, to think, and so on. The breakdown of bonds in energy-rich compounds (sugars and fats, primarily) releases some of the energy that was put there when the bonds were originally formed. This leaves behind simpler, energy-poor molecules. The chemistry going on in all cells forms complex networks of reactions that release the bond energy in small steps, often storing it temporarily in intermediate compounds (one of these is abbreviated ATP—see later) for future use in other reactions. But how does this energy get into your food? It gets there because green plants convert radiant energy from the sun into the chemical energy of sugars and other molecules—food for animals. The number of bonds an atom can make depends upon the number and arrangement of the orbiting electrons.

The most common type of bond in living material is the **covalent bond**. It occurs when two atoms *share electrons* between them. For example, in a molecule of water, one *shared pair* of electrons connects oxygen to each hydrogen. Each connection, a covalent bond, is represented in a structural formula by a short line, as in H-O-H. Hydrogen, with a valence* of *one*, is always connected by *one* covalent bond to other atoms; oxygen, with a valence of *two*, is connected by *two* covalent bonds. Nitrogen atoms always form three covalent bonds and carbon atoms four. Two covalent bonds may be directed toward one atom, as in carbon dioxide (Figure 1). This is called a *double bond*. Readers should be able to draw the structural formulae of molecular oxygen, O_2; methane, CH_4; and ammonia, NH_3.

Another type of connection, weaker than a covalent bond, is also exemplified by water molecules. The oxy-

*Valence refers to the number of electrons that are needed to fill up the outermost "shell" of electrons of an atom.

Table 1 A few isotopes that are useful in biological experimentation

Name of isotope	*Atomic number*	*Mass number*	*Mass number of common isotope*	*Stable or radioactive?*	*Half-life*	*Comment*
Hydrogen-2 (deuterium)	1	2	1	Stable	—	"Heavy hydrogen"
Hydrogen-3 (tritium)	1	3	1	Radioactive	12 years	One of the most useful radioactive tracers
Carbon-14	6	14	12	Radioactive	5,730 years	Another very useful tracer
Nitrogen-15	7	15	14	Stable	—	"Heavy nitrogen"
Phosphorus-32	15	32	31	Radioactive	14 days	Often used to label DNA
Sulfur-35	16	35	32	Radioactive	87 days	Often used to label proteins

<table>
<tr><th>Molecule</th><th>Structural formula</th><th>Shorthand formula</th></tr>
<tr><td>Water</td><td>H—O—H</td><td>H_2O</td></tr>
<tr><td>Carbon dioxide</td><td>O═C═O</td><td>CO_2</td></tr>
<tr><td>Ethyl alcohol</td><td><pre>
   H H
   | |
H—C—C—O—H
   | |
   H H
</pre></td><td>CH_3CH_2OH or C_2H_5OH</td></tr>
</table>

Figure 1 The covalent bonding of some simple substances.

gen in H_2O attracts the shared electrons more strongly than the hydrogens do, so this part of the molecule carries a partial negative charge. Conversely, because the shared electrons are rather far from the hydrogen nuclei, they carry partial positive charges. Although the molecule as a whole is neutral, with no net charge, there is a separation of charge within the molecule. Such molecules are called *polar molecules*. With both positive and negative parts, they tend to attract other polar molecules. In liquid water, for example, the hydrogens tend to orient toward the oxygens of *other* water molecules. These weak attractions *between* molecules of water are called **hydrogen bonds.** Very little energy is needed to make or break such bonds. But there are so many of them in liquid water that its physical and chemical properties are profoundly influenced. For example, water is a good solvent because it wedges between and holds apart the molecules of the dissolved substance with hydrogen bonds. Water is evaporated with more difficulty than most other liquids because it takes a lot of energy to overcome the cohesiveness of the water molecules hydrogen-bonded to one another.

Hydrogen bonds are important in forming the structure of many biological molecules. For example, the helical formation of DNA is stabilized by hydrogen bonding between different parts of the molecule, and proteins owe their three-dimensional conformation (a critically important property) in part to hydrogen bonding. In these cases, some of the hydrogen bonds link a partial positive charge on hydrogen and a partial negative charge on a nitrogen atom.

A third type of bond occurs when an acceptor atom attracts an electron so strongly that the electron's connection with a donor atom is severed completely. In this way, the donor atom loses an electron, becoming a *positively charged ion*; the acceptor atom becomes a *negatively charged ion*. The oppositely charged ions then attract each other to form an **ionic bond**. In table salt, for example, the chlorine atom has pulled an electron away from the sodium atom. Ionic bonds are important in the chemistry of many salts, acids and bases.

ORGANIC MOLECULES

Organic molecules are considered to be those that contain carbon, whether or not they are inside a living organism. Here we categorize some carbon compounds that are important in the structure and function of living things. Readers should note the rules for forming covalent bonds and the tendency for carbon atoms to link covalently into chains.

Hydrocarbons, as the name implies, contain only hydrogen and carbon. These compounds may have been the first organic molecules on Earth, predating the appearance of anything that could be called living. The simplest ones, constituents of natural gas, are

```
Methane        Ethane           Propane

   H            H  H             H  H  H
   |            |  |             |  |  |
H—C—H       H—C—C—H        H—C—C—C—H
   |            |  |             |  |  |
   H            H  H             H  H  H
```

Note that each carbon atom is involved in four covalent bonds, and each hydrogen, one. Other hydrocarbons have longer chains of carbon atoms, sometimes branched, sometimes in circles of six carbons atoms (as in benzene). Compounds with double bonds between adjacent carbon atoms are called *unsaturated* because they contain *fewer hydrogen atoms* than they would if these bonds were single (and the freed electrons were shared with hydrogen instead). Hydrocarbons play a tremendous role in industry and in politics: synthetic and natural rubber are large hydrocarbons; petroleum is a complex mixture of hydrocarbons, from which are manufactured gasoline, fuel oil, asphalt, plastics, and many other petrochemicals.

Functional Groups

Many classes of organic compounds have a hydrocarbon skeleton to which is attached a small group of atoms that enter readily into reactions with other compounds. Such groups are called *functional groups*; a set of compounds with the same functional group engages in similar reactions. About a dozen functional groups are important in living systems, of which three are described here.

1. *Alcohols* are like the hydrocarbons except that one of the hydrogens is replaced by a *hydroxyl group*, —O—H. Examples are

```
Methyl alcohol        Ethyl alcohol
  (a poison)     (in wine, beer, and so on)

    H                  H  H
    |                  |  |
 H—C—O—H           H—C—C—O—H
    |                  |  |
    H                  H  H
```

In addition to being consumed by humans, ethyl alcohol is a major industrial chemical utilized in the manufacture of many other substances.

2. *Organic acids* possess a carboxyl group:

```
     O
    //
 —C
    \
     O
      \
       H
```

The two simplest organic acids are

Formic acid | **Acetic acid (in vinegar)**

```
    Formic acid            Acetic acid
                          (in vinegar)

        O                  H    O
       //                  |   //
   H—C                 H—C—C
       \                   |   \
        O                  H    O
         \                       \
          H                       H
```

Acids tend to ionize to a greater or lesser extent. That is, the hydrogen atom of the carboxyl group completely gives up its electron to the rest of the molecule to become a separate, positively charged *hydrogen ion*, H^+. (Because the hydrogen *atom* consists of just one electron and one proton, the hydrogen *ion* is just a proton.) The ionization of acetic acid is written as follows:

```
  Acetic acid          Acetate ion         Hydrogen ion

         O                    O
        //                   //
  CH3—C       ⟶      CH3—C          +       H+
        \                    \
         O                    O−
          \
           H
```

Note that the OH part of the carboxyl group is not the same as the alcohol functional group; for one thing, the hydroxyl group of alcohols does not ionize.

3. *Amines* possess an *amino group:*

```
      H
     /
 —N
     \
      H
```

The simplest amines are

```
   Methyl amine              Ethyl amine

     H     H                 H  H     H
     |    /                  |  |    /
  H—C—N                  H—C—C—N
     |    \                  |  |    \
     H     H                 H  H     H
```

These compounds are related to ammonia, NH_3, and are classed as *bases*; that is, they tend to ionize in the "opposite" way from the *acids*. The amino functional group *accepts* a hydrogen ion from solution, becoming positively charged.

$$R\text{—}NH_2 + H^+ \longrightarrow R\text{—}NH_3^+$$

The letter R can be used to indicate any hydrocarbon core. Note that an acid is a hydrogen ion donor and a base is a hydrogen ion acceptor.

Many important organic compounds have more than one reactive grouping. Two examples are glycine and glycerol. Glycine carries both an amino group and a carboxyl group:

```
  H      H    O
   \     |   //
    N—C—C
   /     |   \
  H      H    O
               \
                H
```

It is the simplest of about twenty amino acids that join together in long chains to form proteins.

Glycerol, a component of some fats, is a *polyhydroxyl alcohol*:

```
     H  H  H
     |  |  |
  H—C—C—C—H
     |  |  |
     O  O  O
     |  |  |
     H  H  H
```

Large Molecules

Proteins consist of chains of dozens to hundreds of amino acids. The carboxyl group of one amino acid reacts with the amino group of the next one to make the links. Part IV of this text describes some details of their structures and functions. Examples include *hemoglobin*, the oxygen-transporting protein of red blood cells; *antibodies*, the immunity-providing protein of blood plasma and other body fluids; *myosin*, the contractile protein of blood muscles; *keratin*, the structural protein of hair and skin; and most important, *enzymes*, which activate the chemical reactions within an organism. Protein structure is determined by genes. If we know what particular protein is controlled by a gene, we are better able to understand how that gene brings about its characteristic effects.

Nucleic acids also consist of long chains of subunits. These repetitive units are called *nucleotides*, whose structure is discussed in Chapter 13. The genetic material, *DNA*, contains one group of nucleotides; a slightly different group forms the components of the *RNA* molecules that aid in the decoding process. DNA and RNA are called *informational* molecules: the information contained in a sequence of DNA nucleotides is translated by RNA intermediates into a sequence of amino acids—that is, into proteins.

Lipids are fatty substances that are not soluble in water. One type consists of three long-chain organic acids attached one-on-one to the three hydroxyls of glycerol. These *neutral fats* or *triglycerides* pack a caloric wallop. Another type of lipid forms a major component of membranes. A third kind, structurally quite different, is the group of *steroids* containing several six-membered rings of carbon atoms. The steroids include *cholesterol*, *cortisone*, *vitamin D*, and the sex hormones, *estrogen* and *testosterone*.

Carbohydrates include the simple sugars and long chains of these sugars. The breakdown of the six-carbon sugar *glucose* is the primary pathway yielding the chemical energy to drive other reactions. In animal cells, excess glucose may be stored temporarily as *glycogen*, which links together hundreds of glucose subunits. In plant cells, the corresponding storage carbohydrate is called *starch*. The major component of plant cell walls, *cellulose*, is also composed of long chains of glucose subunits. But these are attached to each other differently than in the storage compounds. Wood is about 50% cellulose, and cotton, 100%. In fact, about half of all the carbon atoms in living things are in cellulose. We cannot utilize it as food because we lack the enzymes necessary to break the bonds between the glucose subunits. Cows manage the trick by harboring in their rumens bacteria to do the job for them.

BIOLOGICAL REACTIONS

Many chemical reactions can be generalized by writing:

$$\text{reactant(s)} \longrightarrow \text{product(s)} + \text{energy}$$

In any chemical reaction, energy is conserved; that is, when the chemical bonds joining the atoms of the reactants contain more energy than in the rearranged bonds of the products, this excess energy appears on the product side of the reaction. Part of this extra energy may be picked up as chemical bond energy in other compounds. As an example, consider the breakdown of glucose, which occurs in all cells. Although it involves several dozen separate small steps, the overall reaction is

$$\text{glucose} + \text{oxygen} \longrightarrow \text{carbon dioxide} + \text{water} + \text{energy}$$

The chemical energy within the glucose molecule and oxygen is very much greater than that of the corresponding amounts of carbon dioxide and water. Most of this excess energy appears as heat or is simply unavailable for any kind of useful work in the cell. About a third of it, however, is transferred to the chemical bond energy of a molecule called *adenosine triphosphate, ATP*. The discovery of the role played by ATP was critical in understanding the chemistry of cells. The formation of ATP is *coupled* with each of several of the small steps in the metabolism of glucose. That is, generation of ATP occurs simultaneously with the glucose reactions. Diagrammatically

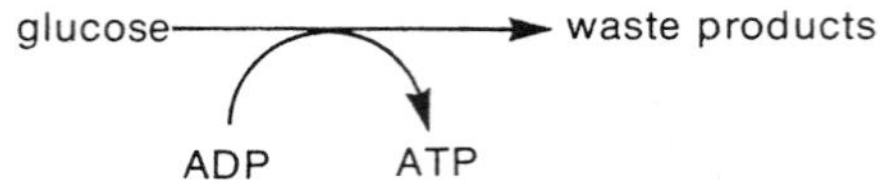

As indicated, the ATP is made from ADP, adenosine *di*phosphate. These two molecules differ by just a single phosphate group (which is present in solution in the cell). ATP is a *high-energy compound* that plays a crucial role in the networks of cellular reactions. ATP is an *energy transfer molecule* that picks up energy from a reaction that has energy to spare (as in the preceding) and delivers it to another reaction that needs an energy input. In financial dealings, money plays an analogous role: dollars made in one transaction are spent in another. Without ATP and molecules with similar functions, metabolism (the totality of chemical reactions in cells) would be as cumbersome as a society that could only barter for its goods and services.

Most chemical reactions in the body are *energy-requiring* and may be written

$$\text{reactant(s)} + \text{energy} \longrightarrow \text{product(s)}$$

All cells must synthesize from simpler substances literally thousands of biochemicals: all DNA and proteins and many of the necessary lipids and carbohydrates. Almost all of these reactions need energy, which is usually supplied by ATP. When ATP loses a phosphate group, becoming ADP, some of the energy that was put into the ATP molecule in its formation becomes available to drive other reactions. Thus the energy transfer circle is completed:

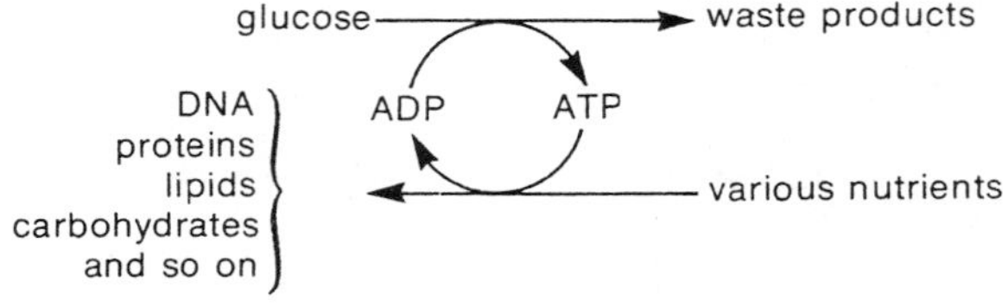

For each arrow, the reader is encouraged to indicate whether energy is produced or required. The reactions are far from 100% efficient. Thus the energy content of the final products—DNA, proteins, and so on—is only a small fraction of the energy available in glucose. Going even farther backward, the energy content of glucose is only a small fraction of the radiant energy of sunlight striking green plants.

Enzymes

Most chemical reactions, regardless of whether they produce energy or need it, will not proceed unless given an initial nudge. For example, paper burns fiercely, yielding large amounts of energy, but only if it is first brought to its kindling temperature. In living things (with fairly constant, low temperatures) activation of reactions is accomplished by **enzymes**. These are large protein molecules that *catalyze* a reaction, that is, speed it up, by bringing the reactants together in one spot and also by straining the molecular bonds that are broken and remade. Without enzymes, random molecular movements would bring the reactants together, but so infrequently as to be negligible. An enzyme is called a *biological catalyst*.

Each enzyme is highly specific, catalyzing only one reaction (or sometimes a few reactions of similar type). For example, the enzyme *hexokinase* catalyzes the first of many small steps in the breakdown of glucose. This is a coupled reaction involving ATP. Hexokinase brings together glucose and ATP and causes a phosphate group from ATP to be transferred to the #6 carbon atom of glucose:

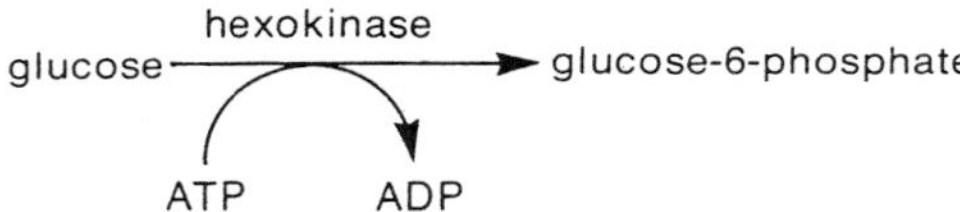

Glucose-6-phosphate has more energy than glucose, the difference (as well as the phosphate group itself) being supplied by the high-energy molecule ATP. Although the total breakdown of glucose to carbon dioxide and water ultimately *yields* considerable energy, this particular starting step *requires* it. (A simple water pump will yield a lot of water after it is first primed with a small amount.)

Enzymes are able to catalyze a reaction partly because of their three-dimensional shape. The folding of the long chain of amino acids leaves an inpocketing of the surface into which the reactants fit. Hydrogen bonding also helps to anchor the reactants to the enzyme and to orient them to each other. After the reaction, the products are released, but the enzyme is unchanged in the process. One enzyme molecule may catalyze tens of thousands of individual reactions per second. Because they are reused, enzymes are needed in only small amounts. Enzymes do eventually wear out, however, and living cells need to continually synthesize those enzymes that are continually used.

Biochemical Pathways

We noted that a sequence of several dozen steps is required for the complete breakdown of glucose. Some steps involve just the splitting or the making of one bond or the transfer of a functional group from one molecule to another. So too are other molecules constructed or degraded, each small reaction step balanced for energy and catalyzed by a specific enzyme. Figure 2 shows how the amino acid *phenylalanine* is converted, in five steps, into the hormone *adrenaline*. Note that each step involves the addition or removal of a single functional group.

One can hardly comprehend the exquisite intricacy of the metabolic pathways operating in a single cell, much less the integration of the reactions occurring in the different cells of the body. Within just one cell, hundreds of enzymes, involved in many dozens of pathways, are at work simultaneously. These reactions are dependent on each other. An intermediate compound in one reaction may be the starting point of another. For example, the intermediate compound DOPA in Figure 2 is the precursor of the melanin pigments found in skin, eyes, and hair. The next two compounds, dopamine and noradrenaline, are themselves neurotransmitters chemically transferring an electric impulse from one nerve cell to the next. The biochemical pathways are thus linked, forming networks of diverging and converging patterns. Each substance appears in the proper concentrations for the correct functioning of the body and its parts. And because the structure of each enzyme depends on one gene (or a few), DNA provides the recipe for all cellular activity. These networks of chemical reactions help determine the final appearance of an organism.

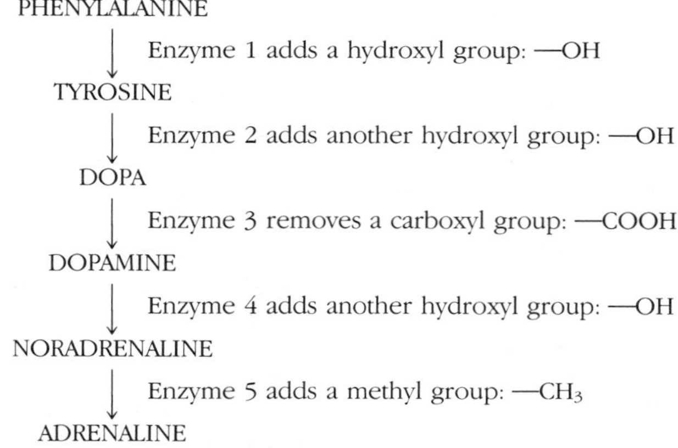

Figure 2 An outline of the synthesis of the hormone adrenaline from the amino acid phenylalanine, which is present in food. (See also Figure 3 in Chapter 16.)

Appendix 4 The Human Gene Map

The human genes listed in this appendix represent about 20% of all mapped genes. The loci of these selected "anchor" genes are shown on the accompanying map. The information is from McKusick (1988).

Gene symbol	*Chromosome location*	*Gene*
ABL	9	Proto-oncogene *ABL*
ABO	9	ABO blood group
ACO1	9	Aconitase, soluble
ACP1	2	Acid phosphatase-1
ACY1	3	Aminoacylase-1
ADHC1	4	ALCOHOL DEHYDROGENASE, CLASS 1, CLUSTER
ADHC1: ADH1	4	Alcohol dehydrogenase-1
ADHC1: ADH2	4	Alcohol dehydrogenase-2
ADHC1: ADH3	4	Alcohol dehydrogenase-3
ADK	10	Adenosine kinase
AFP	4	Alpha-fetoprotein
AK1	9	Adenylate kinase-1, soluble
ALB	4	Albumin
ALDH1	9	Aldehyde dehydrogenase-1
ALPI	2	Alkaline phosphatase, adult intestinal
ALPP	2	Alkaline phosphatase, placental
AMY1	1	Amylase, salivary
AMY2	1	Amylase, pancreatic
ANF	1	Atrial natriuretic factor
APOB	2	Apolipoprotein B
APOLP1	11	APOLIPOPROTEIN I CLUSTER
APOLP1: APOA1	11	Apolipoprotein A-I
APOLP1: APOA4	11	Apolipoprotein A-IV
APOLP1: APOC3	11	Apolipoprotein C-III
APOLP2	19	APOLIPOPROTEIN II CLUSTER
APOLP2: APOC1	19	Apolipoprotein C-I
APOLP2: APOC2	19	Apolipoprotein C-II
APOLP2: APOE	19	Apolipoprotein E
ARAF1	X	Proto-oncogene *ARAF1*
ARG1	6	Arginase, liver
ARSB	5	Arylsulfatase B
AT3	1	Antithrombin III
B2M	15	Beta-2-microglobulin
BCEI	21	Breast cancer estrogen-inducible sequence
BCP	7	Blue cone pigment
C5	9	Complement component-5
CAC	8	CARBONIC ANHYDRASE CLUSTER
CAC: CA1	8	Carbonic anhydrase I
CAC: CA2	8	Carbonic anhydrase II
CAC: CA3	8	Carbonic anhydrase III
CAT	11	Catalase
CBD	X	Deutan colorblindness (green cone pigment)
CBP	X	Protan colorblindness (red cone pigment)
CD13	15	Myeloid membrane antigen, α subunit
CD14	5	Monocyte differentiation antigen CD14
CD4	12	CD4 T-cell antigen
CD8	2	Leu-2 T-cell antigen
CETP	16	Cholesterol ester transfer protein, plasma
CGB	19	CHORIONIC GONADOTROPIN β CHAIN CLUSTER
CHE1	3	Pseudocholinesterase-1
COL1A2	7	Collagen I α-2 chain
COL3A1	2	Collagen III α-1 chain
COL4A1	13	Collagen IV α-1 chain
COL4A2	13	Collagen IV α-2 chain
COL6A1	21	Collagen VI α-1 chain
COL6A2	21	Collagen VI α-2 chain
CP	3	Ceruloplasmin
CPA	7	Carboxypeptidase A
CRH	8	Corticotropin releasing hormone
CRYA1	21	Crystallin α A chain
CRYG	2	Crystallin γ chain
CSF1	5	Macrophage colony stimulating factor
CSF1R	5	Proto-oncogene *FMS*
CSF2	5	Granulocyte-macrophage colony-stimulating factor
CTRB	16	Chymotrypsinogen B

Gene symbol	Chromosome location	Gene
DHFR	5	Dihydrofolate reductase
DHTR	X	Testicular feminization (androgen receptor)
DIA4	16	Diaphorase-4
DNTT	10	Terminal deoxynucleotidyltransferase
DTS	5	Diphtheria toxin sensitivity
ECGF	5	Endothelial cell growth factor
EF2	19	Elongation factor-2
EGF	4	Epidermal growth factor
EGFR	7	Epidermal growth factor receptor
EPO	7	Erythropoietin
ESD	13	Esterase D (s-formylglutathione hydrolase)
ESR	6	Estrogen receptor
ETS1	11	Proto-oncogene *ETS-1*
ETS2	21	Proto-oncogene *ETS-2*
F13A	6	Clotting factor XIII A component
F9	X	Clotting factor IX (hemophilia B)
FES	15	Proto-oncogene *FES*
FGC	4	FIBRINOGEN CLUSTER
FGC: FGA	4	Fibrinogen α chain
FGC: FGB	4	Fibrinogen β chain
FGC: FGG	4	Fibrinogen γ chain
FIM3	3	Homolog of integration site (FMS leukemia virus)
FOS	14	Proto-oncogene *FOS*
FRAXA	X	Fragile site Xq27.3
FSHB	11	Follicle stimulating hormone β-polypeptide
FTL	19	Ferritin light chain
FY	1	Duffy blood group
G6PD	X	Glucose-6-phosphate dehydrogenase
GAPD	12	Glyceraldehyde-3-phosphate dehydrogenase
GC	4	Group-specific component (vitamin-D binding protein)
GCG	2	Glucagon
GLUD	10	Glutamate dehydrogenase
GOT1	10	Glutamate oxaloacetate transaminase, soluble
GRP	18	Gastrin releasing peptide
GSR	8	Glutathione reductase
GST3	11	Glutathione s-transferase-3
GUSB	7	β-Glucuronidase
H	19	Bombay phenotype
HEMA	X	Hemophilia A (clotting factor VIII)

Gene symbol	Chromosome location	Gene
HEXB	5	β-hexosaminidase β chain
HK1	10	Hexokinase-1
HMGCR	5	3-hydroxy-3-methylglutaryl coenzyme A reductase
HMGCS	5	3-hydroxy-3-methylglutaryl coenzyme A synthase
HOX1	7	Homeo box-1
HOX2	17	Homeo box-2
HOX3	12	Homeo box-3
HP	16	Haptoglobin
HPRT	X	Hypoxanthine-guanine phosphoribosyltransferase
HRAS1	11	Proto-oncogene *HRAS1*
IDH2	15	Isocitrate dehydrogenase, mitochondrial
IFNA	9	LEUKOCYTE INTERFERON (α-INTERFERON) CLUSTER
IFNB	9	Fibroblast interferon (β-interferon)
IFNG	12	Interferon, γ or immune type
IGF1R	15	Insulin-like growth factor-1 receptor
IGF2	11	Insulin-like growth factor II (somatomedin A)
IGH	14	IMMUNOGLOBULIN HEAVY CHAIN CLUSTER
IGH: CA1	14	H constant region of IgA1
IGH: CA2	14	H constant region of IgA2
IGH: CD	14	H constant region of IgD
IGH: CE	14	H constant region of IgE
IGH: CEP1	14	H constant region of IgEP1
IGH: CG1	14	H constant region of IgG1
IGH: CG2	14	H constant region of IgG2
IGH: CG3	14	H constant region of IgG3
IGH: CG4	14	H constant region of IgG4
IGH: CM	14	H constant region of IgM
IGH: D	14	H diversity region
IGH: J	14	H joining region
IGH: V	14	H variable region
IGK	2	IMMUNOGLOBULIN κ LIGHT CHAIN CLUSTER
IGK: C	2	κ chain, constant region
IGK: J	2	κ chain, joining region
IGK: V	2	κ chain, variable region
IGL	22	IMMUNOGLOBULIN λ LIGHT CHAIN CLUSTER
IGL: C	22	λ chain, constant region
IGL: J	22	λ chain, joining region
IGL: V	22	λ chain, variable region
IL2	4	Interleukin-2 (T-cell growth factor)
IL2R	10	Interleukin-2 receptor
IL3	5	Interleukin-3

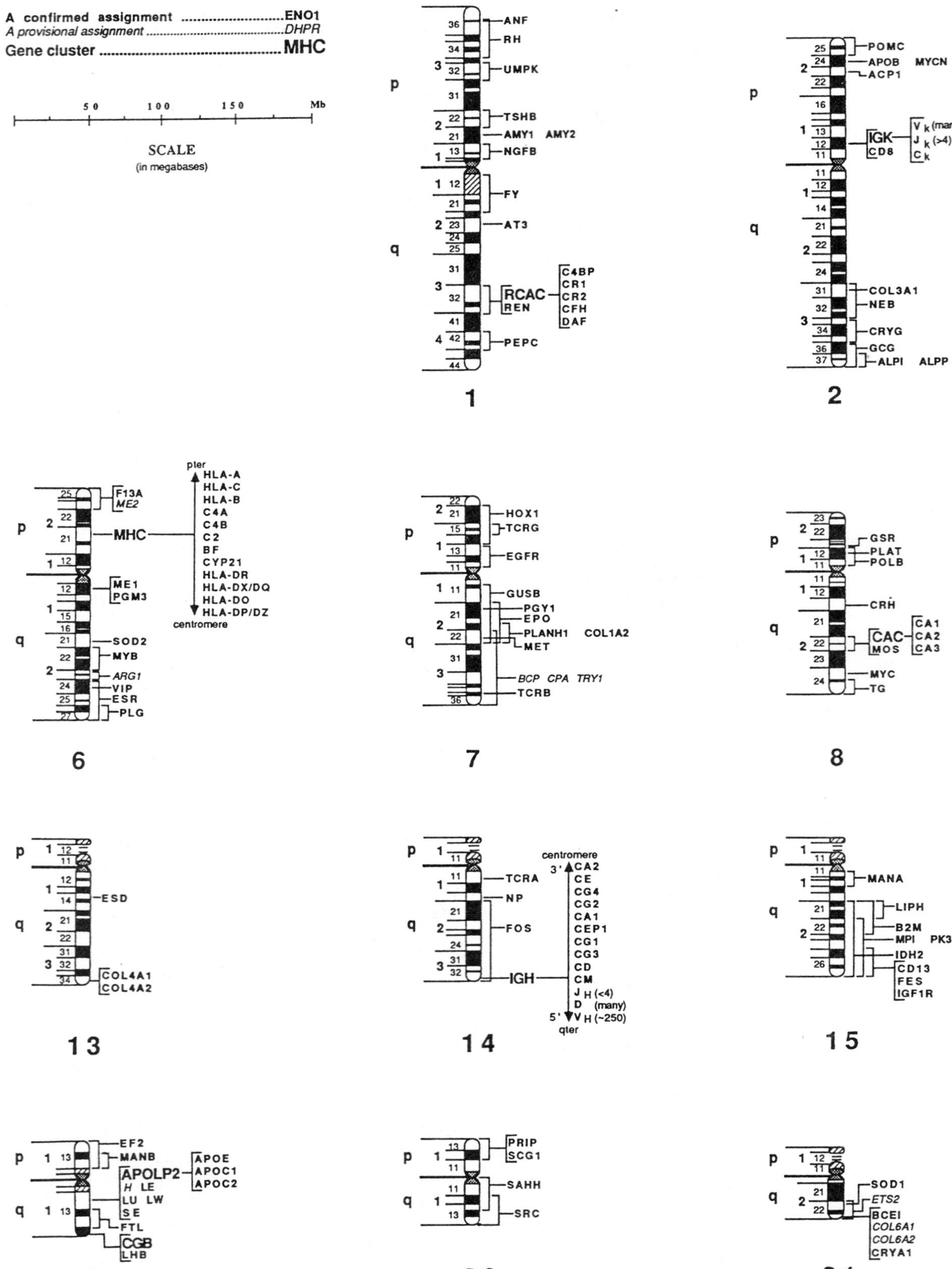

A confirmed assignment ENO1
A provisional assignment ....DHPR
Gene cluster ....MHC
50
100
150
Mb
SCALE
(in megabases)
ANF
RH
UMPK
TSHB
AMY1 AMY2
NGFB
FY
AT3
RCAC
REN
C4BP
CR1
CR2
CFH
DAF
PEPC
1
POMC
APOB MYCN
ACP1
IGK
CD8
V k (many)
J k (>4)
C k
COL3A1
NEB
CRYG
GCG
ALPI ALPP
2
F13A
ME2
MHC
ME1
PGM3
SOD2
MYB
ARG1
VIP
ESR
PLG
pter
HLA-A
HLA-C
HLA-B
C4A
C4B
C2
BF
CYP21
HLA-DR
HLA-DX/DQ
HLA-DO
HLA-DP/DZ
centromere
6
HOX1
TCRG
EGFR
GUSB
PGY1
EPO
PLANH1 COL1A2
MET
BCP CPA TRY1
TCRB
7
GSR
PLAT
POLB
CRH
CAC
MOS
CA1
CA2
CA3
MYC
TG
8
ESD
COL4A1
COL4A2
13
TCRA
NP
FOS
IGH
centromere
3' CA2
CE
CG4
CG2
CA1
CEP1
CG1
CG3
CD
CM
J H (<4)
D (many)
5' V H (~250)
qter
14
MANA
LIPH
B2M
MPI PK3
IDH2
CD13
FES
IGF1R
15
EF2
MANB
APOLP2
H LE
LU LW
SE
FTL
CGB
LHB
APOE
APOC1
APOC2
19
PRIP
SCG1
SAHH
SRC
20
SOD1
ETS2
BCEI
COL6A1
COL6A2
CRYA1
21

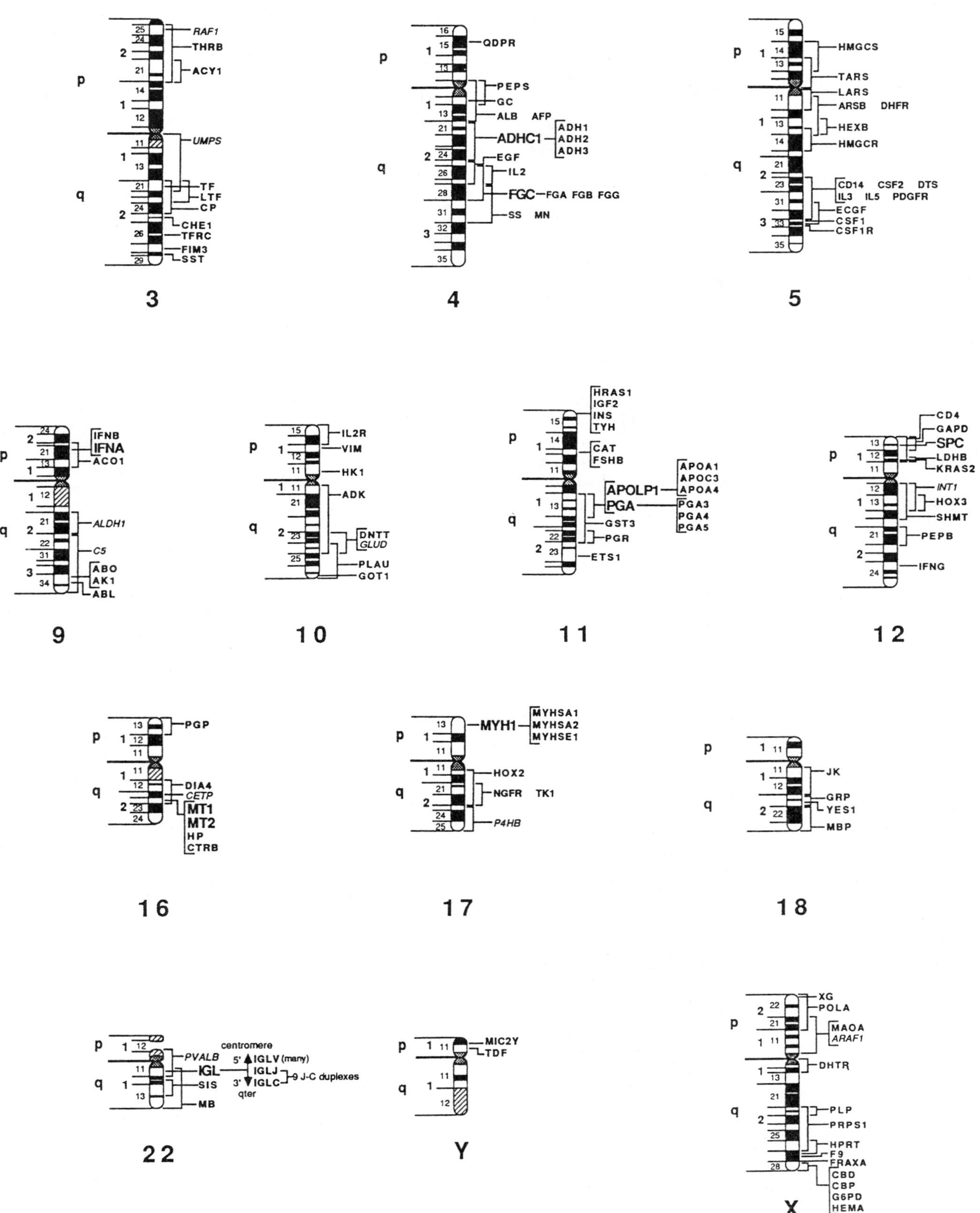
RAF1
THRB
ACY1
UMPS
TF
LTF
CP
CHE1
TFRC
FIM3
SST
3
QDPR
PEPS
GC
ALB AFP
ADHC1
ADH1
ADH2
ADH3
EGF
IL2
FGC
FGA FGB FGG
SS MN
4
HMGCS
TARS
LARS
ARSB DHFR
HEXB
HMGCR
CD14 CSF2 DTS
IL3 IL5 PDGFR
ECGF
CSF1
CSF1R
5
IFNB
IFNA
ACO1
ALDH1
C5
ABO
AK1
ABL
9
IL2R
VIM
HK1
ADK
DNTT
GLUD
PLAU
GOT1
10
HRAS1
IGF2
INS
TYH
CAT
FSHB
APOLP1
APOA1
APOC3
APOA4
PGA
PGA3
PGA4
PGA5
GST3
PGR
ETS1
11
CD4
GAPD
SPC
LDHB
KRAS2
INT1
HOX3
SHMT
PEPB
IFNG
12
PGP
DIA4
CETP
MT1
MT2
HP
CTRB
16
MYH1
MYHSA1
MYHSA2
MYHSE1
HOX2
NGFR TK1
P4HB
17
JK
GRP
YES1
MBP
18
PVALB
IGL
SIS
MB
centromere
5' IGLV (many)
IGLJ
3' IGLC
9 J-C duplexes
qter
22
MIC2Y
TDF
Y
XG
POLA
MAOA
ARAF1
DHTR
PLP
PRPS1
HPRT
F9
FRAXA
CBD
CBP
G6PD
HEMA
X

Gene symbol	Chromosome location	Gene
IL5	5	Interleukin-5
INS	11	Insulin
INT1	12	Proto-oncogene *INT1*
JK	18	Kidd blood group
KRAS2	12	Proto-oncogene *KRAS-2*
LARS	5	Leucyl-tRNA synthetase
LDHB	12	Lactate dehydrogenase B
LE	19	Lewis blood group
LHB	19	Luteinizing hormone β chain
LIPH	15	Hepatic lipase
LTF	3	Lactotransferrin
LU	19	Lutheran blood group
LW	19	LW (Landsteiner-Weiner) blood group
MANA	15	α-Mannosidase, cytoplasmic
MANB	19	α-Mannosidase, lysosomal
MAOA	X	Monoamine oxidase A
MB	22	Myoglobin
MBP	18	Myelin basic protein
ME1	6	Malic enzyme, cytoplasmic
ME2	6	Malic enzyme, mitochondrial
MET	7	Proto-oncogene *MET*
MHC	6	MAJOR HISTOCOMPATIBILITY COMPLEX
MHC: BF	6	MHC, Properdin factor B
MHC: C2	6	MHC, Complement component-2
MHC: C4A	6	MHC, Complement component-4S
MHC: C4B	6	MHC, Complement component-4F
MHC: CYP21	6	MHC, Congenital adrenal hyperplasia
MHC: HLA-A	6	MHC, HLA-A tissue type
MHC: HLA-B	6	MHC, HLA-B tissue type
MHC: HLA-C	6	MHC, HLA-C tissue type
MHC: HLA-DO	6	MHC, HLA-DO tissue type
MHC: HLA-DP/DZ	6	MHC, HLA-DP/DZ tissue type
MHC: HLA-DR	6	MHC, HLA-DR tissue type
MHC: HLA-DX/DQ	6	MHC, HLA-DX/DQ tissue type
MIC2Y	Y	Antigen determined by monoclonal 12E7, Y homolog
MN	4	MN blood group (glycophorin A)
MOS	8	Proto-oncogene *MOS*
MPI	15	Mannosephosphate isomerase
MT1	16	METALLOTHIONEIN I CLUSTER
MT2	16	METALLOTHIONEIN II CLUSTER
MYB	6	Proto-oncogene *MYB*
MYC	8	Proto-oncogene *MYC*
MYCN	2	Proto-oncogene *MYCN*
MYH1	17	MYOSIN HEAVY CHAIN CLUSTER
MYH1: MYHSA1	17	Myosin heavy chain, adult-1
MYH1: MYHSA2	17	Myosin heavy chain, adult-2
MYH1: MYHSE1	17	Myosin heavy chain, embryonic-1
NEB	2	Nebulin
NGFB	1	Nerve growth factor β chain
NGFR	17	Nerve growth factor receptor
NP	14	Nucleoside phosphorylase
P4HB	17	Prolyl-4-hydroxylase β chain
PDGFR	5	Platelet-derived growth factor receptor
PEPB	12	Peptidase B
PEPC	1	Peptidase C
PEPS	4	Peptidase S
PGA	11	PEPSINOGEN A CLUSTER
PGA: PGA3	11	Pepsinogen A3
PGA: PGA4	11	Pepsinogen A4
PGA: PGA5	11	Pepsinogen A5
PGM3	6	Phosphoglucomutase-3
PGP	16	Phosphoglycolate phosphatase
PGR	11	Progesterone receptor
PGY1	7	P-Glycoprotein-1
PK3	15	Pyruvate kinase-3
PLANH1	7	Plasminogen activator inhibitor-1
PLAT	8	Plasminogen activator, tissue type
PLAU	10	Urokinase (plasminogen activator, urinary)
PLG	6	Plasminogen
PLP	X	Myelin proteolipid protein
POLA	X	DNA polymerase, α gene
POLB	8	DNA polymerase, β gene
POMC	2	Proopiomelanocortin
PRIP	20	Prion protein
PRPS1	X	Phosphoribosylpyrophosphate synthetase
PVALB	22	Parvalbumin
QDPR	4	Quinoid dihydropteridine reductase
RAF1	3	Proto-oncogene *RAF1*
RCAC	1	REGULATOR OF COMPLEMENT ACTIVATION CLUSTER
RCAC: C4BP	1	Complement component-4 binding protein
RCAC: CFH	1	Complement factor H
RCAC: CR1	1	Complement component-3b receptor

Gene symbol	Chromosome location	Gene
RCAC: CR2	1	Complement component-3d receptor
RCAC: DAF	1	Decay-accelerating factor of complement
REN	1	Renin
RH	1	Rhesus blood group
SAHH	20	s-Adenosylhomocysteine hydrolase
SCG1	20	Chromogranin B (secretogranin B)
SE	19	Secretor
SHMT	12	Serine hydroxymethyltransferase
SIS	22	Proto-oncogene *SIS*
SOD1	21	Superoxide dismutase-1, soluble
SOD2	6	Superoxide dismutase-1, mitochondrial
SPC	12	SALIVARY PROLINE-RICH PROTEIN COMPLEX
SRC	20	Proto-oncogene *SRC*
SS	4	Ss blood group (glycophorin B)
SST	3	Somatostatin
TARS	5	Threonyl-tRNA synthetase
TCRA	14	T-cell antigen receptor α chain
TCRB	7	T-cell antigen receptor β chain

Gene symbol	Chromosome location	Gene
TCRG	7	T-cell antigen receptor γ chain
TDF	Y	Testis-determining factor
TF	3	Transferrin
TFRC	3	Transferrin receptor
TG	8	Thyroglobulin
THRB	3	Thyroid hormone receptor (Proto-oncogene *ERBA2*)
TK1	17	Thymidine kinase-1
TRY1	7	Trypsin-1
TSHB	1	Thyroid stimulating hormone, β subunit
TYH	11	Tyrosine hydroxylase
UMPK	1	Uridine monophosphate kinase
UMPS	3	Orotate phosphoribosyltransferase
VIM	10	Vimentin
VIP	6	Vasoactive intestinal peptide
XG	X	Xg blood group
YES1	18	Proto-oncogene *YES-1*

Bibliography

Please note: To conserve space, we have (1) cited papers with six or more authors as the first author et al., and (2) used the following journal abbreviations:

AJHG for *The American Journal of Human Genetics*
CSHS for *Cold Spring Harbor Symposia on Quantitative Biology*
NEJM for *The New England Journal of Medicine*
PNAS for *Proceedings of the National Academy of Sciences, U.S.A.*
SA for *Scientific American*
TIG for *Trends in Genetics*

ABELSON, P. H., Editor. 1984. *Biotechnology & Biological Frontiers*. American Association for the Advancement of Science, Washington, D. C.

ADA, G. L. and G. NOSSAL. 1987. The clonal-selection theory. *SA* 257(Aug): 62–69.

ADOLPH, K. W. 1986. Organization of mitotic chromosomes. In *Chromosome Structure and Function*, edited by M. S. Risley, pp. 92–125. Van Nostrand Reinhold, New York.

AGARWAL, D. P. and H. W. GOEDDE. 1987. Genetic variation in alcohol metabolizing enzymes: implications in alcohol use and abuse. In *Genetics and Alcoholism*, edited by H. W. Goedde and D. P. Agarwal, pp. 121–139. Alan R. Liss, New York.

AKAM, M. 1987. The molecular basis for metameric pattern in the *Drosophila* embryo. *Development* 101: 1–22.

ALBERTS, B. et al. 1989. *Molecular Biology of the Cell*, 2nd Edition. Garland Publishing, New York.

ALEIXANDRE, C. et al. 1987. p82H identifies sequences at every human centromere. *Human Genetics* 77: 46–50.

ALFREDSSON, J. H. 1984. Artificial insemination with frozen semen. Sex ratio at birth. *International Journal of Fertility* 29: 152–155.

ALLEN, G. E. 1983. The misuse of biological hierarchies: the American eugenics movement, 1900–1940. *History and Philosophy of the Life Sciences* 5: 106–128.

ALLEN, G. E. 1985a. Thomas Hunt Morgan: materialism in the development of modern genetics. *TIG* 1: 151–154.

ALLEN, G. E. 1985b. Thomas Hunt Morgan: materialism and experimentalism in the development of modern genetics. *TIG* 1: 186–190.

ALLEN, R. D. 1987. The microtubule as an intracellular engine. *SA* 256(Feb): 42–49.

ALLISON, A. C. 1964. Polymorphism and natural selection in human populations. *CSHS* 29: 137–149. [Reprinted in Bajema (1971).]

ALLORE, R. et al. 1988. Gene encoding the β subunit of S100 protein is on chromosome 21: implications for Down syndrome. *Science* 239: 1311–1313.

AMANN, R. P. and G. E. SEIDEL, JR., Editors. 1982. *Prospects for Sexing Mammalian Sperm*. Colorado Associated University Press, Boulder.

AMERICAN FERTILITY SOCIETY. 1986. New guidelines for the use of semen donor insemination: 1986. *Fertility and Sterility* 46(Supplement 2): 95S–110S.

AMERICAN FERTILITY SOCIETY. 1988. Ethical considerations of the new reproductive technologies. *Fertility and Sterility* 49(Supplement 1): 1S–7S.

AMES, B. N. 1979. Identifying environmental chemicals causing mutations and cancer. *Science* 204: 587–593.

AMES, B. N. 1983. Dietary carcinogens and anticarcinogens. *Science* 221: 1256–1264. [See also in this issue the editorial by P. H. Abelson.]

AMES, B. N. 1984. Cancer and diet. *Science* 224: 668ff. [A reply to five responses to Ames (1983).]

AMES, B. N., H. O. KAMMEN and E. YAMASAKI. 1975. Hair dyes are mutagenic: identification of a variety of mutagenic ingredients. *PNAS* 72: 2423–2427.

AMES, B. N., R. MAGAW and L. S. GOLD. 1987. Ranking possible carcinogenic hazards. *Science* 236: 271–280.

ANASTASI, A. 1988. *Psychological Testing*, 6th Edition. Macmillan Publishing Co., New York.

ANDERSON, D. H. 1938. Cystic fibrosis of the pancreas and its relation to celiac disease: a clinical and pathologic study. *American Journal of Diseases of Children* 56: 344–399.

ANDERSON, S. et al. 1981. Sequence and organization of the human mitochondrial genome. *Nature* 290: 457–465. [See other related articles in the same issue of *Nature*.]

ANDERSON, W. F. 1984. Prospects for human gene therapy. *Science* 226: 401–409.

ANDERSON, W. F. and E. G. DIACUMAKOS. 1981. Genetic engineering in mammalian cells. *SA* 245(July): 106–121.

ANDERTON, B. H. 1987a. Alzheimer's disease: progress in molecular pathology. *Nature* 325: 658–659.

ANDERTON, B. H. 1987b. Tangled genes and proteins. *Nature* 329: 106–107.

ANDRES, A. H. and M. S. NAVASCHIN. 1936. Ein Beitrag zür morphologischen Analyse der Chromosomen des Menschen. *Zeitschrift für Zellforschung und mikroskopische Anatomie* 24: 411–426.

ANDREWS, L. B. 1984. *New Conceptions: A Consumer's Guide to the Newest Infertility Treatments*. St. Martin's Press, New York.

ANDREWS, L. B. 1987. The aftermath of Baby M: proposed state laws on surrogate motherhood. *Hastings Center Report* 17(5): 31–40.

ANDREWS, L. B. and H. O. TIEFEL. 1985. When baby's mother is also grandma—and sister. *Hastings Center Report* 15(5): 29–31

ANGIER, N. 1987. Light cast on a darkling gene. *Discover* 8(Mar): 85–96.

ANNAS, G. J. 1981. Righting the wrong of 'wrongful life.' *Hastings Center Report* 11(1): 8–9.

ANNAS, G. J. 1987. Baby M: babies (and justice) for sale. *Hastings Center Report* 17(3): 13–15.

ANNEREN, G. M. et al. 1987. An XXX male resulting from paternal X–Y interchange and maternal X–X nondisjunction. *AJHG* 41: 594–604.

ANONYMOUS. 1833. Westminster Medical Society, 6, 13 and 20 April 1833. *Lancet* II: 146–147.

ANONYMOUS. 1987. Congenital adrenal hyperplasia (1987). *Lancet* II: 663–664.

ANSBACHER, R. 1978. Artificial insemination with frozen spermatozoa. *Fertility and Sterility* 29: 375–379.

ANTONARAKIS, S. E. 1989. Diagnosis of genetic disorders at the DNA level. *NEJM* 320: 153–163.

ANTONARAKIS, S. E. et al. 1985. Hemophilia A: detection of molecular defects and of carriers by DNA analysis. *NEJM* 313: 842–848.

APPEL, S. H. and A. D. ROSES. 1983. The muscular dystrophies. In *The Metabolic Basis of Inherited Disease*, 5th Edition, edited by J. B. Stanbury et al., pp. 1470–1495. McGraw-Hill, New York.

APPLEBAUM, E. G. and S. K. FIRESTEIN. 1983. *A Genetic Counseling Casebook*. The Free Press, New York.

APPLEBY, C. A., J. D. TJEPKEMA and M. J. TRINICK. 1983. Hemoglobin in a nonleguminous plant, *Parasponia*: possible genetic origin and function in nitrogen fixation. *Science* 220: 951–953.

ARPAIA, E. et al. 1988. Identification of an altered splice site in Ashkenazi Tay-Sachs disease. *Nature* 333: 85–86.

ASHTON, G. C. 1980. Mismatches in genetic markers in a large family study. *AJHG* 32: 601–613.

ATALLAH, L. 1976. Report from a test-tube baby. *The New York Times Magazine*, April 18, p. 16ff.

AUFFRAY, C. and J. L. STROMINGER. 1986. Molecular genetics of the human major histocompatibility complex. *Advances in Human Genetics* 15: 197–247.

AUSTIN, C. R. and R. G. EDWARDS, Editors. 1981. *Mechanisms of Sex Differentiation in Animals and Man*. Academic Press, New York.

AUSTIN, C. R. and R. V. SHORT, Editors. 1982. *Germ Cells and Fertilization*, 2nd Edition. Cambridge University Press, Cambridge.

AYALA, F. J. 1977. The question of abortion. In *Heredity and Society: Readings in Social Genetics*, 2nd Edition, edited by A. S. Baer, pp. 323–325. Macmillan Publishing Co., New York.

AYALA, F. J. 1982. *Population and Evolutionary Genetics: A Primer*. Benjamin/Cummings Publishing Co., Menlo Park, California.

BACCETTI, B. 1984. The human spermatozoon. In *Ultrastructure of Reproduction*, edited by J. Van Blerkom and P. M. Motta, pp. 110–126. Martinus Nijhoff Publishers, Boston.

BACCETTI, B., E. S. E. HAFEZ and K. G. GOULD. 1982. Spermatozoa. In *Atlas of Human Reproduction by Scanning Electron Microscopy*, E. S. E. Hafez and P. Kenemans, pp. 197–211. MTP Press, Boston.

BACCI, G. 1965. *Sex Determination*. Pergamon Press, Oxford.

BAILAR, J. C., III and E. M. SMITH. 1986. Progress against cancer? *NEJM* 314: 1226–1232.

BAIRD, P. A. and B. McGILLIVRAY. 1982. Children of incest. *The Journal of Pediatrics* 101: 854–857.

BAJEMA, C. J., Editor. 1971. *Natural Selection in Human Populations*. John Wiley & Sons, New York.

BAJEMA, C. J., Editor. 1976. *Eugenics, Then and Now. Benchmark Papers in Genetics*, Vol. 5. Dowden, Hutchinson & Ross, Stroudsburg, Pennsylvania.

BAKER, B. S., A. T. C. CARPENTER, M. S. ESPOSITO, R. E. ESPOSITO and L. SANDLER. 1976. The genetic control of meiosis. *Annual Review of Genetics* 10: 53–134.

BAKER, B. S., A. T. C. CARPENTER and M. GATTI. 1987. On the biological effects of mutants producing aneuploidy in Drosophila. In *Aneuploidy, Part A: Incidence and Etiology*, edited by B. K. Vig and A. A. Sandberg, pp. 273–296. Alan R. Liss, New York.

BAKKER, E. et al. 1985. Prenatal diagnosis and carrier detection of Duchenne muscular dystrophy with closely linked RFLPs. *Lancet* I: 655–658.

BANDMANN, H.-J. and R. BREIT, Editors. 1984. *Klinefelter's Syndrome*. Springer-Verlag, Berlin.

BARKER, D. et al. 1987. Gene for von Recklinghausen neurofibromatosis is in the pericentromeric region of chromosome 17. *Science* 236: 1100–1102.

BARLOW, P. and C. G. VOSA. 1970. The Y chromosome in human spermatozoa. *Nature* 226: 961–962.

BARNES, D. M. 1987a. Biological issues in schizophrenia. *Science* 235: 430–433.

BARNES, D. M. 1987b. Defect in Alzheimer's is on chromosome 21. *Science* 235: 846–847.

BARNES, D. M. 1988. Orchestrating the sperm–egg summit. *Science* 239: 1091–1092.

BARR, M. L. and E. G. BERTRAM. 1949. A morphological distinction between neurones of the male and female, and the behavior of the nucleolar satellite during accelerated nucleoprotein synthesis. *Nature* 163: 676–677.

BARTON, S. C., M. A. H. SURANI and M. L. NORRIS. 1984. Role of paternal and maternal genomes in mouse development. *Nature* 311: 374–376.

BASERGA, R. 1981. The cell cycle. *NEJM* 304: 453–459.

BATESON, W. 1894. *Materials for the Study of Variation Treated with Especial Regard to Discontinuity in the Origin of the Species*. Macmillan, London.

BATESON, W. and R. C. PUNNETT. 1905–1908. Experimental studies in the physiology of heredity. *Reports to the Evolution Committee of the Royal Society*, Reports 2, 3, and 4. [Reprinted in Peters (1959).]

BAULD, R., G. R. SUTHERLAND and A. D. BAIN. 1974. Chromosome studies in investigation of stillbirths and neonatal deaths. *Archives of Disease in Childhood* 49: 782–788.

BAYES, T. 1763. An essay towards solving a problem in the doctrine of causes. *The Royal Society of London, Philosophical Transactions* 53: 370–418.

BEACH, F. A. 1950. The Snark was a Boojum. *The American Psychologist* 5: 115–124.

BEADLE, G. W. 1967. Mendelism 1965. In *Heritage from Mendel* edited by R. A. Brink, pp. 335–350. University of Wisconsin Press, Madison.

BEADLE, G. W. and E. L. TATUM. 1941. Genetic control of biochemical reactions in *Neurospora*. *PNAS* 27: 499–506.

BEAM, K. G. 1988. Duchenne muscular dystrophy: localizing the gene product. *Nature* 333: 798–799.

BEATTY, R. A. 1970. The genetics of the mammalian gamete. *Biological Reviews of the Cambridge Philosophical Society* 45: 73–119.

BEAUCHAMP, G. K., K. YAMAZAKI and E. A. BOYSE. 1985. The chemosensory recognition of genetic individuality. *SA* 253(July): 86–92.

BEAUDET, A. L. and G. J. BUFFONE. 1987. Prenatal diagnosis of cystic fibrosis. *Journal of Pediatrics* 111: 630–633.

BEGLEITER, H., B. PORJESZ, B. BIHARI and B. KISSIN. 1984. Event-related brain potentials in boys at risk for alcoholism. *Science* 225: 1493–1496.

BEIR [Committee on the Biological Effects of Ionizing Radiation]. 1980. *The Effects on Populations of Exposures to Low Levels of Ionizing Radiation*. National Academy Press, Washington, D. C.

BELL, J. and J. B. S. HALDANE. 1937. The linkage between the genes for colour-blindness and haemophilia in man. *Royal Society of London, Proceedings, Series B* 123: 119–150.

BENDALL, D. S., Editor. 1983. *Evolution From Molecules to Men*. Cambridge University Press, Cambridge.

BENNETT, M. D. et al. 1984. *Chromosomes Today (Chromosomes Today Series*, Vol. VIII). George Allen & Unwin, London.

BENNETT, N. G., Editor. 1983. *Sex Selection of Children*. Academic Press, New York.

BENSON, P. F. and A. H. FENSOM. 1985. *Genetic Biochemical Disorders. Oxford Monographs on Medical Genetics, No. 12*. Oxford University Press, Oxford.

BENTON, W. D. and R. W. DAVIS. 1977. Screening lambda(gt) recombinant clones by hybridization to single plaques in situ. *Science* 196: 180–182.

BERNAL, J. E. 1986. *Human Immunogenetics; Principles and Clinical Applications*. [Translated from Spanish by D. Roberts.] Taylor & Francis, London.

BERRIDGE, M. J. 1985. The molecular basis of communication within a cell. *SA* 253(Oct): 142–152.

BEUTLER, E. 1983. Glucose-6-phosphate dehydrogenase deficiency. In *Metabolic Basis of Inherited Disease*, edited by J. B. Stanbury et al., pp. 1629–1647. McGraw-Hill, New York.

BHENDE, Y. M. et al. 1952. A "new" blood-group character related to the ABO system. *Lancet* I: 903–904.

BIBEL, D. J. 1988. *Milestones in Immunology: A Historical Exploration*. Science Tech Publishers, Madison, Wisconsin.

BICKEL, H. 1985. Differential diagnosis and treatment of hyperphenylalaninaemia. In *Medical Genetics: Past, Present, Future*, edited by K. Berg, pp. 93–107. Alan R. Liss, New York.

BICKEL, H., J. GERARD and E. M. HICKMANS. 1954. The influence of phenylalanine intake on the chemistry and behavior of a phenylketonuric child. *Acta Paediatrica* 43: 64–77.

BICKMORE, W. A. L. and A. T. SUMNER. 1989. Mammalian chromosome banding—an expression of genome organization. *TIG* 5: 144–148.

BIEBER, F. R. et al. 1981. Genetic studies of an acardiac monster: evidence of polar body twinning in man. *Science* 213: 775–777.

BIGGS, R. 1983. Defects in coagulation. In *Principles and Practice of Medical Genetics*, edited by A. E. H. Emery and D. L. Rimoin, pp. 1065–1075. Churchill Livingstone, Edinburgh.

BINGHAM, R. 1981. Outrageous ardor. *Science 81*(Sept): 55–61.

BISHOP, J. A. and L. M. COOK. 1975. Moths, melanism and clean air. *SA* 232(Jan): 90–99.

BISHOP, J. M. 1987. The molecular genetics of cancer. *Science* 235: 305–311.

BITTLES, A. H. 1983. The intensity of human inbreeding depression. *The Behavioral and Brain Sciences* 6: 103–104.

BLAKEMORE, K. J. and M. J. MAHONEY. 1986. Chorionic villus sampling. In *Genetic Disorders and the Fetus; Diagnosis, Prevention, and Treatment*, 2nd Edition, edited by A. Milunsky, pp. 625–660. Plenum Press, New York.

BLEIR, R. 1984. *Science and Gender: A Critique of Biology and Its Theories on Women.* Pergamon Press, New York.

BODMER, W. F., Editor. 1983. *Inheritance of Susceptibility to Cancer in Man.* Oxford University Press, Oxford.

BODMER, W. F. 1986a. Human genetics: the molecular challenge. *CSHS* 51: 1–13.

BODMER, W. F. 1986b. Inherited susceptibility to cancer. In *Introduction to the Cellular and Molecular Biology of Cancer*, edited by L. M. Franks and N. M. Teich, pp. 93–110. Oxford University Press, Oxford.

BODMER, W. F. and L. L. CAVALLI-SFORZA. 1970. Intelligence and race. *SA* 223(Oct): 19–29.

BODMER, W. F. and L. L. CAVALLI-SFORZA. 1976. *Genetics, Evolution, and Man.* W. H. Freeman and Co., San Francisco.

BOEHM, T. L. J. and D. DRAHOVSKY. 1986. Use of a biotinylated DNA probe specific for the human Y chromosome for rapid antenatal sex determination. *Clinical Genetics* 30: 509–514.

BOGENMANN, E., M. A. LOCHRIE and M. I. SIMON. 1988. Cone cell-specific genes expressed in retinoblastoma. *Science* 240: 76–78.

BOHMAN, M. 1978. Some genetic aspects of alcoholism and criminality; a population of adoptees. *Archives of General Psychiatry* 35: 269–276.

BOHMANN, D. et al. 1987. Human proto-oncogene *c-jun* encodes a DNA binding protein with structural and functional properties of transcription factor AP-1. *Science* 238: 1386–1392.

BOIS, E. et al. 1978. Cluster of cystic fibrosis cases in a limited area of Brittany (France). *Clinical Genetics* 14: 73–76.

BOND, D. J. and A. C. CHANDLEY. 1983. *Aneuploidy.* Oxford Monographs on Medical Genetics No. 11. Oxford University Press, Oxford.

BONDY, P. and L. E. ROSENBERG, Editors. 1983. *Duncan's Diseases of Metabolism: Genetics and Metabolism.* W. B. Saunders Co., Philadelphia.

BONGIOVANNI, A. M., W. R. EBERLEIN, A. S. GOLDMAN and M. NEW. 1967. Disorders of adrenal steroid biogenesis. *Recent Progress in Hormone Research* 23: 375–449.

BONNÉ, B. 1963. The Samaritans: a demographic study. *Human Biology* 35: 61–89.

BORGAONKAR, D. S. 1984. *Chromosomal Variation in Man: A Catalog of Chromosomal Variants and Anomalies*, 4th Edition. Alan R. Liss, New York.

BORGAONKAR, D. S. and S. A. SHAH. 1974. The XYY chromosome male—or syndrome? *Progress in Medical Genetics* 10: 135–222.

BORWEIN, J. M. and P. B. BORWEIN. 1988. Ramanujan and pi. *SA* 258(Feb): 112–117.

BOTKIN, J. R. 1988. The legal concept of wrongful life. *JAMA* 259: 1541–1545.

BOTSTEIN, D. 1986. Perspective: the molecular biology of color vision. *Science* 232: 142–143.

BOTSTEIN, D., R. L. WHITE, M. SKOLNICK and R. W. DAVIS. 1980. Construction of a genetic linkage map in man using restriction fragment length polymorphisms. *AJHG* 32: 314–331.

BOUCHARD, T. J., JR. and M. McGUE. 1981. Familial studies of intelligence: a review. *Science* 212: 1055–1059.

BOUÉ, A., J. BOUÉ and A. GROPP. 1985. Cytogenetics of pregnancy wastage. *Advances in Human Genetics* 14: 1–57.

BOUÉ, J., A. BOUÉ and P. LAZAR. 1975. Retrospective and prospective epidemiological studies of 1500 karyotyped spontaneous human abortions. *Teratology* 12: 11–26.

BOVERI, T. 1902. On multipolar mitosis as a means of analysis of the cell nucleus. [Translated by S. Gluecksohn-Waelsch.] Reprinted in *Foundations of Experimental Embryology*, edited by B. H. Willier and J. M. Oppenheimer. Hafner, New York.

BOWEN, P. et al. 1965. Hereditary male pseudohermaphroditism with hypogonadism, hypospadias, and gynecomastia (Reifenstein's syndrome). *Annals of Internal Medicine* 62: 252–270.

BOWLING, F. G. et al. 1987. Monoclonal antibody-based enzyme immunoassay for trypsinogen in neonatal screening for cystic fibrosis. *Lancet* I: 826–827.

BOYCE, A. J. 1983. Computation of inbreeding and kinship coefficients on extended pedigrees. *The Journal of Heredity* 74: 400–404.

BOYD, W. C. 1950. *Genetics and the Races of Man: An Introduction to Modern Physical Anthropology.* Little, Brown and Co., Boston.

BOYER, S. H., IV, Editor. 1963. *Papers on Human Genetics.* Prentice-Hall, Englewood Cliffs, New Jersey

BRACHET, J. 1985. *Molecular Cytology. Vol. 1: The Cell Cycle.* Academic Press, New York.

BRADBURY, E. M., N. MACLEAN and H. R. MATTHEWS. 1981. *DNA, Chromatin, and Chromosomes.* John Wiley & Sons, New York.

BRANDRIFF, B. et al. 1985. Chromosomes of human sperm: variability among normal individuals. *Human Genetics* 70: 18–24.

BRANDRIFF, B. F. et al. 1986. Sex chromosome ratios determined by karyotypic analysis in albumin-isolated human sperm. *Fertility and Sterility* 46: 678–685.

BRAUDE, P., V. BOLTON and S. MOORE. 1988. Human gene expression first occurs between the four- and eight-cell stages of preimplantation development. *Nature* 332: 459–461.

BREG, W. R. 1977. Euploid structural rearrangements in the mentally retarded. In *Population Cytogenetics, Studies in Humans*, edited by E. B. Hook and I. H. Porter, pp. 99–101. Academic Press, New York.

BRENNER, D. A. and J. R. BLOOMER. 1980. The enzymatic defect in variegate porphyria. *NEJM* 302: 765–769.

BRENNER, S. 1974. The genetics of *Caenorhabditis elegans* (1974). *Genetics* 77: 71–94.

BRIDGES, C. 1913. Non-disjunction of the sex chromosomes of Drosophila. *The Journal of Experimental Zoology* 15: 587–605.

BRIDGES, C. 1914. Direct proof through non-disjunction that the sex-linked genes of *Drosophila* are borne by the X-chromosome. *Science* 40: 107–109.

BRILL, A. B., S. J. ADELSTEIN, E. L. SAENGER and E. W. WEBSTER, Editors. 1982. *Low-Level Radiation Effects: A Fact Book.* The Society of Nuclear Medicine, New York.

BRINK, P. A., L. T. STEYN, G. A. COETZEE and D. R. VAN DER WESTHUYZEN. 1987. Familial hypercholesterolemia in South African Afrikaners. *Human Genetics* 77: 32–35.

BROACH, J. R. 1986. New approaches to a genetic analysis of mitosis. *Cell* 44: 3–4.

BROAD, W. J. 1982. Publisher settles suit, says clone book is a fake. *Science* 216: 391.

BROCK, D. J. H., A. E. BOLTON and J. R. SCRIMGEOU. 1974. Prenatal diagnosis of spina bifida and anencephaly through maternal plasma-α-fetoprotein measurement. *Lancet* I: 767–769

BROCK, D. J. H., D. BEDGOOD, L. BARRON and C. HAYWARD. 1985. Prospective prenatal diagnosis of cystic fibrosis. *Lancet* I: 1175–1178.

BROCK, D. J. H. and V. VAN HEYNINGEN. 1986. The facts on cystic fibrosis testing. *Nature* 319: 184.

BRODY, J. E. 1981. In albino studies, scientists search for pigment error. *New York Times* Sept. 15, pp. C1 ff.

BROWDER, L. W., Editor. 1985. *Developmental Biology: A Comprehensive Synthesis. Vol. 1: Oogenesis.* Plenum Press, New York.

BROWN, M. S. and J. L. GOLDSTEIN. 1984. How LDL receptors influence cholesterol and atherosclerosis. *SA* 251(Nov): 58–66.

BROWN, M. S. and J. L. GOLDSTEIN. 1986. A receptor-mediated pathway for cholesterol homeostasis: Nobel lecture, 9 Dec. 1985. *Les Prix Nobel 1985.* Almqvist & Wiksell International, Stockholm. Reprinted in *Science* 232: 34–47.

BROWN, T. R. and C. J. MIGEON. 1986. Androgen receptors in normal and abnormal male sexual differentiation. *Advances in Experimental Biology and Medicine* 196: 227–255.

BROWN, W. T. et al. 1986. The fragile X syndrome. *Annals of the New York Academy of Sciences* 477: 129–150.

BROWNLEE, G. G. and C. RIZZA. 1984. Clotting Factor VIII cloned. *Nature* 312: 307.

BUCKLEY, R. H. 1987. Immunodeficiency diseases. *JAMA* 258: 2841–2850.

BUCKTON, K. E., P. A. JACOBS, L. A. RAE, M. S. NEWTON and R. SANGER. 1971. An inherited X–autosome translocation in man. *Annals of Human Genetics* 35: 171–178.

BUHLER, E. M. 1980. A synopsis of the human Y chromosome. *Human Genetics* 55: 145–175.

BUISSERET, P. D. 1982. Allergy. *SA* 247(Aug): 86–95.

BULL, J. J., D. M. HILLIS and S. O'STEEN. 1988. Mammalian *ZFY* sequences exist in reptiles regardless of sex–determining mechanism. *Science* 242: 567–569.

BULMER, M. G. 1970. *The Biology of Twinning in Man.* Oxford University Press, London.

BURCK, K. B., E. T. LIU and J. W. LARRICK. 1988. *Oncogenes: An Introduction to the Concept of Cancer Genes.* Springer-Verlag, New York.

BURG, J., E. CONZELMANN, K. SANDHOFF, E. SOLOMON and D. M. SWALLOW. 1985. Mapping of the gene coding for the human G_{M2} activator protein to chromosome 5. *Annals of Human Genetics* 49: 41–45.

BURGIO, G. R., M. FRACCARO, L. TIEPOLO and U. WOLF, Editors. 1981. *Trisomy 21. Human Genetics, Supplement 2.* Springer-Verlag, Berlin.

BURKITT, D. P. 1985. The beginnings of the Burkitt's lymphoma story. In *Burkitt's Lymphoma: A Human Cancer Model*, edited by G. M. Lenoir et al., pp. 11–15. International Agency for Research on Cancer, Lyon, France.

BURNET, F. M. 1962. *The Integrity of the Body*. Harvard University Press, Cambridge.

BURNET, F. M. 1968. The impact of ideas on immunology. *CSHS* 32: 1–8.

BURNET, F. M., Editor. 1976. *Immunology: Readings from Scientific American*. W. H. Freeman and Co., San Francisco.

BUSSLINGER, M., J. HURST and R. A. FLAVELL. 1983. DNA methylation and the regulation of globin gene expression. *Cell* 34: 197–206.

BUTLER, S. 1877. *Life and Habit*. Trubner & Co., London. [Reprinted in 1968 by the Ams Press, New York.]

CAIRNS, J. 1985. The treatment of diseases and the war against cancer. *SA* 253(Nov): 51–59.

CALDWELL, J. 1981. Maternal PKU causes new concern. *Boston Globe*, Oct. 14, p. 61ff.

CALLAHAN, D. 1972. New beginnings in life. In *The New Genetics and the Future of Man*, edited by M. P. Hamilton, pp. 90–106. William B. Eerdmans Publishing Co., Grand Rapids, Michigan.

CALNE, R. Y., Editor. 1984. *Transplantation Immunology: Clinical and Experimental*. Oxford University Press, Oxford.

CAMPBELL, J. R. and J. F. LASLEY. 1985. *The Science of Animals that Serve Humanity*, 3rd Edition. McGraw-Hill, New York.

CANADIAN COLLABORATIVE CVS–AMNIOCENTESIS CLINICAL TRIAL GROUP. 1989. Multicentre randomized clinical trial of chorion villus sampling and amniocentesis. *Lancet* I: 1–6.

CANN, R. L., M. STONEKING and A. C. WILSON. 1987. Mitochondrial DNA and human evolution. *Nature* 325: 31–36.

CANTONI, G. L. and A. RIZIN, Editors. 1985. *Biochemistry and Biology of DNA Methylation*. Alan R. Liss, New York.

CARLSON, E. A. 1973. Eugenics revisited: the case for germinal choice. In *Stadler Genetics Symposia,* Vol. 5, edited by G. Kimber and G. P. Rédei, pp. 13–34. University of Missouri, Columbia. [Reprinted in Bajema (1976).]

CAROTHERS, E. E. 1913. The Mendelian ratio in relation to certain Orthopteran chromosomes. *Journal of Morphology* 24: 487–511. [Reprinted in part in Voeller (1968).]

CARPENTER, A. T. C. 1981. EM autoradiographic evidence that DNA synthesis occurs at recombination nodules during meiosis in *Drosophila melanogaster* females. *Chromosoma* 83: 59–80.

CARPENTER, A. T. C. 1984. Genic control of meiosis. *Chromosomes Today* 8: 70–79.

CARR, D. H. and M. GEDEON. 1977. Population cytogenetics of human abortuses. In *Population Cytogenetics: Studies in Humans*, edited by E. B. Hook and I. H. Porter, pp. 1–9. Academic Press, New York.

CARRASCO, A. E., W. McGINNIS, W. J. GEHRING and E. M. DE ROBERTIS. 1984. Cloning of an *X. laevis* gene expressed during early embryogenesis coding for a peptide region homologous to *Drosophila* homeotic genes. *Cell* 37: 409–414.

CARSWELL, H. 1982. "Elephant Man" had more than neurofibromatosis. *JAMA* 248: 1032–1033.

CASKEY, C. T. 1986. Summary: a milestone in human genetics. *CSHS* 51: 1115–1119.

CASKEY, C. T. and R. L. WHITE, Editors. 1983. *Banbury Report 14: Recombinant DNA Applications to Human Disease*. Cold Spring Harbor Laboratory, New York.

CASPERSSON, T., G. LOMAKKA and L. ZECH. 1971. 24 fluorescence patterns of human metaphase chromosomes—distinguishing characters and variability. *Hereditas* 67: 89–102.

CASPERSSON, T., L. ZECH and C. JOHANSSON. 1970. Differential banding of alkylating fluorochromes in human chromosomes. *Experimental Cell Research* 60: 315–319.

CATTANACH, B. M. 1974. Position effect variegation in the mouse. *Genetical Research* 23: 291–306.

CAVALLI-SFORZA, L. L. and W. F. BODMER. 1971. *The Genetics of Human Populations*. W. H. Freeman and Co., San Francisco.

CAVALLI-SFORZA, L. L. et al. 1986. DNA markers and genetic variation in the human species. *CSHS* 51: 411–417.

CEM [COMMITTEE ON CHEMICAL ENVIRONMENTAL MUTAGENS]. 1983. *Identifying and Estimating the Genetic Impact of Chemical Mutagens.* National Academy Press, Washington, D. C.

CHAKRABORTY, R. et al. 1987. Polymorphic DNA haplotypes at the human phenylalanine hydroxylase locus and their relationship with phenylketonuria. *Human Genetics* 76: 40–46.

CHAMBON, P. 1981. Split genes. *SA* 244(May): 60–71.

CHAN, J. Y. H., F. F. BECKER, J. GERMAN and J. H. RAY. 1987. Altered DNA ligase I activity in Bloom's syndrome cells. *Nature* 325: 357–359.

CHAN, S. T. H. and WAI-SUM, O. 1981. Environmental and non-genetic mechanisms in sex determination. In *Mechanisms of Sex Differentiation in Animals and Man*, edited by C. R. Austin and R. G. Edwards, pp. 55–111. Academic Press, New York.

CHANDLEY, A. C. 1979. The chromosomal basis of human infertility. *British Medical Bulletin* 35: 181–186.

CHANDLEY, A. C. 1981. The origin of chromosomal aberrations in man and their potential for survival and reproduction in the adult human population. *Annals of Genetics* 24: 5–11.

CHANDLEY, A. C. 1988. Meiosis in man. *TIG* 4: 79–84.

CHANG, J. C. and Y. W. KAN. 1982. A sensitive new prenatal test for sickle-cell anemia. *NEJM* 307: 30–32.

CHARGAFF, E. 1950. Chemical specificity of nucleic acids and mechanism of their enzymatic degradation. *Experientia* 6: 201–209. [Reprinted in Corwin and Jenkins (1976).]

CHARGAFF, E. 1974. Building the tower of babble. *Nature* 248: 776–779.

CHASE, M. 1987. Sperm banks thrive amid debate over medical and ethical issues. [Plus two other articles.] *The Wall Street Journal*, Apr. 2, page 31.

CHELLY, J., J.-C. KAPLAN, P. MAIRE, S. GAUTRON and A. KAHN. 1988. Transcription of the dystrophin gene in human muscle and non-muscle tissues. *Nature* 333: 858–860.

CHERFAS, J. 1982. *Man-Made Life*. Pantheon Books, New York.

CHILDS, B., J. M. FINUCCI, M. S. PRESTON and A. E. PULVER. 1976. Human behavior genetics. *Advances in Human Genetics* 7: 57–97.

CHILDS, B., N. A. HOLTZMAN, H. H. KAZAZIAN, JR. and D. L. VALLE. 1988. *Molecular Genetics in Medicine* (New Series, Volume 7 of *Progress in Medical Genetics*. Elsevier, New York.

CHILDS, B., W. ZINKHAM, E. A. BROWNE, E. L. KIMBRO and J. V. TORBET. 1958. A genetic study of a defect in glutathione metabolism of the erythrocyte. *Bulletin of the Johns Hopkins Hospital* 102: 21–37.

CHOO, K. H., K. G. GOULD, D. J. G. REES and G. G. BROWNLEE. 1982. Molecular cloning of the gene for human anti-haemophilic factor IX. *Nature*: 299: 178–180.

CHU, G. and E. CHANG 1988. Xeroderma pigmentosum group E cells lack a nuclear factor that binds to damaged DNA. *Science* 242: 564–567.

CHURCH, R. B., F. J. SCHAUFELE and K. MECKLING. 1985. Embryo manipulation and gene transfer in livestock. *Canadian Journal of Animal Science* 65: 527–537.

CLARK, S. M., E. LAI, B. W. BIRREN and L. HOOD. 1988. A novel instrument for separating large DNA molecules with pulsed homogeneous electric fields. *Science* 241: 1203–1205.

CLARKE, C. A. 1968. The prevention of "Rhesus" babies. *SA* 219(Nov): 46–52. [Reprinted in Burnet (1976).]

CLARKE, C. A., Editor. 1975. *Rhesus Hemolytic Disease: Selected Papers and Extracts*. University Park Press, Baltimore.

CLARKSON, B. 1982. The chronic leukemias. In *Cecil Textbook of Medicine*, 16th Edition, edited by J. B. Wyngaarden and L. H. Smith, pp. 926–936. Saunders, Philadelphia.

CLEAVER, J. E. 1983. Xeroderma pigmentosum. In *The Metabolic Basis of Inherited Disease*, 5th Edition, edited by J. B. Stanbury et al., pp.1227–1248. McGraw-Hill, New York.

CLEVELAND, D. W. and K. F. SULLIVAN. 1985. Molecular biology and genetics of tubulin. *Annual Review of Biochemistry* 54: 331–365.

CLONINGER, C. R., M. BOHMAN and S. SIGVARDSSON. 1981. Inheritance of alcohol abuse: cross-fostering analysis of adopted men. *Archives of General Psychiatry* 38: 861–868.

COHEN, L. A. 1987. Diet and cancer. *SA* 257 (Nov): 42–48.

COHEN, M. M., JR. 1983. Craniofacial disorders. In *Principles and Practice of Medical Genetics,* Vol. 1, edited by A. E. H. Emery and D. L. Rimoin, pp. 576–621. Churchill Livingstone, Edinburgh.

COLBURN, Z. 1833. *A Memoir of Zerah Colburn; Written by Himself.* G. and C. Merriam, Springfield, Massachusetts.

COLD SPRING HARBOR SYMPOSIA. 1986. *Molecular Biology of Homo sapiens.* Volume 51 of *CSHS.* Cold Spring Harbor Laboratory, New York.

COLLIER, R. J. and D. A. KAPLAN. 1984. Immunotoxins. *SA* 251(July): 56–64.

COLLINS, F. S. and S. M. WEISSMAN. 1984. The molecular genetics of human hemoglobin. *Progress in Nucleic Acid Research and Molecular Biology* 31: 315–462.
COMINGS, D. E. 1978. Mechanisms of chromosome banding and implications for chromosome structure. *Annual Review of Genetics* 12: 25–46.
COMMITTEE ON GENETICS. 1982. New issues in newborn screening for phenylketonuria and congenital hypothyroidism. *Pediatrics* 69: 104–106.
COMMITTEE ON SCIENCE AND CREATIONISM. 1984. *Science and Creationism: A View from the National Academy of Science*. National Academy Press, Washington, D. C.
CONGREGATION FOR THE DOCTRINE OF THE FAITH. 1987. Instruction on respect for human life. [The complete English text of this Vatican document appears in *The New York Times*, Mar. 11, 1987, pages A14–A17.]
CONLEY, M. E. et al. 1986. Expression of the gene defect in X-linked agammaglobulinemia. *NEJM* 315: 564–567.
CONNEALLY, P. M. 1984. Huntington disease: genetics and epidemiology. *AJHG* 36: 506–526.
CONNEALLY, P. M., A. D. MERRITT and P.-L. YU. 1973. Cystic fibrosis: Population genetics. *Texas Reports on Biology and Medicine* 31: 639–650.
CONNEALLY, P. M. and M. L. RIVAS. 1980. Linkage analysis in man. *Advances in Human Genetics* 10: 209–266.
CONSENSUS CONFERENCE. 1987. Newborn screening for sickle cell disease and other hemoglobinpathies. *National Institutes of Health Consensus Development Conference Statement*, Volume 6, Number 9, Apr. 6–8, 1987. [Reprinted in *JAMA* 258: 1205–1209.]
CONSTANTINI, F., K. CHADA and J. MAGRAM. 1986. Correction of murine β-thalassemia by gene transfer into the germ line. *Science* 233: 1192–1194.
COOPER, B. J. and B. A. VALENTINE. 1988. X-linked muscular dystrophy in the dog. *TIG* 4: 30.
COOPER, D. N. and J. SCHMIDTKE. 1986. Diagnosis of genetic disease using recombinant DNA. *Human Genetics* 73: 1–11.
COOPER, D. N. and J. SCHMIDTKE. 1987. Diagnosis of genetic disease using recombinant DNA. Supplement. *Human Genetics* 77: 66–75.
CORSON, S. L., F. R. BATZER, N. J. ALEXANDER, S. SCHLAFF and C. OTIS. 1984. Sex selection by sperm separation and insemination. *Fertility and Sterility* 42: 756–760.
CORWIN, H. O. and J. B. JENKINS, Editors. 1976. *Conceptual Foundations of Genetics, Selected Readings*. Houghton Mifflin Co., Boston.
COUNCIL OF THE ENVIRONMENTAL MUTAGEN SOCIETY. 1975. Environmental mutagenic hazards. *Science* 187: 503–514.
COUNCIL ON SCIENTIFIC AFFAIRS. 1986. Autologous blood transfusions. *JAMA* 256: 2378–2380.
COURT-BROWN, W. M. 1967. *Human Population Cytogenetics*. North Holland Publishing Co., Amsterdam.
COURT-BROWN, W. M. et al. 1966. *Chromosome Studies on Adults. Eugenics Laboratory Memoirs XLII*. Galton Laboratory, University College London. Cambridge University Press, London.
CRAIG, I. W. and E. TOLLEY. 1986. Steroid sulphatase and the conservation of mammalian X chromosomes. *TIG* 2: 201–204.
CRAWFORD, M. H. and P. L. WORKMAN, Editors. 1973. *Methods and Theories of Anthropological Genetics*. University of New Mexico Press, Albuquerque.
CREA, R., A. KRASZEWSKI, T. HIROSE and K. ITAKURA. 1978. Chemical synthesis of genes for human insulin. *PNAS* 75: 5765–5769.
CREASY, M. R., J. A. CROLLA and E. D. ALBERMAN. 1976. A cytogenetic study of human spontaneous abortions using banding techniques. *Human Genetics* 31: 177–196.
CREEL, D., S. R. GARBER, R. A. KING and C. J. WITKOP, JR. 1980. Auditory brainstem anomalies in human albinos. *Science* 209: 1253–1255.
CREW, F. A. E. 1965. *Sex Determination*, 4th Edition. Dover Publications, New York.
CRICK, F. 1966a. The genetic code—yesterday, today, and tomorrow. *CSHS* 31: 3–9. [Entire volume devoted to the genetic code, including articles by Nobel laureates H. G. Khorana and M. Nirenberg.]
CRICK, F. 1966b. The genetic code: III. *SA* 215 (Oct): 55–62. [Reprinted in Srb et al. (1970).]
CRICK, F. 1974. The double helix: a personal view. *Nature* 248: 766–769.
CRICK, F. 1988. *What Mad Pursuit*. Basic Books, New York.
CRICK, F., L. BARNETT, S. BRENNER and R. J. WATTS-TOBIN. 1961. General nature of the genetic code for proteins. *Nature*: 1227–1232.
CROCE, C. M. and G. KLEIN. 1985. Chromosome translocations and human cancer. *SA* 252(Mar): 54–60.
CRONBACH, L. J. 1975. Five decades of public controversy over mental testing. *American Psychologist* 30: 1–14.
CROW, J. F. 1966. The quality of people: human evolutionary changes. *BioScience* 16: 863–867. [Reprinted in Bajema (1971).]
CROW, J. F. 1982. How well can we assess genetic risks? Not very. In *Critical Issues in Setting Radiation Dose Limits*, pp. 211–235. National Council on Radiation Protection and Measurements, Bethesda, Maryland.
CROW, J. F. 1983. *Genetics Notes, An Introduction to Genetics*, 8th Edition. Burgess Publishing Co., Minneapolis.
CROW, J. F. 1986. *Basic Concepts in Population, Quantitative, and Evolutionary Genetics*. W. H. Freeman and Co., New York.
CROW, J. F. 1987. Population genetics history: a personal view. *Annual Review of Genetics* 21: 1–22.
CROW, J. F. 1988. Sewall Wright (1889–1988). *Genetics* 119: 1–4.
CROW, J. F. and C. DENNISTON. 1985. Mutation in human populations. *Advances in Human Genetics* 14: 59–123.
CROW, J. F. and A. P. MANGE. 1965. Measurement of inbreeding from the frequency of marriages between persons of the same surname. *Eugenics Quarterly* (now *Social Biology*) 12: 199–203. [Reprinted 29: 101–105 (1982).]
CUCKLE, H. S., N. J. WALD and R. H. LINDENBAUM. 1984. Maternal serum α-fetoprotein measurement: a screening test for Down syndrome. *Lancet* I: 926–929.
CUNNINGHAM, C. 1982. *Down's Syndrome: An Introduction for Parents*. Brookline Books, Cambridge, Massachusetts.
CURIE-COHEN, M., L. LUTTRELL and S. SHAPIRO. 1979. Current practice of artificial insemination by donor in the United States. *NEJM* 300: 585–590.
CURRAN, W. J. 1988. Cancer-causing substances in food, drugs, and cosmetics; the *de minimus* rule versus the Delaney clause. *NEJM* 319: 1262–1264
CURTIS, H. 1983. *Biology*, 4th Edition. Worth Publishers, New York.

DALE, B. 1983. *Fertilization in Animals*. Edward Arnold, London.
DALTON, J. 1798. Extraordinary facts relating to the vision of colours, with observation. *Memoirs of the Literary and Philosophical Society of Manchester* 5: 28–45.
DANKS, D. M. 1983. Disorders of copper metabolism. In *Principles and Practice of Medical Genetics,* Vol. 2, edited by A. E. H. Emery and D. L. Rimoin, pp. 1319–1328. Churchill Livingstone, Edinburgh.
DAO, D. D. et al. 1987. Genetic mechanisms of tumor-specific loss of 11p DNA sequences in Wilms tumor. *AJHG* 41: 202–217.
DARNELL, J. E., JR. 1983. The processing of RNA. *SA* 249(Oct): 90–100.
DARNELL, J. E., JR. 1985. RNA. *SA* 253(Oct): 68–78.
DARNELL, J., H. LODISH and D. BALTIMORE. 1986. *Molecular Cell Biology*. Scientific American Books, New York.
DARRAS, B. T., J. F. HARPER and U. FRANCKE. 1987. Prenatal diagnosis and detection of carriers with DNA probes in Duchenne's muscular dystrophy. *NEJM* 316: 985–992.
DARWIN, C. 1859. *The Origin of Species by Means of Natural Selection*. John Murray, London. [A facsimile of the First Edition was reprinted by the Harvard University Press in 1964. The Sixth Edition (1872) is the one more often reprinted.]
DARWIN, C. 1875. *The Variation of Plants and Animals under Domestication*, 2nd Edition, p. 319. John Murray, London.
DARWIN, C. 1969 Reprint. *The Autobiography of Charles Darwin, 1809–1882*. [With original omissions restored; edited with appendix and notes by his granddaughter, Nora Barlow.] W. W. Norton, New York.
DAUSSET, J. 1981. The major histocompatibility complex in man: past, present, and future concepts. *Science* 213: 1469–1474.
DAUTRY-VARSAT, A. and H. F. LODISH. 1984. How receptors bring proteins and particles into cells. *SA* 250(May): 52–58.
DAVENPORT, C. B. 1913. *Heredity of Skin Color in Negro–White Crosses*. Carnegie Institute of Washington, Washington, D. C.
DAVIDSON, E. H. 1986. *Gene Activity in Early Development*, 3rd Edition. Academic Press, Orlando, Florida.

DAVIDSON, R. G., H. M. NITOWSKY and B. CHILDS. 1963. Demonstration of two populations of cells in the human female heterozygous for glucose-6-phosphate dehydrogenase variants. *PNAS* 50: 481–485.

DAVIDSON, R. L., Editor. 1984. *Somatic Cell Genetics (Benchmark Papers in Genetics, Volume 14).* Hutchinson Ross Publishing Co., Stroudsburg, Pennsylvania.

DAVIES, D. A. L. 1984. Non-MHC antigens and their relation to graft rejection. In *Transplantation Immunology* edited by R. Y. Calne, pp. 159–168. Oxford University Press, Oxford.

DAVIS, B. D. 1986. *Storm Over Biology: Essays on Science, Sentiment, and Public Policy.* Prometheus Books, Buffalo, New York.

DAVIS, K. 1974. The migrations of human populations. *SA* 231(Sept): 93–105.

DAVIS, P. 1983. Cystic fibrosis in adults. In *Textbook of Cystic Fibrosis*, edited by J. D. Lloyd-Still, pp. 351–369. John Wright, PSG Inc., Boston.

DAWKINS, R. 1986. *The Blind Watchmaker*. W. W. Norton, New York.

DE CRECCHIO, L. 1865. Sopra un caso di apparenze virile in una donna. *Morgagni* 7: 151–183.

DE DUVE, C. 1984. *A Guided Tour of the Living Cell*, Vol. 2. Scientific American Books, New York.

DE GROUCHY, J. and C. TURLEAU. 1984. *Clinical Atlas of Human Chromosomes*, 2nd Edition. John Wiley & Sons, New York.

DE LA CHAPELLE, A. 1972. Analytic review: Nature and origin of males with XX sex chromosomes. *AJHG* 24: 71–105.

DE LA CHAPELLE, A. 1983. Sex chromosome abnormalities. In *Principles and Practice of Medical Genetics*, edited by A. E. H. Emery and D. L. Rimoin, pp. 193–215. Churchill Livingstone, Edinburgh.

DE LA CHAPELLE, A. 1986a. Sex reversal: genetic and molecular studies on 46,XX and 45,X males. *CSHS* 51: 249–255.

DE LA CHAPELLE, A. 1986b. The use and misuse of sex chromatin screening for gender identification of female athletes. *JAMA* 256: 1920–1923.

DE LA CHAPELLE, A. et al. 1986. The origin of 45,X males. *AJHG* 38: 330–340.

DE LA CRUZ, F. F. and P. S. GERALD, Editors. 1981. *Trisomy 21 (Down Syndrome): Research Perspectives*. University Park Press, Baltimore.

DE MARTINVILLE, B. and U. FRANCKE. 1983. The c-Ha-*ras*1, insulin and β-globin loci map outside the deletion associated with aniridia-Wilms' tumour. *Nature* 305: 641–643.

DEAN, G. 1957. Pursuit of a disease. *SA* 196(3): 133–142.

DELABAR, J.-M. et al. 1987. Beta amyloid gene duplication in Alzheimer's disease and karyotypically normal Down syndrome. *Science* 235: 1390–1392.

DELLARCO, V. L., P. E. VOYTEK and A. HOLLAENDER, Editors. 1985. *Aneuploidy: Etiology and Mechanisms*. Plenum Press, New York.

DEMARINI, D. M. 1983. Genotoxicity of tobacco smoke and tobacco smoke condensate. Mutation Research 114: 59–89.

DENIS, J. 1667. An extract of a letter . . . touching a late cure of an inveterate phrensy by the transfusion of blood. *The Royal Society of London, Philosophical Transactions* 2: 617–623.

DENIS, J. 1668. An extract of a printed letter . . . touching the difference risen about the transfusion of bloud. *The Royal Society of London, Philosophical Transactions* 3: 710–715.

DENNISTON, C. 1982. Low level radiation and genetic risk estimation in man. *Annual Review of Genetics* 16: 329–355.

DESMOND, E. W. 1987. Out in the open; changing attitudes and new research give fresh hope to alcoholics. *Time* 130 (Nov. 30): 80–90.

DESNICK, R. J. and G. A. GRABOWSKI. 1981. Advances in the treatment of inherited metabolic diseases. *Advances in Human Genetics* 11: 281–369.

DESNICK, R. J., G. A. GRABOWSKI and K. HIRSCHHORN. 1985. Prenatal metabolic diagnosis: a compendium. In *Human Prenatal Diagnosis*, edited by K. Filkins and J. F. Russo, pp. 59–108. Marcel Dekker, New York.

D'EUSTACHIO, P. and F. H. RUDDLE. 1983. Somatic cell genetics and gene families. *Science* 220: 919–924.

DEWEY, M. J., D. W. MARTIN, JR., G. R. MARTIN and B. MINTZ. 1977. Mosaic mice with teratocarcinoma-derived mutant cells deficient in hypoxanthine phosphoribosyltransferase. *PNAS* 74: 5564–5568.

DI MAIO, M. S., A. BAUMGARTEN, R. M. GREENSTEIN, H. M. SAAL and M. J. MAHONEY. 1987. Screening for fetal Down's syndrome in pregnancy by measuring maternal serum α-fetoprotein levels. *NEJM* 317: 342–346.

DI SANT' AGNESE, P. A., R. C. DARLING, G. A. PERERA and E. SHEA. 1953. Abnormal electrolyte composition of sweat in cystic fibrosis of the pancreas. Clinical significance and relationship to disease. *Pediatrics* 12: 549–563.

DIAMOND, B. A., D. E. YELTON and M. D. SCHARFF. 1981. Monoclonal antibodies: a new technology for producing serologic reagents. *NEJM* 304: 1344–1349.

DICKERSON, R. E. 1983. The DNA helix and how it is read. *SA* 249(Dec): 94–111.

DICKERSON, R. E. and I. GEIS. 1969. *The Structure and Action of Proteins*. Harper & Row, New York.

DICKERSON, R. E. and I. GEIS. 1983. *Hemoglobin: Structure, Function, Evolution, and Pathology*. Benjamin/Cummings, Menlo Park, California.

DICKMANN, Z., T. H. CHEWE, W. A. BONNEY, JR. and R. W. NOYES. 1965. The human egg in the pronuclear stage. *The Anatomical Record* 152: 293–302.

DIEDEREN, I. 1983. Genetics of schizophrenia. In *Behavior Genetics: Principles and Applications*, edited by J. L. Fuller and E. C. Simmel, pp. 189–215. Lawrence Erlbaum Associates, Hillsdale, New Jersey.

DILELLA, A. G., W.-M. HUANG and S. L. C. WOO. 1988. Screening for phenylketonuria mutations by DNA amplification with the polymerase chain reaction. *Lancet* I: 497–499.

DILELLA, A. G., S. C. M. KWOK, F. D. LEDLEY, J. MARVIT and S. L. C. WOO. 1986. Molecular structure and polymorphic map of the human phenylalanine hydroxylase gene. *Biochemistry* 25: 743–749.

DILELLA, A. G., J. MARVIT, K. BRAYTON and S. L. C. WOO. 1987. An amino-acid substitution involved in phenylketonuria is in linkage disequilibrium with DNA haplotype 2. *Nature* 327: 333–336.

DILELLA, A. G., J. MARVIT, A. S. LIDSKY, F. GUTTLER and S. L. C. WOO. 1986. Tight linkage between a splicing mutation and a specific DNA haplotype in phenylketonuria. *Nature* 322: 799–803.

DIRKSEN, E. R., D. M. PRESCOTT and C. F. FOX, Editors. 1978. *Cell Reproduction. ICN–UCLA Symposia on Molecular and Cellular Biology*, Vol. 12. Academic Press, New York.

DIXON, B. 1984. Of different bloods: 20 discoveries that shaped our lives. *Science 84* (Nov): 65–67.

DOBZHANSKY, TH. 1937. *Genetics and the Origin of Species.* Columbia University Press, New York.

DOBZHANSKY, TH. 1947. N. I. Vavilov, a martyr of genetics. *Journal of Heredity* 38: 227–232.

DOBZHANSKY, TH. 1962. *Mankind Evolving*. Yale University Press, New Haven.

DOETSCHMAN, T. et al. 1987. Targetted correction of a mutant HPRT gene in mouse embryonic stem cells. *Nature* 330: 576–578.

DOLL, R. and R. PETO. 1987. Epidemiology of cancer. In *Oxford Textbook of Medicine*, 2nd Edition, Volume 1, edited by D. J. Weatherall et al., pp. 4.95–4.123. Oxford Medical Publishers, Oxford.

DONAHUE, R. P., W. B. BIAS, J. H. RENWICK and V. A. McKUSICK. 1968. Probable assignment of the Duffy blood group locus to chromosome 1 in man. *PNAS* 61: 949–955.

DOOLITTLE, R. F. 1985. Proteins. *SA* 253(Oct): 88–99.

DOWN, J. L. H. 1866. Observations on an ethnic classification of idiots. *London Hospital Clinical Records and Reports* No. 3: 259–262.

DRAGER, U. C. 1986. Albinism and visual pathways. *NEJM* 314: 1636–1638.

DRESSLER, G. R. and P. GRUSS. 1988. Do multigene families regulate vertebrate development? *TIG* 4: 214–219.

DREYER, W. J. and J. C. BENNETT. 1965. The molecular basis of antibody formation: a paradox. *PNAS* 54: 864–869.

DRIEVER, W. and C. NÜSSLEIN-VOLHARD. 1988a. A gradient of *bicoid* protein in *Drosophila* embryos. *Cell* 54: 83–93.

DRIEVER, W. and C. NÜSSLEIN-VOLHARD. 1988b. The *bicoid* protein determines position in the *Drosophila* embryo in a concentration-dependent manner. *Cell* 54: 95–104.

DRIEVER, W. and C. NÜSSLEIN-VOLHARD. 1989. The bicoid protein is a positive regulator of hunchback transcription in the early Drosophila embryo. *Nature* 337: 138–143.

DROGARI, E., I. SMITH, M. BEASLEY and J. K. LLOYD. 1987. Timing of strict diet in relation to fetal damage in maternal phenylketonuria. *Lancet* II: 927–930.

DRONAMRAJU, K. R. 1964. Mating systems of the Andhra Pradesh people. *CSHS* 29: 81–84.

DUBRIDGE, R. B. and M. P. CALOS. 1987. Molecular approaches to the study of gene mutation in human cells. *TIG* 3: 293–297.

DUCHENNE, G. B. A. 1868. Recherches sur la paralysie musculaire pseudo-hypertrophique ou paralysie myo-sclerosique. *Archives Generales de Medicine* (6th series) 11: 552–588.

DULBECCO, R. 1976. Francis Peyton Rous. *National Academy of Sciences, Biographical Memoirs* 48: 274–306.

DUNCAN, I. 1987. The Bithorax complex. *Annual Review of Genetics* 21: 285–319.

DUNN, L. C. 1962. Cross currents in the history of human genetics. *AJHG* 14: 1–13.

DUNN, L. C. 1965. *A Short History of Genetics*. McGraw-Hill, New York.

DUPRAW, E. J. 1970. *DNA and Chromosomes*. Holt, Rinehart and Winston, New York.

DVOŘÁK, M., J. TESAŘIK, L. PILKA and P. TRÁVNÍK. 1982. Fine structure of human two-cell ova fertilized and cleaved in vitro. *Fertility and Sterility* 37: 661–667.

EBLING, F. J., Editor. 1975. *Racial Variation in Man*. John Wiley & Sons, New York.

EDELMAN, G. M. 1973. Antibody structure and molecular immunology. *Science* 180: 830–840.

EDWARDS v AGUILLARD. 1987. *United States Reports; Cases Adjudged in the Supreme Court* 482: 578–640.

EDWARDS, D. D. 1986. A common medical denominator. *Science News* 129: 60–62.

EDWARDS, R. G. 1974. Fertilization of human eggs in vitro: morals, ethics and the law. *Quarterly Review of Biology* 49: 3–26.

EDWARDS, R. G. 1985. Current status of human conception *in vitro*. *Proceedings of the Royal Society of London, Series B* 223: 417–448.

EDWARDS, R. and P. STEPTOE. 1980. *A Matter of Life*. William Morrow and Co., New York.

EGELAND, J. A. et al. 1987. Bipolar affective disorders linked to DNA markers on chromosome 11. *Nature* 325: 783–787.

EHRHARDT, A. A. and H. F. L. MEYER-BAHLBURG. 1981. Effects of prenatal sex hormones on gender-related behavior. *Science* 211: 1312–1318.

EHRLICH, P. 1900. On immunity with special reference to cell life. *Proceedings of the Royal Scoiety of London* 66: 424–448.

EHRMAN, L. and P. A. PARSONS. 1976. *The Genetics of Behavior*. Sinauer Associates, Sunderland, Massachusetts.

EICHER, E. M. and L. L. WASHBURN. 1986. Genetic control of primary sex determination in mice. *Annual Review of Genetics* 20: 327–360.

EISENBARTH, G. S. 1986. Type I diabetes mellitus; a chronic autoimmune disease. *NEJM* 314: 1360–1368.

EISENBUD, M. 1987. *Environmental Radioactivity From Natural, Industrial, and Military Sources*, 3rd Edition. Academic Press, Orlando, Florida.

EL-HEFNAWI, H., S. M. SMITH and L. S. PENROSE. 1965. Xeroderma pigmentosum—its inheritance and relationships to the ABO blood-group system. *Annals of Human Genetics* 28: 273–290.

ELIAS, S. and G.J. ANNAS. 1987. *Reproductive Genetics and the Law*. Year Book Medical Publishers, Chicago.

ELIAS, S. and J. L. SIMPSON. 1986. Amniocentesis. In *Genetic Disorders and the Fetus; Diagnosis, Prevention, and Treatment*, 2nd Edition, edited by A. Milunsky, pp. 31–52. Plenum Press, New York.

ELLISON, J. W. and L. E. HOOD. 1983. Human antibody genes: evolutionary and molecular genetic perspectives. *Advances in Human Genetics* 13: 113–147.

ELSTON, R. C. 1981. Segregation analysis. *Advances in Human Genetics* 11: 63–120.

ELWOOD, J. M. 1985. Temporal trends in twinning. In *Issues and Reviews in Teratology, Volume 3*, edited by H. Kalter, pp. 65–93. Plenum Press, New York.

EMERY, A. E. H. 1980. Duchenne muscular dystrophy: genetic aspects, carrier detection and antenatal diagnosis. *British Medical Bulletin* 36: 117–122.

EMERY, A. E. H. 1983. The muscular dystrophies. In *Principles and Practice of Medical Genetics*, edited by A. E. H. Emery and D. L. Rimoin, pp. 392–413. Churchill Livingstone, Edinburgh.

EMERY, A. E. H. 1984. *Introduction to Recombinant DNA*. John Wiley & Sons, New York.

EMERY, A. E. H. 1987. *Duchenne Muscular Dystrophy. Oxford University Press (Oxford Monographs in Medical Genetics No. 15).* Oxford University Press, New York.

EMERY, A. E. H. and I. M. PULLEN, Editors. 1984. *Psychological Aspects of Genetic Counselling*. Academic Press, London.

EMERY, A. E. H. and D. L. RIMOIN, Editors. 1983. *Principles and Practice of Medical Genetics* (2 volumes). Churchill Livingstone, Edinburgh.

ENGELHARDT, H. T., JR. 1984. Shattuck Lecture: Allocating scarce medical resources and the availability of organ transplantation. *NEJM* 311: 66–71.

EPEL, D. 1977. The program of fertilization. *SA* 237(Nov): 128–138.

EPEL, D. 1980. Fertilization. *Endeavour* (New Series) 4: 26–31.

EPHRUSSI, B. 1953. *Nucleo-cytoplasmic Relations in Micro-Organisms.* Oxford University Press, New York.

EPHRUSSI, B. 1972. *Hybridization of Somatic Cells*. Princeton University Press, Princeton, New Jersey.

EPSTEIN, C. J. 1986. *The Consequences of Chromosome Imbalance: Principles, Mechanisms and Models*. Cambridge University Press, Cambridge.

EPSTEIN, C. J. 1988. Mechanisms of the effects of aneuploidy in mammals. *Annual Review of Genetics* 22: 51–75.

EPSTEIN, C. J., D. R. COX and L. B. EPSTEIN. 1985. Mouse trisomy 16: an animal model of human trisomy 21 (Down syndrome). In *Molecular Structure of the Number 21 Chromosome and Down Syndrome*, edited by G. F. Smith. *Annals of the New York Academy of Sciences,* Vol. 450.

ERBE, R. W. 1988. Genetic counseling. In *Textbook of Internal Medicine*, edited by W. N. Kelley et al., pp. 2343–2349. J. B. Lippincott, Philadelphia.

ERBE, R. W. and G. R. BOSS. 1983. Newborn genetic screening. In *Principles and Practice of Medical Genetics,* Volume 2, edited by A. E. H. Emery and D. L. Rimoin, pp. 1437–1450. Churchill Livingstone, Edinburgh.

ERICSSON, R. J. and F. BEERNINK. 1987. Sex chromosome ratios in human sperm. *Fertility and Sterility* 47: 531–532.

ERICSSON, R. J., C. N. LANGEVIN and M. NISHINO. 1973. Isolation of fractions rich in human Y sperm. *Nature* 246: 421–424.

ERLICH, H. A., D. H. GELFAND and R. K. SAIKI. 1988. Specific DNA amplification. *Nature* 331: 461–462.

ESTIVILL, X. et al. 1988. Linkage disequilibrium between cystic fibrosis and linked DNA polymorphisms in Italian families: a collaborative study. *AJHG* 43: 23–28.

EVANS, H. J. 1977. Some facts and fancies relating to chromosome structure in man. *Advances in Human Genetics* 8: 347–438.

EVANS, M. J. and M. H. KAUFMAN. 1981. Establishment in culture of pluripotential cells from mouse embryos. *Nature* 292: 154–156.

FALCONER, D. S. 1981. *Introduction to Quantitative Genetics*, 2nd Edition. Longman, London.

FALK, R. 1984. The gene in search of an identity. *Human Genetics* 68: 195–204.

FARRELL, M. et al. 1986. Cystic fibrosis carrier detection using a linked gene probe. *Journal of Medical Genetics* 23: 295–299.

FEASTER, W. W., L. W. KWOK and C. J. EPSTEIN. 1977. Dosage effects for superoxide dismutase-1 in nucleated cells aneuploid for chromosome 21. *AJHG* 29: 563–570.

FEDERMAN, D. D. 1987a. Psychosexual adjustment in congenital adrenal hyperplasia. *NEJM* 316: 209–211.

FEDERMAN, D. D. 1987b. Mapping the X-chromosome: Mining its p's and q's. *NEJM* 317: 161–162.

FEHILLY, C. B., S. M. WILLADSEN and E. M. TUCKER. 1984. Interspecific chimaerism between sheep and goat. *Nature* 307: 634–636.

FELLER, W. 1968. *An Introduction to Probability Theory and Its Applications*, Volume 1, 3rd Edition. John Wiley & Sons, New York.

FELSENFELD, G. 1985. DNA. *SA* 253(Oct): 58–67.

FERGUSON-SMITH, M. A., D. A. AITKEN, C. TURLEAU and J. DE GROUCHY. 1976. Localisation of the human *ABO:Np-1:AK-1* linkage group by regional assignment of *AK-1* to 9q34. *Human Genetics* 34: 35–43.

FERGUSON-SMITH, M. A., B. F. NEWMAN, P. M. ELLIS and D. M. G. THOMSON. 1973. Assignment by deletion of human red cell acid phosphatase gene locus to the short arm of chromosome 2. *Nature New Biology* 243: 271–273.

FERGUSON-SMITH, M. A. and J. R. W. YATES. 1984. Maternal age-specific rates for chromosome aberrations and factors influencing them: report of a collaborative European study on 52,965 amniocenteses. *Prenatal Diagnosis* 4: 5–44.

FESTENSTEIN, H., P. DOYLE and J. HOLMES. 1986. Long-term follow-up in London transplant group recipients of cadaver renal allografts. *NEJM* 314: 7–14.

FICHMAN, K. R., B. R. MIGEON and C. J. MIGEON. 1980. Genetic disorders of male sexual differentiation. *Advances in Human Genetics* 10: 333–377, 387.

FIDDES, J. C. 1977. The nucleotide sequence of a viral DNA. *SA* 237(Dec): 54–67.

FINN, R. 1960. Erythroblastosis (extract of meeting paper). *Lancet* I: 526. [Reprinted in Clarke (1975).]

FISHER, R. A. 1930. *The Genetical Theory of Natural Selection*. The Clarendon Press, Oxford. Reprinted 1958, Dover Publications, New York.

FISHER, R. A. 1936. Has Mendel's work been rediscovered? *Annals of Science* 1: 115–137. [Reprinted in Stern and Sherwood (1966).]

FLEMMING, W. 1879. Contributions to the knowledge of the cell and its life phenomena. *Archiv für Mikroskopische Anatomie* 16: 302–406. [Abridged translation reprinted in Gabriel and Fogel (1955).]

FLETCHER, J. 1974. *The Ethics of Genetic Control: Ending Reproductive Roulette*. Anchor Books, Garden City, New York.

FORD, C. E. and J. L. HAMERTON. 1956. The chromosomes of man. *Nature* 178: 1020–1023.

FORD, C. E., K. W. JONES, P. E. POLANI, J. C. DE ALMEIDA and J. H. BRIGGS. 1959a. A sex-chromosome anomaly in a case of gonadal dysgenesis (Turner's syndrome.) *Lancet* I: 711–713.

FORD, C. E., P. E. POLANI, J. H. BRIGGS and P. M. F. BISHOP. 1959b. A presumptive human XXY/XX mosaic. *Nature* 183: 1030–1032.

FORD, E. H. R. 1973. *Human Chromosomes*. Academic Press, New York.

FORREST, S. M. et al. 1987. Preferential deletion of exons in Duchenne and Becker muscular dystrophies. *Nature* 329: 638–640.

FOWLER, R. E. and R. G. EDWARDS. 1973. The genetics of early human development. *Progress in Medical Genetics* 9: 49–112.

FOX, T. D. 1985. Diverged genetic codes in protozoans and a bacterium. *Nature* 314: 132–133.

FRANCIS, D. P. and J. CHIN. 1987. The prevention of acquired immunodeficiency syndrome in the United States. *JAMA* 257: 1357–1366.

FRANCKE, U. 1983. Gene mapping. In *Principles and Practice of Medical Genetics,* Vol. 1, edited by A. E. H. Emery and D. L. Rimoin, pp. 91–110. Churchill Livingstone, Edinburgh.

FRANCKE, U., M. S. BROWN and J. L. GOLDSTEIN. 1984. Assignment of the human gene for the low density lipoprotein receptor to chromosome 19: synteny of a receptor, a ligand, and a genetic disease. *PNAS* 81: 2826–2830.

FRANCKE, U. et al. 1985. Minor Xp21 chromosome deletion in a male associated with expression of Duchenne muscular dystrophy, chronic granulomatous disease, retinitis pigmentosa, and McLeod syndrome. *AJHG* 37: 250–267

FRANKS, L. M. 1986. What is cancer? In *Introduction to the Cellular and Molecular Biology of Cancer*, edited by L. M. Franks and N. M. Teich, pp. 1–26. Oxford University Press, Oxford.

FRANKS, L. M. and N. M. TEICH, Editors. 1986. *Introduction to the Cellular and Molecular Biology of Cancer*. Oxford University Press, Oxford.

FRASER, F. C. 1973. Survey of Counseling practices. In *Ethical Issues in Human Genetics*, edited by B. Hilton et al. pp. 7–13. Plenum Press, New York.

FRASER, F. C. 1988. Genetic counseling: using the information wisely. *Hospital Practice* 23 (6): 245–266.

FRATANTONI, J. C., C. W. HALL and E. F.NEUFELD. 1968. Hurler and Hunter syndromes: mutual correction of the defect in cultured fibroblasts. *Science* 162: 570–572.

FREIER-MAIA, N. and M. PINHEIRO. 1984. *Ectodermal Dysplasias: A Clinical and Genetic Study*. Alan R. Liss, New York.

FREIFELDER, D. 1978. *The DNA Molecule; Structure and Properties. Original Papers, Analyses, and Problems.* W. H. Freeman and Co., San Francisco.

FRIEDBERG, E. C., Editor. 1986. *Cancer Biology; Readings from Scientific American*. W. H. Freeman and Co., New York.

FRIEDMAN, M. J. and W. TRAGER. 1981. The biochemistry of resistance to malaria. *SA* 244(Mar): 154–164.

FRIZZELL, R. A., G. RECHKEMMER and R. L. SHOEMAKER. 1986. Altered regulation of airway epithelial cell chloride channels in cystic fibrosis. *Science* 233: 558–560.

FRYNS, J. P., A. KLECZKOWSKA, E. KUBIEN, P. PETIT and H. VAN DEN BERGHE. 1984. Cytogenetic survey in couples with recurrent fetal wastage. *Human Genetics* 65: 336–354.

FUCHS, F. 1980. Genetic amniocentesis. *SA* 242 (June): 47–53.

FUDENBERG, H. H., J. R. L. PINK, A.-C. WANG and G. B. FERRARA. 1984. *Basic Immunogenetics*, 3rd Edition. Oxford University Press, Oxford.

FUHRMANN, W. and F. VOGEL. 1983. *Genetic Counseling*, 3rd Edition. [Translated by Sabine Kurth-Scherer.] Springer-Verlag, New York.

FUJITA, T. 1975. Abnormal spermatozoa and infertility (man). In *Scanning Electron Microscopical Atlas of Mammalian Reproduction*, edited by E. S. E. Hafez, pp. 82–87. Springer-Verlag, New York.

FULLER, J. L. and E. C. SIMMEL, Editors. 1983. *Behavior Genetics; Principles and Applications*. Lawrence Erlbaum Associates, Hillsdale, New Jersey.

FULLER, J. L. and W. R. THOMPSON. 1960. *Behavior Genetics*. John Wiley & Sons, New York.

FUNDERBURK, S. J., M. A. SPENCE and R. S. SPARKES. 1977. Mental retardation associated with "balanced" chromosome rearrangements. *AJHG* 29: 136–141.

FUTUYMA, D. J. 1983. *Science on Trial: The Case for Evolution*. Pantheon Books, New York.

FUTUYMA, D. J. 1986. *Evolutionary Biology*, 2nd Edition. Sinauer Associates, Sunderland, Massachusetts.

GABRIEL, M. L. and S. FOGEL, Editors. 1955. *Great Experiments in Biology*. Prentice-Hall, Englewood Cliffs, New Jersey.

GAJDUSEK, D. C. 1963. Kuru. *Royal Society of Tropical Medicine and Hygiene, London, Transactions* 57: 151–166.

GAJDUSEK, D. C. 1977. Unconventional viruses and the origin and disappearance of kuru. *Science* 197: 943–960.

GALE, R. P., Editor. 1986. *Leukemia Therapy*. Blackwell Scientific Publications, Boston.

GALLO, R. C. and L. MONTAGNIER. 1988. AIDS in 1988. *SA* 259(Oct): 40–48. [The entire issue is devoted to AIDS.]

GALTON, F. 1889. *Natural Inheritance*. Macmillan and Co., London.

GALTON, F. 1901. The possible improvement of the human breed under the existing conditions of law and sentiment. *Nature* 64: 659–665. [Reprinted in Bajema (1976) and in Galton (1909).]

GALTON, F. 1905. Eugenics: its definition, scope and aims. In *Sociological Papers*, pp. 45–50. Macmillan & Co., London. [Reprinted in Bajema (1976) and in Galton (1909).]

GALTON, F. 1909. *Essays in Eugenics*. The Eugenics Education Society, London.

GARCIA-BELLIDO, A., P.R. A. LAWRENCE and G. MORATA. 1979. Compartments in animal development. *SA* 241(July): 102–110.

GARDNER, M. 1975. *Mathematical Carnival*. Alfred A. Knopf, New York.

GARDNER, M. 1977. The concept of negative numbers and the difficulty of grasping it. *SA* 236(June): 131–135.

GARROD, A. E. 1902. The incidence of alkaptonuria: a study in chemical individuality. *Lancet* II: 1616–1620.

GARROD, A. E. 1909. *Inborn Errors of Metabolism*. Oxford University Press, Oxford. Reprinted (1963). with a supplement by Harry Harris. Oxford University Press, London.

GARTLER, S. M., K. A. DYER, J. A. M. GRAVES and M. ROCCHI. 1985. A two step model for mammalian X-chromosome inactivation. In *Biochemistry and Biology of DNA Methylation*, edited by G. L. Cantoni and A. Razin, pp. 223–235. Alan R. Liss, New York.

GARTLER, S. M. and A. D. RIGGS. 1983. Mammalian X-chromosome inactivation. *Annual Review of Genetics* 17: 155–190.

GARTLER, S. M., R. C. SCOTT, J. L. GOLDSTEIN, B. CAMPBELL and R. R. SPARKES. 1971. Lesch-Nyhan syndrome: rapid detection of heterozygotes by use of hair follicles. *Science* 172: 572–574.

GARTLER, S. M., S. H. WAXMAN and E. GIBLETT. 1962. An XX/XY human hermaphrodite resulting from double fertilization. *PNAS* 48: 332–335.

GASTON, M. H. et al. 1986. Prophylaxis with oral penicillin in children with sickle cell anemia. *NEJM* 314: 1593–1599.

GATTI, M. et al. 1984. Genetic control of mitotic cell division in *Drosophila melanogaster. Genetics: New Frontiers*, Vol. 3, edited by V. L. Chopra et al., pp. 193–204. Oxford & IBH Publishing Co., New Delhi.

GATTI, R. A. and K. HALL. 1983. Ataxia-telangiectasia: search for a central hypothesis. In *Chromosome Mutation and Neoplasia,* edited by J. German, pp. 23–41. Alan R. Liss, New York.

GATTI, R. A. et al. 1988. Localization of an ataxia-telangiectasia gene to chromosome 11q22–23. *Nature* 336: 577–580.
GEHRING, W. J. 1985. The molecular basis of development. *SA* 253.(Oct): 153–162.
GEHRING, W. J. 1987. Homeo boxes in the study of development. *Science* 236: 1245–1252.
GEHRING, W. J. and Y. HIROMI. 1986. Homeotic genes and the homeobox. *Annual Review of Genetics* 20: 147–173.
GELINAS, D. J. 1983. The persisting negative effects of incest. *Psychiatry* 46: 312–332.
GERALD, P. S. 1980. X-linked mental retardation and an X-chromosome marker. *NEJM* 303: 696–697.
GERMAN, J. 1970. Studying human chromosomes today. *American Scientist* 58: 182–201.
GERMAN, J., Editor. 1983. *Chromosome Mutation and Neoplasia*. Alan R. Liss, New York.
GERRICK, D. J. 1978. *Pharmacogenetics*. Dayton Laboratories.
GHESQUIERE, J., R. D. MARTIN and F. NEWCOMBE, Editors. 1985. *Human Sexual Dimorphism: Symposia of the Society for the Study of Human Biology,* Vol. XXIV. Taylor & Francis, London.
GIBBS, D. A. et al. 1984. First-trimester diagnosis of Lesch-Nyhan syndrome. *Lancet* II: 1180–1183.
GIBBS, R. A. and C. T. CASKEY. 1987. Identification and localization of mutations at the Lesch-Nyhan locus by ribonuclease A cleavage. *Science* 236: 303–305.
GIESER, E. P. and H. F. FALLS. 1961. Hereditary retinoschisis. *American Journal of Ophthalmology* 51: 1193–1200.
GILBERT, F. et al. 1975. Tay-Sachs' and Sandhoff's diseases: the assignment of genes for hexosaminidase A and B to individual human chromosomes. *PNAS* 72: 263–267.
GILBERT, S. F. 1988. *Developmental Biology*, 2nd Edition. Sinauer Associates, Sunderland, Massachusetts.
GILBERT, W. 1985. Genes-in-pieces revisited. *Science* 228: 823–824.
GILBERT, W. S. 1879. *The Pirates of Penzance*. Included in *Plays and Poems of W. S. Gilbert*, Random House, New York, 1932.
GILBOA, E., M. A. EGLITIS, P. W. KANTOFF and W. F. ANDERSON. 1986. Transfer and expression of cloned genes using retroviral vectors. *BioTechniques* 6: 504–512.
GILLESPIE, F. D. 1961. Ocular albinism with report of a family with female carriers. *Archives of Ophthalmology* 66: 774–777.
GITSCHIER, J. et al. 1984. Characterization of the human factor VIII gene. *Nature* 312: 326–330.
GLEDHILL, B. L. 1983. Control of mammalian sex ratio by sexing sperm. *Fertility and Sterility* 40: 572–574.
GLEITMAN, H. 1986. *Psychology*, 2nd Edition. W. W. Norton, New York.
GODDARD, H. H. 1912. *The Kallikak Family: A Study in the Heredity of Feeble-Mindedness.* The Macmillan Co., New York.
GODSON, G. N. 1985. Molecular approaches to malaria vaccines. *SA* 252(May): 52–59.
GOEDDE, H. W. and D. P. AGARWAL, Editors. 1987. *Genetics and Alcoholism*. Alan R. Liss, New York.
GOEDDEL, D. V. et al. 1979. Direct expression in *Escherchia coli* of a DNA sequence coding for human growth hormone. *Nature* 281: 544–548.
GOLDFOOT, D. A. and D. A. NEFF. 1985. On measuring behavioral sex differences in social contexts. In *Handbook of Behavioral Neurobiology*, edited by N. Adler et al., pp. 767–783.
GOLDGABER, D., M. I. LERMAN, O. W. McBRIDE, U. SAFFIOTTI and D. C. GAJDUSEK. 1987. Characterization and chromosomal localization of a cDNA encoding brain amyloid of Alzheimer's disease. *Science* 235: 877–880.
GOLDMAN, M. A. et al. 1987. A chicken transferrin gene in transgenic mice escapes X-chromosome inactivation. *Science* 236: 593–595.
GOLDSTEIN, J. L. and M. S. BROWN. 1979. The LDL receptor locus and the genetics of familial hypercholesterolemia. *Annual Review of Genetics* 13: 259–289.
GOLDSTEIN, J. L. and M. S. BROWN. 1983. Familial hypercholesterolemia. In *The Metabolic Basis of Inherited Disease*, 5th Edition, edited by J. B. Stanbury et al., pp. 672–712.
GOLDSTEIN, J. L., M. S. BROWN, R. G. W. ANDERSON, D. W. RUSSELL and W. J. SCHNEIDER. 1985. Receptor-mediated endocytosis: concepts emerging from the LDL receptor system. *Annual Review of Cell Biology* 1: 1–39.
GOLUB, E. S. 1987a. Somatic mutation: diversity and regulation of the immune repertoire. *Cell* 4: 723–724.
GOLUB, E. S. 1987b. *Immunology: A Synthesis.* Sinauer Associates, Sunderland, Massachusetts.
GONZALEZ, F. J. et al. 1988a. Characterization of the common genetic defect in humans deficient in debrisoquine metabolism. *Nature* 331: 442–446.
GONZALEZ, F. J. et al. 1988b. Human debrisoquine 4-hydroxylase (P450IID1): cDNA and deduced amino acid sequence and mapping of the P450D1 gene to chromosome 22. *Genomics* 2: 147–179.
GOODFELLOW, P. J., S. M. DARLING, N. S. THOMAS and P. N. GOODFELLOW. 1986. A pseudoautosomal gene in man. *Science* 234: 740–743.
GOODFELLOW, P., S. DARLING and J. WOLFE. 1985. The human Y chromosome. *Journal of Medical Genetics* 22: 329–344.
GOODFELLOW, P. N. 1987. From pathological phenotype to candidate gene for Alzheimer's disease. *TIG* 3: 59–60.
GOODFELLOW, P. N., I. W. CRAIG, J. C. SMITH and J. WOLFE, Editors. 1987. *The Mammalian Y Chromosome: Molecular Search for the Sex-Determining Factor*. Supplement to Volume 101 of *Development*. [This volume contains papers presented at the meeting of the British Society for Developmental Biology held at the University of Oxford in March of 1987.]
GOODFIELD, J. 1981. *Reflections on Science and the Media*. American Association for the Advancement of Science, Washington D. C.
GOODWIN, D. W. 1985. Alcoholism and genetics. *Archives of General Psychiatry* 42: 171–174.
GORBSKY, G. J., P. J. SAMMAK and G. G. BORISY. 1987. Chromosomes move poleward in anaphase along stationary microtubules that coordinately disassemble from their kinetochore ends. *The Journal of Cell Biology* 104: 9–18.
GORDON, J. W. and F. H. RUDDLE. 1983. Gene transfer into mouse embryos: production of transgenic mice by pronuclear injection. *Methods in Enzymology* 101: 411–433.
GOSDEN, C. M. and J. R. GOSDEN. 1984. Fetal abnormalities in cystic fibrosis suggest a deficiency in proteolysis of cholecystokinin. *Lancet* II: 541–546.
GOSDEN, J. R. et al. 1984. The use of cloned Y chromosome-specific DNA probes for fetal sex determination in first trimester prenatal diagnosis. *Human Genetics* 66: 347–351.
GOTTLIEB, K., Meeting Organizer. 1983. Surnames as markers of inbreeding and migration. [20 papers] *Human Biology* 55: 209–408.
GOULD, S. J. 1981. *The Mismeasure of Man*. W. W. Norton & Co., New York.
GOULD, S. J. 1986. Reflections from an interior world. *Nature* 320: 647–648.
GOULD, S. J. and N. ELDREDGE. 1977. Punctuated equilibria: the tempo and mode of evolution reconsidered. *Paleobiology* 3: 115–151.
GOWERS, W. 1879. *Pseudohypertrophic Muscular Paralysis*. J. & A. Churchill, London.
GRADWOHL, R. B. H., Editor. 1954. *Legal Medicine*. The C. V. Mosby Co., St. Louis.
GRAHAM, J. B., E. S. BARROW, H. M. REISNER and C.-J. S. EDGELL. 1983. The genetics of blood coagulation. *Advances in Human Genetics* 13: 1–81, 299.
GRANT, S. G. and V. M. CHAPMAN. 1988. Mechanisms of X-chromosome regulation. *Annual Review of Genetics* 22: 199–233.
GRAVEL, R. A., J. A. LOWDEN, J. W. CALLAHAN, L. S. WOLFE and N. M. K. NG YIN KIN. 1979. Infantile sialidosis: a phenocopy of Type 1 G_{M1} gangliosidosis distinguished by genetic complementation and urinary oligosaccharides. *AJHG* 31: 669–679.
GRAVES, J. A. M. 1987. The evolution of mammalian sex chromosomes and dosage compensation: clues from marsupials and monotremes. *TIG* 3: 252–256.
GREENBERG, C. R. et al. 1988. Gene studies in newborn males with Duchenne muscular dystrophy detected by neonatal screening. *Lancet* II: 425–427.
GREENBLATT, A. 1974. *Heredity and You: How You Can Protect Your Family's Future.* Coward, McCann & Geoghegan, New York.
GREENDYKE, R. M. 1980. *Introduction to Blood Banking*, 3rd Edition. Medical Examination Publishing, Garden City, New York.
GREENWALD, I. S., P. W. STERNBERG and H. R. HORVITZ. 1983. The *lin-12* locus specifies cell fates in *Caenorhabditis elegans. Cell* 34: 435–444.
GREGG, R. G. and O. SMITHIES. 1986. Targeted modification of human chromosomal genes. *CSHS* 51: 1093–1099.
GRIPENBERG, U., K. HONGELL, S. KNUUTILA, M. KAHKONEN and J. LEISTI. 1980. A chromosome survey of 1062 mentally retarded

patients. Evaluation of a long-term study at the Rinnekoti Institution, Finland. *Hereditas* 92: 223–228.

GRIVELL, L. A. 1983. Mitochondrial DNA. *SA* 248(Mar): 78–89.

GROBSTEIN, C. 1979. External human fertilization. *SA* 240(June): 57–67.

GROTH, C. G. and O. RINGDÉN. 1984. Transplantation in relation to the treatment of inherited disease. *Transplantation* 38: 319–327.

GRUMBACH, M. M. and F. A. CONTE. 1985. Disorders of sexual differentiation. In *Williams Textbook of Endocrinology*, 7th Edition, edited by J. D. Wilson and D. W. Foster, pp. 312–401. W. B. Saunders, Philadelphia.

GRUNDY, P. et al. 1988. Familial predisposition to Wilms' tumour does not map to the short arm of chromosome 11. *Nature* 336: 374–376. [see also pp. 377–378.]

GRUNDY, S. M. 1986. Cholesterol and coronary heart disease: a new era. *JAMA* 256: 2849–2858.

GUILLERY, R. W. 1974. Visual pathways in albinos. *SA* 230(May): 44–54.

GUSELLA, J. F. 1986. DNA polymorphism and human disease. *Annual Review of Biochemistry* 55: 831–854.

GUSELLA, J. F. et al. 1983. A polymorphic DNA marker genetically linked to Huntington's disease. *Nature* 306: 234–238.

GUSELLA, J. F. et al. 1985. Deletion of Huntington's disease-linked G8(D4S10) locus in Wolf-Hirschhorn syndrome. *Nature* 318: 75–78.

GUSELLA, J. F. et al. 1986. Molecular genetics of Huntington's disease. *CSHS* 51: 359–364.

GUTHRIE, R. and A. SUSI. 1963. A simple phenylalanine method for detecting phenylketonuria in large populations of newborn infants. *Pediatrics* 32: 338–343.

GUTHRIE, R. and S. WHITNEY. 1965. *Phenylketonuria Detection in the Newborn Infant as a Routine Hospital Procedure*. Children's Bureau Publication No. 419–1964, revised 1965. U. S. Department of Health, Education, and Welfare, Washington, D. C.

GÜTTLER, F. et al. 1987. Correlation between polymorphic DNA haplotypes at phenylalanine hydroxylase locus and clinical phenotypes of phenylketonuria. *Journal of Pediatrics* 110: 68–71.

HAAPALA, O. 1983. A prologue to the study of metaphase chromosome structure. In *Kew Chromosome Conference II*, edited by P. E. Brandham and M. D. Bennett, pp. 19–25. George Allen & Unwin, London.

HACKING, I. 1984. Trial by number: Karl Pearson's chi-square test measured the fit between theory and reality, ushering in a new sort of decision making. *Science 84*(Nov): 69–70.

HAFEZ, E. S. E., Editor. 1975. *Scanning Electron Microscopical Atlas of Mammalian Reproduction*. Springer-Verlag, New York.

HAFEZ, E. S. E. and P. KENEMANS, Editors. 1982. *Atlas of Human Reproduction by Scanning Electron Microscopy*. MTP Press, Boston.

HAGERMAN, R. JENSSEN and P. M. McBOGG, Editors. 1983. *The Fragile X Syndrome: Diagnosis, Biochemistry, and Intervention*. Spectra Publishing, Dillon, Colorado.

HAKOMORI, S.-I. 1986. Glycosphingolipids. *SA* 254(May): 44–53.

HALDANE, J. B. S. 1932. *The Causes of Evolution*. Harper & Brothers, London. [Reprinted (1966) by Cornell University Press, Ithaca, New York.]

HALL, S. S. 1987. *Invisible Frontiers: The Race to Synthesize a Human Gene*. Atlantic Monthly Press, New York.

HALUSKA, F. G., Y. TSUJIMOTO and C. M. CROCE. 1987. Oncogene activation by chromosome translocation in human malignancy. *Annual Review of Genetics* 21: 321–345.

HAMERTON, J. L. 1971. *Human Cytogenetics, Volume 1: General Cytogenetics*. Academic Press, New York.

HAMERTON, J. L. 1984. Cytogenetic disorders. *NEJM* 310: 314–316.

HAMERTON, J. L., S. POVEY and N. E. MORTON. 1984. Report of the committee on the genetic constitution of chromosome 1. *Cytogenetics and Cell Genetics* 37: 3–21.

HAMILTON, J. B. 1951. Patterned loss of hair in man: types and incidence. *New York Academy of Sciences, Annals* 53 (Article 3): 708–728.

HAMILTON, M. P., Editor. 1972. *The New Genetics and the Future of Man*. William B. Eerdmans Publishing Co., Grand Rapids, Michigan.

HANAHAN, D. 1988. Dissecting multistep tumorigenesis in transgenic mice. *Annual Review of Genetics* 22: 479–519.

HANAWALT, P. C. and A. SARASIN. 1986. Cancer-prone hereditary diseases with DNA processing abnormalities. *TIG* 2: 124–129.

HANDLER, P. 1979. Basic research in the United States. *Science* 204: 474–479.

HANSEN, M. F. and W. K. CAVENEE. 1988. Retinoblastoma and the progression of tumor genetics. *TIG* 4: 125–128.

HARBOUR, J. W. et al. 1988. Abnormalities in structure and expression of the human retinoblastoma gene in SCLC. *Science* 241: 353–357.

HARDY, G. H. 1908. Mendelian proportions in a mixed population. *Science* 28: 49–50. [Reprinted in Peters (1959), Morris (1971), and Jameson (1977).]

HARNDEN, D. G. 1974. Ataxia telangiectasia syndrome: cytogenetic and cancer aspects. In *Chromosomes and Cancer*, edited by J. German, pp. 619–636. John Wiley & Sons, New York.

HARRIS, A. and M. SUPER. 1987. *Cystic Fibrosis: The Facts*. Oxford University Press, Oxford.

HARRIS, H. 1980. *The Principles of Human Biochemical Genetics*, 3rd, Revised Edition. Elsevier/North-Holland Biomedical Press, Amsterdam.

HARRIS, L. J. 1985. Delicacy of fibres in the brain: early and recent neuropsychological explanations of sex differences in cognition and temperament. In *Human Sexual Dimorphism*, edited by J. Ghesquiere et al., pp. 283–321. Taylor and Francis, London.

HARRIS, R. 1988. Genetic counselling and the new genetics. *TIG* 4: 52–56.

HARRISON, C. J., E. M. JACK, T. D. ALLEN and R. HARRIS. 1983. The fragile X: a scanning electron microscope study. *Journal of Medical Genetics* 20: 280–285.

HARRISON, D. 1970. *Problems in Genetics, with Notes and Examples*. Addison-Wesley Publishers, Reading, Massachusetts.

HARRISON, G. A. and J. J. T. OWEN. 1964. Studies on the inheritance of human skin colour. *Annals of Human Genetics* 28: 27–37. [Reprinted in L. Levine (1971).]

HARTL, D. L. 1983. *Human Genetics*. Harper & Row, New York.

HARTL, D. L. 1985. *Our Uncertain Heritage: Genetics and Human Diversity*, 2nd Edition. Harper & Row, New York.

HARTL, D. L. 1988. *A Primer of Population Genetics*, 2nd Edition. Sinauer Associates, Sunderland, Massachusetts.

HARTL, D. L. and A. G. CLARK. 1989. *Principles of Population Genetics*, 2nd Edition. Sinauer Associates, Sunderland, Massachusetts.

HARTMANN, J. F., Editor. 1983. *Mechanism and Control of Animal Fertilization*. Academic Press, New York.

HARTWELL, L. H. 1978. Cell division from a genetic perspective. *Journal of Cell Biology* 77: 627–637.

HARVEY, J., C. JUDGE and S. WIENER. 1977. Familial X-linked mental retardation with an X chromosome abnormality. *Journal of Medical Genetics* 14: 46–50.

HASSOLD, T. J. 1986. Chromosome abnormalities in human reproductive wastage. *TIG* 2: 105–110.

HASSOLD, T. et al. 1980. A cytogenetic study of 1000 spontaneous abortions. *Annals of Human Genetics* 55: 151–178.

HASSOLD, T. J. and P. A. JACOBS. 1984. Trisomy in man. *Annual Review of Genetics* 18: 69–97.

HAUBRICH, W. S. 1984. *Medical Meanings; A Glossary of Word Origins*. Harcourt Brace Jovanovich, San Diego.

HAUSCHKA, T. S., J. E. HASSON, M. N. GOLDSTEIN, G. F. KOEPF and A. A. SANDBERG. 1962. An XYY man with progeny indicating familial tendency to non-disjunction. *AJHG* 14: 22–30.

HAYDEN, M. R. and J. L. NICHOLS. 1984. Molecular genetic approaches to the study of the nervous system. *Developments in Neuroscience* 6: 189–214.

HAYNES, R. H. and J. D. B. COLLINS. 1984. The mutagenic potential of caffeine. In *Caffeine: Perspectives from Recent Research* edited by P. B. Dews, pp. 221–238. Springer-Verlag, Berlin.

HEARNSHAW, L. S. 1979. *Cyril Burt, Psychologist*. Cornell University Press, Ithaca, New York.

HEATH, C. W. et al. 1984. Cytogenetic findings in persons living near the Love Canal. *JAMA* 251: 1437–1440.

HEDDLE, J. A., Editor. 1982. *Mutagenicity: New Horizons in Genetic Toxicology*. Academic Press, New York.

HEISTERKAMP, N. et al. 1983. Localization of the c–*abl* oncogene adjacent to a translocation break point in chronic myelocytic leukaemia. *Nature* 306: 239–242.

HELMINEN, P., C. EHNHOLM, M.-L. LOKKI, A. JEFFREYS and L. PELTONEN. 1988. Application of DNA "fingerprints" to paternity determinations. *Lancet* I: 574–576.

HENDERSON, N. D. 1982. Human behavior genetics. *Annual Review of Psychology* 33: 403–440.

HENIKOFF, S., M. A. KEENE, K. FECHTEL and J. W. FRISTROM. 1986. Gene within a gene: nested *Drosophila* genes encode unrelated proteins on opposite DNA strands. *Cell* 44: 33–42.

HENLE, W., G. HENLE and E. T. LENNETTE. 1979. The Epstein-Barr virus. *SA* 241 (July): 48–59.

HERTIG, A. T. and J. A. ROCK. 1949. A series of potentially abortive ova recovered from fertile women prior to the first missed menstrual period. *American Journal of Obstetrics and Gynecology* 58: 968–993.

HIGUCHI, R., C. H. VON BEROLDINGEN, G. F. SENSABAUGH and H. A. ERLICH. 1988. DNA typing from single hairs. *Nature* 332: 543–546.

HIRSCHHORN, R. and K. HIRSCHHORN. 1983. Immunodeficiency disorders. In *Principles and Practice of Medical Genetics*, Vol. 2, edited by A. E. H. Emery and D. L. Rimoin, pp 1091–1108. Churchill Livingstone, Edinburgh.

HOBBS, H. H., M. S. BROWN, D. W. RUSSELL, J. DAVIGNON and J. L. GOLDSTEIN. 1987. Deletion in the gene for the low-density-lipoprotein receptor in a majority of French Canadians with familial hypercholesterolemia. *NEJM* 317: 734–737.

HODGKIN, J. 1988. Sex determination: right gene, wrong chromosome. *Nature* 336: 712.

HODSON, M. E., A. P. NORMAN and J. C. BATTEN. 1983. *Cystic Fibrosis.* [Published by Bailliers-Tindall.] Saunders, Philadelphia.

HOFFMAN, E. P., R. H. BROWN, JR. and L. M. KUNKEL. 1987. Dystrophin: the protein product of the Duchenne muscular dystrophy locus. *Cell* 51: 919–928.

HOFFMAN, E. P. et al. 1988. Characterization of dystrophin in muscle-biopsy specimens from patients with Duchenne's or Becker's muscular dystrophy. *NEJM* 318: 1363–1368.

HOLDEN, C. 1987. The genetics of personality. *Science* 237: 598–601.

HOLLIDAY, R. 1987. Ageing: X-chromosome reactivation. *Nature* 327: 661–662.

HOLMES, H. B. 1985. Sex preselection: eugenics for everyone? In *Biomedical Ethics Reviews: 1985*, edited by J. M. Humber and R. F. Almeder. The Humana Press, Clifton, New Jersey.

HOLMQUIST, G. P. 1987. Role of replication time in the control of tissue-specific gene expression. *AJHG* 40: 151–173.

HOLSTEIN, A. F. 1983. Spermatid differentiation in man during senescence. In *The Sperm Cell: Fertilizing Power, Surface Properties, Motility, Nucleus and Acrosome, Evolutionary Aspects*, edited by J. André, pp. 15–18. Martinus Nijhoff Publishers, The Hague.

HOLT, J. T., C. C. MORTON, A.W. NIENHUIS and P. LEDER. 1987. Molecular mechanisms of hematological neoplasms. In *The Molecular Basis of Blood Diseases*, edited by G. Stamatoyannopoulos et al., pp. 347–376. W. B. Saunders, Philadelphia.

HOLTZMAN, N. A. 1983. Pitfalls of newborn screening (with special attention to hypothyroidism): when will we ever learn? *Birth Defects: Original Article Series* 19(5): 111–120.

HOLTZMAN, N. A., R. A. KRONMAL, W. VAN DOORNINCK, C. AZEN and R. KOCH. 1986. Effect of age at loss of dietary control on intellectual performance and behavior of children with phenylketonuria. *NEJM* 314: 593–598.

HOLTZMAN, N. A., C. O. LEONARD and M. R. FARFEL. 1981. Issues in antenatal and neonatal screening and surveillance for hereditary and congenital disorders. *Annual Review of Public Health* 2: 219–251.

HONORE, L. H. 1978. Ageing changes in the human testis: A light-microscopic study. *Gerontology* 24: 58–65.

HOOK, E. B. 1973. Behavioral implications of the human XYY genotype. *Science* 179: 139–151.

HOOK, E. B. 1978. Spontaneous deaths of fetuses with chromosomal abnormalities diagnosed prenatally. *NEJM* 299: 1036–1038.

HOOK, E. B. 1981. Down syndrome: Its frequency in human populations and some factors pertinent to variation in rates. In *Trisomy 21 (Down Syndrome): Research Perspectives*, edited by F. F. de la Cruz and P. S. Gerald, pp. 3–67.

HOOK, E. B. 1985. The impact of aneuploidy upon public health: mortality and morbidity associated with human chromosome abnormalities. In *Aneuploidy: Etiology and Mechanisms*, edited by V. L. Dellarco et al. pp. 7–33. Plenum Press, New York.

HOOK, E. B. and G. M. CHAMBERS. 1977. Estimated rates of Down syndrome in live births by one year maternal age intervals for mothers aged 20–49 in a New York state study—implications of the risk figures for genetic counseling and cost–benefit analysis of prenatal diagnosis programs. *Birth Defects Original Article Series*, Vol. 13, No. 3A, pp. 123–141.

HOOK, E. B. and P. K. CROSS. 1987. Rates of mutant and inherited structural cytogenetic abnormalities detected at amniocenteses: results on about 63,000 fetuses. *Annals of Human Genetics* 51: 27–55.

HOOK, E. B., P. K. CROSS, L. JACKSON, E. PERGAMENT and B. O. BRAMBATI. 1988. Maternal age-specific rates of 47,+21 and other cytogenetic abnormalities diagnosed in the first trimester of pregnancy in chorionic villus biopsy specimens: comparison with rates expected from observations at amniocentesis. *AJHG* 42: 797–807.

HOOK, E. B., P. K. CROSS and R. R. REGAL. 1984. The frequency of 47,+21, 47,+18, and 47,+13 at the uppermost extremes of maternal ages: results on 56,094 fetuses studied prenatally and comparisons with data on livebirths. *Human Genetics* 68: 211–220.

HOOK, E. B., P. K. CROSS and D. M. SCHREINEMACHERS. 1983. Chromosomal abnormality rates at amniocentesis and in liveborn infants. *JAMA* 249: 2034–2038.

HOOK, E. B. and J. L. HAMERTON. 1977. The frequency of chromosome abnormalities detected in consecutive newborn studies—differences between studies—results by sex and by severity of phenotypic involvement. In *Population Cytogenetics: Studies in Humans*, edited by E. B. Hook and I. H. Porter, pp. 63–79. Academic Press, New York.

HOOK, E. B. and I. H. PORTER, Editors. 1977. *Population Cytogenetics: Studies in Humans.* Academic Press, New York.

HOOPER, M., K. HARDY, A. HANDYSIDE, S. HUNTER and M. MONK. 1987. HPRT-deficient (Lesch-Nyhan) mouse embryos derived from germline colonization by cultured cells. *Nature* 326: 292–295.

HOPPE, P. C. and K. ILLMENSEE. 1977. Microsurgically produced homozygous-diploid uniparental mice. *PNAS* 74: 5657–5661.

HOSTETLER, J. A. 1974. *Hutterite Society.* The Johns Hopkins University Press, Baltimore.

HOSTETLER, J. A. 1980. *Amish Society*, 3rd Edition. Johns Hopkins University Press, Baltimore.

HOSTETLER, J. A. 1985. History and relevance of the Hutterite population for genetic studies. *American Journal of Medical Genetics* 22: 453–462.

HOWARD-FLANDERS, P. 1981. Inducible repair of DNA. *SA* 245 (Nov): 72–80.

HOWELL, M. and P. FORD. 1980. *The True History of The Elephant Man.* Allison & Busby Ltd., London.

HOZUMI, N. and S. TONEGAWA. 1976. Evidence for somatic rearrangement of immunoglobulin genes coding for variable and constant regions. *PNAS* 73: 3628–3632.

HRUBEC, Z. and C. D. ROBINETTE. 1984. The study of human twins in medical research. *NEJM* 310: 435–441.

HSIA, Y. E. 1975. Treatment in genetic diseases. In *The Prevention of Genetic Disease and Mental Retardation*, edited by A. Milunsky, pp. 277–305. W. B. Saunders, Philadelphia.

HSU, L. Y. F. 1986. Prenatal diagnosis of chromosome abnormalities. In *Genetic Disorders and the Fetus: Diagnosis, Prevention, and Treatment*, 2nd Edition, edited by A. Milunsky, pp. 115–183. Plenum Press, New York.

HSU, T. C. 1952. Mammalian chromosomes in vitro. I. The karyotype of man. *Journal of Heredity* 43: 167–172.

HSU, T. C. 1979. *Human and Mammalian Cytogenetics: An Historical Perspective.* Springer-Verlag, New York.

HU, F., J. M. HANIFIN, G. H. PRESCOTT and A. C. TONGUE. 1980. Yellow mutant albinism: cytochemical, ultrastructural, and genetic characterization suggesting multiple allelism. *AJHG* 32: 387–395.

HUANG, H.-J. S. et al. 1988. Suppression of the neoplastic phenotype by replacement of the RB gene in human cancer cells. *Science* 242: 1563–1566

HULSE, J. H. 1982. Food science and nutrition: the gulf between rich and poor. *Science* 216: 1291–1294.

HUMAN GENE MAPPING 9. 1987. Ninth International Workshop on Human Gene Mapping. *Cytogenetics and Cell Genetics* 46: 1–762.

HUMPHRIES, S. E. et al. 1985. A common DNA polymorphism of the low-density lipoprotein (LDL) receptor gene and its use in diagnosis. *Lancet* I: 1003–1005.

HUNT, E. 1983. On the nature of intelligence. *Science* 219: 141–146.

HUNTER, T. 1984. The proteins of oncogenes. *SA* 251 (Aug): 70–79.

HUNTINGTON, G. E. 1981. Children of the Hutterites. *Natural History* 90 (Feb): 34–47.

HUNTLEY, O. M. 1984. A mother's perspective. *Hastings Center Report* 14(2): 14–15.

HURVICH, L. M. 1981. *Color Vision*. Sinauer Associates, Sunderland, Massachusetts.

HUXLEY, A. 1932. *Brave New World*. Harper & Row, New York.

HUXLEY, J. 1949. *Soviet Genetics and World Science*. Chatto and Windus, London.

IDLE, J. 1988. Pharmacogenetics: enigmatic variations. *Nature* 331: 391–392.

ILTIS, H. 1924. *Gregor Johann Mendel: Leben, Werk und Wirkung*. Julius Springer, Berlin. [English translation by Eden Paul and Cedar Paul (1932). *Life of Mendel*. George Allen & Unwin, London.]

IMAIZUMI, Y. 1986. A recent survey of consanguineous marriages in Japan. *Clinical Genetics* 30: 230–233.

IMAIZUMI, Y. 1987. Reasons for consanguineous marriages in Japan. *Journal of Biosocial Science* 19: 97–106.

IMPERATO-McGINLEY, J. and T. GAUTIER. 1986. Inherited 5-reductase deficiency in man. *TIG* 2: 130–133.

INGHAM, P. W. 1988. The molecular genetics of embryonic pattern formation in *Drosophila*. *Nature* 335: 25–34.

INGRAM, V. M. 1957. Gene mutations in human haemoglobin: the chemical difference between normal and sickle cell haemoglobin. *Nature* 180: 326–328. [Reprinted in Boyer (1963).]

INOUE, S. 1981. Cell division and the mitotic spindle. *Journal of Cell Biology* 91: 132s–147s.

ISCN. 1985. *An International System for Human Cytogenetic Nomenclature* (1985). S. Karger, Basel. [Also in *Birth Defects: Original Article Series, Volume 21, Number 1*, March of Dimes Birth Defects Foundation, New York.]

ISELIUS, L. and D. A. P. EVANS. 1983. Formal genetics of isoniazid metabolism in man. *Clinical Pharmacokinetics* 8: 541–544.

ISHIHARA, S. 1968. *Tests for Colour Blindness*. Kanehara Shuppan Co., Tokyo.

ITAKURA, K. et al. 1977. Expression in *Escherchia coli* of a chemically synthesized gene for the hormone somatostatin. *Science* 198: 1056–1063.

JACKSON, D. A., R. H. SYMONS and P. BERG. 1972. Biochemical method for inserting new genetic information into DNA of Simian Virus 40: circular SV40 DNA molecules containing lambda phage genes and the galactose operon of *Escherchia coli*. *PNAS* 69: 2904–2909.

JACKSON, R. C. and D. P. HAUBER, Editors. 1983. *Polyploidy*. Benchmark Papers in Genetics. Van Nostrand Reinhold, New York.

JACOBS, G. H. 1981. *Comparative Color Vision*. Academic Press, New York.

JACOBS, P. A. 1982. The William Allen Memorial Award Address: Human Population Cytogenetics: The First Twenty-Five Years. *AJHG* 34: 689–698.

JACOBS, P. A. et al. 1959. Evidence for the existence of the human "super female." *Lancet* II: 423–425.

JACOBS, P. A., M. BRUNTON and M. M. MELVILLE. 1965. Aggressive behavior, mental sub-normality and the XYY male. *Nature* 208: 1351–1352.

JACOBS, P. A., K. E. BUCKTON, C. CUNNINGHAM and M. NEWTON. 1974. An analysis of the break points of structural rearrangements in man. *Journal of Medical Genetics* 11: 50–64.

JACOBS, P. A., J. S. MATSUURA, M. MAYER and I. M. NEWLANDS. 1978. A cytogenetic survey of an institution for the mentally retarded. I. Chromosome abnormalities. *Clinical Genetics* 13: 37–60.

JACOBS, P. A. and J. A. STRONG. 1959. A case of human intersexuality having a possible XXY sex-determining mechanism. *Nature* 183: 302–303.

JACOBS, P. A., C. M. WILSON, J. A. SPRENKLE, N. B. ROSENSHEIN and B. R. MIGEON. 1980. Mechanism of origin of complete hydatidiform moles. *Nature* 286: 714–716.

JACQUARD, A. 1974. *The Genetic Structure of Populations*. Translated by B. and D. Charlesworth. Springer-Verlag, New York.

JAMA: JOURNAL OF THE AMERICAN MEDICAL ASSOCIATION. 1987. *Primer on Allergic and Immunologic Diseases*, 2nd Edition. Volume 258, Nov. 27, 1987 (entire issue).

JAMES, R. and R. A. BRADSHAW. 1984. Polypeptide growth factors. *Annual Review of Biochemistry* 53: 259–292.

JAMES, W. H. 1986a. Hormonal control of sex ratio. *Journal of Theoretical Biology* 118: 427–441.

JAMES, W. H. 1986b. Recent secular trends in dizygotic twinning rates in Europe. *Journal of Biosocial Science* 18: 497–504.

JAMESON, D. L., Editor. 1977. *Evolutionary Genetics. Benchmark Papers in Genetics, Vol. 8*. Dowden, Hutchinson and Ross, Stroudsberg, Pennsylvania.

JANSEN, R. P. S., J. C. ANDERSON and P. D. SUTHERLAND. 1988. Nonoperative embryo transfer to the fallopian tube. *NEJM* 319: 288–291.

JAYNES, J. B. and P. H. O'FARRELL. 1988. Activation and repression of transcription by homoeodomain–containing proteins that bind a common site. *Nature* 336: 744–749.

JEFFREYS, A. J., I. W. CRAIG and U. FRANCKE. 1979. Localization of the $^{G}\gamma$-, $^{A}\gamma$-, δ-, and β-globin genes on the short arm of human chromosome 11. *Nature* 281: 606–608.

JEFFREYS, A. J., V. WILSON and S. L. THEIN. 1985a. Hypervariable "minisatellite" regions in human DNA. *Nature* 314: 67–73.

JEFFREYS, A. J., V. WILSON and S. L. THEIN. 1985b. Individual-specific "fingerprints" of human DNA. *Nature* 316: 76–79.

JENCKS, C. et al. 1972. *Inequality: A Reassessment of the Effect of Family and Schooling in America*. Basic Books, New York.

JENSEN, A. R. 1969. How much can we boost IQ and scholastic achievement? *Harvard Educational Review* 39(1): 1–123. [Reprinted in *Environment, Heredity, and Intelligence* (1969). Reprint Series Number 2 compiled from the Harvard Educational Review.]

JENSEN, A. R. 1980. *Bias in Mental Testing*. The Free Press, New York.

JENSEN, A. R. 1981. *Straight Talk About Mental Tests*. The Free Press, New York.

JERNE, N. K. 1985. The generative grammar of the immune system. *Science* 229: 1057–1059.

JERVIS, G. A. 1947. Phenylpyruvic oligophrenia deficiency of phenylalanine-oxidizing system. *Society for Experimental Biology and Medicine, Proceedings* 82: 514–515.

JOHANNSEN, W. 1903. *Ueber Erblichkeit in Populationen und in reinen Linien*. Gustav Fischer, Jena, (East) Germany. [Abridged translation in Peters (1959).]

JOHN, P. C. L., Editor. 1981. *The Cell Cycle*. Cambridge University Press, Cambridge.

JOHNSEN, D. 1987. A new threat to pregnant women's autonomy. *Hastings Center Report* 17(6): 33–40..

JOHNSON, I. S. 1983. Human insulin from recombinant DNA technology. *Science* 219: 632–637. Reprinted in Abelson (1984).

JOLLY, D. J. et al. 1983. Isolation and characterization of a full-length expressible cDNA for human hypoxanthine phosphoribosyltransferase. *PNAS* 80: 477–481.

JONES, R. E. 1984. *Human Reproduction and Sexual Behavior*. Wiley Interscience, New York.

JOSSO, N. 1981. Differentiation of the genital tract: stimulators and inhibitors. In *Mechanisms of Sex Differentiation in Animals and Man*, edited by C. R. Austin and R. G. Edwards, pp. 165–203. Academic Press, New York.

JOST, A. and S. MAGRE. 1984. Testicular development phases and dual hormonal control of sexual organogenesis. In *Sexual Differentiation: Basic and Clinical Aspects*, edited by M. Serio et al., pp. 1–15. Raven Press, New York.

JOYSEY, V. C. 1984. Tissue typing and clinical kidney grafting. In *Transplantation Immunology*, edited by R. Y. Calne, pp. 101–130. Oxford University Press, Oxford.

JUDSON, H. F. 1979. *The Eighth Day of Creation: Makers of the Revolution in Biology*. Simon & Schuster, New York

KABACK, M. M., Editor. 1977a. *Tay-Sachs Disease: Screening and Prevention. Progress in Clinical and Biological Research*, Vol. 18. Alan R. Liss, New York.

KABACK, M. M., Editor. 1977b. Detection of Tay-Sachs disease carriers: lessons and ramifications. In *Genetic Counseling*, edited by H. A. Lubs and F. de la Cruz, pp. 203–223. Raven Press, New York.

KABACK, M. M. 1983. Heterozygote screening. In *Principles and Practice of Medical Genetics*, Vol. 2, edited by A. E. H. Emery and D. L. Rimoin, pp. 1451–1457. Churchill Livingstone, Edinburgh

KAISER, P. 1984. Pericentric inversions: problems and significance for clinical genetics. *Human Genetics* 68: 1–47.

KAJII, T. et al. 1980. Anatomic and chromosomal anomalies in 639 spontaneous abortuses. *Human Genetics* 55: 87–98.

KAJII, T., K. OHAMA and K. MIKAMO. 1978. Anatomic and chromosomal anomalies in 944 induced abortuses. *Human Genetics* 43: 247–258.

KALMUS, H. 1965. *Diagnosis and Genetics of Defective Colour Vision*. Pergamon Press, Oxford.

KALOW, W., H. W. GOEDDE and D. P. AGARWAL, Editors. 1986. *Ethnic Differences in Reactions to Drugs and Xenobiotics*. Alan R. Liss, New York.

KAMIGUCHI, Y. and K. MIKAMO. 1986. An improved, efficient method for analyzing human sperm chromosomes using zona-free hamster ova. *AJHG* 38: 724–740.

KANG, J. et al. 1987. The precursor of Alzheimer's disease amyloid A4 protein resembles a cell-surface receptor. *Nature* 325: 733–736.

KAO, F.-T. 1983. Somatic Cell Genetics and Gene Mapping. In *International Review of Cytology*, Vol. 85, edited by G. H. Bourne et al., pp. 109–146. Academic Press, New York.

KAPPAS, A., S. SASSA and K. E. ANDERSON. 1983. The porphyrias. In *The Metabolic Basis of Inherited Disease*, edited by J. B. Stanbury et al., pp. 1301–1384. McGraw-Hill, New York.

KASPER, C. K. 1982. Hematology. *JAMA* 247: 2951–2953.

KAUFMAN, S. 1977. Phenylketonuria: biochemical mechanisms. *Advances in Neurochemistry* 2: 1–132.

KAUFMAN, S. 1983. Phenylketonuria and its variants. *Advances in Human Genetics* 13: 217–297.

KEATS, B. 1983. Genetic mapping: X chromosome. *Human Genetics* 64: 28–32.

KEDES, L. H. 1985. The Duchenne dystrophy gene: a Great Leap Forward on the Long March. *TIG* 1: 205–209.

KELLEY, W. N. and J. B. WYNGAARDEN. 1983. Clinical syndromes associated with hypoxanthine-guanine phosphoribosyltransferase deficiency. In *Metabolic Basis of Inherited Disease*, edited by J. B. Stanbury et al., pp. 1115–1143. McGraw-Hill, New York.

KENNEDY, J. L. et al. 1988. Evidence against linkage of schizophrenia to markers on chromosome 5 in a northern Swedish pedigree. *Nature* 336: 167–170.

KERSEY, J. H. et al. 1987. Comparison of autologous and allogeneic bone marrow transplantation for treatment of high-risk refractory acute lymphoblastic leukemia. *NEJM* 317: 461–467.

KEVLES, D. J. 1985. *In the Name of Eugenics: Genetics and the Uses of Human Heredity*. Alfred A. Knopf, New York.

KHAW, K.-T. and E. BARRETT-CONNOR. 1986. Family history of heart attack: a modifiable risk factor? *Circulation* 74: 239–244.

KHOURY, M. J., B. H. COHEN, E. L. DIAMOND, G. A. CHASE and V. A. McKUSICK. 1987a. Inbreeding and prereproductive mortality in the Old Order Amish. I. Genealogic epidemiology of inbreeding. *American Journal of Epidemiology* 125: 453–461.

KHOURY, M. J., B. H. COHEN, G. A. CHASE and E. L. DIAMOND. 1987b. An epidemiologic approach to the evaluation of the effect of inbreeding on prereproductive mortality. *American Journal of Epidemiology* 125: 251–262.

KIDD, K. K. 1987. Phenylketonuria: population genetics of a disease. *Nature* 327: 282–283.

KIMURA, M. 1979. The neutral theory of molecular evolution. *SA* 241(Nov): 98–126.

KIMURA, M. 1983. *The Neutral Theory of Molecular Evolution*. Cambridge University Press, Cambridge.

KING, M.C. and A. C. WILSON. 1975. Evolution at two levels in humans and chimpanzees. *Science* 188: 107–116. [Reprinted in Terzaghi et al. 1984.]

KING, S. 1983. Multiple mechanisms of mitosis? *Nature* 301: 660.

KINGSLEY, C. 1863. *The Water-Babies; A Fairy Tale for a Land-Baby*. Macmillan and Co., London.

KISSIN, B. 1982. Alcohol abuse and alcohol-related illnesses. In *Cecil Textbook of Medicine*, 16th Edition, Vol. 2, edited by J. B. Wyngaarden and L. H. Smith, Jr., pp. 2016–2022. W. B. Saunders, Philadelphia.

KLAT, M. and A. KHUDR. 1986. Religious endogamy and consanguinity in marriage patterns in Beirut, Lebanon. *Social Biology* 33: 139–145.

KLINE, A. H., J. B. SIDBURY, JR. and C. P. RICHTER. 1959. The occurrence of ectodermal dysplasia and corneal dysplasia in one family. *Journal of Pediatrics* 55: 355–366.

KLINE, J., Z. STEIN and M. SUSSER. 1986. Very early pregnancy: fertilization and implantation frequency and cause of loss. In *Perinatal Genetics: Diagnosis and Treatment*, edited by I. Porter et al., pp. 3–22. Academic Press, Orlando, Florida.

KLINEFELTER, H. F., JR., E. C. REIFENSTEIN, JR. and F. ALBRIGHT. 1942. Syndrome characterized by gynecomastia, aspermatogenesis without a-Leydigism, and increased excretion of follicle-stimulating hormone. *Journal of Clinical Endocrinology* 2: 615–627.

KNOWLES, M. R. et al. 1983. Abnormal ion permeation through cystic fibrosis respiratory epithelium. *Science* 221: 1067–1070.

KNOWLTON, R. G. et al. 1985. A polymorphic DNA marker linked to cystic fibrosis is located on chromosome 7. *Nature* 318: 380–382.

KNUDSON, A. G., JR. 1971. Mutation and cancer: statistical study of retinoblastoma. *PNAS* 68: 820–823.

KNUDSON, A. G., JR. 1986. Genetics of human cancer. *Annual Review of Genetics* 20: 231–251.

KOCH, R., P. ACOSTA, N. RAGSDALE and G. N. DONNEL. 1963. Nutrition in the treatment of galactosemia. *Journal of the American Dietetic Association* 43: 216–222.

KÖHLER, G. 1986. Derivation and diversification of monoclonal antibodies. *Science* 233: 1281–1286.

KÖHLER, G. and C. MILSTEIN. 1975. Continuous cultures of fused cells secreting antibody of predefined specificity. *Nature* 256: 495–497.

KOLATA, G. 1985a. Closing in on the muscular dystrophy gene. *Science* 230: 307–308.

KOLATA, G. 1985b. Down syndrome—Alzheimer's linked. *Science* 230: 1152–1153.

KOLATA, G. 1986. Maleness pinpointed on Y chromosome. *Science* 234: 1076–1077.

KOLATA, G. 1987. Remembering a "magical genius." *Science* 236: 1519–1521.

KORENBERG, J. R., M. L. CROYLE and D. R. COX. 1987. Isolation and regional mapping of DNA sequences unique to human chromosome 21. *AJHG* 41: 963–978.

KORENBERG, J. R. and M. C. RYKOWSKI. 1988. Human genome organization: Alu, Lines, and the molecular structure of metaphase chromosome bands. *Cell* 53: 391–400.

KORNBERG, A. 1968. The synthesis of DNA. *SA* 219(Oct): 64–78.

KORNBERG, R. and A. KLUG. 1981. The nucleosome. *SA* 244(Feb): 52–64.

KOSHLAND, D. E., JR. 1987. Nature, nurture, and behavior. *Science* 235: 1445.

KOSHLAND, D. E., T. J. MITCHISON and MARC W. KIRSCHNER. 1988. Polewards chromosome movement driven by microtubule depolymerization in vitro. *Nature* 331: 499–504.

KREMER, J., J. R. H. DIJKHUIS and S. JAGER. 1987. A simplified method for freezing and storage of human semen. *Fertility and Sterility* 47: 838–842.

KRETCHMER, N. 1972. Lactose and lactase *SA* 227(Oct): 70–78.

KRIMSKY, S. 1982. *Genetic Alchemy: The Social History of the Recombinant DNA Controversy*. MIT Press, Cambridge, Massachusetts.

KRUSH, A. J. and K. A. EVANS. 1984. *Family Studies in Genetic Disorders*. Charles C Thomas, Springfield, Illinois.

KUEHN, M. R., A. BRADLEY, E. J. ROBERTSON and M. J. EVANS. 1987. A potential animal model for Lesch-Nyhan syndrome through introduction of HPRT mutations into mice. *Nature* 326: 295–298.

KUHL, P., K. OLEK, P. WARDENBACH and K.-H. GRZESCHIK. 1979. Assignment of a gene for human quinoid-dihydropteridine reductase (QDPR,ec 1.6.5.1) to chromosome 4. *Human Genetics* 53: 47–49.

KULOZIK, A. E. et al. 1986. Geographical survey of β^S-globin gene haplotypes: evidence for an independent Asian origin of the sickle-cell mutation. *AJHG* 39: 239–244.

KUNKEL, L. M. et al. 1986. Analysis of deletions in DNA from patients with Becker and Duchenne muscular dystrophy. *Nature* 322: 73–77.

KWOK, S. C. M., F. D. LEDLEY, A. G. DILELLA, K. J. H. ROBSON and S. L. C. WOO. 1985. Nucleotide sequence of a full-length complementary DNA clone and amino acid sequence of human phenylalanine hydroxylase. *Biochemistry* 24: 556–561.

KWON, B. S., A. K. HAW, S. H. POMERANTZ and R. HALABAN. 1987. Isolation and sequence of a cDNA clone for human tyrosinase that maps at the mouse c-albino locus. *PNAS* 84: 7473–7477.

LA DU, B. N. 1978. Alkaptonuria. In *The Metabolic Basis of Inherited Disease*, 4th Edition, edited by J. B. Stanbury et al., pp. 268–282. McGraw-Hill, New York.

LABIE, D., J. PAGNIER, H. WAJCMAN, M. E. FABRY and R. L. NAGEL. 1986. The genetic origin of the variability of the phenotypic expression of the *HbS* gene. In *Genetic Variation and its Maintenance*, edited by D. F. Roberts and G. F. De Stefano, pp. 149–155. Cambridge University Press, Cambridge.

LAEMMLI, U. K. et al. 1978. Metaphase chromosome structure: the role of nonhistone proteins. *CSHS* 42 (Part 1): 351–360.

LAEMMLI, U. K., C. D. LEWIS and W. C. EARNSHAW. 1983. Long-range order folding of the chromatin fibers in metaphase chromosomes and nuclei. In *Chromosomes and Cancer*, edited by J. D. Rowley and J. E. Ultmann, pp. 15–27. Academic Press, New York.

LAFOND, R. E., Editor. 1988. *Cancer, The Outlaw Cell*, 2nd Edition. American Chemical Society, Washington, D. C.

LAIRD, C. D. 1987. Proposed mechanism of inheritance and expression of the human fragile-X syndrome of mental retardation. *Genetics* 117: 587–599.

LAIRD, C. D., E. JAFFE, G. KARPEN, M. LAMB and R. NELSON. 1987. Fragile sites in human chromosomes as regions of late-replicating DNA. *TIG* 3: 274–281.

LAKE, J. A. 1981. The ribosome. *SA* 245(Aug): 84–97.

LAMM, L. U. et al. 1974. Assignment of the major histocompatibility complex to chromosome no. 6 in a family with a pericentric inversion. *Human Heredity* 24: 273–284.

LAND, J. M., J. A. MORGAN-HUGHES and J. B. CLARK. 1981. Mitochondrial myopathy: biochemical studies revealing a deficiency of NADH-cytochrome *b* reductase activity. *Journal of Neurological Sciences* 50: 1–13.

LANDSTEINER, K. 1901. Über Agglutinationserscheinungen normalen menschlichen Blutes. *Wiener klinische Wochenschrift* 14: 1132–1134. [An English translation is in Boyer (1963).]

LANE, C. 1976. Rabbit hemoglobin from frog eggs. *SA* 235(Aug): 60–71.

LANE, D. and B. STRATFORD, Editors. 1985. *Current Approaches to Down's Syndrome*. Praeger Press, New York.

LANG, K. 1955. Die phenylpyruvische Oligophrenie. *Ergebnisse der Inneren Medizin und Kinderheilkunde* 6: 78–99.

LANGONE, J. 1983. Robert Weinberg: Scientist of the Year. *Discover* 4(Jan): 37–44.

LASKER, G. W. 1985. *Surnames and Genetic Structure*. Cambridge University Press, Cambridge.

LATT, S. A. 1974. Microfluorometric analysis of DNA replication in human X chromosomes. *Experimental Cell Research* 86: 412–415.

LAUGHLIN, H. 1922. Model eugenical sterilization law. In *Eugenical Sterilization in the United States: A Report of the Psychopathic Laboratory of the Municipal Court of Chicago*, pp. 446–461. [Reprinted in Bajema (1976).]

LAURENCE, J. 1985. The immune system in AIDS. *SA* 253(Dec): 84–93.

LAURENCE, K. M. 1983. The genetics and prevention of neural tube defects. In *Principles and Practice of Medical Genetics*, Vol. 1, edited by A. E. H. Emery and D. L. Rimoin, pp. 231–245. Churchill Livingstone, Edinburgh.

LAWN, R. M., E. F. FRITSCH, R. C. PARKER, G. BLAKE and T. MANIATIS. 1978. The isolation and characterization of linked δ- and β-globin genes from a cloned library of human DNA. *Cell* 15: 1157–1174.

LAWN, R. M. and G. A. VEHAR. 1986. The molecular genetics of hemophilia. *SA* 254 (Mar): 48–54.

LAWN, R. M. et al. 1986. Cloned Factor VIII and the molecular genetics of hemophilia. *CSHS* 51: 365–369.

LAWRENCE, R. J. 1985. David the "Bubble Boy" and the boundaries of the human. *JAMA* 253: 74–76.

LAX, E. 1984. *Life and Death on 10 West*. Times Books, New York.

LE BEAU, M. M. and J. D. ROWLEY. 1984. Heritable fragile sites in cancer. *Nature* 308: 607–608.

LE BEAU, M. M. and J. D. ROWLEY. 1986. Chromosomal abnormalities in leukemia and lymphoma: clinical and biological significance. *Advances in Human Genetics* 15: 1–54.

LEBO, R. V. et al. 1979. Assignment of human β-, γ- and δ-globin genes to the short arm of chromosome 11 by chromosome sorting and DNA restriction enzyme analysis. *PNAS* 76: 5804–5808.

LEDBETTER, D. H. et al. 1981. Deletions of chromosome 15 as a cause of the Prader-Willi syndrome. *NEJM* 304: 325–329.

LEDER, P. 1982. The genetics of antibody diversity. *SA* 246(May): 102–115.

LEDERBERG, J. 1970. Genetic engineering and the amelioration of genetic defect. *BioScience* 20: 1307–1310. [Reprinted in Mertens (1975).]

LEDERBERG, J. 1987. Genetic recombination in bacteria: a discovery account. *Annual Review of Genetics* 21: 23–46.

LEDLEY, F. D. 1987. Somatic gene therapy for human disease: background and prospects, parts I and II. *The Journal of Pediatrics* 110: 1–8 and 167–174.

LEDLEY, F. D., A. G. DILELLA and S. L. C. WOO. 1985. Molecular biology of phenylalanine hydroxylase and phenylketonuria. *TIG* 1: 309–313.

LEDLEY, F. D., H. L. LEVY and S. L. C. WOO. 1986. Molecular analysis of the inheritance of phenylketonuria and mild hyperphenylalaninemia in families with both disorders. *NEJM* 314: 1276–1280.

LEE, E. Y.-H. P. et al. 1988. Inactivation of the retinoblastoma susceptibility gene in human breast cancers. *Science* 241: 218–221.

LEE, M. G. and P. NURSE. 1987. Complementation used to clone a human homologue of the fission yeast cell cycle control gene *cdc2*. *Nature* 327: 31–35.

LEE, M. and P. NURSE. 1988. Cell cycle control genes in fission yeast and mannalian cells. *TIG* 4: 287–290.

LEE, W.-H. et al. 1987. Human retinoblastoma susceptibility gene: cloning, identification, and sequence. *Science* 235: 1394–1399.

LEHMANN, A. R. 1983. Xeroderma pigmentosum, Cockayne syndrome and ataxia–telangiectasia: disorders relating DNA repair to carcinogenesis. In *Inheritance of Susceptibility to Cancer in Man*, edited by W. F. Bodmer, pp. 93–118. Oxford University Press, Oxford.

LEHMANN, H. and J. LIDDELL. 1964. Genetical variants of human serum pseudocholinesterase. *Progress in Medical Genetics* 3: 75–105.

LEHNINGER, A. L. 1982. *Biochemistry*. Worth Publishers, New York.

LEHRMAN, M. A. et al. 1987. The Lebanese allele at the low density lipoprotein receptor locus: nonsense mutation produces truncated receptor that is retained in endoplasmic reticulum. *Journal of Biological Chemistry* 262: 401–410.

LEJEUNE, J., M. GAUTIER and R. TURPIN. 1959. Etude des chromosomes somatiques de neuf enfants mongoliens. *Academie des Sciences, Paris: Comptes Rendus Hebdomadaires des Seances* 248: 1721–1722. [An English translation is in Boyer (1963).]

LENKE, R. R. and H. L. LEVY. 1980. Maternal phenylketonuria and hyperphenylalaninemia: An international survey of the outcome of untreated and treated pregnancies. *NEJM* 303: 1202–1208.

LERIDON, H. 1976. Facts and artifacts in the study of intrauterine mortality: a reconsideration from pregnancy histories. *Population Studies* 30: 319–335.

LESCH, M. and W. L. NYHAN. 1964. A familial disorder of uric acid metabolism and central nervous system function. *American Journal of Medicine* 36: 561–570.

LEVENTHAL, A. G., D. J. VITEK and D. J. CREEL. 1985. Abnormal visual pathways in normally pigmented cats that are heterozygous for albinism. *Science* 229: 1395–1397.

LEVIN, R. J. 1987. Human sex pre-selection. In *Oxford Reviews of Reproductive Biology*, Vol. 9, edited by J. R. Clarke, pp. 161–191. Clarendon Press, Oxford.

LEVINE, L. S. 1986. Prenatal detection of congenital adrenal hyperplasia. In *Genetic Disorders and the Fetus: Diagnosis, Prevention and Treatment*, edited by A. Milunsky, pp. 369–385. Plenum Press, New York.

LEVINE, L., Editor. 1971. *Papers on Genetics, A Book of Readings*. C. V. Mosby Co., St. Louis.

LEVINE, M. and T. HOEY. 1988. Homeobox proteins as sequence-specific transcription factors. *Cell* 55: 537–540.

LEVINE, P. 1958. The influence of the ABO system on Rh hemolytic disease. *Human Biology* 30: 14–28.

LEVINE, P., E. ROBINSON, M. CELANO, O. BRIGGS and L. FALKINBURG. 1955. Gene interaction resulting in suppression of blood substance B. *Blood* 10: 1100–1108.

LEVINE, P. and R. E. STETSON. 1939. An unusual case of intra-group agglutination. *JAMA* 113: 126–127.

LEVINSON, A. D. 1986. Normal and activated *ras* oncogenes and their encoded products. *TIG* 2: 81–85.

LEVITAN, M. 1988. *Textbook of Human Genetics*, 3rd Edition. Oxford University Press, New York.

LEVY, H. L. 1973. Genetic screening. *Advances in Human Genetics* 4: 1–104.

LEVY, H. L. 1985. Maternal PKU. In *Medical Genetics: Past, Present, Future*, edited by K. Berg, pp. 109–122. Alan R. Liss, New York.

LEVY, H. L. 1987. Maternal phenoketonuria: review with emphasis on pathogenesis. *Enzyme* 38: 312–320.

LEWIN, B. 1987. *Genes*, 3rd Edition. John Wiley & Sons, New York.

LEWIN, R. 1984. DNA reveals surprises in human family tree. *Science* 226: 1179–1182.

LEWIN, R. 1986. Proposal to sequence the human genome stirs debate. *Science* 232: 1598–1600.

LEWIS, E. B. 1985. Regulation of the genes of the *bithorax* complex in *Drosophila*. *CSHS* 50: 155–164.

LEWIS, R. 1988. DNA fingerprints: witness for the prosecution. *Discover* 9(June): 44–52.

LEWONTIN, R. C. 1974. *The Genetic Basis of Evolutionary Change*. Columbia University Press, New York.

LEWONTIN, R. C. 1985. Population Genetics. *Annual Review of Genetics* 19: 81–102.

LEWONTIN, R. C., S. ROSE and L. J. KAMIN. 1984. *Not in Our Genes*. Pantheon Books, New York.

LIDSKY, A. S. et al. 1984. The PKU locus in man is on chromosome 12. *AJHG* 36: 527–533.

LIDSKY, A. S. et al. 1985. Regional mapping of the phenylalanine hydroxylase gene and the PKU locus in the human genome. *PNAS* 82: 6221–6225.

LINDER, D. and S. M. GARTLER. 1965. Glucose-6-phosphate dehydrogenase mosaicism: utilization as a cell marker in the study of leiomyomas. *Science* 150: 67–69.

LINDGREN, B. W. and G. W. McELRATH. 1978. *Introduction to Probability and Statistics*, 4th Edition. Macmillan Publishing Co., New York.

LINDGREN, V., K. L. LUSKEY, D. W. RUSSELL and U. FRANCKE. 1985. Human genes involved in cholesterol metabolism: Chromosomal mapping of the loci for the low density lipoprotein receptor and 3-hydroxy-3-methylglutaryl-coenzyme A reductase with cDNA probes. *PNAS* 82: 8567–8571.

LINDGREN, V., B. DE MARTINVILLE, A. L. HORWICH, L. E. ROSENBERG and U. FRANCKE. 1984. Human ornithine transcarbamylase locus mapped to band Xp21.1 near the Duchenne muscular dystrophy locus. *Science* 226: 698–700.

LINDSLEY, D. L. et al. 1972. Segmental aneuploidy and the genetic gross structure of the *Drosophila* genome. *Genetics* 71: 157–184.

LINN, S. et al. 1982. No association between coffee consumption and adverse outcomes of pregnancy. *NEJM* 306: 141–145.

LISON, M., S. H. BLONDHEIM and R. N. MELMED. 1980. A polymorphism of the ability to smell urinary metabolites of asparagus. *British Medical Journal* 281: 1676–1678.

LITTLE, P. F. R. 1982. Globin pseudogenes. *Cell* 28: 683–684.

LITTLEFIELD, J. W. 1964. Selection of hybrids from matings of fibroblasts in vitro and their presumed recombinants. *Science* 145: 709 only.

LITTLEFIELD, J. W. 1981. Research on cystic fibrosis. *NEJM* 304: 44–45.

LLOYD, D., R. K. POOLE and S. W. EDWARDS. 1982. *The Cell Division Cycle: Temporal Organization and Control of Cellular Growth and Reproduction*. Academic Press, London.

LLOYD, J. K. and C. R. SCRIVER, Editors. 1985. *Genetic and Metabolic Disease. Butterworth's International Medical Reviews in Pediatrics*. Butterworth, London.

LLOYD-STILL, J. D. 1983. *Textbook of Cystic Fibrosis*. John Wright, PSG Inc., Boston.

LOCKYER, J. et al. 1987. Structure and expression of human dihydropteridine reductase. *PNAS* 84: 3329–3333.

LOEHLIN, J. C. and J. C. DEFRIES. 1987. Genotype–environment correlation and IQ. *Behavior Genetics* 17: 263–277.

LOEHLIN, J. C., G. LINDZEY and J. N. SPUHLER. 1975. *Race Differences in Intelligence*. W. H. Freeman and Co., San Francisco.

LOMBARDO, P. A. 1985. Three generations, no imbeciles: new light on *Buck v. Bell*. *New York University Law Review* 60: 30–62.

LOWITZER, A. C. 1987. Maternal phenylketonuria: cause for concern among women with PKU. *Research in Developmental Disabilities* 8: 1–14.

LUBAHN, D. B. et al. 1988. Cloning of human androgen receptor complementary DNA and localization to the X chromosome. *Science* 240: 327–330.

LUBS, H. A. 1969. A marker X chromosome. *AJHG* 21: 231–244.

LUBS, H. A. 1983. X-linked mental retardation and the marker X. In *Principles and Practice of Medical Genetics*, edited by A. E. H. Emery and D. L. Rimoin, pp. 216–223. Churchill Livingstone, Edinburgh.

LUCARELLI, G. et al. 1987. Marrow transplantation in patients with advanced thalassemia. *NEJM* 316: 1050–1055.

LUCAS, M. 1969. Translocation between both members of chromosome pair number 15, causing recurrent abortions. *Annals of Human Genetics* 32: 347–353.

LUDMAN, M. D., G. A. GRABOWSKI, J. D. GOLDBERG and R. J. DESNICK. 1986. Heterozygote detection and prenatal diagnosis for Tay-Sachs and Type I Gaucher Diseases. In *Genetic Disease, Screening and Management* edited by T. P. Carter and A. M. Willey, pp. 19–48. Alan R. Liss, New York.

LUZZATTO, L. and G. BATTISTUZZI. 1985. Glucose-6-phosphate dehydrogenase. *Advances in Human Genetics* 14: 217–329.

LYNCH, H. T., W. J. KIMBERLING, K. M. BRENNAN and N. W. PAUL, Editors. 1986. *International Directory of Genetic Services*, 8th Edition. March of Dimes Birth Defects Foundation, White Plains, New York.

LYON, M. 1961. Gene action in the X-chromosome of the mouse (*Mus musculus* L.). *Nature* 190: 372–373.

LYON, M. 1968. Chromosomal and sub-chromosomal inactivation. *Annual Review of Genetics* 2: 31–52.

LYON, M. F. 1988. The William Allan Memorial Award address: X-chromosome inactivation and the location and expression of X-linked genes. *AJHG* 42: 8–16.

MABUCHI, H. et al. 1978. Homozygous familial hypercholesterolemia in Japan. *American Journal of Medicine* 65: 290–297.

MACALPINE, I. and R. HUNTER. 1969. Porphyria and King George III. *SA* 221(July): 38–46.

MACCOBY, E. E. and C. N. JACKLIN. 1974. *The Psychology of Sex Differences*. Stanford University Press, Stanford, California.

MACCOBY, E. E. and C. N. JACKLIN. 1980. Sex differences in aggression: a rejoinder and reprise. *Child Development* 51: 964–980.

MACCREADY, R. A. and M. G. HUSSEY. 1963. Newborn phenylketonuria detection program in Massachusetts. *The American Journal of Public Health* 54: 2075–2081.

MACGREGOR, H. C. and J. M. VARLEY. 1988. *Working with Animal Chromosomes*, 2nd Edition. John Wiley & Sons, New York.

MACHIN, G. A. 1974. Chromosome abnormality and perinatal death. *Lancet* I: 549–551.

MACHIN, J. 1732. An uncommon case of a distempered skin. *Royal Society of London Philosophical Transactions* 37: 299–300.

MADDOX, J. 1984. Who will clone a chromosome? *Nature* 312: 306.

MAGENIS, R. E., F. HECHT and E. W. LOVRIEN. 1970. Heritable fragile site on chromosome 16: probable localization of haptoglobin locus in man. *Science* 170: 85–87.

MAGENIS, R. E., K. M. OVERTON, J. CHAMBERLIN, T. BRADY and E. LOVRIEN. 1977. Parental origin of the extra chromosome in Down's syndrome. *Human Genetics* 37: 7–16.

MÄKELÄ, O. 1968. The specificity of antibodies produced by single cells. *CSHS* 32: 423–430.

MAKINO, S. 1975. *Human Chromosomes*. Igaku Shoin, Tokyo.

MANDEL, J. L. et al. 1986. Genetic mapping of the human X chromosome: linkage analysis of the q26–q28 region that includes the fragile X locus and isolation of expressed sequences. *CSHS* 51: 195–203.

MANGE, A. P. 1964. Growth and inbreeding of a human isolate. *Human Biology* 36: 104–133.

MANIATIS, T., E. F. FRITSCH and J. SAMBROOK. 1982. *Molecular Cloning: A Laboratory Manual*. Cold Spring Harbor Laboratory, New York.

MAPLETOFT, R. J. 1984. Embryo transfer technology for the enhancement of animal reproduction. *Bio/technology* 2: 149–160.

MARANTO, G. 1986. Genetic engineering: hype, hubris, and haste. *Discover* 7(June): 50–64.

MARCUS, M. 1985. Mammalian cell cycle mutants. In *Molecular Cell Genetics*, edited by M. M. Gottesman, pp. 591–640. Wiley, New York.

MARCUS, M., A. FAINSOD and G. DIAMOND. 1985. The genetic analysis of mammalian cell-cycle mutants. *Annual Review of Genetics* 19: 389–421.

MARGULIS, L. 1982. *Early Life*. Science Books International, Incorporated, Boston.

MARRACK, P. and J. KAPPLER. 1986. The T cell and its receptor. *SA* 254(Feb): 36–45.

MARSH, W. L. 1978. Chronic granulomatous disease, Kx antigen and the Kell blood groups. In *The Red Cell: Proceedings of the Fourth International Conference on Red Cell Metabolism and Function*, edited by G. J. Brewer, pp. 493–507. Alan R. Liss, New York.

MARSHALL, C. 1985. Human oncogenes. In *RNA Tumor Viruses: Molecular Biology of Tumor Viruses*, 2nd edition (Part 2), edited by R. Weiss et al., pp. 487–558. Cold Spring Harbor Laboratory, New York.

MARTIN, A. O. 1970. The founder effect in a human isolate: evolutionary implications. *American Journal of Physical Anthropology* 32: 351–368.

MARTIN, J. P. and J. BELL. 1943. A pedigree of mental defect showing sex-linkage. *Journal of Neurology and Psychiatry* 6: 154–157.

MARTIN, R. H. et al. 1987. Variation in the frequency and type of sperm chromosomal abnormalities among normal men. *Human Genetics* 77: 108–114.

MARTINI, G. et al. 1986. Structural analysis of the X-linked gene encoding human glucose 6-phosphate dehydrogenase. *Journal of EMBO* 5: 1849–1855.

MARX, J. 1985. Z-DNA: still searching for a function. *Science* 230: 794–796.

MASCOLA, L. and M. F. GUINAN. 1986. Screening to reduce transmission of sexually transmitted diseases in semen used for artificial insemination. *NEJM* 314: 1354–1358.

MASSARIK, F. and M. M. KABACK. 1981. *Genetic Disease Control; A Social Psychological Approach*. Sage Publications, Beverly Hills, California.

MASSIE, R. K. 1985. *Nicholas and Alexandra*. Dell, New York.

MATTEI, M. G., N. SOUIAH and J. F. MATTEI. 1984. Chromosome 15 anomalies and the Prader-Willi syndrome: cytogenetic analysis. *Human Genetics* 66: 313–334.

MAYNARD SMITH, J. 1968. *Mathematical Ideas in Biology*. Cambridge University Press, Cambridge.

MAZIA, D. 1974. The cell cycle. *SA* 230(Jan): 54–64.

MAZIA, D. 1987. The chromosome cycle and the centrosome cycle in the mitotic cycle. *International Review of Cytology* 100: 49–92.

McCARREY, J. R. and U. K. ABBOTT. 1979. Mechanisms of genetic sex determination, gonadal sex differentiation, and germ-cell development in animals. *Advances in Genetics* 20: 217–290.

McCLINTOCK, B. 1984. The significance of responses of the genome to challenge. *Science* 226: 792–801.

McCLURE, H. M., K. H. BELDEN, W. A. PIEPER and C. B. JACOBSON. 1969. Autosomal trisomy in a chimpanzee: resemblance to Down's syndrome. *Science* 165: 1010–1012.

McCRAE, W. M. 1983. Cystic fibrosis. In *Principles and Practice of Medical Genetics,* Vol. 2, edited by A. E. H. Emery and D. L. Rimoin, pp. 899–905. Churchill Livingstone, Edinburgh.

McDONOUGH, P. G., Editor. 1987. Sex chromosome ratios in human sperm. [Letters to the editor.] *Fertility and Sterility* 47: 531–536.

McELHENY, V. K. and S. ABRAHAMSON, Editors. 1979. *Banbury Report I; Assessing Chemical Mutagens: The Risk to Humans.* Cold Spring Harbor Laboratory, New York.

McGHEE, J. D. 1986. The structure of interphase chromatin. In *Chromosome Structure and Function*, edited by M. S. Risley, pp. 1–37.

McGINNIS, W., R. L. GARBER, J. WIRZ, A. KUROIWA and W. J. GEHRING. 1984b. A homologous protein-coding sequence in *Drosophila* homeotic genes and its conservation in other metazoans. *Cell* 37: 403–408.

McGINNIS, W., M. S. LEVINE, E. HAFEN, A. KUROIWA and W. J. GEHRING. 1984a. A conserved DNA sequence in homoeotic genes of the *Drosophila* Antennapedia and bithorax complexes. *Nature* 308: 428–433.

McGRATH, J. and D. SOLTER. 1984. Completion of mouse embryogenesis requires both the maternal and paternal genomes. *Cell* 37: 179–183.

McKEAN, K. 1985. Intelligence: new ways to measure the wisdom of man. *Discover* 6(Oct): 25–41.

McKEE, P. A. 1983. Hemostasis and disorders of blood coagulation. In *The Metabolic Basis of Inherited Disease*, 5th Edition, edited by J. B. Stanbury et al., pp. 1531–1560. McGraw-Hill, New York.

McKINNON, P. J. 1987. Ataxia-telangiectasia: an inherited disorder of ionizing-radiation sensitivity in man. *Human Genetics* 75: 197–208.

McKUSICK, V. A. 1965. The royal hemophilia. *SA* 213(Aug): 88–95.

McKUSICK, V. A. 1966. *Mendelian Inheritance in Man: Catalogs of Autosomal Dominant, Autosomal Recessive, and X-Linked Phenotypes*. The Johns Hopkins University Press, Baltimore.

McKUSICK, V. A. 1975. The growth and development of human genetics as a clinical discipline. *AJHG* 27: 261–273.

McKUSICK, V. A., Editor. 1978. *Medical Genetic Studies of the Amish: Selected Papers*. The Johns Hopkins University Press, Baltimore.

McKUSICK, V. A. 1980. The anatomy of the human genome. *The Journal of Heredity* 71: 370–391.

McKUSICK, V. A. 1988. *Mendelian Inheritance in Man: Catalogs of Autosomal Dominant, Autosomal Recessive, and X-linked Phenotypes*, 8th Edition. The Johns Hopkins University Press, Baltimore.

McKUSICK, V. A. 1989. Mapping and sequencing the human genome. *NEJM* 320: 910–915.

McKUSICK, V. A., J. A. HOSTETLER, J. A. EGELAND and R. ELDRIDGE. 1964. The distribution of certain genes in the Old Order Amish. *CSHS* 29: 99–114. [Reprinted in Morris (1971).]

McKUSICK, V. A. and D. L. RIMOIN. 1967. General Tom Thumb and other midgets. *SA* 217(July): 102–110.

McKUSICK, V. A. and F. H. RUDDLE. 1977. The status of the gene map of the human chromosomes. *Science* 196: 390–405.

McLAREN, A. 1985. Early mammalian development. In *Prevention of Physical and Mental Congenital Defects. Progress in Clinical and Biological Research*, Vol. 163, edited by M. Marois, pp. 29–40. Alan R. Liss, New York.

McLAREN, A. 1988. Sex determination in mammals. *TIG* 4: 153–157.

McLEAN v ARKANSAS BOARD OF EDUCATION. 1982. Opinion of U. S. District Court Judge William R. Overton. [Reprinted in *Science* 215: 934–943.]

McNEILL, C. A. and R. L. BROWN. 1980. Genetic manipulation by means of microcell-mediated transfer of normal human chromosomes into recipient mouse cells. *PNAS* 77: 5394–5398.

MEAD, M. 1949. *Male and Female: A Study of the Sexes in a Changing World*. William Morrow, New York.

MEDAWAR, P. 1986. *Memoir of a Thinking Radish: An Autobiography*. Oxford University Press, New York.

MEDVEDEV, Z. A. 1969. *The Rise and Fall of T. D. Lysenko*. Translated by I. M. Lerner. Columbia University Press, New York.

MENDEL, G. 1865/1866. Versuche über pflanzenhybriden. *Verhandlungen des Naturforschenden Vereines in Brünn* 4: 3–47. [The original paper was published in 1865, but that issue did not appear until 1866. A facsimile version was published in 1983 by Arkana Verlag, Göttingen, Germany. An English translation by W. Bateson is reprinted in Peters (1959) and in Jameson (1977). Another English translation, by E. R. Sherwood, is in Stern and Sherwood (1966).]

MEREDITH, H. V. 1976. Findings from Asia, Australia, Europe, and North America on secular change in mean height of children, youths, and young adults. *American Journal of Physical Anthropology* 44: 315–325.

MERIMEE, T. J., J. ZAPF, B. HEWLETT and L. L. CAVALLI-SFORZA. 1987. Insulin-like growth factors in pygmies: the role of puberty in determining final stature. *NEJM* 316: 906–911.

MERTENS, T. R., Editor. 1975. *Human Genetics: Readings on the Implications of Genetic Engineering*. John Wiley & Sons, New York.

MERTON, R. K. 1973. *The Sociology of Science: Theoretical and Empirical Investigations*. University of Chicago Press, Chicago.

MERZ, B. 1987. Matchmaking scheme solves Tay-Sachs problem. *JAMA* 258: 2636ff.

MERZ, B. 1988. Eye disease linked to mitochondrial gene defect. *JAMA* 260: 894.

MESELSON, M. and F. W. STAHL. 1958. The replication of DNA in *Escherichia coli*. *PNAS* 44: 671–682. [Reprinted in Freifelder (1978) and in Corwin and Jenkins (1976).]

MESSER, A. and I. H. PORTER, Editors. 1983. *Recombinant DNA and Medical Genetics*. Academic Press, New York.

METZ, C. B. and A. MONROY, Editors. 1985a. *Biology of Fertilization, Vol. 1: Model Systems and Oogenesis*. Academic Press, New York.

METZ, C. B. and A. MONROY, Editors. 1985b. *Biology of Fertilization, Vol. 2: Biology of the Sperm*. Academic Press, New York.

METZ, C. B. and A. MONROY, Editors. 1985c. *Biology of Fertilization, Vol. 3: The Fertilization Response of the Egg*. Academic Press, New York.

MEYER, U. A. and R. SCHMID. 1978. The porphyrias. In *The Metabolic Basis of Inherited Disease*, edited by J. B. Stanbury et al., pp. 1166–1220. McGraw-Hill, New York.

MIESCHER, F. 1871. On the chemical composition of pus cells. *Hoppe-Seyler's medizinische-chemische Untersuchungen* 4: 441–460. [Abridged translation in Gabriel and Fogel (1955).]

MIGEON, B. R., T. R. BROWN, J. AXELMAN and C. J. MIGEON. 1981. Studies of the locus for androgen receptor: localization on the human X and evidence for homology with the *Tfm* locus in the mouse. *PNAS* 78: 6339–6343.

MIGEON, C. J. 1980. Male pseudohermaphroditism. *Annales d'Endocrinologie (Paris)* 41: 311–343.

MILKMAN, R., Editor. 1982. *Perspectives on Evolution*. Sinauer Associates, Sunderland, Massachusetts.

MILLER, J. A. 1987. Mammals need moms and dads. *BioScience* 37: 379–383.

MILLER, O. J. 1983. Chromosomal basis of inheritance. In *Principles and Practice of Medical Genetics*, edited by A. E. H. Emory and D. L. Rimoin, pp. 49–64. Churchill Livingstone, Edinburgh.

MILLER, O. L., JR. 1973. The visualization of genes in action. *SA* 228(Mar): 34–42.

MILLER, W. 1988. Gene conversions, deletions, and polymorphisms in congenital adrenal hyperplasia. *AJHG* 42: 4–7.

MILSTEIN, C. 1980. Monoclonal antibodies. *SA* 243(Oct): 66–74.

MILSTEIN, C. 1986. From antibody structure to immunological diversification of immune response. *Science* 231: 1261–1268.

MILUNSKY, A., Editor. 1986. *Genetic Disorders and the Fetus: Diagnosis, Prevention and Treatment*, 2nd Edition. Plenum Press, New York.

MILUNSKY, A. and E. ALPERT. 1984. Results and benefits of a maternal serum α-fetoprotein screening program. *JAMA* 252: 1438–1442.

MILUNSKY, A. and G. J. ANNAS, Editors. 1976, 1980, 1985. *Genetics and the Law, I, II, and III*. Plenum Press, New York.

MINOGUE, B. P., R. TARASZEWSKI, S. ELIAS and G. J. ANNAS. 1988. The whole truth and nothing but the truth? *Hastings Center Report* 18(5): 34–36.

MITCHISON,J. M. 1971. *Biology of the Cell Cycle*. Cambridge University Press, Cambridge.

MITTWOCH, U. 1973. *Genetics of Sex Differentiation*. Academic Press, New York.

MITTWOCH, U. 1981. Whistling maids and crowing hens—hermaphroditism in folklore and biology. *Perspectives in Biology and Medicine* (Summer 1981): 595–606.

MITTWOCH, U. 1986. Males, females and hermaphrodites. *Annals of Human Genetics* 50: 103–121.

MIZEL, S. B. and P. JARET. 1985. *In Self-Defense*. Harcourt Brace Jovanovich, San Diego.

MOENS, P. B., Editor. 1987. *Meiosis*. Academic Press, New York.

MOHANDAS, T. and L. J. SHAPIRO. 1984. DNA methylation and control of X chromosome activity. In *Research Perspectives in Cytogenetics (National Institute of Child Health and Human Development Series)*, edited by R. S. Sparkes and F. F. de la Cruz, pp. 43–52. University Park Press, Baltimore.

MOHR, O. L. and C. WRIEDT. 1919. *A New Type of Hereditary Brachyphalangy in Man*. Carnegie Institute of Washington Publication No. 295. Washington D. C.

MOLLISON, P. L. 1983. *Blood Transfusion in Clinical Medicine*, 7th Edition. Blackwell Scientific Publications, Oxford.

MOLLON, J. D. 1986. Understanding colour vision. *Nature* 321: 12–13.

MOLNAR, S. 1983. *Human Variation: Races, Types, and Ethnic Groups*, 2nd Edition. Prentice-Hall, Englewood Cliffs, New Jersey.

MONACO, A. P., C. J. BERTELSON, C. COLLETTI-FEENER and L. M. KUNKEL. 1987. Localization and cloning of Xp21 deletion breakpoints involved in muscular dystrophy. *Human Genetics* 75: 221–227.

MONACO, A. P. et al. 1985. Detection of deletions spanning the Duchenne muscular dystrophy locus using a tightly linked DNA segment. *Nature* 316: 842–845.

MONACO, A. P. and L. M. KUNKEL. 1987. A giant locus for the Duchenne and Becker muscular dystrophy gene. *TIG* 3: 33–37.

MONACO, A. P. and L. M. KUNKEL. 1988. Cloning of the Duchenne/Becker muscular dystrophy locus. *Advances in Human Genetics* 17: 61–98.

MONEY, J. 1973. Turner's syndrome and parietal lobe functions. *Cortex* 9: 387–393.

MONEY, J. 1981. The development of sexuality and eroticism in humankind. *The Quarterly Review of Biology* 56: 379–404.

MONEY, J. and A. A. EHRHARDT. 1972. *Man & Woman, Boy & Girl: The Differentiation and Dimorphism of Gender Identity from Conception to Maturity*. The Johns Hopkins University Press, Baltimore.

MONEY, J., M. SCHWARTZ and V. G. LEWIS. 1984. Adult erotosexual status and fetal hormonal masculinization and demasculinization: 46,XX congenital virilizing adrenal hyperplasia and 46,XY androgen-insensitivity syndrome compared. *Psychoneuroendocrinology* 9: 405–414.

MONEY, J. and P. TUCKER. 1975. *Sexual Signatures: On Being a Man or a Woman*. Little, Brown and Co., Boston.

MONK, M., A. HANDYSIDE, K. HARDY and D. WHITTINGHAM. 1987. Preimplantation diagnosis of deficiency of hypoxanthine phosphoribosyl transferase in a mouse model for Lesch-Nyhan syndrome. *Lancet* II: 423–425.

MOORE, J. A., Editor. 1972. *Readings in Heredity and Development*. Oxford University Press, New York.

MOORE, K. L., Editor. 1966. *The Sex Chromatin*. W. B. Saunders Company, Philadelphia.

MOORE, K. L. 1988a. *Essentials of Human Embryology*. Saunders, Philadelphia.

MOORE, K. L. 1988b. *The Developing Human: Clinically Oriented Embryology*, 4th Edition. Saunders, Philadelphia.

MOORE, S. et al. 1984. A monoclonal antibody to human blood group B. In *Monoclonal Antibodies: Standardization of their Characterization and Use, Volume 57 of Developments in Biological Standardization*, pp 55–59. S. Karger, Basel.

MØRCH, E. T. 1941. *Chondrodystrophic Dwarfs in Denmark*. Munksgaard, Copenhagen.

MOREL, F. M. M., R. F. BAKER and H. WAYLAND. 1971. Quantitation of human red blood cell fixation by glutaraldehyde. *The Journal of Cell Biology* 48: 91–100.

MORGAN, T. H. 1910. Sex limited inheritance in Drosophila. *Science* 32: 120–122. [Reprinted in Peters (1959) and in Voeller (1968).]

MORGAN, T. H. 1911. Random segregation versus coupling in Mendelian inheritance. *Science* 34: 384. [Reprinted in Voeller (1968).]

MORRIS, L. N., Editor. 1971. *Human Populations, Genetic Variation, and Evolution*. Chandler Publishing, San Francisco.

MORTON, C. C., I. R. KIRSCH, R. TAUB, S. H. ORKIN and J. A. BROWN. 1984. Localization of the β-globin gene by chromosomal in situ hybridization. *AJHG* 36: 576–585.

MORTON, N. E. 1955. Sequential tests for the detection of linkage. *AJHG* 7: 277–318.

MORTON, N. E. 1956. The detection and estimation of linkage between the genes for elliptocytosis and the Rh blood type. *AJHG* 8: 80–96.

MORTON, N. E. 1978. Effect of inbreeding on IQ and mental retardation. *PNAS* 75: 3906–3908.

MORTON, N. E. and G. A. BURNS. 1987. Report of the committee on the genetic constitution of chromosomes 1 and 2. *Cytogenetics and Cell Genetics* 46: 102–130.

MORTON, N. E. and C. S. CHUNG. 1959. Formal genetics of muscular dystrophy. *AJHG* 11: 360–379.

MOSER, H. 1984. Duchenne muscular dystrophy: pathogenic aspects and genetic prevention. *Human Genetics* 66: 17–40.

MOTULSKY, A. G. 1960. Metabolic polymorphisms and the role of infectious diseases in human evolution. *Human Biology* 32: 28–62. Reprinted in Morris (1971).

MOTULSKY, A. G. 1964. Pharmacogenetics. *Progress in Medical Genetics* 3: 49–74.

MOTULSKY, A. G. 1978. Medical and human genetics 1977: trends and directions. *AJHG* 30: 123–131.

MOURANT, A. E. 1983. *Blood Relations: Blood Groups and Anthropology*. Oxford University Press, Oxford.

MOURANT, A. E., A. C. KOPEĆ and K. DOMANIEWSKA-SOBCZAK. 1976. *The Distribution of the Human Blood Groups and Other Polymorphisms*, 2nd Edition. Oxford University Press, London.

MOURANT, A. E., A. C. KOPEĆ and K. DOMANIEWSKA-SOBCZAK. 1978. *Blood Groups and Diseases*. Oxford University Press, London.

MULAIKAL, R. M., C. J. MIGEON and J. A. ROCK. 1987. Fertility rates in female patients with congenital adrenal hyperplasia due to 21-hydroxylase deficiency. *NEJM* 316: 178–182.

MULLER, C. 1938. Xanthomata, hypercholesterolemia, angina pectoris. *Acta Medica Scandinavia* 89(suppl): 75–84.

MULLER, H. J. 1927. Artificial transmutation of the gene. *Science* 66: 84–87. Reprinted in Gabriel and Fogel (1955) and in Peters (1959).

MULLER, H. J. 1935. *Out of the Night: A Biologist's View of the Future*. Vanguard Press, New York.

MULLER, H. J. 1947. Pilgrim Trust Lecture: The gene. *Royal Society of London, Proceedings (Series B)* 134: 1–37. [Reprinted in Moore (1972).]

MULLER, H. J. 1955. Radiation and human mutation. *SA* 193(Nov): 58–68. [Reprinted in Srb et al. (1970).]

MULLER, H. J. 1961. Human evolution by voluntary choice of germ plasm. *Science* 134: 643–649. [Reprinted in Mertens (1975).]

MULLER, H. J. 1965. Radiation genetics: synthesis. In *Genetics Today, Vol. II, Proceedings of the XI International Congress of Genetics*. Pergamon Press, Oxford.

MULLIS, K. et al. 1986. Specific enzymatic amplification of DNA in vitro: the polymerase chain reaction. *CSHS* 51: 263–273.

MURPHY, E. A. 1979. *Probability in Medicine*. The Johns Hopkins University Press, Baltimore.

MURPHY, E. A. and G. A. CHASE. 1975. *Principles of Genetic Counseling*. Year Book Medical Publishers, Chicago.

MURRAY, A. W. 1987. A cycle is a cycle is a cycle. *Nature* 327: 14–15.

MUSTAJOKI, P. and R. J. DESNICK. 1985. Genetic heterogeneity in acute intermittent porphyria: characterisation and frequency of porphobilinogen deaminase mutations in Finland. *British Medical Journal* 291: 505–509.

MUSTAJOKI, P. et al. 1987. Homozygous variegate porphyria: a severe skin disease of infancy. *Clinical Genetics* 32: 300–305.

MYEROWITZ, R. and N. D. HOGIKYAN. 1987. A deletion involving Alu sequences in the β-hexosaminidase α-chain gene of French Canadians with Tay-Sachs disease. *Journal of Biological Chemistry* 262: 15396–15399.

MYERS, R. H. et al. 1985. Maternal factors in onset of Huntington disease. *AJHG* 37: 511–523.

NAFTOLIN, F. and E. BUTZ, Editors. 1981. Sexual dimorphism. *Science* 211: 1263–1324.

NAGEL, R. L. 1984. The origin of the hemoglobin S gene: clinical, genetic and anthropological consequences. *Einstein Quarterly Journal of Biology and Medicine* 2: 53–62.

NANCE, W. E. and F. E. McCONNELL. 1973. Status and prospects of research in hereditary deafness. *Advances in Human Genetics* 4: 173–250.

NATHANS, J. 1989. The genes for color vision. *SA* 260(Feb): 42–49.

NATHANS, J. and D. S. HOGNESS. 1983. Isolation, sequence analysis, and intron–exon arrangement of the gene encoding bovine rhodopsin. *Cell* 34: 807–814.

NATHANS, J., T. P. PIANTANIDA, R. L. EDDY, T. B. SHOWS and D. S. HOGNESS. 1986a. Molecular genetics of inherited variation in human color vision. *Science* 232: 203–210.

NATHANS, J., D. THOMAS and D. S. HOGNESS. 1986b. Molecular genetics of human color vision: the genes encoding blue, green, and red pigments. *Science* 232: 193–202.

NATHANSON, J. A. 1984. Caffeine and related methylxanthines: possible naturally occurring pesticides. *Science* 226: 184–187.

NATIONAL CENTER FOR EDUCATION IN MATERNAL AND CHILD HEALTH. 1985. *Comprehensive Clinical Genetic Services Centers: A National Directory*. U.S. Government Printing Office, Washington, D. C.

NATIONAL RESEARCH COUNCIL, COMMITTEE ON MAPPING AND SEQUENCING THE HUMAN GENOME. 1988. *Mapping and Sequencing the Human Genome*. National Academy Press, Washington, D. C.

NEBERT, D. A., AND F. J. GONZALEZ. 1987. P450 genes: structure, evolution, and regulation. *Annual Review of Biochemistry* 56: 945–993.

NEEL, J. V., M. KODANI, R. BREWER and R. C. ANDERSON. 1949. The incidence of consanguineous matings in Japan. *AJHG* 1: 156–178.

NEEL, J. V. and W. J. SCHULL. 1954. *Human Heredity*. University of Chicago Press, Chicago.

NELSON, K. and L. B. HOLMES. 1989. Malformations due to presumed spontaneous mutations in newborn infants. *NEJM* 320: 19–23.

NEW, M. I., Editor. 1985. *Congenital Adrenal Hyperplasia. Annals of the New York Academy of Sciences*, Vol. 458.

NEW, M. I., B. DUPONT, K. GRUMBACH, and L. S. LEVINE. 1983. Congenital adrenal hyperplasia and related conditions. In *The Metabolic Basis of Inherited Disease*, 5th Edition, edited by J. B. Stanbury et al., pp. 973–1000. McGraw-Hill, New York.

NEWMAN, H. H., F. N. FREEMAN and K. J. HOLZINGER. 1937. *Twins, A Study of Heredity and Environment*. University of Chicago Press, Chicago.

NICHOLS, E. A. and F. H. RUDDLE. 1973. A review of enzyme polymorphism, linkage and electrophoretic conditions for mouse and somatic cell hybrids in starch gels. *Journal of Histochemistry and Cytochemistry* 21: 1066–1081.

NICHOLS, E. E. 1984. Routine antepartum Rh immune globulin administration. *JAMA* 252: 2763.

NICHOLS, E. K. 1988. *Human Gene Therapy*. Harvard University Press, Cambridge.

NICOL, S. E. and I. I. GOTTESMAN. 1983. Clues to the genetics and neurobiology of schizophrenia. *American Scientist* 71: 398–404.

NIEBUHR, E. 1974. Triploidy in man, cytogenetical and clinical aspects. *Humangenetik* 21: 103–125.

NIELSEN, J., A. HOMMA, F. CHRISTIANSEN and K. RASMUSSEN. 1977. Women with tetra-X (48,XXXX). *Hereditas* 85: 151–156.

NIELSEN, J., I. SILLESEN, A. M. SORENSEN and K. SORENSEN. 1979. Follow-up until age 4 to 8 of 25 unselected children with sex chromosome abnormalities, compared with sibs and controls. *Birth Defects: Original Articla Series*, Vol. XV, No. 1, pp. 15–73.

NIGHTINGALE, S. L. 1985. The regulation of α-fetoprotein test kits by the Food and Drug Administration. In *Genetics and the Law III*, edited by A. Milunsky and G. J. Annas, pp. 395–403. Plenum Press, New York.

NIH. 1987. *National Cholesterol Education Program: Report of the Expert Panel on Detection, Evaluation, and Treatment of High Blood Cholesterol in Adults* [draft]. NIH, Bethesda, Maryland.

NIIKAWA, N. and T. KAJII. 1984. The origin of mosaic Down syndrome: four cases with chromosome markers. *AJHG* 36: 123–130.

NILSSON, L. 1975. A portrait of the sperm. In *The Functional Anatomy of the Spermatozoon: Proceedings of the 2nd International Sympopsium, Wenner-Gren Center, Stockholm*, edited by B. A. Afzelius, pp. 79–82. Pergamon Press, Oxford.

NILSSON, L. 1986. *A Child is Born*. Dell Press, New York.

NISONOFF, A. 1984. *Introduction to Molecular Immunology*, 2nd edition. Sinauer Associates, Sunderland, Massachusetts

NOMURA, M. 1984. The control of ribosome synthesis. *SA* 250(Jan): 102–112.

NORMAN, C. 1985. Virus scare halts hormone research. *Science* 228: 1176–1177.

NOSSAL, G. J. V. 1975. *Medical Science and Human Goals*. Edward Arnold, London.

NOSSAL, G. J. V. 1978. *Antibodies and Immunity*, 2nd edition. Basic Books, New York.

NOSSAL, G. J. V. 1985. *Reshaping Life: Key Issues in Genetic Engineering*. Cambridge University Press, Cambridge.

NOSSAL, G. J. V. 1987. The basic components of the immune system. *NEJM* 316: 1320–1325.

NOVITSKI, E. and L. SANDLER. 1956. The relationship between parental age, birth order and the secondary sex ratio in humans. *Annals of Human Genetics* 21(2): 123–131.

NOVOTNY, E. J. et al. 1986. Leber's disease and dystonia: a mitochondrial disease. *Neurology* 36: 1053–1060.

NOWELL, P. C. 1960. Phytohemagglutinin: an initiator of mitosis in cultures of normal human leukocytes. *Cancer Research* 20: 462–466.

NOWELL, P. C. 1976. The clonal evolution of tumor cell populations. *Science* 194: 23–28.

NOWELL, P. C. and D. A. HUNGERFORD. 1960. A minute chromosome in human chronic granulocytic leukemia. *Science* 132: 1497.

NURSE, P. and P. A. FANTES. 1981. Cell cycle controls in fission yeast: a genetic analysis. In *The Cell Cycle*, edited by P. C. . John, pp. 85–98. Cambridge University Press, New York.

NUSBACHER, J. 1984. Routine antepartum Rh immune globulin administration. *JAMA* 252: 2763.

NUSSBAUM, R. L. and D. H. LEDBETTER. 1986. Fragile X syndrome: a unique mutation in man. *Annual Review of Genetics* 20: 109–145.

NÜSSLEIN-VOLHARD, C., H. G. FROHNHOFER and R. LEHMANN. 1987. Determination of anteroposterior polarity in *Drosophila*. *Science* 238: 1675–1681.

NÜSSLEIN-VOLHARD, C. and E. WIESCHAUS. 1980. Mutations affecting segment number and polarity in Drosophila. *Nature* 287: 795–801.

NYHAN, W. L., with E. EDELSON. 1976. *The Heredity Factor: Genes, Chromosomes and You*. Grosset & Dunlap, New York.

O'BRIEN, J. S. 1983. The gangliosidoses. In *The Metabolic Basis of Inherited Disease*, 5th Edition, edited by J. B. Stanbury et al., pp. 945–969. McGraw-Hill, New York.

O'BRIEN, S. J., D. E. WILDT and M. BUSH. 1986. The cheetah in genetic peril. *SA* 254 (May): 84–92.

OFFICE OF TECHNOLOGY ASSESSMENT. 1988a. *Infertility: Medical and Social Choices*, [OTA-BA-358]. U.S. Government Printing Office, Washington, D. C.

OFFICE OF TECHNOLOGY ASSESSMENT. 1988b. *Mapping our Genes: Genome Projects—How Big, How Fast?* The Johns Hopkins University Press, Baltimore.

OHNO, S. 1969. Evolution of sex chromosomes in mammals. *Annual Review of Genetics* 3: 495–524.

OHNO, S. 1973. Conservation of ancient linkage groups in evolution and some insight into the genetic regulatory mechanism of X-inactivation. *CSHS* 38: 155–164.

OKADA, S. and J. S. O'BRIEN. 1969. Tay-Sachs disease: generalized absence of a β-D-nacetylhexosaminidase component. *Science* 165: 698–700.

OLD, R. W. and S. B. PRIMROSE. 1985. *Principles of Gene Manipulation: An Introduction to Genetic Engineering*, 3rd edition. Blackwell Scientific Publications, Oxford.

OLDSTONE, M. B. A. 1987. Molecular mimicry and autoimmune disease. *Cell* 50: 819–820.

OLIVER, C. and A. J. HOLLAND. 1986. Down's syndrome and Alzheimer's disease: a review. *Psychological Medicine* 16(2): 307–322.

OLIVER, G., C. V. E. WRIGHT, J. HARDWICKE and E. M. DE ROBERTIS. 1988. A gradient of homeodomain protein in developing forelimbs of *Xenopus* and mouse embryos. *Cell* 55: 1017–1024.

OLSON, S. B., R. E. MAGENIS and E. W. LOVRIEN. 1986. Human chromosome variation: the discriminatory power of Q-band heteromorphism (variant) analysis in distinguishing between individuals, with specific application to cases of questionable paternity. *AJHG* 38: 235–252.

OREL, V. 1984. *Mendel*. Oxford University Press, Oxford.

ORKIN, S. H. 1983. Controlling the fetal globin switch in man. *Nature* 301: 108–109.

ORKIN, S. H. and S. C. GOFF. 1981. Nonsense and frameshift mutations in β^0-thalassemia detected in cloned β-globin genes. *The Journal of Biological Chemistry* 256: 9782–9784.

ORKIN, S. H. and H. H. KAZAZIAN, JR. 1984. The mutation and polymorphism of the human β-globin gene and its surrounding DNA. *Annual Review of Genetics* 18: 131–171.

ORKIN, S. H., P. F. R. LITTLE, H. H. KAZAZIAN and C. B. BOEHM. 1982. Improved detection of the sickle mutation by DNA analysis. *NEJM* 307: 32–36.

OSBORN, D. 1916. Inheritance of baldness. *Journal of Heredity* 7: 347–355.

OVERZIER, C., Editor. 1963. *Intersexuality*. Academic Press, New York.

OW, D. W. et al. 1986. Transient and stable expression of the firefly luciferase gene in plant cells and transgenic plants. *Science* 234: 856–859.

OWEN, D. R. 1972. The 47,XYY male: a review. *Psychological Bulletin* 78: 209–233.

OWERBACH, D., G. I. BELL, W. J. RUTTER and T. B. SHOWS. 1980. The insulin gene is located on chromosome 11 in humans. *Nature* 286: 82–84.

OWOR, R. and C. OLWENY. 1978. Malignant neoplasms. In *Diseases of Children in the Subtropics and Tropics*, 3rd edition, edited by D. B. Jelliffe and J. P. Stanfield, pp. 605–623. Edward Arnold, London.

OZAR, D. T. 1985. The case against thawing unused frozen embryos. *Hastings Center Report* 15(4): 7–12.

PÄÄBO, S. 1985. Molecular cloning of ancient Egyptian mummy DNA. *Nature* 314: 644–645.

PAGE, D. C. 1986. Sex reversal: deletion mapping the male-determining function of the human Y chromosome. *CSHS* 51: 229–235.

PAGE, D. C., L. G. BROWN and A. DE LA CHAPELLE. 1987a. Exchange of terminal portions of X- and Y-chromosomal short arms in human XX males. *Nature* 328: 437–440.

PAGE, D. C. et al. 1987b. The sex-determining region of the human Y chromosome encodes a finger protein. *Cell* 51: 1091–1104.

PAGNIER, J. et al. 1984. Evidence for the multicentric origin of the sickle cell hemoglobin gene in Africa. *PNAS* 81: 1771–1773.

PAI, G. S., J. A. SPRENKLE, T. T. DO, C. E. MARENI and B. R. MIGEON. 1980. Localization of loci for hypoxanthine phosphate dehydrogenase and glucose-6-phosphate dehydrogenase and biochemical evidence of nonrandom X chromosome expression from studies of a human X–autosome translocation. *PNAS* 77: 2810–2813.

PAINE, R. S. 1964. Phenylketonuria. *Clinical Proceedings* 20(6): 143–152.

PAINTER, T. S. 1923. Studies in mammalian spermatogenesis. II. The spermatogenesis of man. *The Journal of Experimental Zoology* 37: 291–338.

PALMITER, R. D., G. NORSTEDT, R. E. GELINAS, R. E. HAMMER and R. L. BRINSTER. 1983. Metallothionein-human GH fusion genes stimulate growth of mice. *Science* 222: 809–814.

PARKMAN, R. 1986. The application of bone marrow transplantation to the treatment of genetic diseases. *Science* 232: 1373–1378.

PASSARGE, E. 1983. Bloom's syndrome. In *Chromosome Mutation and Neoplasia*, edited by J. German, pp. 11–21. Alan R. Liss, New York.

PATEL, P. I., P. E. FRAMSON,, C. T. CASKEY and A. C CHINAULT. 1986. Fine structure of the human hypoxanthine phosphoribosyltransferase gene. *Molecular and Cellular Biology* 6: 393–403.

PATTERSON, D. 1987. The causes of Down syndrome. *SA* 257(Aug): 52–60.

PAULI, R. M. and E. J. CASSELL. 1978. Case studies in bioethics— nurturing a defective newborn. *Hastings Center Report* 8(1): 13–14.

PAULING, L., H. A. ITANO, S. J. SINGER and I. C. WELLS. 1949. Sickle cell anemia, a molecular disease. *Science* 110: 543–548. [Reprinted in Boyer (1963) and in Schull and Chakraborty (1979).]

PEBLEY, A. R. and C. F. WESTOFF. 1982. Women's sex preferences in the United States: 1970 to 1975. *Demography* 19: 177–189.

PEELE, S. 1986. The implications and limitations of genetic models of alcoholism and other addictions. *Journal of Studies on Alcohol* 47: 63–73.

PENROSE, L. S. 1951. Measurement of pleiotropic effects in phenylketonuria. *Annals of Eugenics* 16: 134–141.

PENROSE, L. S. and C. STERN. 1958. Reconsideration of the Lambert pedigree (ichthyosis hystrix gravior). *Annals of Human Genetics* 22: 258–283.

PERUTZ, M. F. 1978. Hemoglobin structure and respiratory transport. *SA* 239(June): 92–125.

PESTKA, S. 1983. The purification and manufacture of human interferons. *SA* 249(Aug): 36–43.

PETER, K. A. 1987. *The Dynamics of Hutterite Society: An Analytical Approach*. University of Alberta Press, Edmonton.

PETERS, J. A., Editor. 1959. *Classic Papers in Genetics*. Prentice-Hall, Englewood Cliffs, New Jersey.

PETERSON, E. P., N. J. ALEXANDER and K. S. MOGHISSI. 1988. A.I.D. and AIDS—too close for comfort. *Fertility and Sterility* 49: 209–211.

PINES, M. 1984. In the shadow of Huntington's. *Science 84* 5(May): 32–39.

PINSKY, C. M. 1986. Monoclonal antibodies: progress is slow but sure. *NEJM* 315: 704–705.

PINSKY, L. and M. KAUFMAN. 1987. Genetics of steroid receptors and their disorders. *Advances in Human Genetics* 16: 299–472.

PLOMIN, R. and J. C. DEFRIES. 1980. Genetics and intelligence: recent data. *Intelligence* 4: 15–24.

PLOMIN, R., J. C. DEFRIES and G. E. MCCLEARN. 1980. *Behavioral Genetics: A Primer*. W. H. Freeman, San Francisco.

POLANI, P. 1970. Hormonal and clinical aspects of hermaphroditism and the testicular feminizing syndrome in man. *Royal Society of London, Philosophical Transactions, Series B* 259: 187–206.

POLANI, P. E. 1981a. Abnormal sex development in man. I. Anomalies of sex-determining mechanisms. In *Mechanisms of Sex Differentiation in Animals and Man*, edited by C. R. Austin and R. G. Edwards, pp. 465–547. Academic Press, London.

POLANI, P. E. 1981b. Abnormal sex development in man. II. Anomalies of sex-differentiating mechanisms. In *Mechanisms of Sex Differentiation in Animals and Man*, edited by C. R. Austin and R. G. Edwards, pp. 549–590. Academic Press, London.

POLANI, P. E. 1985. The genetic basis of embryonic sexual dimorphism. In *Human Sexual Dimorphism*, edited by J. Ghesquiere et al., pp. 125–150. Taylor & Francis, London.

POLANI, P. E. et al. 1958. Colour vision studies and the X-chromosome constitution of patients with Klinefelter's syndrome. *Nature* 182: 1092–1093.

POLANI, P. E., M. H. LESSOF and P. M. F. BISHOP. 1956. Colour-blindness in ovarian agenesis (gonadal dysplasia). *Lancet* II: 118–120.

PONCZ, M., E. SCHWARTZ, M. BALLANTINE and S. SURREY. 1983. Nucleotide sequence analysis of the δ/β-globin region in humans. *The Journal of Biological Chemistry* 258: 11599–11609.

POPOVSKY, M. 1984. *The Vavilov Affair*. Archon Books, Hamden, Connecticut.

PORTER, I. H. 1982. Control of hereditary disorders. *Annual Review of Public Health* 3: 277–319.

PORTER, I. H., N. HATCHER and A. M. WILLEY,

Editors. 1986. *Perinatal Genetics: Diagnosis and Treatment*. Academic Press, Orlando, Florida.

PORTER, I. H. and E. B. HOOK, Editors. 1980. *Human Embryonic and Fetal Death*. Academic Press, New York.

PORTUGAL, F. H. and J. S. COHEN. 1977. *A Century of DNA: A History of the Discovery of the Structure and Function of the Genetic Substance*. The MIT Press, Cambridge, Massachusetts.

POWELL, J. R. 1987. "In the Air"—Theodosius Dobzhansky's *Genetics and the Origin of Species*. *Genetics* 117: 363–366.

PRENATAL DIAGNOSIS. 1985. Vol. 5: 93–148. [In a special issue devoted to cystic fibrosis.]

PRESCOTT, D. M. 1976. *Reproduction of Eukaryotic Cells*. Academic Press, New York.

PRESCOTT, D. M. 1987. Cell reproduction. *International Review of Cytology* 100: 93–128.

PRESCOTT, D. M. and A. S. FLEXER. 1986. *Cancer: The Misguided Cell*, 2nd edition. Sinauer Associates, Sunderland, Massachusetts.

PRESIDENT'S COMMISSION FOR THE STUDY OF ETHICAL PROBLEMS IN MEDICINE AND BIOMEDICAL AND BEHAVIORAL RESEARCH. 1982. *Splicing Life: A report on the Social and Ethical Issues of Genetic Engineering with Human Beings*. U.S. Government Printing Office, Washington, D. C.

PRESIDENT'S COMMISSION FOR THE STUDY OF ETHICAL PROBLEMS IN MEDICINE AND BIOMEDICAL AND BEHAVIORAL RESEARCH. 1983. *Screening and Counseling for Genetic Conditions*. U.S. Government Printing Office, Washington, D. C.

PROIA, R. L. and E. SORAVIA. 1987. Organization of the gene encoding the human β-hexosaminidase α-chain. *Journal of Biological Chemistry* 262: 5677–5681.

PROVINE, W. B. 1971. *The Origins of Theoretical Population Genetics*. University of Chicago Press, Chicago.

PROVINE, W. B. 1986. *Sewall Wright and Evolutionary Biology*. University of Chicago Press, Chicago.

PRUSINER, S. B. 1984a. Prions. *SA* 251(Oct): 50–59.

PRUSINER, S. B. 1984b. Some speculations about prions, amyloid, and Alzheimer's disease. *NEJM* 310: 661–663.

PUESCHEL, S. M. 1984. *The Young Child with Down Syndrome*. Human Science Press, New York.

PUESCHEL, S. M. 1987. Maternal α-fetoprotein screening for Down's syndrome. *NEJM* 317: 376–378.

PUESCHEL, S. M., C. TINGEY, J. E. RYNDERS, A. C. CROCKER and D. M. CRUTCHER, Editors. 1987. *New Perspectives in Down Syndrome*. Paul H. Brookes Publishing, Baltimore.

PUNNETT, R. C. 1911. Mendelism. In *The Encyclopaedia Britannica*, 11th Edition. The Encyclopaedia Britannica Co., New York.

QUINTON, P. M. and J. BIJMAN. 1983. Higher bioelectric potentials due to decreased chloride absorption in the sweat glands of patients with cystic fibrosis. *NEJM* 308: 1185–1189.

RABBITTS, T. H., T BOEHM and L. MENGLE-GAW. 1988. Chromosomal abnormalities in lymphoid tumours: mechanism and role in tumour pathogenesis. *TIG* 4: 300–304.

RACE, R. R. and R. SANGER. 1975. *Blood Groups in Man*, 6th Edition. Blackwell Scientific Publications, Oxford.

RADMAN, M. and R. WAGNER. 1988. The high fidelity of DNA duplication. *SA* 259(Aug): 40–46.

RAMSAY, M., R. BERNSTEIN, E. ZWANE, D. C. PAGE and T. JENKINS. 1988. XX true hermaphroditism in southern African Blacks: an enigma of primary sexual differentiation. *AJHG* 43: 4–13.

RAMSEY, P. 1970. *Fabricated Man: The Ethics of Genetic Control*. Yale University Press, New Haven, Connecticut.

RATCLIFFE, S. G. and N. PAUL, Editors. 1986. *Prospective Studies on Children with Sex Chromosome Aneuploidy. Birth Defects: Original Article Series*, Vol. 22, No. 3.

RATTAZZI, M. C. and M. M. COHEN. 1972. Further proof of genetic inactivation of the X chromosome in the female mule. *Nature* 237: 393–396.

RATTNER, J. B. and C. C. LIN. 1985. Radial loops and helical coils coexist in metaphase chromosomes. *Cell* 42: 291–296.

RAY, J. H. and J. GERMAN. 1983. The cytogenetics of the "chromosome breakage syndromes." In *Chromosome Mutation and Neoplasia,* edited by J. German, pp. 135–167. Alan R. Liss, New York.

RAY, P. N. et al. 1985. Cloning of the breakpoint of an X:21 translocation associated with Duchenne muscular dystrophy. *Nature* 318: 672–675.

RAYMOND, C. A. 1988. In vitro fertilization enters stormy adolescence as experts debate the odds. *JAMA* 259: 464 ff.

REDDY, P. G. 1987. Effects of consanguineous marriages on fertility among three endogamous groups of Andhra Pradesh. *Social Biology* 34: 68–77.

REED, T. E. 1969. Caucasian genes in American Negroes. *Science* 165: 762–768. Reprinted in Morris (1971).

REICH, T. 1988. Biologic-marker studies in alcoholism. *NEJM* 318: 180–182.

REICH, W. T., Editor. 1978. *Encyclopedia of Bioethics* (4 Volumes). The Free Press, New York.

REILLY, P. 1975. Genetic screening legislation. *Advances in Human Genetics* 5: 319–376.

REILLY, P. R. 1985. Eugenic sterilization in the United States. In *Genetics and the Law III*, edited by A. Milunsky and G. J. Annas, pp. 227–241. Plenum Press, New York.

RETIEF, A. E. et al. 1984. Chromosome studies in 496 infertile males with a sperm count below 10 million/ml. *Human Genetics* 66: 162–164.

RHOADS, G. G. et al. 1989. The safety and efficacy of chorionic villus sampling for early prenatal diagnosis of cytogenetic abnormalities. *NEJM* 320: 609–617.

RICCARDI, V. M. 1981. von Recklinghausen neurofibromatosis. *NEJM* 305: 1617–1627.

RICCARDI, V. M. 1984. High-resolution karyotype–phenotype correlations and focused chromosome analysis. In *Research Perspectives in Cytogenetics*, edited by R. S. Sparkes and F. F. de la Cruz, pp. 53–62.

RICH, A. and S. H. KIM. 1978. The three-dimensional structure of transfer RNA. *SA* 238 (Jan): 52–62.

RICHARDSON, D. W. 1975. Artificial insemination in the human. In *Modern Trends in Human Genetics*, Vol. 2, edited by A. E. H. Emery, pp. 404–448. Butterworths, London.

RIEDER, C. L. 1982. The formation, structure, and composition of the mammalian kinetochore and kinetochore fiber. *International Review of Cytology* 79: 1–58.

RIMOIN, D. L. 1983. Genetic disorders of the pituitary gland. In *Principles and Practice of Medical Genetics*, Vol. 2, edited by A. E. H. Emery and D. L. Rimoin, pp. 1134–1151. Churchill Livingstone, Edinburgh.

RIMOIN, D. L. and R. S. LACHMAN. 1983. The chondrodysplasias. In *Principles and Practice of Medical Genetics*, Vol. 2, edited by A. E. H. Emery and D. L. Rimoin, pp. 703–735. Churchill Livingstone, Edinburgh.

RIORDAN, J. R. and M. BUCHWALD, Editors. 1988. *Genetics and Epithelial Cell Dysfunction in Cystic Fibrosis. Progress in Clinical and Biological Research*, Vol. 254. Alan R. Liss, New York.

RIS, H. and J. KORENBERG. 1979. Chromosome structure and levels of chromosome organization. In *The Structure and Replication of Genetic Material. Cell Biology: A Comprehensive Treatise*, Vol. 2, edited by D. M. Prescott and L. Goldstein, pp. 267–361. Academic Press, New York.

RIS, H. and P. L. WITT. 1981. Structure of the mammalian kinetochore. *Chromosoma* (Berlin) 82: 153–170.

RISLEY, M. S. 1986a. The organization of meiotic chromosomes and synaptonemal complexes. In *Chromosome Structure and Function*, edited by M. S. Risley, pp. 126–151. Van Nostrand Reinhold, New York.

RISLEY, M. S., Editor. 1986b. *Chromosome Structure and Function*. Van Nostrand Reinhold, New York.

ROBBINS, D. C., H. S. TAGER and A. H. RUBENSTEIN. 1984. Biologic and clinical importance of proinsulin. *NEJM* 310: 1165–1175.

ROBERTS, C. J. and C. R. LOWE. 1975. Where have all the conceptions gone? *Lancet* I: 498–499.

ROBERTS, D. F. and G. E. S. AHERNE. 1983. Retinoblastoma. In *Principles and Practice of Medical Genetics*, Vol. 1, edited by A. E. H. Emery and D. L. Rimoin, pp.539–554. Churchill Livingstone, Edinburgh.

ROBERTS, L. 1984. *Cancer Today; Origins, Prevention, and Treatment*. National Academy of Sciences, Washington, D. C.

ROBERTS, L. 1987. Measuring cholesterol is as tricky as lowering it. *Science* 238: 482–483.

ROBERTS, L. 1987. Zeroing in on the sex switch. *Science* 239: 21–23.

ROBERTS, L. 1988a. The race for the cystic fibrosis gene. *Science* 240: 141–144.

ROBERTS, L. 1988b. Race for the cystic fibrosis gene nears end. *Science* 240: 282–285.

ROBERTSON, H. D., A. D. BRANCH and J. E. DAHLBERG. 1985. Focusing on the nature of the scrapie agent. *Cell* 40: 725–727.

ROBERTSON, J. A. and J. D. SCHULMAN. 1987. Pregnancy and prenatal harm to offspring: the case of mothers with PKU. *Hastings Center Report* 17(6): 23–33.

ROBERTSON, M. 1987. Molecular genetics of the mind. *Nature* 325: 755.

ROBINSON, A., H. A. LUBS and D. BERGSMA, Editors. 1979. *Sex Chromosome Aneuploidy: Prospective Studies on Children. Birth Defects: Original Article Series*, Vol. 15, No. 1. Alan R. Liss, New York..

ROBINSON, W. C. 1986. Another look at the Hutterites and natural fertility. *Social Biology* 33: 65–76.

ROBLIN, R. 1975. The Boston XYY case. *Hastings Center Report* 5(4): 5–8.

RODECK, C. H. 1987. Chorion villus biopsy. In *Oxford Reviews of Reproductive Biology*, Vol. 9, edited by J. R. Clarke, pp. 137–160. Oxford University Press, Oxford.

RODGER, J. C. and B. L. DRAKE. 1987. The enigma of the fetal graft. *American Scientist* 75: 51–57.

ROE v. WADE. 1973. *United States Reports: Cases Adjudged in the Supreme Court*. 410: 113–178.

ROGERS, D. B. and L. J. SHAPIRO. 1986. X-linked diseases and disorders of the sex chromosomes. In *Genetic Disorders and the Fetus; Diagnosis, Prevention, and Treatment*, 2nd Edition, edited by A. Milunsky, pp. 341–368. Plenum Press, New York.

ROOT-BERNSTEIN, R. S. 1983. Mendel and methodology. *History of Science* 21: 275–295.

ROSATELLI, C. et al. 1985. Prenatal diagnosis of β-thalassemia with the synthetic-oligomer technique. *Lancet* I: 241–243.

ROSE, N. R. 1981. Autoimmune diseases. *SA* 244(Feb): 80–103.

ROSE, R. J. 1982. Separated twins: data and their limits. *Science* 215: 959–960.

ROSEN, C. M. 1987. The eerie world of reunited twins. *Discover* 8(Sept): 36–46.

ROSEN, F. S., M. D. COOPER and R. J. P. WEDGWOOD. 1984. The primary immunodeficiencies. *NEJM* 311: 300–310.

ROSENBERG, C., Editor. 1985. *Eugenics, Genetics and the Family: Eugenics in Race and State. International Congress of Eugenics, Second Congress*. (History of Hereditarian Thought Series). Reproduction of 1923 Edition. Garland Publishing, New York.

ROSENBERG, L. 1981. On the "Human Life" bill. *Science* 212: 907.

ROSENBERG, L. E. 1985. Can we cure genetic disorders? In *Genetics and the Law III*, edited by A. Milunsky and G. J. Annas, pp. 5–13. Plenum Press, New York.

ROSENBERG, R. N. and H. L. FOWLER. 1981. Autosomal dominant motor system disease of the Portugese: a review. *Neurology* 3l: 1124–1126.

ROSENTHAL, E. 1989. The wolf at the door. *Discover* 10(Feb): 34–37.

ROSNER, F. 1977. *Medicine in the Bible and the Talmud*. Yeshiva University Press, New York.

ROSSANT, J. and R. A. PEDERSEN, Editors. 1986. *Experimental Approaches to Mammalian Embryonic Development*. Cambridge University Press, Cambridge.

ROTHSTEIN, H. 1982. Regulation of the cell cycle by somatomedins. *International Review of Cytology* 78: 127–232.

ROTTER, J. I. and D. L. RIMOIN. 1983. Diabetes mellitus. In *Principles and Practice of Medical Genetics*, Vol. 2, edited by A. E. H. Emery and D. L. Rimoin, pp. 1180–1201. Churchill Livingstone, Edinburgh.

ROUS, P. 1911. Transmission of a malignant new growth by means of a cell-free filtrate. *JAMA* 56: 198.

ROUYER, F. et al. 1986. The pseudoautosomal region of the human sex chromosomes. *CSHS* 51: 221–228.

ROUYER, F., M.-C. SIMMLER, D. C. PAGE and J. WEISSENBACH. 1987. A sex chromosome rearrangement in a human XX male caused by Alu–Alu recombination. *Cell* 51: 417–425.

ROWLAND, L. P. 1988a. Dystrophin: a triumph of reverse genetics and the end of the beginning. *NEJM* 318: 1392–1394.

ROWLAND, L. P. 1988b. Clinical concepts of Duchenne muscular dystrophy. *Brain* 111: 479–495.

ROWLEY, J. D. 1973. A new consistent chromosomal abnormality in chronic myelogenous leukaemia identified by quinacrine fluorescence and Giemsa staining. *Nature* 243: 290–293.

ROWLEY, P. T. 1984. Genetic screening: marvel or menace? *Science* 225: 138–144.

ROYER-POKORA, B. et al. 1986. Cloning the gene for the inherited disorder chronic granulomatous disease on the basis of its chromosomal location. *CSHS* 51: 177–183.

RUDAK, E., P. JACOBS and R. YANAGIMACHI. 1978. Direct analysis of the chromosome constitution of human spermatozoa. *Nature* 274: 911–913.

RUDDLE, F. et al. 1972. Somatic cell genetic assignment of peptidase C and the Rh linkage group to chromosome A-1 in man. *Science* 176: 1429–1431.

RUSHTON, W. A. H. 1975. Visual pigments and color blindness. *SA* 232(Mar): 64–74.

RUSSELL, D. W. et al. 1986. The LDL receptor in familial hypercholesterolemia: use of human mutations to dissect a membrane protein. *CSHS* 51: 811–819.

RUSSELL, P. and P. NURSE. 1987. The mitotic inducer *nim1*+ functions in a regulatory network of protein kinase homologs controlling the initiation of mitosis. *Cell* 49: 569–576

RUSSELL, W. L. 1951. X-ray-induced mutations in mice. *CSHS* 16: 327–336.

SACHS, B. 1887. On arrested cerebral development, with special reference to its cortical pathology. *Journal of Nervous and Mental Disease* 14: 541–553.

SACHS, B. 1896. A family form of idiocy, generally fatal, associated with early blindness (amaurotic family idiocy). *Journal of Nervous and Mental Disease* 23: 475–479.

SAIKI, R. K. et al. 1988a. Primer-directed enzymatic amplification of DNA with a thermostable DNA polymerase. *Science* 239: 487–491.

SAIKI, R. K. et al. 1988b. Diagnosis of sickle-cell anemia and β-thalassemia with enzymatically amplified DNA and nonradioactive allele-specific oligonucleotide probes. *NEJM* 319: 537–541.

SALMON, C., J.-P. CARTRON and P. ROUGER. 1984. *The Human Blood Groups*. Masson Publishing USA, New York.

SAN ROMAN, C. and A. McDERMOTT, Editors. 1984. *Aspects of Human Genetics: With Special Reference to X-Linked Disorders*. S. Karger, New York.

SANDBERG, A. A., Editor. 1983a. *Cytogenetics of the Mammalian X Chromosome. Part A: Basic Mechanisms of X Chromosome Behavior. (Progress and Topics in Cytogenetics, Vol. 3A.* Alan R. Liss, New York.

SANDBERG, A. A., Editor. 1983b. *Cytogenetics of the Mammalian X Chromosome. Part B: X Chromosome Anomalies and Their Clinical Manifestations. (Progress and Topics in Cytogenetics, Vol. 3B).* Alan R. Liss, New York.

SANDBERG, A. A., Editor. 1985. *The Y Chromosome, Part B: Clinical Aspects of Y Chromosome Abnormalities.* Alan R. Liss, New York.

SANDBERG, A. A. 1988. Chromosomal lesions and solid tumors. *Hospital Practice* 23(10): 93–106.

SANDBERG, A. A., G. F. KOEPF, T. ISHIHARA and T. S. HAUSCHKA. 1961. An XYY human male. *Lancet* II: 488–489.

SANDBERG, A. A., C. TURC-CAREL and R. M. GEMMILL. 1988. Chromosomes in solid tumors and beyond. *Cancer Research* 48: 1049–1059.

SANDHOFF, K., K. HARZER, W. WASSLE and H. JATZKEWITZ. 1971. Enzyme alterations and lipid storage in three variants of Tay-Sachs disease. *Journal of Neurochemistry* 18: 2469–2489.

SANDLER, I. and L. SANDLER. 1984. A conceptual ambiguity that contributed to the neglect of Mendel's paper. *History and Philosophy of the Life Sciences* 7: 3–70.

SANDLER, L. and F. HECHT. 1973. Genetic effects of aneuploidy. *AJHG* 25: 332–339.

SANGER, F. 1959. Chemistry of insulin. *Science* 129: 1340–1344.

SANGER, F., S. NICKLEN and A. R. COULSON. 1977. DNA sequencing with chain-terminating inhibitors. *PNAS* 74: 5463–5467.

SAPP, J. 1987. *Beyond the Gene: Cytoplasmic Inheritance and the Struggle for Authority in Genetics*. Oxford University Press, New York.

SASSA, S. and A. KAPPAS. 1981. Genetic, metabolic, and biochemical aspects of the porphyrias. *Advances in Human Genetics* 11: 121–231.

SAYRE, A. 1975. *Rosalind Franklin and DNA*. W. W. Norton, New York.

SCARR, S. 1987. Three cheers for behavior genetics: winning the war and losing our identity. *Behavior Genetics* 17: 219–228.

SCARR, S. and L. CARTER-SALTZMAN. 1983. Genetics and intelligence. In *Behavior Genetics: Principles and Applications*, edited by J. L. Fuller and E. C. Simmel, pp. 217–335. Lawrence Erlbaum Associates, Hillsdale, New Jersey.

SCHINZEL, A. 1984. *Catalogue of Unbalanced Chromosomes in Man*. DeGruyter, Berlin.

SCHLESSELMAN, J. J. 1979. How does one assess the risk of abnormalities from human in vitro fertilization? *American Journal of Obstetrics and Gynecology* 135: 135–148.

SCHNEIDER, E. D., Editor. 1985. *Questions about the Beginning of Life; Christian Appraisals of Seven Bioethical Issues*. Augsburg Publishing House, Minneapolis.

SCHNYDER, U. W. 1970. Inherited ichthyoses. *Archives of Dermatology* 102: 240–252.

SCHREINEMACHERS , D. M., P. K. CROSS and E. B. HOOK. 1982. Rates of trisomies 21, 18, 13 and other chromosome abnormalities in about 20,000 prenatal studies compared with estimated rates in live births. *Human Genetics* 61: 318–324.

SCHULL, W. J. and R. CHAKRABORTY, Editors. 1979. *Human Genetics: A Selection of Insights. Benchmark Papers in Genetics*, Vol. 10. Dowden, Hutchinson & Ross, Stroudsberg, Pennsylvania.

SCHULL, W. J. and J. V. NEEL. 1965. *The Effects of Inbreeding on Japanese Children*. Harper & Row, New York.

SCHULL, W. J. and J. V. NEEL. 1972. The effects of parental consanguinity and inbreeding in Hirado, Japan. V. Summary and interpretation. *AJHG* 24: 425–453.

SCHULL, W. J., M.I OTAKE and J. V. NEEL. 1981. Genetic effects of the atomic bombs: a reappraisal. *Science* 213: 1220–1227.

SCHWARTZ, D. C. and C. R. CANTOR. 1984. Separation of yeast chromosome-sized DNAs by pulsed field gradient gel electrophoresis. *Cell* 37: 67–75.

SCHWARTZ, R. H. 1985. T-lymphocyte recognition of antigen in association with gene products of the major histocompatibility complex. *Annual Review of Immunology* 3: 237–261.

SCIALLI, A. R. 1986. The reproductive toxicity of ovulation induction. *Fertility and Sterility* 45: 315–323.

SCIENTIFIC AMERICAN. 1988. *What Science Knows About AIDS*. Volume 259(Oct) [whole issue].

SCOTT, C. I. 1981. *Achondroplasia*. Human Growth Foundation, Minneapolis. [A free brochure.]

SCOTT, J. A., A. P. WALKER, D L. EUNPU and L. DJURDJINOVIC. 1988. Genetic counselor training: a review and considerations for the future. *AJHG* 42: 191–199.

SCOTT, M. P. 1985. Molecules and puzzles from the antennapedia homoeotic gene complex of *Drosophila*. *TIG* 1: 74–80.

SCOTT, M. P. and A. J. WEINER. 1984. Structural relationships among genes that control development: sequence homology between the *Antennapedia*, *Ultrabithorax*, and *fushi tarazu* loci of *Drosophila*. *PNAS* 81: 4115–4119.

SCRIVER, C. R., A. L. BEAUDET, W.S. SLY and D. VALLE, Editors. 1989. *The Metabolic Basis of Inherited Disease*, 6th Edition. McGraw-Hill, New York.

SCRIVER, C. R. and C. L. CLOW. 1980a. Phenylketonuria and other phenylalanine hydroxylation mutants in man. *Annual Review of Genetics* 14: 179–202.

SCRIVER, C. R. and C. L. CLOW. 1980b. Phenylketonuria: epitome of human biochemical genetics, parts 1 and 2. *NEJM* 303: 1336–1394, 1394–1400.

SCRIVER, C. R., S. KAUFMAN and S. L. C. WOO. 1988. Mendelian hyperphenylalaninemia. *Annual Review of Genetics* 22: 301–321.

SEAGOE, M. V. 1964. *Yesterday Was Tuesday, All Day and All Night*. Little, Brown and Co., Boston.

SEARLE, A. G. 1968. *Comparative Genetics of Coat Color in Mammals*. Academic Press, New York.

SEARLE, T. 1987. Radiation—the genetic risk. *TIG* 3: 152–157.

SEEGER, R. C. et al. 1985. Association of multiple copies of the N-*myc* oncogene with rapid progression of neuroblastomas. *NEJM* 313: 1111–1116.

SEEGMILLER, J. E. 1976. Inherited deficiency of hypoxanthine-guanine phosphoribosyltransferase in X-linked uric aciduria (the Lesch-Nyhan syndrome and its variants). *Advances in Human Genetics* 6: 75–163.

SEEGMILLER, J. E., F. M. ROSENBLOOM and W. N. KELLEY. 1967. Enzyme defect associated with a sex-linked human neurological disorder and excessive purine synthesis. *Science* 155: 1682–1684.

SEEMANOVÁ, E. 1971. A study of children of incestuous matings. *Human Heredity* 21: 108–128.

SEIBEL, M. M. 1988. A new era in reproductive technology: in vitro fertilization, gamete intrafallopian transfer, and donated gametes and embryos. *NEJM* 318: 828–834.

SEIDEL, G. E., JR. 1981. Superovulation and embryo transfer in cattle. *Science* 211: 351–358.

SERIO, M., M. MOTTA, M. ZANISI and L. MARTINI. 1984. *Sexual Differentiation: Basic and Clinical Aspects. Vol. 11, Serono Symposia Publications from Raven Press*. Raven Press, New York.

SEUANEZ, H. N. 1979. *The Phylogeny of Human Chromosomes*. Springer-Verlag, Berlin.

SEX, HORMONES, AND BEHAVIOR: CIBA FOUNDATION SYMPOSIUM 62 (NEW SERIES) 1979. Excerpta Medica, Amsterdam.

SHAPIRO, J. 1981. *Radiation Protection: A Guide for Scientists and Physicians*, 2nd Edition. Harvard University Press, Cambridge.

SHAPIRO, L. J. 1983. Steroid sulfatase deficiency and X-linked ichthyosis. In *The Metabolic Basis of Inherited Disease*, 5th Edition, edited by J. B. Stanbury et al., pp. 1027–1039. McGraw-Hill, New York.

SHAPIRO, L. J. 1985. Steroid sulfatase deficiency and the genetics of the short arm of the human X chromosome. *Advances in Human Genetics* 14: 331–381.

SHARON, N. 1977. Lectins. *SA* 236(June): 108–119.

SHARON, N. 1980. Carbohydrates. *SA* 243 (Nov): 90–116.

SHARP, P. A. 1983. Conversion of RNA to DNA in mammals: *Alu*-like elements and pseudogenes. *Nature* 301: 471–472.

SHARP, P. A. 1988. RNA splicing and genes. *JAMA* 260: 3035–3041.

SHEER, D. 1986. Chromosomes and cancer. In *Introduction to the Cellular and Molecular Biology of Cancer*, edited by L. M. Franks and N. M. Teich, pp. 229–250. Oxford University Press, Oxford.

SHERMAN, S. L. et al. 1985. Further segregation analysis of the fragile X syndrome with special reference to transmitting males. *Human Genetics* 69: 289–299.

SHERRINGTON, R. et al. 1988. Localization of a susceptibility locus for schizophrenia on chromosome 5. *Nature* 336: 164–167.

SHETTLES, L. B. and D. RORVIK. 1984. *How to Choose the Sex of Your Baby*. Doubleday & Company, Garden City, New York.

SHIH, C. and R. A. WEINBERG. 1982. Isolation of a transforming sequence from a human bladder carcinoma cell line. *Cell* 29: 161–169.

SHKLOVSKII, I. S. and C. SAGAN. 1966. *Intelligent Life in the Universe*. Holden-Day, San Francisco.

SHOWS, T. B., A. Y. SAKAGUCHI and S. L. NAYLOR. 1982. Mapping the human genome, cloned genes, DNA polymorphisms, and inherited disease. *Advances in Human Genetics* 12: 341–452.

SHOWS, T. B. et al. 1987. Guidelines for human gene nomenclature; an international system for human gene nomenclature. ISGN, 1987). *Cytogenetics and Cell Genetics* 46: 11–28.

SILVERMAN, L. J., W. N. KELLEY and T. D. PALELLA. 1987. Genetic analysis of human hypoxanthine-guanine phosphoribosyltransferase deficiency. *Enzyme* 38: 36–44.

SIMCHEN, G. 1978. Cell cycle mutants. *Annual Review of Genetics* 12: 161–191.

SIMMONS, K. 1984. Physicians continue to study cause(s) of "bubble" boy's death. *JAMA* 251: 1929–1931.

SIMPSON, E. et al. 1987. Separation of the genetic loci for the H-Y antigen and for testis determination on human Y chromosome. *Nature* 326: 876–878.

SIMPSON, J. L. 1982. Abnormal sexual differentiation in humans. *Annual Review of Genetics 16*: 193–224.

SIMPSON, J. L. 1983. *Genetics in Obstetrics and Gynecology*. Grune & Stratton, Orlando, Florida.

SIMPSON, J. L. 1986. Repetitive spontaneous abortion. In *Perinatal Genetics: Diagnosis and Treatment*, edited by I. Porter et al., pp. 41–69.

SINCLAIR, A. H. et al. 1988. Sequences homolgous to *ZFY*, a candidate human sex-determining gene, are autosomal in marsupials. *Nature* 336: 780–783.

SINEX, F. M. and C. R. MERRIL, Editors. 1982. *Alzheimer's Disease, Down's Syndrome, and Aging*. Kroc Foundation Conference, Oct. 12–16, 1981. Vol. 396, New York Academy of Sciences.

SINISCALCO, M., G. FILIPPI and B. LATTE. 1964. Recombination between protan and deutan genes; data on their relative positions in respect of the G6PD locus. *Nature* 204: 1062–1064.

SIWOLOP, S. and M. MOHS. 1985. The war on Down syndrome. *Discover* 6(Feb): 66–69.

SLEISENGER, M. H. 1982. Management [of malabsorption]. In *Cecil Textbook of Medicine*, 16th Edition, Vol. 1, edited by J. B. Wyngaarden and L. H. Smith, Jr., pp. 690–703. W. B. Saunders, Philadelphia.

SLOAN, H. R. and D. S. FREDRICKSON. 1972. G_{M2} gangliosidosis: Tay-Sachs disease. In *The Metabolic Basis of Inherited Disease*, 3rd Edition, edited by J. B. Stanbury et al., pp. 615–638. McGraw-Hill, New York.

SMITH, G. F., Editor. 1985. *Molecular Structure of the Number 21 Chromosome and Down Syndrome. The Annals of the New York Academy of Sciences*, Vol. 450.

SMITH, L. D., B. S. BAKER and M. GATTI. 1985. Mutations in genes encoding essential mitotic functions in *Drosophila melanogaster*. *Genetics* 110: 647–670.

SMITH, L. M. et al. 1986. Fluorescence detection in automated DNA sequence analysis. *Nature* 321: 674–679.

SMITHIES, O., R. G. GREGG, S. S. BOGGS, M. A. KORALEWSKI and R. S. KUCHERLAPATI. 1985. Insertion of DNA sequences into the human chromosomal β-globin locus by homologous recombination. *Nature* 317: 230–234.

SMOLEV, J. and K. A. FORREST. 1984. Male infertility. In *Men's Reproductive Health*, edited by J. M. Swanson and K. A. Forrest, pp. 162–178. Springer-Verlag, New York.

SOKAL, R. R. and F. J. ROHLF. 1987. *Introduction to Biostatistics*, 2nd Edition. W. H. Freeman, New York.

SOLTER, D. 1988. Differential imprinting and expression of maternal and paternal genomes. *Annual Review of Genetics* 22: 127–146.

SØRENSEN, T. I. A., G. G. NIELSEN, P. K. ANDERSEN and T. W. TEASDALE. 1988. Genetic and environmental influences on premature death in adult adoptees. *NEJM* 318: 727–732.

SOREQ, H., R. ZAMIR, D. ZEVIN-SONKIN and H. ZAKUT. 1987. Human cholinesterase genes localized by hybridization to chromosomes 3 and 16. *Human Genetics* 77: 325–328.

SORSBY, A. 1958. Noah, an albino. *British Medical Journal* 2: 1587.

SOUTHERN, E. M. 1975. Detection of specific sequences among DNA fragments separated by gel electrophoresis. *Journal of Molecular Biology* 98: 503–517.

SOYFER, V. N. 1989. New light on the Lysenko era. *Nature* 339: 415–420.

SPARKES, R. S. and F. F. DE LA CRUZ, Editors. 1984. *Research Perspectives in Cytogenetics (National Institute of Child Health and Human Development Series).* University Park Press, Baltimore.

SPEED, R. M. 1988. The possible role of meiotic pairing anomalies in the atresia of human fetal oocytes. *Human Genetics* 78: 260–266.

SPIESS, E. B. 1977. *Genes in Populations*. John Wiley & Sons, New York.

SRB, A. M., R. D. OWEN and R. S. EDGAR, Editors. 1970. *Facets of Genetics: Readings from Scientific American*. W. H. Freeman, San Francisco.

ST GEORGE-HYSLOP, P. H. et al. 1987. The genetic defect causing familial Alzheimer's disease maps on chromosome 21. *Science* 235: 885–890.

STADLER, L. J. 1954. The gene. *Science* 120: 811–819. [Reprinted in Peters (1959).]

STAFF OF RESEARCH AND EDUCATION ASSOCIATION. 1982. *Behavioral Genetics*. Research and Education Association, New York.

STAHL, F. W. 1987. Genetic recombination. *SA* 256(Feb): 90–101.

STAMATOYANNOPOULOS, G., A. W. NIENHUIS, P. LEDER and P. W. MAJERUS. 1987. *The Molecular Basis of Blood Diseases*. W. B. Saunders, Philadelphia.

STANBURY, J. B., J. B. WYNGAARDEN, D. S. FREDRICKSON, J. L. GOLDSTEIN and M. S. BROWN. 1983. *The Metabolic Basis of Inherited Disease*, 5th Edition. McGraw-Hill, New York.

STANSFIELD, W. D. 1983. *Schaum's Outline of Theory and Problems of Genetics*, 2nd Edition. McGraw-Hill, New York.

STEBBINS, G. L. and F. J. AYALA. 1985. The evolution of Darwinism. *SA* 253(July): 72–82.

STEHELIN, D., H. E. VARMUS, J. M. BISHOP and P. K. VOGT. 1976. DNA related to the transforming gene(s) of avian sarcoma viruses is present in normal avian DNA. *Nature* 260: 170–173.

STEINBROOK, R. 1986. In California, voluntary mass prenatal screening. *Hastings Center Report* 16(5): 5–7.

STEINMETZ, M. 1986. Genes of the immune system. In *Genetic Engineering*, Vol. 5, edited by P. W. J. Rigby, pp 117–158; Academic Press, London.

STEITZ, J. A. 1988. "Snurps." *SA* 258(June): 56–63.

STENT, G. S. 1972. Prematurity and uniqueness in scientific discovery. *SA* 227(Dec): 84–93.

STEPHENS, F. E. and F. H. TYLER. 1951. Studies in disorders of muscle. V. The inheritance of childhood progressive muscular dystrophy in 33 kindreds. *AJHG* 3: 111–125.

STEPTOE, P. C. and R. G. EDWARDS. 1976. Reimplantation of a human embryo with subsequent tubal pregnancy. *Lancet* I: 880–882.

STEPTOE, P. C. and R. G. EDWARDS. 1978. Birth after the reimplantation of a human embryo.. *Lancet* II: 366.

STERN, C. 1965. Mendel and human genetics. *Proceedings of the American Philosophical Society* 109: 216–226.

STERN, C. and E. R. SHERWOOD, Editors. 1966. *The Origin of Genetics: A Mendel Source Book*. W. H. Freeman and Company, San Francisco.

STEWART, D. A., Editor. 1982. *Children with Sex Chromosome Aneuploidy: Follow-up Studies. Birth Defects: Original Article Series*: Vol. 18, No. 4.

STEWART, G. D., T. J. HASSOLD and D. M. KURNIT. 1988. Trisomy 21: molecular and cytogenetic studies of nondisjunction. *Advances in Human Genetics* 17: 99–140.

STINE, G. J. 1977. *Biosocial Genetics.* Macmillan, New York.

STORB, R. and E. D. THOMAS. 1983. Allogeneic bone-marrow transplantation. *Immunological Reviews* 71: 77–102.

STOUT, J. T. and C. T. CASKEY. 1985. HPRT: Gene structure, expression and mutation. *Annual Review of Genetics* 19: 127–148.

STOUT, J. T., H. Y. CHEN, J. BRENNAND, C. T. CASKEY and R. L. BRINSTER. 1985. Expression of human HPRT in the central nervous system of transgenic mice. *Nature* 317: 250–252.

STRAY-GUNDERSON, K., Editor. 1986. *Babies with Down Syndrome: A New Parent's Guide*. Woodbine House, Kensington, Maryland.

STREISSGUTH, A. P. 1986. The behavioral teratology of alcohol: performance, behavioral and intellectual deficits in prenatally exposed children. In *Alcohol and Brain Development*, edited by J. R. West, pp. 3–44. Oxford University Press, Oxford.

STRELKAUSKAS, A. J., Editor. 1987. *Human hybridomas: diagnostic and therapeutic applications.* Marcel Dekker, Inc., New York.

STRICKBERGER, M. W. 1985. *Genetics*, 3rd Edition. Macmillan, New York.

STRYER, L. 1988. *Biochemistry*, 3rd Edition. W. H. Freeman and Co., New York.

STURTEVANT, A. H. 1923. Inheritance of direction of coiling in limnaea. *Science* 58: 269–270.

STURTEVANT, A. H. 1965. *A History of Genetics*. Harper & Row, New York.

SÜDHOF, T. C. et al. 1985a. Cassette of eight exons shared by genes for LDL receptor and EGF precursor. *Science* 228: 893–895.

SÜDHOF, T. C., J. L. GOLDSTEIN, M. S. BROWN and D. W RUSSELL. 1985b. The LDL receptor gene: a mosaic of exons shared with different proteins. *Science* 228: 815–822.

SULSTON, J. E., E.SCHIERENBERG, J. G. WHITE, J. N. THOMSON and G. VON EHRENSTEIN. 1983. The embryonic cell lineage of the nematode *Caenorhabditis elegans. Developmental Biology* 100: 64–119.

SUMMERS, P. M. 1986. Collection, storage and use of mammalian embryos. *International Zoo Yearbook* 24/25: 131–138.

SUMMITT, R. L. 1973. Abnormalities of the autosomes. A *Pediatric Annals* Reprint. Insight Publishing Company, New York.

SUMMITT, R. L. 1981. Abnormalities of the sex chromosomes. In *Genetic Issues in Pediatric and Obstetric Practice*, edited by M. M. Kaback, pp. 63–89. Year Book Medical, Chicago.

SUMNER, A. T. and J. A. ROBINSON. 1976. A difference in dry mass between the heads of X- and Y-bearing human spermatozoa. *Journal of Reproduction and Fertility* 48: 9–15.

SUN, M. 1981. Inmates sue to keep research in prisons. *Science* 212: 650–651.

SUN, M. 1985. Gene-spliced hormone for growth approved. *Science* 230: 523 only.

SURANI, M. A. H. 1986. Evidences and consequences of differences between maternal and paternal genomes during embryogenesis in mouse. In *Experimental Approaches to Mammalian Embryonic Development*, edited by J. Rossant and R. A. Pedersen, pp. 401–435. Cambridge University Press, Cambridge.

SURANI, M. A. H., S. C. BARTON and M. L. NORRIS. 1986. Nuclear transplantation in the mouse: heritable differences between parental genomes after activation of the embryonic genome. *Cell* 45: 127–136.

SUTHERLAND, G. R. 1977. Fragile sites on human chromosomes: demonstration of their dependence on the type of tissue culture medium. *Science* 197: 265–266.

SUTHERLAND, G. R. 1983. The fragile X chromosome. *International Review of Cytology* 81: 107–143.

SUTHERLAND, G. R. and F. HECHT, Editors. 1985. *Fragile Sites on Human Chromosomes.* Oxford University Press, New York.

SUTHERLAND, G. R., A. R. MURCH, A. J. GARDENER, R. F. CARTER and C. WISEMAN. 1976. Cytogenetic survey of a hospital for the mentally retarded. *Human Genetics* 34: 231–245.

SUTTON, W. S. 1902. On the morphology of the chromosome group in Brachystola magna. *Biological Bulletin* 4: 24–39.

SUTTON, W. S. 1903. The chromosomes in heredity. *Biological Bulletin* 4: 231–251. [Reprinted in Peters (1959), in Voeller (1968), and in Gabriel and Fogel (1955).]

SVERDRUP, A. 1922. Postaxial polydactylism in six generations of a Norwegian family. *Journal of Genetics* 12: 217–240.

SWIFT, H. 1974. The organization of genetic material in eukaryotes: progress and prospects. *CSHS* 38: 963–979.

SWIFT, M., P. J. REITNAUER, D. MORRELL and C. L. CHASE. 1987. Breast and other cancers in families with ataxia-telangiectasia. *NEJM* 316: 1289–1294.

SZORADY, I. 1973. *Pharmacogenetics: Principles and Paediatric Aspects.* Akademiai Kiado, Publishing House of the Hungarian Academy of Sciences, Budapest.

TABAKOFF, B. et al. 1988. Differences in platelet enzyme activity between alcoholics and nonalcoholics. *NEJM* 318: 134–139.

TAKIZAWA, T., I.-Y. HUANG, T. IKUTA and A. YOSHIDA. 1986. Human glucose-6-phosphate dehydrogenase: primary structure and cDNA cloning. *PNAS* 83: 4157–4161.

TALAMO, R. C., B. J. ROSENSTEIN and R. W. BERNINGER. 1983. Cystic fibrosis. In *The Metabolic Basis of Inherited Disease*, 5th Edition, edited by J. B. Stanbury et al. McGraw-Hill, New York.

TANZI, R. E. et al. 1987a. Amyloid b protein gene: cDNA, mRNA distribution, and genetic linkage near the Alzheimer locus. *Science* 235: 880–884.

TANZI, R. E. et al. 1987b. The genetic defect in familial Alzheimer's disease is not tightly linked to the amyloid b-protein gene. *Nature* 329: 156–157.

TASSET, D. M., J. A. HARTZ and F.-T. KAO. 1988. Isolation and analysis of DNA markers specific to human chromosome 15. *AJHG* 42: 854–866.

TAUSSIG, L. M. 1984. *Cystic Fibrosis.* Thieme-Stratton Inc., New York.

TAY, W. 1881. Symmetrical changes in the region of the yellow spot in each eye of an infant. *Transactions of the Ophthalmological Society, UK* 1: 55–57.

TAYLOR, S. M. and P. A. JONES. 1982. Changes in phenotypic expression in embryonic and adult cells treated with 5-azacytidine. *Journal of Cellular Physiology* 111: 187–194.

TEICH, N. M. 1986. Oncogenes and cancer. In *Introduction to the Cellular and Molecular Biology of Cancer*, edited by L. M. Franks and N. M. Teich, pp. 200–228. Oxford University Press, Oxford.

TEICHLER-ZALLEN, D. and R. A. DOHERTY. 1980. Amniotic fluid secretor typing: validation for use in prenatal prediction of myotonic dystrophy. *Clinical Genetics* 18: 257–267.

TEITELBAUM, M. S., Editor. 1976. *Sex Differences: Social and Biological Perspectives.* Anchor Books, Garden City, New York.

TEMIN, H. M. 1964. Nature of the provirus of Rous sarcoma. *National Cancer Institute Monographs* 17: 557–570.

TEMIN, H. M. 1972. RNA-directed DNA synthesis. *SA* 226(Jan): 24–33.

TERRY, R. D. and M. WEISS. 1963. Studies in Tay-Sachs disease. II. Ultrastructure of the cerebrum. *Journal of Neuropathology and Experimental Neurology* 22: 18–55.

TERZAGHI, E. A., A. S. WILKINS and D. PENNY, Editors. 1984. *Molecular Evolution, an annotated reader.* Jones and Bartlett Publishers, Boston.

THERMAN, E. 1986. *Human Chromosomes: Structure, Behavior, Effects*, 2nd Edition. Springer-Verlag, New York.

THOMAS, K. R. and M. R. CAPECCHI. 1987. Site-directed mutagenesis by gene targeting in mouse embryo-derived stem cells. *Cell* 51: 503–512.

THOMAS, L. 1984. Foreword to *Cancer Today: Origins, Prevention, and Treatment* by Leslie Roberts. National Academy Press, Washington, D. C.

THOMAS, R. MCG., JR. 1988. Mystery of the Babe. *The New York Times*, June 12, p. S1.

THOMPSON, C. E. 1978. Reproduction in Duchenne dystrophy. *Neurology* 28: 1034–1047.

THOMPSON, M. W. 1980. Genetics of cystic fibrosis. In *Perspectives in Cystic Fibrosis: Proceedings of the 8th International Cystic Fibrosis Congress held in Toronto, Canada May 26–30, 1980*, edited by J. M. Sturgess, pp. 281–291. Canadian Cystic Fibrosis Foundation, Toronto..

THORPE, N. O. 1984. *Cell Biology.* John Wiley & Sons, New York.

TIWARI, J. L. and P. I. TERASAKI. 1985. *HLA and Disease Associations.* Springer-Verlag, New York.

TJIO, J. H. and A. LEVAN. 1956. The chromosome number of man. *Hereditas* 42: 1–6.

TONEGAWA, S. 1983. Somatic generation of antibody diversity. *Nature* 302: 575–581.

TONEGAWA, S. 1985. The molecules of the immune system. *SA* 253(Oct): 122–131.

TONEGAWA, S. 1988. Somatic generation of immune diversity. (Nobel Lecture) *Bioscience Reports* 8: 3–26.

TOURIAN, A. and J. B. SIDBURY. 1983. Phenylketonuria and hyperphenylalaninemia. In *The Metabolic Basis of Inherited Disease*, 5th Edition, edited by J. B. Stanbury et al., pp. 270–286. McGraw-Hill, New York.

TREVES, F. 1885. A case of congenital deformity. *Transactions of the Pathological Society of London* 36: 494–498.

TROUNSON, A. 1986. Preservation of human eggs and embryos. *Fertility and Sterility* 46: 1–12.

TRYON, R. C. 1940. Genetic differences in maze-learning ability in rats. In *The Thirty-ninth Yearbook of the National Society for the Study of Education*, edited by G. M. Whipple, pp. 111–119 and 154–155. Public School Publishing Company, Bloomington, Illinois.

TURNBULL, A. 1988. Woman enough for the games? *New Scientist* 119(15 Sept): 61–64.

TURNER, G. and P. JACOBS. 1983. Marker (X)-linked mental retardation. *Advances in Human Genetics* 13: 83–112, 300.

TURNER, H. H. 1938. A syndrome of infantilism, congenital webbed neck, and cubitus valgus. *Endocrinology* 23: 566–574.

TURPIN, R., J. LEJEUNE, J. LAFOURCADE, P.-L. CHIGOT and C. SALMON. 1961. Presomption de monozygotisme en depit d'un dimorphisme sexuel: sujet masculin XY et sumet neutro haplo X. *Academie des Sciences, Paris: Comptes Rendus Hebdomadaires des Seances* 252: 2945–2946.

UCHIDA, I. A. 1977. Maternal radiation and trisomy 21. In *Population Cytogenetics: Studies in Humans,* edited by E. B. Hook and I. H. Porter, pp. 285–299. Academic Press, New York.

UCHIDA, I. A., R. HOLUNGA and C. LAWLER. 1968. Maternal radiation and chromosomal aberrations. *Lancet* II: 1045–1049.

UCHIDA, I. A. and E. M. JOYCE. 1982. Activity of the fragile X in heterozygous carriers. *AJHG* 34: 286–293.

UCHIDA, I. A. and H. C. SOLTAN. 1975. Dermatoglyphics in medical genetics. In *Endocrine and Genetic Diseases of Childhood and Adolescence*, edited by L. Gardner, pp. 657–674. W. B. Saunders Company, Philadelphia.

UEDA, K. and R. YANAGIMACHI. 1987. Sperm chromosome analysis as a new system to test human X- and Y-sperm separation. *Gamete Research* 17: 221–228.

UNDERWOOD, L. E. 1984. Report of the conference on uses and possible abuses of biosynthetic human growth hormone. *NEJM* 311: 606–608.

UNEP [United Nations Environment Programme]. 1985. *Radiation: Doses, Effects, Risks.* United Nations, New York.

UNSCEAR [United Nations Scientific Committee on the Effects of Atomic Radiation]. 1982. *Ionizing Radiation: Sources and Biological Effects.* United Nations, New York.

UNSCEAR [United Nations Scientific Committee on the Effects of Atomic Radiation]. 1986. *Genetic and Somatic Effects of Ionizing Radiation.* United Nations, New York.

UPTON, A. C. 1982. The biological effects of low-level ionizing radiation. *SA* 246(Feb): 41–49.

VALENTINE, G. H. 1975. *The Chromosome Disorders*, 3rd Edition. J. B. Lippincott, Philadelphia.

VALENTINE, G. H. 1979. The growth and development of six XYY children. *Birth Defects: Original Article Series*, Vol. XV, No. 1, pp. 175–190.

VALENTINE, G. H. 1986. *The Chromosomes and Their Disorders: An Introduction for Clinicians*, 4th Edition. W. Heineman Medical Books, London. [This edition does not include a quote that opens Chapter 9.]

VALLET, H. L. and I. H. PORTER, Editors. 1979. *Genetic Mechanisms of Sexual Development. Birth Defects Institute Symposium Series* No. 7. Academic Press, New York.

VAN BLERKOM, J. and P. M. MOTTA, Editors. 1984. *Ultrastructure of Reproduction: Gametogenesis, Fertilization, and Embryogenesis.* Martinus Nijhoff Publishers, Boston.

VAN BROECKHOVEN, C. et al. 1987. Failure of familial Alzheimer's disease to segregate with the A4-amyloid gene in several European families. *Nature* 329: 153–155.

VAN DEN BERGHE, P. L. 1983. Human inbreeding avoidance: culture in nature. *The Behavioral and Brain Sciences* 6: 91–123.

VAN DER PLOEG, L. H. T. and R. D. FLAVELL. 1980. DNA methylation in the human γ-δ-β-globin gene. *Cell* 19: 947–958.

VAN HEYNINGEN, V. and D. J. PORTEUS. 1986. Mapping a chromosome to find a gene. *TIG* 2: 4–5.
VAN NIEKERK, W. A. 1974. *True hermaphroditism: Clinical, Morphologic and Cytogenetic Aspects*. Harper & Row, Hagerstown, Maryland.
VAN NIEKERK, W. A. and A. E. RETIEF. 1981. The gonads of human true hermaphrodites. *Human Genetics* 58: 117–122.
VAN OMMEN, G. J. B. et al. 1986. A physical map of 4 million bp around the Duchenne muscular dystrophy gene on the human X chromosome. *Cell* 47: 499–504.
VARMUS, H. E. 1988. Retroviruses. *Science* 240: 1427–1435.
VELLUTINO, F. R. 1987. Dyslexia. *SA* 256(Mar): 34–41.
VERP, M. S. and J. L. SIMPSON. 1985. Amniocentesis for cytogenetic studies. In *Human Prenatal Diagnosis*, edited by K. Filkins and J. F. Russo, pp. 13–48. Marcel Dekker, New York.
VIG, B. K. and A. A. SANDBERG, Editors. 1987. *Aneuploidy, Part A: Incidence and Etiology*. Alan R. Liss, new York.
VIRCHOW, R. 1858. *Cellular Pathology as Based upon Physiological and Pathological Histology*. Translated from the 2nd German edition by Frank Chance. Reprinted in 1971 by Dover Publications, New York.
VOELLER, B. R., Editor. 1968. *The Chromosome Theory of Inheritance: Classic Papers in Development and Heredity*. Appleton-Century-Crofts, New York.
VOGEL, F. 1983. Mutation in man. In *Principles and Practice of Medical Genetics*, Vol. 1, edited by A. E. H. Emery and D. L. Rimoin, pp. 26–48. Churchill Livingstone, Edinburgh.
VOGEL, F. and A. G. MOTULSKY. 1986. *Human Genetics: Problems and Approaches*, 2nd Edition. Springer-Verlag, Berlin.
VOGEL, F. and R. RATHENBERG. 1975. Spontaneous mutation in man. *Advances in Human Genetics* 5: 223–318.
VON WETTSTEIN, D., S. W. RASMUSSEN and P. B. HOLM. 1984. The synaptonemal complex in genetic segregation. *Annual Review of Genetics* 18: 331–413.

WABER, D. P. 1985. The search for biological correlates of behavioral sex differences in humans. In *Human Sexual Dimorphism*, edited by J. Ghesquiere et al., pp. 257–282. Taylor and Francis, London.
WAINSCOAT, J. 1987. Human evolution: out of the garden of Eden. *Nature* 325: 13.
WALLACE, D. C. 1986. Mitochondrial genes and disease. *Hospital Practice* 21: 77–92.
WALLACE, D. C. et al. 1988a. Familial mitochondrial encephalomyopathy (MERRF): genetic, pathophysiological and biochemical characterization of a mitochondrial DNA disease. *Cell* 55: 601–610.
WALLACE, D. C. et al. 1988b. Mitochondrial DNA mutation associated with Leber's hereditary optic neuropathy. *Science* 242: 1427–1430.
WALLACE, I. and A. WALLACE. 1978. *The Two*. Simon and Schuster, New York.
WANG, A.-L. et al. 1981. Regional gene assignment of human porphobilinogen deaminase and esterase A(4) to chromosome 11q23-11qter. *PNAS* 78: 5734–5738.
WARBURTON, D. 1987. Reproductive loss: how much is preventable? *NEJM* 316: 158–160.
WARBURTON, D., J. KLINE, Z. STEIN and B. STROBINO. 1986. Cytogenetic abnormalities in spontaneous abortions of recognized conceptions. In *Perinatal Genetics: Diagnosis and Treatment*, edited by I. Porter et al., pp. 23–40. Academic Press, Orlando, Florida.
WAREHAM, K. A., M. F. LYON, P. H. GLENISTER and E. D. WILLIAMS. 1987. Age related reactivation of an X-linked gene. *Nature* 327: 725–727.
WASSARMAN, P. M. 1987. The biology and chemistry of fertilization. *Science* 235: 553–560.
WASSARMAN, P. M. 1988. Fertilization in mammals. *SA* 259(Dec): 78–84.
WATKINS, W. M. 1980. Biochemistry and genetics of the ABO, Lewis, and P blood group systems. *Advances in Human Genetics* 10: 1–136, and 379–385.
WATSON, J. D. 1968. *The Double Helix*. Atheneum, New York.
WATSON, J. D. and F. H. C. CRICK. 1953. Molecular structure of nucleic acids: a structure for deoxyribose nucleic acid. *Nature* 171: 737–738. [Reprinted in Peters (1959) and Freifelder (1978).]
WATSON, J. D., N. H. HOPKINS, J. W. ROBERTS, J. A. STEITZ and A. M. WEINER. 1987. *Molecular Biology of the Gene*, 4th Edition. Volume I: General Principles. Volume II: Specialized Aspects. Benjamin/Cummings Publishing Company, Menlo Park, California.
WATSON, J. D. and J. TOOZE. 1981. *The DNA Story: A Documentary History of Gene Cloning*. W. H. Freeman and Company, New York.
WATSON, J. D., J. TOOZE and D. T. KURTZ. 1983. *Recombinant DNA: A Short Course*. W. H. Freeman and Company, New York.
WEATHERALL, D. J. 1985. *The New Genetics and Clinical Practice*, 2nd Edition. Oxford University Press, Oxford.
WEBSTER v. REPRODUCTIVE HEALTH SERVICES. 1989. *United States Reports: Cases Adjudged in the Supreme Court* (in press).
WEBSTER, C., L. SILBERSTEIN, A. P. HAYS and H. M. BLAU. 1988. Fast muscle fibers are preferentially affected in Duchenne muscular dystrophy. *Cell* 52: 503–513.
WECHSLER, D. 1981. *WAIS-R Manual; Wechsler Adult Intelligence Scale, Revised*. The Psychological Corporation, Cleveland, Ohio.
WECHSLER, R. 1986. Unshackled from diabetes. *Discover* 7(Sept): 77–85.
WEIGALL, A. 1924. *The Life and Times of Cleopatra, Queen of Egypt*. G. P. Putnam's Sons, New York.
WEINBERG, R. A. 1983. A molecular basis of cancer. *SA* 249(Nov): 126–142.
WEINBERG, R. A. 1985. The action of oncogenes in the cytoplasm and nucleus. *Science* 230: 770–776.
WEINBERG, R. A. 1988. Finding the anti-oncogene. *SA* 259(Sept): 44–51.
WEISS, M. C. and H. GREEN. 1967. Human–mouse hybrid cell lines containing partial complements of human chromosomes and functioning human genes. *PNAS* 58: 1104–1111.
WEISSMAN, B. E. et al. 1987. Introduction of a normal human chromosome 11 into a Wilms' tumor cell line controls its tumorigenic expression. *Science* 236: 175–180.
WELLS, R. and M. JENNINGS. 1967. X-linked ichthyosis and ichthyosis vulgaris. Clinical and genetic distinctions in a second series of families. *JAMA* 202: 485–488.
WELSH, M. J. and C. M. LIEDTKE. 1986. Chloride and potassium channels in cystic fibrosis airway epithelia. *Nature* 322: 467–470.
WESTHEIMER, F. H. 1988. Education of the next generation of nonscientists. *Chemical & Engineering News* 66(July 4): 32–38.
WETHERS, D. L. and R. GROVER. 1986. Screening the newborn for sickle cell disease: Is it worth the effort? In *Genetic Disease, Screening and Management*, edited by T. P. Carter and A. M. Willey, pp. 123–136. Alan R. Liss, New York.
WHITE, P. C., M. I. NEW and B. DUPONT. 1987a. Congenital adrenal hyperplasia (first of two parts). *NEJM* 316: 1519–1524.
WHITE, P. C., M. I. NEW and B. DUPONT. 1987b. Congenital adrenal hyperplasia (second of two parts). *NEJM* 316: 1580–1586.
WHITE, R. and J.-M. LALOUEL. 1987. Investigation of genetic linkage in human families. *Advances in Human Genetics* 16: 121–228.
WHITE, R. and J.-M. LALOUEL. 1988. Chromosome mapping with DNA markers. *SA* 258 (Feb): 40–48.
WHYTE, P. et al. 1988. Association between an oncogene and an anti-oncogene: the adenovirus E1A proteins bind to the retinoblastoma gene product. *Nature* 334: 124–129.
WIENER, A. S. 1943. *Blood Groups and Transfusions*, 3rd Edition. Charles C. Thomas, Springfield, Illinois.
WIERINGA, B. et al. 1985. Complex glycerol kinase syndrome explained as X-chromosomal deletion. *Clinical Genetics* 27: 522–523.
WIGGS, J. et al. 1988. Prediction of the risk of hereditary retinoblastoma, using DNA polymorphisms within the retinoblastoma gene. *NEJM* 318: 151–157.
WILCOX, A. J. et al. 1988. Incidence of early loss of pregnancy. *NEJM* 319: 189–194.
WILKINS, A. S. 1986. *Genetic Analysis of Animal Development*. John Wiley & Sons, New York.
WILLIAMS, R. R. 1988. Nature, nurture, and family predisposition. *NEJM* 318: 769–771.
WILLIAMS, S. A., B. E. SLATKO, L. S. MORAN and S. M. DeSIMONE. 1986. Sequencing in the fast lane: a rapid protocol for [$\alpha^{35}S$] dATP dideoxy DNA sequencing. *BioTechniques* 4: 138–147.
WILLIS, A. E. and T. LINDAHL. 1987. DNA ligase I deficiency in Bloom's syndrome. *Nature* 325: 355–357.
WILSON, A. C. 1985. The molecular basis of evolution. *SA* 253(Oct): 164–173.
WILSON, E. B. 1911. The sex chromosomes. *Arch. Mikrobiol. Anat.* 77: 249–271.
WILSON, J. D. and J. E. GRIFFIN. 1985. Mutations that impair androgen action. *TIG* 1: 335–339.

WILSON, J. D., J. E. GRIFFIN, M. LESHIN and P. C. MACDONALD. 1983. The androgen resistance syndromes: 5-reductase deficiency, testicular feminization, and related disorders. In *The Metabolic Basis of Inherited Disease*, 5th Edition, edited by J. B. Stanbury et al., pp. 1001–1026. McGraw-Hill, New York.

WILSON, J., F. W. GEORGE and J. E. GRIFFIN. 1981. The hormonal control of sexual development. *Science* 211: 1278–1284.

WILSON, J. M. et al. 1986. A molecular survey of hypoxanthine-guanine phosphoribosyltransferase deficiency in man. *Journal of Clinical Investigations* 77: 188–195.

WITKIN, H. A. et al. 1976. Criminality in XYY and XXY men. *Science* 193: 547–555.

WITKOP, C. J., JR. 1971. Albinism. *Advances in Human Genetics* 2: 61–142.

WITKOP, C. J. JR., W. C. QUEVEDO, JR. and T. B. FITZPATRICK. 1983. Albinism and other disorders of pigment metabolism. In *The Metabolic Basis of Inherited Disease*, 5th Edition, edited by J. B. Stanbury et al., pp. 301–346. McGraw-Hill, New York.

WITKOWSKI, J. A. 1988. The molecular genetics of Duchenne muscular dystrophy: the beginning of the end? *TIG* 4: 27–30.

WOLF, F. M., L. D. GRUPPEN and J. E. BILLI. 1985. Differential diagnosis and the competing-hypotheses heuristic; a practical approach to judgment under uncertainty and Bayesian probability. *JAMA* 253: 2858–2862.

WOLF, S. F. and B. R. MIGEON. 1982. Studies of X chromosome DNA methylation in normal human cells. *Nature* 295: 667–671.

WOO, S. L. C. 1986. Prenatal diagnosis and carrier detection of classical phenylketonuria. In *Perinatal Genetics: Diagnosis and Treatment*, Edited by I. Porter et al., pp. 79–93. Academic Press, New York.

WOO, S. L. C., A. G. DiLELLA, J. MARVIT and F. D. LEDLEY. 1987. Molecular basis of phenylketonuria and recombinant DNA strategies for its therapy. *Enzyme* 38: 207–213.

WORLD HEALTH ORGANIZATION. 1973. *Pharmacogenetics: Report of a WHO Scientific Group. World Health Organization Technical Report Series* No. 524. United Nations, New York.

WORTON, R. G., P. N. RAY, S. BODRUG and M. W. THOMPSON. 1986. Analysis of an X–autosome translocation responsible for X-linked muscular dystrophy. *CSHS* 51 (Part 1): 345–348.

WORTON, R. G. and M. W. THOMPSON. 1988. Genetics of Duchenne muscular dystrophy. *Annual Review of Genetics* 22: 601–629.

WRAMSBY, H., K. FREDGA and P. LIEDHOLM. 1987. Chromosome analysis of human oocytes recovered from preovulatory follicles in stimulated cycles. *NEJM* 316: 121–124

WRIGHT, S. 1922. Coefficients of inbreeding and relationship. *American Naturalist* 56: 330–338.

WRIGHT, S. 1930. *The Genetical Theory of Natural Selection* [review]. *Journal of Heredity* 21: 349–356.

WRIGHT, S. 1966. Mendel's ratios. In *The Origin of Genetics; A Mendel Sourcebook*, edited by C. Stern and E. Sherwood, pp. 173–175. W. H. Freeman and Company, San Francisco.

WRIGHT, S. 1982. The shifting balance theory and macroevolution. *Annual Review of Genetics* 16: 1–19.

WRIGHT, S. W. and N. E. MORTON. 1968. Genetic studies on cystic fibrosis in Hawaii. *AJHG* 20: 157–169.

WU, T. T. and E. A. KABAT. 1970. An analysis of the sequences of the variable regions of Bence Jones proteins and myeloma light chains and their implications for antibody complementarity. *Journal of Experimental Medicine* 132: 211–250.

WYSOCKI, C. J. and . K. BEAUCHAMP. 1984. Ability to smell androstenone is genetically determined. *PNAS* 81: 4899–4902.

YAMAMOTO, M. and M. WATANABE. 1979. Epidemiology of chromosome anomalies at the early stage of pregnancy. In *Contribution to epidemiology (and biostatistics?)* 1: 101–106.

YEN, P. H. et al. 1987. Cloning and expression of steroid sulfatase cDNA and the frequent occurrence of deletions in STS deficiency: implications for X–Y exchange. *Cell* 49: 443–454.

YOSHIDA, A. 1967. A single amino acid substitution (asparagine to aspartic acid) between normal (B+) and the common Negro variant (A+) of human glocose-6-phosphate dehydrogenase. *PNAS* 57: 835–840.

YOSHIDA, A. 1973. Hemolytic anemia and G6PD deficiency. *Science* 179: 532–537.

YOSHIDA, A. and E. BEUTLER, Editors. 1986. *Glucose-6-Phosphate Dehydrogenase*. Academic Press, Orlando, Florida.

YU, L.-C., J. W. GRAY, R. LANGLOIS, M. A. VAN DILLA and A. V. CARRANO. 1984. Human chromosome karyotyping and molecular biology by flow cytometry. In *Research Perspectives in Cytogenetics*, edited by R. S. Sparkes and F. F. de la Cruz, pp. 63–73. University Park Press, Baltimore.

YUNIS, J. J. 1976. High resolution of human chromosomes. *Science* 191: 1268–1270.

YUNIS, J. J., Editor. 1977. *New Chromosomal Syndromes*. Academic Press, New York.

YUNIS, J. J. 1983. The chromosomal basis of human neoplasia. *Science* 221: 227–236.

YUNIS, J. J., R. D. BRUNNING, R. B. HOWE and M. LOBELL. 1984. High-resolution chromosomes as an independent prognostic indicator in adult acute nonlymphocytic leukemia. *NEJM* 311: 812–818.

YUNIS, J. J. and O. PRAKASH. 1982. The origin of man: a chromosomal pictorial legacy. *Science* 215: 1525–1530.

ZAMBONI, L. 1971. *Fine Morphology of Mammalian Fertilization*. Harper & Row, New York.

ZIEGLER, J. L. 1981. Burkitt's lymphoma. *NEJM* 305: 735–745.

ZIMMERMAN, A. M. and A. FORER, Editors. 1981. *Mitosis/Cytokinesis*. Academic Press, New York.

ZIMMERMAN, D. R. 1973. *Rh, the Intimate History of a Disease and Its Conquest*. Macmillan, New York.

ZOKAEEM, G., J. BAYLERAN, P. KAPLAN, P. HECHTMAN and E. F. NEUFELD. 1987. A shortened β-hexosaminidase α-chain in an Italian patient with infantile Tay-Sachs disease. *AJHG* 40: 537–547.

ZOURLAS, P. A. and H. W. JONES. 1965. Clinical, histologic, and cytogenetic findings in male hermaphroditism. *Obstetrics and Gynecology* 25: 768–778.

ZUBRZYCKA-GAARN, E. E. et al. 1988. The Duchenne muscular dystrophy gene product is localized in sarcolemma of human skeletal muscle. *Nature* 333: 466–469.

ZUCKERMAN, H. 1977. *Scientific Elite: Nobel Laureates in the United States*. Free Press, New York.

Name Index

Entries indicate references in the text itself; citations in the Bibliography beginning on page 542 are not indexed.

Abbott, U. K., 102
Abrahamson, S., 266
Ada, G. L., 376
Adam, 22, 90, 92
Adams, Charles Francis, 205
Adams, Henry, 205
Adams, John, 205
Adams, John Quincy, 205
Adolph, K. W., 66
Agarwal, D. P., 474, 475
Aherne, G. E. S., 392
Akam, M., 85
Alberts, B., 58, 67, 79, 235, 245, 376, 385
Aleixandre, C., 69
Alexandra, Czarina, 21
Alexis, Czarevitch, 21
Alfredsson, J. H., 494
Allen, G. E., 10, 496
Allen, R. D., 60
Allison, A. C., 445
Allore, R., 116
Alpert, E., 510–511
Amann, R. P., 494
Ames, B. N., 253, 259–260, 399
Anastasi, A., 476
Anderson, D. H., 32
Anderson, S., 243
Anderson, W. F., 300, 516
Anderton, B. H., 116, 120
Andres, A. H., 50
Andrews, L. B., 489, 496
Angier, N., 392
Annas, G. J., 485, 488, 496, 521–522
Anneren, G., 135
Ansbacher, R., 485
Antonarakis, S. E., 21, 506
Antony, Mark, 420
Aphrodite, 90
Appel, S. H., 36, 41
Applebaum, E. G., 501
Appleby, C. A., 29
Arber, W., 270
Ardhanarisvara, 90
Arpaia, E., 317
Ashton, G. C., 27
Atallah, L., 485
Auffray, C., 376
Austin, C. R., 79, 102
Avery, O., 6–8
Ayala, F. J., 414, 455, 482

Baby M, 488
Baccetti, B., 64, 71
Bacci, G., 90
Bailar, J. C., III, 378–379, 519
Baird, P. A., 426
Bajema, C. J., 11, 483
Baker, B. S., 77, 79
Bakker, E., 38
Baltimore, D., 381
Bandmann, H.-J., 140
Barker, D., 207
Barlow, P., 494
Barnes, D. M., 63, 120, 469
Barnum, P. T., 185
Barr, M. L., 119, 130–131
Barrett-Connor, E., 461
Barton, S. C., 88
Baserga, R., 79
Bateson, W., 86, 201, 303
Bauld, R., 156, 159
Bayes, T., 179
Beach, F. A., 458
Beadle, G. W., 217, 304
Beam, K. G., 39
Beatty, R. A., 494
Beauchamp, G. K., 41
Beaudet, A. L., 35
Becquerel, Antoine, 261
Beernink, F., 495, 497
Begleiter, H., 475
Bell, J., 136, 199
Bendall, D. S., 455
Bennett, J. C., 363
Bennett, M. D., 53
Bennett, N. G., 492
Benson, P. F., 329
Benton, W. D., 283
Berg, P., 275, 278
Bernal, J. E., 376
Berridge, M. J., 385
Bertram, E. G., 130–131
Beutler, E., 326
Bhende, Y. M., 344
Bibel, D. J., 376
Bickel, H., 306
Bieber, F. R., 477
Biggs, R., 20
Bijman, J., 35
Binet, A., 471
Bingham, R., 220
Bishop, J. A., 441
Bishop, J. M., 383, 400
Bittles, A. H., 426
Blakemore, K. J., 502
Bleier, R., 93, 102
Bloomer, J. R., 327
Bodmer, W. F., 11, 220, 398, 400, 414, 432, 452, 455, 461, 464, 468, 474, 476
Bodycote, J. 261
Boehm, T. L. J. 508
Bogenmann, E., 392
Bohman, M., 475
Bohmann, D., 385
Bois, E., 33–34
Bond, D. J., 120, 140, 164
Bongiovanni, A. M., 98–99
Bonné, B., 433
Borgaonkar, D. S., 120, 124, 128, 129, 144, 153, 164
Borwein, J. M., 470
Borwein, P. B., 470
Boss, G. R., 508
Botkin, J. R., 522
Botstein, D., 3, 19, 23, 286, 454
Boué, A., 156, 158, 159, 160, 164
Boué, J., 156, 158, 159, 160, 164, 442
Bouchard, T. J., Jr., 467, 472
Boveri, T., 45
Bowen, P., 103
Bowling, F. G., 36
Boyce, A. J., 423
Boyd, W. C., 413, 452
Boyer, S. H., IV, 11, 349
Brachet, J., 58, 79
Bradbury, E. M., 58
Bradshaw, R. A., 67
Brandriff, B., 156, 495
Braude, P., 85
Breg, W. R., 160
Breit, R., 140
Brenner, D. A., 327
Brenner, S., 83
Bridges, C., 108–109, 120
Brill, A. B., 266
Brink, P. A., 321
Broach, J. R., 78
Broad, W. J., 494
Brock, D. J. H., 35, 510
Brody, J. E., 313
Browder, L. W., 79
Brown, Louise, 486
Brown, M. S., 41, 317–320, 333, 513
Brown, R. L., 394
Brown, T. R., 96
Brown, W. T., 139, 140
Brownlee, G. G., 23
Buchwald, M., 41
Buck, Carrie, 483
Buckley, R. H., 515
Buckton, K. E., 134
Buffone, G. J., 35
Buhler, E. M., 22
Buisseret, P. D., 373
Bull, J. J., 136
Bulmer, M. G., 466
Bump, Lavinia, 185
Burck, K. B., 400
Burg, J., 317
Burgio, G. R., 120
Burkitt, D. P., 388
Burnet, F. M., 349, 352, 354–355, 376
Burns, G. A., 196–199
Burt, C., 467
Busslinger, M., 89
Butler, S., 60
Butz, E., 102

Caesar, Julius, 420
Cairns, J., 378–379, 400
Caldwell, J., 311
Callahan, D., 4
Calne, R. Y., 370, 376
Calos, M. P., 439–440
Campbell, J. R., 490
Cann, R. L., 23
Cantor, C. R., 277
Capecchi, M. R., 325
Carlson, E. A., 484
Carothers, E. E., 183, 187
Carpenter, A. T. C., 73, 79
Carr, D. H., 156, 158
Carrasco, A. E., 86
Carswell, H., 208
Carter-Saltzman, L., 468, 473, 474
Caskey, C. T., 11, 300, 323
Caspersson, T., 51
Cassell, E. J., 521
Castle, W., 415–416
Cattanach, B. M., 134
Cavalli-Sforza, L. L., 220, 414, 432, 452, 455, 461, 464, 468, 474, 476
Cavenee, W. K., 392
Chakraborty, R., 11, 310
Chambers, G. M., 114
Chambon, P., 237–238, 245
Chan, J. Y. H., 396
Chan, S. T. H., 90
Chandley, A. C., 73–74, 120, 140, 156, 160, 164
Chang, 465
Chang, E., 395
Chang, J. C., 289–290
Chaplin, Charlie, 350
Chapman, V. M., 133, 140
Chargaff, E., 45, 226–227, 245–246
Chase, G. A., 181
Chase, M., 484
Chelly, J., 39
Cherfas, J., 269, 271, 300
Childs, B., 300, 330, 520
Chin, J., 375

Choo, K. H., 21
Chu, G., 395
Chung, C. S., 36
Church, R. B., 490, 492
Clark, A. G., 414, 432
Clark, S. M., 277
Clarke, C. A., 347–349
Clarkson, B., 387
Cleaver, J. E., 394, 395
Cleopatra, 420, 423–424
Cleveland, D. W., 61
Cloninger, C. R., 475
Clow, C. L., 309
Cohen, J. S., 10
Cohen, L. A., 399, 400
Cohen, M. M., Jr., 132, 469
Colburn, Zerah, 470
Collier, R. J., 365, 376
Collins, F. S. 237, 240
Collins, J. D. B., 258
Comings, D. E., 53, 58
Conley, C. L., 172
Conley, M. E., 374
Conneally, P. M., 33, 194–195, 197, 201, 287
Constantini, F., 516
Conte, F. A., 102
Cook, L. M., 441
Cooley, T. B., 256
Cooper, B. J., 39
Cooper, D. N., 500
Corson, S. L., 495
Corwin, H. O., 11
Court-Brown, W. M., 141, 159–160
Craig, I. W., 140
Crawford, M. H., 447
Crea, R., 274
Creasy, M. R., 156
Creel, D., 313
Crew, F. A. E., 90
Crick, F., 4, 7–8, 45, 225–228, 237, 245, 247–249, 254, 258, 266
Croce, C. M. 388–389, 400
Cronbach, L. J., 473
Cross, P. K., 156
Crow, J. F., 181, 264, 414, 419, 427, 431, 432, 435, 438, 439, 443–444, 449, 455, 461, 476
Cuckle, H. S., 115
Cunningham, C., 120
Curie, Marie, 8, 278
Curie-Cohen, M., 484, 496
Curran, W. J., 398
Curtis, H., 93

Dale, B., 79
Dalton, J., 18
Danks, D. M., 513
Dao, D. D., 393
Darnell, J., 79, 376
Darnell, J. E., Jr., 239, 245
Darras, B. T., 40
Darwin, Charles, 7–9, 41, 132, 435–436, 450, 455, 459
Dausset, J., 365–366
Dautry-Varsat, A., 333
Davenport, C. B., 461, 477
David (1971–1984), 370–371, 515
Davidson, E. H., 85, 102
Davidson, R. G., 132
Davidson, R. L., 300
Davies, D. A. L., 369
Davies, K., 38
Davis, B. D., 459
Davis, K., 451
Davis, P., 33
Davis, R. W., 283
Dawkins, R., 436, 450, 454
de Crecchio, L., 98–99
de Duve, C., 58, 79
de Grouchy, J., 120, 140, 144, 164
de la Chapelle, A., 96, 104, 124, 129, 135, 140
de la Cruz, F. F., 120
de Martinville, B., 297
Dean, G., 327
DeFries, J. C., 473
Delabar, J.-M., 116
Dellarco, V. L., 120, 140, 164
DeMarini, D. M., 261
Denis, J., 332
Denniston, C., 264, 438, 439, 455
Desmond, E. W., 474
Desnick, R. J., 327, 504, 513
D'Eustachio, P., 295
Dewey, M. J., 324
Di Maio, M. S., 115
di Sant'Agnese, P. A., 33
Diacumakos, E. G., 300
Diamond, B. A., 363
Dickerson, R. E., 227, 235, 240, 245, 266, 450
Dickman, Z., 65
Diederen, I., 468
DiLella, A. G., 309–311, 509
Dirksen, E. R., 79
Dixon, B., 338
Dobzhansky, Th., 436, 441, 519–520
Doetschman, T., 325
Doherty, R. A., 198
Doll, R., 398
Donahue, R. P., 160–161
Doolittle, R. F., 245
Down, J. L., 107, 112
Drager, U. C., 313
Drahovsky, D., 508
Drake, B. L., 371
Dressler, G. R., 88
Dreyer, W. J., 363
Driever, W., 85
Drogari, E., 308
Dronamraju, K. R., 420
DuBridge, R. B., 439–440
Duchenne, G. B. A., 36
Dulbecco, R., 381
Duncan, I., 86
Dunn, L. C., 10–11

Dvořák, M., 65

Ebling, F. J., 455
Edelman, G. M., 358
Edelson, E., 41
Edwards, D. D., 114, 120
Edwards, R. G., 81, 102, 486, 487, 496
Egeland, J. A., 428, 470
Ehrhardt, A. A., 95, 98–102, 104, 129
Ehrlich, P., 357
Ehrman, L., 469
Eicher, E. M., 92
Einstein, Albert, 8
Eisenbarth, G. S., 373
Eisenbud, M., 262, 263, 265, 266
Eldredge, N., 450
Elias, S., 485, 488, 502
Ellison, J. W. 376
Elston, R. C., 220
Elwood, J. M., 467
El-Hefnawi, H., 395
Emery, A. E. H., 36, 38–41, 266, 300, 329, 376, 501, 521
Eng, 465
Engelhardt, H. T., Jr., 371
Epel, D., 79
Ephrussi, B., 197, 243
Epstein, C. J., 116, 120, 140, 164
Erbe, R. W., 499, 508
Ericsson, R. J., 494–495, 497
Erlich, H. A., 292
Estivill, X., 36
Evans, H. J., 241
Evans, K. A., 27, 41
Evans, M. J., 324, 328
Eve, 22, 91–92

Falconer, D. S., 461, 476
Falk, R., 11
Falls, H. F., 196
Fantes, P. A., 77
Farrell, M., 35
Favorchi, G., 439
Feaster, W. W., 162
Federman, D. D., 100, 128
Fehilly, C. B., 491
Feller, W., 173
Felsenfeld, G., 58, 245
Fensom, A. H., 329
Ferguson-Smith, M. A., 156, 161–163
Festenstein, H., 370
Fichman, K. R., 102
Fiddes, J. C., 266
Finn, R., 348
Firestein, S. K., 501
Fisher, R. A., 8–9, 217, 345, 435–436, 449
Flavell, R. D., 89
Flemming, W., 43
Fletcher, J., 496
Flexer, A. S., 398–400
Følling, A., 306
Ford, C. E., 49, 123, 129, 131
Ford, E. H. R., 130
Ford, P., 220
Forer, A., 77
Forrest, K. A., 486
Forrest, S. M., 40
Fowler, H. L., 27
Fowler, R. E., 81
Fox, T. D., 251
Francis, D. P., 375
Francke, U., 38, 163, 201, 297, 320
Franklin, R., 4, 45, 226, 245
Franks, L. M., 379, 398, 400
Fraser, F. C., 483–484, 498, 500, 521
Fratantoni, J. C., 330
Fredrickson, D. S., 315
Freier-Maia, N., 132
Freud, Sigmund, 8
Friedberg, E. C., 400
Friedman, M. J., 326
Frizzell, R. A., 35
Fryns, J. P., 157
Fuchs, F., 502
Fudenberg, H. H., 349, 376
Fuhrmann, W., 427–428, 521
Fujita, T., 64
Fuller, J. L., 459, 476
Funderburk, S. J., 160
Futuyma, D. J., 436, 455

Gajdusek, D. C., 204, 219, 220
Gale, R. P., 370
Galen, 379
Gallo, R. C., 243, 375
Galton, F., 8, 10, 459, 471, 477, 481, 484–486
Gandhi, Mahatma, 8
Garcia-Bellido, A., 102
Gardner, M., 470
Garrod, A. E., 5–8, 17, 253, 303–304, 312, 313, 328
Gartler, S. M., 95, 133, 140, 141, 324, 380, 400
Gaston, M. H., 512
Gatti, M., 77
Gatti, R. A. 396–397
Gautier, T., 103
Gedeon, M., 156, 158
Gehring, W. J., 84–86, 102
Geis, I., 235, 240, 266, 450
Gelinas, D. J., 419
George III, 327, 469–470
Gerald, P. S., 53, 120
German, J., 58, 395, 400
Gerrick, D. J., 326, 329
Ghesquiere, J., 102
Gibbs, D. A., 324
Gibbs, R. A., 323
Gieser, E. P., 196
Gilbert, F., 317, 320
Gilbert, S. F., 85, 87, 89, 102
Gilbert, W., 239, 278
Gilbert, W. S., 169
Gilboa, E. 516–517
Gillespie, F. D., 196
Gitschier, J., 21
Gledhill, B. L., 494
Gleitman, H., 476
Glisson, J. F., 207
Goddard, H. H., 483
Godson, G. N., 275, 300
Goedde, H. W., 474, 475
Goeddel, D. V., 274
Goff, S. C., 252, 267
Goldfoot, D. A., 93
Goldgaber, D., 115
Goldman, M. A., 133
Goldstein, J. L., 41, 317–320, 333, 513
Golub, E. S., 361, 376
Gonzalez, F. J., 327
Goodfellow, P., 16, 22
Goodfellow, P. J., 102, 135
Goodfellow, P. N., 120, 140
Goodfield, J., 4, 10
Goodwin, D. W., 475
Gorbsky, G. J., 70
Gordon, J. W. 386
Gosden, C. M., 35
Gosden, J. R., 35, 493
Gottesman, I. I., 469
Gottlieb, K., 431

Gould, S. J., 4, 450, 471, 474, 476–477, 483
Gowers, W., 36
Grabowski, G. A., 513
Gradwohl, R. B. H., 350
Graham, J. B., 20
Graham, R. K., 484
Grant, S. G., 133, 140
Granville, H. R., 207
Gravel, R. A., 331
Graves, J. A. M., 135, 140
Green, H., 270, 292
Greenberg, C. R., 40
Greenblatt, A., 41, 220, 266
Greendyke, R. M., 341, 349
Greenwald, I. S., 88
Gregg, R. G., 517
Griffin, J. E., 96
Gripenberg, U., 157
Grivell, L. A., 22, 62
Grobstein, C., 79
Groth, C. G., 515
Grover, R., 512
Grumbach, M. M., 102
Grundy, P., 393
Grundy, S. M., 317, 318
Gruss, P., 88
Guillery, R. W., 313
Guinan, M. F., 485
Gusella, J. F., 287–288, 295, 299, 300
Guthrie, R., 307, 509
Guthrie, Woody, 90, 198, 287
Güttler, F., 310

Haapala, O., 58
Hacking, I., 217
Hafez, E. S. E., 64, 71, 79
Hagerman, R. J., 140
Hakomori, S.-I., 334, 342–343, 349
Haldane, J. B. S., 8–9, 199, 435–436, 439, 445, 449, 455
Hall, K., 396–397
Haluska, F. G., 387–389, 401
Hamerton, J. L., 49, 120–121, 144, 151, 156, 199
Hamilton, J. B., 205
Hanahan, D., 386
Hanawalt, P. C., 394
Handler, P., 4
Hansen, M. F., 392
Harbour, J. W., 393
Hardy, G. H., 405, 408–410, 414–416, 470
Harnden, D. G., 397
Harris, A., 33, 41
Harris, H., 220, 266, 329, 445, 450, 454, 455
Harris, L. J., 102
Harris, R., 501
Harris, S., 253
Harrison, C. J., 137
Harrison, D., 181
Harrison, G. A., 464
Hartl, D. L., 21, 41, 414, 432, 446, 450, 455
Hartmann, J. F., 79
Hartwell, L. H., 77, 79
Harvey, J., 53
Hassold, T. J., 114, 120, 124, 129, 130, 140, 156, 158, 159, 164
Hauber, D. P., 155
Haubrich, W. S., 379
Hauschka, T. S., 126
Hayden, M. R., 41
Haynes, R. H., 258
Hearnshaw, L. S., 467
Heath, C. W., 261
Hecht, F., 53, 110, 137, 139, 140
Heddle, J. A., 266
Heisterkamp, N., 388
Helminen, P., 290
Henderson, N. D., 476
Henikoff, S., 237
Henle, W., 389
Hermaphroditus, 90
Hermes, 90
Hertig, A. T., 158
Higuchi, R., 291
Hippocrates, 379
Hiromi, Y., 86
Hirschhorn, K., 374
Hirschhorn, R., 374
Hitler, Adolf, 481, 482
Hobbs, H. H., 321
Hodgkin, J., 136
Hodson, M. E., 41
Hoey, T., 88
Hoffman, E. P., 39
Hoffmann, J., 334
Hogikyan, N. D., 317
Hogness, D. S., 284
Holden, C., 467, 477
Holland, A. J., 120
Holliday, R., 134
Holm, P. B., 73
Holmes, H. B., 492, 496
Holmes, L. B., 438
Holmes, Oliver Wendell, 483
Holmquist, G. P., 67
Holstein, A. F., 76
Holt, J. T., 388, 389
Holtzman, N. A., 307, 509, 510, 521
Honore, L. H., 76
Hood, L. E., 376
Hook, E. B., 114, 120, 126, 144, 151, 156, 158, 159, 164
Hooper, M., 324
Hoppe, P. C., 88
Hostetler, J. A., 429, 432
Howard-Flanders, P., 266, 394
Howell, M., 220
Hozumi, N., 360
Hrubec, Z., 468
Hsia, Y. E., 513
Hsu, L. Y. F., 156, 164, 503
Hsu, T. C., 50, 58
Hu, F., 314
Huang, H.-J. S., 394
Hubby, J., 450
Hulse, J. H., 490
Humphries, S. E., 321
Hungerford, D. A., 387
Hunt, E., 471
Hunter, R., 327, 469–470
Hunter, T., 384, 385, 400
Huntington, G. E., 432
Huntley, Ola Mae, 498
Hurvich, L. M., 19, 23
Hussey, M. G., 330
Huxley, Aldous, 482
Huxley, J., 519
Idle, J., 327
Illmensee, K., 88
Iltis, H., 10
Imaizumi, Y., 420
Imperato-McGinley, J., 103
Ingham, P. W., 85
Ingram, V. M., 5, 8, 253–254
Inoue, S., 79
Iselius, L., 328
Ishihara, S., 19
Itakura, K., 266, 273–274

Jacklin, C. N., 93, 102
Jackson, D. A., 270
Jackson, R. C., 155
Jacobs, G. H., 23
Jacobs, P. A., 11, 58, 89, 114, 120, 124, 126, 129–131, 137, 140, 148, 157, 159, 164, 165
Jacquard, A., 428
James VI, 327
James, R., 67
James, W. H., 467, 494
Jameson, D. L., 414, 455
Jansen, R. P. S., 486
Jaret, P., 349
Jaynes, J. B., 86–87
Jeffreys, A. J., 289–290, 297
Jencks, C., 472–473
Jenkins, J. B., 11
Jenner, E., 352
Jennings, M., 196
Jensen, A. R., 471, 473–474, 476–477
Jerne, N. K., 354–355
Jervis, G. A., 306
Johannsen, W., 461–462
John, B., 191
John, P. C. L., 79
Johnsen, D., 311
Johnson, I. S., 274
Jolly, D. J., 323
Jones, H. W., 97
Jones, P. A., 89
Jones, R. E., 93
Joseph, Antone, 27
Josso, N., 92
Jost, A., 92
Joyce, E. M., 137
Joysey, V. C., 370
Judson, H. F., 10, 245
Juel-Nielsen, N., 467

Kaback, M. M., 316, 508, 511
Kabat, E. A., 358
Kaiser, P., 153
Kajii, T., 111, 156
Kallikak, M., 483
Kalmus, H., 19
Kalow, W., 329
Kamiguchi, Y., 156, 494
Kan, Y. W., 289–290
Kang, J., 115
Kao, F.-T., 300
Kaplan, D. A., 365, 376
Kappas, A., 327
Kappler, J., 357, 369, 376
Kasper, C. K., 23
Kaufman, M., 96
Kaufman, M. H., 324
Kaufman, S., 306, 308
Kazazian, H. H., Jr., 256
Keats, B., 19, 195
Kedes, L. H., 38
Kell., Mrs., 339
Kelley, W. N., 322
Kenemans, P., 64, 70, 79
Kennedy, J. L., 470
Kersey, J. H., 370
Kevles, D. J., 10, 481, 496
Kezer, J., 191
Khaw, K.-T., 461
Khoury, M. J., 419, 427, 428, 432
Khrushchev, Nikita S., 520
Khudr, A., 428
Kidd, K. K., 310
Kim, S. H., 232, 245
Kimura, M., 439, 450–451, 456
King, M.-C., 200
King, Martin Luther, 8
King, S., 70, 428, 429
Kingsley, Charles, 81
Kinsey, A. C., 100
Kissin, B., 474
Klat, M., 428
Klein, G., 388–389, 400
Kline, A. H., 133
Kline, J., 158
Klinefelter, H. F., Jr., 124
Klug, A., 48
Knowles, M. R., 35
Knowlton, R. G., 35
Knudson, A. G., Jr., 392, 393
Koch, R., 330, 352
Köhler, G., 363, 376
Kolata, G., 37, 92, 113, 470
Korenberg, J. R., 51, 67, 116
Kornberg, A., 229–230, 245
Kornberg, R., 48
Koshland, D. E., 70, 469
Kremer, J., 485
Kretchmer, N., 330
Krimsky, S., 300
Krush, A. J., 27, 41
Kuehn, M. R., 324
Kuhl, P., 309
Kulozik, A. E., 448
Kunkel, L. M., 38, 39, 41
Kwok, S. C. M., 309
Kwon, B. S., 312

La Du, B. N., 17, 303
Labie, D., 448
Lachman, R. S., 438
Laemmli, U. K., 48–49, 66, 70
LaFond, R. E., 400
Laird, C. D., 138, 140
Lake, J. A., 232
Lalouel, J.-M., 298, 300
Lamarck, Jean-Baptiste de, 436
Lambert, E., 26
Lamm, L. U., 160
Land, J. M., 23
Landsteiner, K., 332, 337–338, 343, 349–351
Lane, C., 251, 270
Lane, D., 120
Lang, K., 308
Langone, J., 29
Lasker, G. W., 431, 432
Lasley, J. F., 490
Latt, S. A., 52
Laughlin, H., 483
Laurence, J., 376
Laurence, K. M., 506
Lawn, R. M., 21, 23, 274, 282, 283, 300
Lawrence, R. J., 358, 371
Lax, E., 376
Le Beau, M. M., 54, 387

Lebo, R. V., 297
Ledbetter, D. H., 138, 140, 145
Leder, P., 363, 376
Lederberg, J., 8, 513
Ledley, F. D., 309, 331, 516
Lee, E. Y.-H. P., 393
Lee, M. G., 77–79
Lee, W.-H., 393
Lehmann, A. R., 394
Lehmann, H., 327
Lehninger, A. L., 329
Lehrman, M. A., 321
Lejeune, J., 107, 115
Lenke, R. R., 311
Leridon, H., 158
Lesch, M., 322–325
Levan, A., 49–51, 58
Leventhal, A. G., 313
Levin, R. J., 493–494, 496
Levine, L., 11
Levine, L. S., 99
Levine, M., 88
Levine, P., 338, 344, 347, 348, 350
Levinson, A. D., 384
Levitan, M., 181
Levy, H. L., 311, 510
Lewin, B., 58, 245, 252, 266, 300
Lewin, R., 9, 199
Lewis, E. B., 86
Lewis, R., 290
Lewontin, R. C., 450, 455, 459, 468, 476–477
Liddell, J., 327
Lidsky, A. S., 309
Liedtke, C. M., 35
Lin, C. C., 58
Lincoln, Abraham, 8
Lindahl, T., 396
Linder, D., 141, 380, 400
Lindgren, B. W., 181
Lindgren, V., 38, 320
Lindsley, D. L., 110
Linn, S., 259
Lison, M., 42
Little, P. F. R., 240
Littlefield, J. W., 33, 36, 293
Lloyd, D., 77
Lloyd, J. K., 329
Lloyd-Still, J. D., 41
Lockyer, J., 309
Lodish, H. F., 333
Loehlin, J. C., 461, 473, 474, 476–477
Lombardo, P. A., 483
Lorenz, K., 458
Lowe, C. R., 158
Lowitzer, A. C., 311
Lubahn, D. B., 96
Lubs, H. A., 137, 140
Lucarelli, G., 516
Lucas, M., 165
Ludman, M. D., 511
Lynch, H. T., 499
Lyon, M., 131, 140
Lysenko, Trofim D., 519–520

Mabuchi, H. R., 41
Macalpine, I., 327, 469–470
Maccoby, E. E., 93, 102
MacCready, R. A., 330
Macgregor, H. C., 53, 58
Machin, G. A., 156
Machin, J., 26
Maddox, J., 23
Magenis, R. E., 114, 163
Magre, S., 92
Mahoney, M. J., 502
Mäkelä, O., 354
Makino, S., 58
Mandel, J. L., 138
Mange, A. P., 429, 431
Maniatis, T., 277
Mao Zedong, 8
Mapletoft, R. J., 496
Maranto, G., 276
Marcus, M., 77, 79
Margulis, L., 62, 243
Marrack, P., 357, 369, 376
Marsh, W. L., 196
Marshall, C., 382, 400
Martin, A. O., 430
Martin, J. P., 137
Martin, R. H., 156, 157
Martini, G., 326
Marx, J., 227
Mary, Queen of Scots, 327
Mascola, L., 485
Massarik, F., 511
Massie, R. K., 21
Mattei, M. G., 145
Maynard Smith, J., 181
Mazia, D., 79
McBogg, P. M., 140
McCarrey, J. R., 102
McClintock, B., 8–9, 242
McClure, H. M., 200
McConnell, F. E., 212
McCrae, W. M., 41
McDermott, A., 140
McDonough, P. G., 494
McElheny, V. K. 266
McElrath, G. W., 181
McGhee, J. D., 66
McGillivray, B., 426
McGinnis, W., 86
McGrath, J., 88
McGue, M., 472
McKean, K., 477
McKee, P. A., 20
McKinnon, P. J., 396–397
McKusick, V. A., 11, 23, 24, 29, 30, 36, 41, 94–96, 185, 187, 196–198, 201, 252–253, 255, 280, 300, 305, 325–327, 428, 429, 432, 439, 450, 504
McLaren, A., 91, 102
McNeill, C. A., 394
Mead, M., 93
Medawar, P., 4
Medvedev, Z. A., 520
Mendel, Gregor, 6–18, 23–24, 30, 43, 45, 112, 181, 183–186, 201, 215–217, 221, 405, 419, 435–436, 459, 498, 520
Meredith, H. V., 473
Merimee, T. J., 67
Merrick, Joseph, 208, 220
Merrill, C. R., 120
Merton, R. K., 10
Merz, B., 23, 511
Meselson, M., 228, 245, 260
Messer, A., 300
Metz, C. B., 79
Meyer, U. A., 327
Meyer-Bahlburg, H. F. L., 102
Michurin, I. V., 519
Miescher, F., 7–9, 43
Migeon, B. R., 96, 134
Migeon, C. J., 94, 96
Mikamo, K., 156, 494
Milkman, R., 455
Miller, J. A., 89
Miller, O. J., 58
Miller, O. L., Jr., 237
Miller, W., 99
Milstein, C., 363, 376
Milunsky, A., 496, 510–511, 521
Minogue, B. P., 510
Mitchison, J. M., 79
Mittwoch, U., 90, 94, 102
Mizel, S. B., 349
Moens, P. B., 79
Mohr, O. L., 30–32, 41
Mohs, M., 120
Mollison, P. L., 349
Mollon, J. D., 19
Molnar, S., 455
Monaco, A. P., 38, 41
Money, J., 91, 95, 98–102, 104, 129, 470
Monk, M., 324
Monroy, A., 79
Montagnier, L., 243, 375
Moore, K. L., 82–84, 102, 130
Moore, S., 336
Mørch, E. T., 438
Morel, F. M. M., 333
Morgan, T. H., 7–8, 23, 108, 189, 193
Morton, C. C., 297
Morton, N. E., 33, 36, 194, 196–199, 210, 473
Moser, H., 41
Moses, 417
Motta, P. M., 79
Motulsky, A. G., 11, 326, 328, 329, 440–441, 443–444, 446
Mourant, A. E., 343, 349, 414–415, 452, 453
Mulaikal, R. M., 99–100
Müller, C., 317
Muller, H. J., 8–9, 11, 22, 262, 264, 437, 484–486, 495, 496
Mullis, K., 290–291
Murphy, E. A., 181
Murray, A. W., 77
Mustajoki, P., 327
Myerowitz, R., 317
Myers, R. H., 90

Naftolin, F., 102
Nagel, R. L., 448
Nance, W. E., 212
Napp, Abbott, 6, 12
Nathans, J., 19, 284–286, 300
Nathanson, J. A., 258
Navaschin, M. S., 50
Nebert, D. A., 327
Neel, J. V., 209, 211, 220, 221, 254, 420, 425, 427, 432
Neff, D. A., 93
Nelson, K., 438
New, M. I., 102
Newman, H. H., 467, 468, 477–478
Nicholas, Czar, 21
Nichols, E. A., 296
Nichols, E. E., 347
Nichols, E. K., 516, 518, 521
Nichols, J. L., 41
Nicol, S. E., 469
Niebuhr, E., 155
Nielsen, J., 126, 128, 130
Nightingale, S. L., 511
Niikawa, N., 111
Nilsson, L., 65, 81, 102
Nisonoff, A., 349, 376
Noah, 311
Nomura, M., 232, 245
Norman, C., 274
Nossal, G. J. V., 10, 300, 365, 376
Novitski, E., 174
Novotny, E. J., Jr., 23
Nowell, P. C., 51, 379, 387
Nurse, P., 77–79
Nusbacher, J., 347
Nussbaum, R. L., 138, 140
Nüsslein-Volhard, C., 85
Nyhan, W. L., 41, 322–325

O'Brien, J. S., 316, 317
O'Brien, S. J., 429
O'Farrell, P. H., 86–87
Ohno, S., 39, 200
Okada, S., 316
Old, R. W., 270, 274, 300
Oldstone, M. B. A., 372
Oliver, C., 120
Oliver, G., 87–88
Olson, S. B., 57
Olweny, C., 388
Orel, V., 7, 10, 12
Orkin, S. H., 252, 254, 256, 267, 289
Osborn, D., 205–206
Overzier, C., 96
Ow, D. W., 270
Owen, D. R., 126
Owen, J. J. T., 464
Owerbach, D., 301
Owor, R., 388
Ozar, D. T., 496

Pääbo, S., 241
Page, D. C., 135, 136, 140, 163
Pagnier, J., 448
Pai, G. S., 323,
Paine, R. S., 306
Painter, T. S., 49, 51
Palmiter, R. D., 270, 491
Parkman, R., 371, 515, 521
Parsons, P. A., 469
Passarge, E., 395
Pasteur, Louis, 8, 352, 381
Patel, P. I., 323
Patterson, D., 115, 120
Paul, N., 125, 128, 129, 140
Pauli, R. M., 521
Pauling, L., 8, 253
Pearson, K., 217, 415–416
Pebley, A. R., 492–493
Peele, S., 475
Penrose, L. S., 26, 308
Perutz, M. F., 266
Pestka, S., 300
Peter, K. A., 432
Peters, J. A., 11, 23
Peterson, E. P., 485
Peto, R., 398
Picasso, Pablo, 8
Pines, M., 220, 287
Pinheiro, M., 132
Pinsky, C. M., 365

Pinsky, L., 96
Plomin, R., 473, 476
Plunkett, E., 124, 128
Polani, P. E., 94–95, 140, 191
Poncz, M., 238
Popovsky, M., 519–520
Porter, I. H., 102, 164, 300, 499, 506, 521
Porter, R. R., 358
Porteus, D. J., 163, 297
Portugal, F. H., 10
Powell, J. R., 436
Prakash, O., 200
Prescott, D. M., 66, 79, 398–400
Primrose, S. B., 270, 274, 300
Proia, R. L., 317
Provine, W. B., 410, 414–416, 419, 449, 455
Prusiner, S. B., 219
Ptolemy, 420, 423–424
Pueschel, S. M., 115, 120
Pullen, I. M., 501
Punnett, R. C., 188, 410, 498

Quinton, P. M., 35

Rabbitts, T. H., 401
Race, R. R., 195, 335, 345, 349, 351, 460, 477
Radman, M. 230
Ramanujan, S., 470
Ramsay, M., 96
Ramsey, P., 482, 496
Rasputin, 21
Ratcliffe, S. G., 125, 128, 129, 140
Rathenberg, R., 440
Rattazzi, M. C., 132
Rattner, J. B., 58
Ray, J. H., 395
Ray, P. N., 163
Raymond, C. A., 487
Reddy, P. G., 420
Reed, T. E., 447, 456
Reich, T., 474
Reich, W. T., 496
Reilly, P. R., 482, 483, 509, 512
Retief, A. E., 94, 156–158
Rhoads, G. G., 503
Riccardi, V. M., 145, 207–208
Rich, A., 232, 245
Richardson, D. W., 485
Rieder, C. L., 69
Riggs, A. D., 133, 140
Rimoin, D. L., 184, 185, 207, 266, 329, 376, 438, 469, 521
Ringdén, O., 515
Riordan, J. R., 41
Rios, Elsa and Mario, 496
Ris, H., 67, 69
Risley, M. S., 73, 79
Rivas, M. L., 194–195, 197, 201
Rizza, C., 23
Robbins, D. C., 275
Roberts, C. J., 158
Roberts, D. F., 392
Roberts, L., 35, 135, 140, 321, 400
Robertson, H. D., 219
Robertson, J. A., 311
Robertson, M., 470
Robinette, C. D., 468
Robinson, A., 125
Robinson, J. A., 494
Robinson, W. C., 429
Roblin, R., 128
Rock, J. A., 158
Rodeck, C. H., 502
Rodger, J. C. 371
Roentgen, W., 261
Rogers, D. B., 508
Rohlf, F. J., 220
Roosevelt, Eleanor, 8
Root-Bernstein, R. S., 217
Rorvik, D., 494
Rosatelli, C., 506–507
Rose, N. R., 376
Rose, R. J., 467
Rosen, C. M., 467–468
Rosen, F. S., 374
Rosenberg, C., 11
Rosenberg, L., 482
Rosenberg, L. E., 516, 521
Rosenberg, R. N., 27
Rosenthal, E., 373
Roses, A. D., 36, 41
Rosner, F., 21
Rothstein, H., 67
Rotter, J. I., 207, 469
Rous, P., 6, 381
Rouyer, F., 135, 136, 140
Rowland, L. P., 9, 39, 41
Rowley, J. D., 54, 387
Rowley, P. T., 508
Royer-Pokora, B., 279
Rudak, E., 157
Ruddle, F. H., 201, 295–296, 386
Runk, B. L., 207
Rushton, W. A. H., 19, 23
Russell, D. W., 319
Russell, P., 77
Russell, W. L., 437
Ruth, Babe, 485
Rykowski, M. C., 51

Sachs, B., 314–316
Saiki, R. K., 291–292
Sakti, 90
Salk, J., 8
Salmacis, 90
Salmon, C., 349
San Roman, C., 140
Sandberg, A. A., 102, 126, 140, 390–391, 401
Sandhoff, K., 317
Sandler, I., 10
Sandler, L., 10, 110, 174
Sanger, F., 235, 278
Sanger, R., 195, 335, 345, 349, 351, 460, 477
Sapp, J., 10
Sarasin, A., 394
Sassa, S., 327
Sayre, A., 4, 245
Scarr, S., 468, 473, 474, 477
Schinzel, A., 144
Schleiden, M. J., 14
Schlesselman, J. J., 158
Schmid, R., 327
Schmidtke, J., 500
Schneider, E. D., 488
Schneider, W., 319
Schnyder, U. W., 26
Schreinemachers, D. M., 156
Schull, W. J., 11, 209, 211, 220, 221, 265, 420, 427, 432
Schulman, J. D., 311
Schwann, T., 14
Schwartz, D. C., 277
Schwartz, R. H., 367
Scialli, A. R., 466
Scopes, John, 519
Scott, C. I., 438
Scott, J. A., 499
Scott, M. P., 86
Scriver, C. R., 309, 311, 329
Seagoe, M. V., 120
Searle, A. G., 181
Searle, T., 266
Seeger, R. C., 390
Seegmiller, J. E., 322
Seemanová, E., 426
Seibel, M. M., 486, 487, 496
Seidel, G. E., Jr., 490, 494
Sergovich, F., 146
Serio, M., 102
Seuanez, H. N., 58
Shah, S. A., 128
Shapiro, J., 263, 266
Shapiro, L. J., 134, 140, 508
Sharon, N., 344, 349
Sharp, P. A., 239, 241
Shea, B., 492
Sheer, D., 386–387, 389–391, 400
Sherman, S. L., 138
Sherrington, R., 470
Sherwood, E. R., 11
Shettles, L. B., 494
Shields, J., 467
Shih, C., 383–384
Short, R. V., 79
Shows, T. B., 169, 201, 300
Sidbury, J. B., 307–309
Silverman, L. J., 323
Simchen, G., 79
Simmel, E. C., 476
Simmons, K., 371
Simpson, E., 92, 102
Simpson, J. L., 94, 102, 129, 140, 157, 502, 503
Sinclair, A. H., 136
Sinex, F. M., 120
Siniscalco, M., 195
Siva, 90
Siwolop, S., 120
Sleisenger, M. H., 373
Sloan, H. R., 315
Smith, E. M., 378–379, 519
Smith, G. F., 120
Smith, L. D., 77
Smith, L. M., 280
Smith, R. F., 191
Smithies, O., 324, 517
Smolev, J., 486
Sokal, R. R., 220
Soltan, H. C., 112
Solter, D., 88–90, 102
Soravia, E., 317
Sørensen, T. I. A., 460
Soreq, H., 327
Sorsby, A., 311
Southern, E. M., 285
Soyfer, V. N., 520
Sparrow, A. H., 191
Speck, Richard, 126
Speed, R. M., 74
Spencer, H., 436
Spiess, E. B., 414
Srb, A. M., 11
St George-Hyslop, P. H., 115
Stadler, L. J., 11, 262
Stahl, F. W., 190, 228, 245, 260
Stalin, Joseph, 519–520
Stamatoyannopoulos, G., 256, 266
Stanbury, J. B., 220, 266, 304, 325, 329, 376, 514, 521
Stansfield, W. D., 181
Stebbins, G. L., 455
Stehelin, D., 381
Steinbrook, R., 511
Steinmetz, M., 363, 376
Steitz, J. A., 245
Stent, G. S., 245
Stephens, F. E., 37
Steptoe, P. C., 486, 496
Stern, C., 10–11, 26
Stetson, R. E., 338
Stewart, D. A., 128
Stewart, G. D., 116
Stine, G. J., 220
Storb, R., 370
Stout, J. T., 323, 516
Stratford, B., 120
Stratton, Charles S., 185
Stray-Gunderson, K., 120
Streissguth, A. P., 475
Strelkauskas, A. J., 365
Strickberger, M. W., 181, 220, 245, 266, 396, 414, 444
Strominger, J. L., 376
Strong, J. A., 124, 131
Stryer, L., 329
Sturtevant, A. H., 10, 85, 193
Südhof, T. C., 320
Sullivan, K. F., 61
Sulston, J. E., 83
Summitt, R. L., 119, 140
Sumner, A. T., 494
Sun, M., 274, 482
Super, M., 33, 41
Surani, M. A. H., 88–89
Susi, A., 307
Sutherland, G. R., 53, 137, 139, 140, 157
Sutton, W. S., 45, 183
Sverdrup, A., 209
Swift, H., 47
Swift, M., 398, 401
Szorady, I., 326

Tabakoff, B., 475
Taconis, K., 430
Takizawa, T., 326
Talamo, R. C., 41
Tanzi, R. E., 116
Tasset, D. M., 470
Tatum, E. L., 304
Taussig, L. M., 41
Tay, W., 314
Taylor, S. M., 89
Teich, N. M., 382, 384, 400
Teichler-Zallen, D., 198
Teitelbaum, M. S., 93
Temin, H. M., 245, 381
Terasaki, P. I., 368, 371–372, 376
Terry, R. D., 315
Therman, E., 53, 58, 144, 151, 153, 164
Thomas, E. D., 370
Thomas, K. R., 325

Thomas, L., 378
Thomas, R. M., Jr., 485
Thompson, C. E., 37
Thompson, M. W., 32–33, 151
Thompson, W. R., 459
Thorpe, N. O., 58
Thumb, Tom, 185
Tiefel, H. O., 496
Tinbergen, N., 438
Tiwari, J. L. 368, 371–372, 376
Tjio, J.-H., 49–51, 58
Tolley, E., 140
Tonegawa, S., 357, 360, 362, 376
Tooze, J., 276, 300
Tourian, A., 307–309
Trager, W., 326
Treves, F., 208
Trounson, A., 487
Tryon, R. C., 438–439
Tsui, L.-C., 35
Tucker, P., 91, 102
Turleau, C., 120, 140, 144, 164
Turnbull, A., 104
Turner, G., 137
Turner, H. H., 128, 140
Turpin, R., 140
Tyler, F. H., 37

Uchida, I., 50–52, 112, 117, 137, 145, 146, 155, 161
Ueda, K., 495
Underwood, L. E., 184
Upton, A. C., 265, 266

Valentine, B. A., 39
Valentine, G. H., 119, 120, 124, 128, 143, 145, 146, 164
Vallet, H. L., 102
Van Blerkom, J., 79
Van Broeckhoven, C., 116
van den Berghe, P. L., 419
van der Ploeg, L. H. T., 89
van Heyningen, V., 35, 163, 297
van Mander, K., 439
van Niekerk, W. A., 94, 100, 102
van Ommen, G. J. B., 40, 163
Varley, J. M., 53, 58
Varmus, H. E., 381
Vavilov, N. I., 519–520
Vehar, G. A., 21, 23, 274, 283, 300
Vellutino, F. R., 470
Verp, M. S., 503
Victoria, Queen, 21
Virchow, R., 43
Vogel, F., 427–428, 436, 440–441, 521
von Wettstein, D., 73
Vosa, C. G., 494

Waber, D. P., 102
Wagner, R., 230
Wainscoat, J., 23
Wai-Sum O, 90
Wallace, Alfred Russel, 8–9, 465
Wallace, D. C., 23
Wallace, I., 465
Wang, A.-L., 327
Warburton, D., 156, 160
Wareham, K. A., 134–135
Washburn, L. L., 92
Wassarman, P. M., 79
Watanabe, M., 156
Watkins, W. M., 342, 349
Watson, J. D., 4, 7–8, 45, 58, 66, 79, 102, 190, 225–228, 237, 242, 245, 247, 252, 254, 266, 269–70, 272, 275–276, 280, 300, 381, 385, 396, 400
Weatherall, D. J., 300, 506, 521
Webster, C., 39
Wechsler, D., 471
Wechsler, R., 374
Weigall, A., 420
Weinberg, R. A., 29, 383–384, 393, 394, 400, 402
Weinberg, W., 405, 408, 414–416
Weiner, A. J., 86
Weiss, M. C., 270, 292, 315
Weissman, B. E., 394
Weissman, S. M., 237, 240
Wells, R., 196
Welsh, M. J., 35
Westheimer, F. H., 518
Westoff, C. F., 492–493
Wethers, D. L., 512
Wexler, N., 287
White, P. C., 99
White, R., 35, 298, 300
Whitney, S., 509
Whyte, P. 393
Wiener, A. S., 338, 345, 350
Wieringa, B., 163
Wiggs, J., 393
Wilcox, A. J., 158
Wilkins, A. S., 85, 90, 102
Wilkins, M., 45, 226
Williams, R. R., 459
Williams, S. A., 280
Williamson, R., 35
Willis, A. E., 396
Wilson, A. C., 200, 455
Wilson, E. B., 18
Wilson, J. D., 92, 96, 102
Wilson, J. M., 323
Witkin, H. A., 127
Witkop, C. J., Jr., 312–314, 330
Witkowski, J. A., 39
Witt, P. L., 69
Wolf, F. M., 181
Wolf, S. F., 134
Wolff, S., 261
Woo, S. L. C., 310–311
Workman, P. L., 447
Worton, R. G., 38, 151
Wramsby, H., 156, 158
Wriedt, C., 30–32, 41
Wright, S., 8–9, 33, 217, 419, 423–424, 435–436, 449, 455
Wu, T. T., 358
Wyngaarden, J. B., 322
Wysocki, C. J., 41

Yamamoto, M., 156
Yanagimachi, R., 495
Yates, J. R. W., 156
Yen, P. H., 134, 136
Yoshida, A., 326
Yu, L.-C., 54, 282
Yule, U., 405, 410, 414
Yunis, J. J., 53, 144, 200, 386–387, 390, 400

Zamboni, L., 75, 79
Ziegler, J. L., 388, 390
Zimmerman, A. M., 77
Zimmerman, D. R. 348, 349
Zokaeem, G., 327
Zourlas, P. A., 97
Zubrzycka-Gaarn, E. E., 39
Zuckerman, H., 10

Subject Index

abl (oncogene), 382, 385, 388
ABO blood group(s), 340–345
  alleles, 340–341
  allelic variation, 452–453
  antibodies in plasma, 341, 342
  Bombay phenotype, 344–345, 350
  chemistry of antigens, 342–345
  discovery, 337–338
  epistasis, 210
  frequency calculations, 413, 415
  H antigen, 343–345
  independent assortment with dwarfism, 184–186, 187
  linkage to nail-patella syndrome, 202, 203
  Mendelian transmission of, 7
  polymorphism, 334–335
  protection against Rh-caused hemolytic disease, 348
  transfusion, 332, 341–342
  transplants and, 369
ABO locus, gene mapping, 162
Abortion(s)
  and amniocentesis, 501
  and aneuploidy, 110
  and chromosome aberrations, 156–158, 160, 164, 165, 166
  ethical and legal issues, 489, 493
  following artificial insemination, 485
  frequency, 156–158, 164
  induced, 159
  and polyploidy, 155, 156–157, 164
  and selection, 442
  sex selection and, 493
  and study of development, 81
Acatalasia, 61, 62
Acentric chromosome(s), 144, 152, 153, 154, 164
  double minute, 390, 391
Acetic acid, 533
Achondroplasia, 439, 456
  mutation rate, 438–439, 440
Acid(s), organic, 533
Acid hydrolase, 62
Acid phosphatase, gene mapping, 161
Acquired immune deficiency syndrome, *see* AIDS
Acrocentric chromosome(s), 54–55, 56, 79, 164, 165, 397
  and Robertsonian translocations, 148
Acrosomal cap, 63, 64
Acrosomal reaction, 64
Acrosome, 76
Actin, 60, 61, 70, 233
Acute lymphoblastic leukemia, 370
Acute nonlymphocytic leukemia, 390
ADA deficiency, 374
Additive genetic variance, 461, 462–463
  skin color, 464
Adenine, 45, 46, 47, 51, 226, 227
  metabolism, 322–323
Adenosine deaminase (ADA)
  deficiency, 374
  gene therapy and, 516–518
  isozymes, 296
Adenosine diphosphate (ADP), 533–534
Adenosine triphosphate. *see* ATP
Adenylate cyclase, 475
Adenylate kinase, gene mapping, 162
Adoption studies
  alcoholism, 474–475
  IQ scores, 472, 474
ADP, 533–534
Adrenal glands, 94
  abnormal testosterone production, 97–99, 205
Adrenal hyperplasia. *see* Congenital adrenal hyperplasia
Adrenaline, 307, 309
  biochemistry, 534–535
Adrenogenital syndrome. *see* Congenital adrenal hyperplasia
AFP. *see* α-Fetoprotein
Africa
  Burkitt lymphoma in, 388–389
  Egyptian Pharaohs, 420
  migration and gene flow, 447–448
Agammaglobulinemia, 374
  localization of gene, 500
  treatment, 514
Age. *see also* Maternal age; Paternal age
  gene reactivation and, 134–135
Age of onset, 207–209
Agglutination, of red blood cells, 335–337
Agriculture
  biometric techniques, 10
  recombinant DNA and, 276
  reproductive technology (livestock), 490–491, 492
AID. *see* Artificial insemination
AIDS, 374–375
  retrovirus, 243
  T cells and, 358
AK locus, gene mapping, 162–163
Akaline phosphates, cystic fibrosis and, 35
Alanine, structure, 234
Albinism, 24, 25, 170, 220, 305, 307, 311–314, 328, 330
  ascertainment bias, 215
  biochemistry, 307, 312
  frequency, 313, 314
  frequency calculations, 415
  inbreeding and, 425, 433
  in mice, 134
  ocular, 195, 196, 202, 203
  phenotypes of, 313–314
  probabilities in families, 176–177, 178, 179, 180, 181, 182
  rare types, 314
  yellow variant, 330
Albumin, egg, 233
Albumin-layering technique, 494
Alcohol
  cancer and, 398, 399
  chemistry of, 532–533
Alcohol dehydrogenase, 475
Alcoholism, heritability of, 474–475
Aldehyde dehydrogenase, 475
Aldosterone, 97
Algebra, 172–182, 526–529
Alkapton. *see* Homogentisic acid
Alkaptonuria, 17, 303–304, 307
  ascertainment bias, 213–214
  frequency calculations, 412–413
  inbreeding and, 425
Allele(s), 6, 15
  codominant, 171, 172, 189
  dominant, 170–172
  familial hypercholesterolemia, 321
  hemoglobin, 170, 171–172, 181
  of HLA system, 367–369
  identical by descent, 421–424
  independent, 421
  multiple, 169–170
  mutation rate, 437–440
  overdominant, 171, 172
  of phenylketonuria, 310–311
  recessive, 170–172
  recessive lethal, 442–444
  secretor (*se*), 342–345
  sickle-cell, 289, 290, 444–446, 448
  wildtype, 169
Allele frequencies, 406, 408, 409, 410
  ABO, 413
  alkaptonuria, 412–413
  blood groups, 334–335
  changes with selection, 442–446
  Rh, 346
  sickle-cell, 412, 415
Allelic exclusion, 363
Allergens, 373
Allergy, 373
Alu sequences, 51, 240–241, 243, 317, 321
Alzheimer disease
  Down syndrome and, 113–114, 115–116, 120
  familial, 116, 120
  and kuru, 219
  localization of gene, 116, 500
Ames test, 259–260, 398
Amines, 533
Amino acids, 7, 8, 234–237
  abbreviations, 234
  in antibodies, 359–360, 361, 362
  codons for, 249–250
  essential, 234
  evolutionary change and, 450–451
  genetic code and, 231
  genetic code table, 250
  in insulin, 235
  maple sugar urine disease and, 510
  metabolism, 303–314

protein chemistry and, 533
structure, 234–235
substitutions in hemoglobin, 253–257
in translation, 235–237
visual pigments, 284–285
Amino group, 533
Aminopterin, 294, 295
Amish population, 27, 428–429
genetic drift, 448
Ammonia, 533
Amniocentesis, 21
detection of chromosomal aberrations, 157, 159
procedure, 502–503
sex determination by, 493, 508
for sex selection, 493
sickle-cell anemia, 289, 290
Amnion, 82
Amniotic cavity, 82
Amniotic fluid, 82
analysis of, 502–503
β-Amyloid protein, Down syndrome and, 115–116, 120
Anaphase
meiotic, 72, 74
mitotic, 69, 70
translocation heterozygotes in, 148–149, 164
Anaphylaxis, 373
Ancestry, common. *see* Inbreeding
Androgen(s), 61, 92, 93, 124. *see also* Testosterone
adrenal, 94
and baldness, 204–205
congenital adrenal hyperplasia, 97–99, 102
insensitivity to, 96–97, 102
Androgen insensitivity syndrome, 61, 96–97, 102, 103, 104, 141, 142
Androstendione, 94, 97
Anemia
in G6PD deficiency, 326, 329
sickle-cell. *see* Sickle-cell anemia
Anencephaly, 506–507, 510
Anesthesia, 326–327
Aneuploidy, 110, 111, 119, 120, 144. *see also* Deletion(s); Duplication(s); Inversion(s); Monosomy; Translocation(s); Trisomy
in cancer cells, 379, 380
defined, 107
of sex chromosomes, 123–130, 139
Anhidrotic ectodermal dysplasia, 132, 133, 500, 507
Aniridia, 146
Wilms tumor and, 393
Ankylosing spondylitis, 371–372, 377
Anomaloscope, 19
Antennapedia (*Antp*), 86
Anthropology, human variability, 451–454
Anti-oncogenes, 394
Antibodies
ABO, 337, 338, 341, 342
to centromeres, 69
genes for, 359–363
hypervariable regions, 359, 360
Kell, 338–340
light and heavy chains, 358, 360
MN, 337, 338
monoclonal, 336, 363–365, 376
plasma cells and, 355
as plasma component, 333
production, 354–357
Rh, 337, 338–339, 345–348
source, 335–336
structure, 358–360, 375
variable and constant domains, 358–359, 360
Antibody gene(s), 359–363
Antibody screening, 341–342
Anticodon(s), 232, 235–236
Antigen(s), 334
ABO, 340–345, 342–345. *see also* ABO blood group
antibody production and, 354–357
histocompatibility, 365–371. *see also* HLA system
HLA, 365–371, 515–516
linkage to disease, 371–374
Rh, 345–348. *see also* Rh blood group
transplantation, 365, 368–371. *see also* HLA system
Antigen-antibody reaction, 335–337, 341–342
allergies, 373
Antigen-effect hypothesis, 371–372
Antihemophiliac globulin factor, 21
Antiintellectualism, 518–520
Apoprotein B-100, 318, 319
Arlington Park hemoglobin, 255, 267
Arms, chromosome, 44
Artificial insemination, 484–485
in animals, 490
ethical and legal issues, 489–490
sex selection and, 494
Artificial selection, 441
Ascertainment
of rare chromosomal disorders, 127–128, 130, 141, 142
truncate, 212–215, 221
Asilomar conference, 276
Asparagus, 41, 42
Aspartame, 235
Aspartic acid, 235
structure, 234
Assortative mating, 426
Aster, 68
Ataxia-telangiectasia, 396–398, 401
Atoms, 530–531
bonds, 531
defined, 530
isotopes, 530–531
ATP (adenosine triphosphate), 323
alcoholism and, 475
in chemical reactions, 533–534
Autoantibodies, 373
Autoimmune disorders, 373–374
Autologous blood transfusion, 342
Autoradiography, 52, 279–280, 287, 288, 290
human visual pigment genes, 284–285
Autosomal dominant alleles, mutation rate, 437, 438–440
Autosomal dominant inheritance
brachydactyly, 30–32
characteristics of, 31–32, 40
familial hypercholesterolemia, 317–321
porphyria, 327
retinoblastoma, 392–393, 401
Wilms tumor, 393–394, 401
Autosomal dominant traits
genetic screening for, 198–199
prenatal diagnosis, 505
Autosomal recessive inheritance
albinism, 311–314
ascertainment bias, 212–215
ataxia-telangiectasia, 396–398
Bloom syndrome, 395–396
characteristics of, 33–34, 40
congenital adrenal hyperplasia, 97–99
cystic fibrosis, 32–36
hermaphroditism, 94–96
phenylketonuria, 305–311
Tay-Sachs disease, 314–317
xeroderma pigmentosum, 394–395, 396
Autosomal recessive lethal allele, mutation rate, 439
Autosomal recessive traits
genetic screening for, 199
prenatal diagnosis, 505
Autosome(s), 15, 51, 55
sex-influenced inheritance, 204–206
Azorean neurological disease, 27

B cell(s), 353–354
allergic response and, 373
Burkitt lymphoma, 388–389
gene therapy and, 516–518
somatic mutation, 361–362
B-DNA, 226
Bacteria. *see also Escherichia coli*; *Salmonella*
DNA polymerase from, 291
genetic code and, 250–251
molecular technology and, 8–9
transformation of, 7
Bacteriophage, gene library and, 281–284
Bacteriophage λ, 281–282
Bacteriophage ϕX174, DNA replication, 229–230
Balance hypothesis, of selection, 445–446
Baldness
and ectodermal dysplasia, 41
inheritance of, 30
pattern, 204–206, 220
*Bam*HI, 276–277, 278, 300, 507
Banding patterns, 44
Bands, chromosome. *see* Chromosome bands
Barbiturates, and porphyria, 326
Barr bodies, 130–131. *see also* Sex chromatin bodies
number in various karyotypes, 131, 141, 142
number of, 141, 142
sex determination and, 508
Base analogs, 258, 260–261
Base(s) (DNA, RNA), 45–47, 226–227, 232
Base pairing, 45–46, 227, 228–229
complementarity, 227
gene splicing, 270, 271
reverse, 47
during transcription, 231, 237
during translation, 236–237
Base substitution mutations, 251–252
sickle-cell, 251, 252, 343
Bayes theorem, 178–179, 180, 182
Becker muscular dystrophy, 36, 39
Behavior
Lesch-Nyhan syndrome and, 322
45,X karyotype and, 129
XYY karyotype and, 126–128
Behavioral traits, genetic component, 468–475
Biases, in data collection, 212–215
Bicoid (*bcd*), 85
Bile acids, 333
Binomial coefficient, 175–176, 180–181
Binomial formula, 174–177, 179–180, 213, 529
and chi-square method, 217
polygenic inheritance, 462–463
Binomial theorem, Hardy-Weinberg law and, 410
Biochemical genetics, 3, 7–9, 303–331. *see also* Molecular genetics
of cystic fibrosis, 35–36
of Duchenne muscular dystrophy, 38–39
pedigree analysis and, 29–30
Biochemical pathways, 303–331. *see also* Metabolic pathways
chemical reactions of, 534–535
Biometry, 10, 459

Biotechnology, 269–302
Birds, sex determination in, 136
Birth defects, radiation and, 264–265
Bisexual, 100
Bithorax, 86
Bivalent, 73
Blastocyst, 81–82
Blending inheritance, 7, 10, 14
Blood. *see also* Hemoglobin; Leukemia; Thalassemia
  clotting factors, 333
  composition of, 332–334
Blood clotting, gene linkage, 199
Blood group(s). *see also specific entry:* ABO; Duffy; Kell; MN; Rh; Xg
  codominant alleles, 171
  legal consideration, 350
  racial variation, 452–453
  twins, 477
  variability among races, 452
  variant chromosomes, 160–161
  Xg, 195, 196, 201, 202, 203
Blood grouping techniques, 335–337
Blood plasma, 332–333
Blood transfusion
  ABO typing, 341–342
  antibody screening, 341–342
  autologous, 342
  cross matching, 341–342
  history of, 332
  reactions, 338, 339
  Rh typing, 341–342
Bloom syndrome, 395–396
Bombay phenotype, 344–345, 350
  epistasis, 210
Bonds, 531
Bone marrow
  gene therapy and, 517
  immune response and, 353–354
  transplants, 370–371, 514, 515–516
Bottleneck, 429
Brachydactyly, 30–32, 40, 202, 203
  frequency in populations, 405–416
  probability, 179
Bread mold (*Neurospora*), 304
Breakage cluster region (*bcr*), 387, 388
Breakpoint(s). *see also* Chromosome abnormalities; Translocation breakpoints
  inversions and, 153
Breast cancer, ataxia-telangiectasia and, 398, 401
Breasts
  abnormal development, 96–97, 103, 124, 128, 129
  normal development, 94
Bromodeoxyuridine, 52
5-Bromouracil, 258–260
Bubbles and forks, 228, 229
Burkitt lymphoma, 151, 385, 386, 388–390

C-bands, 52, 53, 57
C (constant) element, antibody gene, 360–361
Caenorhabditis elegans, 83, 85, 90
  homeotic genes, 88
Caffeine, 258–259
Calcium, and Duchenne muscular dystrophy, 38
Calico cats, 131, 132, 134
  male, 141, 142
Calmodulin, 67
Cancer. *see also* Leukemia
  and albinism, 314
  cell cycle and, 68
  cell types, 379–380
  cells, monoclonal antibodies and, 363–364
  characteristics of, 379
  chromosomal abnormalities and, 53, 54, 146, 386–391
  Down syndrome and, 112
  fragile sites and, 54
  incidence of, 378–379
  Mendelian inheritance, 392–398
  oncogenes, 9, 387–390
  polyploidy and, 155
  preventive diet, 399
  and radiation, 263, 264
  single-cell origin of, 379–380
  and smoking, 398, 399
  translocations and, 151, 387–390
Cannibalism, 219
Capacitation, 64
Carbon
  carboxyl group, 533
  isotopes of, 530–531
  hydrocarbons, 532
Carboxyl group, 533
Carcinogen, 259, 398–399
Carcinogenesis, proto-oncogenes to oncogenes, 384–386
Carcinoma, 379, 383
Cardiovascular system, embryonic, 82
Carp mouth, 146
Carrier(s), 19–21. *see also* Genetic screening; Heterozygote(s)
  cystic fibrosis, 35
  fragile X syndrome, 138–139
  frequency calculation, 412–414
  hemophilia, 21
  Lesch-Nyhan syndrome, 323–324
  pedigree symbol, 28, 29
  screening for, 511–512
  Tay-Sachs disease, 316
  translocation Down syndrome, 150, 151, 164, 165
  translocation heterozygote, 149
Carrier proteins, 333
Cat(s)
  albino, 313
  calico, 131, 132, 134, 141, 142
  Siamese, 206, 207
Cat-cry syndrome, 145, 164
Catalase, 62
Catalysis, 534
Cattle
  artifical insemination, 490
  coat color, 177
  cowpox, 352
  embryo splitting, 490–491, 492
  heritability of milk production, 461
  insulin, 235
  reproductive technologies, 490–492
  sex-limited traits, 206
  visual pigments, 284
CDC2 gene, mutation of, 77
Celiac disease, 372–373
Cell(s), 51, 60–64
  cancerous, 379–380
  hybridization, 292–298
  noncycling (G_0), 68
  rods and cones, 284
  transformation by oncogenes, 381–382, 383–384
Cell cycle, 66–68, 78
  cancerous cells, 68
  mutations, 76–78
Cell death, 68
Cell hybridization. *see* Somatic cell fusion
Cell membranes, cystic fibrosis and, 35
Cellular immunity, 354
Cellular metabolism, 62
Cellular oncogenes. *see* Proto-oncogenes
Cellular respiration, 62
Centriole, 63, 66, 67, 68
  in sperm, 76
Centromere(s), 44, 48, 49, 54–55, 56, 69, 72–73, 74
  chiasmata and, 191, 192
  inversions and, 64, 65, 151, 152
  misdivision, 148, 165, 166
  ring chromosomes and, 153, 154
Centromeric index, 54–55
Centrosome, 68
Ceramide, 316
Chain termination codons, 249, 250
Chalones, 68
Chargaff's rule, 245, 226, 245
Checkerboard, 15–16
  Hardy-Weinberg populations, 410, 411
  Mendel's second law, 186–187
  polygenic inheritance, 463
  Punnett square, 15–16
Chediak-Higashi syndrome, 314
Cheetahs, population bottleneck, 429
Chelating agents, 513
Chemical carcinogens, 398–399
Chemical reactions, 533–535
Cherry-red spot, 314, 315
Chi-square test, 216–218
  decision rules, 216–217
  for Hardy-Weinberg equilibrium, 415
  more than two classes, 217–218
  sex selection and, 494
Chiasma (Chiasmata), 72, 73, 74, 190–194, 196
Chickens, immune system, 353
Chimeras, 490
Chimpanzees, kuru and, 219
Cholesterol, 41, 63, 93, 333
  hypercholesterolemia, 513–515
  metabolism, 317–321
  as steroid, 533
Chorion, 82
Chorionic gonadotropin, pregnancy detection and, 158
Chorionic villus sampling (CVS), 21
  detection of chromosome abnormalities, 156–157, 158
  procedure, 502–505
  for sex selection, 493
  uses, 504–507
Christmas disease. *see* Hemophilia B
Chromatid(s), 44, 48
  differential labeling, 260, 261
  involvement in crossing over, 190
  sister, 67, 68, 69, 70
Chromatid exchange, Bloom syndrome and, 395–396. *see also* Sister chromatid exchange (SCE)
Chromatin, 43, 44, 48
  compaction, 43, 47, 48
  replication, 66–67
Chromosomal sex, 91, 102, 104
  of hermaphrodites, 94–95
  reversal of, 135–136
Chromosome(s), 7, 14–15
  bivalent, 73, 74
  bouquet pattern, 71
  cartoon, 458
  coiling, 67
  components, 45–46
  condensation, 66–67, 71, 77
  constrictions, 56, 137
  cytogenetic shorthand, 57

decondensation, 70
defined, 43
double minute, 390, 391
experimental analysis, 50–54
during fertilization, 64, 66
fragile site, 137
gene mapping, 294–298
gene organization, 237–243
gross structure, 44, 46–48, 49
harlequin, 260, 261
homologous, 72–74, 187, 188
Lepore, 257
location of proto-oncogenes, 382
marker, 160–163
meiotic segregation, 72–74
metaphase, 66–67
Philadelphia, 385, 387, 390
primates, 199–200
proteins, 46–48
relative lengths, 55–56
ring, 144, 145, 153, 154, 164, 165
satellites, 56
size of, 47, 48
structure of, 50, 51, 54–57
Chromosome 1
gene map, 197–199
primates, 199–200
Chromosome abnormalities. *see also* specific abnormality
abortions and, 156–157, 158–159, 164, 165, 166
cancer and, 386–391, 401
deletions, 144–147, 164
duplications, 146–147, 164
frequencies, 155–160
gene mapping, 160–163
inversions, 151–153, 164
and maternal age, 125, 156, 159, 160
number of autosomes, 107–119
number of sex chromosomes, 123–142
prenatal diagnosis, 503–504
rearrangements, 143–166, 384
translocations, 147–151, 164
Chromosome arms, 44, 48, 53–57
deletion of, 144, 165
Chromosome bands, 44, 51–53, 54, 57
inversions and, 153, 166
Chromosome groups, 55, 56
Chromosome loss, 110–111, 122. *see also* Deletion(s)
Chromosome number, 44, 49–50, 52, 76
human, 7
mitosis and, 70
Chromosome-specific library, 282
Chromosome theory of heredity, 45
Chromosome walking, 289
Chronic granulomatous disease, 38–39
gene mapping, 163
gene sequencing and, 279
localization of gene, 500
Xk-related, 195, 196
Chronic myelogenous leukemia, 150, 387–388, 389
Cilia, fertilization and, 64
Cis position, 188. *see also* Coupling phase
Class switching, 359, 363
Classical (selectionist) theory of evolution, 450
Cleavage division(s), 66, 70–71, 76, 81, 85, 487
and nondisjunction, 110–111
Cleft lip/palate, 69
recurrence risk, 500
surgical repair, 514, 515
as threshold trait, 468, 469
and trisomy 13, 118, 119
Clitoris, 92, 93, 102
Clonal reproduction, 491
Clonal selection theory, 355
Clone(s)
cancer cells, 379–380
hybrid cells. *see also* Somatic cell fusion
somatic cell fusion, 292–294, 295
Cloning
molecular, 272–276
oncogenes, 383–384
Clotting factor(s), 333
gene sequencing and, 279
recombinant DNA synthesis, 274–275
Club foot, recurrence risk, 500
Coat color
cattle, 177
epistasis, 210
mice, 206, 207, 220, 221
mutation rate and, 437
Coated pits, 318, 319
Code, genetic. *see* Genetic code
Codiscovery, 5–6
Codominant allele(s), 171, 172
Codon(s), 232, 235–236. *see also* Genetic code
deciphering experiments, 247–248
stop, 236, 250
table of, 250
Coefficient, binomial, 175–176, 180–181
Coefficient of inbreeding, 422–424, 432, 433
inter-sibship matings, 430–431
Coenzyme, defined, 513
Cohesive ends, 270, 271
Coin tossing, and hypothesis testing, 215–217
Colchicine, 50, 53, 54
Collagen, 305
Colony-stimulating factor, 67
Color blindness, 18–20. *see also* Green-shift; Red-shift
frequency calculations, 415
linkage relationships, 194–195
molecular biology of, 19, 284–286
and X-chromosome inactivation, 140, 141
Color vision, molecular biology, 19, 284–286
Combining site, antigen, 355
Common ancestor, 417–419. *see also* Inbreeding
inbred, 423–424
Complement (plasma protein), 356
HLA typing and, 367
Complementarity (of DNA), 227
reassociation experiments, 241
Complementary bases. *see* Base pairing
Complementary DNA (cDNA)
library, 283
visual pigment genes, 284
Complementation, 331, 394, 397, 401
Computing ability, 470
Concordance, threshold traits and twins, 468–469
Conditional probability, 177–179, 180, 213–215
Cone cells, 19, 284
retinoblastoma and, 392
Congenital adrenal hyperplasia, 97–99, 100, 102–104
gene mapping, 163
prenatal diagnosis, 505
treatment, 514
Congenital hypothyroidism, 510
Congenital malformations, recurrence risks, 500
Conjoined twins, 465
Consanguinity, 28–29, 419–421. *see also* Inbreeding
and autosomal recessive inheritance, 34
Conservation (genetic)
gene structure and function, 29
histone proteins, 451
of homeobox, 86
of X-linked genes, 39
Constant Spring hemoglobin, 257, 267
Constitutive heterochromatin, 52, 53
Constrictions, chromosomal, 56, 137
Continuous traits, 459–478. *see also* Quantitative variation
defined, 459
Contraception
and eugenics, 484
and infertility, 160
Contractile ring, 68, 70
Cooley's anemia, 256
Core particle, of nucleosome, 47, 48
Corn
hybrid, 171, 520
regulator genes, 9
Cornea transplants, 377
Corpus luteum, 94
Cortical granules, 61, 62, 64
Cortisol, 97, 102
Cortisone, 99, 100
as steroid, 533
Counseling, genetic. *see* Genetic counseling
Coupling phase, 188–189, 193–194
Cousin mating
frequencies, 420, 425
mortality data, 425–427
prognosis, 427–428
Cousins
mating frequencies, 420, 425
terminology, 417–419
Covalent bonding, 531
of amino acids, 234
Covariance, 461
CPK. *see* Creatine phosphokinase
Creatine phosphokinase, Duchenne muscular dystrophy and, 38, 39–40
Creationism, 435, 519
Creutzfeldt-Jakob disease, 219
pituitary transplants and, 274
Criminal behavior, XYY karyotype and, 126–128
Cross-fertilization, 12–14
Cross matching, blood, 341–342
Cross reactivity, 372
Crossing over, 71, 72–74, 74, 78, 187–194
Bloom syndrome and, 395–396
chromatid breakage and rejoining, 190, 192
on chromosome 1, 197–199
color blindness and, 285–286
evolutionary significance, 153, 164
hemoglobin mutations and, 255, 257
illegitimate, 135–136
intergenic vs. intragenic, 285–286
and inversions, 152–153, 164-165
limits, 192–193
multiple, 193
and ring chromosomes, 153, 154
sex differences, 197, 202, 203
sex reversal and, 135–136
thalassemia and, 257
unequal, 146, 148, 257, 285–286, 164, 317, 321
Cumulus cells, 65
CVS. *see* Chorionic villus sampling
Cyanogen bromide, 273
Cysteine, 235, 307
disulfide bonds, 234
structure, 234

Cystic fibrosis (CF), 33–36, 40
gene for, 35
localization of gene, 35, 500
pedigree, 33–34
Cystinosis, prenatal diagnosis, 505
Cytochromes, metabolic disorders and, 327
Cytogenetics, 7, 49–57, 107–119, 123–141, 143–166
of cancer, 396–398
history of, 7
Cytokinesis, 66, 68, 70, 77
Cytoplasm, 44, 60, 61, 70–71
Cytosine, 45, 46, 47, 51, 226, 227
Cytosol, 60

D (diversity) element(s), of antibody gene, 360–363
D–J joining, 361–362
Darwinian fitness. *see* Fitness
Data gathering, 27
ascertainment bias, 212–215
Deafness, 212, 220, 221
Debrisoquine, 327
Decoding, 230. *see also* Transcription; Translation
Deficiencies. *see* Deletions
Degeneracy, of genetic code, 249, 251, 252, 267, 283
Delaney amendment, 398, 401
Deletion(s), 144–147
chromosome 4, 144
chromosome 5, 144–145, 164
chromosome 11, 146
chromosome 13, 15, 145
chromosome 18, 13, 145–146, 146
chromosome 21, 145
and development, 89
Duchenne muscular dystrophy, 38, 40
formation, 144, 146, 149, 150, 152, 164
gene mapping and, 161–163, 164
hypercholesterolemia and, 320–321
retinoblastoma and, 392–393
thalassemia, 255–257
Wilms tumor, 393–394
X-chromosome inactivation and, 162
Deoxyribonucleic acid. *see* DNA
Dementia, 113
Deoxyribose, 45, 226
Dermatoglyphics, 112
Down syndrome, 112, 118
trisomies 13 and 18, 118
Turner syndrome, 129
DES, 398
Deutan gene, 18–19
Deuteranomaly. *see* Green-shift
Deuteranopia, 19, 286
Development, human, 81–83, 84
Developmental genetics, 83–90
Deviations, cause of, 215–217
Diabetes mellitus
age of onset, 207–208
juvenile (Type I), 372, 373–374
pleiotropy, 210, 211
as threshold trait, 468, 469
treatment of, 514, 515
Diakinesis, 71, 73–74
Dicentric chromosome(s), 152, 153, 154, 164, 397
Dideoxy sequencing, 278–280
Diet
cancer and, 398, 399
hypercholesterolemia and, 320, 321
lactose intolerance and, 330, 331
Diet therapy
for galactosemia, 330
for genetic disorders, 513, 514
for phenylketonuria, 306–307, 311, 328, 513, 514
Diethylstilbestrol (DES), 398
Differential imprinting, 89–90
Differentiation, molecular basis, 83, 85–88
Digestion, fats, 318–319
Dihybrid cross(es), 183–187
Dihydrobiopterin synthetase, 309
Dihydropteridine reductase, 309
Dihydrotestosterone, 92, 103, 104
Dilute gene, 206, 220
Diploidy (2n), 44
allelic exclusion and, 363
meiosis and, 71, 75, 76, 78
Diplotene, 71, 72, 73, 74
Disjunction, 74
Disomic gamete(s), 109, 121, 122, 125, 141, 142
Disulfide bonds, 234, 235
insulin synthesis and, 274, 275
Diversity, of antibodies, 359–363
Dizygotic twins, 115, 465–466
DMD locus, deletion mapping, 163
DNA, 45–49, 533
in antibody genes, 359–363
B form, 226
code table, 250
coiling in chromosome, 46–48, 67
complementary (cDNA), 283
defective replication, 396
dideoxy sequencing, 278–280
directionality of strands, 226–227, 229, 230, 231
discovery of, 7
electrophoresis, 276–277, 278, 279–280
gene therapy and, 516–518
homeodomains and, 86–87
late-replicating, 52
left-handed, 227
maternal effects, 84–85
methylation of, 89
mitochondrial (mtDNA), 243–244
prenatal diagnosis, 504–506
promoter, 230, 231
purification and biotechnology, 269
purine metabolism, 322–323
reassociation experiments, 241
recombinant, 270–276. *see also* Recombinant DNA
repetitive, 241–242, 243, 290–292
replication, 45–46, 66–67
replication primer, 291
restriction mapping, 277, 278
satellite, 242
sense and antisense strands, 230–231, 238, 291
snapback sequences, 242
structure, 45–46, 225–227
synthesis, 66, 244
template, 227
terminator, 231
transcription, 230–233. *see also* Transcription
transfection and oncogenes, 382–384
triplets, genetic code table, 250
unique sequence, 241
Z form, 227
DNA fingerprinting, 57, 289–292
DNA–histone complex, 46–48, 66–67, 78
DNA hybridization, Duchenne muscular dystrophy and, 38–39
DNA ligase, 229, 271
defective, 396
DNA polymerase, 228–229, 230, 233
chain reaction, 290–292
DNA probes
genetic counseling and, 499, 500
prenatal diagnosis by, 35, 39, 311, 506–507
DNA repair, cancer and, 394–395, 396
DNA splicing, 8
DNA synthesis, 66–67
Domains
of antibodies, 358–360, 358–363
of chromatin, 66, 68, 78
of LDL receptor, 319–321
Dominance, evolution and, 409–410
Dominance variance, 461
Dominant allele(s), 13, 170–172
evolution and, 170
X-linkage, 20
Dominant inheritance. *see* Autosomal dominant inheritance; X-linked dominant inheritance
Dominant phenotype, frequencies, 406
Dominant traits, inbreeding and, 425–426
DOPA (dihydroxyphenylalanine), 307, 312, 313, 328, 330, 534–535
Dopamine, 307, 309, 534–535
Dopaquinone, 307, 312, 331
Dosage compensation, 131, 140
Dosage effects, 162
Dose-response curves, 264
Double-blind investigations, 127
Double helix, 45, 46, 67. *see also* DNA
Double heterozygote, 183–187, 190, 193, 194
coupling vs. repulsion, 188–189
linkage studies, 195, 197
Double minute chromosomes, 390, 391
Doubling dose (radiation), 265
Down syndrome, 74, 107, 110, 112–118, 120–122, 470
Alzheimer disease and, 113–114, 115–116
animal models, 116
in chimpanzees, 200
and chromosome 21, 55, 57
chromosome number, 7
frequency, 114–115, 143
gene mapping, 162
general features, 112–113
genetic counseling, 501
isochromosome 21, 148
karyotype, 57
and maternal age, 114–115, 120, 150
miscarriage (spontaneous abortion) rate, 114, 157, 159, 160
and mitotic nondisjunction, 117
mosaicism, 118
prenatal diagnosis, 503–504
Robertsonian translocations, 164, 165
and sex chromosome aneuploidy, 117–118
sporadic, 114
translocation, 150–151, 164, 165
and twins, 115, 121, 122
Drift, 429, 448–449
Drosophila
cell cycle mutants, 77
developmental genetics, 83, 84–87
eye color, 108–109
gene mapping and, 7
and inversions, 153
karyotypes (XO, XXX, XXY), 108, 121, 122
nondisjunction in, 108–109, 110, 121, 122
sex determination in, 108
sex reversal in, 96
Drug resistance, cell cycle mutations and, 77
Drugs
and birth defects, 82

and metabolic disorders, 325–328
Duchenne muscular dystrophy (DMD), 36–40, 41, 440
gene mapping, 38–39, 163
linked to chronic granulotamous disease, 279
mutation rate, 440
Duffy blood group
gene mapping, 350
polymorphism, 335
variant chromosomes, 160–161
Duplication(s), 146–147, 149, 164, 165
and development, 89
formation, 150, 152, 164
gene mapping, 162–163
Dwarfism
achondroplastic, 438, 439, 440, 456
Ellis-van Creveld syndrome, 428–429
independent assortment with ABO, 184–186, 187
treatment (pituitary), 514, 515
Dysgenesis, defined, 96
Dyslexia, 470
Dystrophin, 39

Ear pits, 252, 253
*Eco*RI, 270, 271, 276–277, 278, 282, 285, 300
RFLPs and, 286–287
Ectodermal dysplasia, 41, 132, 133, 500, 507
Edwards syndrome. *see* Trisomy 18
Egg(s), 71, 74, 74–76, 75
fertilization of, 64–66
maternal effects, 84–85
mutation rate, 440, 456
structure, 60–61
Egg transfer, 487–488
Electrical charge, on atoms, 530–531
Electron, 530
Electrophoresis, 276–277, 278, 287, 288, 290
detection of variability, 454
enzymes, 294–296
enzyme variation, 330, 331, 450
of fetal DNA, 506, 507
of hemoglobin, 255
sequencing and, 279–280
Elements, of antibody genes, 360–363
Elliptocytosis, linkage with Rh, 210
Ellis-van Creveld syndrome, 428–429, 433, 448
Embryo, four-cell stage, 587
Embryo splitting, 490–491, 492
Embryo transfer, 486–488
in animals, 490
Embryonic development
genetics of, 83–90
human, 81–83, 84
polyploidy and, 155, 164
sex organs, 91–93
Empiric risk, 500, 501
Endocytosis, 63
Endoplasmic reticulum (ER), 61, 62, 69, 319
Energy, chemical reactions and, 533–534
Engrailed (*en*), 86–87
Environment
cancer and, 398–399
IQ and, 472–474
mutation and, 10
phenotype expression and, 206, 207
Environmental variance, 460–464
skin color, 464
twin studies, 467–468
Enzyme(s), 7–8, 170. *see also* Inborn errors of metabolism
alcoholism and, 475
assayed prenatally, 505
chemistry of, 534
congenital adrenal hyperplasia and, 97–99
defects, 303–331
electrophoretic variants, 294–296, 450, 454
fertilization and, 64
gene mapping and, 197–199
in heterozygotes, 170
in mitochondria, 243
protein kinases, 385
replacement of, 514, 515
restriction. *see* Restriction enzymes
somatic cell fusion and, 292–294, 295
variable pigment production and, 206, 207
Enzyme deficiencies, inheritance of, 29
Enzyme therapy, for genetic disorders, 513–514
Epidermal growth factor (EGF), 67, 88
homology to LDL receptor, 319, 320
Epistasis
Bombay phenotype, 344–345
defined, 310
Epstein-Barr virus, 381, 389–390
Equatorial plate, 69, 70, 74
Equatorial region, 74
Equilibrium
Hardy-Weinberg, 408, 409
mutation versus selection, 439
mutational, 437, 439
superior heterozygote, 444–446
Erythroblast, 334
Erythroblastosis fetalis, 340. *see also* Hemolytic disease of the newborn
Erythrocytes. *see* Red blood cells
Escherichia coli
DNA replication, 227–228, 230
gene library and, 281–284
gene transfer experiments, 383–384
genetic code and, 250–251
recombinant DNA, 270–276
Estrogen, 68, 93, 94, 102
in androgen insensitivity syndrome, 97
as steroid, 533
Ethane, 532
Ethical considerations
beginning of life, 482
eugenics, 482
genetic counseling, 501–502
genetic screening, 511, 512, 513
transplants, 371, 377
Euchromatin, 44, 53
Eugenics, 10, 481–487
artifical insemination, 484–485
historical prespective, 481
negative, 481, 483–484
origin of term, 459
positive, 481, 484–486
and recessive traits, 443–444
sterilization, 483–484
Eukaryotes, 60
Euploidy, 107
"Eve", 22–23
Eve principle, 91–92, 95
Evolution
as allele frequency changes, 409–410
antibody genes, 362–363
concepts of, 9
conserved gene sequences, 86, 319, 320
dominance and, 170
drift, 448–449
evidence for, 436
gene flow, 447–449
gene mapping, 199–200
and inversions, 153
maternal inheritance and, 22–23
migration, 447–449
mitochondrial genes, 243–244
molecular, 239–240, 243–244, 362–363, 450–451
principle of, 435
processes, 435–457
pseudogenes and, 240
role of mutation, 437–441
role of selection, 9, 441–446
selectionist vs. neutralist positions, 450–451
shifting balance theory, 449
societal understanding of, 518–519
Exceptional progeny, 108–109, 122
Excision repair, 394–395, 396
Exocytosis, 63
Exons, 237, 238
G6PD gene, 326
HPRT gene, 323
LDL gene, 320, 321, 328
Exponential, defined, 175
Exponential notation, 527–528
Expressed sequences. *see* Exons
Extraembryonic membranes, 82, 88, 89
Eye color
in fruit flies, 108–109
human, 477, 547

F-body, 494
F_1 hybrid, 13–15
F_2 generation, 13
Fabry disease, treatment, 514, 515
Factor VIII, 333, 514, 515
gene sequencing and, 279
recombinant DNA synthesis, 274–275
Factor VIII deficiency. *see* Hemophilia A
Factor IX deficiency. *see* Hemophilia B
Factorial, defined, 175–176
Factoring, 528–529
FAD (familial Alzheimer disease). *see* Alzheimer disease
Fallopian tubes. *see also* Oviducts
in vitro fertilization and, 486–487
Familial hypercholesterolemia (FH), 41, 317–321
frequency, 317
localization of gene, 320, 500
pedigree, 42
receptors and, 61
symptoms, 317–318
treatment of, 513–515
Family linkage studies, 194–195
Favism, 326, 329
treatment, 514
Feedback inhibition
cholesterol and, 319
of hormones, 97
Fertility
chromosome aberration and, 156–157, 158–160, 164, 165, 166
sperm structure and, 63–64
Fertilization, 60, 64–66, 76, 81–82
artificial (AID), 485
double, 95
in vitro, 486–490
Fetal alcohol syndrome, 475
Fetal cells, prenatal analysis, 502–504
Fetal DNA, analysis, 504–506
Fetal membranes, 82, 504
in twin diagnosis, 466
α-Fetoprotein, 506–507
Down syndrome and, 115, 121
genetic screening and, 510–511
Fetoscopy, 21
Fetus, 83, 84
ethical issues, 484

hemolytic disease, 346–348
and hemolytic disease, 338–340
histocompatibility and, 371
viability, 489
Fibrinogen, 333
Fibroblast(s), 295
Fibroblast growth factor, 67
Finger proteins, 88, 135–136
Fingerprint patterns. *see* Dermatoglyphics
Fingerprinting, DNA, 289–292
Fingers
extra. *see* Polydactyly
flexion deformities, 118, 119
shortened, 30–32. *see also* Brachydactyly
Fitness (Darwinian), 436
heterozygote advantage, 444–446
of recessive alleles, 442–444
selection and, 441
Flagellum, sperm, 63, 64
Flow cytometry, 282
Fluoroscopy, 263
Foldback sequences, 47, 242
Follicle cell(s), 61, 63, 64, 76, 94
degeneration, 128
Follicle-stimulating hormone (FSH), 94
Fore population, 204, 218–219
Formic acid, structure, 533
Forward genetics, 29–30
fos (oncogene), 382, 385–386
Founder effect(s), 428–429, 448–449
cystic fibrosis and, 33
hypercholesterolemia, 321
porphyria, 327
Fractions, 526–527
and ratios, 527
Fragile site
cancer and, 54
of X chromosome, 53–54, 123, 323
of chromosome 16, 163
Fragile X site, 123
Lesch-Nyhan syndrome and, 323
Fragile X syndrome, 137–139, 140
mental retardation, 500
Frameshift mutations, 247–248, 258
Ames test and, 259
defined, 251
Fraternal twins. *see* Dizygotic twins
Fraud
cloning, 494
Kallikak book, 483
by Lysenko, 519–520
MZ twin study, 467
FRAX gene, 38–39
Friedreich ataxia, 221
Fructose intolerance, 514
Fruit fly. *see Drosophila*
FSH. *see* Follicle-stimulating hormone
Fushi tarazu (*ftz*), 86–87
Functional group, of atoms, 532–533

G-bands, 37, 52, 53, 57, 67
G (gap) phases
of interphase, 66–67
mutations, 77, 78
G_{M2} ganglioside, 316, 328
activator protein
Galactose, 330, 331
Galactosemia
dietary treatment, 513, 514
genetic screening, 510
lactose and, 330
prenatal diagnosis, 505
Gamete(s), 14, 60–65, 71. *see also* Egg; Sperm
chromosome aberrations, 156, 157–158
disomic, 109, 121, 122, 125, 141, 142
Hardy-Weinberg frequencies, 411
mutation rate, 440–441, 456
nullisomic, 109, 121
number of types, 201
in path diagrams, 423
recombinant and parental, 184–194
and X chromosome activation, 134
Gamete fraction, 15, 16
Gamete × gamete method, 186–187. *see also* Checkerboard
polygenic inheritance, 462–463
Gamete intrafallopian transfer (GIFT), 487, 488
ethical and legal issues, 489–490
Gametogenesis, 71–76, 76. *see also* Oogenesis; Spermatogenesis
and imprinting, 89
Gamma globulin. *see* IgG
Gangliosides, 316–317
Gap genes, 85–87
Garden beans, heritable variation, 460
Garden peas, 12–17
Gel electrophoresis. *see* Electrophoresis
GenBank, 280
Gender identity, 93, 95, 99–101, 102, 103, 104
Gene(s). *see also* DNA; Regulator genes
alleles of, 15
antibody, 359–363
defined, 6, 15
drug resistance, 271–272
families, 239–241
of HLA system, 366, 369
mitochondrial, 243–244
structural, 241
transposable elements of, 242–243
tumor-suppressing, 394
Gene amplification, 384, 390–391
Gene clusters, 199
Gene flow, 447–449
Gene × gene method, 186–187
polygenic inheritance, 462
Gene library, 281–284, 383–384
cDNA and, 283
chromosome-specific, 282
chromosome walking and, 289
color vision genes, 284–285
genomic, 282
Gene linkage
HLA antigens and disease, 371, 372
RFLPs and, 35, 38, 286–289, 309
Gene mapping, 7, 187–200, 536–541
ABO blood group, 187
adding distances, 195 -197
Burkitt lymphoma, 388–389
chromosome 1, 197–199, 200
chromosome aberrations, 160–163
chronic myelogenous leukemia, 387–388
color blindness and, 285–286
hereditary diseases, 500
HLA system, 366
Huntington disease, 287–289
immunoglobulin heavy chain, 361
LDL receptor gene, 319–320
map of human genome (partial), 536–541
oncogenes, 382, 383–384
phenylketonuria, 309–310
pituitary dwarfism, 187
restriction sites, 277–278
retinoblastoma, 392–393
RFLPs and, 35, 38, 286–289, 290, 309
schizophrenia, 470
sex-related differences
sickle-cell anemia, 289, 290
by somatic cell fusion, 292–298
Wilms tumor, 393–394
X chromosome, 194–197
Gene mutation. *see* Mutation
Gene pool, 441
Gene products, replacement of, 514, 515
Gene sequencing, 278–280
Gene splicing, 270–276. *see also* Recombinant DNA
Gene targeting, 517
Lesch-Nyhan syndrome, 324–325
Gene therapy, 516–518
Lesch-Nyhan syndrome, 324–325
somatic vs. germinal, 516
speculation, 522
Gene transfer, oncogene mapping, 382–384
Genetic code, 8, 231–232, 247–251
deciphering codons, 249
degeneracy, 249, 267, 283
as evidence for evolution, 436
properties, 249–251
synonymous substitutions and, 251, 252
table of codons, 250
triplet nature, 247–248
wobble, 249
Genetic counseling, 10, 499–518
cousin matings, 427–428
cystic fibrosis and, 35–36
Duchenne muscular dystrophy and, 39–40
estimating risks, 500–501
ethical concerns, 501–502
fragile X syndrome and, 139
Lesch-Nyhan syndrome and, 324
options, 501
phenylketonuria and, 310–311
prenatal diagnosis, 502–508
procedures, 499–501
screening, 508–512
Tay-Sachs disease and, 316
Genetic diseases. *see also* Inborn errors of metabolism; *specific disease*
candidates for gene therapy, 516–518
diet therapy, 306–307, 311, 328, 330, 513, 514
overall frequencies, 499
prenatal diagnosis, 502–507
prevention vs. treatment, 513
screening for, 508–512, 513–518
treatment, 512–518
Genetic drift, 429, 448–449
Genetic engineering, 269–302, 482, 482–495
artificial insemination, 484–485
defined, 269
ethical issues, 482
farm animals, 490–491
gene therapy, 516–518
in vitro fertilization, 486–490
reproductive technologies, 484–490, 496
sex selection, 491–495
Genetic–environmental interactions, 461, 463
Genetic heterogeneity, defined, 210, 212
Genetic screening, 508–512
Duchenne muscular dystrophy, 40
galactosemia, 510
gene mapping and, 198–199
Huntington disease, 287–289
legal and ethical considerations, 511, 512, 513
neural tube defects, 510–511
phenylketonuria, 307–308, 310–311, 328, 509–510

procedures, 508–511
sickle-cell anemia, 289, 290, 512
Tay-Sachs disease, 316–317, 511–512, 521
47,XYY males, 128
Genetic variance, 460–464
between twins, 468
Genetics, forward vs. reverse, 29–30
Genital tubercule, 91, 92, 101
Genomic library, 282
Genomics, defined, 280
Genotype(s), 15
homozygous vs. heterozygous, 171
Genotype frequencies, 405–407
Hardy-Weinberg, 408–411
Genotype number, for *N* alleles, 169–170
Germ cell(s). *see* Egg; Gamete(s); Sperm
Germinal choice (Muller), 484
Germinal mosaicism, 110
Germinal mutation(s), 251, 262, 264
cancer and, 392
Gibbon, gene mapping and, 199
Giemsa stain, 50, 51–52, 57, 200
for sexing sperm cells, 494
GIFT, 487, 488, 489–490
β-Globin, 237–241
Globin gene(s)
library screening, 283–284
methylation and, 89
organization, 237–241
pseudogenes, 240
β-Globin gene, 245
gene mapping, 297
gene therapy and, 517
sequencing of, 279–280
Glucose, chemical reactions and, 533–534
Glucose-6-phosphate, chemical reactions and, 534
Glucose-6-phosphate dehydrogenase (G6PD), 380
action of, 326
gene, 326
isozymes, 296
linkage relationships, 194–195
X-linkage and evolution, 200
Glucose-6-phosphate dehydrogenase (G6PD) deficiency, 326, 329
biochemistry, 326
and malaria, 326, 329, 446
treatment, 514
variants, 326, 329
and X chromosome inactivation, 132–133, 141, 142, 330, 331
Glutamic acid, sickle-cell mutation and, 251, 252, 254
Glutathione, 326
Glutathione reductase, 326
Gluten, celiac disease and, 372–373
Glycerol, 533
Glycine, structure, 234
Glycolipids, 334, 343
Glycoproteins, 318, 319, 334, 343
Goiters, 514
Golgi complex, 61, 62, 63, 319
in sperm, 76
Gonadal dysgenesis, 96, 103, 104, 141
Gonadal ridges, 82, 91, 95, 101
Gonadoblastoma, 146
Gonadotropic hormone(s), 93, 95
in vitro fertilization and, 486
sex selection and, 494
twinning and, 466
Gonads
development, 91–93, 102. *see also* Ovaries; Testes
errors in development, 94–99
X-rays and, 263
Gorilla, gene map, 199–200
Gout, 322, 323, 514
G6PD. *see* Glucose-6-phosphate dehydrogenase
G6PD (gene), linkage with hemophilia, 21
Gradients, and development, 85, 87–88
Graft-versus-host disease, 370
Grasshoppers
chiasmata, 191
sex chromosomes, 187
Green revolution, 490
Green-shift, 18–20, 24
linkage relationships, 194–195
linkage with hemophilia A, 187–192, 194, 195, 202, 203
Growth factors, 77, 385, 401
cell division and, 67–68
Growth hormone. *see also* Human growth hormone
GTP, proto-oncogenes and, 385
Guanine, 45, 46, 47, 51, 226, 227
metabolism, 322–323
Guthrie test, for phenylketonuria, 509–510

H antigen. *see* ABO blood group
H-Y antigen, 92, 94
Hair bulb incubation test, 313
Hair dyes, mutation and, 260
Hair follicle test, 324
Half-life, 531
Hamster, cell cycle mutation, 77
Hand creases, 112, 115
Haploidy (n), 44
meiosis and, 71, 75, 76, 78
Haplotype(s)
defined, 309, 367–368
HLA system, 367–369, 376–377
LDL receptor gene, 321
of phenylketonuria, 309, 310, 311, 331
of sickle-cell alleles, 448
Haptoglobin (*Hpa*) gene: mapping, 163
Hardy-Weinberg law, 408–411
applications, 412–414
conditions pertaining, 408
extension to inbreeding, 425–426, 433
extension to selection, 442–445
meaning, 409–410
proof, 408–409, 410, 411
X-linkage and, 413–414
Harlequin chromosomes, 52, 260, 261
HAT medium, 293–294, 295, 363–364
Hay fever, 373
HD. *see* Huntington disease
HDN. *see* Hemolytic disease of the newborn
Hearing, in albinos, 313
Heart disease
and cystic fibrosis, 33
and Down syndrome, 112
and Duchenne muscular dystrophy, 36
and hypercholesterolemia, 41
Height
fragile X syndrome, 137
Klinefelter syndrome, 124, 139
Turner syndrome, 128
variation, 454
47,XYY males, 126, 127
Heme group, 239, 240
hemoglobin variants and, 255
Heme synthesis, and porphyria, 326–327
Hemizygote, 19
Hemochromatosis, 500
Hemoglobin, 233, 304
amino acid substitutions, 253–257
Arlington Park, 255, 267
Constant Spring, 257, 267
fetal, 256
fetal vs. adult, 239–240
function, 334
Lepore, 257
molecular evolution and, 239–240
mutations of, 253–257
sickle-cell, 8, 254–255
sickling test, 172
structure, 239–240
thalassemia, 255–257
Wayne, 267
Hemoglobin C disease, 301
Hemoglobin gene(s), 29, 253–257
alleles of β gene, 170, 171–172, 181
base substitutions, 253–257
gene duplication and, 146, 147
gene library, 283–284
gene organization, 237–241
gene therapy, 516–518
molecular evolution and, 239–240
mutations of, 253–257
in plants, 29
prenatal analysis and, 506, 507
sickle-cell alleles, 289, 290
Hemolysis, 338
Hemolytic disease of the newborn, 338, 340
due to ABO, 348
due to Kell, 338–340
due to Rh, 346–348
protection against, 347–348
Hemophilia
affected females, 414
frequency calculation, 413–414
treatment, 514, 515
Hemophilia A (classical), 20–21, 199, 210
gene sequencing, 279
linkage with green-shift, 187–192, 194, 195, 202, 203
localization of gene, 21, 500
mutation rate, 440
recombinant DNA and, 274–275
Hemophilia B (Christmas disease), 20–21, 199, 210
localization of gene, 500
mutation rate, 440
Heredity clinic(s), 499. *see also* Genetic counseling
Heritability, 460–461
of alcoholism, 474–475
broad and narrow sense, 461
IQ, 472–473, 477–478
limitations in meaning, 473
skin color, 464
of skin color, 464
Hermansky-Pudlak syndrome, 314
Hermaphrodite(s), 90, 91, 94–99. *see also* Pseudohermaphrodite(s)
gender identity, 100
true, 94–96, 102, 103, 104
Hernia, 97, 103
Heterochromatin, 44, 52, 53
replication of, 67
Heterokaryon, 394–395, 401
Heterozygosity, inbreeding and, 419
Heterozygote(s), 15. *see also* Hardy-Weinberg law
ascertainment, 212–213
disease resistance and, 255

double, 183–189, 190, 193, 194, 195, 197
genetic screening and, 511–512
intermediate, 171, 172
phenotypes of, 170–172
selection for, 9, 444–446
triple, 201, 202, 203
Heterozygote advantage, 171, 444–446
of sickle-cell allele, 172
Hexokinase, 534
Hexosaminidase A and B, 316–317
HGH (human growth hormone), 274
HGPT. *see* HPRT
High mobility group proteins, 47
High-resolution banding, 52–53
*Hin*dIII, 277, 287, 288, 301
Hindus, inbreeding among, 420
Hiroshima, 264–265, 427
Histamine, 373
Histidine, Ames test and, 260
Histocompatibility antigens. *see also* HLA system
defined, 365
Histone, conservative nature of, 451
Histone proteins, 46, 48, 64, 67, 233
HIV virus, 374–375
HLA locus, gene mapping and, 160–161
HLA system, 365–371
association with diseases, 371–374
BCA loci, 366, 367–369, 377
CA21H gene and, 99
haplotype frequencies, 367, 368
polymorphism, 367–369
pregnancy and, 371
transplants and, 369–371, 515–516
Hoechst stain, 52
Homeobox, 9, 86–88
Homeodomain, 86–88
Homeotic genes, 85–88
Homogeneously staining regions (HSRs), 390
Homogentisic acid, 17, 303–304, 307
Homogentisic acid oxidase, 303–304, 307
Homologous chromosomes, meiotic segregation and, 187, 188
Homologous recombination, 517
Homologous region of X and Y, 18, 135–136
Homologue, 15, 72–74
Homosexual, 100–101
Homozygosity. *see also* Hardy-Weinberg law
inbreeding and, 419, 421
increase due to inbreeding, 422–426
Homozygote(s), 15
Hormone(s)
cell division and, 67–68
gonadotropic, 93, 95
prenatal sex determination and, 508
sexual behavior and, 100
sexual development and, 90, 91, 92–100
"Hot spot," in *DMD* gene, 40
HPRT (hypoxanthine guanine phosphoribosyl transferase)
DNA synthesis and, 294, 295, 322–323
gene, 323
in making monoclonal antibodies, 363–365
Human gene library. *see* Gene library
Human genome, partial map of, 536–541
Human Genome Mapping Organization (HUGO), 280
Human growth hormone, 514, 515
recombinant DNA synthesis, 274
Human leukocyte antigens. *see* Major histocompatibility complex
Humoral immunity, 354
Hunchback (*hb*), 85
Hunter syndrome, 62, 330, 331
localization of gene, 500
prenatal diagnosis, 505, 515
Huntington disease, 90, 208–209
age of onset, 90
gene mapping, 287–289, 296
localization of gene, 500
prenatal diagnosis, 506
screening for, 198–199, 287–289
Hurler syndrome, 62, 330, 331
prenatal diagnosis, 505, 515
Hutterite population, 27, 429–431
path diagram, 419
Hyaluronidase, 64
Hybrid(s)
human populations, 447–448
in Mendel's crosses, 13–15
Hybrid cells. *see* Somatic cell fusion
Hybridization
cell, 292–298
DNA and RNA, 237–238
DNA polymerase chain reaction, 290–292
in situ, 297, 298
Hybridoma, 363–365, 376
Hydatidiform mole, 88
Hydrocarbons, 532
Hydrogen, 531–533
Hydrogen bonds, 532–533
in DNA, 45–56, 226–228
Hydrometrocolpos, 206
Hydroxyl group, 532
Hypercholesterolemia. *see* Familial hypercholesterolemia
Hyperphenylalaninemia(s), 308–309. *see also* Phenylketonuria
maternal, 311
pedigree, 331
Hypervariable region, of antibodies, 359, 360
Hypophosphatemic rickets, 514
Hypothalamus, and sex development, 92–95
Hypothesis testing, 215–219
Hypothyroidism, 510
Hypotonic saline, 50, 54
Hypoxanthine, in DNA synthesis, 293–294, 295
Hypoxanthine guanine phosphoribosyl transferase. *see* HPRT

Ice-minus bacteria, 276
Ichthyosis, 26, 134
linkage to *Xg*, 195, 196, 202, 203
Ichthyosis (hystrix), 26
Identical twins. *see* Monozygotic twins
Identity by descent, 421–422
IgE, allergic reaction and, 373
IgG, 233, 358–361
domains of, 376
rheumatoid arthritis and, 373
"Illegitimate" crossover, sex reversal and, 135–136
Immotile cilia syndrome, 64
Immune deficiency disorders, 374–375
Immune serum, 335, 338
Immunity. *see also* HLA system
allergy, 373
autoimmunity, 373–374
cell types, 353–354
disorders of, 370, 371–375
gene therapy and, 516–518
HLA system, 365–371
humoral vs. cellular, 354
secondary response, 356
transplantation and, 369–371
vaccination, 353
Immunoglobulin(s). *see also* Antibodies
defined, 358
as plasma components, 333
Immunoglobulin classes, 359
Immunoglobulin genes
Burkitt lymphoma and, 389–390
translocation breakpoints, 401
variability, 359–363
Immunoglobulin heavy chain, variability, 359–363
Immunoglobulin light chain, variability, 363
Immunotoxin, 365
Implantation, 82, 158, 486–487
Imprinting
differential, 89–90
X chromosome, 135, 138–139
Inactive-X hypothesis, 131–135, 139–140. *see also* X chromosome inactivation
evidence for, 131–132
and sex chromosome disorders, 134
Inborn errors of metabolism, 7, 303–331
defined, 303
frequency, 305
medical considerations, 325–328
prenatal diagnosis, 504–505
treatment, 513–515
Inbreeding
among Amish, 428–429
brother–sister matings, 419–421, 426
coefficient of, 422–424, 432, 433
complex pedigrees, 423–424
defined, 419
effects, 424–428
Egyptian Pharaohs, 420, 423–424
extension of Hardy-Weinberg law, 425–426, 433
among Hindus, 420
among Hutterites, 429–431
incest, 419–420
increase in recessive traits, 425–426
inter-sibship matings, 430–431
among Japanese, 420, 425, 427
loop, 422–424
measurement from pedigrees, 421–424
measurement of, 430–431
mortality data, 426–427
random and nonrandom components, 431
among Samaritans, 433
self-fertilization, 419, 421
uncle–niece matings, 420, 426
X-linked genes and, 434
Incest, 419–420, 426
Incompatibility, maternal-fetal, 338–340, 346–347. *see also* Blood groups; HLA system
Incomplete androgen insensitivity, 97, 103
Incomplete penetrance, 209–210
Independent assortment, 74, 76
Independent assortment, law of, 6, 184–188. *see also* Mendel's second law
Independent events, probability, 174
Infertility
and artificial insemination, 484–485
blocked oviducts, 81
and chromosome aberrations, 149, 153, 155, 156, 164, 165–166
and congenital adrenal hyperplasia, 99
and in vitro fertilization, 486–490
Klinefelter syndrome, 124, 139, 141, 142
Turner syndrome, 128–129
Informed consent, 483
Inner cell mass, 81–82
Inosine, 232, 249
Insertional mutations, 324
In situ hybridization, 297, 298
Insulin, 67, 233, 514, 515
codons for, 267

deficiency, 210, 211. *see also* Diabetes
gene mapping, 301
HLA antigens and diabetes, 373–374
linkage to β-globin gene, 297
recombinant DNA synthesis, 274, 275
structure, 234–235
Insulin-like growth factor, 67
Intelligence, 471–474. *see also* IQ
Intercalary heterochromatin, 53
Intercourse, timing of, 494
Interleukins, 356–357
Intermediate filaments, 60, 68
Intermediate heterozygote, 171, 172
Interphase, 66–67, 69, 78
Intersex, 90, 94–101, 155. *see also* Hermaphrodite(s); Pseudohermaphrodite(s)
Intervening sequences. *see* Introns
Introns, 237, 238
G6PD gene, 326
HPRT gene, 323
LDL gene, 320, 321
Inverse probability, 179
Inversion(s), 58, 59, 151–153
and crossing over, 152, 164, 165
evolution and, 153, 199
formation, 152
frequencies, 159–160
gene mapping and, 160–161, 164
paracentric, 152
paracentric vs. pericentric, 151, 164, 165
pericentric, 153, 160
rate of incidence, 153
In vitro fertilization (IVF-ET), 81, 486–490
ethical and legal issues, 486–490
In vitro packaging, 282
Ion(s), 530–531, 533
Ionic bond, 531
Ionizing radiation, 261–265
cancer and, 398
IQ
distribution within groups, 473–474
Down syndrome, 113
environmental variables, 472–474
and gene therapy, 522
heritability, 472–473, 477–478
in Lesch-Nyhan syndrome, 322
in phenylketonuria, 306, 307, 308
quantitative variation, 471–474
racial differences, 10, 473–474
tests, 471
twin studies, 472, 477–478
Iron, hemoglobin variants and, 255
Islets of Langerhans, 373–374
Isochromosome(s), 148, 165, 166
Isolates, 428–431
Amish, 428–428
defined, 428
and genetic drift, 448–449
Hutterites, 429–431
Isoniazid, 327–328
Isonymy, 431
Isotopes, 530–531
library probes, 283
in Meselson-Stahl experiment, 228
radioactive, 262–263, 531
Isozyme(s), 294, 295, 296. *see also* Enzyme(s)
IVF-ET. *see* In vitro fertilization

J (joining) element, antibody gene, 360–363
Japanese populations
inbreeding, 420, 425–427
radiation injury, 264–265
Juvenile diabetes (Type I), 372, 373–374. *see also* Diabetes mellitus
treatment of, 514, 515

Kallikak book, 483
Kartagener syndrome, 64
Karyotype(s), 44, 51, 54–57
abortion rate and, 156–157, 158–160
ataxia-telangiectasia, 397
Barr bodies and, 131, 141, 142
of cancer cells, 379, 380, 390, 391
Down syndrome, 115, 150
Klinefelter syndrome (47,XXY), 123, 124–125
multiple X chromosomes, 129–130
polyploid, 155
sex reversal and, 135–136
shorthand notation, 57
translocation gene mapping, 161–162
Turner syndrome (45,X), 129
variant chromosome, 161
47,XYY, 52, 126–128
Karyotyping
prenatal diagnosis and, 508
for sex selection, 493
Kell blood group
discovery, 337, 338–340
hemolytic disease of the newborn, 338–340
polymorphism, 335
Keratin, 233
Keratinocyte, 312
Kidney transplant(s), 369–370
and Fabry disease, 514, 515
Kidneys
polycystic kidney disease, 500
Wilms tumor, 393–394
Kinetochores, 69–70
Klinefelter syndrome (47,XXY), 123–126, 139, 141, 142
Barr bodies, 131
and color blindness, 140, 141
compared to sex reversal, 135
frequency, 124–125
general features, 124–126
karyotype, 59
mosaicism, 125
nondisjunction, 125
and partial X chromosome inactivation, 134
variants, 125
Korea, sex distribution in, 179–180
Krüppel (*Kr*), 85, 87
Kuru, 204, 218–219
genetic hypothesis, 218

Labia, 91, 92, 93, 102
Labioscrotal swellings, 91, 93, 101
Lactase deficiency, 330, 514
Lactation, 94
Lactose, 330, 331
Laminar proteins, 69, 70
Language, and IQ scores, 474
LDL, 317–321, 333
hypercholesterolemia and, 41, 61
LDL receptor, 239
gene for, 319–321
mutations of, 320–321
Leber's hereditary optic neuropathy, 23
Left-handed DNA, 227
Legal considerations, 3–4
artificial insemination, 485
blood groups, 350
cancer and environment, 398–399, 401
eugenics, 482
genetic counseling, 311, 508–509
genetic screening, 511, 512
recombinant DNA, 275–276, 518–519
reproductive technologies, 488–490, 496–497
sterilization, 483–484
transplants, 371, 377

Lepore hemoglobin, 257
Leptotene, 71, 72
Lesch-Nyhan syndrome, 322–325, 469
animal model, 324–325
biochemistry, 322–323
carrier detection, 323–324
frequency, 323
localization of gene, 500
prenatal diagnosis, 505
symptoms, 322
Lethal allele(s), 21–22
kuru, 218–219
Lethal equivalent, 427
Leukemia(s)
bone marrow transplantation and, 370
chromosome aberrations and, 53, 387–388, 389, 390, 401
chronic myelogenous, 150
defined, 379
and Down syndrome, 113
types of, 387
LH (luteinizing hormone), 94
Library. *see* Gene library
Ligands, 61
Limb formation, 83–83, 87–88
Lineage (*lin*) genes, 88
Linkage disequilibrium
HLA system genes, 368, 377
phenylketornuria haplotypes, 310
Linked genes, 187–194
and chiasmata, 189–194
defined, 189
difficulty in detection, 193–194
mapping methods, 194–197
and RFLPs 287–289, 298
Linker, 48
Lipid(s), chemical structure, 533
Lipid metabolism, 314–321, 328
Tay-Sachs disease, 314–317
Lipoprotein metabolism, hypercholesterolemia, 317–321
Liver transplants, 371
Locus, 15
Lod score method, 194
Low-density lipoprotein (LDL). *see* LDL
Lung cancer, 390
Luteinizing hormone, 94
Lymphatic system, 353–354
Lymphocytes, 353, 354
Lymphokines, 353, 357. *see also* Interleukins
Lymphomas, 379
Lymphoproliferative syndrome, 507
Lyon hypothesis (Lyonization), 131–132. *see also* X-chromosome inactivation
X-linked recessive inheritance, 37–38
Lysenkoism, 519–520
Lysine, structure, 234
Lysosomal storage diseases, 62
Tay-Sachs disease, 314–317
treatment, 514, 515
Lysosome(s), 61, 62, 67
cholesterol metabolism and, 318–319

M phase. *see* Mitosis
Macrophages, 353, 354
Major histocompatibility complex (MHC). *see also* HLA system
defined, 366
MHC restriction, 367
proteins, 356, 357
Malaria
geographic distribution, 446

and G6PD deficiency, 326, 446
heterozygote advantage and, 444–446
sickle-cell anemia, 255
and sickle-cell anemia, 9, 172, 444–446
and thalassemia, 256, 446
Malignancy, characteristics of, 379. *see also* Cancer
Mammography, 263
Manic depression, 470, 500
Map units
adding distances, 195 -197
on chromosome 1, 199
and recombination, 193
X-linked genes, 194–197
Maple sugar urine disease
dietary treatment, 510, 512, 514
prenatal diagnosis, 505, 510
Mapping of genes. *see* Gene mapping
Marfan syndrome, 440
Marker chromosome(s), 160–163
in cancer cells, 379
Marker loci. *see also* RFLPs
Marsupials, X-inactivation in, 131, 135, 136
Mass screening
for cystic fibrosis, 35–36
for Duchenne muscular dystrophy, 40
Maternal age
chromosome aberrations and, 156, 159, 160
Down syndrome and, 114–115, 150
trisomies 13 and 18 and, 118
twinning rate and, 466
47,XXX females and, 130
Maternal effect genes, 84–85
Maternal-fetal incompatibility, 338–340, 346–347
Maternal genome, development and, 88–90
Maternal phenylketonuria, 307, 311
Maternal transmission, of mitochondrial genes, 22–23
Mathematical ability, 470
Mathematics, 526–529. *see also* Chi-square test; Probability
Mating(s)
assortative, 426
brother-sister, 419–421, 426
consanguineous. *see* Inbreeding
frequencies, 406–408
inter-sibship, 430–431
random, 325, 407–408
uncle-niece, 420, 426
Mating types, 17
Maturation promotion factor, 78
Maxam-Gilbert method (of sequencing), 278
Measurement, units of, 525
Medullary thyroid carcinoma syndrome, 514
Meiosis, 71–76, 78
color blindness and, 285–286
crossing over, 189
defined, 71
and Mendel's second law, 187, 188
mutations, 77
polyploidy and, 154
translocations, 148–149
Meiotic nondisjunction, 109–110, 119, 120, 121
Melanin(s), 306, 307, 312–313, 328
albinism and, 311–314, 328
Melanism, 409, 441
Melanocytes, 312
Melanosomes, 312, 313
Membranes
G6PD deficiency and, 326
red blood cell, 334
Membranous cytoplasmic bodies, 315, 316
Memory cells, 355, 356
Mendel's first law, 6, 14–15, 169
experimental evidence for, 12–17
Mendel's laws, chi-square and, 217, 221
Mendel's second law, 6, 184–188
cytological evidence, 187
Menkes syndrome, 514
Menstrual cycle, 92, 93, 97, 128, 129
Mental retardation
chromosome aberrations, 144, 145, 146, 157, 160, 164
Down syndrome, 112–118
Duchenne muscular dystrophy, 36
fragile site and, 53–54
hyperphenylalaninemia, 308
hypothyroidism and, 510
Kallikak book, 483
Klinefelter syndrome, 124, 125
Lesch-Nyhan syndrome, 322
metabolic disorders and, 325, 328, 329
phenylketonuria, 306, 307, 308, 328
sterilization and, 483
Wilms tumor and, 393
X-linked, 123
X-linked recessive, 137
47,XXX females, 129
47,XYY males, 126, 127, 139
Meselson-Stahl experiment, 228, 245
Mesoderm, limb formation and, 87–88
Messenger RNA (mRNA), 47, 232
base composition, 245
gene library and, 283
maternal effects and, 85
poly(A) tail, 239, 275
processing of, 237–239, 317
synthesis of, 249
table of codons, 250
in translation, 235–237
wobble and, 249
met (oncogene), 35
Metabolic blocks, 304, 305, 307, 309, 323
Metabolic disorders. *see also* Inborn errors of metabolism; *specific disorders*
detection in newborns, 325
slow metabolizers, 327
Metabolic pathways, 303–331
drug-sensitive, 325–328
gangliosides, 316–317
general scheme, 305
glucose, 326
lipids, 314–317
lipoproteins, 317–321
phenylalanine, 309, 535
phenylalanine–tyrosine, 303, 306-311
purines, 322–323
Metabolism
cellular, 62
errors in, 504–505, 513–515
Metacentric chromosome(s), 54–55, 79
Metaphase
chromosome analysis during, 50–53
meiotic, 72, 74, 75
mitotic, 50, 51, 66, 67, 69, 70
Metastasis, 379
Methane, 532
Methanethiol, 42
Methionine, 235, 236, 273, 274
Method of ascertainment, defined, 127
Methylation, 89
X chromosome inactivation and, 134, 139
Metric system, units of measurement, 525
MHC (major histocompatibility complex), 356, 357, 366, 367. *see also* HLA system
MIC2 gene, 134, 135–136
Mice
cell hybridization, 292–294, 295
coat color, 220, 221
developmental genetics, 83–84, 88–89
Down syndrome model, 116
HPRT-deficient, 324–325
immune system, 353
induced mutations, 265
mdx, 39
monoclonal antibodies, 363–365
mutation rate, 437
nude, 394
tfm gene, 96
transgenic, 324–325, 386, 491
X chromosome inactivation, 131, 134
Microcell transfer, 394
Microfilaments, 60, 68
Microscopes
resolving power, 47
units of measurement, 525
Microtubules, 60, 75
of centrioles, 66
of cytoskeleton, 68
sperm, 63
of spindle, 68, 69, 70
Microvilli, 61, 63
Migration
and evolution, 447–449
Hardy-Weinberg law and, 408
race and, 451
Milk intolerance, 330, 331
Milk production
heritability in cattle, 461
human, 94
as sex-limited trait, 206
Minor histocompatability systems, 369, 377
Miscarriage (spontaneous abortion), 159. *see also* Abortions
Missense mutations, 23
of hemoglobin, 252, 255
Mitochondria, 61, 62
evolution, 62
genes of, 22–23, 243–244
mRNA codons, 250, 251
in sperm, 63, 64, 76
Mitochondrial DNA (mtDNA), 243–244, 245
Mitochondrial myopathy, 23
Mitosis, 66, 68–71, 78
gametogenesis and, 75
mutations, 77, 78
sister chromatid exchange (SCE), 259
stages of, 68–70, 78
Mitotic nondisjunction, 110–111, 122
MN blood group
discovery, 337, 338
frequency calculations, 414–415
polymorphism, 335
Modifier genes, 206
Mole, 155
Molecular cloning, 272–275, 272–276
Molecular disease, defined, 253
Molecular evolution, 239–240, 243–244, 450–451
Molecular genetics, 3. *see also* Biochemical genetics
of cancer, 380–386
of embryonic development, 83–90
evolution and, 9
gene mapping, 298–299
gene therapy, 516–518
history of, 7–9
LDL receptor gene, 319–320
Tay-Sachs disease, 317
Molecular structure, 531–533
Moloney murine leukemia virus, gene therapy and, 517

Mongolism. *see* Down syndrome
Monoclonal antibodies, 336, 363, 365, 376
Monohybrid cross, 13–14
Mononucleosis, 390
Monosomy, 16, 109
  frequencies, 156–157
  gene mapping, 162
  partial, 144, 145, 164
Monosomy 21, 119
Monotremes, 135
Monozygotic twins, 115, 465–466
  raised apart, 467, 477–478
Mosaicism
  defined, 110
  Down syndrome and, 112, 118, 120
  germinal, 110
  HPRT gene, 323–324
  isochromosomes, 165, 166
  Klinefelter syndrome, 124, 1, 142
  and nondisjunction, 110–111, 118, 120
  sex chromosome, 95
  and triploidy, 155
  trisomies 13 and 18, 118
  X chromosome inactivation and, 131–133, 134, 141, 142
  47,XXX females, 129–130
Moths, melanic, 409, 441
mRNA. *see* Messenger RNA
*Mst*II, 289, 301
mtRNA. *see* Mitochondrial RNA
Mucopolysaccharides, metabolism, 330
Mules, and inactive-X hypothesis, 132
Müllerian-inhibiting hormone, 68, 92, 93, 95, 96
Müllerian ducts, 91–93, 95, 101
Multinomial formula, 177
Multiple alleles, 169–170
Multiple crossing over, 193
Multiple gene inheritance, 462–464. *see also* Polygenic inheritance
Multiple myeloma, 358, 376, 380
Multiple sclerosis, 372, 373
Mus musculus. *see* Mice
Muscular dystrophy, 36–40. *see also* Becker muscular dystrophy; Duchenne muscular dystrophy
  localization of gene, 38–39, 500
Mutagens, 398
  Ames test, 259–260
  chemical, 258–261
  defined, 253
  detection, 258–261
  radiation, 261–265
  sister chromatid exchange, 260, 261
Mutation(s), 9, 46, 251–265. *see also* Chromosome aberrations
  albinos, 312
  base substitution, 251–252
  in cancer cells, 379
  of cell cycle, 76–78
  of *Drosophila* development, 85–87
  as experimental tool, 77
  frameshift, 247–248, 251, 258, 259
  of *FRAX*, 138
  germinal, 251, 262, 264
  Hardy-Weinberg law and, 408
  of hemoglobin genes, 253–257
  HPRT gene, 323
  induced, 253
  insertional, 324
  of LDL receptor gene, 320–321
  point, 251
  retinoblastoma, 392–394
  role in evolution, 437–441
  somatic, 251, 262, 264
  somatic cell fusion and, 293–294, 295
  spontaneous, 252–253
  to and from stop codons, 252
  Wilms tumor, 393–394
Mutation rate(s), 437–441
  average value, 440
  direct measurement, 438–439
  Duchenne muscular dystrophy and, 40
  equilibrium, 437–438
  examples, 440
  fruit flies, 437
  for human genes, 439–441
  indirect measurement, 439
  measurement of, 9, 437–439
  mice, 437–438
  microorganisms, 437
  sex-linked differences, 440–441, 456
  X-linkage and, 22
  X-rays and, 22
Mutually exclusive events, probability, 173–174
Myasthenia gravis, 372, 373
myc (oncogene), 382, 386–386, 388–390
Myelin, phenylketonuria and, 306
Myeloma cells, 363–364
Myosin, 70, 233
Myotonic dystrophy, 198

N-acetyl-transferase, 328
Nagasaki, 264–265, 427
Nail-patella syndrome, 202, 203
  gene mapping, 162, 202, 203
Nasopharyngeal carcinoma, 390
Natural selection. *see* Selection
Nature–nurture concept, 10, 459–460
  human behavior and, 468, 471–475
Nazis, 10, 481
Negative eugenics, 481, 483–484
Neonatal death, chromosome aberrations and, 156–157, 159
Neonatal screening
  metabolic disorders, 325
  phenylketonuria, 307–308, 330
Neoplasm, 379. *see also* Cancer
Nerve cell(s), 61
  Lesch-Nyhan syndrome, 322–325
  Tay-Sachs disease, 314–316, 328
Nerve growth factor, 67
Neural fold, 82
Neural tube, 82
Neural tube defects
  genetic screening and, 510–511
  prenatal diagnosis, 506–507, 522
Neuritic plaques, 113, 118
Neuroblastomas, 390
Neurofibrillary tangles, 113, 118
Neurofibromatosis
  mutation rate, 440
  symptoms, 208
Neurospora, 304
Neurotransmitters, 307, 309
Neutral genes, 450–451
Neutralist position (evolution), 450–451
Neutron, 530
Nitrogen-15, 228
Nitrous acid, 258
Noncycling (G_0) cells, 68
Nondisjunction, 7
  of autosomes, 112–121
  defined, 74, 107, 108, 119
  Down syndrome, 115–118
  as evidence that genes are on chromosomes, 108–109, 120
  in fruit flies, 108–109, 110, 121, 122
  genes causing, 118
  Klinefelter syndrome, 125, 140, 141–142
  meiotic, 109–110, 119, 120, 121
  mitotic, 110–111
  and mosaicism, 110–111
  primary, 116
  primary vs. secondary, 109, 119
  rates of, 116–118
  secondary, 109, 118, 121, 122
  of sex chromosomes, 108–110, 124–130, 140–142
  trisomy 13, 118, 119
  trisomy 18, 118, 119
  Turner syndrome, 129
  47,XXX females, 130
  47,XYY, 126–128
Nonhistone proteins, 46–48, 49
Nonsense codons, 250
Nonsense mutation, 321
Noradrenaline, 307, 309, 534–535
NORs. *see* Nucleolar-organizing regions
Notch gene, 88
Notochord, 82
Nuclear envelope, 60, 61, 64, 67, 69, 70, 73, 74
Nucleic acids, chemical structure, 533. *see also* DNA; RNA
Nuclein, 7, 43
Nucleolar-organizing chromosomes, 70
  and Robertsonian translocations, 148
Nucleolar-organizing regions (NORs), 54, 56, 232
Nucleolus, 44, 130, 315
  in cell division, 68, 69, 70, 74
  origin, 56
Nucleosome(s), 46–48, 66, 67
Nucleotide(s), 45, 228, 533. *see also* Base pairing
  defined, 227
  dideoxy sequencing of, 278–280
  metabolism, 322–323
  polymerase chain reaction, 290–291
  reassociation experiments, 241
Nucleus, 44, 60, 61, 63, 65, 315
  sperm, 63
Nullisomic gamete(s), 109, 121
Nutrition, 304–305
Nystagmus, albinism, 313

Octamer, 88
Ocular albinism (OA), 313–314
  linkage to *Xg*, 195, 196, 202, 203
Oculocutaneous albinism (OCA), 313–314. *see also* Albinism
Okazaki fragments, 229
Oligonucleotide probe(s), 283
  prenatal diagnosis by, 311, 506–507
Olympic games, sex testing at, 102, 104
Oncogenes, 9, 68, 115
  anti-oncogenes, 394
  isolation and cloning, 383–384
  number of, 384, 401
  protein products, 384–386
  retroviruses and, 243, 381–382, 401
  transgenic mice and, 386
One-gene-one-polypeptide hypothesis, 8, 304
Oocyte(s), 64, 71, 72–75, 76, 78
  chromosome aberrations, 156, 158
  maturation, 74, 79, 80
Oogenesis, 71–76, 76, 79, 80
  and Mendel's second law, 188
Oogonium, 71, 72, 75
Opsin, 19

Orangutan, gene map, 199–200
Orcein stain, 51
Organic acids, 533
Origin of Species, 9
Orotic aciduria, 513, 514
Ovalbumin gene, 238
  methylation and, 89
Ovaries, 76
  cancer of, 391
  development, 92, 93, 101
  in hermaphrodites, 94
  X-rays and, 263
Overdominant alleles, 171, 172
Oviducts, 76
  blocked, 81
  development, 92, 93
  sterilization and, 484
  in vitro fertilization and, 486
Ovulation, 71, 76, 94
  sex selection and, 494
  twinning and, 466

Pachytene, 71, 72, 73
PAH gene, 308, 309–310
Pair-rule gene, 85–87
Palindrome, 47, 300
Papain, 376
Parental gametes, 184–194
Parental generation, 13–14
Parental isonymy, 431
Parthenogenesis, 154–155
  defined, 88
Particulate inheritance, 10
Pascal's triangle, 177
Patau syndrome. *see* Trisomy 13
Paternal age. *see also* Maternal age
  and Down syndrome, 114–115, 120
  and Klinefelter syndrome, 125
Paternal genome, development and, 88–90
Path diagrams, 417–419, 421, 422, 424, 433
Pattern baldness, 204–206, 220
Pattern formation, 85–88
Pedigree(s)
  ABO, 345, 350
  agammaglobulinemia, 374
  autosomal dominant, 28–29, 31, 209
  autosomal recessive, 33–36, 41
  biochemical analysis, 29–30
  consanguineous, 418–420
  construction of, 27–29
  Duchenne muscular dystrophy, 37
  Duffy blood group, 161
  Egyptian Pharoahs, 420
  familial hypercholesterolemia, 42
  fragile X syndrome, 138–139
  inter-sibship marriages, 430–431
  linkage studies, 194–195
  linked red- and green-shift, 194, 195
  pattern baldness, 206
  phenylketonuria, 331
  polydactyly, 209
  sex-influenced trait, 206
  translocation Down syndrome, 151
  X-linked recessive, 37–38
Pedigree analysis, genetic counseling, 499
Penetrance, 31, 209–210
  mutation rate and, 438–439, 456
Penicillamine, 513
Penis, development, 92, 93
  in congenital adrenal hyperplasia, 98, 99
  in hermaphrodite, 100
Peptidase C, isozymes, 295, 296
Peptide bond, 235, 236
Permease, 313
Peroxisomes, 62
Phage. *see also* Bacteriophage
  gene transfer and, 383–384
Phagocytosis, 63
Pharmacogenetics, 325–328
Phase (of double heterozygote)
  defined, 188–189
  difficulties, 193–194
Phenocopy
  defined, 210
  mutations and, 258
Phenotype(s), 15
  behavioral, 468–475
  color vision abnormalities, 284–286
  effects of mutations, 251, 252
  frequencies, 406
  genetic vs. environmental causation, 218–219
  of hemoglobin gene, 171–172
  of heterozygote, 170–172
  intelligence, 471–474
  threshold traits, 468–469
Phenotypic variance, 460–464
  IQ, 473–474
Phenotypic variation, 204–212
  age of onset, 207–209
  genetic heterogeneity, 210, 212
  incomplete penetrance, 209–210
  pleiotropy, 210–212
  sex differences, 204–206
  variable expressivity, 206–207
Phenylalanine, 235
  genetic screening and, 509, 510
  metabolism, 303, 306–311, 534–535
  phenylketonuria and, 306, 307–309, 311, 328
  structure, 234
Phenylalanine hydroxylase (PAH), 306, 307, 308–309, 328
  gene for, 308, 309–310
  phenylketonuria and, 306, 307–309
Phenylketonuria, 305–311
  animal models, 306, 311
  biochemistry, 306, 307
  carriers, 306
  frequencies, 306, 330, 331
  localization of gene, 309, 500
  maternal, 307, 311
  pedigree, 331
  prenatal diagnosis, 504
  screening for, 330, 509–510
  symptoms, 307
  treatment, 306–307, 311, 328, 513, 514
  variants, 307
Phenylpyruvic acid, 306, 307
Phenylthiocarbamide (PTC) tasting, 415
Philadelphia chromosome, 151, 385, 387, 390
Phosphate (in DNA), 45, 46, 226, 228
Phosphoric acid, 226
Phosphorus-32, 279, 283
Phosphorylation, 67, 68, 385
Phytohemagglutinin, 51, 54
Pigment(s)
  color blindness and, 19
  color vision, 284–285
Pigmentation, 171
  albinism and, 312–314, 330
  fur, 131, 132, 134
  mutation rate and, 437
  variability of skin color, 464
  mammals, 206, 207
Pituitary dwarfism, 183–186, 187
  treatment, 514, 515
Pituitary gland, hormones, 94, 95
PKU. *see* Phenylketonuria
Placenta, 82, 88, 504
  polyploid, 155
Plant breeding, 519–520
Plants
  allergies to, 373
  bean, 244, 460
  flower color, 171
  hybrid corn, 171, 520
  Mendel's peas, 12–17
  mitochondrial and chloroplast DNA, 239, 240
  polyploidy, 154
  strawberry, 276
  trillium, 191
Plaques
  atherosclerotic, 317, 319
  Down syndrome, 113, 118
  gene library and, 281–284
Plasma, blood, 332–333
Plasma cells
  immune response and, 353–354
  in making monoclonal antibodies, 363–364, 376
Plasma membrane, 61, 63, 69
  fertilization and, 64
Plasmids, 271–272. *see also Escherichia coli*
Platelet-derived growth factor (PDGF), 67
  oncogenes and, 385
Pleiotropy, 210–212
  of ataxia-telangiectasia, 396–397
  defined, 210
Ploidy, 154–155, 164, 165
Point mutation, defined, 251
Polar bod(ies), 61, 63, 64–65, 72, 74, 74–75, 78, 79
  genes in, 201, 203
  and triple-X karyotype, 130
Poly(A) tail, 239, 275
Polycystic kidney disease, 500
Polydactyly, 202, 203, 209, 210
  and Ellis-van Creveld syndrome, 428–429
  linked to computing ability, 470
  and trisomy 13, 118, 119
Polyethylene glycol, 292
Polygenic inheritance, 462–464
  additive model, 462–463
  eye color, 477
  gamete types, 463
  skin color, 464
Polymerase chain reaction, 290–292
Polymorphism(s)
  blood group systems, 334–335
  and concept of race, 451–454
  defined, 334
  enzyme, 450, 454
  HLA, 367–369
  neutralist vs. selectionist positions, 450–451
  phenylketonuria, 309–310
  RFLPs, 286–287
  sickle-cell, 444–446
  stable, 445–446
  in twin diagnosis, 466
Polypeptide(s), 7–9, 233–237
  in antibodies, 359–360
  antibody genes, 359–363
  in hemoglobin, 237–241, 253–257
  of HLA system, 366–367
  in insulin, 234–235, 274, 275
  peptide bonding, 235, 236
  in various proteins, 233
Polyploidy, 154–155, 164, 165
  in abortions, 156–157, 164
  in animals, 154–155, 164

colchicine and, 50
frequencies, 156–157
in plants, 154
Polyposis of colon, 514
Polyuracil, 249
Population
defined, 405
Hardy-Weinberg conditions, 408
Population genetics, 3, 405–416
bottlenecks, 429
and concept of race, 451–454
defined, 405
evolutionary processes, 435–457
founder effect, 429–430, 448–449
genetic drift, 429, 448–449
Hardy-Weinberg law, 408–411
history of, 9–10
inbreeding, 417–428
isolates, 428–431
quantitative variation, 458–478
Porcupine men, 26, 27
Pore complex, 70
Porphobilinogen deaminase, 327
Porphyria, 326–327, 329, 469–470
localization of gene, 327, 500
prenatal diagnosis, 505
Porphyrin molecules, 327
Positive eugenics, 481, 484–486
Posterior probability, 178–179
Prader-Willi syndrome, 145, 164
Pregnancy
histocompatibility and, 371
phenylketonuria and, 311
prenatal diagnosis and, 504
Rh and, 338, 339, 346–348
technological aids to, 486–490
Pregnancy detection, 158
Premature discovery, 6
Premutation, 138
Prenatal diagnosis, 10, 502–508
abortions and, 156–157, 158–160
of cystic fibrosis, 35
of Down syndrome, 114–115
of Duchenne muscular dystrophy, 39–40
ethical considerations, 522
of fragile X syndrome, 139
gene mapping and, 198
of hypercholesterolemia, 321
of Lesch-Nyhan syndrome, 323–324
of phenylketonuria, 310–311
reasons for, 503–507
of sickle-cell anemia, 512
of Tay-Sachs disease, 316–317
Prenatal treatment, Rh, 347
Preproinsulin, 274, 275
Primaquine sensitivity, 326
Primates, gene maps, 199–200
Primer, DNA replication and, 291
Primordial germ cell(s), 71, 72, 82
Prion, 219
Prior probability, 178–179
Probability, 172–182
addition rule, 173–174
binomial, 174–177, 179–180
compared to frequency, 405
conditional, 177–179, 180, 213–215
defined, 173
in hypothesis testing, 215–217
inbreeding and, 425–426
independent events, 174
inverse, 179
multinomial, 177
multiplication rule, 174
mutually exclusive events, 173–174
Pascal's triangle, 177
and ratios, 177–178
prior vs. posterior, 178–179
rules, 173–179
sex distributions, 174–176, 179–180, 181
in truncate ascertainment, 213–215
Proband (propositus), 28–29
genetic hypothesis and, 218
Probe(s)
gene transfer experiments, 383–384
Huntington disease gene mapping, 287–289
in situ hybridization, 297, 298
library screening and, 283–284
muscular dystrophy gene mapping, 39
prenatal diagnosis, 506–507
Proflavine, 258
Progeny testing, 16–17
Progesterone, 94, 99, 100
Proinsulin, 274, 275, 456
Prokaryotes, 60
Prolactin, 68, 94
Prometaphase, 69
chromosome bands, 53
Promoter (DNA), 230, 231, 316
Pronucleus, 64–65, 76, 88
Proofreading, of DNA, 230
Propane, 532
Prophase
meiotic, 71–74, 763
mitotic, 68–69, 78
Propositus (proband), 28–29
genetic hypothesis and, 218
Prospective studies, 117
47,XYY males, 128
Prostate gland, 92, 103
Protamines, 46, 63, 64, 76
Protan gene, 18–19
Protanomaly. *see* Red-shift
Protanopia, 19
Protein(s)
carrier, 333
in chromosomes, 46–48
defects in metabolism, 303–331
electrophoretic variants, 294–296. *see also* Enzyme(s), electrophoretic variants
gene sequencing and, 279
high mobility group, 47
histone, 46, 48
homeobox, 87–88
human vs. mouse, 295, 296
laminar, 69, 70
MHC, 356, 357
of oncogenes, 384–386
receptor, 61, 63, 67
recognition, 334
steroids, 533
structure of, 7
synthesis by recombinant DNA, 272–275
three dimensional shape, 234–235
types, 233, 533
in viruses and prions, 219
Protein gradients, 87–88
Protein kinases, oncogenes and, 385, 401
Protein phenotypes, inheritance of, 29
Protein synthesis, 233–237
in vitro, 249
initiation, 235
mechanism, 235–237
termination signal, 236
Proto-oncogenes, 381–382
Proton, 530
Protonopia, 286
Protoporphyrinogen oxidase, 327
Provirus, 243, 375
Pseudoautosomal region, 76, 135–136
Pseudocholinesterase, 327
Pseudogenes, 240
Pseudohermaphrodite(s), 94, 96–99, 100, 103, 104, 141
Pseudomonas, ice-minus strain, 276
Psoriasis, 68, 372
Puberty, 100, 103, 104
gamete production, 71, 74, 76
in hermaphrodites, 96–97
precocious, 98
sex hormones, 93–94, 95, 97, 102
Pulsed-field electrophoresis, 277
Punnett square. *see* Checkerboard
Pure-breeding lines, 12–14
Purine(s), 45. *see also* Adenine; Guanine
Purine metabolism, 322–323, 329
Pyrimidine(s), 45. *see also* Cytosine; Thymine; Uracil
Pyrimidine dimer, 394, 396

Q-bands, 52, 57, 66
Quantitative variation, 10, 458–478
defined, 459
genetic and environmental components, 459–461
IQ, 471–474
polygenes, 462–464
skin color, 464
twin studies, 467–468
Quinacrine stain, 51–52, 56, 57
for sexing sperm cells, 494

ras (oncogene), 382, 384–385
R-bands, 52, 53, 67
R (restriction) point, of interphase, 66
Rabbits
Californian, 207
Himalayan, 206, 207, 220
Race(s), 451–454
allele frequency variation, 452–454
and IQ, 10, 473–474
variation in twinning rates, 466
Radiation
acute and chronic, 263
and birth defects, 82
and cancer, 263, 264
as carcinogen, 398–399
dosage, 262–265
effects, 263–265
exposure guidelines, 265
germinal effects, 264–265
medical and dental, 262, 263, 265
mutagenesis, 261–265
sources, 262–265
units of measurement, 263
Radioactive probe, 283
Radioactive isotopes, 262–263, 531
Radioactivity, 262–263
Radon, 263
Random genetic drift, 429, 448–449
Random mating, 407–408. *see also* Hardy-Weinberg law
compared to inbreeding, 425
Ratio(s), 527. *see also* Sex ratio
and probabilities, 177–178
Rats, behavioral experiments, 458–459
Reading frame, 248
Reassociation experiments, 241
Receptor proteins, 61, 63, 67. *see also* LDL receptor
androgen insensitivity syndrome and, 96, 102
growth factor, 77
Tay-Sachs disease, 316

Recessive allele(s), 13, 170–172
X-linkage, 20
Recessive detrimental allele(s), selection against, 444
Recessive inheritance. *see* Autosomal recessive inheritance; X-linked recessive inheritance
Recessive lethal allele, selection against, 442–444, 456
Recessive phenotype, frequencies, 406
Recessive traits, inbreeding and, 425–426
Reciprocal crosses, green-shift and, 20
Reciprocal mating, 13, 17–18
Reciprocal translocation(s), 147, 164, 165
gene mapping, 162
meiotic segregation, 149
Recognition proteins, 334
Recombinant DNA, 270–276
benefits and risks, 275–276
cartoon, 5
containment, 275–276
controversy, 518–519
Factor VIII from, 515
guidelines, 275–276
protein synthesis by, 272–275
Recombinant gametes, 184–194
Recombination, 183–194
antibody diversity and, 360–361
and chiasmata, 72–74, 189–194
defined, 198
gamete types, 184–194
homologous, 517
linked genes, 187–194
mapping methods, 194–197
mutations of, 77
unlinked genes, 184–187
within *DMD* gene, 40
Recombination distances, 298
Recombination percentages, 189–194
limits, 192–193
sex differences, 197, 202, 203
Red blood cells, 333–351. *see also* Hemoglobin
agglutination, 335–337
function, 334
hemolysis, 338
membrane, 326, 334
sickling, 172, 254–257
structure, 333–334
Red-shift, 19
linkage relationships, 194, 195, 202, 203
Refsum syndrome, 514
Regulator gene(s), 9
Reifenstein syndrome, 97
Reiter disease, 372
Relatives. *see also* Inbreeding
terminology, 417–419
Repetitive DNA, 241–242
Alu family, 51
DNA fingerprinting and, 290–292
L1 family, 51
terminal repeats, 243
Replication (of DNA), 45–46, 227–230
bubbles and forks, 228–229
Meselson-Stahl experiment, 228
origin, 66
primer, 291
proofreading mechanism, 230
Replication forks, defects in, 396
Replicons, 66
Reproductive technologies
animal, 490–491
ethical and legal issues, 488–490, 496–497
human, 486–490
Reptiles, sex determination in, 136
Repulsion phase, 188–189, 193–194
Resistance (R) factors, 271–272
Respiration, cellular, 62
Restriction point, 66
Restriction enzymes, 270–271, 273
electrophoresis and, 276–277, 278
gene mapping and, 197–199, 309
making gene library, 282
RFLPs, 286–289
use in prenatal diagnosis, 506, 507
Restriction fragment length polymorphisms. *see* RFLPs
Restriction fragments, 277, 278
Restriction mapping, 277, 278
Restriction nucleases, 8
Restriction sites, M13 virus, 280
Retina, in Tay-Sachs disease, 314, 315
Retinitis pigmentosa, 500
Retinoblastoma, 146, 392–393
chromosomal location, 393, 500
mutation rate, 440
Retinoschisis, linkage to *Xg*, 195, 196, 202, 203
Retrospective studies, 117
47,XYY males, 126–127
Retrovirus(es), 243, 375
as oncogene carriers, 381–382, 401
used in gene therapy, 517
Reverse banding, 52, 53
Reverse genetics, 30
DMD and, 39
Reverse transcriptase, 243, 381
RFLPs, 21, 286–289, 290
CF gene and, 34
on chromosome 21 (Down syndrome), 116
defined, 34, 287
DMD gene and, 38–39
Klinefelter syndrome and, 125
LDL receptor gene and, 321
mapping hereditary disease and, 500, 506
molecular gene mapping, 298–299
and phenylketonuria, 309–311
prenatal diagnosis and, 506
sickle-cell allele and, 448
TDF locus and, 135
Xg locus and, 135
Rh blood group, 24, 25, 345–348
allele frequencies, 346, 452
alleles, 345–346
discovery, 337, 338
linkage with elliptocystosis, 210
frequency calculations, 415, 416
gene flow, 447–448
hemolytic disease of the newborn, 338, 346–348
incompatibility, 346–347
measurement of gene flow, 447
notational systems, 345–346
polymorphism, 335
transfusion reaction, 332, 339
Rhesus monkey, Rh blood group, 337, 338
Rheumatoid arthritis, 372, 373
Rhodopsin, gene for, 119, 284–285
Ribonucleic acid. *see* RNA
Ribose, 47, 226
Ribosomal RNA (rRNA), 47, 54, 232–233
repetitive DNA and, 242
Ribosome(s), 61, 62, 232–233, 319
Rickets, 514
Ring chromosomes, 144, 145, 153, 154, 164, 165
Risk taking, 253, 398–399, 501
RNA, 230–233, 235–237
primer, 229
purine metabolism, 322–323
retroviruses and oncogenes, 381
structure of, 47, 230
transcription, 230–233. *see also* Transcription
translation, 233–237
types, 47, 232–233
RNA polymerase, 229, 230, 231
RNA viruses. *see* Retrovirus(es)
Robertsonian translocation(s), 148, 164, 165
Down syndrome and, 164, 165
evolution and, 199–200
meiotic segregation, 150
Rod cells, 284
Rous sarcoma virus (RSV), 381
rRNA. *see* Ribosomal RNA

S (synthesis) phase
of interphase, 66, 67, 68, 70
mutations, 77
premeiotic, 70, 71, 77
Saccharomyces (yeast), 77–78, 275
Salamanders, chiasmata, 191
Salmonella
in Ames test, 259–260
biosynthetic pathways, 329–330
Salt balance, congenital adrenal hyperplasia, 97–98
Samaritan population, 433
Sandhoff disease, 317
Sanger method, dideoxy sequencing, 278–280
Sarcoma, 379
Satellite, 56
Satellite DNA, 242
SCE (sister chromatid exchange), 260, 261
Schizophrenia, 470
as threshold trait, 468, 469
Schmiedenleut, 429–431
Scholastic Aptitude Tests (SATs), 471
SCID *see* Severe combined immunodeficiency disease
Science
nature of, 4–6
and society, 3–4, 518–520
subversion of, 519–520
Scientists
communication with public, 473, 518
life spans of, 8
public image of, 4–5
Scleroderma, 69
Scrapie, 219
Screening. *see* Genetic screening
Scrotum, 92–93, 103
Second messenger, 475
Secondary nondisjunction, 109, 118, 121, 122
Secretor trait, 342–345
Segment polarity genes, 85–87
Segmentation genes, 85–87
Segregation, meiotic, 149, 164, 165
Segregation, law of. *see* Mendel's first law
Selection, 9, 441–446
artificial, 441
balance hypothesis, 445–446
Hardy-Weinberg law and, 408, 442–445
heterozygote advantage, 444–446
intensification, 441
models of, 442
in modern society, 441–442
against a recessive lethal, 442–444, 456
relaxation, 441
and sickle-cell anemia, 444–446
Selectionist position, 450
Selective techniques
somatic cell fusion, 293–294, 295
tetracycline method, 272

Self-fertilization, 12–15, 419, 421
Self-mutilation, Lesch-Nyhan syndrome, 322
Seminiferous tubules, 76
Semisterility, and chromosome aberrations, 149, 153, 164, 165, 166
Sendai virus, 292
Sequencing, dideoxy, 278–280
Serine, 249
Serotonin, 309
Serum, 333
  immune, 335, 338
Severe combined immunodefiency disease (SCID), 370–371, 374
  gene therapy, 516–518
  treatment, 514, 515
Sex
  chromosomal, 91, 94, 95
  evolutionary value, 90
  gonadal, 94–99
Sex assignment, 93, 95
Sex chromatin
  detection of, 131, 139
  and number of X chromosomes, 131
  sex determination and, 508
Sex chromatin bodies, defined, 130. *see also* Barr bodies
Sex chromosome(s), 15, 18, 55. *see also* X chromosome; Y chromosome
  evolution, 135
  meiotic segregation and, 187, 188
  mosaicism, 95
  nondisjunction, 108–110
Sex determination, 18, 90–91, 101, 507–508
  in *Drosophila*, 108, 121, 122
  and polyploidy, 154–155
  prenatal, 493, 507–508
  and sex chromosome aberrations, 123
  Y chromosome and, 22
Sex development
  Down syndrome, 112
  environmental and cultural factors, 93, 95, 102, 104
  errors, 94–101, 102
  physiological factors, 93–95
  postnatal, 93–94
  prenatal, 91–93, 95
Sex distributions, probability, 174–176, 179–180, 181, 527
Sex hormones, 93–94, 95, 204–206. *see also* Estrogen; Testosterone
Sex identity, 93, 95, 99–101
Sex-influenced inheritance, 204–206
Sex-limited traits, 206
Sex-linked regions, of X and Y chromosomes, 135–136
Sex-linked traits. *see* X-linkage
Sex organs, development, 91–93, 101, 102
Sex preferences, of prospective parents, 492–493
Sex ratio, 18, 221, 492, 527. *see also* Sex distribution
  ascertainment bias, 212, 221
  and lethal genes, 20
  sex selection and, 494
Sex reversal, 96, 135–136, 140, 141
  gene mapping, 163
Sex roles, 93, 95, 96, 99–101
Sex selection, 491–495, 497
  postconception techniques, 493
  preconception techniques, 493–495
Sheep, scrapie, 219
Sherman paradox, 138–139
Shifting balance theory, 449
Shotgun technique, 282
Siamese twins, 465
Sibship(s), 28–29
  inter-sibship matings, 430–431
  sex distributions, 174–176, 179–180, 181
Sibship size, and ascertainment bias, 213–215, 221
Sickle-cell allele
  gene flow and, 448
  mutations of, 251, 252
Sickle-cell anemia
  cascade of effects, 254–255
  confusion with trait, 512
  frequency calculations, 412, 415
  gene mapping, 289, 290, 500
  gene therapy and, 516
  levels of observation, 171–172
  and malaria, 9, 172, 255, 444–446
  as molecular disease, 254–255
  prenatal diagnosis, 512
  screening for heterozygotes, 512
Sickle-cell hemoglobin, 8
Sickle-cell trait, 27, 254–255
Side chains, 234, 235
Silver staining, 54, 56
Single-copy DNA, 241
Single-gene inheritance, 29–38
Sister chromatid exchange (SCE), mutagenicity test, 260, 261
Sister chromatids, 44, 48, 52, 54, 67, 69
  in meiosis, 72–74, 78
  in nondisjunction, 107, 111, 119
Skin color, 10, 312
  variation, 464, 477
Slow metabolizers, 327
Small-cell lung carcinoma, 390
Smallpox, 352
Smell, genetics of, 42
Smoking
  cancer and, 399
  pregnancy and, 259
  spontaneous abortion rate and, 159
Snapback sequences, 242
Socioeconomic status, and IQ scores, 474
Somatic cell fusion, 292–298, 363–365
  methods, 292–294, 295
Somatic cell genetics, 197
Somatic cells, cancer and, 398
Somatic mutation, 251, 262, 264, 361–362
Somatomedin, 67
Somatostatin, 266, 273–274
Somatotropin, 67, 68
Somites, 82
Southern blotting, 295, 296
  methodology, 285
  visual pigment genes, 285
Species, defined, 451
Sperm, 71, 75, 76, 92, 94, 103
  abnormal, 63–64
  artificial insemination, 484–485
  chromosome aberrations, 156, 157–158
  frozen, 485, 490
  mutation rate, 440–441, 456
  separating X and Y, 490–491, 493–495
  sexing of, 494
  structure, 63–64, 65
Sperm banks, 484, 485
Sperm production, Y-linkage, 22
Spermatid, 75, 76
Spermatocyte(s), 72, 75, 76, 78, 187, 188
Spermatogenesis, 71–76
  and Mendel's second law, 187, 188
  and nondisjunction, 109–110
  Y-linkage, 22
Spermatogonium, 75, 76
Spermiogenesis, 75, 76
Spherocytosis, 514
Sphingosine, 316
Spina bifida, 506–507, 510
  recurrence risk, 500
Spindle, 66, 68, 69, 74–77
  fibers, 68, 69
  meiotic, 60, 61
Spontaneous abortion. *see* Abortions
Squashing technique, 51, 54
Stable polymorphisms, 445–446
Stanford-Binet tests, 471, 477
Stearic acid, 316
Stem cells, gene therapy and, 324–325, 517
Sterility, male, 63–64
Sterilization, 483–484
Steroids, 533
Sticky ends, 270, 271
Stillbirth(s). *see also* Abortions
  chromosome aberrations and, 156–157, 159, 164
  defined, 159
Stop codons, 236, 250
  mutation to, 252
Structural genes, defined, 241
Submetacentric chromosome(s), 54–55, 56
Sugars, in DNA and RNA. *see* Deoxyribose; Ribose
Superoxide dismutase, 115, 116, 162
Suppressor genes, 206
Surgery, for genetic defects, 98–99, 514, 515
Surnames, analysis of (isonymy), 431
Surrogate motherhood, 488
  ethical and legal issues, 489–490
Suxamethonium, 327
Sweat glands
  ectodermal dysplasia, 132, 133
  lack of, 41
Symbiosis, molecular evolution and, 243
Synapsis, 71–73, 76, 77
  chiasmata, 73 -74, 190, 191
  disjunction and, 74
  and inversions, 152, 164, 165
  misaligned, 257
  translocation heterozygote, 148, 164, 165
Synaptonemal complex, 73
Syndactyly, 202, 203
Syndrome, defined, 112
Synonymous substitutions, 251, 252
Synteny, 295, 301
Systemic lupus erythematosus, 372, 373

T cell(s), 354–355, 357–358
  AIDS and, 375
  allergic response and, 373
  cytotoxic, 358
  gene therapy and, 516–518
  helper, 356, 357–358
  nude mice and, 394
  suppressor, 358
T cell receptors, 357, 358, 363
T4 (virus), frameshift mutations, 247–248, 258
Tangles, Down syndrome, 113, 118
TATA box, 230
Tay-Sachs disease, 62, 314–317, 328
  biochemistry, 314, 316
  frequencies, 314
  inbreeding and, 425
  prenatal diagnosis, 505
  screening for heterozygotes, 511–512
  symptoms, 314, 316
  treatment, 514, 515
  variants, 317

TDF gene, 92, 94, 95, 101
  hermaphroditism and, 96
  mapping, 163
  sex reversal and, 135–136, 140
Technology. *see also* Recombinant DNA; *specific technology*
  molecular, 8–9
  reproductive, 486–491, 496–497
  scientific discovery and, 6
Teeth, lack of, 41. *see also* Ectodermal dysplasia
Telomeres, 67, 71, 73
Telophase
  meiotic, 72, 74
  mitotic, 69, 70
Temperature-sensitive mutation, 77
Template
  dideoxy sequencing and, 278–279
  DNA, 227
Termination codon, 236
Terminator (DNA), 230, 231, 273
Test tube baby, 486
Testcross, 16–17
Testes
  absence in males, 96
  in androgen insensitivity syndrome, 96–97
  in congenital adrenal hyperplasia, 97–98
  development, 92, 94, 95, 101
  development in hermaphrodites, 95–96
  in Klinefelter syndrome, 124
Testicular feminizing syndrome. *see* Androgen insensitivity syndrome
Testis-determining factor, 92, 94, 95
  gene mapping, 163
  hermaphroditism and, 96
  sex reversal and, 135–136
Testosterone, 68, 92–95, 101
  in androgen insensitivity syndrome, 97, 102
  in congenital adrenal hyperplasia, 97–98
  prenatal sex determination and, 508
  as steroid, 533
Tetracycline, as selective agent, 272
Tetrad(s), 72, 190, 191–194
Tetraploidy (4n), 154, 155
Thalassemia(s), 255–257
  alpha (α), 257, 267
  beta (β), 256–257, 267
  bone marrow transplants and, 514, 515–516
  gene therapy, 516
  localization of gene, 500
  and malaria, 256, 446
  prenatal diagnosis, 506–507
  stop codon mutation and, 252
Therapy, for genetic disorders, 512–518
Thermophilic bacteria, 291
Threshold (radiation), 264
Threshold traits, 468–469
Thymidine
  in DNA synthesis, 293–294, 295
  in fragile site, 54
Thymidine kinase (TK), 292–294, 295
Thymine, 45, 46, 51, 226, 227
Thymine dimer, 394, 396
Thymus, 353
Thyroid-stimulating hormone, hypothyroidism and, 510
Thyroxin, 68, 307
Tissue culture, 50, 324–325
Tissue typing, 365–367. *see also* HLA system
TK. *see* Thymidine kinase
Toad. *see Xenopus*
Tobacco, carcinogen, 398–399
Trans position, 189. *see also* Repulsion phase
Transcription, 230–233, 237
  base pairing, 231
  promoter and terminator, 230–231
Transfection, oncogene mapping, 382–384
Transfer RNA (tRNA), 47, 232
  repetitive DNA and, 242
  in translation, 235–237
  wobble and, 249
Transferrin, 333
Transformation, oncogenes and, 381–382, 383–384
Transfusion. *see* Blood transfusion
Transgenic farm animals, 491
Transgenic mice, 386
  gene therapy and, 516
  Lesch-Nyhan syndrome and, 324–325
Translation, 230, 235–237. *see also* Protein synthesis
Translocation(s)
  aneuploidy and, 110
  balanced, 148, 164
  chromosomes 11 and 22, 151
  chromosomes 14 and 21, 150–151
  development and, 89
  Duchenne muscular dystrophy and, 38
  evolution and, 199–200
  formation, 147–148
  frequencies, 159–160
  gene mapping and, 161–163, 297
  gene inactivation, 134–135
  meiotic segregation, 149, 164, 165
  reciprocal, 147, 164
  Robertsonian, 148, 164, 165
  X;autosome, 151
Translocation breakpoints
  leukemia and, 387–389
  proto-oncogenes and, 385, 387–389
Translocation Down syndrome, 150–151, 164, 165
Translocation heterozygotes, 149, 164, 165
Transmission, laws of, 6
Transmission genetics, 3, 12–40, 183–203
  history of, 6–7
Transmitter males, fragile X syndrome, 138–139
Transplantation
  antigens, 365, 368–371
  genetic disease treatment, 514, 515–516
  HLA system and, 365–367, 369–371
  legal and ethical concerns, 371, 377
  of organs, 369–371, 376, 377
Transposable elements, 9, 242–243
Transposons, 242–243
Transsexual, 101
Transvestite, 101
Triglycerides, 318, 533
Trillium, chiasmata in, 191
Trinomials, Hardy-Weinberg equilibrium, 410
Triple-X females, 129–130
Triplets
  genetic code, 247–251
  table of, 250
Triploidy (3n), 154, 165
  mosaicism, 155
Triradius. *see* Dermatoglyphics
Trisomy, 109–120
  and abortion, 156–157
  double, 117
  Down syndrome, 115
  frequencies, 156–157
  gene mapping, 162
  partial, 145
Trisomy 13, 118, 119
Trisomy 18, 118, 119, 521
  miscarriages and stillbirths, 159
Trisomy 21, 115–116. *see also* Down syndrome
Trisomy 22, 118–119
Tritiated thymidine, 52
tRNA. *see* Transfer RNA
Truncate ascertainment, 212–215
  complete vs. incomplete, 213–215
  in four-child families, 213, 214
  and sibship size, 213–214, 221
Trypsin, 233, 245
  cystic fibrosis and, 35–36
Tryptophan
  frequency, 249
  metabolism, 309
Ts mutation, 77
TSD gene, 317
Tubal ligation, 484
Tuberculosis, isoniazid metabolism, 327–328
Tubulin, 60, 61, 68, 70
Tumor-suppressing genes, 394
Tumors. *see also* Cancer
  types of, 379–380
Turner syndrome (45,X), 123, 128–129, 139, 470
  Barr bodies, 131
  and color blindness, 140, 141
  karyotype, 57, 59
  and partial X inactivation, 134
Twinning, rates, 466, 467
Twins, 465–468
  Bayes theorem and probability, 178–179, 182
  blood groups, 477
  concordance of threshold traits, 468–469
  conjoined (Siamese), 465
  Down syndrome and, 115, 121, 122
  ectodermal dysplasia, 133
  embryo splitting (cattle), 490–491
  gender identity, 101
  in genetic studies, 10, 467–468, 477–478
  IQ studies, 472, 477–478
  MZ vs. DZ probability, 178–179, 182
  pedigree symbol, 28–29
  phenotypic differences, 141, 142
  and Turner syndrome, 140, 141
  types, 465, 466
  and X-chromosome inactivation, 132, 133
Two-genes-one-polypeptide hypothesis, 359–360, 363
Tyrosinase, 307
  albinism and, 312–314
  gene, 312
  hair bulb test, 313, 314, 330
Tyrosine
  metabolism, 303–304, 306, 307, 309, 311, 313, 330, 534–535
  protein kinases and, 385
  structure, 234

Ubiquitin, 46–47
Ulex, 244
Ultrabithorax (*Ubx*), 86
Ultrasound scanning, 507
  neural tube defects and, 510
  sex determination and, 508
Ultraviolet (UV) radiation, 262
  albinisim and, 312
  cancer and, 394–395, 396, 398
Unequal crossing over, 317, 321
  color blindness and, 285–286
  DNA fingerprinting and, 398
  Hex-A gene, 317
  LDL gene, 321
  Lepore chromosome and, 257

Unique sequence DNA, 241
Universal donor and recipient, 342
Uracil, 47, 227, 230, 231
  single-base sequence, 249
Urethal folds, 91, 92, 93, 101
Uric acid, Lesch-Nyhan syndrome, 322–323
Uridine, orotic aciduria and, 513, 514
Uterus, 76
  development, 92, 93, 103
  and implantation of embryo, 81–82

V-D-J-C gene, 360–361, 376
V (variable) element, antibody gene, 360–363
Vaccination, 352
Vaccines, recombinant DNA synthesis, 275
Vagina
  in androgen insensitivity syndrome, 96–97
  development, 92, 93, 102
Valence, 531
Valine
  sickle-cell mutation and, 251, 252, 254
  structure, 234
Variability
  among human populations, 452–454
  antibody diversity and, 359–363
Variable age of onset, 90, 207–209
Variable expressivity, defined, 206, 207
Variance, genetic and environmental, 460–464
Variance (statistic)
  IQ, 473
  between twins, 467–468
Variant chromosomes, 160–161
Variation
  appreciation of, 519–520
  phenotypic, 204–212. *see also* Phenotypic variation
  quantitative, 458–478. *see also* Quantitative variation
Vas deferens, 92
Vasectomy, 484
Vector
  gene library and, 281
  for gene therapy, 517
  recombinant DNA and, 271–272
Vesicle, 61
Vinculin, 385
Virus(es)
  bacteriophage , 281–282
  and birth defects, 82
  as carcinogen, 380–382, 389–390
  gene therapy and, 516–518
  HIV, 374–375
  M13, sequencing gel, 280
  molecular technology and, 8
  oncogenes and, 381–382
  retroviruses, 243
  slow, 219
  vaccines and, 275
Vision
  in albinos, 313
  color, 284–286
  Tay-Sachs disease, 314
Visual pigments, retinoblastoma and, 392
von Recklinghausen neurofibromatosis. *see* Neurofibromatosis
von Willebrand disease, 20

WAGR syndrome, 146
Wechsler scales, 471
Weight
  units of measurement, 525
  variation in beans, 460
White blood cells, 54
  HLA system and, 366–367
  immune response, 353–354
  malignancies of, 379
Wildtype alleles, defined, 169
Wilms tumor, 146, 393–394
  chromosomal location, 393
  treatment, 394
Wilson disease, 500, 513, 514
Wiskott-Aldrich syndrome, 507
Wobble, 249
Wolffian ducts, 91–93, 95, 101

X chromosome, 18, 52, 55
  banding, 57
  comparison among primates, 199–200
  fragile site, 137, 140
  gene mapping, 162, 163, 187–190, 194–197
  genes on, 20
  inactivation and reactivation, 131–135
  isochromosomes, 165, 166
  map length, 196–197
  multiple, 129–130
  one lacking (45,X), 128–129
  pseudoautosomal region, 135–136, 140
  sex-linked region, 22, 135–136
  structure, 22, 44
  visual pigment genes and, 284–286
X chromosome inactivation
  agammaglobulinemia and, 374
  cancer origin and, 379–380
  deletions and, 162
  ectodermal dysplasia, 132, 133
  enzyme activity and, 330, 331
  germ cells and, 134
  Lesch-Nyhan syndrome and, 323
  mosaicism and, 141, 142
  partial, 134, 140
  prenatal diagnosis and, 507–508
  steps, 133–134
X chromosome reactivation, 134–135
  fragile X syndrome and, 138
X-linkage, 18–22, 36–40
  crossing over, 187–189, 194–197
  frequency calculations, 413–414, 415
  gene conservation and, 39
  gene mutation rate, 440, 441
  Hardy-Weinberg law and, 413–414
  in inbred populations, 434
  lethal genes, 21–22
X-linked disorders. *see also* X-linked recessive inheritance
  color blindness, 18–20
  frequencies of affected males and females, 20, 21
  gene mapping, 163
  hemophilia, 21
  prenatal diagnosis, 507
X-linked dominant inheritance, characteristics of, 30, 41, 42
X-linked mental retardation, 53, 123, 136–139
X-linked recessive inheritance, 267
  agammaglobulinemia, 374
  characteristics of, 37–38, 40
  G6PD deficiency, 326
  Lesch-Nyhan syndrome, 322–325
  muscular dystrophy, 36–40
  prenatal diagnosis, 505, 507–508
X-ray(s). *see also* Autoradiography
  ataxia-telangiectasia and, 396–398
  chest, 263, 265
  and chromosome breaks, 143
  and Down syndrome, 117
  effect on mutation rate, 22
  as mutagen, 261–265
  mutation rate and, 9
ϕX174 (virus)
  DNA replication, 229–230
  overlapping genes, 266
X–Y pairing region, 135–136, 140
Xenopus
  pattern formation in, 87–88
  recombinant DNA, 270
Xeroderma pigmentosum, 394–395, 396, 401
Xg blood group
  and inactive-X hypothesis, 134
  and Klinefelter syndrome, 125
  linkage relationships, 195, 196
  phenotypes, 201
  polymorphism, 335
Xm serum protein, 195
47,XYY males, 123, 126–128, 139, 501–502
  and criminal behavior, 126–127
  Danish study, 127–128
  fluorescent Y chromosome, 52
  frequency, 126–127

Y chromosome, 18, 55, 92, 101
  comparison among primates, 199
  extra (47,XYY karyotype), 126–128
  fluorescence, 52, 494, 508
  homology with X chromosome, 22
  length variation, 159–160
  probes, 508
  pseudoautosomal region, 135–136, 140
  and sex determination, 18, 22, 91, 95, 123
  sex-linked region, 22, 135–136
  structure, 44
  variants, 57
  zinc-finger protein on, 135–136
Y-linkage, 22, 41, 42
Yeast
  cell cycle mutants, 77–78
  use in recombinant DNA, 275
Yellow mutant, albinisim, 314
Yolk sac, 82, 89

Z-DNA, 227
Zero, as exponent, 527–528
Zinc-finger protein, 88, 135–136
Zona pellucida, 61, 63, 64–65, 75, 76, 487
Zygote, 64–65, 71, 78, 81, 101
  defined, 60
Zygotene, 71, 72
Zygotic genes, 83–90
  time of activation, 85

ABOUT THE BOOK

Book and Cover Design: Rodelinde Albrecht
Artwork: Katherine Doktor-Sargent
Production Coordinator: Joseph J. Vesely
Editorial Coordinator: Carol Wigg
Copy Editor: Jodi Simpson
Composition: The text was set in Linotron 202 Garamond by DEKR Corporation
Cover Manufacture: New England Book Components, Inc.
Book Manufacture: The Murray Printing Company